GEOMICROBIOLOGY
Fifth Edition

GEOMICROBIOLOGY

Fifth Edition

Henry Lutz Ehrlich
Dianne K. Newman

CRC Press
Taylor & Francis Group
Boca Raton   London   New York

CRC Press is an imprint of the
Taylor & Francis Group, an **informa** business

CRC Press
Taylor & Francis Group
6000 Broken Sound Parkway NW, Suite 300
Boca Raton, FL 33487-2742

© 2009 by Taylor & Francis Group, LLC
CRC Press is an imprint of Taylor & Francis Group, an Informa business

International Standard Book Number-13: 978-0-8493-7906-2 (Hardcover)

Library of Congress Cataloging-in-Publication Data

Ehrlich, Henry Lutz, 1925-
   Geomicrobiology / Henry Lutz Ehrlich. -- 5th ed. / and Dianne K. Newman.
      p. cm.
   Includes bibliographical references and index.
   ISBN 978-0-8493-7906-2 (alk. paper)
   1. Geomicrobiology. I. Newman, Dianne K. II. Title.

QR103.E437 2009
551.9--dc22   2008029570

Visit the Taylor & Francis Web site at
http://www.taylorandfrancis.com

and the CRC Press Web site at
http://www.crcpress.com

Dedication

We dedicate this edition to Terry Beveridge:
dear friend, inspiring mentor, and geomicrobiologist par excellence.

Contents

Preface

Several important advances have occurred in the field of geomicrobiology since the last edition of this book, including a number of observations made possible by the introduction of genetic and molecular biological techniques that make revision and updating of the previous edition of *Geomicrobiology* timely.

Henry Lutz Ehrlich, author of the earlier four editions, has been joined by Dianne K. Newman for this fifth edition to lend her expertise in the area of molecular geomicrobiology. This has resulted in a new chapter (Chapter 8) in this edition, which is entitled "Molecular Methods in Geomicrobiology." The techniques described in this chapter illuminate the processes by which bacteria catalyze important geomicrobial reactions. For example, we are beginning to understand the molecular details whereby some gram-negative bacteria export electrons to mineral oxides with which they are in physical contact in their respiratory metabolism. Such electron transfer is enabled by respiratory enzymes in the outer membrane and periplasm of such organisms. Molecular techniques have also demonstrated that at least one gram-negative bacterium can import electrons donated by an electron donor, ferrous iron, in contact with the outer surface of the outer membrane of this organism. In some cases, electron shuttles have been shown to facilitate electron transfer. Further important advances in this area are anticipated. Collectively, these mechanistic observations make clear that microbes play a much more direct role in the transformation of oxidizable and reducible minerals than had been previously believed by many researchers in this field. We anticipate that as mechanistic molecular approaches are increasingly applied to diverse problems in geomicrobiology, exciting discoveries will be made about how life sustains itself even in seemingly inhospitable environments such as the deep subsurface.

Just as in the case of the previous editions of *Geomicrobiology*, the chief aim of the fifth edition is to serve as an introduction to the subject and an up-to-date reference. To continue to provide a broad perspective of the development of the field, discussion of the older literature that appeared in earlier editions of this book has been retained. Changes in understanding and viewpoints are pointed out where necessary. Although we do not claim that the reference citations at the end of each chapter are exhaustive, cross-referencing should reveal other pertinent literature. As before, a glossary of terms that may be unfamiliar to some readers has been added. All chapters have been updated where necessary by introducing the findings of recent research.

We are continuing to retain some of the drawings prepared by Stephen Chiang for the first edition. Other illustrations from the fourth edition have been retained in the current edition, with appropriate acknowledgments to their source when not originating from us, and some new illustrations have been added. We are very grateful to Andreas Kappler for allowing us to use the photomicrograph of *Chlorobium ferrooxidans* for the book cover illustration of this edition.

We owe special thanks to Martin Polz, Victoria Orphan, and Alex Sessions for stimulating discussions that shaped the content of Chapter 8; and we gratefully acknowledge Alexandre Poulain for his help in preparing the figures for this chapter. We also owe sincere thanks to Jon Price for his assistance in obtaining the photograph of the sample of basalt from the rock collection at Rensselaer Polytechnic Institute.

We appreciate the encouragement and editorial assistance of Judith Spiegel, Barbara Norwitz, and Patricia Roberson of Taylor & Francis Group LLC.

Responsibility for the presentation and interpretation of the subject matter in this edition rests entirely with the authors.

Henry Lutz Ehrlich
Dianne K. Newman

Authors

Dr. Henry Lutz Ehrlich earned a BS degree from Harvard College (major: biochemical sciences) in 1948, an MS degree in 1949 (major: agricultural bacteriology), and a PhD degree in 1951 (major: agricultural bacteriology; minor: biochemistry); both of the latter degrees from the University of Wisconsin, Madison. He joined the faculty of the Biology Department of Rensselaer Polytechnic Institute as an assistant professor in the fall of 1951, attaining the rank of full professor in 1964. Dr. Ehrlich became professor emeritus in 1994 but continues to be active in the department in pursuit of some scholarly work. He began teaching a course in geomicrobiology in the spring semester of 1966.

Dr. Ehrlich is a fellow of the American Academy of Microbiology, American Association for the Advancement of Science, the International Union of Pure and Applied Chemistry, and the International Symposia on Environmental Biogeochemistry. He is a member of the Interdisciplinary Committee of the World Cultural Council (Consejo Cultural Mundial) and an honoree of the 11th International Symposium on Water/Rock held in 1994 in Saratoga Springs, New York. Dr. Ehrlich has been a consultant at various times for a number of different companies. He was editor-in-chief of *Geomicrobiology Journal* (1983–1995) and has since continued as co-editor-in-chief. He is a member of the editorial boards of *Applied and Environmental Microbiology* and *Applied Microbiology and Biotechnology*. He is also emeritus member of American Association for the Advancement of Science, American Institute of Biological Sciences, American Society for Microbiology, and the Society of Industrial Microbiology.

Dr. Ehrlich's research interests have resided in bacterial oxidation of Mn(II) and reduction of Mn(IV) associated with marine ferromanganese concretions, marine hydrothermal vent communities, and some freshwater environments; bacterial oxidation of arsenic(III); bacterial reduction of Cr(VI); bacterial interaction with bauxite; and bioleaching of ores including metal sulfides, bauxite, and others. He is author or coauthor of more than 100 articles dealing with various topics in geomicrobiology.

Dr. Dianne K. Newman earned a BA degree from Stanford University (major: German studies) in 1993, and a PhD degree in 1997 (major: environmental engineering with an emphasis on microbiology) from the Massachusetts Institute of Technology (MIT). She spent two years as an exchange scholar at Princeton University in the Geosciences department from 1995 to 1997. Dr. Newman was a postdoctoral fellow in the Department of Microbiology and Molecular Genetics at Harvard Medical School from 1998 to 2000. She joined the faculty of the California Institute of Technology in 2000, where she was jointly appointed in the divisions of Geological and Planetary Sciences and Biology. In 2007, she returned to MIT, where she is currently the John and Dorothy Wilson Professor of Biology and Geobiology, with a joint appointment in the departments of Biology and Earth, Atmospheric and Planetary Sciences. Dr. Newman is also an Investigator of the Howard Hughes Medical Institute.

Dr. Newman's honors include being a Clare Boothe Luce assistant professor, an Office of Naval Research young investigator, a David and Lucille Packard Fellow in science and engineering, an Investigator of the Howard Hughes Medical Institute, and a fellow of the American Academy of Microbiology. She was the 2008 recipient of the Eli Lily and Company Research Award from the American Society for Microbiology. She is an editor of the *Geobiology Journal*, and is on the editorial board of the *Annual Review of Earth and Planetary Science*. She is on the scientific advisory board of Mascoma Corporation, and is a member of the American Society of Microbiology and the American Geophysical Union.

Dr. Newman's laboratory seeks to gain insights into the evolution of metabolism as recorded in ancient rocks by studying how modern bacteria catalyze geochemically significant reactions. Specifically, she focuses on putatively ancient forms of photosynthesis and respiration, with a specific interest in the cellular mechanisms that enable these complex processes to work.

1 Introduction

Geomicrobiology deals with the role that microbes play at present on Earth in a number of fundamental geologic processes and have played in the past since the beginning of life. These processes include the cycling of organic and some forms of inorganic matter at the surface and in the subsurface of Earth, the weathering of rocks, soil and sediment formation and transformation, and the genesis and degradation of various minerals and fossil fuels.

Geomicrobiology should not be equated with microbial ecology or microbial biogeochemistry. *Microbial ecology* is the study of interrelationships between different microorganisms; among microorganisms, plants, and animals; and between microorganisms and their environment. *Microbial biogeochemistry* is the study of microbially influenced geochemical reactions, enzymatically catalyzed or not, and their kinetics. These reactions are often studied in the context of cycling of inorganic and organic matter with an emphasis on environmental mass transfer and energy flow. These subjects overlap to some degree, as shown in Figure 1.1.

It is unclear as to when the term *geomicrobiology* was first introduced into the scientific vocabulary. This term is obviously derived from the term *geological microbiology*. Beerstecher (1954) defined geomicrobiology as "the study of the relationship between the history of the Earth and microbial life upon it." Kuznetsov et al. (1963) defined it as "the study of microbial processes currently taking place in the modern sediments of various bodies of water, in ground waters circulating through sedimentary and igneous rocks, and in weathered Earth crust [and also] the physiology of specific microorganisms taking part in presently occurring geochemical processes." Neither author traced the history of the term, but they pointed to the important roles that scientists such as S. Winogradsky, S. A. Waksman, and C. E. ZoBell played in the development of the field.

Geomicrobiology is not a new scientific discipline, although until the 1980s it did not receive much specialized attention. A unified concept of geomicrobiology and the biosphere can be said to have been pioneered in Russia under the leadership of V. I. Vernadsky (1863–1945) (see Ivanov, 1967; Lapo, 1987; Bailes, 1990; Vernadsky, 1998, for insights and discussions of early Russian geomicrobiology and its practitioners).

Certain early investigators in soil and aquatic microbiology may not have thought of themselves as geomicrobiologists, but they nevertheless exerted an important influence on the subject. One of the first contributors to geomicrobiology was Ehrenberg (1836, 1838), who discovered the association of *Gallionella ferruginea* with ochreous deposits of bog iron in the second quarter of the nineteenth century. He believed that this organism, which he classified as an infusorian (protozoan), but which we now recognize as a stalked bacterium (see Chapter 16), played a role in the formation of such deposits. Another important early contributor to geomicrobiology was S. Winogradsky, who discovered that *Beggiatoa*, a filamentous bacterium (see Chapter 19), could oxidize H_2S to elemental sulfur (Winogradsky, 1887) and that *Leptothrix ochracea*, a sheathed bacterium (see Chapter 16), promoted oxidation of $FeCO_3$ to ferric oxide (Winogradsky, 1888). He believed that each of these organisms gained energy from the corresponding processes. Still other important early contributors to geomicrobiology were Harder (1919), a researcher trained as a geologist and microbiologist, who studied the significance of microbial iron oxidation and precipitation in relation to the formation of sedimentary iron deposits, and Stutzer (1912) and others, whose studies led to the recognition of the significance of microbial oxidation of H_2S to elemental sulfur in the formation of sedimentary sulfur deposits. Our early understanding of the role of bacteria in sulfur deposition in nature received a further boost from the discovery of bacterial sulfate reduction by Beijerinck (1895) and van Delden (1903).

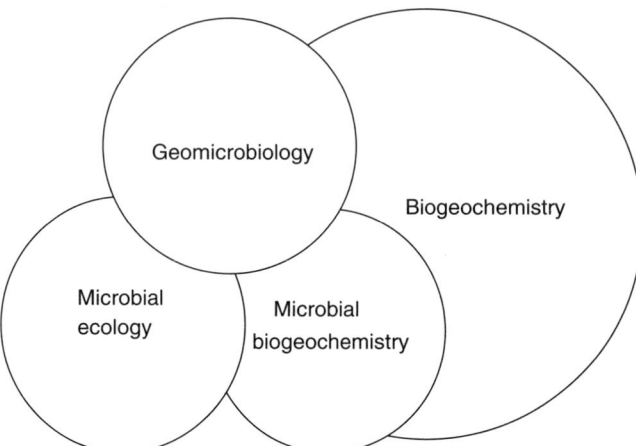

FIGURE 1.1 Interrelationships between geomicrobiology, microbial ecology, microbial biogeochemistry, and biogeochemistry.

Starting with the Russian investigator Nadson (1903, 1928) at the end of the nineteenth century, and continuing with such investigators as Bavendamm (1932), the important role of microbes in some forms of $CaCO_3$ precipitation began to be noted. Microbial participation in manganese oxidation and precipitation in nature was first recognized by Beijerinck (1913), Soehngen (1914), Lieske (1919), and Thiel (1925). Zappfe (1931) later related this activity to the formation of sedimentary manganese ore (see Chapter 17). A microbial role in methane formation (methanogenesis) became apparent through the observations and studies of Béchamp (1868), Tappeiner (1882), Popoff (1875), Hoppe-Seyler (1886), Omeliansky (1906), Soehngen (1906), and Barker (1956). The role of bacteria in rock weathering was first suggested by Muentz (1890) and Merrill (1895). Later, the involvement of acid-producing microorganisms, such as nitrifiers, and crustose lichens and fungi in such weathering was suggested (see Waksman, 1932). Thus by the beginning of the twentieth century, many important areas of study of geomicrobial processes had begun to receive serious attention from microbiologists. In general it may be said that most of the early geomicrobially important discoveries were made through physiological studies in the laboratory, which revealed the capacity of specific organisms to promote geomicrobially important transformations, causing later workers to study the extent of the occurrence of such processes in nature.

In the United States, geomicrobiology can be said to have begun with the work on iron-depositing bacteria by Harder (1919). Other early American investigators of geomicrobial phenomena include J. Lipman, S. A. Waksman, R. L. Starkey, and H. O. Halvorson, all prominent in soil microbiology, and G. A. Thiel, C. Zappfe, and C. E. ZoBell, all prominent in aquatic microbiology. ZoBell was a pioneer in marine microbiology (see Ehrlich, 2000).

Very fundamental discoveries in geomicrobiology continue to be made, some having been made as the twentieth century progressed and others very recently. For instance, the concept of environmental limits of pH and E_h for microbes in natural habitats was first introduced by Baas-Becking et al. (1960) (see Chapter 6). The pH limits as these authors defined them have since been extended at both the acidic and alkaline ends of the pH range (pH 0 and 13) as a result of new observations. Life at high temperature was systematically studied for the first time in the 1970s by Brock (1978) and associates in Yellowstone National Park in the United States. A specific acidophilic, iron-oxidizing bacterium, originally named *Thiobacillus ferrooxidans* and later renamed *Acidithiobacillus ferrooxidans*, was discovered by Colmer et al. (1950) in acid coal mine drainage in the late 1940s and thought by these investigators and others to be directly involved in its formation by promoting oxidation of pyrite occurring as inclusions in bituminous coal seams (see also Chapters 16 and 20).

The subsequent demonstration of the presence of *A. ferrooxidans* in acid mine drainage from an ore body with sulfidic copper as chief constituent, located in Utah, United States (Bingham Canyon open pit mine), and the experimental finding that *A. ferrooxidans* can promote the leaching (mobilization by dissolution) of metals from various metal sulfide ores (Bryner et al., 1954) led to the first industrial application of a geomicrobially active organism to ore extraction (Zimmerley et al., 1958; Ehrlich, 2001, 2004). After these pioneering studies on microbial participation of *A. ferrooxidans* in the formation of acid mine drainage, other organisms with iron-oxidizing capacity have been discovered in acid mine drainage from different sources and implicated in its formation, as have other microorganisms associated in consortia with the iron oxidizers (see review by Ehrlich, 2004).

The first attempt at visual detection of Precambrian prokaryotic fossils in sedimentary rocks was made by Tyler and Barghoorn (1954), Schopf et al. (1965), and Barghoorn and Schopf (1965) (see Chapter 3). These paleontological discoveries have had a profound influence on current theories about the origin and evolution of life on Earth (Schopf, 1983). The discovery of geomicrobially active microorganisms around submarine hydrothermal vents (Jannasch and Mottl, 1985; Tunnicliffe, 1992) and the demonstration of a significant viable microflora with a potential for geomicrobially important activity in the deep subsurface of the Earth's continents at depths of hundreds and thousands of meters below the surface (Ghiorse and Wilson, 1988; Sinclair and Ghiorse, 1989; Fredrickson et al., 1989; Pedersen, 1993) and deep beneath the surface of the ocean floor (Parkes et al., 1994) have revealed previously unsuspected regions of Earth where microbes are geomicrobiologically active. These discoveries have also had a major impact on the development of the field of astrobiology.

As this book will show, many areas of geomicrobiology remain to be fully explored or developed further.

REFERENCES

Baas-Becking LGM, Kaplan IR, Moore D. 1960. Limits of the environment in terms of pH and oxidation and reduction potentials. *J Geol* 68:243–284.

Bailes KE. 1990. *Science and Russian Culture in the Age of Revolution. V. I. Vernadsky and His Scientific School. 1863–1945.* Bloomington, MN: Indiana University Press.

Barghoorn ES, Schopf JW. 1965. Microorganisms from the late Precambrian of Central Australia. *Science* 150:337–339.

Barker HA. 1956. *Bacterial Fermentations. CIBA Lectures in Microbial Biochemistry.* New York: Wiley.

Bavendamm W. 1932. Die mikrobiologische Kalkfällung in der tropischen See. *Arch Mikrobiol* 3:205–276.

Béchamp E. 1868. Lettre de M.A. Béchamp á M. Dumas *Ann Chim Phys* 13:103 (as cited by Barker, 1956).

Beerstecher E. 1954. *Petroleum Microbiology.* New York: Elsevier.

Beijerinck MW. 1895. Über *Spirillum desulfuricans* als Ursache der Sulfatreduktion. *Zentralbl Bakteriol Parasitenk Infektionskr Hyg Abt I Orig* 1:1–9, 49–59, 104–114.

Beijerinck MW. 1913. Oxydation des Mangancarbonates durch Bakterien und Schimmelpilzen. *Folia Microbiol* (Delft) 2:123–134.

Brock TD. 1978. *Thermophilic Microorganisms and Life at High Temperatures.* New York: Springer.

Bryner LC, Beck JV, Davis DB, Wilson DG. 1954. Microorganisms in leaching sulfide minerals. *Ind Eng Chem* 46:2587–2592.

Colmer AR, Temple KL, Hinkle HE. 1950. An iron-oxidizing bacterium from the acid drainage of some bituminous coal mines. *J Bacteriol* 59:317–328.

Ehrenberg CG. 1836. Vorläufige Mitteilungen über das wirkliche Vorkommen fossiler Infusorien und ihre grosse Verbreitung. *Poggendorfs Ann* 38:213–227.

Ehrenberg CG. 1838. *Die Infusionsthierchen als vollkommene Organismen.* Leipzig, Germany: L. Voss.

Ehrlich HL. 2000. ZoBell and his contributions to the geosciences. In: Bell CR, Brylinsky M, Johnson-Green P, eds. *Microbial Biosystems: New Frontiers. Proceedings of the 8th Symposium on Microbial Ecology.* Atlantic Canada Society for Microbial Ecology, Halifax, Canada, Vol. 1, pp. 57–62.

Ehrlich HL. 2001. Past, present and future of biohydrometallurgy. *Hydrometallurgy* 59:127–134.

Ehrlich HL. 2004. Beginnings of rational bioleaching and highlights in the development of biohydrometallurgy: A brief history. *Eur J Miner Process Environ Protect* 4:102–112.

Fredrickson JK, Garland TR, Hicks RJ, Thomas JM, Li SW, McFadden K. 1989. Lithotrophic and heterotrophic bacteria in deep subsurface sediments and their relation to sediment properties. *Geomicrobiol J* 7:53–66.

Ghiorse WC, Wilson JT. 1988. Microbial ecology of the terrestrial subsurface. *Adv Appl Microbiol* 33:107–172.

Harder EC. 1919. Iron depositing bacteria and their geologic relations. US Geol Surv Prof Pap 113.

Hoppe-Seyler FZ. 1886. Ueber Gährung der Cellulose mit Bildung von Methan und Kohlensäure. *Physiol Chem* 10:201, 401 (as cited by Barker, 1956).

Ivanov MV. 1967. The development of geological microbiology in the U.S.S.R. *Mikrobiologiya* 31:795–799.

Jannasch HW, Mottl MJ. 1985. Geomicrobiology of the deep sea hydrothermal vents. *Science* 229:717–725.

Kuznetsov SI, Ivanov MV, Lyalikova NN. 1963. *Introduction to Geological Microbiology* (Engl. Transl.). New York: McGraw-Hill.

Lapo AV. 1987. *Traces of Bygone Biospheres*. Moscow: Mir Publishers.

Lieske R. 1919. Zur Ernährungsphysiologie der Eisenbakterien. *Zentralbl Bakteriol Parasitenk Infektionskr Hyg Abt II* 49:413–425.

Merrill GP. 1895. Disintegration of granite in the District of Columbia. *Geol Soc Am Bull* 6:321–332 (as cited by Waksman, 1932).

Muentz A. 1890. Sur la décomposition des roches et la formation de la terre arable. *CR Acad Sci (Hebd Séances)* (Paris) 110:1370–1372.

Nadson GA. 1903. *Microorganisms as Geologic Agents*. I. Tr Komisii Isslect Min Vodg, St. Petersburg: Slavyanska, 1903.

Nadson GA. 1928. Beitrag zur Kenntnis der bakteriogenen Kalkablagerung. *Arch Hydrobiol* 19:154–164.

Omeliansky W. 1906. About methanogenesis in nature by biological processes (Engl. Transl.) *Zentralbl Bakteriol Parasitenk Infektionskr Hyg Abt II* 15:673 (as cited by Barker, 1956).

Parkes RJ, Cragg BA, Bale SJ, Getliff JM, Goodman K, Rochelle PA, Fry JC, Weightman AJ, Harvey SM. 1994. Deep bacterial biosphere in Pacific Ocean sediments. *Nature* (London) 371:410–413.

Pedersen K. 1993. The deep subterranean biosphere. *Earth Sci Rev* 34:243–260.

Popoff L. 1875. *Arch Ges Physiol* 10:142 (as cited by Barker, 1956).

Schopf JW. ed. 1983. *Earth's Earliest Biosphere. Its Origin and Evolution*. Princeton, NJ: Princeton University Press.

Schopf JW, Barghoorn ES, Maser MD, Gordon RO. 1965. Electron microscopy of fossil bacteria two billion years old. *Science* 149:1365–1367.

Sinclair JL, Ghiorse WC. 1989. Distribution of aerobic bacteria, protozoa, algae, and fungi in deep subsurface sediments. *Geomicrobiol J* 7:15–31.

Soehngen NL. 1906. Het oustaan en verdwijnen van waterstof en methaan ouder invloed van het organische leven. Thesis. Technical University Delft. Delft, The Netherlands.

Soehngen NL. 1914. Umwandlung von Manganverbindungen unter dem Einfluß mikrobiologischer Prozesse. *Zentralbl Bakteriol Parasitenk Infektsionskr Hyg Abt II* 40:545–554.

Stutzer O. 1912. Origin of sulfur deposits. *Econ Geol* 7:733–743.

Tappeiner W. 1882. Über Celluloseverdauung. *Ber Deut Chem Ges* 15:999 (as cited by Barker, 1956).

Thiel GA. 1925. Manganese precipitated by microorganisms. *Econ Geol* 20:301–310.

Tunnicliffe V. 1992. Hydrothermal-vent communities of the deep sea. *Am Sci* 80:336–349.

Tyler SA, Barghoorn ES. 1954. Occurrence of structurally preserved plants in Precambrian rocks of the Canadian Shield. *Science* 119:606–608.

van Delden A. 1903. Beitrag zur Kenntnis der Sulfatreduktion durch Bakterien. *Zentralbl Bakteriol Parasitenk Infektsionskr Hyg Abt II* 11:81–94.

Vernadsky VI. 1998. *The Biosphere*. New York: Springer (Engl. Transl. of Vernadsky's *Biosfera*, first published in Russian in 1926).

Waksman SA. 1932. *Principles of Soil Microbiology*. 2nd ed. rev. Baltimore, MD: William & Wilkins.

Winogradsky S. 1887. Über Schwefelbakterien. *Bot Ztg* 45:489–600.

Winogradsky S. 1888. Über Eisenbakterien. *Bot Ztg* 46:261–276.

Zappfe C. 1931. Deposition of manganese. *Econ Geol* 26:799–832.

Zimmerley SR, Wilson DG, Prater JD. 1958. Cyclic leaching process employing iron oxidizing bacteria. U.S. Patent 2,829,964.

2 Earth as a Microbial Habitat

2.1 GEOLOGICALLY IMPORTANT FEATURES

The interior of the planet Earth consists of three successive concentric regions (Figure 2.1), the innermost being the *core*. It is surrounded by the *mantle*, which, in turn, is surrounded by the outermost region, the *crust*. The crust is surrounded by a gaseous envelope, the *atmosphere*.

The core, whose radius is estimated to be ~3450 km, is believed to consist of a Fe–Ni alloy with an admixture of small amounts of the siderophile elements cobalt, rhenium, and osmium, probably some sulfur and phosphorus, and perhaps even hydrogen (Mercy, 1972; Anderson, 1992; Wood, 1997). The inner portion of the core, which has an estimated radius of ~1250 km, is solid, has a density of 13 g cm^{-3} and is subjected to a pressure of 3.7×10^{12} dyn cm^{-2}. The outer portion of the core has a thickness of ~2200 km and is molten, owing to a higher temperature but lower pressure than at the central core (1.3–3.2×10^{12} dyn cm^{-2}). The density of this portion is 9.7–12.5 g cm^{-3}.

The mantle, which has a thickness of ~2865 km, has a very different composition from the core and is separated from it by the Wickert–Gutenberg discontinuity (Madon, 1992). Seismic measurements of the mantle regions have revealed distinctive layers called the upper mantle (365 km thick), the asthenosphere or transition zone (270 km thick), and lower mantle (1230 km thick) (Madon, 1992). The mantle rock is dominated by the elements O, Mg, and Si with lesser amounts of Fe, Al, Ca, and Na (Mercy, 1972). The consistency of the rock in the upper mantle, although not truly molten, is thought to be plastic, especially in the region called the *asthenosphere*, situated 100–220 km below the Earth's surface (Madon, 1992). Upper mantle rock penetrates the crust on rare occasions and may be recognized as an outcropping, as in the case of some ultramafic rock on the bottom of the western Indian Ocean (Bonatti and Hamlyn, 1978).

The crust is separated from the mantle by the Mohorovičić discontinuity. The thickness of the crust varies from as little as 5 km under ocean basins to as great as 70 km under continental mountain ranges. The average crustal thickness is 45 km (Madon, 1992; Skinner et al., 1999). The rock of the crust is dominated by O, Si, Al, Fe, Mg, Na, and K (Mercy, 1972). These elements make up 98.03% of the weight of the crust (Skinner et al., 1999) and occur predominantly in the rocks and sediments. The bedrock under the oceans is generally basaltic, whereas that of the continents is granitic to an average crustal depth of 25 km. Below this depth it is basaltic to the Mohorovičić discontinuity (Ronov and Yaroshevsky, 1972, p. 243). Sediment covers most of the bedrock under the oceans. In thickness, it ranges from 0 to 4 km. Sedimentary rock and sediment (soil in a nonaquatic context) cover the bedrock of the continents; their thickness may exceed that of marine sediments (Kay, 1955, p. 655). The continents make up 64% of the crustal volume; oceanic crust, 21%; and the shelf and subcontinental, the remaining 15% (Ronov and Yaroshevsky, 1972).

Although until the 1960s the Earth's crust was usually viewed as a coherent structure that rests on the mantle, it is now seen to consist of a series of moving and interacting *plates* of varying sizes and shapes. Some plates support the continents and parts of the ocean floor, whereas others support only parts of the ocean floor. The estimate of the number of major plates is still not fully agreed upon but ranges between 10 and 12 according to Keary (1993) and 10–16 according to the National Geographic Society (1995, 1998). Figure 2.2 shows the outlines of some of the major plates and adjacent continents. The plates float on the asthenosphere of the mantle. The crust and the upper mantle above the asthenosphere is sometimes referred to as the lithosphere by geologists.

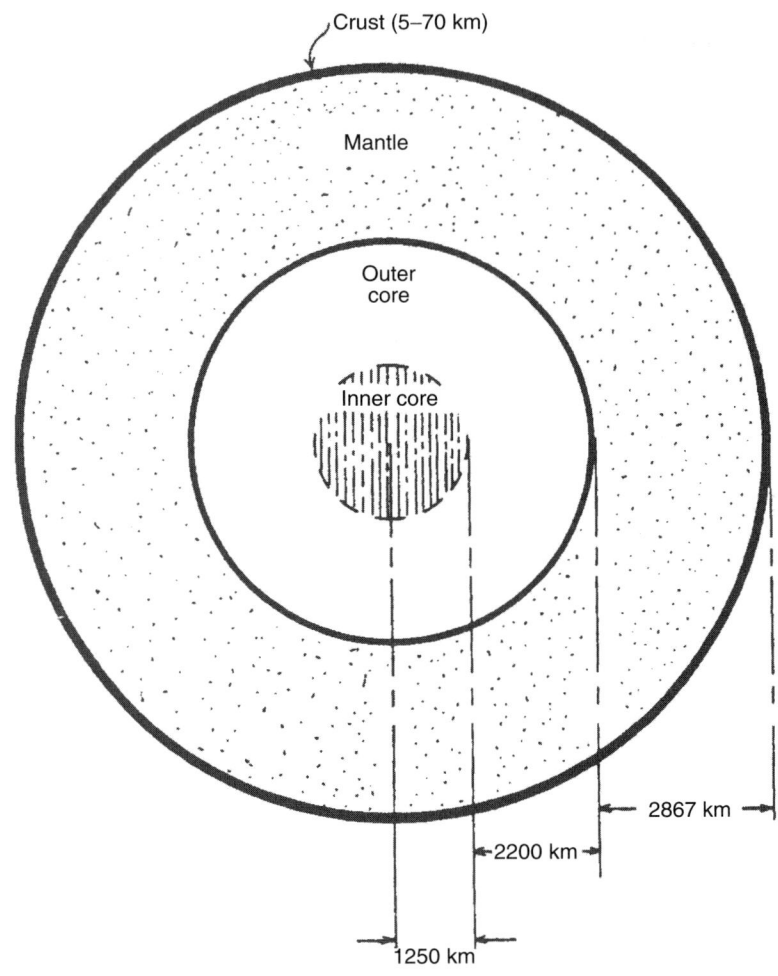

FIGURE 2.1   Diagrammatic cross section of the Earth. Radii of core and mantle drawn to scale.

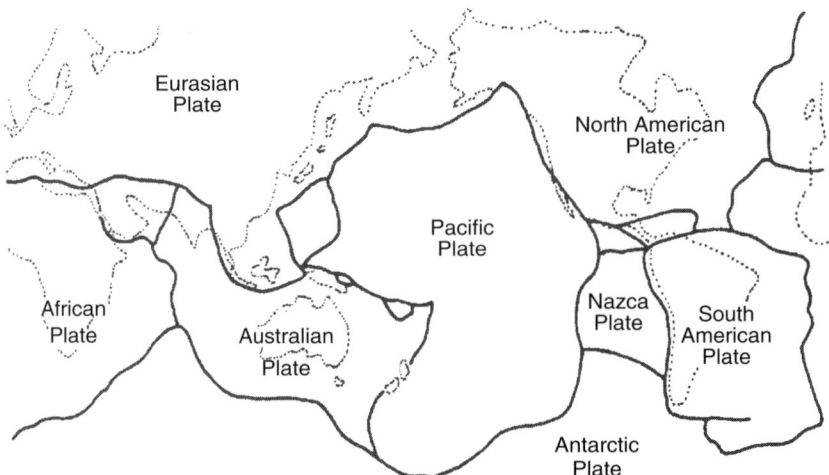

FIGURE 2.2   Major crustal plates of the Earth.

Convection resulting from the thermal gradients in the plastic rock of the asthenosphere is believed to be the cause of movement of the crustal plates (Kerr, 1995; Wysession, 1995; Ritter, 1999). In some locations this movement may manifest itself in a collision of plates and in other locations in plates of nearly equal density sliding past one another along transform faults. In still others, interacting plates may partially slide over one another in a process of crustal convergence called *subduction* where a denser oceanic plate slides below a lighter continental plate. Either of the last two processes may lead to formation of a trench–volcanic island arc system. Island arc systems result from a sedimentary wedge formed by the oceanic plate. In subduction, the resulting arc system may eventually accrete to the continental margin as a result of the movement of the subducting oceanic plate in the direction of the continental plate (Van Andel, 1992; Gurnis, 1992).

Oceanic plates grow along oceanic ridges, the sites of *crustal divergence*. Two examples of such divergence are represented by the Mid-Atlantic Ridge and the East Pacific Rise (Figure 2.3). The older portions of growing oceanic plates are destroyed through subduction with the formation of deep-sea trenches, such as the Marianas, Kurile, and Phillipine trenches in the Pacific Ocean and the Puerto Rico Trench in the Atlantic Ocean. Growth of the oceanic plates at the midocean ridges is the result of submarine volcanic eruptions of *magma* (molten rock from the deep crust or upper mantle). This magma gets added to opposing plate margins along a midocean ridge, causing adjacent parts of the plates to be pushed away from the ridge in opposite directions (Figure 2.4). The oldest portions of the interacting oceanic plates are consumed by subduction more or less in proportion to the formation of new oceanic plate at the midocean ridges, thereby maintaining a fairly constant plate size.

Volcanism occurs not only at midocean ridges but also in the regions of subduction where the sinking crustal rock undergoes melting as it descends toward the upper mantle. The molten rock may then erupt through fissures in the crust and contribute to mountain building at the continental

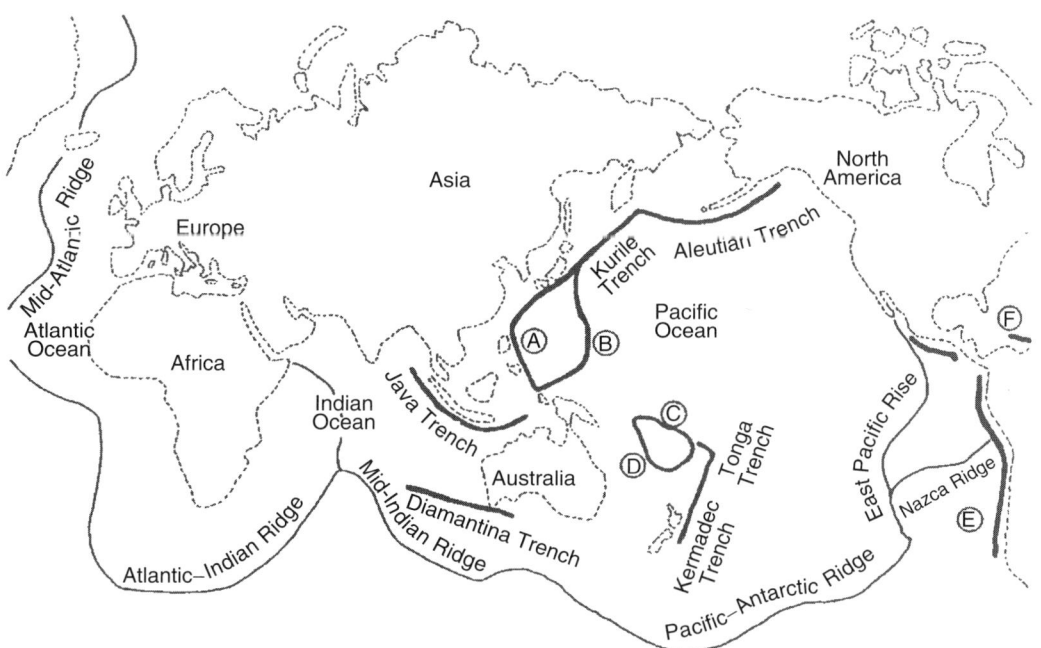

FIGURE 2.3 Major midocean rift systems (thin continuous lines) and ocean trenches (heavy continuous lines) (A, Philippine Trench; B, Marianas Trench; C, Vityaz Trench; D, New Hebrides Trench; E, Peru–Chile Trench; F, Puerto Rico Trench). The East Pacific Ridge is also known as the East Pacific Rise.

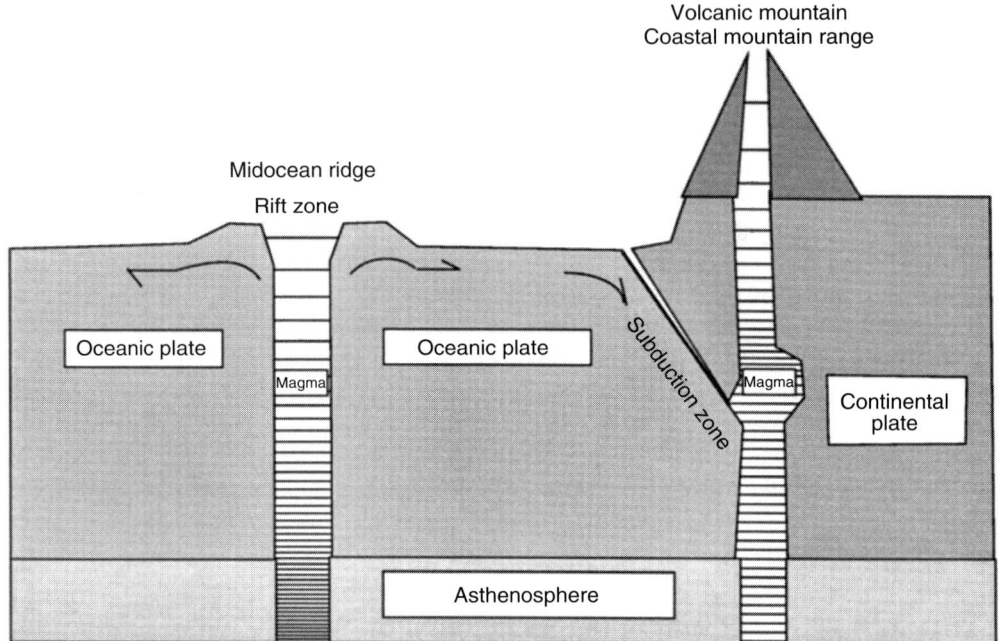

FIGURE 2.4 Schematic representation of sea floor spreading and plate subduction. New oceanic crust is formed at the rift zone of the midocean ridge. Old oceanic crust is consumed in the subduction zone near a continental margin or island arc.

margins (*orogeny*). It is plate collision and volcanic activity associated with subduction at continental margins that explain the existence of coastal mountain ranges. The origin of the Rocky Mountains and the Andes on the North- and South American continent, respectively, is associated with subduction activity, whereas Himalayas are the result of collision of the plate bearing the Indian subcontinent with that bearing the Asian continent.

Volcanic activity may also occur away from crustal plate margins, at the so-called hot spots. In the Pacific Ocean, one such hot spot is represented by the island of Hawaii with its active volcanoes. The remainder of the Hawaiian island chain had its origin at the same spot where the island of Hawaii is presently located. Crustal movement of the Pacific Ocean plate westward caused the remaining islands to be moved away from the hot spot so that they are no longer volcanically active.

The continents as they exist today are thought to have derived from a single continental mass, *Pangaea*, which broke apart less than 200 million years ago as a result of crustal movement. Initially this separation gave rise to *Laurasia* (which included present-day North America, Europe, and most of Asia) and *Gondwana* (which included present-day Africa, South America, Australia, Antarctica, and the Indian subcontinent). These continents separated subsequently into the continents we know today, except for the Indian subcontinent, which did not join the Asian continent until some time after this breakup (Figure 2.5) (Dietz and Holden, 1970; Fooden, 1972; Matthews, 1973; Palmer, 1974; Hoffman, 1991; Smith, 1992). The continents that evolved became modified by accretion of small landmasses through collision with plates bearing them. Pangaea itself is thought to have originated 250–260 million years ago from an aggregation of crustal plates bearing continental landmasses including Baltica (consisting of Russia, west of the Ural Mountains; Scandinavia; Poland; and Northern Germany), China, Gondwana, Kazakhstania (consisting of present-day Kazakhstan), Laurentia (consisting of most of North America, Greenland, Scotland, and the Chukotski Peninsula of eastern Russia), and Siberia (Bambach et al., 1980). Mobile continental plates are believed to have existed as long as 3.5 billion years ago (Kroener and Layer, 1992). The Earth seems to have had

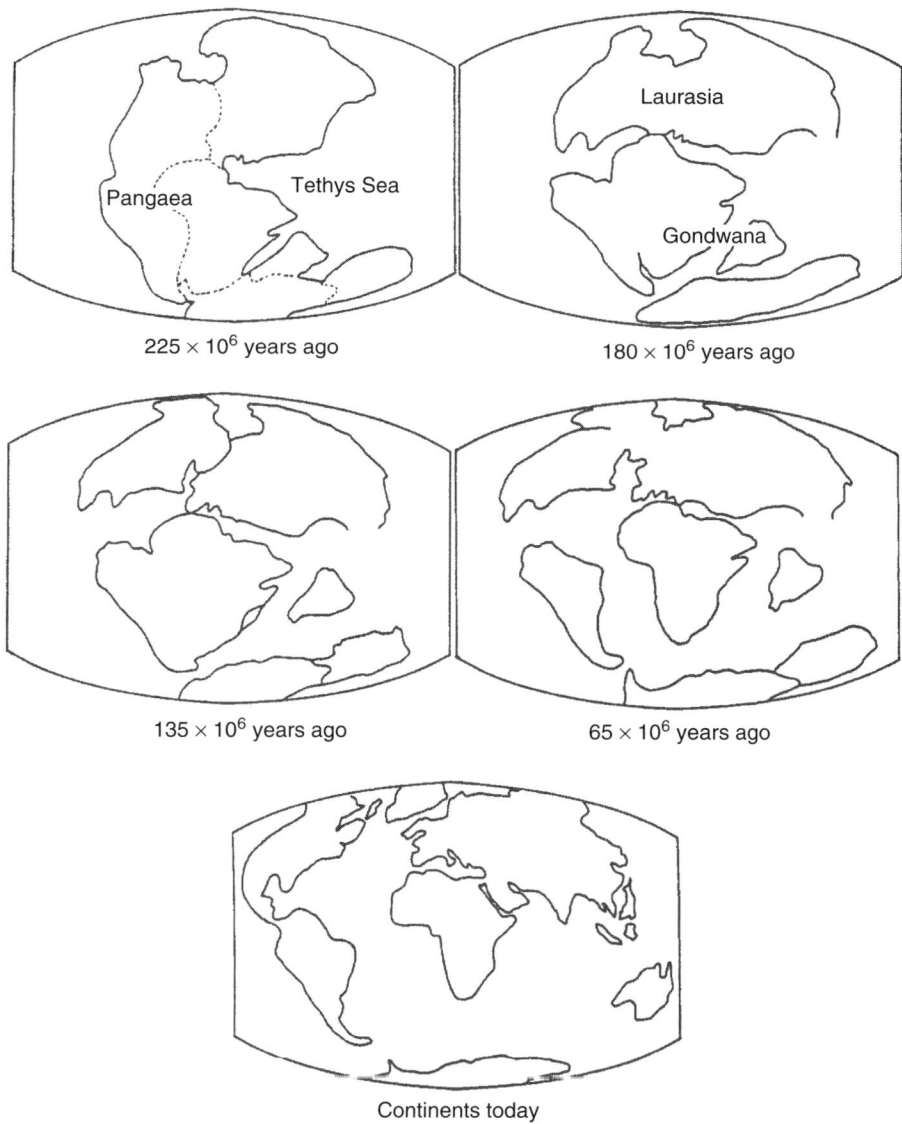

FIGURE 2.5 Continental drift. Simplified representation of the breakup of Pangaea to present time. (Reproduced from Dietz RS, Holden JC, *J. Geophys. Res.*, 75, 4939–4956, 1970. With permission.)

a crust as early as 4.35–4.4 eons ago—the age of the Earth being 4.65 eons (Amelin, 2005; Harrison et al., 2005; Watson and Harrison, 2005; Wilde et al., 2001).

The evidence for the origin and movement of the present-day continents has been obtained from at least three kinds of studies: (1) paleomagnetic and seismic examinations of the Earth's crust; (2) comparative sedimentary analyses of deep-ocean cores obtained from drillings by the *Glomar Challenger*, an ocean-going research vessel; and (3) paleoclimatic studies (Bambach et al., 1980; Nierenberg, 1978; Vine, 1970; Ritter, 1999). Although the separation of the present-day continents with the breakup of Pangaea had probably no significant effect on the evolution of prokaryotes (they had pretty much evolved to their present complexity by this time), it did have a profound effect on the evolution of metaphytes and metazoans (McKenna, 1972; Raven and Axelrod, 1972). Flowering plants, birds, and mammals, for example, had yet to establish themselves.

2.2 BIOSPHERE

The *biosphere*, the portion of the Earth that supports life, is restricted to the uppermost part of the crust and to a certain degree the lowermost part of the atmosphere. It includes the land surface, that is, the exposed sediment or soil and rock and the subsurface to a depth of 1 km and more, and the sediment surface and subsurface on the ocean floor (Ghiorse and Wilson, 1988; Parkes et al., 1994; Pedersen, 1993; Pokrovskiy, 1961; van Waasbergen et al., 2000; Wellsbury et al., 2002). The sediment, soil, and rock at and near the surface of the crust are sometimes referred to as the *lithosphere* by ecologists (however, see Section. 2.1 for geologists' definition of this term). The biosphere also includes the *hydrosphere*, the freshwater and especially the marine water that cover a major portion of the Earth's crust. The presence of living microorganisms has been demonstrated in groundwater samples taken at a depth of 3500 m from a borehole in granitic rock in the Siljan Ring in central Sweden (Szewzky et al., 1994). The water from this depth contained thermophilic, anaerobic fermenting bacteria related to *Thermoanaerobacter* and *Thermoanaerobium* species and one strain related to *Clostridium thermohydrosulfuricum* but no sulfate-reducing or methanogenic bacteria. The bacteria that were cultured grew in a temperature range of 45–75°C (65°C optimum) at atmospheric pressure in the laboratory. In continental crust, the temperature has been estimated to increase by ~25°C km^{-1} of depth (Fredrickson and Onstott, 1996). Using this constant, the *in situ* temperature at a depth of 3500 m should be ~87.5°C, which is higher than the maximum temperature tolerated by the cultures isolated by Szewzky et al. (1994) when grown under laboratory conditions, but well within the temperature range of hyperthermophilic bacteria (recently found maximum growth temperature was ~121°C; Kashefi and Lovley, 2003). Within a very limited range, elevated hydrostatic pressure to which microbes would be subjected at great depths may increase their temperature tolerance slightly, as suggested by the observations of Haight and Morita (1962) and Morita and Haight (1962). Clearly, temperature and hydrostatic pressure are important determinants of the depth limit at which life can exist within the crust. Other limiting factors are porosity and the availability of moisture (Colwell et al., 1997).

Unlike the lithosphere, the hydrosphere is inhabited by life at all water depths, some as great as 11,000 m—the depth of the Marianas Trench. In marine sediments, microbial life has now been detected at depths of >500 mbsf (meters below sea floor) (Parkes et al., 1994; Cragg et al., 1996). Bacterial alteration of the glass in ocean basalts has been seen to decreasing extents for 250–500 mbsf (Torsvik et al., 1998; Furnes and Staudigel, 1999). In some parts of the hydrosphere, some special *ecosystems* have evolved whose primary energy source is geothermal rather than radiant energy from the sun (Jannasch, 1983). These ecosystems occur around hydrothermal vents at midocean rift zones. Here heat from magma chambers in the lower crust and upper mantle diffuses upward into overlying basalt, causing seawater that has penetrated deep into the basalt to react with it (see Figure 17.17 for diagrammatic representation of this process). This seawater–basalt interaction results in the formation of hydrogen sulfide and in the mobilization of some metals, particularly iron and manganese and in some cases some other metals such as copper and zinc. The altered seawater (now a *hydrothermal solution*) charged with these dissolved metals is eventually forced up through cracks and fissures in the basalt to enter the overlying ocean through hydrothermal vents. Autotrophic bacteria living free around the vents or in symbiotic association with some metazoa at these sites use the hydrogen sulfide as an energy source for converting carbon dioxide into organic matter. Some of this organic matter is used as food by heterotrophic microorganisms and metazoa at these locations (Jannasch, 1983; Tunnicliffe, 1992). The hydrogen sulfide–oxidizing bacteria are the chief primary producers in these ecosystems, taking the place of photosynthesizers such as anoxygenic photosynthesizing bacteria, cyanobacteria, algae, and plants—the usual primary producers of Earth. Photosynthesizers cannot operate in the location of hydrothermal vent communities because of the perpetual darkness that prevails at these sites (see also Section 19.8).

Not all submarine communities featuring chemosynthetic hydrogen sulfide oxidizers as primary producers are based on hydrothermal discharge. On the Florida Escarpment in the Gulf of Mexico, ventlike biological communities have been found at abyssal depths around hydrogen sulfide seeps whose discharge is at ambient temperature. The sulfide in this instance may originate from an adjacent carbonate platform containing fluids with 250‰ dissolved solids and temperatures up to 115°C (Paul et al., 1984).

In some other locations, such as at the Oregon subduction zone or at some sites of the Florida Escarpment, methane of undetermined origin expelled from the pore fluids of the sediments, rather than hydrogen sulfide, is the basis for primary production on the seafloor. Metazoa share in the carbon fixed by free-living or symbiotic methane-oxidizing bacteria (Kuhn et al., 1986; Childress et al., 1986; Cavanaugh et al., 1987) (see also Chapter 22).

Finally the biosphere includes the lower portion of the atmosphere. Living microbes have been recovered from it at heights as great as 48–77 km above the Earth's surface (Imshenetsky et al., 1978; Lysenko, 1979).

Whether the atmosphere constitutes a true microbial habitat is very debatable. Although it harbors viable vegetative cells and spores, it is generally not capable of sustaining growth and multiplication of the organisms because of lack of sufficient moisture and nutrients and because of lethal radiation, especially at higher elevations. At high humidity in the physiological temperature range, some bacteria may, however, propagate to a limited extent (Dimmick et al., 1979; Straat et al., 1977). The residence time of microbes in air may also be limited, owing to their eventual fallout. In the case of microbes associated with solid particles suspended in still air, the fallout rate may range from 10^{-3} cm s^{-1} for particles in a 0.5 μm size range to 2 cm s^{-1} for particles in a 10 μm size range (Brock, 1974, p. 541). Even if it is not a true habitat, the atmosphere is nevertheless important to microbes. It is a vehicle for spreading microbes from one site to another; it is a source of oxygen for strict and facultative aerobes; it is a source of nitrogen for nitrogen-fixing microbes; and its ozone layer screens out most of the harmful ultraviolet radiation from the sun.

Although the biosphere is restricted to the upper crust and the atmosphere, the core of the Earth does exert an influence on some forms of life. The core, with its solid center and molten outer portion, acts like a dynamo in generating the magnetic field surrounding the Earth (Strahler, 1976, p. 36; Gubbins and Bloxham, 1987; Su et al., 1996; Glatzmaier and Roberts, 1996). Magnetotactic bacteria can align themselves with respect to the Earth's magnetic field because they form magnetite (Fe_3O_4) or greigite crystals (Fe_3S_4) in special membrane vesicles, magnetosomes, in their cells that behave like compasses. Although it has been thought that their ability to sense the Earth's magnetic field enables the cells to seek their preferred habitat, which is a partially reduced environment (Blakemore, 1982; DeLong et al., 1993), this interpretation appears to be too simplistic (Simmons et al., 2006) (see also Chapter 16).

2.3 SUMMARY

The surface of the Earth includes the lithosphere, hydrosphere, and atmosphere; all of which are habitable by microbes to a greater or lesser extent and constitute the biosphere of the Earth.

The structure of the Earth can be separated into the core, the mantle, and the crust. Of these, only the upper part of the crust is habitable by living organisms. The crust is not a continuous solid layer over the mantle but consists of a number of crustal plates afloat on the mantle, or more specifically on the asthenosphere of the mantle. Some of the plates lie entirely under the oceans. Others carry parts of a continent and an ocean. Oceanic plates are growing along midocean spreading centers, whereas old portions of these plates are being destroyed by subduction under or by collision with continental plates. The crustal plates are in constant, albeit slow, motion owing to the action of convection cells in the underlying mantle. This plate motion accounts for continental drift.

REFERENCES

Amelin Y. 2005. A tale of early Earth told in zircons. *Science* 310:1914–1915.
Anderson DL. 1992. The earth's interior. In: Brown CG, Hawksworth CJ, Wilson RCL, eds. *Understanding of the Earth*. Cambridge, U.K.: Cambridge University Press, pp. 44–66.

Bambach RK, Scotese CR, Ziegler AF. 1980. Before Pangaea: The geography of the Paleozoic world. *Am Sci* 68:26–38.

Blakemore RP. 1982. Magnetotactic bacteria. *Annu Rev Microbiol* 36:217–238.

Bonatti E, Hamlyn PR. 1978. Mantle uplifted block in western Indian Ocean. *Science* 201:249–251.

Brock TD. 1974. *Biology of Microorganisms*. 2nd ed. Englewood Cliffs, NJ: Prentice Hall.

Cavanaugh CM, Levering PR, Maki JS, Mitchell R, Lidstrom ME. 1987. Symbiosis of methylotrophic bacteria and deep-sea mussels. *Nature* (London) 325:346–348.

Childress JJ, Fischer CR, Brooks JM, Kennecutt MC II, Bidigare R, Anderson AE. 1986. A methanotrophic marine molluscan (Bivalvia, Mytilidae) symbiosis: Mussels fueled by gas. *Science* 233:1306–1308.

Colwell FS, Onstott TC, Delwiche ME, Chandler D, Fredrickson JK, Yao Q-J, McKinley JP, Boone DR, Griffiths R, Phelps TJ, Ringelberg D, White DC, LaFreniere L, Balkwill D, Lehman RM, Konisky J, Long PE. 1997. Microorganisms from deep, high temperature sandstones: Constraints on microbial colonization. *FEMS Microbiol Rev* 20:425–435.

Cragg BA, Parkes RJ, Fry JC, Weightman AJ, Rochelle PA, Maxwell JR. 1996. Bacterial populations and processes in sediments containing gas hydrates (OPD Leg 146: Cascadia Margin). *Earth Planet Sci Lett* 139:497–507.

DeLong EF, Frankel RB, Bazylinski DA. 1993. Multiple evolutionary origin of magnetotaxis. *Science* 259:803–806.

Dietz RS, Holden JC. 1970. Reconstruction of Pangaea: Breakup and dispersion of continents, Permian to present. *J Geophys Res* 75:4939–4956.

Dimmick RL, Wolochow H, Chatigny MA. 1979. Evidence that bacteria can form new cells in air-borne particles. *Appl Environ Microbiol* 37:924–927.

Fooden J. 1972. Breakup of Pangaea and isolation of relict mammals in Australia, South America, and Madagascar. *Science* 175:894–898.

Fredrickson JK, Onstott TC. 1996. Microbes deep inside the Earth. *Sci Am* 275:68–83.

Furnes H, Staudigel H. 1999. Biological mediation in ocean crust alteration: How deep is the deep biosphere? *Earth Planet Sci Lett* 166:97–103.

Ghiorse WC, Wilson JT. 1988. Microbial ecology of the terrestrial subsurface. *Adv Appl Microbiol* 33:107–171.

Glatzmaier GA, Roberts PH. 1996. Rotation and magnetism of Earth's inner core. *Science* 274:1887–1891.

Gubbins D, Bloxham J. 1987. Morphology of the geomagnetic field and implications for the geodynamo. *Nature* (London) 325:509–511.

Gurnis M. 1992. Rapid continental subsidence following initiation and evolution of subduction. *Science* 255:1556–1558.

Haight RD, Morita RY. 1962. Interaction between the parameters of hydrostatic pressure and temperature on aspartase of *Escherichia coli*. *J Bacteriol* 83:112–120.

Harrison TM, Blichert-Toft J, Müller W, Albarede F, Holden P, Mojzsis SJ. 2005. Heterogeneous Hadean hafnium: Evidence of continental crust at 4.4 to 4.5 Ga. *Science* 310:1947–1050.

Hoffman PF. 1991. Did the breakup of Laurentia turn Gondwanaland inside out? *Science* 252:1409–1412.

Imshenetsky AA, Lysenko SV, Kazatov GA. 1978. Upper boundary of the biosphere. *Appl Environ Microbiol* 35:1–5.

Jannasch HW. 1983. Microbial processes at deep sea hydrothermal vents. In: Rona PA, Bostrom K, Laubier L, Smith KL Jr., eds. *Hydrothermal Processes at Sea Floor Spreading Centers*. New York: Plenum Press, pp. 677–710.

Kashefi K, Lovley DR. 2003. Extending the upper temperature limit for life. *Science* 301:934.

Kay M. 1955. Sediments and subsidence through time. In: Poldervaart A, ed. *Crust of the Earth: A Symposium*. Spec Pap 62A. New York: Geological Society of America.

Keary P, ed. 1993. *The Encyclopedia of the Solid Earth Sciences*. Oxford, U.K.: Blackwell, p. 472.

Kerr RA. 1995. Earth's surface may move itself. *Science* 269:1214–1215.

Kroener A, Layer PW. 1992. Crust formation and plate motion in the Early Archean. *Science* 256:1405–1411.

Kuhn LD, Suess E, Moore JC, Carson B, Lewis BT, Ritger SD, Kadko DC, Thornburg TM, Ebley RW, Rugh WD, Maasoth GJ, Lagseth MG, Cochrane GR, Scamman RL. 1986. Oregon subduction zone: Venting, fauna, and carbonates. *Science* 231:561–566.

Kuznetsov SI, Ivanov MV, Lyalikova NN. 1963. *Introduction to Geological Microbiology* (Engl. transl.). New York: McGraw-Hill.

Lysenko SV. 1979. Microorganisms in the upper atmospheric layers. *Mikrobiologiya* 48:1066–1074 (Engl. transl. 871–877).

Madon M. 1992. Mantle. In: Nierenberg WA, ed. *Encyclopedia of Earth System Science, Vol. 3*. San Diego, CA: Academic Press, pp. 85–99.

Matthews SW. 1973. This changing Earth. *Natl Geogr* 143:1–37.

McKenna MC. 1972. Possible biological consequences of plate tectonics. *BioScience* 22:519–525.

Mercy E. 1972. Mantle geochemistry. In: Fairbridge RW, ed. *The Encyclopedia of Geochemistry and Environmental Sciences. Encycl Earth Sci Ser, Vol. IVA*. New York: Van Nostrand Reinhold, pp. 677–683.

Morita RY, Haight RD. 1962. Malic dehydrogenase activity at 101°C under hydrostatic pressure. *J Bacteriol* 83:1341–1346.

National Geographic Society. 1995. The Earth's fractured surface (map supplement). *Natl Geogr* 187(4).

National Geographic Society. 1998. Millennium in maps. Physical Earth (map supplement). *Natl Geogr* 193(5) (May 1998 issue).

Nierenberg WA. 1978. The deep sea drilling project after 10 years. *Am Sci* 66:20–29.

Palmer AR. 1974. Search for the Cambrian world. *Am Sci* 62:216–224.

Parkes RJ, Cragg BA, Bale SJ, Gettliff JM, Goodman K, Rochelle PA, Fry JC, Weightman AJ, Harvey SM. 1994. Deep bacterial biosphere in Pacific Ocean sediments. *Nature* (London) 371:410–413.

Paul CK, Hecker B, Commeau R, Freeman-Lynde RP, Neumann C, Corso WP, Golubic S, Hook JE, Sikes E, Curray J. 1984. Biological communities at the Florida Escarpment resemble hydrothermal vent taxa. *Science* 226:965–967.

Pedersen K. 1993. The deep subterranean biosphere. *Earth Sci Rev* 34:243–260.

Pokrovskiy VA. 1961. On the lower boundary of the biosphere in the European part of the USSR, on the basis of regional geothermal investigations. In: *The Geological Activity of Microorganisms. A Symposium*. Trudy In-ta Mikrobiol AN SSR, No. 9 (as cited in Kuznetsov et al., 1963).

Raven PH, Axelrod DI. 1972. Plate tectonics and Australian paleogeography. *Science* 176:1379–1386.

Ritter JRR. 1999. Rising through Earth's mantle. *Science* 286:1865–1866.

Ronov AB, Yaroshevsky AA. 1972. Earth's crust and geochemistry. In: Fairbridge RW, ed. *The Encyclopedia of Geochemistry and Environmental Sciences. Encycl Earth Sci Ser, Vol. IVA*. New York: Van Nostrand Reinhold, pp. 243–254.

Simmons SL, Bazylinski DA, Edwards KJ. 2006. South-seeking magnetotactic bacteria in the northern hemisphere. *Science* 311:371–374.

Skinner BJ, Porter SC, Botkin DB. 1999. *The Blue Planet. An Introduction to Earth Systems Science*. 2nd ed. New York: Wiley.

Smith AG. 1992. Plate tectonics and continental drift. In: Brown G, Hawkesworth C, Wilson C, eds. *Understanding the Earth*. Cambridge, U.K.: Cambridge University Press, pp. 187–203.

Straat PA, Woodrow H, Dimmick RL, Chatigny MA. 1977. Evidence for incorporation of thymidine into deoxyribonucleic acid in air-borne bacterial cells. *Appl Environ Microbiol* 34:292–296.

Strahler N. 1976. *Principles of Physical Geology*. New York: Harper & Row.

Su W-j, Dziewonski AM, Jeanloz R. 1996. Planet within a planet: Rotation of the inner core of the Earth. *Science* 274:1883–1887.

Szewzky U, Szewzky R, Stenström T-A. 1994. Thermophilic, anaerobic bacteria isolated from a deep borehole in granite in Sweden. *Proc Natl Acad Sci USA* 91:1810–1813.

Torsvik T, Furness H, Muehlenbachs K, Torseth IH, Tumyr O. 1998. Evidence for microbial activity at the glass-alteration interface in oceanic basalts. *Earth Planet Sci Lett* 162:165–176.

Tunnicliffe V. 1992. Hydrothermal-vent communities. *Am Sci* 80:336–349.

Van Andel TH. 1992. Seafloor spreading and plate tectonics. In: Brown G, Hawkesworth C, Wilson C, eds. *Understanding the Earth*. Cambridge, U.K.: Cambridge University Press, pp. 167–186.

Van Waasbergen LG, Balkwill DL, Crocker FH, Bjornstad BN, Miller RV. 2000. Genetic diversity among *Arthrobacter* species collected across a heterogeneous series of deep-subsurface sediments as determined on the basis of 16S rRNA and *recA* gene sequences. *Appl Environ Microbiol* 66:3454–3463.

Vine FJ. 1970. Spreading of the ocean floor: New evidence. *Science* 154:1405–1415.

Watson EM, Harrison TM. 2005. Zircon thermometer reveals minimum melting conditions on earliest Earth. *Science* 308:841–844.

Wellsbury P, Mather I, Parkes RJ. 2002. Geomicrobiology of deep, low organic carbon sediments in the Woodlark Basin, Pacific Ocean. *FEMS Microbiol Ecol* 42:59–70.

Wilde SA, Valley JW, Peck WH, Graham CM. 2001. Evidence from detrital zircons for the existence of continental crust and oceans on the Earth 4.4 Gyr ago. *Nature* 409:175–178.

Wood BJ. 1997. Hydrogen: An important constituent of the core? *Science* 278:1727.

Wysession M. 1995. The inner workings of the Earth. *Am Sci* 83:134–147.

3 Origin of Life and Its Early History

3.1 BEGINNINGS

The Earth is thought to be ~4.54 × 10^9 years old (~4.6 eons) (Jacobsen, 2003). One accepted view holds that it was derived from an accretion disk that resulted from gravitational collapse of interstellar matter. A major portion of the matter condensed to form the Sun, a star. Other components in the disk subsequently accreted to form planetesimals of various sizes. These in turn accreted to form our Earth and the other three inner planets of our solar system, namely, Mercury, Venus, and Mars. All four of these planets are rocky. As accretion of the Earth proceeded, its internal temperature could have risen sufficiently to result ultimately in separation of silicates and iron, leading to a differentiation into mantle and core. Alternatively, and more likely, a primordial rocky core could have been displaced by a liquid iron shell that surrounded it. Displacement of the rocky core would have been made possible if it fragmented as a result of nonhydrostatic pressures, causing the inner core to become surrounded by a hot, well-mixed mantle or rock material in a *catastrophic process*. Whichever process actually took place, much heat must have been released during this formational process, resulting in outgassing from the mantle to form a primordial atmosphere and, possibly, hydrosphere. It has been suggested recently that bombardment of the early Earth by giant comets that consisted of water ice and cosmic dust introduced much of the water on the Earth's surface (see, for instance, Delsemme, 2001; Broad, 1997; Robert, 2001). All of this is thought to have occurred in a span of ~10^8 years. Recent evidence suggests the presence of liquid water at the Earth's surface as long ago as 4.3 eons before the present (BP) (Mojzsis et al., 2001).

As the planet cooled, segregation of the mantle components is thought to have occurred and a thin crust to have formed by 4.0–3.8 eons ago. Accretion by meteoritic (bolide) bombardment is believed to have become insignificant by this time. Results from very recent geophysical investigations involving zircon thermometry suggest that the Earth developed a crust as early as 4.35 eons ago and that the process of plate tectonics originated in less than 100 million years (Myr) thereafter (Watson and Harrison, 2006; Wilde et al., 2001). Previous estimates of the origin of crustal plates ranged from 3.8 to 2.7 eons ago. Protocontinents may have emerged at this time to be subsequently followed by the development of true continents. (For earlier views on the details about these early steps in the formation of the Earth, see Stevenson, 1983; Ernst, 1983; Taylor, 1992.) How and when did life originate on this newly formed Earth?

3.1.1 ORIGIN OF LIFE ON EARTH: PANSPERMIA

According to the *panspermia* hypothesis, life arrived on the planet as one or more kinds of spores from another world. This view finds some support in laboratory studies published by Weber and Greenberg (1985). Their studies employed spores of *Bacillus subtilis*, a common soil bacterium, enveloped in a mantle of 0.5 μm thickness or greater derived from equal parts of H_2O, CH_4, NH_3, and CO (presumed interstellar conditions). The mantle shielded the spores from short ultraviolet (UV) radiation (100–200 μm wavelength) in ultrahigh vacuum (<1 × 10^6 torr) at 10 K, but not from long UV radiation (200–300 μm). From experimentally determined survival rates of the spores, the investigators calculated that if spores were enveloped in a mantle of 0.9 μm thickness having a refractive index of 0.5, which would protect them from short- and long-wavelength UV radiation, they could survive in sufficient numbers over a period of 4.5–45 Myr in outer space to allow them to travel from one solar system to another. Spores could have entered outer space in high-speed ejecta as a result of collisions between a life-bearing planet and a meteorite or comet (Weber and Greenberg, 1985).

Instead of individual spores coated in a mantle of H_2O, CH_4, NH_3, and CO arriving on the Earth's surface, it is possible that spores were carried inside ejecta of rock fragments generated by a meteorite impact on another planet that harbored life (Cohen, 1995; Nicholson et al., 2000; Nisbet and Sleep, 2001). As shown in other chapters of this book (e.g., Chapter 9), microbial life is known to exist inside some rocks on the Earth, and thus the idea of viable spores inside ejecta of rock fragments is not preposterous. If such rock fragments are large enough, shock-induced heating and pressure through meteorite impact and the acceleration that an ejected rock fragment would undergo immediately after meteorite impact could be survived by bacterial spores inside the rock fragment (for more details see Nicholson et al., 2000). Enclosure in a protective film or in a salt crystal is thought to enable spores to survive the dehydrating effect of high vacuum of space (see Weber and Greenberg, 1985; Nicholson et al., 2000). Enclosure in a rock fragment is thought to protect spores sufficiently not only from UV radiation but also from cosmic ionizing radiation to survive interplanetary travel (Nicholson et al., 2000; Fajardo-Cavazos and Nicholson, 2006). Furthermore, spores in a large rock fragment should be able to survive entry into and penetration of the Earth's atmosphere and subsequent impact on the Earth. Breakup of the entering rock fragment due to aerodynamic drag in the lower atmosphere would ensure scattering of the inoculum at the Earth's surface (see Nicholson et al., 2000 for more detail).

Despite the possibility that life on Earth could have originated elsewhere in the universe, a more widely held view is that life began *de novo* on Earth.

3.1.2 ORIGIN OF LIFE ON EARTH: *DE NOVO* APPEARANCE

For life to have originated *de novo* on Earth, the existence of a primordial nonoxidizing atmosphere was of primary importance. There is still no common agreement as to whether Earth's primordial atmosphere was reducing or nonreducing. Its constituents may have included H_2O, H_2, CO_2, CO, CH_4, N_2, and NH_3 (see Table 4.3 in Chang et al., 1983), and HCN (Chang et al., 1983). The exact composition of Earth's early atmosphere will have changed as time progressed. Photochemical reactions and reactions driven by electric discharge (lightening) in the atmosphere, interaction of some gases with mineral constituents at high temperature, and escape of the lightest gases (e.g., hydrogen) into space (Chang et al., 1983; Schopf et al., 1983) could be the causes of this change. Two opposing views have been expressed on how life may have arisen *de novo* on Earth, the *organic soup theory* and the *surface metabolism theory* (Bada, 2004).

3.1.3 LIFE FROM ABIOTICALLY FORMED ORGANIC MOLECULES
IN AQUEOUS SOLUTION (ORGANIC SOUP THEORY)

An older view, and one that is still much favored, is that life arose in a dilute *aqueous, organic soup* (*broth*) that covered the surface of the planet. This view arose from the proposals of Haldane (1929) and Oparin (1938) (see also Nisbet and Sleep, 2001; Bada and Lazcano, 2003). According to this view, the biologically important organic molecules in the soup were synthesized by abiotic chemical interactions among some of the atmospheric gases, driven by heat, electric discharge, and light energy (see, for instance, discussion by Chang et al., 1983). If, as Bada et al. (1994) have theorized, the surface of the early Earth was frozen because the sun was less luminous than it was to become later, bolide impacts could have caused episodic melting, during which time the abiotic reactions took place. Alternatively, it is possible that few or none of the early organic molecules in the organic soup were formed on Earth, but were mostly or entirely introduced on the Earth's surface by collision with giant comets. Whatever the origin of these molecules, special polymeric molecules that had an ability to self-reproduce (the beginning of true life) arose abiotically at the expense of certain organic molecules (building blocks) that continued to be abiotically synthesized or introduced on the Earth by comet bombardment. Clays could have played an important role as catalysts and templates in the assembly of the polymeric molecules (Cairns-Smith and Hartman, 1986). Ribonucleic acid (RNA) may have been the most important original polymeric molecule (Gilbert, 1986; Joyce, 1991) that was able to self-assemble autocatalytically from abiotically formed nucleotides, according to the findings of Cech (1986), Doudna and Szostak (1989), and others. As this self-reproducing RNA evolved, it acquired new functions through mutations and recombinations, with the result that an *RNA world* emerged. In time, a form of RNA (template RNA) arose that assumed a direct role in the assembly of proteins from constituent amino acids. Many of the proteins were enzymes (biocatalysts), and among these proteins were some that assumed a catalytic role in RNA synthesis. The protein catalysts were more efficient than RNA catalysts (Gilbert, 1986). Still later, deoxyribonucleic acid (DNA), which may have arisen independently of RNA, acquired information stored in RNA related to protein structure and resultant function by a process of *reverse transcription*, a process in which information stored in RNA was transcribed into DNA (Gilbert, 1986). This speculative scenario has been proposed as a result of studies in the past two to three decades in which some RNAs were discovered in living cells that can modify themselves by self-splicing through catalysis of phosphoester cleavage and phosphoester transfer reactions (*ribozyme* activity) (Kruger et al., 1982; Guerrier-Takada et al., 1983; Cech, 1986; Doudna and Szostak, 1989).

The ability of certain RNAs to transform themselves catalytically is not unique to them. Some proteins are also known to catalyze their own transformation. Thus in considering the origin of life on Earth, it cannot be ruled out that proteins with self-reproducing properties arose spontaneously from abiotically formed amino acids (Doebler, 2000). Among these proteins may have been some that were able to catalyze polymerization of abiotically formed building blocks of RNA, the ribonucleotides, into RNAs. Some of these RNAs may subsequently have developed an ability to serve as templates in protein synthesis, making synthesis of specific proteins more orderly. Other RNAs may have evolved into reactants (transfer RNAs) in the protein assembly reactions in which amino acids are linked to each other in a specific sequence by peptide bonds, making the polymerization more efficient. As template RNA became more diverse through mutation and recombination, the diversity of catalytic proteins increased. This resulted in controlled accelerated synthesis of the building blocks (amino acids, fatty acids, sugars, nucleotides, etc.) from which vital polymers (proteins, lipids, polysaccharides, nucleic acids, etc.) could be synthesized by other newly evolved catalytic proteins. Enzyme-catalyzed synthesis was much more efficient than abiotic synthesis.

We may assume that to optimize the various biochemical processes that had become interdependent or had a potential for it, they became encapsulated in a structure we now recognize as a cell. The encapsulation is thought to have involved enclosure in a lipid membrane vesicle, whose interior provided an environment in which vital syntheses could proceed at optimal rates. Whether the first membranes were like the bilayered lipid membranes of cells today remains unknown but seems likely. A model for a primitive form of encapsulation may be a present-day observation of enzyme-catalyzed RNA synthesis from nucleotides in artificially formed lipid bilayer membrane vesicles of dimyristoyl phosphatidylcholine whose interior contained a template-independent polymerase protein. Adenosine diphosphate substrate penetrated such vesicles readily from the exterior solution and was transformed into long-chain RNA polymers in the vesicles with the help of the template-independent RNA polymerase (Chakrabarti et al., 1994).

As the primitive cells evolved, special proteins (transport proteins) became introduced into their membranes. These proteins exerted positive or negative control over the passage of specific substances into and out of a cell. In time, the membrane of some cells also acquired an energy-transducing system, the electron transport or respiratory chain involving electron carriers and enzymes, which made possible the use of externally available terminal electron acceptors such as O_2, Fe^{3+}, and CO_2 that made metabolic energy conservation more efficient than strictly intracellular processes that were independent of externally supplied terminal electron acceptors (fermentation).

According to the *organic soup* scenario, the first primitive cells that arose in the evolution of life were *heterotrophs*, which depended on abiotically formed organic building blocks in the organic soup. As time went on, supply of abiotically formed organic molecules must have become progressively more limiting. This happened because of changes in environmental conditions at the Earth's surface. Conditions for abiotic synthesis became less and less favorable but the demand for building blocks increased exponentially. The emergence of *autotrophs*, which had acquired an ability to form their own organic building blocks from inorganic constituents in their surrounding environment by using chemical or radiant energy as the driving force for these reactions, made the heterotrophs independent of the supply of abiotically formed building blocks. They were now able to feed on secretions of excess organic synthate formed by the autotrophs, or on their dead remains.

3.1.4   SURFACE METABOLISM THEORY

An alternative to the organic soup theory of how life may have originated on the Earth is the more recently proposed *surface metabolism theory* formulated by Wächtershäuser (1988). According to it, building blocks were synthesized and polymerized starting with key inorganic constituents (carbon dioxide, phosphate, and ammonia) on the surface of minerals with a positive (anodic) surface charge (Wächtershäuser favors iron pyrite, FeS_2). It is axiomatic for Wächtershäuser (1988) that polymerizations of surface-bound molecular building blocks are thermodynamically favorable, whereas polymerizations of the same molecular building blocks in solution in an organic soup scenario are thermodynamically unfavorable. In the latter case, the water in which the molecular building blocks and the polymers formed from them are dissolved favors hydrolytic cleavage of the polymers. In Wächtershäuser's surface metabolism scenario, building-block molecules were synthesized autocatalytically on a pyrite surface and, because of their ability to self-replicate, constituted the first life forms, which were two-dimensional (*surface metabolists* in Wächtershäuser's terminology). With the emergence of synthesis of isoprenoid lipids, lipid membranes with amphoteric properties formed, which could detach from the mineral surface to cover surface metabolists to form half-cells. In time, these membranes completely enclosed surface metabolists, which thus became the first cells. The cells featured a membrane-enclosed cytosol in which initially the vital chemistry still occurred on the surface of a mineral grain. However, with the passage of time and the appearance of some critical molecules, the vital chemistry became progressively independent of the mineral grain and assumed a distinct existence in the aqueous phase of the cytosol. In the original cells, the mineral grain could have consisted of pyrite (FeS_2), which may have been the product of early energy and reducing power generation according to Wächtershäuser (1988) (see Section 3.2.2). The heteropolymers DNA and RNA, which eventually became key components of the genetic apparatus and assumed firm control of the cell's metabolic behavior and its perpetuation, originated independent of other surface metabolism processes in the precellular stage and did not exert control over them. The evolution of the genetic apparatus after the cellular stage had been attained involved, among other processes, the encoding in DNA via RNA of structural information of specific proteins and the development of a mechanism for deciphering this information to enable the synthesis of proteins, most of which had the ability to serve as enzymes. After cellular metabolism became independent of the mineral grain, these enzymes took over as catalysts of the metabolic processes in the cytosol. The expression of the genetic determinants of the enzymes in the DNA regulated their formation and function as well as the timing of their appearance in the cell.

The appearance of enzymes in the cytosol made possible for the first time the utilization of accumulated, surface-detached organic substances in the cytosol, a process called salvaging action by Wächtershäuser (1988). Surface metabolists had been completely unable to use these substances.

In contrast to the organic soup theory, the surface metabolist theory proposes that the first life in the attached and detached states was *autotrophic*, that is, it depended on CO_2, CO, NH_3, H_2S, and H_2O to form organic molecules. Driving energy and reducing power for converting inorganic into organic carbon in autotrophic metabolism may have come from an interaction of ferrous iron and hydrogen sulfide (Wächtershäuser, 1988).

$$Fe^{2+} + 2H_2S \rightarrow FeS_2 + 4H^+ + 2e \quad (\Delta G°, -2.62 \text{ kcal or } -11 \text{ kJ}) \tag{3.1}$$

Both ferrous iron and H_2S should have been plentiful on the primitive Earth. The driving energy and reducing power could also have come from iron pyrite (FeS_2) and H_2 formation in a reaction between iron monosulfide and hydrogen sulfide under strictly anaerobic conditions. Such a reaction was experimentally demonstrated by Drobner et al. (1990):

$$FeS + H_2S \rightarrow FeS_2 + H_2 \quad (\Delta G°, -6.26 \text{ kcal or } -26.17 \text{ kJ}) \tag{3.2}$$

H_2 could then serve as an energy source and a reductant in further metabolism.

Evolution of a membrane-bound respiratory chain can be assumed to have followed the formation of the cell membrane. According to Wächtershäuser (1988), the respiratory chain, once it had arisen, liberated organisms that had developed it from having to rely on reactions involving ferrous iron and H_2S as a source of energy and reducing power by enabling them to use reactions like those in which H_2 reduces elemental sulfur (S^0) or sulfate to H_2S (see Chapter 19). The surface metabolist theory also proposes that substrate-level phosphorylation (see Chapter 6) originated with autotrophic catabolism as an alternative means of conserving energy.

Heterotrophy, in which energy needed by the cell is generated in the oxidation of reduced carbon compounds, is believed, by Wächtershäuser (1988), to have evolved from autotrophic catabolism in some cell lines. This required a transport mechanism across the cell membrane for importing dissolved organic molecules from the environment surrounding the cells. The first prokaryotic anoxygenic photosynthesizers probably appeared some time before the first heterotrophs appeared. Thus in the surface metabolism scenario, autotrophy preceded heterotrophy, and anaerobic chemosynthetic autotrophy preceded anoxygenic photosynthetic autotrophy.

3.1.5 ORIGIN OF LIFE THROUGH IRON MONOSULFIDE BUBBLES IN HADEAN OCEAN AT THE INTERFACE OF SULFIDE-BEARING HYDROTHERMAL SOLUTION AND IRON-BEARING OCEAN WATER

In an even newer proposal by Russell and Hall (1997), bubbles coated by iron monosulfide are postulated to have formed on the surface of sulfide mounds resulting from hydrothermal seepage on the ocean floor. More specifically, they are thought to have been formed at the solution interface that existed between hot (~150°C), extremely reduced, alkaline, bisulfide-bearing solution issuing from these mounds and warm (~90°C), iron-bearing, acidic, Hadean ocean water ~4.2 eons ago. The iron monosulfide coating of the bubbles included some nickel and acted like a semipermeable membrane. Its catalytic property depended on the redox potential and pH gradient (acid outside) across it with a ΔE of ~300 mV. Organic anions, formed by the membrane from reactants in the seawater, accumulated inside the bubbles and generated osmotic pressure that kept the bubbles from collapsing. Indeed, continuing solute accumulation inside a bubble could have led to its budding and splitting in two. The simple organic molecules inside the bubbles eventually polymerized with the help of pyrophosphate hydrolysis. The pyrophosphate resulted from a primitive form of chemiosmosis (see Chapter 6). Eventually the iron monosulfide membrane was replaced by an organic (phospholipid) membrane.

3.2 EVOLUTION OF LIFE THROUGH THE PRECAMBRIAN: BIOLOGICAL AND BIOCHEMICAL BENCHMARKS

The geologic time scale since the origin of the Earth is divided into the Precambrian, which extends from ~4.5 to 0.54 eons or billion years BP and the Phanerozoic, which extends from 0.54 eons to the present. The Precambrian may be divided into two *eras*: the Archean (4.5–2.5 eons BP) and the Proterozoic (2.5–0.57 eons BP). The Archean may be further subdivided into three *periods*: the Hadean (4.5–3.9 eons BP), the Early Archean (3.9–2.9 eons BP), and the Late Archean (2.9–2.5 eons BP). The Proterozoic may be subdivided into three *periods*: the Early Proterozoic (2.5–1.6 eons BP), the Middle Proterozoic (1.6–0.9 eons BP), and the Late Proterozoic (0.9–0.57 eons BP). The complete geologic time scale is summarized in Table 3.1.

Although until very recently it was believed that during the Hadean era no solid rock existed at the Earth's surface, recent discoveries suggest that this assumption was incorrect. The Earth seems to have had a crust as early as 4.35–4.4 eons BP (see Chapter 2). This finding affects speculation about when life may have first appeared on Earth.

Recent thinking based on some fossil finds and recent geochemical evidence is that the first life, that is, the first living entities, whatever their form, could have originated as early as 3.8 eons ago or earlier (e.g., 4 eons ago) (Schopf et al., 1983; Mojzsis et al., 1996; Holland, 1997). Remnants of a *stromatolite* (in this instance a fossilized mat of filamentous microorganisms) have been found in chert of the Warrawoona Group in the Pilbara Block of Western Australia (Figure 3.1). Its age has been determined to be ~3.5 billion years (Lowe, 1980; Walter et al., 1980; see also a critique by Brasier et al., 2002). Slightly younger microfossils (3.3–3.5 billion years old) having a recognizable cell type that resembles cyanobacteria have been found in the Early Archean Apex Basalt and Tower

TABLE 3.1
Geologic Time Scale

		Years Before the Present
Eon	Precambrian	
Era	Archean	$4.5–2.5 \times 10^9$
Period	Hadean	$4.5–3.9 \times 10^9$
	Early Archean	$3.9–2.9 \times 10^9$
	Late Archean	$2.9–2.5 \times 10^9$
Era	Proterozoic	$2.5–0.57 \times 10^9$
Period	Early Proterozoic	$2.5–1.6 \times 10^9$
	Middle Proterozoic	$1.6–0.9 \times 10^9$
	Late Proterozoic	$0.9–0.57 \times 10^9$
Eon	Phanerozoic	
Era	Paleozoic	$570–225 \times 10^6$
Period	Cambrian	$570–500 \times 10^6$
	Ordovician	$500–430 \times 10^6$
	Silurian	$430–395 \times 10^6$
	Devonian	$395–345 \times 10^6$
	Carboniferous	$345–280 \times 10^6$
	Permian	$280–225 \times 10^6$
Era	Mesozoic	$225–65 \times 10^6$
Period	Triassic	$225–190 \times 10^6$
	Jurassic	$190–136 \times 10^6$
	Cretaceous	$136–65 \times 10^6$
Era	Cenozoic	
Period	Tertiary	$65–1 \times 10^6$
	Quaternary	1×10^6 to present

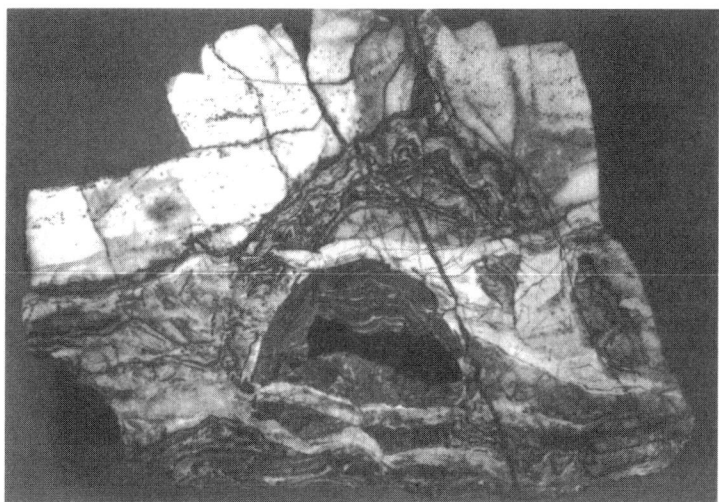

FIGURE 3.1 Fossil remnant of ancient life: domical stromatolite (×0.35) from a stratum of the 3.5 billion-year-old Warrawoona Group in the North Pole Dome region of northwestern Australia. A stromatolite is formed from fossilization of a mat of filamentous microorganisms such as cyanobacteria. (Reproduced from the frontispiece of Schopf JW, *Earth's Earliest Biosphere*, Princeton University Press, Princeton, NJ, 1983. With permission.)

Formation of the Warrawoona Group (Figure 3.2) (Schopf and Packer, 1987). Some other stromatolites of approximately similar age have been reported from limestone in the Fort Victoria greenstone belt of the Rhodesian Archean Craton within Zimbabwe in Africa (Orpen and Wilson, 1981).

The discovery of fossils of once-living organisms as old as 3.5 billion years leads to the conclusion that noncellular life and single-celled life must have preceded the emergence of stromatolites by a span of 500 Myr or more. Indeed, Mojzsis et al. (1996) found carbon isotopic evidence in carbonaceous inclusions (graphitized carbon) within grains of apatite (basic calcium phosphate) from the oldest known sediment sequences that supports the existence of biotic activity. These sediment sequences are the ~3.8 billion-year-old banded iron formation (BIF) of the Isua supracrustal belt of West Greenland and a similar ~3.85 billion-year-old sedimentary formation on the nearby Akilia Island. The isotopic signature in this case is represented by a significant enrichment of the graphitized carbon in the light, stable isotope of carbon, ^{12}C, relative to a graphite reference standard (see Chapter 6 for an explanation of isotope enrichment and its significance). The observed magnitude of the enrichment is best explained on a biological basis. An abiotic process is deemed unlikely in this instance (Mojzsis et al., 1996; Holland, 1997). The measurements were made with an ion microprobe.

3.2.1 EARLY EVOLUTION ACCORDING TO ORGANIC SOUP SCENARIO

In the organic soup scenario, the precursor molecules such as amino acids, purine and pyrimidine bases, and sugars from which life originated must have appeared in sufficient quantities by the middle Hadean as a result of abiotic synthesis. As mentioned in Section 3.1, these monomers must subsequently have polymerized into heteropolymers such as proteins and nucleic acids. The sources of the energy driving the abiotic syntheses of the monomers and their polymerizations were heat, sunlight, and electric discharge.

If we accept the ability to self-reproduce as the basic definition of life, then the appearance of the first proteins and/or nucleic acids with this ability marked the beginning of life. Clays, as already mentioned, may have played an important role in the production of self-reproducing molecules, especially proteins (Cairns-Smith and Hartman, 1986). Clays may have served as templates and catalysts in these early syntheses. If proteins were the first living molecules, they would have developed independent of abiotically produced nucleic acids at this stage. If RNAs were the first self-reproducing entities, they may have given rise to templates on which the first proteins were assembled. Regardless of whether the emergence of self-reproducing RNA preceded the emergence of proteins or whether the emergence of self-reproducing proteins preceded the emergence of RNA, semipermeable lipid vesicles are assumed to have subsequently enclosed these self-reproducing entities to better cope with adverse environmental influences and make the self-reproduction process more efficient. These primitive cells must then have evolved systems for intracellular production of the monomers from which the polymers were formed.

If proteins were the first self-reproducing molecules, they and independently evolved RNA must have given rise to a replication system in the primitive cells in which some of the RNA replaced the clays as templates in protein synthesis. DNA evolved to become the repository for structural information of the different proteins that resided in the RNA templates. Many of these proteins were catalysts needed for the synthesis of monomers and for assembly of polymers from the monomers.

If RNAs were the first living molecules, then their enclosure in primitive cells would have led, through the evolution of the template RNA, to the synthesis of proteins from abiotically formed amino acids absorbed by the primitive cells. Many of these proteins would have possessed catalytic functions (Haldane, 1929; Joyce, 1991; Miller and Orgel, 1974; Oparin, 1938; Schopf et al., 1983). DNA, which is thought to have evolved subsequently, became a more stable repository for the information in the RNA templates and, therefore, protein structure. Whatever the exact sequence of events, they probably occurred in the middle-to-late Hadean.

As the primitive cells evolved, they soon must have developed the traits that we associate with modern *prokaryotic cells*, as suggested by micropaleontological evidence. This implies that they possessed a cell envelope or wall surrounding a plasma membrane and enclosing an interior featuring a large DNA strand (repository of genetic information), nucleoprotein granules (ribosomes), and other proteins and smaller polymers and monomers. At least in the early beginnings, cell multiplication is likely to have been by binary fission, a process involving replication of all vital cell components and their equal partitioning between daughter cells. To ensure equal partitioning, a *cytoskeleton* was evolved. This cellular apparatus also assisted in maintenance of cell shape (Gitai, 2005; Møller-Jensen and Löwe, 2005).

The plasma membrane of the cell must have soon acquired mechanisms for transporting externally available organic and inorganic molecules across it. These first prokaryotes must have been anaerobic *heterotrophs* that were able to live without free

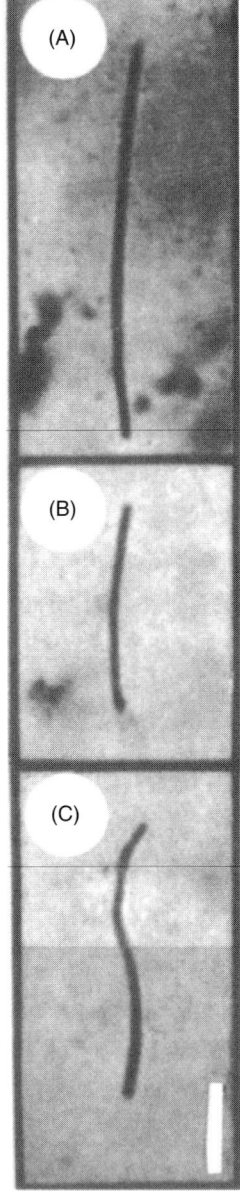

FIGURE 3.2   Rod-shaped, thread-like, juvenile forms of apparently nonseptate bacteria in petrographic thin section of stromatolitic black chert from the 3.5 billion-year-old Warrawoona Group (Pilbara Supergroup) of Western Australia. Scale mark in panel (C) is 5 μm and also applies to panels (A) and (B). (Reproduced from photo 9-4 C, D, and E in Schopf JW, *Earth's Earliest Biosphere*, Princeton University Press, Princeton, NJ, 1983. With permission.)

oxygen, because the Earth was surrounded at this time by an atmosphere devoid or nearly devoid of oxygen. In the beginning, these cells may have fed on externally available organic molecules that probably were mostly or entirely abiotically synthesized monomers. However, as the Archean progressed, an evolutionary step was required that would make these cells independent of abiotic syntheses. Such a step was needed because abiotic synthesis was rate-limited so that as cell populations increased exponentially, abiotically synthesized products could no longer meet cellular demand for them. In any case, conditions needed to sustain abiotic synthesis gradually disappeared. To sustain life, *autotrophy* emerged. Autotrophy is a process in which an organism obtains the energy it needs either from the oxidation of an inorganic compound (*chemosynthetic autotrophy* or *chemolithoautotrophy*) or from the transformation of radiant energy from sunlight into chemical energy (*photosynthetic autotrophy* or *photolithoautotrophy*). In these processes, CO_2 is converted into a form of reduced (organic) carbon. Such autotrophs not only completely satisfied their own needs for reduced carbon monomers from inorganic matter but could also feed already existing heterotrophs. They did this by excreting some of the reduced carbon monomers that they had synthesized and that were in excess of their needs. Furthermore, upon their death, their remains became food for heterotrophic scavengers.

Whether the first autotrophs were chemosynthetic or photosynthetic is still a matter of debate. One school of thought favors chemosynthetic autotrophs in the form of *methanogens* (e.g., Kasting, 2004; Kasting and Siefert, 2002), which formed methane according to the reaction

$$4H_2 + CO_2 \rightarrow CH_4 + 2H_2O \quad (\Delta G° = -33.2\,\text{kcal or} -138.8\,\text{kJ}) \tag{3.3}$$

Low Archean sulfate concentrations, resulting from volcanic SO_2 outgassing, and low rates of sulfate reduction were favorable to the evolution of methanogenesis (Habicht et al., 2002).

Methanogens exist today. They are strict anaerobes. Because a number of the extant methanogens live at relatively high temperatures ($+60$ to $+90°C$), not unlike temperatures that are likely to have prevailed on the Earth at the time of the Archean (Schopf et al., 1983), and because the large majority of them can use hydrogen, a gas thought to have been sufficiently abundant in the primordial atmosphere, as an energy source, they could have been the first autotrophs to evolve. Analyses by the techniques of molecular biology have shown that methanogens have a very ancient origin and that they should be grouped in a special domain, the Archaea (initially called Archaebacteria) (Fox et al., 1977; Woese and Fox, 1977). If they were the first autotrophs, they were, however, soon displaced by photosynthetic autotrophs as *primary producers.*

The other school of thought favors photosynthetic prokaryotes in the domain Bacteria as the first autotrophs. This notion is supported by the existence of the 3.5 billion-year-old Warrawoona stromatolite (Schopf et al., 1983; Schopf and Packer, 1987). This microfossil has been interpreted, on the basis of comparison with modern counterparts, to have been formed from a mat of cyanobacteria (once called blue-green algae). However, modern cyanobacteria are aerobes, which usually produce O_2 when photosynthesizing, whereas the primordial atmosphere, 3.5 billion years ago when this stromatolite formed, is thought to have been lacking in oxygen. Anoxygenic photosynthetic bacteria, of which modern purple and green bacteria are a counterpart, are believed to have preceded the emergence of cyanobacteria. Indeed, recent molecular analysis of photosynthesis genes suggests that purple bacteria represent the oldest photosynthetic prokaryotes and that cyanobacteria were relative latecomers evolutionarily (Xiong et al., 2000; Des Marais, 2000). Like their modern counterparts, the anaerobic photosynthesizing bacteria photosynthesized without producing oxygen. They transformed radiant energy from sunlight into chemical energy that enabled them to reduce CO_2 to organic carbon with H_2 or H_2S instead of H_2O as the reductants. Elemental sulfur or sulfate was produced if H_2S was the reductant of CO_2, for example,

$$2H_2S + CO_2 \rightarrow (CH_2O) + H_2O + 2S \tag{3.4}$$

where (CH_2O) represents reduced carbon with a ratio of C to H to O of 1:2:1. Such photosynthetic organisms would have kept the Archean atmosphere oxygen-free.

The emergence of anoxygenic photosynthesis required the appearance of chlorophyll, the light-harvesting and energy-transducing pigment (Chapter 6). A structural modification of this chlorophyll was required to enable the emergence of oxygenic phototrophs, that is, cyanobacterialike organisms. This modification enabled the substitution of H_2O for H_2 or H_2S as reductant of CO_2, leading to the introduction of O_2 into the atmosphere.

$$H_2O + CO_2 \rightarrow (CH_2O) + O_2 \tag{3.5}$$

A few modern cyanobacteria that have the capacity to carry out not only oxygenic photosynthesis but also anoxygenic photosynthesis that requires anaerobic conditions and the presence of H_2S are known. In the latter instance they photosynthesize like purple or green bacteria, using the H_2S rather than H_2O to reduce CO_2 (Cohen et al., 1975). This means that the microfossils that Schopf and Packer (1987) found in 3.3–3.5 billion-year-old chert from the Warrawoona Group in Western Australia could have been cyanobacteria that were photosynthesizing anoxygenically or were just beginning to develop an oxygenic photosynthetic system. Buick (1992) inferred from stromatolites in the Tumbiana Formation of the late Archean located in Western Australia that oxygenic photosynthesis occurred ~2.7 billion years ago.

Stromatolites became common in the Proterozoic (2.5–0.57 billion years BP) as the cyanobacteria achieved dominance as carbon-fixing and oxygen-evolving microorganisms. These structures formed as a result of the aggregation of some filamentous forms of cyanobacteria into mats that trapped siliceous and carbonaceous sediment, which in many instances contributed to their ultimate preservation by silicification and transformation into stromatolites. Environmental conditions at this time seemed to favor the mat-forming growth habit of cyanobacteria: continental emergence, development of shallow seas, and climatic and atmospheric changes resulting from oxygenic photosynthesis probably exerted selective pressure favoring this growth habit (Knoll and Awramik, 1983). Most microfossil finds representing this period have been stromatolites, possibly because they are among the more easily recognized microfossils. Unicellular microfossils would be much harder to find and identify, so one should not draw the conclusion that mat-forming cyanobacteria were necessarily the only common form of life at that time.

Based on current geologic and geochemical evidence, oxygen began to accumulate in the Earth's atmosphere beginning ~2.3–2.4 billion years ago (e.g., Kump et al., 2001; Kasting, 2001; Catling et al., 2001; Holland, 2002; Yang et al., 2002; Bekker et al., 2004; Claire, 2005), but no common agreement on the cause of this oxygen accumulation has been reached yet. The emergence of oxygenic photosynthesis has been viewed by many as the primary source of the oxygen (Schopf, 1978; Claire 2005). According to that view, oxygen that was photosynthetically evolved initially probably reacted with oxidizable inorganic matter such as iron [Fe(II)], forming iron oxides such as magnetite (Fe_3O_4) and hematite (Fe_2O_3) (see Chapter 16). But by ~2.3 billion years ago after the oxidizable inorganic matter was sufficiently depleted, free oxygen began to accumulate in the atmosphere, gradually changing it from a nonoxidizing atmosphere to a distinctly oxidizing one.

Two recent proposals oppose the idea that the rise of oxygen concentration in the Earth's atmosphere beginning ~2.3 billion years ago had a biological cause, that is, the emergence of oxygenic photosynthesis. These proposals invoke instead abiologic geophysical/geochemical processes as the primary cause. One of these proposals attributes the rise in oxygen to an abrupt mantle overturn or plume activity close to the Archean–Proterozoic transition (Kump et al., 2001), and the other to a gradual qualitative and quantitative change in emissions of volatiles from active volcanoes that by some time near 2.3 billion years ago had attained a biologically critical composition (Holland, 2002). Either proposal, however, still relies on oxygenic photosynthesis by relevant microbes to build up the O_2 concentration in the atmosphere. The chief difference between the biological and abiological proposals is that in the biological proposal evolutionary emergence of oxygenic photosynthesis is the trigger for the sudden rise in oxygen concentration in the atmosphere whereas in the abiological proposals geologic change is the trigger (selective agent) to which organisms with an oxygenic photosynthesizing potential respond.

As the concentration of oxygen in the atmosphere increased, some organisms began to evolve biochemical catalysts, and other molecules capable of electron transfer that included *cytochromes*, proteins containing iron porphyrins, could have arisen from the magnesium porphyrins of the chlorophylls of photosynthetic prokaryotes. The cytochromes were incorporated into electron transport systems in the plasma membrane and, in some instances, in the cell envelope covering the plasma membrane (e.g., the outer membrane of gram-negative bacteria), thereby enabling cells to dispose of excess reducing power to molecular oxygen or other externally available electron acceptors (respiration) instead of partially reduced organic compounds inside the cell (fermentation) (see Chapter 6). The cytochrome system made possible a reaction sequence of discrete steps (respiratory chain); in some of the steps metabolic energy that was released could be conserved in special chemical (anhydride) bonds for subsequent utilization in energy-requiring reactions (syntheses and polymerizations) (see Chapter 6).

Because the biochemical oxygen-reduction process leads to the formation of very toxic superoxide radicals (O_2^-) (Fridovich, 1977), the oxygen-utilizing organisms had to evolve a special protective system against them. Such a system included superoxide dismutase and catalase, metalloenzymes that together catalyze the reduction of superoxide to water and oxygen. Superoxide dismutase catalyzes the reaction

$$2O_2^- + 2H^+ \rightarrow H_2O_2 + O_2 \qquad (3.6)$$

and catalase catalyzes the reaction

$$2H_2O_2 \rightarrow 2H_2O + O_2 \qquad (3.7)$$

Peroxidase may replace catalase as the enzyme for disposing of the toxic hydrogen peroxide,

$$2RH + H_2O_2 \rightarrow 2H_2O + 2R \qquad (3.8)$$

where RH represents an oxidizable organic molecule.

Schopf et al. (1983) suggested that the first oxygen-utilizing prokaryotes were *amphiaerobic*, that is, they retained the ability to live anaerobically although they had acquired the ability to metabolize aerobically. (Present-day amphiaerobes are called *facultative* organisms.) From them, according to Schopf et al. (1983), obligate aerobes evolved more than 2 billion years ago. Towe (1990), however, has proposed that aerobes could have first appeared in the Early Archean. In any case, aerobes probably became well established only ~2 billion years ago. The evolutionary sequence leading from anaerobes to aerobes was probably complex because present-day facultative aerobes include some that have a respiratory system that can use nitrate, ferric iron, manganese(IV) oxides, and some other oxidized inorganic species as terminal electron acceptors, some only in the absence of atmospheric oxygen, others in the presence or absence of oxygen. Respiratory processes using such oxidized forms of inorganic ions could not have existed before the atmosphere became oxidizing because, except for ferric iron, most of these electron acceptors would not have occurred in sufficient quantities in a reducing atmosphere.

Amphiaerobes, which evolved in the Late Archean to Early Proterozoic, are best defined as organisms that have the ability to *respire*[*] aerobically in the presence of oxygen or to *ferment*[†] in its absence. The methanogens, which must have appeared well before the amphiaerobes, are an

---

[*] Respiration refers to a physiological process in which a substrate is oxidized. In this oxidation, electrons are removed from the substrate and transferred to an externally supplied terminal electron acceptor such as O_2, NO_3^-, Fe^{3+}, Mn(IV), SO_4^{2-}, and CO_2. During the electron transfer, some of the free energy liberated is conserved for use by energy-consuming reactions such as biosyntheses and polymerizations.

[†] Fermentation refers to disproportionation of a substrate in which a part of it is oxidized by transfer of electrons from it to the remainder of the substrate, which serves as terminal electron acceptor. Some of the free energy liberated in the process is conserved for energy-consuming reactions in the cell. In fermentation, unlike respiration, no externally supplied electron acceptor is involved. See Chapter 6 for further details.

exception. Although strict anaerobes, as a group, the methanogens, like their modern counterparts, must have respired anaerobically using CO_2 as the terminal electron acceptor (among modern methanogens, a few are found that can produce methane by fermenting acetate). The sulfate reducers are another exception. They are obligate anaerobes that respire on SO_4^{2-} (some modern sulfate reducers are known that can reduce sulfate aerobically as well but cannot grow in this mode; see Chapter 19 for details).

Some strict anaerobes probably also evolved subsequent to the appearance of oxygen in the atmosphere. These were organisms with a capacity to respire anaerobically using oxidized inorganic species as terminal electron acceptors that became available in sufficient quantity only after the appearance of O_2 in the atmosphere. Examples of such species are nitrate, sulfate, and Mn(IV) oxide. Sulfate reducers in the domain Bacteria have been viewed as having made their first appearance in the Late Archean (Ripley and Nichol, 1981). They have been important ever since their first appearance in the reductive segment of the sulfur cycle in the mesophilic temperature range (~15–40°C) (Schidlowski et al., 1983). At ambient temperature and pressure, bacterial sulfate reduction is the only process whereby sulfate can be reduced to hydrogen sulfide. A group of extremely thermophilic, archaeal sulfate reducers were isolated more recently from marine hydrothermal systems in Italy (Stetter et al., 1987), which in the laboratory grew in a temperature range of 64–92°C with an optimum near 83°C. They reduce sulfate, thiosulfate, and sulfite with H_2, formate, formamide, lactate, and pyruvate. Stetter et al. (1987) suggested that these types of bacteria may have inhabited Early Archean hydrothermal systems containing significant amounts of sulfate of magmatic origin.

The accumulation of oxygen in the atmosphere led to the buildup of an ozone (O_3) shield that screened out UV components of sunlight. The ozone results from the action of short-wavelength UV on O_2:

$$3O_2 \rightarrow 2O_3 \tag{3.9}$$

The ozone screen would have stopped any abiotic synthesis dependent on UV irradiation and at the same time would have allowed the emergence of life forms onto land surfaces of the Earth, where they were directly exposed to sunlight. This emergence would have been impossible earlier because of the lethality of UV radiation (Schopf et al., 1983).

With the appearance of oxygen-producing cyanobacteria and aerobic heterotrophs, the stage was set for extensive cellular compartmentalization of such vital processes as photosynthesis and respiration. New types of photosynthetic cells evolved, where photosynthesis was carried on in special organelles, the *chloroplasts*, and new types of respiring cells evolved in which respiration was carried on in other special organelles, the *mitochondria*. Based on molecular biological evidence, summarized below, it is now generally accepted that these organelles arose by endosymbiosis, a process in which primitive cells that were incapable of photosynthesis or respiration were invaded by cyanobacterialike organisms[*] to evolve into chloroplasts and by aerobically respiring bacteria to evolve into mitochondria. These invading organisms established a permanent symbiosis with their host (see discussion by Margulis, 1970; Dodson, 1979). The genetic apparatus in the host cell became enveloped in a double membrane by an as-yet-unknown mechanism. Eventually the relationship of some of the host cells with their endosymbionts became one of absolute interdependence, the endosymbionts having lost their capacity for an independent existence. The result was the appearance of the first *eukaryotic cells*, possibly as late as 1 billion years ago (Schopf et al., 1983), but more likely as early as 2.1 billion years ago (Han and Runnegar, 1992). The highly compartmentalized organization of these cells contrasts with the much less compartmentalized *prokaryotic cells* (Bacteria including cyanobacteria, and Archaea).

---

[*] Prochloron, which contains both chlorophyll a and b, like most chloroplasts, has been considered a possible candidate. Most other cyanobacteria contain only chlorophyll a.

Supporting evidence for an endosymbiotic origin of chloroplasts and mitochondria in eukaryotic cells is the discovery in the organelles of genetic substance—DNA—and cell particles called ribosomes, each of a type that is otherwise found only in prokaryotes. Furthermore, in the case of chloroplasts, molecular comparisons of the highly conserved 16S RNA fraction in some cyanobacterial and chloroplast ribosomes have revealed a close relatedness to bacteria (Giovannoni et al., 1988). The emergence of eukaryotic cells was evidently needed for the evolution of more complex forms of life, such as protozoa, fungi, plants, and animals.

3.2.2 EARLY EVOLUTION ACCORDING TO SURFACE METABOLIST SCENARIO

If surface metabolism as described by Wächtershäuser (1988) preceded cellular metabolism, the description of the beginning of life and its early evolution requires an extensive revision of the sequence of evolutionary steps as they are visualized in the organic soup theory scenario. Wächtershäuser's surface metabolist scenario is an autocatalytic process that does not involve enzymes or templates for the metabolist formation or reproduction. Any dissolved organic molecules synthesized abiotically with heat, electric, or radiant energy did not play a role in surface metabolism as conceived by him. Metabolist synthesis involving chemical combination of select inorganic species as well as metabolist reproduction occurred on positively charged mineral surfaces, pyrite (FeS_2) surfaces being favored by Wächtershäuser. The metabolism was *autotrophic* in that the starting molecules (reactants) were CO_2, CO, NH_3 (or N_2 after emergence of nitrogen fixation), H_2S, H_2, and H_2O. Detachment of metabolists from the positively charged mineral surface took place after they became enveloped in a lipid membrane. The membrane formed on the positively charged mineral surface before partial detachment to enable the envelopment of surface-bound metabolists. After complete metabolist envelopment, each membrane enclosed a cytosol in which enzymes and templates for regulating them evolved gradually with the development of a genetic apparatus. But until the advent of enzymes and templates in these membrane vesicles, surface metabolism was still central to continued life. As in the precellular state, it took place on a mineral grain with a positively charged surface, but now located in the cytosol. The grains were still very likely iron pyrite, according to Wächtershäuser, formed in energy metabolism involving the interaction of ferrous iron with hydrogen sulfide (see Section 3.1.4 and Equation 3.1). After the appearance of enzymes in the cytosol, cells gradually dispensed with surface metabolism on the intracellular mineral particle, probably by shifting to energy-yielding reactions that did not form the particles (see below).

In Wächtershäuser's view (1988), the cytosol and the appearance of enzymes in it made possible for the first time the transformation and degradation of intracellularly accumulated, surface-detached reaction products (salvaging) that were not needed for biosynthesis. Because the first cells were autotrophs, this represents a form of autotrophic *catabolism*. The evolution of a membrane-bound respiratory system, which followed, was a consequence of the development of the membrane. According to Wächtershäuser (1988), the respiratory chain freed organisms that developed it from the need to use the reaction of ferrous iron with H_2S as a source of energy and reducing power. They were now able to use reactions such as the one in which H_2 reduces elemental sulfur (S^0) or sulfate to hydrogen sulfide. Wächtershäuser also saw autotrophic catabolism as a point of departure for the evolution of substrate-level phosphorylation (see Chapter 6) as an alternative means of conserving energy. In his view, heterotrophy evolved from autotrophic catabolism in cells, which at that point developed a transport mechanism for importing dissolved organic molecules from the surrounding environment. He believes that this occurred at a later stage in the evolution of early cells. The first prokaryotic anoxygenic photosynthesizers probably appeared some time before the first heterotrophs appeared. Thus, in the surface metabolism scenario, autotrophy preceded heterotrophy, and anaerobic chemosynthetic autotrophy preceded anoxygenic photosynthetic autotrophy.

3.3   EVIDENCE

The scenarios for the origin and early evolution of life outlined in the previous section are at best based on educated guesses. Scenarios for the evolution of cellular life have been constructed in part on the basis of observations in the geologic record, in part on the basis of comparisons between fossilized microorganisms and their present-day counterparts, and in part on molecular biological studies. The geologic record has contributed supporting evidence in the form of microfossils and geochemical data. Morphological similarities between microfossils and certain present-day organisms as well as some geochemical data relating to the site where the fossilized organisms were found have permitted inferences to be drawn concerning likely physiological and biochemical activities of the fossilized organisms when they were alive. Molecular biological analysis of present-day organisms has permitted the construction of evolutionary trees that reflect biochemical evolution and permit an estimate of the time of first appearance of phylogenetically distinct groups of organisms using as a basis the conservation of certain genetic information over geologic periods of time (Woese, 1987; Olsen et al., 1994).

The identification of Precambrian microfossils is difficult because structures that appear to be fossilized microorganisms may actually be modern contaminants or abiotically formed structures resembling the microfossils in appearance. A true microfossil should meet the following criteria (Schopf and Walter, 1983):

1. The sedimentary rock in which the microfossil was found must be of a scientifically established age.
2. The purported fossil must be indigenous to the rock sample and not a modern contaminant.
3. The purported fossil must have been formed at the same time as the enclosing rock, that is, *syngenetically.*
4. The purported fossil must have had a true biogenic origin.

On the basis of these criteria, some previously identified microfossils have had to be rejected as bona fide and reclassified as *dubiofossils* (uncertainty about biogenicity), *nonfossils* (may be present-day microbes that invaded a particular rock *in situ*), or *contaminants* introduced during sampling. Walter (1983) and Hofmann and Schopf (1983) reclassified Precambrian fossil finds before 1983 into the aforementioned categories based on the criteria of Schopf and Walter (1983).

As already implied in referring to the origin of stromatolites, microfossils in general arose as a result of entrapment of microorganisms in siliceous sediment. This was followed by impregnation of cell structures, for instance those of cyanobacteria, with dissolved silica from mineral diagenesis. Subsequent dewatering under conditions of moderately elevated temperature and pressure resulted in precipitation of silica in the cells. This process has been reproduced in the laboratory on a time scale obviously compressed over many orders of magnitude (Oehler and Schopf, 1971). Fossilization of bacteria may also have involved the concentration of certain metallic ion species in their cell envelope and subsequent crystallization of specific sulfide, phosphate, oxide, carbonate, silicate, or other mineral from the accumulated metal (Beveridge et al., 1983; Ferris et al., 1986). Observations by Ferris et al. (1986) suggest that microfossils formed as a result of mineral precipitation were probably best preserved if they had been previously embedded in a fibrous silica matrix (*premineralization*). Growth of crystals of metal-containing minerals would otherwise have caused rupture of the cells. Microfossils are visualized by thin-sectioning sedimentary rock containing them and examining such sections, after suitable treatment, by light or electron microscopy.

Bona fide Archean microfossils are represented by the Warrawoona specimens (~3.5 billion years old) from the North Pole Dome region of the Pilbara Block in Western Australia (Figure 3.1) in which examples of *filamentous bacteria* (Figure 3.2) (see Schopf and Walter, 1983) and microfossiliferous stromatolites (Walter, 1983) have been found. Unicell-like spheroids, which are currently classified as dubiofossils, have also been seen in these specimens (see Schopf and Walter, 1983). Spheroid structures of an age similar to that of the Warrawoona specimens, classified as dubiofossils, have been found in the Onverwacht Group of the Swaziland System, eastern Transvaal, South Africa. Filamentous fossil bacteria, ~2.7 billion years old, have been identified in specimens of the Fortescue Group, Tumbiana Formation in Western Australia (see Schopf and Walter, 1983).

Bona fide Proterozoic microfossils have been identified in greater numbers than Archean microfossils. They include filamentous organisms named *Gunflintia minuta* Barghoorn in Pokegama quartzite (1.8–2.1 billion years old), coccoid, septate filamentous, and tubular unbranched and budding bacterialike microfossils in the Gunflint Formation of northwestern Ontario, Canada (1.8–2.1 billion years old); coccoid, septate filamentous, and tubular unbranched microfossils in the Tyler Formation of Gogebic County, northern Michigan, United States (1.6–2.5 billion years old); spheroid, planktonic, organic-walled microfossils in the Krivoirg Series of the Ukranian Shield; and a number of other examples from other parts of the world (Figure 3.3). Many of these microfossils were associated with stromatolites, which abounded 1.7–0.57 billion years ago in the Proterozoic (Krylov and Semikhatov, 1976). These stromatolites were formed chiefly by cyanobacteria. Their abundance in the Proterozoic is attributed to environmental conditions that favored formation of cyanobacterial mats, that is, the absence of organisms capable of grazing on such mats (Knoll and Awramik, 1983). With the evolutionary emergence of grazers in the Phanerozoic, stromatolites became rare and have remained so to the present day. Modern stromatolites are confined to very special locations.

The finding of microfossils in Precambrian sedimentary rock formations is evidence of the presence of life at that time. Because of the resemblance of some of these microfossils to present-day cyanobacteria, inferences can be drawn concerning the physiology and biochemistry of these microfossils. It can be inferred, for instance, that photosynthesis, possibly oxygenic photosynthesis, occurred at the time corresponding to the geologic age of these particular fossils. However, if independent geochemical data indicate that a reducing or at least oxygen-lacking atmosphere prevailed at that time, the fossils may represent anoxic precursors of oxygenic cyanobacteria, or they may represent the time of emergence of oxygenic cyanobacteria.

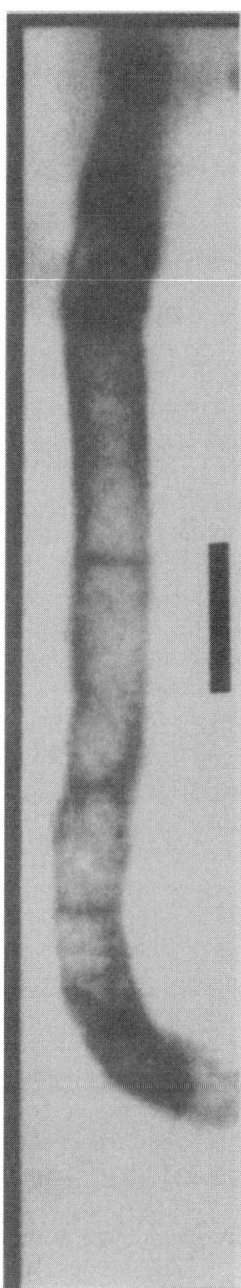

FIGURE 3.3 *Gunflintia*? sp.: a septate filament with unusually elongated cells, preserved in dark-brown organic matter from Duck Creek Dolomite, Mount Stuart area, Western Australia: ~2.02 billion years old. Scale mark on right is 10 µm. (Reproduced from photo 14-3 J in Schopf JW, *Earth's Earliest Biosphere*, Princeton University Press, Princeton, NJ, 1983. With permission.)

No microfossils of the earliest life forms have been found, nor are they ever likely to be found on the Earth because of the weathering or diagenetic processes to which sedimentary rock on our planet has been subjected from the start and to which it is continuing to be subjected. This weathering was originally a physicochemical process involving water and various reactive substances in the planet's atmosphere. With the emergence of prokaryotic life, microbes also became important agents of weathering. Their present-day weathering activities are discussed in Chapters 4 and 9.

Geochemical studies of Precambrian rocks can also tell us something about early life. For instance, measurements of stable isotope ratios of major elements important to life, namely, C, H, N, and S (Schidlowski et al., 1983) and perhaps even Fe (Beard et al., 1999; but see also Mandernack et al., 1999 and Anbar et al., 2000), can give an indication of whether a biological agent was involved in their formation or transformation. Such interpretation rests on observations that some present-day organisms can discriminate among stable isotopes of an element by metabolizing the lighter species faster than the heavier species. Thus under appropriate conditions, they attack ^{12}C more readily than ^{13}C, hydrogen more readily than deuterium, ^{14}N more readily than ^{15}N, ^{32}S more readily than ^{34}S, or ^{54}Fe more readily than ^{56}Fe. Abiotic reactions do not discriminate among stable isotopes to this extent. As a result, products of microbial isotope fractionation reactions will show an enrichment with respect to the lighter isotope. The fractionation is most noticeable in the initial stages of a reaction in a closed system or in an open system with a low rate of substrate consumption. Residual substrates will show an enrichment in the heavier isotope in a closed system (Chapter 6). Carbon isotope studies of many sediment samples of Archean age indicate that life played a dominant role in the carbon cycle as far back as 3.5 billion years ago (Schidlowski et al., 1983). Sulfur isotopic data for Early Archean sediments indicate the likely activity of photosynthetic bacteria. For instance, barites ($BaSO_4$) of this time were only slightly enriched in ^{34}S compared to sulfides from the same sequence (Schidlowski et al., 1983). They also lend support to the notion that sulfate respiration by prokaryotes may not have occurred to a significant extent before 2.7 billion years ago. Some more recent geochemical evidence indicates, however, that some sulfate reduction occurred as long ago as 3.4 billion years (Ohmoto et al., 1993) or even slightly earlier (Moizsis et al., 2003).

Organic geochemistry provides another approach to seeking clues to early life on the Earth. Organic matter trapped in sediment subjected to abiological transformation due to heat and pressure may be changed into products that are stable *in situ* over geologic time. Organic matter that underwent this kind of transformation is likely to have been in a form that was not rapidly degraded by biological means as would, for instance, carbohydrates, nucleic acids, and nucleotides and most proteins. Nevertheless, amino acids, fatty acids, porphyrins, *n*-alkanes, and isoprenoid hydrocarbons have been identified in sediments of Archean age (Kvenvolden, cited by Schopf, 1977; Hodgson and Whiteley, 1980). If the source compound of any stable organic product identified in an ancient sample of sedimentary rock is known, the latter can be used as an indicator or *biological marker* of the source compound. If the source compound such as porphyrin is a key compound in a particular process, it indicates that a process such as photosynthesis or respiration or both was occurring when the source material became trapped in sediment. *Kerogen* is an example of a stabilized substance formed from ancient organic matter. Its presence in an ancient sedimentary rock suggests the existence of life contemporaneous with the age of that rock.

Studies in molecular biology have revealed that the proportion and sequence of certain monomers in some bioheteropolymers such as ribosomal RNAs are highly conserved in various organisms, that is, they have not become significantly modified over very long times due to extremely slow mutation rates. Such conserved sequences can be used to study the degree of relatedness among different groups of organisms (see Fox et al., 1980; Woese, 1987; Olsen et al., 1994) and can also be used to estimate the geologic time at which they first appeared. Such studies have led to the conclusion that Archaea (formerly Archaebacteria or Archaeobacteria) and Bacteria (formerly called Eubacteria) diverged early in Archean times from a common prokaryotic ancestor and have evolved ever since along independent parallel lines. They also indicate that gram-positive bacteria most likely had a photosynthetic ancestry (Woese et al., 1985). The fact that certain physiological processes such as protein synthesis, energy conservation by chemiosmosis, and some biodegradative as well as biosynthetic pathways are held in common by the Bacterial and Archaean domains, although differing in some details, suggests that these pathways may have existed in a common ancestor but became modified during *divergent* evolution. *Convergent* evolution cannot, however, be ruled out in all cases.

Combining several lines of paleontological evidence can lend strong support to a model of an ancient biological process or microbe responsible for it. Summons and Powell (1986) found in a certain Canadian petroleum deposit of Silurian age (~400 Myr ago) the presence of (1) characteristic biological markers indicating an ancient presence of aromatic carotenoids from green sulfur bacteria (Chlorobiaceae) and (2) enrichment in ^{13}C in these markers relative to the saturated oils. Relating these findings to the paleoenvironmental setting of the oil deposit, the investigators deduced that microbial communities that included Chlorobiaceae must have existed in the ancient restricted sea in which the source material from which the oil derived was emplaced.

3.4 SUMMARY

The Earth is ~4.6 eons old. Primitive life probably arose *de novo*, first appearing 0.5–0.7 eons after formation of the planet. Initially the Earth was surrounded by an atmosphere that lacked oxygen or contained at most small traces and may have been reducing or nonreducing. Oxygen did not begin to accumulate in the atmosphere until oxygen-generating, photosynthetic microorganisms, the cyanobacteria, had evolved and become established, and oxygen-scavenging substances like ferrous iron had been depleted by reacting with evolved oxygen. The time at which oxygen began to accumulate is not yet precisely known. A conservative estimate is 2.3 eons ago, but it may have been earlier.

The earliest forms of cellular life were anaerobic prokaryotes. Except for cyanobacteria, aerobic prokaryotes did not evolve until free oxygen began to accumulate in the atmosphere, and eukaryotic forms did not appear until the accumulated oxygen in the atmosphere attained significant levels.

The evolutionary sequence of prokaryotes in terms of physiological types according to the organic soup scenario started with fermenting heterotrophs and progressed with the development of anaerobic photo- and chemoautotrophs, some anaerobic respirers, oxygenic photoautotrophs, aerobically respiring heterotrophs and chemoautotrophs, and other anaerobic respirers. Eukaryotes evolved by endosymbiosis involving anaerobic heterotrophic prokaryotes as host cells and oxygenically photosynthetic and aerobically respiring prokaryotes as intracellular symbionts.

According to the surface metabolist theory, the sequence of emergence of prokaryotic physiological types was autotrophic surface metabolists (precellular); semicellular surface metabolists; membrane-bound, detached chemosynthetically autotrophic primitive cells; anoxygenic photosynthetic autotrophs; oxygenic photosynthetic autotrophs, and aerobically respiring heterotrophs; and followed by the emergence of eukaryotes. Major steps in Precambrian evolution according to the organic soup scenario and the surface metabolist scenario are compared in Table 3.2.

Aspects of currently held views of how life evolved on Earth are supported by the microfossil record in geologically dated sedimentary rock, by inorganic and organic geochemical studies of Precambrian rocks, and by comparative molecular biological analysis of highly conserved biopolymers such as portions of DNA, RNA, and proteins from living cells.

TABLE 3.2
Milestones in Precambrian Evolution of Life

Event	Years Before the Present[a]
Organic soup scenario	
Origin of the Earth	4.6×10^9
First self-reproducing molecules	$4.3–4.0 \times 10^9$ (?)
First primitive heterotrophic cells	$4.0–3.8 \times 10^9$ (?)
First autotrophs (methanogens and anoxygenic photosynthesizers)	$~3.8 \times 10^9$ (?)
Warrawoona stromatolite	$~3.5 \times 10^9$
First oxygenic photosynthesizers	$3.5–3.0 \times 10^9$
First anaerobic respirers (sulfate reducers)[b]	$~2.7 \times 10^9$
Fully oxidizing atmosphere	$~2.1 \times 10^9$
First aerobic respirers[c]	$~2.0 \times 10^9$
First eukaryotic cells	$~1.4 \times 10^9$
Surface metabolism scenario[d]	
Origin of the Earth	4.6×10^9
First surface metabolists (chemoautotrophic)	$~4.3 \times 10^9$ (?)
First primitive cells (chemoautotrophic)	$~4.1 \times 10^9$ (?)
First anaerobic S and SO_4 respirers (autotrophic)	$~4.0 \times 10^9$ (?)
First autotrophs	$~3.8 \times 10^9$ (?)
First heterotrophs	$~3.5 \times 10^9$ (?)
Warrawoona stromatolite	3.5×10^9
First oxygenic photosynthesizers	$~3.5–3.0 \times 10^9$ (?)
Fully oxidizing atmosphere	$~2.1 \times 10^9$
First aerobic respirers	2.0×10^9
First eukaryotic cells	1.4×10^9

[a] Dates followed by (?) represent guesses without any paleontological or molecular evolutionary backup.

[b] Most known sulfate reducers belong to the domain Bacteria. The discovery of sulfate-reducing Archaea (e.g., *Archeoglobus fulgidus*) suggests that some bacterial sulfate reduction may have occurred earlier (e.g., see Shen et al., 2001).

[c] According to some students of early evolution, aerobic respirers could have evolved earlier than this.

[d] Sequence in the surface metabolism scenario is based on evolutionary description by Wächtershäuser (1988).

REFERENCES

Anbar AD, Roe JE, Barling, J, Nealson KH. 2000. Nonbiological fractionation of isotopes. *Science* 288:126–128.

Bada JL. 2004. How life began on Earth: A status report. *Earth Planet Sci Lett* 226:1–15.

Bada JL, Bingham C, Miller SL. 1994. Impact melting of frozen oceans on the early Earth: Implications for the origin of life. *Proc Natl Acad Sci USA* 91:1248–1250.

Bada JL, Lazcano A. 2003. Prebiotic soup—revisiting the miller experiment. *Science* 300:745–746.

Beard BL, Johnson CM, Cox L, Sun H, Nealson KH, Anguillar C. 1999. Iron isotope biosignatures. *Science* 285:1889–1892.

Bekker A, Holland HD, Wang P-L, Rumble D III, Stein HJ, Hannah JL, Coetze LL, Beukes NJ. 2004. Dating the rise of atmospheric oxygen. *Nature* 427:117–120.

Beveridge TJ, Meloche JD, Fyfe WS, Murray RGE. 1983. Diagenesis of metals chemically complexed to bacteria: Laboratory formulation of metal phosphates, sulfides and organic condensates in artificial sediments. *Appl Environ Microbiol* 45:1094–1108.

Brasier MD, Green OR, Jephcoat AP, Kleppe AK, Van Krankendonk MJ, Lindsay JF, Steele A, Grassineau NV. 2002. Questioning the evidence for Earth's oldest fossils. *Nature* 416:76–81.

Broad WJ. 1997. Spotlight on comets in shaping of Earth, Science Section, *New York Times*, June 3, 1997.

Buick R. 1992. The antiquity of oxygenic photosynthesis: Evidence from stromatolites in sulfate-deficient Archean lakes. *Science* 255:74–77.

Cairns-Smith AG, Hartman H, eds. 1986. *Clay Minerals and the Origin of Life*. Cambridge, U.K.: Cambridge University Press.

Catling DC, Zhanle KJ, McKay CP. 2001. Biogenic methane, hydrogen escape, and the irreversible oxidation of early Earth. *Science* 293:839–843.

Cech TR. 1986. A model for the RNA-catalyzed replication of RNA. *Proc Natl Acad Sci USA* 83:4360–4363.

Chakrabarti AC, Breaker RR, Joyce GF, Deamer DW. 1994. Production of RNA by a polymerase protein encapsulated within phospholipids vesicles. *J Mol Evol* 39:555–559.

Chang S, Des Marais D, Mack R, Miller SL, Strathearn GE. 1983. Prebiotic organic syntheses and the origin of life. In: Schopf JW, ed. *Earth's Earliest Biosphere. Its Origin and Evolution*. Princeton, NJ: Princeton University Press, pp. 53–92.

Claire MW. 2005. Modeling the rise of atmospheric oxygen. Earth System Processes 2, abstract, paper no. 16-5, The Geological Society of America meeting, 8–11 August 2005.

Cohen J. 1995. Getting all turned around over the origins of life on Earth. *Science* 267:1265–1266.

Cohen Y, Padan E, Shilo M. 1975. Facultative anoxygenic photosynthesis in the cyanobacterium *Oscillatoria limnetica*. *J Bacteriol* 123:855–861.

Delsemme AH. 2001. An argument for the cometary origin of the biosphere. *Am Sci* 89:432–442.

Des Marais DJ. 2000. When did photosynthesis emerge on Earth? *Science* 289:1703–1705.

Dodson EO. 1979. Crossing the prokaryote–eukaryote border: Endosymbiosis or continuous development. *Can J Microbiol* 25:651–674.

Doebler SA. 2000. The dawn of the protein era. *Bioscience* 50:15–20.

Doudna JA, Szostak JW. 1989. RNA-catalyzed synthesis of complementary-strand RNA. *Nature* (London) 339:519–522.

Drobner E, Huber H, Wächtershäuser G, Rose D, Stetter KO. 1990. Pyrite formation linked with hydrogen evolution under anaerobic conditions. *Nature* (London) 346:742–744.

Ernst WG. 1983. The early Earth and the Archean rock record. In: Schopf JW, ed. *Earth's Earliest Biosphere. Its Origin and Evolution*. Princeton, NJ: Princeton University Press, pp. 41–52.

Fajardo-Cavazos P, Nicholson W. 2006. *Bacillus* endospores isolated from granite: Close molecular relationships to globally distributed *Bacillus* spp. from endolithic and extreme environments. *Appl Environ Microbiol* 72:2856–2863.

Ferris FG, Beveridge TJ, Fyfe WS. 1986. Iron–silica crystalline nucleation by bacteria in a geothermal sediment. *Nature* (London) 320:609–611.

Fox GE, Magrum LJ, Balch WE, Wolfe RS, Woese CR. 1977. Classification of methanogenic bacteria by 16S ribosomal RNA characterization. *Proc Natl Acad Sci USA* 74:4537–4541.

Fox GE, Stackebrandt E, Hespell RB, Gibson J, Maniloff T, Dyer A, Wolfe RS, Balch WE, Tanner RS, Magrum LJ, Zablen LB, Blakemore R, Gupta R, Bonen L, Lewis BJ, Stahl DFA, Luehrsen KR, Chen KN, Woese CR. 1980. The phylogeny of prokaryotes. *Science* 209:457–463.

Fridovich I. 1977. Oxygen is toxic! *Bioscience* 27:462–466.

Gilbert W. 1986. The RNA world. *Nature* (London) 319:618.

Giovannoni SJ, Turner S, Olsen GJ, Barns S, Land DJ, Pace NR. 1988. Evolutionary relationships among cyanobacteria and green chloroplasts. *J Bacteriol* 170:3584–3592.

Gitai Z. 2005. The new bacterial cell biology: Moving parts and subcellular architecture. *Cell* 120:577–586.

Guerrier-Takada C, Gardiner K, Marsh T, Pace N, Altman S. 1983. The RNA moiety of ribonuclease P is the catalytic subunit of the enzyme. *Cell* 35:849–857.

Habicht KS, Gade M, Thamdrup B, Berg P, Canfield DE. 2002. Calibration of sulfate levels in the Archean Ocean. *Science* 298:2372–2374.

Haldane JBS. 1929. The origin of life. *Rationalist Annu* 1929:148–169 (Reprinted, Smith JM, ed. 1985. *On Being the Right Size and Other Essays*. Oxford, U.K.: Oxford University Press, pp. 101–112).

Han T-M, Runnegar B. 1992. Megascopic eukaryotic algae from 2.1 billion-year-old Negaunee iron-formation, Michigan. *Science* 257:232–235.

Hodgson GW, Whiteley CG. 1980. The universe of porphyrins. In: Trudinger PA, Walter MR, Ralph BJ, eds. *Biogeochemistry of Ancient and Modern Environments*. Canberra, Australia: Australian Academy of Science, pp. 35–44.

Hofmann HJ, Schopf JW. 1983. Early Proterozoic microfossils. In: Schopf WJ, ed. *Earth's Earliest Biosphere. Its Origin and Evolution*. Princeton, NJ: Princeton University Press, pp. 321–360.

Holland HD. 1997. Evidence for life on Earth more than 3850 million years ago. *Science* 275:38–39.

Holland HD. 2002. Volcanic gases, black smokers, and the Great Oxidation Event. *Geochim Cosmochim Acta* 66:3811–3826.

Jacobsen SB. 2003. How old is the planet Earth? *Science* 300:1513–1514.

Joyce GF. 1991. The rise and fall of the RNA world. *New Biol* 3:399–407.

Kasting JF. 2001. The rise of atmospheric oxygen. *Science* 293:819–820.

Kasting JF. 2004. Today methane-producing microbes are confined to oxygen-free settings, such as the guts of cows, but Earth's distant past, they ruled the world: When methane made climate. *Sci Am* 291:78–81, 83–85.

Kasting JF, Siefert JL. 2002. Life and evolution of Earth's atmosphere. *Science* 296:1066–1068.

Knoll AN, Awramik SM. 1983. Ancient microbial ecosystems. In: Krumbein WE, ed. *Microbial Geochemistry*. Oxford, U.K.: Blackwell Scientific, pp. 287–315.

Kruger K, Grabowski PJ, Zaug AJ, Sands J, Gottschling DE, Cech TR. 1982. Self-splicing RNA: Autoexcision and autocatalyzation of the ribosomal RNA intervening sequence of *Tetrahymena*. *Cell* 31:147–157.

Krylov IN, Semikhatov MA. 1976. Appendix II. Table of time ranges of the principal groups of Precambrian stromatolites. In: Walter MR, ed. *Stromatolites*. Amsterdam: Elsevier, pp. 693–694.

Kump LR, Kasting JF, Barley ME. 2001. Rise of atmospheric oxygen and the "upside-down" Archean mantle. *Geochem Geophy Geosy* 2(2). American Geophysical Union. Online computer file: http://g-cubed.org/gs2001/2000GC000114/article2000GC000114.pdf.

Lowe DR. 1980. Stromatolites 3,400 Myr old from the Archean of Western Australia. *Nature* (London) 284:441–443.

Mandernack KW, Bazylinski DA, Shanks WC III, Bullen TD. 1999. Oxygen and iron isotope studies of magnetite produced by magnetotactic bacteria. *Science* 285:1892–1896.

Margulis L. 1970. *Origin of Eukaryotic Cells*. New Haven, CT: Yale University Press.

Miller SL, Orgel LE. 1974. *The Origin of Life on Earth*. Englewood Cliffs, NJ: Prentice-Hall.

Moizsis SJ, Coath CD, Greenwood JP, McKeegan KD, Harrison TM. 2003. Mass-independent isotope effect in Archean (2.5 to 3.8 Ga) sedimentary sulfides determined by ion microprobe analysis. *Geochim Cosmochim Acta* 67:1635–1658.

Mojzsis SJ, Arrhenius G, McKeegan KD, Harrison TM, Nutman AP, Friend CRL. 1996. Evidence for life on Earth before 3800 million years ago. *Nature* (London) 384:55–59.

Mojzsis SJ, Harrison TM, Pidgeon RT. 2001. Oxygen-isotope evidence from ancient zircons for liquid water at the Earth's surface 4,300 Myr ago. *Nature* 409:178–181.

Møller-Jensen J, Löwe J. 2005. Increasing complexity of the bacterial cytoskeleton. *Curr Opin Cell Biol* 17:75–81.

Nicholson WL, Munakata N, Horneck G, Melosh HJ, Setlow P. 2000. Resistance of *Bacillus* endospores to extreme terrestrial and extraterrestrial environments. *Microbiol Mol Biol Rev* 64:548–572.

Nisbet EG, Sleep NH. 2001. The habitat and nature of early life. *Nature* 409:1083–1091.

Oehler JH, Schopf JW. 1971. Artificial microfossils: Experimental studies of permineralization of blue-green algae in silica. *Science* 174:1229–1231.

Ohmoto H, Kagegawa T, Lowe DR. 1993. 3.4 Billion-year-old biogenic pyrites from Barberton, South Africa: Sulfur isotope evidence. *Science* 262:555–557.

Olsen GJ, Woese CR, Overbeek R. 1994. The winds of evolutionary change: Breathing new life into microbiology. *J Bacteriol* 176:1–6.

Oparin AI. 1938. *The Origin of Life*. New York: Macmillan. (Reprinted by Dover Publications, Inc., New York, 1953.)

Orpen JL, Wilson JF. 1981. Stromatolites at ~3500 Myr and a greenstone-granite unconformity in the Zimbabwean Archean. *Nature* (London) 291:218–220.

Ripley EM, Nicol DL. 1981. Sulfur isotopic studies of Archean slate and greywacke from northern Minnesota: Evidence for the existence of sulfate-reducing bacteria. *Geochim Cosmochim Acta* 45:839–846.

Robert F. 2001. The origin of water on Earth. *Science* 293:1056–1058.

Russell MJ, Hall AJ. 1997. The emergence of life from iron monosulfide bubbles at a submarine hydrothermal redox and pH front. *J Geol Soc London* 154:377–402.

Schidlowski M, Hayes JM, Kaplan IR. 1983. Isotopic inferences of ancient biochemistries: Carbon, sulfur, hydrogen, and nitrogen. In: Schopf JW, ed. *Earth's Earliest Biosphere. Its Origin and Evolution.* Princeton, NJ: Princeton University Press, pp. 149–186.

Schopf JW. 1977. Evidences of Archean life. In: Ponnamperuma C, ed. *Chemical Evolution of the Early Precambrian.* New York: Academic Press, pp. 101–105.

Schopf JW. 1978. The evolution of the earliest cells. *Am Sci* 238:110–139.

Schopf JW, ed. 1983. *Earth's Earliest Biosphere. Its Origin and Evolution.* Princeton, NJ: Princeton University Press.

Schopf JW, Hayes JM, Walter MR. 1983. Evolution of Earth's earliest ecosystem: Recent progress and unsolved problems. In: Schopf JW, ed. *Earth's Earliest Biosphere. Its Origin and Evolution.* Princeton, NJ: Princeton University Press, pp. 360–384.

Schopf JW, Packer BM. 1987. Early Archean (3.3-billion-year-old) microfossils from Warrawoona Group, Australia. *Science* 237:70–73.

Schopf JW, Walter MR. 1983. Archean microfossils: New evidence of ancient microbes. In: Schopf JW, ed. *Earth's Earliest Biosphere. Its Origin and Evolution.* Princeton, NJ: Princeton University Press, pp. 214–239.

Shen YA, Buick R, Canfield DE. 2001. Isotopic evidence for microbial sulphate reduction in the early Archean era. *Nature* (London) 410:77–81.

Stetter KO, Lauerer G, Thomm M, Neuner A. 1987. Isolation of extremely thermophilic sulfate reducers: Evidence for a novel branch of Archaebacteria. *Science* 236:822–823.

Stevenson DJ. 1983. The nature of the Earth prior to the oldest known rock record. The Hadean Earth: In: Schopf JW, ed. *Earth's Earliest Biosphere. Its Origin and Evolution.* Princeton, NJ: Princeton University Press, pp. 32–40.

Summons RE, Powell TG. 1986. Chlorobiaceae in Paleozoic seas revealed by biological markers, isotopes, and geology. *Nature* (London) 319:763–765.

Taylor SR. 1992. The origin of the Earth. In: Brown G, Hawkesworth C, Wilson C, eds. *Understanding the Earth.* Cambridge, U.K.: Cambridge University Press, pp. 25–43.

Towe KM. 1990. Aerobic respiration in the Archean. *Nature* (London) 348:54–56.

Wächtershäuser G. 1988. Before enzymes and templates: Theory of surface metabolism. *Microbiol Rev* 52:452–484.

Walter MR. 1983. Archean stromatolites: Evidence of the Earth's earliest benthos. In: Schopf JW, ed. *Earth's Earliest Biosphere. Its Origin and Evolution.* Princeton, NJ: Princeton University Press, pp. 187–213.

Walter MR, Buici R, Dunlop JSR. 1980. Stromatolites 3,400–3,500 Myr old from the North Pole area, Western Australia. *Nature* (London) 284:443–445.

Watson EB, Harrison TM. 2006. Zircon thermometer reveals minimum melting conditions on earliest Earth. *Science* 308:841–844.

Weber P, Greenberg JM. 1985. Can spores survive in interstellar space? *Nature* (London) 316: 403–407.

Wilde SA, Valley JW, Peck WH, Graham CM. 2001. Evidence from detrital zircons for the existence of continental crust and oceans on the Earth 4.4 Gyr ago. *Nature* 409.175–178.

Woese CR. 1987. Bacterial evolution. *Microbiol Rev* 51:221–271.

Woese CR, Debrukner-Vossbrinck BA, Oyaizu H, Stackebrandt E, Ludwig W. 1985. Gram-positive bacteria: Possible photosynthetic ancestry. *Science* 229:762–765.

Woese CR, Fox GE. 1977. Phylogenetic structure of the prokaryotic domain: The primary kingdoms. *Proc Natl Acad Sci USA* 74:5088–5090.

Xiong J, Fischer WM, Inoue K, Nakahara M, Bauer CE. 2000. Molecular evidence for the early evolution of photosynthesis. *Science* 289:1724–1730.

Yang W, Holland HD, Rye R. 2002. Evidence for low or no oxygen in the late Archean atmosphere from the ~2.76 Ga Mt. Roe #2 paleosol, Western Australia: Part 3. *Geochim Cosmochim Acta* 66:3707–3718.

4 Lithosphere as Microbial Habitat

4.1 ROCK AND MINERALS

To understand how the lithosphere supports the existence of microbes on and in it and how microbes influence the formation and transformation of some of its constituent rocks and minerals, we must review some of the general chemical and physical features of the lithosphere components.

Geologically, the term *rock* refers to massive, solid inorganic matter consisting usually of two or more intergrown minerals. Rock may be igneous in origin; that is, it may arise by cooling of *magma* (molten rock material) from the interior of the Earth (crust and asthenosphere). The cooling may be a slow or fast process. In slow cooling, different minerals begin to crystallize at different times, owing to their different melting points, leading to the formation of rock with a visually distinguishable mixture of intergrown crystals of which granite is a typical example (Figure 4.1A). In fast cooling, rapid crystallization occurs, leading to the formation of rock containing only tiny crystals that are not visible to the naked eye. Basalt is an example of rock formed in this way (Figure 4.1B).

Rock may also be *sedimentary* in origin; that is, it may arise through the accumulation and compaction of sediment that consists mainly of mineral matter derived from breakdown of other rocks. In other instances, sedimentary rock may arise as a result of cementation of accumulated inorganic sediment by carbonate, silicate, aluminum oxide, ferric oxide, or a combination thereof. The cementing substance may result from microbial activity. These transformations of loose sediment into sedimentary rock are termed *lithification*. Sedimentary rock facies often exhibit a layered structure in vertical section, reflecting changes in composition as the sediment is accumulated. Analysis of the different layers may tell about the environmental conditions during which they accumulated. Examples of sedimentary rock are limestone, shale, and sandstone.

Finally, rock may be *metamorphic* in origin; that is, it may have formed through alteration of igneous or sedimentary rock by the action of heat and pressure. Examples of metamorphic rock are marble, derived from limestone; slate, derived from shale; quartzite, derived from sandstone; and gneiss, derived from granitic rock.

Geochemically, minerals are usually defined as inorganic compounds, usually crystalline but occasionally amorphous, of specific chemical composition and structure.

Sometimes the term *mineral* is also applied to certain organic compounds in nature, such as asphalt and coal. Inorganic minerals may be very simple in composition, such as elemental sulfur (S^0) or quartz (SiO_2), or very complex, as in the case of the igneous mineral biotite [$K(Mg,Fe,Mn)_3AlSi_3O_{10}(OH)_2$]. Minerals that result from crystallization during the cooling of magma are *primary* or *igneous* minerals whereas those that result from chemical alteration (weathering or diagenesis) of primary minerals are known as *secondary* minerals. Microbes may play a role in the transformation of primary to secondary minerals (Chapter 9). Examples of primary and secondary mineral groups are listed in Table 4.1. In addition, minerals can result from precipitation from solutions, in which case they are called *authigenic* minerals. Microbes may also play a role in their formation (e.g., ferromanganese concretions; see Chapter 17).

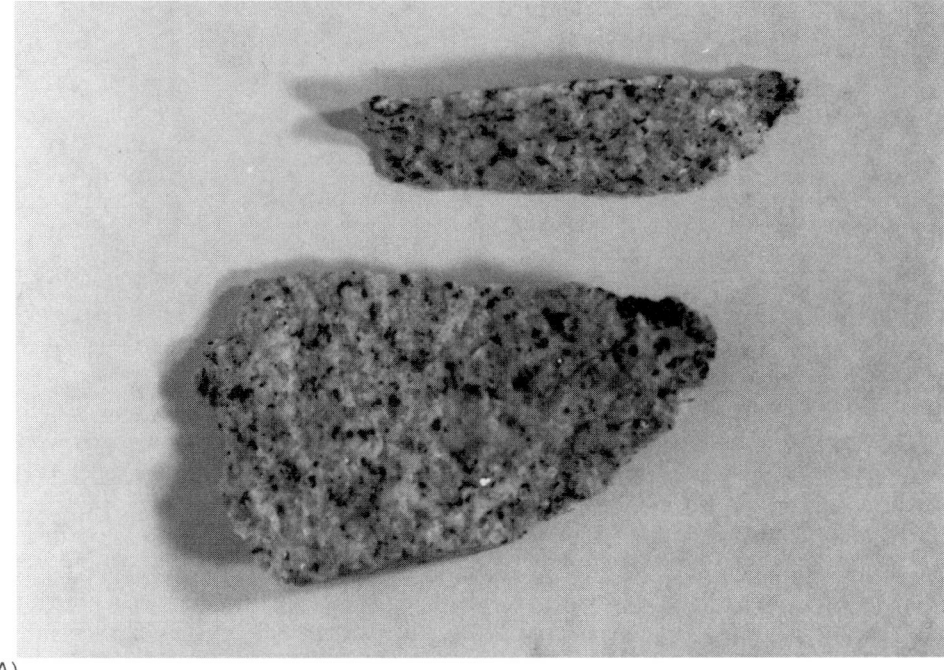

(A)

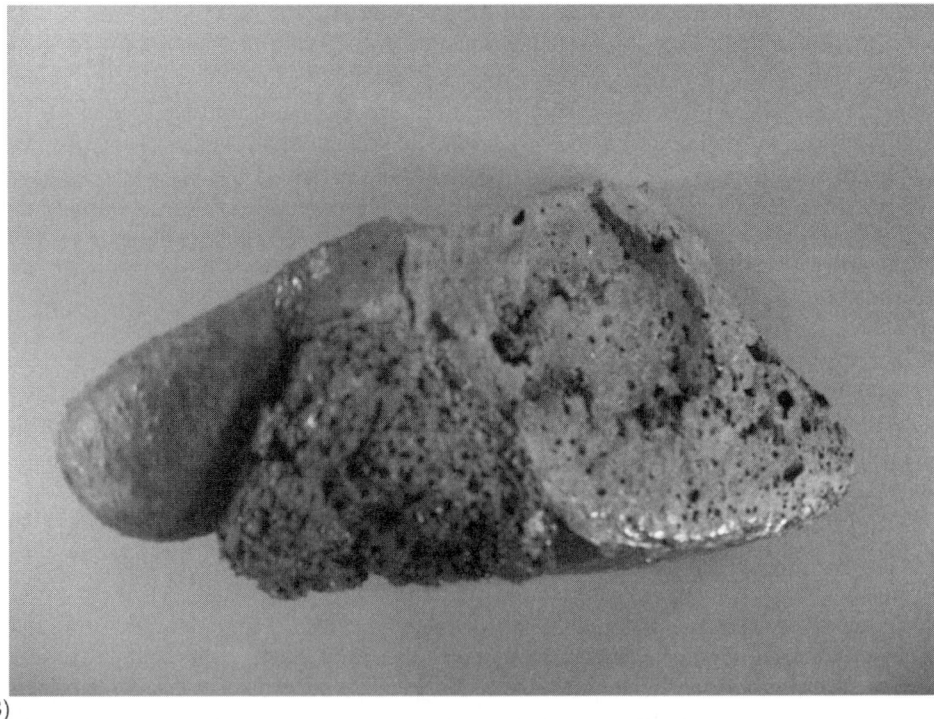

(B)

FIGURE 4.1 (A) Pieces of granite showing phenocrysts, that is, visible crystals of mineral in a fine crystalline ground mass: an igneous rock. The inset fragment is 5 cm long. (B) Pahoehoe basalt from Kilauea, Hawaii. Dark spots represent holes left by outgassing as the molten rock solidified. Note the absence of visible crystals. (Courtesy of rock collection of the Department of Earth and Environmental Sciences, Rensselaer Polytechnic Institute.)

TABLE 4.1
Minerals Classified Based on Mode of Formation

Primary minerals
Feldspars
Pyroxenes and amphiboles
Olivines
Micas
Silica

Secondary minerals
Clay minerals
   Kaolinites
   Montmorillonites
   Illites
Hydrated iron and aluminum oxides
Carbonates

Source: Based on Lawton K. in *Chemistry of the Soil*, Van Nostrand Reinhold, New York, 1955.

4.2 MINERAL SOIL

4.2.1 ORIGIN OF MINERAL SOIL

The mineral constituents of mineral soil are ultimately derived from rock that underwent weathering. Weathering, which leads to soil formation, is a process in which rock is eroded or broken down into ever smaller particles and finally into constituent minerals. Some or all of these minerals may become chemically altered. Some forms of rock weathering involve physical processes. For example, freezing and thawing of water in cracks and fissures of a rock may cause expansion of the cracks and fissures and ultimate splitting of the rock because the ice formed from the water that originally filled the cracks and fissures occupies a larger volume than the water from which it formed. Sand carried by wind may cause sandblasting of rock surfaces. Alternate heating by the sun's rays by day and cooling at night may cause expansion and contraction of rock, leading to widening of cracks and fissures. Waterborne abrasives or rock collisions may cause rock to break. Seismic activity may cause rock to crumble. Evaporation of hard water in cracks and fissures of rock and resultant formation of crystals formed from the solutes in the hard water may cause rock to break because the crystals occupy a larger volume than the original water solution from which they formed, thereby widening the cracks and fissures through the pressure they exert. Mere alternate wetting and drying may itself cause rock breakup.

Rock weathering processes may also be chemical when the weathering agents are of nonbiological origin. Examples are the solvent action of water; CO_2 of volcanic origin; and mineral acids such as H_2SO_3, HNO_2, and HNO_3 formed from gases of nonbiological origin, such as SO_2, NO, and NO_2, respectively. Chemical weathering may also be caused by redox reagents of nonbiological origin, such as H_2S of volcanic origin or nitrate of atmospheric origin.

Finally, rock weathering may be the result of biological activity. Some of this activity may be physical, as when roots of plants penetrate cracks and fissures in rock, forcing it apart. However, much of it is biochemical, resulting from the activity of algae, fungi, lichens, and bacteria frequently residing on rock surfaces and in the interior of porous rock. The microorganisms on the surface of rocks may exist in biofilms, especially in a moist and wet environment. In biofilms containing a mixed microbial population, the different organisms may arrange themselves in distinct zones where conditions are most favorable for their existence (Costerton et al., 1994). Some microorganisms, the so-called boring organisms, may form cavities in limestone rock that they occupy by causing dissolution of $CaCO_3$ (Golubic et al., 1975). In other cases, opportunistic microorganisms invade preformed cavities in rock (chasmolithic organisms) (Friedmann, 1982). Invertebrates, snails in particular, may feed on boring organisms (Golubic and Schneider, 1979; Shachak et al., 1987) or chasmolithic microorganisms by grinding away the superficial rock to expose the former to be consumed. The rock debris that the snails generate becomes part of a soil (Shachak et al., 1987; Jones and Shachak, 1990).

Microbes dissolve rock minerals through the corrosive action of metabolic products, such as NH_3, HNO_3, and CO_2 (forming H_2CO_3 in water), and oxalic, citric, and gluconic acids they excrete. Studies using scanning electron microscopy have shown that organic compounds formed by microorganisms such as lichens cause distinct weathering (Jones et al., 1981). Waksman and Starkey (1931) cited the following reactions as examples of how microbes can affect weathering of minerals:

$$\underset{\text{Orthoclase}}{2KAlSi_3O_8} + 2H_2O + CO_2 \rightarrow \underset{\text{Kaolinite}}{H_4Al_2Si_2O_9} + K_2CO_3 + 4SiO_2 \qquad (4.1)$$

$$\underset{\text{Olivine}}{12MgFeSiO_4} + 26H_2O + 3O_2 \rightarrow \underset{\text{Serpentine}}{4H_4Mg_3Si_2O_9} + 4SiO_2 + 6Fe_2O_3 \cdot 3H_2O \qquad (4.2)$$

Reaction 4.1 is promoted by CO_2 production in the metabolism of heterotrophic microorganisms, and Reaction 4.2 by O_2 production in oxygenic photosynthesis by cyanobacteria, algae, and lichens inhabiting the surface of rocks. Further investigations have extended these observations. In recent studies, reactions were examined in which organic acids that are excreted by microorganisms promote weathering of primary minerals such as feldspars and secondary minerals such as clays (Browne and Driscoll, 1992; Lucas et al., 1993; Hiebert and Bennett, 1992; Welch and Ullman, 1993; Brady and Carroll, 1994; Oelkers et al., 1994; Ullman et al., 1996; Bennett et al., 1996; Barker and Banfield, 1996, 1998). Some current weathering models favor protonation as a means of displacing cationic components from the crystal lattice followed by cleaving of Si–O and Al–O bonds (Berner et Holdren, 1977; Chou and Wollast, 1984). Others favor complexation, for instance, of Al and Si in aluminosilicates, as a primary mechanism of dissolution (Wieland and Stumm, 1992; Welch and Vandevivere, 1995).

Mineral soil may derive from aquatic sediment or *alluvium* left behind after the water that carried it from its place of origin to its final site of deposition has receded. Mineral soil can also form in place as a result of progressive weathering of parent rock and subsequent differentiation of weathering products. Soils originating by either mechanism undergo eluviation (removal of some products by washing out) or alluviation (addition of new material by water transport). Any soil, once formed, undergoes further gradual transformation due to the biological activity it supports (Buol et al., 1980).

4.2.2 SOME STRUCTURAL FEATURES OF MINERAL SOIL

Mineral soil varies in composition, depending on the source of the parent material, the extent of weathering, the amount of organic matter introduced into or generated in the soil, and the amount of moisture it holds. Its texture is affected by the particle sizes of its inorganic constituents (stones, >2 mm; sand grains, 0.05–2 mm; silt, 0.002–0.05 mm; clay particles, <0.002 mm), which determine its porosity and thus its permeability to water and gases.

Many, but not all, mineral soils tend to be more or less obviously stratified. As many as three or four major strata or *horizons* may be recognizable in agricultural and forest *soil profiles*. A soil profile is a vertical section through soil (Figure 4.2). The strata are labeled O, A, B, and C horizons. The O horizon represents the litter zone, consisting of much undecomposed and partially decomposed organic matter. Some soil profiles may lack an O horizon. The A and B horizons represent the true

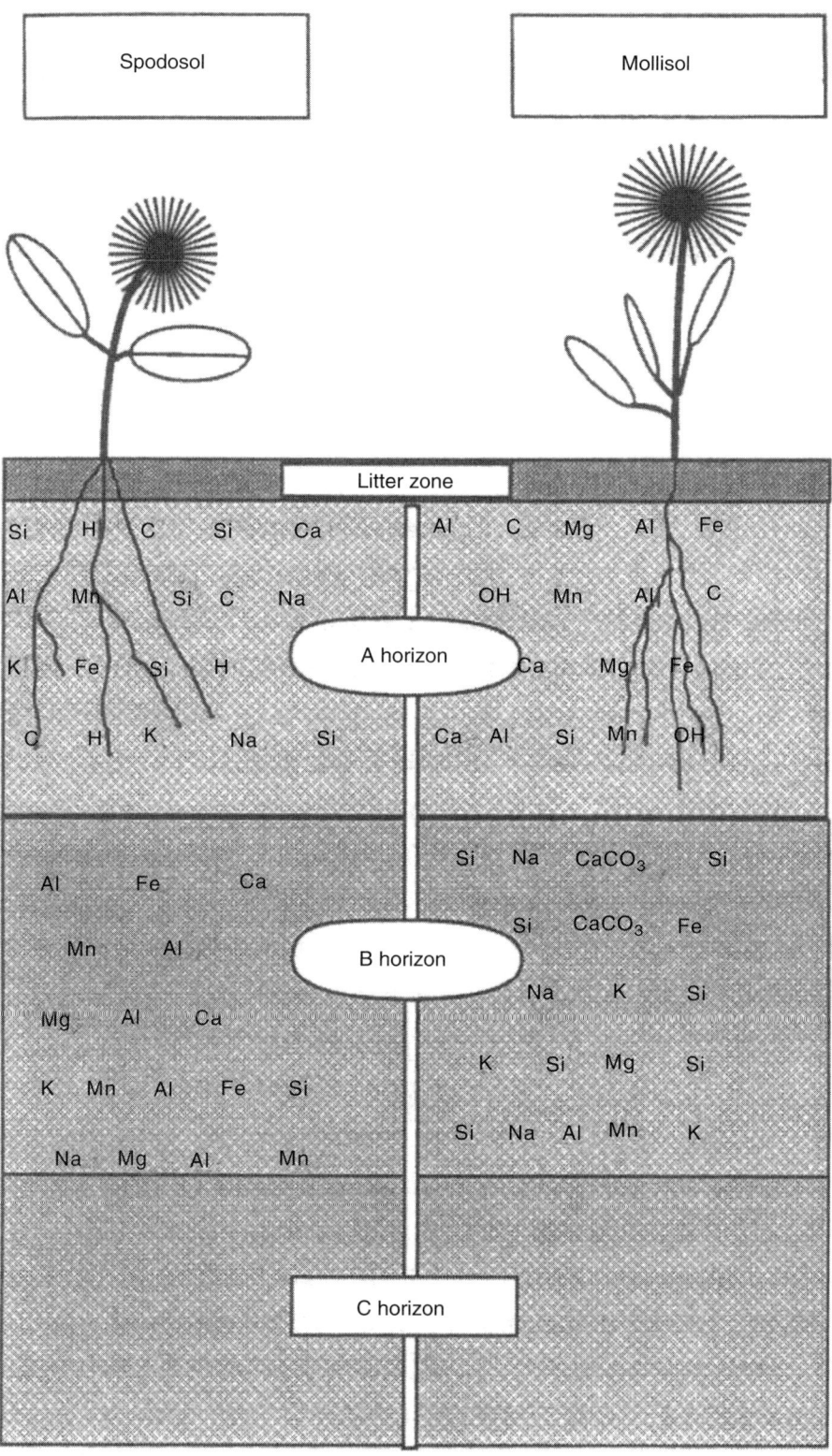

FIGURE 4.2   Schematic representation of the major soil horizons of spodosol and mollisol. The litter zone is also called the O horizon. The A and B horizons may be further subdivided on the basis of soil chemistry.

soil. The C horizon represents the parent material from which the soil was formed. It may be the bedrock or an earlier soil. The A and B horizons are often further subdivided, although these divisions are somewhat arbitrary. The A horizon is the biologically most active zone, containing most of the root systems of plants growing on it and the microbes and other life forms that inhabit soil. As is to be expected, the carbon content in this horizon is also greater. The biological activity in the A horizon may cause solubilization of organic and inorganic matter, some or all of which, especially the inorganic matter, is carried by soil water into the B horizon. At times, the A horizon is therefore known as the *leached layer*, and the B horizon as the *enriched layer*. Both biological and abiological factors play a role in soil profile formation.

4.2.3   Effects of Plants and Animals on Soil Evolution

Plants assist in soil evolution by contributing organic matter through excretions from their root systems and as dead organic matter. The plant excretions may react directly with some soil mineral constituents, or they may first be modified together with dead plant matter by microbes, resulting in products that then react with soil mineral constituents. During their lifetime, plants remove some minerals from soil and contribute to water movement through the soil by water absorption via their roots and transpiration from their leaves. Their root system may also help prevent destruction of the soil through wind and water erosion by anchoring it.

Burrowing invertebrates, from small mites to large earthworms, help to break up soil, keep it porous, and redistribute organic matter. The habitat of some of these invertebrates is restricted to specific regions in the soil profile.

4.2.4   Effects of Microbes on Soil Evolution

Microbes contribute to soil evolution by mineralizing some or all of any added organic matter during the decay process. Some of the metabolic products from this decay, such as organic and inorganic acids, CO_2, and NH_3, interact slowly with soil minerals and cause their alteration or dissolution, which is an important step in soil profile formation (Berthelin, 1977; Welch and Ullman, 1993; Ullman et al., 1996; Barker and Banfield, 1996, 1998). For instance, the mineral chlorite has been reported to be bacterially altered in this manner through loss of Fe and Mg and an increase in Si. The mineral vermiculite has been reported to be bacterially altered through mobilization by dissolution of Si, Al, Fe, and Mg; thereby forming montmorillonite (Berthelin and Boymond, 1978). Certain microbes may interact directly (i.e., enzymatically) with certain inorganic soil minerals by oxidizing or reducing them or their constituents (see Chapters 13, 14, and Chapters 16 through 19; Ehrlich, 2001), resulting in their mobilization by dissolution, or in the formation of new minerals (Berthelin, 1977). Microbes may also play an important role in *humus* formation.

Humus is an important constituent of soil, consisting of humic and fulvic acids, humins and amino acids, lignin, amino sugars, and other compounds of biological origin (Paul and Clark, 1996, pp. 148–152; Stevenson, 1994). Humic and fulvic acids are dispersible in solutions of NaOH or sodium pyrophosphate whereas humin is not. Humic acids are precipitated at acid pH whereas fulvic acids are not. The humus constituents humins and humic and fulvic acids represent components of soil organic matter that are only slowly decomposed. They are mostly formed by microbial attack of plant organic matter introduced into the litter zone (O horizon) and in the A horizon. Humus gives proper texture to soil and plays a significant role in regulating the availability of the mineral elements that are important in plant nutrition and in detoxifying those that are harmful to plants by complexing them. Humus also contributes to the water-holding capacity of soil. Some microorganisms in soil can use humic substances as terminal electron acceptors in anaerobic oxidation of various other organic compounds and H_2, and as electron shuttles in the anaerobic reduction of Fe(III) oxides (Lovley et al., 1996, 1998; Newman and Kolter, 2000; Hernandez and Newman, 2001; Hernandez et al., 2004).

4.2.5 EFFECTS OF WATER ON SOIL EROSION

Water from rain or melting snow may mobilize and transport some soluble soil components and cause precipitation of others. This can contribute to horizon development as the water permeates the soil. Precipitates, especially inorganic ones, that form in the soil water may lead to soil clumping when formed in sufficient quantity. Water may also affect the distribution of soil gases by displacing the rather insoluble ones, such as nitrogen and oxygen, and absorbing the more soluble and potentially corrosive ones, such as CO_2, NH_3, and H_2S.

4.2.6 WATER DISTRIBUTION IN MINERAL SOIL

Only ~50% of the volume of mineral soil is solid matter. The other 50% is pore space occupied by water and gases such as CO_2, N_2, and O_2. As might be expected, owing to the biological activity in soil and the slow gas exchange with the external atmosphere, the CO_2 concentration in the gas space in soil usually exceeds that in air, whereas the O_2 concentration is less than that in air. According to Lebedev (see Kuznetsov et al., 1963), soil water may be distributed in distinct layers around soil particles (Figures 4.3A and 4.3B). Surrounding a soil particle is *hygroscopic water*, a thin film of

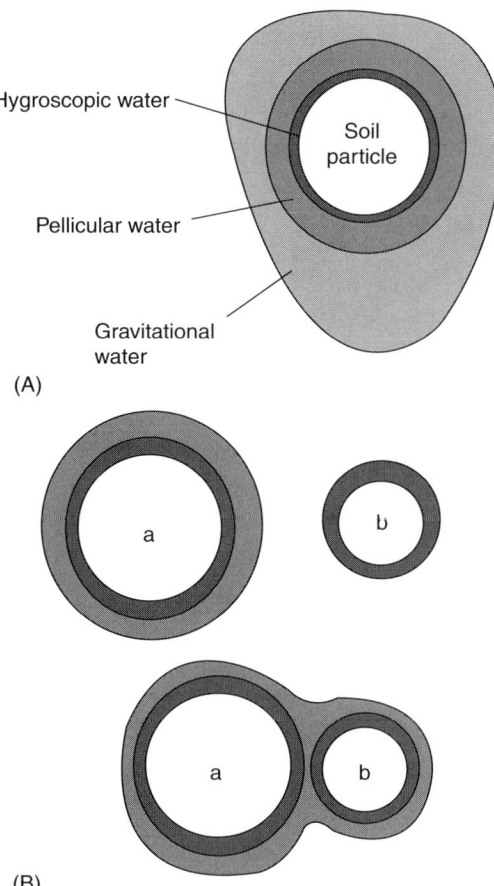

FIGURE 4.3 Diagrammatic representation of soil water distribution according to Lebedev. (A) Water layers around a soil particle when soil moisture is in excess of what the soil atmosphere can hold. (B) Movement of pellicular water from soil particle (a), which is surrounded by it, to soil particle (b) lacking it; particles (a) and (b) are in very close proximity. (Adapted from Kuznetsov SI, Ivanov MV, Lyalikova NN, *Introduction to Geological Microbiology*, McGraw-Hill, New York, 1963.)

3×10^{-2} μm in thickness when surrounding a 25 mm diameter particle. This water never freezes and never moves as a liquid. It is adsorbed by soil particles from water vapor in the atmosphere. In a water-saturated atmosphere, *pellicular water* surrounds the hygroscopic water. Pellicular water may move from one soil particle to another by intermolecular attraction, but not by gravity (Figure 4.3B). It may contain dissolved salts, which may depress the freezing point to −1.5°C. *Gravitational water* in Lebedev's model surrounds pellicular water when moisture in excess of what the soil atmosphere can hold is present. It moves by gravity and responds to hydrostatic pressure, unlike hygroscopic water and pellicular water. So far, it is unclear as to which of these forms of water is available to microorganisms. A reasonable guess is that gravitational water and probably pellicular water can be used by them, but not hygroscopic water. The water requirement of microorganisms is usually studied in terms of moisture content, water activity, or water potential without regard to the form of soil water (Dommergues and Mangenot, 1970).

Water activity of a soil sample is a measure of the degree of water saturation of the vapor phase in the soil and is expressed in terms of relative humidity, but as a fractional number instead of percentage. Pure water has a water activity of 1. Except for extreme halophiles, bacteria have a higher minimum water activity requirement (>0.85) than many fungi (>0.60) (Brock et al., 1984).

Water potential of soil is a measure of water availability in terms of the difference between the free energy of the combined matrix and osmotic potentials of soil water and pure water at the same temperature. Matric effects on water availability have to do with the effect of water adsorption to solid surfaces such as soil particles, which lower water availability. Osmotic effects have to do with the effect of dissolved solutes on water availability; their presence lowers it. As matric and osmotic effects lower the free energy of water, water potential values are negative. The more negative the water potential value, the lower the water availability. A zero potential is equivalent to pure water. Osmotic water potential can be calculated from the effect of solute on the freezing point of water by using the following formula of Lang (1967):

$$\text{Water potential (J kg}^{-1}) = 1.322 \times \text{freezing point depression} \tag{4.3}$$

where 100 J kg^{-1} is equal to 1 bar. Matric water requirements can be determined by the method of Harris et al. (1970). In this method, NaCl or glycerol solutions of desired water potentials, solidified with agar, are used to equilibrate with matrix material on which microbial growth is to occur. (For further discussion on water potential, see Brock et al., 1984; Brown, 1976.)

The water potential requirement for two strains of the acidophilic iron oxidizer *Acidihiobacillus* (formerly *Thiobacillus*) *ferrooxidans* has been determined by Brock (1975). Using NaCl as osmotic agent, strain 57-5 exhibited a minimum water potential requirement at −18 to −32 bar, whereas strain 59-1 exhibited it at −18 to −20 bar. Using glycerol as osmotic agent, strain 57-5 exhibited a minimum water potential at −8.8 bar, whereas strain 59-1 exhibited it at −6 bar (Table 4.2). The same study showed that significant amounts of CO_2 were assimilated by *A. ferrooxidans* on coal refuse material with water potentials between −8 and −29 bar, whereas none was assimilated when the water potential of the refuse was less than −90 bar.

4.2.7 NUTRIENT AVAILABILITY IN MINERAL SOIL

Organic or inorganic nutrients required by soil microbes are distributed between the soil solution and the surface of mineral particles. Partitioning effects determine their relative concentrations in the two phases. Their presence on the surface of soil particles may be the result of adsorption or ion exchange. Nonionizable molecules tend to be adsorbed, whereas ionizable ones will bind as a result of charges of opposite sign and may involve ion exchange. Microbial utilization of surface-bound molecules that are metabolized intracellularly may require either their displacement from the mineral surface to be taken into the cells or, in the case of some polymeric organic molecules, direct attack, for example, by hydrolysis, of the portion of the molecule that is not bound to the surface. If displacement or direct attack at the mineral surface cannot be effected, such nutrient will be unavailable.

TABLE 4.2
Effect of Osmotic Water Potential (Glycerol) on Growth of *T. ferrooxidans*

Glycerol	Total Water Potential (bar)	Strain 57-5	Strain 59-1
184	−61	−	−
147	−49	−	−
92	−32	−	−
74	−26	−	−
55	−20	−	−
37	−15	−	−
18.4	−8.8	+	−
9.2	−6	+	+
3.7	−4.2	+	+
0	−3	+	+

Note: All experiments were done in replicate tubes, which showed the same results. The incubation period was 2 weeks. Iron concentration in the medium, 10 g of $FeSO_4 \cdot 7H_2O$ per liter; +, visible iron oxidation and microscopically visible growth; −, no iron oxidation nor microscopically visible growth.

Source: Reproduced from Brock TD, *Appl. Microbiol.*, 29, 495–501, 1975. With permission.

Clay particles are especially important in ionic binding of organic or inorganic cationic solutes (those having a positive charge). Such particles exhibit mostly negative charges except at their edges, where positive charges may appear. Their capacity for ion exchange depends on their crystal structure. The partitioning of solutes between soil solution and mineral surfaces often results in greater concentration of solutes on mineral surfaces than in the soil solution, and as a result the mineral surfaces may be the preferred habitat of soil microbes that require these solutes in more concentrated form. However, ionically bound solutes on clay or other soil particles may be less available to soil microbes because the microbes may not be able to dislodge them from the particle surface. In that instance, soil solution may be the preferred habitat for microbes that have a requirement for such solutes. Ionic binding to soil particles may be beneficial if a solute subject to such binding is toxic and not readily dislodged (see Chapter 10).

4.2.8 Some Major Soil Types

Distinctive soil types may be identified by and correlated with climatic conditions and with the vegetation they support (Bunting, 1967; Buol et al., 1980). Climatic conditions determine the kind of vegetation that may develop. Thus, in the high northern latitudes, *tundra soil*, a type of *inceptisol*, prevails, which in cold climate is often frozen and therefore supports only limited plant and microbial development. It has a poorly developed profile and may be slightly alkaline. Examples of tundra soil are Arctic brown soil and bog soil. In the cool (i.e., temperate), humid zones at midlatitudes, *spodosols* (Figures 4.2 and 4.4) prevail, which support extensive forests, particularly of the coniferous type. Spodosols tend to be acidic and have a strongly leached, grayish A horizon depleted in colloids and compounds of iron and aluminum and a brown B horizon enriched in colloids and compounds of iron and aluminum that are leached from the A horizon. In regions of moderate rainfall in temperate climates at midlatitudes, *mollisols* (Figures 4.2 and 4.5) prevail. These are soils that support grasslands (i.e., they are prairie soils). They exhibit rich black topsoil and show lime accumulation in the B horizon because they have neutral to alkaline pH. *Oxisols* are found at low latitudes in tropical, humid climates. They are poorly zoned, highly weathered jungle soils with a B horizon rich in sesquioxides or clays. Owing to the hot, humid climate conditions under which

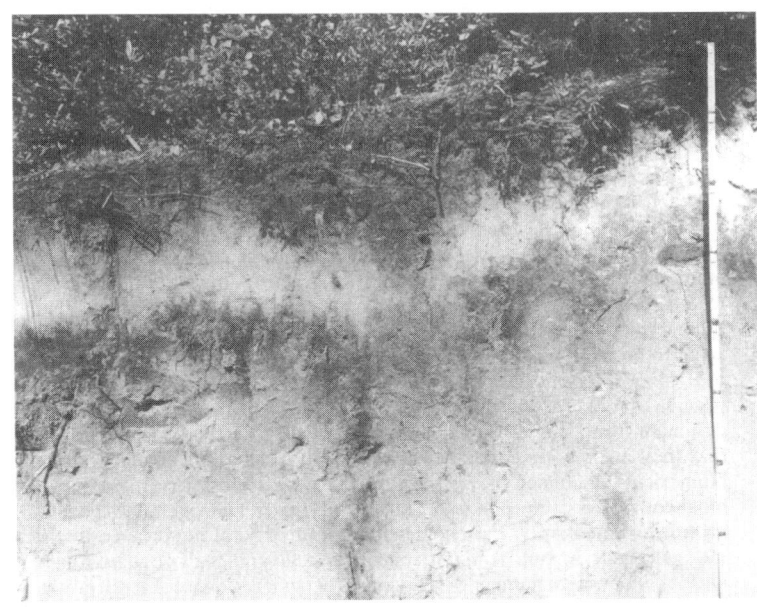

FIGURE 4.4 Soil profile of spodosol (podsol). (Courtesy of US Department of Agriculture (U.S.D.A.) Soil Conservation Service, Washington, DC, USA.)

FIGURE 4.5 Soil profile of mollisol (chernozem). (Courtesy of US Department of Agriculture (U.S.D.A.) Soil Conservation Service, Washington, DC, USA.)

they exist, these soils are intensely active microbiologically and require constant replenishment of organic matter by the vegetation growing on them and from animal excretions and remains to stay fertile. The neutral to alkaline pH conditions of oxisols promote leaching of silicate and precipitation of iron and aluminum as sesquioxides. When oxisols are denuded of the arboreal vegetation, as in slash-and-burn agriculture, they quickly lose their fertility as a result of intense microbial activity, which rapidly destroys soil organic matter. Because little organic matter is returned to the soil in its agricultural exploitation, conditions favor *laterization*, a process in which iron and aluminum oxides, silica, and carbonates are precipitated that cement the soil particles together and greatly reduce the porosity and water-holding capacity of the soil and make it generally unfavorable for plant growth.

Aridisols and *entisols* are desert soils that are present mostly in hot, arid climates at low latitudes. Aridisols feature an ochreous surface soil and may show one or more subsurface horizons, such as argillic horizon (a layer with silica and clay minerals dominating), cambic horizon (an altered, light-colored layer, low in organic matter, with carbonates usually present), natric horizon (dominant presence of sodium in exchangeable cation fraction), salic horizon (enriched in water-soluble salts), calcic horizon (secondarily enriched in $CaCO_3$), gypsic horizon (secondarily enriched in $CaSO_4 \cdot 2H_2O$), and duripan horizon (primarily cemented by silica and secondarily by iron oxides and carbonates) (Fuller, 1974; Buol et al., 1980). Entisols are poorly developed immature desert soils without subsurface development. They may arise from recent alluvial deposits or from rock erosion (Fuller, 1974; Buol et al., 1980).

Desert soils are not fertile due to the lack of sufficient moisture that prevents the development of lush vegetation. However, insufficient nitrogen as major nutrient; and zinc, iron, and sometimes copper, molybdenum, or manganese as minor nutrients may also limit plant growth. Desert soils support a specially adapted macroflora and fauna that cope with the stressful conditions in such an environment. They also harbor a characteristic microflora of bacteria, fungi, algae, and lichens. Actinomycetes and lichens may sometimes be dominant. Cyanobacteria seem to be more important in nitrogen fixation in desert soils than other bacteria. Desert soils can sometimes be converted to productive agricultural soils by irrigation. Such watering often results in extensive solubilization of salts from the subhorizons where they have accumulated during soil-forming episodes. As a consequence, the salt level in the available groundwater in the growth zone of the soil may increase to a concentration that becomes inhibitory to plant growth. The drainage water from such irrigated soil will also become increasingly salty and present a disposal and reuse problem.

4.2.9 Types of Microbes and Their Distribution in Mineral Soil

Microorganisms found in mineral soil include prokaryotes (Bacteria and Archaea), fungi, protozoa, and algae, and also viruses associated with these groups. Members of the prokaryotes may inhabit the soil water and the surface of mineral soil particles. Some may be restricted to or be dominant in soil water, whereas others may be restricted or dominant on the surface of soil particles. Prokaryotes living on soil particles may form or inhabit biofilm, which may contribute to soil clumping. Mycelial fungi may inhabit soil pores and spread on the surface of soil particles. Nonmycelial fungi, protozoa, and algae will dominate wherever their source of nutrients is most readily available. Viral particles may exist in soil water or be adsorbed to soil particles.

A great variety of prokaryotes may be encountered in soil. A major portion of these have not yet been cultured but are known to exist through analysis of DNA extracted from soil samples. Morphological types of cultured Bacteria include gram-positive rods and cocci, gram-negative rods and spirals, sheathed bacteria, stalked bacteria, mycelial bacteria (*actinomycetes*), budding bacteria, and others. In terms of oxygen requirements, they may be aerobic, facultative, or anaerobic. Physiological types include cellulolytic, pectinolytic, saccharolytic, proteolytic, ammonifying, nitrifying, denitrifying, nitrogen-fixing, sulfate-reducing, iron-oxidizing and iron-reducing, manganese-oxidizing and manganese-reducing, and other types. Morphologically the dominant forms of culturable Bacteria seem to be gram-positive cocci, probably representing the coccoid phase of *Arthrobacter* or possibly microaerophilic cocci related to *Mycococcus* (Casida, 1965). At one time, non-spore-forming rods were held to be the dominant form. Spore-forming rods are not very prevalent despite the fact that they are readily encountered when culturing the flora in soil in the laboratory. Numerical dominance of a given type as determined by classical enumeration techniques employing selective culture methods does not necessarily speak of its biochemical importance in soil. Thus nitrifying and nitrogen-fixing bacteria, although less numerous than some other physiological types, are of vital importance to the nitrogen cycle in soil. A given soil under a given set of conditions will harbor an optimum number of individuals of each resident microbial group. These numbers will change with modification in prevailing physical and chemical conditions. Total viable counts of members of the domain Bacteria in soils generally range from 10^5 g^{-1} in poor soil to 10^8 g^{-1} in garden soil.

Certain members of the Bacteria in soil are primarily responsible for mineralization of organic matter, nitrogen fixation, nitrification, denitrification (Campbell and Lees, 1967), and some other geochemically important processes such as sulfate reduction, which cannot proceed in soil without intervention of sulfate-reducing bacteria. Members of the Bacteria and certain Archaea play a significant role in mineral mobilization and immobilization. Some types of Bacteria, especially *copiotrophs* (microorganisms requiring a nutrient-rich environment), often reside in microcolonies or biofilms on soil particles because the optimum nutritional and other requirements for their existence are found there (see Section 4.2.6; Flemming et al., 2007). Conditions of nutrient supply, oxygen supply, moisture availability, pH, and E_h may vary widely from one particle to another, owing in part to the activity of different bacteria or other micro- or macroorganisms. Thus soil may contain many different microenvironments. The colonization of soil particles by bacteria, especially those forming exopolysaccharide (EPS), may cause some particles to adhere to one another (Martin and Waksman, 1940, 1941), which means that bacteria can affect soil texture.

Fungi reside mainly in the O horizon and the upper A horizon of soil because they are, for the most part, strict aerobes and find their richest food supply at these sites (Atlas and Bartha, 1997). They are of great importance in the degradation of natural polymers such as cellulose and lignin, which are the chief constituents of wood and which most kinds of bacteria are unable to attack. Fungi share the degradation products from these polymers with bacteria, which then mineralize them. Some fungi are predaceous and help control the protozoan (Alexander, 1977, p. 67) and nematode populations (Pramer, 1964) in soil. Their mycelial growth habit causes them to grow over soil particles and penetrate the pore space of soil. They may also cause clumping of soil particles. The soil fungi include members of all the major groups: Phycomycetes, Ascomycetes, Basidiomycetes, and Deuteromycetes, and also slime molds. The last ones are usually classified separately from fungi and protozoa, although possessing attributes of both. In numbers, the fungi represent a much smaller fraction of the total microbial population in soil than bacteria. Total numbers of fungi, expressed as propagules (spores, hyphae, hyphal fragments), may range from 10^4 to 10^6 g^{-1} of soil.

Protozoa are also found in soil. They inhabit mainly the upper portion of soil where their food source (prey) is abundant. They are represented by flagellates (Mastigophora), amoebae (Sarcodina), and ciliates (Ciliata). Like fungi, they are less numerous than the bacteria, typically ranging from 7×10^3 to 4×10^5 g^{-1} of soil (Alexander, 1977; Atlas and Bartha, 1997). The types and numbers of protozoa in a given soil depend on soil type and soil condition. Although both *saprozoic* and *holozoic* types occur, it is the latter that are of ecological importance in soil. Being predators, the holozoic forms help to keep the bacteria, and, to a much lesser extent, other protozoa, fungi, and algae in check (Paul and Clark, 1996).

In the study of soils, cyanobacteria and algae have been considered as a single group labeled algae, although the cyanobacteria are prokaryotes and the algae are eukaryotes. Although both groups are associated mostly with aquatic environments, they occur in significant numbers in the O horizon and the uppermost portion of the A horizon in soils (Alexander, 1977; Atlas and Bartha, 1997) where sufficient light penetrates through translucent minerals and pore space. Overall, they

TABLE 4.3
Bacterial Densities at Different Depths of Different Soil Types on July 7, 1915

Soil Depth (in.)	Bacterial Densities (Bacteria per Gram of Air-Dried Soil $\times$ 10^6)			
	Garden Soil	Orchard Soil	Meadow Soil	Forest Soil
1	7.76	6.23	6.34	1.00
4	6.22	3.70	5.20	0.34
8	2.81	1.01	3.80	0.27
12	0.80	0.82	1.11	0.060
20	0.31	0.075	0.10	0.040
30	0.30	0.052	0.70	0.023

Source: Waksman SA, *Soil. Sci.*, 1, 363–380, 1916.

are the least numerous group of the various microbial groups, ranging from 10^2 to 10^4 g^{-1} of soil just below the soil surface (Alexander, 1977), but as many as 10^6 g^{-1} have been reported (Atlas and Bartha, 1997). The most important true algal groups represented in soil are the green algae (chlorophytes) and diatoms (chrysophytes). Xanthophytes may also be present. Although cyanobacteria and algae are photosynthetic and therefore grow mostly at or just below the soil surface where light can penetrate, they may also be found below the light zone, where they seem to grow heterotrophically. Some cyanobacteria, green algae, and diatoms have been shown to be capable of heterotrophic growth in the dark. Some of the cyanobacteria in soil are capable of nitrogen fixation and may in some cases be more important in enriching the soil in fixed nitrogen than other bacteria. The growth of algae in soil is dependent on adequate moisture and CO_2 supply. The latter is rarely limiting. The pH will influence which algae will predominate. Cyanobacteria prefer neutral to alkaline soil, whereas green algae will also grow in acid soil. When growing photosynthetically, cyanobacteria and algae are primary producers, generating at least some of the reduced carbon in soil on which heterotrophs depend. Because of their role as primary producers, cyanobacteria and algae are the pioneers in the formation of new soils, for instance, in volcanic areas.

Bacterial distribution based on viable counts in mineral soil is given in Table 4.3. Based on such counts, the largest number of organisms has been found characteristically in the upper A horizon and the smallest in the B horizon. Generally, aerobic bacteria have been found to be more numerous than anaerobic bacteria, actinomycetes, fungi, or algae. In enumerations as those in Table 4.3, anaerobes were found to decrease with depth to about the same degree as aerobes. This seems contradictory but may reflect the fact that most of the anaerobes that were enumerated by the methodology used in the year 1915 were facultative. Special techniques for cultivating anaerobes were not developed until much later (Chung and Bryant, 1997; Levett, 1992).

Determination of microbial distribution in soil, when done by culturing as in the study summarized in Table 4.3, never yields an absolute estimate because no universal culture medium exists on which all living microbes can grow. A somewhat better estimate can be obtained through direct counts using fluorescence microscopy with soil preparations treated with special fluorescent reagents (see Chapter 7), provided viable cells can be distinguished from dead cells.

4.3 ORGANIC SOILS

In some special locations, *organic soils* or *histosols* are found. They form from rapid accumulation and slow decomposition of organic matter, especially plant matter, as a result of the displacement of air by water, which prevents rapid and extensive microbial decomposition of the organic matter.

These soils are thus sedimentary in origin and are never a result of rock weathering. They consist of 20% or more of organic matter (Lawton, 1955; Buol et al., 1989; Atlas and Bartha, 1997). Their formation is associated with the evolution of swamps, tidal marshes, bogs, and even shallow lakes. An organic soil such as peat may have an ash content of 2–50% and contain cellulose, hemicellulose, lignin, and derivatives; heterogeneous complexes; fats; waxes; resins; and water-soluble substances such as polysaccharides, sugars, amino acids, and humus (Lawton, 1955). The pH of organic soils may range from 3 to 8.5. Examples of such soils are peat and *mucks*. They accumulate to depths ranging from >1 m to <8 m (Lawton, 1955) and are not stratified like mineral soils. Their occurrence is rare and some are agriculturally very productive.

4.4   THE DEEP SUBSURFACE

For ~100 years, the methods available for the study of microbial distribution in the lithosphere (mainly soils) depended mainly on microbial detection and enumeration by cultivation from samples that were collected without regard to whether they were exposed to air or not. This approach seemed to indicate that viable microbes decrease rapidly to negligible numbers with depth in soil (see Table 4.3). The existence of microbial life below 50 m or more in the lithosphere was at that time considered very unlikely or at best rare. More recent investigations have shown that living microbes exist in significant numbers in the lithosphere at depths as great as 3500 m (Pedersen, 1993; Szewzyk et al., 1994; Moser et al., 2003; Onstott et al., 2003), and even 4000–5000 m (Moser et al., 2005). These studies used newly developed techniques for sampling and specialized methods for aerobic and anaerobic culture enrichment preparation, relying at first on the old enumeration methods involving cultivation but later also on methods that did not involve cultivation, that is, molecular biological techniques. The new sampling techniques involved special drilling techniques and included controls to ensure that the organisms that were recovered were not contaminants introduced during the drilling and sampling process (Pedersen, 1993; Griffin et al., 1997; Russell, 1997, Moser et al., 2003). By this new approach, bacterial population densities in groundwater from depths of 50 to 3500 m were found to range from 10 to 10^8 mL^{-1}, depending on the sampling site (Pedersen, 1993). Examples of deep subsurface formations whose groundwater was examined included sediments and porous sedimentary rock strata (Sargent and Fliermans, 1989; Colwell et al., 1997) or cracked or fissured igneous rock, for example, granite (Pedersen, 1993). Stratigraphically, bacteria have been found only in permeable rock and sedimentary strata, which may be separated by over- and underlying impermeable rock and sedimentary strata (Sinclair and Ghiorse, 1989). Besides having been detected in groundwater itself, bacteria in the deep subsurface may also reside as microcolonies and biofilms on the surface of sediment particles and the walls of rock channels (Pedersen, 1993). Mineral surfaces appear to be very important habitats for subsurface microbes.

The advent of molecular techniques in the 1990s for detecting bacterial DNA and ribosomal RNA (Kowalchuk et al., 2004) made techniques available that are much more sensitive than cultural techniques for detecting microbes. An additional advantage of applying these techniques is that they do not require growth of the organisms for detection. The application of these molecular techniques has clearly established that the majority of the microbes that are detected in this way are not presently culturable, in most if not all instances, because special physical and chemical growth requirements for their growth are as yet unknown. The molecular methods can be used qualitatively to answer the question of "who is there," and quantitatively "how many" and in some instances can even reveal some key physiological attributes of some of the molecularly characterized individuals.

Culturable and unculturable bacteria that have been detected to date represent fairly diverse physiological groups and include members of the domains Bacteria and Archaea, representing mesophiles and thermophiles, and autotrophs and heterotrophs. Their physiological attributes enable them to cope with their special environment (Balkwill, 1989; Balkwill and Boone, 1997; Pedersen, 1993; Colwell et al., 1997; Kotelnikova and Pedersen, 1998). Aerobic bacteria along with fungi, protozoa, and algae have been found restricted to shallow subsurface depths (Sinclair and Ghiorse, 1989). Anaerobic Bacteria and Archaea range over most of the subsurface with respect to depth.

In the absence of circulating, nutrient-charged groundwaters, organic nutrients are relatively sparse in the deep-subsurface habitats. Anaerobic chemolithotrophic bacteria capable of using H_2, mostly abiotic in origin, or formate, if available, as an energy source are thought to be important among primary producers of reduced forms of carbon. Such organisms include methanogens (Archaea), which form methane, and homoacetogens (Bacteria), which form acetate, in both instances by reducing CO_2 with H_2 from which they derive the energy needed for their growth (Kieft and Phelps, 1997). Important decomposers (consumers of organic matter as energy and carbon source) include anaerobic methanotrophic consortia capable of oxidizing methane with sulfate (see Chapter 22) and acetotrophs (Archaea and Bacteria) capable of anaerobically oxidizing acetate with various inorganic electron acceptors such as NO_3^-, Fe(III), Mn(III, IV), S^0, SO_4^{2-}, UO_2^{2+}, and others (Chapters 16, 17, 19, and 22) (Kieft and Phelps, 1997). These organisms in turn become a source of organic nutrients for other heterotrophic microbes through excretion of metabolites, predation, or scavenging of dead biomass. Rock weathering by subsurface microbes may result in mobilization of suitable electron acceptors for their energy metabolism and needed trace elements.

The rate of microbial metabolism in the deep subsurface has been reported to be several orders of magnitude slower (Chapelle and Lovley, 1990; Phelps et al., 1994) than the metabolic rates measured in soil. Very low *in situ* rates at specific deep subsurface sites were estimated by applying geochemical models. In one model, microbial CO_2 production from organic carbon along an aquifer flow path was estimated by taking into consideration CO_2 production from available organic matter; interaction of the produced CO_2 with carbonate in the aquifer matrix; and the resultant changes in the groundwater chemistry, the groundwater flow rate, and the length of the flow path (Chapelle and Lovley, 1990). By comparison, metabolic rate measurements obtained by direct application of small amounts of radiolabeled test substrates to subsurface samples from the same sites led to apparent overestimation of *in situ* rates by two orders of magnitude or more. In dismissing metabolic rate measurements obtained by the application of small amounts of test substrates as overestimations, the following needs to be remembered, however. Although *in situ* nutrient supply in the deep subsurface is very limited, fresh supplies that stimulate metabolism may be introduced very episodically in cases where replenishment by groundwater movement is possible. Hence metabolic rates in the subsurface may rise briefly at widely spaced time intervals. Because of the relatively short duration of these increased *in situ* rates, they have little influence on the average rates determined by applying geochemical models. Metabolic rate estimates obtained by the addition of radiolabeled substrates to deep subsurface samples will give an indication of *metabolic potential*.

Factors other than carbon or energy source availability may limit *in situ* microbial metabolism in the deep subsurface. Kieft et al. (2005) have concluded that groundwater consisting of 2.0 Ga-old saline fracture water admixed with meteoritic water in a fracture zone of the Klooft 7-shaft mine at 3.1 kmbls (kilometers below the surface) has a residence time estimated to be 20–160 Ma. *In situ* growth of naturally present sulfate-reducing bacteria, detectable at a concentration in the order of 10^3 mL^{-1}, appears to be limited, despite the availability of significant concentrations of H_2, CH_4, CO, ethane, propane, butane, and acetate of abiogenic origin, because of a limitation in available inorganic nutrients such as Fe and phosphate. Much remains to be learned about the microbiology of the deep subsurface.

4.5 SUMMARY

The lithosphere of the Earth consists of rock, which may be igneous, metamorphic, or sedimentary. Rock is composed of intergrown minerals. The rock surface and, in the case of porous rock, the interior of rock may be habitats for microbes. Rock may be broken down by weathering, which may ultimately lead to the formation of mineral soil. Some of the rock minerals are chemically altered in this process. Weathering may be biological, especially microbiological, as well as chemical and physical.

Progress of mineral soil development is recognizable in a soil profile. A vertical section through mineral soil may reveal more or less well-developed horizons. Typical horizons of spodosols and mollisols include the litter zone (O horizon), a leached layer (A horizon), an enriched layer (B horizon), and the parent material (C horizon). The aspect of the horizons varies with soil type. Climate is one of the several important determinants of soil type. The horizons are the result of intense biological activity in the litter zone and A horizon. Much of the organic matter in the litter zone is microbially solubilized and at least partly degraded. Soluble components are washed into the A horizon or transported there by some invertebrates, where they may be further metabolized and they contribute directly or indirectly to the transformation of some of the mineral matter. Soluble products, especially inorganic ones, formed in the A horizon may be washed into the B horizon. The more refractory organic matter in the soil accumulates as humus, which contributes to the soil's texture, water-holding capacity, and general fertility. Mineral soil may have 50% solid matter and 50% pore space. The pore space is occupied by gases, such as N_2, CO_2, and O_2, and water. Water also surrounds soil particles to varying degrees. Microbes, including bacteria, fungi, protozoa, and algae, may inhabit the soil pores or live on the surface of soil particles. They are most numerous in the upper layer of soil.

Not all soils can be classified as mineral soils. A few soils are organic and have a different origin. They arise from the slow decomposition of organic matter, mainly plant residues, which accumulates by sedimentation as in swamps, marshes, and shallow lakes. They are not stratified and usually have low mineral content.

Soil is not the only important microbial habitat of the lithosphere. Microbes have been detected in the deep subsurface of the lithosphere, at depths in excess of 3500 m. Although aerobic bacteria, fungi, protozoa, and algae are found at shallower depths, anaerobic bacteria predominate in deeper zones. These organisms can be found in permeable strata formed by sediments, sedimentary rock, and cracked or fissured igneous rock. They inhabit the pore water in these strata and also the exposed mineral surfaces of mineral particles or rock on which they may form microcolonies or biofilms.

The bacteria in the lithosphere exhibit great diversity morphologically and physiologically. Their average *in situ* metabolic rates in the deep subsurface appear to be very low owing to limitations in major or essential minor nutrients. Much remains to be discovered about life in the deep subsurface of the lithosphere.

REFERENCES

Alexander M. 1977. *Introduction to Soil Microbiology*. 2nd ed. New York: Wiley.

Atlas RM, Bartha R. 1997. *Microbial Ecology. Fundamentals and Applications*, 4th ed. Menlo Park, CA: Addison Wesley Longman.

Balkwill DL. 1989. Numbers, diversity, and morphological characteristics of aerobic, chemoheterotrophic bacteria in deep subsurface sediments from a site in South Carolina. *Geomicrobiol J* 7:33–52.

Balkwill DL, Boone DR. 1997. Identity and diversity of microorganisms cultured from subsurface environments. In: Amy PS, Haldeman DL, eds. *The Microbiology of the Terrestrial Deep Subsurface*. Boca Raton, FL: CRC Lewis, pp. 105–117.

Barker WW, Banfield JF. 1996. Biological versus inorganically mediated weathering reactions: Relationships between minerals and extracellular microbial polymers in lithobiontic communities. *Chem Geol* 132:55–69.

Barker WW, Banfield JF. 1998. Zones of chemical and physical interaction at interfaces between microbial communities and minerals: A model. *Geomicrobiol J* 15:223–224.

Bennett PC, Hiebert FK, Choi WJ. 1996. Microbial colonization and weathering of silicates in petroleum-contaminated groundwater. *Chem Geol* 132:45–53.

Berner RA, Holdren GR Jr. 1977. Mechanism of feldspar weathering: Some observational evidence. *Geology* 5:369–372.

Berthelin J. 1977. Quelques aspects des mécanismes de transformation des minéraux des sols par les microorganismes hétérotrophes. *Sci Bull AFES* 1:13–24.

Berthelin J, Boymond D. 1978. Some aspects of the role of heterotrophic microorganisms in the degradation of waterlogged soils. In: Krumbein WE, ed. *Environmental Biogeochemistry and Geomicrobiology*. Ann Arbor, MI: Ann Arbor Science, pp. 659–673.

Brady PV, Carroll SA. 1994. Direct effects of CO_2 and temperature on silicate weathering. Possible implications for climate control. *Geochim Cosmochim Acta* 58:1853–1856.

Brock TD. 1975. Effect of water potential on growth and iron oxidation by *Thiobacillus ferrooxidans*. *Appl Microbiol* 29:495–501.

Brock TD, Smith DW, Madigan MT. 1984. *Biology of Microorganisms*. 4th ed. Englewood Cliffs, NJ: Prentice Hall.

Brown AD. 1976. Microbial water stress. *Bacteriol Rev* 40:803–846.

Browne BA, Driscoll CT. 1992. Soluble aluminum silicates: Stoichiometry, stability, and implications for environmental geochemistry. *Science* 256:1667–1670.

Bunting BT. 1967. *The Geography of Soil*. London: Hutchison University Library.

Buol SW, Hole FD, McCracken RJ. 1980. *Soil Genesis and Classification*. 2nd ed. Ames, IA: Iowa State University Press.

Buol SW, Hole FD, McCracken RJ. 1989. *Soil Genesis and Classification*. 3rd ed. Ames, IA: Iowa State University Press.

Campbell NER, Lees H. 1967. The nitrogen cycle. In: McLaren AD, Petersen GH, eds. *Soil Biochemistry*. Vol. 1. New York: Marcel Dekker, pp. 194–215.

Casida EL. 1965. Abundant microorganisms in soil. *Appl Microbiol* 13:327–334.

Chapelle FH, Lovley DR. 1990. Rates of microbial metabolism in deep coastal plain aquifers. *Appl Environ Microbiol* 56:1865–1874.

Chou L, Wollast R. 1984. Study of weathering of albite at room temperature and pressure with a fluidized bed reactor. *Geochim Cosmochim Acta* 48:2205–2218.

Chung K-T, Bryant MP. 1997. Robert E. Hungate: Pioneer of anaerobic microbial ecology. *Anaerobe* 3:213–217.

Colwell FS, Onstott TC, Delwiche ME, Chandler D, Fredrickson JK, Yao Q-J, McKinley JP, Boone DR, Griffiths R, Phelps TJ, Ringelberg D, White DC, LaFreniere L, Balkwill D, Lehman RM, Konnisky J, Long PE. 1997. Microorganisms from deep, high temperature sandstones: Constraints on microbial colonization. *FEMS Microbiol Rev* 20:425–435.

Costerton JW, Lewandowski Z, deBeer D, Caldwell D, Korber D, James G. 1994. Biofilms, the customized microniche. *J Bacteriol* 176:2137–2142.

Dommergues Y, Mangenot F. 1970. *Ecologie Microbienne du Sol*. Paris: Masson.

Ehrlich HL. 2001. Interactions between microorganisms and minerals under anaerobic conditions. In: Huang PM, Bollag J-M, Senesi N, eds. *Interactions Between Soil Particles and Microorganisms. Impact on the Terrestrial Environment. IUPAC Series on Analytical Physical Chemisty of Environmental Systems*. Vol. 8. New York: Wiley, Chap. 11, pp. 459–494.

Flemming H-C, Neu TR, Wozniak DJ. 2007. The EPS matrix: The "house of biofilm cells." *J Bacteriol* 189:7945–7947.

Friedmann EI. 1982. Endolithic microorganisms in the Antarctic cold desert. *Science* 215:1045–1053.

Fuller WH. 1974. Desert soils. In: Brown GW, ed. *Desert Biology*. Vol. II. New York: Academic Press, pp. 31–101.

Golubic S, Perkins DD, Lukas KJ. 1975. Boring microorganisms and microborings in carbonate substrates. In: Frey RW, ed. *The Study of Trace Fossils*. New York: Springer, pp. 229–259.

Golubic S, Schneider J. 1979. Carbonate dissolution. In: Trudinger PA, Swaine DJ, eds. *Biogeochemical Cycling of Mineral Forming Elements*. Amsterdam: Elsevier, pp. 107–129.

Griffin WT, Phelps TJ, Colwell FS, Fredrickson JK. 1997. Methods for obtaining deep subsurface microbiological samples by drilling. In: Amy PS, Haldeman DL, eds. *The Microbiology of the Terrestrial Deep Subsurface*. Boca Raton, FL: CRC Lewis, pp. 23–44.

Harris RF, Gardner WR, Adebayo AA, Sommers LE. 1970. Agar dish isopiestic equilibration method for controlling the water potential of solid substrates. *Appl Microbiol* 19:536–537.

Hernandez ME, Kappler A, Newman DK. 2004. Phenazines and other redox-active antibiotics promote microbial mineral reduction. *Appl Environ Microbiol* 70:921–928.

Hernandez ME, Newman DK. 2001. Extracellular electron transfer. *CMLS, Cell Mol Life Sci* 58:1562–1571.

Hiebert FK, Bennett PC. 1992. Microbial control of silicate weathering in organic-rich groundwater. *Science* 258:278–281.

Jones CG, Shachak M. 1990. Fertilization of the desert soil by rock-eating snails. *Nature* (London) 346:839–841.

Jones D, Wilson MJ, McHardy WJ. 1981. Lichen weathering of rock-forming minerals: Application of scanning electron microscopy and microprobe analysis. *J Microsc* 124:95–104.

Kieft TL, McCuddy SM, Onstott TC, Davidson M, Lin L-H, Mislowack B, Pratt L, Boice E, Lollar BS, Lippmann-Pipke J, Pfiffner SM, Phelps TJ, Gihring T, Moser D, van Heerden A. 2005. Geochemically generated, energy-rich substrates and indigenous microorganisms in deep, ancient groundwater. *Geomicrobiol J* 22:325–335.

Kieft TL, Phelps TJ. 1997. Life in the slow lane: Activities of microorganisms in the subsurface. In: Amy PS, Haldeman DL, eds. *The Microbiology of the Terrestrial Deep Subsurface*. Boca Raton, FL: CRC Lewis, pp. 137–163.

Kotelnikova S, Pedersen K. 1998. Distribution and activity of methanogens and homo-acetogens in deep granitic aquifers at Äspö Hard Rock Laboratory, Sweden. *FEMS Microbiol Ecol* 26:121–134.

Kowalchuk GA, de Bruijn FJ, Head IM, Akkermans ADL, van Elsas JD, eds. 2004. *Molecular Microbial Ecology*. 2nd ed. Dordrecht, The Netherlands: Kluwer Academic.

Kuznetsov SI, Ivanov MV, Lyalikova NN. 1963. *Introduction to Geological Microbiology*. Engl. transl. New York: McGraw-Hill, pp. 33–41.

Lang ARG. 1967. Osmotic coefficients and water potentials of chloride solutions from 0–40°C. *Aust J Chem* 20:2017–2023.

Lawton K. 1955. Chemical composition of soils. In: Bear FE, ed. *Chemistry of the Soil*. New York: Van Nostrand Reinhold, pp. 53–84.

Levett PN. 1992. *Anaerobic Microbiology: A Practical Approach*. Oxford, UK: Oxford University Press.

Lovley DR, Coates JD, Blunt-Harris EL, Phillips EJP, Woodward JC. 1996. Humic substances as electron acceptors for microbial respiration. *Nature* (London) 382:445–448.

Lovley DR, Fraga JL, Blunt-Harris EL, Hayes LA, Phillips EJP, Coates JD. 1998. Humic substances as a mediator for microbially catalyzed metal reduction. *Acta Hydrochimica et Hydrobiologica* 26:152–157.

Lucas Y, Luizao FJ, Chauvel A, Rouiller J, Nahon D. 1993. The relation between biological activity of the rain forest and mineral composition of soils. *Science* 260:521–523.

Martin JP, Waksman SA. 1940. Influence of microorganisms on soil aggregation and erosion. *Soil Sci* 50:29–47.

Martin JP, Waksman SA. 1941. Influence of microorganisms on soil aggregation and erosion. II. *Soil Sci* 52:381–394.

Moser DP, Gihring TM, Brockman FJ, Fredrickson JK, Balkwill DL, Dollhopf ME, Lollar BS, Pratt LM, Boice E, Southam G, Wanger G, Baker BJ, Pfiffner SM, Lin L-H, Onstott TC. 2005. *Desulfotomaculum* and *Methanobacterium* spp. dominate a 4- to 5-kilometer-deep fault. *Appl Environ Microbiol* 71:8773–8783.

Moser DP, Onstott TC, Fredrickson JK, Brockman FJ, Balkwill DL, Drake GR, Pfiffner SM, White DC, Takai K, Pratt LM, Fong J, Lollar BS, Slater G, Phelps TJ, Spoelstra N, Deflaun M, Southam G, Welty AT, Baker BJ, Hoek J. 2003. Temporal shifts in the geochemistry and microbial community structure of an ultradeep mine borehole following isolation. *Geomicrobiol J* 20:517–548.

Newman DK, Kolter R. 2000. A role for excreted quinines in extracellular electron transfer. *Nature* (London) 405:94–97.

Oelkers EH, Schott J, Devidal J-L. 1994. The effect of aluminum, pH, and chemical affinity on the rates of aluminosilicate dissolution reactions. *Geochim Cosmochim Acta* 58:2011–2024.

Onstott TC, Moser DP, Pfiffner SM, Fredrickson JK, Brockman FJ, Phelps TJ, White DC, Peacock A, Balkwill D, Hoover R, Krumholz LR, Borscik M, Kieft TL, Wilson R. 2003. Indigenous and contaminant microbes in ultradeep mines. *Environ Microbiol* 5:1168–1191.

Paul EA, Clark FE. 1996. *Soil Microbiology and Biochemistry*. 2nd ed. San Diego, CA: Academic Press.

Pedersen K. 1993. The deep subterranean biosphere. *Earth Sci Rev* 34:243–260.

Phelps TJ, Murphy EM, Pfiffner SM, White DC. 1994. Comparison between geochemical and biological estimates of subsurface microbial activities. *Microb Ecol* 28:335–349.

Pramer D. 1964. Nematode trapping fungi. *Science* 144:382–388.

Rock collection of the Department of Earth and Environmental Sciences, Rensselaer Polytechnic Institute.

Russell CE. 1997. The collection of subsurface samples by mining. In: Amy PS, Haldeman DL, eds. *The Microbiology of the Deep Subsurface*. Boca Raton, FL: CRC Lewis, pp. 45–59.

Sargent KA, Fliermans CB. 1989. Geology and hydrology of the deep subsurface microbiology sampling sites at the Savannah River Plant, South Carolina. *Geomicrobiol J* 7:3–13.

Shachak M, Jones CG, Granot Y. 1987. Herbivory in rocks and the weathering of a desert. *Science* 236:1098–1099.

Sinclair JL, Ghiorse WC. 1989. Distribution of aerobic bacteria, protozoa, algae, and fungi in deep subsurface sediments. *Geomicrobiol J* 7:15–31.

Stevenson FJ. 1994. *Humus Chemistry. Genesis, Composition, Reactions.* New York: Wiley.

Szewzyk U, Szewzyk R, Senström T-A. 1994. Thermophilic, anaerobic bacteria islated from a deep borehole in granite in Sweden. *Proc Natl Acad Sci USA* 91:1810–1813.

Ullman WJ, Kirchman DL, Welch SA, Vandevivere P. 1996. Laboratory evidence for microbially mediated silicate mineral dissolution in nature. *Chem Geol* 132:11–17.

US Department of Agriculture (U.S.D.A.) Soil Conservation Service, Washington, DC, USA.

Waksman SA. 1916. Bacterial numbers in soils at different depths and in different seasons of the year. *Soil Sci* 1:363–380.

Waksman SA, Starkey RL. 1931. *The Soil and the Microbe.* New York: Wiley.

Welch SA, Ullman WJ. 1993. The effect of organic acids on plagioclase dissolution rates and stoichiometry. *Geochim Cosmochim Acta* 57:2725–2736.

Welch SA, Vandevivere P. 1995. Effect of microbial and other naturally occurring polymers in mineral dissolution. *Geomicrobiol J* 12:227–238.

Wieland E, Stumm W. 1992. Dissolution kinetics of kaolinite in acidic aqueous solutions at 25°C. *Geochim Cosmochim Acta* 56:3339–3355.

5 The Hydrosphere as Microbial Habitat

5.1 THE OCEANS

5.1.1 PHYSICAL ATTRIBUTES

The oceans are a habitat for various forms of life, ranging from the largest organisms anywhere on Earth to the smallest. The *fauna* includes various vertebrates (mammals, birds, reptiles, and fish) as well as a wide range of invertebrates. The *flora* includes the algae—from the macroscopic kelps to the small unicellular forms. The *plankton* includes the floating biota. The *phytoplankton* includes free-floating algae such as diatoms and dinoflagellates, and the *zooplankton* includes the free-floating microscopic protozoa and invertebrates. The *bacterioplankton* consists of free, unattached bacterial forms that include some kinds of cyanobacteria. In considering the ecology of the oceans, cyanobacteria, which are prokaryotes, have often been considered as the true algae, which, however, are eukaryotes.

The oceans cover ~70% of the Earth's surface, occupying an area of 3.6×10^8 km^2 and a volume of 1.37×10^9 km^3, which amounts to 1.41×10^{21} kg of water. By comparison, the total mass of the Earth is estimated to be 5.98×10^{24} kg. Because of the unequal distribution of the contents between the northern and southern hemispheres of the present Earth, only 60.7% of the northern hemisphere is covered by oceans, whereas 80.9% of the southern hemisphere is covered by them. The world's major oceans include the Atlantic Ocean, 16.2%; the Pacific Ocean, 32.4%; the Indian Ocean, 14.4%; and the Arctic Ocean, 2.8% of the Earth's surface. The average depth of all oceans is 3795 m. The average depth of the Atlantic Ocean is 3296 m, that of the Pacific Ocean is 4282 m, that of the Indian Ocean is 3693 m, and that of the Arctic Ocean is 1205 m. The greatest depths in the oceans occur in the *ocean trenches*. For instance, in the Pacific Ocean, the water depth of the Marianas Trench is 10,500 m; in the Atlantic Ocean, the water depth of the Puerto Rico Trench is 8650 m; and in the Indian Ocean, the water depth of the Java Trench is 7450 m. Shallow ocean depths are encountered in the marginal seas along the coasts of continents. These are usually <2000 m and frequently <1000 m deep. Figure 5.1 shows the oceans of the world (Williams, 1962; Bowden, 1975).

An ocean includes a basin with several structural features. Its walls are formed by the *continental margin*. Projecting from each continental shore is the *continental shelf*; all shelves together encompassing ~7.5% of the ocean area. Each shelf slopes gently downward in the direction of the ocean to a water depth of ~130 m at an average angle of 7'. The average shelf width is ~65 km, but may range from 0 to 1290 km, the greatest width being represented by the shelf projecting from the coast of Siberia into the North Polar Sea. The waters over the continental shelves are a biologically important part of the oceans. They are sites of high biological productivity because they receive a significant contribution of nutrients in general runoff from the adjacent land, particularly from rivers emptying into the waters over a shelf.

At the edge of the continental shelf, the ocean floor drops sharply at an average angle of 4° (range 1–10°) to abyssal depths of ~3000 m. This is the region of the *continental slope* and constitutes ~12% of the ocean area. In some places slopes are cut by deep canyons, which often occur at the mouths of large rivers (e.g., Hudson River, Amazon River). Many canyons were formed by *turbidity currents* over geologic time. Such currents consist of strong water movements that carry high sediment load, picked up as a river flows oceanward. As the river meets the sea, the sediment is

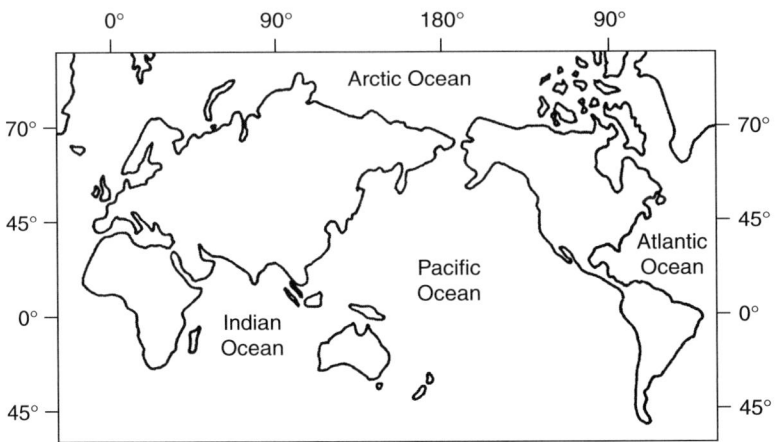

FIGURE 5.1   Oceans of the world.

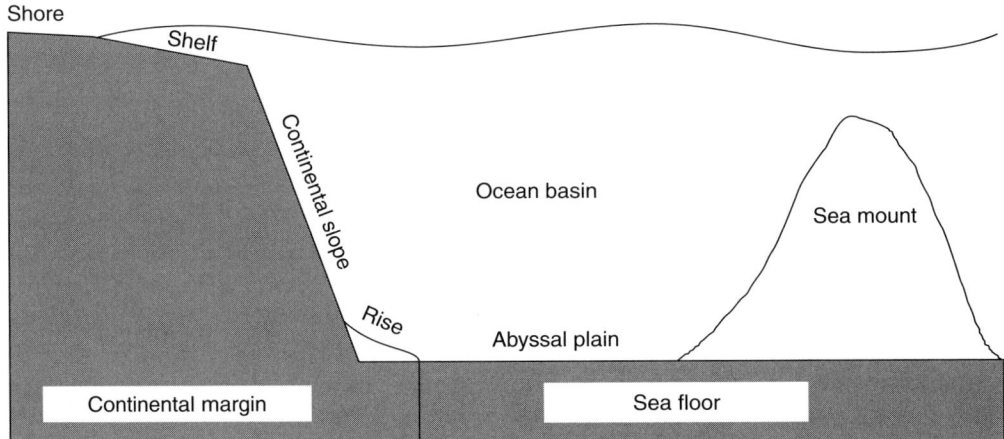

FIGURE 5.2   Schematic representation of a profile of an ocean basin.

dropped, and when settling it abrades the slope. Marine canyons may also be cut by slumping of an unstable sediment deposit on a portion of the continental slope and the consequent abrasion of the slope. Occasionally, the continental slope may be interrupted by a terraced region, as in the case of Blake Plateau off the southern Atlantic coast of the United States. This particular shelf is ~302 km wide and drops gradually from a depth of 732 to 1100 m over this distance. It was gouged out of the continental slope by the northward flowing Gulf Stream.

At the foot of the continental slope lies the *continental rise*, consisting of accumulations of sediment carried downslope by turbidity currents. Such deposits may extend for 100 km or more from the foot of the continental slope. The continental rise may form fanlike structures in some places and wedges in others. An idealized profile of a continental margin is shown in Figure 5.2.

The *ocean basin* takes up 80% of the ocean area. Its floor, far from being a flat expanse, as once believed by some, often exhibits a rugged topography. Submarine mountain ranges cut by fracture zones and rift valleys stretch over thousands of kilometers as the midocean ridge systems where the new ocean floor is created (see Chapter 2). Elsewhere, somewhat more isolated submarine mountains, some of which are active and others dormant volcanoes, dot the ocean floor. Some of the seamounts have flattened tops and have been given the special name of *guyots*. Some of the flattened tops of seamounts, especially in the Pacific Ocean, reach surface waters at depths of 50–100 m where the temperature is ~21°C. In these positions, the flat tops may serve as substratum for colonization by corals (coelenterates) and coralline algae, which then form atolls and reefs.

Covering the ocean floor almost everywhere are *sediments*. They range in thickness from 0 to 4 km, with an average thickness of 300 m. Their rate of accumulation varies, being slowest in midocean (<1 cm per 10^3 years) and fastest on continental shelves (10 cm per 10^3 years). These rates may be even greater in some inland seas and gulfs (e.g., 1 cm per 10–15 years in the Gulf of California and 1 cm in 50 years in the Black Sea). In some regions of the deep ocean, the sediments consist mainly of deposits of siliceous and calcareous remains of marine organisms. The siliceous remains are derived from the frustules of diatoms (algae) and the support skeleton and spines of radiolarians (protozoa). The calcareous remains are derived from the tests of foraminifera (protozoa), carbonate platelets from the walls of coccolithophores (algae), and shells from pteropods (mollusks). *Diatomaceous oozes* predominate in colder waters (e.g., in the North Pacific between 40° and 70° north latitude and 140° west to 145° east longitude; Horn et al., 1972). *Radiolarian oozes* predominate in warmer waters (e.g., in the North Pacific between 5° and 20° north latitude and 90° and 180° west longitude; Horn et al., 1972). *Calcareous oozes* are found mainly in warmer waters on ocean bottoms no deeper than 4550–5000 m (e.g., in the North Pacific between 0° and 10° north latitude and 80° and 180° west longitude; Horn et al., 1972). At greater depths, the CO_2 concentration in the water is high enough to cause dissolution of carbonate unless the structures are enclosed in a protective membrane.

Other vast areas of the ocean floor are covered by clays (*red clay* or *brown mud*), which are probably of terrigenous origin and washed into the sea by rivers and general runoff from continents and islands and carried into the ocean basins by ocean currents, mudflows, and turbidity currents. At high latitudes in both hemispheres, particularly on and near continental shelves, ice-rafted sediments are found. They were dropped into the ocean by melting icebergs, which had previously separated from glacier fronts that had picked up terrigenous debris during glacial progression. Except for ice-rafted detritus, only the fine portion of terrigenous debris (clays and silts) is carried out to sea. The clay particles are defined as having a diameter <0.004 mm, and the silt particles as having a size range of 0.004–0.1 mm in diameter. Figure 5.3 shows the appearance of some Pacific Ocean sediments under the microscope.

5.1.2 OCEAN IN MOTION

A significant portion of the ocean is in motion at all times (Williams, 1962; Bowden, 1975). The causes of this motion are (1) wind stress on the surface waters, (2) Coriolis force arising from the rotation of the Earth, (3) density variation of seawater resulting from temperature and salinity changes, and (4) tidal movement due to gravitational influences on the water exerted by the sun and the moon. Surface currents (Figure 5.4) are prominent in regions of prevailing winds, such as the trade winds, which blow from east to west at ~20° north and south latitudes; the westerlies, which blow from west to east between 40° and 60° north and south latitude; and the easterly polar winds, which blow in a westerly direction south of the Arctic Ocean. The effect of these winds, together with the deflecting influence of the continents and the Coriolis force, is to set up surface circulations in the form of *gyrals* between the north and south poles in each major ocean. They are the North Subtropical Gyral (large), the North Tropical Gyral (small), the South Tropical Gyral (small), the South Subtropical Gyral (large), and the Antarctic Current that circulates around Antarctica from west to east (Figure 5.4A). Thus the Gulf Stream, together with the Canary Current and the North Equatorial Current, is part of the North Subtropical Gyral of the Atlantic Ocean (Figure 5.4B). The flow rates of the waters in these gyrals and segments of them are different. The flow rate of the water in the Gulf Stream is fastest of any surface current—250 cm s^{-1}. Other currents have flow rates that are mostly in the range of 25–65 cm s^{-1}.

Meanders in the Gulf Stream in the Atlantic and the Kuroshio Current in the Pacific Ocean may give rise to the so-called rings—small closed current systems that may measure as much as

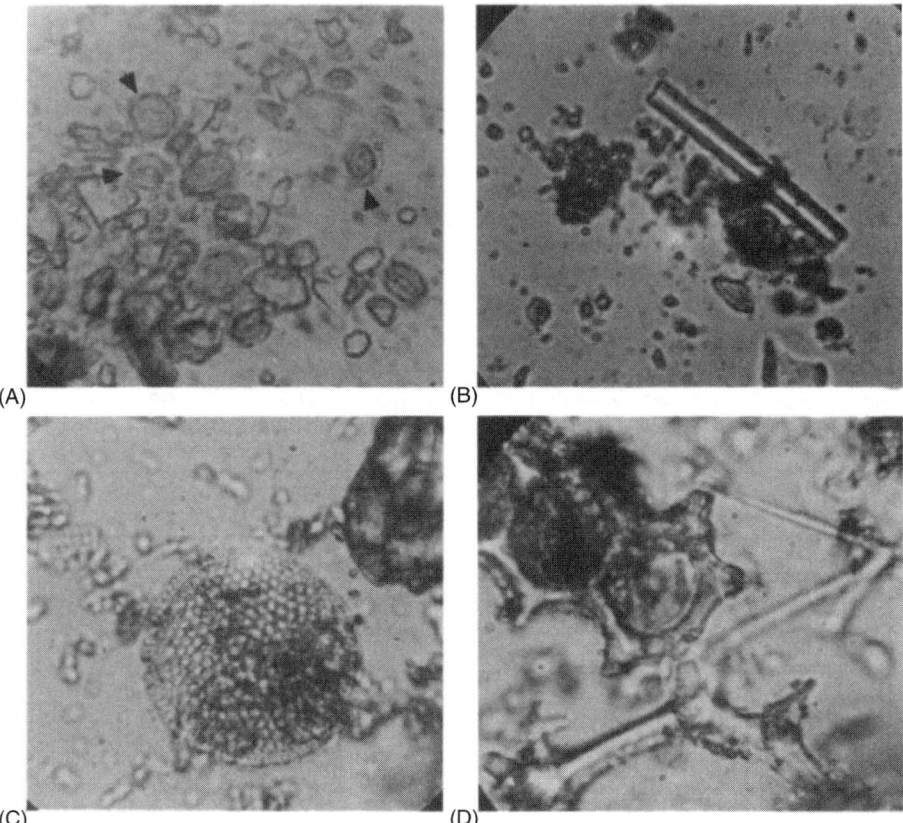

FIGURE 5.3 Microscopic appearance of marine sediments (×1750). (A) Atlantic sediment showing coccoliths (CaCO$_3$) (arrows) and clay particles. (B) Atlantic sediment showing diatom frustules (SiO$_2$) and other debris. (C) Pacific sediment showing a centric diatom frustule (SiO$_2$) and other debris. (D) Pacific sediment showing fragments of radiolarian tests (SiO$_2$).

300 km in diameter and may have a depth as great as 2 km. Such rings may move 5–10 km per day. The chemical, physical, and biological characteristics of the water enclosed in a ring may be significantly different from those of the surrounding water. A slow exchange of solutes and biota as well as heat transfer may take place across the boundary of a ring. Rings thus constitute the means of nutrient transport from ocean currents. The rings may ultimately rejoin the current that spawned them (Gross, 1982; Richardson, 1993; Ring Group, 1981). More recently, *anticyclones* have been reported to arise from the Gulf Stream in addition to the rings, and *meddies* from the north of the Strait of Gibraltar (Richardson, 1993). Although rings have a cold water core surrounded by a warm water layer and counterclockwise rotation, anticyclones have a warm water core surrounded by colder water and clockwise rotation. Meddies have a core that is more saline than the surrounding ocean water and a clockwise rotation. Collectively, these formations are known as *ocean eddies*.

Deep water is also in motion. Its movement appears to be caused by slow diffusion resulting from density differences of water masses through broad zones in the oceans. Some of the deep currents that have been measured in the Atlantic Ocean have a velocity between 1 and 2 cm s^{-1} (Dietrich and Kalle, 1965, pp. 399, 407; Gross, 1982). Occasional short bursts in velocity may occur. The bottom current movement is influenced by bottom topography.

Deep water may rise toward the surface in a process called *upwelling*. This results from the moving apart of two surface water masses, causing the deep water to rise to take the place of the divergent waters. The moving apart of the water masses is called *divergence* (Williams, 1962).

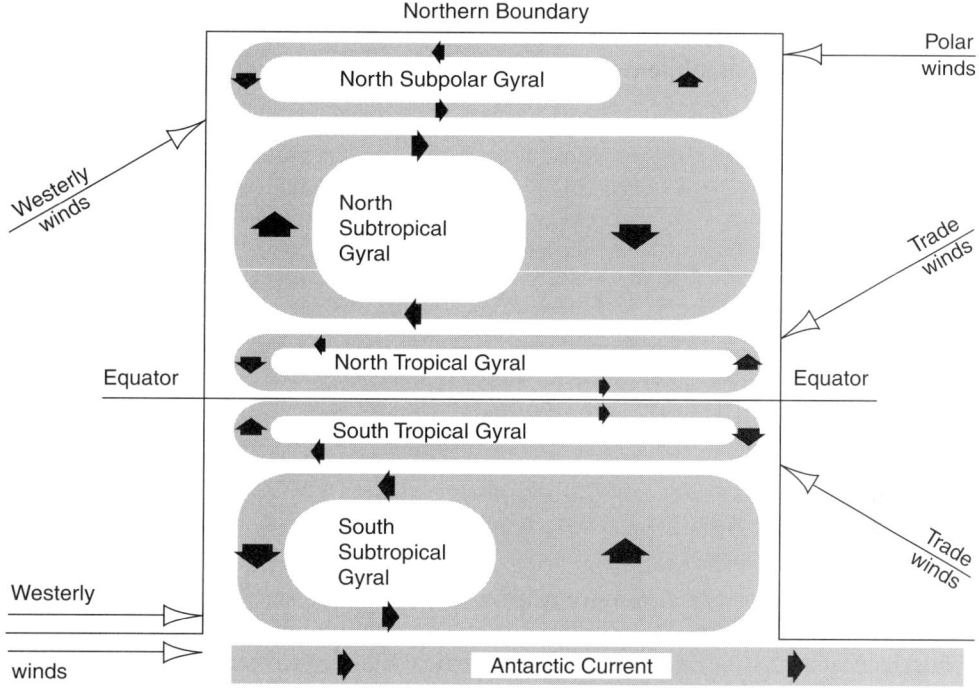

(A)

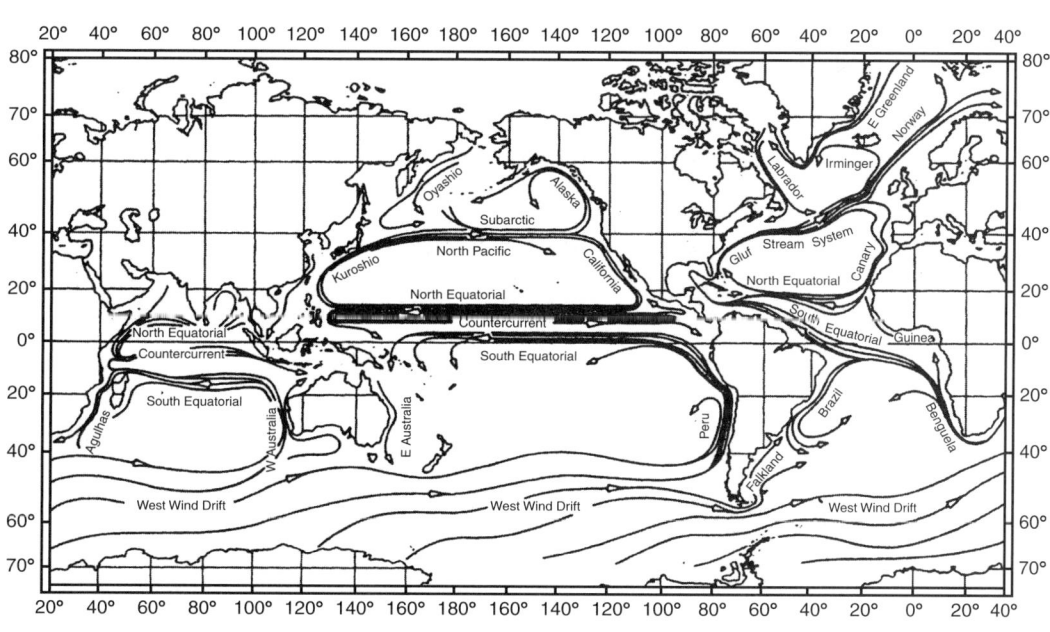

(B)

FIGURE 5.4 Oceanic surface currents. (A) Schematic representation of the prevailing winds and their effects on the surface currents of an imaginary rectangular ocean. (B) Average surface currents of the world's oceans. (From Williams J, *Oceanography*, Little, Brown, Boston, MA, 1962. With permission. [Jerome Williams is deceased.])

TABLE 5.1
Some Constituents of Seawater (μg L^{-1})

Major Constituents		Minor Constituents			
Cl	1.9×10^7	Si	3×10^3	Cu	3
Na	1.1×10^7	N	6.7×10^2	Fe	3
Mg	1.3×10^6	Li	1.7×10^2	U	3
S (SO$_4$)	9.0×10^5	Rb	1.2×10^2	As	2.6
Ca	4.1×10^5	P	90	Mn	2
K	3.9×10^5	I	60	Al	1
Br	6.7×10^4	Ba	20	Co	0.4
C (CO$_3$, HCO$_3$)	2.8×10^4	Mo	10	Se	9×10^{-2}
B	4.5×10^3	Zn	10	Pb	3×10^{-2}
		Ni	7	Ra	1×10^{-7}

Source: Marine Chemistry. *A Report to the Marine Chemistry Panel of the Committee
of Oceanography*, National Academy of Sciences, Washington, DC, 1971.

Upwelling of deep water may also result when winds blow large surface water masses away from coastal regions (Smith, 1968). Deep water is relatively rich in mineral nutrients, including nitrate and phosphate, and thus, upwelling is of great ecological significance because it replenishes biologically depleted nutrients in the surface waters. Regions of upwelling are therefore very fertile. An important region of upwelling in the eastern Pacific Ocean occurs off the coast of Peru. A disturbance in the surface water circulation in the southern Pacific can result in failure of upwelling in this region (El Niño) and can spell temporary disaster for the fisheries of the area.

When a dense surface water mass meets a lighter water mass, a *convergence* occurs, and the heavier water will sink to a level where it meets water of the same density. This phenomenon is important because the denser, sinking surface water will carry oxygen to the deep waters. Important convergences in the oceans occur at high latitude in both the hemispheres.

5.1.3   Chemical and Physical Properties of Seawater

Seawater is saline. Some important chemical components of seawater, listed in decreasing order of concentration, are presented in Table 5.1 (Marine Chemistry, 1971). Of these components, chloride (55.2%), sodium (30.4%), sulfate (7.7%), magnesium (3.7%), calcium (1.16%), potassium (1.1%), bromide (0.1%), strontium (0.04%), and borate (0.07%) account for 99.5% of the total salts in solution (percent, in weight per volume). Because these components generally occur in constant proportions relative to one another in true ocean waters, it has been possible to estimate salt concentrations in seawater samples by merely measuring chloride concentration. The chloride concentration in grams per kilogram (chlorinity, Cl) is related to the total salt concentration (salinity, S) in grams per kilogram by the relationship

$$S(\text{‰}) = 0.030 + 1.8050\text{Cl} \ (\text{‰})* \tag{5.1}$$

The salinity so determined is an estimate of the total amount of solid material in a unit mass of seawater in which all carbonate salts have been converted to oxides and all bromide and iodide have been replaced by chloride, and in which all organic matter has been completely oxidized.

---

* The symbol "‰" represents parts per thousand or grams per kilogram. Equation 5.1 was amended to S(‰) = 0.030 + 1.80655Cl (‰) by UNESCO in 1969.

For reference purposes, the salinity of standard seawater has been taken as 34.3‰. The actual salinity of different parts of the world oceans can vary from <33.4‰ to almost 36‰ (Dietrich and Kalle, 1965, p. 156).

Accurate titrimetric chlorinity determinations for determining seawater salinity are cumbersome and require skill. Nowadays, determinations of salinity are made using a salinometer, which measures electrical conductivity of seawater. Such measurements are translated into *practical salinity* (*S*) using the following relationship formulated and adopted by UNESCO/CES/SCOR/IAPSO Joint Panel on Oceanographic Tables and Standards in 1978:

$$S = 0.008 - 0.1692K_{15}^{1/2} + 25.3851K_{15} + 14.0941K_{15}^{3/2} - 7.0261K_{15}^2 + 2.7081K_{15}^{5/2} \quad (5.2)$$

where K_{15} represents the ratio of electrical conductivity of a seawater sample at 15°C and 1 standard atmosphere of pressure to that of a standard KCl solution containing 35.4356 g of KCl in 1 kg of solution (see http://www.salinometry.com/content/view/18/31/). The value of K_{15} for seawater having a salinity of 35‰ is exactly unity. Practical salinity as presently defined is dimensionless.

Table 5.2 lists the salinities, determined from chlorinity measurements, of some different marine waters as well as those of some saline lakes. It must be pointed out that whereas the Great Salt Lake in Utah has a salt composition that is qualitatively similar to that of oceans, some other inland hypersaline water bodies, such as the Dead Sea at the mouth of the Jordan River, have a different salt composition. In the Dead Sea, the dominant cations and their respective concentrations are in descending order: Mg (~44 g L^{-1}), Na (40 g L^{-1}), Ca (17 g L^{-1}), and K (7.5 g L^{-1}), and the dominant anions and their respective concentrations are Cl (225 g L^{-1}) and Br (5.5 g L^{-1}) (Nissenbaum, 1979).

Although some portions of the salts in seawater derive from the runoff from the continents and the weathering of minerals in the surficial sediments, a most important contribution to seawater solutes is made by hydrothermal discharges from vents at the midocean spreading centers. These discharges are the consequence of seawater penetration deep into the porous basalt (down to depths of 1–3 km) beneath the ocean floor, where the seawater then reacts with constituents of the basalt in various ways. The reactions include the reduction of seawater sulfate to hydrogen sulfide by ferrous iron in the basalt. They also include the removal of magnesium from seawater as magnesium hydroxide and the incorporation of seawater calcium into minerals such as plagioclase to form new aluminosilicate minerals such as clinozoisite or Ca-rich amphibole, accompanied by the generation of acidity. In the case of calcium incorporation into plagioclase, this can be illustrated by the reaction

$$3CaAl_2Si_2O_8 + Ca^{2+} + 2H_2O \rightarrow 2Ca_2Al_3Si_3O_{12}(OH) + 2H^+ \quad (5.3)$$

TABLE 5.2
Salinities (‰) of Some Marine Waters and Salt Lakes

Water Body	Salinity(‰)	References
Gulf of Bothnia	2–6	Smith (1974)
Baltic Sea	6–17	Smith (1974)
Black Sea	16–18	Smith (1974)
Mediterranean Sea	37–39	Bowden (1975)
Red Sea	40–41	Bowden (1975)
Dead Sea	320	ZoBell (1946)
Great Salt Lake	320	Zobell (1946)
Ocean bottoms	34.66–34.92	Defant (1961)

The resultant acidity is the cause of leaching of some other components from the basalt, such as hydrogen sulfide from pyrrhotite and base metals from some other basalt minerals. All these reactions are possible because of high *hydrostatic pressure* (HP) exerted on the solution in the basalt at these depths and because of the high temperature (~350°C) from heat diffusion from the underlying magma chambers into the reaction zone of the basalt. The resultant chemically altered seawater is forced upward by HP through porous channels and fissures in the basalt. It is ultimately discharged as a hydrothermal solution from vents at the spreading centers into the overlying seawater and mixed with it (Bischoff and Rosenbauer, 1983; Edmond et al., 1982; Seyfried and Janecky, 1985; Shanks et al., 1981) (see also Chapters 2, 17, and 20). These reactions contribute significantly to the stability of seawater composition.

Seawater contains a pH-buffering system that consists of bicarbonate and carbonate ions, borates, and silicates. The carbonate plus bicarbonate ions constitute 0.35% of the solutes in seawater. Together, these buffers keep the pH of seawater in the range of 7.5–8.5. Surface seawater pH tends to fall into a narrow range of 8.0–8.5. At depth, seawater pH may approach 7.5. To a major extent, the variation in pH of seawater with depth may be related to oxygen utilization during respiration by marine organisms, which results in CO_2 production from organic carbon. To a lesser extent, it may be related to carbonate minerals (e.g., $CaCO_3$) dissolution (Park, 1968). Figure 5.5 illustrates the changes in pH with depth at one particular station in the Pacific Ocean.

Because of the alkaline pH and elevated E_h of seawater, nutritionally available iron appears to be a limiting micronutrient for bacterioplankton and phytoplankton (Tortell et al., 1999; Hutchins et al., 1999; Gelder, 1999). This is because iron under these conditions will be ferric and, unless complexed, will predominate in the form of insoluble hydroxide, oxyhydroxides, and oxides. Various bacteria and algae release ligands (siderophores) that complex with ferric iron and thus keep it in solution at the alkaline pH of seawater and make it nutritionally available (Tortell et al., 1999; Hutchins et al., 1999; Martinez et al., 2000). Growth by some siderophore-producing marine bacteria in iron-limited waters can be stimulated by exogenous siderophores (Guan et al., 2001). In at least some parts of the oceans, it is possible to stimulate growth of phytoplankton and bacterioplankton by fertilizing the ocean water with iron (Coale et al., 1996; Church et al., 2000; Arrieta et al., 2004). Growth stimulation of phytoplankton in the ocean by iron fertilization might offer a means of significant enhancement of CO_2 sequestration in the ocean and, thus, have a positive

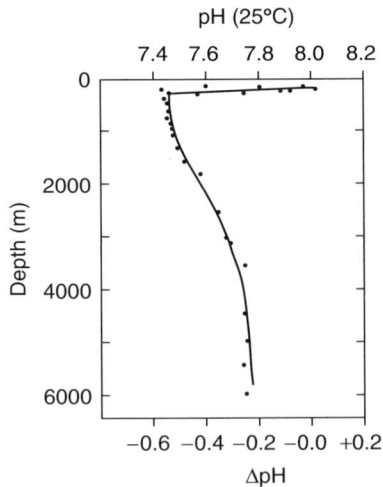

FIGURE 5.5 Vertical profile of pH at station 54°46′N, 138°36′W in the northeastern Pacific Ocean. (Adapted from Park PK, *Science*, 162, 357–368, © 1968 by the American Association for the Advancement of Science. With permission.)

effect on climate control. It might even enhance the bioproduction of silica by growth stimulation of marine diatoms, an important food source for some plankton feeders, for instance, in the seas around Antarctica (Southern Ocean). However, theoretical considerations and results from field experiments examining these possibilities suggest that the extent of sequestration of carbon and its transfer to the ocean floor as particulate organic carbon (POC) is not great enough to have a significant impact on climate control (Liss et al., 2005; Zeebe and Archer, 2005; Brzezinski et al., 2005; Arrieta et al., 2004; Buessler et al., 2004; Buessler and Boyd, 2003; Bakker, 2002).

The salts dissolved in seawater impart a special osmotic property to it. The *osmotic pressure* of seawater is of the order of magnitude of the internal pressure of bacterial cells or the cell sap of eukaryotic cells. At a salinity of 35‰ and a temperature of 0°C, seawater has an osmotic pressure of 23.37 bar (23.07 atm), whereas at the same salinity but at 20°C it has an osmotic pressure of 25.01 bar (24.96 atm). Clearly then, the osmotic pressure of seawater is not deleterious to living cells.

With increasing depth in the water column, HP becomes a significant factor in the life of microbes and other forms of life in the sea. On average, the HP in the open ocean increases ~1 atm (1.013 bar) for every 10 m of depth. Related to the weight of overlying water at a given depth, HP in the oceans ranges from 0 to more than 1000 atm (1013 bar). Thus, the highest pressures occur in the deep ocean trenches. Among the marine fauna, some members are adapted to live only in surface waters, others at intermediate depths, and still others at abyssal depths. Generally, none are known to live over the entire depth range of the open ocean. Although microorganisms such as bacteria appear to be more adaptable to changes in HP, facultative (pressure tolerant) and obligately barophilic (pressure-requiring) bacteria are known (see also Section 5.1.5).

Salinity and temperature affect the *density* of seawater. At 0°C, seawater having a salinity of 30–37‰ has a corresponding density range of 1.024–1.030 g cm^{-3}. A variation in seawater density due to variation in salinity is one cause of water movement in the ocean, because denser water will sink below lighter water (convergence), or conversely, lighter water will rise above denser water (upwelling). The following processes may cause changes in salinity and, therefore, density: (1) dilution of seawater by runoff or less-saline water; (2) dilution by rain or snow; (3) concentration through surface evaporation; (4) freezing, which excludes salts from ice and thus leaves any residual, unfrozen water more saline; or (5) thawing of ice, which dilutes the already existent saline water.

As already stated, variation in salinity of seawater is not the sole cause of variation in density. The other important cause of density variation of seawater is temperature. Unlike freshwater, whose density is greatest at 4°C (Figure 5.6B), seawater with a salinity of 24.7‰ or greater has maximum density at its freezing point, that is, 0°C (Figure 5.6A). A body of freshwater thus freezes from its surface downward because freshwater at its freezing point is lighter than at a temperature of 4°C. Ocean water in the Arctic or Antarctic seas also freezes from the surface downward, but in this instance because ice, which excludes salts as it forms from seawater, is lighter than the seawater and will thus float on it.

The temperature of seawater ranges from about −2°C (the freezing point at 36‰ salinity) to +30°C, in contrast to the temperature of air over the ocean, which can range from −65 to +65°C. The narrower temperature range for seawater can be related to (1) its heat capacity, (2) its latent heat of evaporation, and (3) heat transfer from lower to higher latitudes by surface currents in both hemispheres. The major source of heat in the ocean is solar radiation. More than half the surface waters of the ocean are at 15–30°C. Only 27% of the surface waters are below 10°C. From ~50° north latitude to 50° south latitude, the ocean is thermally stratified. In this range of latitudes, the seawater temperature below the depth of ~1000 m is below 4°C (deep water). At depths from ~300 to 1000 m, the temperature drops rapidly with increasing depth. The zone of this rapid temperature change is called the *thermocline*. Its thickness and position vary with geographic location and season of the year. Above the thermocline lies the warm surface water, the *mixed layer*, which is extensively agitated by wind and water currents and thus exhibits relatively little temperature change with increasing depth.

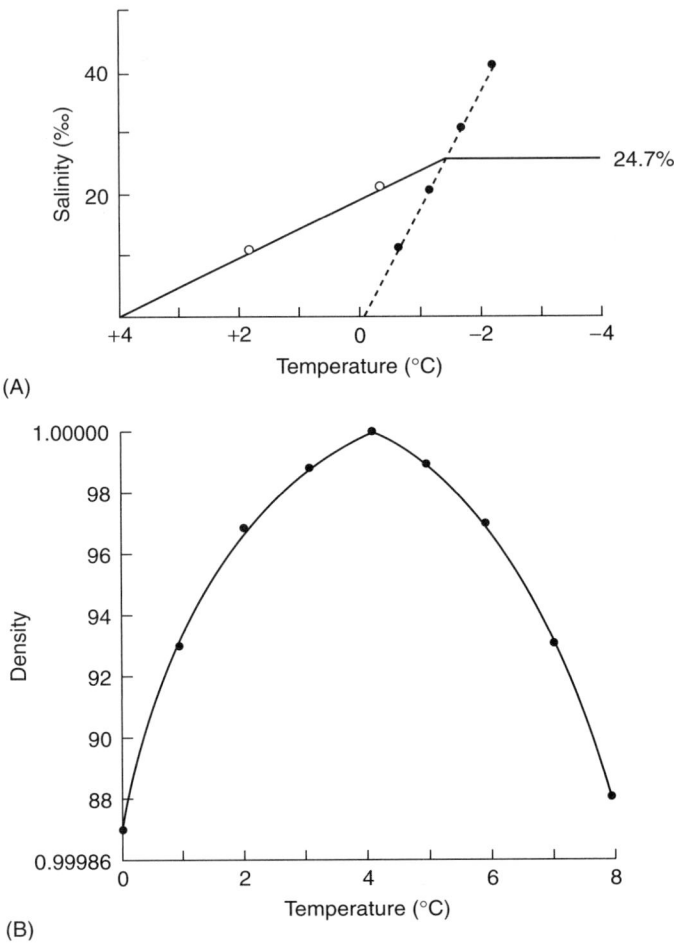

FIGURE 5.6 Density relationships in seawater and freshwater. (A) Relationship of seawater salinity to freezing point. Symbols: open circles, temperature of maximum density at a given salinity; closed circles, freezing-point temperature at a given salinity. Note that above a salinity of 24.7‰ seawater freezes at its maximum density as its temperature at maximum density cannot be lower than its freezing point. (B) Relationship of freshwater density (in g cc^{-1}) to temperature. Data points from chemically pure water are shown. Note that in the case of freshwater, its density at its freezing point is lower than its density at 4°C.

At latitudes higher than 50°N and 50°S, seawater is not thermally stratified. The waters around Antarctica, being cold (−1.9°C) and hypersaline (34.82‰) owing to ice formation, are hyperdense and thus sink below warmer, less-dense water to the north and flow northward along the bottom of the ocean basin. This is an example of convergence. Similarly, Atlantic waters from the subarctic region, having a temperature in the range of 2.8–3.3°C and a salinity in the range of 34.9–34.96‰, sink and flow southward at near-bottom or bottom levels of the ocean. Because the Arctic Ocean bottom is separated from the other oceans by barriers such as the shallow Bering Strait in the case of the Pacific Ocean and a shallow ridge in the case of the Atlantic Ocean, it does not influence the water masses of the Pacific and Atlantic Oceans directly. Other convergences occur in the world's oceans in both hemispheres because of the interaction of waters of different densities. In these instances, the heavier waters sink to lesser depths because they have lower densities than the heavier waters at high latitudes.

The water convergences alluded to above help to explain why generally ocean water is oxygenated at all depths (Figure 5.7; Kester, 1975). Of all ocean waters, only some coastal and near-coastal

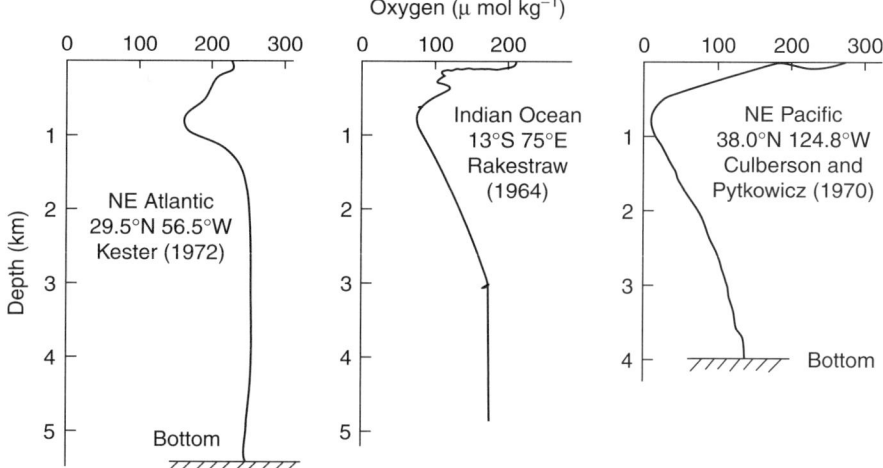

FIGURE 5.7   Vertical distribution of oxygen in the ocean. Profiles from three ocean basins. (From Kester DR, *Chemical Oceanography*, Academic Press, New York, 1975. With permission. Copyright by Elsevier.)

waters (e.g., estuarine waters; Cariaco Trench) may, as a result of intense biological activity, be devoid of oxygen at depth. At some sites, intense biological activity is sometimes the direct result of pollution caused by human beings. Surface waters of the open ocean tend to be saturated with oxygen because of oxygenation by the atmosphere and, equally important, by the photosynthetic activity of the phytoplankton. Oxygenation by phytoplankton can occur to depths of ~100 m (200 m in exceptional cases), where light penetration is 1% of the surface illumination. Seawater salinity of 34.352‰ is saturated at 5.86 mL or 8.40 mg of oxygen per liter at 760 mm Hg and 15°C. The higher the salinity and the higher the temperature, the lower the solubility of oxygen in seawater.

Starting at the top of the water column, the oxygen concentration in seawater will at first decrease with depth, owing mainly to oxygen consumption by the respiration of living organisms (Figure 5.7). Because many life forms in the oceans tend to be concentrated in the upper waters, oxygen concentration will fall to a minimum at ~600–900 m of depth, where respiration (oxygen consumption) by zooplankton and other animal forms as well as bacterioplankton occurs but not photosynthesis (oxygen production) by phytoplankton. Below this depth, because of rapidly decreasing biological activity, the oxygen concentration may at first increase once more and then slowly decrease again toward the bottom. Bottom water, however, may still be half-saturated with oxygen relative to surface water. This oxygen is not supplied by *in situ* photosynthesis, which cannot occur in the absence of light at these depths, nor is it the result of significant oxygen diffusion from the atmosphere to these depths. As previously indicated, the oxygenated waters at depth derive from the antarctic and subarctic convergences. The oxygen-carrying waters from the antarctic convergences flow northward along the bottom and at intermediate depths of the ocean basins, whereas the waters from subarctic convergence in the Atlantic flow southward at more intermediate depths. The oxygen content of these waters is only slowly depleted because of the low numbers of oxygen-consuming organisms in these deep regions of the oceans and the low rate of oxygen consumption in the upper sediments.

Photosynthetic activity of the phytoplankton is dependent on the penetration of sunlight into the water column because phytoplankton derives its energy almost exclusively from sunlight. It has been shown that light absorption by pure water in the visible range between 400 and 700 μm increases greatly toward the red end of the spectrum. It has also been shown that 60% of the light that penetrates transparent water is absorbed at a depth of 1 m. And 80 and 99% of the same light is absorbed at depths of 10 and 140 m, respectively. In less transparent coastal water, 95% of the light may have been absorbed at 10 m. Although the photosynthetic process of phytoplankters can use light over the entire visible spectrum, action spectra show peaks in the red and blue ends of the spectrum, where chlorophylls absorb optimally. Accessory pigments, such as carotenoids, absorb light at intermediate visible wavelengths. Clearly, light penetration limits the depth at which phytoplankton can grow. This depth is ~80–100 m on average (200 m maximally) and often much less in less transparent waters. Two exceptions have been noted, however. One was seen off the northern border of San Salvador Island in the Bahamas, where crustose coralline algae (*Rhodophyta*) were growing attached to rock at a depth of 268 m, observed from a submersible. At this location, the light intensity was only ~0.0005% of that at the surface (Littler et al., 1985). The other exception was noted in the Black Sea. Here, the photosynthetic sulfur bacterium *Chlorobium phaeobacteroides* was found to grow in a chemocline at a depth of 80 m, where light transmission from surface irradiance has been calculated to be 0.0005% (Overmann et al., 1992), as at the station at San Salvador Island.

The water layer from the ocean surface to the depth below which photosynthesis cannot take place constitutes the *euphotic zone*. Zooplankton and bacteria, except for cyanobacteria, may abound to somewhat lower depths than phytoplankton (~750 m), being scavengers and able to feed on dying and dead phytoplankters and their remains in the process of settling.

5.1.4 Microbial Distribution in Water Column and Sediments

Microbial distribution in the open oceans is not uniform throughout the water column (Figure 5.8). Factors affecting this distribution are energy, carbon, nitrogen, and phosphorus limitations (Wu et al., 2000) and also temperature, HP, and salinity. Accessory growth factors, such as vitamins, may

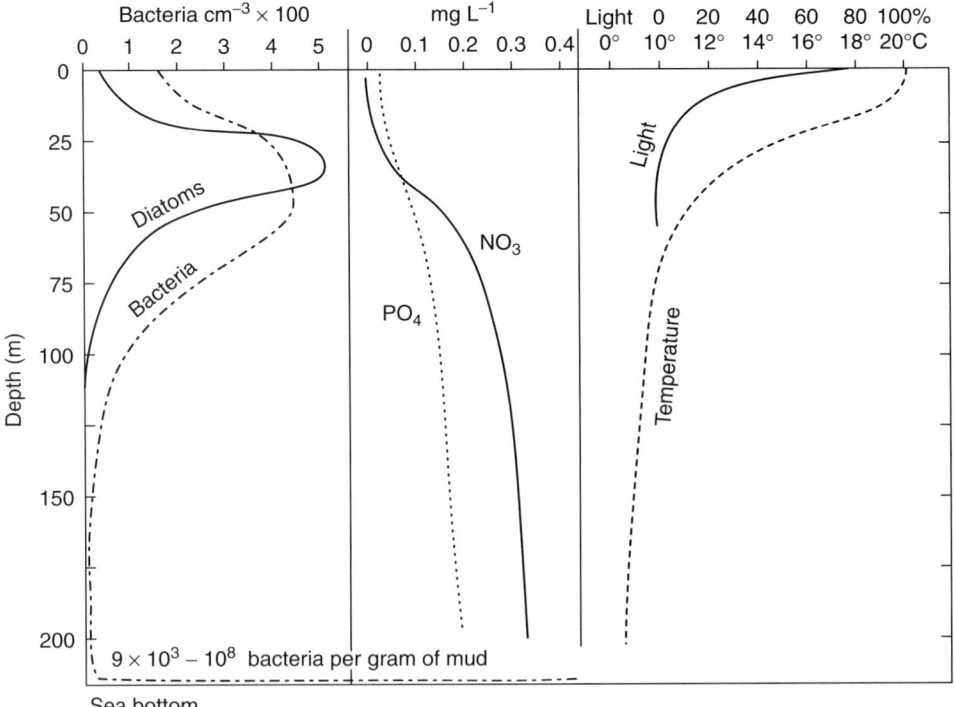

FIGURE 5.8 Vertical distribution of bacteria (number per cubic centimeter of water), diatoms (number per liter of water), PO_4, NO_3 (milligrams per liter), light, and temperature in the sea based on average results at several different stations off the coast of Southern California. (Reproduced from ZoBell CE, *Sci Mon.*, 55, 320–330, © 1942 by the American Association for the Advancement of Science. With permission.)

also be limiting to those microbes that cannot synthesize them by themselves. Phytoplankton distribution is limited to the euphotic zone of the water column primarily by available sunlight—the energy source for the organisms. However, phytoplankton distribution in the euphotic zone may also be limited by temperature and to some extent by salinity as well as by available dissolved nitrogen (nitrate) and phosphorus (phosphate).

Nonphotosynthetic microorganisms are limited to certain zones in the water column of the oceans by nutrient availability and in addition by temperature, salinity, and HP. The nonphotosynthetic microorganisms include predators (zooplankton), scavengers (zooplankton, fungi), and decomposers (bacteria and fungi and, possibly to a small extent, zooplankton). Zooplankters therefore dominate the euphotic zone, where they can feed optimally on phytoplankton, zooplankton, and bacteria. Bacteria and fungi are also prevalent here, because they find sufficient sources of nutrients produced by the phytoplankton and zooplankton in secretions and excretions and in dead remains.

Very high bacterial populations occur at the air–water interface of the ocean as a result of concentrations of organic carbon in the surface film, which may exceed concentrations in the waters below by three orders of magnitude. The bacterial population in this film is known as *bacterioneuston* (Sieburth, 1976; Sieburth et al., 1976; Wangersky, 1976). A recent comparison of 16S rRNA clone libraries constructed from DNA in samples from a surface microlayer of the North Sea off the coast of Northumbria with 16S rRNA clone libraries from the underlying pelagic seawater indicated a much more limited bacterial diversity in the surface microlayer than in the underlying pelagic seawater (Franklin et al., 2005). The bacterioneuston was dominated by *Vibrio* spp. (57%) and *Pseudoalteromonas* (32%), whereas in the pelagic water these two groups represented only 13 and 8%, respectively, suggesting that the bacterial community in the bacterioneuston is specialized.

Marine sediments contain significant numbers of living bacteria, fungi, and other benthic microorganisms as well as higher forms of life. In 1940, Rittenberg reported the recovery of viable bacteria from 350 cm below the sediment surface in the Pacific Ocean (Rittenberg, 1940), a finding, which at that time must have seemed remarkable and which ZoBell considered very significant (ZoBell, 1946; Ehrlich, 2000a). In 1994, Parkes and colleagues reported finding bacteria at depths >500 m below the sediment surface at five Pacific Ocean sites (Parkes et al., 1994). The bacterial numbers decrease with depth in a sediment column.

Fungi seem commonly to be restricted to shallow water depths, whereas bacteria, protozoa, and metazoa are found associated with sediments of shallow as well as abyssal water depths. The chief function of the microbes is to aid in scavenging and decomposing organic matter that settled undecomposed or partially decomposed from the overlying regions of biological productivity in the water column. Most of the organic matter from the euphotic zone settles to the bottom in the form of fecal pellets from metazoa. It should be pointed out that not all settled organic matter in deep-sea sediments is utilizable by microbes, for reasons that are not yet clearly understood. The unutilizable organic matter constitutes a significant part of the *sedimentary humus*.

The metabolic activity of the free-living bacteria of deep-sea sediment has been observed to be at least 50 times lower than that of microorganisms in shallow waters or on sediments at shallow depths (Jannasch et al., 1971; Jannasch and Wirsen, 1973; Wirsen and Jannasch, 1974). Environmental factors contributing to this slow rate of bacterial metabolism seem to be low temperature (<5°C) and, especially, elevated HP. Turner (1973) observed that pieces of wood left for 104 days on sediment at a station in the Atlantic Ocean at a depth of 1830 m was rapidly attacked by two species of wood borers (mollusks). This observation led to the suggestion that the primary attackers of organic matter in the deep sea, including sediment, are metazoans. Bacteria and other microbes harbored in the digestive tract of these metazoans decompose this organic matter only after ingestion by the metazoa (see, for instance, Jannasch, 1979). The intestinal bacteria appear essential to the digestion of cellulose to enable these metazoans to assimilate it.

Schwartz and Colwell (1976) were the first to report that the bacterial flora of the intestines of amphipods (crustaceans) that they collected in the Pacific Ocean at a depth of 7050 m was able to grow and metabolize nearly as rapidly at 780 atm and 3°C as at 1 atm and 3°C in laboratory experiments. Their study suggested that these types of gut bacteria behave very differently in response to HP from free-living bacteria from the same depths. These findings have been extended by other observations on amphipod microflora. Deming and Collwell (1981) reported finding barophilic and barotolerant bacteria in the intestinal flora of amphipods living at depths of 5200–5900 m. Yayanos et al. (1979) isolated a barophile from a decomposing amphipod that grew optimally at ~500 bar and 2–4°C and poorly at atmospheric pressure in this temperature range. The same workers (Yayanos et al., 1981) isolated an obligately barophilic bacterium from an amphipod recovered from 10,746 m in the Marianas Trench.

The generally low rate of metabolism of the biological community (benthos) on deep-sea sediments is also reflected by respiratory measurements carried out at 1850 m. The measurements revealed a rate of oxygen consumption that was orders of magnitude less than in sediments at shallow shelf depths (Smith and Teal, 1973).

Phytoplankton, zooplankton, bacteria, and fungi are not found in significant numbers at intermediate depths in the water column of the oceans. The main reason for this is the lack of adequate nutrient supply, including a suitable source of energy, but low temperature can also be a factor in the case of some organisms. Kriss (1970), having examined water samples from a north–south transect in the Atlantic Ocean, concluded that an uneven distribution of bacteria at intermediate water depths was attributable to the different origins and characteristics of particular water masses. He drew his conclusions on the basis of available metabolizable nutrients in the different water masses, claiming that higher concentrations occur in water masses of equatorial-tropical origin, owing to *autochthonous* (of native, e.g., planktonic, origin) and *allochthonous* (from runoff from continents and islands) contributions, than in water masses of Arctic and Antarctic origin.

Growth and reproduction of bacteria and fungi in ocean water also occur on surfaces of some living organisms and on the surface of suspended organic and inorganic particles (epiphytes), because at these sites essential nutrients may be very concentrated (Sieburth, 1975, 1976; Hermansson and Marshall, 1985). The microorganisms may form microcolonies or a biofilm on these surfaces. Detritus, although not a nutrient by itself, usually has adsorptive capacity, which helps to concentrate nutrients on its surface and thus makes for a preferred microbial habitat. The beneficial effect that the buildup of nutrients by adsorption to particle surfaces has on microbial growth is great, because the concentration of the nutrients in solution in seawater is very low ($0.35-0.7$ mgL^{-1}) (Menzel and Rhyter, 1970). ZoBell (1946) long ago showed a significant increase in the bacterial population in natural seawater during 24 h of storage in an Erlenmeyer flask. He attributed this to the adsorption of essential nutrients in the seawater to the walls of the flask, where the bacteria actually grew.

5.1.5 EFFECTS OF TEMPERATURE, HYDROSTATIC PRESSURE, AND SALINITY ON MICROBIAL DISTRIBUTION IN OCEANS

Temperature and pressure may have a profound influence on where a given nonphotosynthetic microbe may live in the ocean. Some will grow only in the temperature range of 15–45°C (*mesophiles*) with an optimum near 30°C, others only in the range from 0°C or slightly below to 20°C with an optimum at 15°C or below (*psychrophiles*), and still others in the range of 0–30°C or even higher (37°C) (Ehrlich, 1983) with an optimum near 25°C (*psychrotrophs*) (Morita, 1975). The mesophiles would be expected to grow only in waters of the mixed zone and near active hydrothermal vents, whereas psychrophiles would grow only below the thermocline and in polar seas. Psychrotrophs would be expected to grow above, in, and below the thermocline and the polar seas, although they might do better in and above the thermocline. Mesophiles can be recovered from cold waters and deep sediments, where they are able to survive but cannot grow (i.e., they are *psychrotolerant*).

Many bacteria that normally grow at atmospheric pressure are not inhibited by HPs up to ~200 atm (202.6 bar), but their growth is retarded at 300 atm (303.9 bar), and they will not grow above 400 atm (405.2 bar). One effect that increased HP has on these bacteria is on cell morphology and structure (Kaletunç et al., 2004). Many bacteria isolated from waters at 500 atm (506.5 bar) and 600 atm (607.8 bar) were found to grow better at these pressures under laboratory conditions than at atmospheric pressure. Such organisms are called *barophiles*. Some organisms, described in a pioneering study by ZoBell and Morita (1957), which had been recovered from extreme depths (10,000 m) were suspected of having been obligate barophiles. Since that time, an obligately barophilic bacterium has actually been isolated from an amphipod taken at 10,476 m in the Marianas Trench and studied (Yayanos et al., 1981). It exhibited an optimal growth rate (generation time of 25 h) at 2°C and 690 bar (681 atm) of HP. As already mentioned, Yayanos et al. (1979) also isolated a facultatively barophilic spirillum from 5700 m depth that grows fastest at ~500 bar (493.5 atm) and 2–4°C, with a generation time of 4–13 h.

In prokaryotes, the most pressure-sensitive biochemical process is protein synthesis. It determines the degree of barotolerance and limits growth under pressure. Other processes, including nucleic acid synthesis, in the same cells are less pressure sensitive (Pope and Berger, 1973). The step most sensitive to pressure in protein synthesis is translation, resulting from the effect of HP on the 30S ribosomal subunit (Smith et al., 1975). (For a more complete discussion on ecological implications of temperature and HP in the marine environment, see Marquis, 1982; Morita, 1967, 1980; and Jannasch and Wirsen, 1977.)

Marine microorganisms, especially bacteria, vary in their salinity requirements. Those that can grow only in a narrow range of salinities are said to be *stenohaline*, and those that can grow in wide range of salinities are called *euryhaline*. Both types are found in the open ocean. Their salt requirement is not explained on the basis of osmotic pressure but by a specific requirement for one or more of the following ions: Na^+, K^+, Mg^{2+}, and Ca^{2+}. These ions may affect cell permeability or specific enzyme activities (MacLeod, 1965). They may also be needed to maintain cell integrity. The tolerance of a wide range of salinities by some euryhaline bacteria may be explainable in part by genetic control of cytosolic and membrane phospholipids and protein patterns in these organisms in response to different salinities and HPs. In four strains of *Halomonas*, changes in salinities and HP appeared to work synergistically under some conditions (e.g., high HP, low temperature) and antagonistically under other conditions with respect to phospholipids and protein synthesis in the cytosol and membranes (Kaye and Baross, 2004).

5.1.6 Dominant Phytoplankters and Zooplankters in Oceans

Diatoms, dinoflagellates, coccolithophores, and other flagellates are the dominant phytoplankters of the sea (Figure 5.9; Sieburth, 1979). The first two are the chief source of food for herbivorous marine organisms. Diatoms are also important agents in the control of Si and Al concentrations in seawater (Mackenzie et al., 1978). Kelps and other sessile algae are mostly restricted to the shelf areas of the seas because they cannot grow at depths below ~30 m. A few kelps, however, are able to float in the open sea (e.g., Sargasso weed). The dominant members of the zooplankton not only include protozoa but also invertebrates such as coelenterates, pteropods, and crustaceans; some of which are not found free-floating as adults but have planktonic larval stages. Among protozoa of the zooplankton, dominant forms include foraminifera and radiolarians. Their place in the ecology of the oceans is chiefly as predators on bacteria and some other members of the plankton population. Some of these forms are also the food for higher predatory animals. The phytoplankters are the principal *primary producers* (i.e., the chief synthesizers of organic carbon by photosynthesis) in the euphotic zone of the oceans. (For further discussion, see Sieburth, 1979.)

At special sites at abyssal depths around hydrothermal vents or seeps from which H_2S or methane is discharged, primary production is the result of chemoautotrophic bacteria that obtain energy from the oxidation of hydrogen sulfide (Paull et al., 1984; Jannasch and Taylor, 1984; Stein, 1984; Tunnicliffe, 1992) or methane (Jannasch and Mottl, 1985; Kulm et al., 1986) (see Chapters 17, 19, and 22). This primary production is the basis for existence of biological communities including

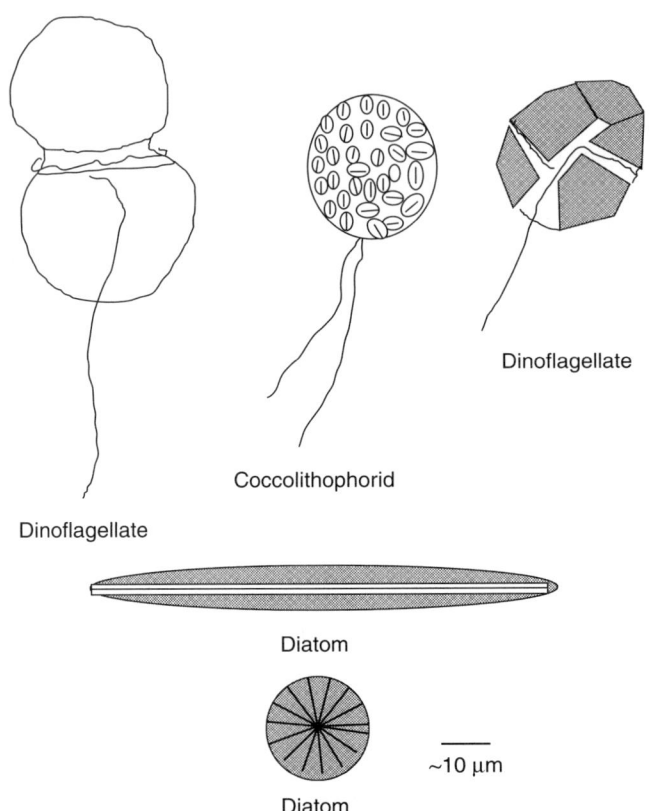

Dinoflagellate

Coccolithophorid

Dinoflagellate

Diatom

~10 μm

Diatom

FIGURE 5.9 Sketches of important marine phytoplankters. (Adapted from Wood EJF, *Marine Microbial Ecology*, Reinhold, New York, 1965; Sieburth JMcN, *Sea Microbes*, Oxford University Press, New York, 1979.)

metazoa and even vertebrates that are spatially restricted to the site of activity of the chemoautotrophic primary producers and their energy sources.

5.1.7 PLANKTERS OF GEOMICROBIAL INTEREST

The phytoplankters of special geomicrobial interest include the diatoms, coccolithophores, and silicoflagellates. The zooplankters of special geomicrobial interest include the foraminifers and radiolarians. It is these organisms that precipitate much of the $CaCO_3$ and SiO_2 in the open sea. Their calcareous and siliceous remains, respectively, settle out and become incorporated into the sediments (see Chapters 9 and 10).

5.1.8 BACTERIAL FLORA IN OCEANS

The bacterial flora of the seas is primarily represented by members in the domain Bacteria that are aerobic, facultative, and gram negative (Dworkin, 1999) and members in the domain Archaea, including *Crenarcheota* and *Euryarcheota*, which mostly have not yet been cultured (Massana et al., 2000). Members of the genera *Shewanella*, *Pseudomonas*, *Alteromonas*, *Vibrio*, and *Oceanospirillum* are common, although many other gram-negative genera are also found, albeit in lesser numbers and in special niches. Gram-positive bacteria such as *Arthrobacter* or spore-forming rods are relatively less common and are encountered frequently in regions directly influenced by runoff from land. Most bacteria in the sea are aerobic or facultative. Strict anaerobes, such as sulfate reducers, are encountered mainly where organic matter accumulates in significant quantities, such as salt marshes, estuaries, and some near-shore waters. Although heterotrophic and mixotrophic bacteria are the most numerous, autotrophic bacteria are encountered in significant numbers, particularly in certain niches. Chemosynthetic autotrophs are the primary producers around deep-sea hydrothermal vents and may be free-living or occur as facultative or strict symbionts of some invertebrates. The latter reside, for instance, in *trophosomes* in the coelomic cavity of vestimentiferan worms and on the gills of certain mollusks. The cyanobacterium *Trichodesmium*, a photosynthetic autotroph, is widespread in the euphotic zone of the open ocean. Apart from being a photosynthetic primary producer, it is also an important nitrogen fixer in the sea (Zehr, 1992; Capone et al., 1997). Other cyanobacteria, such as *Oscillatoria*, *Lyngbia*, *Plectonema*, and *Spirulina*, are also encountered in the oceans. Nonsymbiotic heterotrophic, mixotrophic, or autotrophic bacteria in the marine environment may be free-living or attached to living organisms or inert particles suspended in the seawater column (e.g., inorganic matter such as clay particles and fecal pellets; for more information on bacterial attachment to fecal pellets, see Turner and Ferrante, 1979). They can also be detected in the sediment column in all parts of the ocean.

Marine bacteria are usually defined as types that will not grow in media prepared in freshwater because they need one or more of the salt constituents of seawater. Bacteria that do not require seawater for growth can, however, be readily isolated from seawater and marine sediment samples far from shore. Many of these organisms can grow readily in media prepared in seawater. They may represent terrestrial forms. The active bacteria in the marine environment are important as decomposers and in special marine niches as primary producers. Some may also play a role in mineral formation and mineral diagenesis (e.g., ferromanganese deposits; see Chapter 17; for further discussion, see Ehrlich, 1975, 2000b; Jannasch, 1984; Kulm et al., 1986; and Sieburth, 1979).

5.2 FRESHWATER LAKES

Freshwater amounts to <3% of the total water on Earth. Like the oceans, it furnishes important habitats for certain life forms. Among these habitats are lakes, which are part of the *lentic* environments—the standing waters. Other lentic environments are ponds and swamps. Lakes represent only 0.009% of the total water in land areas (van der Leeden et al., 1990), most of the freshwater being tied up in ice caps and glaciers (2.14%) and in groundwater (0.61%) (van der Leeden et al., 1990).

Lakes have arisen in various ways. Some resulted from past glacial action. An advancing glacier gouged out a basin that, when the glacier retreated, filled with water from the melting ice and was later kept filled by runoff from the surrounding *watershed*. Other basins resulted from landslides that obstructed valleys and blocked the outflow from their watershed. Still others resulted from crustal up-and-down movement (dip-slip faulting) that formed dammed basins for the collection of runoff water. Some resulted from the solution of underlying rock, especially limestone, which led to the formation of basins in which water collected. Lakes have also been formed by the collection of water from glacier melts in craters of extinct volcanoes and by the obstruction of river flow or changes in river channels (Welch, 1952; Strahler and Strahler, 1974; Skinner et al., 1999).

Lakes vary greatly in size. The combined Great Lakes in the United States cover an area of 328,000 km^2, an unusually great expanse. More commonly, lakes cover areas of 26–520 km^2, but many are smaller. Most lakes are <30 m deep. However, the deepest lake in the world, Lake Baikal in southern Siberia (Russia), has a depth of 1700 m. The average depth of the Great Lakes is 700 m, and that of Lake Tahoe on the California–Nevada border is 487 m. The elevation of lakes ranges from below sea level (e.g., the Dead Sea at the mouth of the Jordan River) to as high as 3600 m (Lake Titicaca in the Andes on the border between Bolivia and Peru).

5.2.1 SOME PHYSICAL AND CHEMICAL FEATURES OF LAKES

Some of the water of lakes may be in motion, at least intermittently. Most prevalent are horizontal currents, which result from wind action and the deflecting action of shorelines. Vertical movements are rare in lakes of average or small size. They may result from thermal, morphological, or hydrostatic influences. Thermal influence can result in changes of water density such that heavier (denser) water sinks below lighter water. Morphological influence can result from rugged bottom topography, which may deflect horizontal water flow downward or upward. Hydrostatic influence can result from springs at the lake bottom that force water upward into the lake. Besides horizontal and vertical movement, return currents may occur as a result of water being forced against a shore by wind and piling up. Depending on the type of lake and the season of the year, only a portion of the total water mass of a lake, or all of it, may be circulated by the wind. (For further discussion, see Welch, 1952 and Strahler and Strahler, 1974.)

The waters of lakes may vary in composition from very low salt content (e.g., Lake Baikal) to a very high salt content (e.g., Dead Sea between Israel and Jordan and Lake Natron in Africa), and from low organic content to high organic content. Salt accumulation in lakes is the result of input from runoff from the watershed, including stream flow, very gradual solution of sediment components and rock minerals in the lake bed, and evaporation.

The waters of lakes may or may not be thermally stratified, depending on various factors: geographic location, the season of the year, and lake depth and size. Thermal stratification, when it occurs, may or may not be permanent. Absence of stratification may be the result of complete mixing or *turnover* as a result of wind action. When waters are thermally stratified into a warmer layer in the upper portion of a lake (*epilimnion*) and a cooler layer in the lower portion (*hypolimnion*), complete mixing does not occur because of a density difference between the two layers. A thermocline forms between the epilimnion and the hypolimnion, which is a relatively thin layer of water in which a temperature gradient exists ranging from the temperature corresponding to the bottom of the epilimnion to that of the top of the hypolimnion. Lakes may be classified according to whether and when they turn over (Reid, 1961). The categories can be defined as follows:

Amictic lakes are bodies of water that never turn over, being permanently covered by ice. Such lakes are found in Antarctica and at high altitudes in mountains.

Cold monomictic lakes are bodies of water that contain waters never exceeding 4°C, which turn over once during the summer, being thermally stratified the rest of the year.

Dimictic lakes turn over twice a year, in spring and fall. They are thermally stratified at other times. These are typically found in temperate climates and at higher altitudes in subtropical regions.

Warm monomictic lakes have water that is never colder than 4°C. They turn over once a year in winter and are thermally stratified the rest of the year.

Oligomictic lakes contain water that is significantly warmer than 4°C and turn over irregularly. Such lakes are found mostly in tropical zones.

Polymictic lakes have water just over 4°C and turn over continually. Such lakes occur at high altitude in equatorial regions.

Meromictic lakes are deep, narrow lakes whose bottom waters never mix with the waters above. The bottom waters generally have a relatively high concentration of dissolved salts, which makes them dense and separates them from the overlying waters by a *chemocline*. The upper waters in temperate climates may be thermally stratified in summer and winter and may undergo turnover in spring and fall.

A dimictic lake in a temperate zone during spring thaw accumulates water near 0°C, which because of its lower density, floats on the remaining denser water, which is near 4°C. As the season progresses, the colder surface water is slowly warmed by the sun to near 4°C. At this point, all water has a more or less uniform temperature and thus uniform density. This allows the water to be completely mixed or turned over by wind agitation if the lake is not excessively deep like a meromictic lake. As the surface water undergoes further warming by the sun, segregation of water masses occurs as warmer, lighter water comes to lie over the colder, denser water. A thermocline is established between the two water masses, separating them into epilimnion and hypolimnion. The temperature of the water in the epilimnion may be higher than 10°C and vary little with depth (perhaps 1°C m^{-1}). However, the water in the thermocline will show a rapid drop in temperature with depth. This drop may be as drastic as 18°C m^{-1} but is more usually ~8°C m^{-1}. The thickness of the thermocline varies with position in the lake and between different lakes—an average thickness being ~1 m. The water in the hypolimnion will have a temperature well below that in the epilimnion and show a small drop in temperature with depth, usually <1°C m^{-1}.

The water in the epilimnion but not in the hypolimnion is subject to wind agitation and is thus fairly well mixed at all times. It is the greater density of the water in the hypolimnion that prevents it from getting mixed by the wind action. Continual warming by the sun and mixing by the wind produces horizontal currents and, in larger lakes, return currents over the thermocline, resulting in some exchange with water of the thermocline. This water exchange progressively increases the volume of the epilimnion and causes a progressive drop in position of the thermocline.

At fall turnover, the thermocline will have touched the bottom in the lake and disappeared, the water now having a uniform warm temperature. With the approach of winter, the lake water will cool. Once the surface water has cooled below 4°C, a thermocline will be reestablished, but at this time the water in the epilimnion will be colder than the water in the hypolimnion. Ice may form on the epilimnion if the water temperature reaches the freezing point. The winter thermocline will usually remain near the lower surface of any ice cover on the lake. Figure 5.10 shows in idealized form the seasonal cycle of thermal stratification of a dimictic lake.

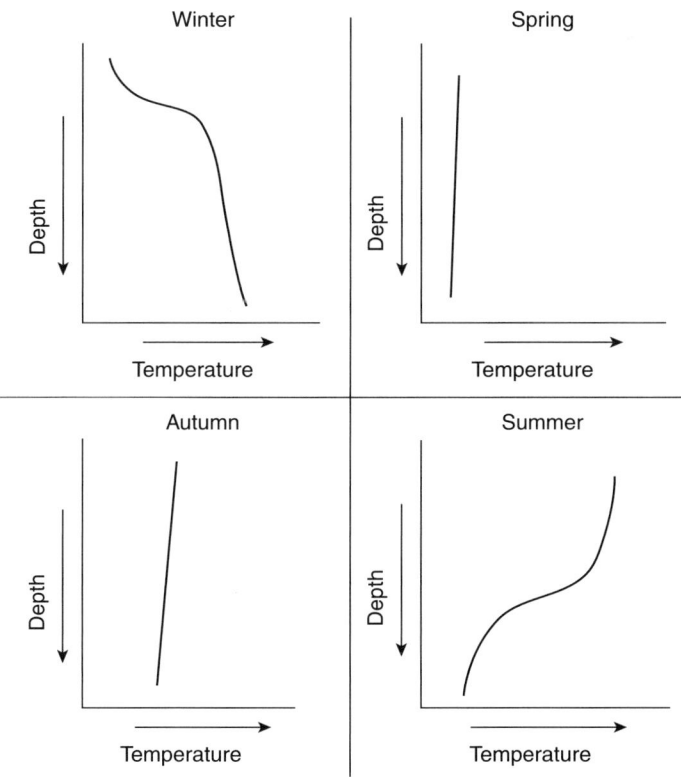

FIGURE 5.10 Schematic representation of thermal stratification in a dimictic lake at different seasons of the year.

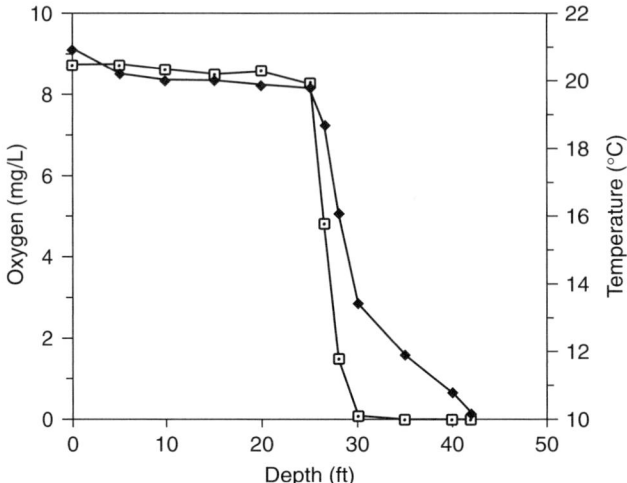

FIGURE 5.11   Oxygen profile in Tomhannock Reservoir (near Troy, New York) on September 5, 1967. (□) Oxygen, mgL^{-1}. (■) Temperature, °C. (From LaRock PA, The bacterial oxidation of manganese in a fresh water lake. PhD Thesis, Rensselaer Polytechnic Institute, Troy, New York, 1969. With permission.)

The thermocline of a lake acts as a barrier between the epilimnion and the hypolimnion. It prevents easy exchange of salts, dissolved organic matter, and gases because the two water masses that it separates do not readily mix owing to their difference in density and the very slow diffusion across the thermocline. The oxygen content in the epilimnion is usually around the saturation level. At times of intense photosynthetic activity of phytoplankton, oxygen supersaturation may be achieved. The source of oxygen in the epilimnion is photosynthesis and aeration. The latter process is especially significant during wind agitation. The oxygen concentration in freshwater at saturation at 0°C is 14.62 ppm or 10.23 mL L^{-1}; at 15°C is 10.5 ppm or 7.10 mL L^{-1}; and at 20°C is 9.2 ppm or 6.5 mL L^{-1}. Optimal conditions of light and oxygen concentration together with adequate nutrient supply permit phytoplankton, zooplankton, bacteria, fungi, and other life to attain their greatest numbers in the waters of the epilimnion. The phytoplankters are the primary producers on which the remaining life forms depend for food, directly or indirectly.

Oxygen that was distributed into all parts of the lake during spring turnover will be gradually depleted from the hypolimnion of a fertile (eutrophic) lake after reemergence of thermal stratification. This is a consequence of biological activity, especially on and in the sediment. Thus, the hypolimnion may be anoxic during a shorter or longer period before fall turnover. Only anaerobic or facultative organisms will carry on active life processes under these anoxic conditions. Such organisms include bacteria and protozoa as well as certain nematodes, annelids, immature stages of certain insects, mollusks, and some fishes (Welch, 1952; Strahler and Strahler, 1974). Figure 5.11 illustrates the oxygen distribution measured in a dimictic lake during summer stratification.

5.2.2   LAKE BOTTOMS

The nature of lake bottoms is highly variable, depending on the location and history of the lake. The basins of many smaller lakes are flat expanses of sediment overlying bedrock. In contrast, the basins of larger lakes (e.g., the Great Lakes) have a more rugged topography in many places. The bottom of lakes may be dominated by sand and grit; clay; a brown, mud-rich humus; diatom oozes; ochreous mud rich in limonitic iron oxide; or calcareous deposits. The organic components of any sediment may derive from dead or dying plankters that have settled to the bottom or from plant or animal remains. Some inorganic and organic components may have been introduced into the lake sediment by the wind. Much silt and clay is washed into lakes by runoff. Some is also contributed by wind erosion of the shoreline. The sediments are a major habitat of microbes.

5.2.3 LAKE FERTILITY

Lakes may be classified in terms of their fertility or their nutritional status (i.e., their ability to support flora and fauna). *Oligotrophic* lakes have an impoverished nutrient supply in which phosphorus and nitrogen are in short supply and oxygen concentration is high at all depths. *Eutrophic* lakes, in contrast, are fertile lakes in which phosphorus and nitrogen are significantly more plentiful than in oligotrophic lakes. *Mesotrophic* lakes are intermediate in nutritional status between oligotrophic and eutrophic lakes. *Dystrophic* lakes are defined as having an oversupply of organic matter that cannot be completely decomposed because of an insufficiency of oxygen and alternative terminal electron acceptors such as ferric iron, nitrate, or sulfate. They are also sometimes deficient in assimilable nitrogen or phosphorus. The waters of such lakes are turbid and often acid. The origin of dystrophic conditions may be encroachment of the shoreline by plants, including reeds, shrubs, and trees.

5.2.4 LAKE EVOLUTION

Lakes have an evolutionary history. Once fully matured, they age progressively. Their basin slowly fills with sediment, partly contributed by feed streams, by the surrounding land through erosion and runoff, and partly by the biological activity in the lake. The size of the contribution that each process makes depends on the fertility of the lake. Changes in climate may also contribute to lake evolution (e.g., through lessened rainfall, which can cause a drop in water level, or through warming of the climate, which can cause more rapid water evaporation). These effects usually make themselves felt slowly. Ultimately, a lake may change into a swamp.

5.2.5 MICROBIAL POPULATIONS IN LAKES

The microbial population in eutrophic lakes tends to be orders of magnitude greater than in the seas. Numbers of culturable bacteria may range from 10^2 to 10^5 mL^{-1} of lake water and be of the order of 10^6 g^{-1} of lake sediment. The size of the bacterial population may be affected by runoff, which contributes soil bacteria. The culturable bacterial population of lakes consists predominantly of gram-negative rods (Wood, 1965, p. 36), although gram-positive spore-forming bacteria and actinomycetes can be readily isolated, especially from sediments.

Like marine bodies of water, bodies of freshwater, such as lakes and ponds feature surface films at the air–water interface that harbor a number of microorganisms that are greater than in the underlying water (reviewed by Wotton and Preston, 2005). The organisms in such surface films include bacteria, protozoa, and even invertebrate larvae (e.g., mosquito larvae). Some of the bacteria may occur in microcolonies, being embedded in an exopolymer (see Figure 4 in Wotton and Preston, 2005). Accumulations of organic matter in the surface films help to sustain the heterotrophic microbial population in these films.

Few, if any, of the types of bacteria found in lakes seem to be exclusively limnetic organisms. The main activity of the bacteria, other than the cyanobacteria, is that of decomposers. The cyanobacteria along with algae serve as the primary producers. Fungi and protozoa are also found. Important functions of the former are as scavengers and decomposers and of the latter as predators.

Cyanobacteria and algae are abundant in eutrophic lakes. The algae include green forms as well as diatoms and pyrrhophytes. *Cyanobacterial* and *algal blooms* may occur at certain times when one species suddenly multiplies explosively and becomes the dominant phytoplankton member temporarily, often forming a carpetlike layer or mat on the water surface. After having reached a population peak, most of the cells in the phytoplankton bloom die off and are attacked by scavengers and decomposers. Especially favorable growth conditions appear to be the stimulus for such blooms.

5.3 RIVERS

Rivers, which account for only 0.0001% of the total water in the world (van der Leeden et al., 1990), are part of the *lotic* environment in which waters move in channels on the land surface. Such flowing water may start as a brook, then widen into a stream, and ultimately become a river. The source of this water is surface runoff and groundwater reaching the surface through springs or, more important, through general seepage. A riverbed is shaped and reshaped by the flowing water that scours the bottom and sides with the help of suspended particles ranging in size from clay particles to small stones. Young rivers may feature rapids and steep valley slopes. Mature rivers lack rapids and feature more uniform stream flow, owing to a smoothly graded river bottom and an ever-widening riverbed. Old rivers may develop meanders in wide, flat floodplains. The flow of the water is caused by gravity because the head of a river always lies above its mouth. Average flow of rivers range from 0 to 9 m s^{-1}. However when viewed in cross section, the flow of water in a river is not uniform. The water in some portions in such a cross section flows much faster than in other portions. This can be attributed to frictional effects related to the riverbed topography as well as to density differences of different parts of the water mass. Density variation may arise from temperature differences and from solute concentration differences between parts of the river. Portions of river water may exhibit strong turbulence caused in part by certain features of the river topography. Water velocity, turbulence, and terrain determine the size of particles a river may sweep along (see Strahler and Strahler, 1974; Stanley, 1985; Skinner et al., 1999).

Most river water is ultimately discharged into an ocean, but exceptions exist. The Jordan River, whose headwaters originate in the mountains of Syria and Lebanon, empties into a lake called the Dead Sea, which has no connection with any ocean. The Dead Sea does not overflow because it loses its water by evaporation, which accounts, in part, for its high salt accumulation. (Nowadays, actual shrinkage in the size of the Dead Sea is observed that is attributable to commercial exploitation for recovery of minerals from its waters, and to a decrease in water inflow from the Jordan river as a result of water diversion as a source of potable and irrigation water.) The waters in the Dead Sea are nearly saturated with salts, which makes life impossible except for specially adapted organisms.

When river water is discharged into an ocean, an estuary is frequently formed where the less dense river water will flow over the denser saline water from the sea with incomplete mixing. Tidal effects of the sea may alter the water level of the discharging river, sometimes to a considerable distance upstream. Estuaries form special habitats for microbes and higher forms of life, which must cope with periodic changes in salinity, water temperature, nutrients, oxygen availability, and so forth, engendered by tidal movement.

Because of the relatively constant water movement, the water temperature of rivers tends to be rather uniform (i.e., rivers generally are not thermally stratified the way lakes are, when examined in cross section). Only where a tributary with a different water temperature enters a river there may be local temperature stratification. Different segments of a river may, however, differ in temperature. The pH of river water can range from very acidic (e.g., pH 3), for instance, in streams receiving acid mine drainage, which is the result of microbial activity (see Chapter 20), to alkaline (e.g., pH 8.6) (Welch, 1952, p. 413). Unless heavily polluted by human activity, rivers are generally well aerated. It has been thought that in unpolluted rivers most organic and inorganic nutrients supporting microbial as well as higher forms of life are largely introduced by runoff (allochthonous). It is now believed that a significant portion of fixed carbon in such streams and rivers may be contributed by photosynthetic autotrophs, mainly algae growing in parts of a river with quiet waters (autochthonous) (Minshall, 1978). Pollution may cause overloading, which because of excessive oxygen demand, will result in anoxic conditions with the consequent elimination of many micro- as well as macroorganisms.

Planktonic organisms tend to be found in greater numbers in the more stagnant or slower-flowing waters of a river than in fast-flowing portions. The plankters include algae such as diatoms, cyanobacteria, green algae, protozoa, and rotifers. The proportions depend on the environmental conditions of a particular river and its sections. Sessile plants or algae tend to develop to significant extents only in sluggish streams or in the backwaters of otherwise rapidly flowing streams. Bacteria are present in significant numbers where physical and chemical conditions favor them. Rheinheimer (1980) reported total bacterial numbers, estimated by culture methods, in the River Elbe in Germany to range from 4.7×10^9 to 6.9×10^9 L^{-1} and bacterial biomass to range from 0.55 to 0.71 mg L^{-1} in an unspecified year. As in lakes, no unique microflora occurs in unpolluted rivers.

5.4 GROUNDWATERS

Water that collects below the land surface in soil, sediment, and permeable rock strata is called *groundwater*. It represents 0.61% of all water in the world (van der Leeden et al., 1990). Groundwater derives mainly from *surface water* whose origin is meteoritic precipitation such as rain and melted snow. Surface water includes the water of rivers, lakes, and the like (Figure 5.12). A minor amount of groundwater derives from *connate water, water of dehydration*, or *juvenile water*. Connate water, often of marine origin and therefore saline, is water that got trapped in rock strata in the geologic past by up or down warping or faulting and as a consequence became isolated as a stagnant reservoir. Its salt composition frequently has become highly altered from that of the original water from which it is derived as a result of long-term interaction with the enclosing rock. Connate waters are often associated with oil formations. Waters of dehydration are derived from waters of crystallization, which are part of the structure of certain crystalline minerals. They are released as a result of the action of heat and pressure in the lithosphere. Juvenile waters are associated with magmatism that causes them to escape from the interior of the Earth. They are termed juvenile because they are assumed to have never reached the Earth's surface before.

Surface water slowly infiltrates permeable ground as long as the ground is not already saturated. It passes through a zone of aeration or unsaturation called the *vadose zone* to the zone of saturation or *aquifer*, which lies over an impermeable stratum (Strahler and Strahler, 1974; Chapelle, 1993; Skinner et al., 1999). The vadose zone may include the soil, an intermediate zone, and the capillary fringe, and can range in thickness from a few centimeters to 100 m or more. The resident water in

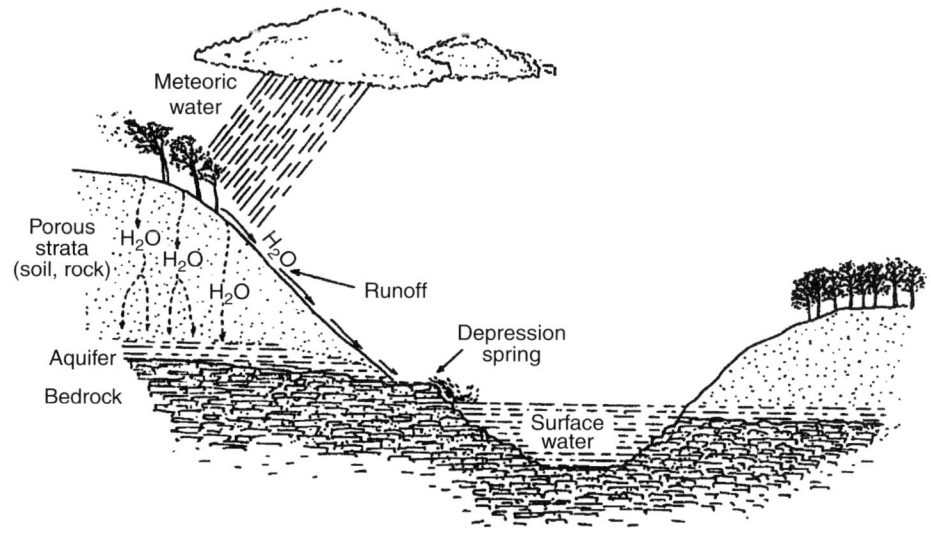

FIGURE 5.12 Interrelation of meteoritic, surface, and groundwaters.

the vadose zone is pore water (pellicular water, see Chapter 4) that is under less than atmospheric pressure and held there by capillarity. In soil, this water supports the soil micro- and macroflora and fauna and plant growth. The water in the aquifer can be extracted by human intervention, but that in the vadose zone cannot (Hackett, 1972). The rate of infiltration of permeable strata depends not only on the surface water supply but also on the porosity of the permeable strata. Similarly, the water holding capacity of the aquifer depends on the porosity of the matrix rock. The cause of water movement below ground level is not only gravity but also intermolecular attraction between water molecules, capillary action, and hydrostatic head (see Chapter 4).

At a given location, two or more aquifers may occur, one over the other, separated by one or more impervious strata. An example of such multiple aquifers is found in the Upper Atlantic Coastal Plain Province in South Carolina, United States (Sargent and Fliermans, 1989). In such a sequence of aquifers, the uppermost may be directly rechargeable with surface water from above and is then called an *unconfined aquifer*, or it may be separated from a larger aquifer just below by a thin lens of material with low or no porosity, called an *aquiclude*, in which case the uppermost aquifer is called a *perched aquifer*. The material composing an aquiclude may be clay or shale. Aquifers underlying a perched aquifer are called *confined aquifers* and are rechargeable only where the aquiclude is absent. The matrix of aquifers may consist of sand and gravel, fractured limestone and other soluble rock, basalt and fractured volcanic rocks, sandstone and conglomerate, crystalline and metamorphic rocks, or other porous but poorly permeable materials (Hackett, 1972).

Groundwater may escape to the surface or into the atmosphere through *springs* or by evaporation with or without the mediation of plants (transpiration). Some water will be accumulated by the vegetation itself. Depending on the relative rates of water infiltration and water loss, the level of the *water table* in the ground may rise, fall, or remain constant. Groundwater that reaches the surface through springs may do so under the influence of gravity, which may create sufficient head to force the water to the surface through a channel, as in an *artesian spring*. Groundwater may also reach the surface as a result of an intersection of the water table with a sloping land surface, as in a *depression spring*. Finally, groundwater may reach the surface in springs under the influence of thermal energy applied to reservoirs of water deep underground. Such *hot springs* in their most spectacular form are *geysers* from which hot water spurts forth intermittently. Some hot springs emit not only water derived from infiltration of surface water but also juvenile water.

As it infiltrates permeable soil and rock strata, surface water will undergo changes in the composition of dissolved and suspended organic and inorganic matter. These changes are the result of adsorption and ion exchange by surfaces of soil and rock particles. They are also the result of biochemical action of microbes, including bacteria, fungi, and protozoa, which exist mainly in biofilms on the surface of many of the rock particles and metabolize the adsorbed organic and (to a limited extent) inorganic matter (Cullimore, 1992; Costerton et al., 1994; Flemming et al., 2007). Plant roots and microbes associated with them in the rhizosphere affect the solute composition through absorption or excretion and metabolism. Polluted water infiltrating the ground may become thoroughly purified, provided that it moves through a sufficient depth and does not encounter major cracks and fissures, which, because of reduced surface area, would exert only limited filtering action. Under some circumstances, the groundwater may also become highly mineralized during filtration or after reaching the water table. If such mineralized water reaches the land surface, it may leave extensive deposits of calcium carbonate, iron oxide, or other material as it evaporates.

Systematic studies of microbes in groundwater have been undertaken. As many as 10^6 bacteria per gram have been recovered by viable and direct counts from the vadose zone and some shallow water table aquifers. They included gram-positive and gram-negative types, the former in apparently greater numbers (Wilson et al., 1983). Evidence of fungal spores and yeast cells was also seen in one instance (Ghiorse and Balkwill, 1983), but eukaryotic microbes were generally thought to be absent from subsurface samples associated with groundwater. Sinclair and Ghiorse (1987) showed that the number of protozoa, mostly flagellates and amoebae, decreased sharply to 28 g^{-1} (dry weight) with increasing depth in the vadose zone at the bottom of a clay loam subsoil material taken from a site in

Lula, Oklahoma, United States. They were absent from the saturated zone except in gravelly, loamy sand matrix at a depth of 7.5 m, which also contained significant numbers of bacteria.

Special drilling methods that minimize the possibility of microbial contamination have been developed (see Fredrickson et al., 1993; Russell et al., 1992) for studying the microbiology of vadose zones and aquifers in the deep subsurface. Fredrickson et al. (1991) reported as many as 10^4–10^6 colony-forming heterotrophic bacteria per gram of Middendorf (~366–416 m deep) and Cape Fear (~457–470 m deep) sediments of Cretaceous age, which were obtained by drilling into the Atlantic Coastal Plain at Savannah River Plant, South Carolina. The isolates from the two sediments were physiologically distinct. Contrary to what they expected from their study of a shallow aquifer in Lula, Oklahoma (see earlier), Sinclair and Ghiorse (1989) found significant numbers of fungi and protozoa in samples from deep aquifer material from the Upper Atlantic Coastal Plain Province. The numbers of these organisms were highest where the prokaryotic population was high. The bacteria in the samples included members of the Bacteria and Archaea. Diverse physiological groups were represented, including autotrophs such as sulfur oxidizers, nitrifiers, and methanogens (Fredrickson et al., 1989; Jones et al., 1989) as well as heterotrophs such as aerobic and anaerobic mineralizers. The latter included denitrifiers, sulfur reducers, and iron(III) reducers (Balkwill, 1989; Hicks and Fredrickson, 1989; Phelps et al., 1989; Madsen and Bollag, 1989; Francis et al., 1989; Chapelle and Lovley, 1992). Hydrogen- and acetate-oxidizing methanogens were also found (Jones et al., 1989). Rates of metabolism in deep aquifers of the Atlantic Coastal Plain have been estimated to range from 10^{-4} to 10^{-6} mmol of CO_2 per liter per year, using a method of geochemical modeling of groundwater chemistry (see also Chapter 4) (Chapelle and Lovley, 1990).

Significant numbers of autotrophic methanogens and homoacetogens have been detected in groundwater from the deep granite aquifers (112–446 m) at the Äspö Hard Rock Laboratory in Sweden (Kotelnikova and Pedersen, 1998). According to the investigators, these organisms may be the primary producers in this aquifer, using available hydrogen as their energy source. However, only 0–4.5×10^{-1} autotrophic methanogens and 0–2.2×10^1 homoacetogens per milliliter were found in groundwater samples from deep igneous rock aquifers in Finland (Haveman et al., 1999). These samples did contain 0 to $>1.6 \times 10^4$ sulfate reducers and 7.0×10^0 to 1.6×10^4 iron(III) reducers. A subsurface paleosol sample from a depth of 188 m in the Yakima Borehole at the U.S. Department of Energy's Hanford Site in the state of Washington yielded a variety of bacteria, including members of the Bacteria and Archaea. The members of the Bacteria included relatives of *Pseudomonas*, *Bacillus*, *Micrococcus*, *Clavibacter*, *Nocardioides*, *Burkholderia*, *Comamonas*, and *Erythromicrobium* in addition to six novel types with some affinity to the *Chloroflexaceae*. The members of the Archaea showed an affinity to the *Crenarchaea* branch (Chandler et al., 1998). The paleosol in which these bacteria were detected is part of a sedimentary formation overlying the Columbia River Basalt. It has been proposed that methanogens and homoacetogens along with sulfate and iron(III) reducers, present in groundwater from aquifers in the Columbia River Basalt, form a biocenosis that uses H_2 generated in an interaction between water and ferromagnesian silicates in basalt as its energy source (Stevens and McKinley, 1995; Stevens, 1997). However, the occurrence of a reaction between ferromagnesian silicates in basalt and water has been questioned (Madsen et al., 1996; Lovley and Chapelle, 1996; Stevens and McKinley, 1996; Anderson et al., 1998). Indeed, although it can be shown on a thermodynamic basis that H_2 can be formed in a reaction between Fe^{2+} and water at pH 7.0,

$$3Fe^{2+} + 4H_2O \rightarrow H_2 + Fe_3O_4 + 6H^+ \quad (\Delta G_r^{0'} = -50.33\,\text{kJ}) \tag{5.4}$$

it cannot be formed in a reaction between ferrous silicate and water at pH 7.0, for example,

$$3FeSiO_3 + H_2O \rightarrow H_2 + Fe_3O_4 + SiO_2 \quad (\Delta G_r^{0'} = +12.59\,\text{Kcal or} +52.63\,\text{kJ}) \tag{5.5}$$

The hydrogen in the Columbia River Basalt may come from other sources, as pointed out by Madsen (1996) and Lovley and Chapelle (1996).

Much remains to be learned about the microbial populations of pristine aquifers and their associated vadose zones and the response of this population to environmental stresses (pollution).

5.5 SUMMARY

The hydrosphere is mainly marine. It occupies more than 70% of the Earth's surface. The world's oceans reside in basins whose walls arise from the margins of continental landmasses that project into the sea by way of the continental shelf, the continental slope, and the continental rise, bottoming out at the ocean floor. The ocean floor is traversed by mountain ranges cut by fracture zones and rift valleys—the midocean spreading centers—where new ocean floor is being formed. The ocean floor is also cut by deep trenches representing zones of subduction, where the margin of an oceanic crustal plate slips beneath a continental crustal plate. Parts of the ocean floor also feature isolated mountains that are live or extinct volcanoes and may project above sea level as islands. The average world ocean depth is 3975 m; the greatest depth is ~11,000 m in the Marianas Trench.

Most of the ocean floor is covered by sediment of 300 m average thickness, accumulating at rates <1 to >10 cm per 1000 years. Ocean sediments may consist of sand, silt, and clays of terrigenous origin and of oozes of biogenic origin, such as diatomaceous, radiolarian, or calcareous oozes.

Different parts of the ocean are in motion at all times, driven by wind stress, the Earth's rotation, density variations, and gravitational effects exerted by the sun and moon. Surface, subsurface, and bottom currents have been found in various geographic locations.

Where water masses diverge, upwelling occurs, which replenishes nutrients for plankton in the surface waters. Where the water masses converge, surface water sinks and carries oxygen to deeper levels of the ocean, ensuring some degree of oxygenation at all levels.

Seawater is saline (average salinity ~35‰) owing to the presence of chloride, sodium, sulfate, magnesium, calcium, potassium, and some other ions. Variations in total salt concentrations affect the density of seawater, as does variation in water temperature. The ocean is thermally stratified between 50° north and 50° south latitude into a mixed layer (to a depth of ~300 m), with water at more or less uniform temperature between 15 and 30°C depending on latitude; a thermocline (from a depth of ~300 m to ~1000 m) in which the temperature drops to ~4°C with depth; and the deep water (from the thermocline to the bottom), where the temperature is more or less uniform, that is, between <0 and 4°C. HP in the water column increases by ~1 atm (1.013 bar) for every 10 m of increase in depth. Light penetrates to an average depth of ~100 m, which restricts phytoplankton to shallow depths. Zooplankton and bacterioplankton can exist at all depths but are found in greatest numbers at the seawater–air interface, near where the phytoplankton abounds, and on the ocean sediment. Intermediate depths are at most sparsely inhabited because of limited nutrient supply.

Marine phytoplankton is constituted of algal forms, mainly diatoms, dinoflagellates, and coccolithophorids, whereas marine zooplankton is constituted mainly of flagellated and amoeboid protozoa as well as some small invertebrates. Bacterioplankton is composed of members of Bacteria, chiefly heterotrophs, and of members of Archaea, whose traits we know little so far. Phytoplankton organisms are the primary producers, zooplankton organisms the predators and scavengers, and the heterotrophic bacteria the decomposers. The metabolic rate of free-living microorganisms decreases markedly with depth, probably as a result of the effects of high HP and low temperature. Different life forms in the ocean show different tolerances to salinity, temperature, and HP.

Freshwater is found in lakes and streams above ground and in saturated and unsaturated strata below ground. Lakes are standing bodies of water, usually of low salinity, which may be thermally stratified into epilimnion, thermocline, and hypolimnion. The degrees of stratification may vary with the season of the year. Water below the thermocline (i.e., the hypolimnion) may develop anoxia because the thermocline is an effective barrier to diffusion of oxygen into it. Only after the disappearance of the thermocline do these waters become reoxygenated due to total mixing by wind agitation. Lakes vary in nutrient quality. Phosphorus is usually the most limiting element to lake life. Phytoplankton, zooplankton, and bacterioplankton are important life forms in lakes.

Phytoplankton is restricted to the epilimnion, whereas zooplankton and bacterioplankton together with fungi may be distributed throughout the water column and in the sediment.

Rivers constitute moving freshwater. They are generally not thermally stratified. Abundant life forms, such as phytoplankton, zooplankton, and bacterioplankton are concentrated mainly in the quieter potions of streams, especially those that are not polluted.

Groundwaters are derived from surface waters that seep into the ground and accumulate in aquifers above impervious rock strata. Water from an aquifer may come to the surface again by way of springs or seepage. In passing through the ground, water is purified. Microorganisms as well as organic and inorganic chemicals are removed by adsorption to rock and soil particles. Organic matter may be mineralized by the microbial decomposers. Sediment samples from shallow as well as deep aquifers have revealed the presence of a significant microbial population with very diverse physiological potentials.

REFERENCES

Anderson RT, Chapelle FN, Lovley DR. 1998. Evidence against hydrogen-based microbial ecosystems in basalt aquifers. *Science* 281:976–977.

Arrieta JM, Weinbauer MG, Lute C, Herndl GJ. 2004. Response of bacterioplankton to iron fertilization in the Southern Ocean. *Limnol Oceanogr* 49:799–808.

Bakker DCE. 2002. The marine carbon cycle and climate. *Recent Research Developments in Biotechnology and Bioengineering, Special Issue: Biotechnology and Bioengineering of CO_2 fixation.* Kerala, India: Research Signpost, pp. 23–28.

Balkwill DL. 1989. Numbers, diversity, and morphological characteristics of aerobic, chemoheterotrophic bacteria in deep subsurface sediments from a site in South Carolina. *Geomicrobiol J* 7:32–52.

Bischoff JL, Rosenbauer RJ. 1983. A note on the chemistry of seawater in the range of 350°–500°C. *Geochim Cosmochim Acta* 47:139–144.

Bowden KF. 1975. Oceanic and estuarine mixing processes. In: Riley JP, Skirrow G, eds. *Chemical Oceanography*, Vol. 1, 2nd ed. London: Academic Press, pp. 43–72.

Brzezinski MA, Jones JL, Demarest MS. 2005. Control of silica production by iron and silicic acid during the southern ocean iron experiment (SOFeX). *Limnol Oceanogr* 50:810–824.

Buessler KO, Andrews JE, Pike SM, Charette MA. 2004. The effects of iron fertilization on carbon sequestration in the Southern Ocean. *Science* 304:414–417.

Buessler KO, Boyd PW. 2003. Climate change. Will ocean fertilization work? *Science* 300:67–68.

Capone DG, Zehr JP, Paerl HW, Bergman B, Carpenter EJ. 1997. *Trichodesmium*, a globally significant marine cyanobacterium. *Science* 276:1221–1230.

Chandler DP, Brockman FJ, Bailey TJ, Fredrickson JK. 1998. Phylogenetic diversity of archaea and bacteria in a deep subsurface paleosol. *Microb Ecol* 36:37–50.

Chapelle FH. 1993. *Groundwater Microbiology and Geochemistry*. New York: Wiley.

Chapelle FH, Lovley DR. 1990. Rates of microbial metabolism in deep coastal plain aquifers. *Appl Environ Microbiol* 56:1865–1874.

Chapelle FH, Lovley DR. 1992. Competitive exclusion of sulfate reduction by Fe(III) reducing bacteria: A mechanism for producing discrete zones of high-iron groundwater. *Ground Water* 30:29–36.

Church MJ, Hutchins DA, Ducklow HW. 2000. Limitation of bacterial growth by dissolved organic matter and iron in the southern ocean. *Appl Environ Microbiol* 66:455–466.

Coale KH, Johnson KS, Fitzwater SE, Gordon RM, Tanner S, Chavez FP, Ferioli L, Sakamoto C, Rogers P, Millero F, Steinberg P, Nightingale P, Cooper D, Cochlan WP, Landry MR, Constantinou J, Rollwagen G, Trasvina A, Kudela R. 1996. A massive phytoplankton bloom induced by an ecosystem-scale iron fertilization experiment in the equatorial Pacific Ocean. *Nature* (London) 383:495–501.

Costerton JW, Lewandowski Z, DeBeer D, Caldwell D, Korber D, James G. 1994. Biofilms, the customized microniche. *J Bacteriol* 176:2137–2142.

Cullimore DR. 1992. *Practical Groundwater Microbiology*. Chelsea, MI: Lewis.

Defant A. 1961. *Physical Oceanography*, Vol. 1. Oxford, U.K.: Pergamon Press.

Deming JW, Collwell RR. 1981. Barophilic bacteria associated with deep-sea animals. *BioScience* 31:507–511.

Dietrich G, Kalle K. 1965. *Allgemeine Meereskunde. Eine Einführung in die Ozeanographie*. Berlin: Gebrüder Bornträger.

Dworkin M. 1999. *The Prokaryotes. An Evolving Electronic Resource for the Microbiological Community.* Online Edition, New York: Springer.

Edmond JM, Von Damm KL, McDuff RE, Measures CI. 1982. Chemistry of hot springs on the East Pacific Rise and their effluent dispersal. *Nature* (London) 297:187–191.

Ehrlich HL. 1975. The formation of ores in the sedimentary environments of the deep sea with microbial participation: The case for ferromanganese concretions. *Soil Sci* 119:36–41.

Ehrlich HL. 1983. Manganese oxidizing bacteria from a hydrothermally active area on the Galapagos Rift. *Ecol Bull* (Stockholm) 35:357–366.

Ehrlich HL. 2000a. ZoBell and his contributions to the geosciences. In: Bell CR, Brylinsky M, Johnson-Green P, eds. *Microbial Biosystems: New Frontiers.* Proceedings of the 8th International Symposium on Microbial Ecology, Halifax, NS: Canada Society for Microbial Ecology, pp. 57–62.

Ehrlich HL. 2000b. Ocean manganese nodules: Biogenesis and bioleaching possibilities. *Miner Metall Process* 17:121–128.

Flemming H-C, Neu TR, Wozniak DJ. 2007. The EPS matrix: The "house of biofilm cells." *J Bacteriol* 189:7945–7947.

Francis AJ, Slater JM, Dodge CJ. 1989. Denitrification in deep subsurface sediments. *Geomicrobiol. J* 7:103–116.

Franklin MP, McDonald IR, Bourne DG, Owens NJP, Upstill-Goddard RC, Murell JC. 2005. Bacterial diversity in the bacterioneuston (sea surface microlayer): the bacterioneuston through the looking glass. *Environ Microbiol* 7:723–736.

Fredrickson JK, Balkwill DL, Zachara JM, Li S-MW, Brockman FJ, Simmons MA. 1991. Physiological diversity and distributions of heterotrophic bacteria in deep Cretaceous sediments in the Atlantic Coastal Plain. *Appl Environ Microbiol* 57:402–411.

Fredrickson JK, Brockman FJ, Bjornstad BN, Long PE, Li SW, McKinley JP, Wright JV, Conca JL, Kieft TL, Balkwill DL. 1993. Microbial characteristics of pristine and contaminated deep vadose sediments from an arid region. *Geomicrobiol J* 11:95–107.

Fredrickson JK, Garland TR, Hicks RJ, Thomas JM, Li SW, McFadden KM. 1989. Lithotrophic and heterotrophic bacteria in deep subsurface sediments and the relationship to sediment properties. *Geomicrobiol J* 7:53–66.

Gelder RJ, 1999. Complex lessons of iron uptake. *Nature* (London) 400:815–816.

Ghiorse WC, Balkwill DL. 1983. Enumeration and morphological characterization of bacteria indigenous to subsurface environments. *Dev Ind Microbiol* 24:213–224.

Gross MG. 1982. *Oceanography: A View of the Earth.* 3rd ed. Englewood Cliffs, NJ: Prentice-Hall.

Guan LL, Kanoh K, Kamino K. 2001. Effect of exogenous siderophores on iron uptake activity of marine bacteria under iron-limited conditions. *Appl Environ Microbiol* 67:1710–1717.

Hackett OM. 1972. Groundwater. In: Fairbridge RW, ed. *The Encyclopedia of Geochemistry and Environmental Sciences.* Encyclopedia of Earth Science Series, Vol. IVA. New York: Van Nostrand Reinhold, pp. 470–478.

Haveman SA, Pedersen K, Ruotsalainen P. 1999. Distribution and metabolic diversity of microorganisms in deep igneous rock aquifers of Finland. *Geomicrobiol J* 16:277–294.

Hermansson M, Marshall KC. 1985. Utilization of surface localized substrate by non-adhesive marine bacteria. *Microb Ecol* 11:95–105.

Hicks RJ, Fredrickson JK. 1989. Aerobic metabolic potentials of microbial populations indigenous to deep subsurface environments. *Geomicrobiol J* 7:67–77.

Horn DR, Horn BM, Delach MN. 1972. *Sedimentary Provinces, North Pacific (Map).* Palisades, New York: Lamont-Doherty Observatory of Columbia University.

Hutchins DA, Winter AE, Butler A, Luther GW III. 1999. Competition among marine phytoplankton for different chelated iron species. *Nature* (London) 400:855–861.

Jannasch HW. 1979. Microbial turnover of organic matter in the deep sea. *BioScience* 29:228–232.

Jannasch HW. 1984. Microbial processes at deep sea hydrothermal vents. In: Rona PA, Bostrom K, Laubier L, Smith KL Jr., eds. *Hydrothermal Processes at Seafloor Spreading Centers.* New York: Plenum Press, pp. 677–709.

Jannasch HW, Eimhjellen K, Wirsen CO, Farmanfarmaian A. 1971. Matter in the deep sea. *Science* 171:672–675.

Jannasch HW, Mottl MJ. 1985. Geomicrobiology of the deep-sea hydrothermal vents. *Science* 229:717–725.

Jannasch HW, Taylor CD. 1984. Deep-sea microbiology. *Annu Rev Microbiol* 38:487–514.

Jannasch HW, Wirsen CO. 1973. Deep-sea microorganisms: In situ response to nutrient enrichment. *Science* 180:641–643.

Jannasch HW, Wirsen CO. 1977. Microbial life in the deep sea. *Sci Am* 236:2–12.

Jones RE, Beeman RE, Suflita JM. 1989. Anaerobic metabolic processes in deep terrestrial subsurface. *Geomicrobiol J* 7:117–130.

Kaletunç G, Lee J, Alpas H, Bozoglu F. 2004. Evaluation of structural changes induced by hydrostatic pressure in *Leuconostoc mesenteroides. Appl Environ Microbiol* 70:1116–1122.

Kaye JZ, Baross JA. 2004. Synchronous effects of temperature, hydrostatic pressure, and salinity on growth, phospholipids profiles, and protein patterns of four *Halomonas* species isolated from deep-sea hydrothermal vent and sea surface environments. *Appl Environ Microbiol* 70:6220–6229.

Kester DR. 1975. Dissolved gases other than CO_2. In: Riley JP, Skirrow G, eds. *Chemical Oceanography*, Vol. 1, 2nd ed. New York: Academic Press, pp. 497–547.

Kotelnikova S, Pedersen K. 1998. Distribution and activity of methanogens and homoacetogens in deep granitic aquifers at Äspö Hard Rock Laboratory, Sweden. *FEMS Microbiol Ecol* 26:121–134.

Kriss AE. 1970. Ecological-geographic patterns in the distribution of heterotrophic bacteria in the Atlantic Ocean. *Mikrobiologiya* 39:362–371 (English translation pp. 313–320).

Kulm LD, Suess E, Moore JC, Carson B, Lewis BT, Ritger SD, Kado DC, Thornburg TM, Embley RW, Rugh WD, Massoth GJ, Langseth MC, Cochrane GR, Scamman RL. 1986. Oregon subduction zone: Venting, fauna, and carbonates. *Science* 231:561–566.

LaRock PA. 1969. The bacterial oxidation of manganese in a fresh water lake. PhD Thesis. Rensselaer Polytechnic Institute, Troy, New York.

Liss P, Chuck A, Bakker D, Turner S. 2005. Ocean fertilization with iron: Effect on climate and air quality. *Tellus B Chem Phys Meteorol* 57B:269–270.

Littler MM, Littler DS, Blair SM, Norris JN. 1985. Deepest known plant life discovered on an uncharted seamount. *Science* 227:57–59.

Lovley DR, Chapelle FH. 1996. Technical comment. Hydrogen-based microbial ecosystems in the Earth. *Science* 272:896.

Mackenzie FT, Stoffym M, Wollast R. 1978. Aluminum in seawater: Control by biological activity. *Science* 199:680–682.

MacLeod RA. 1965. The question of the existence of specific marine bacteria. *Bacteriol Rev* 29:9–23.

Madsen EL, Stevens T, Lovley DR, Chapelle H, McKinley JP. 1996. Technical comment. Hydrogen-based microbial ecosystems in the Earth. *Science* 272:896–897.

Madsen EL, Bollag JM. 1989. Aerobic and anaerobic microbial activity in deep subsurface sediments from the Savannah River Plant. *Geomicrobiol J* 7:93–101.

Marine Chemistry. 1971. *A Report to the Marine Chemistry Panel of the Committee of Oceanography.* Washington, DC: National Academy of Sciences.

Marquis RE. 1982. Microbial barobiology. *BioScience* 32:267–271.

Martinez JS, Zhang GP, Holt PD, Jung H-T, Carrano CJ, Haygood MG, Butler A. 2000. Self-assembling amphiphilic siderophores from marine bacteria. *Science* 287:1245–1247.

Massana R, DeLong EF, Pedrós-Alió C. 2000. A few cosmopolitan phylotypes dominate planktonic archaeal assemblages in widely different oceanic provinces. *Appl Environ Microbiol* 66.1777–1787.

Menzel DW, Rhyter JH. 1970. Distribution and cycling of organic matter in the ocean. In: Wood DE, ed. *Organic Matter in Natural Waters.* Institute of Marine Science Occasional Publ No. 1. Fairbanks, AK: University of Alaska, pp. 31–54.

Minshall GW. 1978. Autotrophy in stream ecosystems. *BioScience* 28:767–771.

Morita RY. 1967. Effects of hydrostatic pressure on marine bacteria. *Oceanogr Marine Biol Annu Rev* 5:187–203.

Morita RY. 1975. Psychrophilic bacteria. *Bacteriol Rev* 39:144–167.

Morita RY. 1980. Microbial life in the deep sea. *Can J Microbiol* 26:1375–1385.

Nissenbaum A. 1979. Life in the Dead Sea—Fables, allegories, and scientific research. *BioScience* 29:153–157.

Overmann J, Cypionka H, Pfennig N. 1992. An extremely low light-adapted phototrophic sulfur bacterium from the Black Sea. *Limnol Oceanogr* 37:150–155.

Park PK. 1968. Seawater hydrogen-ion concentration: Vertical distribution. *Science* 162:357–358.

Parkes RJ, Cragg BA, Bale SJ, Getliff JM, Goodman K, Rochelle PA, Fry JC, Weightman AJ, Harvey SM. 1994. Deep bacterial biosphere in Pacific Ocean sediments. *Nature* (London) 371:410–413.

Paull CK, Hecker B, Commeau R, Freeman-Lynde RP, Neumann C, Corso WP, Golubic S, Hook JE, Sikes E, Curray J. 1984. Biological communities at the Florida Escarpment resemble hydrothermal vent taxa. *Science* 226:965–967.

Phelps TJ, Raione EG, White DC, Fliermans CB. 1989. Microbioal activities in deep subsurface environments. *Geomicrobiol J* 7:79–91.

Pope DH, Berger LR. 1973. Inhibition of metabolism by hydrostatic pressure: What limits microbial growth? *Arch Mikrobiol* 93:367–370.

Reid GK. 1961. *Ecology of Inland Waters and Estuaries*. New York: Reinhold.

Rheinheimer G. 1980. *Aquatic Microbiology*. 2nd ed. Chichester, U.K.: Wiley.

Richardson PL. 1993. Tracking ocean eddies. *Am Sci* 81:261–271.

Ring Group. 1981. Gulf-Stream cold-core rings: Their physics, chemistry, and biology. *Science* 212:1091–1100.

Rittenberg SC. 1940. Bacteriological analysis of some long cores of marine sediments. *J Mar Res* 3:191–201.

Russell BF, Phelps TJ, Griffin WT, Sargent KA. 1992. Procedures for sampling deep subsurface microbial communities in unconsolidated sediments. *Ground Water Monitor Rev* 12:96–104.

Sargent KA, Fliermans CB. 1989. Geology and hydrology of the deep subsurface microbiology sampling site at the Savannah River Plant, South Carolina. *Geomicrobiol J* 7:3–13.

Schwartz JR, Colwell RR. 1976. Microbial activities under deep-ocean conditions. *Dev Indust Microbiol* 17:299–310.

Seyfried WE Jr., Janecky DR. 1985. Heavy metal and sulfur transport during subcritical hydrothermal alteration of basalt: Influence of fluid pressure and basalt composition and crystallinity. *Geochim Cosmochim Acta* 49:2545–2560.

Shanks WC III, Bischoff JL, Rosenbauer RJ. 1981. Seawater sulfate reduction and sulfur isotope fractionation in basaltic systems: Interaction of seawater with fayalite and magnetite at 200–350°C. *Geochim Cosmochim Acta* 45:1977–1995.

Sieburth J McN. 1975. *Microbial Seascape. A Pictorial Essay on Marine Microorganisms and Their Environments*. Baltimore, MD: University Park Press.

Sieburth J McN. 1976. Bacterial substrates and productivity in marine ecosystems. *Annu Rev Ecol Syst* 7:259–285.

Sieburth J McN. 1979. *Sea Microbes*. New York: Oxford University Press.

Sieburth J McN, Willis P-J, Johnson KM, Burney CM, Lavoie DM, Hinga KR, Caron DA, French FW III, Johnson PW, Davis PG. 1976. Dissolved organic matter and heterotrophic microneuston in the surface microlayers of the North Atlantic. *Science* 194:1415–1418.

Sinclair JL, Ghiorse WC. 1987. Distribution of protozoa in subsurface sediments of a pristine groundwater study site in Oklahoma. *Appl Environ Microbiol* 53:1157–1163.

Sinclair JL, Ghiorse WC. 1989. Distribution of aerobic bacteria, protozoa, algae, and fungi in deep subsurface sediments. *Geomicrobiol J* 7:15–31.

Skinner BJ, Porter SC, Botkin DB. 1999. *The Blue Planet. An Introduction to Earth System Science*. 2nd ed. New York: Wiley.

Smith FGW, ed. 1974. *Handbook of Marine Science*, Vol. 1. Cleveland, OH: CRC Press, p. 617.

Smith KL, Teal JM. 1973. Deep-sea benthic community respiration: An in situ study at 1850 meters. *Science* 179:282–283.

Smith RL. 1968. Upwelling. *Oceanogr Marine Biol Annu Rev* 6:11–46.

Smith W, Pope D, Landau JV. 1975. Role of bacterial ribosome subunits in barotolerance. *J Bacteriol* 124:582–584.

Stanley SM. 1985. *Earth and Life Through Time*. New York: WH Freeman.

Stein JL. 1984. Subtidal gastropods consume sulfur-oxidizing bacteria: Evidence from coastal hydrothermal vents. *Science* 223:696–698.

Stevens TO. 1997. Lithoautotrophy in the subsurface. *FEMS Microbiol Rev* 20:327–337.

Stevens TO, McKinley JP. 1995. Lithoautotrophic microbial ecosystems in deep basalt aquifers. *Science* 270:450–454.

Stevens TO, McKinley JP. 1996. Response to technical comment. Hydrogen-based microbial ecosystems in the Earth. *Science* 272:896–897.

Strahler AN, Strahler AH. 1974. *Introduction to Environmental Science*. Santa Barbara, CA: Hamilton.

Tortell PD, Maldonado MT, Granger J, Price NM. 1999. Marine bacteria and biogeochemical cycling of iron in the oceans. *FEMS Microbiol Ecol* 29:1–11.

Tunnicliffe V. 1992. Hydrothermal-vent communities in the deep sea. *Am Sci* 80:336–349.

Turner JT, Ferrante JG. 1979. Zooplankton fecal pellets in aquatic ecosystems. *BioScience* 29:670–677.

Turner RD. 1973. Wood-boring bivalves, opportunistic species in the deep-sea. *Science* 180:1377–1379.

van der Leeden F, Troise FL, Todd DK. 1990. *The Water Encyclopedia*. 2nd ed. Chelsea, MI: Lewis.

Wangersky PJ. 1976. The surface film as a physical environment. *Annu Rev Ecol Syst* 7:161–176.

Welch PH. 1952. *Limnology*. 2nd ed. New York: McGraw-Hill.

Williams J. 1962. *Oceanography*. Boston, MA: Little, Brown.

Wilson JT, McNabb JF, Balkwill DL, Ghiorse WC. 1983. Enumeration and characterization of bacteria indigenous to a shallow water-table aquifer. *Ground Water* 21:134–142.

Wirsen CO, Jannasch HW. 1974. Microbial transformation of some ^{14}C-labeled substances in coastal water and sediment. *Microb Ecol* 1:25–37.

Wood EJF. 1965. *Marine Microbial Ecology*. New York: Reinhold.

Wotton RS, Preston TM. 2005. Surface films: Areas of water bodies that are often overlooked. *BioScience* 55:137–145.

Wu J, Sunda W, Boyle EA, Karl DM. 2000. Phosphate depletion in the western North Atlantic Ocean. *Science* 289:759–762.

Yayanos AA, Dietz AS, van Boxtel R. 1979. Isolation of a deep-sea barophilic bacterium and some of its growth characteristics. *Science* 205:808–810.

Yayanos AA, Dietz AS, van Boxtel R. 1981. Obligately barophilic bacterium from the Marianas Trench. *Proc Natl Acad Sci USA* 78:5212–5215.

Zeebe RE, Archer D. 2005. Feasibility of ocean fertilization and its impact on future atmospheric CO_2 levels. *Geophys Res Lett* 32:L09703/1–L09703/5.

Zehr JP. 1992. Molecular biology of nitrogen fixation in natural populations of marine cyanobacteria. In: Carpenter EJ, Capone DG, Rueter JG, eds. *Marine Pelagic Cyanobacteria: Trichodesma and Other Diazotrophs*. Dordrecht: Kluwer Academic, pp. 249–264.

ZoBell CE. 1942. Bacteria in the marine world. *Sci Mon* 55:320–330.

ZoBell CE. 1946. *Marine Microbiology*. Waltham, MA: Chronica Botanica.

ZoBell CE, Morita RY. 1957. Barophilic bacteria in some deep sea samples. *J Bacteriol* 73:563–568.

6 Geomicrobial Processes: Physiological and Biochemical Overview

6.1 TYPES OF GEOMICROBIAL AGENTS

Various microorganisms, including *prokaryotes* and *eukaryotes*, contribute actively to certain geological processes, a fact that until not too long ago seems not always to have been sufficiently appreciated by some microbiologists and geologists. Geomicrobially active prokaryotes include members of the domain Bacteria (formerly designated Eubacteria) and the domain Archaea (formerly known as Archaebacteria or Archaeobacteria). Both Bacteria and Archaea are prokaryotes because they lack a true nucleus. Each has its genetic information encoded in a large circular polymeric molecule of *deoxyribonucleic acid* (DNA). This structure is often called the bacterial chromosome; but unlike the chromosomes of eukaryotic cells, it does not contain structural protein such as histone, nor is it surrounded by a nuclear membrane. The molecular size of a prokaryotic chromosome measures on the order of 10^9 Da. Some genetic information in prokaryotes may also be located in one or more extrachromosomal circularized DNA molecules, called *plasmids*. The exact molecular sizes of different plasmids vary, depending on the amount of genetic information they carry but generally range ~10^7 Da.

Another reason for classifying Bacteria and Archaea as prokaryotes is that they lack *mitochondria* and *chloroplasts*, organelles that carry out respiration and photosynthesis, respectively, in eukaryotic cells. In prokaryotes, respiratory activity is carried out in the plasma membrane, and in some gram-negative, anaerobic and facultative bacteria, this activity may also involve the periplasm and outer membrane. Photosynthetic activity in members of the domain Bacteria is carried out either by internal membranes derived from the plasma membrane (purple bacteria, cyanobacteria) or by special internal membranes (green bacteria). Similar photosynthetic activity is so far unknown in the domain Archaea.

Archaea are not distinguishable from Bacteria when seen as intact cells with a light microscope. At a submicroscopic level, however, Archaea exhibit distinct differences from Bacteria in the structure and composition of their cell envelope and plasma membrane and in the structure of their ribosomes, which are the sites of protein synthesis in Bacteria. They also differ in key enzymes involved in nucleic acid and protein synthesis (Brock and Madigan, 1988; Atlas, 1997; Schaechter et al., 2006).

Eukaryotic microorganisms include algae, fungi, protozoa, and slime molds. They differ from prokaryotic microorganisms in possessing a true nucleus, which is an organelle enclosed in a double membrane in which the chromosomes, the bearers of genetic information, and the nucleolus, the center for ribonucleic acid (RNA) synthesis, are located. Eukaryotic cells also feature mitochondria, chloroplasts, and vacuoles, all of which are membrane-bound organelles. The structure and mode of operation of their *flagella*, organs of locomotion, if present, also differ from those of the flagella of prokaryotic cells, when present. In eukaryotic cells, some key metabolic processes are highly compartmentalized, unlike those in prokaryotes. Figure 6.1 shows the phylogenetic interrelationship of the prokaryotic domains of Bacteria and Archaea and the Eukaryotes.

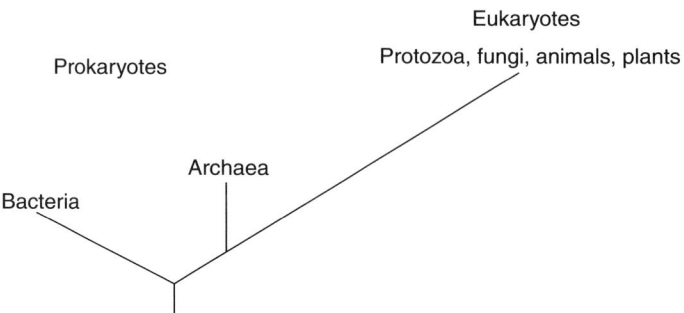

FIGURE 6.1 Phylogenetic relationships of the Prokaryotes, domains of the Bacteria and the Archaea, and the Eukaryotes. (Based on Woese CR, Kandler O, Wheelis ML, *Proc. Natl. Acad. Sci. USA*, 87, 4576–4579, 1990.)

Examples of geomicrobially important Archaea include *methanogens* (methane-forming bacteria), oxidizers of reduced forms of sulfur, extreme *halophiles*, and *thermoacidophiles*. Examples of geomicrobially important Bacteria include some aerobic and anaerobic hydrogen-metabolizing bacteria, iron-oxidizing and iron-reducing bacteria, manganese-oxidizing and manganese-reducing bacteria, nitrifying and denitrifying bacteria, sulfate-reducing bacteria, sulfur-oxidizing and sulfur-reducing bacteria, anaerobic photosynthetic sulfur bacteria, oxygen-producing cyanobacteria, and many others. Examples of geomicrobially important eukaryotes include fungi that can attack silicate, carbonate, and phosphate minerals, among others. They are also important in initiating degradation of somewhat recalcitrant natural organic polymers such as lignin, cellulose, and chitin, as in the O and A horizons of soil (see Chapter 4), or on and in surface sediments. Other geomicrobially important eukaryotes are algae, which together with cyanobacteria (prokaryotes), are a major source of oxygen in the atmosphere. Some algae promote calcium carbonate precipitation or dissolution, and others precipitate silica as frustules. Still other geomicrobially important eukaryotes include protozoa, some of which lay down siliceous, calcium carbonate, strontium sulfate, or manganese oxide tests, and others may accumulate preformed iron oxide on their cell surface.

6.2 GEOMICROBIALLY IMPORTANT PHYSIOLOGICAL GROUPS OF PROKARYOTES

Prokaryotes can be divided into various physiological groups such as chemolithoautotrophs, photolithoautotrophs, mixotrophs, photoheterotrophs, and heterotrophs (Figure 6.2). Each of these groups includes some geomicrobially important organisms. *Chemolithoautotrophs* (chemosynthetic autotrophs) include members of both the Bacteria and the Archaea. They are microorganisms that derive energy for doing metabolic work from the oxidation of inorganic compounds and assimilate carbon as CO_2, HCO_3^-, or CO_3^{2-} (see Wood, 1988). *Photolithoautotrophs* (photosynthetic autotrophs) include various Bacteria but no known Archaea. They are microorganisms that derive energy for doing metabolic work by converting radiant energy from the sun into chemical energy and use it, in part, in the assimilation of carbon as CO_2, HCO_3^-, or CO_3^{2-} as their carbon source (photosynthesis). Some of these microbes are anoxygenic (do not produce oxygen from photosynthesis), whereas others are oxygenic (produce oxygen from photosynthesis). *Mixotrophs* include some members of the Bacteria and the Archaea. They may derive energy simultaneously from the oxidation of reduced carbon compounds and oxidizable inorganic compounds and their carbon simultaneously from organic carbon and CO_2; or they may derive their energy totally from the oxidation of an inorganic compound but their carbon from organic compounds. *Photoheterotrophs* include mostly Bacteria but also a few Archaea (extreme halophiles). They derive all or part of their energy from sunlight

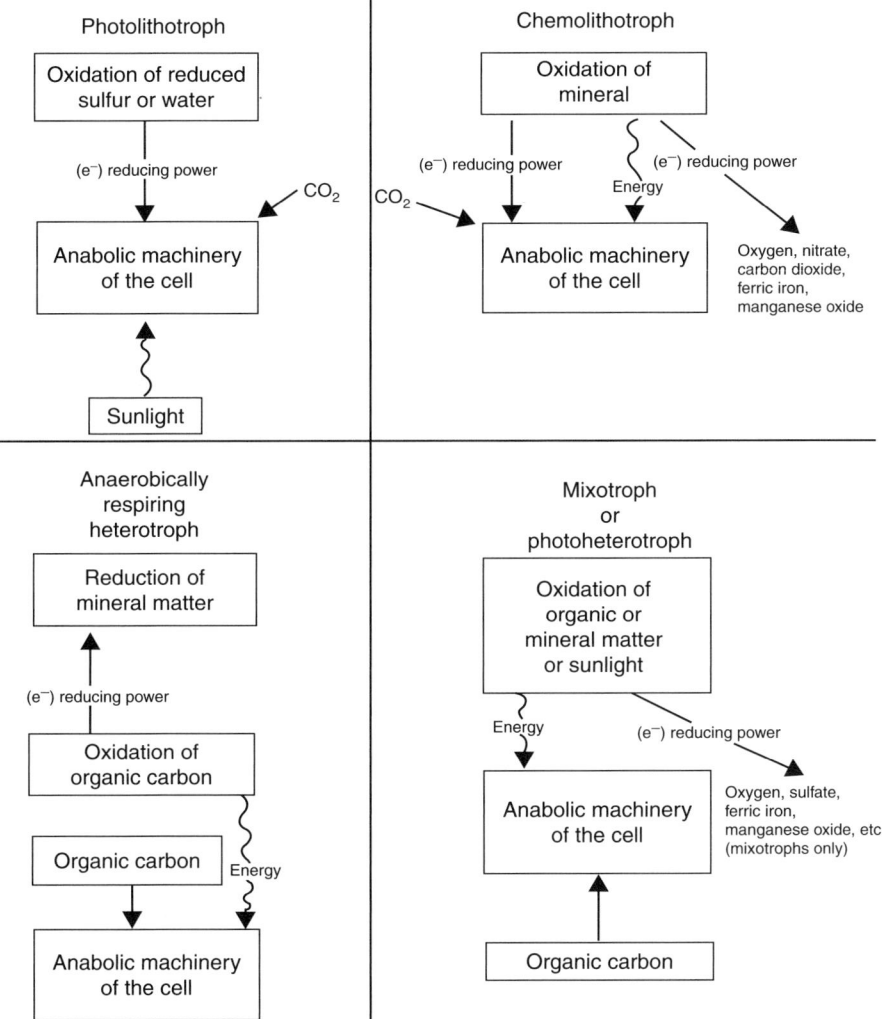

FIGURE 6.2   Geomicrobially important physiological groups among prokaryotes.

but their carbon by assimilating organic compounds. *Heterotrophs* include members of both the Bacteria and the Archaea. They derive all of their energy from the oxidation of organic compounds and most or all of their carbon from the assimilation of organic compounds. They may respire (oxidize their energy source) aerobically or anaerobically, or they may ferment their energy source by disproportionation (see Sections 6.5.1 through 6.5.4).

6.3   ROLE OF MICROBES IN INORGANIC CONVERSIONS IN LITHOSPHERE AND HYDROSPHERE

A number of microbes in the biosphere can be considered to be *geologic agents*. They may serve as agents of concentration, dispersion, or fractionation of geologically important matter. As *agents of concentration*, they cause localized accumulation of inorganic matter by (1) depositing inorganic products of metabolism in or on special parts of the cell, (2) passive accumulations involving surface adsorption or ion exchange, or (3) promoting precipitation of insoluble compounds external to the cell (Ehrlich, 1999). An example of mineral accumulation of an inorganic metabolic product inside a cell is the deposition of polyphosphate (volutin) in the cytoplasm of bacteria, such as *Spirillum volutans*, lactobacilli, rhizobia, and some others. An example of metabolic product accumulation in the bacterial cell envelope is the deposition of elemental sulfur granules in the periplasm (region between the plasma membrane and the outer membrane of gram-negative bacteria) by *Beggiatoa* and *Thiothrix* (Strohl et al., 1981; Smith and Strohl, 1991). An example of metabolic product accumulation at the cell surface is the formation of silica frustules by diatoms (algae), the frustules being their cell walls (de Vrind-de Jong and de Vrind, 1997) (see also Chapter 10).

Examples of passive accumulation of inorganic matter by adsorption or ion exchange are the binding of specific metallic cations by carboxyl groups of peptidoglycan or phosphate groups of teichoic or teichuronic acids in the cell wall of gram-positive bacteria (e.g., *Bacillus subtilis*), or by the lipopolysaccharide phosphoryl groups of outer membranes of gram-negative bacteria (e.g., *Escherichia coli*). The bound cations may subsequently react with certain anions, such as sulfide, carbonate, or phosphate, and form insoluble salts that may serve as nuclei in the formation of corresponding minerals (Beveridge et al., 1983; Beveridge, 1989; Beveridge and Doyle, 1989; Doyle, 1989; Ferris, 1989; Geesey and Jang, 1989; Macaskie et al., 1987, 1992).

An example of extracellular inorganic accumulation is the precipitation of metal cations in the cellular surround (bulk phase) by sulfides produced in sulfate reduction by sulfate-reducing bacteria. Many such sulfides are insoluble and fairly stable in the absence of oxygen (anoxic condition) (see Chapter 20).

As *agents of dispersion*, microbes promote dissolution of insoluble mineral matter as, for example, in the dissolution of $CaCO_3$ by respiratory CO_2 (see Chapter 9), or in the biochemical reduction of insoluble ferric oxide or manganese(IV) oxide to corresponding soluble compounds (see Chapters 16 and 17).

As *agents of fractionation*, microbes may act on a mixture of insoluble inorganic compounds (minerals) by promoting selective mobilization involving one or a few compounds in the mixture. One example is the oxidation of arsenopyrite (FeAsS) in pyritic gold ore by *Acidithiobacillus* (formerly *Thiobacillus*) *ferrooxidans* (Ehrlich, 1964) (see also Chapter 14). In this process, some of the iron solubilized by oxidation reacts with arsenic, which is simultaneously mobilized from the mineral, to precipitate in the bulk phase as a new compound—ferric arsenate. Another example is the preferential solubilization by reduction of Mn(IV) over Fe(III) contained in ferromanganese nodules by bacteria (Ehrlich et al., 1973; Ehrlich, 2000) (see also Chapter 17).

Microbes may also cause fractionation by preferentially attacking the light isotope in a mixture of stable heavy and light isotopes of an element in a compound in preference to the heavier isotope(s). Examples are the reduction of $^{32}SO_4^{2-}$ in preference to $^{34}SO_4^{2-}$ by some sulfate-reducing bacteria and the assimilation of $^{12}CO_2$ in preference to $^{13}CO_2$ by some autotrophs, in either instance under conditions of slow growth (see discussion by Doetsch and Cook, 1973). Other isotope mixtures that may be fractionated by microbes include hydrogen/deuterium (Estep and Hoering, 1980), $^{6}Li/^{7}Li$ (Sakaguchi and Tomita, 2000), $^{14}N/^{15}N$ (Wada and Hattori, 1978), $^{16}O/^{18}O$ (Duplessy et al., 1981), $^{28}Si/^{30}Si$ (De La Rocha et al., 1997), and $^{54}Fe/^{56}Fe$ (Beard et al., 1999). In the laboratory, the magnitude of these fractionations may be relatively large and may involve significant changes in isotopic ratios in a relatively short time. In some natural settings, corresponding microbial isotope fractionations are also readily detectable but may be of somewhat smaller magnitude. Studies so far lead to the impression that only a few mostly unrelated organisms have the capacity to fractionate stable isotope mixtures.

6.4 TYPES OF MICROBIAL ACTIVITIES INFLUENCING GEOLOGICAL PROCESSES

Microbes influence a number of geologic processes at the Earth's surface and in the uppermost crust (deep subsurface). *Lithification* is a type of geological process in which microbes may produce the cementing substance that binds inorganic sedimentary particles together to form sedimentary rock.

The microbially produced cementing substance may be calcium carbonate, iron- or aluminum-oxide, or silicate.

Some types of mineral formation may be the result of microbial activity. Iron sulfides such as pyrite (FeS_2), iron oxides such as magnetite (Fe_3O_4) or goethite (FeOOH), manganese oxides such as vernadite (MnO_2) or psilomelane (Ba, $Mn^{2+}Mn^{4+}O_{16}(OH)_4$), calcium carbonates such as calcite and aragonite ($CaCO_3$), and silica (SiO_2) may be generated authigenically by microbes (for a more extensive survey, see Lowenstamm, 1981).

In some instances, microbes may be responsible for mineral *diagenesis* in which microbes may cause alteration of rock structure and transformation of primary into secondary minerals, as in the conversion of orthoclase to kaolinite (Chapter 4).

Rock weathering may be promoted by microbes through production and excretion of metabolic products, which attack the rock and cause solubilization or diagenesis of some mineral constituents of the rock. Rock weathering may also involve direct enzymatic attack of certain oxidizable or reducible rock minerals by microbes, thereby causing their solubilization or their diagenesis.

Microbes may contribute to *sediment accumulation* in the form of calcium carbonate tests as those from coccolithophores or foraminifera, silica frustules from diatoms, or silica tests from radiolaria or actinopods in oceans and lakes. The aging of lakes may be influenced by microbes through their rock weathering activity or their generation of organic debris from incomplete decomposition of organic matter (see Chapter 5).

Geologic processes that are not influenced by microbes include *magmatic activity* or *volcanism*, *rock metamorphism* resulting from heat and pressure, *tectonic activity* related to crustal formation and transformation, and the allied processes of *orogeny* or mountain building. *Wind* and *water erosion* should also be included, although these processes may be facilitated by prior or concurrent microbial weathering activity. Although microbes do not influence these geologic processes, microbes may be influenced by them because these processes may create new environments that may be more or less favorable for microbial growth and activities than their previous occurrence.

6.5 MICROBES AS CATALYSTS OF GEOCHEMICAL PROCESSES

Most of the influence that microbes exert on geological processes is physiological. They may act as *catalysts* in some geochemical processes or as producers or consumers of certain geochemically active substances and thereby influence the rate of a geochemical reaction in which such substances are reactants or products (see Ehrlich, 1996). In either case, the microbes act through their *metabolism*, which has two components. One of these components is *catabolism*, which provides the cell with needed energy through *energy conservation* and may also yield to the cell some compounds that can serve as building blocks for polymers. A key reaction in energy conservation is the oxidation of a suitable nutrient or *metabolite* (a compound metabolically derived from a nutrient). The other component of metabolism is *anabolism*. It deals with assimilation (synthesis and polymerization) and leads to the formation of organic polymers such as nucleic acids, proteins, polysaccharides, lipids, and others. It also deals with the synthesis of inorganic polymers such as the polysilicates in diatom frustules and radiolarian tests and the polyphosphate granules that are formed by some bacteria and yeasts as energy storage compounds within their cells. Anabolism, by contributing to an increase in cellular mass and duplication of vital molecules, makes growth and reproduction possible. Catabolism and anabolism are linked to one another in such a way that catabolism provides the energy and some or all of the building blocks that make anabolism, which overall is an energy-consuming process, possible. Both catabolism and anabolism may play a geomicrobial role. Catabolism is involved, for instance, in large-scale oxidation that brings about transformation of inorganic substances and degradation of organic molecules, whereas anabolism is involved, for instance, in the synthesis of organic compounds from which fossil fuels (peat, coal, and petroleum) are generated. Anabolism is also the process by which the diatom frustules and radiolarian tests that accumulate in siliceous oozes are formed.

Catabolism may take the form of aerobic or anaerobic respiration, both of which are oxidation processes, or fermentation. Catabolism may thus be carried on in the presence or absence of oxygen in air. Oxygen is used as terminal electron acceptor. Indeed, microorganisms can be grouped as *aerobes* (oxygen-requiring organisms), *anaerobes* (oxygen-shunning organisms), *microaerophilic organisms* (requiring low concentrations of oxygen), or *facultative organisms* (can adapt their catabolism to operate in the presence or absence of oxygen in air). Facultative organisms use oxygen as terminal electron acceptor when it is available. When oxygen is not available, they use a reducible inorganic (e.g., nitrate or ferric iron) or an organic (e.g., fumarate) compound as a substitute terminal electron acceptor, or they ferment.

6.5.1 CATABOLIC REACTIONS: AEROBIC RESPIRATION

In *aerobic respiration*, hydrogen atoms or electrons are removed in the oxidation of organic compounds and electrons in the oxidation of inorganic entities by various biochemical reactions and conveyed by an *electron transport system* (ETS) to oxygen to form water. Among these biochemical reactions, an important reaction sequence in which reducing power (hydrogen atoms, electrons) is generated as part of aerobic respiration is the Krebs tricarboxylic acid cycle (Figure 6.3). By this reaction sequence, organic substances are completely oxidized to CO_2 and H_2O (Stryer, 1995; Schaechter et al., 2006). The reaction sequence is initiated when acetyl~SCoA, produced enzymatically in the oxidative degradation of a large variety of organic nutrients, is enzymatically combined with oxalacetate to form citrate with the release of CoASH. Citrate is converted stepwise to isocitrate, α-ketoglutarate, succinate, fumarate, malate, and back to oxalacetate. One turn of this cycle produces four hydrogen pairs and two CO_2 as well as one adenosine 5′-triphosphate (ATP) by *substrate-level phosphorylation*. The hydrogen pairs are the source of reducing power that is fed into the ETS and transported to oxygen as part of aerobic respiration to form water. In the transfer of the reducing power via the ETS, some of the energy that is liberated is conserved in special phosphate anhydride bonds of ATP by a chemiosmotic process called *oxidative phosphorylation* (see Section 6.5.5). Upon hydrolysis, these bonds (Figure 6.4) yield 7.3 kcal mol^{-1} (30.5 kJ mol^{-1}) of free energy at pH 7 and 25°C (Stryer, 1995), as opposed to ordinary phosphate ester bonds, which release only ~2 kcal mol^{-1} (8.4 kJ mol^{-1}) of energy under these conditions. The energy in high-energy bonds is used by cells for driving energy-consuming reactions such as syntheses or polymerizations.

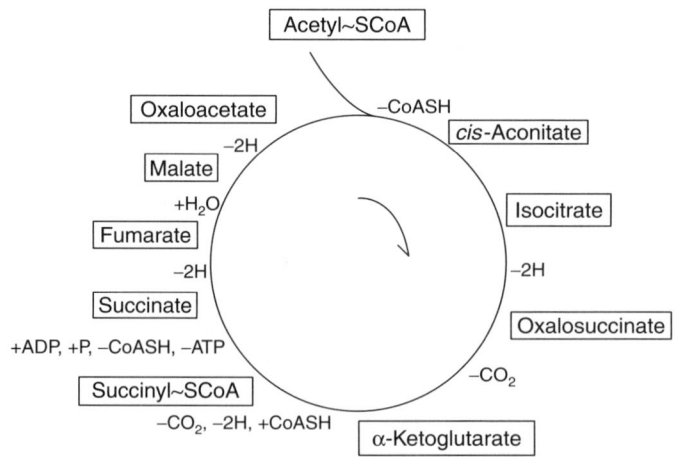

FIGURE 6.3 Krebs tricarboxylic acid cycle. One turn of the cycle converts one molecule of acetate to two molecules of CO_2 and four hydrogen pairs (2H), with the formation of one molecule of ATP by substrate-level phosphorylation. An additional 11 ATP can be formed when the four hydrogen pairs are oxidized to H_2O with oxygen as terminal electron acceptor.

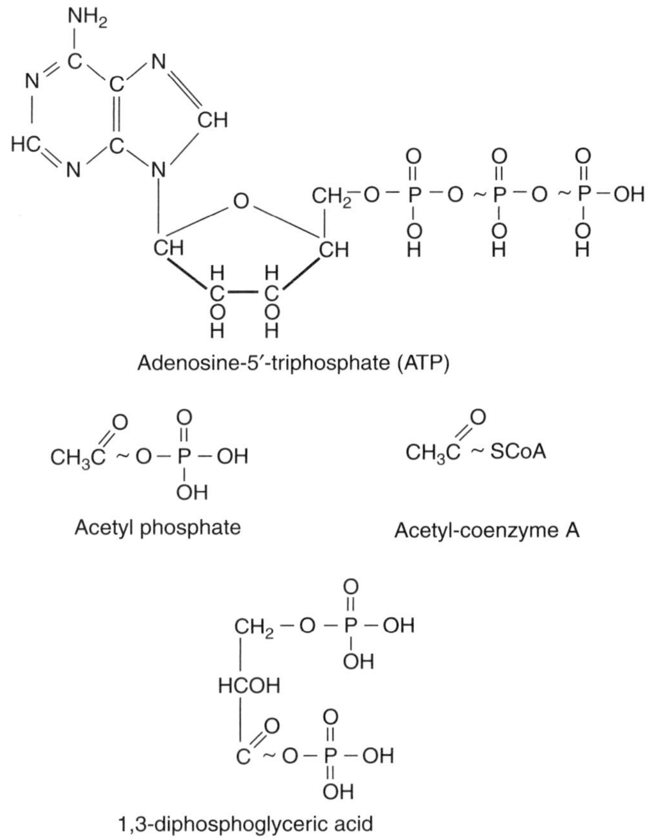

FIGURE 6.4 Examples of compounds containing one or more high-energy phosphate bonds ($\sim$).

Typical components of the ETS include nicotinamide adenine dinucleotide (NAD), flavoproteins (FP), iron–sulfur protein (Fe–S), quinone (CoQ), cytochromes (cyt Fe), and cytochrome oxidase (cyt oxid). They are arranged in complexes in the plasma membrane of aerobically respiring bacteria, as, for instance, in *Paracoccus denitrificans* (Payne et al., 1987; Onishi et al., 1987) and marine bacterial strain SSW_{22} (Graham, 1987) (Figure 6.5), and in the inner mitochondrial membrane eukaryotes. The types of electron carriers and enzymes and their arrangement in complexes, if any, differ among different kinds of bacteria. Indeed, in the same bacterium, the carriers may vary quantitatively and qualitatively, depending on growth conditions. Whatever may be the makeup of the assemblage of electron carriers, they interact in a specific sequence such as the one shown in Figure 6.6. Hydrogen or electrons enter the ETS where the E_h of the half-reaction by which they are removed from a substrate is near or below the E_h of the half-reaction of the appropriate hydrogen- or electron-accepting component of the system. For example, electrons from the oxidation of H_2 or pyruvate may enter the transport system at the level of complex I via NAD^+ as carrier and are transferred to complex III via CoQ and thence to complex IV via cytochrome *c*. Complex IV transfers the electrons that it receives to O_2, which is then transformed to form H_2O (Stryer, 1995; Schaechter et al., 2006). Electrons from the oxidation of succinate enter the transport system via complex II and are then transferred to complex III and so on to O_2. Electrons from the oxidation of ferrous iron enter the ETS at the level of complex IV. Table 6.1 lists the E_h values of some geomicrobially important enzyme-catalyzed oxidations, the level at which their hydrogens or electrons are fed into the ETS upon their oxidation, and also the estimated maximum number of high-energy phosphate bonds (ATP) that may be generated in the transfer of hydrogen or electron pairs to oxygen.

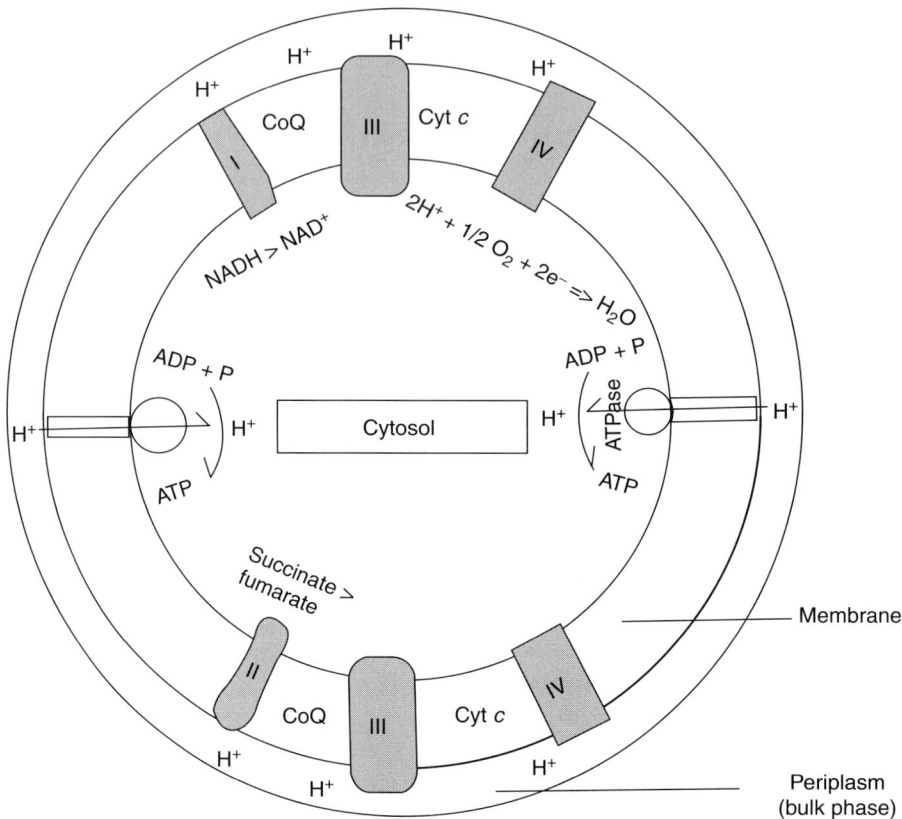

FIGURE 6.5 Schematic display of the bioenergetic machinery in a prokaryotic (domain Bacteria) cell enve-lope. Structures labeled I, II, III, and IV represent specific electron transport complexes involved in some prokaryotes. Complex I, reactive with NADH, includes a flavoprotein and an Fe–S protein; complex II, reac-tive with succinate, includes succinic dehydrogenase (another flavoprotein) and an Fe–S protein; complex III includes cytochromes b and c_1 and an Fe–S protein; and complex IV includes cytochrome oxidase (e.g., cyto-chrome $a + a_3$). Coenzyme Q (CoQ) and cytochrome c (cyt c) shuttle electrons between respective complexes. Proton translocation from the cytosol to the periplasm involves complex I or II, CoQ and complex III, and often complex IV. Oxygen reduction to water occurs on the inner surface of the plasma membrane. ATPase (ATP synthase) is the site of ATP synthesis.

6.5.2 CATABOLIC REACTIONS: ANAEROBIC RESPIRATION

In aerobic respiration, oxygen is always the terminal electron acceptor, whereas in anaerobic respi-ration other reducible substrates species such as nitrate, Fe^{3+}, sulfate, carbon dioxide, or an organic compound such as fumarate serve as terminal electron acceptors. Anaerobic respiration is per-formed by some Bacteria and some Archaea. Microorganisms performing such respiration may be facultative (e.g., nitrate reducers, some iron(III) and manganese(IV) reducers) or obligately respir-ing anaerobes (e.g., sulfate reducers, methanogens, homoacetogens, some other Fe(III) and Mn(IV) reducers). Some fermenters (see Section 6.5.4), although not anaerobic respirers, may also be facultative—that is, aerobically they respire (see Section 6.5.1) whereas anaerobically they ferment. Certain facultative respirers may reduce O_2 and concurrently another inorganic electron acceptor (e.g., certain nitrate, chromate, and MnO_2 reducers; see Chapters 13, 17, and 18, respectively). In most cases of anaerobic respiration by facultative organisms, oxygen competes with the other pos-sible terminal electron acceptors and thus must be absent or present at significantly lower concen-tration than in normal air for anaerobic respiration to occur. Anaerobic respiration usually employs

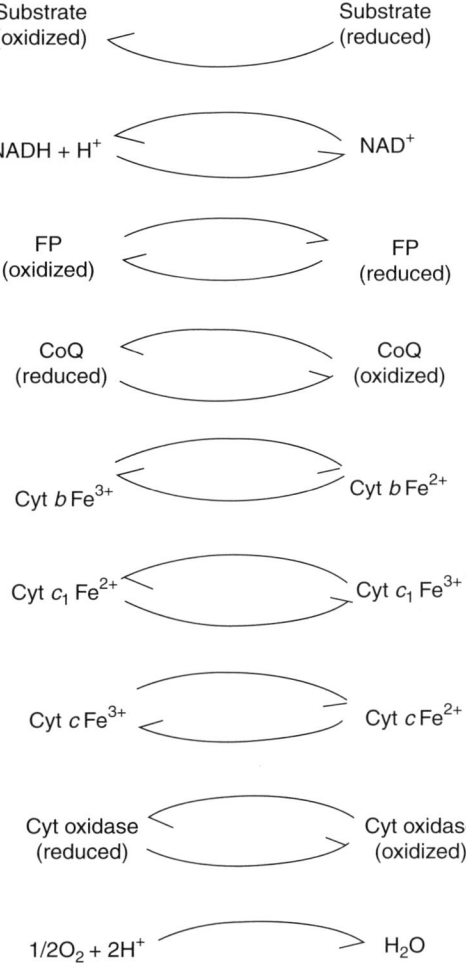

FIGURE 6.6 Schematic representation of the sequence of interactions of components of an ETS in a bacterial membrane by which reducing power is transferred from a substrate to oxygen.

TABLE 6.1
Microbially Catalyzed Oxidations of Geological Significance and Some Characteristics of Their Interaction with the ETS

Reaction	E_h at pH 7 (V)	Entrance Level into ETS	ATP/2e⁻ or 2H
$Fe^{2+} \rightarrow Fe^{3+} + e^-$	+0.77	Complex IV	1
$S^0 + 4H_2O \rightarrow SO_4^{2-} + 8H^+ + 6e^-$	−0.20	Complex III or IV	2 or 1[a]
$H_2S \rightarrow S^0 + 2H^+ + 2e^-$	−0.27	Complex I or III	3 or 2
$H_2 \rightarrow 2H^+ + 2e^-$	−0.42	Complex I or III	3 or 2
$Mn^{2+} + 2H_2O \rightarrow MnO_2 + 4H^+ + 2e^-$	+0.46	Complex IV (?)	1

[a] Add 0.5 mol of ATP per mol of SO_3^{2-} oxidized to SO_4^{2-} if substrate-level phosphorylation is part of the oxidation process.

some of the hydrogen and electron carriers of aerobic respiration but usually substitutes a suitable terminal reductase for cytochrome oxidase to convey electrons to the terminal electron acceptor that replaces oxygen. If the organic substrate being consumed anaerobically is oxidized completely, the tricarboxylic acid cycle may be involved; but other pathways may be used instead. Among the best characterized pathways of these anaerobic respiratory systems are those in which sulfate and nitrate are reduced as terminal electron acceptors.

6.5.3 CATABOLIC REACTIONS: RESPIRATION INVOLVING INSOLUBLE INORGANIC SUBSTRATES AS ELECTRON DONORS OR ACCEPTORS

It is important to recognize that in prokaryotic cells the ETS is located in the plasma membrane (Figure 6.7), and sometimes parts of it are located in the cell envelope (Figures 6.8A and 6.8B). By contrast, in eukaryotic cells the ETS is located internally in special organelles called mitochondria (Figure 6.9). As a result, the prokaryotes that are endowed with appropriate oxidoreductases

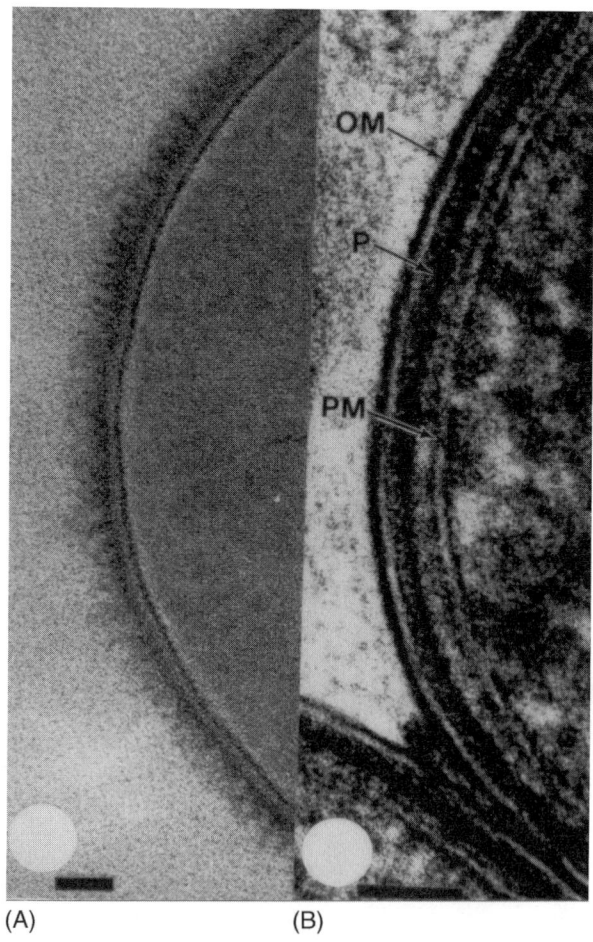

(A)   (B)

FIGURE 6.7 Location of ETS in typical prokaryotes. Thin sections of (A) the gram-positive cell wall of *Bacillus subtilis* and (B) the gram-negative cell wall of *Escherichia coli*. Both sections were prepared by freeze substitution. OM, outer membrane; PM, plasma membrane; P, periplasmic gel containing peptidoglycan located between the outer and plasma membranes. In both types of cells, the ETS is located in the plasma membrane. The bars in (A) and (B) equal 25 nm. (From Beveridge TJ, Doyle RJ, *Metal Ions and Bacteria*, Wiley, New York, 1989. With permission.)

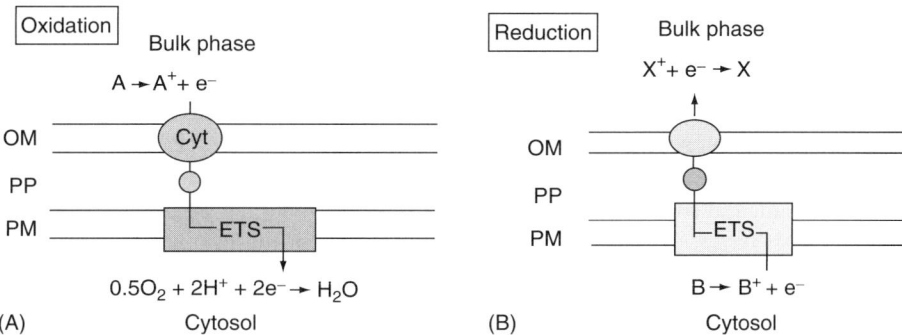

FIGURE 6.8 Generalized diagrams of the ETS of gram-negative prokaryotes capable of oxidizing or reducing oxidizable or reducible constituents of insoluble minerals or dissolved electron donors or acceptors at their cell surface–bulk phase interface by electron import or export, respectively. (A) Electron import (oxidation). (B) Electron export (reduction). OM, outer membrane; PP, periplasm; PM, plasma membrane; ETS, electron transport system in plasma membrane.

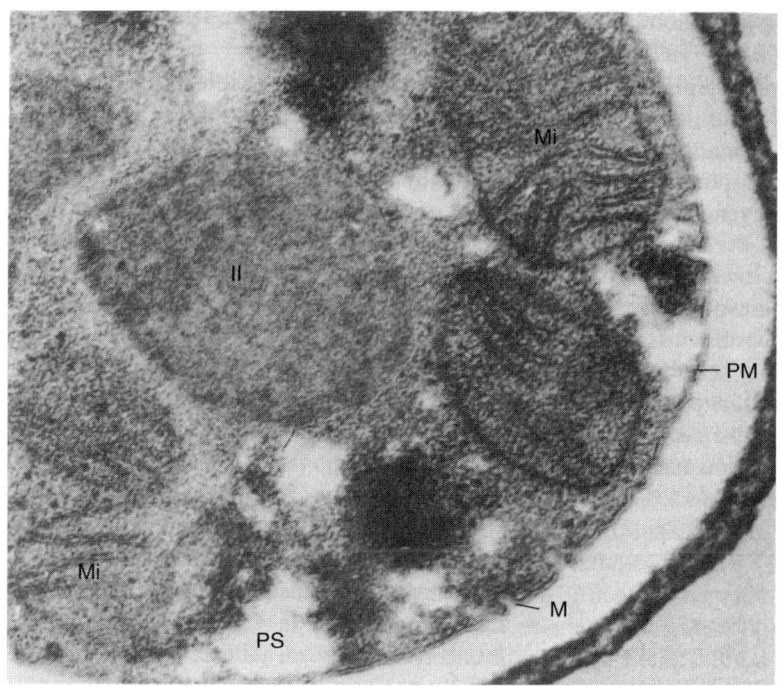

FIGURE 6.9 Location of ETS in eukaryotes. Cross section of a dormant conidium (spore) of *Aspergillus fumigatus*, a fungus (×64,000). ETS is located in the mitochondria. Mi, mitochondria; PM, plasma membrane; M, thin layer of electron-dense material; PS, polysaccharide storage material; II, membrane-bound storage body. (Courtesy of Ghiorse WC, Department of Microbiology, Cornell University, Ithaca, NY, USA.)

(enzymes that transfer hydrogen atoms or electrons) in their cell surface are able to oxidize or reduce insoluble inorganic substrates when these are in physical contact with the cell surface (Figures 6.8A and 6.8B); in other words, these organisms can use insoluble, oxidizable or reducible inorganic substrates (e.g., minerals) as electron donors or terminal electron acceptors, respectively, in their respiration by importing or exporting electrons, respectively. In at least one instance, a bacterium known as *Geobacter sulfurreducens* has been shown to facilitate electron transfer to the surface of Fe(III) oxides via special pilli—protein filaments projecting from the cell surface (Reguera et al., 2005).

Oxidoreductases in the cell envelope of some prokaryotes also enable these organisms to oxidize dissolved electron donors or acceptors without first taking them into the cytosol. This avoids any possible intracellular toxic effects from these dissolved electron donors or acceptors, or from the intracellular accumulation of insoluble products resulting from the oxidation or reduction of these electron donors or acceptors that cannot be readily expelled from the cell. Examples of insoluble inorganic substrates that can serve as electron donors or acceptors at the cell surface of some prokaryotes are elemental sulfur, iron sulfide, iron(III) oxide, and manganese(IV) oxide.

In gram-negative bacteria, enzymes and electron carriers in the periplasmic space of the cell envelope participate in the transfer of electrons in the appropriate direction between catalytic sites in the outer membrane and the ETS in the plasma membrane. The details of the mechanism of electron transfer in gram-positive Bacteria and Archaea that oxidize or reduce electron donors or acceptors at their cell surface remain to be elucidated. Pham et al. (2008) found that the gram-positive *Brevibacillus* sp. strain PTH1 in culture was able to export electrons to the anode of a microbial fuel cell using acetate as electron donor in the presence of purified phenazine-1-carboxamide (PCN) from *Pseudomonas* sp. CMR12a and rhamnolipids as biosurfactants. PCN appeared to serve as an electron shuttle. *Brevibacillus* also appeared to be able to reduce goethite (FeOOH) under these conditions, but only to a limited extent. The ability of phenazines to act as an electron shuttle in electron export in gram-negative *Pseudomonas chlororaphis* PCL1391 and some other gram-negative bacteria was first shown by Hernandez et al. (2004).

Although most experimental evidence in support of electron transfer across the cell envelope has so far been gathered from studies of gram-negative bacteria, such as *Shewanella oneidensis* MR-1 and *Geobacter* spp., which are able to respire anaerobically using ferric oxide or Mn(IV) oxide as terminal electron acceptors (Myers and Myers, 1992; Lovley, 2000; review by Ehrlich, 2002), evidence for the presence of c-type cytochrome Cyc2 in the outer membrane of iron-grown cells of *A. ferrooxidans* strain 33020 indicates that during Fe(II) oxidation by this organism, electrons are transferred from Fe(II) via this outer membrane cytochrome and rusticyanin and another cytochrome in the periplasmic space to the ETS in the plasma membrane (Yarzábal et al., 2002 a,b, 2004; Appia-Ayme et al., 1999; see also Chapter 20). This electron transport mechanism probably operates as well when these organisms oxidize insoluble metal sulfide minerals such as chalcocite (Cu_2S) and covellite (CuS) (see discussion in Chapter 20).

Eukaryotic cells are unable to enzymatically oxidize or reduce insoluble, oxidizable or reducible substrates because their ETS is located on the inner membrane of the mitochondria (Figure 6.9), which reside in the cytoplasm of these cells. Thus, the mitochondrial ETS, which is spatially removed from the cell surface, lacks direct access to insoluble substrates (Ehrlich, 1978; Stryer, 1995; Schaechter et al., 2006).

6.5.4   Catabolic Reactions: Fermentation

Fermentation is a catabolic process that involves energy conservation from a disproportionation process in which part of the energy-yielding substrate is oxidized by the reduction of the remainder of the consumed substrate. No externally supplied terminal electron acceptor is involved in the redox process. Glucose fermentation to lactic acid by the Embden–Meyerhoff pathway is a typical example (Figure 6.10). Pairs of hydrogen atoms are removed in an oxidation step from an intermediate metabolic product, *glyceraldehyde 3-phosphate*, resulting in the formation of *1,3-diphosphoglycerate*. The hydrogen pairs that are removed are transferred to *pyruvate*, thereby reducing it to *lactic acid*. The source of pyruvate is the stepwise enzymatic transformation of the previously formed 1,3-diphosphoglycerate. A variant of this pathway leads to the formation of ethanol and CO_2. Other mechanisms of glucose fermentation that can lead to the formation of acetate include the Entner–Doudoroff pathway, the pentose phosphate pathway, and the pentose phosphoketolase pathway (Stryer, 1995; Schaechter et al., 2006). Recently, a new glycolytic pathway leading to the formation of acetate and formate was discovered in the archeon *Thermococcus zilligii* (Xavier et al., 2000).

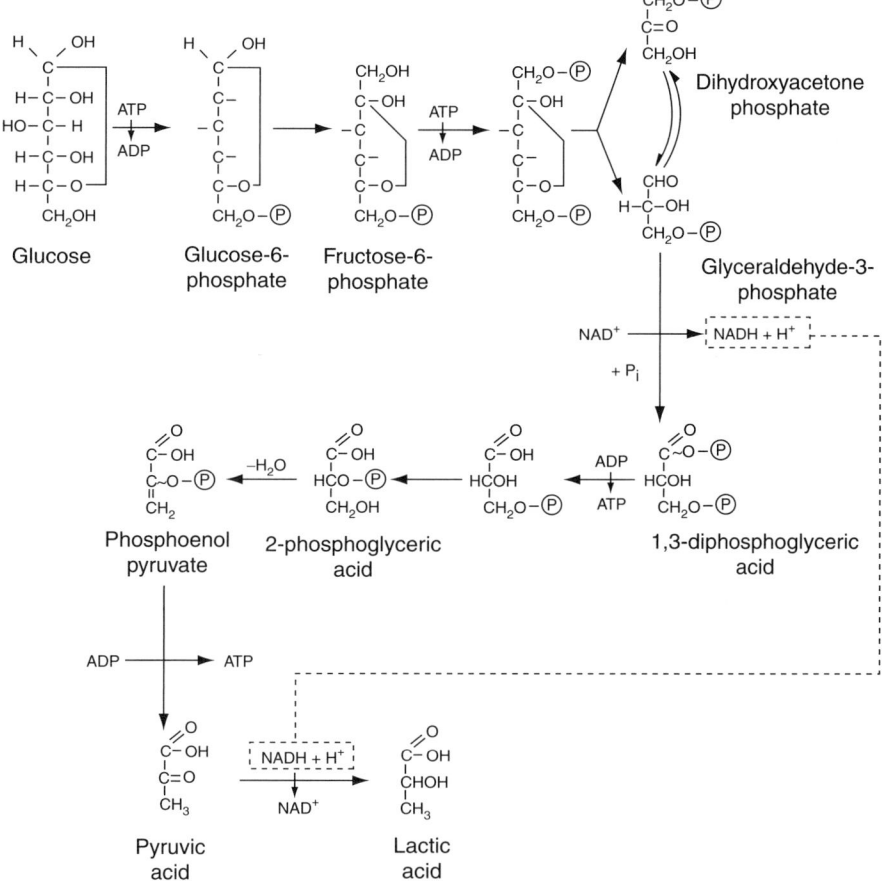

FIGURE 6.10 An example of fermentation: conversion of glucose to lactic acid by glycolysis.

Energy conservation in fermentation is by substrate-level phosphorylation (see Section 6.5.5), which almost never involves a plasma membrane–bound ETS. Major exceptions are the fermentation of acetate to methane and CO_2 (see Chapter 22) and the anaerobic disproportionation of thiosulfate to sulfide and sulfate (Bak and Cypionka, 1987; Bak and Pfennig, 1987; Finster et al., 1998; Janssen et al., 1996; Jackson and McInerney, 2000; see Chapter 19). Fermentation always occurs in the cytoplasm of a cell. A number of prokaryotes, both facultative and anaerobic, are capable of performing fermentation; but it is relatively rare among eukaryotic microorganisms. Certain fungi, such as the yeast *Saccharomyces cerevisiae*, are exceptions.

6.5.5 How Energy Is Generated by Aerobic and Anaerobic Respirers and Fermenters During Catabolism

In aerobic and anaerobic respiration, most useful energy is trapped in high-energy phosphate bonds conserved in ATP as a result of oxidative phosphorylation. The reaction leading to the formation of a high-energy bond may be summarized as*

$$ADP + P_i \rightarrow ATP + H_2O \qquad (6.1)$$

---

* ADP, adenosine 5'-diphosphate; ATP, adenosine 5'-triphosphate; P_i, inorganic phosphate

Reaction 6.1 is energy-consuming and is made possible by charge separation between the inside and the outside of the plasma membrane in prokaryotes. In most aerobically respiring prokaryotes, this charge separation results from the passage of electrons down the electron transport chain to oxygen, which is coupled concurrently with the pumping of protons across the plasma membrane from the cell interior to the periplasm or its functional equivalent in the cell envelope. In eukaryotes, the passage of electrons down the electron transport chain is coupled with the pumping of protons from the interior (matrix) to the outside of the inner membrane of the mitochondria. The cytoplasm of actively respiring prokaryotic cells or the matrix of the active mitochondria of eukaryotic cells is alkaline relative to the outside of the respective membranes enclosing them and therefore more electronegative. The plasma membrane and the inner mitochondrial membrane are impermeable to protons except at the sites where a protein complex, F_1F_0-ATP synthase/ATPase,* is located. The F_1F_0-ATP synthase/ATPase, which has been shown to behave like a nanomotor (see discussion in Weber and Senior, 2003), is anchored in the plasma membrane and projects into the cytoplasm in bacteria and the matrix in mitochondria. It permits the reentry of protons into the cytosol of a prokaryotic cell or the matrix of the mitochondria in a eukaryotic cell through a proton channel (Figure 6.5). It couples this proton reentry with ATP synthesis (Reaction 6.1). Proton reentry via F_1F_0-ATPase is facilitated in aerobes by the consumption of protons in the reduction of O_2 to water catalyzed by cytochrome oxidase on the inside of the plasma membrane or the inner mitochondrial membrane. In anaerobically respiring bacteria, protons may be consumed in the reduction of an electron acceptor that replaces oxygen; this reduction is catalyzed by an enzyme other than cytochrome oxidase. The energy that drives Reaction 6.1 comes from the proton gradient and the membrane potential

$$PMF = \Delta\psi - 2.3RT(\Delta pH/F) \tag{6.2}$$

where PMF is the proton motive force, $\Delta\psi$ the transmembrane potential, ΔpH the pH gradient across the membrane, R the universal gas constant, T absolute temperature, and F the Faraday constant. The overall process by which energy is generated and conserved in aerobic and anaerobic respiration is called chemiosmosis (see Hinkle and McCarty, 1978; Stryer, 1995; Weber and Senior, 2003; Schaechter et al., 2006, for further discussion of the process). As many as three molecules of ATP may be formed per electron pair transferred from donor to terminal electron acceptor in aerobic respiration, and a probable maximum of two in anaerobic respiration.

In methanogens, the ETS is as yet incompletely characterized. However, it is generally agreed that they use a chemiosmotic mechanism for the production of ATP. Many methanogens achieve charge separation in the form of a pH gradient and transmembrane potential by generating protons in the oxidation of H_2 in the periplasmic space outside the plasma membrane and by conducting electrons to the cell interior for use in CO_2 assimilation. This model is supported by evidences developed by Blaut and Gottschalk (1984), Butsch and Bachofen (1984), Mountford (1978), and Sprott et al. (1985). Because the interior of methanogens has a pH near neutrality, H_2 oxidation leading to H^+ formation in the periplasmic space equivalent generates a pH gradient across the plasma membrane in actively metabolizing cells. This pH gradient appears to be utilized in the generation of ATP by ATP synthase/ATPase complexes projecting from the inner surface of the plasma membrane. The electrons removed in the oxidation of hydrogen are conveyed to the cell interior, probably by a membrane-bound hydrogenase, and used in the reduction of CO_2 to methane (see Chapter 22). The charge separation mechanism of methanogens for generating ATP by oxidative phosphorylation is probably illustrative of the earliest chemiosmotic mechanisms from which the more elaborate systems utilizing various membrane-bound electron carriers and enzymes found in modern aerobic and anaerobic respirers evolved. Some methanogens appear to use Na^+ rather than H^+ to achieve charge separation (see Chapter 22; also Ruppert et al., 1998), as do homoacetogens (Müller, 2003).

---

* ATP synthase refers to the protein complex when it catalyzes ATP synthesis from ATP + P_i. ATPase refers to the same protein complex when it catalyzes the hydrolysis of ATP to ADP + P_i.

In fermentation, useful energy is conserved by substrate-level phosphorylation—a process in which a high-energy bond, which traps some of the total energy released during oxidation—is formed on the substrate molecule (metabolite) that is being oxidized. An example is the oxidation of glyceraldehyde 3-phosphate to 1,3-diphosphoglycerate in glucose fermentation illustrated in Figure 6.10. Substrate-level phosphorylation may also occur during aerobic and anaerobic respiration, but it contributes only a small portion of the total energy conserved in high-energy bonds by cells. Clearly, aerobic and anaerobic respirations are more efficient energy-yielding processes in a cell than fermentation. It takes less substrate to satisfy a fixed energy demand by a cell if the substrate is oxidized by aerobic or anaerobic respiration than by fermentation. If the energy-yielding substrate is organic, the greater efficiency may also result from the fact that respirers may oxidize a substrate completely to CO_2 and H_2O, whereas fermenters cannot.

Although many of the microbes that oxidize inorganic substrates to obtain energy are aerobes, a few are not. All autotrophically growing methanogens (domain Archaea) oxidize hydrogen gas (H_2) by transferring electron from H_2 to CO_2 to form methane (CH_4), generating ATP by oxidative phosphorylation in the process. Homoacetogens (domain Bacteria) carry out a similar reduction of CO_2 by hydrogen but form acetate instead of methane (Eden and Fuchs, 1983).

$$4H_2 + 2CO_2 \rightarrow CH_3COOH + 2H_2O \ (\Delta G^0, -25 \text{ kcal or } -104.8 \text{ kJ}) \tag{6.3}$$

Some oxidizers of sulfur compounds can transfer electrons from a reduced sulfur substrate, such as thiosulfate or elemental sulfur, to nitrate in the absence of oxygen. In the presence of oxygen, these sulfur-oxidizing organisms transfer electrons from the reduced sulfur compounds to oxygen. The maximum ATP yield in methane formation from H_2 reduction of CO_2 and in the oxidation of reduced sulfur by nitrate—two examples of anaerobic respiration—has not yet been established.

6.5.6 HOW CHEMOLITHOAUTOTROPHIC BACTERIA (CHEMOSYNTHETIC AUTOTROPHS) GENERATE REDUCING POWER FOR ASSIMILATING CO_2 AND CONVERTING IT INTO ORGANIC CARBON

Unlike chemoheterotrophs, most chemolithoautotrophs have a special problem in generating NADPH when reducing CO_2 with NADPH. These chemolithoautotrophs, which possess an electron transport chain containing Fe–S proteins, quinones, and cytochromes in their plasma membrane, whether they are aerobic or anaerobic respirers, depend on reverse electron transport to reduce $NADP^+$ to NADPH. In reverse electron transport, electrons must travel against a redox gradient with the expenditure of energy contained in high-energy bonds of ATP. The source of the electrons, which is also the source of energy, usually has a midpoint potential (E_h) that is significantly higher than that of the $NADP^+$/NADPH couple. Methanogens, which do not use NADPH for reducing fixed CO_2 to organic carbon but instead employ unique hydrogen carriers such as factor F_{420} (8-OH-deazaflavin) and carbon dioxide–reducing (CDR) factor, appear not to consume ATP in CO_2 reduction by hydrogen (Ferris, 1993). Homoacetogens are another exception. They employ ferredoxin and other Fe–S proteins, whose reduction by H_2 does not require the expenditure of energy (Pezacka and Wood, 1984).

6.5.7 HOW PHOTOSYNTHETIC MICROBES GENERATE ENERGY AND REDUCING POWER

Anoxygenic bacteria, such as the purple and green sulfur bacteria, purple nonsulfur bacteria, and some cyanobacteria that have the capacity to grow anaerobically in the presence of H_2S, generate their ATP by transducing light energy of appropriate wavelengths into chemical energy, which they conserve in high-energy phosphate bonds in ATP via cyclic or noncyclic photophosphorylation (Gottschalk, 1985). They operate under a chemiosmotic principle analogous to respiration. In cyclic photophosphorylation, as in purple sulfur bacteria, electrons pass from a reduced, low-potential Fe–S protein (E_h, –530 mV) to bacteriochlorophyll along an electron transport pathway, which includes membrane-bound quinones and cytochromes. High-energy phosphate bonds are generated in this phase of electron passage and conserved in ATP. The passage of electrons is coupled with proton pumping and a resultant proton gradient as in respiration, except that in this case the proton pumping is in a direction opposite to that in respiration, that is, from the outside of the membrane barrier to the inside. The PMF thus generated causes ATP synthase in the photosynthetic membrane to generate ATP from ADP plus P_i. For the electrons to return to the low-potential Fe–S protein from the high-potential bacteriochlorophyll in cyclic photophosphorylation, they have to be energized by light absorption at an appropriate wavelength (Figure 6.11). Therefore, it is light that drives the movement of the electrons in the cycle.

In green sulfur bacteria, photophosphorylation is noncyclic as well as cyclic (Figure 6.11). The cyclic mechanism is similar to that of the purple sulfur bacteria. In the noncyclic photophosphorylation process, ATP is synthesized in a reaction sequence in which an external electron donor such as H_2S, S^0, or $S_2O_3^2$ reduces chlorobium chlorophyll. The electrons that have reduced the chlorophyll are then used to reduce $NADP^+$ to NADPH. This requires input of light energy because the midpoint potential for the chlorophyll reduction is much higher (approximately +440 mV) than that for the $NADP^+$/NADPH couple (approximately –350 mV). As in most known chemolithoautotrophs, the NADPH is needed for CO_2 assimilation. The ATP-synthesizing mechanism in both cyclic

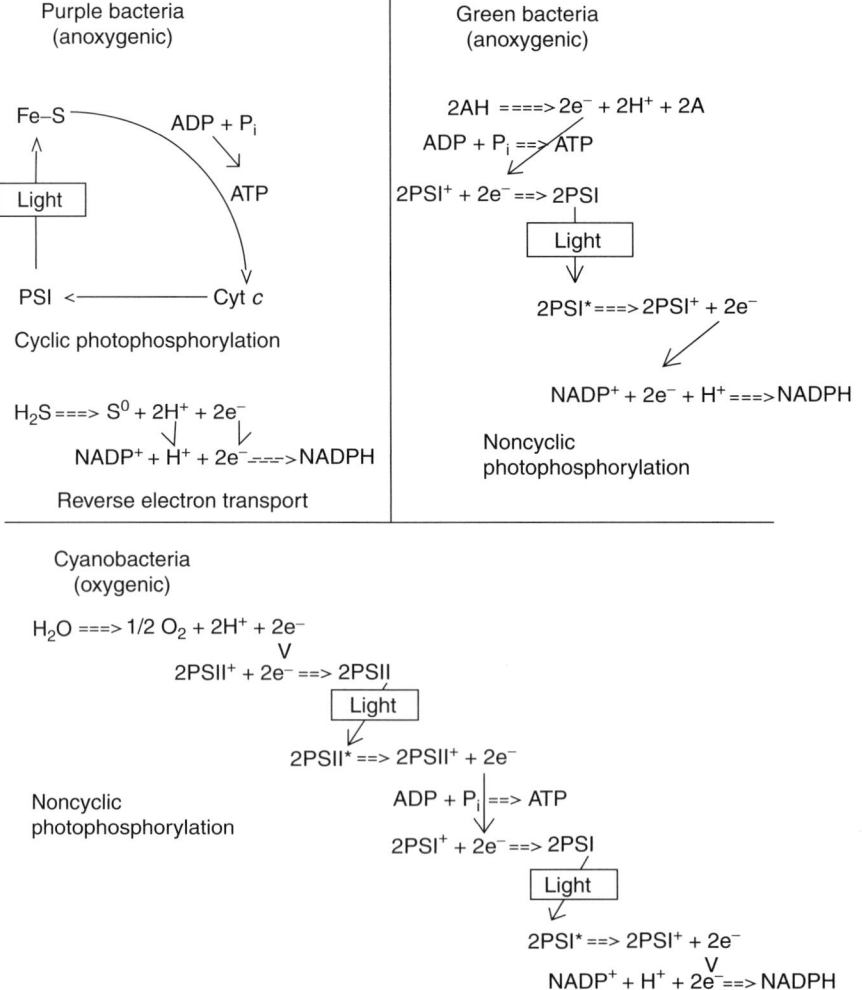

FIGURE 6.11 Diagrammatic representation of the mechanisms of photophosphorylation and generation of reducing power (NADPH) in purple and green photosynthetic bacteria and cyanobacteria. PSI, photosystem I; PSII, photosystem II. (Adapted from Stanier RY et al., *The Microbial World*, Prentice-Hall, Englewood Cliffs, NJ, 1986.)

and noncyclic photophosphorylation of green sulfur bacteria involves chemiosmosis. The extent to which the cyclic and noncyclic mechanisms contribute to ATP and NADPH production depends on the cellular demands for these products.

Cyanobacteria also use a noncyclic photophosphorylation process for generating ATP, but they use a more complex pathway than the green sulfur bacteria (Figure 6.11). Their photosynthetic machinery, which is normally oxygenic because it uses H_2O as a source of reducing power, unlike that of the purple or green sulfur bacteria, involves two major components: photosystems I and II. These are linked to one another by a reaction sequence involving a series of electron carriers that transfer electrons from photosystem II to I and promote proton pumping, which permits chemiosmotic ATP synthesis. Photosystem II generates electrons by the photolysis of water with transduced light energy. Photosystem I sends the electrons that it receives from photosystem II to $NADP^+$ to form NADPH with the help of another boost from transduced light energy. As in green sulfur bacteria, NADPH is required for reducing fixed CO_2. Besides the noncyclic photophosphorylation, oxygen-producing cyanobacteria perform cyclic photophosphorylation that involves only photosystem I to generate additional ATP.

In purple sulfur bacteria, the need for NADPH to reduce CO_2 to organic carbon is met by reverse electron transport that is not directly dependent on light energy input (Figure 6.11). They transfer electrons from a high-potential membrane carrier to low-potential membrane carriers and finally to $NADP^+$ in a dark reaction by the consumption of ATP. The ATP consumption is needed because the electrons travel against a redox gradient. For further discussion of these photosynthetic processes, the reader is referred to Stanier et al. (1986), Atlas (1997), and Schaechter et al. (2006).

6.5.8 Anabolism: How Microbes Use Energy Trapped in High-Energy Bonds to Drive Energy-Consuming Reactions

As an example of how aerobic chemolithoautotrophs couple ATP formation with CO_2 assimilation and reduction to organic carbon, the process in *A. ferrooxidans*, an acidophilic iron oxidizer, will be considered. This organism oxidizes ferrous to ferric iron at acid pH

$$2Fe^{2+} \rightarrow 2Fe^{3+} + 2e^- \tag{6.4}$$

Some of the reducing power (e^-) generated in this way is transferred to oxygen

$$0.5O_2 + 2H^+ + 2e^- \rightarrow H_2O \tag{6.5}$$

with simultaneous chemiosmotic production of ATP

$$ADP + P_i \rightarrow ATP + H_2O \tag{6.6}$$

At maximum efficiency, one ATP is formed for every electron pair ($2e^-$) transferred to oxygen. The remaining reducing power from the oxidation of the ferrous iron is used to reduce pyridine nucleotide (NAD^+, $NADP^+$). Because the electrons in this case have to travel against a redox gradient from +800 (high-potential Q-cycle intermediate in *A. ferrooxidans* at pH 2) (see Ingledew, 1982; Ehrlich et al., 1991, Elbehti et al., 2000) to −305 mV (E_m at pH 6.5 for $NADP^+$/NADPH), energy conserved in high-energy phosphate bonds has to be consumed

$$NAD^+ + 2H^+ + 2e^- + 2ATP \rightarrow NADH + H^+ + 2ADP + P_i \tag{6.7}$$

$$NADH + H^+ + NADP^+ \rightarrow NAD^+ + NADPH + H^+ \tag{6.8}$$

The $NADPH + H^+$, together with some ATP, is used in the assimilation of CO_2 and its reduction to organic carbon,

$$\text{Ribulose 5-phosphate} + ATP \rightarrow \text{ribulose 1,5-diphosphate} + ADP \tag{6.9}$$

$$\text{Ribulose 1,5-diphosphate} + CO_2 \rightarrow 2(\text{3-phosphoglycerate}) + ATP \qquad (6.10)$$

$$2(\text{3-phosphoglycerate}) + 2NADPH + 2H^+ + 2ATP$$
$$\rightarrow 2(\text{glyceraldehyde 3-phosphate}) + 2NADP^+ + 2ADP + 2P_i \qquad (6.11)$$

From glyceraldehyde 3-phosphate, the various organic constituents of the cell are then manufactured, including the building blocks for polymers such as proteins, nucleic acids, lipids, and polysaccharides. These are subsequently combined into the corresponding polymers with the expenditure of additional ATPs because some of the steps in the synthesis of the building blocks and the polymerizations are energy-requiring reactions. Also, some ribulose 5-phosphate is regenerated to permit continued CO_2 fixations (Reactions 6.10 and 6.11). Although generally chemolithoautotrophs can grow in the complete absence of organic matter under laboratory conditions, many, if not all, of the organisms can assimilate some types of organic compounds such as amino acids and vitamins. Some chemolithoautotrophs are able to use organic carbon as a sole energy source under some conditions (facultative chemolithoautotrophs), but others cannot (obligate chemolithoautotrophs).

Anaerobic chemolithoautotrophs such as methanogens, homoacetogens, and some sulfate reducers use a mechanism of CO_2 fixation that differs from that used by *A. ferrooxidans*. They use a mechanism in which they form acetate from two molecules of CO_2. This involves the stepwise reduction of one CO_2 to a methyl carbon (CH_3-) and the other CO_2 to carbon monoxide (CO) that is subsequently transformed into a carboxyl carbon ($-COOH$). The methyl and carboxyl carbons are then joined to form acetate (CH_3COOH) in an ATP-consuming process. Acetate is subsequently carboxylated by combining with another molecule of CO_2 in another ATP-consuming process to form pyruvate ($CH_3CH(OH)COOH$) (Figure 6.12). Pyruvate is a key precursor for the formation of all other monomeric building blocks from which the various polymers are formed by ATP-consuming processes.

Anoxygenic photolithoautotrophs assimilate CO_2 by one of the several different mechanisms. Purple sulfur bacteria usually fix CO_2 and reduce it to organic carbon by the same set of Reactions 6.9 through 6.11, as aerobic chemolithoautotrophs such as *A. ferrooxidans*. They obtain needed NADPH through reduction of $NADP^+$ by reverse electron transport with H_2, H_2S, S^0, or $S_2O_3^{2-}$ as electron donors (see Figure 6.11).

The filamentous green photolithoautotroph *Chloroflexus aurantiacus* fixes CO_2 by a 3-hydroxypropionate cycle (Ivanovsky et al., 1993; Strauss and Fuchs, 1993; Eisenreich et al., 1993; Herter et al., 2001, 2002; Friedmann et al., 2006). In the cycle as proposed by Herter et al. (2001, 2002), CO_2 is incorporated into acetyl~SCoA to form malonyl~SCoA, which in turn is transformed into 3-hydroxypropionate, from which propionyl~SCoA is formed. Propionyl~SCoA is transformed into succinyl~SCoA via methylmalonyl~SCoA by further CO_2 fixation. Succinyl~SCoA is transformed into malonyl~SCoA, which is then cleaved into acetyl~SCoA and glyoxalate, completing the cycle

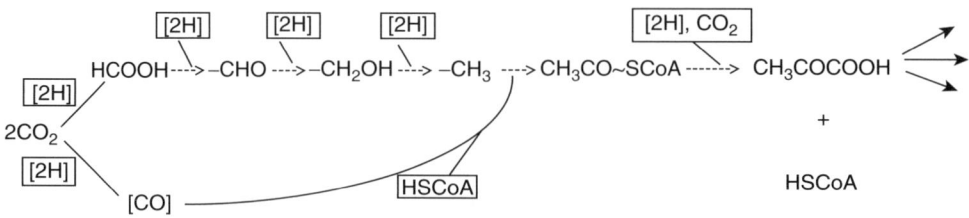

FIGURE 6.12 Pathway of carbon assimilation in methanogens (the activated acetate [$CH_3CO~SCoA$] pathway). Pyruvate ($CH_3COCOOH$) is a key intermediate for forming various building blocks for the cell, including sugars, amino acids, fatty acids, and so on.

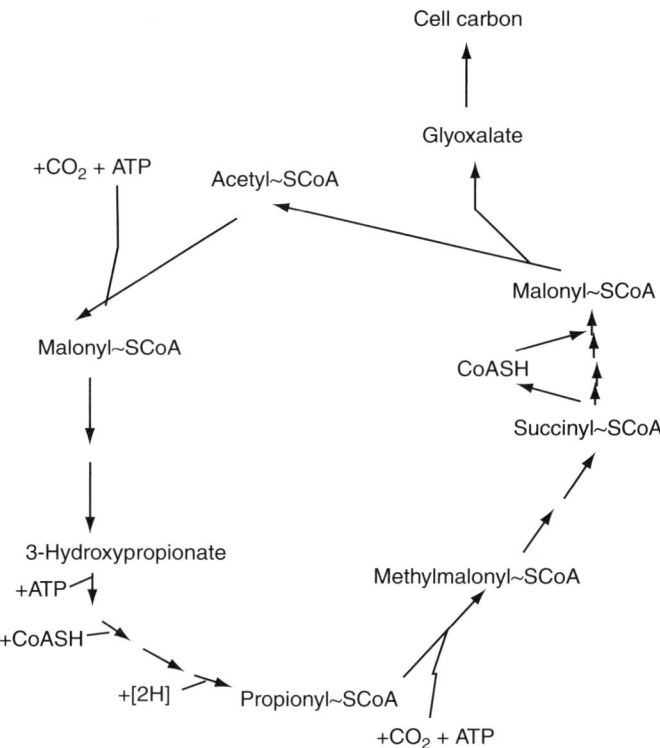

FIGURE 6.13 Pathway of autotrophic CO_2 fixation by *Chloroflexus aurantiacus* (3-OH-propionate cycle) as proposed by Herter et al. (2001). (Adapted from Herter S et al., *J. Bacteriol.*, 183, 4305–4316, 2001.)

(Figure 6.13). Glyoxalate is coupled to propionyl~CoA and converted to pyruvate, which is used in the synthesis of cellular carbon compounds (Herter et al., 2002).

Green sulfur bacteria of the genus *Chlorobium* fix CO_2 and reduce it by a reverse tricarboxylic acid cycle (Figure 6.14). In this process, CO_2 is combined with pyruvate in an ATP-consuming process to form oxalate, which is then converted via malate, fumarate, and succinate to 2-ketoglutarate, the last step requiring consumption of ATP. The 2-ketoglutarate is a key precursor in amino acid synthesis as well as citrate synthesis. Formation of citrate involves fixation of another CO_2. Citrate is cleaved to form oxaloacetate and acetate. Acetate serves as a precursor in the synthesis of pyruvate by CO_2 fixation and ATP consumption, thus completing the reverse tricarboxylic acid cycle. Pyruvate is a key precursor for the synthesis of other biochemical building blocks. NADH and NADPH needed for the operation of this cycle are generated by a noncyclic photoreduction mechanism (see Figure 6.11). Although once thought unique to *Chlorobium*, the reverse tricarboxylic acid cycle has since been found to operate as a mechanism of CO_2 assimilation in some characteristic autotrophs, for example, *Aquifex pyrophilus*, a chemolithoautotrophic, H_2-oxidizing member of the domain Bacteria; and *Thermoproteus neutrophilus*, a chemolithoautotrophic, thermophilic, H_2-oxidizing and S^0-reducing archeon (Beh et al., 1993). Most recently, the reverse tricarboxylic acid cycle has been reported in the ε-proteobacteria (e.g., *Thiomicrospira denitrificans*, *Candidatus* Arcobacter sulfidicus) (Hügler et al., 2005).

Oxygenic photolithoautotrophs fix CO_2 and reduce it to organic carbon by a reaction sequence similar to Reactions 6.9 through 6.11, which is also called the Calvin–Benson cycle. They produce NADPH for this process via noncyclic photophosphorylation and form ATP by both noncyclic and cyclic photophosporylation.

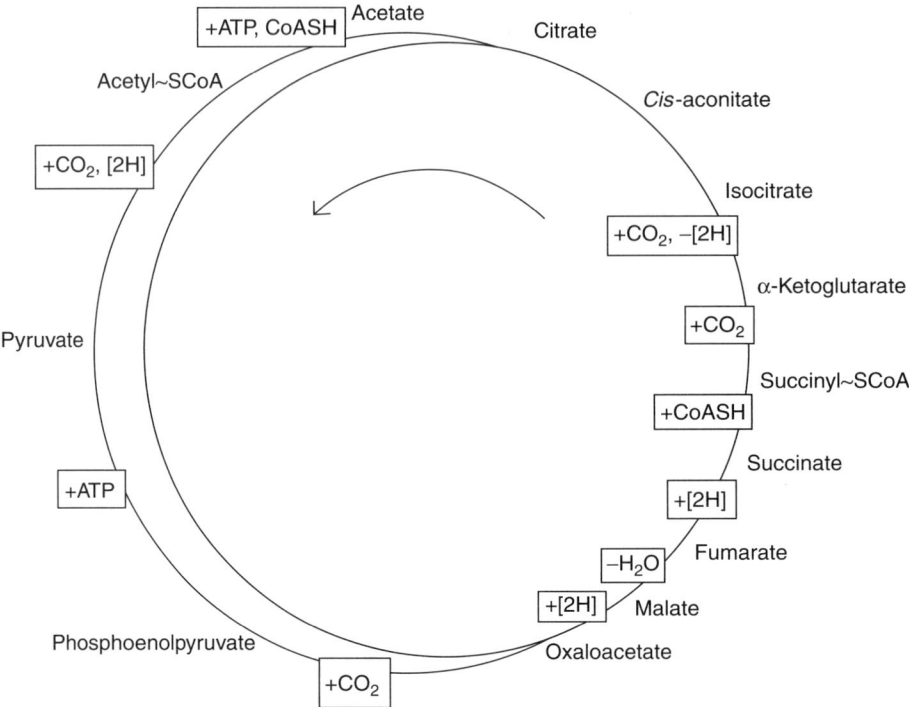

FIGURE 6.14 The reverse tricarboxylic acid cycle used by green sulfur bacteria for carbon assimilation. (Modified from Stanier RY et al., *The Microbial World*, Prentice-Hall, Englewood Cliffs, NJ, 1986; scheme originally presented by Evans MCW, Buchanan BB, Arnon DI, *Proc. Natl. Acad. Sci. USA*, 55, 928–934, 1966.)

6.5.9 CARBON ASSIMILATION BY MIXOTROPHS, PHOTOHETEROTROPHS, AND HETEROTROPHS

Because heterotrophs fashion some monomeric building blocks by catabolism and acquire others preformed from the environment external to the cell, they use much of the ATP, which they generate catabolically by respiration or fermentation, for polymerization reactions, as in the formation of proteins, polysaccharides, nucleotides and nucleic acids, lipids, and others. Mixotrophs and photoheterotrophs perform anabolic reactions that are similar to those performed by chemolithoautotrophs, photolithoautotrophs, and heterotrophs.

So far, nothing is known of the detailed biochemical steps used by algae or protozoa in forming inorganic polymers such as polysilicate. Such polymerization is expected to involve the consumption of ATP or its equivalent.

6.6 MICROBIAL MINERALIZATION OF ORGANIC MATTER

Microbes play a major role in the transformation of organic matter in the upper lithosphere (soils, sediments, deep subsurface) and in the hydrosphere (oceans and bodies of freshwater). Because biological availability of carbon and other nutritionally vital inorganic elements in the biosphere is limited, it is essential that these elements be recycled for the continuation of life. This recycling involves mostly complete degradation of dead organic matter into inorganic matter, whether the dead organic matter is in the form of remains of dead organisms or metabolic wastes. However, some incompletely degraded organic matter may also be recycled in some instances. When the organic matter is completely degraded, the process is called *mineralization*. It may proceed aerobically or anaerobically.

In *aerobic mineralization*, organic matter is completely degraded (oxidized) to CO_2 and H_2O, and if N, S, and P are present in the original organic matter, to NO_3^-, SO_4^{2-}, and PO_4^{3-}. The process by which this mineralization occurs is aerobic respiration, that is, oxygen serves as terminal electron acceptor in the oxidations. In many instances of aerobic mineralization, a single microorganism may be responsible for the complete degradation of a compound. In others, however, consortia of two or more organisms may collaborate in the process, especially in the degradation of polymers because only one or a few consortium members produce the enzymes for depolymerization, which in general occurs extracellularly.

In *anaerobic mineralization*, products of complete degradation of organic matter are CH_4 or CO_2, H_2, NH_3, H_2S, PO_4^{3-}, etc. The process by which this degradation is accomplished may be anaerobic respiration in which NO_3^-, Fe(III), Mn(IV), SO_4^{2-}, and CO_2, or other reducible entities serve as terminal electron acceptors instead of O_2. It may involve a single microorganism or a succession of different ones. In a succession of microorganisms, the key contribution to the overall process by one group may be extracellular hydrolytic enzymes to depolymerize proteins, carbohydrates, nucleic acids, lipids, and so forth. In the presence of sufficient NO_3^-, Mn(IV), Fe(III), SO_4^{2-}, or other suitable electron acceptors, denitrifiers, manganese(IV), iron(III), sulfate, or other anaerobic respirers, respectively, may then oxidize the products of hydrolysis to CO_2, H_2O, etc. In the absence of sufficient NO_3^-, Mn(IV), Fe(III), or SO_4^{2-}, fermenters may convert the products of polymer hydrolysis and other soluble organic carbon to acetate, CO_2, and H_2. Methanogens may then convert acetate and CO_2 and H_2 to CH_4 (see also Chapter 22). Thus, in the absence of a suitable terminal electron acceptor to carry on anaerobic respiration, methanogenesis may be an important process of organic carbon mineralization in soils, sediments, and the deep subsurface. The carbon in methane will be recycled when methane enters an aerobic environment and is oxidized to CO_2 by methanotrophs. In some sedimentary marine environments, methane may be anaerobically oxidized to CO_2 by the collaboration of a special kind of methanogen and a special kind of sulfate reducer (see Chapter 22).

In some environments, mineralization of organic matter may lag or be absent, resulting in the accumulation of organic matter. This may be because the organic matter that accumulates is difficult for many microbes to degrade, for instance, lignin in wood. In other cases when the organic matter is unsuitable for fermentation, the accumulation may be the result of limited availability of O_2, NO_3^-, Mn(IV), Fe(III), SO_4^{2-}, or CO_2 as terminal electron acceptors. In still other instances, incomplete degradation of organic matter may be the result of enzymatic deficiencies in the members of the microbial community in the environment where the organic matter is accumulating.

In soil, incomplete oxidation of some of the organic matter results in the formation of an important soil constituent, *soil humus*, a mixture of polymeric substances derived from the partial decomposition of plant, animal, and microbial remains and from microbial syntheses. It is usually recognizable as a brownish-black organic complex, only portions of which are soluble in water; a larger fraction is soluble in alkali. Soil humus includes aromatic molecules, often in polymerized form, whose origin is mostly lignin; bound and free amino acids; uronic polymers; free and polymerized purines and pyrimidines; and other forms of bound phosphorus.

Organic matter accumulating in marine sediments has been called *marine humus* because of a similarity in its C/N ratio to that of soil humus (Waksman, 1933) and because of its relative resistance to aerobic microbial decomposition (Waksman and Hotchkiss, 1937; Anderson, 1940). However, more detailed chemical analysis of marine humus indicates differences from soil humus, which is not surprising in view of the differences in their origin (Jackson, 1975; Moore, 1969, p. 271). Soil humus is formed mainly from plant remains, whereas typical marine humus in sediments far from land derives mainly from phytoplankton remains and fecal residues. Marine humus from three northern Pacific sediments was found to contain 0.14–0.34% organic matter, including 20–1145 ppm alkali-soluble humic acids, 40–55% of benzene-soluble bitumen, and 50–180 ppm of amino acids. The remainder was kerogen—a material insoluble in aqueous and nonpolar solvents (Palacas et al., 1966). The organic matter of deep-sea sediments contains a fraction that, although refractory to microbial attack *in situ*, is readily attacked by microbes when brought to the surface (Ehrlich et al., 1972). Presumably hydrostatic pressure (>300 atm or 303.9 bar) and low temperature (<4°C) prevent rapid microbial *in situ* decomposition (Jannasch and Wirsen, 1973; Wirsen and Jannasch, 1975). Metabolizable organic matter in shallow-water sediments will undergo more complete decomposition provided that it does not accumulate too rapidly.

6.7   MICROBIAL PRODUCTS OF METABOLISM THAT CAN CAUSE GEOMICROBIAL TRANSFORMATIONS

Many heterotrophic bacteria, whether aerobic, facultative, or anaerobic, form significant quantities of organic acids among the products from their catabolism in addition to CO_2. At least some of the CO_2 will react with water to form carbonic acid (H_2CO_3) in aqueous solution. Some chemolithoautotrophs and photolithoautotrophs form significant amounts of sulfuric or nitric acid, depending on the substrate they use as their source of energy and reducing power. These acids may react chemically with certain minerals, resulting in their partial or complete dissolution or alteration (diagenesis) (see Chapters 9 and 10). Other heterotrophs, when growing at the expense of nitrogenous carbon and energy sources such as proteins or peptides, generate ammonia, which forms NH_4OH, a base, in aqueous solution. This base can solubilize some silicates.

Various prokaryotes and some eukaryotes form ligands that can complex inorganic ions. Some ligands, such as siderophores, are very specific to the ion they complex. In the case of minerals, when the ion complex formed by the ligand is more stable than the source of the ion, namely the mineral, the ligand is able to withdraw the ion from the mineral, resulting in its diagenesis or dissolution (see Chapters 10, 12, and 16).

Some bacteria can form strong reductants, for example, Fe^{2+} by reducing FeOOH, or H_2S by reducing SO_4^{2-}. If the reductant formed is Fe^{2+}, it may then react nonbiologically with a reducible mineral such as pyrolusite (MnO_2), dissolving it by reducing the $Mn(IV)$ in MnO_2 to water-soluble Mn^{2+} (Chapter 17). This reaction is favored by acid pH. If the reductant formed is H_2S, it may react with FeOOH by reducing it to FeS (see Chapter 16). Acidophilic iron oxidizers such as *A. ferrooxidans* produce Fe^{3+} from Fe^{2+}. Fe^{3+} may chemically oxidize a metal sulfide such as CuS, dissolving it by forming Cu^{2+} and SO_4^{2-} (see Chapter 20).

6.8   PHYSICAL PARAMETERS THAT INFLUENCE GEOMICROBIAL ACTIVITY

Temperature is an important parameter that influences geomicrobial activity. In fact, it influences biological activity in general. This is because biochemical reaction rates, like all chemical reaction rates, increase with a rise in temperature, except that with enzyme-catalyzed reactions, the positive response is confined to a relatively narrow temperature range because of the limited heat stability of enzyme proteins. Proteins denature, that is, they become structurally randomized above a maximum temperature. If they are enzymes, this means that they lose their catalytic activity. Denaturation of some enzymes at temperatures slightly above the maximum temperature can be prevented if accompanied by a moderate increase in hydrostatic pressure (Haight and Morita, 1962).

The lipid phase of cell membranes also responds to temperature. It is more fluid at higher temperature than at lower temperature. A certain degree of membrane fluidity is essential for proper cell functioning. Cells can control this fluidity by adjusting the degree of saturation of the fatty acids in their membrane lipids. The more saturated the fatty acids of a given chain length in membrane lipid, the higher the temperature required for a desirable degree of fluidity, and conversely, the more unsaturated these fatty acids, the lower the temperature required to maintain a similar degree of fluidity.

At present, life is known to exist in a temperature range from slightly below 0°C to as high as 121°C, with survival at +130°C (Kashefi and Lovley, 2003). However, no organism exists that spans this entire range. This is because proteins and some other structural components of cells serving a particular function require somewhat different compositions and structures for stability and activity at different temperature intervals within the overall temperature range in which life exists. No organisms are known to exist that are genetically endowed to produce respective components to cover all these different temperature intervals. The heat-stability range of the enzymes and critical cell structures, including cell membranes, of a microorganism reflects the temperature range in which it is able to grow. In other words, key molecules in organisms with different temperature requirements have different heat liabilities (Brock, 1967; Tansey and Brock, 1972; Morgan-Kiss et al., 2006). *Psychrophiles* grow in a range from slightly below 0 to ~20°C, with an optimum at 15°C or lower (Morita, 1975). *Psychrotrophs* grow over a wider temperature range than do psychrophiles (e.g., 0–30°C), with an optimum near 25°C. *Mesophiles* are microbes that grow in the range of 10–45°C, with an optimum range for some of about 25–30°C and for others of about 37–40°C. *Thermophiles* are microbes that live in a temperature range of 42–121°C, but the range for any given thermophile is considerably narrower. The temperature optimum for any one thermophilic organism depends on its identity and usually corresponds to the predominant temperature of its normal habitat. Extreme- or hyperthermophiles, those growing optimally >60°C, seem to be mostly archaea. Generally, thermophilic photosynthetic prokaryotes cannot grow at temperatures >73°C. In contrast, thermophilic eukaryotic algae cannot grow at temperatures >56°C (Brock, 1967, 1974, 1978). Thermophilic fungi generally exhibit temperature maxima ~60°C, and thermophilic protozoa, ~50°C. Only nonphotosynthetic, thermophilic prokaryotes exhibit temperature maxima that may be as high as 121°C (Kashefi and Lovley, 2003). For growth at temperatures at and above the boiling point of water, elevated hydrostatic pressure has to prevail to keep the water liquid, as liquid water is a requirement for life.

The parameters of pH and E_h also exert important influences on geomicrobial activity, as they do on biological activity in general. Each enzyme has its characteristic pH optimum, and E_h optimum in the case of redox enzymes, at which it catalyzes most efficiently. That is not to say that in a cell or, in the case of extracellular enzymes, outside a cell, an enzyme necessarily operates at its optimal pH and E_h. The interior of living cells tends to have a pH around neutrality and an E_h that may be lower or higher than its external environment. Enzymes with higher or lower pH optima will operate at less than optimal efficiency. This helps a cell to modulate and integrate individual enzyme reactions in a sequence so that no shortage or unneeded buildup of metabolic intermediates occurs in such a sequence. Changes in external pH that are within the physiological range of a microorganism do not affect its internal pH because of its plasma membrane barrier and its ability to control internal pH. However, extreme changes will have adverse effects.

Environmental pH and E_h control the range of distribution of microorganisms (see, however, Ehrlich, 1993 for environmental significance of E_h), as recognized by Baas Becking et al. (1960). As shown in Figure 6.15, Bass Becking et al. (1960) gave recognition to the prevalence of iron-oxidizing bacteria and, to some extent, thiobacteria in environments of relatively reduced potential and elevated pH. More recent studies have extended the environmental pH limits. For instance, iron-oxidizing *Ferroplasma acidiphilum* has been found to grow in a pH range of 1.3–2.2 (Golyshina et al., 2000) and iron-oxidizing *Ferroplasma acidarmanus* in a pH range of 0–2.5 (Edwards et al., 2000) (see also Chapter 16).

As mentioned in Chapter 5, hydrostatic pressure in excess of 400 atm (405 bar) at a fixed physiologically permissive temperature below the boiling point of water generally prevents the growth of *nonbarophilic* microbes. Pressure between 200 and 400 atm (203 and 405 bar) at such a temperature tends to interfere reversibly with the cell division of bacteria (ZoBell and Oppenheimer, 1950). *Barophilic* organisms can grow at pressures >400 atm (405 bar) at physiologically permissive temperatures. *Facultative barophiles* grow progressively more slowly with increasing pressure, whereas *obligate barophiles* grow best at or near the pressure and temperature of the native environment and grow progressively more slowly with decreasing pressure and usually not at all at atmospheric pressure at the same temperature (Yayanos et al., 1982). The growth-inhibiting effect of hydrostatic

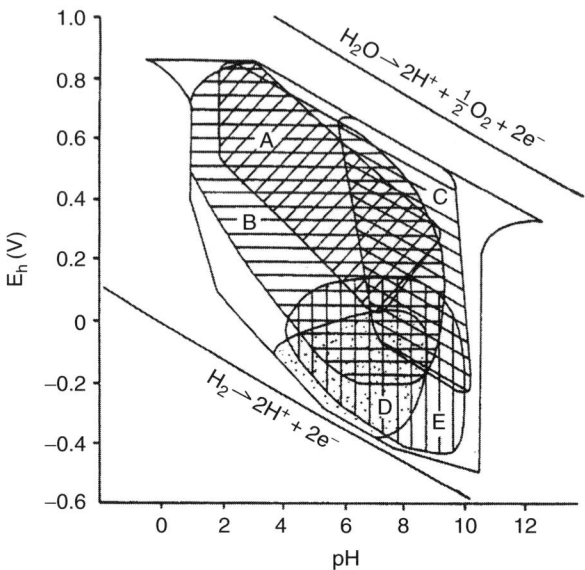

FIGURE 6.15 Environmental limits of E_h and pH for some bacteria. A, iron bacteria; B, thiobacteria; C, denitrifying bacteria; D, facultative and anaerobic heterotrophic bacteria and methanogens; E, sulfate-reducing bacteria. (Adapted from Baas Becking LGM, Kaplan IR, Moore D, *J. Geol.*, 68, 243–284, 1960. With permission from the University of Chicago Press.)

pressure is attributable to its effect on protein synthesis (Schwarz and Landau, 1972a,b; Pope et al., 1975; Smith et al., 1975). Many other biochemical reactions are much less pressure-sensitive (Pope and Berger, 1973) (see also Chapter 5).

6.9   SUMMARY

Microbes may make a geologically significant contribution to lithification, mineral formation, mineral diagenesis, and sedimentation, but not to volcanism, tectonic activity, orogeny, or wind and water erosion. They may act as agents of concentration, dispersion, or fractionation of mineral matter. Their influence may be direct, through action of their enzymes, or indirect, through chemical action of their metabolic products, passive concentration of insoluble substances on their cell surface, and alteration of pH and E_h conditions in their environment. Their metabolic influence may involve anabolism or catabolism under aerobic or anaerobic conditions. Respiratory activity of prokaryotes may cause oxidation or reduction of certain inorganic compounds, resulting in their precipitation, often as minerals, or their solubilization. Chemolithoautotrophic and some mixotrophic bacteria can obtain useful energy from the oxidation of some inorganic substances such as H_2, Fe(II), Mn(II), H_2S, S^0, and so on. Photolithoautotrophic bacteria can use H_2S as a source of reducing power in the assimilation of CO_2 and, in this process, deposit sulfur. Anaerobically respiring organisms, which use any of the various oxidized inorganic substances as terminal electron acceptors, are important in the mineralization of organic matter in environments devoid of atmospheric oxygen. Mineralization of organic matter by microbes under aerobic conditions in soil, freshwater, and marine environments leads to the formation of CO_2, H_2O, NO_3^-, SO_4^{2-}, PO_4^{3-}, etc., and under anaerobic conditions to the formation of CH_4, CO_2, NH_3, H_2S, PO_4^{3-}, etc. Under some special conditions in the marine environment, CH_4 may be oxidized anaerobically to CO_2 and H_2O by collaboration of certain specific methanogens and sulfate reducers. Some organic matter is refractory to mineralization under anaerobic conditions and is microbially converted to humus. All microbial activities are greatly influenced by temperature, pH, and E_h conditions in an environment.

REFERENCES

Anderson DQ. 1940. Distribution of organic matter in marine sediments and its availability to further decomposition. *J Mar Res* 2:225–235.

Appia-Ayme C, Guilliani N, Ratouchniak J, Bonnefoy V. 1999. Characterization of an operon encoding two *c*-type cytochromes, an *aa3*-type cytochrome oxidase, and rusticyanin in *Thiobacillus ferrooxidans* ATCC 33020. *Appl Environ Microbiol* 65:4781–4787.

Atlas RM. 1997. *Principles of Microbiology*. Boston, MA: WCB/McGraw-Hill.

Baas Becking LGM, Kaplan IR, Moore D. 1960. Limits of the natural environment in terms of pH and oxidation-reduction potentials. *J Geol* 68:243–284.

Bak F, Cypionka H. 1987. A novel type of energy metabolism involving fermentation of inorganic sulfur compounds. *Nature* (London) 326:891–892.

Bak F, Pfennig N. 1987. Chemolithotrophic growth of *Desulfovibrio sulfodismutans* sp. nov. by disproportionation of inorganic sulfur compounds. *Arch Microbiol* 147:184–189.

Beard BL, Johnson CM, Cox L, Sun H, Nealson KH, Aguilar C. 1999. Iron isotope biosignatures. *Science* 285:1889–1892.

Beh M, Strauss G, Huber R, Stetter K-O, Fuchs G. 1993. Enzymes of the reductive citric acid cycle in the autotrophic eubacterium *Aquifex pyrophilus* and in the archaebaterium *Thermoproteus neutrophilus*. *Arch Microbiol* 160:306–311.

Beveridge TJ. 1989. Metal ions and bacteria. In: Beveridge TJ, Doyle RJ, eds. *Metal Ions and Bacteria*. New York: Wiley, pp. 1–29.

Beveridge TJ, Doyle RJ, eds. 1989. *Metal Ions and Bacteria*. New York: Wiley.

Beveridge TJ, Meloche JD, Fyfe WS, Murray RGE. 1983. Diagenesis of metals chemically complexed by bacteria: Laboratory formation of metal phosphates, sulfides and organic condensates in artificial sediments. *Appl Environ Microbiol* 45:1094–1108.

Blaut M, Gottschalk G. 1984. Proton motive force-driven synthesis of ATP during methane formation from molecular hydrogen and formaldehyde or carbon dioxide in *Methanosarcina barkeri*. *FEMS Microbiol Lett* 24:103–107.

Brock TD. 1967. Life at high temperatures. *Science* 158:1012–1019.

Brock TD. 1974. *Biology of Microorganisms*. 2nd ed. Englewood Cliffs, NJ: Prentice-Hall.

Brock TD. 1978. *Thermophilic Microorganisms and Life at High Temperatures*. New York: Springer.

Brock TD, Madigan MT. 1988. *Biology of Microorganisms*. 5th ed. Englewood Cliffs, NJ: Prentice-Hall.

Butsch BM, Bachofen R. 1984. The membrane potential in whole cells of *Methanobacterium thermoautotrophicum*. *Arch Microbiol* 138:293–298.

De La Rocha CL, Brzezinski MA, DeNiro MJ. 1997. Fractionation of silicon isotopes by marine diatoms during biogenic silica formation. *Geochim Cosmochim Acta* 61:5051–5056.

de Vrind-de Jong WW, de Vrind JPM. 1997. Algal deposition of carbonates and silicates. In: Banfield JF, Nealson KH, eds. *Geomicrobiology: Interactions Between Microbes and Minerals. Rev Mineral*, Vol. 35, Washington, DC: Mineralogical Society of America, pp. 267–307.

Doetsch RN, Cook TM. 1973. *Introduction to Bacteria and Their Ecobiology*. Baltimore, MD: University Park Press.

Doyle RJ. 1989. How cell walls of gram-positive bacteria interact with metal ions. In: Beveridge TJ, Doyle RJ, eds. *Metal Ions and Bacteria*. New York: Wiley, pp. 275–293.

Duplessy J-C, Blanc P-L, Be AWH. 1981. Oxygen-18 enrichment of planktonic foraminifera due to gametogenic calcification below the euphotic zone. *Science* 213:1247–1250.

Eden G, Fuchs G. 1983. Autotrophic CO_2 fixation in *Acetobacterium woodii*. II. Demonstration of enzymes involved. *Arch Microbiol* 135:68–73.

Edwards KJ, Bond PL, Gihring TM, Banfield JF. 2000. An archaeal iron-oxidizing extreme acidophile important in acid mine drainage. *Science* 287:1796–1799.

Ehrlich HL. 1964. Bacterial oxidation of arsenopyrite and enargite. *Econ Geol* 59:1306–1312.

Ehrlich HL. 1978. Inorganic energy sources for chemolithotrophic and mixotrophic bacteria. *Geomicrobiol J* 1:65–83.

Ehrlich HL. 1993. Bacterial mineralization of organic carbon under anaerobic conditions. In: Bollag J-M, Stotzky G, eds. *Soil Biochemistry*, Vol. 8, New York: Marcel Dekker, pp. 219–247.

Ehrlich HL. 1996. How microbes influence mineral growth and dissolution. *Chem Geol* 132:5–9.

Ehrlich HL. 1999. Microbes as geologic agents: Their role in mineral formation. *Geomicrobiol J* 16:135–153.

Ehrlich HL. 2000. Ocean manganese nodules: Biogenesis and bioleaching possibilities. *Miner Metall Process* 17:121–128.

Ehrlich HL. 2002. How microbes mobilize metals in ores: A review of current understandings and proposals for further research. *Miner Metall Process* 19:220–224.

Ehrlich HL, Ghiorse WC, Johnson GL II. 1972. Distribution of microbes in manganese nodules from the Atlantic and Pacific Oceans. *Dev Ind Microbiol* 13:57–65.

Ehrlich HL, Ingledew WJ, Salerno JC. 1991. Iron- and manganese-oxidizing bacteria. In: Shively JM, Barton LL, eds. *Variations in Autotrophic Life*. London: Academic Press, pp. 147–170.

Ehrlich HL, Yang SH, Mainwaring JD Jr. 1973. Bacteriology of manganese nodules. VI. Fate of copper, nickel, cobalt, and iron during bacterial and chemical reduction of the manganese(IV). *Z Allgem Mikrobiol* 13:39–48.

Eisenreich W, Strauss G, Werz U, Fuchs G, Bacher A. 1993. Retrobiosynthetic analysis of carbon fixation in the phototrophic eubacterium *Chloroflexus aurantiacus*. *Eur J Biochem* 215:619–632.

Elbehti A, Brasseur G, Lemesle-Meunier D. 2000. First evidence for existence of an uphill transfer through the bc1 and NADH-Q oxidoreductase complexes of the acidophilic obligate chemolithotrophic ferrous ion-oxidizing bacterium *Thiobacillus ferrooxidans*. *J Bacteriol* 182:3602–3606.

Estep MF, Hoering TC. 1980. Biogeochemistry of the stable hydrogen isotopes. *Geochim Cosmochim Acta* 44:1197–1206.

Evans MCW, Buchanan BB, Arnon DI. 1966. A new ferredoxin-dependent carbon reduction cycle in a photosynthetic bacterium. *Proc Natl Acad Sci USA* 55:928–934.

Ferris FG. 1989. Metallic ion interactions with the outer membrane of gram-negative bacteria. In: Beveridge TJ, Doyle RJ, eds. *Metal Ions and Bacteria*. New York: Wiley, pp. 295–323.

Ferris JG. 1993. *Methanogenesis. Ecology, Physiology, Biochemistry and Genetics*. New York: Chapman & Hall.

Finster K, Liesack W, Thamdrup B. 1998. Elemental sulfur and thiosulfate disproportionation by *Desulfocapsa sulfoexigens* sp. nov., a new anaerobic bacterium isolated from marine subsurface sediment. *Appl Environ Microbiol* 64:119–125.

Friedmann S, Steindorf A, Alber BE, Fuchs G. 2006. Properties of succinyl-coenzyme A:L-malate coenzyme a transferase and its role in the autotrophic 3-hydroxypropionate cycle of *Chloroflexus aurantiacus*. *J Bacteriol* 188:2646–2655.

Geesey GG, Jang L. 1989. Interactions between metal ions and capsular polymer. In: Beveridge TJ, Doyle RG, eds. *Metal Ions and Bacteria*. New York: Wiley, pp. 325–357.

Ghiorse WC. Department of Microbiology, Cornell University, Ithaca, NY, USA.

Golyshina OV, Pivovarova TA, Karavaiko GI, Kondrat'eva TF, Moore ERB, Abraham WR, Lunsdorf H, Timmis KN, Yakimov MM, Golyshin PM. 2000. *Ferroplasma acidiphilum* gen. nov., spec. nov., an acidophilic, autotrophic, ferrous-iron oxidizing, cell-wall lacking, mesophilic member of the *Ferroplasmaceae* fam. nov., comprising a distinct lineage of the Archaea. *Int J Syst Evol Microbiol* 50:997–1006.

Gottschalk G. 1985. *Bacterial Metabolism*, 2nd ed. New York: Springer.

Graham LA. 1987. Biochemistry and electron transport of a manganese-oxidizing bacterium. PhD Dissertation. Rensselaer Polytechnic Institute, Troy, NY.

Haight RD, Morita RY. 1962. Interaction between the parameters of hydrostatic pressure and temperature on aspartase of *Escherichia coli*. *J Bacteriol* 83:112–120.

Hernandez ME, Kappler A, Newman DK. 2004. Phenazines and other redox-active antibiotics promote microbial mineral reduction. *Appl Environ Microbiol* 70:921–928.

Herter S, Farfsing J, Gad'on N, Rieder C, Eisenreich W, Bacher A, Fuchs G. 2001. Autotrophic CO_2 fixation by *Chloroflexus aurantiacus*: Study of glyoxalate formation and assimilation via the 3-hydroxypropionate cycle. *J Bacteriol* 183:4305–4316.

Herter S, Fuchs G, Bacher A, Eisenreich W. 2002. A bicyclic autotrophic CO_2 fixation pathway in *Chloroflexus aurantiacus*. *J Biol Chem* 277: 20277–20283.

Hinkle PC, McCarty RE. 1978. How cells make ATP. *Sci Am* 238:104–123.

Hügler M, Winsen Co, Fuchs G, Taylor CD, Sievert SM. 2005. Evidence for autotrophic CO_2 fixation via the reductive tricarboxylic acid cycle by members of the ε subdivision of proteobacteria. *J Bacteriol* 187: 3020–3027.

Ingledew WJ. 1982. *Thiobacillus ferrooxidans*. The bioenergetics of an acidophilic chemolithotroph. *Biochim Biophys Acta* 683:89–117.

Ivanovsky RN, Krasilnikova EN, Fal YI. 1993. A pathway of the autotrophic CO_2 fixation in *Chloroflexus aurantiacus*. *Arch Microbiol* 159:257–264.

Jackson BE, McInerney MJ. 2000. Thiosulfate disproportionation by *Desulfotomaculum thermobenzoicum*. *Appl Environ Microbiol* 66:3650–3653.

Jackson TA. 1975. Humic matter in natural waters and sediments. *Soil Sci* 119:56–64.

Jannasch HW, Wirsen CO. 1973. Deep-sea microorganisms: In situ response to nutrient enrichment. *Science* 180:641–643.

Janssen PH, Schuhmann A, Bak F, Liesack W. 1996. Disproportionation of inorganic sulfur compounds by the sulfate-reducing bacterium *Desulfocapsa thiozymogenes* gen. nov., spec. nov. *Arch Microbiol* 166:184–192.

Kashefi K, Lovley DR. 2003. Extending the upper temperature limit for life. *Science* 301:934.

Lovley DR. 2000. Fe(III) and Mn(IV) reduction. In: Lovley DR, ed. *Environmental Microbe–Metal Interactions*. Washington, DC: ASM Press, pp. 3–30.

Lowenstamm HA. 1981. Minerals formed by microorganisms. *Science* 211:1126–1131.

Macaskie LE, Dean ACR, Cheetham AK, Jakeman RJB, Skarnulis AJ. 1987. Cadmium accumulation by a *Citrobacter* sp.: The chemical nature of the accumulated metal precipitate and its location in the bacterial cells. *J Gen Microbiol* 133:539–544.

Macaskie LE, Empson RM, Cheetham AK, Grey CP, Skarnulis AJ. 1992. Uranium bioaccumulation by a *Citrobacter* sp. as a result of enzymatically mediated growth of polycrystalline HUO_2PO_4. *Science* 257:782–784.

Moore LR. 1969. Geomicrobiology and geomicrobial attack on sediment organic matter. In: Eglinton G, Murphy MTJ, eds. *Organic Geochemistry: Methods and Results*. New York: Springer, pp. 264–303.

Morgan-Kiss RM, Priscu JC, Pocock T, Gudynaite-Savitch L, Huner NPA. 2006. Adaptation and acclimation of photosynthetic microorganisms to permanently cold environments. *Microbiol Mol Biol Rev* 70:222–252.

Morita RY. 1975. Psychrophilic bacteria. *Bacteriol Rev* 39:144–167.

Mountford DO. 1978. Evidence for ATP synthesis driven by a proton gradient in *Methanobacterium barkeri*. *Biochem Biophys Res Commun* 85:1346–1351.

Müller V. 2003. Energy conservation in acetogenic bacteria. *Appl Env Microbiol* 69:6345–6353.

Myers CR, Myers JM. 1992. Localization of cytochromes to the outer membrane of anaerobically grown *Shewanella putrefaciens* MR-1. *J Bacteriol* 174:3429–3438.

Onishi T, Meinhardt SW, Yagi T, Oshima T. 1987. Comparative studies on the NADH-Q oxido-reductase segment of the bacterial respiratory chain. In: Kim CH, Teschi H, Diwan JJ, Salerno JC, eds. *Advances in Membrane Biochemistry and Bioenergetics*. New York: Plenum Press, pp. 237–248.

Palacas JG, Swanson VE, Moore GW. 1966. Organic geochemistry of three North Pacific deep sea sediment samples. *U.S. Geol Surv Prof Pap* 550C, pp. C102–C107.

Payne WE, Yang X, Trumpower BL. 1987. Biochemical and genetic approaches to elucidating the mechanism of respiration and energy transduction in *Paracoccus denitrificans*. In: Kim CH, Tedeschi H, Diwan JJ, Salerno JC, eds. *Advances in Membrane Biochemistry and Bioenergetics*. New York: Plenum Press, pp. 273–284.

Pezacka E, Wood HG. 1984. The synthesis of acetyl-CoA by *Clostridium thermoaceticum* from carbon dioxide, hydrogen, coenzyme A and methyltetrahydrofolate. *Arch Microbiol* 137:63–69.

Pham TH, Boon N, Aelterman P, Clauwaert P, De Schamphelaire L, Vanhaecke L, De Mayer K, Hötte M, Verstraete W, Rabaey K. 2008. Metabolites produced by *Pseudomonas* sp. enable a Gram-positive bacterium to achieve extracellular electron transfer. *Appl Microbiol Biotechnol* 77:1119–1129.

Pope DH, Berger LR. 1973. Inhibition of metabolism by hydrostatic pressure: What limits microbial growth? *Arch Mikrobiol* 93:367–370.

Pope DH, Smith WP, Swartz RW, Landau JV. 1975. Role of bacterial ribosomes in barotolerance. *J Bacteriol* 121:664–669.

Reguera G, McCarthy KD, Mehta T, Nicoll JS, Tuominen MT, Lovley DR. 2005. Extracellular electron transfer via microbial nanowires. *Nature* 435:1098–1101.

Ruppert C, Sönke W, Lemker T, Müller V. 1998. The A_1A_0 ATPase from *Methanosarcina mazei*: Cloning of the 5' end of the *aha* operon encoding the membrane domain and expression of the proteolipid in a membrane-bound form in *Escherichia coli*. *J Bacteriol* 180:3448–3452.

Sakaguchi T, Tomita O. 2000. Bioseparation of lithium isotopes by using microorganisms. *Resource Environ Biotechnol* 3:173–182.

Schaechter M, Ingraham JL, Neidhardt FC. 2006. *Microbe*. Washington, DC: ASM Press.

Schwarz JR, Landau JV. 1972a. Hydrostatic pressure effects on *Escherichia coli*: Site of inhibition of protein synthesis. *J Bacteriol* 109:945–948.

Schwarz JR, Landua JV. 1972b. Inhibition of cell-free protein synthesis by hydrostatic pressure. *J Bacteriol* 112:1222–1227.

Smith DW, Strohl WR. 1991. Sulfur oxidizing bacteria. In: Shively JM, Barton LL, eds. *Variations in Autotrophic Life*. London: Academic Press, pp. 121–146.

Smith W, Pope D, Landau JV. 1975. Role of bacterial ribosome subunits in barotolerance. *J Bacteriol* 124:582–584.

Sprott GD, Bird SE, McDonald IJ. 1985. Proton motive force as a function of cell pH at which *Methanobacterium bryantii* is grown. *Can J Microbiol* 31:1031–1034.

Stanier RY, Ingraham JL, Wheelis ML, Painter PR. 1986. *The Microbial World*. 5th ed. Englewood Cliffs, NJ: Prentice-Hall.

Strauss G, Fuchs G. 1993. Enzymes of a novel autotrophic CO_2 fixation pathway in the phototrophic bacterium *Chloroflexus aurantiacus*, the 3-hydroxypropionate cycle. *Eur J Biochem* 215:633–643.

Strohl WR, Geffers I, Larkin JM. 1981. Structure of the sulfur inclusion envelopes from four beggiatoas. *Curr Microbiol* 6:75–79.

Stryer L. 1995. *Biochemistry*. 4th ed. New York: WH Freeman.

Tansey MR, Brock TD. 1972. The upper temperature limit for eukaryotic organisms. *Proc Natl Acad Sci USA* 69:2426–2428.

Wada E, Hattori A. 1978. Nitrogen isotope effects in the assimilation of inorganic nitrogen compounds by marine diatoms. *Geomicrobiol J* 1:85–101.

Waksman SA. 1933. On the distribution of organic matter in the sea and chemical nature and origin of marine humus. *Soil Sci* 36:125–147.

Waksman SA, Hotchkiss M. 1937. On the oxidation of organic matter in marine sediments by bacteria. *J Mar Sci* 36:101–118.

Weber J, Senior AE. 2003. ATP synthesis driven by proton transport in F_1F_0-ATP synthase. *FEBS Lett* 545:61–70.

Wirsen CO, Jannasch HW. 1975. Activity of marine psychrophilic bacteria at elevated hydrostatic pressure and low temperature. *Mar Biol* 31:201–208.

Woese CR, Kandler O, Wheelis ML. 1990. Towards a natural system of organisms: Proposal for the domains Archaea, Bacteria, and Eucarya. *Proc Natl Acad Sci USA* 87:4576–4579.

Wood P. 1988. Chemolithotrophy. In: Anthony C, ed. *Bacterial Energy Transduction*. London: Academic Press, pp. 183–230.

Xavier KB, Da Costa MS, Santos H. 2000. Demonstration of a novel glycolytic pathway in the hyperthermophilic archeon *Thermococcus zilligii* by ^{13}C-labeling experiments and nuclear magnetic resonance analysis. *J Bacteriol* 182:4632–4636.

Yarzábal A, Appia-Ayme C, Ratouchniak J, Bonnefoy V. 2004. Regulation of the expression of the *Acidithiobacillus ferrooxidans rus* operon encoding two cytochromes c, a cytochrome oxidase and rusticyanin. *Microbiology* 150:2113–2123.

Yarzábal A, Brasseur G, Bonnefoy V. 2002a. Cytochromes *c* of *Acidithiobacillus ferrooxidans*. *FEMS Microbiol Lett* 209:189–195.

Yarzábal A, Brasseur G, Ratouchniak J, Lund K, Lemesle-Meunier D, DeMoss JA, Bonnefoy V. 2002b. The high-molecular-weight cytochrome *c* Cyc2 of *Acidithiobacillus ferrooxidans* is an outer membrane protein. *J Bacteriol* 184:313–317.

Yayanos AA, Dietz AS, Van Boxtel R. 1982. Dependence of reproduction rate on pressure as a hallmark of deep-sea bacteria. *Appl Environ Microbiol* 44:1356–1361.

ZoBell CE, Oppenheimer CH. 1950. Some effects of hydrostatic pressure on the multiplication and morphology of marine bacteria. *J Bacteriol* 60:771–781.

7 Nonmolecular Methods in Geomicrobiology

7.1 INTRODUCTION

Geomicrobial phenomena can be studied in the field (*in situ*) and in the laboratory (*in vitro*) in microcosms or as isolated reactions. Field study of a given geomicrobial phenomenon should ideally involve identification and enumeration of the geomicrobially active agents and *in situ* measurements of their average growth rate. It should also involve chemical and physical identification of the substrates, that is, the reactants (e.g., minerals [insoluble] and dissolved inorganic or organic substances) and the products that are formed in the geomicrobial process. Furthermore, it should involve measurement of the overall rate at which the process occurs and assessment of the impact of different environmental factors on it. In practice, however, it may happen that a suspected geomicrobial process is no longer operating at a given site but took place in the geologic past. In such instances, the role of the microorganisms in the process has to be reconstructed from microscopic observations (e.g., searching for microfossils associated with the starting materials, if still present, and especially the products of the process). It may also be reconstructed from geochemical observations such as biomarker (*fingerprint*) compounds in sedimentary rock, which indicate the past existence of an organism or a group of organisms that could have been the geochemical agents responsible. If applicable, evidence of isotopic fractionation of a key element relevant to the geomicrobial process should be sought.

In situ observations of an ongoing geomicrobial process should include the study of the setting in which it occurs in nature. In a terrestrial environment, the kinds of rocks, soil, or sediment, whichever are involved, and their constituent minerals ought to be characterized, and the prevailing temperature, pH, oxidation–reduction potential (E_h), sunlight intensity, seasonal cycles, and the source and availability of moisture, oxygen or other terminal electron acceptors, and nutrients ought to be identified. In an aqueous environment, water depth, availability of oxygen or other terminal electron acceptors, turbidity, light penetration, thermal stratification, pH, E_h, chemical composition of the solutes in the water, nature of the sediment if part of the habitat, and nutrient sources and availability should be examined.

In the laboratory, a geochemical process can be studied in a *microcosm*. For this, a large sample of water, soil, sediment, or rock on or in which the process is occurring is collected. It is placed in a suitable vessel, which may be a flow-through chamber, a glass or plastic column, a battery jar, or another kind of suitable vessel. Filter-sterilized water from the site at which a solid sample was collected or a synthetic nutrient solution of a composition that approximates qualitatively and quantitatively the nutrient supply available at the sampling site is added intermittently or continuously. The added nutrient solution should displace an equivalent volume of spent solution from the culture vessel. The experimental setup may be placed in the same environment from which the sample was taken, or it may be incubated in the laboratory at the temperature with the illumination and access to air, or lack of it, and humidity to which the sample was exposed at the sampling site. Measurement of the concentration of nutrients and products in the influent and effluent critical to the process under study will give a measure of the process rate. Solid products that are not recoverable in the effluent can be identified and measured in representative samples taken from the microcosm. Continuous or intermittent measurement of temperature, pH, E_h, and oxygen availability in the microcosm will give information about any changes in these parameters, some of which may be a result of microbial activity in the microcosm.

The microcosm will probably contain a *mixed population* of bacteria, not all of which are likely to play a role in the geochemical process of interest. Manipulation of the microcosm through qualitative or quantitative changes in nutrient supply, adjustment of pH or temperature, or a combination of these factors may cause selective increases of the organisms directly responsible for the geomicrobial process of interest and intensify the process. Because of enrichment biases, it is possible that a minor member of the indigenous population may grow to dominate the microcosm, and therefore, care must be taken before concluding that this member is the causative agent of the geomicrobial process *in situ*.

In vitro laboratory study of a geomicrobial process may be done by isolating the responsible microorganism(s) in *pure culture*, if possible, from a representative sample from the geomicrobially active site. The process originally observed in the field is then recreated with the isolate(s) in batch or continuous culture. Characterization of the process mechanism will involve qualitative and quantitative measurements of the biogeochemical transformation(s). It may include genetic and biochemical studies (see Chapter 8 for more details), where appropriate, as well as an assessment of environmental effects on the *in vitro* process. *In vitro* laboratory study may be important in lending support to field interpretation of geomicrobial processes that are occurring at present or have occurred in the past.

7.2 DETECTION, ISOLATION, AND IDENTIFICATION OF GEOMICROBIALLY ACTIVE ORGANISMS

A geomicrobial process may be the result of a single microbial species or an association of two or more. An association of microbial species is often called a *consortium*. The basis for the association may be *synergism*, in which one type of organism is not capable of carrying out the complex process but in which each member of the consortium carries out part of the process in a sequential set of interactions. It is also possible that not all members of an association of microbes contribute directly to an overall geomicrobial process but instead carry out reactions that create environmental conditions relating, for instance, to pH or E_h that facilitate the geomicrobial process under consideration.

Even if a geomicrobial process is the result of the action of a single organism, that organism rarely occurs as a pure culture in the field. It will usually be accompanied by other organisms, which may not play a direct role in the geomicrobial process under study although they may compete with the geomicrobially active agent for living space and nutrients and may even produce metabolites that stimulate or inhibit the geomicrobial agent to a degree. Three types of microorganisms may be found in a geomicrobial sample taken in the field: (1) *indigenous organisms*, whose normal habitat is being examined and which include the geomicrobially active organism(s); (2) *adventitious organisms*, which were introduced by chance into the habitat by natural circumstances and which may or may not grow in the new environment but do survive in it; and (3) *contaminants*, which were introduced in manipulating the environment during *in situ* geomicrobial study or sampling. Distinctions among these groups are frequently difficult to make experimentally.

A criterion for identifying indigenous organisms may be their frequency of occurrence in a given habitat and in similar habitats at different sites. A criterion for identifying adventitious organisms may be their inability to grow successfully in the habitat under study and their lower frequency of occurrence than in their normal habitat. Neither of these criteria is absolute, however. Identification of a contaminant may simply be based on the knowledge about the organism concerned that would make its natural existence in the habitat under study unlikely.

7.2.1 *In Situ* Observation of Geomicrobial Agents

To detect geomicrobially active microorganisms *in situ*, visual approaches including direct observation with the naked eye, light microscopy, or transmission electron microscopy (TEM) or scanning electron microscopy (SEM), especially environmental SEM (ESEM), are possible. Direct visual observation is possible only in rare instances, namely, when the microbes occur so massively as to be easily seen as, for instance, algal or bacterial mats in hot springs (e.g., in Yellowstone National Park; see Brock, 1978), or lichen growth on rocks. In most instances, visual observations of microbes in their natural habitat requires magnification. In soil or sediment, such observation may be made by the buried slide method or the capillary technique of Perfil'ev (Perfil'ev and Gabe, 1969). In the buried slide method, a clean microscope slide is inserted in soil or sediment and left undisturbed for a number of days. It is then withdrawn, washed, and suitably stained. A choice of stains includes those that discriminate between living and dead cells and those that do not. Examination with a light microscope, equipped for fluorescence microscopy if fluorescent stains were used (Lawrence et al., 1997), will then reveal microorganisms, especially bacteria and fungi, which became attached to the glass surface during burial (Figure 7.1).

In the capillary technique, one or more glass capillaries with optically flat sides are inserted into soil (pedoscope) or sediment (peloscope). Each capillary takes up soil or sediment solution and minute soil or sediment particles that become the culture medium for microbes, which entered the capillary lumen at the moment of emplacement or subsequently. The capillaries can be periodically withdrawn and their contents examined under a light microscope. The capillaries may also be perfused with special nutrient solutions. The capillaries permit observations of trapped microbes in a living or nonliving state (Figure 7.2). Using this technique, Perfil'ev and Gabe (1965) discovered several previously unknown bacteria in soil and sediment, including *Metallogenium*, *Kuznetsovia*, and *Caulococcus*.

Although the buried slide and capillary methods give an indication of some of the organisms present in soil or sediment where the slides or capillaries were inserted, they do not indicate whether the organisms that are seen resided preferentially on the soil or sediment particles or in the pore fluid. To determine this, direct observation of samples of pore fluid and of soil or sediment particles is necessary. To observe organisms in pore fluids, axenically collected samples of fluid can be filtered and any microbial cells in the fluid deposited on suitable filter membranes, which are then stained and subsequently examined microscopically (e.g., Clesceri et al., 1989).

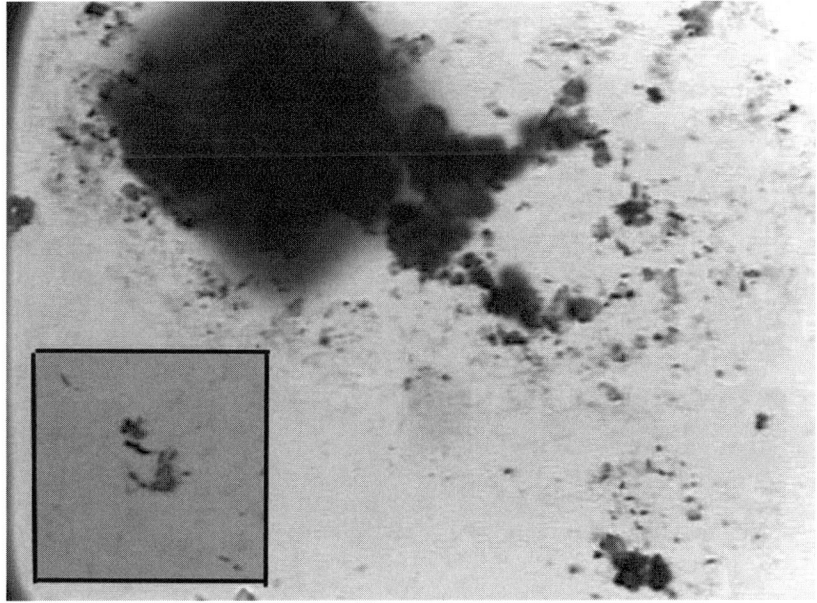

FIGURE 7.1   Demonstration of microbes in soil by the buried slide method. Slide was buried for 1 week. After withdrawal from soil and gentle washing, it was stained with crystal violet. Main view is of isolated shorter and longer rods, and some cocci near soil particles. Inset shows a clump of rods in slime (biofilm?) (×2200).

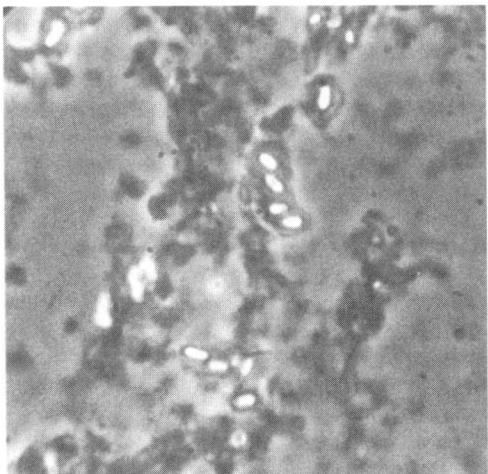

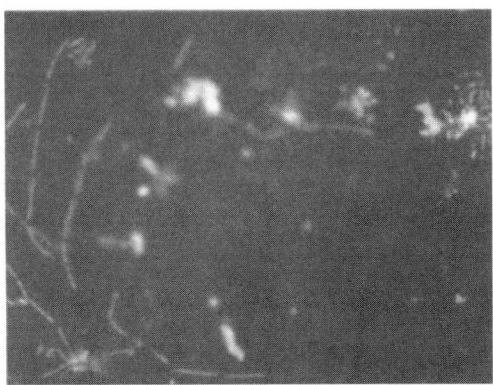

FIGURE 7.2   Microbial development in a capillary tube inserted into lake sediment contained in a beaker and incubated at ambient temperature (×5720). The oval, refractile structures are bacterial spores.

FIGURE 7.3   Bacteria growing on the surface of 304L stainless steel immersed in tap water (Troy, New York) for 3 days, stained with fluorescein isothiocyanate. (Courtesy of Pope DH)

To observe microbes directly on soil or sediment particles or on rock fragments, fluorescence microscopy may be used in conjunction with staining with fluorescent dyes such as 4′,6-diamidino-2-phenylindole (DAPI), acridine orange, or the live/dead stain syto 9/propidium iodide, or with fluorescently labeled antibodies if specific microbes are being sought (Figure 7.3) (Bohlool and Brock, 1974; Casida, 1962, 1965, 1971; Edwards et al., 1999; Eren and Pramer, 1966; Huber et al., 1985; Kepner and Pratt, 1994; Muyzer et al., 1987; Schmidt and Bankole, 1965).

Another approach is to examine sectioned samples by TEM, which is also very useful in detecting fossilized microbes (Barker and Banfield, 1998; Ghiorse and Balkwill, 1983; Jannasch and Mottl, 1985; Jannasch and Wirsen, 1981; Schopf, 1983). Another approach is the use of SEM (e.g., Figure 17.16) (Jannasch and Wirsen, 1981; Sieburth, 1975; LaRock and Ehrlich, 1975; Edwards et al., 1999).

7.2.2   IDENTIFICATION BY APPLICATION OF MOLECULAR BIOLOGICAL TECHNIQUES

Recent advances in the techniques of molecular biology have led to the development of powerful methods for identifying microorganisms and studying their phylogeny. These methods have been adapted to locate and enumerate microorganisms in environmental samples, even microorganisms that, though viable, have not yet been cultured in the laboratory (Ward et al., 1990; Stahl, 1997; Pace, 1997).

7.3   SAMPLING

To determine the nature of a geomicrobially active organism in terms of its morphology, physiology, and the particular geomicrobial process for which it is responsible, it should be isolated and cultivated in the laboratory if possible. Some geomicrobial cultures have to be studied in the laboratory in enrichment cultures because of unsuccessful attempts to obtain them in pure culture, or because the geomicrobial activity of interest is the result of a consortium of microorganisms. In the case of a consortium, its members should be identified by the application of molecular techniques, isolated, and characterized, if possible, to determine their particular contribution to the overall process. Samples, whatever their nature, brought to the laboratory must be obtained under conditions as aseptic as possible, that is, with very little or preferably no contamination. Working surfaces of sampling tools should be thoroughly washed and alcohol-flamed. If a rock outcrop is to be sampled, use of a rock hammer or chisel may be required. If the interior of a small rock specimen is to be sampled in the laboratory, the surface of the specimen should be sterilized if possible. This can be accomplished by using a disinfectant such as carbol-gentian violet-methanol spray (Bien and Schwartz, 1965) or by flaming briefly with a propane torch, or, if done in the laboratory, for 1 min with a Bunsen flame (Weirich and Schweisfurth, 1985). If a sampling device cannot be sterilized, the sample it gathers should be subsampled to obtain a portion that is least likely to have been contaminated. Rock chips should be collected in sterile containers. If the collection is done manually, the hands should be covered by sterile surgical gloves. Weirich and Schweisfurth (1985) devised a special method for obtaining an undisturbed core of rock with a hollow drill under sterile conditions and with cooling by sterile tap water. The extracted rock core was aseptically cut into sections with a flamed chisel, and each section was then aseptically crushed in a flamed mortar mill in sterile dispensing solution and microbiologically tested.

7.3.1 TERRESTRIAL SURFACE/SUBSURFACE SAMPLING

To sample the terrestrial subsurface down to 3000–4000 m or more, drilling methods that depend on the use of special drilling equipment have been devised, which causes minimal if any contamination during the collection of samples (cores) (Phelps and Russell, 1990; Pedersen, 1993; Griffin et al., 1997). One method uses modified wireline coring tools, with cores collected in Lexan- or polyvinyl chloride (PVC)-lined barrels. The drill rig, rods, and tools are steam-cleaned. The drilling fluid system includes a recirculation tank, the drilling fluid being chlorinated water. Tracers such as potassium bromide, the dye rhodamine T, fluorescent beads (~2 μm diameter), and perfluorocarbons added to the drilling fluid aid in determining to what extent, if any, cores were contaminated during drilling. The assessment is made by measuring the extent of tracer presence, if any, in the cores. The extent of bacterial contamination can be determined by quantitative enumeration of bacteria such as coliforms, which were not expected as part of the normal flora of the core, the enumeration being done on the drilling fluid and core samples (Beeman and Suflita, 1989). If anaerobes as well as aerobes are sought in subsurface samples, the cores should be kept away from air and be processed in an oxygen-free atmosphere. Subsamples may then be tested for aerobes and anaerobes by appropriate culture techniques.

Soil and sediment samples from shallow depths on land surfaces may be collected manually with an auger or other coring device under aseptic conditions. Cores should be subdivided aseptically for sampling at different depths. If the cores cannot be obtained with a sterilized sampling device, they should be subsampled so as to obtain the least contaminated sample.

7.3.2 AQUATIC SAMPLING

To obtain aquatic samples, special gear may be required. Water samples at any given depth below the surface, including deepwater samples, may be obtained with a Van Dorn sampler. It consists of a piece of large-diameter plastic tube fitted with rubber closures, which can be kept in an open or closed position at both ends. The plastic tube is mounted on a cable or rope in such a way that it can be lowered vertically into the water column with the rubber closures held in an open position by a spring mechanism. While the device is being lowered, water will pass through it. When a desired depth has been reached, the rubber closures at both ends of the tube are caused to block the openings of the tube by tripping the spring mechanism by means of a messenger (brass weight), which slides down the wire or rope to which the sampler is attached. Below-surface water samples can also be collected with a Niskin sampler, which consists of a

FIGURE 7.4   Ekman dredge. The brass messenger on the rope is 5 cm long.

collapsible, sterile plastic bag with a tubelike opening mounted between two hinged metal plates. The sampler, with the bag in a collapsed state between the metal plates, is lowered on a cable or rope to a desired depth. A spring mechanism of the sampler is then activated by a messenger, as with the Van Dorn sampler, causing the hinged metal plates to open and expanding the bag, which now draws water into it.

Aquatic sediment samples may be obtained with dredging or coring devices. Lake sediment can be collected with an Ekman (Figure 7.4) or a Peterson dredge (Clesceri et al., 1989, pp. 10–100) if surface sediment is desired. A corer needs to be used if different depths of a sediment column are to be examined. Ocean surface sediment may be collected by dragging a bucket dredge over a desired area of the ocean floor. Such a sample will, however, consist of combined, mixed surface sediment encompassing the total surface area sampled. To obtain samples representing different sediment depths at a given location, a gravity (Figure 7.5) or a box corer (Figure 7.6) has to be used. Such devices are rammed into the sediment. Box corers of sufficient cross section provide the least disturbed cores. All cores need to be subsampled to obtain representative, minimally contaminated samples that are representative of the sediment at a given depth. Large rock fragments or concretions on the sediment surface may be collected with a chain dredge or similar device dragged over the sea bottom in a desired area (Figure 7.7).

7.3.3   SAMPLE STORAGE

If samples cannot be examined immediately after collection, they should be stored so as to minimize microbial multiplication or loss of viability. Cooling a sample is usually the best way to preserve it temporarily in its native state, but the extent of cooling may be critical. Freezing may be destructive to at least some of the microbes. However, icing may not prevent growth of psychrophiles or psychrotrophs. The duration of storage before examination should not be longer than absolutely necessary.

FIGURE 7.5 Gravity corer. This is simply a hollow pipe containing a removable plastic liner and having a cutting edge at the lower end with a *core catcher* to retain the sediment core when the device is pulled out of the sediment. A heavy lead weight at the top helps to ram the corer into the sediment when allowed to free-fall just above the sediment surface.

FIGURE 7.6 Box corer. After the frame hits bottom, the coring device is forced into the sediment mechanically. (Courtesy of Sand M)

FIGURE 7.7   Seafloor samplers. (A) Chain dredge. (B) Dredge for collecting manganese nodules from the ocean floor. A conical bag of nylon netting is attached to a pyramidal frame.

7.3.4   CULTURE ISOLATION AND CHARACTERIZATION OF ACTIVE AGENTS FROM ENVIRONMENTAL SAMPLES

To study agents active in a geomicrobial process of interest, culture enrichments and pure culture isolations should be attempted as far as possible. Not all cultures obtained by these procedures may be geomicrobially active. Each isolate must be tested for its ability to perform the particular geomicrobial activity, or a part of it, that is under investigation. Because some geomicrobial processes are the result of the activity of a microbial consortium, no one of the pure culture isolates may exhibit all of the desired activity but may have to be tested with others of the isolates in different combinations. Examples of geomicrobial cooperation among microorganisms in microbial manganese oxidation that have been described include (1) the bacterium *Metallogenium symbioticum* in association with the fungus *Coniothyrium carpaticum* (Zavarzin, 1961; Dubinina, 1970; but see also, Schweisfurth, 1969); (2) the bacteria *Corynebacterium* sp. and *Chromobacterium* sp. (Bromfield and Skerman, 1950; Bromfield, 1956); and (3) two strains of *Pseudomonas* (Zavarzin, 1962). More recently, it was discovered that anaerobic methane oxidation depended on the action of a consortium consisting of a methanogen and a sulfate reducer (e.g., Boetius et al., 2000) (for more detail see Chapter 22).

Enrichment of and isolation from a mixed culture require selective conditions. If a microbial agent with a specific geomicrobial attribute is sought, the selective culture medium should have ingredients incorporated that favor the geomicrobial activity of interest. Apart from special nutrients, special pH, E_h, and temperature conditions may also have to be chosen to favor selective growth of the geomicrobial agents.

Isolation and characterization of pure cultures from enrichments should follow standard bacteriological technique, including determination of the molecular phylogeny of the agent(s) (for details, see for instance, Gerhardt et al., 1981, 1993; Hurst et al., 1997; and Skerman, 1967).

7.4 *IN SITU* STUDY OF PAST GEOMICROBIAL ACTIVITY

Past *in situ* geomicrobial activity in sedimentary deposits that ceased as far back as the early Precambrian or during the Phanerozoic up to modern times can sometimes be inferred through the detection of specific organic biomarkers in samples from such deposits. These biomarkers represent preserved organic derivatives formed from some characteristic cellular constituent, for example, chlorophylls from photosynthetic prokaryotes (bacteria, cyanobacteria); phototrophic eukaryotic microbes (Brocks and Pearson, 2005; Brocks et al., 2005; Roselle-Melé and Koç, 1997); carotenoids of anoxygenic photosynthetic bacteria and the oxygenic cyanobacteria (Hebting et al., 2006); and lipid cell membrane components from Archaea or Bacteria (Brocks and Pearson, 2005; Hebting et al., 2006). The process by which these cell constituents were preserved appears to have involved mostly or exclusively chemical reduction (abiotic) under anaerobic conditions. *In situ*, the agents of preservation in the case of preservation of carotenoids such as β- (phytoplankton) and γ-carotene (cyanobacteria and green nonsulfur bacteria, i.e., Chloroflexaceae), and okenone (purple sulfur bacteria, i.e., Chromatiaceae appear to have been reducing agents in the form of reduced sulfur compounds such as H_2S. The corresponding products of preservation from these pigments were β- and γ-carotane and okenane or perhydrookenone (Brocks et al., 2005; Hebting et al., 2006). The detection of the biomarker okenane, derived naturally from okenane of photosynthetic purple nonsulfur bacteria (Bradyrhizobiaceae) and chlorobactane from green sulfur bacteria (Chlorobiacea) in sediments of the 1.64-Gyr-old Barney Creek Formation (BCF) of the McArthur Group in northern Australia, has been used to argue that oxygen levels at this site at that time were well below modern levels although such levels had been approached or reached elsewhere on Earth (Brocks et al., 2005). This interpretation was supported by the simultaneous finding of extremely low concentrations of 2α-methylhopanes, biomarkers from cyanobacteria at the BCF site, which are normally considered to be oxygenic photosynthesizers (Brocks et al., 2005). However, Rashby et al. (2007) have demonstrated that a contemporary strain of the phototroph *Rhodopseudomonas palustris* (purple nonsulfur bacteria) is capable of synthesizing 2-methylbacteriohopanepolyols from which 2-methylhopane biomarkers could be formed during fossilization, indicating that this type of biomarker may not be a unique indicator for cyanobacteria.

Under certain circumstances, the occurrence of past geomicrobial activity can also be identified in terms of isotopic fractionation. Certain prokaryotic and eukaryotic microbes have been shown to distinguish between stable isotopes of elements such as C, H, O, N, S, Si, Li, and Fe. These microbes prefer to metabolize molecules containing the lighter isotopes of these elements (^{12}C in preference to ^{13}C, H in preference to D, ^{16}O in preference to ^{18}O, ^{14}N in preference to ^{15}N, ^{32}S in preference to ^{34}S, ^{28}Si in preference to ^{30}Si, ^{6}Li in preference to ^{7}Li, ^{54}Fe in preference to ^{56}Fe), especially under conditions of slow growth (see Jones and Starkey, 1957; Emiliani et al., 1978; Mortimer and Coleman, 1997; Mandernack et al., 1999; Wellman et al., 1968; Estep and Hoering, 1980; De La Rocha et al., 2000; Sakaguchi and Tomita, 2000; Croal et al., 2004; Crosby et al., 2007; Beard et al., 1999). They distinguish kinetically between different stable isotopes of the same element in a substrate on the basis of a difference in the rate of reaction the substrate undergoes in a specific biochemical step, the reaction rate involving the substrate with the lighter isotope being faster than that with heavier stable isotope. The isotope fractionation may occur during a specific enzyme-catalyzed chemical reaction into which the isotope-containing substrate enters, or it may occur during membrane transport of the isotope-containing substrate, or both (see Hoefs, 1997). Thus, the products of metabolism will be enriched in the lighter isotope compared to the starting compound or to some reference standard that has not been subjected to isotope fractionation. In practice, isotope fractionation is measured by determining isotope ratios of the heavier isotope of an element to the lighter, using mass spectrometry, and then calculating the amount of enrichment from the relationship

$$\delta\,\text{Isotope}\,(\permil) = \frac{\text{isotopic ratio of sample} - \text{isotope ratio of standard}}{\text{isotopic ratio of standard}} \times 1000 \qquad (7.1)$$

If the enrichment value (δ) is negative, the sample tested was enriched in the lighter isotope relative to a reference standard, and if the value is positive, the sample tested was enriched in the heavier isotope relative to a reference standard. Thus, for instance, to determine if a certain metal sulfide mineral deposit is of biogenic origin, various parts of the deposit are sampled and δ^{34}S values of the sulfide are determined. If the values are generally negative (although the magnitude of the δ^{34}S may vary among the samples and fall in the range of –5 to –50‰), the deposit can be viewed as of biogenic origin because a chemical explanation for such ^{32}S enrichment under natural conditions is not likely. If the δ^{34}S values are positive and fall in a narrow range, the deposit is viewed as being abiogenic in origin.

7.5 *IN SITU* STUDY OF ONGOING GEOMICROBIAL ACTIVITY

Ongoing geomicrobial activity may be measurable *in situ*. Such activity may be followed by the use of radioisotopes. For instance, bacterial sulfate-reducing activity may be determined by adding a small quantity of $Na^{35}SO_4$ to water, soil, or sediment sample of known sulfate content in a closed vessel. After incubation under *in situ* conditions, the sample is analyzed for loss of $^{35}SO_4^{2-}$ and buildup of $^{35}S^{2-}$ by separating these two entities and measuring their quantity in terms of their radioactivity. In the case of a water sample, incubation of the reaction mixture in a closed vessel in the water column may be at the depth from which the sample was taken. An example of a direct application of this method is that of Ivanov (1968). It allows the estimation of the rate of sulfate reduction in the sample without having any knowledge of the number of physiologically active organisms present in it. A modified method is that of Sand et al. (1975). Their method allows an estimation of the sulfate-reducing activity in terms of the number of physiologically active bacteria in the sample as distinct from an estimation of the sum of physiologically active and inactive bacteria. The assay for the estimation of active bacteria can be set up either to measure percentage of sulfate reduced in a fixed amount of time that is proportional to the logarithm of the concentration of active cells, or to measure the length of time required to reduce a fixed amount of sulfate to sulfide, which is related to the concentration of physiologically active sulfate-reducing bacteria in the sample. Ivanov's (1968, pp. 30–32) method can be adapted to measure the formation of elemental sulfur and sulfates from sulfide by adding $^{35}S^{2-}$ to a sample and after incubation in a reaction vessel *in situ*, separating ^{35}S and $^{35}SO_4^{2-}$ and measuring their quantity in terms of their radioactivity.

Microbial action on manganese (Mn^{2+} fixation by biomass; Mn^{2+} oxidation) *in situ* is another example of a geomicrobial process, which can be followed by the use of a radioisotope, $^{54}Mn^{2+}$ in this case (LaRock, 1969; Emerson et al., 1982; Burdige and Kepkay, 1983). One approach is to measure manganese oxidation in terms of the decrease in dissolved $^{54}Mn^{2+}$. It assumes that the oxidized manganese is insoluble. Decreases in dissolved $^{54}Mn^{2+}$ are measured on acidified samples of the reaction mixture from which the oxidized manganese has been removed by centrifugation. Acidification of samples before centrifugation ensures resolubilization of any adsorbed Mn^{2+}. The difference in radioactivity counts between a zero-time sample and a sample taken at a subsequent time is a measure of the amount of Mn^{2+} oxidized by a combination of biological and chemical processes over this time interval. Manganese oxidation due to biological activity alone can be estimated by subtracting chemical oxidation from the total Mn^{2+} oxidation for a corresponding time interval. An estimate of the amount of chemical oxidation can be obtained from a separate reaction mixture in which the biological activity is inhibited by use of autoclaved instead of active cells, by adding one or more chemical inhibitors to the oxidation mixture containing active cells, or by excluding air (if the enzymatic manganese oxidation is oxygen-dependent). This experimental approach has been used to assess the manganese-oxidizing activity in lake sediment (LaRock, 1969). It makes no assumption about binding of unoxidized or oxidized manganese to the bacterial cells.

Another approach is to measure manganese oxidation in terms of product formation. Manganese binding by metabolically active bacteria in a water sample may be followed in terms of ^{54}Mn accumulation by the cells. A manganese-oxidizing culture is incubated in a suitable reaction mixture containing added $^{54}Mn^{2+}$. After an appropriate length of incubation, the amount of radiomanganese bound by the cells is determined. For this determination, a measured sample of the bacterial suspension is filtered through a 0.2 μm membrane and washed, and the radioactivity retained on the filter membrane (assumed to be bound to the cells) is determined on a suitable counter. The results of this experiment are then compared to those of a parallel experiment in which the cells were inhibited by a poison such as formaldehyde or a mixture of sodium azide, penicillin G, and tetracycline HCl. The difference in radioactivity between these two experiments represents manganese bound by actively metabolizing cells. It includes manganese that was bound as Mn^{2+}, presumably a very minor amount, and that which was bound due to its oxidation by the cells (Emerson et al., 1982). Manganese binding by bacteria in sediment can be followed by a modified form of this method performed in a special reaction vessel called a peeper (Burdige and Kepkay, 1983). $^{54}Mn^{2+}$ adsorbed by the cells deposited on a filter membrane is displaced by washing with $CuSO_4$ solution. The radioactivity recovered in the wash is then counted. Residual ^{54}Mn associated with the cells (taken as oxidized manganese) is dissolved by washing with hydroxylamine-HCl solution followed by washing with $CuSO_4$ solution, and the radioactivity in the resultant solution is determined. Hydroxylamine-HCl is a reducing agent, which converts Mn(IV) to Mn(II).

Biological manganese binding in lake sediments may be studied by a method that requires controls in which biological inhibitors are used to account for abiological manganese binding by sediment constituents in the overall manganese budget (Burdige and Kepkay, 1983).

One advantage of using radioisotopes in the quantitative assessment of a specific geomicrobial transformation in nature is that their detection is extremely sensitive so that only minute amounts of radiolabeled substrate, which do not significantly change the naturally occurring concentration of the substrate, need to be added. Another advantage is that in cases where the rate of transformation is very slow although the natural substrate concentration is high, *spiking* the reaction with radiolabeled substrate allows analysis after a relatively brief reaction time because of the sensitivity of radioisotope detection.

The use of radioisotopes is, however, not essential for quantitative assessment in all instances of biogeochemical transformation in a natural environment. Other analytical methods with sufficient sensitivity may be applicable (see, for instance, Jones et al., 1983, 1984; Hornor, 1984; Kieft and Caldwell, 1984; Tuovila and LaRock, 1987).

To fully appreciate *in situ* geomicrobial activity, knowledge of local vertical chemical, pH, and redox gradients over relatively narrow depth intervals (e.g., millimeters or less) is important. The application of specific microelectrodes has made such determinations possible, as the following examples, a few among many, show.

Thus, use of an Au/Hg voltammetric microelectrode made it possible to measure simultaneously and with a spatial resolution in millimeters the vertical three-dimensional distribution of O_2, Mn^{2+}, Fe^{2+}, HS^-, and I^- in pore water of undisturbed sediment cores from the Canadian continental shelf and slope, and in the sediment surrounding an actively irrigated worm burrow in a mesocosm (Luther et al., 1998). Solid-state gold amalgam voltammetric microelectrodes have also made the monitoring of sulfur speciation *in situ* in sediments, microbial mats, and hydrothermal vent waters possible (Luther et al., 2001). Use of a fiber-optic scalar irradiance microsensor and oxygen microelectrode spaced 120 μm apart made it possible to measure scalar irradiance and oxygenic photosynthesis with 100 μm spatial resolution in marine microbial mats dominated by cyanobacteria with a surface layer of pinnate diatoms on sandy sediment along the coast of Limfjorden, Denmark (Lassen et al., 1992). Microelectrodes have also been employed in measuring O_2, H_2S, and pH microgradients at depth increments of 50 μm in *Beggiatoa* mats from marine sediments and *Thiovulum* films above the sediment to determine the microbial response to O_2 and H_2S (Jørgensen and Revsbech, 1983). Freshly prepared nitrate-selective microelectrodes with a liquid ion exchanger have been employed in determining nitrate gradients in sediments from a mesotrophic lake in the Netherlands (De Beer and Swearts, 1989). Although all the examples cited here involve sediments in freshwater and marine systems, microelectrodes are equally useful in studying terrestrial systems (e.g., soil).

7.6   LABORATORY RECONSTRUCTION OF GEOMICROBIAL PROCESSES IN NATURE

It is often important to reconstruct a naturally occurring geomicrobial process in the laboratory to investigate the mechanism whereby the process operates. Laboratory reconstruction can permit optimization of a process through the application of more favorable conditions than in nature. Examples are the use of a pure culture or a *purified* consortium to eliminate interference by competing microorganisms and the optimization of substrate availability, temperature, pH, E_h, and oxygen and carbon dioxide supply.

The activity of organisms growing on the surface of solid substrates such as soils, sediments, rocks, and ore may be investigated in batch culture, in air-lift columns, in percolation columns, or in a chemostat. A batch culture represents a closed system in which an experiment is started with a finite amount of substrate that is continually depleted during the growth of the culture. Cell population and metabolic products buildup and changes in pH and E_h are likely to occur. Conditions within the culture are thus continually changing and becoming progressively less favorable. Batch experiments may be least representative of a natural process, which usually occurs in an open system with continual or intermittent replenishment of substrate and removal of at least some of the metabolic wastes. A culture in an *air-lift column* (Figure 7.8) is a partially open system, where the microbes grow and carry on their biogeochemical activity on a mineral charge in the column. They are continually fed with recirculated nutrient solution from which nutrients are depleted by the organisms. The recirculation removes metabolic products from the solid substrate charge in the column. *Percolation columns* (Figure 7.9) are even more open systems than air-lift columns. In them, microbes grow on the solid substrate charge in the columns, but they are fed with nutrient solution that is not recirculated. This fresh nutrient solution is added continually or at intervals, and wastes are removed at the same time in the effluent without recirculation, while pH, E_h, and temperature are held constant or nearly so. Steady-state conditions such as in a *chemostat* idealize the open

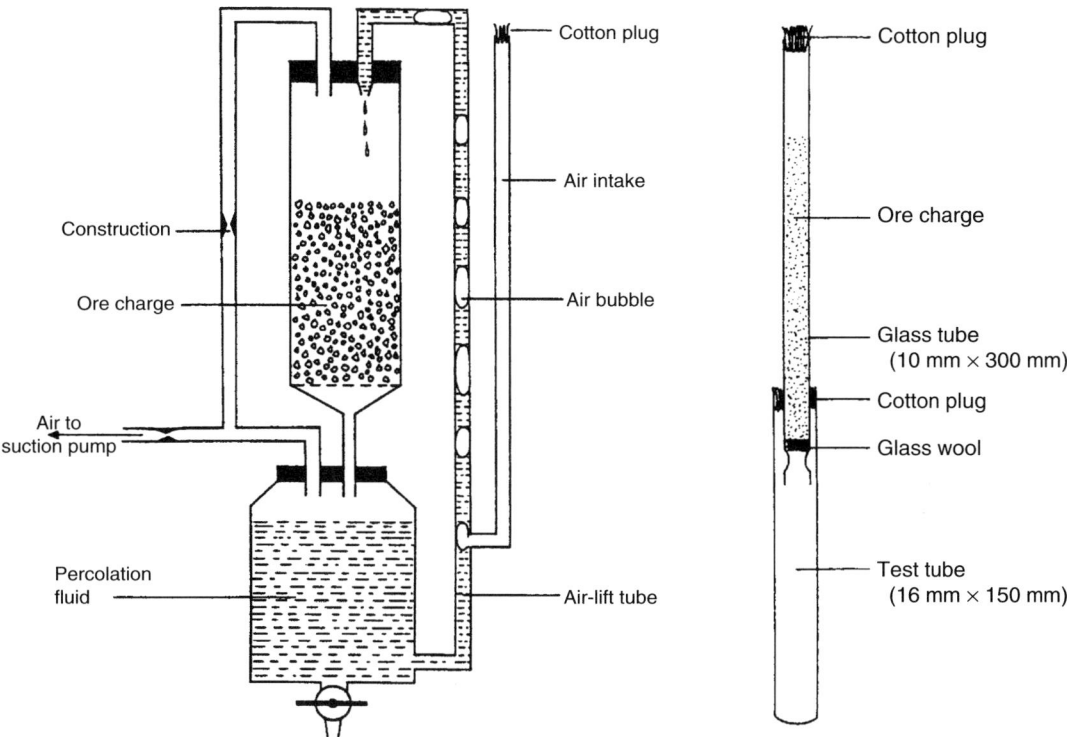

FIGURE 7.8 Air-lift column for ore leaching.

FIGURE 7.9 Percolation column for ore leaching.

culture system. They do not imitate nature because conditions are too constant. In open systems in nature, some fluctuation in various environmental parameters occurs over time.

The chemostat is a liquid culture system of constant volume. The continuous introduction of fresh nutrient solution at a constant rate keeps the actively growing cell population and the concentration of accumulating metabolic products constant in the culture vessel through medium displacement to maintain constant volume. In other words, steady-state conditions are maintained. This can be expressed mathematically as

$$\frac{dx}{dt} = \mu x - Dx \tag{7.2}$$

where

dx/dt = rate of cell population change in the chemostat
μ = instantaneous growth rate constant
D = dilution rate
x = cell concentration or cell number in the chemostat

The dilution rate is defined as the flow rate (f) of the influent feed or the effluent waste divided by the liquid volume (V) of the culture in the chemostat. Under steady-state conditions, $dx/dt = 0$, and therefore, $\mu x = Dx$ (i.e., instantaneous growth rate equals dilution rate, $\mu = D$). Under conditions where $D > \mu$, the cell population in the chemostat will decrease over time and may ultimately be washed out. Conversely, if $\mu > D$, the cell population in the chemostat will increase until a new steady state is reached, which is determined by the growth-limiting concentration of an essential substrate.

The steady state in the chemostat can also be expressed in terms of the rate change of growth-limiting substrate concentration (ds/dt). This is based on the principle that the rate of change in substrate concentration is dependent on the rate of substrate addition to the chemostat, the rate of washout from the chemostat, and the rate of substrate consumption by the growing organism:

$$\frac{ds}{dt} = D(S_{\text{inflow}} - S_{\text{outflow}}) - \mu(S_{\text{inflow}} - S_{\text{outflow}}) \tag{7.3}$$

where
   D = dilution rate
   S_{inflow} = substrate concentration entering the chemostat
   S_{outflow} = concentration of unconsumed substrate
   μ = instantaneous growth rate constant

At steady state, $ds/dt = 0$. The substrate consumed, that is, $S_{\text{inflow}} - S_{\text{outflow}}$, is related to the cell mass produced (x) according to the relationship

$$x = y(S_{\text{inflow}} - S_{\text{outflow}}) \tag{7.4}$$

where y is the growth yield constant (mass of cells produced per mass of substrate consumed). These relationships require modification if a solid substrate is included in the chemostat (see, for instance, Section 7.6).

The chemostat can be used, for example, to determine limiting substrate concentrations for the growth of bacteria under simulated natural conditions. Thus, the limiting concentration of lactate, glycerol, and glucose required for growth at different relative growth rates (D/μ_m) of *Achromobacter aquamarinus* (strain 208) and *Spirillum lunatum* (strain 102) in seawater have been determined by this method (Table 7.1) (Jannasch, 1967). The chemostat principle can also be applied to a study of growth rates of microbes in their natural environment by laboratory simulation (Jannasch, 1969) or directly in the natural habitat. For instance, the size of the algal population in an algal mat of a hot spring in Yellowstone National Park, Montana, was found to be relatively constant, implying that the algal growth rate equaled its washout rate from the spring pool. When a portion of the algal mat was darkened by blocking access of sunlight, thereby stopping photosynthesis and thus algal growth

TABLE 7.1
Threshold Concentrations of Three Growth-Limiting Substrates (mg L^{-1}) in Seawater at Several Relative Growth Rates of Six Strains of Marine Bacteria and the Corresponding Maximum Growth Rates (μ_m h^{-1})

Strain	D/μ_m	Lactate	Glycerol	Glucose
208	0.5	0.5	1.0	0.5
	0.1	0.5	1.0	0.5
	0.005	1.0	5.0	1.0
μ_m		0.15	0.20	0.34
102	0.3	0.5	No growth	0.5
	0.1	1.0		5.0
	0.05	1.0		10.0
μ_m		0.45		0.25

Source: Excerpted from Table 2 in Jannasch HW, *Limnol. Oceanogr.*, 12, 264–271, 1967. With permission.

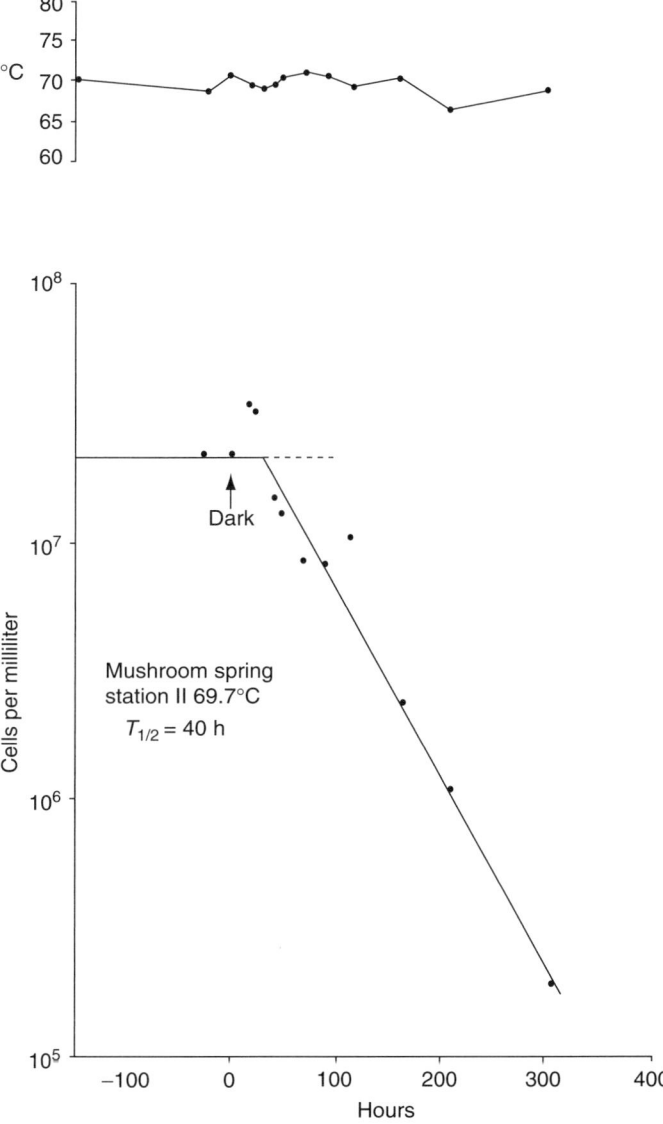

FIGURE 7.10 Washout rate of part of an algal mat at a station at Mushroom Spring in Yellowstone National Park, Montana, after it was shaded experimentally. (From Brock TC, Brock ML, *J. Bacteriol.*, 95, 811–815, 1968. With permission.)

and multiplication in that part of the mat, the algal cells were washed out from it at a constant rate after a short lag (Figure 7.10). The washout rate under these conditions equaled the growth rate in the illuminated part of the mat. This follows from Equation 7.2 when $dx/dt = 0$ (Brock and Brock, 1968). Similarly, the population of the sulfur-oxidizing thermophile *Sulfolobus acidocaldarius* has been found to be in steady state in hot springs in Yellowstone National Park, implying, as in the case of the algae in the mat, that the growth rate of the organism equals its washout rate from the spring. In this instance, the washout rate was measured by following the water turnover rate in terms of dilution rate of a small measured amount of NaCl added to the spring pool (Figure 7.11). The dilution rate was then translated into the growth rate of *S. acidocaldarius* in the spring (Mosser et al., 1974).

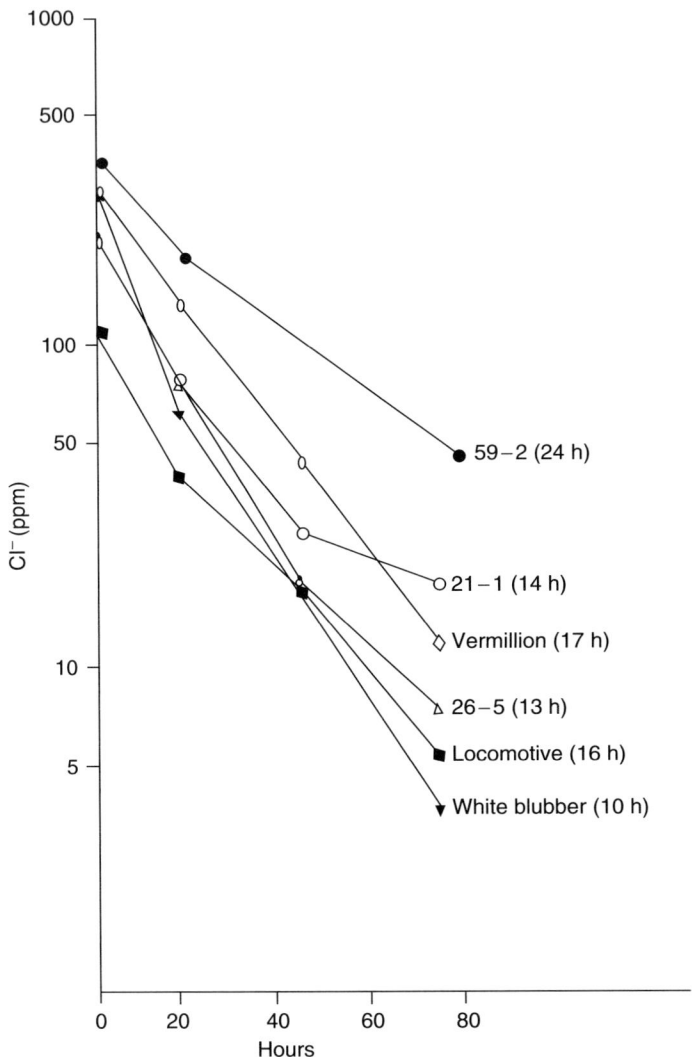

FIGURE 7.11 Chloride dilution in several small springs in Yellowstone National Park, Montana. Estimated half-times for chloride dilution are given in parentheses. At site 21–1, chloride concentration had reached the natural background level by the final sampling time, and the dilution rate was estimated from the data from the first three sampling times. (From Mosser JL, Bohlool BB, Brock TD, *J. Bacteriol.*, 118, 1075–1081, 1974. With permission.)

7.7 QUANTITATIVE STUDY OF GROWTH ON SURFACES

The chemostat or its principle of operation is not applicable to all culture situations. In the study of geomicrobial phenomena, the central microbial activity often occurs on the surface of inorganic or organic solids. Indeed, the solid may be the growth-limiting substrate upon which the organism acts. Under these conditions, it may be assumed that the microbial population as it colonizes the surface will increase geometrically once it has settled on it, approximating the relationship

$$\log N = \log N_0 + \frac{\Delta t}{g} \log 2 \tag{7.5}$$

where

> N = final cell concentration per unit area
> N_0 = initial cell concentration per unit area
> Δt = time required for N_0 to multiply to N cells
> g = average doubling time (generation time)

It is also assumed that the cell multiplication rate is significantly slower than the rate of attachment of cells in the bulk phase to the solid surface. Once all available space on the surface has been occupied, however, the cell population on it will remain constant (provided that the surface area does not decrease significantly because of solid substrate consumption or dissolution). The cell population in the liquid in contact with the solid will then show an arithmetic increase in cell numbers according to the relationship

$$N_{final,liquid} = N_{initial,liquid} + zN_{solid} \qquad (7.6)$$

where

> z = number of cell doublings on the solid
> N_{solid} = N cells on the solid when all attachment sites are occupied

Equation 7.6 states that for every cell doubling on the surface of the cell-saturated solid, one daughter cell will be displaced into the liquid medium for lack of space on the solid. This model assumes that the bulk phase, liquid medium by itself cannot support growth of the organism. As long as there is no significant change in the surface area of the solid phase, the model applies. An example to which the model can be applied is the growth of autotrophic thiobacteria on water-insoluble metal sulfide, which serves as their sole energy source in a mineral salts solution that satisfies all other nutrient requirements (Chapter 20).

To introduce the concept of time into Equation 8.6, the following relationship should be considered:

$$g = \frac{\Delta t}{z} \qquad (7.7)$$

where

> g = doubling time
> Δt = time interval between $N_{initial,liquid}$ and $N_{final,liquid}$ determinations

If Equations 7.6 and 7.7 are combined, we get

$$N_{final,liquid} = N_{initial,liquid} + \frac{\Delta t}{g} N_{solid} \qquad (7.8)$$

when solving for g, we get

$$g = \frac{\Delta t\, N_{solid}}{N_{final,liquid} - N_{initial,liquid}} \qquad (7.9)$$

Espejo and Romero (1987) took a different mathematical approach to bacterial growth on a solid surface. They developed the following relationship to describe growth before surface saturation was reached:

$$\frac{dN_a}{dt} = \frac{\mu N_a (N_s - N_a)}{N_s} \qquad (7.10)$$

where

    N_a = number of attached bacteria

    N_s = limit value of bacteria that can attach to the surface under consideration

    μ = specific growth rate

After surface saturation by the bacteria, they propose the relationship

$$\frac{dN_f}{dt} = \mu N_s \qquad (7.11)$$

where N_f represents the number of free (unattached) bacteria in the liquid phase with which the surface is in contact. From Equation 7.11, the following relationship for the specific growth rate of the culture after surface saturation is derived:

$$\mu = \frac{\Delta N_f}{\Delta t\, N_s} \qquad (7.12)$$

These relationships were tested by the researchers in a real experiment growing *Acidithiobacillus ferrooxidans* on elemental sulfur in which the sulfur was the only available energy source for the bacterium. They observed logarithmic population increase before surface saturation and a linear increase thereafter. They concluded, however, that the value for N_s was not constant in their experiment but changed gradually after linear growth had begun.

Another mathematical model was presented by Konishi et al. (1994), in which the change in particle size of the solid substrate with time as well as cell adsorption to the solid substrate was taken into account.

7.8 TEST FOR DISTINGUISHING BETWEEN ENZYMATIC AND NONENZYMATIC GEOMICROBIAL ACTIVITY

To determine if a geomicrobial transformation is an enzymatic or a nonenzymatic process, attempts should be made to reproduce the phenomenon with cell-free extract, especially if a single organism is involved. If catalysis is observed with the cell-free extract, identification of the responsible enzyme system should be undertaken by standard techniques. Spent culture medium from which all cells have been removed, for example, by filtration or by inactivation by heating, should also be tested for activity. If equal levels of activity are observed with untreated and treated spent medium, operation of a nonenzymatic process may be inferred. If the level of activity in treated cell-free extract is lower than in untreated cell-free spent medium, extracellular enzyme activity may be present in the untreated cell-free spent medium. This can be verified by testing activity in unheated cell-free medium over a range of temperatures. If a temperature optimum is observed, at least part of the activity in the cell-free medium can be attributed to extracellular enzymes.

7.9 STUDY OF REACTION PRODUCTS OF GEOMICROBIAL TRANSFORMATION

Ideally, the products of geomicrobial transformation, if they are precipitates, should be studied not only with respect to chemical composition but also with respect to mineralogical properties through one or more of the following techniques: SEM and TEM, including energy dispersive x-ray measurements (EDX), microprobe examination, x-ray photoelectron spectroscopy (XPS), infrared spectroscopy, x-ray diffraction, x-ray absorption fine structure (XFAS), and x-ray absorption near-edge structure spectroscopy (XANES), and mineralogical examination with a polarizing microscope. Similar studies should ideally be undertaken on the substrate if it is an insoluble mineral complex to be able to detect any mineralogical changes it may have undergone with time during geomicrobial transformation. Raman spectroscopy can be applied to study organic compounds associated with ancient rocks.

Studies of geomicrobial phenomena require ingenuity in the application of standard microbiological, chemical, and physical techniques and often require collaboration among microbiologists, biochemists, geochemists, mineralogists, and other specialists to unravel a problem.

7.10 SUMMARY

Geomicrobial phenomena can be studied in the field and in the laboratory. Direct observation may involve microscopic examination and chemical and physical measurements. Very specific and sensitive molecular biological techniques should be applied to determine the number and kind of microbes in field samples. Laboratory study may involve an artificial reconstruction of a geomicrobial process. Special methods have been devised for sampling, direct observation, and laboratory manipulation. The latter two categories include the use of fluorescence microscopy, radioactive tracers, and mass spectrometry for observing microorganisms *in situ*, measuring process rates, and measuring microbial isotope fractionation, respectively. The chemostat principle has been applied in the field to measure natural growth rates of geomicrobial agents. It has also been used under simulated conditions for determining limiting substrate concentrations. The study of growth on particle surfaces may require special experimental approaches.

REFERENCES

Barker WW, Banfield JF. 1998. Zones of chemical and physical interaction at interfaces between microbial communities and minerals: A model. *Geomicrobiol J* 15:223–244.

Beard BL, Johnson CM, Cox L, Sun H, Nealson KH, Aguilar C. 1999. Iron isotope biosignatures. *Science* 285:1889–1892.

Beeman RE, Suflita JM. 1989. Evaluation of deep subsurface sampling procedures using serendipitous microbial contaminants as tracer organisms. *Geomicrobiol J* 7:223–233.

Bien E, Schwartz W. 1965. Geomikrobiologische Untersuchungen. IV. Über das Vorkommen konservierter toter und lebender Bakterienzellen in Salzgestein. *Z Allg Mikrobiol* 5:185–205.

Boetius A, Ravenschlag K, Schubert CJ, Rickert D, Widdel F, Gieseke A, Amann R, Jørgensen BB, Witte U, Pfannkuche O. 2000. A marine microbial consortium apparently mediating anaerobic oxidation of methane. *Nature* (London) 407:623–626.

Bohlool BB, Brock TD. 1974. Immunofluorescence approach to the study of the ecology of *Thermoplasma acidophilium* in a coal refuse material. *Appl Environ Microbiol* 28:11–16.

Brock TC, Brock ML. 1968. Measurement of steady-state growth rates of a thermophilic alga directly in nature. *J Bacteriol* 95:811–815.

Brock TD. 1978. *Thermophilic Microorganisms and Life at High Temperatures*. New York: Springer.

Brocks JJ, Love GD, Summons RE, Knoll AH, Logan GA, Bowden SA. 2005. Biomarker evidence for green and purple sulphur bacteria in a stratified Paleoproterozoic sea. *Nature* (London) 437:866–870.

Brocks JJ, Pearson A. 2005. Building the biomarker tree of life. *Rev Mineral Geochem* 59:233–258.

Bromfield SM. 1956. Oxidation of manganese by soil microorganisms. *Aust J Biol Sci* 9:238–252.

Bromfield SM, Skerman VBD. 1950. Biological oxidation of manganese in soils. *Soil Sci* 69:337–348.

Burdige DJ, Kepkay PE. 1983. Determination of bacterial manganese oxidation rates in sediments using an in-situ dialysis technique. I. Laboratory studies. *Geochim Cosmochim Acta* 47:1907–1916.

Casida LE. 1962. On the isolation and growth of individual microbial cells from soil. *Can J Microbiol* 8:115–119.

Casida LE. 1965. Abundant microorganisms in soil. *Appl Microbiol* 13:327–334.

Casida LE. 1971. Microorganisms in unamended soil as observed by various forms of microscopy and staining. *Appl Microbiol* 21:1040–1045.

Clesceri LS, Greenberg AE, Trussell RR, eds. 1989. *Standard Methods for the Examination of Water and Wastewater*. 17th ed. Washington, DC: American Public Health Association, pp. 9-65 to 9-66.

Croal LR, Johnson CM, Beard BL, Newman DK. 2004. Iron isotope fractionation by Fe(II)-oxidizing photoautotrophic bacteria. *Geochim Cosmochim Acta* 68:1227–1242.

Crosby HA, Roden EE, Johnson CM, Beard BL. 2007. The mechanism of iron isotope fractionation produced during dissimilatory Fe(III) reduction by *Shewanella putrefaciens* and *Geobacter sulfurreducens*. *Geobiology* 5:169–189.

De Beer D, Sweerts JPRA. 1989. Measurement of nitrate gradients with an ion-selective microelectrode. *Anal Chim Acta* 219:351–356.

De La Rocha CL, Brzczinski MA, DeNiro MJ. 2000. A first look at the distribution of the stable isotopes of silicon in natural waters. *Geochim Cosmochim Acta* 64:2467–2477.

Dubinina GA. 1970. Untersuchungen über die Morphologie und die Beziehung zu *Mycoplasma*. *Z Allg Mikrobiol* 10:309–320.

Edwards KJ, Goebel BM, Rodgers TM, Schrenk MO, Gihring TM, Cardona MM, Hu B, McGuire MM, Hamers RJ, Pace NR, Banfield JF. 1999. Geomicrobiology of pyrite (FeS$_2$) dissolution: Case study at Iron Mountain, CA. *Geomicrobiol J* 16:155–179.

Emerson S, Kalhorn S, Jacobs L, Tebo BM, Nealson KH, Rosson RA. 1982. Environmental oxidation rate of manganese(II): Bacterial catalysis. *Geochim Cosmochim Acta* 46:1073–1079.

Emiliani C, Hudson J, Shinn A, George RY. 1978. Oxygen and carbon isotope growth record in a reef coral from the Florida Keys and a deep-sea coral from the Blake Plateau. *Science* 202:627–629.

Eren J, Pramer D. 1966. Application of immunofluorescent staining to studies of the ecology of soil microorganisms. *Soil Sci* 101:39–49.

Espejo RT, Romero P. 1987. Growth of *Thiobacillus ferrooxidans* on elemental sulfur. *Appl Environ Microbiol* 53:1907–1912.

Estep MF, Hoering TC. 1980. Biogeochemistry of the stable hydrogen isotopes. *Geochim Cosmochim Acta* 44:1197–1206.

Gerhardt P, Murray RGE, Costilow RN, Nester EW, Wood WA, Krieg NR, Phillips GB, eds. 1981. *Manual of Methods for General Bacteriology*. Washington, DC: The American Society for Microbiology.

Gerhardt P, Murray RGE, Wood WA, Krieg NR. 1993. *Methods for General and Molecular Bacteriology*. Washington, DC: ASM Press.

Ghiorse WC, Balkwill DL. 1983. Enumeration and morphological characterization of bacteria indigenous to subsurface environments. *Dev Ind Microbiol* 24:213–224.

Griffin WT, Phelps TJ, Colwell FS, Fredrickson JK. 1997. Methods for obtaining deep subsurface microbiological samples by drilling. In: Amy PS, Haldeman DL, eds. *The Microbiology of the Terrestrial Deep Subsurface*. Boca Raton, FL: CRC Lewis, pp. 23–44.

Hebting Y, Schaeffer P, Behrens A, Adam P, Schmitt G, Schneckenburger P, Bernasconi SM, Albrecht P. 2006. Biomarker evidence for a major preservation pathway of sedimentary organic carbon. *Science* 312:1627–1630.

Hoefs J. 1997. *Stable Isotope Geochemistry*. 4th ed. Berlin: Springer, pp. 5, 41–42, 59–60.

Hornor SG. 1984. Microbial leaching of zinc concentrate in fresh-water microcosms: Comparison between aerobic and oxygen-limited conditions. *Geomicrobiol J* 3:359–371.

Huber H, Huber G, Stetter KO. 1985. A modified DAPI fluorescence staining procedure suitable for the visualization of lithotrophic bacteria. *Syst Appl Microbiol* 6:105–106.

Hurst CJ, Knudsen GR, McInerney MJ, Stetzenbach LD, Walter MV. 1997. *Manual of Environmental Microbiology*. Washington, DC: ASM Press.

Ivanov MV. 1968. *Microbiological Processes in the Formation of Sulfur Deposits*. Washington, DC: U.S. Department of Agriculture and the National Science Foundation (Israel Program for Scientific Translations).

Jannasch HW. 1967. Growth of marine bacteria at limiting concentrations of organic carbon in seawater. *Limnol Oceanogr* 12:264–271.

Jannasch HW. 1969. Estimation of bacterial growth in natural waters. *J Bacteriol* 99:156–160.

Jannasch HW, Mottl MJ. 1985. Geomicrobiology of the deep-sea hydrothermal vents. *Science* 229:717–725.

Jannasch HW, Wirsen CO. 1981. Morphological survey of microbial mats near deep-sea thermal vents. *Appl Environ Microbiol* 41:528–538.

Jørgensen BB, Revsbech NP. 1983. Colorless sulfur bacteria, *Beggiatoa* spp. and *Thiovulum* spp. in oxygen and hydrogen sulfide microgradients. *Appl Environ Microbiol* 45:1261–1270.

Jones GE, Starkey RL. 1957. Fractionation of stable isotopes of sulfur by microorganisms and their role in deposition of native sulfur. *Appl Microbiol* 5:111–118.

Jones JG, Gardener S, Simon BM. 1983. Bacterial reduction of ferric iron in a stratified eutrophic lake. *J Gen Microbiol* 129:131–129.

Jones JG, Gardener S, Simon BM. 1984. Reduction of ferric iron by heterotrophic bacteria in lake sediments. *J Gen Microbiol* 130:45–51.

Kepner RL Jr., Pratt JR. 1994. Use of fluorochromes for direct enumeration of total bacteria in environmental samples: Past and present. *Microbiol Rev* 58:603–615.

Kieft TL, Caldwell DE. 1984. Weathering of calcite, pyrite, and sulfur by *Thermothrix thiopara* in a thermal spring. *Geomicrobiol J* 3:201–215.

Konishi YY, Takasaka Y, Asai S. 1994. Kinetics of growth and elemental sulfur oxidation in batch culture of *Thiobacillus ferrooxidans*. *Biotechnol Bioeng* 44:667–673.

LaRock PA. 1969. The bacterial oxidation of manganese in a freshwater lake. Ph.D. Thesis. Rensselaer Polytechnic Institute, Troy, New York.

LaRock, PA, Ehrlich HL. 1975. Observations of bacterial microcolonies on the surface of ferromanganese nodules from Blake Plateau by scanning electronmicroscopy. *Microb Ecol* 2:84–96.

Lassen C, Ploug H, Jørgensen BB. 1992. Microalgal photosynthesis and spectral scalar irradiance in coastal marine sediments of Limfjorden, Denmark. *Limnol Oceanogr* 37:760–772.

Lawrence JR, Korber DR, Wolfaardt GM, Caldwell DE. 1997. Analytical imaging and microscopy techniques. In: Hurst CJ, Knudsen GR, McInerney MJ, Stetzenbach LD, Walter MV, eds. *Manual of Environmental Microbiology*. Washington, DC: ASM Press, pp. 29–51.

Luther GW III, Brendel PJ, Lewis BL, Sundby F, Lefrançois L, Silverberg N, Nuzzio DB. 1998. Simultaneous measurement of O_2, Mn, Fe, I and S(-II) in marine pore waters with a solid-state voltammetric microelectrode. *Limnol Oceanogr* 43:325–333.

Luther GW III, Glazer BT, Hohmann L, Popp JI, Taillefert M, Rozan TF, Brendel PJ, Theberge SM, Nuzzio DB. 2001. Sulfur speciation monitored in situ with solid state gold amalgam voltammetric microelectrodes: Polysulfides as a special case in sediments, microbial mats and hydrothermal vent waters. *J Environ Monit* 3:61–66.

Mandernack KW, Bazilinski DA, Shanks WC III, Bullen TD. 1999. Oxygen isotope studies of magnetite produced by magnetotactic bacteria. *Science* 285:1892–1896.

Mortimer RJG, Coleman ML. 1997. Microbial influence on the oxygen isotopic composition of diagentic siderite. *Geochim Cosmochim Acta* 61:1705–1711.

Mosser JL, Bohlool BB, Brock TD. 1974. Growth rates of *Sulfolobus acidocaldarius* in nature. *J Bacteriol* 118:1075–1081.

Muyzer G, de Bruyn AC, Schmedding DJM, Bos P, Westbroek P, Kuenen GJ. 1987. A combined immuno-fluorescence-DNA fluorescence staining technique for enumeration of *Thiobacillus ferrooxidans* in a population of acidophilic bacteria. *Appl Environ Microbiol* 53:660–664.

Pace NR. 1997. A molecular view of microbial diversity and the biosphere. *Science* 276:734–740.

Pedersen K. 1993. The deep subterranean biosphere. *Earth-Sci Rev* 34:243–260.

Perfil'ev BV, Gabe DR. 1965. The use of the microbial-landscape method to investigate bacteria which concentrate manganese and iron in bottom deposits. In: Perfil'ev BV, Gabe DR, Gal'perina AM, Rabinovich VA, Sapotnitskii AA, Sherman EE, Troshanov EP, eds. *Applied Capillary Microscopy: The Role of Microorganisms in the Formation of Iron and Manganese Deposits*. New York: Consultants Bureau, pp. 9–54.

Perfil'ev BV, Gabe DR. 1969. *Capillary Methods for Studying Microorganisms*. English Translation by Shewan J. Toronto, Canada: University Toronto Press.

Phelps TJ, Russell BF. 1990. Drilling and coring deep subsurface sediments for microbiological investigations. In: Fliermans CB, Hazen TC, eds. *Proceedings of the First International Symposium on Microbiology of the Deep Subsurface*. Aiken, SC: WSRC Infor Services Sec Publ Group, pp. 2–35 to 2–47.

Pope DH,

Rashby SE, Sessions AL, Summons RE, Newman DK. 2007. Biosynthesis of 2-methylobacteriohopanepolyols by an anoxygenic phototroph. *Proc Nat Acad Sci USA*, 104:15099–15104.

Roselle-Melé A, Koç N. 1997. Paleoclimatic significance of the stratigraphic occurrence of photosynthetic biomarker pigments in the Nordic seas. *Geology* 25:49–52.

Sakaguchi T, Tomita O. 2000. Bioseparation of lithium isotopes by using microorganisms. *Resource Environ Biotechnol* 3:173–182.

Sand M,

Sand MD, LaRock PA, Hodson RE. 1975. Radioisotope assay for the quantification of sulfate reducing bacteria in sediment and water. *Appl Microbiol* 29:626–634.

Schmidt EL, Bankole RO. 1965. Specificity of immunofluorescent staining for study of *Aspergillus flavus* in soil. *Appl Microbiol* 13:673–679.

Schopf JW, ed. 1983. *Earth's Earliest Biosphere: Its Origin and Evolution*. Princeton, NJ: Princeton University Press.

Schweisfurth R. 1969. Mangan-oxydierende Pilze. *Z Parasitenk, Infektionskr Hyg Abt 1, Orig* 212:486–491.

Sieburth JMcN. 1975. *Microbial Seascapes. A Pictorial Essay on Marine Microorganism and Their Environments*. Baltimore, MD: University Park Press.

Skerman VBD. 1967. *A Guide to the Identification of the Genera of Bacteria. With Methods and Digest of Generic Characteristics*. 2nd ed. Baltimore, MD: Williams and Wilkins.

Stahl DA. 1997. Molecular approaches for the measurement of density, diversity and phylogeny. In: Hurst CJ, Knudson GR, McInerney MJ, Stotzenbach LD, Walter MV, eds. Manual of Environmental Microbiology. Washington, DC: ASM Press, pp. 102–114.

Tuovila BJ, LaRock PA. 1987. Occurrence and preservation of ATP in Antarctic rocks and its implication in biomass determination. *Geomicrobiol J* 5:105–118.

Ward DM, Weller R, Bateson MM, 1990. 16SrRNA sequences reveal numerous uncultured microorganisms in a natural community. *Nature* (London) 345: 63–65.

Weirich G, Schweisfurth R. 1985. Extraction and culture of microorganisms from rock. *Geomicrobiol J* 4:1–20.

Wellman RP, Cook FD, Krouse HR. 1968. Nitrogen-15: Microbiological alterations in abundance. *Science* 161:269–270.

Zavarzin GA. 1961. Symbiotic culture of a new microorganism oxidizing manganese. *Mikrobiologiya* 30:393–395 (English translation pp. 343–345).

Zavarzin GA. 1962. Symbiotic oxidation of manganese by two species of *Pseudomonas*. *Mikrobiologiya* 31:586–588 (English translation pp. 481–482).

8 Molecular Methods in Geomicrobiology

8.1 INTRODUCTION

In addition to the classical techniques used in geomicrobiology (described in Chapter 7), in the past three decades, molecular tools have become increasingly important in the study of the presence, activity, and mechanism(s) of catalysis by geomicrobial organisms. Today, various molecules (deoxyribonucleic acid [DNA], ribonucleic acid [RNA], protein, and constituents of lipids) are used to detect specific geomicrobial agents *in situ*, and by coupling measurements of their isotopic composition to the identification of these molecules, it is possible not only to identify geomicrobial agents but also to make inferences about their metabolic activity. Rapid advances in whole genome sequencing and biophysical sorting are greatly expanding our appreciation of the genetic potential of isolated geomicrobial organisms, as well as uncultured species from the environment. Finally, the application of molecular genetic, cell biological, and biochemical techniques to study the genes and gene products that catalyze geochemically significant reactions is permitting geomicrobial processes to be understood mechanistically. Together, these molecular tools can provide us with the information, we need to be able to predict when a geomicrobial process will occur and estimate its impact on the geochemistry of the environment.

8.2 WHO IS THERE? IDENTIFICATION OF GEOMICROBIAL ORGANISMS

Molecular approaches to identify pure cultures of microorganisms include DNA/DNA hybridization, DNA/RNA hybridization, DNA fingerprinting, and 16S ribosomal RNA (16S rRNA) analysis (Amann et al., 1990; Saylor and Layton, 1990; Stahl, 1997; Ward et al., 1992). DNA fingerprinting involves extracting DNA from a pure culture of the unknown organism, digesting the DNA with specific restriction enzymes, which cleave it into polynucleotide fragments, subjecting the digest to electrophoresis, and comparing the resultant banding pattern of the different nucleotide fragments with patterns obtained in the same way from DNA extracted from pure cultures of known organisms. Because the vast majority of geomicrobial organisms have not yet been cultured, culture-independent methods for their identification have become commonplace.

8.2.1 CULTURE-INDEPENDENT METHODS

Although, the first widely applied culture-independent methods to document the presence of diverse organisms in the environment were based on comparisons of phospholipid fatty acid (PLFA) profiles (Bobbie and White, 1980; Tunlid and White, 1990), today, DNA and RNA sequences are more commonly used to identify microorganisms and study their phylogeny. These methods have been adapted to locate and enumerate microorganisms in environmental samples, even microorganisms that, though viable, have not been cultured in the laboratory (Ward et al., 1990; Stahl, 1997).

The recognition that a 16S rRNA molecule possesses both highly conserved nucleotide sequences and sequences that are variable to different extents has provided a handle for distinguishing between the domains of the Bacteria, Archaea, and Eucarya and between groups of organisms within each these three domains (Woese and Fox, 1977). Thus, based on the level of variability, distinctions can be made at the phylogenetic, generic, and species levels (e.g., Ward et al., 1992). In practice, species-level differentiation of microorganisms is commonly assigned a 97% identify cutoff; this is based on a weak correlation of whole genome DNA–DNA hybridization results with 16S rRNA sequence similarity (Stackebrandt and Goebel, 1994). In whole genome hybridization, strains that are >70% similar are considered to be part of the same species and it was shown that above this cutoff no 16S rRNA sequence similarity <97% could be detected. Because any 16S rRNA sequence of a new organism once established is entered into a databank (e.g., Ribosomal Database Project II: http://rdp.cme.msu.edu/or GenBank: http://www.ncbi.nlm.nih.gov/Genbank/), it is possible to compare 16S rRNA sequences from a new isolate with those in the database to determine if it has been previously reported and to identify its closest relatives. Although 16S rRNA sequences have proven to be a fairly robust phylogenetic marker, we refer to the reader to a review by Olsen and Woese that discusses the importance of comparing results from multiple molecules, both as a method for testing the overall reliability of the organismal phylogeny and as a method for more broadly exploring the history of the genome (Olsen and Woese, 1993). Because sequencing whole microbial genomes is a routine now, it is also possible to consider phylogenetic relationships based on whole genomes. This is sometimes referred to as *phylogenomics* and we refer the reader to a review by Delsuc et al. (2005) that discusses this approach.

Probes based on specific 16S rRNA nucleotide sequences can be applied directly to intact cells that have been treated to make them permeable to the probes and allow hybridization between probes and the corresponding nucleotide sequence in the ribosomal 16S rRNA. When such probes are labeled with a fluorescent dye or with a radioactive isotope, cells that have reacted with the labeled probe can be readily located by fluorescence microscopy or autoradiography, respectively (Giovannoni et al., 1988; DeLong et al., 1989; Amann et al., 1990, 1992, 1995; Tsien et al., 1990; Ward et al., 1990; Braun-Howland et al., 1992,1993; Jurtshuk et al., 1992; Schrenk et al., 1998; Orphan et al., 2001; Michaelis et al., 2002). This fluorescence microscopy technique is called *fluorescence in situ hybridization* (FISH). The 16S rRNA probes can be group-, genus-, and species-specific. Caveats of this method are that it requires genes to be actively transcribed (the metabolic rates of some important geomicrobial organisms may be so slow that insufficient rRNA is present to detect), it can only be used with sequences that are already known, and there is a chance that related but nontarget genes will interfere with meaningful identification. One approach to circumvent the first of these issues is to amplify the signal using catalyzed reporter deposition (*CARD-FISH*) (Pernthaler et al., 2002); this technique has been used to identify organisms in biofilm communities in soil and marine environments (Fazi et al., 2005; Ferrari et al., 2006). Alternatively, *chromosomal painting* can be used to directly target genomic DNA with fluorescently labeled nucleic acid probes to identify specific sequences without the need for gene expression (Lanoil and Giovannoni, 1997).

Beyond probing for organisms with identifying sequences that are already known, various culture-independent molecular approaches can be used to detect (new) organisms in the environment, and estimate the diversity of a microbial community. By using the *polymerase chain reaction* (PCR) technique (Guyer and Koshland, 1989) with primers targeting 16S rRNA genes (16S rDNA), it is possible to amplify 16S rRNA genes in the DNA extracted from the community. The resultant mixture of 16S rDNA must then be resolved. One approach employs cloning. The 16S rDNA in each clone is reamplified by PCR and the nucleotide sequence of the rRNA is determined. This is sometimes referred to as making a *clone library*. Another approach to resolving the 16S rDNA mixture derived from amplification of the DNA extract is to use *denaturing gradient gel electrophoresis* (DGGE), which can separate fragments of the same length that differ by only one or two nucleotides. Attempts can then be made to hybridize rDNA bands in the electrophoretic patterns with oligonucleotide probes from known organisms and thus identify the organisms corresponding to the band that successfully hybridized with a probe. The bands from the electrophoresis gels can also be eluted, purified if needed, reamplified, and their nucleotide sequence determined (Ward, 1992; Muyzer et al., 1993; Stahl, 1997; Ward et al., 1998; Madigan et al., 2000). Such sequences can then be compared to sequences in an appropriate database to determine if they were previously reported. The second major technique for fingerprinting microbial communities is *terminal restriction fragment length polymorphism* (t-RFLP) (Liu et al., 1997; Clement et al., 1998). t-RFLP relies on the discrimination of sequences by restriction enzymes, which fractionate DNA molecules according to the location of specific restriction sites. The resulting fragments are separated by size. Individual sequences are differentiated because one of the PCR primers carries a fluorescent label so that the fragment adjacent to this primer can be detected when it migrates past a laser detector on a DNA sequencer. Although a convenient way to compare the diversity of different samples, t-RFLP is a poor method to accurately resolve phylogenetic differences (Dubar et al., 2001). Increasingly, *DNA microarrays* are being used to rapidly assess the microbial diversity of an environmental sample (Wu et al., 2006).

An important caveat in the estimation of microbial diversity by 16S-based approaches is that 16S rRNA genes can occur in multiple, and in some cases, divergent, copies per genome (Klappenbach et al., 2001). Other potential complications to accurate diversity assessment through PCR-based 16S techniques include potential biases in nucleic acid recovery due to differences in the ability of cells to lyse based on differential membrane durability; sorption of cells onto particles or burial within a complex matrix making them inaccessible; and co-extraction of contaminants that may hamper downstream processing, shearing of DNA leading to chimeric products, and errors in the PCR reaction itself (Acinas et al., 2005). For a more complete discussion of these issues, we refer the reader to von Wintzingerode et al. (1997).

8.2.2 New Culturing Techniques

Although molecular approaches have revolutionized our ability to detect organisms in nature, ultimately, if we are to understand how they operate in any detail (see Section 8.4), we must be able to study them in culture. For many years, the *great plate count anomaly* (i.e., the failure of the majority of cells from nature to form colonies on plates, despite being visualized by microscopy as viable) vexed microbiologists interested in studying dominant organisms from the environment (Eilers et al., 2000). In recent years, however, a number of important advances have been made toward this end. These include using diffusion chambers in media that simulate the composition of the natural environment (Kaeberlein et al., 2002), high-throughput procedures for isolating cell cultures through the dilution of natural microbial communities into very low nutrient media (Rappe et al., 2002), and encapsulating cells in gel microdroplets for massively parallel cultivation under low nutrient flux conditions, followed by flow cytometry to detect microdroplets containing microcolonies (Zengler et al., 2002). Two key concepts underlying the success of these methods are (1) the composition of designed media must approximate the composition of the natural environment as closely as possible, and (2) organisms often need to be cultured in combination with others, as cross-feeding between cells may be essential for their growth and not readily simulated in the laboratory. Increasing attention on the part of geochemists to characterize the composition of the environments in which microorganisms live is helping geomicrobiologists design more appropriate media for their cultivation (Svensson et al., 2004).

8.3 WHAT ARE THEY DOING? DEDUCING ACTIVITIES OF GEOMICROBIAL ORGANISMS

Although the use of molecular signatures to identify specific organisms in the environment has revolutionized microbial ecology, these molecules can only tell us *who is there* but cannot inform us about their metabolic activities. To infer what geomicrobial organisms are doing in any particular locale, the classic techniques described in Chapter 7 have great value. In addition, several techniques have been developed in recent years that permit concomitant molecular identification of geomicrobial organisms and their metabolic activity (Figure 8.1). Some of these techniques are even operable at the single-cell level. In general, they all rely on three steps: (1) identifying the

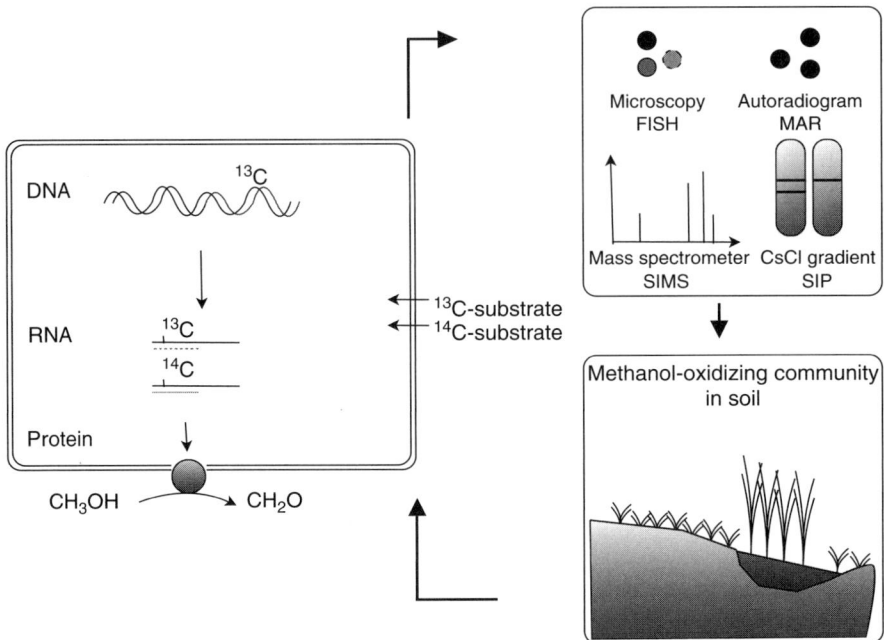

FIGURE 8.1   An illustration of some of the techniques described in Section 8.3. Starting with an environmental sample (e.g., a methanol oxidizing soil community), a labeled substrate (labeled either with stable isotopes or radioisotopes) is provided to the cells within the sample. Cells that can metabolize this substrate incorporate the label into DNA, RNA, or other cellular constituents. Depending on the method used, cells that incorporate this label are identified using MAR or secondary iron mass spectrometry (SIMS), and their phylogenetic identify is confirmed using fluorescent probes that bind to their DNA or RNA (FISH). Before attempting FISH-MAR or FISH-SIMS, SIP can be used to identify the organisms that take up the substrate of interest. See text for details.

geomicrobial organism(s) of interest using a molecular label; (2) separating this population from other organisms in the sample; and (3) analyzing this population with respect to its isotopic composition and its genomic content (to infer potential metabolic functions).

8.3.1   SINGLE-CELL ISOTOPIC TECHNIQUES

The first attempts to combine molecular identification (using 16S rRNA) with isotopic analysis of single cells were made in 1999 by two different groups (Lee et al., 1999; Ouverney and Fuhrman, 1999). Both employed FISH (described in Section 8.2) with microautoradiography (MAR), and enabled direct microscopic observation of whether particular cells had consumed a radioactively labeled substrate under specific incubation conditions or not. In the past several years, this technique has been perfected and employed in many different contexts, from single cells to biofilms (for a recent review, see Wagner et al., 2006). Recently, a quantitative FISH-MAR technique was developed and used to measure the substrate affinity, K_s, of uncultured target organisms in activated sludge (Nielsen et al., 2003). FISH-MAR only measures assimilation of the radiolabeled substrates into macromolecules; unincorporated labeled compounds are not retained by the FISH fixation process. Nevertheless, FISH-MAR is a very sensitive technique because radiotracers are incorporated into *all* macromolecules, not just nucleic acids. An advantageous consequence of this is that FISH-MAR requires relatively short incubation times (generally a few hours), which minimizes the possibility of substrate breakdown and cross-feeding of secondary organisms.

The main limitations of FISH-MAR for geomicrobial studies are (1) it depends on *a priori* knowledge about the phylogenetic affiliation of the studied organism(s); (2) the number of populations that can be specifically detected in a single FISH experiment is limited by the number of available fluorophores that can be applied simultaneously (currently <10); (3) some geomicrobial samples may not be well suited for FISH analysis, as sediment/soil particles might obscure the resident bacteria; and (4) the radioactive substrates of geomicrobial interest may not be available. The first, second, and third limitations can be circumvented by molecular techniques that cast a broader net, such as stable isotope probing (SIP) or the *isotope array* (described later); the fourth limitation may be circumvented in some cases by simultaneously incubating the sample with unlabeled complex substrate and labeled $^{14}CO_2$. This technique, called *HetCO$_2$-MAR*, takes advantage of the fact that most heterotrophs assimilate CO_2 during biosynthesis in various carboxylation reactions. In addition, HetCO$_2$-MAR may permit better differentiation between substrate incorporation without growth (e.g., into storage compounds) and actual growth than traditional FISH-MAR because growth is thought to be required for active carboxylation reactions. An important caveat to this approach, however, is that autotrophs can rapidly consume the labeled $^{14}CO_2$ substrate, so if heterotrophs are the target, specific autotrophic inhibitors must be added to avoid false positives. Another potential limitation of FISH-MAR is that it requires active cellular incorporation of the substrate (important organisms could be missed if populations of interest are metabolically quiescent at the time of sampling). The more general caveats to FISH discussed in Section 8.2 earlier, also apply.

A different single-cell approach that directly couples isotopic and phylogenetic analysis is FISH-secondary ion mass spectrometry (FISH-SIMS) (Orphan et al., 2001). In this technique, fluorescent rRNA-targeted probes are applied to identify the microbial cells of interest in an environmental sample (FISH). The precise location of hybridized cells, which are fixed on a glass or silicon slide round, is mapped, and then the slide is transferred to an ion microprobe (e.g., SIMS, nanoSIMS, time-of-flight-SIMS [ToF-SIMS]) to measure their stable isotopic composition. The SIMS works by using a focused primary ion beam (typically Cs^+) to sputter the surface of a specimen, thereby ejecting secondary ions that are measured with a mass spectrometer to determine the elemental, isotopic, or molecular composition of the specimen. SIMS is a highly sensitive surface analysis technique, able to detect elements in the parts per billion range from samples with a primary beam spot size ranging from tens of nanometers to tens of micrometers. Depending upon the ion microprobe instrument used, data may be acquired with a static primary beam, resulting in either an average isotopic value for the specimen, or if analyzing a cell aggregation or biofilm, time-dependent isotopic profiles may be obtained. Alternatively, ion imaging using nanoSIMS or multi-isotope imaging mass spectrometry (MIMS) also shows promise, enabling the visualization of biologically relevant ions (elemental or isotopic) within microorganisms, structured microbial assemblages, or tissues (Lechene et al., 2007; Li et al., 2008). ToF-SIMS coupled to a gallium beam has also been used successfully for micron-scale ion imaging (Cliff et al., 2002) and, recently, has been extended to the characterization of archaeal lipid biomarkers in environmental samples with a spatial resolution approaching 5 μm (Thiel et al., 2007).

Although, in principle, it is possible to measure a variety of biologically interesting stable isotopes and elements with SIMS, including isotopes of C, N, H, O, S, Ca, and Fe, to date, FISH-SIMS has only been used to measure natural abundance $\delta^{13}C$ in single cells collected from the environment. Like FISH-MAR, this method is compatible with stable isotope labeling experiments and has been used successfully with both ^{13}C and ^{15}N labeling. For example, Orphan et al. (2001) used FISH-SIMS to measure the $\delta^{13}C$ profiles from the periphery to the interior of individual syntrophic consortia from marine sediments at methane seeps. These studies revealed extremely low values of $\delta^{13}C$ in the inner cores of layered cell aggregates, identified as belonging to an uncultured group of archaea, known as the ANME-2. Surrounding these cells, were sulfate-reducing bacteria of the *Desulfosarcina-Desulfococcus* clade. This work provided strong evidence of metabolic syntrophy between these organisms catalyzing the anaerobic oxidation of methane. Although the light ^{13}C signal affiliated with methane-oxidization is diagnostic and readily interpreted by SIMS analysis, other forms of C metabolism may be more difficult to interpret without the use of labeled substrate additions. Additional variables to be considered for natural abundance stable isotope measurements in single cells include the isotopic effect of exogenous carbon added during cell fixation and during the hybridization with the fluorescently labeled oligonucleotide probe.

The major limitation of single-cell isotopic techniques is that they can only provide information about the metabolism of substrates that can be labeled and traced isotopically. This handicaps our ability to infer an integrated picture of metabolism for a single cell. Toward this end, two new PCR-based methods have been developed that permit the isolation of the genomic content of single cells from complex environmental samples. The first such technique is termed *microfluidic digital PCR*, and couples 16S rDNA identification of single, uncultivated cells with the simultaneous detection of other genes in these cells. Cells are diluted and separated from a complex mixture using a microfluidic device. This separation step is critical, as single, partitioned cells then serve as templates for individual multiplex PCR reactions. Although to date, this technique has been used to link the presence of a particular metabolic gene of interest with single bacterial species residing in the hindguts of wood-feeding termites (Ottesen et al., 2006), in principle, it can be applied to link any number of genes to 16S rDNA–identified single cells. This technique operates independently of the physiological state of the cell at the time of harvest, gene expression or position on the genome. Moreover, the use of single-cell PCR avoids PCR artifacts such as amplification biases and unresolvable chimeric products (described in Section 8.2) (Ottesen et al., 2006). Subsequent to the termite hindgut study, similar microfluidic techniques were used to isolate and identify organisms from the human subgingival crevice. In this work, genomes were amplified from individual cells of interest, permitting the sequencing and assembly of >1000 of their genes (Marcy et al., 2007). As genome-sequencing methods improve for single cells (Hutchinson and Venter, 2006), we can anticipate even better genomic recovery in the future. The strength of this approach is that it allows specific organisms that have the potential to catalyze interesting reactions (as measured by the presence of particular genes encoding the function[s] of interest) to be identified from complex environmental samples, assuming no *a priori* knowledge of which organisms are involved. In this respect, it is similar to the isotope array or SIP, only with microfluidic digital PCR, the cell as a distinct informational entity is preserved.

Another technique that has been developed recently (Pernthaler et al., 2008) is called *magneto-FISH*. Magneto-FISH combines 16S rRNA–targeted FISH and a tyramide-based amplification reaction (CARD-FISH) (Pernthaler et al., 2002), with immunomagnetic capture of hybridized cells using paramagnetic beads coated with an antibody targeting the fluorochrome applied in the CARD-FISH procedure. The selective capture of target organisms from complex environmental samples, combined with metagenomic analysis and PCR surveys of diagnostic metabolic genes, can provide information regarding the metabolic potential of uncultured microorganisms/microbial assemblages recovered directly from the environment. Pernthaler et al. (2008) used this method to purify a specific lineage of syntrophic anaerobic methane oxidizing archaea from marine sediment, along with their associated partner bacteria. Subsequent metagenome sequencing, PCR, and microscopy of these captured consortia revealed unexpected diversity in the associated bacteria, and yielded new insights into the versatility and metabolism of this syntrophic association. A unique advantage of the magneto-FISH method is that it enables the recovery of the complete genetic content of uncultured target organisms, and allows native partnerships to be identified by physical coassociation. A caveat is that spurious microorganisms may potentially be copurified along with the true consortia. Although physical association alone cannot be used as exclusive proof of a syntrophic relationship, inferred associations and metabolic potential revealed by magneto-FISH and metagenomics identify new areas for follow-up investigations.

8.3.2 SINGLE-CELL METABOLITE TECHNIQUES

So far, we have discussed single-cell techniques that employ fluorescent probes to identify organisms phylogenetically, providing a handle for subsequent isotopic or genomic analysis to infer function.

A large number of fluorescent probes also exist that can potentially be used to measure other parameters of interest including pH, membrane potential, enzyme activity, reactive oxygen and nitrogen species, toxins, and viability. A commercial supplier of such probes is Molecular Probes®, which specializes in producing fluorescent reagents to measure a variety of attributes of living cells. For example, the popular LIVE/DEAD stain (or *Bac*Light kit) is used to measure the viability of cells in environmental samples, and respiratory activity can be visualized using the dye 5-cyano-2,3-ditolyl tetrazolium chloride (CTC). We note, however, that often these probes are developed in yeast or *Escherichia coli*, and their application to geomicrobial organisms has not been calibrated. This means one must be cautious when using these probes, as conventional interpretations of what they signify may not apply (e.g., see Teal et al., 2006 and for a discussion of their use in biofilm systems, see a recent review by Stewart and Franklin, 2008). Nevertheless, the application of fluorescent probes developed in other systems to geomicrobiology has the potential to greatly expand our appreciation of cell biological processes in diverse organisms (see Section 8.4). For a probe to be successful, however, it must be able to localize to a proper place in the cell (or microbial community), where it can measure the attribute of interest. A need exists for probe development to make cell biological studies of this kind possible for diverse geomicrobial organisms. In the future, it is possible that endogenous metabolites that exhibit a diagnostic and measurable fluorescent signal, may even be detected directly inside living cells.

8.3.3 COMMUNITY TECHNIQUES INVOLVING ISOTOPES

Until now, this section has dealt with techniques that are performed at the single-cell level. Although this can be advantageous, there are times when a coarser degree of resolution is more helpful in approaching a problem in geomicrobiology. For example, if one is interested in a particular metabolic process, but does not know which organisms are involved, methods that employ FISH are not as helpful as techniques that do not require *a priori* knowledge of identifying sequences. Two primary methods exist for this purpose: *SIP* and the *isotope array*.

SIP was first applied to the analysis of PLFAs. Because groups of microorganisms often have diagnostic PLFA molecules, it is possible to identify those that are enriched in naturally abundant stable isotopes such as ^{13}C or ^{15}N. If one knows the source of the label, one may conclude that the organism (identified by its PLFA molecules) metabolized the substrate. Petsch et al. used this technique to make a case for the involvement of a complex microbial community in the weathering of organic material from late Devonian black shales (Petsch et al., 2003). However, because nothing is known about PLFA patterns for microorganisms for which there are no cultured representatives, PLFA molecules are not as useful biomarkers as nucleic acids. Accordingly, most SIP studies today employ isotopically labeled DNA or RNA as the phylogenetic hook. Because the buoyant density of DNA varies with its guanine and cytosine (GC) content, the incorporation of a heavy isotope (e.g., ^{13}C) into DNA enhances the density of labeled DNA compared to unlabeled (e.g., ^{12}C) DNA. A band of heavy DNA can thus be separated from the unlabeled population, cloned and sequenced to identify organisms involved in the metabolism of the labeled substrate using PCR-primers targeting 16S rDNA. Alternatively, PCR can amplify other genes that might play a role in catalyzing the metabolism of the substrate. DNA-SIP was first made use of to identify organisms involved in methanol consumption in forest soil (Radajewski et al., 2000). These initial studies involved artificially high substrate concentrations and relatively long incubation times (~40 days), which raised the issue that such exposure might also result in the detection of organisms that used labeled intermediates or by-products generated by the primary consumer organisms. If one is interested in identifying (syn)trophic interactions, however, this may be useful. RNA-based SIP can potentially mitigate this problem. Regardless of whether primary consumers or downstream cross-feeders are the target, one powerful attribute of SIP is that ^{13}C-labeled DNA can be cloned to generate a metagenomic library. Unlink magneto-FISH, which enriches for genomes of interest based on 16S phylogeny, DNA-SIP can enrich for genomes based on the incorporation of specific substrates into the DNA itself.

For thorough recent reviews of DNA and RNA SIP, with more complete discussions of their advantages and limitations see (Dumont and Murrell, 2005) and (Whiteley et al., 2006).

One potential drawback to SIP is that it requires high levels of ^{13}C incorporation (>50%) to enable separation of the nucleic acids of interest using standard techniques, which makes it insensitive to slowly growing organisms (or organisms using multiple substrates). A new technique that permits precise isotopic analysis of nanogram amounts of samples has recently been developed (Sessions et al., 2005) that may be useful in this context. Known as *spooling wire microcombustion* (SWiM), this technique has been used to analyze both rRNA, specifically captured to target particular phylogenetic groups (Pearson et al., 2008) and whole cells sorted by flow cytometry (i.e., *fluorescence-activated cell sorting* [FACS]) (Eek et al., 2006). The method is compatible both with substrates containing a natural ^{13}C abundance, and with artificially ^{13}C-enriched tracers. In brief, a single droplet of liquid sample is placed onto a continuously, slowly spooling wire that is carried into a combustion reactor where nonvolatile organic matter is quantitatively oxidized to CO_2, H_2O, and NO_x. CO_2 is then carried in a helium gas stream to an isotope ratio mass spectrometer (IRMS) for measurement of the $^{13}C/^{12}C$ ratio. Although remarkably sensitive, an important practical limitation of this FACS-SWiM marriage is the difficulty of separating and accumulating sufficient biomass (~10^7 bacterial cells or 10^4 eukaryotic cells, Eek et al., 2006) to make the measurement, particularly for cells that are not naturally autofluorescent. To detect these cells, exogenous amplified fluorophores (such as those used in CARD-FISH reactions) must be applied, but this appears to add too much exogenous carbon to the cell to yield an accurate $\delta^{13}C$ value. The development of brighter fluorescent labels for whole cells that do not add appreciable carbon is needed to make FACS-SWiM more broadly applicable.

An alternative to using stable isotopes to identify organisms involved in the uptake of particular substrates is to use radioisotopes. This was described earlier for single cells (FISH-MAR), but microarrays can also be used to screen microbial communities. In the isotope array, radiolabeled RNA (typically containing ^{14}C) is extracted from a sample, labeled with a fluorophore and probed against an oligonucleotide microarray that targets 16S rRNA of different microorganisms. In this manner, organisms that can take up the label can be rapidly identified, without *a priori* knowledge of the community's composition (Adamczyk et al., 2003). The isotope array is a high-throughput method and requires only a short incubation time. But unlike SIP, the isotope array can only detect organisms whose sequences are represented on the array, so in this regard, it is more limited. As noted previously in the discussion of limitations to FISH-MAR, organisms that are metabolically quiescent (or simply growing slowly) at the time of sampling would be missed, and the technique is limited to substrates for which there is a radiolabel.

8.3.4 COMMUNITY TECHNIQUES INVOLVING GENOMICS

Although isotopic labeling can be a powerful way to link specific metabolic capabilities with particular organisms in the environment, it is limited to metabolisms for which there are isotopic tracers. In the absence of these, one can use genomes to infer potential metabolic capabilities or the ability to catalyze geochemically significant reactions. We emphasize the word *potential*, as genomes, at best, can only hint at what might be possible for a given organism. This type of analysis (often called *metagenomics*; Riesenfeld et al., 2004; DeLong, 2007) relies heavily upon computational analyses to compare different sequences to each other and to identify motifs in the DNA or the gene products that are predicted to have a specific function. Hypotheses can be generated about what types of reactions a given protein might catalyze, or the conditions under which the gene that encodes it might be expressed; sometimes, genomic analysis can even be used to make predictions about the behavior of entire microbial communities (Tyson et al., 2004; DeLong et al., 2006). Ultimately, these hypotheses must be tested through classical genetic and biochemical analyses to prove that the connection between the presence of a particular gene and an inferred metabolism/predicted geochemical effect is actually causal as opposed to correlative (see Section 8.4 and Martinez et al., 2007).

Various methods exist that enable genomic information to be acquired from specific organisms in the environment (e.g., magneto-FISH, digital PCR). In cases where the cellular host has been lost (e.g., SIP or simply DNA-purified in bulk from the environment without isotopic selection and cloned into *cosmid, fosmid* or *bacterial artificial chromosome* [BAC] libraries), metagenomic information can still be assigned to specific organismal entities if sufficient DNA is cloned to permit the assembly of putative whole genomes. We underscore the word *putative*, as it remains very challenging, despite advances in bioinformatics, to reconstruct original genomes from environmental DNA. Despite these challenges, environmental genomic data can be extremely powerful. Not only does it permit gene expression in communities to be monitored *in situ* (Ram et al., 2005), but it allows the tracking of specific genotypes and their close relatives in their natural environment (Rich et al., 2008) and may guide the design of media to enable the isolation/study of particular geomicrobial organisms (Tyson et al., 2005; Sabehi et al., 2005). For a more complete discussion of the opportunities and challenges of environmental genomics in the context of geomicrobial problems, see Banfield et al. (2005).

8.3.5 Probing for Expression of Metabolic Genes or Their Gene Products

If one is interested in the expression of a particular metabolic gene or its gene product, specific probes can be designed to detect it. For example, PCR primers designed to target archaeal *amoA* (the gene encoding ammonia monooxygenase alpha-subunit) revealed the widespread presence of ammonia-oxidizing archaea in marine water columns and sediments, suggesting that these organisms may play a significant role in the global nitrogen cycle (Francis et al., 2005). Another example is the use of PCR primers targeting the *arrA* gene, which encodes one of the subunits of the respiratory arsenate reductase common to most arsenate-respiring bacteria, to detect the expression of this gene in arsenic-contaminated sediments (Malasarn et al., 2004). In these studies, bulk mRNA was extracted from sediments and reverse transcribed to DNA prior to PCR analysis. Although these studies looked for the expression of genes in the environment qualitatively, it is possible to quantify gene expression using *real-time PCR* or *quantitative PCR* (q-PCR) (Walker, 2002). Such an approach was used to quantify niche partitioning among *Prochlorococcus* ecotypes along ocean-scale environmental gradients (Johnson et al., 2006).

A caveat to PCR-based approaches is that, in the case of proteins, posttranscriptional controls may affect the abundance (integrated production and decay) of the gene products of interest; accordingly, detection of a DNA or RNA sequence in a given environment cannot prove that its product is present in the environment. A better check for this is to use antibodies targeted against the protein of interest. One example of where this has been used is in the area of (per)chlorate reduction, where immunoprobes were raised against the enzyme chlorite dismutase (CD) (O'Connor and Coates, 2002). CD is highly conserved among the dominant (per)chlorate-reducing bacteria in the environment, thus CD appears to be a good target for a probe specific to this process. If one seeks spatial resolution of gene/protein expression within microbial communities, FISH can be used, with the caveat that signal amplification is usually needed to detect low-abundance mRNA or protein (Stewart and Franklin, 2008).

8.4 HOW ARE THEY DOING IT? UNRAVELING THE MECHANISMS OF GEOMICROBIAL ORGANISMS

As discussed in Sections 8.2 and 8.3, molecular biology has found wide application in geomicrobiology by helping investigators determine which microorganisms are present in a given environment and what their (potential) metabolic activities are in that environment. Underlying many of these techniques are assumptions about the biological function of specific molecules in the cell (e.g., rDNA is a good phylogenetic marker because it is known to be an essential, highly conserved, and slowly evolving constituent of the replication machinery common to all life; measuring the expression of specific metabolic genes in the environment presupposes we know which genes to look for, and what their gene products do). How does one elucidate these genes in the first place, and determine their cellular function? This is the province of classical genetics, biochemistry, and cell biology—fields whose tools have been developed and honed over many decades in a variety of model systems ranging from *Escherichia coli* to yeast to fruit-flies to mice.

Because geomicrobially significant organisms have often been assumed to be difficult to tame in the laboratory (although this may simply reflect a lack of trying rather than an inherent problem), less is currently understood about how they catalyze geobiologically significant processes. But this is changing rapidly. The past decade has seen heightened interest in exploring the genetics, biochemistry, and cell biology of geomicrobial organisms. Owing to advances in culturing, the plummeting costs of sequencing microbial genomes, and the entrance into the field of students with training in genetics, biochemistry, and cell biology, geomicrobiology is now primed for exciting discoveries on the mechanistic level. Why does mechanism matter? Because it permits a variety of interesting questions to be asked, including What genes are required for a process of interest? What are the products of these genes? Where do they reside in the cell? How do they function (e.g., what molecular partners—cofactors, other proteins—are required)? If they are enzymes, what are their catalytic rates? And under what conditions are they expressed? Ultimately, the answers to these type of questions will permit geomicrobiologists to return to the environment to search for the presence and expression of particular genes/proteins once their function is known (e.g., Karkhoff-Schweizer et al., 1995; Malasarn et al., 2004; Sabehi et al., 2005).

8.4.1 GENETIC APPROACHES

There are many ways one can use genetics to study living cells. Here, we restrict our discussion to the genetic approaches that are routinely applied in geomicrobiology: the construction of mutants, the expression of DNA in a foreign host, and the labeling of specific genes to visualize their expression or the cellular location of their products.

Mutagenesis is used to identify genes essential for catalyzing a process of interest. Various methods for mutagenesis exist, offering the potential to eliminate/delete genes entirely, introduce point mutations into specific genes, or introduce genes into an organism. The effects of these different types of mutations can be far ranging, from altering the amino acid composition of a protein and thereby affecting its substrate specificity, to changing the regulation of an entire network of genes, to eliminating the ability to make a set of proteins. The latter type of mutagenesis represents a *loss-of-function* approach, and is commonly used to identify genes essential for a given function. Examples of where this has been used in geomicrobiology include work on phototrophic iron oxidation (Jiao et al., 2005; Jiao and Newman, 2007), manganese oxidation (van Waasbergen et al., 1996), iron reduction (Coppi et al., 2001; DiChristina et al., 2002; Myers and Myers, 2002), arsenate reduction (Saltikov and Newman, 2003; Murphy and Saltikov, 2007), magnetosome formation (Komeili et al., 2004; Scheffel et al., 2008), and methanogenesis (Pritchett and Metcalf, 2005). Many other examples are provided elsewhere in this book.

Mutagenesis can either be random or directed. To make random mutations, *transposons* are often used. Transposons are mobile genetic elements that (usually) insert in an unbiased fashion into the genome of the organism of interest. (*Note:* some transposons such as Tn7, insert in a neutral chromosomal site, and thus can be used to introduce genes in single-copy—see Koch et al., 2001.) They disrupt the sequence of the gene into which they land, thereby destroying the gene product and rendering it dysfunctional. Commonly, transposons carry a constitutively expressed antibiotic resistance marker, so that their insertion may be selected by plating cells on medium containing the antibiotic. Mating between a donor strain (that carries the transposon) and a recipient strain (the one to be mutagenized) is often used for transposon delivery. Transposon libraries must be screened to identify the subset of mutants that have a defect in the *phenotype* (i.e., property or function) of interest. Although chemical mutagens or UV can also be used to make random mutations throughout

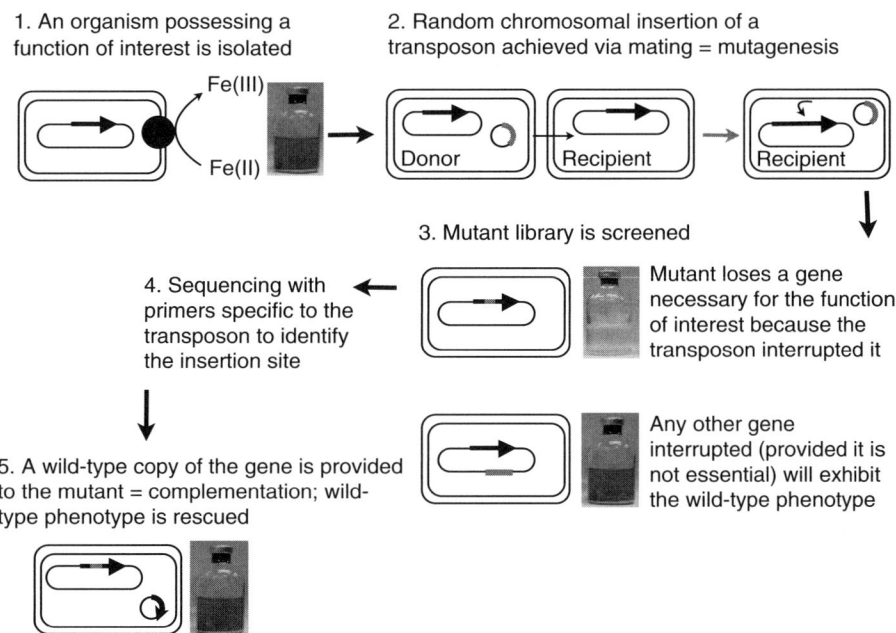

1. An organism possessing a function of interest is isolated

Fe(III)

Fe(II)

2. Random chromosomal insertion of a transposon achieved via mating = mutagenesis

Donor

Recipient

Recipient

3. Mutant library is screened

Mutant loses a gene necessary for the function of interest because the transposon interrupted it

4. Sequencing with primers specific to the transposon to identify the insertion site

Any other gene interrupted (provided it is not essential) will exhibit the wild-type phenotype

5. A wild-type copy of the gene is provided to the mutant = complementation; wild-type phenotype is rescued

FIGURE 8.2 Steps involved in transposon mutagenesis (i.e., a *loss-of-function* assay). As an example, Fe(II) oxidation to Fe(III) is provided as the phenotype of interest. The blue fragment represents the part of the chromosome that contains the gene(s) that encodes the function of interest. The yellow fragment represents the transposon. Qualitatively, bottles turn a rusty brown color when iron oxidation occurs. See text for details.

the genome, the advantage of transposons is that their site of insertion can be readily mapped using molecular techniques, permitting identification of the genes into which they insert. For many applications in geomicrobiology, this is sufficient to determine which genes control a process of interest (Figure 8.2). If one seeks to learn more about the importance of particular residues within a given gene, however, transposons are not appropriate; in this case, *site-directed mutagenesis* (a controlled technique for generating mutations, including point mutations) or *chemical mutagenesis* (the addition of chemicals that promote modifications of DNA at the scale of single residues) must be used. For example, these more highly resolved mutagenic techniques would be appropriate if one wanted to understand the function of a specific protein in greater detail (e.g., which sites are necessary for it to bind its substrate, which are required for the binding of cofactors, and which control its ability to associate with other molecules in the cell).

After mutagenesis is performed, mutants are identified either through a selection or a screen. A *selection* permits only those mutants that have the desired properties to grow, whereas a *screen* requires characterizing the behavior of thousands of mutants to identify those that have the phenotype of interest. Depending on the manner in which one has identified candidate mutants, secondary screens may be required to narrow the pool of candidates down to only those that are interesting. For example, if one performs a screen to find genes that control various steps in a biochemical reaction and if the assay for mutant identification involves looking at the rate at which a reaction proceeds, then *false* mutants could be identified by the screen that are simply slow growers but which do not have a specific defect in the reaction of interest. These mutants could be sorted out by measuring the growth rate of all candidates and only continuing to study those whose growth were normal with respect to the *wild-type* (i.e., parent) strain. Once interesting mutants are identified, the nature of the mutation must be determined through sequencing and genetic verification. Often this includes a *complementation* experiment, where a wild-type copy of the gene of interest is put back into the mutagenized strain (often on a plasmid), to demonstrate that it can restore the original phenotype.

In the event that the wild-type phenotype is not rescued by this experiment, it means that the mutation responsible for the phenotype of interest lies elsewhere on the chromosome. Sequence analysis can help generate hypotheses to explain why the mutant behaves the way it does (see the discussion on bioinformatics in Section 8.4.2), and thus infer what affects the process of interest. To test these hypotheses, however, physiological, biochemical, or cell biological experiments are required.

However, not every organism is amenable to mutagenesis. For a practical discussion of this, see Newman and Gralnick (2005). An alternative strategy to identify genes from these organisms (or from environmental genomic sequences for which an organismal host may or may not be known) that are necessary and sufficient to catalyze a particular reaction is to clone and express them in a foreign host (i.e., *heterologous complementation*). This is a *gain-of-function* strategy, in contrast to the loss-of-function mutagenesis approach described above (Figure 8.3). Toward this end, the host must have the metabolic machinery necessary to process the gene product(s) one wishes to express. For example, if one seeks to express genes encoding multiheme cytochromes, such as those found in *Geobacter* and *Shewanella*, in a genetically tractable foreign host such as *Escherichia coli*, *E. coli* must be capable of processing these cytochromes in addition to its own cytochromes; this may be accomplished by engineering *E. coli* to express extra versions of the machinery for cytochrome

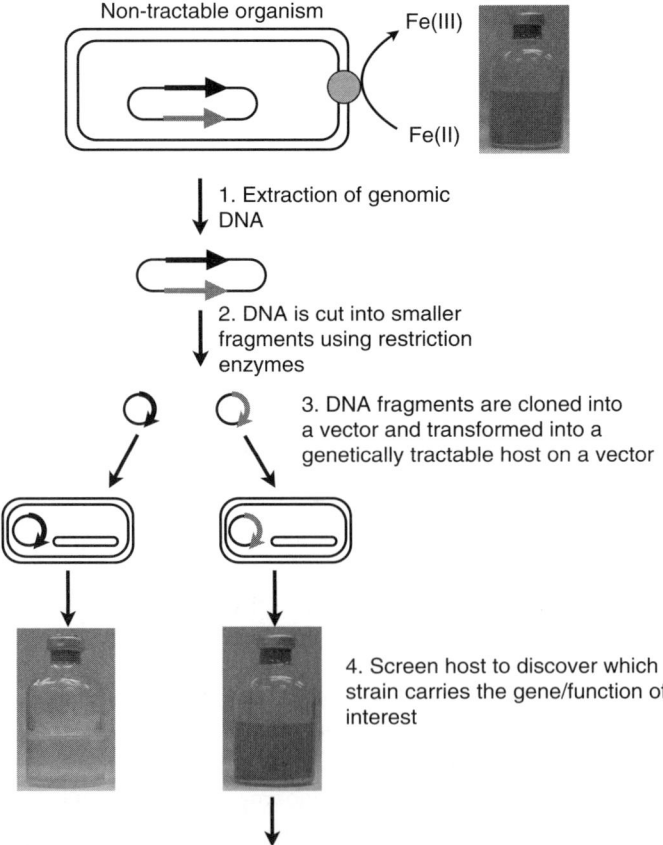

FIGURE 8.3 Steps involved in heterologous complementation (i.e., a *gain-of-function* assay). As an example, Fe(II) oxidation to Fe(III) is provided as the phenotype of interest. The red, blue, and green fragments symbolize different parts of the chromosome. The green fragment represents the one that contains the gene(s) that encodes the function of interest. Qualitatively, bottles turn a rusty brown color when iron oxidation occurs. See text for details.

assembly (Londer et al., 2006). Examples where a gain-of-function approach has been used to identify and elucidate the function of genes of geomicrobial interest include work on phototrophic iron oxidation (Croal et al., 2007) and marine proteorhodopsin (Beja et al., 2000; Martinez et al., 2007). The latter example was also discussed previously in the context of community genomics in Section 8.3.

8.4.2 Bioinformatic Approaches

Help in understanding the specific function of genes identified through mutagenesis can come from analyzing the amino acid sequence encoded by the gene. A tool that can provide significant clues is the *basic local alignment searchtTool* (BLAST) search engine available at National Center for Biotechnology Information (NCBI—http://www.ncbi.nlm.nih.gov). BLAST can be used to infer functional and evolutionary relationships between amino acid or nucleotide sequences. This program compares sequences entered by the user to a selected database, which can include all known sequences. The program assigns a statistical significance to matches within the database. If the gene of interest encodes a protein with a significant match to a protein of known function in the database, this may suggest it has a similar activity. If the protein has no significant match or is only similar to other proteins of unknown function (a common occurrence), there are several additional analyses that can be performed on the sequence to gain insight into its function (although these do not always help). Other useful types of searches are motif, posttranslational modification, and topology. Several programs in each category can be found on the *Expert Protein Analysis System* (ExPASy) web server (http://www.expasy.org/tools), *SoftBerry* (http://linux1.softberry.com) and *BioProspector* (http://ai.stanford-edu/~Xsliu/Bioprospector). Programs found on these sites will allow further analysis of the protein sequence of interest to predict characteristics such as subcellular localization, cofactor binding, and transmembrane domains. For example, if the protein is predicted to bind a redox-active cofactor such as heme, we may hypothesize that it plays a role in electron transfer. It is important to remember that database predictions are only suggestive, and must be verified experimentally; however, programs such as these greatly help one formulate testable hypotheses.

8.4.3 Follow-Up Studies

Once mutants have been identified and the involvement of particular genes in the process of interest verified, many avenues become open for exploration. By identifying a variety of mutants with similar phenotypes, one can begin to construct a model for how a process works at the molecular level. Studies can be initiated to determine how genes that catalyze a process of interest are regulated by environmentally relevant factors (e.g., Saltikov et al., 2005). Putative marker genes for this activity may be identified, potentially leading to the design of molecular tools (e.g., PCR probes, antibodies) to monitor when this process is occurring in a given environment (see the discussion at the end of Section 8.3). Once regulatory elements (e.g., promoters) are known for a gene, a strain can be engineered to *report* when it is expressing that gene. For example, a promoter from a gene of interest can be cloned into a plasmid and used to drive expression of a protein that can be detected by fluorescence or colormetric assay (e.g., the expression of *gfp*, which encodes the green fluorescent protein [GFP] or *lacZ*, which encodes beta-galactosidase). Manipulating various environmental conditions in the laboratory can provide precise information regarding when the engineered strain is expressing the gene of interest by quantifying fluorescence or β-galactosidase activity. This is an alternative to quantifying gene expression using q-PCR, discussed in Section 8.3. While q-PCR can only be used to monitor gene transcription, gene fusions can be used to monitor both transcription and translation. In the latter case, the reporter gene is fused to the gene that encodes the protein of interest, such that when it is expressed, it carries the reporter tag. Depending on the nature of this tag, the translated fused product can be visualized directly within living cells (e.g., if GFP-fusions are made; see Komeili et al., 2004), or detected via antibodies to a particular epitope (e.g., hemagglutinin [HA-tag]; see Gralnick et al., 2006). Finally, the biochemical properties of the gene product can be studied, either within the host strain, or by cloning and over-expressing the gene that encodes it in another organism. Many interesting mechanistic questions can be answered at this stage. For example, if the protein is an enzyme, one can determine its substrate specificity and kinetic parameters. Once a method for purifying the protein has been developed, one can study it in great depth—from solving its crystal structure to determining what other proteins it interacts with in the cell.

8.5   SUMMARY

The application of molecular biology to geomicrobiology has greatly expanded our appreciation of microbial diversity. Thanks to the wealth of phylogenetic information now available, identification of novel organisms through their 16S rDNA sequence has become routine. Not only is it possible to detect organisms of geomicrobial interest in a wide range of habitats, but the coupling of phylogenetic analysis to isotopic probing and metagenomic sequencing (both at the level of communities and single cells) can provide clues into what these organisms may be doing *in situ*. Advances in culturing methods, and media development informed by geochemical analysis, promise to improve the cultivation success rate of an increasing number of geomicrobially significant organisms. Finally, the application of classical genetic, biochemical, and cell biological methods to the study of these organisms (or even the DNA of uncultivated organisms) can reveal important information on how they function, and guide the development of strategies to detect and quantify their activities in nature.

REFERENCES

Acinas SG, Sarma-Rupavtarm R, Klepac-Ceraj V, Polz MF. 2005. PCR-induced sequence artifacts and bias: Insights from comparison of two 16S rRNA cone libraries constructed from the same sample. *Appl Environ Microbiol* 71:8966–8969.

Adamczyk J, Hesselsoe M, Iversen N, Horn M, Lehner A, Nielsen P, Scholter M, Roslev P, Wagner M. 2003. The isotope array, a new tool that employs substrate-mediated labeling of rRNA for determination of microbial community structure and function. *Appl Environ Microbiol* 69:6875–6887.

Amann RI, Krumholz L, Stahl, DA. 1990. Fluorescent-oligonucleotide probing of whole cells for determinative, phylogenetic, and environmental studies. *J Bacteriol* 172:762–770.

Amann RI, Ludwig W, Schleifer K-H. 1995. Phylogenetic identification and in situ detection of individual microbial cells without cultivation. *Microbiol Rev* 59:143–169.

Amann RI, Zarda B, Stahl DA, Schleifer K-H. 1992. Identification of individual prokaryotic cells using enzyme-labeled, rRNA-targeted oligonucleotide probes. *Appl Environ Microbiol* 58:3007–3011.

Banfield JF, Verberkmoes NC, Hettich RL, Thelen MP. 2005. Proteogenomic approaches for the molecular characterization of natural microbial communities. *OMICS-J Integr Biol* 9:301–333.

Beja O, Aravind L, Koonin EV, Suzuki MT, Hadd A, Nguyen LP, Jovanovich SB, Gates CM, Feldman RA, Spudich JL, Spudich EN, DeLong EF. 2000. Bacterial rhodopsin: Evidence for a new type of phototrophy in the sea. *Science* 289:1902–1906.

Bobbie RJ, White CD. 1980. Characterization of benthid microbial community structure by high-resolution gas chromatography of fatty acid methyl esters. *Appl Environ Microbiol* 39:1212–1222.

Braun-Howland EB, Danielsen SA, Nierzwicki-Bauer SA. 1992. Development of a rapid method for detecting bacterial cells in situ using 16S rRNA-targeted probes. *Biotechniques* 13:928–934.

Braun-Howland EB, Vescio PA, Nierzwicki-Bauer SA. 1993. Use of a simplified cell blot technique and 16S rRNA-directed probes for identification of common environmental isolates. *Appl Environ Microbiol* 59:3219–3224.

Clement BG, Kehl LE, Debord KL, Kitts CL. 1998. Terminal restriction fragment patterns (TRFPs), a rapid, PCR-based method for the comparison of complex bacterial communities. *J Microbiol Method* 31:135–142.

Cliff JB, Gaspar DJ, Bottomley PJ, Myrold DD. 2002. Exploration of inorganic C and N assimilation by soil microbes with time-of-flight secondary ion mass spectrometry. *Appl Environ Microbiol* 68:4067–4073

Coppi MV, Leang C, Sandler SJ, Lovley DR. 2001. Development of a genetic system for *Geobacter sulfurreducens*. *Appl Environ Microbiol* 67:3180–3187.

Croal LR, Jiao YQ, Newman DK. 2007. The *fox* operon from *Rhodobacter* strain SW2 promotes phototrophic Fe(II) oxidation in *Rhodobacter capsulatus* SB1003. *J Bacteriol* 189:1774–1782.

DeLong EF. 2007. Metagenomics defined. *Tech Rev* 110:26–27.

DeLong EF, Preston CM, Mincer T, Rich V, Hallam SJ, Frigaard NU, Martinez A, Sullivan MB, Edwards R, Brito BR, Chisholm SW, Karl DM. 2006. Community genomics among stratified microbial assemblages in the ocean's interior. *Science* 311:496–503.

DeLong EF, Wickham GS, Pace NR. 1989. Phylogenetic stains: Ribosomal RNA-based probes for the identification of single cells. *Science* 243:1360–1363.

Delsuc F, Brinkmann H, Philippe H. 2005. Phylogenomics and the reconstruction of the tree of life. *Nat Rev Genet* 6:361–375.

DiChristina TJ, Moore CM, Haller CA. 2002. Dissimilatory Fe(III) and Mn(IV) reduction by *Shewanella putrefaciens* requires ferE, a homolog of the pulE (gspE) type II protein secretion gene. *J Bacteriol* 184:142–151.

Dubar J, Ticknor LO, Kuske CR. 2001. Phylogenetic specificity and reproducibility and a new method for analysis of terminal restriction fragment profiles of 16S rRNA genes from bacterial communities. *Appl Environ Microbiol* 69:2555–2562.

Dumont MG, Murrell JC. 2005. Stable isotope probing—linking microbial identity to function. *Nat Rev Microbiol* 3:499–504.

Eek KM, Sessions AL, Lies DP. 2006. Carbon-isotopic analysis of microbial cells sorted by flow cytometry. *Geobiology* 5:85–95.

Eilers H, Pernthaler J, Glockner FO, Amann R. 2000. Culturability and in situ abundance of pelagic bacteria from the North Sea. *Appl Environ Microbiol* 66:3044–3051.

Fazi S, Amalfitano S, Pernthaler J, Puddu A. 2005. Bacterial communities associated with benthic organic matter in headwater stream microhabitats. *Environ Microbiol* 7:1633–1640.

Ferrari BC, Tujula N, Stoner K, Kjelleberg S. 2006. Catalyzed reporter deposition-fluorescence *in situ* hybridization for enrichment-independent detection of microcolony-forming soil bacteria. *Appl Environ Microbiol* 72:918–922.

Francis CA, Roberts KJ, Beman JM, Santoro AE, Oakley BB. 2005. Ubiquity and diversity of ammonia-oxidizing archaea in water columns and sediments of the ocean. *Proc Natl Acad Sci USA* 102:14683–14688.

Giovannoni SJ, DeLong EF, Olson GJ, Pace NR. 1988. Phylogenetic group-specific oligonucleotide probes for identification of single microbial cells. *J Bacteriol* 170:720–726.

Gralnick JA, Vali J, Lies DP, Newman DK. 2006. Extracellular respiration of dimethyl sulfoxide by *Shewanella oneidensis* strain MR-1. *Proc Natl Acad Sci* 103:4669–4674.

Guyer RL, Koshland DE. 1989. The molecule of the year. *Science* 246:1543–1546.

Hutchinson CA, III, Venter C. 2006. Single-cell genomics. *Nat Biotechnol* 24:657–658.

Jiao YQ, Kappler A, Croal LR, Newman DK. 2005. Isolation and characterization of a genetically tractable photoautotrophic Fe(II)-oxidizing bacterium, *Rhodopseudomonas palustris* strain TIE-1. *Appl Environ Microbiol* 71:4487–4496.

Jiao YQ, Newman DK. 2007. The *pio* operon is essential for phototrophic Fe(II) oxidation in *Rhodopseudomonas palustris* TIE-1. *J Bacteriol* 189:1765–1773.

Johnson Zi, Zinser ER, Coe A, McNulty NP, Woodward EMS, Chisholm SW. 2006. Niche partitioning among *Prochlorococcus* ecotypes along ocean-scale environmental gradients. *Science* 311:1737–1740.

Jurtshuk RJ, Blick M, Bresser J, Fox GE, Jurtshuk PJ. 1992. Rapid in situ hybridization technique using 16S rRNA segments for detecting and differentiating the closely related gram-positive organisms *Bacillus polymyxa* and *Bacillus macerans*. *Appl Environ Microbiol* 58:2571–2578.

Kaeberlein T, Lewis K, Epstein SS. 2002. Isolating uncultivable microorganisms in pure culture in a simulated natural environment. *Science* 296:1127–1129.

Karkhoff-Schweizer RR, Huber DPW, Voordouw G. 1995. Conservation of the genes for dissimilatory sulfite reductase from *Desulfovibrio vulgaris* and *Archaeoglobus fulgidus* allows their detection by PCR. *Appl Environ Microbiol* 61:290–296.

Klappenbach JA, Saxman PR, Cole JR, Schmidt TM. 2001. rrndb: The Ribosomal RNA Operon Copy Number Database. *Nucleic Acids Res* 29:181–184.

Koch B, Jensen LE, Nybroe O. 2001. A panel of Tn7-based vectors for insertion of the *gfp* marker gene or for delivery of cloned DNA into gram-negative bacteria at a neutral chromosomal site. *J Microbiol Method* 45:187–195.

Komeili A, Vali H, Beveridge TJ, Newman DK. 2004. Magnetosome vesicles are present before magnetite formation, and MamA is required for their activation. *Proc Natl Acad Sci USA* 101:3839–3844.

Lanoil BD, Giovannoni SJ. 1997. Identification of bacterial cells by chromosomal painting. *Appl Environ Microbiol* 63:1118–1123.

Lechene CP, Luyten Y, McMahon G, Distel DL. 2007. Quantitative imaging of nitrogen fixation by individual bacteria within animal cells. *Science* 317: 1563–1566.

Lee N, Nielsen P, Andreasen K, Juretschko S, Nielsen J, Schleifer K-H, Wagner M. 1999. Combination of fluorescent *in situ* hybridization and microautoradiography—a new tool for structure-function analyses in microbial ecology. *Appl Environ Microbiol* 65:1289–1297.

Li T, Wu T-D, Mazeas L, Toffin L, Guerquin-Kern J-L, Leblon G, Bouchez T. 2008. Simultaneous analysis of microbial identity and function using NanoSIMS. *Environ Microbiol* 10:580–588.

Liu W-T, Marsh TL, Cheng H, Forney LJ. 1997. Characterization of microbial diversity by determining terminal restriction fragment length polymorphism of genes encoding 16S rRNA. *Appl Environ Microbiol* 63:4516–4522.

Londer YY, Pokkuluri PR, Orshonsky V, Orshonsky L, Schiffer M. 2006. Heterologous expression of dodeca-heme nanowire cytochromes c from *Geobacter sulfurreducens*. *Protein Expr Purif* 47:241–248.

Madigan MR, Matinko AJM, Parker J, Brock TD. 2000. *Biology of Microorganisms*, 9th ed. Upper Saddle River, NJ: Prentice Hall, pp. 432–438.

Malasarn D, Saltikov CW, Campbell KM, Santini JM, Hering JG, Newman DK. 2004. *arrA* is a reliable marker for As(V) respiration. *Science* 306:455.

Marcy Y, Ouverney C, Mik EM, Losekann T, Ivanova N, Martin HG, Szeto E, Platt D, Hugenholtz P, Relman DA, Quake SR. 2007. Dissecting biological *dark matter* with single-cell genetic analysis of rare and uncultivated TM7 microbes from the human mouth. *Proc Natl Acad Sci* 104:11889–11894.

Martinez A, Bradley AS, Waldbauer JR, Summons RE, DeLong EF. 2007. Proteorhodopsin photosystem gene expression enables photophosphorylation in a heterologous host. *Proc Natl Acad Sci USA* 104:5590–5595.

Michaelis W, Seifert R, Nauhaus K, Treude T, Thiel V, Blumenberg M, Knittel K, Gieseke A, Peterknecht K, Pape T, Boetius A, Amann R, Jorgensen BB, Widdel F, Peckmann J, Pimenov NV, Gulin MB. 2002. Microbial reefs in the Black Sea fueled by anaerobic oxidation of methane. *Science* 297:1013–1015.

Murphy JN, Saltikov CW. 2007. The *cymA* gene, encoding a tetraheme c-type cytochrome, is required for arsenate respiration in *Shewanella* species. *J Bacteriol* 189:2283–2290.

Muyzer G, de Waal EC, Uitterlinden AG. 1993. Profiling of complex microbial populations by denaturing gel electrophoresis analysis of polymerase chain-reaction amplified genes coding for 16S rRNA. *Appl Environ Microbiol* 59:695–700.

Myers JM, Myers CR. 2002. Genetic complementation of an outer membrane cytochrome *omcB* mutant of *Shewanella putrefaciens* MR-1 requires *omcB* plus downstream DNA. *Appl Environ Microbiol* 68:2781–2793.

Newman DK, Gralnick JA. 2005. What genetics offers geobiology. In: Banfield JF et al., ed. *Reviews in Mineralogy and Geochemistry*. Chantilly, VA: Mineralogical Society of America Vol. 59, pp. 9–26.

Nielsen J, Chirstensen D, Kloppenborg M, Nielsen P. 2003. Quantification of cell-specific substrate uptake by probe-defined bacteria under *in situ* conditions by microautoradiography and fluorescence *in situ* hybridization. *Environ Microbiol* 5:202–211.

O'Connor SM, Coates JD. 2002. Universal immunoprobe for (per)chlorate-reducing bacteria. *Appl Environ Microbiol* 68:3108–3113.

Olsen GJ, Woese CR. 1993. Ribosomal RNA: A key to phylogeny. *FASEB J* 7:113–123.

Orphan VJ, House CH, Hinrichs K-U, McKeegan KD, DeLong EF. 2001. Methane-consuming archaea revealed by directly coupled isotopic and phylogenetic analysis. *Science* 293:484–487.

Ottesen EA, Hong JW, Quake SR, Leadbetter JR. 2006. Microfluidic digital PCR enables multigene analysis of individual environmental bacteria. *Science* 314:1464–1467.

Ouverney C, Fuhrman J. 1999. Combined microautoradiography-16S rRNA probe technique for determination of radioisotope uptake by specific microbial cell types *in situ*. *Appl Environ Microbiol* 65:1746–1752.

Pearson A, Kraunz KS, Sessions AL, Dekas AE, Leavitt WR, Edwards KJ. 2008. Quantifying microbial utilization of petroleum hydrocarbons in salt-marsh sediments using the ^{13}C content of bacterial rRNA. *Appl Environ Microbiol* 74: 1157–1166.

Pernthaler A, Brown CT, Goffredi S, Dekas A, Embaye T, Orphan VJ. 2008. Diverse syntrophic partnerships from deep-sea methane vents revealed by direct cell capture and metagenomics. *Proc Natl Acad Sci* 105: 7052–7057.

Pernthaler A, Pernthaler J, Amann R. 2002. Fluorescence in situ hybridization and catalyzed reporter deposition (CARD) for the identification of marine bacteria. *Appl Environ Microbiol* 68:3094–3101.

Petsch ST, Edwards KJ, Eglinton TI. 2003. Abundance, distribution and d13C analysis of microbial phospho-lipid-derived fatty acids in a black shale weathering profile. *Org Geochem* 34:731–743.

Pritchett MA, Metcalf WM. 2005. Genetic, physiological and biochemical characterization of multiple metha-nol methyltrasferase isozymes in *Methanosarcina acetivorans* C2A. *Mol Micro* 56:1183–1194.

Radajewski S, Ineson P, Parekh NR, Murrell JC. 2000. Stable-isotope probing as a tool in microbial ecology. *Nature* 403:646–649.

Ram RJ, VerBerkmoes NC, Thelen MP, Tyson GW, Baker BJ, Blake RC, Shah M, Hettich RL, Banfield JF. 2005. Community proteomics of a natural microbial biofilm. *Science* 308:1915–1920.

Rappe MS, Connon SA, Vergin KL, Giovannoni SJ. 2002. Cultivation of the ubiquitous SAR11 marine bacte-rioplankton clade. *Nature* 418:630–633.

Rich VI, Konstantinidis K, DeLong EF. 2008. Design and testing of genome-proxy microarrays to profile marine microbial communities. *Environ Microbiol* 10:506–521.

Riesenfeld CS, Schloss PD, Handelsman J. 2004. Metagenomics: Genomic analysis of microbial communities. *Annu Rev Genet* 38:525–552.

Sabehi G, Loy A, Jung K-H, Partha R, Spudich JL, Isaacson T, Hirschberg J, Wagner M, Beja O. 2005. New insights into metabolic properties of marine bacteria encoding proteorhodopsins. *PLoS Biology* 3:1409–1417.

Saltikov CW, Newman DK. 2003. Genetic identification of a respiratory arsenate reductase. *Proc Natl Acad Sci USA* 100:10983–10988.

Saltikov CW, Wildman RA, Newman DK. 2005. Expression dynamics of arsenic respiration and detoxification in *Shewanella* sp. strain ANA-1. *J Bacteriol* 187:7390–7396.

Saylor G, Layton AC. 1990. Environmental application of nucleic acid hybridization. *Annu Rev Microbiol* 44:625–648.

Scheffel A, Gardes A, Grunberg K, Wanner G, Schuler D. 2008. The major magnetosome proteins MamGFDC are not essential for magnetite biomineralization in Magnetospirillum gryphiswaldense but regulate the size of magnetosome crystals. *J Bacteriol* 190:377–386.

Schrenk MO, Edwards KJ, Goodman RM, Hamers RJ, Banfield JF. 1998. Distribution of *Thiobacillus fer-rooxidans* and *Leptospirillum ferrooxidans*: Implications for generation of acid mine drainage. *Science* 279:1519–1522.

Sessions AL, Sylva SP, Hayes JM. 2005. A moving-wire device for carbon-isotopic analyses of nanogram quantities of nonvolatile organic carbon. *Anal Chem* 77:6519–6527.

Stackebrandt E, Goebel BM. 1994. Taxomonic note: A place for DNA-DNA reassociation kinetics and sequence analysis in the present species definition in bacteriology. *Int J Syst Bacteriol* 44:846–849.

Stahl D. 1997. Molecular approaches for the measurement of density, diversity and phylogeny. In: Hurst CJ, Knudsen GR, McInerney MJ, Stetzenbach LD, Walter MV, eds. *Manual of Environmental Microbiology*. Washington: ASM Press, pp. 102–114.

Stewart PS, Franklin MJ. 2008. Physiological heterogeneity in biofilms. *Nat Rev Microbiol* 6:199–210.

Svensson E, Skoog A, Amend JP. 2004. Concentration and distribution of dissolved amino acids in a shallow hydrothermal system, Vulcano Island (Italy). *Org Geochem* 35:1001–1014.

Teal TK, Lies DP, Wold BJ, Newman DK. 2006. Spatiometabolic stratification of *Shewanella oneidensis* bio-films. *Appl Environ Microbiol* 72:7324–7330.

Thiel V, Toporski J, Schumann G, Sjövall P, Lausmaa J. 2007. Analysis of archaeal core ether lipids using Time of Flight—Secondary Ion Mass Spectrometry (ToF-SIMS): Exploring a new prospect for the study of biomarkers in geobiology. *Geobiology* 5:75–83.

Tsien HC, Bratina BJ, Tsuji K, Hanson RS. 1990. Use of oligonucleotide signature probes for identification of physiological groups of methylotrophic bacteria. *Appl Environ Microbiol* 56:2858–2865.

Tunlid A, White DC. 1990. Use of lipid biomarkers in environmental samples. In Fox A, ed. *Analytical Microbiological Methods*. New York: Plenum Press, pp. 259–274.

Tyson GW, Chapman J, Hugenholtz P, Allen EE, Ram RJ, Richardwon PM, Solovyev VV, Rubin EM, Rokhsar DS, Banfield JF. 2004. Community structure and metabolism through reconstruction of microbial genomes from the environment. *Nature* 428:37–43.

Tyson GW, LO I, Baker BJ, Allen EE, Hugenholtz P, Banfield JF. 2005. Genome-directed isolation of the key nitrogen fixer *Leptospirillum ferrodiazotrophum* sp nov from an acidophilic microbial community. *Appl Environ Microbiol* 71:6319–6324.

van Waasbergen LG, Hildebrand M, Tebo BM. 1996. Identification and characterization of a gene clus-ter involved in manganese oxidation by spores of the marine *Bacillus* sp. strain SG-1. *J Bacteriol.* 178:3517–3530.

von Wintzingerode F, Gobel UB, Stackebrandt E. 1997. Determination of microbial diversity in environmental samples: Pitfalls of PCR-based rRNA analysis. *FEMS Microbiol Rev* 21:213–229.

Wagner MP, Nielsen P, Loy A, Nielsen J, Daims H. 2006. Linking microbial community structure with function: Fluorescence *in situ* hybridization-microautoradiography and isotope arrays. *Curr Opin Biotechnol* 17:83–91.

Walker NJ. 2002. A technique whose time has come. *Science* 296:557–559.

Ward DM, Ferris MJ, Nold SC, Bateson MM. 1998. A natural view of microbial biodiversity within hot spring cyanobacterial mat communities. *Microbiol Mol Biol Rev* 62:1353–1370.

Ward DM, Weller R, Bateson MM. 1990. 16S rRNA sequences reveal numerous uncultured microorganisms in a natural community. *Nature* 345:63–65.

Ward DMB, Weller MM, Ruff-Roberts R, Alyson L. 1992. Ribosomal RNA analysis of microorganisms as they occur in nature. *Adv Microb Ecol* 12:219–286.

Whiteley AS, Manefield M, Lueders T. 2006. Unlocking the *microbial black box* using RNA-based stable isotope probing technologies. *Curr Opin Biotechnol* 17:67–71.

Woese CR, Fox GE. 1977. Phylogenetic structure of the prokaryotic domain: The primary kingdoms. *Proc Natl Acad Sci USA* 74:5088–5090.

Wu L, Liu X, Schadt CW, Zhou J. 2006. Microarray-based analysis of subnanogram quantities of microbial community DNAs by using whole-community genome amplification. *Appl Environ Microbiol* 72:4931–4941.

Zengler K, Toledo G, Rappé M, Elkins J, Mathur EJ, Short JM, Keller M. 2002. Cultivating the uncultured. *Proc Natl Acad Sci USA* 99: 15681–15686.

9 Microbial Formation and Degradation of Carbonates

9.1 DISTRIBUTION OF CARBON IN EARTH'S CRUST

Carbon is an element central to all life on Earth. Although it is one of the less abundant elements in the crust (320 ppm) (Weast and Astle, 1982), it is widely but unevenly distributed (Figure 9.1). In some places it occurs at high concentrations in nonliving matter. Much of the carbon on the surface of the Earth is tied up inorganically in the form of carbonates such as limestone and dolomite, amounting to $\sim 1.8 \times 10^{22}$ g of carbon. Much is also trapped as aged organic matter, such as bitumen and kerogen, coal, organic matter in shale, and natural gas and petroleum. This carbon amounts to $\sim 2.5 \times 10^{22}$ g as compared with $\sim 3.5 \times 10^{18}$ g of carbon in unaged, dead organic matter in soils and sediments and $\sim 8.3 \times 10^{17}$ g of carbon in living matter (estimates from Fenchel and Blackburn, 1979; Bowen, 1979). The atmosphere around the Earth holds $\sim 6.4 \times 10^{17}$ g of carbon as CO_2 (Bolin, 1970; Fenchel and Blackburn, 1979). From the quantities of carbon in each of these compartments, it is seen that the carbon in living matter represents only a small fraction of the total carbon, as does the carbon in unaged, dead organic matter and atmospheric carbon. The carbon in limestone and dolomite, and in aged organic matter, insofar as it is not mined as fossil fuel and combusted by human beings, is not readily available for assimilation by living organisms. Therefore, living systems have to depend on unaged, dead organic matter and atmospheric carbon as ultimate sources of assimilable carbon. In order not to exhaust this carbon, it has to be recycled by biological mineralization of organic matter (see Chapter 6). In the recycling process, some of the carbon released by mineralization of the organic matter enters the atmosphere as CO_2, some is trapped in carbonate deposits, and some is reassimilated by living organisms. In the absence of human interference, *homeostasis* is assumed to operate insofar as transfer of carbon among compartments representing living and dead organic matter and the atmosphere is concerned. Present fears are that human interference by burning fossil fuels is increasing the size of the atmospheric reservoir of carbon and that the remaining reservoirs cannot accommodate the extra CO_2 from this combustion. The consequence of this CO_2 buildup in the atmosphere is blockage of heat radiation into space and an overall warming of the Earth's climate.

9.2 BIOLOGICAL CARBONATE DEPOSITION

Some of the CO_2 generated in energy metabolism of living organisms can be fixed in insoluble carbonates. Indeed, a significant portion of the insoluble carbonate at the Earth's surface is of biogenic origin, but another portion is the result of magmatic and metamorphic activity (see Bonatti, 1966; Skirrow, 1975; Berg, 1986). Direct biological incorporation of carbon in carbonates involves some bacteria, fungi, and algae as well as some metazoa. These carbonates can be deposited extra- and intracellularly. The bacteria, including cyanobacteria, as well as some fungi that are involved, deposit calcium carbonate extracellularly (Bavendamm, 1932; Monty, 1972; Krumbein, 1974, 1979; Morita, 1980; Verrecchia et al., 1990; Chafetz and Buczynski, 1992). The bacterium *Achromatium oxaliferum* seems to be an exception. It has been reported to deposit calcium carbonate intracellularly (Buchanan and Gibbons, 1974; De Boer et al., 1971). Some algae, including certain green, brown, and red algae, and chrysophytes, such as coccolithophores (Lewin, 1965), deposit calcium carbonate as surface structures of their cells, and some protozoa lay it down as tests or shells (foraminifera). Calcium carbonate is also incorporated into the skeletal support structures of certain

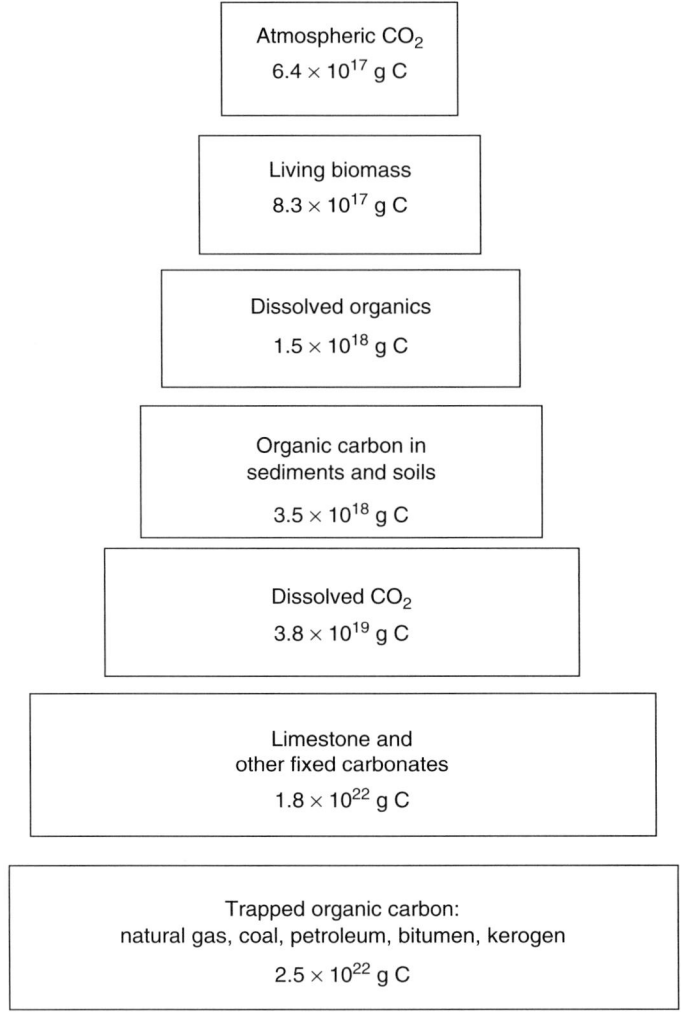

FIGURE 9.1 Distribution of carbon in the lithosphere of the Earth. (Estimates from Fenchel T, Blackburn TH, *Bacteria and Mineral Cycling*, Academic Press, London, 1979; Bowen, 1979.)

sponges and invertebrates such as coelenterates (corals), echinoderms, bryozoans, brachiopods, and mollusks. In arthropods it is associated with their chitinous exoskeleton. The function of the structural calcium carbonate in each of these organisms is to provide support and protection. In all these cases, calcium and some magnesium ions are combined with carbonate ions of biogenic origin (Lewin, 1965). Figure 9.2 illustrates a massive biogenic carbonate deposit in the form of chalk, the White Cliffs of Dover, England.

9.2.1 Historical Perspective of Study of Carbonate Deposition

The study of microbial carbonate precipitation began in the late nineteenth century when Murray and Irvine (1889/1890) and Steinmann (1899/1901) reported formation of $CaCO_3$ in conjunction with urea decomposition and putrefaction caused by microbial activity in seawater culture media (as cited by Bavendamm, 1932). These investigations and others have elicited some controversy.

In 1899 and 1903, Nadson (1903, 1928) presented the first extensive evidence that bacteria could precipitate $CaCO_3$. Nadson studied this process in Lake Veisovo in Kharkov, southern Ukraine.

FIGURE 9.2   White cliffs of Dover, England: a foraminiferal chalk deposit. (Courtesy of the British Tourist Authority, New York.)

He described this lake as a shallow one with a funnellike depression to 18 m near its center, and as resembling the Black Sea physicochemically and biologically. He found that the lake bottom was covered by black sediment with a slight admixture of calcium carbonate. He also noted that the lake water contained $CaSO_4$, with which it was saturated at the bottom. He found that in winter the lake water was clear to a depth of 15 m. From there to the bottom it was turbid, owing to a suspension of elemental sulfur (S^0) and $CaCO_3$. The clear water revealed a varied flora and fauna and a microbial population resembling that of the Black Sea. Gorlenko et al. (1974) demonstrated that in the lake, a significant part of the suspended S^0 in water was the result of oxidation of H_2S by photosynthetic bacteria, especially the green sulfur bacterium *Pelodictyon phaeum*, along with thiobacilli and other colorless sulfur bacteria. These authors reported that the H_2S used by these bacteria arose from bacterial sulfate reduction in the lake.

When Nadson incubated black sediment from Lake Veisovo covered with water from the lake in a test tube open to air, he noted the appearance of crystalline $CaCO_3$ with time in the water above the sediment. He also noted the development of a rust color on the sediment surface owing to the appearance of oxidized iron (Figure 9.3). These changes did not occur with sterilized sediment unless it was inoculated with a small portion of unsterilized sediment. When access to air was blocked by a stopper after the development of $CaCO_3$ and ferric oxide and the test tube was then reincubated, the reactions were reversed. The $CaCO_3$ got dissolved, and the sediment again turned black (Figure 9.3). Nadson isolated a number of organisms from the black lake sediment, including *Proteus vulgaris*, *Bacillus mycoides*, *B. salinus* n. sp., *Actinomyces albus* Gasper, *A. verrucosis* n. sp., and *A. roseolus* n. sp. He did not report the presence of sulfate-reducing bacteria, which were still unknown at that time.

When in a separate experiment Nadson incubated a pure culture of *Proteus vulgaris* in a test tube containing sterilized, dried lake sediment and a 2% peptone solution prepared in distilled water, he noted the development of a pellicle of silicic acid that yellowed in time and then became brown and opaque from ferric hydroxide deposition. At the same time he noted that $CaCO_3$ was deposited in the pellicle. None of these changes occurred in sterile, uninoculated medium.

In other experiments, Nadson observed $CaCO_3$ formation by *Bacillus mycoides* in nutrient broth, nutrient agar, and gelatin medium. He did not identify the source of calcium in these instances, but it must have been a component of each of these media. He also noted that bacterial decomposition of dead algae and invertebrates in seawater led to the precipitation of $CaCO_3$. He further reported the formation of ooliths (i.e., shell-like deposits around normal or involuted cells) by *Bacterium*

Aerobic

Anaerobic

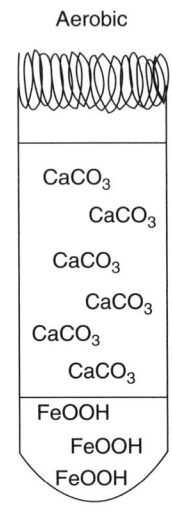

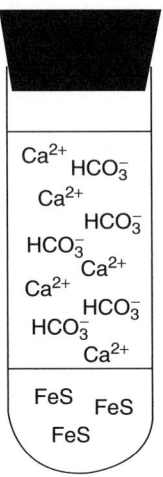

$$2FeS + 1.5O_2 + H_2O = 2S + 2FeOOH$$

Bacteria
$$Protein ======> CO_2, H_2O, NH_3, SO_4^{2-}$$

$$CO_2 + H_2O ==> H_2CO_3$$
$$H_2CO_3 + 2NH_3 ==> CO_3^{2-} + 2NH_4^+$$
$$Ca^{2+} + CO_3^{2-} ==> CaCO_3$$

Bacteria
$$Protein ======> CO_2, NH_3,$$
$$Organic\ products$$

Bacteria
$$SO_4^{2-} + 2C + 2H^+ ====> H_2S + 2CO_2$$
$$2FeOOH + 3H_2S ==> 2FeS + S^0 + 4H_2O$$
$$CO_2 + H_2O ==> H_2CO_3$$
$$CaCO_3 + H_2CO_3 ==> Ca^{2+} + 2HCO_3^-$$

FIGURE 9.3 Diagrammatic representation of Nadson's experiments with mud from Lake Veisovo. Contents of the tubes represent the final chemical state after microbial development. Chemical equations describe important reactions leading to the final chemical state.

helgolandium in 1% peptone seawater broth. He had isolated this bacterium from decomposing *Alcyonidium*, an alga. Finally, Nadson claimed to have observed precipitation of magnesian calcite or dolomite in two experiments. One involved 3 1/2 years of incubation of black mud from a salt lake near Kharkov. The second involved 1 1/2 years of incubation of a culture of *Proteus vulgaris* growing in a mixture of sterilized sediment from Lake Veisovo and seawater enriched with 2% peptone.

The pioneering experiments of Nadson showed that $CaCO_3$ precipitation was not brought about by any special group of bacteria but depended on appropriate environmental conditions. However, Drew (1911, 1914), who was apparently not aware of Nadson's studies, concluded from laboratory experiments that in the tropical seas around the Bahamas, denitrifying bacteria were responsible for $CaCO_3$ precipitation. He found that water samples from this geographic location contained denitrifying bacteria in significant numbers. He isolated a denitrifying, $CaCO_3$-precipitating culture from the water samples, which he named *Bacterium calcis*. He apparently held it responsible for the $CaCO_3$ precipitation in the sea (see Bavendamm, 1932). Although many geologists of his day appeared to accept Drew's explanation of the origin of $CaCO_3$ in the seas around the Bahamas, it is not accepted today.

Lipman (1929) rejected Drew's conclusion on the basis of some studies of seawater off the island of Tutuila (American Samoa) in the Pacific Ocean and the Tortugas in the Gulf of Mexico. In his work, Lipman confined himself mainly to an examination of water samples and only a few sediment samples. Using various culture media, he was able to demonstrate $CaCO_3$ precipitation in the laboratory by various bacteria, not just *Bacterium calcis* as Drew had suggested. However, because the number of viable organisms in the water samples and the few sediment samples that Lipman examined seemed small to him, he felt that these organisms were not important in $CaCO_3$ precipitation in the sea. He probably should have examined the sediments more extensively.

Bavendamm (1932) reintroduced Nadson's concept of microbial $CaCO_3$ precipitation as an important geochemical process that is not a specific activity of any special group of bacteria. As a result of his microbiological investigations in the Bahamas, he concluded that this precipitation occurs chiefly in sediments of shallow bays, lagoons, and swamps. He isolated and enumerated heterotrophic and autotrophic bacteria, including sulfur bacteria; photosynthetic bacteria; agar-, cellulose-, and urea-hydrolyzing bacteria; nitrogen-fixing bacteria; and sulfate-reducing bacteria. According to him, all of these bacteria had an ability to precipitate $CaCO_3$. He rejected the idea of significant participation of cyanobacteria in $CaCO_3$ precipitation. As we know now, however, some cyanobacteria can cause significant $CaCO_3$ deposition (see Golubic, 1973).

There is still no complete consensus about the origin of calcium carbonate suspended in the seas around the Bahamas. Some geochemists favor an inorganic mechanism of formation (Skirrow, 1975). They feel that a finding of supersaturation of the waters with $CaCO_3$ around the Bahamas and an abundant presence of nuclei for $CaCO_3$ deposition explains the $CaCO_3$ precipitation in the simplest way. Others, including Lowenstamm and Epstein (1957), favor algal involvement based on a study of $^{18}O/^{16}O$ ratios of the precipitated carbonates. McCallum and Guhathakurta (1970) isolated a number of different bacteria from the sediments of Bimini, Brown's Cay, and Andros Island in the Bahamas, which under laboratory conditions had the capacity to precipitate calcium carbonate, in most cases as aragonite. They concluded that the naturally formed calcium carbonate in the Bahamas is the result of a combination of biological and chemical factors. This also appears from the work of Buczynski and Chafetz (1991), who illustrated the ability of marine bacteria to induce calcium carbonate precipitation in the form of calcite and aragonite in elegant scanning electron photomicrographs (Figure 9.4). Even more recent investigations of the origin of calcium carbonate suspended in the waters around the Great Bahama Banks, west of Andros Island, suggest coupling on a biophysicochemical level among the microbial community, physical circulation on the Bank, and water chemistry; but further study is needed to fully resolve this phenomenon (Thompson, 2000).

Despite the numerous reports of calcium carbonate precipitation by bacteria other than cyanobacteria (see Krumbein, 1974, 1979; Shinano and Sakai, 1969; Shinano 1972a,b; Ashirov and Sazonova, 1962; Greenfield, 1963; Abd-el-Malek and Rizk, 1963a; Roemer and Schwartz, 1965; McCallum and Guhathakurta, 1970; Morita, 1980; del Moral et al., 1987; Ferrer et al., 1988a,b; Thompson and Ferris, 1990; Rivadeneyra et al., 1991; Buczynski and Chafetz, 1991; Chafetz and Buczynski, 1992), the significance of many of these reports has been questioned by Novitsky (1981). He feels that, in most instances, *in situ* environmental conditions do not meet requirements essential for calcium carbonate precipitation in laboratory experiments (especially pH values >8.3). He was unable to demonstrate $CaCO_3$ precipitation with active laboratory isolates at *in situ* pH in water samples from around Bermuda and from the Halifax, Nova Scotia, harbor. However, Novitsky did not acknowledge the laboratory observation by McCallum and Guhathakurta (1970) of bacterial $CaCO_3$ precipitation by a number of different organisms in calcium acetate–containing seawater medium with added KNO_3 in which the pH fell to ~6. The authors of this study stressed that the mineral form of the calcium carbonate that precipitated in the presence of the bacteria was aragonite, as is generally observed *in situ* in marine environments. When calcium carbonate precipitated in any of their sterile controls, it seemed to be in a different mineral form. The authors did not appear to consider that the bacterial cell surface might play a role in the nucleation of calcium carbonate formation.

9.2.2 BASIS FOR MICROBIAL CARBONATE DEPOSITION

Carbonate compounds are relatively insoluble in water. Table 9.1 lists the solubility constants of several geologically important carbonates. Because of the relative insolubility of carbonates, they are readily precipitated from aqueous solution at relatively low carbonate and counterion concentrations. The following example will illustrate the fact.

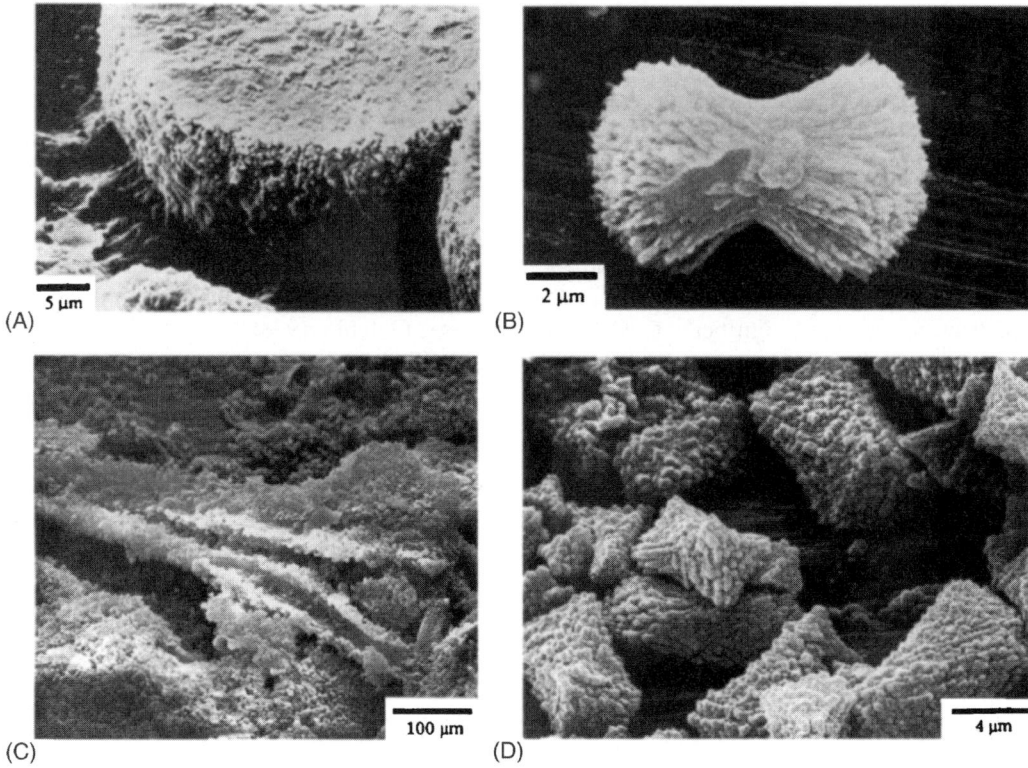

FIGURE 9.4 Bacterial precipitation of calcium carbonate as aragonite and calcite. Scanning electron micrographs of (A) a crust of a hemisphere of aragonite precipitate formed by bacteria in liquid culture in the laboratory; (B) an aragonite dumbbell precipitated by bacteria in liquid medium in the laboratory ([A] and [B]: From Buczynski C, Chafetz HS, *J. Sedim. Petrol.*, 61, 226–233, 1991. With permission.); and (C) crystal bundles of calcite encrusting dead cyanobacterial filaments that were placed in gelatinous medium inoculated with bacteria from Baffin Bay. The specimen was treated with 30% H_2O_2 to remove organic matter. It should be noted that the crystal aggregates have been cemented to form a rigid crust that does not depend on the organic matter for support. (D) Higher magnification view of some of the crystal aggregates from (C) (area close-up not from field of view in [C]). These aggregates sometimes resemble rhombohedra, tetragonal disphenoids, or tetragonal pyramids. ([C] and [D]: From Chafetz HS, Buczynski C, *Palaios*, 7, 277–293, 1992. With permission.)

TABLE 9.1
Solubility Products of Some Carbonates

Compound	Solubility Constant (K_{sol})	Reference
$CaCO_3$	$10^{-8.32}$	Latimer and Hildebrand (1942)[a]
$MgCO_3 \cdot 3H_2O$	10^{-5}	Latimer and Hildebrand (1942)
$MgCO_3$	$10^{-4.59}$	Weast and Astle (1982)
$CaMg(CO_3)_2$	$10^{-16.7}$	Stumm and Morgan (1981)

[a] Note that Stumm and Morgan (1981) give a solubility product in freshwater of $10^{-8.42}$ (25°C) for calcite and $10^{-8.22}$ (25°C) for aragonite.

In an aqueous solution containing $10^{-4.16}$ M Ca^{2+},* the calcium will be precipitated by CO_3^{2-} at a concentration in excess of $10^{-4.16}$ M. This is because the product of the concentrations of these two ions will exceed the solubility product (K_{sol}) of $CaCO_3$, which is

$$[Ca^{2+}][CO_3^{2-}] = K_{sol} = 10^{-8.32} \tag{9.1}$$

In general, Ca^{2+} will be precipitated as $CaCO_3$ when the carbonate ion concentration is in excess of the ratio $10^{-8.32}/[Ca^{2+}]$ (see Equation 9.1).

Carbonate ion in an unbuffered aqueous solution undergoes hydrolysis. This process causes a solution of 0.1 M Na_2CO_3 to exhibit a hydroxyl ion concentration of 0.004 M. The following reactions explain this phenomenon:

$$Na_2CO_3 \Leftrightarrow 2Na^+ + CO_3^{2-} \tag{9.2}$$

$$CO_3^{2-} + H_2O \Leftrightarrow HCO_3^- + OH^- \tag{9.3}$$

$$HCO_3^- + H_2O \Leftrightarrow H_2CO_3 + OH^- \tag{9.4}$$

Of these three reactions, the third can be considered negligible.

As bicarbonate dissociates according to the reaction

$$HCO_3^- \Leftrightarrow CO_3^{2-} + H^+ \tag{9.5}$$

whose dissociation constant (K_2) is

$$[CO_3^{2-}][H^+]/[HCO_3^-] = 10^{-10.33} \tag{9.6}$$

and as the ionization constant of water (K_w) at 25°C is

$$[H^+][OH^-] = 10^{-14} \tag{9.7}$$

the equilibrium constant ($K_{equilib}$) for the hydrolysis of CO_3^{2-} (Reaction 9.3) is

$$[HCO_3^-][OH^-]/[CO_3^{2-}] = 10^{-14}/10^{-10.33} = 10^{-3.67} \tag{9.8}$$

At pH 7, where the hydroxyl concentration is 10^{-7} (see Equation 9.7), the ratio of bicarbonate to carbonate is therefore

$$[HCO_3^-]/[CO_3^{2-}] = 10^{-3.67}/10^{-7} = 10^{3.33} \tag{9.9}$$

This means that at pH 7 the bicarbonate concentration is $10^{3.33}$ times greater than the carbonate concentration in an aqueous solution, assuming that an equilibrium exists between the CO_2 of the atmosphere in contact with the solution and the CO_2 in solution.

From Equation 9.1, we can predict that at a freshwater concentration of 0.03 g of Ca^{2+} per liter of solution (i.e., at $10^{-3.14}$ M Ca^{2+}), an excess of $10^{-5.18}$ M carbonate ion would be required to precipitate the calcium as calcium carbonate, because from Equation 9.1 it follows that

$$[CO_3^{2-}] = 10^{-8.32}/10^{-3.14} = 10^{-5.18} \tag{9.10}$$

Now, at pH 7, $10^{-5.18}$ M carbonate would be in equilibrium with $10^{-1.85}$ M bicarbonate ($10^{-5.18}$ multiplied by $10^{3.33}$ according to Equation 9.9). This amount of bicarbonate and carbonate is equivalent to ~0.6 g of CO_2 per liter of solution.

---

* The ion concentrations here refer really to ion activities.

Similarly, from Equation 9.1 we can predict that at the calcium concentration in normal seawater, which is 10^{-2} M, a carbonate concentration in excess of $10^{-6.32}$ M would be required to precipitate it. At pH 8, this amount of carbonate ion would be expected to be in equilibrium with $\sim 10^{-3.99}$ M bicarbonate ion, which is equivalent to ~ 0.0045 g of CO_2 per liter of solution. Assuming the combined concentration of carbonate and bicarbonate ions in seawater to be 2.8×10^4 µg of carbon per liter (Marine Chemistry, 1971), we can calculate that the carbonate concentration at pH 8 must be $\sim 10^{-4.97}$ M and the bicarbonate concentration, $\sim 10^{-2.64}$ M. Because the product of the carbonate ($10^{-4.97}$ M) and calcium concentrations (10^{-2} M) is $10^{-6.97}$, which is greater than the solubility product of $CaCO_3$ ($10^{-8.32}$), seawater is saturated with respect to calcium carbonate. In reality, this seems to be true only for marine surface waters (Schmalz, 1972). (Mg^{2+} in seawater is not readily precipitated as $MgCO_3$ because of the relatively high solubility of $MgCO_3$.)

A quantity of 0.6 g of CO_2 can be derived from the complete oxidation of 0.41 g of glucose according to the following equation:

$$C_6H_{12}O_6 + 6CO_2 \rightarrow 6CO_2 + 6H_2O \tag{9.11}$$

Similarly, 0.0045 g of CO_2 can be derived from the complete oxidation of 0.003 g of glucose. Such quantities of glucose are readily oxidized in a relatively short time by appropriate populations of bacteria or fungi.

9.2.3 CONDITIONS FOR EXTRACELLULAR MICROBIAL CARBONATE PRECIPITATION

As already mentioned, some bacteria, including cyanobacteria, and fungi precipitate $CaCO_3$ or other insoluble carbonates extracellularly under various conditions. The following points define the conditions under which this precipitation can take place.

1. *Aerobic or anaerobic oxidation of carbon compounds consisting of carbon and hydrogen with or without oxygen, for example, carbohydrates, organic acids, and hydrocarbons.* If such oxidations occur in a well-buffered neutral or alkaline environment containing adequate amounts of calcium ions or other appropriate cations, at least some of the CO_2 that is generated will be transformed into carbonate, which will then precipitate with appropriate cations.

$$CO_2 + H_2O \Leftrightarrow H_2CO_3 \tag{9.12}$$

$$H_2CO_3 + OH^- \Leftrightarrow HCO_3^- + H_2O \tag{9.13}$$

$$HCO_3^- + OH^- \Leftrightarrow CO_3^{2-} + H_2O \tag{9.14}$$

Calcium carbonate precipitation under these conditions has been demonstrated by the formation of aragonite and other calcium carbonates by bacteria and fungi in seawater media containing organic matter at concentrations of 0.01 and 0.1% (Krumbein, 1974). The organic matter in different experiments consisted of glucose, sodium acetate, or sodium lactate. The aragonite precipitated on the surface of the bacteria or fungi after 36 h of incubation. Between 36 and 90 h, the cells in the $CaCO_3$ precipitate were still viable (although deformed), but after 4–7 days they were nonviable. Phosphate above a critical concentration can interfere with calcite formation by soil bacteria (Rivadeneyra et al., 1985).

Verrecchia et al. (1990) suggested that in semiarid regions, the role of fungi is to immobilize Ca^{2+} with oxalate, which is a product of their metabolism. Upon death of the fungi, bacteria convert the calcium oxalate (whewellite) to secondary calcium carbonate by mineralizing the oxalate.

Luff and Wallmann (2003) developed a model to "quantify biogeochemical processes and methane turnover in gas-hydrate-bearing surface sediment (at) a cold vent site at Hydrate Ridge … in the Cascadia Margin subduction zone." In this model they predicted that a significant portion of CO_2 produced in anaerobic biooxidation of methane that escapes via vents from the gas hydrate, known to occur at this site (see Boetius et al., 2000), is transformed authigenically into calcite and aragonite. Sulfate is the terminal electron acceptor in the anaerobic methane biooxidation. The carbonate deposit that develops at the vents from this process ultimately interferes with the replenishment of methane from the underlying gas hydrate at the biologically active sites and thereby limits anaerobic methane oxidation (Luff et al., 2004). The biogenic carbonate thus helps in the natural preservation of the gas hydrate.

2. *Aerobic or anaerobic oxidation of organic nitrogen compounds with the release of NH_3 and CO_2 in unbuffered environments containing sufficient amounts of calcium, magnesium, or other appropriate cations.* NH_3 is formed in the deamination of amines, amino acids, purines, pyrimidines, and other nitrogen-containing compounds, especially by bacteria. In water, NH_3 hydrolyzes to NH_4OH, which dissociates partially into NH_4^+ and OH^-, thereby raising the pH of the environment to the point where at least some of the CO_2 produced may be transformed into carbonate. $CaCO_3$ precipitation under these conditions has been observed by the formation of aragonite and other calcium carbonates by bacteria and fungi in seawater media containing nutrients such as asparagines or peptone in a concentration of 0.01–0.1%, or homogenized cyanobacteria (Krumbein, 1974). Moderately halophilic bacteria have been shown to precipitate $CaCO_3$ under laboratory conditions as calcite, aragonite, or vaterite, depending on culture conditions (del Moral et al., 1987; Ferrer et al., 1988a; Rivadeneyra et al., 1991, 2006). Other examples are the precipitation of calcium carbonate by various species of *Micrococcus* and a gram-negative rod in peptone media made up of natural seawater and Lyman's artificial seawater (Shinano and Sakai, 1969; Shinano, 1972a,b). The organisms in this case came from inland seas of the North Pacific and from the western Indian Ocean. Lithification of beach rock along the shores of the Gulf of Aqaba (Sinai) is an example of *in situ* formation of $CaCO_3$ by heterotrophic bacteria in their mineralization of cyanobacterial remains (Krumbein, 1979).

3. *The reduction of $CaSO_4$ to CaS by sulfate-reducing bacteria such as Desulfovibrio spp., Desulfotomaculum spp., and so on, using organic carbon.* For the purpose of this discussion, organic carbon is indicated by the formula (CH_2O) in Equation 9.15. It serves as the source of reducing power. The CaS formed by these organisms hydrolyzes readily to H_2S, which has a small dissociation constant ($K_1 = 10^{-6.95}$, $K_2 = 10^{-15}$). The Ca^{2+} then reacts with CO_3^{2-} derived from the CO_2 produced in the oxidation of the organic matter by the sulfate-reducing bacteria. A reaction sequence describing the reduction of $CaSO_4$ to H_2S and $CaCO_3$ may be written as follows:

$$CaSO_4 + 2(CH_2O) \xrightarrow{\text{Sulfate-reducing bacteria}} CaS + 2CO_2 + 2H_2O \qquad (9.15)$$

$$CaS + 2H_2O \rightarrow Ca(OH)_2 + H_2S \qquad (9.16)$$

$$CO_2 + H_2O \rightarrow H_2CO_3 \qquad (9.17)$$

$$Ca(OH)_2 + H_2CO_3 \rightarrow CaCO_3 + 2H_2O \qquad (9.18)$$

It should be noted that in Reaction 9.15, 2 mol of CO_2 is formed for every mole of SO_4^{2-} reduced, yet only 1 mol of CO_2 is required to precipitate the Ca^{2+}. Hence, $CaCO_3$ precipitation under these circumstances depends on one of the three conditions, namely, the loss of CO_2 from the environment by volatilization, the presence of a suitable buffer system, or the development of alkaline conditions.

Evidence of $CaCO_3$ deposition linked to bacterial sulfate reduction is found in the work of Abd-el-Malek and Rizk (1963a). They demonstrated the formation of $CaCO_3$ during bacterial sulfate reduction in experiments using fertile clay loam soil enriched with starch and sulfate or sandy soil enriched with sulfate and plant matter. Other evidences of microbial carbonate formation during sulfate reduction are found in the works of Ashirov and Sazonova (1962) and Roemer and Schwartz (1965). Ashirov and Sazonova showed that secondary calcite was deposited when an enrichment of sulfate-reducing bacteria was grown in quartz sand bathed in Shturm's medium: $(NH_4)SO_4$, 4 g; $NaHPO_4$, 3.5 g; KH_2PO_4, 1.5 g; $CaSO_4$, 0.5 g; $MgSO_4 \cdot 7H_2O$, 1 g; $NaCl$, 20 g; $(NH_4)_2Fe(SO_4)_2 \cdot 6H_2O$, 0.5 g; Na_2S, 0.03 g; $NaHCO_3$, 0.5 g; and distilled water, 1 L. Electron donors were hydrogen (H_2), calcium lactate and acetate (serving as carbon source), or petroleum. The petroleum may have first been broken down to usable hydrogen donors for sulfate reduction by other organisms in the mixed culture (see Nazina et al., 1985), which the investigators used as inoculum, before the sulfate reducers carried out their activity. The results from these experiments have lent support to the notion that incidents of sealing of some oil deposits by $CaCO_3$ may be due to the activity of sulfate-reducing bacteria at the petroleum–water interface of an oil reservoir.

Roemer and Schwartz (1965) showed that sulfate reducers were able to form calcite $(CaCO_3)$ from gypsum $(CaSO_4 \cdot 2H_2O)$ and anhydrite $(CaSO_4)$. Their cultures also formed strontianite $(SrCO_3)$ from celestite $(SrSO_4)$ and witherite $(BaO \cdot CO_2)$ from barite $(BaSO_4)$, but they formed aluminum hydroxide rather than aluminum carbonate from aluminum sulfate.

Still another example of calcium carbonate formation as a result of bacterial sulfate reduction is the deposition of secondary calcite in cap rock of salt domes. This activity has been inferred from a study of $^{13}C/^{12}C$ ratios of samples taken from these deposits (Feeley and Kulp, 1957; see also Chapter 19).

4. *The hydrolysis of urea leading to the formation of ammonium carbonate.* Urea hydrolysis can be summarized by the following equation:

$$NH_2(CO)NH_2 + H_2O \rightarrow (NH_4)_2CO_3 \quad\quad (9.19)$$

This reaction causes precipitation of Ca^{2+}, Mg^{2+}, or other appropriate cations when present at suitable concentrations. Urea is an excretory product of ureotelic animals, including adult amphibians and mammals. The hydrolysis of urea was first observed in experiments by Murray and Irvine (see Bavendamm, 1932). They believed it to be important in the marine environment. However, they did not implicate bacteria in urea hydrolysis, whereas Steinmann (1899/1901), working independently, did (as cited by Bavendamm, 1932). Bavendamm (1932) observed extensive $CaCO_3$ precipitation by urea-hydrolyzing bacteria from the Bahamas. Most recently, Fujita et al. (2000) demonstrated the presence of urea-hydrolyzing organisms in groundwater samples from the Eastern Snake River Plain, United States, that precipitated calcite rapidly in a medium containing urea and calcium. Of the three isolates, two belonged to the genus *Pseudomonas* and one to the genus *Variovorax*. Mitchell and Ferris (2006) observed that although *Bacillus pasteurii* did not affect the morphology or mineralogy of the $CaCO_3$ formed as a result of urea hydrolysis in artificial groundwater in a laboratory microcosm, it significantly enhanced the rate of crystal growth of the $CaCO_3$ formed.

As it is now perceived, urea hydrolysis is the least important mechanism of microbial carbonate deposition because urea is not a widely distributed compound in nature.

5. *Removal of CO_2 from a bicarbonate-containing solution.* Such removal causes an increase in carbonate ion concentration according to the following relationship:

$$2HCO_3^- \Leftrightarrow CO_2\uparrow + H_2O + CO_3^{2-} \quad\quad (9.20)$$

In the presence of an adequate supply of Ca^{2+}, $CaCO_3$ will precipitate.

An important process of CO_2 removal is photosynthesis in which CO_2 is assimilated as the chief source of carbon for the photosynthesizing organism. Some chemolithotrophs, as long as they do not generate acids in the oxidation of their inorganic energy source, can also promote $CaCO_3$ precipitation through assimilation of CO_2 as their sole carbon source. Examples of microbes that precipitate $CaCO_3$ around them as a result of their photosynthetic activity include certain filamentous cyanobacteria associated with stromatolites (Monty, 1972; Golubic, 1973; Walter, 1976; Krumbein and Giele, 1979; Wharton et al., 1982; Nekrasova et al., 1984) (see also Section 9.2). In Flathead Lake delta, Montana, cyanobacteria and algae deposit calcareous nodules and crusts on subaqueous levees. The calcium carbonate deposition in the outer portion of the nodules and concretions may result in a local rise in pH, which promotes the dissolution of silica of diatom frustules, which is also found on the nodules and concretions. The dissolved silica is reprecipitated with calcium carbonate in the interior zones of the concretions. The source of the calcium and organic carbon from which CO_2 is generated by mineralization is not the lake, which is oligotrophic, but the Flathead River feeding into the lake at the site of deposition. Deposition of the concretions and crusts coincides with periods of high productivity (Moore, 1983).

This mechanism can also function in the deposition of structural carbonate. Examples are found among some of the green, brown, and red algae and some of the chrysophytes, which are all known to deposit calcium carbonate in their walls (see Lewin, 1965; Friedmann et al., 1972). Not all calcareous algae form $CaCO_3$ as a result of photosynthetic removal of CO_2; however, some form it from respiratory CO_2. In any case, photosynthetic removal of CO_2 is probably one of the most important mechanisms of biogenic $CaCO_3$ formation in the open, aerobic environment.

9.2.4 CARBONATE DEPOSITION BY CYANOBACTERIA

Carbonate deposition by cyanobacteria has been described by Golubic (1973), Pentecost (1978), and Pentecost and Bauld (1988), among others. In this process a distinction must be made between the cyanobacteria that entrap and agglutinate preformed calcium carbonate in their thalli and those that precipitate it in their thalli as a result of their photosynthetic activity. Preformed calcium carbonate used in entrapment and agglutination processes is formed at a site other than the site of deposition, whereas the calcium carbonate deposited as a result of photosynthetic activity of the cyanobacteria is being formed at the site of deposition (see Krumbein and Potts, 1979; Pentecost and Bauld, 1988). Calcium carbonate associated with stromatolite structures, originating from special types of cyanobacterial mats, may be the result of deposition by entrapment and agglutination or by cyanobacterial photosynthesis. In the cases of *Homeothrix crustacea* (Pentecost, 1988), *Lyngbya aestuarii*, and *Scytonema myochrous*, it is due to photosynthesis (Pentecost and Bauld, 1988).

Calcium carbonate associated with travertine and lacustrine carbonate crusts and nodules can result from the photosynthetic activity of cyanobacteria in freshwater environments. Travertine, a porous limestone, is formed from rapid calcium carbonate precipitation resulting, in part, from cyanobacterial photosynthesis in waterfalls and streambeds of fast-flowing rivers in which cyanobacteria tend to be buried. The cyanobacteria avoid being trapped by outward growth movement, which contributes to the porosity of the deposit (Golubic, 1973). On the basis of $^{14}CO_2$ photoassimilation, Pentecost (1978, 1995) has estimated that the contribution of cyanobacteria to the calcification process in travertine formation may amount to no more than ~10%; the rest of the $CaCO_3$ is formed abiotically as a result of degassing of stream water (loss of CO_2 to the atmosphere). At Waterfall Beck in Yorkshire, England, Spiro and Pentecost (1991) observed that the calcite deposited by cyanobacteria was richer in ^{13}C than nearby travertine, suggesting that the travertine was formed by a different mechanism than the calcite formed on the cyanobacteria. Degassing of CO_2 from the stream water may have played a role in that case of travertine formation (Pentecost and Spiro, 1990).

Lacustrine carbonate crusts are formed by benthic cyanobacteria attached to rocks or sediment, which deposit calcium carbonate through their photosynthetic activity in shallow portions of lakes with carbonate-saturated waters (Golubic, 1973). Calcareous nodules are formed around rounded rocks and pebbles or shells to which calcium carbonate–depositing cyanobacteria are attached. The nodules are rolled by water currents, thus exposing different parts of their surface to sunlight at different times and promoting photosynthetic activity and calcium carbonate precipitation by the attached cyanobacteria (Golubic, 1973). High-magnesium calcite precipitated in the sheaths of certain cyanobacteria such as *Scytonema* may be related to the ability of the sheaths to concentrate magnesium three to five times over the concentrations of magnesium in seawater (Monty, 1967; see also discussion by Golubic, 1973).

Thompson and Ferris (1990) demonstrated the ability of *Synechococcus* from Green Lake, Lafayette, New York, to precipitate calcite, gypsum, and probably magnesite from filter-sterilized water from the lake in laboratory simulations (Figure 9.5). This lake has an average depth of ~28 m (52.5 m maximum) and is meromictic with a distinct, permanent chemocline at a depth of ~18 m (Brunskill and Ludlam, 1969). Its water is naturally alkaline (pH ~7.95) and contains on the order of 11 mM Ca^{2+}, 2.8 mM Mg^{2+}, and 10 mM SO_4^{2-}. It has an ionic strength of ~54.1 and an alkalinity of ~3.24 (Thompson and Ferris, 1990). Gypsum crystals developed on the surface of *Synechococcus* cells before calcite crystals, but the calcite deposit became more massive and less prone to being shed by the cells on division than the gypsum deposit. Calcite deposition coincided with a rise in pH in the immediate surround of the cells that was related to their photosynthetic activity. Gypsum deposition occurred in the dark as well as the light and hence was not driven by photosynthesis, as was calcium deposition. Indeed, Thompson and Ferris suggested that calcite may replace gypsum in the developing bioherm (a microbialite) in the lake. Calcite is deposited on the *Synechococcus* cells as a result of the interaction between calcium ions bound at the cell surface and carbonate ions generated as a result of the photosynthetic activity of the cells. The cell-bound calcium ions also capture sulfate ions to form gypsum. This activity explains the origin of the marl and the calcified bioherm that are found in Green Lake (Thompson et al., 1990).

There are some instances where the calcium carbonate that formed in the lithification of some cyanobacterial mats did not originate during the photosynthesis of the cyanobacteria. It originated from the activity of bacteria associated with cyanobacteria (Chafetz and Buczynski, 1992). According to Chafetz and Buczynski, cyanobacterial stromatolites may thus owe their existence to bacterial $CaCO_3$ precipitation rather than cyanobacterial photosynthesis.

9.2.5 POSSIBLE MODEL FOR OOLITE FORMATION

A process that may serve as a model for oolite formation is that involving the deposition of carbonate on the cell surface of a marine pseudomonad, strain MB-1. Living or dead cells of this bacterium adsorb calcium and magnesium ions on the cell surface (cell wall–membrane complex) (Greenfield, 1963), which can then react with carbonate ions in solution. This process of $CaCO_3$ deposition is therefore not directly dependent on the living state of the organism. The dead cells adsorb calcium ions more extensively than magnesium ions. Carbonate in the medium derives mostly from respiratory CO_2 produced by the living cells of the organism. The conversion of CO_2 to carbonate is brought about by hydrolysis of ammonia produced from organic nitrogen compounds by actively metabolizing cells. The $CaCO_3$ deposited on the cells has the form of aragonite. The cells with $CaCO_3$ deposited on them serve as nuclei for further calcium carbonate precipitation (Figure 9.6). Observations by Buczynski and Chafetz (1991) strongly support this model. A similar phenomenon was also shown with a marine yeast (Buck and Greenfield, 1964; as cited by McCallum and Guhathakurta, 1970).

9.2.6 STRUCTURAL OR INTRACELLULAR CARBONATE DEPOSITION BY MICROBES

In general, morphological and physiological studies have shown that bacteria, including cyanobacteria, and some algae cause precipitation of $CaCO_3$ mostly in the bulk phase close to or at their cell surface. In contrast, some other algae and protozoa form $CaCO_3$ intracellularly and then export it to the cell surface to become support structures. Examples of eukaryotic algae that form $CaCO_3$

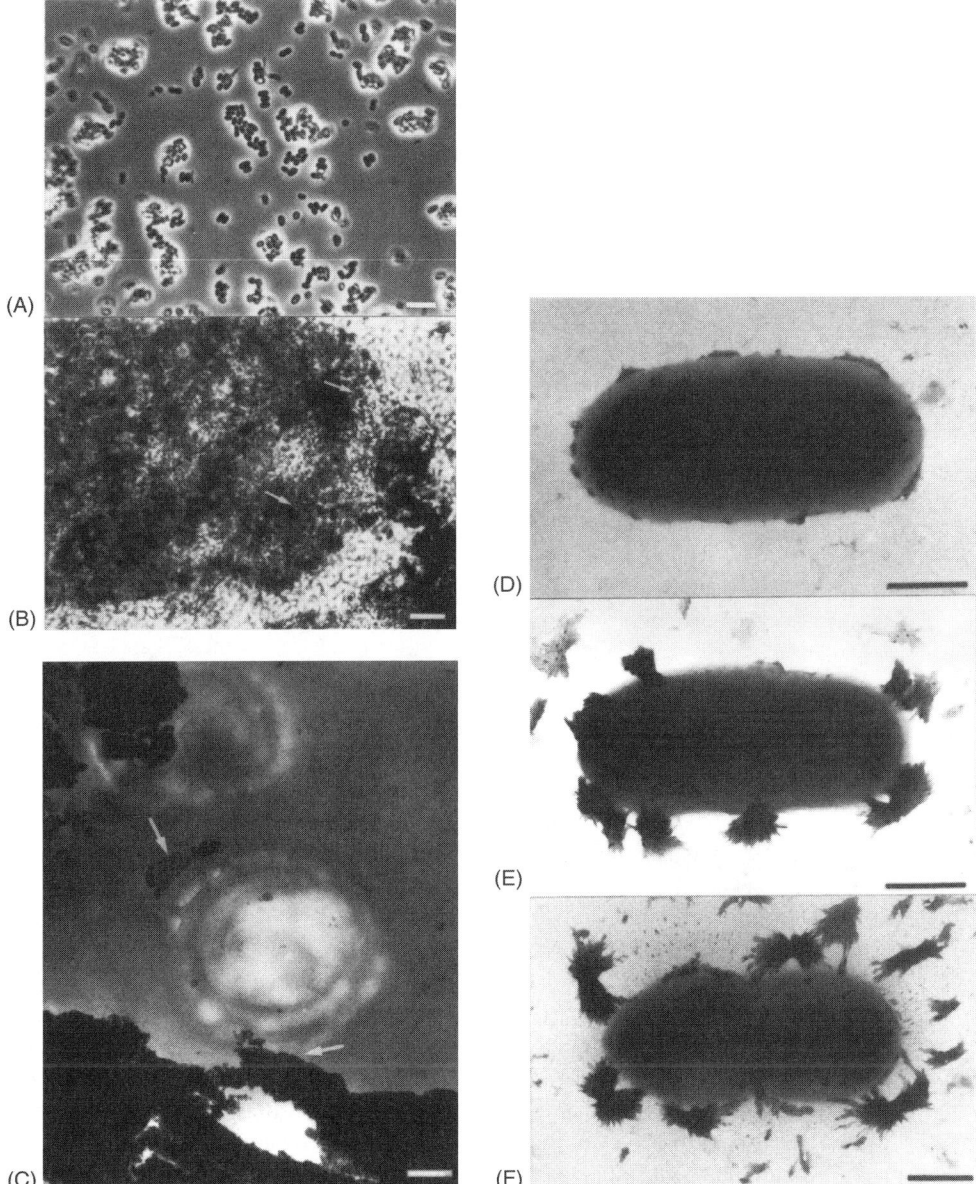

FIGURE 9.5 Calcite and gypsum precipitation by *Synechococcus* sp. isolated from Fayetteville Green Lake, New York. (A) Phase contrast photomicrograph of *Synechococcus* laboratory culture. (B) Petrographic thin-section photomicrograph of calcite crystal from Green Lake showing evidence of occlusion of numerous small bacterial cells within calcite grain (arrows). Note similar size of *Synechococcus* in (A) and occluded bacterial cells in (B). Scale bars in (A) and (B) equal 5 μm. (C) Thin-section transmission electron micrograph of two *Synechococcus* cells and calcite from a 72 h culture (cell represented by white oval area between arrows). Arrows point to calcite (electron-dense material) on the cell surface. Cells are unstained to avoid dissolution by heavy metal stains that are used to provide contrast to biological specimens. Scale bar equals 200 nm. (D)–(F) Series of transmission electron micrographs showing progression of gypsum precipitation on the cell wall of *Synechococcus* (whole mounts). (D) Initiation of numerous nucleation sites on the cell surface. (E) Gypsum precipitation spreading away from the cell. Gypsum still appears to be covered by a thin layer of bacterial slime. (F) Dividing *Synechococcus* cell shedding some of the precipitated gypsum. Scale bars equal 500 nm. (Courtesy of Thompson JB, Ferris FG, *Geology*, 18, 995–998, 1990. With permission.)

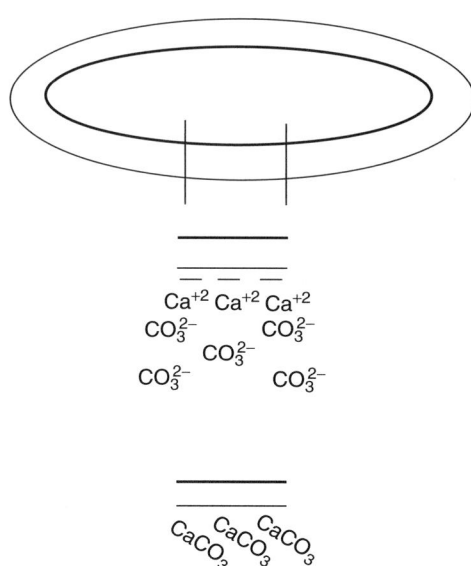

FIGURE 9.6 Schematic diagram of calcium carbonate deposition on the cell surface of a prokaryote (domain Bacteria), resulting in calcite or aragonite formation. See text for details.

external to their cells include the members of Chlorophyceae such as *Chara* and *Halimeda* (de Vrind-de Jong and de Vrind, 1997) and probably *Acetabularia*, *Nitella*, *Penicillus*, and *Padina* and coralline algae such as *Arthrocardia silvae* and *Amphiroa beauvoisii* (Figure 9.7). Organisms that form $CaCO_3$ intracellularly and then export it to the cell surface include the coccolithophores (green algae) (Figure 9.8), such as *Scyphosphaera*, *Rhabdosphaera*, and *Calciococcus*, and foraminifera (Protozoa, Sarcodinae), such as *Heterostegina* and *Globigerina* (Figure 9.9). The mineral form of the calcium carbonate that is deposited is either calcite or aragonite (Lewin, 1965). The amount of carbon incorporated into carbonate by algae as a result of photosynthesis may be a significant portion of the total carbon assimilated (Jensen et al., 1985). Wefer (1980) measured *in situ* $CaCO_3$ production by *Halimeda incrassata*, *Penicillus capitatus*, and *Padina sanctae-crucis* in Harrington Sound, Bermuda, that amounted to approximately 50, 30, and 240 g m^{-2} per year, respectively.

Chara and *Halimeda* induce calcification by creating conditions of supersaturation at specific sites at the cell surface. These sites involve internodal cells in *Chara* and intercellular spaces in *Halimeda* (de Vrind-de Jong and de Vrind, 1997). Gelatinous or mucilaginous substances in association with the cell walls of these algae may be involved in the deposition and organization of the crystalline $CaCO_3$. If the CO_2 for forming the algal carbonate comes from seawater and is transformed to carbonate as a result of photosynthesis, it is likely to be enriched in ^{13}C relative to the carbon in seawater CO_2, whereas if the CO_2 from which algal carbonate is formed has a respiratory origin, the algal carbonate is likely to be enriched in ^{12}C relative to the carbon in seawater CO_2 (Lewin, 1965). The basis for the morphogenesis of structures as intricate as those found associated with calcareous foraminifera remains to be explained in detail. Involvement of the Golgi apparatus in the cytoplasm is likely.

Some insight into intracellular coccolith assembly has been gained. In the coccolithophore *Pleurochrysis carterae*, coccolith formation is associated with coccolithosome-containing vesicles and cisternae-containing scales in the Golgi apparatus, which is located in the cytoplasm of the cells. Once formed, the coccoliths are exported to the cell surface (van der Wal et al., 1983; de Jong et al., 1983; de Vrind-de Jong and de Vrind, 1997). By contrast, coccoliths of *Emiliania huxleyi* are formed in special vesicles called the coccolith-production compartments, which may be derived by the fusion of Golgi vesicles (de Vrind-de Jong and de Vrind, 1997). Acidic polysaccharides appear to play an organizing role (template?) in $CaCO_3$ deposition. In strains A92, L92, and 92D of

(A)

(B)

FIGURE 9.7 Articulated coralline (calcareous) algae. (A) *Arthrocardia silvae*, Johansen, California. Scale bar equals 1.5 cm. (B) *Amphiroa beauvoisii*, Lamouroux, Gulf of California. Scale bar equals 1.5 cm. (Courtesy of Johansen HW, Department of Biology, Clark University, Worcester, MA, USA.)

E. huxleyi, acidic Ca^{2+}-binding polysaccharides have been found to inhibit precipitation of $CaCO_3$ *in vitro* (Borman et al., 1982, 1987) and are thought to regulate $CaCO_3$ precipitation in the intracellular coccolith-forming vesicles (see de Jong et al., 1983). *E. huxleyi* derives its carbonate mainly from photosynthesis, whereas *P. carterae* derives it from photosynthesis in the light and respiration in the dark (de Vrind-de Jong and de Vrind, 1997).

9.2.7 MODELS FOR SKELETAL CARBONATE FORMATION

A possible clue to the biochemical mechanism of $CaCO_3$ deposition in cell structures, especially protozoa (heterotrophs), was suggested in an observation by Berner (1968). He noted that during bacterial decomposition of butterfish and smelts in sweater in a sealed jar, calcium was precipitated

FIGURE 9.8 Coccoliths, the calcareous skeletons of an alga belonging to the class Chrysophyceae (see sketch in Figure 5.9). These specimens were found residing on the surface of a ferromanganese nodule from Blake Plateau of the Atlantic coast of the United States. (From LaRock PA, Ehrlich HL, *Microb. Ecol.*, 2, 84–96, 1975. With permission from Springer Science and Business Media.)

not as $CaCO_3$ but as calcium soaps or adipocere (calcium salts of fatty acids) despite the presence of HCO_3^- and CO_3^{2-} species and an alkaline pH in the reaction mixture. The prevailing fatty acid concentration favored calcium soap formation over calcium carbonate formation. Berner suggested that in nature such soaps could later be transformed into $CaCO_3$ through mineralization of the fatty acid ligand. However, an actual analog of such reactions in calcification in cells has not been reported to date.

McConnaughey (1991) has proposed that *Chara corallina*, a calcareous alga that lays down $CaCO_3$ pericellularly, forms it at the cell surface by a process involving an ATP-driven H^+/Ca^{2+} exchange. In this process, protons produced in the reaction

$$Ca^{2+} + CO_2 + H_2O \rightarrow CaCO_3 + 2H^+ \qquad (9.21)$$

which is assumed to occur on the $CaCO_3$ surface facing the cell, are exchanged for intracellular Ca^{2+}. The protons then react with HCO_3^- in the cell to generate CO_2 for photosynthesis.

$$HCO_3^- + H^+ \rightarrow CO_2 + H_2O \qquad (9.22)$$

In other organisms that form $CaCO_3$, calcium intended for calcium carbonate structures may first be localized by the formation of an organic calcium salt at the site of calcium deposition (e.g., the plasma membrane, Golgi apparatus) (see de Vrind-de Jong and de Vrind, 1997). It may then be converted to $CaCO_3$ in the presence of carbonic anhydrase, an enzyme that can promote the conversion of dissolved metabolic CO_2 to bicarbonate and carbonate in a reversible reaction. Carbonic anhydrase has been detected in some microalgae and cyanobacteria (Aizawa and Miyachi, 1986). Although Aizawa and Miyachi considered the activity of caronic anhydrase from the standpoint of its role in CO_2 assimilation in photosynthesis, it may very well also play a role in CO_2 conversion to carbonate under different physiological conditions. *Hymenomonas*, a coccolithophorid, has been

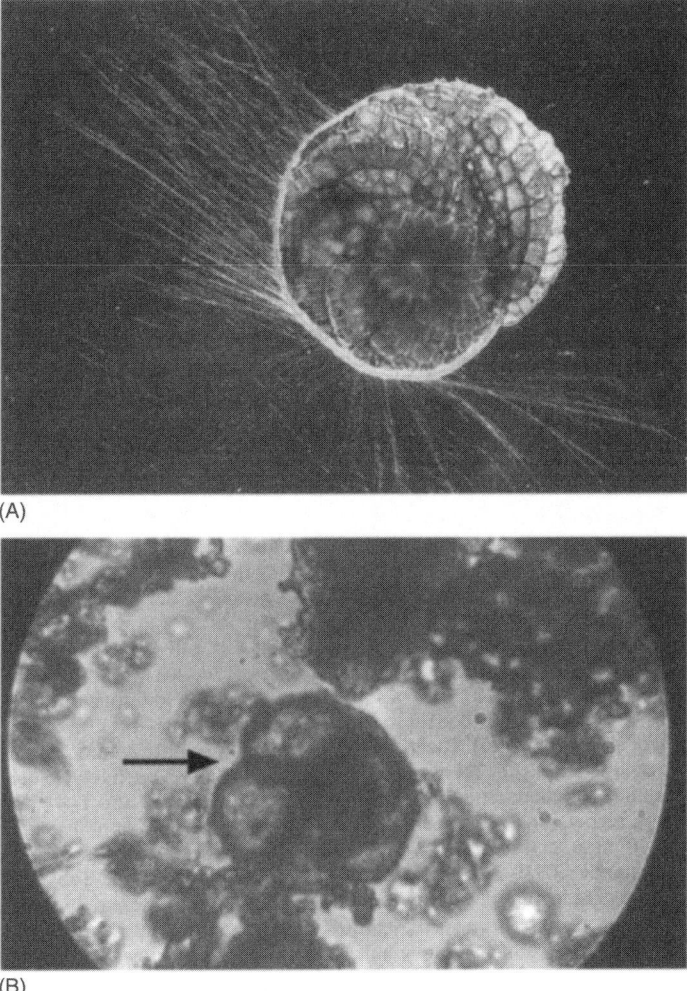

(A)

(B)

FIGURE 9.9   Foraminifera. (A) A living foraminiferan specimen, *Heterostegina depressa*, from laboratory cultivation. Note the multichambered test and the projection of fine protoplasmic threads from the test. Test diameter is 3 mm. (Courtesy of Röttger R, *Marine Biol.* (Berlin), 26, 5–12, 1974.) (B) Foraminiferan test (arrow) in Pacific sediment: *Globigerina* (?) (×2430).

reported to utilize "a hydroxyproline–proline-rich peptide and sulfated polysaccharide moieties" in $CaCO_3$ deposition (Isenberg and Lavine, 1973).

9.2.8   MICROBIAL FORMATION OF CARBONATES OTHER THAN THOSE OF CALCIUM

9.2.8.1   Sodium Carbonate

Carbonate may occur in solid phases not only as calcium or calcium and magnesium salts but also as a sodium salt (natron, $Na_2CO_3 \cdot 10H_2O$). In at least one instance, the Wadi Natrun in the Libyan Desert, such a deposit has been clearly associated with the activity of sulfate-reducing bacteria (Abd-el-Malek and Rizk, 1963b). As the authors described it, the Wadi (the channel of a water-course that is dried up except during periods of rainfall, an arroyo) in this case contains a chain of small lakes 23 m below sea level. The smallest of the lakes dries up almost completely during summer, and the larger ones dry out partially at that time. Natron is in solution in the water of all of the lakes and in solid form at the bottom of some of the lakes. The water feeding the lakes is supplied by springs and partly by streamlets, which probably derive their water from the nearby Rosetta branch of the Nile. On its way to the lakes, the surface water passes through cypress swamps. The authors found sulfate and carbonate at high concentration in the lakes (189–204 meq of carbonate per liter and 324–1107 meq of sulfate per liter) and at low concentration in the cypress swamps (0 meq to traces of carbonate per liter and 2–13 meq of sulfate per liter). Bicarbonates were present in significant amounts in the lakes and swamps (11–294 and 11–16 meq L^{-1}, respectively). Soluble sulfides predominated in the lakes and swamps (7–13 and 1–4 meq L^{-1}, respectively). Considerable numbers of sulfate-reducing bacteria (1×10^6 to 8×10^6 mL^{-1}) were detected in the swamps and in the soil at a distance of 150 m from the lakes, but not elsewhere (see Table 9.2 for more detailed data). Sulfate reduction was inferred to occur chiefly in the cypress swamps because of the significant presence of sulfate reducers and the readily available organic nutrient supply at those sites. The sulfate reduction leads to the production of bicarbonate as follows:

$$SO_4^{2-} + 2(CH_2O) \rightarrow H_2S + 2CO_2 + 2OH^- \qquad (9.23)$$

$$2CO_2 + 2OH^- \rightarrow 2HCO_3^- \qquad (9.24)$$

Most of the soluble products of sulfate reduction, HCO_3^- and HS^-, were found to be washed into the lakes, where, upon evaporation, they were concentrated and partially precipitated as carbonate, including natron, and sulfide salts. Some of the sulfide produced in the swamps was found to combine with iron to form ferrous sulfide, which imparts a characteristically black color to the swamps. The carbonate in the lakes resulted from the loss of CO_2 from the water to the atmosphere, which was promoted by the warm water temperatures, especially in summer (CO_2 solubility in water decreases with increase in temperature) (Equation 9.20).

9.2.8.2 Manganous Carbonate

Carbonate may also combine with manganous manganese to form rhodochrosite ($MnCO_3$) in nature. The origin of at least some $MnCO_3$ deposits has been ascribed to microbial activity. An example is the occurrence of rhodochrosite together with siderite ($FeCO_3$) in Punnus-Ioki Bay of Lake Punnus-Yarvi on the Karelian Peninsula in Russia (Sokolova-Dubinina and Deryugina, 1967). Lake Punnus-Yarvi is 7 km long and 1.5 km wide at its broadest part. It has a greatest depth of 14 m and is only slightly stratified thermally. The authors reported that the oxygen concentration in its surface water ranges from 11.8 to 12.1 mg L^{-1} and in its bottom water from 5.7 to 6.6 mg L^{-1}. The pH of its water was given as slightly acidic (6.3–6.6). The Mn^{2+} concentration ranged from 0.09 to 0.02–0.2 mg L^{-1} in its bottom water (1.4 mg L^{-1} in winter). The lake is fed by the Suantaka-Ioki and Rennel Rivers and 24 small streams that drain surrounding swampland. The lake in turn feeds into the Punnus-Yarvi River. It is estimated that 48% of the water in the lake is exchanged every year. The manganese and the iron in the lake are derived from surface and underground drainage of Cambrian and Quaternary glacial deposits, carrying 0.2–0.8 mg of manganese per liter and 0.4–2 mg of iron per liter. The oxidized forms of manganese and iron are incorporated into silt, where they are reduced and subsequently concentrated by upward migration to the sediment–water interface and reoxidation into lake ore. This occurs mostly in Punnus-Ioki Bay, which has oxide deposits on the sediment at water depth down to 5–7 m. The oxide layer has a thickness of 5–7 cm.

All of the sediment and ore samples taken from the lake (mainly Punnus-Ioki Bay) contained manganese-reducing bacteria. They were concentrated chiefly in the upper sediment layer. They included an unidentified, non-spore-forming bacillus in addition to *Bacillus circulans* and *B. polymyxa*. Only limited numbers of sulfate-reducing bacteria, which the investigators (Sokolova-Dubinina and Deryugina, 1967) attributed to the lack of extensive accumulation of organic matter,

TABLE 9.2
Chemical and Bacteriological Analyses of Water and Soil Samples from Wadi Natrun

Type and Source of Sample	pH Range	Milliequivalents[a] of					Total Soluble Salts (g L^{-1})	Organic Matter (%)	Viable Counts of Sulfate Reducers[b] (10^6 mL^{-1})
		CO$_3^{2-}$	HCO$_3^-$	SO$_4^{2-}$	Cl$^-$	S^{2-}			
Water									
Artesian wells	7.4–7.8	0	2–5	9–13	18–30	0.2–0.5	2–3		—[c]
Burdi swamps	6.8–7.2	0–trace	11–61	2–13	1–7	1–4	1–8		1–5
Lakes	9.5–10.1	189–240	22–294	324–1107	107–210	7–13	180–239		—[c]
Soil									
Newly cultivated uplands	7.0–7.6	0	1–2	14–18	11–19[d]	—[d]	—[d]	0.2–0.5	—[c]
About 150 m from lakes	7.2–7.5	0	2–3	12–23	1–6[d]	—[d]	—[d]	0.1–5.2	5–8
Swamps	7.4–7.8	Trace	3–11	4–7	3–8[d]	—[c]	—[d]	3.4–7.8	0.8–2

[a] Milliequivalents per liter of water and per 100 g of soil.
[b] Counts per milliliter of water and per gram of soil.
[c] Not detected.
[d] Not determined.

Source: From Abd-el-Malek Y, Rizk SG, *J. Appl. Microbiol.*, 26, 20–26, 1963b. With permission.

were recovered from the ore. They were associated with hydrotroilite (FeS·nH$_2$O). Carbonates of calcium and manganese at most stations in the lake and bay were reported of low occurrence (0.1%, calculated on the basis of CO$_2$). However, in a limited area near the center of Punnus-Ioki Bay, ore contained as much as 4.7% calcite, 5.96% siderite, and 4.99% rhodochrosite, together with 15.8% hydrogoethite and 38.9% barium psilomelanes and wads (complex oxides of manganese) (Sokolova-Dubinina and Deryugina, 1967). The investigators related the relatively localized concentration of carbonates to the localized availability of organic matter and its attack by heterotrophs. The organic matter was the ultimate source of CO$_2$–CO$_3^{2-}$ and the cause of the essentially low E$_h$. It was noted that the decaying remains of the plant life on the lakeshore accumulated in sufficient quantities in the Punnus-Ioki Bay area only where extensive carbonate ores were found. The weak sulfate-reducing activity at this location may explain the low iron and manganese sulfide formation and the significant carbonate formation. In view of much more recent discoveries of various new types of sulfate-reducing bacteria and the ability of a number of them to use ferric iron and Mn(IV) oxides in place of sulfate as terminal electron acceptor (see Chapters 16, 17, and 19), this site should be reinvestigated using phylogenetic probes and physiological tests.

Mixed deposits including manganous and ferrous carbonates, manganous and ferrous sulfides, and manganous and calcium-iron phosphates have been found in the sediments of Landsort Deep in the central Baltic Sea, which is an anoxic site. The minerals are thought to have formed authigenically as a result of microbial mineralization of organic matter (Suess, 1979). Metal carbonate and sulfide deposition appeared to be compatible there, perhaps because iron and manganese were available in nonlimiting supply.

Bacterial rhodochrosite (manganous carbonate) formation in the reduction of Mn(IV) oxide has been observed in pure culture under laboratory conditions using isolate GS-15, which is now known as *Geobacter metallireducens* (Lovley and Phillips, 1988).

9.2.8.3 Ferrous Carbonate

Siderite (FeCO$_3$) beds in the Yorkshire Lias in England are thought to have resulted from Fe$_2$O$_3$ reduction and subsequent reaction with HCO$_3^-$. The bicarbonate could have resulted from microbial mineralization of organic matter. At the time of this study, formation of siderite at this spot was explained on the basis of an exclusion of sulfate by an overlying clay layer, which blocked the entry of sulfate to the site of siderite deposition. If sulfate had entered the site, it would have been expected to be bacterially reduced to sulfide, leading to the preferential formation of iron sulfide instead of siderite (see Sellwood, 1971).

Microbial precipitation in siderite formation is supported by recent observations in rapidly accreting tidal marsh sediments at very shallow depths on the Norfolk coast, England. Extensive bacterial decay of organic matter is occurring there at low interstitial sulfate and sulfide concentrations (Pye, 1984; Pye et al., 1990). Despite the low interstitial sulfate concentration, sulfate-reducing bacteria were detected. Scanning electron microscopic examination and x-ray powder diffraction analysis of siderite concretions from this site revealed that siderite formed a void-filling cement and coating around quartz grains. Traces of greigite (Fe$_3$S$_4$), iron monosulfide, and calcite were also detected (Pye, 1984). Carbon isotope fractionation studies supported a microbiological role in the formation of siderite (Pye et al., 1990). The investigators explained the simultaneous formation of siderite and ferrous sulfide as a result of a faster rate of reduction of ferric iron than sulfate. At that time, the investigators believed that sulfate and ferric iron reductions were always caused by two different types of organisms. Since then, it has been shown that some sulfate reducers can use ferric iron as an alternative terminal electron acceptor in place of sulfate (Coleman et al., 1993; Lovley et al., 1993).

Coleman et al. (1993) demonstrated that *Desulfovibrio* in salt marsh sediment from the Norfolk site could reduce ferric to ferrous iron with H$_2$ as electron donor. When neither sulfate as electron acceptor nor H$_2$ as electron donor was limiting, both sulfate and ferric iron were reduced. However, when H_2 was limiting but sulfate was not, ferric iron seemed to be the preferential electron acceptor. Interestingly, significantly more of the carbonate (75%) in the siderite nodules in the marsh derived from seawater than from the degradation of organic matter (25%). Thus in the case of siderite formation in the Norfolk coastal marsh, bacterial Fe(III) reduction made a more important contribution to siderite formation than microbial mineralization of organic carbon.

Ehrlich and Wickert (1997) observed siderite formation by bacteria resident on crushed bauxite in a column experiment in the laboratory. The columns containing bauxite ore were fed with a sucrose–mineral salts solution. The siderite was detected by x-ray diffraction analysis of ore residue taken from the column at the end of the experiment. Concurrent iron sulfide formation was noted in this experiment. The source of the sulfide was the reduction of sulfate in the mineral salts solution. Sulfate-reducing bacteria as well as other kinds of bacteria were detected in the column.

9.2.8.4 Strontium Carbonate

Strontium carbonate is little more insoluble ($K_{sol} = 10^{-8.8}$) than calcium carbonate ($K_{sol} = 10^{-8.07}$), so it should not be surprising that under appropriate conditions Sr^{2+} can be precipitated by biogenic carbonate. Three cases of microbial strontium carbonate formation have been reported so far—all under laboratory conditions. In the first instance, Roemer and Schwartz (1965) showed that sulfate-reducing bacteria could form strontianite ($SrCO_3$) from celestite ($SrSO_4$). In the second instance, Anderson and Appanna (1994) showed that a soil strain of *Pseudomonas fluorescens* could form crystalline strontium carbonate from 5 mM strontium citrate. The strontium carbonate was identified by x-ray fluorescence spectroscopy, x-ray diffraction spectrometry, Fourier transform infrared spectroscopy, and acid treatment. The source of the carbonate was the mineralization of citrate by the bacterium. In the third instance, Roden and Ferris (2000) demonstrated the formation of strontium carbonate during amorphous hydrous ferric oxide reduction by *Shewanella putrefaciens* CN32 in a defined bicarbonate-buffered, Ca-free, and Ca-amended medium. The $SrCO_3$ was incorporated in to the ferrous carbonate (siderite) produced in the ferric oxide reduction at initial Sr concentrations of 0.01 and 0.1 M. According to the authors, the aqueous phase at these two Sr concentrations was undersaturated with respect to $SrCO_3$ at all times. It was not reported as to whether any Sr was incorporated into calcite in experiments with added calcium. Similar observations of strontium immobilization during hydrous ferric oxide reduction were made with *Shewanella alga* BrY (Parmar et al., 2000).

In nature, Ferris et al. (1995) observed precipitation of strontium calcite on a serpentine outcrop in a groundwater discharge zone near Rock Creek, British Columbia, Canada, as a result of the photosynthetic activity of epilithic cyanobacteria, including *Calothrix*, *Synechococcus*, and *Gloeocapsa*. The strontium content of calcite was up to 1 wt%. The cyanobacteria served as nucleation sites in the formation of calcite. The Ca concentration in water samples from the study site was 32–36 ppm and the Sr concentration was 5.8–6.6 ppm. The pH of the water ~2 m above the outcrop was 8.5 and at its base (0.2 m level) 8.8. This pH difference was attributed to cyanobacterial photosynthesis. The existence of strontium in calcite was explained as a homogenous solid solution of $SrCO_3$ in calcite. Such solid solution of $SrCO_3$ was also observed when calcite was precipitated by *Bacillus pasteurii* in a urea medium containing Ca^{2+} and small amounts of Sr^{2+} (Warren et al., 2001). The Sr^{2+} apparently substitutes for Ca^{2+} in some structural sites of calcite.

9.2.8.5 Magnesium Carbonate

Microbial communities in which actinomycetes belonging to the genus *Streptomyces* predominate may play a role in the formation of hydromagnesite ($Mg_5(CO_3)_4(OH)_2 \cdot 4H_2O$) (Cañaveras et al., 1999). These organisms were found associated with moonmilk deposits containing hydromagnesite and needle-fiber aragonite in Altamira Cave in northern Spain.

9.3 BIODEGRADATION OF CARBONATES

Carbonates in nature may be readily degraded as a direct or indirect result of biological activity, especially microbial activity (Golubic and Schneider, 1979). A chemical basis for this decomposition is the instability of carbonates in acid solution. For instance

$$CaCO_3 + H^+ \rightarrow Ca^{2+} + HCO_3^- \tag{9.25}$$

$$HCO_3^- + H^+ \rightarrow H_2CO_3 \tag{9.26}$$

$$H_2CO_3 \rightarrow H_2O + CO_2\uparrow \tag{9.27}$$

Because $Ca(HCO_3)_2$ is highly soluble compared with $CaCO_3$, the latter begins to dissolve even in weak acid solutions. In stronger acid solutions, $CaCO_3$ dissolves more rapidly because, as Equation 9.27 shows, the CO_2 is likely to be lost from solution due to degassing. Therefore, from a biochemical standpoint, any microorganism that generates acid metabolic wastes is capable of dissolving insoluble carbonates. Even the mere metabolic generation of CO_2 during respiration can have this effect, because

$$CO_2 + H_2O \rightarrow H_2CO_3 \tag{9.28}$$

and

$$H_2CO_3 + CaCO_3 \rightarrow Ca^{2+} + 2HCO_3^- \tag{9.29}$$

Thus it is not surprising that various kinds of CO_2- and acid-producing microbes have been implicated in the breakdown of insoluble carbonates in nature.

9.3.1 BIODEGRADATION OF LIMESTONE

Breakdown of lime in the cement of reservoir walls and docksides was attributed in part to bacterial action as long ago as 1899 (Stutzer and Hartleb, 1899). However, extensive investigation into microbial decay of limestone was first undertaken three decades later by Paine et al. (1933). These workers showed that both sound and decaying limestones usually carried a sizeable bacterial flora, the numbers ranging from 0 to over 8×10^6 g^{-1}. The size of the population in a particular sample seemed to depend in part on the environment around the stone. As one might expect, the surface of the limestone was generally more densely populated than its interior. The authors suggested that in many limestones the bacteria were unevenly distributed, inhabiting pockets and interstices in the limestone structures. The kinds of bacteria found in limestones that they examined included gram-variable and gram-negative rods and cocci. Spore formers, such as *Bacillus mycoides*, *B. megaterium*, and *B. mesentericus*, appeared to have been rare or absent. The investigators performed an experiment to estimate the rate of limestone decay through bacterial action under controlled conditions. They employed a special apparatus in which evolved CO_2 was trapped in barium hydroxide solution. They found 0.18 mg of CO_2 per hour per 350 g of stone to be evolved in one case, and 59 mg of CO_2 per hour per 350 g of stone in another. In the latter instance, they calculated from the data that 28 g of CO_2 would have been evolved from 1 kg of stone in 1 year. As expected, the rate of CO_2 evolution from decaying stone was greater than from sound stone. Although organic acids and CO_2 from heterotrophic metabolism of organic matter were the cause of the observed limestone decay in these experiments, autotrophic nitrifying and sulfur-oxidizing bacteria were also shown to be able to promote limestone decay. These latter organisms accomplished it through the production of nitric and sulfuric acids from ammonia and reduced forms of sulfur, respectively. Nitrifying and sulfur-oxidizing bacteria were detected by Paine et al. (1933) in some limestone samples.

In a much later study, variable numbers of fungi, algae, and ammonifying, nitrifying, and sulfur-oxidizing bacteria were found on the surface of some limestones in Germany (Krumbein, 1968). Krumbein found that the number of detectable organisms depended on the type of stone, the length of elapsed time since the collection of the stone from a natural site, the surface structure of the stone (i.e., the degree of weathering), the cleanliness of the stone, and the climate and microclimatic conditions prevailing at the collection site. In the case of strongly weathered stone, the bacteria had sometimes penetrated the stone to a depth of 10 cm. Ammonifying bacteria were generally most numerous. Nitrogen-fixing bacteria were few, and denitrifiers were absent. The number of bacteria was not directly related to the presence of lichens or algae. The greatest number of ammonifiers was found on freshly collected and weathered stone. This was related to the pH of the stone surface (pH 8.1–8.3 in water extract). Sulfur oxidizers were more numerous in city environments, where the atmosphere contains more oxidizable sulfur compounds than in the countryside. The concentration of nitrifiers on limestone could not be correlated with city and country atmospheres. The number of ammonifiers on limestone was also found to be greater in stones exposed to city air than in stones exposed to country air. This can be explained on the basis that city air contains more organic pollutants that serve as nutrients for these bacteria than does country air. Laboratory experiments by Krumbein confirmed the weathering of limestone by its natural flora. In the observations, the ammonifiers were less directly responsible for the weathering of limestone than they were in generating ammonia from which the nitrifying bacteria could form nitric acid, which then corroded the limestone.

In yet another important study of the decomposition of limestone, numerous bacteria and fungi were isolated from a number of samples (Wagner and Schwartz, 1965). The active organisms appeared to weather the stone through the production of oxalic and gluconic acids. The investigators also noted the presence of nitrifying bacteria and thiobacilli in their samples and that the corresponding mineral acids produced by these organisms also contributed to the weathering of the stone.

Marble, a metamorphic type of $CaCO_3$ rock, can also be attacked by microorganisms. Figure 9.10 shows the corrosive effect of microcolonial black yeast on marble from the Dionysos Theater of the

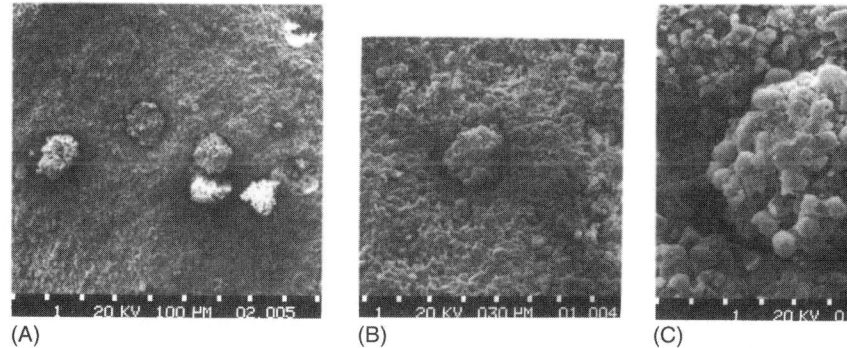

(A)   (B)   (C)

FIGURE 9.10   (A)–(C) SEM photomicrographs showing a section of marble from the Dionysos Theater, Acropolis, Athens, Greece, at three different magnifications (note scale and scale marks on the bottom of each photograph). The marble has been extensively corroded by *biopitting*. Deep holes of different sizes (between 2 and 5 mm in diameter and depth) were incised chemically (etched by metabolically produced substances) and physically (mechanical action) by black yeasts and meristematically growing yeastlike fungi. The microcolonial fungi can be confused with algae in SEM micrographs. The fungi have a cell size similar to that of the marble grain. The chemical and physical corrosive actions of these fungi have been demonstrated in laboratory experiments. (Courtesy of Krumbein WE, from Anagnostidis K, Gehrmann K, Gross M, Krumbein WE, Lisi S, Panasidou A, Urzì C, Zagari M. Biodeterioration of marbles of the Parthenon and Propylea, Acropolis, Athens—Associated organisms, decay, and treatment suggestions, in Decrouez D, Chamay J, Zezza F, eds. II International Symposium for the Conservation of Monuments in Mediterranean Basin. Musée d'Histoire Naturelle, Genève, Switzerland, 1992.)

Acropolis in Athens, Greece (Anagnostidis et al., 1992). *Micrococcus halobius* was shown to colonize the surfaces of Carrara marble slabs, forming biofilm and producing gluconic, lactic, pyruvic, and succinic acids from glucose that etched the marble surfaces (Urzì et al., 1991). This organism also caused a discoloration of the marble surface, suggesting that natural patina on marble structures may have a microbial origin.

The microbial weathering of rock surfaces such as marble, basalt, and granite, the latter two of which are not $CaCO_3$ rock, may involve not only dissolution but also precipitation of new secondary minerals. The new minerals appear as surface crusts and may include calcite, apatite, and wilkeite (Urzì et al., 1999). Such secondary $CaCO_3$ precipitation caused by microbes may be exploitable in the preservation of monuments and statuary made from carbonate rock, as studied under laboratory conditions with *Myxococcus xanthus* as the inducer of $CaCO_3$ (Rodriguez-Navarro et al., 2003).

Black fungi of the Dematiaceae have been implicated in the destruction of some marble and limestone (Gorbushina et al., 1994). However, instead of attacking the marble through the formation of corrosive metabolic products such as acids, the fungi appear to attack the marble by exerting physical pressure in pores and crevices in which they grow and by changing the water activity in superficial polymers of the cells and surrounding the cells in the stone. The melanin pigment produced by the fungi has been implicated in the blackening of the surfaces of marble structures.

Cave formation in a limestone region (karst) is under the influence of various microbes. Some microbes, including autotrophs and heterotrophs, generate the corrosive agents that attack the $CaCO_3$ of limestone and dissolve it. Others may reduce oxidized sulfur minerals such as gypsum or oxides of iron or manganese (see Herman, 1994, and other papers cited therein). Some caves, such as Movile Cave in Romania, which receives little input from the surface environment, have developed light-independent ecosystems in which the primary producers are chemosynthetic instead of photosynthetic autotrophs (Sarbu et al., 1994, 1996).

9.3.2 CYANOBACTERIA, ALGAE, AND FUNGI THAT BORE INTO LIMESTONE

Endolithic cyanobacteria, algae, and fungi have been found to cause local dissolution of limestone, thereby forming tubular passages in which they can grow (Figure 9.11; Golubic et al., 1975). The kinds of limestone they attack in nature include coral reefs, beach rock, and other types. Active algae include some green, brown, and red algae (Golubic, 1969). The mechanism by which any of these organisms bore into limestone is not understood. Some filamentous boring cyanobacteria possess a terminal cell that is directly responsible for the boring action, presumably dissolution of the carbonate (Golubic, 1969). Different boring microorganisms form tunnels of characteristic morphology (Golubic et al., 1975). In a pure mineral such as Iceland spar, boring follows the planes of crystal twinning, diagonal to the main cleavage planes (Golubic et al., 1975). The depth to which cyanobacteria and algae bore into limestone is limited by light penetration in the rock, because they need light for photosynthesis. Boring cyanobacteria may have unusually high concentrations of phycocyanin, an accessory pigment of the photosynthetic apparatus, to compensate for the low light intensity in the limestone. In contrast, boring fungi are not limited by light penetration. Being incapable of photosynthesis, they have no need for light.

Endolithic fungi and cyanobacteria or algae in limestone and sandstone may form a special relationship in the form of lichens (Figure 9.12). The cyanobacteria and algae in these associations share the carbon they fix with their fungal partner, whereas the fungi share minerals that they mobilize and some other less well-defined functions with their cyanobacterial or algal partners. These lichens, although growing within limestone, may serve as food to some snails, as has been reported from the Negev Desert in Israel (Shachak et al., 1987). The snails have a tonguelike organ in their mouth, the *radula*, which has toothlike structures embedded in it that are useful for abrading. The toothlike structures frequently consist of the iron mineral hematite. The snails scrape the surface of the limestone beneath which the lichens grow with their radula to feed on the lichens, which are their preferred food. They ingest some of the pulverized limestone with the lichens and leave behind a trail of this limestone powder as they move over the limestone surface. It has been

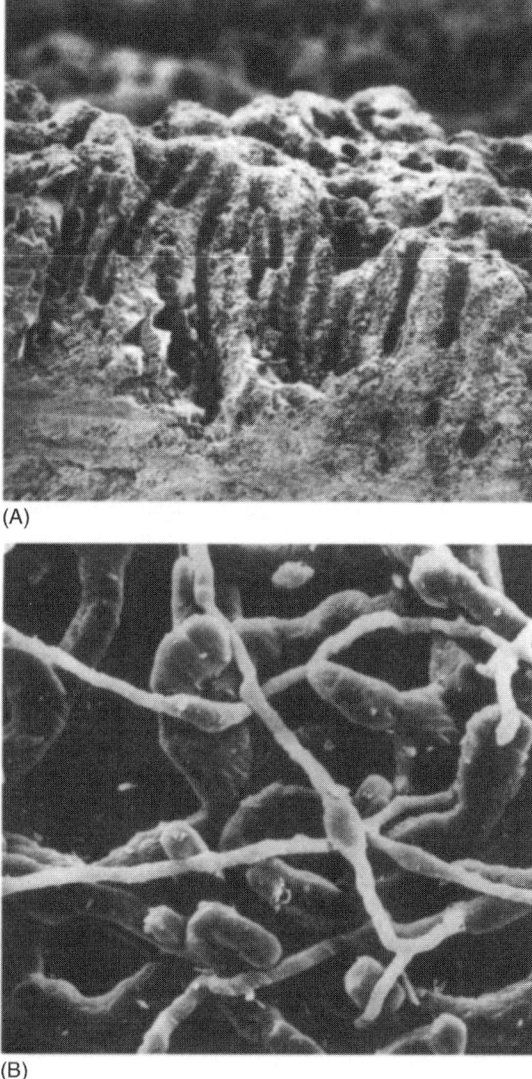

(A)

(B)

FIGURE 9.11 Microorganisms that bore into limestone. (A) Limestone sample experimentally recolonized by the cyanobacterium *Hyella balani* Lehman ($\times 234$). The exposed tunnels are the result of boring by the cyanobacterium. (B) Casts of the boring of the green alga *Eugamantia sacculata* Kormann (larger filaments) and the fungus *Ostracoblabe implexa* Bornet and Flahault in calcite spar ($\times 2000$). The casts were made by infiltrating a sample of fixed and dehydrated bore mineral with synthetic resin and the etching sections of the embedded material (e.g., with dilute mineral acid) to expose the casts of the organism. (Courtesy of Golubic S.)

estimated that in the Negev Desert this weathering affects 0.7–1.1 metric tons of rock per hectare per year (Shachak et al., 1987). A similar biological weathering phenomenon was previously noted on some reef structures in Bermuda (see Golubic and Schneider, 1979).

Cyanobacteria and fungi inhabit not only limestone but also other porous rock. To distinguish among rock-inhabiting microbes, the term *euendoliths* has been coined for true boring microbes in limestone. Opportunists that invade preexistent pores of rock where they may cause alteration of the rock in the area that they inhabit are classified as *cryptoendoliths* whereas those that invade preexistent cracks and fissures without altering the rock structure are classified as *chasmoendoliths* (Golubic et al., 1981).

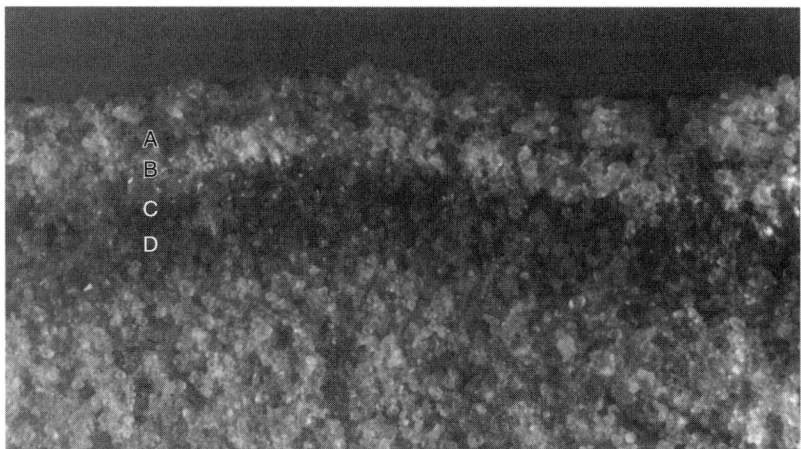

FIGURE 9.12 Cryptoendolithic microorganisms in vertically fractured Beacon sandstone. (A) Lichen (small black bodies between rock particles). (B) Zone of fungus filaments. (C) Yellowish-green zone of unicellular cyanobacterium. (D) Blue-green zone of unicellular cyanobacterium. The color difference between (C) and (D) is not apparent in this black-and-white photograph. Sample A76-77/36, north of Mount Dido, elevation 1750 m, magnification ×4.5. (Courtesy of Friedmann EI, *Antarctic J. U.S.*, 12, 26–30, 1977.)

In some environments, such as Antarctic dry valleys, cryptoendolithic cyanobacteria (Friedmann and Ocampo, 1976) and lichens (in this instance symbiotic associations of a green alga and a filamentous fungus) (Friedmann, 1982; Friedmann and Ocampo-Friedmann, 1984) inhabit superficial cavities in sandstone, 1–2 mm below the surface (Figure 9.12). The near-surface locale in the sandstone (orthoquartzite) is the major habitat for microorganisms in this inhospitable environment (Friedmann, 1982). These cryptoendolithic microorganisms appear to differ from the boring microbes in that they are not directly responsible for cavity formation in the sandstone but instead invade preexistent pores. However, they promote exfoliation of the sandstone surface—a physical process facilitated through solubilization of the cementing substance ($CaCO_3$?) that holds the quartz grains together (Friedmann and Weed, 1987). The lichen activity may manifest itself visibly by mobilization of iron in the region of lichen activity. Exfoliation may expose the lichens to the external environment from which, if they survive the exposure, they may reinvade the sandstone through pores. Fissures in granite and grandiorite in the Antarctic dry valleys may also be inhabited by lichens and coccoid cyanobacteria (Friedmann, 1982). These organisms are clearly not boring microorganisms.

The presence of cryptoendoliths has also been recently reported from rock outcroppings (sandstone) in the Canadian High Arctic, that is, Ellesmere Island, Nunavut, Canada, where the temperatures are warmer and the precipitation is more extensive than in the Antarctic dry valleys (Omelon et al., 2006a,b). The cryptoendoliths were found to reside in small spaces, 0.5–5 mm below the sandstone surface. The organisms were represented by cyanobacterial, algal, fungal, and heterotrophic bacteria (Omelon et al., 2006a,b). The authors implicated extensive exopolysaccharide (EPS) produced by the cyanobacteria as an important source of nutrients for the heterotrophic bacteria.

Cryptoendoliths are not unique to the Antarctic dry valleys but are also found in hot desert environments (e.g., the Mojave Desert, California; Sonoran Desert, Mexico; Negev Desert, Israel) (Friedmann, 1980). Because the hot deserts, such as the cold Antarctic deserts, represent extreme environments in which moisture is a limiting factor for survival, rock interiors afford protection and permit life to persist.

Walker and Pace (2007) have recently reported on the phylogenetic makeup of endolithic communities that they found in sandstone, limestone, and granite cliffs in semiarid montane zones in

Colorado and Wyoming, United States, using culture-independent methodology involving determination of the rRNA gene sequence content of these communities. Their results indicated phylogenetic affiliation to the Bacteria (13 different groups of which actinobacterial, cyanobacterial, and proteobacterial affiliations were the most common), the Archaea (Crenarchota), and the Eucarya (Chlorophytes). Interestingly, they did not report the detection of rRNA genes from members of the Fungi. The authors concluded that the enodolithic communities were very stable and exhibited limited diversity compared with the other types of microbial communities and proposed that they are seeded by members from a global metacommunity that are uniquely adapted to the endolithic niche, as previously proposed by Friedmann and Ocampo-Friedmann (1984). The authors also inferred from their observations that the endolithic communities they studied were under biogeographical controls, which included not only rock type, but also climate and water chemistry.

9.4 BIOLOGICAL CARBONATE FORMATION AND DEGRADATION AND THE CARBON CYCLE

The biological aspect of the carbon cycle involves chiefly the fixation of inorganic carbon (CO_2 or its equivalent) to form organic carbon and the remineralization of some of this organic carbon to inorganic carbon (Figure 9.13). The carbon cycle operates in such a way that a certain portion of the biologically fixed carbon is transitionally immobilized as *standing crop* (see Smith, 1981). However, the standing crop undergoes continual turnover at such a rate that its size does not markedly change unless it is subjected to some major environmental change.

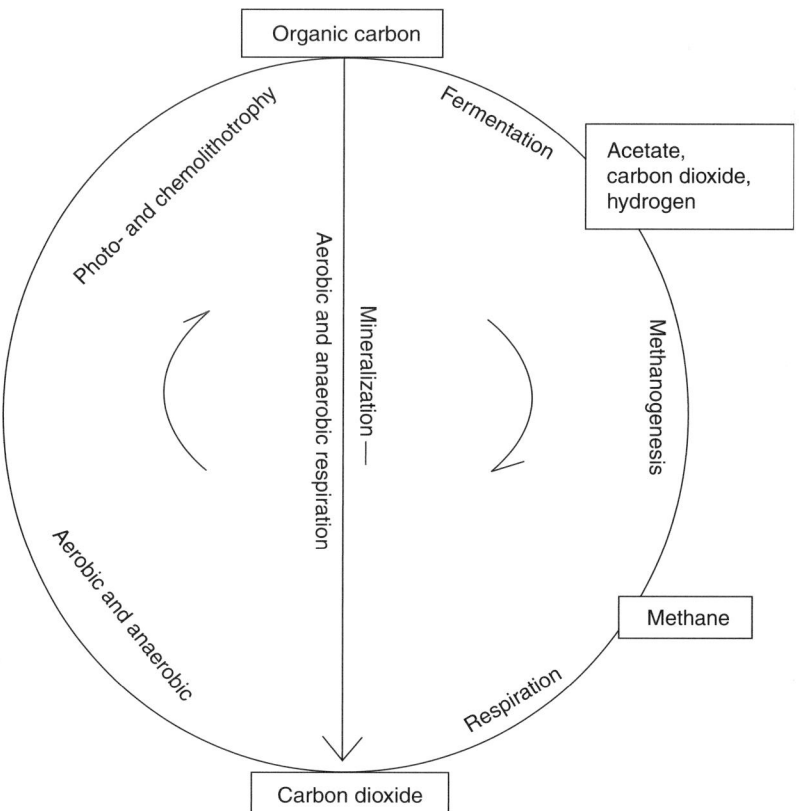

FIGURE 9.13   The carbon cycle.

Some sediments become a permanent sink of organic matter trapped in them. Through burial, such organic matter is removed from the carbon cycle and may be gradually transformed into coal, petroleum, bitumen, kerogen, and natural gas (see Chapter 22), or, very rarely, graphite or diamonds. The carbon in these substances reenters the carbon cycle with few exceptions only through human intervention (the exploitation of fossil fuels).

As this chapter has demonstrated, inorganic carbon may also be removed from the carbon cycle through biotransformation into inorganic carbonates. Such carbon when trapped in sediments may not reenter the active carbon cycle for extended geologic time intervals. Reentry of carbon into the cycle, as this chapter has shown, is under extensive biological control. Indeed, the biological component of the carbon cycle as a whole exerts the chief directing influence over it.

For a more detailed discussion of the carbon cycle on Earth, the following references should be examined: Bolin (1970), Golubic et al. (1979), Krumbein and Swart (1983), and Post et al. (1990).

9.5   SUMMARY

Carbon dioxide is trapped on Earth mainly as calcium and calcium–magnesium carbonates but also to a much lesser extent as carbonates of iron, manganese, sodium, magnesium, and strontium. In many cases, these carbonates are of biogenic origin. Calcium or calcium–magnesium carbonates may be precipitated by bacteria, cyanobacteria, and fungi by (1) aerobic or anaerobic oxidation of organic carbon compounds devoid of nitrogen in neutral or alkaline environments with a supply of calcium and magnesium counterions, (2) aerobic or anaerobic oxidation of organic nitrogen compounds in unbuffered environments with a supply of calcium and magnesium counterions, (3) $CaSO_4$ reduction by sulfate-reducing bacteria, (4) the hydrolysis of urea in the presence of Ca and Mg counterions, or (5) photo- and chemosynthetic autotrophy in the presence of Ca and Mg counterions.

The complete mechanisms of calcium carbonate precipitation by living organisms is not understood in many cases, although it always depends on metabolic CO_2 production or CO_2 consumption and the relative insolubility of calcium carbonate. Bacteria and fungi deposit it extracellularly. Bacterial cells may also serve as nuclei around which calcium carbonate is laid down. Some algae and protozoa form structural calcium carbonate either extracellularly or intracellularly. Structural $CaCO_3$ formation may involve localized fixation of calcium prior to its reaction with carbonate, regardless of whether the carbonate is of photosynthetic or respiratory origin.

Microbial calcium carbonate precipitation has been observed in soil and in freshwater and marine environments. Sodium carbonate (natron) deposition associated with microbial sulfate-reducing activity has been noted in the hot climate of the Wadi Natrun in the Libyan Desert. Ferrous carbonate deposition associated with microbial activity has been noted in some special coastal marine environments. Manganese carbonate deposition associated with microbial activity has been noted in a freshwater lake. Both ferrous and manganous carbonate formation has been observed in the laboratory with pure bacterial cultures.

Insoluble carbonates may be broken down by microbial attack. This is usually the result of organic and inorganic acid formation by the microbes but may also involve physical processes. Various bacteria, fungi, and even algae have been implicated. Such activity is evident on limestones and marble used in building construction, but it is also evident in natural limestone formations such as coral reefs, where limestone-boring cyanobacteria, algae, and fungi are active in the breakdown process. Bacteria and fungi contribute to the discoloration of structural limestone and marble and may be the cause of patina.

REFERENCES

Abd-el-Malek Y, Rizk SG. 1963a. Bacterial sulfate reduction and the development of alkalinity. II. Laboratory experiments with soils. *J Appl Bacteriol* 26:14–19.

Abd-el-Malek Y, Rizk SG. 1963b. Bacterial sulfate reduction and the development of alkalinity. III. Experiments under natural conditions in the Wadi Natrun. *J Appl Microbiol* 26:20–26.

Aizawa K, Miyachi S. 1986. Carbonic anhydrase and CO_2 concentration mechanism in microalgae and cyanobacteria. *FEMS Microbiol Rev* 39:215–233.

Anagnostidis K, Gehrmann K, Gross M, Krumbein WE, Lisi S, Panasidou A, Urzì C, Zagari M. 1992. Biodeterioration of marbles of the Parthenon and Propylea, Acropolis, Athens—Associated organisms, decay, and treatment suggestions. In: Decrouez D, Chamay J, Zezza F, eds. *II International Symposium for the Conservation of Monuments in Mediterranean Basin*. Genève, Switzerland: Musée d'Histoire Naturelle, pp. 305–325.

Anderson S, Appanna VD. 1994. Microbial formation of crystalline strontium carbonate. *FEMS Microbiol Lett* 116:43–48.

Ashirov KB, Sazonova IV. 1962. Biogenic sealing of oil deposits in carbonate reservoirs. *Mikrobiologiya* 31:680–683 (Engl transl pp 555–557).

Bavendamm W. 1932. Die mikrobiologische Kalkfällung in der tropischen See. *Arch Mikrobiol* 3:205–276.

Berg GW. 1986. Evidence for carbonate in the mantle. *Nature* (London) 324:50–51.

Berner RA. 1968. Calcium carbonate concretions formed by the decomposition of organic matter. *Science* 159:195–197.

Boetius A, Ravenschlag K, Schubert CJ, Rickert D, Widdel F, Gieske A, Amann R, Jorgensen BB, Witte U, Pfannkuche O. 2000. A marine microbial consortium apparently mediating anaerobic oxidation of methane. *Nature* (London) 407:623–626.

Bolin R. 1970. The carbon cycle. *Sci Am* 223:124–132.

Bonatti E. 1966. Deep-sea authigenic calcite and dolomite. *Science* 152:534–537.

Borman AH, de Jong EW, Huizinga M, Kok DJ, Bosch L. 1982. The role of $CaCO_3$ crystallization of an acid Ca^{2+}-binding polysaccharide associated with coccoliths of *Emiliania huxleyi*. *Eur J Biochem* 129:179–183.

Borman AH, de Jong EW, Thierry R, Westbroek P, Bosch L. 1987. Coccolith-associated polysaccharides from cells of *Emiliania huxleyi* (Haptophyceae). *J Phycol* 23:118–123.

Bowen HJM, 1979. *Environmental Chemistry of the Elements*. London: Academic Press.

British Tourist Authority, 551 Fifth Avenue, New York, USA.

Brunskill GJ, Ludlam SD. 1969. Fayetteville Green Lake, New York. I. Physical and chemical limnology. *Limnol Oceanogr* 14:817–829.

Buchanan RE, Gibbons NE, eds. 1974. *Bergey's Manual of Determinative Bacteriology*. 8th ed. Baltimore, MD: Williams and Wilkins.

Buck JD, Greenfield LJ. 1964. Calcification in marine-occurring yeast. *Bull Mar Sci Gulf Carib* 14:239–245.

Buczynski C, Chafetz HS. 1991. Habit of bacterially induced precipitates of calcium carbonate and the influence of medium viscosity on mineralogy. *J Sedim Petrol* 61:226–233.

Cañaveras JC, Hoyos M, Sanchez-Moral S, Sanz-Rubino E, Bedoya J, Soller V, Groth I, Schumann P, Laiz L, Gonzalez I, Sais-Jimenez C. 1999. Microbial communities associated with hydromagnesite and needle-fiber aragonite deposits in a karstic cave (Altamira, northern Spain). *Geomicrobiol J* 16:9–25.

Chafetz HS, Buczynski C. 1992. Bacterially induced lithification of microbial mats. *Palaios* 7:277–293.

Coleman ML, Hedrick DF, Lovley DR, White DC, Pye K. 1993. Reduction of Fe(III) in sediments by sulfate-reducing bacteria. *Nature* (London) 361:436–438.

De Boer W, la Rivière J, Schmidt K. 1971. Some properties of *Achromobacter oxaliferum*. *Leeuwenhoek* 37:553–556.

de Jong EW, van der Wal P, Borman AH, de Vrind JPM, Van Emburg P, Westbroek P, Bosch L. 1983. Calcification in coccolithophorids. In: Westbroek P, de Jong WE, eds. *Biomineralization and Biological Metal Accumulation*. Dodrecht, Holland: Reidel, pp. 291–301.

de Vrind-de Jong EW, de Vrind JPM. 1997. Algal deposition of carbonates and silicates. In: Banfield JG, Nealson KH, eds. *Geomicrobiology: Interactions Between Microbes and Minerals. Rev Mineral*, Vol. 35, Washington, DC: Mineralological Society of America, pp. 267–307.

del Moral A, Roldan E, Navarro J, Monteoliva-Sanchez M, Ramos-Cormenzana A. 1987. Formation of calcium carbonate crystals by moderately halophilic bacteria. *Geomicrobiol J* 5:79–87.

Drew GH. 1911. The action of some denitrifying bacteria in tropical and temperate seas, and the bacterial precipitation of calcium carbonate in the sea. *J Mar Biol Assoc UK* 9:142–155.

Drew GH. 1914. On the precipitation of calcium carbonate in the sea by marine bacteria, and on the action of denitrifying bacteria in tropical and temperate seas. *Carnegie Inst Washington Publ 182* 5:7–45.

Ehrlich HL, Wickert LM. 1997. Bacterial action on bauxites in columns fed with full-strength and dilute sucrose-mineral salts medium. In: Lortie L, Bédard P, Gould WD, eds. *Biotechnology and the Mining Environment. Proc 13th Annu Gen Meet Biominet* Sp 97-1. Ottawa, Canada: CANMET, Natural Resources Canada, pp. 74–89.

Feeley HW, Kulp JL. 1957. Origin of Gulf Coast salt-dome sulfur deposits. *Bull Am Assoc Petrol Geol* 41:1802–1853.

Fenchel T, Blackburn TH. 1979. *Bacteria and Mineral Cycling*. London: Academic Press.

Ferrer MR, Quevedo-Sarmiento J, Bejar V, Delgado R, Ramos-Cormenzana A, Rivadeneyra MA. 1988a. Calcium carbonate formation by *Deleya halophila*: Effect of salt concentration and incubation temperature. *Geomicrobiol J* 6:49–57.

Ferrer MR, Quevedo-Sarmiento J, Rivadeneyra MA, Bejar V, Delgado R, Ramos-Cormenzana A. 1988b. Calcium carbonate precipitation by two groups of moderately halophilic microorganisms at different temperatures and salt concentrations. *Curr Microbiol* 17:221–227.

Ferris FG, Fratton CM, Gerits JP, Schultze-Lam S, Lollar BS. 1995. Microbial precipitation of a strontium calcite phase at a groundwater discharge zone near Rock Creek, British Columbia, Canada. *Geomicrobiol J* 13:57–67.

Friedmann EI. 1977. Microorganisms in Antarctic desert rocks from dry valleys and Dufek Massif. *Antarctic J U.S.* 12:26–30.

Friedmann EI. 1980. Endolithic microbial life in hot and cold deserts. *Orig Life* 10:223–235.

Friedmann EI. 1982. Endolithic microorganisms in the Antarctic Cold Desert. *Science* 215:1045–1053.

Friedmann EI, Ocampo R. 1976. Endolithic blue-green algae in the dry valleys: Producers in the Antarctic desert ecosystem. *Science* 193:1247–1249.

Friedmann EI, Ocampo-Friedmann R. 1984. Endolithic microorganisms in extreme dry environments: Analysis of a lithobiontic microbial habitat. In: Klug MJ, Reddy CA, eds. *Current Perspectives in Microbial Ecology. Proc 3rd Int Symp Microb Ecol*. Washington, DC. American Society of Microbiology, pp. 177–185.

Friedmann EI, Roth WC, Turner JB, McEwin RS. 1972. Calcium oxalate crystals in the aragonite-producing green alga *Penicillus* and related genera. *Science* 177:891–893.

Friedmann EI, Weed R. 1987. Microbial trace-fossil formation, biogenous and abiotic weathering in the Antarctic Cold Desert. *Science* 236:703–705.

Fujita Y, Ferris FG, Lawson RD, Colwell FS, Smith RW. 2000. Calcium carbonate precipitation by ureolytic subsurface bacteria. *Geomicrobiol J* 17:305–318.

Golubic S. 1969. Distribution, taxonomy, and boring patterns of marine endolithic algae. *Am Zool* 9:747–751.

Golubic S. 1973. The relationship between blue-green algae and carbonate deposits. In: Carr NG, Whitton BA, eds. *The Biology of Blue Green Algae*. Oxford, U.K.: Blackwell Scientific, pp. 434–472.

Golubic S, Friedmann I, Schneider J. 1981. The lithobiontic ecological niche, with special reference to microorganisms. *J Sediment Petrol* 51:475–478.

Golubic S, Krumbein WE, Schneider J. 1979. The carbon cycle. In: Trudinger PA, Swaine DJ, eds. *Biogeochemical Cycling of Mineral Forming Elements*. Amsterdam: Elsevier, pp. 29–45.

Golubic S, Perkins RD, Lukas KJ. 1975. Boring microorganisms and microborings in carbonate substrates. In: Frey RW, ed. *The Study of True Fossils*. New York: Springer, pp. 229–259.

Golubic S, Schneider J. 1979. Carbonate dissolution. In: Trudinger PA, Swaine DJ, eds. *Biogeochemical Cycling of Mineral-Forming Elements*. Amsterdam: Elsevier, pp. 107–129.

Gorbushina AA, Krumbein WE, Hamman CH, Panina L. Solukharjevski S, Wollenzien U. 1994. Role of black fungi in color change and biodeterioration of antique marbles. *Geomicrobiol J* 11:205–211.

Gorlenko VM, Chebotarev EN, Kachalkin VI. 1974. Microbial oxidation of hydrogen sulfide in Lake Veisovo (Slavyansk Lake). *Mikrobiologiya* 43:530–534 (Engl transl pp. 450–453).

Greenfield LJ. 1963. Metabolism and concentration of calcium and magnesium and precipitation of calcium carbonate by a marine bacterium. *Ann NY Acad Sci* 109:23–45.

Herman JS. 1994. Karst geomicrobiology and redox geochemistry: State of the science. *Geomicrobiol J* 12:137–140.

Isenberg HD, Lavine LS. 1973. Protozoan calcification. In: Zipkin I, ed. *Biological Mineralization*. New York: Wiley, pp. 649–686.

Jensen PR, Gibson RA, Littler MM, Littler DS. 1985. Photosynthesis and calcification in four deep-water *Halimeda* species (Chlorophyceae, Caulerpales). *Deep Sea Res* 32:451–464.

Johansen HW. Department of Biology, Clark University, Worcester, MA, USA.

Krumbein WE. 1968. Zur Frage der biologischen Verwitterung: Einfluss der Mikroflora auf die Bausteinverwitterung und ihre Abhängigkeit von adaphischen Faktoren. *Z Allg Mikrobiol* 8:107–117.

Krumbein WE. 1974. On the precipitation of aragonite on the surface of marine bacteria. *Naturwissenschaften* 61:167.

Krumbein WE. 1979. Photolithotrophic and chemoorganotrophic activity of bacteria and algae as related to beachrock formation and degradation (Gulf of Aqaba, Sinai). *Geomicrobiol J* 1:139–203.

Krumbein WE, Giele C. 1979. Calcification in a coccoid cyanobacterium associated with the formation of desert stromatolites. *Sedimentology* 26:593–604.

Krumbein WE, Potts M. 1979. *Givanella*-like structures formed by *Plectonema gloeophilum* (Cyanophyta) from the Borrego Desert in Southern California. *Geomicrobiol J* 1:211–217.

Krumbein WE, Swart PK. 1983. The microbial carbon cycle. In: Krumbein WE, ed. *Microbial Geochemistry*. Oxford, U.K.: Blackwell, pp. 5–62.

LaRock PA, Ehrlich HL. 1975. Observations of bacterial microcolonies on the surface of ferromanganese nodules from Blake Plateau by scanning electron microscopy. *Microb Ecol* 2:84–96.

Latimer WM, Hildebrand JH. 1942. *Reference Book of Inorganic Chemistry*. Rev ed. New York: Macmillan.

Lewin J. 1965. Calcification. In: Lewin RA, ed. *Physiology and Biochemistry of Algae*. New York: Academic Press, pp. 457–465.

Lipman CB. 1929. Further studies on marine bacteria with special reference to the Drew hypothesis on $CaCO_3$ precipitation in the sea. *Carnegie Inst Washington Publ 391* 26:231–248.

Lovley DR, Phillips EJP. 1988. Novel mode of microbial energy metabolism: Organic carbon oxidation coupled to dissimilatory reduction of iron and manganese. *Appl Environ Microbiol* 54:1427–1480.

Lovley DR, Roden EE, Phillips EJP, Woodward JC. 1993. Enzymatic iron and uranium reduction by sulfate-reducing bacteria. *Marine Geol* 113:41–53.

Lowenstamm HA, Epstein S. 1957. On the origin of sedimentary aragonite needles of the Great Bahama Bank. *J Geol* 65:364–375.

Luff R, Wallmann K. 2003. Fluid flow, methane fluxes, carbonate precipitation and biogeochemical turnover in gas hydrate-bearing sediments at Hydrate Ridge, Cascadia Margin: Numerical modeling and mass balances. *Geochim Cosmochim Acta* 67:3403–3421.

Luff R, Wallmann K, Aloisi G. 2004. Numerical modeling of carbonate crust formation at cold vent sites: Significance for fluid and methane budgets and chemosynthetic biological communities. *Earth Planet Sci Lett* 221:337–353.

Marine Chemistry. 1971. A report of the Marine Chemistry Panel of the Committee of Oceanography. National Academy of Sciences, Washington, DC.

McCallum MF, Guhathakurta K. 1970. The precipitation of calcium carbonate from seawater by bacteria isolated from Bahama Bank sediments. *J Appl Bacteriol* 33:649–655.

McConnaughey T. 1991. Calcification in *Chara corallina*: CO_2 hydroxylation generates protons for bicarbonate assimilation. *Limnol Oceanogr* 36:619–628.

Mitchell AC, Ferris FG. 2006. The influence of *Bacillus pasteurii* on the nucleation and growth of calcium carbonate. *Geomicrobiol J* 23:213–226.

Monty CLV. 1967. Distribution and structure of recent stromatolitic algal mats, Eastern Andros Island, Bahamas. *Ann Soc Geol Belgi* 90:55–99.

Monty CLV. 1972. Recent algal stromatolitic deposits, Andros Island, Bahamas. Preliminary Report. *Geol Rundschau* 61:742–783.

Moore JN. 1983. The origin of calcium carbonate nodules forming on Flathead Lake delta, northwestern Montana. *Limnol Oceanogr* 28:646–654.

Morita RY. 1980. Calcite precipitation by marine bacteria. *Geomicrobiol J* 2:63–82.

Murray J, Irvine R. 1889/1890. On coral reefs and other carbonate of lime formations in modern seas. *Proc Roy Soc Lond A* 17:79–109.

Nadson GA. 1903. Die Mikroorganismen als geologische Faktoren: I. über die Schwefelwasserstoffgährung und über die Beteiligung der Mikroorganismen bei der Bildung des schwarzen Schlammes (Heilsschlammes). Arb d Comm Erf D Mineralseen bei Slawansk, St Petersburg.

Nadson GA. 1928. Beitrag zur Kenntniss der bakteriogenen Kalkablagerung. *Arch hydrobiol* 191:154–164.

Nazina TN, Rozanova EP, Kuznetsov SI. 1985. Microbial oil transformation processes accompanied by methane and hydrogen sulfide formation. *Geomicrobiol J* 4:103–130.

Nekrasova VK, Gerasimenko LM, Romanova AK. 1984. Precipitation of calcium carbonate in the presence of cyanobacteria. *Mikrobiologiya* 53:833–836 (Engl transl pp. 691–694).

Novitsky JA. 1981. Calcium carbonate precipitation by marine bacteria. *Geomicrobiol J* 2:375–388.

Omelon CR, Pollard WH, Ferris FG. 2006a. Chemical and ultrastructural characterization of high Arctic cryptoendolithic habitats. *Geomicrobiol J* 23:189–200.

Omelon CR, Pollard WH, Ferris FG. 2006b. Environmental controls on microbial colonization of high Arctic cryptoendolithic habitats. *Polar Biol* (in press).

Paine SG, Lingood FV, Schimmer F, Thrupp TC. 1933. IV. The relationship of microorganisms to the decay of stone. *Roy Soc Phil Trans* 222B:97–127.

Parmar N, Warren LA, Roden EE, Ferris FG. 2000. Solid phase capture of strontium by the iron reducing bacteria *Shewanella alga* strain BrY. *Chem Geol* 169:281–288.

Pentecost A. 1978. Blue-green algae and freshwater carbonate deposits. *Proc Roy Soc Lond* B 200:43–61.

Pentecost A. 1988. Growth and calcification of the cyanobacterium *Homeothrix crustacea*. *J Gen Microbiol* 134:2665–2671.

Pentecost A. 1995. Significance of the biomineralizing microniche in a *Lyngbya* (Cyanobacterium) travertine. *Geomicrobiol J* 13:213–222.

Pentecost A, Bauld J. 1988. Nucleation of calcite on the sheaths of cyanobacteria using a simple diffusion cell. *Geomicrobiol J* 6:129–135.

Pentecost A, Spiro B. 1990. Stable carbon and oxygen isotope composition of calcites associated with modern freshwater cyanobacteria and algae. *Geomicrobiol J* 8:17–26.

Post WH, Peng T-H, Emanuael WE, King AW, Dale H, DeAngellis DL. 1990. The global carbon cycle. *Am Sci* 78:310–326.

Pye K. 1984. SEM analysis of siderite cements in intertidal marsh sediments, Norfolk, England. *Mar Geol* 56:1–12.

Pye K, Dickson AD, Schiavon N, Coleman ML, Cox M. 1990. Formation of siderite-Mg-calcite-iron sulfide concentrations in intertidal marsh and sandflat sediments, north Norfolk, England. *Sedimentology* 37:325–343.

Rivadeneyra MA, Delgado R, Quesada E, Ramos-Cormenzana A. 1991. Precipitation of calcium carbonate by *Deleya halophila* in media containing NaCl as sole salt. *Curr Microbiol* 22:185–190.

Rivadeneyra MA, Martín-Algarra A, Sánchez-Navas A, Martín-Ramos D. 2006. Carbonate and phosphate precipitation by *Chromohalobacter marismortui*. *Geomicrobiol J* 23:1–13.

Rivadeneyra MA, Perez-Garcia I, Salmeron V, Ramos-Cormenzana A. 1985. Bacterial precipitation of calcium carbonate in the presence of phosphate. *Soil Biol Biochem* 17:171–172.

Roden EE, Ferris FG. 2000. Immobilization of aqueous strontium during carbonate formation coupled to microbial Fe(III) oxide reduction. Preprints Ext Abstract ACS Nat Meet, *Am Chem Soc, Div Environ Chem* 40(2):400–403.

Rodriguez-Navarro C, Rodriguez-Gallego M, Chekroun KB, Gonzalez-Muñoz MT. 2003. Conservation of ornamental stone by *Myxococcus xanthus*-induced carbonate mineralization. *Appl Environ Microbiol* 69:2182–2193.

Roemer R, Schwartz E. 1965. Geomikrobiologische Untersuchungen. V. Verwertung von Sulfatmineralien and Schwermetallen. Tolleranz bei Desulfizierern. *Z Allg Mikrobiol* 5:122–135.

Röttger R. 1974. Large foraminifera: Reproduction and early stages of development in *Heterostigina depressa*. *Marine Biol* (Berlin) 26:5–12.

Sarbu SM, Kane TC, Kinkle BK. 1996. A chemoautotrophically based cave ecosystem. *Science* 272:1953–1955.

Sarbu SM, Kinkle BK, Vlasceanu L, Kane TC, Popa R. 1994. Microbiological characterization of a sulfide-rich groundwater ecosystem. *Geomicrobiol J* 12:175–182.

Schmalz RF. 1972. Calcium carbonate: Geochemistry. In: Fairbanks RW, ed. *The Encyclopedia of Geochemistry and Environmental Sciences*. *Encycl Earth Sci Ser*, Vol IVA. New York: Van Nostrand Reinhold, p. 110.

Sellwood RW. 1971. The genesis of some siderite beds in the Yorkshire Lias (England). *J Sediment Petrol* 41:854–858.

Shachak M, Jones CG, Granot Y. 1987. Herbivory in rocks in the weathering of a desert. *Science* 236:1098–1099.

Shinano H. 1972a. Studies on marine microorganisms taking part in the precipitation of calcium carbonate. II. Detection and grouping of the microorganisms taking part in the precipitation of calcium carbonate. *Bull Jpn Soc Sci Fisheries* 38:717–725.

Shinano H. 1972b. Microorganisms taking part in the precipitation of calcium carbonate. III. A taxonomic study of marine bacteria taking part in the precipitation of calcium carbonate. *Bull Jpn Soc Sci Fisheries* 38:727–732.

Shinano H, Sakai M. 1969. Studies on marine bacteria taking part in the precipitation of calcium carbonate. I. Calcium carbonate deposited in peptone medium prepared with natural seawater and artificial seawater. *Bull Jpn Soc Sci Fisheries* 35:1001–1005.

Skirrow G. 1975. The dissolved gases—Carbon dioxide. In: Riley JR, Skirrow G, eds. *Chemical Oceanography*, Vol. 2, 2nd ed. London: Academic Press, pp. 144–152.

Smith SV. 1981. Marine macrophytes as a global carbon sink. *Science* 211:838–840.

Sokolova-Dubinina GA, Deryugina ZP. 1967. On the role of microorganisms in the formation of rhodochrosite in Punnus-Yarvi Lake. *Mikrobiologiya* 36:535–542.

Spiro B, Pentecost A. 1991. One day in the life of a stream—A diurnal inorganic carbon mass balance for the travertine depositing stream (Waterfall Beck, Yorkshire). *Geomicrobiol J* 9:1–11.

Steinmann G. 1899/1901. Über die Bildungsweise des dunklen Pigments bei den Mollusken nebst Bemerkung über die Entstehung von Kalkcarbonat. *Ber Naturf Ges Freiburg i B* 11:40–45.

Stumm W, Morgan JJ. 1981. *Aquatic Chemistry. An Introduction Emphasizing Chemical Equilibria in Natural Waters*. New York: Wiley.

Stutzer A, Hartleb R. 1899. Die Zersetzung von Cement unter dem Einfluss von Bakterien. *Z Angew Chem* 12:402 (cited by Paine et al., 1933).

Suess E. 1979. Mineral phases formed in anoxic sediments by microbial decomposition of organic matter. *Geochim Cosmochim Acta* 43:339–352.

Thompson JB. 2000. Microbial whitings. In: Riding RE, Awaramik SM, eds. *Microbial Sediments*. Berlin: Springer, pp. 250–260.

Thompson JB, Ferris FG. 1990. Cyanobacterial precipitation of gypsum, calcite and magnesite from natural alkaline lake waters. *Geology* 18:995–998.

Thompson JB, Ferris FG, Smith DA. 1990. Geomicrobiology and sedimentology of the mixolimnion and chemocline in Fayetteville Green Lake, New York. *Palaios* 5:52–75.

Urzì C, Garcia-Valles M, Vendrell M, Pernice A. 1999. Biomineralization processes on rock and monument surfaces observed in field and laboratory conditions. *Geomicrobiol J* 16:39–54.

Urzì C, Lisi S, Criseo G, Penrice A. 1991. Adhesion to and decomposition of marble by a *Micrococcus* strain isolated from it. *Geomicrobiol J* 9:81–90.

van der Wal P, de Jong EW, Westbroek P, de Bruijn WC, Mulder-Stapel AA. 1983. Polysaccharide localization, coccolith formation, and Golgi dynamics in the coccolithophorid *Hymenomonas carterae*. *J Ultrastruct Res* 85:139–158.

Verrecchia EP, Dumant J-L, Collins KE. 1990. Do fungi building limestone exist in semi-arid regions? *Naturwissenschaften* 77:584–586.

Wagner E, Schwartz W. 1965. Geomikrobiologische Untersuchungen. VIII. Über das Verhalten von Bakterien auf der Oberfläche von Gesteinen und Mineralien und ihre Rolle bei der Verwitterung. *Z Allg Mikrobiol* 7:33–52.

Walker JJ, Pace NR. 2007. Phylogenetic composition of Rocky Mountain endolithic microbial ecosystems. *Appl Environ Microbiol* 73:3497–3504.

Walter MR, ed. 1976. *Stromatolites. Developments in Sedimentology 20*. Amsterdam: Elsevier.

Warren LA, Maurice PA, Parmar N, Ferris FG. 2001. Microbially mediated calcium carbonate precipitation: Implications for interpreting calcite precipitation and for solid-phase capture of inorganic contaminants. *Geomicrobiol J* 18:93–115.

Weast RC, Astle MJ, eds. 1982. *CRC Handbook of Chemistry and Physics*. 63rd ed. Boca Raton, FL: CRC Press.

Wefer G. 1980. Carbonate production by algae *Halimeda, Penicillus*, and *Padina*. *Nature* (London) 285:323–324.

Wharton RA Jr, Parker BC, Simmons GM Jr, Seaburg KS, Love FG. 1982. Biogenic calcite structures in Lake Fryxell, Antarctica. *Nature* (London) 29:403–405.

10 Geomicrobial Interactions with Silicon

10.1 DISTRIBUTION AND SOME CHEMICAL PROPERTIES

The element silicon is one of the most abundant in the Earth's crust, ranking second only to oxygen. Its estimated crustal abundance is 27.7% (w/w) whereas that of oxygen is 46.6% (Mitchell, 1955). The concentration of silicon in various components of the Earth's surface is listed in Table 10.1.

In nature, silicon occurs generally in the form of silicates, including aluminosilicates and silicon dioxide (silica). It is found in primary and secondary minerals. It can be viewed as an important part of the backbone of silicate rock structure. In silicates and aluminosilicates, silicon is usually surrounded by four oxygen atoms in tetrahedral fashion (Kretz, 1972). In aluminosilicates the aluminum is coordinated with oxygen in tetrahedral or octahedral fashion, depending on the mineral (see Tan, 1986). In clays, which result from the weathering of primary aluminosilicates, silica tetrahedral sheets and aluminum hydroxide octahedral sheets are layered in different ways depending on the clay type. In *montmorillonite*-type clays, structural units consisting of single aluminum hydroxide octahedral sheets are sandwiched between silica tetrahedral sheets. The units are interspaced with layers of water molecules of variable thickness into which other polar molecules, including some organic ones, can enter. This variable water layer allows montmorillonite-type clays to swell. The structural units of *illite*-type clays resemble those of montmorillonite-type clays but differ from them in that Al replaces some of the Si in the silica tetrahedral sheets. These substituting Al atoms impart extra charges, which are neutralized by potassium ions between the silica sheets between two successive units. The potassium ions act as bridges that prevent the swelling exhibited by montmorillonite in water. In *kaolinite* clays, structural units consist of silica tetrahedral sheets alternating with aluminum hydroxide octahedral sheets joined to one another by oxygen bridges (see Toth, 1955).

Silicon–oxygen bonds of *siloxane linkages* (Si–O–Si) in silicates and aluminosilicates are very strong (their energy of formation ranges from 3110 to 3142 kcal mol^{-1} or 13,031 to 13,165 kJ mol^{-1}), whereas Al–O bonds are weaker (their energy of formation ranges from 1793 to 1878 kcal mol^{-1} or 7531 to 7869 kJ mol^{-1}). Bonds between nonframework cations and oxygen are the weakest (energy of formation ranges from 299 to 919 kcal mol^{-1} or 1252 to 3851 kJ mol^{-1}) (values cited by Tan, 1986). The strength of these bonds determines their susceptibility to weathering. Thus Si–O bonds are relatively resistant to acid hydrolysis (Karavaiko et al., 1984) unlike Al–O bonds. Bonds between cations and oxygen are readily broken by protonation or cation exchange.

Silicate in solution at pH 2–9 exists in the form of undissociated monosilicic acid (H_4SiO_4), whereas at pH 9 and above it transforms into silicate ions (see Hall, 1972). Monosilicic acid polymerizes at a concentration of 2×10^{-3} M and above, forming oligomers of polysilicic acids (Iler, 1979). This polymerization reaction appears to be favored around neutral pH (Avakyan et al., 1985). Polymerization of monosilicate can be viewed as a removal of water from between adjacent silicates to form a siloxane linkage. *Silica* can be viewed as an anhydride of *silicic acid*:

$$H_4SiO_4 \rightarrow SiO_2 + 2H_2O \qquad (10.1)$$

TABLE 10.1
Abundances of Silicon on the Earth's Surface

Phases	Concentration	Reference
Granite	336,000 ppm	Bowen (1979)
Basalt	240,000 ppm	Bowen (1979)
Shale	275,000 ppm	Bowen (1979)
Limestone	32,000 ppm	Bowen (1979)
Sandstone	327,000 ppm	Bowen (1979)
Soils	330,000 ppm	Bowen (1979)
Seawater	3×10^3 μg L^{-1}	Marine Chemistry (1971)
Freshwater	7 ppm	Bowen (1979)

Dissociation constants for silicic acid are as follows (see Anderson, 1972):

$$H_4SiO_4 \rightarrow H^+ + H_3SiO_4^- \quad (K_1 = 10^{-9.5}) \tag{10.2}$$

$$H_3SiO_4^- \rightarrow H^+ + H_2SiO_4^{2-} \quad (K_2 = 10^{-12.7}) \tag{10.3}$$

Silica can exist in partially hydrated form called *metasilicic acid* (H_3SiO_3) or in a fully hydrated form called orthosilicic acid (H_4SiO_4). Each of these forms can be polymerized, the ortho acid forming, for instance, $H_3SiO_4 \cdot H_2SiO_3 \cdot H_3SiO_3$ (Latimer and Hildebrand, 1940; Liebau, 1985). The polymers may exhibit colloidal properties, depending on size and other factors. Colloidal forms of silica tend to exist locally at high silica concentrations or at saturation levels and are favored by acid conditions (Hall, 1972).

Common silicon-containing minerals include quartz (SiO_2), olivine [$(Mg,Fe)_2SiO_4$], orthopyroxene ($Mg,FeSiO_3$), biotite [$K(Mg,Fe)_3AlSi_3O_{10}(OH)_2$], orthoclase ($KAlSi_3O_8$), plagioclase [$(Ca,Na)(Al,Si)AlSi_3O_8$], kaolinite [$Al_4Si_4O_{10}(OH)_8$], and others.

Silica and silicates form an important buffer system in the oceans (Garrels, 1965), together with the $CO_2/HCO_3^-/CO_3^{2-}$ CO_3^{2-} system. The latter is a rapidly reacting system, whereas the system based on reaction with silica and silicates is slow (Garrels, 1965; Sillén, 1967).

Aluminosilicates in the form of clay perform a buffering function in mineral soils. This is because of their ion-exchange capacity, net electronegative charge, and adsorption powers. Their ion-exchange capacity and adsorption power, moreover, make them important reservoirs of cations and organic molecules. Montmorillonite exhibits the greatest ion-exchange capacity, illite less, and kaolinite the least (Dommergues and Mangenot, 1970, p. 469).

10.2 BIOLOGICALLY IMPORTANT PROPERTIES OF SILICON AND ITS COMPOUNDS

Silicon is taken up and concentrated in significant quantities by certain forms of life. These include microbial forms such as diatoms and other chrysophytes; silicoflagellates and some xanthophytes; radiolarians and actinopods; some plants such as horsetails, ferns, grasses, and some flowers and trees; and also some animals such as sponges, insects, and even vertebrates. Some bacteria (Heinen, 1960) and fungi (Heinen, 1960; Holzapfel and Engel, 1954a,b) have also been reported to take up silicon to a limited extent. According to Bowen (1966), diatoms may contain from 1,500 to 20,000 ppm silicon, land plants from 200 to 5,000 ppm, and marine animals from 120 to 6,000 ppm.

Although the function of silicon in higher forms of life, animals and plants, is not presently apparent, it is clearly structural in some microorganisms such as diatoms, actinopods, and radiolarians.

In diatoms, silicon also seems to play a metabolic role in the synthesis of chlorophyll (Werner, 1966, 1967), DNA (Darley and Volcani, 1969; Reeves and Volcani, 1984), and DNA polymerase and thymidylate kinase (Sullivan, 1971; Sullivan and Volcani, 1973).

Silicon compounds in the form of clays (aluminosilicates) may exert an effect on microbes in soil. They may stimulate or inhibit microbial metabolism, depending on the conditions (Marshman and Marshall, 1981a,b; Weaver and Dugan, 1972; see also discussion by Marshall, 1971). These effects of clays are mostly indirect, that is, clays tend to modify the microbial habitat physicochemically, thereby eliciting a physiological response by the microbes (Stotzky, 1986). For beneficial effect, clays may buffer the soil environment and help maintain a favorable pH, thereby promoting growth and metabolism of some microorganisms that might otherwise be slowed or stopped if the pH became unfavorable (Stotzky, 1986). Certain clays have been found to enable some bacteria that were isolated from marine ferromanganese nodules or associated sediments to oxidize Mn^{2+}. Intact cells of these organisms can oxidize Mn^{2+} if it is bound to bentonite (montmorillonite-type clay) or kaolinite but not illite if each has been pretreated with ferric iron. They cannot oxidize Mn^{2+} that is free in solution (Ehrlich, 1982). Cell-free preparations of these bacteria oxidize Mn^{2+} bound to bentonite and kaolinite without ferric iron pretreatment, although manganese-oxidizing activity of the cell extracts is greater when the clays with Mn^{2+} bound to them are pretreated with ferric iron (Ehrlich, 1982). Like intact cells, the cell-free extract cannot oxidize dissolved Mn^{2+} (Ehrlich, 1982). Clays may also enhance the activity of some enzymes such as catalase when the enzymes are bound to their surface (see Stotzky, 1986, p. 404).

By contrast, clays may suppress microbial growth and metabolism by adsorbing organic nutrients, thereby making them less available to microbes. Clays may also adsorb microbial antibiotics and thereby lower the inhibitory activity of these agents (see Stotzky, 1986). In soils, the results may be that an antibiotic producer is outgrown by organisms that *in vitro* it keeps in check with the help of the antibiotic it excretes. These effects of clay can be explained, at least in part, by the strength of binding to a negatively charged clay surface and the inability of many microbes to attack adsorbed nutrients, or by the inability of adsorbed antibiotics to inhibit susceptible microbes (see Dashman and Stotzky, 1986). High concentrations of clay may interfere with diffusion of oxygen by increasing the viscosity of a solution, which can have a negative effect on aerobic microbial respiration (see Stotzky, 1986). Clays may also modulate other interactions between different microbes and between microbes and viruses in soil, and they may affect the pathogenicity of these disease-causing soil microbes (see Stotzky, 1986).

Although clay-bound organic molecules may be less available or unavailable to organisms in the bulk phase or even attached to the mineral surface, this cannot be a universal phenomenon. Portions of attached large organic polymers may be attacked by appropriate extracellular enzymes, producing smaller unattached units that can be taken up by microbes. Electrostatically bound organic molecules that are potential nutrients may be dislodged by exchange with protons excreted as acids in the catabolism of some microbes. These processes of remobilization must also apply to mineral sorbents other than clays.

10.3   BIOCONCENTRATION OF SILICON

10.3.1   Bacteria

Some bacteria have been shown to accumulate silicon. A soil bacterium, strain B_2 (Heinen, 1960), and a strain of *Proteus mirabilis* (Heinen, 1968) have been found to take up limited amounts of silicon when it is furnished in the form of silica gel, quartz, or sodium silicate. Sodium silicate was taken up most easily, and quartz least easily. The silicon seemed to substitute partially for phosphorus in phosphorus-deficient media (Figure 10.1). This substitution reaction was reversible (Heinen, 1962). The silicon taken up by the bacteria appeared to become organically bound in a metabolizable form (Heinen, 1962). Sulfide, sulfite, and sulfate were found to affect phosphate–silicate exchange

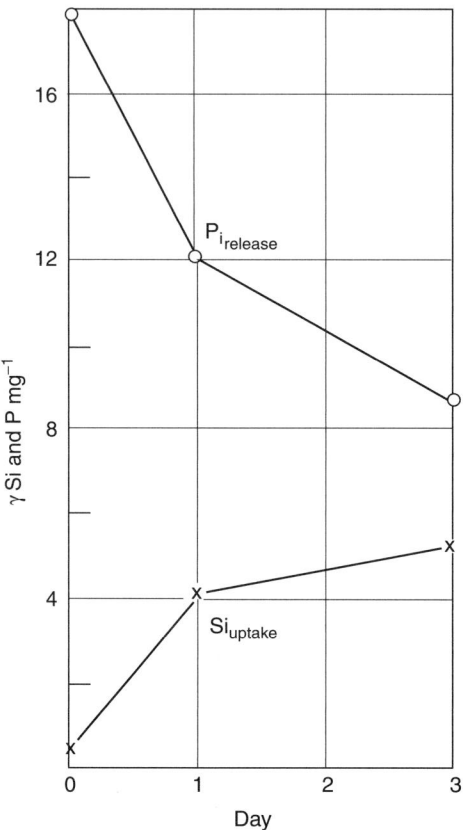

FIGURE 10.1   Relationship between Si uptake and P_i release during incubation of resting cells of strain B_2 in silicate solution (80 μg Si mL^{-1}). (From Heinen W, *Arch. Mikrobiol.*, 41, 229–246, 1962. With permission of Springer Science and Business Media.)

in different ways, depending on concentration, whereas KCl and NaCl were without effect. NH_4Cl, NH_4NO_3, and $NaNO_3$ stimulated the formation of adaptive (inducible) enzymes involved in the phosphate–silicate exchange (Heinen, 1963a). The presence of sugars such as glucose, fructose, or sucrose and of amino acids such as alanine, cysteine, glutamine, methionine, asparagines, and citrulline as well as of metabolic intermediates pyruvate, succinate, and citrate stimulated silicon uptake. On the contrary, acetate, lactate, phenylalanine, peptone, and wheat germ oil inhibited silicon uptake. Glucose at an initial concentration of 1.2 mg mL^{-1} of medium stimulated silicon uptake maximally (Figure 10.2). Higher concentrations of glucose caused the formation of particles of protein, carbohydrates, and silicon outside the cell. $CdCl_2$ inhibited the stimulatory effect of glucose on silicon uptake, but 2,4-dinitrophenol was without effect. The simultaneous presence of $NaNO_3$ and KH_2PO_4 lowered the stimulatory effect of glucose but did not eliminate it (Heinen, 1963b). The silicon that was fixed in the bacteria was readily displaced by phosphate in the absence of external glucose. In cells that were preincubated in a glucose–silicate solution, only a small portion of the silicon was released by glucose–phosphate, but all of the silicon was releasable on incubation in glucose–carbonate solution.

Some of the silicon taken up by the bacterial cells appeared to be tied up in labile ester bonds (C–O–Si), whereas other silicon appeared to be tied up in more stable bonds (C–Si) (Heinen, 1963c, 1965). Studies of intact cells and cell extracts of *Proteus mirabilis* after the cells were incubated in the presence of silicate suggested that the silicate taken up was first accumulated in the cell walls and then slowly transferred to the interior of the cell (Heinen, 1965). The silicon was organically

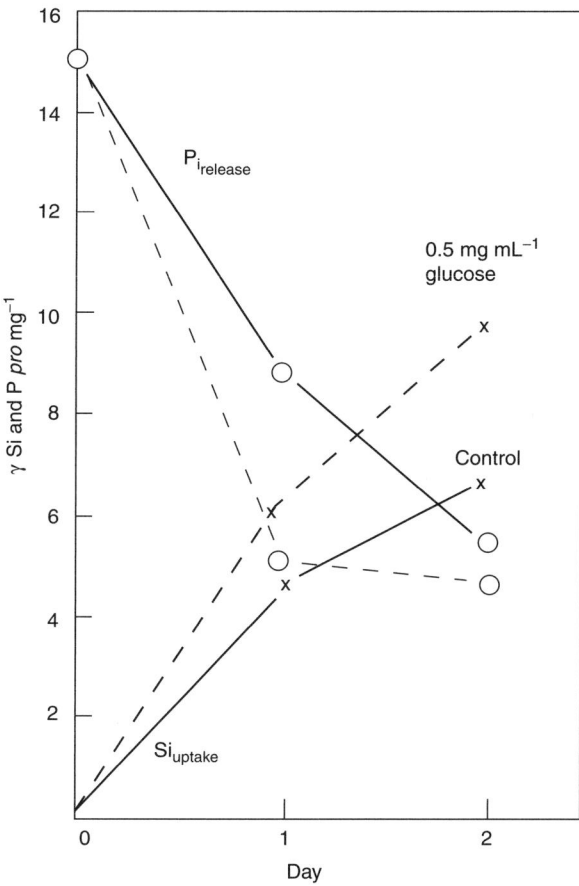

FIGURE 10.2 Influence of glucose and silicate uptake (x—x, without glucose, control; x---x, with glucose), and phosphate release (o—o, without glucose; o---o, with glucose). (From Heinen W, *Arch. Mikrobiol.*, 45, 162–171, 1963. With permission of Springer Science and Business Media.)

bound in the wall and within the cell. A particulate fraction from *P. mirabilis* bound silicate organically in an oxygen-dependent process (Heinen, 1967).

10.3.2 FUNGI

Some fungi have also been reported to accumulate silicon (Holzapfel and Engel, 1954a,b). When growing on a silicate-containing agar medium, the vegetative mycelium of such fungi exhibits an induction period of 5–7 days before taking up silicon. When the silicon in the medium is in the form of galactose–quartz or glucose–quartz complexes, silicon uptake by vegetative mycelium can occur within 12–18 h (Holzapfel, 1951). Evidently, inorganic silicates have to be transformed into organic complexes during the prolonged induction period before silicon is taken into the fungal cells.

10.3.3 DIATOMS

Among eukaryotic microorganisms that take up silicon, diatoms have been most extensively studied with respect to this process (Figure 10.3) (Lewin, 1965; de Vrind-de Jong and de Vrind, 1997). Their silicon uptake ability affects redistribution of silica between fresh and marine waters. In the Amazon River estuary, for instance, diatoms remove 25% of the dissolved silica from the river water.

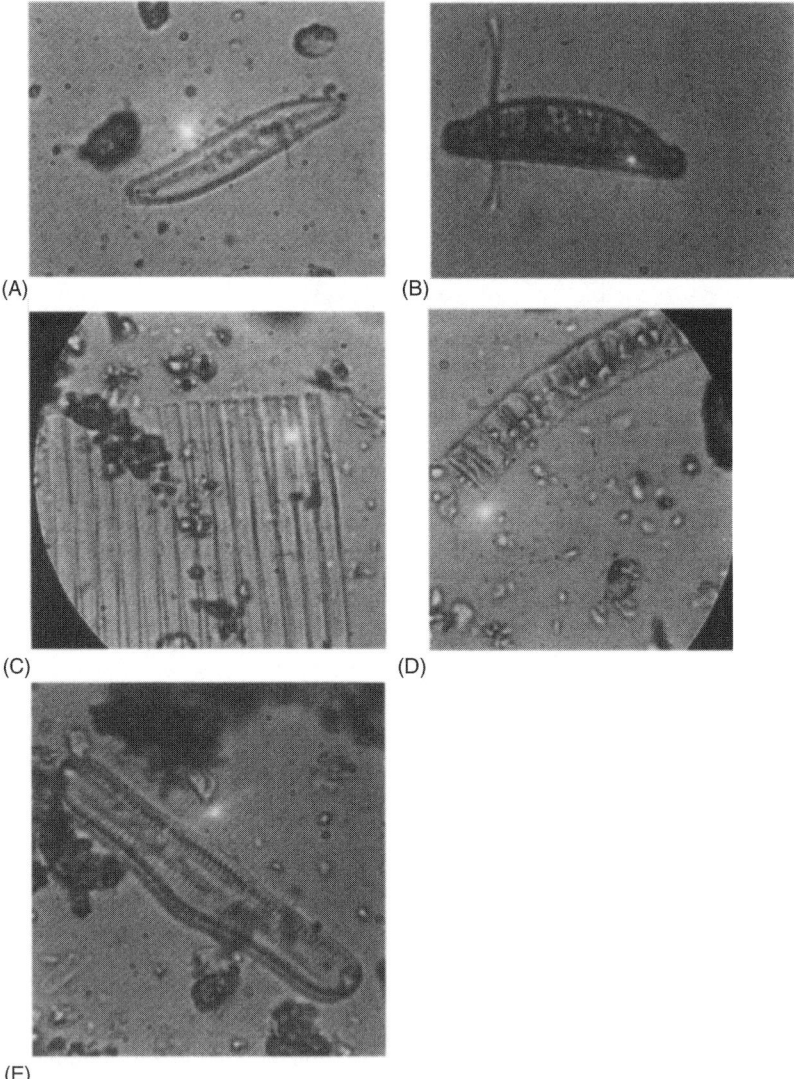

FIGURE 10.3 Diatoms. (A) *Gyrosigma*, from freshwater (×1944); (B) *Cymbella*, from freshwater (×1944); (C) *Fragellaria*, from freshwater (×1864); (D) ribbon of diatoms from freshwater (×1990); (E) marine diatom frustule, from Pacific Ocean sediment (×1944).

Their frustules are not swept oceanward upon their death but are transported coastward and incorporated into dunes, mud, and sandbars (Millman and Boyle, 1975).

Diatoms are unicellular algae enclosed in a wall of silica consisting of two valves, an epivalve and a hypovalve, in pillbox arrangement. One or more girdle bands are loosely connected to the epivalve. The valves are usually perforated plates, which may have thickened ribs. Their shape may be pennate or centric. The pores serve as sites of gas and nutrient exchange (see de Vrind-de Jong and de Vrind, 1997). In cell division, each daughter cell receives either the epivalve or hypovalve of the mother cell and synthesizes the other valve *de novo* to fit into the one already present. To prevent excessive reduction in size of the daughter diatoms that receive the hypovalve upon each cell division, a special reproductive step called *auxospore formation* returns these daughter cells to maximum size. It occurs when a progeny cell that has received a hypovalve has reached minimum size after repeated divisions. Auxospore formation is a sexual reproductive process in which the cells escape from their frustules and increase in size in their zygote membrane, which may become weakly silicified. After a time, the protoplast in the zygote membrane contracts and forms the typical frustules of the parent cell (Lewin, 1965).

The silica walls of the diatoms consist of hydrated amorphous silica, a polymerized silicic acid (Lewin, 1965). The walls of marine diatoms may contain as much as 96.5% SiO_2, but only 1.5% Al_2O_3 or Fe_2O_3 and 1.9% water (Rogall, 1939). In clean, dried frustules of freshwater *Navicula pelliculosa*, 9.6% water has been found (Lewin, 1957). Thin parts of diatom frustules reveal a foamlike substructure when viewed under the electron microscope, suggesting silica gel (Helmcke, 1954), which may account for the adsorptive power of such frustules. The silica gel may be viewed as arranged in small spherical particles about 22 µm in diameter (Lewin, 1965). Because of the low solubility of amorphous silica at the pH of most natural waters, frustules of living diatoms do not dissolve readily (Lewin, 1965). At pH 8, however, it has been found that 5% of the silica in the walls of *Thalassiosira nana* and *Nitschia linearis* dissolves. Moreover, at pH 10, 20% of the silica in the frustules of *N. linearis* and all of the silica in the frustules of *T. nana* dissolves (Jorgensen, 1955). This silica dissolution may reflect the state of integration of newly assimilated silica in the diatom frustule. Some bacteria naturally associated with diatoms have been shown to accelerate dissolution of silica in frustules by an unknown mechanism (Patrick and Holding, 1985). Frustules of living diatoms are to some extent protected against dissolution by an organic film, when present, and their rate of dissolution has been shown to exhibit temperature dependence (Katami, 1982). After the death of diatoms, their frustules may dehydrate to form more crystalline SiO_2 that is much less soluble in alkali than that in living diatoms. This may account for the accumulation of diatomaceous ooze.

Rates of silicic acid uptake and incorporation by diatoms can be easily measured with radioactive [^{65}Ge]germanic acid as tracer (Azam et al., 1973; Azam, 1974; Chisholm et al., 1978). At low concentration (Ge/Si molar ratio of 0.01), germanium, which is chemically similar to silicon, is apparently incorporated together with silicon into the silicic acid polymer of the frustule. At higher concentrations (Ge/Si molar ratio of 0.1), germanium is toxic to diatoms (Azam et al., 1973). Genetic control of silicic acid transport into diatoms has begun to be studied on a molecular level (see review by Martin-Jézéquel et al., 2000).

Diatoms are able to discriminate between ^{28}Si and ^{30}Si by assimilating the lighter isotope preferentially. The fraction (α) for each of the three diatom species *Skeletonema costatum*, *Thalassiosira weissflogii*, and *Thalassiosira* sp. was 0.9989 ± 0.004. It was independent of temperature between 12 and 22°C and thus independent of growth rate (De La Rocha et al., 1997). This fractionation ability appears to be usable as a signature in identifying biogenic silica (De La Rocha et al., 2000).

Diatoms take up silica in the form of orthosilicate. More highly polymerized forms of silicate are not taken up unless first depolymerized, as by some bacteria (Lauwers and Heinen, 1974). Organic silicates are also not available to them. Ge, C, Sn, Pb, P, As, B, Al, Mg, and Fe do not replace silicon extensively if at all (Lewin, 1965). The concentration of silicon accumulated by a diatom depends to some extent on its concentration in the growth medium and on the rate of cell division (the faster the cells divide, the thinner their frustules). Silicon is essential for cell division, but resting cells in a medium in which silica is not at a limiting concentration continue to take up silica (Lewin, 1965). Synchronously growing cells of *Navicula pelliculosa* take up silicon at a constant rate during the cell division cycle (Lewin, 1965). Silica uptake appears dependent on energy-yielding processes (Lewin, 1965; Azam et al., 1974; Azam and Volcani, 1974; review by Martin-Jézéquel et al., 2000) and seems to involve intracellular receptor sites (Blank and Sullivan, 1979). Uncoupling of oxidative phosphorylation stops silica uptake by *Navicula pelliculosa* and *Nitzchia angularis*. Starved cells of *Navicula pelliculosa* show an enhanced silica uptake rate when fed glucose or lactate in the dark or when returned to the light, where they can photosynthesize (Healy et al., 1967). Respiratory inhibitors prevent Si and Ge uptake by *Nitzchia alba* (Azam et al., 1974; Azam and Volcani, 1974).

Total uptake of phosphorus and carbon is decreased during silica starvation of *Navicula pelliculosa*. Upon restoration of silica to the medium, the total uptake of phosphorus is again increased (Coombs and Volcani, 1968). Sulfhydryl groups (–SH) appear to be involved in silica uptake (Lewin, 1965).

Some progress has been made in understanding how diatoms form their siliceous cell walls (see de Vrind-de Jong and de Vrind, 1997). Valve and girdle-band assembly takes place inside the cell and happens late in the cell cycle during the last part of mitosis. For this assembly, silicate is taken into the cell and polymerized in special membrane-bound *silica deposition vesicles* (SDVs), leading to the formation of the girdle bands and valves. The SDV seems to arise from the Golgi apparatus, a special membrane system within the cell. The endoplasmic reticulum, a membrane network within the cell that is connected to the plasma membrane and the nuclear membrane, may participate in SDV development. The active SDVs are located adjacent to the plasma membrane. The shape of the SDV may be determined by interaction with various cell components such as the plasma membrane, actin filaments, microtubules, and cell organelles. The SDV is believed to help determine the morphology of the valves. Frustule buildup in the SDV appears to start along the future raphe, which appears as a longitudinal slot in each mature pennate frustule. The raphe has a central thickening called a nodule. Completed valves are exocytosed by the cell, that is, they are exported to the cell surface. When the valves are in place at the cell surface of the diatom, the raphe plays a role in its motility. Mature frustules have glycoprotein associated with them, which may have played a role in silica assembly during valve formation. It may help in determining valve morphology and in the export of assembled valves to the cell surface. For additional information, the reader is referred to de Vrind-de Jong and de Vrind (1997) and references cited therein.

10.4 BIOMOBILIZATION OF SILICON AND OTHER CONSTITUENTS OF SILICATES (BIOWEATHERING)

Some bacteria and fungi play an important role in mobilization of silica and silicates in nature. Part of this microbial involvement is manifested in the weathering of rock silicates and aluminosilicates. The solubilizing action may involve the cleavage of Si–O–Si (siloxane) or Al–O framework bonds or the removal of cations from the crystal lattice of silicate, causing subsequent collapse of the silicate lattice structure. The mode of attack may be by (1) microbially produced ligands of cations; (2) microbially produced organic or inorganic acids, which are a source of protons; (3) microbially produced alkali (ammonia or amines); or (4) microbially produced extracellular polysaccharides acting at acidic pH. The source of the polysaccharides may be the glycocalyx of some bacteria.

Bioweathering action of silica or silicates seems not only to be restricted to corrosive agents that have been excreted by appropriate microorganisms into the bulk phase but also to involve microbes attached to the surface of silica or silicates (Bennett et al., 1996, 2001). Because they are attached, their excreted metabolic products can attack the mineral surface in more concentrated form. Such attack may be manifested in etch marks. Some of the polysaccharides by which the microbes attach to the mineral surface may themselves be corrosive.

Bioweathering, like abiotic weathering, can lead to the formation of new minerals. This is the result of reprecipitation and crystallization of some of the mobilized constituents from the mineral that is weathered (Barker and Banfield, 1996; Adamo and Violante, 2000). New, secondary minerals may form on the surface of the weathered mineral. Microbes attached to the surface of minerals that are weathered may serve as nucleating agents in mineral neoformation (Schultze-Lam et al., 1996; Macaskie et al., 1992).

10.4.1 SOLUBILIZATION BY LIGANDS

Microbially produced ligands of divalent cations have been shown to cause dissolution of calcium-containing silicates. For instance, a soil strain of *Pseudomonas* that produced 2-ketogluconic acid from glucose dissolved synthetic silicates of calcium, zinc, and magnesium and the minerals wollastonite ($CaSiO_3$), apophyllite [$KCa_4Si_8O_{20}(F,OH) \cdot 8H_2O$], and olivine [$(Mg,Fe)_2SiO_4$] (Webley et al., 1960). The demonstration consisted of culturing the organism for 4 days at 25°C on separate agar

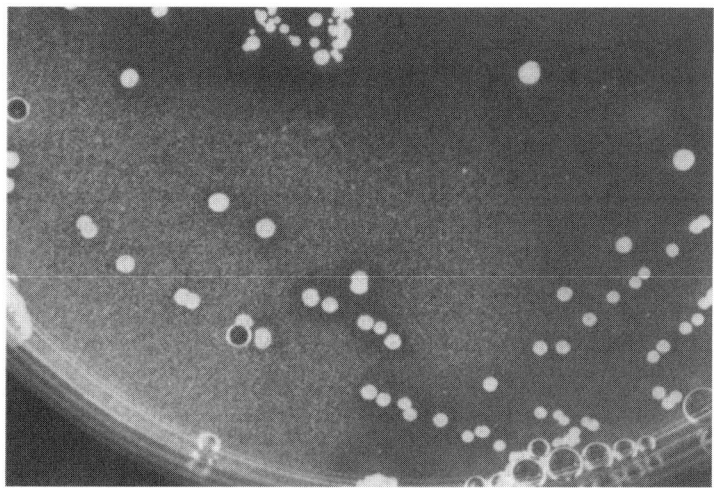

FIGURE 10.4 Colonies of bacterial isolate C-2 from a sample of weathered rock on synthetic calcium silicate selection medium showing evidence of calcium silicate dissolution around colonies. Basal medium was prepared by aseptically mixing 10 mL of sterile solution A (3 g dextrose or 3 g levulose in 100 mL distilled water), 10 mL of sterile solution B (0.5 g $(NH_4)_2SO_4$, 0.5 g $MgSO_4 \cdot 7H_2O$, 0.5 g Na_2HPO_4, 0.5 g KH_2PO_4, 2 g yeast extract, 0.05 g $MnSO_4 \cdot 2H_2O$ in 500 mL distilled water), and 20 mL of sterile 3% agar and distributing the mixture in Petri plates. Capping agar was prepared by mixing 10 mL of sterile synthetic $CaSiO_3$ suspension with 7.5 mL of sterile solution A, 7.5 mL of sterile solution B, and 15 mL of sterile 3% agar and distributing 3 mL of this mixture aseptically over the surface of the solidified basal agar in the plates.

media, each containing 0.25% (w/v) of one of the synthetic or natural silicates, which rendered the medium turbid. A clear zone was observed around the bacterial colonies when silicate was dissolved (Figure 10.4). A similar silicate-dissolving action was also shown with a gram-negative bacterium, strain D_{11}, which resembled *Erwinia* spp., and with *Bacterium* (now *Erwinia*) *herbicola* or with some *Pseudomonas* strains, all of which were able to produce 2-ketogluconate from glucose (Duff et al., 1963). The action of these bacteria was tested in glucose-containing basal medium: KH_2PO_4, 0.54 g; $MgSO_4 \cdot 7H_2O$, 0.25 g; $(NH_4)_2SO_4$, 0.75 g; $FeCl_3$, trace; Difoc yeast extract, 2 g; glucose 40 g; distilled water, 1 L; and 5–500 mg pulverized mineral per 5–10 mL of medium. It was found that dissolution of silicates in these cases resulted from the complexation of the cationic components of the silicates by 2-ketogluconate. The complexes were apparently more stable than the silicate. For example,

$$CaSiO_3 \Leftrightarrow Ca^{2+} + SiO_3^{2-} \tag{10.4}$$

$$Ca^{2+} + \text{2-ketogluconate} \rightarrow Ca(\text{2-ketogluconate}) \tag{10.5}$$

The structure of 2-ketogluconate is

The silicon that was liberated or released in these experiments and subsequently transformed took three forms: (1) low-molecular-weight or ammonium molybdate-reactive silicate (monomeric?);

(2) a colloidal polymeric silicate of higher molecular weight, which did not react with dilute hydrofluoric acid; and (3) an amorphous form that could be removed from solution by centrifugation and dissolved in cold 5% aqueous carbonate (Duff et al., 1963). Polymerized silicate can be transformed by bacteria into monomeric silicate, as has been shown in studies with *Proteus mirabilis* and *Bacillus caldolyticus* (Lauwers and Heinen, 1974). The *Proteus* culture was able to assimilate some of the monomeric silicate. The mechanism of depolymerization has not been elucidated. It may involve an extracellular enzyme.

Gluconic acid produced from glucose by several different types of bacteria has been shown to solubilize bytownite, albite, kaolinite, and quartz at near-neutral pH (Vandevivere et al., 1994). The activity around neutral pH suggests that the mechanism of action of gluconate involves chelation.

Quartz (SiO_2) has been shown to be subject to slow dissolution by organic acids such as citric, oxalic, pyruvic, and humic acids (Bennett et al., 1988), all of which can be formed by fungi or bacteria. In a pH range of 3–7, the effect was greatest at pH 7, indicating that the mechanism of action was not protonation but chelation. Bennett et al. (1988) suggest that the chelation involves an electron donor–acceptor system. Acetate, fumarate, and tartrate were ineffective in dissolving silica as a complex.

10.4.2 SOLUBILIZATION BY ACIDS

The effect of acids in solubilizing silicates has been noted in various studies. Waksman and Starkey (1931) cited the action of CO_2 on orthoclase

$$2KAlSi_3O_8 + 2H_2O + CO_2 \rightarrow H_4Al_2Si_2O_9 + K_2CO_3 + 4SiO_2 \qquad (10.6)$$

The CO_2 is, of course, very likely to be a product of respiration or fermentation.

Its weathering action is best viewed as based on the formation of the weak acid H_2CO_3 through hydration of CO_2. Another example of a silicate attack by acid is that involving spodumene ($LiAlSi_2O_6$) (Karavaiko et al., 1979). In this instance, an *in situ* correlation was observed between the extent of alteration of a spodumene sample and the acidity it produced when in aqueous suspension. Unweathered spodumene generated a pH in the range of 5.1–7.5, whereas altered spodumene generated a pH in the range of 4.2–6.4. Non-spore-forming heterotrophs were found to predominate in weathered spodumene. They included bacteria such as *Athrobacter pascens*, *A. globiformis*, and *A. simplex* as well as *Nocardia globerula*, *Pseudomonas fluorescens*, *Ps. putida*, and *Ps. testeronii* and fungi such as *Trichoderma lignorum*, *Cephalosporum atrum*, and *Penicillium decumbens*.

Acid decomposition of spodumene may be formulated as follows (Karavaiko et al., 1979):

$$4LiAlSi_2O_6 + 6H_2O + 2H_2SO_4 \rightarrow 2Li_2SO_4 + Al_4(Si_4O_{10})(OH)_8 + 4H_2SiO_3 \qquad (10.7)$$

The aluminosilicate product in this reaction is kaolinite.

Further investigation into microbial spodumene degradation revealed that among the most active organisms are the fungi *Penicillium notatum* and *Aspergillus niger*, thionic bacteria like *Thiobacillus thiooxidans*, and the slime-producing bacterium *Bacillus mucilaginosus* var. *siliceous* (Karavaiko et al., 1980; Avakyan et al., 1986). The fungi and *T. thiooxidans*, which produce acid, were most effective in solubilizing Li and Al. *B. mucilaginosus* was effective in solubilizing Si in addition to Li and Al by reaction of its extracellular polysaccharide with the silicate in spodumene.

Solubilization of silicon along with other constituents in the primary minerals amphibolite, biotite, and orthoclase by acids (presumably citric and oxalic acids) formed by several fungi and yeasts at the expense of glucose has also been demonstrated (Eckhardt, 1980; see also Barker et al., 1997). These findings of silicon mobilization are similar to those in earlier studies on the action of the fungi *Botritis*, *Mucor*, *Penicillium*, and *Trichoderma* isolated from rock surfaces and weathered stone. In these experiments, citric and oxalic acids produced by the fungi solubilized Ca, Mg, and Zn silicates (Webley et al., 1963). In studies by Henderson and Duff (1963), *Aspergillus niger* has been shown to release Si from apophyllite, olivine, saponite, vermiculite, and wollastonite, but not augite, garnet, heulandite, hornblende, hypersthene, illite, kaolinite, labradorite, orthoclase, or talc. However, Berner et al. (1980) found in laboratory experiments that augite, hypersthene, hornblende, and diopside in soil samples were subject to weathering by soil acids, presumably of biological origin. Organisms different from those used by Henderson and Duff (1963) and, as a result, different metabolic products were probably involved. *Penicillium simplicissimus* released Si from basalt, granite, grandiorite, rhyolite, andesite, peridotite, dunite, and quartzite with metabolically produced citric acid (Silverman and Munoz, 1970). Acid formed by *Pen. notatum* and *Pseudomonas* sp. release Si from plagioclase and nepheline (Aristovskaya and Kutuzova, 1968; Kutuzova, 1969).

In a study of weathering by organic and inorganic acids of three different plagioclase specimens (Ca–Na feldspars), it was found that steady-state dissolution rates were highest at approximately pH 3 and decreased as the pH was increased toward neutrality (Welch and Ullman, 1993). The organic and inorganic acids whose weathering action was studied are representative of some end products of microbial metabolism. Polyfunctional acids, including oxalate, citrate, succinate, pyruvate, and 2-ketoglutarate, were the most effective, whereas acetate and propionate were less effective. However, all organic acids were more effective than the inorganic acids HCl and HNO_3. The polyfunctional acids acted mostly as acidulants at very acidic pH and mainly as chelators at near-neutral pH. Ullman et al. (1996) found that in some instances the combined effect of protonation and chelation enhanced the solubilizing action of some polyfunctional acids on feldspars by a factor of 10 above the expected proton-promoted rate. Ca and Na were rapidly released in these experiments. The chelate attack appeared to be at the Al sites. Those organic acids that preferentially chelated Al were the most effective in enhancing plagioclase dissolution. Although the products of dissolution of feldspars are usually considered to include separate aluminum and silicate species, soluble aluminosilicate complexes may be intermediates (Browne and Driscoll, 1992).

The practical effect of acid attack of aluminosilicates can be seen in the corrosion of concrete sewer pipes. Concrete is formed from a mixture of cement (heated limestone, clay, and gypsum) and sand. On setting, the cement includes the compounds Ca_2SiO_4, Ca_3SiO_5, and $Ca(AlO_3)_2$, which hold the sand in their matrix. H_2S produced by microbial mineralization of organic sulfur compounds and by bacterial sulfate reduction of sulfate in sewage can itself corrode concrete. But corrosion is enhanced if the H_2S is first oxidized to sulfuric acid by thiobacilli *Thiobacillus neapolitanus* (now renamed *Halothiobacillus neapolitanus*), *T. intermedius*, *T. novellus*, and *T. thiooxidans* (now renamed *Acidithiobacillus thiooxidans*) (Parker, 1947; Milde et al., 1983; Sand and Bock, 1984; Okabe et al., 2007).

Groundwater pollution with biodegradable substances has been found to result in silicate weathering of aquifer rock. Products of microbial degradation of the substances cause the weathering. This was observed in an oil-polluted aquifer near Bemidji, Minnesota (Hiebert and Bennett, 1992). Microcosm experiments of 14 months' duration in the aquifer with a mixture of crystals such as albite, anorthite, anorthoclase, and microcline, each of which is a feldspar mineral, and quartz revealed microbial colonization of the mineral surfaces by individual cells and small clusters. Intense etching of the feldspar minerals and light etching of the quartz occurred at or near where the bacteria were seen. Such aquifer rock weathering can affect water quality.

10.4.3 SOLUBILIZATION BY ALKALI

Alkaline conditions are very conducive for mobilizing silicon, whether from silicates, aluminosilicates, or even quartz (Karavaiko et al., 1984). This is attributable to the significant lability of the Al–O and Si–O bonds under these conditions, because both types of bonds are susceptible to nucleophilic attack (see discussion by Karavaiko et al., 1984). *Sarcina ureae* growing in peptone–urea broth released silicon readily from nepheline, plagioclase, and quartz (Aristovskaya and Kutuzova, 1968; Kutuzova, 1969). In this instance, ammonia resulting from the hydrolysis of urea was the source of the alkali. In microbial spodumene degradation, alkaline pH also favors silicon release (Karavaiko et al., 1980).

Pseudomonas mendocina was able to enhance mobilization of Al, Si, and Fe impurities from kaolinite in a succinate-mineral salts medium in which the pH rose from ~7.7 to 9.2 in 4 days of growth under aerobic conditions (Maurice et al., 2001).

10.4.4 SOLUBILIZATION BY EXTRACELLULAR POLYSACCHARIDE

Extracellular polysaccharide has been claimed to play an important role in silicon release, especially in the case of quartz. Such polysaccharide is able to react with siloxanes to form organic siloxanes. It can be of bacterial origin (e.g., from *Bacillus mucilaginosus* var. *siliceous* [Avakyan et al., 1986] or unnamed organisms in a microbial mat on the rock around a hot spring [Heinen and Lauwers, 1988]) or fungal origin (e.g., from *Aspergillus niger*; Holzapfel and Engel, 1954a). The reaction appears not to be enzyme-catalyzed because polysaccharide from which the cells have been removed is reactive. Indeed, such organic silicon-containing compounds can be formed with reagent-grade organics (Holzapfel, 1951; Weiss et al., 1961) and have been isolated from various biological sources other than microbes (Schwarz, 1973). With polysaccharide from *B. mucilaginosus*, the reaction appears to be favored by acid metabolites (Malinovskaya et al., 1990). It should be noted that Welch and Vandevivere (1995) found that polysaccharides from different sources either had no effect or interfered with solubilization of plagioclase by gluconate at a pH between 6.5 and 7.

Barker and Banfield (1996) described the weathering by bacteria and lichens of amphibole syenite associated with the Stettin complex near Wausau, Wisconsin. The process involved penetration of grain boundaries, cleavages, and cracks. Mineral surfaces were coated with acid mucopolysaccharides (biofilm formation?). In the weathering, dissolution by metabolically produced corrosive agents and selective transport of mobilized constituents, probably mediated by acid mucopolysaccharides, occurred. Some mobilized constituents reprecipitated, leading to the formation of clay minerals.

A more detailed study revealed that the site of bioweathering by lichens (in this instance a symbiotic consortium of a fungus and an alga) could be divided into four zones (Barker and Banfield, 1998). The authors concluded that in the uppermost zone (zone 1), represented by the upper lichen thallus, no weathering occurs. This is the photosynthetic zone. In zone 2, involving the lower lichen thallus, active weathering due to interaction with lichen products occurs. Mineral fragments coated with organic polymers of incipient secondary minerals that resulted from the weathering may appear in the thallus. In zone 3, reactions occur, which are an indirect consequence of lichen action. In zone 4, any weathering, if it occurs, is due to abiotic processes.

10.4.5 DEPOLYMERIZATION OF POLYSILICATES

Because silicate can exist in monomeric form as well as polymeric form (metasilicates, siloxanes) and because silicon uptake by microbes depends on the monomeric form (orthosilicate), depolymerization of siloxanes is important. *Proteus mirabilis* and *Bacillus caldolyticus* have the capacity to promote this process. In the case of *B. caldolyticus*, it appears to be growth-dependent, although the organism does not assimilate Si (Lauwers and Heinen, 1974). In the weathering of quartz, the degradation of the mineral to the monomeric stage appears to proceed through an intermediate oligomeric stage. Organosilicates may also be formed transitionally (Avakyan et al., 1985). The detailed mechanism by which these transformations proceed is not known. It is clear, however, that these biodegradative processes are of fundamental importance to the biological silica cycle.

10.5 ROLE OF MICROBES IN THE SILICA CYCLE

As the foregoing discussion shows, some microbes (even some plants and animals; Drever, 1994) have a significant influence on the distribution and form of silicon in the biosphere. Those organisms that assimilate silicon clearly act as concentrators of it. Those that degrade silica, silicates, or aluminosilicates act as agents of silicon dispersion. They are an important source of orthosilicate on which the concentrators depend, and they are also important agents of rock weathering. Comparative electron microscopic studies provide clues to the extent of microbial weathering action (see Berner et al., 1980; Callot et al., 1987; Barker and Banfield, 1998).

It has been argued that silicate-mobilizing reactions by microbially produced acids and complexing agents under laboratory conditions occurred at glucose concentrations that may not be encountered in nature and may therefore be laboratory artifacts. A counterargument can be made, however, that microenvironments exist in soil and sediment that have appropriately high concentrations of utilizable carbohydrates, nitrogenous compounds, and other needed nutrients. They originate from the excretory products and dead remains of organisms from which appropriate bacteria and fungi can form the compounds that can promote breakdown of quartz, silicates, and aluminosilicates. Indeed, fungal hyphae in the litter zone and A horizon of several different soils have been shown by scanning electron microscopy to carry calcium oxalate crystals attached to them. This is evidence for extensive *in situ* production of oxalate by the fungus (Graustein et al., 1977). The basidiomycete *Hysterangium crassum* was shown to weather clay *in situ* with the oxalic acid it produced (Cromack et al., 1979). Lichens show evidence, observable *in situ*, of extensive rock weathering activity (Jones et al., 1981). Although Ahmadjian (1967) and Hale (1967) cast doubt on this ability of lichens, current evidence strongly supports the rock weathering activity of these organisms. Biodegradation of silica, silicates, and aluminosilicates is usually a slower process in nature than in the laboratory because the conditions in natural environments are usually less favorable. If this were not so, rock in the biosphere would be very unstable.

Thus, silicon in nature may follow a series of cyclic biogeochemical transformations (Kuznetsov, 1975; Harriss, 1972; Lauwers and Heinen, 1974). Silica, silicates, and aluminosilicates in rocks are subject to the weathering action of biological, chemical, and physical agents. The extent of the contribution of each of these agents must depend on the particular environmental circumstances. Silicon liberated in these processes as soluble silicate may be leached away by surface water or groundwater, and it may be removed from these waters by chemical and biological precipitation at new sites, or it may be swept into bodies of freshwater or the sea. There, silicate will tend to be removed by biological agents. Upon their death, other biological agents will release this silicon back into the solution or the siliceous remains will be incorporated into the sediment (Allison, 1981; Patrick and Holding, 1985; Weaver and Wise, 1974), where some or all of the silicon may later be returned to solution by weathering. The sediments of the ocean appear to be a sink for excess silica swept into the oceans because the silica concentrations of seawater tend to remain relatively constant. But over geologic time, this silicon in the sediments is not permanently immobilized. Plate tectonics will ultimately cause it to be recycled.

10.6 SUMMARY

The environmental distribution of silicon is significantly influenced by microbial activity. Certain microorganisms assimilate it and build it into cell-support structures. They include diatoms, some chrysophytes, silicoflagellates, some xanthophytes, radiolarians, and actinopods. Silicon uptake rates by diatoms have been measured, but the mechanism by which silicon is assimilated is still only partially understood. Certain silicon-incorporating bacteria may provide a biochemical model for some aspects of silicon assimilation. Silicon-assimilating microorganisms such as diatoms and radiolaria are important in the formation of siliceous oozes in the oceans, and diatoms are important in forming such oozes in lakes.

Some bacteria, fungi, and lichens are able to solubilize silicates and silica. They accomplish this by forming chelators, acids, bases, and exopolysaccharides that react with silica and silicates. These reactions are important in weathering of rock and cycling silicon in nature.

REFERENCES

Adamo P, Violante P. 2000. Weathering of rocks and neogenesis of minerals associated with lichen activity. *Appl Clay Sci* 16:229–256.

Ahmadjian V. 1967. *The Lichen Symbiosis*. Waltham, MA: Blaidsell.

Allison CW. 1981. Siliceous microfossils from the Lower Cambrian of northwest Canada: Possible source for biogenic chert. *Science* 211:53–55.

Anderson GM. 1972. Silica solubility. In: Fairbridge RW, ed. *The Encyclopedia of Geochemistry and Environmental Sciences*. Encyclopedia of Earth Science Series, Vol. IVA. New York: Van Nostrand Reinhold, pp. 1085–1088.

Aristovskaya TV, Kutuzova RS. 1968. Microbiological factors in the mobilization of silicon from poorly soluble natural compounds. *Pochvovedenie* 12:59–66.

Avakyan ZA, Belkanova NP, Karavaiko GI, Piskunov VP. 1985. Silicon compounds in solution during bacterial quartz degradation. *Mikrobiologiya* 54:301–307 (English translation pp. 250–256).

Avakyan ZA, Pivovarova TA, Karavaiko GI. 1986. Properties of a new species, *Bacillus mucilaginosus*. *Mikrobiologiya* 55:477–482 (English translation pp. 477–482).

Azam F. 1974. Silicic-acid uptake in diatoms studied with [^{68}Ge]germanic acid as tracer. *Planta* 121:205–212.

Azam F, Hemmingsen B, Volcani BE. 1973. Germanium incorporation into the silica of diatom cell walls. *Arch Microbiol* 92:11–20.

Azam F, Hemmingsen B, Volcani BE. 1974. Role of silicon in diatom metabolism. V. Silicic acid transport and metabolism in the heterotrophic diatom *Nitschia alba*. *Arch Mikrobiol* 97:103–114.

Azam F, Volcani B. 1974. Role of silicon in diatom metabolism. VI. Active transport of germanic acid in the heterotrophic diatom *Nitschia alba*. *Arch Microbiol* 101:1–8.

Barker WW, Banfield JF. 1996. Biologically versus inorganically mediated weathering reactions: Relationships between minerals and extracellular microbial polymers in lithiobiontic communities. *Chem Geol* 132:55–69.

Barker WW, Banfield JF. 1998. Zones of chemical and physical interactions at interfaces between microbial communities and minerals. A model. *Geomicrobiol J* 15:223–244.

Barker WW, Welch SA, Banfield JF. 1997. Biogeochemical weathering of silicate minerals. In: Banfield JF, Nealson KH, eds. *Geomicrobiology: Interactions Between Microbes and Minerals*. Reviews in Mineralogy, Vol. 35. Washington, DC: Mineralogical Society of America, pp. 391–428.

Bennett PC, Hiebert FK, Choi WJ. 1996. Microbial colonization and weathering of silicates in a petroleum-contaminated aquifer. *Chem Geol* 132:45–53.

Bennett PC, Melcer ME, Siegel DI, Hassett JP. 1988. The dissolution of quartz in dilute aqueous solutions of organic acids at 25°C. *Geochim Cosmochim Acta* 52:1521–1530.

Bennett PC, Rogers JR, Choi WJ, Hiebert FK. 2001. Silicates, silicate weathering, and microbial ecology. *Geomicrobiol J* 18:3–19.

Berner RA, Sjoeberg EL, Velbel MA, Krom MD. 1980. Dissolution of pyroxenes and amphiboles during weathering. *Science* 207:1205–1206.

Blank GS, Sullivan CW. 1979. Diatom mineralization of silicic acid. III. Si(OH)$_4$ binding and light dependent transport in *Nitschia angularis*. *Arch Microbiol* 123:17–164.

Bowen HJM. 1966. *Trace Elements in Biochemistry*. London: Academic Press.

Bowen HJM. 1979. *Environmental Chemistry of the Elements*. London: Academic Press.

Browne BA, Driscoll CT. 1992. Soluble aluminum silicates: Stoichiometry, stability, and implications for environmental geochemistry. *Science* 256:1667–1670.

Callot B, Maurette M, Pottier L, Dubois A. 1987. Biogenic etching of microfractures in amorphous and crystalline silicates. *Nature* (London) 328:147–149.

Chisholm SW, Azam F, Epply RW. 1978. Silicic acid incorporation in marine diatoms on light-dark cycles: Use as an assay for phased cell division. *Limnol Oceanogr* 23:518–529.

Coombs J, Volcani BE. 1968. The biochemistry and fine structure of silica shell formation in diatoms. Silicon induced metabolic transients in *Navicula pelliculosa*. *Planta* 80:264–279.

Cromack K, Jr., Sollins P, Graustein WC, Speidel K, Todd AW, Spycher G, Li CY, Todd RL. 1979. Calcium oxalate accumulation and soil weathering in mats of the hypogenous fungus *Hysterangium crassum*. *Soil Biol Biochem* 11:463–468.

Darley WM, Volcani BE. 1969. Role of silicon in diatom metabolism. A silicon requirement for deoxyribonucleic acid synthesis in the diatom *Cylindrotheca fusiformis*, Reiman and Lewin. *Exp Cell Res* 58:334–343.

Dashman T, Stotzky G. 1986. Microbial utilization of amino acids and a peptide bond on homoionic montmorillonite and kaolinite. *Soil Biol Biochem* 18:5–14.

De La Rocha CL, Brzezinski MA, DeNiro MJ. 1997. Fractionation of silicon isotopes by marine diatoms during biogenic silica formation. *Geochim Cosmochim Acta* 61:5051–5056.

De La Rocha Cl, Brzezinski MA, DeNiro MJ. 2000. A first look at the distribution of the stable isotopes of silicon in natural waters. *Geochim Cosmochim Acta* 64:2467–2477.

de Vrind-de Jong EW, de Vrind JPM. 1997. Algal deposition of carbonates and silicates. In: Banfield JF, Nealson KH, eds. *Geomicrobiology: Interactions Between Microbes and Minerals. Reviews in Mineralology*, Vol. 35. Washington, DC: Mineralogical Society of America, pp. 267–307.

Dommergues Y, Mangenot F. 1970. *Écologie Microbienne du Sol*. Paris: Masson.

Drever JL. 1994. The effect of land plants on weathering of silicate minerals. *Geochim Cosmochim Acta* 58:2325–2332.

Duff RB, Webley DM, Scott RO. 1963. Solubilization of minerals and related minerals by 2-ketogluconic acid-producing bacteria. *Soil Sci* 95:105–114.

Eckhardt FEW. 1980. Microbial degradation of silicates. Release of cations from aluminosilicate minerals by yeast and filamentous fungi. In: Oxley TA, Becker G, Allsopp D, eds. *Biodeterioration. Proceedings of the 4th International Biodeterioration Symposium*, Berlin. London: Pitman, pp. 107–116.

Ehrlich HL. 1982. Enhanced removal of Mn^{2+} from seawater by marine sediment and clay minerals in the presence of bacteria. *Can J Microbiol* 28:1389–1395.

Garrels RM. 1965. Silica: Role in the buffering of natural waters. *Science* 148:69.

Graustein WC, Cromack K, Jr., Sollins P. 1977. Calcium oxalate: Occurrence in soils and effect on nutrient and geochemical cycles. *Science* 198:1252–1254.

Hale ME, Jr. 1967. *The Biology of Lichens*. London: Edward Arnold.

Hall FR. 1972. Silica cycle. In: Fairbridge RW, ed. *The Encyclopedia of Geochemistry and Environmental Sciences*. Encyclopedia of Earth Science Series, Vol. IVA. New York: Van Nostrand Reinhold, pp. 1082–1085.

Harriss RC. 1972. Silica–biogeochemical cycle. In: Fairbridge RW, ed. *The Encyclopedia of Geochemistry and Environmental Sciences*. Encyclopedia of Earth Science Series, Vol. IVA. New York: Van Nostrand Reinhold, pp. 1080–1082.

Healy FP, Combs J, Volcani BE. 1967. Changes in pigment content of the diatom *Navicula pelliculosa* in silicon starvation synchrony. *Arch Mikrobiol* 59:131–142.

Heinen W. 1960. Silicium-Stoffwechsel bei Mikroorganismen. I. Mitteilung. Aufnahme von Silicium durch Bakterien. *Arch Mikrobiol* 37:199–210.

Heinen W. 1962. Silicium-Stoffwechsel bei Mikroorganismen. II. Mitteilung. Beziehung zwischen dem Silicat- und Phosphat-Stoffwechsel bei Bakterien. *Arch Mikrobiol* 41:229–246.

Heinen W. 1963a. Silicium-Stoffwechsel bei Mikroorganismen. III. Mitteilung. Einfluss verschiedener Anionen auf den bakeriellen Si-Stoffwechsel. *Arch Mikrobiol* 45:145–161.

Heinen W. 1963b. Silicium-Stoffwechsel bei Mikroorganismen. IV. Mitteilung. Die Wirkung organischer Verbindungen, insbesondere Glucose, auf den Silicium-Stoffwechsel bei Bakterien. *Arch Mikrobiol* 45:162–171.

Heinen W. 1963c. Silicium-Stoffwechsel bei Mikroorganismen. V. Mitteilung. Untersuchungen zur Mobilität der inkorporierten Kieselsäure. *Arch Mikrobiol* 45:172–178.

Heinen W. 1965. Time-dependent distribution of silicon in intact cells and cell-free extracts of *Proteus mirabilis* as a model of bacterial silicon transport. *Arch Biochem Biophys* 110:137–149.

Heinen W. 1967. Ion accumulation in bacterial systems. III. Respiration-dependent accumulation of silicate by a particulate fraction from *Proteus mirabilis* cell-free extract. *Arch Biochem Biophys* 120:101–107.

Heinen W. 1968. The distribution and some properties of accumulated silicate in cell-free bacterial extracts. *Acta Bot Neerl* 17:105–113.

Heinen W, Lauwers AM. 1988. Leaching of silica and uranium and other quantitative aspects of the lithobiontic colonization in a radioactive thermal spring. *Microb Ecol* 15:135–149.

Helmcke J-G. 1954. Die Feinstruktur der Kieselsäure und ihre physiologische Bedeutung in Diatomschalen. *Naturwissenschaften* 11:254–255.

Henderson MEK, Duff RB. 1963. The release of metallic and silicate ions from minerals, rocks and soils by fungal activity. *J Soil Sci* 14:236–246.

Hiebert FK, Bennett PC. 1992. Microbial control of silicate weathering in organic-rich ground water. *Science* 258:278–281.

Holzapfel L. 1951. Siliziumverbindung in biologischen Systemen. Organ. Kieselsäureverbindungen. XX. Mitteilung. *Z Elektrochem* 55:577–580.

Holzapfel L, Engel W. 1954a. Der Einfluss organischer Kieselsäureverbindungen auf das Wachstum von *Aspergillus niger* und *Triticum. Z Naturforsch* 9b:602–606.

Holzapfel L, Engel W. 1954b. Über die Abhängigkeit der Wachstumsgeschwindigkeit von *Aspergillus niger* in Kieselsäurelösungen bei O_2-Belüftung. *Naturwissenschaften* 41:191–192.

Iler R. 1979. *Chemistry of Silica*. New York: Wiley.

Jones D, Wilson MJ, McHardy WJ. 1981. Lichen weathering of rock-forming minerals: Application of scanning electron microscopy and microprobe analysis. *J Microsc* 124(1):95–104.

Jorgensen EG. 1955. Solubility of silica in diatoms. *Physiol Plant* 8:864–851.

Karavaiko GI, Avakyan ZA, Krutsko VS, Mel'nikova EO, Zhdanov V, Piskunov VP. 1979. Microbiological investigations on a spodumene deposit. *Mikrobiologiya* 48:502–508 (English translation pp. 383–398).

Karavaiko GI, Belkanova NP, Eroshchev-Shak VA, Avakyan ZA. 1984. Role of microorganisms and some physicochemical factors of the medium in quartz destruction. *Mikrobiologiya* 53:976–981 (English translation pp. 795–800).

Karavaiko GI, Krutsko VS, Mel'nikova EO, Avakyan ZA, Ostroushko YI. 1980. Role of microorganisms in spodumene degradation. *Mikrobiologiya* 49:547–551 (English translation pp. 402–406).

Katami A. 1982. Dissolution rates of silica from diatoms decomposing at various temperatures. *Marine Biol* (Berlin) 68:91–96.

Kretz R. 1972. Silicon: Element and geochemistry. In: Fairbridge RW, ed. *The Encyclopedia of Geochemistry and Environmental Sciences*. Encyclopedia of Earth Science Series, Vol. IVA. New York: Van Nostrand Reinhold, pp. 1091–1092.

Kutuzova RS. 1969. Release of silica from minerals as result of microbial activity. *Mikrobiologiya* 38:714–721 (English translation pp. 596–602).

Kuznetsov SI. 1975. The role of microorganisms in the formation of lake bottom deposits and their diagenesis. *Soil Sci* 119:81–88.

Latimer WM, Hildebrand JH. 1940. *Reference Book of Inorganic Chemistry*. Revised edition. New York: Macmillan.

Lauwers AM, Heinen W. 1974. Bio-degradation and utilization of silica and quartz. *Arch Microbiol* 95:67–78.

Lewin JC. 1957. Silicon metabolism in diatoms. IV. Growth and frustule formation in *Navicula pelliculosa. J Gen Physiol* 39:1–10.

Lewin JC. 1965. Silicification. In: Lewin RA, ed. *Physiology and Biochemistry of Algae*. New York: Academic Press, pp. 445–455.

Liebau F. 1985. *Structural Chemistry of Silicates. Structure, Bonding and Classification*. Berlin, Germany: Springer.

Macaskie LE, Empson RM, Cheetham AK, Gey CP, Skarnulis AJ. 1992. Uranium bioaccumulation by a *Citrobacter* sp. as a result of enzymatically mediated growth of polycrystalline HUO_2PO_4. *Science* 257:782–784.

Malinovskaya IM, Kosenko LV, Votselko SK, Podgorskii VS. 1990. Role of *Bacillus mucilaginosus* polysaccharide in degradation of silicate minerals. *Mikrobiologiya* 59:70–78 (English translation pp. 49–55).

Marine Chemistry. 1971. *A Report on the Marine Chemistry Panel of the Committee of Oceanography*. Washington, DC: National Academy of Sciences.

Marshall KC, 1971. Sorption interactions between soil particles and microorganisms. In: McLaren AD, Skujins J, eds. *Soil Biochemistry*, Vol. 2. New York: Marcel Dekker, pp. 409–445.

Marshman NA, Marshall KC. 1981a. Bacterial growth on proteins in the presence of clay minerals. *Soil Biol Biochem* 13:127–134.

Marshman NA, Marshall KC. 1981b. Some effect of montmorillonite on the growth of mixed microbial cultures. *Soil Biol Biochem* 13:135–141.

Martin-Jézéquel V, Hildebrand M, Brezinski MA. 2000. Silicon metabolism in diatoms: Implications for growth. *J Phycol* 36:821–824.

Maurice PA, Vierkoen MA, Hersman LE, Fulghum JE, Ferryman A. 2001. Enhancement of kaolinite dissolution by an aerobic *Pseudomonas mendocina* bacterium. *Geomicrobiol J* 18:21–35.

Milde K, Sand W, Wolff W, Bock E. 1983. Thiobacilli of the corroded concrete walls of the Hamburg sewer system. *J Gen Microbiol* 129:1327–1333.

Millman JD, Boyle E. 1975. Biological uptake of dissolved silica in the Amazon estuary. *Science* 189:995–997.

Mitchell RL. 1955. Trace elements. In: Bear FE, ed. *Chemistry of the Soil*. New York: Van Nostrand Reinhold, pp. 235–285.

Okabe S, Odagiri M, Ito T, Satoh H. 2007. Succession of sulfur-oxidizing bacteria in the microbial community on corroding concrete in sewer systems. *Appl Environ Microbiol* 73:971–978.

Parker CD. 1947. Species of sulfur bacteria associated with corrosion of concrete. *Nature* (London) 159:439.

Patrick S, Holding AJ. 1985. The effect of bacteria on the solubilization of silica in diatom frustules. *J Appl Microbiol* 59:7–16.

Reeves CD, Volcani BE. 1984. Role of silicon in diatom metabolism. Patterns of protein phosphorylation in *Cylindrotheca fusiformis* during recovery from silicon starvation. *Arch Microbiol* 137:291–294.

Rogall E. 1939. Über den Feinbau der Kieselmembran der Diatomeen. *Planta* 29:279–291.

Sand W, Bock E. 1984. Concrete corrosion in the Hamburg sewer system. *Environ Technol Lett* 5:517–528.

Schultze-Lam S, Fortin D, Davis BS, Beveridge TJ. 1996. Mineralization of bacterial surfaces. *Chem Geol* 132:171–181.

Schwarz K. 1973. A bound form of silicon in glycosaminoglycans and polyuronides. *Proc Natl Acad Sci USA* 70:1608–1612.

Sillén LG. 1967. The ocean as chemical system. *Science* 156:1189–1197.

Silverman MP, Munoz EF. 1970. Fungal attack on rock: Solubilization and altered infrared spectra. *Science* 169:985–987.

Stotzky G. 1986. Influences of soil mineral colloids on metabolic processes, growth, adhesion, and ecology of microbes and viruses. In: Huang PM, Schnitzer M, eds. *Interaction of Soil Minerals with Organics and Microbes*. SSSA Spec. Publ. no. 17. Madison, WI: Soil Science Society of America, pp. 305–428.

Sullivan CW. 1971. A silicic acid requirement for DNA polymerase, thymidylate kinase, and DNA synthesis in the marine diatom *Cylindrica fusiformis*. PhD Thesis, University of California.

Sullivan CW, Volcani EB. 1973. Role of silicon in diatom metabolism. The effects of silicic acid on DNA polymerase, TMP kinase and DNA synthesis in *Cyclotheca fusiformis*. *Biochim Biophys Acta* 308:212–229.

Tan KH. 1986. Degradation of soil minerals by organic acids. In: Huang PM, Schnitzer M, eds. *Interaction of Soil Minerals with Organics and Microbes*. SSSA Spec. Publ. no. 17. Madison, WI: Soil Science Society of America, pp. 1–27.

Toth SJ. 1955. Colloidal chemistry of soils. In: Bear FE, ed. *Chemistry of the Soil*. New York: Van Nostrand Reinhold, pp. 85–106.

Ullman WJ, Kirchman DL, Welch SA, Vandevivere P. 1996. Laboratory evidence for microbially mediated silicate mineral dissolution in nature. *Chem Geol* 132:11–17.

Vandevivere P, Welch SA, Ullman WJ, Kirchman DL. 1994. Enhanced dissolution of silicate minerals by bacteria at near-neutral pH. *Microb Ecol* 27:241–251.

Waksman SA, Starkey RL. 1931. *The Soil and the Microbe*. New York: Wiley.

Weaver FM, Wise SW, Jr. 1974. Opaline sediments of the southeastern coastal plain and horizon A: Biogenic origin. *Science* 184:899–901.

Weaver TL, Dugan PR. 1972. Enhancement of bacteria methane oxidation by clay minerals. *Nature* (London) 237:518.

Webley DM, Duff RB, Mitchell WA. 1960. A plate method for studying the breakdown of synthetic and natural silicates by soil bacteria. *Nature* (London) 188:766–767.

Webley DM, Henderson MEF, Taylor IF. 1963. The microbiology of rocks and weathered stones. *J Soil Sci* 14:102–112.

Weiss A, Reiff G, Weiss A. 1961. Zur Kenntnis wasserbeständiger Kieselsäureesster. *Z Anorg Chem* 311:151–179.

Welch SA, Ullman WJ. 1993. The effect of organic acids on plagioclase dissolution rates and stoichiometry. *Geochim Cosmochim Acta* 57:2725–2736.

Welch SA, Vandevivere P. 1995. Effect of microbial and other naturally occurring polymers on mineral dissolution. *Geomicrobiol J* 12:227–238.

Werner D. 1966. Die Kieselsäure im Stoffwechsel von *Cyclotella cryptica* Reimann, Lewin and Guillard. *Arch Mikrobiol* 55:278–308.

Werner D. 1967. Hemmung der Chlorophyllsynthese und des NADP$^+$-abhängigen Glyzeraldehyd-3-phosphat-dehydrogenase durch Germaniumsäure bei *Cyclotella cryptica*. *Arch Mikrobiol* 57:51–60.

11 Geomicrobiology of Aluminum: Microbes and Bauxite

11.1 INTRODUCTION

Aluminum is the most plentiful element in the Earth's crust after silicon and oxygen. Its crustal abundance has been estimated to be 8.19% (Gornitz, 1972; Strahler, 1976). Its concentration in rocks and shales amounts to 82,000 ppm, in sandstones 25,000 ppm, and in limestones 4200 ppm. Its average concentration in freshwater has been given as 0.24 ppm and in seawater 0.001 ppm (Bowen, 1966; Marine Chemistry, 1971). The aluminum concentration in surface seawater may be under biological control according to the studies with the diatom *Skeletonema costatum* (Mackenzie et al., 1978; Stoffyn, 1979). However, inorganic control of aluminum in seawater has also been suggested (Hydes, 1979).

Although aluminum is a metal, it does not occur in a metallic state in nature. This is best explained by the fact that the reduction of aluminum salts or oxides to Al^0 is an extremely endothermic process (for the reaction $Al_2O_3 \rightarrow 2Al^0 + 1.5O_2$, ΔG^0 is approximately $+377$ kcal mol^{-1} or $+1578$ kJ mol^{-1}). Commercial production of aluminum metal involves electrolysis of molten Al_2O_3 in cryolite (sodium aluminum fluoride) at high temperature.

Aluminum is a constituent of igneous minerals, including feldspars, micas, some pyroxenes and amphiboles, and secondary minerals, especially clays (see Chapter 10) and aluminum hydroxides and oxides such as gibbsite (γ-$Al(OH)_3$), boehmite (γ-$AlO(OH)$), diaspore (α-$AlO(OH)$), and corundum (Al_2O_3) (Gornitz, 1972).

Apart from its presence as a constituent of specific minerals, aluminum exists as Al^{3+} in aqueous solution below pH ~5 and mainly as a cationic complex, $Al(OH)_4^-$, above pH 7.4. The aluminum ion tends to form hydroxyl complexes ($Al(OH)^{2+}$ and $Al(OH)_2^+$) and precipitates as $Al(OH)_3$ at neutral pH (Macdonald and Martin, 1988; Nordstrom and May, 1995; Garcidueñas Piña and Cervantes, 1996). Alumina (Al_2O_3) is more insoluble in water than silica (SiO_2) (Gornitz, 1972). Al^{3+} can be complexed by various organic compounds including hydroxamate siderophores, sugar acids, phenols, phenolic acids, and polyphenols (Vance et al., 1995). It can also form sulfato and phosphato complexes (Nordstrom and May, 1995).

The Al^{3+} is toxic to most forms of life because it can react with negatively charged groups on proteins, including enzymes, and with other vital polymers (Harris, 1972; Garcidueñas Piña and Cervantes, 1996). However, a few plants not only tolerate it but may have a requirement for it. The tree *Orites excelsa* contains basic aluminum succinate in its woody tissue. The ashes of the club moss *Lycopodium alponium* (Pteridophyta) may contain as much as 33% Al_2O_3 (Gornitz, 1972). The bacterium *Pseudomonas fluorescens* can detoxify Al^{3+} in the bulk phase by producing an extracellular phospholipid that sequesters the aluminum (Appanna et al., 1994; Appanna and St Pierre, 1994; Appanna and Hamel, 1996). The cyanobacterium *Anaboena cylindrica* can detoxify limited amounts of aluminum taken into the cell with the help of intracellular inorganic polyphosphate granules, which sequester Al^{3+} (Petterson et al., 1985).

A number of bacteria, fungi, and lichens are known to participate in the formation of secondary aluminum-containing minerals through their ability to weather aluminum-containing rock minerals (see Chapters 4 and 10).

11.2   MICROBIAL ROLE IN BAUXITE FORMATION

11.2.1   NATURE OF BAUXITE

Bauxite is an ore that contains 45–50% Al_2O_3 in the form of gibbsite, boehmite, and/or diaspore and not more than 20% Fe_2O_3 as hematite, goethite, or aluminian goethite. Such ore also contains a combination of 3–5% silica and silicates (Valeton, 1972). The silicates in the ore are chiefly in the form of kaolinite.

Bauxite is a product of surficial weathering of aluminosilicate minerals in rock (Butty and Chapalaz, 1984). Warm and humid climatic conditions with wet and dry seasons favor the weathering process. The parent material from which bauxite arises may be volcanic and other aluminosilicate rocks, limestone associated with karsts, and alluvium (Butty and Chapalaz, 1984). Weathering that leads to bauxite formation begins at the surface of an appropriate exposed or buried rock formation and in cracks and fissures. This process includes breakdown of the aluminosilicates in the parent substance with gradual solubilization of Al, Si, Fe, and other constituents, starting at the mineral surface. The biological contributions to the weathering are favored by warm temperatures and humid conditions (see Section 11.2.2). In the case of limestone as parent substance, an important part of the weathering process is the solubilization of the $CaCO_3$. The solubilized products reprecipitate when their concentration and the environmental pH and E_h are favorable. Groundwater flow may transport some of the solubilized constituents away from the site of weathering, contributing to an enrichment of the constituents left behind. The initial stages of weathering produce materials that could serve as precursors in the formation of laterite or bauxite. A key difference between laterite and bauxite is that the former is richer in iron relative to aluminum, whereas the latter is richer in aluminum. Biotic and abiotic environmental conditions determine whether laterite or bauxite will accumulate (Schellman, 1994).

Bauxite formation in nature is a slow process and impacted by vegetation at the site of formation and tectonic movement in addition to climate. Vegetation provides cover that protects against erosion of weathered rocks, limits water evaporation, and may generate weathering agents (Butty and Chapalaz, 1984). It is also the source of nutrients for microbiota that participate in rock weathering and some other aspects of bauxite formation. Tectonic movement contributes to topography and geomorphology in the area of bauxite formation. Alterations in topography as well as variation in climate can affect the groundwater level. Alternating moist and dry conditions are needed during weathering of host rock for the formation and buildup of the secondary minerals that make up bauxite.

11.2.2   BIOLOGICAL ROLE IN WEATHERING OF THE PARENT ROCK MATERIAL

Biological participation in bauxite formation has been suggested in the past. Butty and Chapalaz (1984) invoked microbial activity in controlling pH and E_h. They viewed rock weathering as being promoted by microbes through the generation of acids or ligands for mobilizing rock components and direct participation in redox reactions affecting iron, manganese, and sulfur compounds.

A more detailed proposal of biogenic bauxite formation is that of Taylor and Hughes (1975). They concluded that bauxite deposits in Rennell Island in the south Solomon Sea near Guadalcanal were the result of biodegradation of volcanic ash that originated in eruptions on Guadalcanal, 180 km distant, and was deposited in pockets of karstic limestone in the lagoons in Rennell Island in the Plio-Pleistocene. The authors established that the bauxite deposit, enclosed in dolomitic limestone from a reef, was not derived from the residues left after the dolomite had weathered away. They speculated that sulfate-reducing bacteria generated CO_2, which caused weathering of aluminosilicates and ferromagnesian minerals in the volcanic ash, giving rise to transient kaolin that would dissolve at low pH to yield Al^{3+} and silicic acid. According to the authors, bacterial pyrite formation by sulfate-reducing bacteria would create pH and E_h conditions that favor weathering of the minerals in the volcanic ash. As initial microbial activity subsided owing to nutrient depletion, pH was predicted to rise, resulting in the formation of iron and aluminum oxides—the chief constituents of bauxite. The authors speculated that bacteria played a role in the formation of a gel of oxides. Uplift in the northwestern part of the island, groundwater flow, and oxidation of the pyrite were seen to play a role in the maturation of the bauxite.

Natarajan et al. (1997) inferred from the presence of the members of the bacterial genera *Thiobacillus*, *Bacillus*, and *Pseudomonas* in the Jamnagar bauxite mines in Gujarat, India, that these microorganisms are involved in bauxite formation. They based their inference on the known ability of these organisms to weather aluminosilicates, precipitate oxyhydroxides of iron, dissolve and transform alkaline metal species, and form alumina, silica, and calcite minerals. On the same basis, they also implicated the fungi of the genus *Cladosporium*, which they suggest can reduce ferric iron and dissolve aluminosilicates.

On the basis of the published discussions of bauxite formation (bauxitization) (see Valeton, 1972; Butty and Chapalaz, 1984), the process can be divided into two stages, which may overlap to some extent. The first stage is the weathering of the parent rock or alluvium that leads to the liberation of Al, Fe, and Si from primary and secondary minerals that contain aluminum. The second stage consists of the formation of bauxite from the weathering products. Each of the two stages is thought to be aided by microorganisms.

11.2.3 WEATHERING PHASE

Extensive evidence exists that bacteria, fungi, and lichens have the ability to weather rock minerals (see Chapter 10). This evidence was amassed in laboratory experiments and by *in situ* observation. The rock weathering resulted from the excretion of corrosive metabolic products by various microbes. These products included inorganic and organic acids, bases, or organic ligands. In instances where oxidizable or reducible rock components are present, enzymatic redox reactions may also have come into play. Most studies of microbial weathering have involved aerobic bacteria and fungi. However, anaerobic bacteria must also be considered in some cases of weathering. Many of them are actually a better source of corrosive organic acids needed for rock weathering than aerobic bacteria. Moreover, bacterial reduction of ferric iron, whether in solution or in minerals, requires anaerobic conditions.

The products of rock weathering may be soluble or insoluble. In the latter case, they may accumulate as secondary minerals. In bauxitization, Al, Si, and Fe will be mobilized. Control of pH, mostly by microorganisms, helps to segregate these products to some extent by affecting their respective solubilities. Vegetative cover over the site of bauxitization is a source of nutrients required by the microorganisms to grow and form the weathering agents. Warm temperature enhances microbial growth and activity. Infiltrating surface water (rain) and groundwater help to separate the soluble weathering products from the insoluble ones derived from the source material, leaving behind a mineral mixture that will include aluminum and iron oxides, silica, and silicates (especially kaolinite, formed secondarily from the interaction of Al^{3+} and silicate ions). The mineral conglomerate is *protobauxite*.

11.2.4 BAUXITE MATURATION PHASE

In this stage, the protobauxite becomes enriched in aluminum oxides (gibbsite, diaspore, or boehmite) by selective removal of iron oxides, silica, and silicates. Such enrichment has been shown to occur in laboratory experiments under anaerobic conditions (Ehrlich et al., 1995; Ehrlich and Wickert, 1997). Unsterilized Australian ore was placed in presterilized glass or Lexan™ columns. The ore in the columns was then completely immersed in a sterile sucrose–mineral salts solution. After the outgrowth of bacteria resident on the ore had taken place over 3–5 days at 37°C, the columns were fed daily with fresh sterile medium from the bottom over a time interval of 20–30 min, depending on the size of the columns. The medium was not deaerated before it is introduced into the columns. Control columns in which the outgrowth of bacteria was suppressed by 0.1% or 0.05% thymol added to the nutrient solution fed to these columns were run in parallel. The effluent of spent medium displaced by each addition of fresh medium was collected and analyzed by measuring pH; determining the concentration of solubilized ferrous and total iron, Si, and Al; and examining the morphology of the bacteria displaced in the effluents.

The content of the columns quickly turned anaerobic as bacteria grew out from the ore. This was indicated by strong foaming and outgassing in the headspace of the columns and by detection of significant numbers of *Clostridia* among the bacteria in the displaced medium in successive effluents from the columns. The evolved gas probably consisted of CO_2 and H_2, both of which are known to be produced from sucrose by clostridial fermentation. Analyses of successive effluents showed that the bacteria solubilized iron in the bauxite, which was in the form of hematite, goethite, or aluminian goethite, by reducing it to ferrous iron. As expected, solubilization from unground Australian pisolitic bauxite was slower and less extensive than the same ore preground to a mesh size of −10 (particle size of 2 mm or less). In an experiment with the unground pisolitic bauxite* (Ehrlich et al., 1995), 25% of the iron was mobilized in 106 days. In the same time, the bacteria also solubilized 2.2% of the SiO_2 or kaolinite in the ore. Al was solubilized over this time interval to the extent of 5.9%, whereas Fe and Si were solubilized at a fairly constant rate once the bacteria had grown out from the ore. Al was not solubilized until the pH in the bulk phase in the column had dropped gradually from ~6.5 initially to ~4.5 after ~20 days. At the start of the experiments, the ore contained 50% Al (calculated as Al_2O_3), 20% Fe (calculated as Fe_2O_3), and 6.5% Si (calculated as SiO_2) by weight. No measurable Fe, Si, or Al solubilization took place in the control columns.

Results from column experiments with bauxite samples from different geographic locations (Ehrlich et al., 1995; Ehrlich and Wickert, 1997) support the notion that for optimal aluminum enrichment of protobauxite, the pH in the bulk phase should remain >4.5. In most cases in the field, the pH probably rarely, if ever, drops <4.5. This is because in nature bauxite maturation occurs in an open system where the bacterial activity will be much slower and acidic metabolites are more readily diluted and carried away by moving groundwater than in the column experiments. However, an exception seems to be a deposit in northern Brazil in which bauxite weathering has given rise to a kaolin deposit as a result of iron mobilization (deferritization) and apparently some aluminum mobilization (Kotschoubey et al., 1999). The experimental results described earlier clearly showed that the action of the bacteria that grew from the ore in the columns enriched the ore in aluminum.

The column effluents contained fermentation acids such as acetic and butyric acids and, sometimes, neutral solvents such as butanol, acetone or isopropanol, and ethanol, detectable by spot confirmation using high performance liquid chromatography (HPLC) analysis and by odor. With unground pisolitic ore, Fe and Si but not Al solubilization leveled off after ~60 days. With the same ore ground to −2 mm particle size, Fe, Si, and Al solubilization continued at a steady rate over the entire experimental period, which in some cases exceeded 100 days.

Some pisolites taken from the active and control columns at the end of the experiment with Australian pisolitic ore described earlier were cross-sectioned, surface polished, and examined microscopically by reflected light and scanning electron microscopy (SEM) coupled with energy-dispersive x-ray (EDX) analysis (Ehrlich et al., 1995). Color images of the cross sections of pisolites that had been subjected to bacterial action in the active column showed a distinct bleached rim surrounding a reddish-brown core. Comparable sections of pisolites from the control column showed only a faintly bleached zone surrounding a reddish-brown core. A SEM–EDX image of a cross section of a pisolite that had been acted upon by bacteria showed a significant depletion of iron in the bleached zone around the core, whereas a similar cross section of a pisolite from a control column showed no significant iron depletion. No depletion of Si or Al was visible in either of the cross sections, probably because the percentages of these elements that were mobilized were too small (see earlier). Interestingly, cross

---

* A form of bauxite consisting of pea size (e.g., 0.2 to ~4 mm in diameter), quasi-spherical concretions, reddish-brown in color, called pisolites, or pisoliths if irregular in shape (see Valeton, 1972).

sections of the pisolites collected at the Weipa bauxite deposit in Queensland, Australia (Rintoul and Fredericks, 1995), resemble the cross section of the microbially attacked pisolites in the previously described experiments. The finding of Rintoul and Fredericks supports the idea that what happened to the pisolites in the experimental columns is representative of a natural process.

The iron-depleted bleached zone around the undepleted core in the cross sections of pisolites that had been acted upon by bacteria presents an enigma. The pisolites are not significantly porous, as has been shown by placing untreated pisolites in boiling water and noting a lack of effervescence originating from the pisolitic surface, indicating the absence of air entrapped in the pores in the pisolites. Thus, bacteria cannot penetrate the pisolites to effect iron mobilization by Fe(III) reduction below the pisolite surface. It is proposed that the bacteria bring about iron mobilization by enzymatic reduction of Fe(III) because daily replacement of a major portion of the medium in the columns, which would dilute any chemical reductant, Fe(III)-complexing agent, or extracellular Fe(III)-reducing enzyme in the bulk phase, did not significantly change the rate of iron reduction. Reduction of Fe(III) below the pisolite surface must therefore depend on a nonenzymatic redox mechanism. Such a mechanism may involve Fe^{2+} produced microbially at the surface by the enzymatic reduction of Fe_2O_3 with a suitable electron donor (the CH_2O in Reaction 11.1 representing an unspecified organic electron donor):

$$2Fe_2O_{3 \text{ surface}} + (CH_2O) + 8H^+ \rightarrow 4Fe^{2+}_{\text{surface}} + CO_2 + 5H_2O \qquad (11.1)$$

This Fe^{2+} reacts somehow with Fe_2O_3 below the surface.

$$2Fe^{2+}_{\text{surface}} + Fe_2O_{3 \text{ interior}} + 6H^+ \rightarrow 2Fe^{3+}_{\text{surface}} + 2Fe^{2+}_{\text{interior}} + 3H_2O \qquad (11.2)$$

The Fe^{3+} produced at the pisolite surface in Reaction 11.2 is also reduced microbially to Fe^{2+}.

$$4Fe^{3+}_{\text{surface}} + (CH_2O) + H_2O \rightarrow 4Fe^{2+}_{\text{surface}} + CO_2 + 4H^+ \qquad (11.3)$$

Reaction 11.2 is best visualized as involving the conduction of an electron from an Fe^{2+}_{surface} (Reaction 11.4a) to the interior, where it reacts with Fe(III) of $Fe_2O_{3 \text{ interior}}$ (Reaction 11.4b):

$$2Fe^{2+}_{\text{surface}} \rightarrow 2Fe^{3+}_{\text{surface}} + 2e \qquad (11.4a)$$

$$Fe_2O_{3 \text{ interior}} + 2e + 6H^+ \rightarrow 2Fe^{2+}_{\text{interior}} + 3H_2O \qquad (11.4b)$$

The Fe^{3+} generated at the pisolite surface in Reaction 11.2 is immediately reduced to Fe^{2+} by the bacteria (Reaction 11.3). The Fe^{2+} generated in the interior of the pisolite in Reaction 11.4b escapes to the exterior through passages created by the solubilization of Fe and Si, and later Al, if the pH drops below 4.5 in the interior. Because Reaction 11.2 is thermodynamically unfavorable ($\Delta G_r^0 = +1.99$ kcal or $+8.32$ kJ), it is the rapid, bacterially catalyzed reduction of Fe(III) at the surface of the pisolites (Reactions 11.1 and 11.3) that provide the energy that drives Reaction 11.2. If, for instance, H_2 instead of CH_2O was the reductant in Reaction 11.1, the value of ΔG_r^0 would be -35.56 kcal (-112.4 kJ); and for Reaction 11.3 it would be -71.12 kcal (-297.3 kJ). If acetate was the electron donor, the value of ΔG_r^0 would be -26.9 kcal (-112.4 kJ) for Reaction 11.1; and for Reaction 11.3 it would be -62.6 kcal (-261.7 kJ) for the reduction of 4 mol Fe^{3+}. Yan et al. (2004) obtained some evidence using x-ray fluorescence and x-ray absorption near-edge structure spectroscopy (XANES) image analysis that is consistent with this microbe-dependent mechanism of Fe(III) reduction below the surface of pisolites.

The conduction of electrons to the interior of the pisolite to reduce $Fe_2O_{3 \text{ interior}}$ to $Fe^{2+}_{\text{interior}}$ may be similar to that in a reaction of anaerobic microbial reduction of structural iron(III) within a ferruginous smectite by *Shewanella putrefaciens* MR-1 with formate or lactate as electron donor

(Kostka et al., 1996). Stucki et al. (1987) previously showed that a bacterium indigenous to the clay reduced the structural iron in ferruginous smectites including the smectite used by Kostka et al. (1996). The difference between the reaction with smectites and the one postulated for bauxitic pisolites is that the reduced iron formed in the smectites was not mobilized by the bacteria unlike that formed in the reaction with the pisolites. Structural iron(II) produced in smectite remained in place and could be reoxidized, thus making ferrigenous smectite a renewable terminal electron acceptor (Ernstsen et al., 1998). A simulation of electron transfer on hematite surfaces by computer may also have a direct bearing on iron mobilization from pisolites (Kerisit and Rosso, 2006). The recent observation of Fe(III) reduction in kaolinite by *Geobacter pickeringii* sp. nov., *G. argillaceus* sp. nov., and *Pelosinus fermentans* gen. nov., sp. nov. may be a related phenomenon (Shelobolina et al., 2007). Interestingly, the first two of these organisms are Fe(III) respirers, whereas the third is a fermentor, which though able to reduce Fe(III), does not respire it but uses it as an electron sink.

Bacterial reduction of hematite in oxidized samples from the Central Plateau of Brazil incubated in the presence of sucrose at 25°C was reported by Macedo and Bryant (1989). They observed preferential attack of hematite over aluminian goethite. Initial outgrowth of the bacteria from the soil in these experiments required 3–9 weeks. Microbial activity was correlated with a decrease in redness of the soil.

11.2.5 BACTERIAL REDUCTION OF FE(III) IN BAUXITES FROM DIFFERENT LOCATIONS

Physiologically similar bacterial cultures grew from bauxite from the pisolitic deposit in Australia and from deposits in the Amazon in Brazil and the island of Jamaica in the Caribbean Sea (Ehrlich and Wickert, 1997). When the Amazonian and Jamaican ores in columns were fed with sucrose–mineral salts medium, behavior similar to that observed with the Australian bauxite was noted with respect to Fe, Si, and Al solubilization and pH changes in successive effluents. *Clostridia* were among the bacteria that were first noted in column effluents. The *Clostridia* from these two ores showed a close phylogenetic relationship to that from the Australian ore (B. Methé, unpublished results). These similarities suggest that the natural flora associated with the bauxites may play a role in bauxite maturation, that is, its enrichment in Al over time. A caveat is, however, that none of the ore samples was collected under controlled conditions that would have minimized or prevented contamination of the ore in the collection process or in subsequent storage. However, the probability that all ore samples were heavily contaminated during collection or storage so that very similar mixed anaerobic bacterial assemblages would arise only after 3–4 days of incubation in experimental columns seems small.

11.2.6 OTHER OBSERVATIONS OF BACTERIAL INTERACTION WITH BAUXITE

Others have demonstrated bacterial interaction with bauxite, mainly for the purpose of testing whether the ore could be made industrially more attractive (biobeneficiation). These interactions occurred generally under aerobic conditions. Anand et al. (1996) found that *Bacillus polymyxa* strain NCIM 2539 was able to mobilize in shake culture all the calcium and ~45% of the iron from a bauxite ground to 53–74 μm particle size. The bacterial treatment occurred at 30°C in Bromfield medium containing 2% sucrose. The change in composition of the bauxite was attributed to the direct action of the cells and the action of cellular products such as exopolysaccharides and organic acids. The oxidation state of the mobilized iron was not determined.

Groudev (1987) reported silicon removal by *Bacillus circulans* and *B. mucilaginosus* from low-grade bauxites. The Si mobilization was attributed to the action of exopolysaccharides that were elaborated by the bacteria. Some Al was also mobilized in these experiments.

Orgutsova et al. (1989) reported variations in the ability of the strains of the fungi *Aspergillus niger* and *Penicillium chrysogenum* and various yeasts and pseudomonads to mobilize Al, Fe, and Si from a ground bauxite of which 70% had a –74 μm particle size. The oxidation state of the mobilized iron was not reported. The mobilization of Al, Fe, and Si was attributed to the action of metabolic products formed by the test organisms.

In another study, Karavaiko et al. (1989) found that a strain of *Bacillus mucilaginosus* removed Si from bauxite ground to −0.074 mm particle size and incubated in a sucrose–mineral salts medium with a 10% inoculum in shake-flask culture at 30°C. The Si removal was attributed to the selective adherence of fine particles of ore rich in Si to the exopolysaccharides at the surface of the bacterial cells and not to dissolution. The mycelial fungi *Aspergillus niger* and *A. pullulans*, on the contrary, were able to mobilize varying amounts of Fe, Al, and Si from the same bauxites by dissolution with acids, which are formed them metabolically in a sucrose–mineral salts medium.

Bandyopadhay and Banik (1995) were able to mobilize 39.9% silica and 46.4% iron from a bauxite ore with a mutated strain of *Aspergillus niger* in a laboratory experiment in which the fungus was allowed to grow at the surface of 80 mL of culture liquid at an initial pH of 4 in a flask at 30°C. The culture medium contained glucose as the energy source and $NaNO_3$ as the nitrogen source. The ore was ground to a mesh size of −170 to +200 and then added to the medium at a concentration of 0.3%. The mobilization of Si and Fe was attributed to action of the organic acids, probably citric and oxalic, produced by the fungus.

11.3 SUMMARY

Aluminum is the third most abundant element in the Earth's crust, silicon and oxygen being more abundant. Of these three elements, aluminum is the only one for which a physiological function has not been found, although a very small number of higher organisms are known to accumulate it. Al^{3+} is generally toxic. At least one known cyanobacterium and a strain of *Escherichia coli* have each developed a different mechanism of resistance to it. Various microbes are known to participate in the formation of some aluminum-containing minerals through weathering action.

The formation of bauxite (bauxitization), whose major constituents are Al_2O_3 in the form of gibbsite, boehmite, or diaspore; Fe_2O_3 in the form of hematite, goethite, or aluminian goethite; and SiO_2 or aluminosilicate in the form of silica or kaolinite, can be visualized as involving two stages that may overlap to some extent. Evidence to date suggests that microbes are involved in both stages. The source material in bauxitization may be volcanic and other aluminosilicate rocks, limestone associated with karsts, and alluvium. The first stage of bauxitization involves weathering of source rock and the formation of protobauxite, and the second stage the maturation of protobauxite to bauxite. The first stage, if aerobic, may be promoted by bacteria and fungi, and if anaerobic, by facultative and anaerobic bacteria. The second stage is promoted by iron-reducing and fermentative bacteria under anaerobic conditions. The first stage involves the mobilization of Al, Fe, and Si from host rock and the subsequent precipitation of these rock constituents as oxides, silica, and silicate minerals. The second stage involves the selective mobilization of iron oxides and silica or silicate, enriching the solid residue in aluminum. The process is favored in warm, humid climates with alternate wet and dry seasons. The site of formation must be associated with vegetation that can serve as a source of nutrients to the microorganisms and may yield weathering agents as a result of microbial attack of plant residues. Microbes are expected to play a significant role in the control of pH *in situ* during bauxite maturation, which ensures that little of the aluminum oxide is mobilized.

REFERENCES

Anand P, Modak JM, Natarajan KA. 1996. Biobeneficiation of bauxite using *Bacillus polymyxa*: Calcium and iron removal. *Int J Mineral Process* 48:51–60.

Appanna VD, Hamel R. 1996. Aluminum detoxification mechanism in *Pseudomonas fluorescens* is dependent on iron. *FEMS Microbiol Lett* 143:223–228.

Appanna VD, Kepes M, Rochon P. 1994. Aluminum tolerance in *Pseudomonas fluorescens* ATCC 13525: Involvement of a gelatinous lipid-rich residue. *FEMS Microbiol Lett* 119:295–302.

Appanna VD, St Pierre M. 1994. Influence of phosphate on aluminum tolerance in *Pseudomonas fluorescens*. *FEMS Microbiol Lett* 124:327–332.

Bandyopadhay N, Banik AK. 1995. Optimization of physical factors for bioleaching of silica and iron from bauxite ore by a mutant strain of *Aspergillus niger*. *Res Ind* 40:14–17.

Bowen HJM. 1966. *Trace Elements in Biochemistry*. London: Academic Press.

Butty DL, Chapalaz CA. 1984. Bauxite genesis. In: Jacob LE, Jr, ed. *Bauxite: Proceedings of the 1984 Bauxite Symposium*, Los Angeles, CA. New York: Society of Mining Engineers of the American Institute of Mining, Metallurgical, Petroleum Engineers, pp. 11–151.

Ehrlich HL, Wickert LM. 1997. Bacterial action on bauxites in columns fed with full-strength and dilute sucrose-mineral salts medium. In: Lortie L, Bédard P, Gould WD, eds. *Biotechnology and the Mining Environment. SP 97–1*. Ottawa, Canada: CANMET Natural Resources Canada, pp. 74–89.

Ehrlich HL, Wickert LM, Noteboom D, Doucet J. 1995. Weathering pisolitic bauxite by heterotrophic bacteria. In: Vargas T, Jerez CA, Wiertz JV, Toledo H, eds. *Biohydrometalllurgical Processing*, Vol. I, Santiago: University of Chile, pp. 395–403.

Ernstsen V, Gates WP, Stucki JW. 1998. Microbial reduction of structural iron in clays: A renewable resource of reduction capacity. *J Environ Qual* 27:761–766.

Garcidueñas Piña R, Cervantes C. 1996. Microbial interaction with aluminum. *BioMetals* 9:311–316.

Gornitz V. 1972. Aluminum: Element and geochemistry. In: Fairbridge RW, ed. *Encyclopedia of Geochemistry and Environmental Sciences. Encyclopedia of Earth Science Series*, Vol. IVA, New York: Van Nostrand Reinhold, pp. 21–23.

Groudev S. 1987. Use of heterotrophic microorganisms in mineral biotechnology. *Acta Biotechnol* 7:299–306.

Harris SA. 1972. Aluminum toxicity. In: Fairbridge RW, ed. *Encyclopedia of Geochemistry and Environmental Sciences. Encyclopedia of Earth Science Series*, Vol. IVA, New York: Van Nostrand Reinhold, pp. 27–29.

Hydes DJ. 1979. Aluminum in seawater: Control by inorganic processes. *Science* 205:1260–1262.

Karavaiko GI, Avakyan ZA, Ogurtsova LV, Safonova OF. 1989. Microbiological processing of bauxite. In: Salley J, McCready RGL, Wichlacz PL, eds. *Biohydrometallurgy. Proceedings of the International Symposium*, Jackson Hole, WY. CANMET SP89–10. Ottawa, Canada: CANMRT, Canada Centre for Mineral and Energy Technology, pp. 93–102.

Kerisit S, Rosso KM. 2006. Computer simulation of electron transfer at hematite surfaces. *Geochim Cosmochim Acta* 70:1888–1903.

Kostka JE, Stucki JW, Nealson KH, Wu J. 1996. Reduction of structural Fe(III) in smectite by a pure culture of *Shewanella putrefaciens* strain MR-1. *Clays Clay Min* 44:522–529.

Kotschoubey B, De Souza Duarte AL, Truckenbrodt W. 1999. The bauxitic mantle and origin of the kaolin deposit of Morro do Felipe, lower course of the Jari River, Amapa State, northern Brazil. *Rev Bras Geoscience* 29:331–338.

Macdonald TL, Martin RB. 1988. Aluminum ion in biological systems. *Trends Biochem Sci* 13:15–19.

Macedo J, Bryant RB. 1989. Preferential microbial reduction of hematite over goethite in a Brazilian oxisol. *Soil Sci Soc Am J* 53:1114–1118.

Mackenzie FT, Stoffyn M, Wollast R. 1978. Aluminum in seawater: Control by biological activity. *Science* 199:680–682.

Marine Chemistry. 1971. Report of Marine Chemistry Panel of the Committee of Oceanography. Washington, DC: National Academy of Sciences.

Natarajan KA, Modak JM, Anand P. 1997. Some mineralogical aspects of bauxite mineralization and beneficiation. *Min Metall Process* 14:47–53.

Nordstrom DK, May HM. 1995. Aqueous equilibrium data for mononuclear aluminum species. In: Esposito G, ed. *The Environmental Chemistry of Aluminum*. Boca Raton, FL: CRC Lewis, pp. 39–80.

Orgutsova LV, Karavaiko GI, Avakyan ZA, Korenevskii AA. 1989. Activity of various microorganisms in extracting elements from bauxite. *Mikrobiologiya* 58:956–962 (Engl transl pp. 774–780).

Petterson A, Kunst L, Berman B, Roomans GM. 1985. Accumulation of aluminum by *Anaboena cylindrica* into polyphosphate granules and cell walls: An x-ray energy-dispersive microanalysis study. *J Gen Microbiol* 131:2545–2548.

Rintoul L, Fredericks PM. 1995. Infrared microspectroscopy of bauxite pisoliths. *Appl Spectrosc* 49:1608–1616.

Schellman W. 1994. Geochemical differentiation in laterite and bauxite formation. *Catena* 21:131–143.

Shelobolina ES, Nevin KP, Blakeney-Hayward JD, Johnsen CV, Plaia TW, Krader P. Woodard T, Holmes DE, Gaw VanPraagh C, Lovley DR. 2007. *Geobacter pickeringii* sp. nov., *Geobacter argillaceus* sp. nov., and *Pelosinus fermentans* gen. nov., sp. nov., isolated from subsurface kaolin lenses. *Int J Syst Evol Microbiol* 57:126–135.

Stoffyn M. 1979. Biological control of dissolved aluminum in seawater: Experimental evidence. *Science* 203:651–653.

Strahler AN. 1976. *Principles of Earth Science*. New York: Harper & Row, p. 5.

Stucki JW, Komadel P, Wilkinson HT. 1987. Microbial reduction of structural iron(III) in smectites. *Soil Sci Soc Am J* 51:1663–1665.

Taylor GR, Hughes GW. 1975. Biogenesis of the Rennell bauxite. *Econ Geol* 70:542–546.

Valeton I. 1972. *Bauxites*. Amsterdam: Elsevier.

Vance GF, Stevenson FJ, Sikora FJ. 1995. Environmental chemistry of aluminum-organic complexes. In: Esposito G, ed. *The Environmental Chemistry of Aluminum*. Boca Raton, FL: CRC Lewis, pp. 169–220.

Yan B, Abrajano T, Newville M, Sutton S, Sturchio NC, Ehrlich H. 2004. Anaerobic bacterial reduction of ferric iron in pisolites. In: Wanty RB, Seal, RR II, eds. *Water–Rock Interaction: 11th International Symposium on Water–Rock Interaction*, Vol. 2, Leiden, Netherlands: AA Balkema Publishers, pp. 1165–1169.

12 Geomicrobial Interactions with Phosphorus

12.1 BIOLOGICAL IMPORTANCE OF PHOSPHORUS

Phosphorus is an element fundamental to life, and a structural and functional component in all organisms. It is found universally in such vital cell constituents as nucleic acids, nucleotides, phosphoproteins, and phospholipids. It occurs in teichoic and teichuronic acids in the walls of gram-positive bacteria, and in phytins (also known as inositol phosphates) in plants. In many types of bacteria and some yeasts it may also be present intracellularly as polyphosphate granules. Simple phosphates (orthophosphate) can form anhydrides with other phosphates, as in organic and inorganic pyrophosphates (see Figure 6.4) and polyphosphate. Phosphate is also capable of forming anhydrides with carboxyl groups of organic acids, with amino groups of amines, and with sulfate (as in adenosine $5'$-phosphosulfate). The phosphate anhydride bond serves to store biochemically useful energy. For example, a standard free energy change ($\Delta G°$) of -7.3 kcal (30.6 kJ) per mole is associated with the hydrolysis of the terminal anhydride bond of adenosine $5'$-triphosphate (ATP), yielding adenosine $5'$-diphosphate (ADP) + P_i. Unlike many anhydrides, some of those involving phosphates such as ATP are unusually resistant to hydrolysis in the aqueous environment (Westheimer, 1987). Chemical hydrolysis of these bonds requires 7 min of heating in dilute acid (e.g., 1 N HCl) at the temperature of boiling water (Lehninger, 1970, p. 290). At more neutral pH and physiological temperature, hydrolysis proceeds at an optimal rate only in the presence of appropriate enzymes (e.g., ATPase). The relative resistance of phosphate anhydride bonds to hydrolysis is attributable to the negative charges on the phosphates at neutral pH (Westheimer, 1987). It is the probable reason why ATP got selected in the evolution of life as a repository and universal transfer agent of chemical energy in biological systems.

12.2 OCCURRENCE IN EARTH'S CRUST

Phosphorus is found in all parts of the biosphere. Its gross abundance at the surface of the Earth has been cited by Fuller (1972) to be 0.10–0.12% (w/w). It occurs mostly in the form of inorganic phosphates and organic phosphate derivatives. The organic derivatives in soil are mostly phytins (Paul and Clark, 1996). Total phosphorus concentrations in mineral soil range from 35 to 5300 mg kg^{-1} (average 800 mg kg^{-1}; Bowen, 1979). An average concentration in freshwater is 0.02 mg kg^{-1} (Bowen, 1979) and in seawater 0.09 mg L^{-1} (Marine Chemistry, 1971). The ratio of organic to inorganic phosphorus (P_{org}/P_i) varies widely in these environments. In mineral soil, P_{org}/P_i may range from 1:1 to 2:1 (Cosgrove, 1967, 1977). In lake water, as much as 50% of the organic faction may be phytin and releasable as inorganic phosphorus through hydrolysis catalyzed by phytase (Herbes et al., 1975). The organic phosphorus in lake water may constitute 80–99% of the total soluble phosphorus. In the particular examples cited by Herbes et al. (1975), the total organic phosphorus rarely exceeded 40 µg phosphate per liter. They speculated that hydrolyzable phosphate compounds other than phytins were largely absent because they are much more labile than phytins. Readily measurable phosphatase activity was detected in Sagima and Suruga Bays, Tokyo, by Kobori and Taga (1979) and Taga and Kobori (1978).

12.3 CONVERSION OF ORGANIC INTO INORGANIC PHOSPHORUS AND SYNTHESIS OF PHOSPHATE ESTERS

An important source of free organic phosphorus compounds in the biosphere is the breakdown of animal and vegetable matter. However, living microbes such as *Escherichia coli* and organisms from activated sludge have been found to excrete aerobically assimilated phosphorus as inorganic phosphate when incubated anaerobically (Shapiro, 1967). Organically bound phosphorus is for the most part not directly available to living organisms because it cannot be taken into the cell in this form. To be taken up, it must be freed from organic combination through mineralization. This is accomplished through hydrolytic cleavage catalyzed by phosphatases. In the soil, as much as 70–80% of the microbial population are able to participate in this process (Dommergues and Mangenot, 1970, p. 266). Active organisms include bacteria such as *Bacillus megaterium*, *B. subtilis*, *B. malabarensis*, *Serratia* sp., *Proteus* spp., *Arthrobacter* spp., *Streptomyces* spp., and fungi such as *Aspergillus* sp. *Penicillium* sp., *Rhizopus* sp., and *Cunninghamella* sp. (Dommergues and Mangenot, 1970, p. 266; Paul and Clark, 1996). These organisms secrete, or liberate upon their death, phosphatases with greater or lesser substrate specificity (Skujins, 1967). Such activity has also been noted in the marine environment (Ayyakkannu and Chandramohan, 1971).

Phosphate liberation from phytin generally requires the enzyme phytase:

$$\text{Phytin} + 6\text{H}_2\text{O} \rightarrow \text{inositol} + 6\text{P}_i \tag{12.1}$$

Phosphate liberation from nucleic acid requires the action of nucleases, which yield nucleotides, followed by the action of nucleotidases, which yield nucleosides and inorganic phosphate:

$$\text{Nucleic acid} \xrightarrow[+\text{H}_2\text{O}]{\text{nucleases}} \text{nucleotides} \xrightarrow[+\text{H}_2\text{O}]{\text{nucleotidase}} \text{nucleosides} + \text{P}_i \tag{12.2}$$

Phosphate liberation from phosphoproteins, phospholipids, ribitol, and glycerol phosphates requires phosphomono- and phosphodiesterases. Phosphodiesterases attack phosphodiesters at either the 3′ or 5′ carbon linkage, whereas phosphomonoesterases (phosphatases) attack monoester linkages (Lehninger, 1975, pp. 184, 323–325), for example,

$$\underset{\substack{3' \quad\quad 5' \\ \\ |\\ \text{OH}}}{\text{R}-\text{O}-\overset{\overset{\text{O}}{\|}}{\text{P}}-\text{O}-\text{R}'} \xrightarrow[+\text{H}_2\text{O}]{\text{diesterase}} \text{ROH}$$

$$+\,\text{HO}-\underset{\substack{|\\\text{OH}}}{\overset{\overset{\text{O}}{\|}}{\text{P}}}-\overset{5'}{\text{O}}-\text{R}' \xrightarrow[+\text{H}_2\text{O}]{\text{phosphatase}} \text{HO}-\underset{\substack{|\\\text{OH}}}{\overset{\overset{\text{O}}{\|}}{\text{P}}}-\text{OH} + \text{HOR}' \tag{12.3}$$

Synthesis of organic phosphates (monomeric phosphate esters) is an intracellular process and normally proceeds through reaction between the hydroxyl (OH) of a carbinol group (CHOH), as for instance in glucose, and ATP in the presence of an appropriate kinase. For example,

$$\text{Glucose} + \text{ATP} \xrightarrow{\text{glucokinase}} \text{glucose 6-phosphate} + \text{ADP} \tag{12.4}$$

Phosphate esters in cells may also arise through phosphorolysis of certain polysaccharide polymers such as starch or glycogen [(glucose)$_n$]:

$$(\text{Glucose})_n + H_3PO_4 \xrightarrow{\text{phosphorylase}} (\text{glucose})_{n-1} + \text{glucose 1-phosphate} \qquad (12.5)$$

$$\text{Glucose 1-phosphate} \xrightarrow{\text{phosphoglucomutase}} \text{glucose 6-phosphate} \qquad (12.6)$$

12.4 ASSIMILATION OF PHOSPHORUS

ATP may be generated from ADP by *adenylate kinase*,

$$2\text{ADP} \rightarrow \text{ATP} + \text{AMP} \qquad (12.7)$$

or by *substrate level phosphorylation*, as in the reaction sequence:

$$\text{3-Phosphoglyceraldehyde} + NAD^+ + P_i \xrightarrow[\text{dehydrogenase}]{\text{triosephosphate}} \text{1,3-diphosphoglycerate} + NADH + H^+ \qquad (12.8)$$

$$\text{1,3-Diphosphoglycerate} + \text{ADP} \xrightarrow{\text{ADP kinase}} \text{3-phosphoglycerate} + \text{ATP} \qquad (12.9)$$

It may also be generated by *oxidative phosphorylation*:

$$\text{ADP} + P_i \xrightarrow[\text{ATPase}]{\text{electron transport system}} \text{ATP} \qquad (12.10)$$

or by photophosphorylation:

$$\text{ADP} + P_i \xrightarrow[\text{light, ATPase}]{\text{photosynthetic system}} \text{ATP} \qquad (12.11)$$

Phosphate polymers are generally produced through reactions such as

$$(\text{Polynucleotide})_{n-1} + \text{nucleotide triphosphate} \xrightarrow{\text{polymerase}} (\text{polynucleotide})_n + P \sim P \qquad (12.12)$$

In many organisms, inorganic pyrophosphate (P~P) is enzymatically hydrolyzed,

$$P \sim P \xrightarrow{\text{pyrophosphatase}} 2P_i \qquad (12.13)$$

without conservation of the energy released by cleavage of the anhydride bond.

However, in a few bacteria, inorganic pyrophosphate has been reported to be able to serve as an energy source (Liu et al., 1982; Varma et al., 1983). Although it is easy to understand that this ability can be of great importance for energy conservation from intracellularly formed pyrophosphate in bacteria, it remains to be clarified how important it may be for extracellularly available pyrophosphate in nature. Liu et al. (1982) found gram-positive and gram-negative motile and nonmotile bacteria in pyrophosphate enrichments from freshwater anaerobic environments, which grew at the expense of the pyrophosphate as energy source. Nothing appears to be known about the mechanism of pyrophosphate uptake in these organisms.

Like pyrophosphate, intracellular inorganic polyphosphate granules formed by some microbes (Friedberg and Avigad, 1968) are a form of metaphosphate and can represent an energy storage compound (van Groenestijn et al., 1987) as well as a phosphate reserve. In the case of the cyanobacterium *Anabaena cylindrica*, it may also play a role as detoxifying agent by combining with aluminum ions that are taken into the cell (Petterson et al., 1985) (see also Chapter 11).

12.5 MICROBIAL SOLUBILIZATION OF PHOSPHATE MINERALS

Inorganic phosphorus may occur in soluble and insoluble forms in nature. The most common inorganic form is orthophosphate (e.g., H_3PO_4). As an ionic species, the concentration of phosphate is controlled by its solubility in the presence of an alkaline earth cation such as Ca^{2+} or Mg^{2+} or in the presence of metal cations such as Fe^{2+}, Fe^{3+}, or Al^{3+} at appropriate pH values (see Table 12.1). In seawater, for instance, the soluble phosphate concentration (~3×10^{-6} M, maximum) is controlled by Ca^{2+} ions (4.1×10^2 mg L^{-1}), which form hydroxyapatites with phosphate in a prevailing pH range of ~7.9–8.1.

Insoluble phosphate occurs most commonly in the form of apatite $[Ca_5(PO_4)_3(F, Cl, OH)]$ in which the (F, Cl, OH) radical may be represented exclusively by F, Cl, or OH or any combination of these. In soil, insoluble phosphate may also occur as an aluminum salt (e.g., variscite, $AlPO_4 \cdot 2H_2O$) or as iron salts vivianite ($Fe_3(PO_4)_2 \cdot 8H_2O$) and strengite ($FePO_4 \cdot 2H_2O$).

Insoluble forms of inorganic phosphorus (calcium, aluminum, and iron phosphates) may be solubilized through microbial action. The mechanism by which the microbes accomplish this solubilization varies. The first mechanism may be the production of inorganic or organic acids that attack the insoluble phosphates. A second mechanism may be the production of chelators such as gluconate and 2-ketogluconate (Duff and Webley, 1959; Banik and Dey, 1983; Babu-Khan et al., 1995) (see also Chapter 10), citrate, oxalate, and lactate. All of these chelators can complex the cationic portion of the insoluble phosphate salts and thus force the dissociation of the salts. A third mechanism of phosphate solubilization may be the reduction of iron in ferric phosphate (e.g., strengite) to ferrous iron by enzymes and metabolic products of nitrate reducers such as *Pseudomonas fluorescens* and *Alcaligenes* sp. in sediment (Jansson, 1987). A fourth mechanism may be the production of hydrogen sulfide (H_2S), which can react with the iron in iron phosphate and precipitate it as iron sulfide, thereby mobilizing phosphate, as in the reaction

$$2FePO_4 + 3H_2S \rightarrow 2FeS + 2H_3PO_4 + S^0 \qquad (12.14)$$

Table 12.2 lists some of the organisms active in phosphate solubilization.

Solubilization of phosphate minerals has been noted directly in soil (Alexander, 1977; Babenko et al., 1984; Chatterjee and Nandi, 1964; Dommergues and Mangenot, 1970; Patrick et al., 1973). Indeed, soil containing significant amounts of immobilized calcium, aluminum, or iron phosphates has been thought to benefit from inoculation with phosphate-mobilizing bacteria (see discussion by Dommergues and Mangenot, 1970, p. 262). Important microbial phosphate-solubilizing activity in soil occurs in rhizospheres (Alexander, 1977), probably because root secretions allow phosphate-solubilizing bacteria to generate sufficient acid or ligands to effect dissolution of calcium and other

TABLE 12.1
Solubility Products of Some Phosphate Compounds

Compound	K_s	Reference
$CaHPO_4 \cdot 2H_2O$	2.18×10^{-7}	Kardos (1955, p. 185)
$Ca_{10}(PO_4)_6(OH)_2$	1.53×10^{-112}	Kardos (1955, p. 188)
$Al(OH)_2HPO_4$	2.8×10^{-29}	Kardos (1955, p. 184)
$FePO_4$	1.35×10^{-18}	From ΔG of formation

TABLE 12.2
Some Organisms Active in Phosphate Solubilization

Organism	Mechanism of Solubilization	References
Bacillus megaterium	H_2S production, FeS precipitation	Sperber (1958a); Swaby and Sperber (1958)
Thiobacillus sp.	H_2SO_4 production from sulfur	Lipman and McLean (1916)
Nitrifying bacteria	NH_3 oxidation to HNO_3	Dommergues and Mangenot (1970, p. 263)
Pseudomonads, Arthrobacter, and *Erwinia*-like bacterium	Chelate production; glucose converted to gluconate or 2-ketogluconate	Duff et al. (1963); Sperber (1958b); Dommergues and Mangenot (1970, p. 262); Babu-Khan et al. (1995)
Sclerotium	?	Dommergues and Mangenot (1970, p. 262)
Aspergillus niger, A. flavus, Sclerotium rolfsii, Fusarium oxysporum, Cylindrosporium sp., and *Penicillium* sp.	Organic acid production (e.g., citric acid)	Agnihotri (1970)

insoluble phosphates. Phosphate-deficient soil may be beneficially fertilized with insoluble inorganic phosphate rather than soluble phosphate salts because the former will be solubilized slowly and thus will be better conserved than soluble phosphate salts, which can be readily leached. Soluble phosphate in soil may consist not only of orthophosphate but also of pyrophosphate (metaphosphate). The latter is readily hydrolyzed by pyrophosphatase, especially in flooded soil (Racz and Savant, 1972).

Recent experiments with *Bacillus megaterium* found that phosphate mobilization when the organism is in direct contact with apatite was 3–10 times slower than when it is not in direct contact with the mineral (Hutchens et al., 2006). The authors suggest that when in direct contact, the organism may block reactive sites at the mineral surface.

12.6   MICROBIAL PHOSPHATE IMMOBILIZATION

Microorganisms can cause fixation or immobilization of phosphate, either by promoting the formation of inorganic precipitates or by assimilating the phosphate into organic cell constituents or intracellular polyphosphate granules. The latter two processes have been called *transitory phosphate immobilization* by Dommergues and Mangenot (1970) because of the ready solubilization of phosphate through mineralization upon death of the cell. In soil and freshwater environments, transitory phosphate immobilization is often more important, although fixation of phosphate by Ca^{2+}, Al^{3+}, and Fe^{3+} is recognized. In a few marine environments (coastal waters or shallow seas) where phosphorite deposits occur, the precipitation mechanism may be more important (McConnell, 1965).

12.6.1   PHOSPHORITE DEPOSITION

Phosphorite (Figure 12.1) in nature may form authigenically or diagenetically. In authigenesis, the phosphorite forms as a result of a reaction of soluble phosphate with calcium ions forming corresponding insoluble calcium phosphate compounds. In diagenesis, phosphate may replace carbonate in calcareous concretions. The role of microbes in these processes may be one or more of the following: (1) making reactive phosphate available, (2) making reactive calcium available, or (3) generating or maintaining the pH and redox conditions that favor phosphate precipitation.

FIGURE 12.1 Micronodules of phosphorite (phosphatic pellets) from the Peru shelf. The average diameter of such pellets is 0.25 mm. According to Burnett (personal communication), such pellets are more representative of what is found in the geologic record than the larger phosphorite nodules. (Courtesy of Burnett WC, Institute for International Cooperative Environmental Research, Florida State University, Tallahassee FL.)

12.6.1.1 Authigenic Formations

Models of authigenic phosphorite genesis assume the occurrence of mineralization of organic phosphorus in biologically productive waters, such as at ocean margins, that is, at shallow depths on continental slopes, shelf areas, or plateaus. Here detrital accumulations may be mineralized at the sediment–water interface and in interstitial pore waters, liberating phosphate, some of which may then interact chemically with calcium in seawater to form phosphorite grains. These grains may subsequently be redistributed within the sediment units (Riggs, 1984; Mullins and Rasch, 1985). Dissolution of fish debris (bones) has also been considered an important source of phosphate in authigenic phosphorite genesis (Suess, 1981). Upwelling probably plays an important role in many cases of authigenic phosphorite formation on western continental margins at latitudes in both the northern and southern hemispheres, where prevailing winds (e.g., trade winds) cause upwelling (see, e.g., discussion by Burnett et al., 1982; Jahnke et al., 1983; Riggs, 1984). Nathan et al. (1993) cite evidence that in the southern Bengula upwelling system (Cape Peninsula, western coast of South Africa) during nonupwelling periods in winter, phosphate-sequestering bacteria of the oxidative genera *Pseudomonas* and *Acinetobacter* become dominant in the water column. Fermentative Vibrios and Enterobacteriaceae are dominant during upwelling in summer. It has been suggested that *Pseudomonas* and *Acinetobacter*, which sequester phosphate as polyphosphate under aerobic conditions and hydrolyze the polyphosphate under anaerobic conditions to obtain energy of maintenance and to sequester volatile fatty acids for polyhydroxybutyrate formation, contribute to authigenic phosphorite formation. Locally elevated, excreted orthophosphate becomes available for precipitation as phosphorite by reacting with seawater calcium. In the northern Bengula upwelling system off the coast of Namibia, where upwelling occurs year-round, Nathan et al. (1993) found that phosphate-sequestering cocci occurred in the water column. They suggested that these organisms such as *Pseudomonas* and *Acinetobacter* may release sequestered phosphate when they reach waters with low oxygen concentration below 10 m water depth and thereby contribute to phosphorite formation.

Authigenic phosphorite at some eastern continental margins, where upwelling, if it occurs at all, is a weak and intermittent process, may have been formed more directly as a result of intracellular bacterial phosphate accumulation, which became transformed into carbonate fluorapatite upon death of the cells and accumulated in sediments in areas where the sedimentation rate was very low (O'Brien and Veeh, 1980; O'Brien et al., 1981). Ruttenberg and Berner (1993) concluded that carbonate fluorapatite accumulations in Long Island Sound and Mississippi Delta sediments are the result of mineralization of organic phosphorus. These accumulations increased as organic phosphorus concentrations decreased with depth. Thus, important phosphorus sinks occur in sediments of continental margins outside upwelling regions.

Youssef (1965) proposed that phosphorite could be formed in a marine setting through mineralization of phytoplankton remains that have settled into a depression on the seafloor, leading to localized accumulation of dissolved phosphate. According to him, this phosphate could then precipitate as a result of reaction with calcium in seawater. Piper and Codespoti (1975) proposed that carbonate fluorapatite $[Ca_{10}(PO_4,CO_3)_6F_{2-3}]$ precipitation in the marine environment may be dependent on bacterial denitrification in the oxygen minimum layer of the ocean as it intersects with the ocean floor. A loss of nitrogen due to denitrification means lowered biological activity and can lead to excess accumulation of phosphate in this zone. The lower pH (7.4–7.9) in the deeper waters compared to the surface waters keeps phosphate dissolved and allows for its transport by upwelling to regions where phosphate precipitation is favored (pH >8) (Figure 12.2). This model takes into account the conditions of marine apatite formation described by Gulbrandsen (1969) and helps to explain the occurrence of probable contemporary formation of phosphorite in regions of upwelling

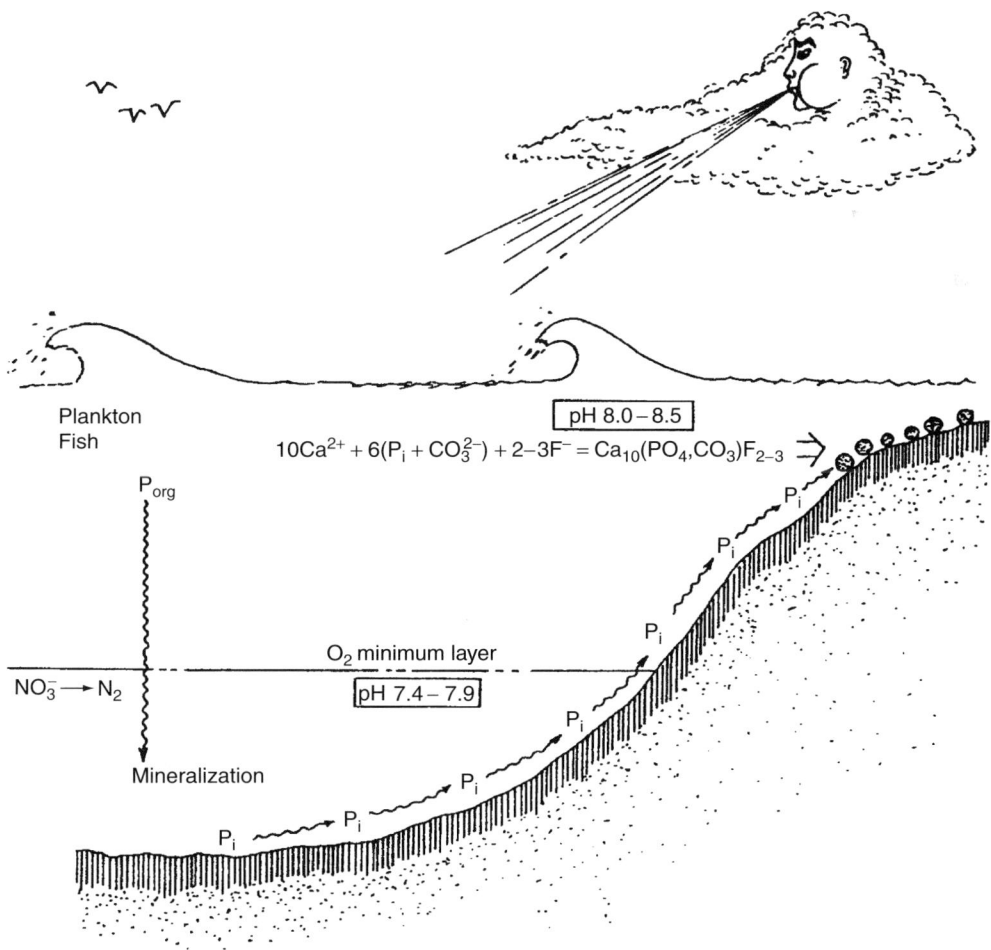

FIGURE 12.2 Schematic presentation of phosphorite formation in the marine environment. Note that the rising P_i upslope is due to upwelling. (Based on Piper DZ, Codespoti LA, *Science*, 179, 564–565, 1975.)

such as the continental margin off Peru (Veeh et al., 1973; see also, however, Suess, 1981) and on the continental shelves of southwestern Africa (Baturin, 1973; Baturin et al., 1969). To explain the more extensive ancient phosphorite deposits, a periodic warming of the ocean can be invoked, which would reduce oxygen solubility and favor more intense denitrification in deeper waters, resulting in temporarily lessened biological activity and a consequent increase in dissolved phosphate concentration that would lead to phosphate precipitation (Piper and Codespoti, 1975). Mullins and Rasch (1985) proposed an oxygen-depleted sedimentary environment for biogenic apatite formation along the continental margin of central California during the Miocene. They view oolitic phosphorite as having resulted from organic matter mineralization by sulfate reducers in sediments in which dolomite was concurrently precipitated. The phosphate, according to their model, then tended to precipitate interstitially as phosphorite, in part around bacterial nuclei. O'Brien et al., 1981 had previously reported the discovery of fossilized bacteria in a phosphorite deposit on the East Australian margin.

12.6.1.2 Diagenetic Formation

Models of phosphorite formation through diagenesis generally assume an exchange of phosphate for carbonate in accretions that have the form of calcite and aragonite. The role of bacteria in this process is to mobilize phosphate by mineralizing detrital organic matter. This has been demonstrated in marine and freshwater environments under laboratory conditions (Lucas and Prévot, 1984; Hirschler et al., 1990a,b). Adams and Burkhart (1967) propose that diagenesis of calcite to form apatite explains the origin of some deposits in the North Atlantic. Phosphorite deposits off Baja California and in a core from the eastern Pacific Ocean seem to have formed as a result of partial diagenesis (d'Anglejan, 1967, 1968).

12.6.2 OCCURRENCES OF PHOSPHORITE DEPOSITS

Sizable phosphorite deposits are associated with only six brief geological intervals: the Cambrian, Ordovician, Devonian-Mississippian, Permian, Cretaceous, and Cenozoic eras. Because in many instances these phosphorite deposits are associated with black shales and contain uranium in reduced form (Altschuler et al., 1958), they are presumed to have accumulated under reducing conditions. Nowadays, apatite appears to be forming in the sediments at the Mexican continental margin (Jahnke et al., 1983) and in the deposits off the coast of Peru (Burnett et al., 1982; Suess, 1981; Veeh et al., 1973).

12.6.3 DEPOSITION OF OTHER PHOSPHATE MINERALS

Microbes can also play a role in the authigenic or diagenetic formation of other phosphate minerals such as vivianite, strengite, and variscite. In these instances, the bacteria may contribute orthophosphate to the mineral formation by degrading organic phosphate in detrital matter and they may contribute iron or aluminum by mobilizing these metals from other minerals. Authigenic formation of such phosphate minerals is probably most common in soil. A case of diagenetic formation of vivianite from siderite ($FeCO_3$) in the North Atlantic coastal plain has been proposed by Adams and Burkhart (1967). Microbial control of pH and E_h can influence the stability of these phosphate minerals (Patrick et al., 1973; Williams and Patrick, 1971).

Citrobacter sp. has been reported to form metal phosphate precipitates, for example, cadmium phosphate ($CdHPO_4$) and uranium phosphate (UO_2HPO_4), which encrust the cells (Macaskie et al., 1987, 1992). The precipitates form as a result of the action of a cell-bound, metal-resistant phosphatase on organophosphates such as glycerol-2-phosphate, liberating orthophosphate (HPO_4^{2-}) that reacts in the immediate surroundings of the cells with metal cations to form corresponding metal phosphates. The metal phosphates form deposits on the cell surface.

Some microbes, such as certain strains of *Arthrobacter* sp., *Chromohalobacter marismortui*, *Flavobacterium* sp., *Listeria* sp., and *Pseudomonas* sp., can cause struvite ($MgNH_4PO_4 \cdot 6H_2O$) to form, at least under laboratory conditions (Rivadeneyra et al., 1983, 1992b, 2006). A major, but not necessarily exclusive, microbial contribution to this process appears to be ammonium formation (Rivadeneyra et al., 1992a). Struvite is formed when orthophosphate is added at pH 8 to seawater solutions in which the NH_4^+ concentration is 0.01 M (Handschuh and Orgel, 1973). The presence of calcium ion in sufficient quantity can suppress struvite formation and promote apatite formation instead (Rivadeneyra et al., 1983). Although struvite formation is probably of little significance in nature today, it may have been significant in the primitive world of Precambrian times if NH_4^+ was present in concentrations as high as 10^{-2} M (Handschuh and Orgel, 1973).

12.7   MICROBIAL REDUCTION OF OXIDIZED FORMS OF PHOSPHORUS

Phosphorus may also undergo redox reactions; some or all of which may be catalyzed by microbes. Of biogenic interest are the +5, +3, +1, and −3 oxidation states, as in orthophosphate (H_3PO_4), orthophosphite (H_3PO_3), hypophosphite (H_3PO_2), and phosphine (PH_3), respectively. Reduction of phosphate to phosphine by soil bacteria has been reported (Rudakov, 1927; Tsubota, 1959; Devai et al., 1988). Mannitol appeared to be a suitable electron donor in the reaction described by Rudakov (1927) and glucose in the experiments described by Tsubota (1959). Phosphite and hypophosphite were claimed to be intermediates in the reduction process (Rudakov, 1927; Tsubota, 1959). Devai et al. (1988) detected phosphine evolution in anaerobic sewage treatment in Imhoff tanks in Hungary and confirmed the observation in anaerobic laboratory experiments. Gassmann and Schorn (1993) detected phosphine in surface sediments in Hamburg Harbor. The phosphine was most readily detected in porous sediments in which the porosity was due to gas bubbles. Iron phosphide (Fe_3P_2) is reported to have been formed when a cell-free preparation of *Desulfovibrio* was incubated in the presence of steel in a yeast extract broth under hydrogen gas (Iverson, 1968). Inositol hexaphosphate, a product of plants and present in yeast extract, may be a substrate for phosphine formation (Iverson, 1998). Hydrogenase from *Desulfovibrio* may have been responsible for the formation of phosphine from the inositol phosphate in the yeast extract, using the hydrogen in the system as the reductant in Iverson's (1968) experiment. The phosphine could then have reacted with ferrous iron from the steel corrosion to form Fe_3P_2 (Iverson, 1968).

Questions have been raised about the ability of microbes to reduce phosphate. Liebert (1927) showed that on the basis of thermodynamic calculations using heats of formation, the reduction of phosphate to phosphite by mannitol is an energy-consuming process and could therefore not serve a respiratory function. He calculated a heat of reaction value (ΔH) of +20 kcal on the basis of the following equation:

$$\underset{316\,kcal}{C_6H_{14}O_6} + \underset{5390\,kcal}{13Na_2HPO_4} \rightarrow \underset{4460\,kcal}{13Na_2HPO_3} + \underset{566\,kcal}{6CO_2} + \underset{478\,kcal}{7H_2O} \tag{12.15}$$

He also calculated a ΔH of +438 kcal for the reduction of phosphate to hypophosphite and a ΔH of +1147 kcal for the reduction of phosphate to phosphine by mannitol. These conclusions can also be reached when the free energy of reaction (ΔG) is considered instead of heats of reaction (ΔH). Woolfolk and Whiteley (1962) reported that phosphate was not reduced by hydrogen in the presence of an extract of *Veillonella alcalescens* (formerly *Micrococcul lactilyticus*), although this extract could catalyze the reduction of some other oxyanions with hydrogen. Skinner (1968) also questioned the ability of bacteria to reduce phosphate. He could not find such organisms in soils he tested. Burford and Bremner (1972), while unable to demonstrate phosphine evolution from water-logged soils, were not able to rule out microbial phosphine genesis because they found that soil constituents can adsorb phosphine. Thus, unless bacteria form phosphine in excess of the adsorption capacity of a soil, phosphine detection in the gas phase may not be possible.

Interestingly, Barrenscheen and Beckh-Widmanstetter (1923) reported the production of hydrogen phosphide (phosphine, PH_3) from organically bound phosphate during putrefaction of beef blood. Much more recently, Metcalf and Wanner (1991) presented evidence supporting the existence of a C–P lyase in *Escherichia coli* that catalyzes the reductive cleavage of compounds such as methyl phosphonate to phosphite and methane:

$$HO-\underset{\underset{OH}{|}}{\overset{\overset{O}{\|}}{P}}-CH_3 + 2(H) \xrightarrow[\text{C–P lyase}]{} HO-\underset{\underset{OH}{|}}{\overset{\overset{O}{\|}}{P}}-H + CH_4 \qquad (12.16)$$

$$P(+5) \hspace{5cm} P(+3)$$

This enzyme activity was previously studied in *Agrobacterium radiobacter* (Wackett et al., 1987), although it was described by these authors as a hydrolytic reaction. Phosphonolipids are known to exist in organisms from bacteria to mammals (Hilderbrand and Henderson, 1989, cited by Metcalf and Wanner, 1991). Thus, biochemical mechanisms for synthesizing orthophosphonates exist, and it is therefore highly likely that an organophosphate such as methyl- or ethylphosphonate that requires a C–P lyase to release the phosphorus as orthophosphite (Metcalf and Wanner, 1991) is an intermediate in the conversion of orthophosphate to orthophosphite and that C–P lyase activity represents the reductive step in this transformation. This needs further investigation.

12.8   MICROBIAL OXIDATION OF REDUCED FORMS OF PHOSPHORUS

Reduced forms of phosphorus can be aerobically and anaerobically oxidized by bacteria. Thus, *Bacillus caldolyticus*, a moderate thermophile, can oxidize hypophosphite to phosphate aerobically (Heinen and Lauwers, 1974). The active enzyme system consists of an $(NH_4)_2SO_4$-precipitable protein fraction, nicotinamide adenine dinucleotide (NAD), and respiratory chain components. The enzyme system does not oxidize phosphite. Adams and Conrad (1953) first reported the aerobic oxidation of phosphite by bacteria and fungi from soil. All phosphite that was oxidized by these strains was assimilated. None of the oxidized phosphite, that is, phosphate, was released into the medium before the organisms died. Phosphate added to the medium inhibited phosphite oxidation. Active organisms included the bacteria *Pseudomonas fluorescens*, *P. lachrymans*, *Aerobacter* (now known at *Enterobacter*) *aerogenes*, *Erwinia amylovora*; fungi *Alternaria*, *Aspergillus niger*, *Chaetomium*, *Penicillium notatum*; and some actinomycetes. In later studies, Casida (1960) found that a culture of *P. fluorescens* strain 195 formed orthophosphate aerobically from orthophosphite in excess of its needs and released phosphate into the medium. The culture was heterotrophic, and its phosphite-oxidizing activity was inducible and stimulated by yeast extract. The enzyme system involved in phosphite oxidation was an orthophosphite-NAD oxidoreductase, which was inactive on arsenite, hypophosphite, nitrite, selenite, and tellurite, and was inhibited by sulfite (Malacinski and Konetzka, 1966, 1967).

Oxidation of reduced phosphorus compounds can also occur anaerobically. A soil bacillus has been isolated that is capable of anaerobic oxidation of hypophosphite and phosphite to phosphate (Foster et al., 1978). In a mixture of phosphite and hypophosphite, phosphite was oxidized first. Phosphate inhibited the oxidation of either phosphite or hypophosphite. The organism did not release phosphate into the medium.

Because phosphite and hypophosphite have not been reported in detectable quantities in natural environments, it has been suggested that microbial ability to utilize these compounds, especially anaerobically, may be a vestigial property that originated at a time when the Earth had a reducing atmosphere surrounding it that favored the occurrence of phosphite (Foster et al., 1978).

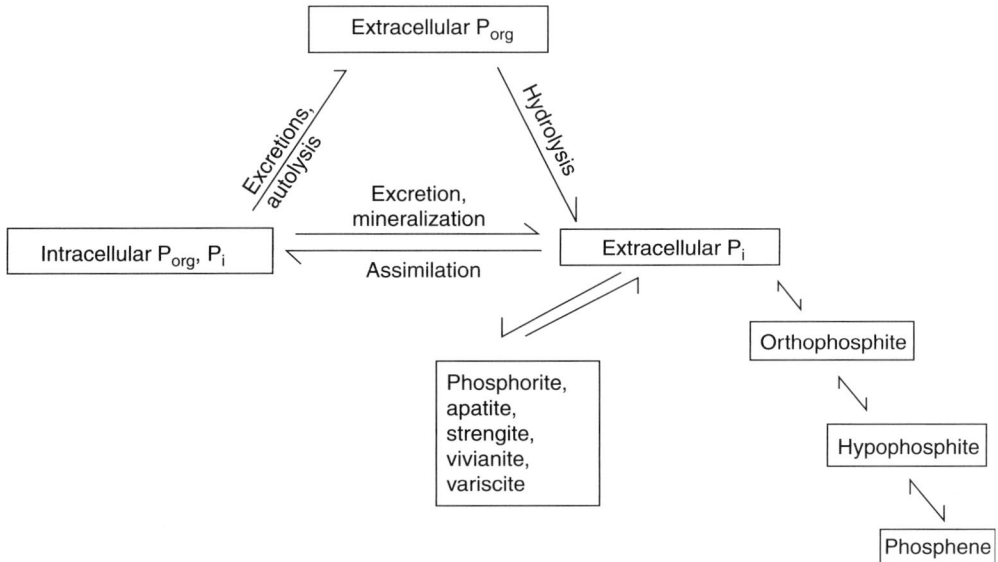

FIGURE 12.3 The phosphorus cycle.

12.9 MICROBIAL ROLE IN THE PHOSPHORUS CYCLE

In many ecosystems, phosphorus availability may determine the extent of microbial growth and activity. The element follows cycles in which it finds itself alternately outside and inside the living cells, in organic and inorganic form, free or fixed, dissolved or precipitated. Microbes play a major role in these changes of state, as outlined in Figure 12.3 and as discussed in this chapter.

12.10 SUMMARY

Phosphorus is a very important element for all forms of life. It is used in cell structure as well as cell function. It plays a role in transducing biochemically useful energy. When free in the environment, it occurs primarily as organic phosphate esters and as inorganic phosphates. Some of the latter, such as calcium, aluminum, and iron phosphates, are insoluble at neutral or alkaline pH. To be nutritionally available, organic phosphates have to be enzymatically hydrolyzed to liberate orthophosphate. Microbes play a central role in this process. Microbes may also free orthophosphate from insoluble inorganic phosphates, by producing organic or mineral acids or chelators, or in the case of iron phosphates by producing H_2S. Under some conditions, microbes may promote the formation of insoluble inorganic phosphates such as those of calcium, aluminum, and iron. They have been implicated in phosphorite formation in the marine environment.

Microbes have been implicated in the reduction of pentavalent phosphorus to lower valence states. The experimental evidence for this is somewhat equivocal, however. It is likely that organic phosphorus compounds are intermediates in these reductions. Microbes have also been implicated in the aerobic and anaerobic oxidation of reduced forms of phosphorus to phosphate. The experimental evidence in this case is strong. It includes demonstration of enzymatic involvement. The geomicrobial significance of these redox reactions in nature is not clearly understood.

REFERENCES

Adams JK, Burkhart B. 1967. Diagenetic phosphates from the northern Atlantic coastal plain. *Abstr Annu Geophys Soc Am Assoc Soc Joint Meeting*. New Orleans, LA. November 20–22, 1967. Program, p. 2.
Adams F, Conrad JP. 1953. Transition of phosphite to phosphate in soil. *Soil Sci* 75:361–371.

Agnihotri VP. 1970. Solubilization of insoluble phosphates by some soil fungi isolated from nursery seedbeds. *Can J Microbiol* 16:877–880.

Alexander M. 1977. *Introduction to Soil Microbiology*. 2nd ed. New York: Wiley.

Altschuler S, Clarke RS, Jr, Young EJ. 1958. *Geochemistry of Uranium in Apatite and Phosphorite*. U.S. Geol Surv Prof Pap 314D.

Ayyakkannu K, Chandramohan D. 1971. Occurrence and distribution of phosphate solubilizing bacteria and phosphatase in marine sediments at Porto Nova. *Mar Biol* 11:201–205.

Babenko YS, Tyrygina GI, Grigor'ev EF, Dolgikh LM, Borisova TI. 1984. Biological activity and physiological-biochemical properties of phosphate dissolving bacteria. *Mikrobiologiya* 53:533–539 (English translation pp. 427–433).

Babu-Khan S, Yeo TC, Martin WL, Duran MR, Rogers RD, Goldstein AH. 1995. Cloning of a mineral phosphate-solubilizing gene from *Pseudomonas capacia*. *Appl Environ Microbiol* 61:972–978.

Banik S, Dey BK. 1983. Phosphate solubilizing potentiality of the microorganisms capable of utilizing aluminum phosphate as a sole phosphate source. *Z Mikrobiol* 138:17–23.

Barrenscheen HK, Beckh-Widmanstetter HA. 1923. Über bakterielle Reduktion organisch gebundener Phosphorsäure. *Biochem Z* 140:279–283.

Baturin GN. 1973. Genesis of phosphorites of the southwest African shelf. *Tr Inst Okeanol Akad Nauk SSSR* 95:262–267.

Baturin GN, Merkhulova KI, Chalov PI. 1969. Radiometric evidence for recent formation of phosphatic nodules in marine shelf sediments. *Mar Geol* 13:M37–M41.

Bowen HJM. 1979. *Environmental Chemistry of the Elements*. London: Academic Press.

Burford JR, Bremner JM. 1972. Is phosphate reduced to phosphine in waterlogged soils? *Soil Biol Biochem* 4:489–495.

Burnett WC. Institute for International Cooperative Environmental Research, Florida State University, Tallahassee FL.

Burnett WC, Beers MJ, Roe KK. 1982. Growth rates of phosphate nodules from the continental margin off Peru. *Science* 215:1616–1618.

Casida LE, Jr. 1960. Microbial oxidation and utilization of orthophosphite during growth. *J Bacteriol* 80:237–241.

Chatterjee AK, Nandi P. 1964. Solubilization of insoluble phosphates by rhizosphere fungi of legumes (*Arachis cyanopsis, Desmodium*). *Trans Bose Res Inst* (Calcutta) 27:115–120.

Cosgrove DJ. 1967. Metabolism of organic phosphates in soil. In: McLaren AD, Petersen GH, eds. *Soil Biochemistry*, Vol. 1. New York: Marcel Dekker, pp. 216–228.

Cosgrove DJ. 1977. Microbial transformations in the phosphorus cycle. *Adv Microb Ecol* 1:95–134.

d'Anglejan BF. 1967. Origin of marine phosphates off Baja California, Mexico. *Mar Geol* 5:15–44.

d'Anglejan BF. 1968. Phosphate diagenesis of carbonate sediments as a mode of in situ formation of marine phosphorites: Observations in a core from the Eastern Pacific. *Can J Earth Sci* 5:81–87.

Devai I, Feldoeldy L, Wittner I, Plosz S. 1988. Detection of phosphine: New aspects of the phosphorus cycle in the hydrosphere. *Nature* (London) 333:343–345.

Dommergues Y, Mangenot F. 1970. *Écologie Microbienne du Sol*. Paris: Masson.

Duff, RB, Webley DM. 1959. 2-Ketogluconic acid as a natural chelator produced by soil bacteria. *Chem Ind* (London) 1376–1377.

Duff RB, Webley DM, Scott RO. 1963. Solubilization of minerals and related materials by 2-ketogluconic acid-producing bacteria. *Soil Sci* 95:105–114.

Foster TL, Winans L, Jr., Helms JS. 1978. Anaerobic utilization of phosphite and hypophosphite by *Bacillus* sp. *Appl Environ Microbiol* 35:937–944.

Friedberg I, Avigad G. 1968. Structures containing polyphosphate in *Micrococcus lysodeikticus*. *J Bacteriol* 96:544–553.

Fuller WH. 1972. Phosphorus: Element and geochemistry. In: Fairbridge WR, ed. *The Encyclopedia of Geochemistry and Environmental Sciences*. Encyclopedia of Earth Science Series, Vol. IVA. New York: Van Nostrand Reinhold, pp. 942–946.

Gassmann G, Schorn F. 1993. Phosphine from harbor surface sediments. *Naturwissenschaften* 80:78–80.

Gulbrandsen RA. 1969. Physical and chemical factors in the formation of marine apatite. *Econ Geol* 64:365–382.

Handschuh GJ, Orgel LE. 1973. Struvite and prebiotic phosphorylation. *Science* 179:483–484.

Heinen W, Lauwers AM. 1974. Hypophosphite oxidase from *Bacillus caldolyticus*. *Arch Microbiol* 95:267–274.

Herbes SE, Allen HE, Mancy KH. 1975. Enzymatic characterization of soluble organic phosphorus in lake water. *Science* 187:432–434.

Hilderbrand RL, Henderson TO. 1989. Phosphonic acids in nature. In: Hilderbrand RL, ed. *The Role of Phosphonates in Living Systems.* Boca Raton, FL: CRC, pp. 6–25.

Hirschler A, Lucas J, Hubert J-C. 1990a. Apatite genesis: A biologically induced or biologically controlled mineral formation process? *Geomicrobiol J* 7:47–57.

Hirschler A, Lucas J, Hubert J-C. 1990b. Bacterial involvement in apatite genesis. *FEMS Microbiol Ecol* 73:211–220.

Hutchens E, Valsami-Jones E, Harouiya N, Chaïrat C, Oelkers EH, McEldoney S. 2006. An experimental investigation of the effect of *Bacillus megaterium* on apatite dissolution. *Geomicrobiol J* 23:177–182.

Iverson WP. 1968. Corrosion of iron and formation of iron phosphide by *Desulfovibrio desulfuricans. Nature* (London) 217:1265.

Iverson WP. 1998. Possible source of a phosphorus compound produced by sulfate-reducing bacteria that cause anaerobic corrosion of iron. *Mater Perform* 37:46–49.

Jahnke RA, Emerson SR, Roe KK, Burnett WC. 1983. The present day formation of apatite in Mexican continental margin sediments. *Geochim Cosmochim Acta* 47:259–266.

Jansson M. 1987. Anaerobic dissolution of iron-phosphorus complexes in sediment due to the activity of nitrate-reducing bacteria. *Microb Ecol* 14:81–89.

Kardos LT. 1955. Soil fixation of plant nutrients. In: Bear FE, ed. *Chemistry of the Soil.* New York: Van Nostrand Reinhold, pp. 177–199.

Kobori H, Taga N. 1979. Phosphatase activity and its role in mineralization of organic phosphorus in coastal sea water. *J Exp Mar Biol Ecol* 36:23–39.

Lehninger AL. 1970. *Biochemistry.* New York: Worth.

Lehninger AL. 1975. *Biochemistry.* 2nd ed. New York: Worth.

Liebert F. 1927. Reduzieren Mikroben Phosphat? *Z Bakt Parasitenk Infektionskr Hyg* 72:369–374.

Lipman JG, McLean H. 1916. The oxidation of sulfur in soils as a means of increasing the availability of mineral phosphates. *Soil Sci* 1:533–539.

Liu CL, Hart N, Peck HD, Jr. 1982. The utilization of inorganic pyrophosphate as a source of energy for the growth of sulfate reducing bacteria, *Desulfotomaculum* sp. *Science* 217:363–364.

Lucas J, Prévot L. 1984. Synthèse de l'apatite par voie bactérienne á partir de matière organique phosphatée et de divers carbonates de calcium dans des eaux douce et marine naturelles. *Chem Geol* 42:101–118.

Macaskie LE, Dean ACR, Cheetham AK, Jakeman RJB, Skarnulis AJ. 1987. Cadmium accumulation by a *Citrobacter* sp.: The chemical nature of the accumulated metal precipitate and its location in the bacterial cells. *J Gen Microbiol* 133:539–544.

Macaskie LE, Empson RM, Cheetham AK, Grey CP, Skarnulis AJ. 1992. Uranium bioaccumulation by a *Citrobacter* sp. as a result of enzymatically mediated growth of polycrystalline HUO_2PO_4. *Science* 277:782–784.

Malacinski G, Konetzka WA. 1966. Bacterial oxidation of orthophosphite. *J Bacteriol* 91:578–582.

Malacinski G, Konetzka WA. 1967. Orthophosphite-nicotinamide adenine dinucleotide oxidoreductase from *Pseudomonas fluorescens. J Bacteriol* 93:1906–1910.

Marine Chemistry. 1971. *A Report of the Marine Chemistry Panel of the Committee of Oceanography.* Washington, DC: National Academy of Sciences.

McConnell D. 1965. Precipitation of phosphates in sea water. *Econ Geol* 60:1059–1062.

Metcalf WW, Wanner BL. 1991. Involvement of the *Escherichia coli* phn (φD) gene cluster in assimilation of phosphorus in the form of phosphonates, phosphite, P_i esters, and P_i. *J Bacteriol* 173:587–600.

Mullins HT, Rasch RF. 1985. Sea-floor phosphorites along the Central California continental margin. *Econ Geol* 80:696–715.

Nathan Y, Bremner JM, Loewenthal RE, Monteiro P. 1993. Role of bacteria in phosphorite genesis. *Geomicrobiol J* 11:69–76.

O'Brien GWO, Veeh HH. 1980. Holocene phosphorite on the East Australian margin. *Nature* (London) 288:690–692.

O'Brien GWO, Harris JR, Milnes AR, Veeh HH. 1981. Bacterial origin of the East Australian continental margin phosphorites. *Nature* (London) 294:442–444.

Paul EA, Clark FE. 1996. *Soil Microbiology and Biochemistry.* 2nd ed. San Diego, CA: Academic Press.

Patrick WH, Jr., Gotoh S, Williams BG. 1973. Strengite dissolution in flooded soils and sediment. *Science* 179:564–565.

Petterson A, Kunst L, Bergman B, Roomans GM. 1985. Accumulations of aluminum by *Anabaena cylindrica* into polyphosphate granules and cell walls: An x-ray energy dispersive microanalysis study. *J Gen Microbiol* 131:2545–2548.

Piper DZ, Codespoti LA. 1975. Marine phosphorite deposits and the nitrogen cycle. *Science* 179:564–565.

Racz GJ, Savant NK. 1972. Pyrophosphate hydrolysis in soil as influenced by flooding and fixation. *Soil Sci Soc Am Proc* 36:678–682.

Riggs SR. 1984. Paleooceanographic model of neogene phosphorite deposition, U.S. Atlantic Continental Margin. *Science* 223:123–131.

Rivadeneyra MA, Martín-Algarra A, Sánchez-Navas A, Martín-Ramos D. 2006. Carbonate and phosphate precipitation by *Chromohalobacter marismortui*. *Geomicrobiol J* 23:1–13.

Rivadeneyra MA, Perez-Garcia I, Ramos-Cormenzana A. 1992a. Influence of ammonium ion on bacterial struvite production. *Geomicrobiol J* 10:125–137.

Rivadeneyra MA, Perez-Garcia I, Ramos-Cormenzana A. 1992b. Struvite precipitation by soil and fresh water bacteria. *Curr Microbiol* 24:343–347.

Rivadeneyra MA, Ramos-Cormenzana A, Garcia-Cervigon A. 1983. Bacterial formation of struvite. *Geomicrobiol J* 3:151–163.

Rudakov KJ. 1927. Die Reduktion der mineralischen Phosphate auf biologischem Wege. II. Mitteilung. *Z Bakt Parasitenk Infektionskr Hyg Abt II* 79:229–245.

Ruttenberg KC, Berner RA. 1993. Authigenic apatite formation and burial in sediments from non-upwelling, continental margin environments. *Geochim Cosmochim Acta* 57:991–1007.

Shapiro J. 1967. Induced rapid release and uptake of phosphate by microorganisms. *Science* 155:1269–1271.

Skinner FA. 1968. The anaerobic bacteria in soil. In: Gray TGR, Parkinson D, eds. *The Ecology of Soil Bacteria.* Toronto, Ontario, Canada: University of Toronto Press, pp. 573–592.

Skujins JJ. 1967. Enzymes in soil. In: McLaren AD, Peterson GH, eds. *Soil Biochemistry,* Vol. 1. New York: Marcel Dekker, pp. 371–414.

Sperber JI. 1958a. Release of phosphates from soil minerals by hydrogen sulfide. *Nature* (London) 181:934.

Sperber JI. 1958b. The incidence of apatite-solubilizing organisms in the rhizosphere and soil. *Aust J Agric Res* 9:778–781.

Suess E. 1981. Phosphate regeneration from sediments of the Peru continental margin by dissolution of fish debris. *Geochim Cosmochim Acta* 45:577–588.

Swaby RJ, Sperber J. 1958. Phosphate-dissolving microorganisms in the rhizosphere of legumes. In: Hallsworth EG, ed. *Nutrition of Legumes. Proc Univ Nottingham Easter School Agric Sci,* 5th ed. London: Academic Press, pp. 289–294.

Taga N, Kobori H. 1978. Phosphatase activity in eutrophic Tokyo Bay. *Marine Biol* (Berlin) 49:223–229.

Tsubota G. 1959. Phosphate reduction in the paddy field. I. *Soil Plant Food* 5:10–15.

van Groenestijn JW, Deinema MN, Zehnder AJB. 1987. ATP production from polyphosphate in *Acinetobacter* strain 210A. *Arch Microbiol* 148:14–19.

Varma AK, Rigsby W, Jordan DC. 1983. A new inorganic pyrophosphate utilizing bacterium from a stagnant lake. *Can J Microbiol* 29:1470–1474.

Veeh HH, Brunett WC, Soutar A. 1973. Contemporary phosphorites on the continental margin of Peru. *Science* 181:844–845.

Wackett LP, Shames SL, Venditti CP, Walsh CT. 1987. Bacterial carbon-phosphorus lyase: Products, rates, and regulation of phosphonic and phosphinic acid metabolism. *J Bacteriol* 169:710–717.

Westheimer FH. 1987. Why nature chose phosphates. *Science* 235:1173–1178.

Williams BG, Patrick WH, Jr. 1971. Effect of E_h and pH on the dissolution of strengite. *Nature* (*Phys Sci*) 234:16–17.

Woolfolk CA, Whiteley HR. 1962. Reduction of inorganic compounds with molecular hydrogen by *Micrococcus lactilyticus*. I. Stoichiometry with compounds of arsenic, selenium, tellurium, transition and other elements. *J Bacteriol* 84:647–658.

Youssef MI. 1965. Genesis of bedded phosphates. *Econ Geol* 60:590–600.

13 Geomicrobially Important Interactions with Nitrogen

13.1 NITROGEN IN BIOSPHERE

Nitrogen is an element essential to life. It is abundant in the atmosphere, mostly as dinitrogen (N_2), representing roughly four-fifths by volume or three-fourths by weight of the total gas, the rest being mainly oxygen and small amounts of carbon dioxide and trace amounts of other common gases. Small amounts of oxides of nitrogen are also present. In soil, sediment, freshwater, and ocean water, nitrogen exists in both inorganic and organic forms. Geomicrobially important inorganic forms include ammonia and ammonium ion, nitrate, nitrite, and gaseous oxides of nitrogen. Table 13.1 (Fenchel and Blackburn, 1979; Brock and Madigan, 1988) lists the estimated abundance of some of these forms as well as that of organic nitrogen in the environments (see also Stevenson, 1972). Geomicrobially important organic nitrogen compounds include humic and fulvic acids, proteins, peptides, amino acids, purines, pyrimidines, pyridines, other amines, and amides.

Chemically, nitrogen occurs in the oxidation states $-3(NH_3)$, $-2(N_2H_4)$, $-1(NH_2OH)$, $0(N_2)$, $+1(N_2O)$, $+2(NO)$, $+3(HNO_2)$, $+4(N_2O_4)$, and $+5(HNO_3)$. Of these, the -3, -1, 0, $+1$, $+2$, $+3$, and $+5$ oxidation states have biological significance because they can be altered enzymatically. Although nitrogen compounds with nitrogen in the oxidation state of $+4$ are not metabolized by microbes, nitrite formed from the disproportionation of NO_2 after it is absorbed by soil can be oxidized to nitrate by as yet unidentified organisms (Ghiorse and Alexander, 1978).

Generally, inorganic nitrogen compounds exist in nature either as gases in the atmosphere and dissolved in water, or as compounds in aqueous solution. Exceptions are small deposits of nitrates of sodium, potassium, calcium, magnesium, and ammonium known as guano or cave playa, or caliche nitrates (Lewis, 1965). In some cases these nitrate deposits were apparently formed by bacteriological transformation of organic nitrogen produced as a result of nitrogen fixation by bacteria, including cyanobacteria. In other cases the deposits arose from bacterial transformation of organic nitrogen contained in animal droppings such as those of birds and bats. The organic nitrogen was released as ammonia and then oxidized to nitrate by a consortium of bacteria (Ericksen, 1983; see Section 13.2 for discussion of the reactions). In Chile, deposits of this type, which were probably formed mainly from the microbiota in playa lakes (Ericksen, 1983), are commercially exploited.

Organic nitrogen in nature may exist dissolved in an aqueous phase or in an insoluble state, in the latter case usually in polymers (e.g., certain proteins such as keratin). Insofar as is known, organic nitrogen is usually metabolizable by microbes.

Nitric acid formed through bacterial nitrification can be an important agent in the weathering of rocks and minerals (Chapters 4 and 6, and Chapters 9 through 11).

13.2 MICROBIAL INTERACTIONS WITH NITROGEN

13.2.1 AMMONIFICATION

Most plants derive the nitrogen that they assimilate from soil. This nitrogen is in most instances in the form of nitrate. The nitrate anion is much less readily bound by mineral soil particles, especially clays, which have a net negative charge, than the ammonium cation. The nitrate supply of the soil depends on recycling of spent organic nitrogen (plant, animal, and microbial excretions and

TABLE 13.1
Abundance of Nitrogen in the Biosphere

Biosphere Compartment	Form of Nitrogen	Estimated Quantity of Nitrogen (kg)
Atmosphere	N_2	3.9×10^{18}
Oceans	Organic N	9×10^{14}
	NH_4^+, NO_2^-, NO_3^-	10^{14}
Land	Organic N	8×10^{14}
	NH_4^+, NO_2^-, NO_3^-	1.4×10^{14}
Sediments	Total N	4×10^{17}
Rocks	Total N	1.9×10^{20}
Living biomass	Total N	1.3×10^{13}

Source: Extracted from Fenchel T, Blackburn TH, *Bacteria and Mineral Cycling*, Academic Press, London, 1979; Brock TD, Madigan MT, *Biology of Microorganisms*. 5th ed., Prentice-Hall, Englewood Cliffs, NJ, 1988.

remains). The first step in the recycling process is *ammonification*, in which the organic nitrogen is transformed into ammonia. An example of ammonification is the deamination of amino acids:

$$\begin{array}{c} \text{RCHCOOH} + \text{NAD}^+ \rightarrow \text{RCCOOH} + \text{NADH} + \text{H}^+ \\ | \qquad\qquad\qquad\qquad\quad || \\ \text{NH}_2 \qquad\qquad\qquad\qquad \text{NH} \end{array}$$

$$\begin{array}{c} \text{RCCOOH} + \text{H}_2\text{O} \rightarrow \text{RCCOOH} + \text{NH}_3 \\ || \qquad\qquad\qquad\qquad || \\ \text{NH} \qquad\qquad\qquad\quad \text{O} \end{array} \qquad (13.1)$$

The NH_3 reacts with water to form ammonium hydroxide, which dissociates as follows:

$$NH_3 + H_2O \rightarrow NH_4OH \rightarrow NH_4^+ + OH^- \qquad (13.2)$$

In the laboratory it is commonly observed that when heterotrophic bacteria grow in proteinaceous medium, such as nutrient broth consisting of peptone and beef extract, in which the organic nitrogen serves as the source of energy, carbon, and nitrogen, the pH rises over time owing to the liberation of ammonia and its hydrolysis to ammonium ion. Indeed, ammonification is always an essential first step when an amino compound such as an amino acid serves as an energy source.

Ammonia is also formed as a result of urea hydrolysis catalyzed by the enzyme urease.

$$NH_2CONH_2 + 2H_2O \rightarrow 2NH_4^+ + CO_3^{2-} \qquad (13.3)$$

Urea is a nitrogen waste product excreted in the urine of many mammals. Although urease is produced by a variety of prokaryotic and eukaryotic microbes, a few prokaryotic soil microbes, for example, *Bacillus pasteurii* and *B. freudenreichii*, seem to be specialists in degrading urea. They prefer to grow at an alkaline pH such as that generated when urea is hydrolyzed (see Alexander, 1977).

13.2.2 Nitrification

Plants can readily assimilate ammonia. But the ammonia produced in ammonification in aqueous systems at neutral pH exists as a positively charged ammonium ion (NH_4^+) owing to protonation. It is adsorbed by clays and then not readily available to plants. Thus it is important that ammonia be converted into an anionic nitrogen species, which is not readily adsorbed to the negatively charged clays and is thus more readily available to plants. *Nitrifying bacteria* play a central role in this conversion. The majority of nitrifying bacteria are autotrophs and can be divided into two groups; one includes those bacteria that oxidize ammonia to nitrous acid (e.g., *Nitrosomonas*, *Nitrosocystis*) and the second includes those bacteria that oxidize nitrite to nitrate (e.g., *Nitrobacter*, *Nitrococcus*). All the members of these two groups of nitrifying bacteria are obligate autotrophs except for *Nitrobacter winogradskyi* strains, which appear to be facultative. They are all strict aerobes. Representatives are found in soil, freshwater, and seawater (for further characterization see Paul and Clark, 1996; Balows, 1992).

Although ammonification is the major source of ammonia in soil and sediments, a special anaerobic respiratory process may also be a significant source of ammonia in some environments. In this process, nitrate is reduced to ammonia via nitrite (Jørgensen and Sørensen, 1985, 1988; Binnerup et al., 1992). The overall process can be summarized in the equation

$$NO_3^- + 8(H) + H^+ \rightarrow NH_3 + 3H_2O \tag{13.4}$$

This process is known as *nitrate ammonification* and can be carried on by a number of different facultative and strictly anaerobic bacteria (see Cruz-García et al., 2007; Dannenberg et al., 1992; Sørensen, 1987; review by Ehrlich, 1993, pp. 232–233). A variety of organic compounds, H_2, and inorganic sulfur compounds can serve as electron donors in this reaction (Dannenberg et al., 1992).

13.2.3 Ammonia Oxidation

Oxidation of ammonia by ammonia-oxidizing bacteria involves hydroxylamine (NH_2OH) as an intermediate (see review by Wood, 1988). The formation of hydroxylamine is catalyzed by an oxygenase.

$$NH_3 + 0.5O_2 \rightarrow NH_2OH \quad (\Delta G^\circ = +0.77 \text{ kcal}; +3.21 \text{ kJ}) \tag{13.5}$$

This reaction does not yield biochemically useful energy. Indeed, it is slightly endothermic and proceeds in the direction of hydroxylamine because it is coupled to the oxidation of hydroxylamine to nitrous acid, which is strongly exothermic. The overall reaction of the oxidation of hydroxylamine to HNO_2 can be summarized as

$$NH_2OH + O_2 \rightarrow HNO_2 + H_2O \quad (\Delta G^\circ = -62.42 \text{ kcal}; -260.9 \text{ kJ}) \tag{13.6}$$

It is Reaction 13.6 from which chemoautotrophic ammonia oxidizers obtain their energy, by using chemiosmotic coupling, that is, oxidative phosphorylation. The conversion of hydroxylamine to nitrous acid involves some intermediate steps (Hooper, 1984).

The enzyme that catalyzes ammonia oxidation (Reaction 13.5) is a nonspecific oxygenase. It can also catalyze the oxygenation of methane to methanol (Jones and Morita, 1983).

$$CH_4 + 0.5O_2 \rightarrow CH_3OH \quad (\Delta G^\circ = -29.74 \text{ kcal}; -124.3 \text{ kJ}) \tag{13.7}$$

Under standard conditions, this reaction is thermodynamically more favorable than Reaction 13.5. This does not mean, however, that ammonia oxidizers can grow on methane. They lack the ability to oxidize methanol.

Ammonia oxidizers can also form some NO and N_2O in side reactions of ammonia oxidation under oxygen limitation in which nitrite replaces oxygen as terminal electron acceptor (see discussion by Knowles, 1985; Bock et al., 1991; Davidson, 1993). This is an important observation because it means that biogenically formed N_2O and NO are not solely the result of denitrification (see Section 13.2.7).

13.2.4 Nitrite Oxidation

The nitrite oxidizers convert nitrite to nitrate:

$$NO_2^- + 0.5O_2 \rightarrow NO_3^- \quad (\Delta G^\circ = -18.18 \text{ kcal}; -76.0 \text{ kJ}) \tag{13.8}$$

They obtain useful energy from this process by coupling it chemiosmotically to ATP generation (Aleem and Sewell, 1984; Wood, 1988).

13.2.5 Heterotrophic Nitrification

Ammonia is also converted to nitrate by certain heterotrophic microorganisms, but the process is probably of minor importance in nature in most instances. Rates of heterotrophic nitrification measured under laboratory conditions so far are significantly slower than those of autotrophic nitrification. The organisms capable of heterotrophic nitrification include both bacteria, such as *Arthrobacter* sp., and fungi, such as *Aspergillus flavus*. They gain no energy from the conversion. The pathway from ammonia to nitrate may involve intermediates such as hydroxylamine, nitrite, and 1-nitrosoethanol in the case of bacteria; and 3-nitropropionic acid in the case of fungi (see Alexander, 1977; Paul and Clark, 1996).

13.2.6 Anaerobic Ammonia Oxidation (Anammox)

Within the last decade, some *Planctomycetes* (domain Bacteria) have been found to be involved in the *anaerobic* oxidation of ammonium (NH_4^+) with nitrite (NO_2^-) to form dinitrogen (N_2) (see van de Graaf et al., 1995; and reviews by Jetten et al., 2003; Kuenen and Jetten, 2001; Strous and Jetten, 2004). An equation summarizing the overall reaction is as follows:

$$NH_4^+ + NO_2^- \rightarrow N_2 + 2H_2O \quad (\Delta G^\circ = -85.51 \text{ kcal}; -357.79 \text{ kJ}) \tag{13.9}$$

Hydrazine (N_2H_4) and hydroxylamine (NH_2OH) are intermediates. Acetylene and methanol have been found to inhibit different reactions in anammox (Jensen et al., 2007). Anammox was first discovered in an anaerobic sewage treatment process but has since been demonstrated to occur in nature in both freshwater and marine environments (Jetten et al., 2003; Trimmer et al., 2003; Tal et al., 2005). In the oceans it may account for 50% of the loss of fixed nitrogen (Dalsgaard et al., 2005). The free energy from this reaction (Equation 13.9) supports CO_2 fixation. The anammox reaction appears to be carried out in a special intracytoplasmic organelle in active bacteria. The organelle is called an anammoxosome. It is bounded by a membrane that contains ladderane lipids (see van Niftrik et al., 2004). The bacteria capable of the anammox reaction belong to the phylum Planctomycetes. *Candidatus* Brocadia anammoxidans is one member that has been described

(Strous et al., 1999; Kuenen and Jetten, 2001). Two others are *Candidatus* Kuenenia stuttgartiensis and *Candidatus* Scalindua sorokinii (Jetten et al., 2003).

13.2.7 DENITRIFICATION

Nitrate, nitrite, and nitrous and nitric oxides can serve as electron acceptors in microbial respiration, usually under anaerobic conditions. The transformation of nitrate to nitrite as an anaerobic respiratory process is called *dissimilatory nitrate reduction*, and the reduction of nitrate to nitric oxide (NO), nitrous oxide (N_2O), and dinitrogen is called *denitrification*. *Assimilatory nitrate reduction* is the first step in a process in which nitrate is reduced to ammonia for the purpose of assimilation. Only as much nitrate is consumed in this process as is needed for assimilation. It is not a form of respiration and is performed by many organisms that cannot use nitrate for respiration. Some nitrate-respiring bacteria are capable of only nitrate reduction, lacking the enzymes for reduction of nitrite to dinitrogen, whereas others are capable of reducing nitrite to ammonia instead of dinitrogen in a process that has been called nitrate ammonification by Sørensen (1987) (see also Section 13.2.6). All nitrate respiratory processes have been found to operate to varying degrees in terrestrial, freshwater, and marine environments and represent an important part of the nitrogen cycle favored by anaerobic conditions.

Nitrate reduction is described by the half-reaction

$$NO_3^- + 2H^+ + 2e \rightarrow NO_2^- + H_2O \qquad (13.10)$$

The electron donor may be any one of a variety of organic metabolites or reduced forms of sulfur such as H_2S or S^0. The enzyme catalyzing reaction 13.10 is called *nitrate reductase* and is an iron molybdoprotein. It is not only capable of catalyzing nitrate reduction but may also catalyze reduction of ferric to ferrous iron (see Chapter 16) and reduction of chlorate to chlorite. Nitrate can competitively affect ferric iron reduction by nitrate reductase (Ottow, 1969).

The reduction of nitrite to dinitrogen is illustrated by the following series of half-reactions, with organic metabolites or reduced sulfur acting as electron donor:

$$NO_2^- + 2H^+ + e \rightarrow NO + H_2O \qquad (13.11)$$

$$2NO + 2H^+ + 2e \rightarrow N_2O + H_2O \qquad (13.12)$$

$$N_2O + 2H^+ + 2e \rightarrow N_2 + H_2O \qquad (13.13)$$

The reduction of nitrite to ammonia may be summarized by the equation

$$NO_2^- + 7H^+ + 6e \rightarrow NH_3 + 2H_2O \qquad (13.14)$$

The electron donor may be one of a variety of organic metabolites.

Although it was previously believed that these reactions can only occur at low oxygen tension or in the absence of oxygen, evidence now indicates that in some cases organisms such as *Thiosphaera pantotropha* can perform the reactions at near-normal oxygen tension (Robertson and Kuenen, 1984a,b; but see also disagreement by Thomsen et al., 1993). This organism can actually use oxygen and nitrate simultaneously as terminal electron acceptors. The explanation is that in *Tsa. pantotropha* the enzymes of denitrification are produced aerobically as well as anaerobically, whereas in oxygen-sensitive denitrifiers they are produced only at low oxygen tension or anaerobically. Nitrate reductase in *Tsa. pantotropha* is constitutive, whereas in many anaerobic denitrifiers it is inducible. Moreover, nitrate reductase in *Tsa. pantotropha* is not inactivated by oxygen, unlike in some anaerobic denitrifiers. Finally, oxygen does not repress the formation of the denitrifying enzymes in *Tsa. pantotropha*, whereas it does in some anaerobic denitrifiers. Aerobic denitrification in *Tsa. pantotropha* appears to be linked to heterotrophic nitrification (Robertson et al., 1988; Robertson and Kuenen, 1990). The organism seems to use denitrification as a means of disposing of excess reducing power because its cytochrome system is insufficient for this purpose. Oxygen tolerance in denitrification has also been observed with some other bacteria (Hochstein et al., 1984; Davies et al., 1989; Bonin et al., 1989).

For a more complete discussion of denitrification, the reader is referred to a monograph by Payne (1981) and a review article by the same author (Payne, 1983).

13.2.8 NITROGEN FIXATION

If nature had not provided for microbial *nitrogen fixation* to reverse the effect of microbial depletion of fixed nitrogen from soil or water as volatile nitrogen oxides or dinitrogen as a result of denitrification and anammox, life on Earth would not have long continued after the processes of denitrification and anammox first evolved. Nitrogen fixation is dependent on a special enzyme, *nitrogenase*, which is found only in prokaryotic organisms, including aerobic and anaerobic photosynthetic and non-photosynthetic Bacteria and Archaea. Nitrogenase is an oxygen-sensitive enzyme, usually a combination of an iron protein and a molybdoprotein (Eady and Postgate, 1974; Orme-Johnson, 1992), but in some cases (e.g., *Azotobacter chroococcum*) may also be a combination of an iron protein and a vanadoprotein (Robson et al., 1986; Eady et al., 1987), and in another case (*A. vinelandii*) may be a combination of two iron proteins (Chiswell et al., 1988). Nitrogenase catalyzes the reaction

$$N_2 + 6H^+ + 6e \rightarrow 2NH_3 \tag{13.15}$$

The enzyme is not specific for dinitrogen. It can also catalyze the reduction of acetylene (CHCH), as well as of hydrogen cyanide (HCN), cyanogen (NCCN), hydrogen azide (HN_3), hydrogen thiocyanate (HCNS), protons (H^+), carbon monoxide (CO), and some other compounds (Smith, 1983).

The reducing power (the term "6e" in Equation 13.15) needed for dinitrogen reduction is provided by reduced ferredoxin. Reduced ferredoxin can be formed in a reaction in which pyruvate is oxidatively decarboxylated (Lehninger, 1975),

$$CH_3COCOOH + NAD^+ + CoASH \rightarrow CH_3CO \sim SCoA + CO_2 + NADH + H^+ \tag{13.16}$$

$$NADH + (ferredoxin)_{ox} \rightarrow NAD^+ + (ferredoxin)_{red} + H^+ \tag{13.17}$$

In phototrophs, the reduced ferrodoxin is produced as part of the photophosphorylation mechanism (see Chapter 6).

Nitrogen fixation is a very energy-intensive reaction, consuming as many as 16 moles of ATP in the reduction of 1 mole of dinitrogen to ammonia (Newton and Burgess, 1983).

Nitrogen fixation may proceed symbiotically or nonsymbiotically. *Symbiotic nitrogen fixation* requires that the nitrogen-fixing bacterium associates with a specific host plant (e.g., a legume), one of several nonleguminous angiosperms, the water fern *Azolla*, fungi (certain lichens), or, in rare cases, with an animal host to carry out nitrogen fixation. Even then, dinitrogen will be fixed only if the fixed-nitrogen level in the surrounding environment of the host plant is low or the diet of the animal host is nitrogen-deficient. In some plants (legumes or alder), the nitrogen fixer may be localized in the cells of the cortical root tissue that are transformed into nodules. Invasion of the plant tissue may have occurred via root hairs. In some other plants, the nitrogen fixers may be localized in special leaf structures (e.g., in *Azolla*). In animals, the nitrogen fixer may be found to be a member of the flora of the digestive tract (Knowles, 1978). The plant host in symbiotic nitrogen fixation provides the energy source required by the nitrogen fixer for generating ATP. The energy source may take the form of compounds such as succinate, malate, and fumarate (Paul and Clark, 1996). The plant host also provides an environment in which access to oxygen is controlled so that nitrogenase in the nitrogen fixer is not inactivated. In root nodules of legumes, leghemoglobin is involved in the control of oxygen. The nitrogen fixer shares the ammonia that it forms from dinitrogen with its plant or animal host.

Symbiotic nitrogen-fixing bacteria include members of the genera *Rhizobium*, *Bradyrhizobium*, *Frankia*, and *Anabaena*. Some strains of *B. japonicum* have been found to be able to grow autotrophically on hydrogen as energy source because they possess uptake hydrogenase. They can couple hydrogen oxidation to ATP synthesis, which they can use in CO_2 assimilation via the ribulose bisphosphate carboxylase/oxidase system. In nitrogen fixation, the ability to couple hydrogen oxidation to ATP synthesis may represent an energy recovery system because nitrogenase generates a significant amount of hydrogen during nitrogen fixation, the energy content of which would otherwise be lost to the system.

About 30 years ago, a special symbiotic nitrogen-fixing relationship was discovered in Brazil between certain cereal grasses, such as maize, and nitrogen-fixing spirilla such as *Azospirillum lipoferum* (Von Bülow and Döbereiner, 1975; Day et al., 1975; Smith et al., 1976). In these symbioses, the nitrogen-fixing bacterium does not invade the host plant roots or any other part of the plant but lives in close association with the roots in the *rhizosphere*. Apparently, the plants excrete compounds via their roots that the nitrogen fixer can use as energy sources and that enable it to fix nitrogen that it can share with the plant if the soil is otherwise deficient in fixed nitrogen.

In *nonsymbiotic nitrogen fixation*, the active organisms are free-living in soil or water and fix nitrogen if fixed nitrogen is limiting. Their nitrogenase is not distinctly different from that of symbiotic nitrogen fixers. Unlike the symbiotic nitrogen fixers, aerobic nonsymbiotic nitrogen fixers appear to be able to maintain an intracellular environment in which nitrogenase is not inactivated by oxidizing conditions. The capacity for nonsymbiotic nitrogen fixation is widespread among prokaryotes. The best-known and the most efficient examples include the aerobes *Azotobacter* and *Beijerinckia* and the anaerobe *Clostridium pasteurianum*, but many other aerobic and anaerobic genera include species with nitrogen-fixing capacity, even some photo- and chemolithotrophs. Most of the nitrogen fixers are active only at environmental pH values between 5 and 9, but some strains of the acidophile *Acidithiobacillus ferrooxidans* have been shown to fix nitrogen at a pH as low as 2.5.

For a more detailed discussion of nitrogen fixation, the reader is referred to Alexander (1984), Broughton (1983), Newton and Orme-Johnson (1980), and Quispel (1974).

13.3 MICROBIAL ROLE IN THE NITROGEN CYCLE

Owing to their special capacities of transforming inorganic compounds, which plants and animals lack, microbes, especially prokaryotes and certain fungi, play a central role in the nitrogen cycle (Figure 13.1). Many reactions of the cycle are entirely dependent on them; nitrogen fixation is restricted to prokaryotes. The direction of transformations in the cycle is determined by environmental conditions, especially the availability of oxygen, but also by the supply of particular nitrogen compounds. Anaerobic conditions may encourage denitrification and anammox and thus cause nitrogen limitation unless the process is counteracted by anaerobic nitrogen fixation. Limitations in the supply of organic or ammonia nitrogen affect the availability of nitrate. Availability of fixed nitrogen has been viewed as a growth-limiting factor in the marine environment but infrequently in unpolluted freshwater, in which phosphate is more likely to limit productivity. Fixed nitrogen can be a limiting factor in soil, especially in agriculturally exploited soils.

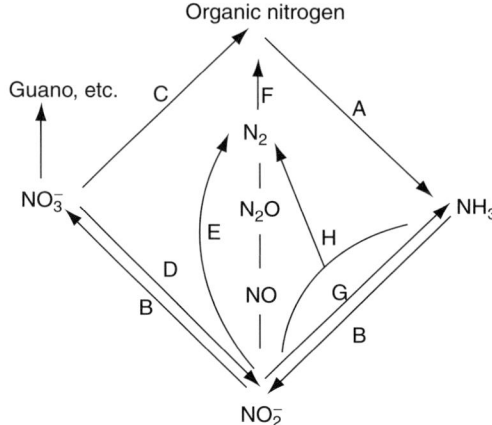

FIGURE 13.1 The nitrogen cycle. (A) Ammonification, aerobic and anaerobic; (B) autotrophic nitrification, strictly aerobic; (C) nitrate assimilation, aerobic and anaerobic; (D) nitrate reduction, usually anaerobic, but see (D + E); (D + E) denitrification, usually anaerobic but sometimes aerobic, see text; (F) nitrogen fixation, aerobic and anaerobic; (D + G) nitrate ammonification, anaerobic; and (H) anammox, anaerobic.

13.4 SUMMARY

Nitrogen is essential to all forms of life. It is assimilated by cells in the form of ammonia or nitrate, and in some instances as amino-nitrogen, mostly in the form of one or more amino acids. It is released from organic combination in the form of ammonia. The latter process is called ammonification. It occurs both aerobically and anaerobically. Ammonia is an energy-rich compound and can be oxidized to nitrate by way of nitrite by some aerobic, autotrophic bacteria (nitrifiers). It can also be converted to nitrate by some heterotrophic bacteria and some fungi in a nonenergy yielding process, but this is much less common. The conversion of ammonia to nitrate is important in soil and sediments because negatively charged clay particles can adsorb ammonia, making it unavailable to plants. Under reducing conditions, nitrate can be transformed by anaerobic respiration to nitrite, nitric and nitrous oxide, and dinitrogen, or ammonia by appropriate bacteria. In soil, the reduction of nitrate to dinitrogen can have the effect of lowering its fertility, as can the anaerobic oxidation of ammonia to dinitrogen by nitrite (anammox). Depletion of soil nitrogen through dinitrogen evolution can, however, be reversed by symbiotic and nonsymbiotic nitrogen-fixation by bacteria, which are able to reduce dinitrogen to ammonia. These various biological interactions are part of a cycle that is essential to the sustenance of life on Earth.

REFERENCES

Aleem MIH, Sewell DL. 1984. Oxidoreductase system in *Nitrobacter agilis*. In: Strohl WR, Tuovinen OH, eds. *Microbial Chemoautotrophy*. Columbus, OH: Ohio State University Press, pp. 185–210.

Alexander M. 1977. *Introduction to Soil Microbiology*. 2nd ed. New York: Wiley.

Alexander M, ed. 1984. *Biological Nitrogen Fixation: Ecology, Technology, and Physiology*. New York: Plenum Press.

Balows A, ed. 1992. *The Prokaryotes: A Handbook on the Biology of Bacteria: Ecophysiology, Isolation, Identification, Applications*. 2nd ed. New York: Springer.

Binnerup SJ, Jensen K, Revsbech NP, Jensen MH, Sørensen J. 1992. Denitrification, dissimilatory reduction of nitrate to ammonium, and nitrification in a bioturbated estuarine sediment as measured with ^{15}N and microsensor techniques. *Appl Environ Microbiol* 58:303–313.

Bock E, Koops H-P, Harms H, Ahlers B. 1991. The biochemistry of nitrifying organisms. In: Shively JM, Barton LL, eds. *Variations in Autotrophic Life*. London: Academic Press, pp. 171–200.

Bonin P, Gilewicz M, Bertrand JC. 1989. Effects of oxygen on each step of denitrification on *Pseudomonas nautica*. *Can J Microbiol* 35:1061–1064.

Brock TD, Madigan MT. 1988. *Biology of Microorganisms*. 5th ed. Englewood Cliffs, NJ: Prentice-Hall.

Broughton WJ, ed. 1983. *Nitrogen fixation, Vol I, Ecology; Vol 2, Rhizobium; Vol 3, Legumes*. Oxford, U.K.: Clarendon Press.

Chiswell JR, Premarkumar R, Bishop PE. 1988. Purification of a second alternative nitrogenase from a *nif* HDK deletion strain of *Azotobacter vinelandii*. *J Bacteriol* 170:27–33.

Cruz-García C, Murray AE, Klappenbach JA, Stewart V, Tiedje JM. 2007. Respiratory nitrate ammonification by *Shewanella oneidensis* MR-1. *J Bacteriol* 189:656–662.

Dalsgaard T, Thamdrup B, Canfield DE. 2005. Anaerobic ammonium oxidation (anammox) in the marine environment. *Res Microbiol* 156:457–464.

Dannenberg S, Kroder M, Dilling W, Cypionka H. 1992. Oxidation of H_2, organic compounds and inorganic sulfur compounds coupled to reduction of O_2 or nitrate by sulfate-reducing bacteria. *Arch Microbiol* 158:93–99.

Davies KJP, Lloyd D, Boddy L. 1989. The effect of oxygen on denitrification in *Paracoccus denitrificans* and *Pseudomonas aeruginosa*. *J Gen Microbiol* 135:2445–2451.

Davidson EA. 1993. Soil water content and the ratio of nitrous to nitric oxide emitted from soil. In: Oremland RS, ed. *Biogeochemistry of Global Change. Radiatively Active Trace Gases*. New York: Chapman and Hall, pp. 369–386.

Day JM, Neves MCP, Döbereiner J. 1975. Nitrogenase activity on the roots of tropical forage grasses. *Soil Biol Biochem* 7:107–112.

Eady RR, Postgate JR. 1974. Nitrogenase. *Nature* (London) 249:805–810.

Eady RR, Robson RL, Richardson TH, Miller RW, Hawkins M. 1987. The vanadium nitrogenase of *Azotobacter chroococcum*. Purification and properties of the vanadium iron protein. *Biochem J* 244:197–207.

Ehrlich HL. 1993. Bacterial mineralization of organic matter under anaerobic conditions. In: Bollag J-M, Stotzky G, eds. *Soil Biochemistry*, Vol. 8. New York: Marcel Dekker, pp. 219–238.

Ericksen GE. 1983. The Chilean nitrate deposits. *Am Sci* 71:366–374.

Fenchel T, Blackburn TH. 1979. *Bacteria and Mineral Cycling*. London: Academic Press.

Ghiorse WC, Alexander M. 1978. Nitrifying populations and the destruction of nitrogen dioxide in soil. *Microb Ecol* 4:233–240.

Hochstein LI, Betlach M, Kritikos K. 1984. The effect of oxygen on denitrification during steady-state growth of *Paracoccus halodenitrificans*. *Arch Microbiol* 137:74–78.

Hooper AB. 1984. Ammonia oxidation and energy transduction in the nitrifying bacteria. In: Strohl WR, Tuovinen OH, eds. *Microbial Chemoautotrophy*. Columbus, OH: Ohio State University Press, pp. 133–167.

Jensen MM, Thamdrup B, Dalsgaard T. 2007. Effects of specific inhibitors on anammox and denitrification in marine sediments. *Appl Environ Microbiol* 73:3151–3158.

Jetten MSM, Slickers O, Kuypers M, Dalsgaard T, van Niftrik L, Cirpus I, van de Pas-Schoonen K, Lvik G, Thamdrup B, Le Paslier D, Op den Camp HJM, Hulth SK, Nielsen LP, Abma W, Third K, Engström P, Kuenen JG, Jørgensen BB, Canfield DE, Sinninghe Damsté JS, Revsbech NP, Fuerst J, Weissenbach J, Wagner M, Schmidt I, Schmid M, Strous M. 2003. Anaerobic ammonium oxidation by marine and freshwater planctomycete-like bacteria. *Appl Microbiol Biotechnol* 63:107–114.

Jones RD, Morita RY. 1983. Methane oxidation by *Nitrosococcus oceanus* and *Nitrosomonas europaea*. *Appl Environ Microbiol* 45:401–410.

Jørgensen BB, Sørensen J. 1985. Seasonal cycles of O_2, NO_3^- and SO_4^{2-} reduction in estuarine sediments. The significance of a NO_3 maximum in the spring. *Mar Ecol Progr Ser* 24:65–74.

Jørgensen BB, Sørensen J. 1988. Two annual maxima of nitrate reduction and denitrification in estuarine sediment (Norsminde Fjord, Denmark). *Mar Ecol Progr Ser* 48:147–154.

Knowles R. 1978. Free-living bacteria. In: Döbereiner R, Burris H, Hollaender A, Franco AA, Neyra CA, Scott DB, eds. *Limitations and Potentials for Biological Nitrogen Fixation in the Tropics*. New York: Plenum Press, pp. 25–40.

Knowles R. 1985. Microbial transformations as sources and sinks of nitrogen oxides. In: Caldwell DE, Brierley JA, Brierley CL, eds. *Planetary Ecology*. New York: Van Nostrand Reinhold, pp. 411–426.

Kuenen JG, Jetten MSM. 2001. Extraordinary anaerobic ammonium oxidizing bacteria. *Am Soc Microbiol News* 67:456–463.

Lehninger AL. 1975. *Biochemistry*. 2nd ed. New York: Worth.

Lewis RW. 1965. Nitrogen. In: *Mineral Facts and Problems, Bull 630*. Washington, DC: US Bur Miners, pp. 621–629.

Newton WE, Burgess BK. 1983. Nitrogen fixation: Its scope and importance. In: Mueller A, Newtan WE, eds. *Nitrogen Fixation. The Chemical–Biochemical–Genetic Interface*. New York: Plenum Press, pp. 1–19.

Newton WE, Orme-Johnson WH, eds. 1980. *Nitrogen Fixation*. Baltimore, MD: University Park Press.

Orme-Johnson WH. 1992. Nitrogenase structure: Where to now? *Science* 257:1639–1640.

Ottow JCG. 1969. Der Einfluss von Nitrat, Chlorat, Sulfat, Eisenoxyd-form und Wachstumsbedingungen auf das Ausmass der Bakteriellen Eisenreduktion. *Z Pflanzenernähr Düngung Bodenk* 124:238–253.

Paul EA, Clark PF. 1996. *Soil Microbiology and Biochemistry*. San Diego, CA: Academic Press.

Payne WJ. 1981. *Denitrification*. New York: Wiley.

Payne WJ. 1983. Bacterial denitrification: Asset or defect. *BioScience* 33:319–325.

Quispel A, ed. 1974. *The Biology of Nitrogen Fixation*. Amsterdam: North-Holland.

Robertson LA, Kuenen JG. 1984a. Aerobic denitrification: A controversy revived. *Arch Microbiol* 139:351–354.

Robertson LA, Kuenen JG. 1984b. Aerobic denitrification: Old wine in new bottles? Antonie van Leeuwenhoek 50:525–544.

Robertson LA, Kuenen JG. 1990. Combined heterotrophic nitrification and aerobic denitrification in *Thiosphaera pantotropha* and other bacteria. *Antonie van Leeuwenhoek* 57:139–152.

Robertson LA, van Niel EWJ, Torremans RAM, Kuenen JG. 1988. Simultaneous nitrification and denitrification in aerobic chemostat cultures of *Thiosphaera pantotropha*. *Appl Environ Microbiol* 54:2812–2818.

Robson RL, Eadky RR, Richardson TH, Miller RW, Hawkins M, Postgate JR. 1986. The alternative nitrogenase of *Azotobacter chroococcum* is a vanadium enzyme. *Nature* (London) 322:388–390.

Smith BE. 1983. Reactions and physicochemical properties of the nitrogenase MoFe proteins. In: Mueller A, Newton WE, eds. *Nitrogen Fixation. The Chemical-Biochemical-Genetic Interface*. New York: Plenum Press, pp. 23–62.

Smith RL, Bouton JH, Schank SC, Queensberry KH, Tyler ME, Milam JR, Gaskin MH, Littell RC. 1976. Nitrogen fixation in grasses inoculated with *Spirillum lipoferum*. *Science* 193:1003–1005.

Sørensen J. 1987. Nitrate reduction in marine sediment: Pathway and interactions with iron and sulfur recycling. *Geomicrobiol J* 5:401–421.

Stevenson FJ. 1972. Nitrogen: Element and geochemistry. In: Fairbridge RW, ed. *The Encyclopedia of Geochemistry and Environmental Sciences. Encyclopedia of Earth Science Series, Vol. IVA*. New York: Van Nostrand Reinhold, pp. 795–801.

Strous, M, Fuerst JA, Kramer EHM, Logemann S, Muyzer G, Van De Pas-Schoonen KT, Webb R, Kuenen JG, Jetten MSM. 1999. Missing lithotroph identified as new planctomycete. *Nature* (London) 400:446–449.

Strous M, Jetten MSM. 2004. Anaerobic oxidation of methane and ammonium. *Annu Rev Microbiol* 58:99–117.

Tal Y, Watts JEM, Schreier HJ. 2005. Anaerobic ammonia-oxidizing bacteria and related activity in Baltimore Inner Harbor sediment. *Appl Environ Microbiol* 71:1816–1821.

Thomsen JK, Iversen JJL, Cox RP. 1993. Interactions between respiration and denitrification during growth of *Thiosphaera pantotropha* in continuous culture. *FEMS Microbiol Lett* 110:319–324.

Trimmer M, Nicholls JC, Deflandre B. 2003. Anaerobic ammonium oxidation measured in sediments along the Thames Estuary, United Kingdom. *Appl Environ Microbiol* 69:6447–6454.

van de Graaf AA, Mulder A, de Bruijn P, Jetten MSM, Robertson LA, Kuenen JG. 1995. Anaerobic oxidation of ammonium is a biologically mediated process. *Appl Environ Microbiol* 61:1246–1251.

van Niftrik LA, Fuerst JA, Sinninghe Damsté JS, Kuenen JG, Jetten MSM, Strous M. 2004. The anammoxosome: An intracytoplasmic compartment in anammox bacteria. *FEMS Microbiol Lett* 233:7–13.

Von Bülow JFW, Döbereiner J. 1975. Potential for nitrogen fixation in maize genotypes in Brazil. *Proc Natl Acad Sci USA* 72:2389–2393.

Wood P. 1988. Chemolithotrophy. In: Anthony C, ed. *Bacterial Energy Transduction*. London: Academic Press, pp. 183–230.

14 Geomicrobial Interactions with Arsenic and Antimony

14.1 INTRODUCTION

Although arsenic and antimony are generally toxic to life, some microorganisms exist that can metabolize some forms of these elements. Some can use arsenite or stibnite as partial or sole energy sources whereas others can use arsenate as terminal electron acceptors. Still other microbes can metabolize arsenic and antimony compounds to detoxify them. These reactions are important from a geomicrobial standpoint because they indicate that some microbes contribute to arsenic and antimony mobilization or immobilization in the environment and play a role in arsenic and antimony cycles.

14.2 ARSENIC

14.2.1 DISTRIBUTION

Arsenic is widely distributed in the upper crust of the Earth, mostly at very low concentrations. Its average abundance in igneous rocks has been estimated to be of the order of 5 g tn^{-1} (Carapella, 1972). It rarely occurs in elemental form. More often it is found to be combined with sulfur, as in orpiment (As_2S_3), realgar (As_2S_2 or AsS), or arsenopyrite ($FeAsS$); with selenium, as in As_2Se_3; with tellurium, as in As_2Te; or as a sulfosalt, as in enargite (Cu_3AsS_4). It is also found in arsenides of heavy metals such as iron (loellingite, $FeAs_2$), copper (domeykite, Cu_3As), nickel (nicolite, $NiAs$), and cobalt (Co_2As). Sometimes it occurs in the form of arsenite minerals (arsenolite or claudetite, As_2O_3) or arsenate minerals (erythrite, $Co_3(AsO_4)_2 \cdot 8H_2O$; scorodite, $FeAsO_4 \cdot 2H_2O$; olivenite, $Cu_2(AsO_4)(OH)$). Arsenopyrite is the most common and widespread mineral form of arsenic, but orpiment and realgar are also fairly common. The ultimate source of arsenic on the Earth's surface is igneous activity. On weathering of arsenic-containing rocks, which may harbor as much as 1.8 ppm of the element, the arsenic is dispersed through the upper lithosphere and the hydrosphere.

Arsenic concentration in soil may range from 0.1 to more than 1000 ppm. The average concentration in seawater has been reported to be 3.7 μg L^{-1} and in freshwater, 1.5–53 ng m^{-3} (Bowen, 1979). However, in groundwater in the Munshiganj District of Bangladesh, the As concentration approaches a maximum of 640 mg m^{-3} at a depth of 30–40 m but decreases to 58 mg m^{-3} at 107 m (Swartz et al., 2004). Some living organisms may concentrate arsenic many fold over its level in the environment. For example, some algae have been found to accumulate arsenic 200–3000 times in excess of its concentration in the growth medium (Lunde, 1973). Humans may artificially raise the arsenic concentrations in soil and water through the introduction of sodium arsenite ($NaAsO_3$) or cacodylic acid (($CH_3)_2AsO \cdot OH$) as herbicides.

14.2.2 SOME CHEMICAL CHARACTERISTICS

In nature, arsenic is usually encountered in the oxidation states of 0, +3, and +5. Its coordination numbers are in the range of 3–6 (Cullen and Reimer, 1989). Except for the oxidation states of As in arsenate and arsenite, the oxidation state of other compounds, whether organic or inorganic, is often unclear and depends on its definition (Cullen and Reimer, 1989, p. 715). According to Cullen and Reimer (1989, p. 715), arsenious acid and its salts in aqueous solution exist in the ortho form (H_3AsO_3) but not in the meta ($HAsO_2$) form. Environmentally, arsenite is more mobile than arsenate, but it can be significantly adsorbed under certain conditions. This is because of the tendency of arsenate to become strongly adsorbed to mineral surfaces such as those of ferrihydrite. However, an ability of ferrihydrite and schwertmannite to adsorb both As(V) and As(III) has been observed by Raven et al. (1998) and Carlson et al. (2002). The ability of disordered mackinawite (FeS), which can be formed in sediments where bacterial sulfate reduction occurs, to adsorb As(V) and As(III) has been reported by Wolthers et al. (2005). As(III) adsorption by amorphous iron oxide and goethite has also been studied by Dixit and Hering (2003, 2006), who noted preferential uptake of As(V) in a pH range of 5–6 and As(III) in a pH range of 7–8. They also noted that with goethite, current single-sorbate models were able to predict the adsorption of Fe(II) and As(III) more satisfactorily than double-sorbate models. Overall, because of the lesser mobility of arsenate compared with arsenite under common environmental conditions, reduction of arsenate to arsenite in the environment, whether chemical or biological, leads to an increase in arsenic toxicity (Cullen and Reimer, 1989).

14.2.3 TOXICITY

Arsenic compounds are toxic for most living organisms. Arsenite (AsO_3^{3-}) has been shown to inhibit dehydrogenases such as pyruvate, α-ketoglutarate-, and dihydrolipoate-dehydrogenases (Mahler and Cordes, 1966). Arsenate (AsO_4^{3-}) uncouples oxidative phosphorylation, that is, it inhibits chemiosmotic ATP synthesis (Da Costa, 1972).

Both the uptake and the inhibitory effect of arsenate on metabolism can be modified by phosphate (Button et al., 1973; Da Costa, 1971, 1972). This is because of the existence of a common transport mechanism for phosphate and arsenate in the membranes of some organisms. However, a separate transport mechanism for phosphate may also exist (Bennett and Malamy, 1970). In the latter case, phosphate uptake does not affect arsenate uptake, nor does arsenate uptake affect phosphate uptake. In one reported case of a fungus, *Cladosporium herbarium*, arsenite toxicity could also be ameliorated by the presence of phosphate. In that instance, prior oxidation of the arsenite to arsenate by the fungus appeared to be the basis for the effect (Da Costa, 1972). In growing cultures of *Candida humicola*, phosphate can inhibit the formation of trimethylarsine from arsenate, arsenite, and monomethylarsonate, but not from dimethylarsinate (Cox and Alexander, 1973). In similar cultures, phosphite can suppress the formation of trimethylarsine from monomethylarsonate, but not from arsenate or dimethylarsinate. Hypophosphite can cause temporary inhibition of the conversion of arsenate, monomethylarsonate, and dimethylarsinate (Cox and Alexander, 1973). High antimonite concentrations lower the rate of conversion of arsenate to trimethylarsine by resting cells of *C. humicola* (Cox and Alexander, 1973).

Bacteria can develop genetically determined resistance to arsenic (Ji and Silver, 1992; Ji et al., 1993). The gene locus for this resistance may reside on a plasmid as, for example, in *Staphylococcus aureus* (Dyke et al., 1970) and *Escherichia coli* (Hedges and Baumberg, 1973). The mechanism of resistance in these bacterial species is a special pumping mechanism that expels the arsenic taken up as arsenate by the cells (Silver and Keach, 1982). It involves intracellular reduction of arsenate to arsenite followed by efflux of the arsenite promoted by an oxyanion-translocating ATPase (Broeer et al., 1993; Ji et al., 1993). Some of the resistant organisms have the capacity to oxidize reduced forms of arsenic to arsenate and others to reduce oxidized forms (see Sections 14.2.4 and 14.2.6). In *Shewanella* sp. strain ANA-3, the traits of arsenate resistance and arsenate respiration are encoded in two distinct genetic loci, operons *ars* and *arrs*, respectively (Saltikov and Newman, 2003; Saltikov et al., 2003; see also Section 14.2.7).

14.2.4 MICROBIAL OXIDATION OF REDUCED FORMS OF ARSENIC

14.2.4.1 Aerobic Oxidation of Dissolved Arsenic

Bacterial oxidation of arsenite to arsenate was first reported by Green (1918). He discovered bacteria with this ability in arsenical cattle-dipping solution used for protection against insect bites. He isolated an organism with this trait, which was named *Bacillus arsenoxydans* by him. Quastel and Scholefield (1953) observed arsenite oxidation in perfusion experiments in which they passed 2.5×10^{-3} M sodium arsenite solution through columns charged with Cardiff soil. They did not isolate the organism or organisms responsible for the oxidation but showed that a 0.1% solution of NaN_3 inhibited oxidation. The onset of arsenite oxidation in their experiments occurred after a lag. The length of this lag was increased when sulfanilamide was added with the arsenite. A control of pH was found to be important for sustained bacterial activity. They observed an almost stoichiometric O_2 consumption during arsenite oxidation.

Further investigation of arsenical cattle-dipping solutions led to the isolation of 15 aerobic arsenite-oxidizing strains of bacteria (Turner, 1949, 1954). These organsims were assigned to the genera *Pseudomonas*, *Xanthomonas*, and *Achromobacter*. *Achromobacter arsenoxydans-tres* was later considered synonymous with *Alcaligenes faecalis* (Hendrie et al., 1974). This organism was described in the eighth edition of *Bergey's Manual of Determinative Bacteriology* (Buchanan and Gibbons, 1974) as frequently having the capacity of arsenite oxidation.

Of Turner's 15 isolates, *Pseudomonas arsenoxydans-quinque* was studied in detail with respect to arsenite oxidation. Resting cells of this culture oxidized arsenite at an optimum pH of 6.4 and an optimum temperature of 40°C (Turner and Legge, 1954). Cyanide, azide, fluoride, and pyrophosphate inhibited the activity. Under anaerobic conditions, 2,6-dichlorophenol indophenol, *m*-carboxyphenolindo-2,6-dibromophenol, and ferricyanide could act as electron acceptors. Pretreating cells with toluene and acetone or desiccating them rendered them incapable of oxidizing arsenite in air. The arsenite-oxidizing enzyme was described as *adaptable*. Examination of cell-free extracts of *P. arsenoxydans-quinque* suggested the presence of soluble *dehydrogenase* activity, which under anaerobic conditions conveyed electrons from arsenite to 2,6-dichlorophenol indophenol (Legge and Turner, 1954). This activity was partly inhibited by 10^{-3} M *p*-chloromercurybenzoate. The entire arsenite-oxidizing system was believed to consist of dehydrogenase and an oxidase (Legge, 1954).

A soil strain of *Alcaligenes faecalis* was isolated by Osborne (1973) whose arsenite-oxidizing ability was found to be inducible by arsenite and arsenate (Osborne and Ehrlich, 1976). Very recent phylogenetic study of this strain by 16S rDNA analysis showed that it should be reassigned to the genus *Achromobacter* (Santini, 2001, personal communication). It was shown to oxidize arsenite stoichiometrically to arsenate (Table 14.1):*

$$AsO_2^- + H_2O + 0.5O_2 \rightarrow AsO_4^{3-} + 2H^+ \tag{14.1}$$

Inhibitor and spectrophotometric studies suggested that the oxidation involved an oxidoreductase with a bound flavin that passed electrons from arsenite to oxygen by way of a c-type cytochrome and cytochrome oxidase (Osborne, 1973; Osborne and Ehrlich, 1976).

Anderson et al. (1992) isolated an inducible arsenite-oxidizing enzyme that was located on the outer surface of the plasma membrane of *Alcaligenes faecalis* strain NCIB 8687. The enzyme location suggests that in intact cells of this organism, arsenite oxidation occurs in the periplasmic space. Biochemical characterization showed that this enzyme is a molybdenum-containing hydroxylase consisting of a monomeric 85 kDa peptide with a pterin cofactor and one atom of molybdenum, five

---

* The formula for *meta*-arsenite is used in Equation (14.1) although *ortho*-arsenite is the form of As(III) in aqueous solution according to Cullen and Reimer (1989). This equation as shown conveys that, if unbuffered, the reaction mixture turns acid as the oxidation progresses. The increase in acidity results from the fact that arsenious acid is a much weaker acid ($K = 10^{-9.2}$) than arsenic acid ($K_1 = 10^{-2.25}$; $K_2 = 10^{-6.77}$; $K_3 = 10^{-11.4}$) (Weast and Astle, 1982).

TABLE 14.1

Stoichiometry of Oxygen Uptake by *Alcaligenes faecalis* on Arsenite Based on the Reaction of $AsO_2^- + H_2O + 1/2O_2 \rightarrow AsO_4^{3-} + 2H^+$

| NaAsO$_2$ Added (μmol) | Oxygen Uptake | | |
	Theoretical	Experimental	Percent of Theoretical
19.25	9.63	8.79	91.3
38.50	19.25	18.48	96.0
57.75	28.88	27.05	93.7
77.00	38.50	37.05	96.2

Source: Courtesy of Osborne FH, Arsenite oxidation by a soil isolate of *Alcaligenes*, PhD Thesis, Rensselaer Polytechnic Institute, Troy, NY, 1973. With permission.

or six atoms of iron, and inorganic sulfide. Both azurin and cytochrome c from *A. faecalis* served as electron acceptors in arsenite oxidation catalyzed by this enzyme. A more detailed discussion of arsenite oxidase and its phylogenetic distribution can be found in Silver and Phung (2005).

A strain of *Alcaligenes faecalis* similar to the one isolated by Osborne (1973) (see also Osborne and Ehrlich, 1976) was independently isolated and characterized by Philips and Taylor (1976). Neither their strain nor that of Osborne oxidized arsenite strongly until late exponential or stationary phase of growth was reached in batch culture (Philips and Taylor, 1976; Ehrlich, 1978). Other heterotrophic arsenite-oxidizing bacteria that have been identified more recently include *Pseudomonas putida* strain 18 and *Alcaligenes eutrophus* strain 280, both of which were isolated from gold–arsenic deposits (Abdrashitova et al., 1981), and a strain that appears to belong to the genus *Zoogloea* (Weeger et al., 1999).

The observation reported by Osborne and Ehrlich (1976) that their strain passes electrons from arsenite to oxygen via an electron transport system that involves c-type cytochrome and cytochrome oxidase suggested that their organism is able to conserve energy from this oxidation. Indeed, indirect evidence indicated that the organism may be able to derive maintenance energy from arsenite oxidation (Ehrlich, 1978). Donahoe-Christiansen et al. (2004) found that a strain of *Hydrogenobaculum*, an obligate chemolithoautotroph using H$_2$ as sole energy source, was able to oxidize arsenite but not use it as a sole energy source. They had isolated their culture from a geothermal spring in Yellowstone National Park. Arsenite-oxidizing bacteria were also found in Hot Creek, California, by Salmassi et al. (2002, 2006). Salmassi et al. (2002) concluded that arsenite oxidation by their organism, *Agrobacterium albertimagni*, from Hot Creek was a detoxification process, whereas Christiansen et al. (2004) concluded that arsenite oxidation by *Hydrogenobaculum* was not by means of detoxification because this organism was more sensitive to As(V) than As(III).

Strong evidence that arsenite can be used as a sole energy source by some arsenite-oxidizing bacteria was presented by Ilyaletdinov and Abdrashitova (1981), who isolated a culture from a gold-arsenic ore deposit and named it *Pseudomonas arsenitoxidans*. This culture was an obligate autotroph that grew on arsenite as sole energy source. It also oxidized arsenic in arsenopyrite. Santini et al. (2000) reported the isolation of two new strains of autotrophic arsenite oxidizers, N-25 and N-26, from a gold mine in the Northern Territory of Australia. Both strains belong to the *Agrobacterium–Rhizobium* branch of the α-Proteobacteria. Growth of strain N-26 was accelerated by the addition of a trace of yeast extract to an arsenite–mineral salts growth medium. In addition to growing on arsenite as energy source, strain N-26 was also able to grow heterotrophically on a wide range of organic carbon or energy sources including glucose, fructose, succinate, fumarate, pyruvate, and several others. This strain thus appears to be facultatively autotrophic, being also able to grow mixotrophically and heterotrophically. The arsenite oxidase of strain N-26 appears to be periplasmic. The oxidase was shown to have a molecular mass of 219 kDa and consist of two heterologous subunits, AroA (98 kDa) and AroB (14 kDa) in an $\alpha_2\beta_2$ configuration containing 2 Mo and 9 or 10 Fe atoms per $\alpha_2\beta_2$ unit. Gene sequence analysis revealed similarities when compared with the arsenite oxidase of *Alcaligenes faecalis* (Santini and vanden Hoven, 2004).

Recently an arsenite-oxidizing chemolithotroph, phylogenetically related to *Acidicaldus*, was isolated from a microbial mat in an acid-sulfate-chloride spring in Yellowstone National Park, United States (D'Imperio et al., 2007). The ability of this organism to oxidize arsenite was inhibited by H_2S, which affected its distribution in the spring. The inhibition appeared to be noncompetitive.

Arsenite (As(III)) may under some conditions also be subject to abiotic oxidation by Mn(IV), but apparently to a much lesser extent or not at all by Fe(III) (Oscarson et al., 1981).

14.2.4.2 Anaerobic Oxidation of Dissolved Arsenic

Oremland et al. (2002) isolated a facultative chemoautotrophic bacterium, strain MLHE-1, from arsenite-enriched bottom water from Mono Lake, California, United States, that oxidized arsenite anaerobically to arsenate using nitrate as terminal electron acceptor. In laboratory culture experiments, H_2 and sulfide were able to replace arsenite as electron donor in a nitrate–mineral salts medium. This organism was shown to possess the gene for ribulose-1,5-bisphosphate carboxylase/oxidase consistent with the observation that it was able to fix CO_2 in the dark. It was also able to grow heterotrophically with acetate as carbon and energy sources and oxygen (aerobic growth) or nitrate (anaerobic growth) as terminal electron acceptors. Phylogenetic analysis based on its 16S rDNA places this organism with the haloalkaliphilic *Ectothiorhodospira* of the γ-Proteobacteria.

14.2.5 Interaction with Arsenic-Containing Minerals

Arsenic in arsenic-containing minerals that also contain iron, copper, and sulfur may be mobilized by bacteria. This mobilization may or may not involve bacterial oxidation of the arsenic if it exists in reduced form in the mineral. The simplest of these compounds, orpiment (As_2O_3), was found to be oxidized by *Thiobacillus* (now renamed *Acidithiobacillus*) *ferrooxidans* TM in a mineral salts solution (9K medium without iron; for formulation see Silverman and Lundgren, 1959) to which

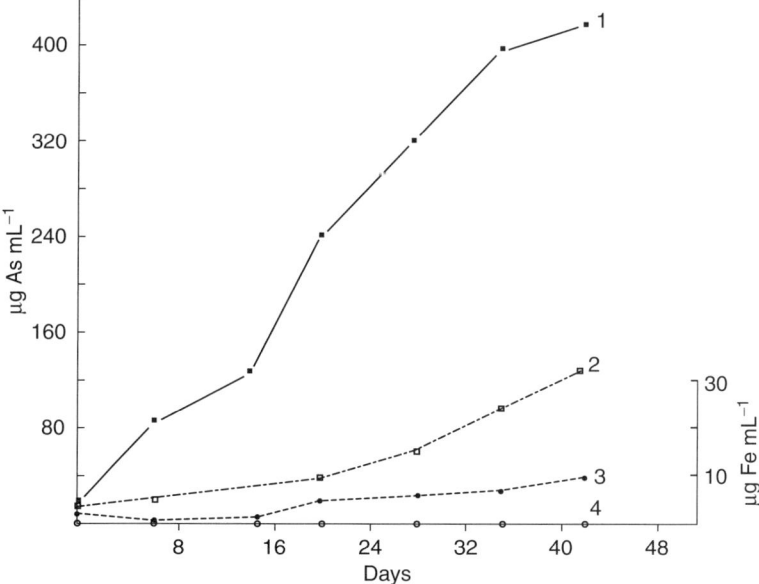

FIGURE 14.1 Bacterial solubilization of orpiment. 1, Total arsenic mobilized with bacteria (*Thiobacillus ferrooxidans*); 2, total arsenic mobilized without bacteria; 3, total iron mobilized with bacteria (*T. ferrooxidans*); 4, total iron mobilized without bacteria. (From Ehrlich HL, Society of Economic Geologists, *Economic Geology,* Fig.1, p. 993, 1963. With permission.)

the mineral had been added in pulverized form (Ehrlich, 1963) (Figure 14.1). Arsenite and arsenate accumulated in the medium over time. Because *T. ferrooxidans* does not oxidize arsenite to arsenate, the arsenate that appeared in the inoculated flasks probably resulted from the autoxidation of arsenite produced by this organism. Chemical oxidation by traces of ferric iron from the mineral may also have contributed to the formation of arsenate. The initial pH in the inoculated flasks was 3.5, which was dropped to 2 in 35 days. In contrast, in an uninoculated control in which orpiment autoxidized, but only slowly, the pH rose from 3.5 to 5, suggesting differences between the reactions in the experimental and control flasks. Realgar (As_2S_2) was not attacked by *T. ferrooxidans* TM.

Arsenopyrite (FeAsS) and enargite (Cu_3AsS_4) were also oxidized by an iron-oxidizing acidophilic *Thiobacillus* culture under the same test conditions as those used with orpiment (Ehrlich, 1964). During growth on arsenopyrite, the arsenic in the ore was transformed to arsenite and arsenate. The iron in the ore appeared ultimately as mobilized ferric iron. It precipitated extensively as ferric arsenite and arsenate. Because acidophilic *Thiobacillus ferrooxidans* did not oxidize arsenite to arsenate, the arsenate in the culture was probably formed through the oxidation by ferric iron, although some autoxidation cannot be ruled out (Wakao et al., 1988; Braddock et al., 1984; Monroy-Fernandez et al., 1995). The pH dropped from 3.5 to 2.5 in both the inoculated and uninoculated flasks in the first 21 days. However, in the uninoculated controls it rose to 4.3 by the 40th day whereas it remained at 2.5 in the inoculated flasks. Oxidation of arsenopyrite in the absence of bacteria was significantly slower

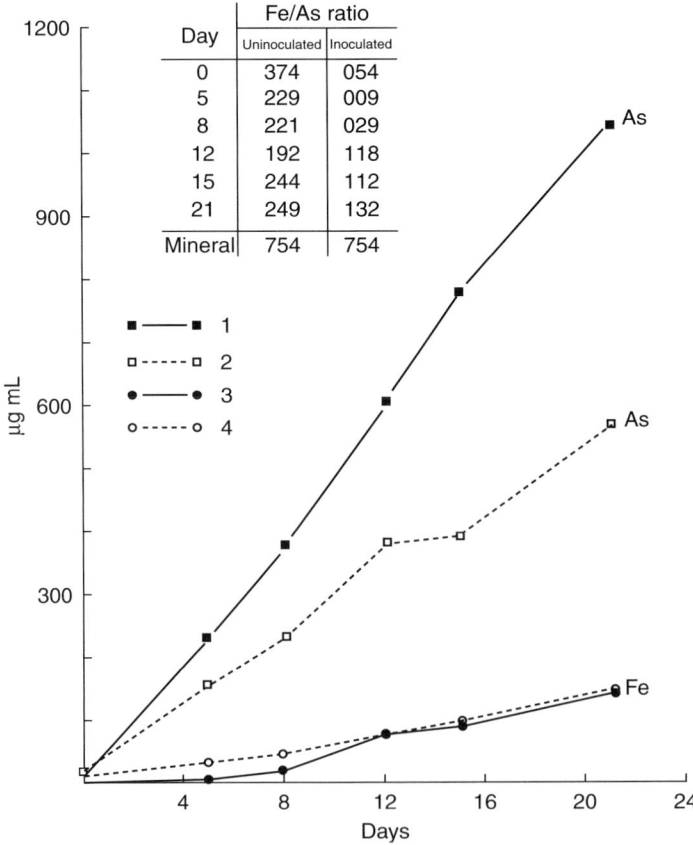

Day	Fe/As ratio	
	Uninoculated	Inoculated
0	374	054
5	229	009
8	221	029
12	192	118
15	244	112
21	249	132
Mineral	754	754

FIGURE 14.2 Oxidation of arsenopyrite by *Thiobacillus ferrooxidans*. Curves 1 and 3 represent changes in inoculated reaction mixture in flasks. Curves 2 and 4 represent changes in uninoculated reaction mixture in flasks. (From Ehrlich HL, Society of Economic Geologists, *Economic Geology*, Fig.1, p. 1308, 1964. With permission.)

than that in their presence (Figure 14.2). In a later study, Carlson et al. (1992) identified the minerals jarosite and scorodite among the solid products formed in the oxidation of arsenic-containing pyrite by a mixed culture of moderately thermophilic acidophiles. In another study, with *T. ferrooxidans* at 22°C and with a moderately thermophilic mixed culture at 45°C on arsenopyrite, Tuovinen et al. (1994) found that the mixed culture oxidized the mineral nearly completely whereas acidophilic *T. ferrooxidans* did not. Jarosite was a major sink for ferric iron in both cultures, but some amorphous ferric arsenate and S^0 were also formed. The absence of scorodite formation in this case was attributed to an insufficient drop in pH in the experiments. Groudeva et al. (1986) and Ngubane and Baecker (1990) reported that the extremely thermophilic archeon *Sulfolobus* sp. oxidizes arsenopyrite.

Although acidophilic *Thiobacillus ferrooxidans* appears not to be able to oxidize arsenite to arsenate, the thermophilic archeon *Sulfolobus acidocaldarius* strain BC is able to do so (Sehlin and Lindström, 1992).

Exposure of sterilized arsenopyrite particles for ~2 months in an underground area of the Richmond Mine at Iron Mountain, northern California (see Figure 20.5), revealed a very extensive surface attachment of rod-shaped bacteria, probably *Sulfobacillus* sp., accompanied by extensive pitting of the mineral particles and the deposition of elemental sulfur on the particle surface (Figure 14.3). The S^0 represented more than 50% of the sulfur in the arsenopyrite.

Ehrlich (1964) found that a strain of *Thiobacillus ferrooxidans*, when attacking enargite (Cu_3AsS_4) released cupric copper and arsenate into the bulk phase together with some iron that was present as an impurity in the mineral. The culture medium in which he performed the experiments contained (in g L^{-1}) $(NH_4)_2SO_4$, 3.0; KCl, 0.1; K_2HPO_4, 0.5; $MgSO_4 \cdot 7H_2O$, 0.5; $Ca(NO_3)_2$, 0.01, and was acidified to pH 3.5. To this salts solution, 0.5 g enargite ground to a mesh size of +63 μm was added. He found that in the presence of active bacteria, the pH of the reaction mixture dropped from 3.5 to between 2 and 2.5. In the absence of bacteria, the pH usually rose to 4.5. Some mobilized copper and arsenic precipitated in the experiments. The rate of enargite oxidation without bacteria was significantly slower and may have followed a different course of reaction on the basis of slower rates of Cu and As solubilization and the difference in pH changes.

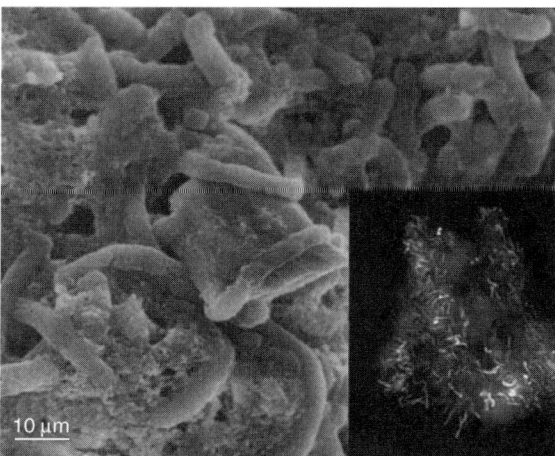

FIGURE 14.3 Arsenopyrite particle with bacteria colonizing its surface. The particle was sterilized and exposed in an underground stream in the Richmond Mine at Iron Mountain, northern California, for ~ 2 months (see Figure 20.5) prior to examination by SEM. The image reveals a highly pitted surface covered with elongate cells and some secondary minerals. Based on microbial ecological studies at the site, the cells are probably *Sulfobacillus* sp. Inset shows an epifluorescence optical microscopic image in which the cells on an arsenopyrite grain are labeled with a fluorescent dye that binds to DNA. Parallel experimental studies by McGuire et al. (2001) demonstrated that ~50% of the sulfur in the arsenopyrite ended up as elemental sulfur on the surface. (Images courtesy of Banfield JF, Department of Earth and Planetary Science, Environmental Science, Policy, Management, University of California, Berkeley, CA.)

Escobar et al. (1997) reported on the action of *Thiobacillus* (now *Acidithiobacillus*) *ferrooxidans* strain ATCC 19859 on enargite, ground to a mesh size of $-147 + 104$ μm and added to a salts solution containing (in g L^{-1}) $(NH_4)_2SO_4$, 0.4; $MgSO_4 \cdot 7H_2O$, 0.4; and KH_2PO_4, 0.056, acidified to pH 1.6. In the absence of added ferric iron, the mobilization of copper in the enargite was more rapid and sustained in the presence of the bacteria than in the sterile medium. In the presence of ferric iron (3.0 g L^{-1}), the mobilization of copper in the enargite by the bacteria was more rapid and sustained than that without added ferric iron. It was also more rapid than that in sterile medium with ferric iron. Copper mobilization without bacteria but with added ferric iron was significantly faster than that without ferric iron and the bacteria, but in both of these instances much less sustained than with the bacteria.

The ability of *Acidithiobacillus* (*Thiobacillus*) *ferrooxidans* and other acidophilic, iron-oxidizing bacteria to oxidize arsenopyrite is now being industrially exploited in beneficiation of sulfidic ores containing precious metals (Brierley and Brierley, 1999; Dew et al., 1997; Olson, 1994; Livesey-Goldblatt et al., 1983; Pol'kin et al., 1973; Pol'kin and Tanzhnyanskaya, 1968). This process involves the selective removal of arsenopyrite as well as pyrite by bacterial oxidation (bioleaching) before gold or silver is chemically (nonbiologically) extracted by cyanidation (complexing with cyanide). Arsenopyrite and pyrite interfere with cyanidation for several reasons. They encapsulate finely disseminated gold and silver, blocking access of cyanide reagent used to solubilize gold and silver. They consume oxygen that is needed to convert metallic gold to aurous (Au^+) and auric (Au^{3+}) gold before it can be complexed by cyanide. Finally, the oxidation products of arsenopyrite and pyrite consume cyanide by forming thiocyanate and iron–cyanide complexes from which cyanide cannot be regenerated (Livesey-Goldblatt et al., 1983; Karavaiko et al., 1986). This increases the amount of cyanide reagent required for precious metal recovery.

As of 1997, Chile began to consider microbial removal of enargite from sulfidic gold ores (Acevedo et al., 1997; Escobar et al., 1997).

Beneficiation of carbonaceous gold ores containing as much as 6% arsenic by the action of *Thiobacillus ferrooxidans* was reported by Kamalov et al. (1973). They succeeded in removing as much as 90% of the arsenic in the ore in 10 days under some circumstances. In another study, it was found to be possible to accelerate leaching of arsenic from a copper–tin–arsenic concentrate by six- to seven-fold, using an adapted strain of *T. ferrooxidans* (Kulibakin and Laptev, 1971).

14.2.6 MICROBIAL REDUCTION OF OXIDIZED ARSENIC SPECIES

Some bacteria, fungi, and algae are able to reduce arsenic compounds. One of the first reports on arsenite reduction involves fungi (Gosio, 1897). Although originally the product of this reduction was thought to be diethylarsine (Bignelli, 1901), it was later shown to be trimethylarsine (Challenger et al., 1933, 1951). A bacterium from cattle-dipping tanks that reduced arsenate to arsenite was also described early in the twentieth century (Green, 1918). Much more recently, it was reported that a strain of *Chlorella* was able to reduce a part of the arsenate it absorbs from the medium to arsenite (Blasco et al., 1971). Woolfolk and Whiteley (1962) demonstrated arsenate reduction to arsenite by H_2 with cell extracts of *Micrococcus* (*Veillonella*) *lactilyticus* and intact cells of *M. aerogenes*. The enzyme hydrogenase was involved. Arsine (AsH_3) was not formed in these reactions. *Pseudomonas* sp. and *Alcaligenes* sp. have been reported to be able to reduce arsenate and arsenite anaerobically to arsine (Cheng and Focht, 1979).

Intact cells and cell extracts of the strict anaerobe *Methanobacterium* M.o.H. were shown to produce dimethylarsine from arsenate with H_2 (McBride and Wolfe, 1971). The extracts of this organism used methylcobalamine (CH_3B_{12}) as methyl donor. The reaction required consumption of ATP. The compounds 5-CH_3-tetrahydrofolate or serine could not replace CH_3B_{12}, although CO_2 could when tested by isotopic tracer technique. The reaction sequence as described by McBride and Wolfe (1971) was

$$AsO_4^{3-} \xrightarrow{2e^-} AsO_3^{3-} \xrightarrow{CH_3B_{12}}$$

$$CH_3 - \underset{\underset{OH}{\overset{\displaystyle \|}{|}}}{\overset{\displaystyle O}{As}} - OH \xrightarrow{CH_3B_{12}} CH_3 - \underset{\underset{CH_3}{\overset{\displaystyle \|}{|}}}{\overset{\displaystyle O}{As}} - OH \longrightarrow CH_3 - \underset{\underset{CH_3}{\overset{\displaystyle \|}{|}}}{\overset{\displaystyle O}{As}} - H \qquad (14.2)$$

Methylarsonic   Dimethylarsinic   Dimethylarsine

acid   acid

In an excess of arsenite, methylarsonic acid was the final product, the supply of CH_3B_{12} being limiting. In an excess of CH_3B_{12}, a second methylation step yielding dimethylarsinic acid (cacodylic acid) followed the first. The last step was shown to occur in the absence of CH_3B_{12}. All of these steps were enzymatic. Reaction sequence (Equation 14.2) was later found to be more complex (see below), and the natural methyl donor in *Methanobacterium* M.o.H (now *M. bryantii*) was shown to be methyl-coenzyme M (see discussion by Cullen and Reimer, 1989). Cell extracts of *Desulfovibrio vulgaris* were also found to produce a volatile arsinic derivative—presumably an arsine (McBride and Wolfe, 1971).

The fungus *Scopulariopsis brevicaulis* is able to convert arsenite to trimethylarsine. The reaction sequence originally proposed by Challenger (1951) has been modified and includes the following steps (Cullen and Reimer, 1989, p. 720):

$$As(III)O_3^{3-} \xrightarrow{RCH_3} CH_3As(V)O(OH)_2 \xrightarrow{2e^-} CH_3As(III)(OH)_2 \xrightarrow{RCH_3} (CH_3)_2As(V)O(OH)$$

Methylarsonic acid   Methylarsinous acid   Dimethylarsinic acid

or cacodylic acid

$$\xrightarrow{2e^-} ((CH_3)_2As(III)OH) \xrightarrow{RCH_3} (CH_3)_3As(V)O \xrightarrow{2e^-} (CH_3)_3As(III)$$

Unknown intermediate   Trimethylarsine oxide   Trimethylarsine

$$(14.3)$$

It should be noted that in this reaction sequence the oxidation state of the arsenic changes successively from +3 to +5 to +3. The reaction sequence leading to dimethylarsinic acid is similar to that observed in arsenic detoxification by methylation in human beings (Petrick et al., 2000; Aposhian et al., 2000). The methyl donors (RCH_3) in Reaction 14.3 can be a form of methionine (see Cullen and Reimer, 1989). Besides *S. brevicaulis*, other fungi such as *Aspergillus, Mucor, Fusarium, Paecilomyces,* and *Candida humicola* have also been found to be active in such reductions (Alexander, 1977; Cox and Alexander, 1973; Pickett et al., 1981). Trimethylarsine oxide was found to be an intermediate in trimethylarsine formation by *C. humicola* (Pickett et al., 1981). A review of biomethylation of arsenic can be found in Bentley and Chasteen (2002). The reduction of arsenate and arsenite to volatile arsines is a means of detoxification.

14.2.7 ARSENIC RESPIRATION

Some bacteria reduce arsenate only as far as arsenite. These organisms use arsenate as terminal electron acceptor in anaerobic respiration. A number of them have been characterized (Table 14.2) (Stolz and Oremland, 1999; Oremland and Stolz, 2000) and they include the members of the domains Bacteria and Archaea. Ahmann et al. (1994) isolated a comma-shaped motile rod, strain MIT-13, from arsenic-contaminated sediment from the Aberjona watershed in Massachusetts, United States.

TABLE 14.2
Arsenate-Respiring Bacteria and Archaea

Organism	References
Bacillus arsenicoselenatis	Switzer Blum et al. (1998)
Bacillus selenitireducens	Switzer Blum et al. (1998) and Oremland et al. (2000)
Chrysiogenes arsenatis	Macy et al. (1996) and Krafft and Macy (1998)
Desulfomaculum auripigmentum	Newman et al. (1997a,b)
Desulfomicrobium strain Ben-RB	Macy et al. (2000)
Pyrobaculum aerophilum[a]	Huber et al. (2000)
Pyrobaculum arsenaticlum[a]	Huber et al. (2000)
Shewanella sp. strain ANA-3	Saltikov et al. (2003)
Sulfurihydrogenibium subterraneum strain HGMK1[T]	Takai et al. (2003)
Sulfurospirillum arsenophilum	Stolz et al. (1999)
Sulfurospirillum barnesii	Lavermann et al. (1995) and Zobrist et al. (2000)
Strain MLMS-1	Hoeft et al. (2004)

[a] *Pyrobaculum* belongs to the domain Archaea.

This organism was later named *Sulfospirillum arsenophilum*. It belongs to the domain Bacteria (Stolz et al., 1999). Anaerobically growing cultures of *S. arsenophilum* MIT-13 respired on arsenate using lactate but not acetate as electron donor and produced stoichiometric amounts of arsenite.

Chrysiogenes arsenatis, isolated from gold mine wastewater in Australia, is the only arsenate respirer isolated so far that is capable of using acetate as electron donor (Macy et al., 1996). Its arsenate reductase, a periplasmic enzyme, has been analyzed and found to consist of two subunits. One of these subunits has a molecular mass of 87 kDa, and the other 29 kDa. The enzyme structure includes Zn, Fe, Mo, and acid-labile S (Krafft and Macy, 1998).

Sulfurospirillum barnesii strain SES-3 is another member of the domain Bacteria that can respire anaerobically on lactate using arsenate to arsenite reduction for terminal electron disposal (Laverman et al., 1995). Besides arsenate, it can also use selenate (see Chapter 21), Fe(III), thiosulfate, elemental sulfur, Mn(IV), nitrate, nitrite, trimethylamine *N*-oxide, and fumarate as terminal electron acceptors (Oremland et al., 1994). It can also grow microaerophilically, albeit sluggishly (15 O_2, optimum) (Laverman et al., 1995).

Newman et al. (1997a,b) isolated *Desulfotomaculum auripigmentum* strain OREX-4 from sediment of upper Mystic Lake, Massachusetts, United States. It is a gram-positive rod belonging to the domain Bacteria. It anaerobically reduces arsenate to arsenite and sulfate to sulfide with lactate as electron donor. When both arsenate and sulfate are present together, this organism precipitates orpiment (As_2S_3). The reactions can be summarized as follows (Newman et al., 1997b):

$$CH_3CHOHCOO^- + 2HASO_4^- + 4H^+ \rightarrow CH_3COO^- + HAsO_2 + CO_2 + 3H_2O$$

$$(\Delta G^{0'} = -41.1\,kcal\,mol^{-1}\ or\ -172\,kJ\,mol^{-1}\ lactate) \tag{14.4}$$

$$CH_3CHOHCOO^- + 0.5SO_4^{2-} + 0.5H^+ \rightarrow CH_3COO^- + 0.5HS^- + CO_2 + H_2O$$

$$(\Delta G^{0'} = -21.3\,kcal\,mol^{-1}\ or\ -89\,kJ\,mol^{-1}\ lactate) \tag{14.5}$$

$$2HAsO_2 + 3HS^- + 3H^+ \rightarrow As_2S_3 + 4H_2O \tag{14.6}$$

Reactions 14.4 and 14.5 are catalyzed by the organism, whereas Reaction 14.6 is not.

The microbial reduction of arsenate and sulfate when both are present together is sequential at rates that avoid the accumulation of excess sulfide. Orpiment did not form when arsenate and sulfide were added to the culture medium in the absence of the bacteria. Maintenance of a pH of ~6.8 was essential for orpiment stability.

Shewanella sp. strain HN-41 was recently shown to form arsenic-sulfide nanotubes (20–100 nm $\times$ 30 µm) when growing anaerobically in a lactate medium with a mixture of thiosulfate and As(V) as electron acceptors at circumneutral pH. The nanotubes initially consisted of amorphous As_2S_3 but with continued incubation developed polycrystalline phases of realgar (AsS) and duranusite (As_4S). Mature nanotubes exhibited metallic properties in terms of electrical and photoconductive properties (Lee et al., 2007).

Macy et al. (2000) isolated two sulfate reducers belonging to the domain Bacteria that had the capacity to reduce arsenate to arsenite when lactate was used as the source of reductant. The lactate was oxidized to acetate. One of these organisms was *Desulfomicrobium* strain Ben-RB, which was able to reduce arsenate and sulfate concurrently and respired on arsenate as sole terminal electron acceptor in the absence of sulfate. The other strain was *Desulfovibrio* strain Ben-RA, which was also able to reduce arsenate and sulfate concurrently. However, it did not grow with arsenate as sole terminal electron acceptor. In this organism, arsenate reduction may be part of an arsenate-resistance mechanism (see Silver, 1998).

Switzer Blum et al. (1998) discovered two species of the genus *Bacillus* in Mono Lake, California, United States, that were able to reduce not only arsenate but also selenium oxyanions (see Chapter 21). They are *Bacillus arsenicoselenatis* and *B. selenitireducens* belonging to the domain Bacteria. Both organisms are moderate halophiles and alkaliphiles, growing maximally at pH between 9 and 11. Each reduces arsenate to arsenite.

Saltikov et al. (2003) isolated *Shewanella* sp. strain ANA-3 from an arsenic-treated wooden pier located in Eel Pond, Woods Hole, Massachusetts, which was found to respire anaerobically on lactate using arsenate as terminal electron acceptor. The arsenate could be replaced by soluble ferric iron, oxides of iron and manganese, nitrate, fumarate, thiosulfate, and the humic acid functional analog 2,6-anthraquinone disulfonate as terminal electron acceptors. This organism is also able to respire using oxygen. Like other arsenate respirers, it produces arsenite from the arsenate it consumes. Besides reducing arsenate anaerobically, it is also able to reduce it aerobically to arsenite. It is resistant to arsenite concentrations as high as 10 mM aerobically and anaerobically, enabled by an *ars* operon—*arsDABC*. Although the *ars* operon is not required for arsenate respiration, its presence in the genome enhances the strain's growth rate and cell yield on lactate. The arsenate reductase of the strain is encoded in two gene clusters—*arrAB*. The *arrA* cluster codes for a 95.2 kDa molybdoprotein, and the *arrB* cluster codes for a 25.7 kDa iron–sulfur protein (Saltikov and Newman, 2003). Malasarn et al. (2008) determined kinetic properties of this enzyme.

The *arrA* gene has been shown to be a reliable marker for As(V) respiration in bacteria (Malasarn et al., 2004). The arsenate resistance operon, *ars*, is expressed both aerobically and anaerobically, whereas the arsenate reductase gene cluster, *arrAB*, is expressed only anaerobically and is repressed in the presence of oxygen and nitrate. Arsenate respiration and resistance are both induced specifically by arsenite. Arsenate respiration is induced at an arsenite concentration 1000 times lower than arsenate resistance. Arsenate respiration but not arsenite resistance is inducible by low micromolar concentrations of arsenate (Saltikov et al., 2005).

Takai et al. (2003) described a new member of the Aquificales in the domain Bacteria, isolated from the subsurface hot aquifer water of the Hishikari gold mine, Kagoshima Prefecture, Japan, which was capable of reducing arsenate to arsenite. They named the organism *Sulfurihydrogenibium subterraneum* strain HGMK1^T. The organism, which is a motile, straight to slightly curved rod (1.5–2.5 $\times$ 0.3–0.5 µm) is a thermophile with optimum growth at 65°C in a pH range of 6.4–8.8. It grows chemolithoautotrophically using H_2, S^0, or $S_2O_3^{2-}$ as an energy source and O_2, NO_3^-, Fe(III), SeO_4^{2-}, and SeO_3^{2-} besides As(V) as alternative electron acceptors. Nakagawa et al. (2005) subsequently found that this organism is facultatively heterotrophic, being able to use acetate as an organic carbon source. A comparison of *S. subterraneum* HGMK1[T] with *S. azorense* Az-Fu1[T] and *S. yellowstonense* SS-S[T], of which *S. yellowstonense* cannot use As(V) as terminal electron acceptor, has been presented by Nakagawa et al. (2005).

Hoeft et al. (2004) isolated a bacterial strain, MLMS-1, from Mono Lake bottom water that was able to reduce arsenate to arsenite using sulfide as electron donor. This organism is a chemoautotrophic, gram-negative, motile curved rod belonging to the δ-Proteobacteria distantly related to *Desulfobulbus*. It assimilated CO_2 by an enzyme system that did not seem to involve ribulose-1,5-bisphosphate carboxylase/oxidese (Hoeft et al., 2004).

Huber et al. (2000) isolated a member of the domain Archaea, *Pyrobaculum arsenaticum*, from a hot spring at Pisciarelli Solfataras at Naples, Italy, capable of reducing arsenate as terminal electron acceptor using H_2 as electron donor. This organism is a thermophilic, facultative autotroph that reduces arsenate to arsenite in its anaerobic respiration. In addition to arsenate, it can also use thiosulfate and elemental sulfur as alternative electron acceptors for autotrophic growth. Organotrophically, this organism produces realgar (As_2S_2) when using arsenate and thiosulfate or arsenate and cysteine as terminal electron acceptors. It can also reduce selenate but not selenite organotrophically to elemental selenium in a respiratory process.

Another species of *Pyrobaculum*, *P. aerophilum*, was found to reduce arsenate and selenate as terminal electron acceptors in autotrophic growth on H_2 as electron donor (Huber et al., 2000). Organotrophically, this organism can use arsenate as terminal electron acceptor when growing anaerobically. However, it does not form a visible precipitate of realgar on arsenate and thiosulfate or arsenate and cysteine. Selenate and selenite can be alternative electron acceptors. The selenium compounds are reduced to elemental selenium. A further discussion of respiratory arsenate reductase can be found in Silver and Phung (2005).

14.2.8 DIRECT OBSERVATIONS OF ARSENITE OXIDATION AND ARSENATE REDUCTION *IN SITU*

Newman et al. (1998) reviewed environmental microbial arsenate respiration with emphasis on non-marine systems, that is, freshwater as well as hypersaline systems. They compared and contrasted relevant physiological traits of four different respirers.

Oremland et al. (2000) developed extensive evidence of arsenate reduction in meromictic, alkaline, hypersaline Mono Lake, California, United States, which contains ~ 200 μM total arsenic. The oxidation state of the dissolved arsenic changed predominantly from +5 in the oxic surface waters (mixolimnion) to +3 in the anoxic bottom waters (monimolimnion). No significant bacterial reduction was noted in the oxic waters, but it was significant in the anoxic waters. A rate of ~5.9 μmol L[-1] per day was measured at depths between 18 and 21 m. Sulfate reduction occurred at depths below 21 m; the highest rate of ~2.3 μmol L[-1] per day was measured at a depth of 28 m. Oremland et al. (2000) estimated that arsenate respiration is second in importance after sulfate respiration in Mono Lake and may account for mineralization of 14.2% of the photosynthetically fixed carbon annually in the lake. Oremland et al. (2002) subsequently demonstrated anaerobic oxidation of arsenite by a facultative, chemoautotrophic isolate, strain MLHE-1 (recently named *Alkalilimnicola ehrlichii* by Hoeft et al., 2007), from Mono Lake using nitrate as terminal electron acceptor. The nitrate could be replaced by Fe(III) and Mn(IV), which also occur in Mono Lake at low concentrations (10 and 0.4 μM, respectively). Thus in the anaerobic hypolimnion of Mono Lake, a bacterially promoted arsenic cycle seems to occur (Oremland et al., 2002, 2004). The microbial diversity with respect to the domain Bacteria at different depths in Mono Lake has been examined by Humayoun et al. (2003). They found greater diversity in the anoxic bottom waters of the lake than in the oxic surface waters, despite the presence of toxic chemical species including sulfide, arsenate, and arsenite.

Searles Lake in the Mojave Desert in California, ~270 km south–southeast of Mono Lake, featuring a salt-saturated, alkaline brine with an unusually elevated arsenic content (total As ~3.9 mM), is another lake studied by Oremland et al. (2005). They found that the brine contained traces of sulfide, methane, and ammonia, but they did not detect any nitrate. The sulfide and methane concentrations in the brine of Searles Lake were 0.1–0.01 of those in anoxic bottom waters of Mono Lake, and methane concentration in the sediment of Searles Lake was 0.1–0.01 of its concentration in sediments of Mono Lake. They found that in the sediment of Searles Lake, the oxidation state of arsenic changed from As(V) to As(III) with depth. In sediment slurries, As(V) reduction was stimulated by H_2 or sulfide under anaerobic conditions, whereas As(III) was rapidly oxidized under aerobic conditions. Some of the As(III) formed in the As(V) reduction formed thioarsenites, especially when sulfide was the electron donor. As(V)-reducing and As(III)-oxidizing activities were abolished when the sediment slurries were autoclaved. From the lake sediment, they isolated an anaerobic, extremely alkihalophilic, facultatively lithotrophic As(V)-respiring bacterium, strain SLAS-1, which is a curved, motile rod. This organism oxidized lactate to acetate while reducing As(V) to As(III). It also could use H_2 or sulfide as electron donor for As(V) reduction. It fixed CO_2 when growing lithoautotrophically with sulfide as electron donor. Phylogenetically, strain SLAS-1 is affiliated with the Halanaerobacteriales and appears to represent a new species. The findings of the authors clearly show that a complete arsenic cycle is operative in the sediments of Searles Lake.

Islam et al. (2004), using a microcosm approach, showed that metal-reducing bacteria played a key role in mobilizing arsenic in sediments from a contaminated aquifer in West Bengal. In this instance, soluble As(III) was generated only after Fe(III) was reduced to Fe(II) by the members of the Geobacteraceae in acetate-fed microcosms, indicating that As(V) bound to iron oxyhydroxides was reduced after its release due to bacterial Fe(III) reduction; but the agent responsible for the reduction of As(V) was not identified. A subsequent study by Islam et al. (2005) with *Geobacter sulfurreducens* showed that this organism was unable to reduce As(V) to As(III) with or without energy conservation. When respiring Fe(III) in a suitable culture medium in the presence of As(V), *G. sulfurreducens* generated vivianite ($Fe_3(PO_4)_2 \cdot 8H_2O$) accompanied by arsenic immobilization, 85% as As(V) and 15% as As(III). Similar arsenic removal occurred when the organism formed siderite ($FeCO_3$) under different culture conditions, with 50% of the immobilized arsenic as As(V) and 50% as As(III), or magnetite (Fe_3O_4) with 100% of the immobilized arsenic as As(V).

Johnson (1972) made direct observations relating to microbial reduction of arsenic in the ocean. He found that bacteria from phytoplankton samples in Narragansett Bay and from Sargasso Sea water were able to reduce added arsenate. An arsenate reduction rate of 10^{-11} μmol cell^{-1} min^{-1} was measured over a 12 h period of incubation at 20–22°C. Arsenic was not accumulated by the bacteria, and none was lost from the medium through volatilization. Scudlark and Johnson (1982) presented evidence of bacterial oxidation in aged seawater from Narragansett Bay and some other marine sites. Depending on the geographic location of the collected samples, initial oxidation rates ranged from 3 to 1280 nmol L^{-1} per day. These observations may help to explain why the observed As(V)/As(III) ratio in the seawater was 10^{-1}:10 instead of 10^{26}:1, as predicted under equilibrium conditions in oxygenated seawater at pH 8.1 (Johnson, 1972; Andreae, 1979).

Cheng and Focht (1979) observed that bacteria are able to reduce arsenate and arsenite to arsine under anaerobic conditions in soils. They indicated that, unlike fungi, the bacteria produce mono- and dimethylarsine only when methylarsonate or dimethylarsinate is available.

Oxidation of arsenite to arsenate in activated sludge under aerobic conditions was reported by Myers et al. (1973). Under anaerobic conditions, arsenate was found to be reduced stepwise to lower oxidation states over an extended period in the activated sludge. *Pseudomonas fluorescens* was an active reducer in this system under overall aerobic conditions (Myers et al., 1973). Reduction may actually have occurred in anaerobic microenvironments in the floc.

Figure 14.4 summarizes the reactions involving arsenic compounds in nature that are catalyzed by microorganisms. The oxidation of methylated arsine, although not indicated in this figure, has been suggested by Cheng and Focht (1979). The arsenic cycle in natural freshwaters has been discussed by Ferguson and Gavis (1972) and Scudlark and Johnson (1982).

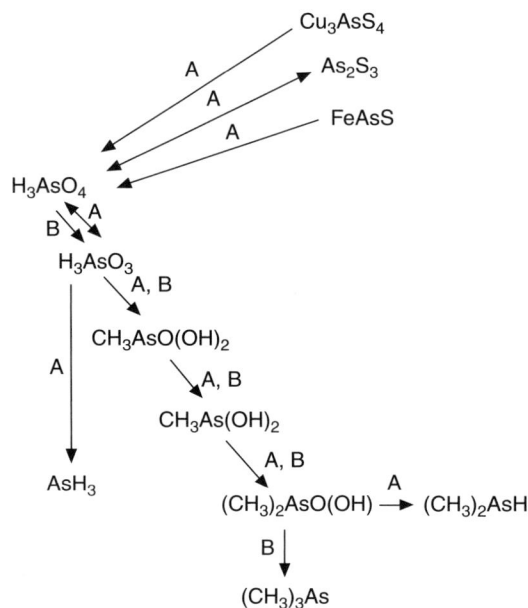

FIGURE 14.4 Summary of observed microbial interactions with arsenic compounds performed by (A) bacteria and (B) fungi.

14.3 ANTIMONY

14.3.1 Antimony Distribution in Earth's Crust

Antimony is a rare element. Its abundance in the Earth's crust when normalized to its abundances in chondritic stone meteorites and adjusted to Si equal to 1.00 has been given as 0.1 (Winchester, 1972). Its average concentration in igneous rocks is 0.2 ppm, in shales 1.5 ppm, in limestone 0.3 ppm, in sandstonze 0.05 ppm, and in soil 1 ppm. Its average concentration in seawater is 0.124 ppb and in freshwater 0.2 ppb (Bowen, 1979). As a mineral, it may occur as stibnite (Sb_2S_3), kermesite (Sb_2S_2O), senarmontite (Sb_2O_3), jamesonite ($2PbS \cdot Sb_2S_3$), boulangerite ($5PbS \cdot 2Sb_2S_3$), and tetrahedrite ($Cu_{12}Sb_4S_{13}$). It may also occur as sulfantimonides of silver and nickel, and sometimes as elemental antimony (Gornitz, 1972). Stibnite is the most common antimony mineral.

Antimony can exist in the oxidation states -3, 0, $+3$, and $+5$. Like arsenic compounds, antimony compounds tend to be toxic for most living organisms. The basis for this toxicity has not been established.

Elevated antimony concentrations have been measured in soils surrounding mining sites and smelters dealing with antimony-containing ores, which fall in the range of 100–1489 mg kg^{-1} (Li and Thornton, 1993; Ainsworth et al., 1990). By using bacterial biosensors, the polluting antimony at five former British sites with levels up to 700 mg kg^{-1} was shown not to be bioavailable (Flynn et al., 2003).

14.3.2 Microbial Oxidation of Antimony Compounds

The report by Bryner et al. (1954) is among the earliest reports of the biooxidation of antimony-containing minerals in which the oxidation of tetrahedrite ($Cu_{12}Sb_4S_{13}$) by *Thiobacillus* (now *Acidothiobacillus*) *ferrooxidans* is recorded. Lyalikova (1961) reported the oxidation of antimony trisulfide (Sb_2S_3) by *T. ferrooxidans*. In both of these instances, the oxidation proceeded under acidic conditions (pH 2.45) (Kuznetsov et al., 1963). The oxidation state of the solubilized antimony was not determined in either case. More recently, Silver and Torma (1974) reported on the oxidation of synthetic antimony sulfides by *T. ferrooxidans*. Torma and Gabra (1977) examined the oxidation of stibnite (Sb_2S_3) by the same organism. The latter authors suspected that *T. ferrooxidans* oxidized trivalent antimony (Sb(III)) to pentavalent antimony (Sb(V)) but offered no proof. Lyalikova et al. (1972) reported on the oxidation of Sb–Pb sulfides, Sb–Pb–Te sulfides, and Sb–Pb–As sulfides, resulting in the formation of minerals such as anglesite ($PbSO_4$) and valentinite (Sb_2O_3). The formation of valentinite suggests that antimony in the minerals was solubilized but not oxidized. Ehrlich (1986, 1988) reported on the oxidation of a mixed sulfide ore that included tetrahedrite by *T. ferrooxidans* strain 19759. Although he followed mobilization of silver, copper, and zinc from the ore in these experiments, he did not follow the mobilization of antimony. In all the foregoing studies, most or all of the oxidation apparently involved sulfide in the minerals.

Trivalent antimony is susceptible to direct oxidation by *Stibiobacter senarmontii* (Lyalikova, 1972, 1974; Lyalikova et al., 1976). This organism, which was isolated from a sample of mine drainage in the former Yugoslavia, oxidizes senarmontite (Sb_2O_3) or Sb_2O_4 to Sb_2O_5, deriving useful energy from this process. It is a gram-positive rod (0.5–1.8 × 0.5 μm) with a single polar flagellum and has the ability to form rudimentary mycelia in certain stages of development. It grows at neutral pH and generates acid when oxidizing Sb_2O_3 (the pH can drop from 7.5 to 5.5). When grown on reduced antimony oxide (senarmontite), this organism possesses the enzyme ribulose bisphosphate carboxylase, indicating that it has chemolithotrophic propensity (Lyalikova et al., 1976). Antimony sulfide ores can thus be oxidized to antimony pentoxide in the following two steps:

$$Sb_2S_3 \xrightarrow{\ O_2\ } Sb_2O_3 \xrightarrow{\ O_2\ } Sb_2O_5 \qquad (14.7)$$
$$\phantom{Sb_2S_3 \xrightarrow{\ }} (1) \phantom{Sb_2O_3 \xrightarrow{\ }} (2)$$

The first step is catalyzed by an organism such as *Thiobacillus* Y or *T. ferrooxidans* (see [1] in Reaction 14.7) and the second by *S. senarmontii* (see [2] in Reaction 14.7) (Lyalikova et al., 1974).

14.3.3 MICROBIAL REDUCTION OF OXIDIZED ANTIMONY MINERALS

No strong evidence exists to date that microbes are able to reduce oxidized antimony species as terminal electron acceptor in anaerobic respiration, as is the case for oxidized arsenic species (see Iverson and Brinckman, 1978). However, bioreduction of Sb(V) to Sb(III) is suggested from antimony methylation studies (Bentley and Chasteen, 2002). This limited evidence should not be taken to mean that microbial antimony reduction is of rare occurrence, but rather that it has so far not been thoroughly studied.

14.4 SUMMARY

Although arsenic and antimony compounds are toxic to most forms of life, some microbes metabolize them. Arsenite has been found to be enzymatically oxidized by several bacteria. The enzyme system has been shown to be inducible in at least some microbes. In laboratory experiments, *Alcaligenes faecalis* oxidized arsenite most intensely only after having gone through active growth. This organism probably can derive maintenance energy from arsenite oxidation. *Pseudomonas arsenitoxidans* and two as yet unnamed isolates from Australia can grow autotrophically with arsenite as their sole source of energy. Simple and complex arsenic sulfides are oxidized by *Thiobacillus* (now *Acidithiobacillus*) *ferrooxidans*. However, no evidence has been obtained that this organism can oxidize trivalent to pentavalent arsenic enzymatically.

Arsenate and arsenite have been shown to be reduced by certain bacteria and fungi. When used as terminal electron acceptor, arsenate is reduced to arsenite. In some forms of arsenic detoxification, bacteria can reduce arsenate or arsenite to arsine or dimethylarsine, whereas fungi produce trimethylarsine. All of these arsines are volatile.

Antimony compounds have also been shown to be microbially oxidized. *Thiobacillus* (now *Acidithiobacillus*) *ferrooxidans* has been shown to attack various antimony-containing sulfides. Although enzymatic oxidation by *T. ferrooxidans* of Sb(III) to Sb(V) has been claimed in the case of Sb_2S_3, clear proof is lacking. Generally only the sulfide moiety and ferrous iron, if present, in an antimony mineral are oxidized by this organism. *Stibiobacter senarmontii*, an autotroph that was isolated from an ore deposit in the former Yugoslavia, oxidizes trivalent antimony in Sb_2O_3 or Sb_2O_4 to pentavalent antimony (Sb_2O_5). Microbial reduction of oxidized antimony compounds has so far been reported only in connection with antimony methylation.

REFERENCES

Abdrashitova SA, Mynbaeva BN, Ilyaletdinov AN. 1981. Oxidation of arsenic by the heterotrophic bacteria *Pseudomonas putida* and *Alcaligenes eutrophus*. *Mikrobiologiya* 50:41–45 (Engl transl pp. 28–31).

Acevedo F, Gentina JC, Alegre C, Arévalo P. 1997. Biooxidation of a gold bearing enargite concentrate. In: *International Biohydrometallurgy Symposium IBS97 Biomine 97, "Biotechnology Comes of Age." Conference Proceedings*, Glenside, South Australia: Australian Mineral Foundation, pp. M3.2.1–M3.2.9.

Ahmann D, Roberts AL, Krumholz LR, Morel FMM. 1994. Microbe grows by reducing arsenic. *Nature* (London) 371:750.

Ainsworth N, Cooke JA, Johnson MS. 1990. Distribution of antimony in contaminated grassland: 1–vegetation and soils. *Environ Poll* 65:65–77.

Alexander M. 1977. *Introduction to Soil Microbiology*. New York: Wiley.

Anderson GL, Williams J, Hille R. 1992. The purification and characterization of arsenite oxidase from *Alcaligenes faecalis*, a molybdenum-containing hydroxylase. *J Biol Chem* 267:23674–23682.

Andreae MO. 1979. Arsenic speciation in seawater and interstitial waters: The influence of biological–chemical interactions on the chemistry of a trace element. *Limnol Oceanogr* 24:440–452.

Aposhian HV, Zheng B, Aposhian MM, Le XC, Cebrian ME, Cullen W, Zakharian RA, Ma M, Dart RC, Cheng Z, Andrewes P, Yip L, O'Malley GF, Maiorino RM, Van Voorhies W, Healy SM, Titcomb A. 2000. DMPS: Arsenic challenge test. II. Modulation of arsenic species, including monoethylarsonous acid (MMA[III]), excreted in human urine. *Toxicol Appl Pharm* 165:74–83.

Banfield JF. Department of Earth and Planetary Science, Environmental Science, Policy, Management, University of California, Berkeley, CA.

Bennett RL, Malamy MH. 1970. Arsenate-resistant mutants of *Escherichia coli* and phosphate transport. *Biochem Biophys Res Commun* 40:496–503.

Bentley R, Chasteen TG. 2002. Microbial methylation of metalloids: Arsenic, antimony, and bismuth. *Microbiol Molec Biol Rev* 66:250–271.

Bignelli P. 1901. *Gazz Chim Ital* 31:58 (as cited by Challenger, 1951).

Blasco F, Gaudin C, Jeanjean R. 1971. Absorption of arsenate ions by *Chlorella*. Partial reduction of arsenate to arsenite. *CR Acad Sci (Paris) Ser D* 273:812–815.

Bowen HJM. 1979. *Environmental Chemistry of the Elements*. London: Academic Press.

Braddock JG, Luong HV, Brown EJ. 1984. Growth kinetics of *Thiobacillus ferrooxidans* isolated from mine drainage. *Appl Environ Microbiol* 48:48–55.

Brierley JA, Brierley CL. 1999. Present and future commercial applications of biohydrometallurgy. In: Amils R, Ballester A, eds. *Biohydrometallurgy and the Environment Toward the Mining of the 21st Century. Part A*. Amsterdam: Elsevier, pp. 81–89.

Broeer S, Ji G, Broeer A, Silver S. 1993. Arsenic efflux governed by the arsenic resistance plasmid pI258. *J Bacteriol* 175:3480–3485.

Bryner LC, Beck JV, Davis BB, Wilson DG. 1954. Microorganisms in leaching sulfide minerals. *Ind Eng Chem* 46:2587–2592.

Buchanan RE, Gibbons NE. 1974. *Bergey's Manual of Determinative Bacteriology*. 8th ed. Baltimore, MD: Williams & Wilkins.

Button DK, Dunker SS, Moore ML. 1973. Continuous culture of *Rhodotorula rubra*: Kinetics of phosphate–arsenate uptake, inhibition and phosphate limited growth. *J Bacteriol* 113:599–611.

Carapella SC, Jr. 1972. Arsenic: Element and geochemistry. In: Fairbridge RW, ed. *The Encyclopedia of Geochemistry and Environmental Sciences. Encyclopedia of Earth Science, Series IVA*. New York: Van Nostrand Reinhold, pp. 41–42.

Carlson L, Bigham JM, Schwertmann U, Kyek A, Wagner F. 2002. Scavenging of As from acid mine drainage by schertmannite and ferrihydrite: A comparison with synthetic analogues. *Environ Sci Technol* 36:1712–1719.

Carlson L, Lindström EB, Hallberg KB, Tuovinen OH. 1992. Solid-phase production of bacterial oxidation of arsenic/pyrite. *Appl Environ Microbiol* 58:1046–1049.

Challenger F. 1933. The formation of organo-metalloid compounds by microorganisms. Part II. Trimethylarsine and dimethylarsine. *J Chem Soc* 95:101.

Challenger F. 1951. Biological methylation. *Adv Enzymol* 12:429–491.

Cheng C-N, Focht DD. 1979. Production of arsine and methylarsines in soil and in culture. *Appl Environ Microbiol* 38:494–498.

Cox DP, Alexander M. 1973. Effect of phosphate and other anions on trimethylarsine formation by *Candida humicola*. *Appl Environ Microbiol* 25:408–413.

Cullen WR, Reimer KJ. 1989. Arsenic speciation in the environment. *Chem Rev* 89: 713–784.

Da Costa EWB. 1971. Suppression of the inhibitory effects of arsenic compounds by phosphate. *Nature (New Biol)* (London) 231:32.

Da Costa EWB. 1972. Variation in toxicity of arsenic compounds to microorganisms and the suppression of the inhibitory effects of phosphate. *Appl Environ Microbiol* 23:46–53.

Dew DW, Lawson EN, Broadhurst JL. 1997. The BIOX® process for biooxidation of gold-bearing ores and concentrates. In: Rawlings DE, ed. *Biomining. Theory, Microbes and Industrial Processes*. Berlin, Germany: Springer, pp. 45–80.

D'Imperio S, Lehr CR, Breary M, McDermott TR. 2007. Autecology of an arsenite chemolithotroph: Sulfide constraints on function and distribution in a geothermal spring. *Appl Environ Microbiol* 73:7067–7074.

Dixit S, Hering JC. 2003. Effects of arsenate reduction and iron oxide transformation on arsenic mobility. *Env Sci Technol* 37:4182–4189.

Dixit S, Hering JC. 2006. Sorption of Fe(II) and As(III) on goethite in single- and dual-sorbate systems. *Chem Geol* 228:6–15.

Donahoe-Christiansen J, D'Imperio S, Jackson CR, Innskeep WP, McDermott TR. 2004. Arsenite-oxidizing *Hydrogenobaculum* strain isolated from an acid-sulfate-chloride geothermal spring in Yellowstone National Park. *Appl Environ Microbiol* 70:1865–1868.

Dyke KGH, Parker MT, Richmond MH. 1970. Penicillinase production and metal-ion resistance in *Staphylococcus aureus* cultures isolated from hospital patients. *J Med Microbiol* 3:125–136.

Ehrlich HL. 1963. Bacterial action on orpiment. *Econ Geol* 58:991–994.

Ehrlich HL. 1964. Bacterial oxidation of arsenopyrite and enargite. *Econ Geol* 59:1306–1312.

Ehrlich HL. 1978. Inorganic energy sources for chemolithotrophic and mixotrophic bacteria. *Geomicrobiol J* 1:65–83.

Ehrlich HL. 1986. Bacterial leaching of silver from silver-containing mixed sulfide ore by a continuous process. In: Lawrence RW, Branion RMR, Ebneer HG, eds. *Fundamental and Applied Biohydrometallurgy*. Amsterdam: Elsevier, pp. 77–88.

Ehrlich HL, 1988. Bioleaching of silver from a mixed sulfide ore in a stirred reactor. In: Norris PR, Kelly DP, eds. *Biohydrometallurgy*. Kew Surrey, U.K.: Science & Technology Letters, pp. 223–231.

Escobar B, Huenupi E, Wirtz J. 1997. Chemical and biological leaching of enargite. *Biotechnol Lett* 19:719–722.

Ferguson JF, Gavis J. 1972. A review of the arsenic cycle in natural waters. *Water Res* 6:1259–1274.

Flynn HC, Meharg AA, Bowyer PK, Paton GI. 2003. Antimony bioavailability in mine soils. *Environ Poll* 124:93–100.

Gornitz V. 1972. Antimony: Element and geochemistry. In: Fairbridge RW, ed. *The Encyclopedia of Geochemistry and Environmental Sciences. Encyclopedia Earth Science Series IVA*. New York: Van Nostrand Reinhold, pp. 33–36.

Gosio B. 1897. Zur Frage, Wodurch die Giftigkeit arsenhaltiger Tapeten bedingt wird. *Ber Deut Chem Ges* 30: 1024–1027 (as cited by Challenger, 1951).

Green HH. 1918. Description of a bacterium which oxidizes arsenite to arsenate, and of one which reduces arsenate to arsenite, isolated from a cattle-dipping tank. *S Afr J Sci* 14:465–467.

Groudeva VI, Groudev SN, Markov KI. 1986. A comparison between mesophilic and thermophilic bacteria with respect to their ability to leach sulfide minerals. In: Lawrence RW, Branion RMR, Ebner HG, eds. *Fundamental and Applied Biohydormetallurgy*. Amsterdam: Elsevier, pp. 484–485.

Hedges RW, Baumberg S. 1973. Resistance to arsenic compounds conferred by a plasmid transmissible between strains of *Escherichia coli*. *J Bacteriol* 115:459–460.

Hendrie MS, Holding AJ, Shewan JM. 1974. Emended description of the genus *Alcaligenes* and of *Alcaligenes faecalis* and a proposal that the generic name of *Achromobacter* be rejected: Status of the named species *Alcaligenes* and *Achromobacter*. *Int J Syst Bacteriol* 24:534–550.

Hoeft SE, Kulp TR, Stolz JF, Hollibaugh JT, Oremland RS. 2004. Dissimilatory arsenate reduction with sulfide as electron donor: Experiments with Mono Lake water and isolation of strain MLMS-1, a chemoautotrophic arsenate respirer. *Appl Environ Microbiol* 70:2741–2747.

Hoeft SE, Switzer Blum J, Stolz JF, Tabita R, Witte B, King GM, Santini JM, Oremland RS. 2007. *Alkalilimnicola ehrlichii* sp. nov., a novel, arsenite oxidizing haloalkaliphilic gammaproteobacterium capable of chemoautotrophic or heterotrophic growth with nitrate or oxygen as the electron acceptor. *Int J Syst Evol Microbiol* 57:504–512.

Huber R, Sacher M, Vollman A, Huber H, Rose D. 2000. Respiration of arsenate and selenate by hyperthermophilic Archaea. *Syst Appl Microbiol* 23:305–314.

Humayoun SB, Banno N, Hollibaugh JT. 2003. Depth distribution of microbial diversity in Mono Lake, a meromictic soda lake in California. *Appl Environ Microbiol* 69:1030–1042.

Ilyaletdinov AN, Abdrashitova SA. 1981. Autotrophic oxidation of arsenic by a culture of *Pseudomonas arsenitoxidans*. *Mikrobiologiya* 50:197–204 (Engl transl pp. 135–140).

Islam FS, Gault AG, Boothman C, Polya DA, Charnock JM, Chatterjee D, Lloyd JR. 2004. Role of metal-reducing bacteria in arsenic release from Bengal delta sediments. *Nature* (London) 430:68–71.

Islam FS, Pederick RL, Gault AG, Adams LK, Polya DA, Charnock JM, Lloyd JR. 2005. Interactions between the Fe(III)-reducing bacterium *Geobacter sulfurreducens* and arsenate, and capture of the metalloid by biogenic Fe(II). *Appl Environ Microbiol* 71:8642–8648.

Iverson WP, Brinckman FE. 1978. Microbial metabolism of heavy elements. In: Mitchell R, ed. *Water Pollution Microbiology*, Vol. 2. New York: Wiley, pp. 201–232.

Ji G, Silver S. 1992. Regulation and expression of the arsenic resistance operon from *Staphylococcus aureus* plasmid pI258. *J Bacteriol* 174:3684–3694.

Ji G, Silver S. Garber EAE, Ohtake H, Cervantes C, Corbisier P. 1993. Bacterial molecular genetics and enzymatic transformations of arsenate, arsenite, and chromate. In: Torma AE, Apel ML, Brierley CL, eds. *Biohydrometallurgical Technologies, Vol 2. Fossil Energy Materials Bioremediation, Microbial Physiology*. Warrendale, PA: The Minerals, Metals, and Materials Society, pp. 529–539.

Johnson DL. 1972. Bacterial reduction of arsenite in seawater. *Nature* (London) 240:44–45.

Kamalov MR, Karavaiko GI, Ilyaletdinov AN, Abrashitova SA. 1973. The role of *Thiobacillus ferrooxidans* in leaching arsenic from a concentrate of carbonaceous gold-containing ore. *Izv Akad Nauk Kaz SSR Ser Biol* 11:37–44.

Karavaiko GI, Chuchalin LK, Pivovarova TA, Yemel'yanov BA, Forofeyev AG. 1986. Microbiological leaching of metals from arsenopyrite containing concentrates. In: Lawrence RW, Branion RMR, Ebner HG, eds. *Fundamental and Applied Biohydrometallurgy*. Amsterdam: Elsevier, pp. 125–126.

Krafft T, Macy JM. 1998. Purification and characterization of the respiratory arsenate reductase of *Chrysiogenes arsenatis*. *Eur J Biochem* 255:647–653.

Kulibakin VG, Laptev SF. 1971. Effect of adaptation of *Thiobacillus ferrooxidans* to a copper–arsenic–tin concentrate on the arsenic leach rate. *Sb Tr Tsent Nauk-Issled Inst Olomyan Prom* 1:75–76.

Kuznetsov SI, Ivanov MV, Lyalikova NN. 1963. *Introduction to Geological Microbiology*. Engl transl. New York: McGraw-Hill.

Laverman AM, Switzer Blum J, Schaefer JK, Phillips EJP, Lovley DR, Oremland RS. 1995. Growth of strain SES-3 with arsenate and other diverse electron acceptors. *Appl Environ Microbiol* 61:3556–3561.

Lee J-H, Kim M-G, Yoo B, Myung NV, Maeng J, Lee T, Dohnalkova AC, Fredrickson JK, Sadowsky MJ, Hur H-G. 2007. Biogenic formation of photoactive arsenic-sulfide nanotubes by *Shewanella* sp. strain HN-41. *Proc Natl Acad Sci* (Washington, DC) 104:20410–20415.

Legge JW. 1954. Bacterial oxidation of arsenite. IV. Some properties of the bacterial cytochromes. *Aust J Biol Sci* 7:504–514.

Legge JW, Turner AW. 1954. Bacterial oxidation of arsenite. III. Cell-free arsenite dehydrogenase. *Aust J Biol Sci* 7:496–503.

Li X, Thornton I. 1993. Arsenic, antimony, and bismuth in soil and pasture herbage in some old metalliferous mining areas in England. *Environ Geochem Health* 15:135–144.

Livesey-Goldblatt E, Norman P, Livesey-Goldblatt DR. 1983. Gold recovery from arsenopyrite/pyrite ore by bacterial leaching and cyanidation. In: Rossi G, Torma AE, eds. *Recent Progress in Biohydrometallurgy*. Iglesias, Italy: Assoc Mineraria Sarda, pp. 627–641.

Lunde G. 1973. Synthesis of fat and water soluble arseno organic compounds in marine and limnetic algae. *Acta Chem Scand* 27:1586–1594.

Lyalikova NN. 1961. The role of bacteria in the oxidation of sulfide ores. *Tr In-ta Mikrobiol AN SSR*, No. 9 (as cited by Kuznetsov et al., 1963).

Lyalikova NN. 1972. Oxidation of trivalent antimony to higher oxides as an energy source for the development of a new autotrophic organism *Stibiobacter* gen. n. *Dokl Akad Nauk SSSR Ser Biol* 205:1228–1229.

Lyalikova NN. 1974. *Stibiobacter senarmontii*: A new antimony-oxidizing microorganisms. *Mikrobiologiya* 43:941–948 (Engl transl pp. 799–805).

Lyalikova NN, Shlain LB, Trofimov VG. 1974. Formation of minerals of antimony(V) under the effect of bacteria. *Izv Akad Nauk SSSR Ser Biol* 3:440–444.

Lyalikova NN, Shlain LB, Unanova OG, Anisimova LS. 1972. Transformation of products of compound antimony and lead sulfides under the effect of bacteria. *Izv Akad Nauk SSSR Ser Biol* 4:564–567.

Lyalikova NN, Vedenina IYa, Romanova AK. 1976. Assimilation of carbon dioxide by a culture *of Stibiobacter senarmontii*. *Mikrobiologiya* 45:552–554 (Engl transl pp. 476–477).

Macy JM, Nunan K, Hagen KD, Dixon DR, Harbour PJ, Cahill M, Sly L. 1996. *Chrysiogenes arsenatis* gen. nov., spec. nov., a new arsenate-respiring bacterium isolated from gold mine wastewater. *Int J Syst Bacteriol* 46:1153–1157.

Macy JM, Santini JM, Pauling BV, O'Neill AH, Sly LI. 2000. Two new arsenate/sulfate-reducing bacteria: Mechanism of arsenate reduction. *Arch Microbiol* 173:49–57.

Mahler HR, Cordes EH. 1966. *Biological Chemistry*. New York: Harper & Row.

Malasarn D, Keeffe JR, Newman DK. 2008. Characterization of the arsenate respiratory reductase from *Shewanella* sp. strain ANA-3. *J Bacteriol* 190:135–142.

Malasarn D, Saltikov CW, Campbell KM, Santini JM, Hering JG, Newman DK. 2004. *arrA* is a reliable marker for As(V) respiration. *Science* 306:455.

McBride BS, Wolfe RS. 1971. Biosynthesis of dimethylarsine by *Methanobacterium*. *Biochemistry* 10:4312–4317.

McGuire MM, Banfield JF, Hamers RJ. 2001. *Quantitative Determination of Elemental Sulfur at the Arsenopyrite Surface After Oxidation by Ferric Iron: Mechanistic Implications. Geochemical Transactions*, Vol. 4; Digital Object Identifier: 10.1039/b104111h.

Monroy-Fernández MG, Mustin C, de Donato P, Berthelin J, Barion P. 1995. Bacterial behavior and evolution of surface oxidized phases during arsenopyrite oxidation by *Thiobacillus ferrooxidans*. In: Vargas T, Jerez JV, Toledo H, eds. *Biohydrometallurgical Processing*. Vol. 1. Santiago: University of Chile, pp. 57–66.

Myers DJ, Heimbrook ME, Osteryoung J, Morrison SM. 1973. Arsenic oxidation state in the presence of microorganisms. Examination by differential pulse polarography. *Environ Lett* 5:53–61.

Nakagawa S, Shtaih Z, Banta A, Beveridge TJ, Sako Y, Reysenbach A-L. 2005. *Sulfurihdyrogenibium yellowstonense* sp. nov., an extremely thermophilic, facultative heterotrophic, sulfur-oxidizing bacterium from Yellowstone National Park, and emeneded descriptions of the genus *Sulfurihydrogenibium, Sulfurihydrogenibium subterraneum* and *Sulfurihydroenibium azorense*. *Int J Syst Evol Microbiol* 55:2263–2268.

Newman DK, Ahmann D, Morel FMM. 1998. A brief review of microbial arsenate respiration. *Geomicrobiol J* 15:255–268.

Newman DK, Beveridge TJ, Morel FM. 1997a. Precipitation of arsenic trisulfide by *Desulfotomaculum auripigmentum*. *Appl Environ Microbiol* 63:2022–2028.

Newman DK, Kennedy EK, Coates JD, Ahmann D, Ellis DJ, Lovley DR, Morel FM. 1997b. Dissimilatory arsenate and sulfate reduction in *Desulfotomaculum auripigmentum*, sp. nov. *Arch Microbiol* 168:380–388.

Ngubane WT, Baecker AAW. 1990. Oxidation of gold-bearing pyrite and arsenopyrite by *Sulfolobus acidocaldarius* and *Sulfolobus* BC in airlift bioreactors. *Biorecovery* 1:225–269.

Olson GJ. 1994. Microbiological oxidation of gold ores and gold bioleaching. *FEMS Microbiol Lett* 119:1–6.

Oremland RS, Dowdle PR, Hoeft S, Sharp JO, Schaefer JK, Miller LG, Switzer Blum J, Smith RL, Bloom NS, Wallschlaeger D. 2000. Bacterial dissimilatory reduction of arsenate and sulfate in meromictic Mono Lake, California. *Geochim Cosmochim Acta* 64:3073–3084.

Oremland RS, Hoeft SE, Santini JM, Bano N, Hollibaugh RA, Hollibaugh JT. 2002. Anaerobic oxidation of arsenite in Mono Lake water and by a facultative, arsenite-oxidizing chemoautotroph, strain MLHE-1. *Appl Environ Microbiol* 68:4795–4802.

Oremland RS, Kulp TR, Switzer Blum J, Hoeft SE, Baesman S, Miller LG, Stolz JF. 2005. A microbial arsenic cycle in a salt-saturated, extreme environment. *Science* 308:1305–1308.

Oremland RS, Stolz J. 2000. Dissimilatory reduction of selenate and arsenate in nature. In: Lovley DR, ed. *Environmental Microbial–Metal Interactions*. Washington, DC: ASM Press, pp. 199–224.

Oremland RS, Stolz JF, Hollibaugh JT. 2004. The microbial arsenic cycle in Mono Lake, California. *FEMS Microbiol Ecol* 48:15–27.

Oremland RS, Switzer Blum J, Culbertson CW, Visscher PT, Miller LG, Dowdle P, Strohmaier FE. 1994. Isolation, growth, and metabolism of an obligately anaerobic, selenate-respiring bacterium, strain SES-3. *Appl Environ Microbiol* 60:3011–3019.

Osborne FH. 1973. Arsenite oxidation by a soil isolate of *Alcaligenes*. PhD Thesis. Rensselaer Polytechnic Institute, Troy, NY.

Osborne FH, Ehrlich HL. 1976. Oxidation of arsenite by a soil isolate of *Alcaligenes*. *J Appl Bacteriol* 41:295–305.

Oscarson DW, Huang PM, Defosse C, Herbillon A. 1981. Oxidative power of Mn(IV) and Fe(III) oxides with respect to As(III) in terrestrial and aquatic environments. *Nature* (London) 291:50–51.

Petrick JS, Ayala-Fierro F, Cullen WR, Carter DE, Aposhian HV. 2000. Monomethylarsonous acid (MMA^{III}) is more toxic than arsenite in Chang human erythrocytes. *Toxixol Appl Pharmacol* 163:203–207.

Philips SE, Taylor ML. 1976. Oxidation of arsenite to arsenate by *Alcaligenes faecalis*. *Appl Environ Microbiol* 32:392–399.

Pickett AW, McBride BC, Cullen WR, Manji H. 1981. The reduction of trimethylarsine oxide by *Candida humicola*. *Can J Microbiol* 27:773–778.

Pol'kin SI, Tanzhnyanskaya ZA. 1968. Use of bacterial leaching for ore enrichment. *Izv Vyssch Ucheb Zaved Tsvet Met* 11:115–121.

Pol'kin SI, Yudina N, Nanin VV, Kim DKh. 1973. Bacterial leaching of arsenic from an arsenopyrite gold-containing concentrate in a thick pulp. *Nauchno-Issled Geologorazved Inst Tsvetn Blagorodn Metall* 107:34–41.

Quastel JH, Scholefield PG. 1953. Arsenite oxidation in soil. *Soil* 75:279–285.

Raven KP, Jain A, Loeppert RH. 1998. Arsenite and arsenate adsorption on ferrihydrite: Kinetics, equilibrium, and adsorption envelopes. *Environ Sci Technol* 32:344–349.

Salmassi TM, Venkateswaren K, Satomi M, Nealson KH, Newman DK, Hering JG. 2002. Oxidation of arsenite by *Agrobacterium albertimagni*, AOL15, sp. nov., isolated from Hot Creek, California. *Geomicrobiol J* 19:53–66.

Salmassi TM, Walker JJ, Newman DK, Leadbetter JR, Pace NR, Hering JG. 2006. Community and cultivation analysis of arsenite oxidizing biofilms at Hot Creek. *Environ Microbiol* 8:50–59.

Saltikov CW, Cifuentes A, Venkateswaran K, Newman DK. 2003. The *ars* detoxification system is advantageous but not required for As(V) respiration by the genetically tractable *Shewanella* species strain ANA-3. *Appl Environ Microbiol* 69:2800–2809.

Saltikov CW, Newman DK. 2003. Genetic identification of a respiratory arsenate reductase. *Proc Natl Acad Sci USA* 100:10983–10988.

Saltikov CW, Wildman RA, Jr, Newman DK. 2005. Expression dynamics of arsenic respiration and detoxification in *Shewanella* sp. strain ANA-3. *J Bacteriol* 187:7390–7396.

Santini JM, Sly LI, Schnagl RD, Macy JM. 2000. A new chemolithotrophic arsenite-oxidizing bacterium isolated from a gold mine: Phylogenetic, physiological, and preliminary biochemical studies. *Appl Environ Microbiol* 66:92–97.

Santini JM, vanden Hoven RN. 2004. Molybdenum-containing arsenite oxidase of the chemolithoautotrophic arsenite oxidizer NT-26. *J Bacteriol* 186:1614–1619.

Scudlark JR, Johnson DL. 1982. Biological oxidation of arsenite in seawater. *Estuarine, Coastal and Shelf Science* 14:693–706.

Sehlin HM, Lindström EB. 1992. Oxidation and reduction of arsenic by *Sulfolobus acidocaldarius* strain BC. *FEMS Microbiol Lett* 93:87–92.

Silver M, Torma AE. 1974. Oxidation of metal sulfides by *Thiobacillus ferrooxidans* grown on different substrates. *Can J Microbiol* 20:141–147.

Silver S. 1998. Genes for all metals: A bacterial view of the periodic table. *J Ind Microbiol Biotechnol* 20:1–12.

Silver S, Keach D. 1982. Energy-dependent arsenate flux: The mechanism of plasmid-mediated resistance. *Proc Natl Acad Sci USA* 79:6114–6118.

Silver S, Phung LT. 2005. Genes and enzymes involved in bacterial oxidation and reduction of inorganic arsenic. *Appl Environ Microbiol* 71:599–608.

Silverman MP, Lundgren DG. 1959. Studies on the chemoautotrophic iron bacterium *Ferrobacillus ferrooxidans*. 1. An improved medium and a harvesting procedure for securing high cell yields. *J Bacteriol* 77:642–647.

Stolz JF, Ellis DJ, Switzer Blum J, Ahmann D, Lovley DR, Oremland RS. 1999. *Sulfurospirillum barnesii* sp. nov., *Sulfurospirillum arsinophilus* sp. nov., and the *Sulfurospirillum* clade in the Epsilon Proteobacteria. *Int J Syst Bacteriol* 49:1177–1180.

Stolz JF, Oremland RS. 1999. Bacterial respiration of arsenic and selenium. *FEMS Microbiol Rev* 23:615–627.

Swartz CH, Blute NK, Badruzzman B, Ali A, Brabander D, Jay J, Besancon J, Islam S, Hemond HF, Harvey CF. 2004. Mobility of arsenic in a Bangladesh aquifer: Inferences from geochemical profiles, leaching data, and mineralogical characterization. *Geochim Cosmochim Acta* 68:4539–4557.

Switzer Blum J, Burns Bindi A, Buzzelli J, Stolz JF, Oremland RS. 1998. *Bacillus arsenicoselenatis*, sp. nov., and *Bacillus selenitireducens*, sp. nov.: Two haloalkaliphiles from Mono Lake, California that respire oxyanions of selenium and arsenic. *Arch Microbiol* 171:19–30.

Takai K, Kobayashi H, Nealson KH, Horikoshi K. 2003. *Sulfurihydrogenibium subterraneum* gen. nov., sp. nov., from a subsurface hot aquifer. *Int J Syst Evol Microbiol* 53:823–827.

Torma AE, Gabra GG. 1977. Oxidation of stibnite by *Thiobacillus ferrooxidans*. *Antonie v Leeuwenhoek* 43:1–6.

Tuovinen OH, Bhatti TM, Bigham JM, Hallberg KB, Garcia O, Jr, Lindström EB. 1994. Oxidative dissolution of arsenopyrite by mesophilic and moderately thermophilic acidophiles. *Appl Environ Microbiol* 60:3268–3274.

Turner AW. 1949. Bacterial oxidation of arsenite. *Nature* (London) 164:76–77.

Turner AW. 1954. Bacterial oxidation of arsenite. I. Description of bacteria isolated from arsenical cattle-dipping fluids. *Aust J Biol Sci* 7:452–478.

Turner AW, Legge JW. 1954. Bacterial oxidation of arsenite. II. The activity of washed suspensions. *Aust J Biol Sci* 7:479–495.

Wakao N, Koyatsu H, Komai Y, Shimokawara H, Sakurai Y, Shibota H. 1988. Microbial oxidation of arsenite and occurrence of arsenite-oxidizing bacteria in acid mine water from a sulfur-pyrite mine. *Geomicrobiol J* 6:11–24.

Weast RC, Astle MJ. 1982. *CRC Handbook of Chemistry and Physics*. 63rd ed. Boca Raton, FL: CRC Press.

Weeger W, Lièvremont D, Perret M, Lagarde F, Hubert J-C, Leroy M, Lett M-C. 1999. Oxidation of arsenite to arsenate by a bacterium from an aquatic environment. *BioMetals* 12:141–149.

Winchester JW. 1972. Geochemistry. In: Fairbridge RW, ed. *The Encyclopedia of Geochemistry and Environmental Sciences. Encyclopedia Earth Science Series IVA*, Vol. IVA. New York: Van Nostrand Reinhold, pp. 402–410.

Wolthers M, Charlet L, van der Weijden CH, van der Linde PR, Rickard D. 2005. Arsenic mobility in the ambient sulfidic environment: Sorption of arsenic(V) and arsenic(III) onto disordered mackinawite. *Geochim Cosmochim Acta* 69:3483–3492.

Woolfolk CA, Whiteley HR. 1962. Reduction of inorganic compounds with molecular hydrogen by *Micrococcus lactilyticus*. I. Stoichiometry with compounds of arsenic, selenium, tellurium, transition and other elements. *J Bacteriol* 84:647–658.

Zobrist J, Dowdle PR, Davis JA, Oremland RS. 2000. Mobilization of arsenate by dissimilatory reduction of adsorbed arsenate. *Environ Sci Technol* 34:4747–4753.

15 Geomicrobiology of Mercury

15.1 INTRODUCTION

The element mercury has been known as a specific chemical from at least as far back as 1500 BC. The physician Paracelsus (AD 1493–1541) attempted to cure syphilis by administering metallic mercury to sufferers of the disease. His treatment was probably based on intuitive or empirical knowledge that at an appropriate dosage, mercury was more toxic to the cause of the disease than to the patient. The true etiology of syphilis was, however, unknown to him. The extent of mercury toxicity for human beings and other animals became very apparent only in recent times as a consequence of environmental pollution by mercury compounds. The toxicity manifests itself in major physical impairments and death from the intake of the compounds in food and water. Incidents of mercury poisoning in Japan (Minimata disease), Iraq, Pakistan, Guatemala, and the United States drew special attention to the problem. In some cases, food was consumed, which had been made from seed grain treated with mercury compounds to inhibit fungal damage before planting. The seed grain had not been intended for food use. In other cases, food such as meat had become tainted because the animals yielding the meat drank water that had become polluted by mercury compounds or they had eaten mercury-tainted feed. Tracing the fate of mercury introduced into the environment has revealed an intimate role of microbes in the interconversion of some mercury compounds.

15.2 DISTRIBUTION OF MERCURY IN EARTH'S CRUST

The abundance of mercury in the Earth's crust has been reported as 0.08 ppm (Jonasson, 1970). Its concentration in freshwater may range from 0.01 to 10 ppb, although concentrations as high as 1600 ppb have been measured in waters in contact with copper deposits in the southern Urals (Jonasson, 1970). The maximum permissible level in potable waters was set at 1 ppb as of 1990 by the World Health Organization (WHO) (Cotruvo and Vogt, 1990). The average mercury concentration in seawater has been reported as 0.2 ppb (Marine Chemistry, 1971).

In nature, mercury can exist as metal or in inorganic and organic compounds. The metal is liquid at ambient temperature and has a significant vapor pressure of 1.2×10^{-3} mm Hg at 20°C and heat of vaporization of 14.7 cal mol^{-1} at 25°C (Vostal, 1972). The most prevalent mineral of mercury is cinnabar (HgS). It is found in highest concentration in volcanically active zones such as the circumpacific volcanic belt, the East Pacific Rise, and the Mid-Atlantic Ridge. The occurrence of mercury in its metallic state is rare. In water, inorganic mercury may exist as aquo, hydroxo, halido, and bicarbonate complexes of mercuric ion, but the mercuric ion may also be adsorbed to particulate or colloidal materials in suspension (Jonasson, 1970). In soil, inorganic mercury may exist in the form of elemental mercury vapor that may be adsorbed to soil matter, at least in part. It can also exist as mercury humate complexes at pH 3–6, or as $Hg(OH)^+$ and $Hg(OH)_2$ in the pH range 7.5–8. The latter two species may be adsorbed to soil particles (Jonasson, 1970). Mercury in soil and water may also exist as methylmercury [$(CH_3)Hg^+$], which may be adsorbed by negatively charged particles such as clays.

Mercuric ions (Hg^{2+}) are toxic because they bind readily to exposed sulfhydryl (–SH) groups of enzyme proteins and are therefore potent nonspecific enzyme inhibitors. Their toxicity can be modulated by various organic solutes that can form mercury(II) complexes. This is of special significance because it can affect the determination of mercury toxicity for microbes in growth assays (Farrell et al., 1993).

15.3 ANTHROPOGENIC MERCURY

The local mercury level in the environment may be affected by human activity. These activities include industrial operations such as the synthesis of certain chemicals such as vinyl chloride and acetaldehyde, which employ inorganic mercury compounds as catalysts. They also include the electrolytic production of chlorine gas and caustic soda, which employs mercury electrodes, and they also include the manufacture of paper pulp, which makes use of phenylmercuric acetate as a slimicide (Jonasson, 1970). Some of the mercury used in these processes may accidentally pollute the environment. In agriculture, organic compounds used as fungicides to prevent fungal attack of seeds may pollute the soil. In mining, the exposure of mercury ore deposits and other deposits in which mercury is only a trace component leads to weathering and resultant solubilization introduces some of the mercury into the environment.

15.4 MERCURY IN ENVIRONMENT

As Jonasson (1970) has pointed out, in the past inorganic mercury compounds were considered less toxic than organic mercury compounds, but since the discovery that inorganic mercury compounds can be converted into organic ones (e.g., methylmercury), this is no longer considered to be true. Living tissue has a high affinity for methylmercury [$(CH_3)Hg^+$]. Fish have been found to concentrate it up to 3000 times the concentration found in water. This is because methylmercury is fat- as well as water-soluble and is taken up more readily by living cells than mercuric ion. Owing to the lipid solubility of methylmercury, nervous tissue, especially the brain, has a high affinity for it. It is also bound by inert matter, especially negatively charged particles such as clays.

Dimethylmercury [$(CH_3)_2Hg$] is volatile. It can thus enter the atmosphere from soil or water phases. The ultraviolet component of sunlight can, however, dissociate dimethylmercury into volatile elemental mercury, methane, and ethane.

Microorganisms have been shown to be intimately involved in the interconversions of inorganic and organic mercury compounds (Trevors, 1986). The initial discoveries of such microbial activities were those of Jensen and Jernelöv (1969) who demonstrated the production of methylmercury from mercuric chloride ($HgCl_2$) added to lake sediment samples and incubated for several days in the laboratory. They also noted the production of dimethylmercury from decomposing fish tissue containing methylmercury or supplemented with Hg^{2+} and incubated for several weeks. Later work established that methylation was brought about by bacteria and fungi (see Section 15.5). Mercury methylation is inhibited at acidic pH (Steffan et al., 1988).

Microbial action on mercury compounds is a means of *detoxification*. By forming volatile elemental mercury (see Sections 15.5.1, 15.5.2, and 15.5.4) or dimethylmercury, neither of which is water-soluble, the microbes ensure removal of mercury from their environment into the atmosphere. Even the microbial formation of methylmercury can be a form of mercury detoxification, because the methylmercury can be immobilized by adsorption to negatively charged clay particles in sediment or soil, which removes it as a toxicant from the microbial environment. Similarly, the precipitation of HgS by reaction of Hg^{2+} with biogenic H_2S is a type of mercury detoxification, because the solubility of HgS is very low ($K_{sol} = 10^{-49}$). Of all these detoxification mechanisms, formation of volatile metallic mercury has been thought to be the predominant microbial mercury detoxification mechanism (Robinson and Tuovinen, 1984). Baldi et al. (1987) demonstrated the presence of a significant number of mercury-resistant bacteria that could reduce Hg^{2+} to Hg^0 but not methylate it at sites surrounding natural mercury deposits situated in Tuscany, Italy. Baldi et al. (1989) also isolated 37 strains of aerobic, mercury-resistant bacteria from the Fiora River in southern Tuscany, which receives drainage from a cinnabar mine. All 37 strains were able to reduce Hg^{2+} to Hg^0, and three were also able to degrade methylmercury. None was able to generate methylmercury.

15.5 SPECIFIC MICROBIAL INTERACTIONS WITH MERCURY

15.5.1 NONENZYMATIC METHYLATION OF MERCURY BY MICROBES

An early study of the biochemistry of microbial methylation of mercury involved the use of a cell extract of a methanogenic culture in the presence of low concentrations of Hg^{2+}. This extract caused the formation of $(CH_3)_2Hg$ but little amount of methane through preferential interaction between methylcobalamin and Hg^{2+} (Wood et al., 1968). Although the production of methylcobalamin in this instance depended on enzymatic catalysis, the production of $(CH_3)_2Hg$ from the reaction of Hg^{2+} with methylcobalamin did not. This nonenzymatic nature of mercury methylation by methylcobalamin was confirmed by Bertilsson and Neujahr (1971), Imura et al. (1971), and Schrauzer et al. (1971). The nonenzymatic mechanism of mercury methylation by methylcobalamin was explained as follows (DeSimone et al., 1973):

$$Hg^{2+} \xrightarrow{\quad CH_3B_{12} \quad} (CH_3)Hg^+ \xrightarrow{\quad CH_3B_{12} \quad} (CH_3)_2Hg \qquad (15.1)$$

According to Wood (1974), the initial methylation of Hg^{2+} in this reaction sequence proceeds 6000 times as fast as the second one. However, more recent study of these reactions has indicated that the methylation of methylmercury can proceed as fast as the initial methylation of mercury (Filippelli and Baldi, 1993; Baldi, 1997). The methylation rate of methylmercury is affected by the counter ion associated with methylmercuric ion. The rate is faster if sulfate is frequently the counter ion than when it is halogen ion such as chloride or iodide. The halogens have a greater tendency than sulfate to bind covalently to Hg^{2+}.

Dimethylmercury can also arise from a reaction of methylmercury with hydrogen sulfide (Craig and Bartlett, 1978; Baldi, 1997). In nature, hydrogen sulfide is frequently of biogenic origin, formed anaerobically by sulfate-reducing bacteria. The transformation of methyl- to dimethylmercury can be summarized by the reactions

$$2(CH_3)Hg^+ + H_2S \rightarrow (CH_3)Hg-S-Hg(CH_3) + 2H^+ \qquad (15.2a)$$

$$(CH_3)Hg-S-Hg(CH_3) \rightarrow (CH_3)_2Hg + HgS \qquad (15.2b)$$

Reaction 15.2b is slow and appears to be the rate-controlling reaction (Baldi et al., 1993; Baldi, 1997).

Other studies have revealed that mercury can be nonenzymatically methylated by microbes other than methanogens, including both aerobes and anaerobes (see review by Robinson and Tuovinen, 1984). Among anaerobes, *Clostridium cochlearium* was shown to methylate mercury contained in HgO, $HgCl_2$, $Hg(NO_3)_2$, $Hg(CN)_2$, $Hg(SCN)_2$, and $Hg(OOCH_3)_2$ (Yamada and Tonomura, 1972a,b).

Among aerobes, *Pseudomonas* spp., *Bacillus megaterium*, *Escherichia coli*, *Enterobacter aerogenes*, and others have been implicated (see summary by Robinson and Tuovinen, 1984). Even fungi such as *Aspergillus niger*, *Scopularis brevicaulis*, and *Saccharomyces cerevisiae* have been found capable of mercury methylation (see Robinson and Tuovinen, 1984).

Bacteria and other microbes have been found to be able to methylate metals other than mercury. Among these metals are cadmium, lead, tin, and thallium and the metalloids selenium and tellurium (Brinckman et al., 1976; Guard et al., 1981; Schedlbauer and Heumann, 2000; Wong et al., 1975; review by Summers and Silver, 1978). Methylation of some metals may occur as a result of nonbiological transmethylation by methylated donor compounds of biogenic origin such as trimethyltin and methyl iodide (Brinckman and Olson, 1986). Methyl halides, including methyl iodides (White, 1982;

Brinckman and Olson, 1986), are produced by some marine algae and also by microorganisms associated with them (White, 1982; Manley and Dastoor, 1988) and by fungi (Harper, 1985). They can readily react nonenzymatically with some metal salts (Brinckman and Olson, 1986). Trimethyltin yields cabanions that can methylate metal ions such as those of palladium, tallium, platinum, and gold, forming unstable methylated intermediates that undergo reductive demethylation to yield the metal in the elemental state. Mercuric ion reacting with trimethyltin, however, forms stable methylmercury (Brinckman and Olson, 1986).

Although in the laboratory nonenzymatic methylation appears to be favored by anaerobic conditions, partially aerobic conditions are needed in nature. This is because under *in situ* anaerobic conditions, biogenic H_2S may prevail, and as a result, mercuric mercury will exist most probably as HgS (Fagerstrom and Jernelöv, 1971; Vostal, 1972). HgS cannot be methylated without prior conversion to a soluble Hg^{2+} salt or HgO (Yamada and Tonomura, 1972c).

15.5.2 Enzymatic Methylation of Mercury by Microbes

Sulfate reducers such as *Desulfovibrio desulfuricans* appear to be the principal methylators of mercury in some anoxic estuarine sediments when sulfate is limiting and fermentable organic energy sources are available (Compeau and Bartha, 1985; Benoit et al., 2001). Some other sulfate reducers also have this capacity (Benoit et al., 2001). The methyl group was shown to originate from carbon-3 of the amino acid serine, which is transferred to tetrahydrofolate (THF) to form methylene-THF catalyzed by serine hydroxymethyl transferase. The methylene-THF is then reduced to methyl-THF by reduced ferrodoxin. In the methylation of mercury, the methyl group of methyl-THF is transferred to mercury via a cobalamin–protein complex, the cobalamin being the transfer agent and the protein being the catalyst (enzyme) of the methyl transfer from methyl-THF to Hg^{2+} (Berman et al., 1990; Choi and Bartha, 1993; Choi et al., 1994a,b). Methyl-THF may also arise from formate via formyl-THF, methenyl-THF, and methylene-THF. The normal role of methylcobalamin–protein complex in *D. desulfuricans* is to provide the methyl group in acetate synthesis from CO_2 by the acetyl-CoA pathway (see Chapters 6 and 19). Hg^{2+} evidently acts as a competing methyl acceptor in acetate formation in the organism (Choi et al., 1994b). The methyl transfer from the methylcobalamin–protein complex to mercury follows Michaelis–Menten kinetics and is 600 times as fast as the uncatalyzed transfer from methylcobalamin at pH 7. These findings raise a question about the abiotic mercury methylation described in Section 15.5.1. These mercury methylation reactions should be reexamined to check whether a transmethylase is never involved in these instances. Dissimilatory iron-reducing bacteria belonging to the genera *Geobacter* and *Desulfuromonas* showed ability to methylate mercury while reducing Fe(III), whereas *Shewanella* strains were not able to do so (Kerin et al., 2006). As Kerin et al. (2006) pointed out, *Geobacter* and *Desulfuromonas* are closely related to mercury-methylating sulfate reducers in the *Deltaproteobacteria* whereas *Shewanella* is not.

The fungus *Neurospora crassa* uses a different mechanism of methylating mercury (Landner, 1971). In this organism, mercury forms a complex with homocysteine or cysteine nonenzymatically, and then, with the help of a methyl donor and the enzyme transmethylase, methylmercury is cleaved from this complex. The following equation illustrates this reaction with cysteine:

$$\underset{\substack{\\ \text{COOH}}}{\overset{\substack{\text{SH} \\ | \\ \text{CH}_2 \\ | }}{Hg^{2+} + \text{CHNH}_2}} \quad \longrightarrow \quad \underset{\substack{\\ \text{COOH}}}{\overset{\substack{\text{SHg}^+ \\ | \\ \text{CH}_2 \\ | }}{\text{CHNH}_2}} \quad \xrightarrow[\text{transmethylase}]{\text{Methyl donor}} \quad \underset{\substack{\\ \text{COOH}}}{\overset{\substack{\text{SH} \\ | \\ \text{CH}_2 \\ | }}{\text{CHNH}_2}} + (\text{CH}_3)\text{Hg}^+ \qquad (15.3)$$

Suitable methyl donors are betaine or choline but not methylcobalamin.

15.5.3 MICROBIAL DIPHENYLMERCURY FORMATION

A case of conversion of phenylmercuric acetate to diphenylmercury has been reported (Matsumara et al., 1971). The reaction can be summarized as follows (where φ represents the phenyl moiety):

$$2\phi Hg^+ \rightarrow \phi\text{--}Hg\text{--}\phi + \text{unknown Hg compound and a trace of } Hg^{2+} \qquad (15.4)$$

15.5.4 MICROBIAL REDUCTION OF MERCURIC ION

Mercuric ion is reduced to volatile metallic mercury by various members of the domain Bacteria. Examples of these include strains of *Pseudomonas* spp., enteric bacteria, *Staphylococcus aureus*, *Acidithiobacillus ferrooxidans*, group B *Streptococcus*, *Bacillus*, *Vibrio*, coryneform bacteria, *Cytophaga*, *Flavobacterium*, *Achromobacter*, *Alcaligenes*, and *Acinetobacter* and *Streptomyces* (Komura et al., 1970; Nelson et al., 1973; Nelson and Colwell, 1974; Olson et al., 1981; Nakahara et al., 1985; Summers and Lewis, 1973). Bacteria that are able to reduce Hg^{2+} to Hg^0 are mercury-resistant. They possess the *mer* operon that may be located on a plasmid, transposon, or bacterial chromosome. Its genetic components include the gene sequence *merR*, *merT*, *merP*, *merC*, *merD*, *merA*, and *merB* (Nascimento and Chartone-Souza, 2003). The gene (*merA*) codes for the enzyme mercuric reductase (*merA*), which is responsible for Hg^{2+} reduction to Hg^0. Gene (*merP*) codes for a protein that scavenges Hg^{2+}, and gene (*merT*) codes for a membrane protein involved in transporting the Hg^{2+} from the periplasm into the cytoplasm where the reduction takes place. The expression of the *mer* operon is controlled by a product of *merR* (reviewed in Barkey et al., 2003; Schelert et al., 2004).

Information on the ability of Archaea to reduce Hg^{2+} to Hg^0 is limited to date. Schelert et al. (2004) detected the presence of mercury-resistant *Crenarcheota* in Coso Hot Springs of Yellow Stone National Park, United States, whose DNA included a *merA* gene. Study of the *merA* gene in *Sulfolobus solfataricus* revealed that it was constitutive, unlike *merA* in Bacteria. Nevertheless, expression of *merR* was shown to be essential for *S. solfataricus* to be resistant to Hg^{2+} (Simbahan et al., 2005).

At least two species of the yeast *Cryptococcus* (Fungi, Eukaryota), *C. albidus* and *C. neoformans*, have been shown to contain a *merA* gene (see Tax Blast report: *Cryptococcus*,%20Hg/927499… 928899. fa_tax.html). Brunker and Bott (1974) found that *C. albidus* reduced Hg^{2+} to Hg^0. However, the Hg^0 appeared to accumulate in the cell wall, membrane, and vacuoles of the yeast. Hg^0 volatilization was not reported.

The enzyme that catalyzes Hg^{2+} reduction to Hg^0 involves a soluble NADPH-dependent cytoplasmic flavoprotein that is active in the presence of an excess of exogenously supplied thiols (RSH). The thiols react with Hg^{2+} to form a thiol complex (RS–Hg–SR). The thiols (in the laboratory they may be mercaptoethanol, dithiothreitol, glutathione, or cysteine) ensure the reduced state of mercuric reductase and the formation of Hg^0. The reaction catalyzed by mercuric reductase may be summarized as follows (Fox and Walsh, 1982; Foster, 1987):

$$NADPH + H^+ + RS\text{--}Hg\text{--}SR \rightarrow NADP^+ + Hg^0\uparrow + 2RSH \qquad (15.5)$$

In some reactions, the dimercaptal derivative of Hg^{2+} may be replaced by a monomercaptal or an ethylenediamine (EDTA) derivative. NADPH can be replaced by NADH with enzyme preparations from some organisms, but the preparation is then less active. The kinetics for the purified enzyme is biphasic. (See review by Robinson and Tuovinen, 1984, for a more detailed discussion.) Although the reaction occurs under reducing conditions, it is performed by many obligate and facultative aerobes (Nelson et al., 1973; Spangler et al., 1973a,b).

Mercuric reductase activity is not entirely substrate-specific. Besides mercuric ion, the enzyme can also reduce ionic silver and ionic gold to corresponding metal colloids (Summers and Sugarman, 1974; Summers, 1986). Silver and gold resistance in bacteria is not, however, related to this enzyme activity (Summers, 1986).

Not all reduction of mercury observed in nature is biological (Nelson and Colwell, 1975). Chemical reduction may occur as a result of interaction with humic acid (Alberts et al., 1974).

15.5.5   FORMATION OF META-CINNABAR (ß-HgS) FROM Hg(II) BY CYANOBACTERIA

The cyanobacteria *Limnothrix planctonica* (Lemm.), *Synecococcus leopoldensis* (Racib.) Komarek, and *Phormidium limnetica* (Lemm.) were recently shown to form mercuric sulfide from mercuric ion and to accumulate it intracellularly (Lefebvre et al., 2007). The sulfide with which Hg^{2+} taken into the cells was combined was formed by the cells from an intracellular thiol pool, as demonstrated by use of the inhibitors dimethylfumarate and iodoacetamide. On initial exposure to Hg(II) in the medium in laboratory experiments, some Hg^0 was formed as well as ß-HgS, but the rate of Hg^0 formation decreased rapidly. Increase in growth temperature enhanced ß-HgS formation and decreased Hg^0 evolution.

15.5.6   MICROBIAL DECOMPOSITION OF ORGANOMERCURIALS

Phenyl- and methylmercury can be microbially converted to volatile Hg^0 by bacteria in lake and estuarine sediments and in soil (Nelson et al., 1973; Spangler et al., 1973a,b; Tonomura et al, 1968). The bacteria that seem most frequently involved are mercury-resistant strains of *Pseudomonas*. Although mercury-resistant strains of other genera, such as *Escherichia coli* and *Staphylococcus aureus*, also exhibit this activity, they seem to be much less active (Nelson and Colwell, 1975). The removal of phenyl or methyl groups linked to mercuric mercury is catalyzed by a class of enzymes called mercuric lyases, encoded by the gene *merB* in the *mer* operon (see Section 15.5.4). They catalyze the cleavage of carbon–mercury bonds and in laboratory demonstrations require the presence of an excess of reducing agent such as L-cysteine. They release Hg^{2+}, which is then reduced to Hg^0 by mercuric reductase (Furukawa and Tonomura, 1971, 1972a,b; Tezuka and Tonomura, 1976, 1978; Robinson and Tuovinen, 1984). Phenylmercuric lyase can be inducible (Nelson et al., 1973; Robinson and Tuovinen, 1984). The overall reactions the lyases catalyze can be summarized as follows:

$$\phi Hg^+ + H^+ + 2e \rightarrow Hg^0 + H\phi \qquad (15.6)$$

$$(CH_3)Hg^+ + H^+ + 2e \rightarrow Hg^0 + CH_4 \qquad (15.7)$$

Besides methyl- and phenylmercury compounds, some bacteria are able to decompose one or more of the following organomercurials: ethylmercuric chloride (EMC), fluorescine mercuric acetate, *para*-mercuribenzoate (pHMB), thimerosal, and merbromin (see review by Robinson and Tuovinen, 1984). The dephenylation of triphenyltin (TPT) to diphenyltin by *Pseudomonas chloraphis* ATCC 9446, *P. fluorescens* ATCC 13525, and *P. aeruginosa* ATCC 15962 does not involve mercuric lyase but a TPT-degrading factor, which has a low molecular mass of ~1000 Da and resembles pyoverdin (Inoue et al., 2000).

15.5.7   OXIDATION OF METALLIC MERCURY

Elemental mercury (Hg^0) has been reported to be oxidizable to mercuric ion in the presence of certain bacteria (Holm and Cox, 1975). Although strains of *Pseudomonas aeruginosa*, *P. fluorescens*, *Escherichia coli*, and *Citrobacter* oxidized only small amounts of Hg^0, strains of *Bacillus subtilis* and *B. megaterium* oxidized more significant amounts. In none of these cases was methylmercury formed. The observed oxidation was reported not to have been enzymatic but due to reaction with metabolic products, which acted as oxidants. Even yeast extract was found to be able to oxidize Hg^0.

By contrast, Smith et al. (1998) demonstrated enzymatic oxidation of metallic mercury in vapor form (monatomic) by hydroperoxidase-catalase, KatG, in growing *Escherichia coli*. The hydroperoxidase-catalase active in stationary cells, Kat E, was found proportionately less active. *Bacillus subtilis* PY79 exhibiting strong catalase activity also promoted Hg^0 oxidation, as did *Streptomyces venezuelae* with weaker catalase activity (Smith et al., 1998). Because of the great toxicity of Hg^{2+}, it is unclear how this activity benefits organisms capable of this activity unless it occurs in environments where Hg^{2+} is rapidly immobilized. The oxidation of monatomic mercury by catalase was previously observed in mammals and plants (see references cited by Smith et al., 1998). An assessment is needed of the relative contributions of biotic and abiotic Hg^0 oxidation in different environments.

15.6   GENETIC CONTROL OF MERCURY TRANSFORMATIONS

In general, resistance to the toxicity of inorganic mercury compounds in bacteria is attributable to the ability to form mercuric reductase and, for certain organomercurials, mercuric lyase. However, in a strain of *Enterobacter aerogenes*, bacterial resistance to some organomercurials may be due to membrane impermeability (Pan Hou et al., 1981), and in *Clostridium cochlearium* it is due to demethylation followed by precipitation with H_2S generated by the organism (Pan Hou and Imura, 1981).

As previously noted, the genes encoding mercuric reductase and mercuric lyase are part of the *mer* operon, which may occur on a plasmid, that is, it is R- or sex-factor linked (Belliveau and Trevors, 1990; Komura and Izaki, 1971; Loutit, 1970; Novick 1967; Richmond and John, 1964; Smith, 1967; Schottel et al., 1974; Summers and Silver, 1972; Silver and Phung, 1996; Silver, 1997), on a transposon, or even on the bacterial chromosome (Foster, 1987; Nascimento and Chartone-Souza, 2003; Barkey et al., 2003). Some needed components of the mercury resistance gene complex (operon) on a gram-negative transposon Tn*21* from *Shigella* plasmid R100 were determined by Hamlett et al. (1992). The mercury resistance determinant (gene complex) in *Bacillus cereus* RC607 was mapped by Gupta et al. (1999). Except in *Acidithiobacillus ferrooxidans*, the mercury resistance genes (*mer*) in all bacteria so far tested are expressed only in the presence of mercury compounds; that is, the gene products for which they code are *inducible* (Robinson and Tuovinen, 1984). Depending on the organism, induction of the two enzymes mercuric lyase and mercuric reductase may be *coordinated*; that is, the two genes are under common regulatory control. In such an instance, an organomercurial induces both the lyase and the mercuric reductase (see Robinson and Tuovinen, 1984). As previously indicated, in *A. ferrooxidans* (formerly *T. ferrooxidans*), the mercuric resistance (Hg^r) trait appears to be constitutive (Olson et al., 1982). In *Clostridium cochlearium*, which lacks the *mer* operon, the genetic determinants for demethylation of methylmercury are also plasmid-encoded (Pan Hou and Imura, 1981). The demethylated mercury may be precipitated as HgS with biogenic H_2S to render it nontoxic.

Many gram-positive bacteria in the environment that are sensitive to mercury have nevertheless been found to possess the regulatory gene, *merR*, and the gene *merA*, which codes for the mercuric reductase enzyme and is inducible in these organisms (Bogdanova et al., 1992). Mercuric reductase activity in these strains was tested in extracts of induced cells. These organisms appear to lack a transport function (e.g., *merT* and *merP*) needed by the mercury resistance mechanism, leaving these cultures sensitive to mercury. In other words, they have a cryptic *mer* operon (Bogdanova et al., 1992).

In nature, plasmid-determined mercury resistance in bacteria can be transferred from resistant to susceptible cells through conjugation or phage transduction among gram-negative organisms and through phage transduction among gram-positive organisms (Summers and Silver, 1972). In other words, a mercury-sensitive strain can become mercury resistant by acquiring a plasmid containing the *mer* genes from an Hg^r bacterium.

In the fungus *Neurospora crassa*, in which Hg^{2+} methylation is the basis for mercury tolerance, the mercury-resistant strain isolated by Landner (1971) appears to be a mutant that has lost control over one of the last enzymes in methionine biosynthesis, so that methylation of Hg^{2+} no longer competes with methionine biosynthesis, as it does in the wild-type parent strain. The resistant strain could tolerate 225 mg Hg L^{-1}.

15.7   ENVIRONMENTAL SIGNIFICANCE OF MICROBIAL MERCURY TRANSFORMATIONS

The enzymatic attack of mercury compounds is not for the dissipation of excess reducing power or respiration, nor is it for the production of useful metabolites. Its function is detoxification. This is well illustrated in experiments with mercury-sensitive and mercury-resistant strains of *Thiobacillus* (now *Acidithiobacillus*) *ferrooxidans* (Baldi and Olson, 1987). The mercury-sensitive strain could not oxidize pyrite admixed with cinnabar (HgS) in oxidation columns, whereas two resistant strains could. Yet even the resistant strains could not use cinnabar as an energy source. The mercury-resistant strains volatilized the mercury as Hg^0.

Extensive mercury methylation has been observed in mercury-polluted Clear Lake in California (Macalady et al., 2000). Analysis of the community structure in sediment samples from various locations in the lake indicated that *Desulfobacter* sp. was the dominant methylator in the lower arm of the lake, where the organic carbon content was highest.

Volatile mercury (Hg^0) and dimethylmercury, owing to their high volatility and low water solubility, are readily lost to the atmosphere from the normal habitat of the microbes and other creatures. Methylmercury, because of its positive charge, can become immobilized by adsorption to negatively charged clay particles in soil and sediment. The metabolic transformation to volatile forms of mercury or to methylmercury, which may become fixed in soil or sediment, protects not only the organisms actively involved in the conversion but also the coinhabitants that lack this ability and are susceptible to mercury poisoning. By contrast, development of mercury resistance in microbes that is due to lessened permeability to a mercury compound benefits only those organisms that acquired this trait.

15.8   MERCURY CYCLE

On the basis of the interactions of microbes with mercury compounds described in the foregoing sections, it is apparent that microbes play an important role in the movement of mercury in nature, that is, the soil, sediment, and aqueous environments. One of the main results of microbial action on mercury, whether in the form of mercuric ion or alkyl- or arylmercury ions, seems to be its volatilization as Hg^0. Methylmercury ion may also be an important end product of microbial action on mercuric ion, but it also can be an intermediate in the formation of volatile dimethylmercury. Methylmercury is more toxic to susceptible forms of life than mercuric ion, owing to the greater lipid solubility combined with the positive charge of the former compared to the latter. Baldi (1997) suggests that microbial activity is an important factor in homeostatic control on the level of available mercury in the environment.

A mercury cycle is outlined in Figure 15.1. Mercuric sulfide of volcanic origin slowly autoxidizes to mercuric sulfate on exposure to air and moisture and may become disseminated in soil and water through groundwater movement. Bacteria and fungi may reduce the Hg^{2+} to Hg^0, as may humic substances. Hydroperoxidase-catalase of some bacteria and other organisms, and organic matter may catalyze oxidation of monatomic Hg^0 to Hg^{2+}, but some organic matter may also serve as oxidant of Hg^0.

The volatile mercury (Hg^0) may be adsorbed by soil, sediment, and humic substances or lost to the atmosphere. Some of the Hg^{2+} may also become methylated through the action of bacteria

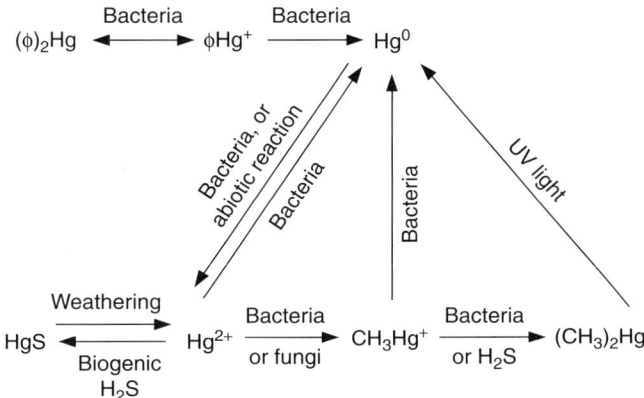

FIGURE 15.1 Mercury transformations by microbes and by chemical and physical agents.

and fungi. Some of the positively charged mercuric ions, methylmercury ions as well as phenyl-mercury ions, may be fixed by negatively charged soil and sediment particles and by humic matter and thereby become immobilized. This may explain why mercury concentrations in soil are often higher in topsoil than in subsoil (Anderson, 1967). Those ions that are not fixed can be further disseminated by water movement.

Mercury methylation and demethylation have been observed to occur seasonally in the surficial sediments of the deep parts of an oligotrophic lake, with measurable amounts of the methylmercury passing into the overlying water column to be subsequently demethylated and reduced to volatile metallic mercury (Korthals and Winfrey, 1987). Methylmercury was also demethylated in the surficial sediments. During the year, methylation was most intense from mid-July through September, whereas demethylation was most active from spring to late summer (Korthals and Winfrey, 1987).

Methylation and demethylation of mercury together with mercury reduction have also been observed in estuarine sediments (Compeau and Bartha, 1984), sulfate reducers being the principal agents of methylation when the supply of sulfate is limiting (Compeau and Bartha, 1985).

Some methylmercury ions may be further methylated to volatile dimethylmercury, which readily escapes into the atmosphere from soil and water. In the atmosphere it may be decomposed into Hg^0, methane, and ethane by solar ultraviolet radiation. Phenylmercury, which usually originates from human activities, may be reduced to volatile mercury by bacteria in soil, but it may also be converted to diphenylmercury. Mercuric ion may be converted to mercuric sulfide by bacterially generated H_2S and thereby be immobilized. Although the formation of mercuric sulfide is strictly dependent on anaerobiosis, the methylation or reduction of mercuric mercury can occur aerobically and anaerobically.

15.9 SUMMARY

Mercuric ion (Hg^{2+}) may be methylated by bacteria and fungi to methylmercury [(CH_3)Hg^+]. At least some fungi use a methylation mechanism different from that of bacteria. Methylmercury is lipid- and water-soluble and more toxic than mercuric ion. In nature, some methylmercury may, however, be adsorbed by soil or sediment, which removes it as a toxicant as long as it remains adsorbed. Some bacteria can methylate methylmercury, forming volatile dimethylmercury. This may be a form of detoxification when it occurs in soil or sediment, because this compound is water-insoluble and can escape into the atmosphere. Methylmercury as well as phenylmercury can be enzymatically reduced to volatile metallic mercury (Hg^0) by some bacteria. Phenylmercury can also be microbially converted to diphenylmercury. Most important, mercuric ion can be enzymatically reduced to metallic mercury by bacteria and fungi. This also is a form of detoxification, because the product of the reduction is volatile. Metallic mercury may also be oxidized to mercuric mercury by bacteria. However, this reaction is not enzymatic but rather the result of interaction of metallic mercury with metabolic by-products. Microbes capable of metabolizing mercury are generally resistant to its toxic effects.

The mercury cycle in nature is under the influence of microorganisms.

REFERENCES

Alberts JJ, Schindler JE, Miller RW, Nutter DE, Jr. 1974. Elementary mercury evolution mediated by humic acid. *Science* 184:895–897.

Anderson A. 1967. Mercury in soil. *Grundförbättring* 20:95–105.

Baldi F. 1997. Microbial transformation of mercury species and their importance in the biogeochemical cycle of mercury. In: Sigel A, Sigel H, eds. *Metal Ions in Biological Systems*. New York: Marcel Dekker, pp. 213–257.

Baldi F, Filippelli M, Olson GJ. 1989. Biotransformation of mercury by bacteria isolated from a river collecting cinnabar mine waters. *Microb Ecol* 17:263–274.

Baldi F, Olson GJ. 1987. Effects of cinnabar on pyrite oxidation by *Thiobacillus ferrooxidans* and cinnabar mobilization by a mercury-resistant strain. *Appl Environ Microbiol* 53:772–776.

Baldi F, Olson GJ, Brinckman FE. 1987. Mercury transformation by heterotrophic bacteria isolated from cinnabar and other metal sulfide deposits in Italy. *Geomicrobiol J* 5:1–16.

Baldi F, Pepi M, Filippelli M. 1993. Methylmercury resistance in *Desulfovibrio desulfuricans* strains in relation to methylmercury degradation. *Appl Environ Microbiol* 59:2479–2485.

Barkey T, Miller SM, Summers AO. 2003. Bacterial mercury resistance from atoms to ecosystems. *FEMS Microbiol Rev* 27:355–384.

Belliveau BH, Trevors JT. 1990. Mercury resistance determined by a self-transmissible plasmid in *Bacillus cereus* 5. *Biol Metals* 3:188–196.

Benoit JM, Gilmour CC, Mason RP. 2001. Aspects of bioavailability of mercury for methylation in pure cultures of *Desulfobulbus propionicus* (1pr3). *J Bacteriol* 183:51–58.

Berman M, Chase T, Jr., Bartha R. 1990. Carbon flow in mercury biomethylation by *Desulfovibrio desulfuricans*. *Appl Environ Microbiol* 56:298–300.

Bertilsson L, Neujahr HY. 1971. Methylation of mercury compounds by methylcobalamin. *Biochemistry* 10:2805–2808.

Bogdanova ES, Mindlin SZ, Pakrova E, Kocur M, Rouch DA. 1992. Mercuric reductase in environmental gram-positive bacteria sensitive to mercury. *FEMS Microbiol Lett* 97:95–100.

Brinckman FE, Iverson WP, Blair W. 1976. Approaches to the study of microbial transformation of metals. In: Sharpley JM, Kaplan AM, eds. *Proceedings of the Third International Biodegradation Symposium*. London: Applied Science, pp. 916–936.

Brinckman FE, Olson GJ. 1986. Chemical principles underlying bioleaching of metals from ores and solid wastes, and bioaccumulation of metals from solution. In: Ehrlich HL, Holmes DE, eds. *Workshop on Biotechnology for the Mining, Metal-Refining, and Fossil Fuel Processing Industries. Biotechnology and Bioengineering Symposium 16*. New York: Wiley, pp. 35–44.

Brunker RL, Bott TL. 1974. Reduction of mercury to the elemental state by a yeast. *Appl Microbiol* 27:870–873.

Choi S-C, Bartha R. 1993. Cobalamin-mediated mercury methylation by *Desulfovibrio desulfuricans* LS. *Appl Environ Microbiol* 59:290–295.

Choi S-C, Chase T, Jr., Bartha R. 1994a. Enzymatic catalysis of mercury methylation by *Desulfovibrio desulfuricans* LS. *Appl Environ Microbiol* 60:1342–1346.

Choi S-C, Chase T, Jr., Bartha R. 1994b. Metabolic pathways leading to mercury methylation in *Desulfovibrio desulfuricans* LS. *Appl Environ Microbiol* 60:4072–4077.

Compeau G, Bartha R. 1984. Methylation and demethylation of mercury under controlled redox, pH, and salinity conditions. *Appl Environ Microbiol* 48:1203–1207.

Compeau G, Bartha R. 1985. Sulfate-reducing bacteria: Principal methylators of mercury in anoxic estuarine sediments. *Appl Environ Microbiol* 50:498–502.

Cotruvo JA, Vogt CD. 1990. Rationale of water quality standards and controls. In: *American Water Works Association. Water Quality and Treatment. A Handbook of Community Water Supplies*, 4th ed. New York: McGraw-Hill, pp. 1–62.

Craig PJ, Bartlett PD. 1978. The role of hydrogen sulfide in the environmental transport of mercury. *Nature* (London) 275:635–637.

DeSimone RE, Penley MW, Charbonneau L, Smith SG, Wood JKM, Hill HAO, Pratt JM, Ridsale S, Williams JP. 1973. The kinetics and mechanism of cobalamin-dependent methyl and ethyl transfer to mercuric ion. *Biochim Biophys Acta* 304:851–863.

Fagerstrom T, Jernelöv A. 1971. Formation of methylmercury from pure mercuric sulfide in aerobic organic sediment. *Water Res* 5:121–122.

Farrell RE, Germida JJ, Huang PM. 1993. Effects of chemical speciation in growth media on the toxicity of mercury(II). *Appl Environ Microbiol* 59:1507–1514.

Filippelli M, Baldi F. 1993. Alkylation of ionic mercury to methylmercury and dimethylmercury by methylcobalamin: Simultaneous determination by purge-and-trap GC in line with FTIR. *Appl Organometal Chem* 7:487–493.

Foster TJ. 1987. The genetics and biochemistry of mercury resistance. *CRC Crit Rev Microbiol* 15:117–140.

Fox B, Walsh CT. 1982. Mercuric reductase: Purification and characterization of a transposon-encoded flavoprotein containing an oxidation-reduction-active disulfide. *J Biol Chem* 257:2498–2503.

Furukawa K, Tonomura K. 1971. Enzyme system involved in decomposition of phenyl mercuric acetate by mercury-resistant *Pseudomonas*. *Agric Biol Chem* 35:604–610.

Furukawa K, Tonomura K. 1972a. Induction of metallic-mercury-releasing enzyme in mercury-resistant Pseudomonas. *Agric Biol Chem* 36(Suppl. 13):2441–2448.

Furukawa K, Tonomura K. 1972b. Metallic mercury-releasing enzyme in mercury-resistant *Pseudomonas*. *Agric Biol Chem* 36:217–226.

Guard HE, Cobet AB, Coleman WM III. 1981. Methylation of trimethyltin compounds by estuarine sediments. *Science* 213:770–771.

Gupta A, Phung LT, Chakravarty L, Silver S. 1999. Mercury resistance in *Bacillus cereus* RC607: Transcriptional organization and two new open reading frames. *J Bacteriol* 181:7080–7086.

Hamlett NV, Landsdale EC, Davis BH, Summers AO. 1992. Roles of the Tn21 *merT*, *merP*, and *merC* gene products in mercury resistance and mercury binding. *J Bacteriol* 174:6377–6385.

Harper DB. 1985. Halomethane from halide ion: A highly effective fungal conversion of environmental significance. *Nature* (London) 315:55.

Holm HW, Cox MF. 1975. Transformation of elemental mercury by bacteria. *Appl Microbiol* 29:491–494.

Imura N, Sukegawa E, Pan S-K, Nagao K, Kim J-Y, Kwan T, Ukita T. 1971. Chemical methylation of inorganic mercury with methyl-cobalamin, a vitamin B_{12} analog. *Science* 172:1248–1249.

Inoue H, Takimura H, Fuse H, Marakami K, Kamimura K, Yamaoka Y. 2000. Degradation of triphenyltin by a fluorescent pseudomonad. *Appl Environ Microbiol* 66:3492–3498.

Jensen S, Jernelöv A. 1969. Biological methylation of mercury in aquatic organisms. *Nature* (London) 223:753–754.

Jonasson IP. 1970. Mercury in the natural environment. A review of recent work. *Geol Surv Can*, Paper 70-57, 1970.

Kerin EJ, Gilmour CC, Roden E, Suzuki MR, Coates JD, Mason RP. 2006. Mercury methylation by dissimilatory iron-reducing bacteria. *Appl Environ Microbiol* 72:7919–7921.

Komura I, Izaki K. 1971. Mechanism of mercuric chloride resistance in microorganisms. I. Vaporization of a mercury compound from mercuric chloride by multiple drug resistant strains of *Escherichia coli*. *J Biochem* (Tokyo) 70:885–893.

Komura I, Izaki K, Takahashi H. 1970. Vaporization of inorganic mercury by cell-free extracts of drug-resistant *Escherichia coli*. *Agric Biol Chem* 34:480–482.

Korthals ET, Winfrey MR. 1987. Seasonal and spatial variations in mercury methylation and demethylation in an oligotrophic lake. *Appl Environ Microbiol* 53:2397–2404.

Landner L. 1971. Biochemical model for the biological methylation of mercury suggested from methylation studies in vivo with *Neurospora crassa*. *Nature* (London) 230:452–454.

Lefebvre D, Kelly D, Budd K. 2007. Biotransformation of Hg(II) by cyanobacteria. *Appl Environ Microbiol* 73:243–249.

Loutit JS. 1970. Mating systems of *Pseudomonas aeruginosa* strain I. VI. Mercury resistance associated with the sex factor (FP). *Genet Res* 16:179–184.

Macalady JL, Mack EE, Nelson CD, Scow KM. 2000. Sediment microbial community structure and mercury methylation in mercury-polluted Clear Lake, California. *Appl Environ Microbiol* 66:1479–1488.

Manley SL, Dastoor MN. 1988. Methyl iodide (CH_3I) production by kelp and associated microbes. *Mar Biol* (Berlin) 98:477–482.

Marine Chemistry. 1971. *A Report of the Marine Chemistry Panel of the Committee of Oceanography.* Washington, DC: National Academy of Sciences.

Matsumara F, Gotoh Y, Brush GM. 1971. Phenyl mercuric acetate: Metabolic conversion by microorganisms. *Science* 173:49–51.

Nascimento AMA, Chartone-Souza E. 2003. Operon *mer*: Bacterial resistance to mercury and potential for bioremediation of contaminated environments. *Genet Mol Res* 2:92–101.

Nakahara H, Schottle JL, Yamada T, Miyagawa Y, Asakawa M, Harville J, Silver S. 1985. Mercuric reductase enzymes from *Streptomyces* species and group B *Streptococcus*. *J Gen Microbiol* 131:1053–1059.

Nelson JD, Blair W, Brinckman FE, Colwell RR, Iverson WP. 1973. Biodegradation of phenylmercuric acetate by mercury-resistant bacteria. *Appl Microbiol* 26:321–326.

Nelson JD, Colwell RR. 1974. Metabolism of mercury compounds by bacteria in Chesapeake Bay. In: Acker RF, Brown BF, De Palma JR, eds. *Proceedings of the Third International Congress on Marine Corrosion Fouling*, 1972. Evanston, IL: Northwestern University Press, pp. 767–777.

Nelson JD, Colwell RR. 1975. The ecology of mercury-resistant bacteria in Chesapeake Bay. *Microb Ecol* 1:191–218.

Novick RP. 1967. Penicillinase plasmids of *Staphylococcus aureus*. *Fed Proc* 26:29–38.

Olson GJ, Iverson WP, Brinckman FE. 1981. Volatilization of mercury by *Thiobacillus ferrooxidans*. *Curr Microbiol* 5:115–118.

Olson GJ, Porter FD, Rubinstein J, Silver S. 1982, Mercuric reductase enzyme from a mercury-volatilizing strain of *Thiobacillus ferrooxidans*. *J Bacteriol* 151:1230–1236.

Pan Hou HS, Imura N. 1981. Role of hydrogen sulfide in mercury resistance determined by plasmid of *Clostridium cochlearium* T-2. *Arch Microbiol* 129:49–52.

Pan Hou HS, Nishimoto M, Imura N. 1981. Possible role of membrane proteins in mercury resistance of *Enterobacter aerogenes*. *Arch Microbiol* 130:93–95.

Richmond MH, John M. 1964. Cotransduction by a staphylococcal phage of the genes responsible for penicillinase synthesis and resistance to mercury salts. *Nature* (London) 202:1360–1361.

Robinson JB, Tuovinen OH. 1984. Mechanism of microbial resistance and detoxification of mercury and organomercury compounds: Physiological, biochemical and genetic analyses. *Microbiol Rev* 48:95–124.

Schedlbauer OF, Heumann KG. 2000. Biomethylation of thallium by bacteria and first determination of biogenic dimethylthallium in the ocean. *Appl Organometal Chem* 14:330–340.

Schelert J, Dixit V, Hoang V, Simbahan J, Drozda M, Blum P. 2004. Occurrence and characterization of mercury resistance in the hyperthermophilic Archeon *Sulfolobus solfataricus* by use of gene disruption. *J Bacteriol* 186:427–437.

Schottel JA, Mandel D, Clark S, Silver S, Hodges RW. 1974. Volatilization of mercury and organomercurials determined by inducible R-factor systems in enteric bacteria. *Nature* (London) 251:335–337.

Schrauzer GN, Weber JH, Beckham TM, Ho RKY. 1971. Alkyl group transfer from cobalt to mercury: Reaction of alkyl cobalamins, alkylcobaloximes, and of related compounds with mercury acetate. *Tetrahedron Lett* 3:275–277.

Silver S. 1997. The bacterial view of the periodic table: Specific functions for all elements. In: Banfield J, Nealson KH, eds. *Geomicrobiology: Interactions Between Microbes and Minerals*. Reviews in Mineralalogy, Vol. 35. Washington, DC: Mineralogical Society of America, pp. 345–360.

Silver S, Phung LT. 1996. Bacterial heavy metal resistance: New surprises. *Ann Rev Microbiol* 50:753–789.

Simbahan J, Kurth E, Schelert J, Dillman A, Moriyama E, Javanovich S, Blum P. 2005. Community analysis of a mercury hot spring supports occurrence of domain-specific forms of mercuric reductase. *Appl Environ Microbiol* 71:8836–8845.

Smith DH. 1967. R factors mediate resistance to mercury, nickel, and cobalt. *Science* 156:1114–1116.

Smith T, Pitts K, McGarvey JA, Summers AO. 1998. Bacterial oxidation of mercury metal vapor, Hg(O). *Appl Environ Microbial* 64:1328–1332.

Spangler WA, Spigarelli JL, Rose JM, Fillipin RS, Miller HM. 1973a. Degradation of methylmercury by bacteria isolated from environmental samples. *Appl Microbiol* 25:488–493.

Spangler WA, Spigarelli JL, Rose JM, Miller HM. 1973b. Methylmercury: Bacterial degradation in lake sediments. *Science* 180:192–193.

Steffan RJ, Korthals ET, Winfrey MR. 1988. Effects of acidification on mercury methylation, demethylation, and volatilization in sediments from an acid-susceptible lake. *Appl Environ Microbiol* 54:2003–2009.

Summers AO, 1986. Genetic mechanisms of heavy-metal and antibiotic resistances. In: Carlisle D, Berry WI, Kaplan IR, Watterson JR, eds. *Mineral Exploration: Biological Systems and Organic Matter*. Rubey Vol. 5. Englewood Cliffs, NJ: Prentice Hall, pp. 265–281.

Summers AO, Lewis E. 1973. Volatilization of mercuric chloride by mercury-resistant plasmid-bearing strains of *Escherichia coli*, *Staphylococcus aureus*, and *Pseudomonas aeruginosa*. *J Bacteriol* 113:1070–1072.

Summers AO, Silver S. 1972. Mercury resistance in a plasmid-bearing strain of *Escherichia coli*. *J Bacteriol* 112:1228–1236.

Summers AO, Silver S. 1978. Microbial transformations of metals. *Annu Rev Microbiol* 32:637–672.

Summers AO, Sugarman LI. 1974. Cell-free mercury(II)-reducing activity in a plasmid-bearing strain of *Escherichia coli*. *J Bacteriol* 119:242–249.

Tezuka T, Tonomura K. 1976. Purification and properties of an enzyme catalyzing the splitting of carbon-mercury linkages from mercury resistant *Pseudomonas* K 62-strain. *J Biochem* (Tokyo) 80:779–787.

Tezuka T, Tonomura K. 1978. Purification and properties of a second enzyme catalyzing the splitting of carbon-mercury linkages from mercury-resistant *Pseudomonas* K-62. *J Bacteriol* 135:138–143.

Tonomura K, Makagami T, Futai F, Maeda D. 1968. Studies on the action of mercury-resistant microorganisms on mercurials. I. The isolation of a mercury-resistant bacterium and the binding of mercurials to cells. *J Ferment Technol* 46:506–512.

Trevors JR. 1986. Mercury methylation by bacteria. *J Basic Microbiol* 26:499–504.

Vostal J. 1972. Transport and transformation of mercury in nature and possible routes of exposure. In: Friberg L, Vostal J, eds. *Mercury in the Environment. An Epidemiological and Toxicological Appraisal.* Cleveland, OH: CRC Press, pp. 15–27.

White RH. 1982. Analysis of dimethyl sulfonium compounds in marine algae. *J Mar Res* 40:529–536.

Wong PTS, Chau YK, Luxon PL. 1975. Methylation of lead in the environment. *Nature* (London) 253:263–264.

Wood JM. 1974. Biological cycles of toxic elements in the environment. *Science* 183:1049–1052.

Wood JM, Kennedy FS, Rosen CG. 1968. Synthesis of methylmercury compounds by extracts of a methanogenic bacterium. *Nature* (London) 220:173–174.

Yamada M, Tonomura K. 1972a. Formation of methylmercury compounds from inorganic mercury by *Clostridium cochlearium*. *J Ferment Technol* 50:159–166.

Yamada M, Tonomura K. 1972b. Further study of formation of methylmercury from inorganic mercury by *Clostridium cochlearium* T-2. *J Ferment Technol* 50:893–900.

Yamada M, Tonomura K. 1972c. Microbial methylation of mercury in hydrogen sulfide-evolving environments. *J Ferment Technol* 50:901–909.

16 Geomicrobiology of Iron

16.1 IRON DISTRIBUTION IN EARTH'S CRUST

Iron is the fourth most abundant element in the Earth's crust and the most abundant element in the Earth as a whole (see Chapter 2). Its average concentration in the crust has been estimated to be 5% (Rankama and Sahama, 1950). It is found in a number of minerals in rocks, soil, and sediments. Table 16.1 lists mineral types in which iron is a major or minor structural component.

The primary source of iron accumulated at the Earth's surface is volcanic activity. Weathering of iron-containing rocks and minerals is often an important phase in the formation of local iron accumulations including sedimentary ore deposits.

16.2 GEOCHEMICALLY IMPORTANT PROPERTIES

Iron is a very reactive element. Its common oxidation states are 0, +2, and +3. In a moist environment exposed to air or in an aerated solution at pH values >5, its ferrous form (+2) readily autoxidizes to the ferric form (+3). In the presence of an appropriate reducing agent, that is, under reducing conditions, ferric iron is readily reduced to the ferrous state. In dilute acid solution, hydrogen ions (protons) readily oxidize metallic iron to ferrous iron.

$$Fe^0 + 2H^+ \rightarrow Fe^{2+} + H_2 \tag{16.1}$$

Ferric iron may exist as a hydroxide ($Fe(OH)_3$), oxyhydroxide ($FeO(OH)$), or oxide (e.g., Fe_2O_3, $Fe_2O_3 \cdot H_2O$) in neutral to slightly alkaline solution but as Fe^{3+} in acid solution. Ferric hydroxide dissolves in strongly alkaline solution because of its amphoteric nature, which allows it to exist as an oxyanion (e.g., $Fe(OH)_2O^-$). In an aqueous environment, a mixture of Fe^{3+} and metallic iron undergoes a disproportionation (dismutation) reaction, resulting in the formation of ferrous iron.

$$Fe^0 + 2Fe^{3+} \rightarrow 3Fe^{2+} \tag{16.2}$$

Hydrogen sulfide reduces ferric iron readily to ferrous iron and, when present in excess, precipitates the ferrous iron as a sulfide or disulfide.

$$2Fe^{3+} + H_2S \rightarrow 2Fe^{2+} + 2H^+ + S^0 \tag{16.3}$$

$$Fe^{2+} + H_2S \rightarrow FeS + 2H^+ \tag{16.4}$$

$$2Fe^{3+} + 2H_2S \rightarrow FeS_2 + Fe^{2+} + 4H^+ \tag{16.5}$$

Reactions 16.3 through 16.5 are microbiologically important (see also Chapter 20).

TABLE 16.1
Iron-Containing Minerals

Igneous (Primary) Minerals	Secondary Minerals	Sedimentary Minerals
Pyroxenes	Montmorillonite[a] $(OH_4Si_8Al_4O_{20} \cdot nH_2O)$	Siderite $(FeCO_3)$
Amphiboles		Goethite $(Fe_2O_3 \cdot H_2O)$
Olivines	Illite $(OH)_4K_y(Al_4Fe_4Mg_4Mg_6)–(Si_{8-y}Al_y)O_{20}$	Limonite $(Fe_2O_3 \cdot nH_2O$ or $FeOOH)$
Micas		Hematite (Fe_2O_3)
		Magnetite (Fe_3O_4)
		Pyrite, marcasite (FeS_2)
		Pyrrhotite (Fe_nS_{n+1}); $n = 5-6$
		Ilmenite $(FeO \cdot TiO_2)$

[a] Montmorillonite contains iron by lattice substitution for aluminum.

16.3   BIOLOGICAL IMPORTANCE OF IRON

16.3.1   FUNCTION OF IRON IN CELLS

All organisms, prokaryotic and eukaryotic, single-celled and multicellular, require iron nutritionally. A small group of homolactic bacteria is the only known exception (Pandey et al., 1994). The other organisms need iron in some enzymatic processes that involve the transfer of electrons. Examples of such processes are aerobic and anaerobic respiration in which cytochromes, special proteins that bear iron-containing heme groups, and nonheme iron–sulfur proteins play a role in the transfer of electrons to a terminal electron acceptor. Phototrophic organisms also need iron for ferredoxin, a nonheme iron–sulfur protein, and some cytochromes that are part of the photosynthetic system. Some cells also employ iron in certain superoxide dismutases, which convert superoxide to hydrogen peroxide by a disproportionation reaction (see Chapter 3). Most aerobic cells produce the iron-containing enzymes catalase and peroxidase for catalyzing the decomposition of toxic hydrogen peroxide to water and oxygen. Prokaryotes that are capable of nitrogen fixation employ ferredoxin and another nonheme iron–sulfur protein (component II of nitrogenase) as well as an iron–smolybdo-, iron–vanado-, or iron–iron-protein (component I) of nitrogenase (see Chapter 13).

As will be discussed in Section 16.4, ferrous iron may serve as a major energy source or source of reducing power for certain bacteria. Ferric iron may serve as a terminal electron acceptor for a number of different bacteria, usually under anaerobic conditions. As discussed in Chapter 3, ferrous iron may have served as an important reductant during evolutionary emergence of oxygenic photosynthesis by scavenging the toxic oxygen produced in the process until the appearance of oxygen-protective superoxide dismutase. Cloud (1973) believed that the banded iron formations (BIFs) that appeared in the sedimentary record from 3.3 to 2 billion years ago are evidence for the oxygen-scavenging action by ferrous iron, but modifications of this view have been offered more recently (see discussion in Section 16.9). BIFs consist of cherty magnetite (Fe_3O_4) and hematite (Fe_2O_3). From a biogeochemical viewpoint, large-scale microbial iron oxidation is important because it may lead to extensive iron precipitation (immobilization), and microbial iron reduction is important because it may lead to extensive iron solubilization (mobilization). In some anaerobic environments, microbial iron reduction plays an important role in the mineralization of organic carbon (iron respiration).

16.3.2   IRON ASSIMILATION BY MICROBES

In aerobic environments of neutral pH, ferric iron precipitates readily from solution as hydroxide, oxyhydroxide, and oxide. At circumneutral pH, these compounds, especially the latter two, are insoluble to various degrees. This is a special problem in seawater, where iron is a growth-limiting nutrient (Gelder, 1999; also Chapter 5). As previously stated, iron is an essential trace nutrient for nearly all forms of cellular life. Because iron can be taken into a cell only in soluble form, a number of microorganisms have acquired the ability to synthesize chelators (ligands) that help to keep ferric iron in solution at circumneutral pH or that can return it to solution in sufficient quantities from an insoluble state to meet nutritional demands. Collectively, such chelators are known as *siderophores*. They have an extremely high affinity for ferric iron but very low affinity for ferrous iron. They exhibit proton-independent binding constants for Fe^{3+} ranging from $10^{22.5}$ to 10^{49}–10^{53} (Reid et al., 1993) compared with ~10^8 for Fe^{2+} (for additional data, see Dhungana and Crumbliss, 2005). Examples of siderophores are enterobactin or enterochelin, a catechol derivative from *Salmonella typhimurium* (Pollack and Neilands, 1970); aerobactin, a hydroxamate derivative produced by *Enterobacter* (formerly *Aerobacter*) *aerogenes* (Gibson and Magrath, 1969); alterobactin from *Alteromonas luteoviolacea* (Reid et al., 1993); ferrioxamine E (nocardamine) and ferrioxamines G, D_2, and X_1 from *Pseudomonas stutzeri* (Essén et al., 2007); and rhodotorulic acid, a hydroxamate derivative produced by the yeast *Rhodotorula* (Neilands, 1974) (Figure 16.1). Citrate, produced by some fungi, can also serve as a siderophore for some bacteria (Rosenberg and Young, 1974). Iron mobilization action of siderophores under circumneutral, aerobic conditions may be assisted by an extracellular reductant. Hersman et al. (2000) reported on mobilization of iron from hematite (Fe_2O_3) by a siderophore secreted by *P. mendocina* and extracellular iron-reducing activity. The reducing activity was maximal under conditions of extreme iron deprivation in the absence of hematite or FeEDTA. The siderophore mobilized the iron in the form of a ferric chelate whereas the iron-reducing activity mobilized the iron in ferrous form. The authors speculated that the iron-reducing activity was due to

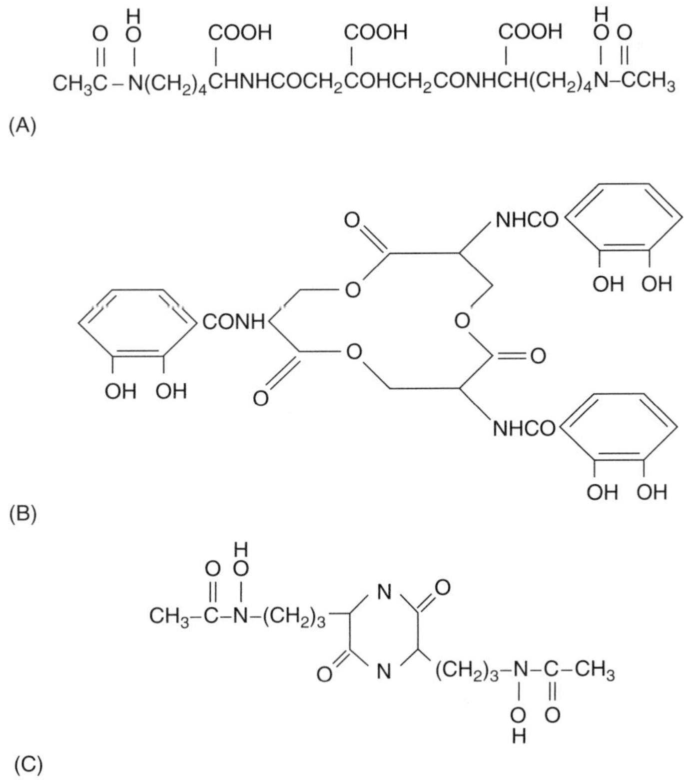

FIGURE 16.1 Examples of siderophores. (A) Aerobactin (bacterial). (B) Enterochelin (bacterial). (C) Rhodotorulic acid (fungal).

an extracellular enzyme produced by *P. mendocina* that was similar to extracellular iron reductases produced by *Escherichia coli* and *P. aeruginosa* (Vartanian and Cowart, 1999). Alternatively, small redox-active metabolites (such as phenazine antibiotics) released by *Pseudomonas* species may also facilitate iron acquisition via electron shuttling (Wang and Newman, 2008).

Chelated ferric iron is usually taken up by first binding to ferrisiderophore-specific receptors at the cell surface of the microbial species that produces the siderophore. In some cases, it may also bind to surface receptors of certain species of microorganisms that are incapable of synthesizing the desferrisiderophore (iron-free siderophore). After the chelated ferric ion is being transported into the cell, it is usually enzymatically reduced to ferrous iron and rapidly released by the sidero-phore because it has only low affinity for Fe^{2+} (Brown and Ratledge, 1975; Cox, 1980; Ernst and Winkelmann, 1977; Tait, 1975; Wandersman and Delepelaire, 2004). The desferrisiderophore may then be excreted again for further scavenging of iron. In some instances, the ferrisiderophore taken up by the cell is degraded first before the ferric iron is reduced to ferrous iron. In that instance, the desferrisiderophore is not recyclable. In still another few instances, the chelated ferric iron is exchanged with another ligand at the initial binding site at the cell surface before being transported into the cell and reduced to ferrous iron. The ferrous iron, by whatever mechanism it is generated in the cell, is immediately assimilated into heme protein or nonheme iron–sulfur protein.

16.4 IRON AS ENERGY SOURCE FOR BACTERIA

16.4.1 Acidophiles

Microbes can promote iron oxidation, but this does not mean that the oxidation is always enzymatic. Because ferrous iron has the tendency to autoxidize in aerated solution at pH values >5, it is difficult to demonstrate enzyme-catalyzed iron oxidation in near-neutral, air-saturated solutions. To date, much of the evidence for enzymatic ferrous iron oxidation has been amassed at pH values <5. The bacteria with this capacity include members of the domains Bacteria and Archaea.

16.4.2 Domain Bacteria: Mesophiles

Various acidophilic iron-oxidizing bacteria have been recognized and characterized in various acidic environments (Pronk and Johnson, 1992). They include autotrophs, mixotrophs, and heterotrophs. A more detailed characterization of some of them is given in the follwing subsections.

16.4.2.1 *Acidithiobacillus* (Formerly *Thiobacillus*) *ferrooxidans; See Kelly and Wood, 2000*

16.4.2.1.1 General Traits

The most widely studied acidophilic, ferrous iron-oxidizing bacterium is *Acidithiobacillus ferrooxidans* belonging to the γ-subclass of the Proteobacteria. It is a gram-negative, motile rod (0.5×1 μm) (Figure 16.2). Under phosphate starvation in the laboratory, the rod-shaped cells become filamentous cells and exhibit some qualitative and quantitative changes in their protein complement (Seeger and Jerez, 1993). This organism can derive energy and reducing power from the oxidation of ferrous iron, reduced forms of sulfur (H_2S, S^0, $S_2O_3^{2-}$), metal sulfides, H_2, and formate. It gets its carbon from CO_2 and derives its nitrogen preferentially from NH_3-N but can also use NO_3-nitrogen (Temple and Colmer, 1951; Lundgren et al., 1964). Some strains can fix N_2 (Mackintosh, 1978; Stevens et al., 1986).

Morphologically, the cells of *Acidithiobacillus ferrooxidans* exhibit the multilayered cell wall (outer membrane, periplasm, thin peptidoglycan layer) typical of gram-negative bacteria (Figure 16.3) (Avakyan and Karavaiko, 1970; Remsen and Lundgren, 1966). They do not contain special internal membranes like those found in nitrifiers and methylotrophs. Cell division is mostly by constriction but occasionally also by partitioning (Karavaiko and Avakyan, 1970).

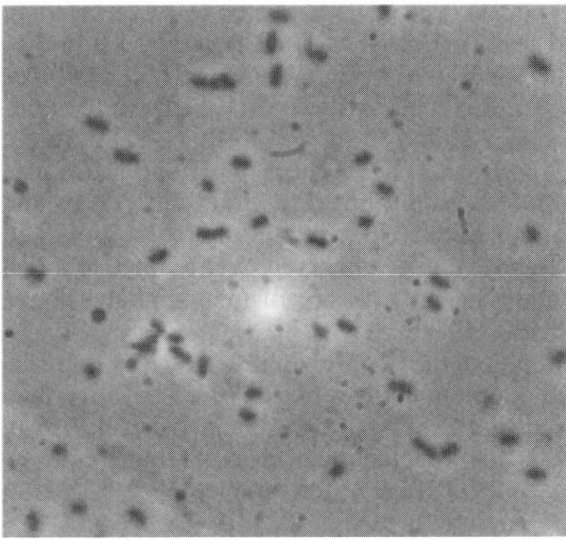

FIGURE 16.2   *Acidithiobacillus ferrooxidans* (×5170). Cell suspension viewed by phase contrast microscope.

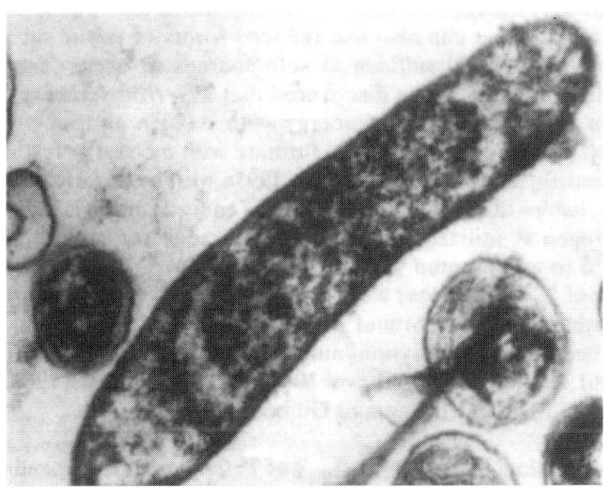

FIGURE 16.3   *Acidithiobacillus ferrooxidans* (×30,000). Transmission electron micrograph of a thin section. (Courtesy of Lundgren DG, Department of Biology, Syracuse University, Syracuse, NY, USA.)

Acidithiobacillus ferrooxidans is acidophilic. The original isolate by Temple and Colmer (1951) grew and oxidized ferrous iron in a pH range of 2–2.5, but these investigators did not report an overall pH range for growth. The TM strain used by Silverman and Lundgren (1959a) oxidized ferrous iron optimally in a pH range of 3–3.6, whereas strain NCIB No. 9490 used by Landesman et al. (1966) oxidized ferrous iron optimally at pH 1.6. Razzell and Trussell (1963) gave an optimum pH for growth on ferrous iron as 2.5, and Jones and Kelly (1983) gave the pH range for growth on ferrous iron as 1–4.5. Differences in pH optima for ferrous iron oxidation and growth on ferrous iron can be attributed to strain differences and differences in experimental conditions. In general, all strains of *A. ferrooxidans* grow and oxidize iron at ~pH 2. Growth can be observed over an initial pH range of 1.5–6. Ferric iron produced by the organism precipitates above pH 1.9 (Buchanan and Gibbons, 1974; Starr et al., 1981).

Acidithiobacillus ferrooxidans is mesophilic, that is, it grows in a temperature range of 15–42°C with an optimum in the range of 30–35°C (Silverman and Lundgren, 1959a; Ahonen and Tuovinen, 1989; Niemelä et al., 1994). Ahonen and Tuovinen (1989) found that acidophilic iron-oxidizing cultures resembling *A. ferrooxidans* obtained from water samples from a Finnish copper mine grew at incubation temperatures of 4, 7, 10, 13, 16, 19, 28, and 37°C but not at 46°C. Optimum rates of iron oxidation by resting cells of *A. ferrooxidans* have been observed at 40°C with at least some strains (Silverman and Lundgren, 1959b; Tuttle and Dugan, 1976). A linear increase in ferrous oxidation rates was detected between 4 and 28°C with the Finnish cultures (Ahonen and Tuovinen, 1989).

As already noted, ferrous iron is not the only energy source for *Acidithiobacillus ferrooxidans*. It can also use reduced forms of sulfur, such as H_2S, S^0, $S_2O_3^{2-}$, and some metal sulfides as sole sources of energy (see Chapters 19 and 20). In addition, it has now been shown that *A. ferrooxidans* can use H_2 as sole source of energy with oxygen as terminal electron acceptor (Drobner et al., 1990) and formate with oxygen or ferric iron as terminal electron acceptor (Pronk et al., 1991a,b). The key enzyme hydrogenase that enables utilization of H_2 as energy source is inducible. The hydrogenase from *A. ferrooxidans* strain 19859 has been purified and recovered as 100 and 200 kDa units containing 6.02 mol of Fe and 0.72 mol of Ni per mole of enzyme. This enzyme was made up of two subunits of 64 and 34 kDa molecular mass, respectively, occurring in a 1:1 ratio in the enzyme (Fischer et al., 1996). When H_2 is utilized, the optimum pH range for growth was 3–5.8 as opposed to an optimum of pH 2 on ferrous iron.

Some strains of *Acidithiobacillus ferrooxidans* were originally named as *Ferrobacillus ferrooxidans* (Leathen et al., 1956) and *F. sulfooxidans* (Kinsel, 1960). These strains were subsequently considered synonymous with *Thiobacillus* (now *Acidithiobacillus*) *ferrooxidans* (Unz and Lundgren, 1961; Ivanov and Lyalikova, 1962; Hutchinson et al., 1966; Kelly and Tuovinen, 1972; Buchanan and Gibbons, 1974).

16.4.2.1.2 Laboratory Cultivation

Laboratory study of *Acidithiobacillus ferrooxidans* depends on the ease of culturing. Table 16.2 lists the ingredients of four liquid media of which the 9K and T&K media are among the widely

TABLE 16.2
Media for Cultivating *Acidithiobacillus ferrooxidans*

Ingredient	Quantity of Ingredients (g L^{-1})			
	9K[a]	T&K[b]	L[c]	T and C[d]
$(NH_4)_2SO_4$	3.0	0.4	0.15	0.5
KCl	0.1	—	—	—
K_2HPO_4	0.5	0.4	0.05	—
$MgSO_4 \cdot 7H_2O$	0.5	0.4	0.5	1.0
$Ca(NO_3)_2$	0.01	—	0.01	—
$FeSO4 \cdot 7H_2O$	44.22	33.3	1.0	129.1

[a] Medium of Silverman and Lundgren (1959a). $FeSO_4$ is dissolved in 300 mL of distilled water and filter-sterilized. The remaining salts are dissolved in 700 mL of distilled water and autoclaved after adjustment of the pH with 1 mL of 10 N H_2SO_4.
[b] Medium of Tuovinen and Kelly (1973). The salts are dissolved in 0.11 N H_2SO_4. This medium is sterilized by filtration.
[c] Medium of Leathen et al. (1956).
[d] Medium of Temple and Colmer (1951). The pH of this medium is adjusted to 2–2.5 with H_2SO_4. This medium is sterilized by filtration.

used ones. For purification, cloning, and enumerating viable cells in a culture, gelled (solid) medium is essential. Use of standard agar or silica gel has not been successful. Tuovinen and Kelly (1973) suggested that galactose from the standard agar is inhibitory to the organism. Modified 9K medium gelled with agarose or purified agar will, however, support the growth of most strains. On such media, the organism forms rust-colored colonies with varying morphology, depending on the strain. The colonies usually develop in 1–2 weeks (Manning, 1975; Mishra and Roy, 1979; Mishra et al., 1983; Holmes et al., 1983). Another approach is to grow *A. ferrooxidans* on membrane filters (Sartorius or Millipore types) on T&K medium solidified with 0.4% Japanese agar at an initial pH of 1.55 and incubating at 20°C (Tuovinen and Kelly, 1973). Rust-colored colonies of 1–1.5 mm diameter develop after 2 weeks.

Johnson et al. (1987) used the following gelled medium for concurrent isolation and enumeration of *Acidithiobacillus ferrooxidans* and acidophilic heterotrophic bacteria that usually accompany *A. ferrooxidans* in nature. This medium contains 1 part by volume of 20% ferrous sulfate solution adjusted to pH 2 or 2.5; 14 parts by volume of a basal salts–tryptone soy broth solution—consisting of (in g L^{-1}) $(NH_4)_2SO_4$, 1.8; $MgSO_4 \cdot 7H_2O$, 0.7; tryptone soy broth (TSB, Oxoid Ltd., Basingstoke, U.K.), 0.35—adjusted to pH 2, 2.5, or 3; and 5 parts by volume of 0.7% agarose type 1 (Sigma Ltd., St. Louis, Minnesota). The best final pH values for optimal recovery of *A. ferrooxidans* were 2.7 and 2.3. A modification of this gelled medium, which also supported the growth of moderately thermophilic, acidophilic iron-oxidizing bacteria, was developed by Johnson and McGinness (1991). In this medium, the final concentration of iron was lowered from 1% (33 mM) to 0.76% (25 mM), and the concentration of agarose type 1 was lowered from 0.7 to 0.5%. Addition of some trace elements to the medium was found to be beneficial for some organisms. For culturing, the final medium was dispensed in two layers in Petri dishes. The bottom layer (~20 mL) consisted of medium that had been preinoculated with a desired acidophilic heterotroph, and a thin top layer (~10 mL) consisted of sterile medium. The purpose of the bacteria in the bottom layer was to have them consume as much as possible of any sugars that are contributed by the agarose and that might suppress the growth of iron oxidizers.

16.4.2.1.3   Facultative Heterotrophy

A number of claims have appeared in the literature that some strains of *Thiobacillus* (now *Acidithiobacillus*) *ferrooxidans* can be adapted to grow heterotrophically if glucose is used instead of ferrous iron as the sole source of energy and carbon (Lundgren et al., 1964; Shafia and Wilkinson, 1969; Shafia et al., 1972; Tabita and Lundgren, 1971a; Tuovinen and Nicholas, 1977; Sugio et al., 1981). Some of these strains could be reverted to ferrous iron-dependent autotrophy, whereas others could not. The discovery that some cultures of *T. ferrooxidans* (now *A. ferrooxidans*) contained heterotrophic satellite organisms (Zavarzin, 1972; Guay and Silver, 1975; Mackintosh, 1978; Arkesteyn, 1979; Harrison, 1981, 1984; Johnson and Kelso, 1983; Lobos et al., 1986; Wichlacz et al., 1986) cast serious doubt on the existence of any *A. ferrooxidans* strains that can grow heterotrophically.

16.4.2.1.4   Facultative Mixotrophy

At least some strains of *Acidithiobacillus ferrooxidans* seem to be capable of facultative mixotrophy. Barros et al. (1984) found that ferrous iron oxidation by strain FD1 was faster in the absence of glucose than in its presence and increased as the ratio of iron to glucose increased. At an initial iron/glucose concentration ratio of 5:9, neither iron nor glucose was utilized. Even 0.5 g of glucose per liter in a medium with 9 g of iron per liter inhibited iron oxidation, but lowering the initial iron concentration from 9 g L^{-1} resulted in simultaneous use of Fe and glucose. At an initial concentration ratio of 7:5, a slight lag occurred during which only iron was oxidized, but when the initial concentrations of ferrous iron and glucose were 5 g L^{-1} each, iron and glucose were utilized concurrently from the start by strain FD1. The authors showed that on the basis of mole percent GC ratios, the cultures of FD1 grown autotrophically on iron and mixotrophically on iron and glucose were identical.

16.4.2.1.5 Consortia with Acidithiobacillus ferrooxidans

The existence of satellite organisms that appear to live in close association with *Acidithiobacillus ferrooxidans* was first reported by Zavarzin (1972). As noted earlier, confirmation of their existence soon followed. Indeed, taxonomically different organisms were isolated from different *A. ferrooxidans* consortia. Zavarzin (1972) isolated his organisms in modified Leathen's medium in which yeast extract (0.01–0.02%) replaced $(NH_4)_2SO_4$ and the pH was adjusted to 3–4. The organism was morphologically distinct from *A. ferrooxidans*, being a single rod. It could not oxidize iron, although it required its presence at high concentration in the medium. It was acidophilic (optimum pH range 2–3). It required yeast extract but did not grow in an excess of it. Glucose, fructose, ribose, maltose, xylose, mannitol, ethanol, and citric, succinic, and fumaric acids could serve as energy sources. This organism was originally isolated from peat in an acid bog. It resembled *Acetobacter xylinum* but grew at lower pH than the latter and exhibited only a weak ability to form acetic acid and ethanol.

Guay and Silver (1975) derived a satellite organism from a culture of *Thiobacillus* (now *Acidithiobacillus*) *ferrooxidans* strain TM by successive subculturing in 9K medium containing increasing amounts of glucose from 0.1 to 1% and concomitant decreases in Fe^{2+} from 9000 to 10 ppm in four steps. They isolated a pure culture of the satellite organism, which they named *T. acidophilus* (now *Acidiphilium acidophilum*; Harrison, 1983) and assigned to the α-subclass of the Proteobacteria, using 9K basal salts–glucose agar that contained 1 ppm ferrous iron at pH 4.5 and incubating at 25°C. The pure culture consisted of a gram-negative, motile rod (0.5–0.8 μm × 1–1.5 μm) and was morphologically not very distinct from the *T. ferrooxidans* strain. Physiological study of *T. acidophilus* by Guay and Silver (1975) and subsequently by Norris et al. (1986), Mason et al. (1987), and Mason and Kelly (1988) showed the organism to be capable of heterotrophic growth in suitable medium and also capable of mixotrophic growth on tetrathionate and glucose. In addition, it was found to grow autotrophically in media with elemental sulfur (S^0), thiosulfate, trithionite, or tetrathionate as energy sources and CO_2 as carbon source. This organism is incapable of oxidizing Fe^{2+} or metal sulfides. It is able to use D-ribose, D-xylose, L-arabinose, D-glucose, D-fructose, D-galactose, D-mannitol, sucrose, citrate, malate, DL-aspartate, and DL-glutamate as carbon and energy sources in 9K basal medium without Fe^{2+}. It cannot use D-mannose, L-sorbose, L-rhamnose, ascorbic acid, lactose, D-maltose, cellobiose, trehalose, D-melobiose, raffinose, acetate, lactate, pyruvate, glyoxalate, fumarate, succinate, mandelate, cinnamate, phenylacetate, salicylate, phenol, benzoate phenylamine, tryptophane, tyrosine, or proline.

Like *Acidithiobacillus ferrooxidans*, *Thiobacillus acidophilus* is acidophilic (pH range 1.5–6, optimum 3). Its DNA has a GC ratio of 62.9–63.2 mol%, which is distinctly different from that of *A. ferrooxidans*, whose GC ratio is 56.1 mol%. Differences in key enzymes such as glucose 6-phosphate dehydrogenase, 6-phosphogluconate dehydrogenase, fructose 1,6-diphosphate aldolase, isocitric dehydrogenase, 2-ketoglutarate dehydrogenase, NADH:acceptor oxidoreductase, thiosulfate-oxidizing enzyme, and rhodanase between *A. ferrooxidans* and *T. acidophilus* were also noted. The ribulose 1,5-bisphosphate carboxylase level for CO_2 fixation in *T. acidophilus* depends on whether the organism is growing mixotrophically or autotrophically. Glucose in mixotrophic cultures can partially repress the enzyme (Mason and Kelly, 1988). What is puzzling about *T. acidophilus* is that it was carried for years as a satellite of *A. ferrooxidans* TM in 9K medium without an organic supplement. The explanation may be that *T. acidophilus* is oligotrophic as given by Arkesteyn and DeBont (1980). In coculture with *A. ferrooxidans*, it lives at the expense of organic excretions (organic acids, alcohols, amino acids) from *A. ferrooxidans* that inhibit the growth of *A. ferrooxidans* if they accumulate in the growth medium.

Other satellite organisms that have been found to be associated with some *Acidithiobacillus ferrooxidans* cultures include *Acidiphilium (Aphil.) cryptum* (Harrison, 1981), *Aphil. organovorum* (Lobos et al., 1986), *Aphil. angustum*, *Aphil. facilis*, and *Aphil. rubrum* (Wichlacz et al., 1986). *Aphil. cryptum* is a heterotrophic, motile, gram-negative rod (0.3–0.4 μm × 0.5–0.8 μm) that grows in very dilute organic media in a pH range of 1.9–5.9. This organism is an oligotroph whose growth is inhibited at high organic nutrient concentrations. Its GC content is 68–70 mol%, which clearly distinguishes it from *A. ferrooxidans* and *Thiobacillus acidophilus*. It cannot use reduced forms of sulfur or ferrous iron as an energy source and is inhibited by as little as 0.01% acetate. *Aphil. organovorum* differs from *Aphil. cryptum* by its lower GC ratio (64 mol%) and especially by its facultative oligotrophy. The *Acidiphilium* species isolated by Wichlacz et al. (1986) differs from *Aphil. cryptum* in cell size, pigment production, nutritional traits, and genome homology. In support of the theory of Arkesteyn and DeBont (1980) concerning the role of *T. acidophilus* in nature in relation to its association with *A. ferrooxidans*, Harrison (1984) showed that the growth of *A. ferrooxidans* is enhanced in the presence of *Aphil. cryptum* when 0.004% pyruvate is present in the growth medium. Pyruvate can be used by *Aphil. cryptum* as carbon and energy source but inhibits the growth of *A. ferrooxidans* at a critical concentration, although *A. ferrooxidans* excretes pyruvate during its growth (Schnaitman and Lundgren, 1965).

16.4.2.1.6   *Genetics*

Understanding of the genetics of *Acidithiobacillus ferrooxidans* lagged behind that of many other bacteria. One reason for this was the past difficulty of culturing these organisms on solid media to obtain pure cultures when selecting wild-type and mutant strains from a mixed culture. Another reason was that the acidophilic nature of the organism made the application of standard molecular biological techniques difficult.

With the development of new culture techniques, significant advances in unraveling the genetics of *Acidithiobacillus ferrooxidans* have been made in the last 25 years or so (Croal et al., 2004). For instance, in the older literature, reports appeared of *training A. ferrooxidans* to tolerate high concentrations of toxic metal ions such as those of Cu, Ni, Co, and Pb by serial subculture in the presence of increasing amounts of the toxicants (Marchlewitz et al., 1961; Tuovinen et al., 1971; Kamalov, 1972; Tuovinen and Kelly, 1974). Many interpreted this phenomenon on the basis of spontaneous mutation and selection. However, considering the relatively rapid rate at which these cultures developed tolerance for these metals and their limited growth yields in appropriate media, some questioned this interpretation. Pure cultures of rapidly growing heterotrophs acquire resistance to a toxicant more gradually when a spontaneous mutation and selection process is involved. The discovery of mobile elements in the genome of many *A. ferrooxidans* strains offers a more plausible explanation for their rapid acquisition of metal resistance and phenotypic switching with respect to iron oxidation (Holmes et al., 1989; Yates and Holmes, 1987; Schrader and Holmes, 1988; Holmes et al., 2001).

Like most bacteria, *Acidithiobacillus ferrooxidans* contains most of its genetic information on a *chromosome*, and like many bacteria, most strains of *A. ferrooxidans* also contain some nonessential genetic information on extrachromosomal DNA called *plasmids* (Mao et al., 1980; Martin et al., 1981; Rawlings et al., 1986). Some of the chromosomal and plasmid genes have been mapped and characterized (Rawlings and Kusano, 1994). These include genes of nitrogen metabolism, such as glutamine synthetase (see Rawlings and Kusano, 1994) and nitrogenase (Pretorius et al., 1986, 1987; Rawlings and Kusano, 1994), genes involved in CO_2 fixation and energy conservation (see Rawlings and Kusano, 1994), genes responsible for resistance to mercury, and other genes (see Rawlings and Kusano, 1994).

The genomes of different strains of *Acidithiobacillus ferrooxidans* and *A. thiooxidans* have been compared (Harrison, 1982, 1986). The 13 strains of *A. ferrooxidans* that were examined fell into different DNA homology groups. Although genetically related, they were not identical. None showed significant DNA homology with six strains of *A. thiooxidans*, indicating little genetic relationship. The *A. thiooxidans* strains fell into two different DNA homology groups. On the basis of 16S rRNA relationships, *A. ferrooxidans* belongs to the Proteobacteria, most to the β-subgroup, but at least one strain to the γ-subgroup (see Rawlings and Kusano, 1994). A gapped genome sequence has now been determined for *A. ferrooxidans* ATCC 23270 (Selkov et al., 2000).

After many failures, it finally became possible under laboratory conditions to introduce genes into strains of *Acidithiobacillus ferrooxidans* that lacked them, using bacterial transformation via electroporation and bacterial conjugation, and having them expressed by recipient strains. Thus, the mercury resistance gene complex (*mer* operon) was introduced into a mercury-sensitive strain of *A. ferrooxidans* (Kusano et al., 1992). Heterogeneous arsenic resistance genes on plasmids were transferred from a strain of *Escherichia coli* (heterotroph) to an arsenic-sensitive strain of *A. ferrooxidans* by filter mating (Peng et al., 1994). To what extent gene transfer occurs among *A. ferrooxidans* strains in nature is still unknown.

16.4.2.1.7   The Energetics of Ferrous Iron Oxidation

Oxidation of ferrous iron does not furnish much energy on a molar basis, for instance, when compared to the oxidation of glucose. In the past, estimates of free energy change (ΔG^0) for iron oxidation ranged ~10 kcal mol^{-1} (41.8 kJ mol^{-1}), as calculated, for example, by Baas Becking and Parks (1927) from the reaction

$$4FeCO_3 + O_2 + 6H_2O \rightarrow 4Fe(OH)_3 + 4CO_2$$

$$(\Delta G^0_{r\,298} = -40\,\text{kcal or } -167.2\,\text{kJ})$$

(16.6)

Another examination of the question of free energy yield from ferrous iron oxidation by Lees et al. (1969), who assumed that the reaction was

$$Fe^{2+} + H^+ + 0.25O_2 \rightarrow Fe^{3+} + 0.5H_2O \tag{16.7}$$

and who took into account the effects of pH and ferric iron solubility on the reaction, led to the following equation for estimating ΔG^0_r between pH 1.5 and 3:

$$\Delta G^0_r = 1.3(7.7 - \text{pH} - 0.17) \tag{16.8}$$

From this equation it may be calculated that the ΔG^0_r at pH 2.5 is -6.5 kcal mol^{-1} or -27.2 kJ mol^{-1}, barely enough for the synthesis of 1 mol of adenosine 5'-triphosphate (ATP) (which requires ~7 kcal mol^{-1} or 29.3 kJ mol^{-1}). If we assume that it takes 120 kcal of energy to assimilate 1 mol of carbon at 100% efficiency (Silverman and Lundgren, 1959b), then ~18.5 mol of Fe^{2+} would have to be oxidized to incorporate this much carbon. However, *Acidithiobacillus ferrooxidans* is not 100% efficient in using energy available from iron oxidation. An early experimental estimate of true efficiency of carbon assimilation at the expense of Fe^{2+} oxidation was 3.2% (Table 16.3) (Temple and Colmer, 1951). At that efficiency level, it would take the oxidation of 577 mol of ferrous iron to assimilate 1 mol of carbon. Taking the efficiency determination of Silverman and Lundgren (1959b) given in Table 16.3, oxidation of 90.1 mol of ferrous iron to assimilate 1 mol of CO$_2$ would be predicted. This is greater than the consumption of 50 mol observed by Silverman and Lundgren (1959b) and raises

TABLE 16.3
Estimates of Free Energy Efficiency of Carbon Assimilation by Iron-Oxidizing *Thiobacilli*

Efficiency (%)	Age of Cells	Reference
3.2	17 days	Temple and Colmer (1951)
30	?	Lyalikova (1958)
4.8–10.6	?	Beck and Elsden (1958)
20.5 ± 4.3	Late log phase	Silverman and Lundgren (1959b)

the question as to whether the assumption that it takes 120 kcal in these cells to assimilate 1 mol of carbon is correct. However, the estimate of 90.1 mol of Fe^{2+} per mole of carbon approaches the results of another experiment (Beck, 1960) in which ~100 mol of Fe^{2+} had to be oxidized to fix 1 mol of CO_2 by a strain of *A. ferrooxidans*. No matter what the actual efficiency of energy utilization, the observed ratios of moles of Fe^{2+} oxidized to moles of CO_2 assimilated illustrate that large amounts of ferrous iron have to be oxidized to satisfy the energy requirements for the growth of these organisms.

16.4.2.1.8   Iron-Oxidizing Enzyme System in Acidithiobacillus ferrooxidans

Significant advances have been made in elucidating the enzymatic mechanism of iron oxidation in *A. ferrooxidans*. Kinetic studies with whole cells of the organism yielded apparent K_m values of 2.2 mM and for cell-free preparations, 5.6 mM (pH optimum 3.5) (Ingledew, 1982).*

Ferric iron resulting from Fe^{2+} oxidation causes product inhibition and limits the growth of *Acidithiobacillus ferrooxidans* as it builds up in the medium (Kelly and Jones, 1978; Kovalenko et al., 1982). The inhibitory effect can be modulated by a change in temperature. Increasing temperature decreases the inhibitory effect. Physiological age of the cells also has an effect on iron susceptibility as the lag-phase cells are more sensitive to inhibition than log (exponential)-phase cells (Kovalenko et al., 1982). Ferrous iron may cause substrate inhibition in chemostat growth (Jones and Kelly, 1983).

A requirement for sulfate ions by the ferrous iron-oxidizing system of *Acidithiobacillus ferrooxidans* was established by Lazaroff (1963). Changes in SO_4^{2-} concentration as well as pH affect the values of V_{max} but not K_m in the Michaelis–Menten relationship. Even when cells were adapted to the presence of Cl^-, SO_4^{2-} could not be fully replaced by Cl^- (Lazaroff, 1963; see also Kamalov, 1967; Vorreiter and Madgwick, 1982). In at least one instance, SO_4^{2-} could be partially replaced by HPO_4^{2-} or $HAsO_4^{2-}$, but not by BO_3^-, MoO_4^{2-}, NO_3^-, or Cl^- ions. Formate and MoO_4^{2-} ions inhibited iron oxidation (Schnaitman et al., 1969). Selenate could replace sulfate in iron oxidation by *A. ferrooxidans* but did not permit its growth. In the presence of sulfate or selenate, iron oxidation was enhanced by tellurate, tungstate, arsenate, or phosphate (Lazaroff, 1977). A role of sulfate in iron oxidation by *A. ferrooxidans* appears to be the stabilization of the hexa-aquated complex of Fe^{2+}, which serves as substrate for its iron-oxidizing system. Selenate can replace sulfate completely as an anionic stabilizer, and tellurate, tungstate, arsenate, and phosphate can replace it partially (Lazaroff, 1983). The extensive formation of jarosite, a crystalline basic ferric sulfate, in the presence but not in the absence of *A. ferrooxidans* suggests that in abiological oxidation of ferrous iron, water displaces sulfate in the ferric product (Lazaroff et al., 1982, 1985).

An early model for the iron-oxidizing system in *Acidithiobacillus ferrooxidans*, proposed by Ingledew (1986), is illustrated in Figure 16.4A. According to this model, bulk-phase Fe^{2+} is oxidized at the outer surface of the outer membrane of the cell envelope by transfer of an electron to a structurally bound iron in the outer membrane, that is, in the +3 oxidation state, and is reduced to the +2 state by the transfer. The structurally bound iron was called polynuclear iron (Ingledew, 1986). The resultant bound ferrous iron in the outer membrane gives up its acquired electron to the copper protein rusticyanin catalyzed by an unidentified enzyme (X). The rusticyanin in turn transfers its acquired electron to periplasmic *c*-type cytochrome. The reduced *c*-type cytochrome then binds to the outer surface of the plasma membrane, allowing for transfer of the electrons across the membrane to a cytochrome oxidase (cytochrome a_1) located on the inside surface of the plasma membrane. The reduced cytochrome oxidase then reacts with O_2, leading to the formation of water. The inclusion of the catalytic component (X) in the model in Figure 16.4A is necessary because the kinetics of electron transfer from Fe(II) to periplasmic *c*-type cytochrome is otherwise too slow to explain the observed rate of iron oxidation by intact cells (Ingledew, 1986; Cox and Boxer, 1986;

---

* K_m is defined by the Michaelis–Menten equation $v = V_{max}[S]/([S] + K_m)$, where v is the initial reaction velocity, V_{max} the maximal velocity, $[S]$ the initial substrate concentration, and K_m a constant (Segel, 1975).

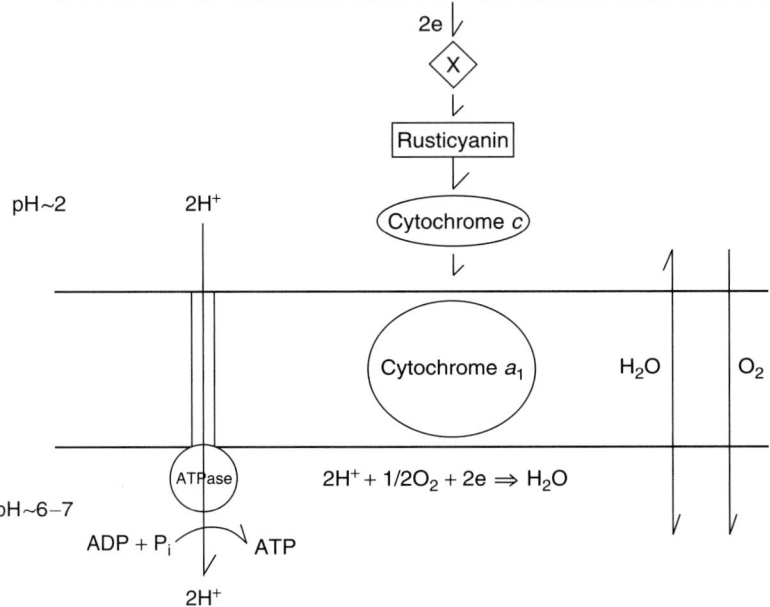

(A)

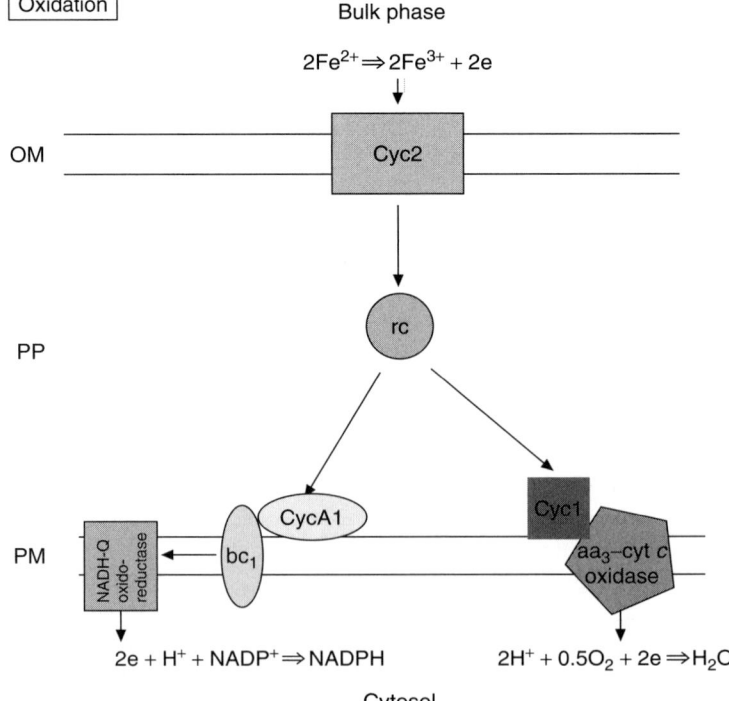

(B)

Blake and Shute, 1987). Fry et al. (1986) claimed that they found evidence for such an enzyme in the form of a nonheme iron protein. Ingledew's suggestion that outer membrane-bound iron as proposed in his model may be the initial electron acceptor is based on a proposal by Dugan and Lundgren (1965) that some iron is bound in the cell envelope of *A. ferrooxidans*. It is also based on the finding by Agate and Vishniac (1970) that phosphatidyl serine in the membrane may be the site of this binding.

Alternative models for the iron-oxidizing system in *Acidithiobacillus ferrooxidans* have been proposed. The most plausible subsequent model of the alternative models was one based on direct observations by Blake et al. (1992) of rapid electron transfer from a molar excess of Fe^{2+} to rusticyanin catalyzed by a partially purified iron rusticyanin oxidoreductase that appeared to consist of some form of *c*-type cytochrome. The reaction depended on the presence of sulfate ions. The kinetics of ferrous iron oxidation was consistent with that observed with intact cells. Polynuclear iron in the outer membrane and the postulated enzyme (X) were not required in this model. The seminal observations by Blake et al. (1992) and Blake and Shute (1994) did not establish the location of the rusticyanin-reducing *c*-type cytochrome, because they used solubilized preparations of rusticyanin and cytochrome c_{552} obtained from cell-free extracts of *A. ferrooxidans* ATCC 23270. Their findings left unclear as to how electrons from Fe^{2+} would enter the periplasm—the previously established location of rusticyanin. It could have been by the uptake of Fe^{2+} into the periplasm or, if Fe^{2+} was oxidized at the cell surface, by an electron transfer agent in the outer membrane for transporting electrons from bulk-phase Fe^{2+} into the periplasm. This question has now been resolved.

Yarzábal et al. (2002a,b) demonstrated the presence of a high-molecular-weight *c*-type cytochrome, Cyc2, in the outer membrane of *Acidithiobacillus ferrooxidans* strains 23270 and 33020 by molecular biological analysis. It was previously shown that the gene for Cyc2 was part of a gene complex, called *rus* operon (Appia-Ayme et al., 1999), which includes the genes for all the proteins involved in the passage of electrons from Fe^{2+} to O_2, and their expression is highly regulated (Yarzábal et al., 2004). The proteins expressed by the genes in the *rus* operon, in the order in which they pass electrons to the ultimate electron acceptor, O_2, are cytochrome Cyc2, rusticyanin, cytochrome Cyc1, and cytochrome oxidase (aa_3) (Figure 16.4B). The outer-membrane location of cytochrome Cyc2, which is directly involved in the removal of an electron from each Fe^{2+} that is oxidized, makes clear that the iron that is used as energy source by *A. ferrooxidans* never enters the cell. The electron transfer from Fe^{2+} to cytochrome oxidase on the plasma membrane thus involves, in sequence, cytochrome Cyc2, rusticyanin, cytochrome Cyc1, and cytochrome oxidase (aa_3) (Appia-Ayme et al., 1999; Figure 16.4B). The involvement of cytochrome oxidase (aa_3) in the *rus* operon contrasts with earlier claims of involvement of cytochrome *a*, which is not based on molecular biological analysis.

The presence of cytochrome Cyc2 in the outer membrane of *Acidithiobacillus ferrooxidans* is the first report of an enzyme located in the outer membrane of a gram-negative bacterium that is able to remove electrons from a substrate at the cell exterior (bulk phase) for transfer into the cell for

FIGURE 16.4 Alternative models of bioenergetic mechanisms of iron oxidation in *Acidithiobacillus ferrooxidans*. (A) Model as proposed by Ingledew et al. (1977); for discussion see text. (B) Diagram of most recent model showing path to O_2 of electrons removed from Fe^{2+} during its oxidation by *A. ferrooxidans* (based on Appia-Ayme C, Guiliani N, Ratouchniak J, Bonnefoy V, *Appl. Environ. Microbiol.*, 65, 4781–4787, 1999; Yarzábal A, Brasseur G, Ratouchniak J, Lund K, Lemesle-Meunier D, DeMoss JA, Bonnefoy V, *J. Bacteriol.*, 184, 313–317, 2002a; Yarzábal A, Brasseur G, Bonnefoy V, *FEMS Microbiol. Lett.*, 209, 189–195, 2002b; Yarzábal A, Appia-Ayme C, Ratouchniak J, Bonnefoy V, *Microbiology* (Reading), 150, 2113–2123, 2004; Elbehti A, Brasseur G, Lemesle-Meunier D, *J. Bacteriol.*, 182, 3602–3606, 2000; Bruscella P, Appia-Ayme C, Levcán G, Ratouchniak J, Jedlicki E, Holmes DS, Bonnefoy V, *Microbiology* (Reading), 153, 102–110, 2007). OM, outer membrane; PP, periplasm; PM, plasma membrane; Cyc1, cytochrome c_1; Cyc2, cytochrome c_2; rc, rusticyanin; bc_1, cytochrome bc_1 complex; aa_3, cytochrome aa_3. The arrows indicate directions of electron flow.

energy generation by respiration. Electron-translocating enzyme systems that transport electrons from the cell interior to the cell exterior in the anaerobic reduction of MnO_2 and Fe(III) oxide were previously discovered in some other gram-negative bacteria. Such systems have been found, for instance, in the outer membrane of *Shewanella putrefaciens* MR-1 (now *S. oneidensis* MR-1) and *Geobacter sulfurreducens* when grown anaerobically in the presence of MnO_2 and Fe(III) oxide, respectively (see discussion in Chapter 17).

Historically, evidence for the involvement of the cytochrome system in the oxidation of iron by *Acidithiobacillus ferrooxidans* was first presented by Vernon et al. (1960) and subsequently by Tikhonova et al. (1967) in a more detailed analysis. The biochemical basis for the succeeding models of iron oxidation by *A. ferrooxidans* emerged from the discovery of the copper protein rusticyanin in its periplasm (Cobley and Haddock, 1975; Cox and Boxer, 1978, 1986; Ingledew et al., 1977; Ingledew and Houston, 1986). Blaylock and Nason (1963) separated an iron–cytochrome c reductase from a particulate iron oxidase preparation from *Ferrobacillus* (now *Acidithiobacillus*) *ferrooxidans*. Yates and Nason (1966a,b) believed the reductase to be a DNA-containing enzyme protein, but Din et al. (1967a) reported it to be an RNA-containing enzyme. The location and significance of this enzyme in the intact cell remains unclear. Fukumori et al. (1988) isolated an Fe(II)-oxidizing enzyme from an *A. ferrooxidans* strain that conveyed electrons from Fe(II) to cytochrome c_{552} but not to rusticyanin. Neither of these enzymes appears to play a role in the most recent model of respiratory Fe^{2+} reduction by *A. ferrooxidans* presented by Yarzábal et al. (2002a,b, 2004).

Mansch and Sand (1992) described a membrane-bound iron-oxidizing system in *Acidithiobacillus ferrooxidans* strain F427 that includes cytochromes of the a_1, b, and c type (compare Yarzábal et al., 2002b). The c-type cytochrome consisted of at least three different acid-stable forms with M_r values of 60,000, 30,000, and 25,000, respectively, an acid-stable protein with noncovalently bound heme with an M_r value of 20,000, and an acid-stable protein with an M_r value of 18,000, which probably was rusticyanin. Sulfur-grown cells of this strain contained aa$_3$-type cytochrome in addition to the others, but cytochromes b and aa$_3$ were acid labile. The investigators proposed that ferrous iron oxidation in *A. ferrooxidans* involves more than one key enzyme. It is unclear as to how this F427 strain of *A. ferrooxidans* is compared genetically to the more extensively studied strains 23270 and 33020 used by Yarzábal et al. in their studies. It remains to be determined if strain F427 possesses a *rus* operon comparable to that in strains 23270 and 33020 described by Yarzábal et al. (2004).

16.4.2.1.9 *Energy Coupling in Iron Oxidation by* Acidithiobacillus ferrooxidans

Energy coupling in iron oxidation is best understood in terms of a chemiosmotic mechanism (Ingledew, 1982; Ingledew et al., 1977). Such a mechanism implies that a proton-motive force is set up across the plasma membrane of *A. ferrooxidans*, owing to charge separation across the two sides of the plasma membrane. The proton-motive force results from a pH gradient generated from the higher proton concentration in the acid periplasm relative to the near-neutral cytoplasm of an active *A. ferrooxidans* cell and from a transmembrane electric potential. The transfer of electrons to O_2 via the electron transport system results in pumping of protons from the cytoplasm into the periplasm, which together with the protons formed in the hydrolysis of ferric iron generated in the oxidation at the cell surface is a cause of the proton gradient. Energy coupling, that is, ATP synthesis, results from the fact that the plasma membrane is impermeable to protons in the periplasm except at the sites where adenosine 5′-triphosphate synthase (ATPase) is anchored in the membrane. ATPase contains a proton channel that allows the passage of protons from the periplasm in the direction of the cytoplasm. As a result of this proton movement, the ATPase causes the synthesis of ATP through the following reaction:

$$ADP + P_i \rightarrow ATP + H_2O \qquad (16.9)$$

Stoichiometrically, only one molecule of ATP can be synthesized per electron pair originating from the oxidation of two Fe^{2+} ions when the electrons are passed to oxygen at 100% efficiency. Because, as has already been discussed, the observed efficiency is much less than 100%, much ferrous iron needs to be oxidized to meet the energy demand of the *A. ferrooxidans* cell.

For a discussion of molecular biochemical analysis of ATP synthase of *Acidithiobacillus ferrooxidans*, the reader is referred to Rawlings and Kusano (1994).

16.4.2.1.10   Reverse Electron Transport

The assimilation of CO_2 by any autotroph requires a source of reducing power. When *Acidithiobacillus ferrooxidans* grows at the expense of ferrous iron oxidation, this source is Fe^{2+}. Ferrous iron thus has a dual function in the nutrition of this organism, namely, as sources of energy and reducing power. The electrons needed to reduce fixed CO_2 originate from the oxidation of ferrous ions and are transferred across the periplasmic space with the help of rusticyanin, just as in energy conservation from iron oxidation. However, unlike in energy conservation, in which the electrons from rusticyanin are transferred to cytochrome Cyc1 at the outer surface of the plasma membrane, in the case of NAD(P)H generation for CO_2 assimilation, they are transferred from rusticyanin to cytochrome CycA1 at the outer surface of the plasma membrane and thence to a bc_1 complex in the plasma membrane (Yarzábal et al., 2004; Bruscella et al., 2007). From the bc_1 complex, the electrons are transferred to $NAD(P)^+$ against an electropotential gradient by expenditure of energy (consumption of ATP) in a process called *reverse electron transport* (Aleem et al., 1963; Ingledew, 1982). The expenditure of energy is necessary because the *c*-type cytochrome$_{oxd}$/*c*-type cytocrome$_{red}$ redox couple (i.e., the level at which electrons removed from Fe^{2+} in its oxidation enter the plasma membrane-bound electron transport system) has a much higher $E_h^{0'}$ (e.g., +245 mV) than does the NAD/NADH redox couple ($E_h^{0'} = -320$ mV) or the NADP/NADPH couple ($E_h^{0'} = -324$ mV). The involvement of a cytochrome *c*, cytochromes c_1 and *b* (bc_1 complex), and a flavin as participants in the reverse electron transport system in *A. ferrooxidans* was first identified by Tikhonova et al. (1967) and Elbehti et al. (2000). As expected, arsenate, an uncoupler of oxidative phosphorylation, and amytal, an inhibitor of flavin reduction, blocked this system.

16.4.2.1.11   Carbon Assimilation

The major mechanism of CO_2 assimilation in *Acidithiobacillus ferrooxidans* involves the Calvin–Benson cycle (Din et al., 1967b; Gale and Beck, 1967; Maciag and Lundgren, 1964). A minor CO_2-fixation mechanism involving phosphoenol pyruvate (PEP) carboxylase also exists in this organism (Din et al., 1967b). The latter enzyme is needed for the formation of certain amino acids. In the Calvin–Benson cycle, CO_2 is fixed by ribulose 1,5-bisphosphate, generated from ribulose 5-phosphate, as follows (see also Chapter 6):

$$\text{Ribulose 5-phosphate} + \text{ATP} \xrightarrow{\text{Phosphoribulokinase}} \text{ribulose 1,5-bisphosphate} + \text{ADP} \qquad (16.10)$$

$$\text{Ribulose 1,5-biphosphate} + CO_2 \xrightarrow{\text{Ribulose bisphosphate carboxylase}} 2(\text{3-phosphoglycerate}) \qquad (16.11)$$

Each molecule of 3-phosphoglycerate is then reduced to 3-phosphoglyceraldehyde (PGA).

$$\text{3-Phosphoglycerate} + \text{NADPH} + H^+ + \text{ATP} \xrightarrow{\text{3-PGA dehydrogenase}} \text{3-phosphoglyceraldehyde}$$
$$+ \text{NADP} + + \text{ADP} + P_i \qquad (16.12)$$

In a complex series of steps, the 3-PGA is converted to various cell constituents as well as catalytic amounts of ribulose 5-phosphate to keep the Calvin–Benson cycle operating.

PEP carboxylase catalyzes the fixation of CO_2 by PEP, which is formed from 3-phosphoglycerate as follows:

$$3\text{-Phosphoglycerate} \xrightarrow{\text{Phosphoglyceromutase}} 2\text{-phosphoglycerate} \qquad (16.13)$$

$$2\text{-Phosphoglycerate} \xrightarrow{\text{Enolase}} \text{phosphoenol pyruvate} \qquad (16.14)$$

The PEP is then combined with CO_2 to form oxalacetate.

$$\text{Phosphoenolpyruvate} + CO_2 \xrightarrow{\text{PEP carboxylase}} \text{oxalacetate} + P_i \qquad (16.15)$$

No evidence for a functional tricarboxylic acid (TCA) cycle, whether operating in a forward or reverse direction, has been obtained to date in *Acidithiobacillus ferrooxidans* when growing autotrophically on iron. Although Anderson and Lundgren (1969) reported experimental evidence for such a cycle in iron-grown cells, Tabita and Lundgren (1971b) found it only in glucose-grown cells. The strain used in these studies may have been mixed with satellite organisms.

16.4.2.2 *Thiobacillus prosperus*

The mesophilic organism *Thiobacillus* (now *Acidithiobacillus*) *prosperus* resembles *A. ferrooxidans* closely except that it can grow in the presence of up to 6% NaCl (Huber and Stetter, 1989). Chloride ion is toxic to *A. ferrooxidans*.

16.4.2.3 *Leptospirillum ferrooxidans*

General traits. *Leptospirillum ferrooxidans* was first isolated by Markosyan (1972) and was further studied by Balashova et al. (1974) and Pivovarova et al. (1981). A comparison with other acidophiles from the same habitat as *L. ferrooxidans* was first published by Harrison (1984). The original isolation was from a copper deposit in Armenia. *L. ferrooxidans* is an acidophile. Its cells are vibrioid in shape with a polar flagellum ~25 nm in diameter. Involution cells may have a spiral shape. Tori form at pH values <2. In the laboratory, organisms have been grown in Leten's (Leathen's) medium at pH 2–3 (Kuznetsov and Romanenko, 1963) and in 9K medium (Silverman and Lundgren, 1959a) at pH 1.5. This organism oxidizes ferrous iron for energy conservation but cannot oxidize reduced forms of sulfur and is incapable of growth in organic media, that is, it is an obligate autotroph. At least one strain of *L. ferrooxidans*, CF12, oxidizes ferrous iron much more effectively at pH 1.25 and exhibits a lower temperature optimum (≤25°C) than *Acidithiobacillus ferrooxidans* ATCC 23270 (~30°C) (Gómez et al., 1999). However, two other strains of *L. ferrooxidans*, DSM 2705 and BC, have been shown to grow optimally at ~32 and ~35°C, respectively (Norris, 1990). The mechanism by which *L. ferrooxidans* oxidizes ferrous iron also differs from that of *A. ferrooxidans*. Its iron-oxidizing system lacks rusticyanin. Strain DSM 2705 of *L. ferrooxidans* contains a novel red cytochrome that is soluble, acid stable, and reducible by ferrous iron and seems to lack *a*-type cytochrome (Hart et al., 1991). The red cytochrome consists of a single polypeptide having a molecular mass of 17,900 Da and a standard reduction potential of +680 mV at pH 3.5. It contains one equivalent each of Fe and Zn (Hart et al., 1991). Strain P3A, on the contrary, contains a red cytochrome that has a smaller apparent molecular mass—12,000 Da. This is in keeping with the conclusion by Harrison and Norris (1985) that not all morphologically similar acidophilic iron oxidizers called *Leptospirillum* belong to the same species. A similar conclusion was reached by Johnson et al. (1989) and González-Toril et al. (1999), who used a different experimental approach. Ferrous iron oxidation by *L. ferrooxidans*-like organisms is less susceptible to product inhibition by Fe^{3+} than that by *A. ferrooxidans* (Norris et al., 1988).

The cells of *Leptospirillum ferrooxidans* have been shown to contain an active ribulose 1,5-bisphosphate carboxylase, typical of many chemolithotrophs (autotrophs) that use the Calvin–Benson cycle for CO_2 assimilation (Balashova et al., 1974).

16.4.2.4 *Metallogenium*

An acid-tolerant organism, named *Metallogenium*, was reported from mesoacidic, iron-bearing groundwaters (Walsh and Mitchell, 1972). It is a filamentous organism consisting of branching filaments (0.1–0.4 μm and >1 μm long) that are usually encrusted with iron oxide. This organism tolerates a pH range of 3.5–6.8, with an optimum at 4.1, and is thus an intermediate between acidophiles and neutrophiles. In the laboratory, the addition of 0.024 M phthalate at pH 4.1 was important for observing iron oxidation and growth at initial Fe^{2+} concentrations >100 mg L^{-1}. Acetate, citrate, or phosphate could not replace phthalate in the medium. The growth medium contained $(NH_4)_2SO_4$, 0.1%; KH_2PO_4, 0.001%; $CaCO_3$, 0.01%; $MgSO_4 \cdot 7H_2O$, 0.02%; and $FeSO_4 \cdot 7H_2O$, 9%. These ingredients were dissolved in distilled water, and the solution was adjusted to pH 4.1. It is not clear as to whether this organism is an autotroph or a heterotroph. Care has to be taken in identifying it. Some inorganic phosphates may resemble it morphologically (Ivarson and Sojak, 1978).

16.4.2.5 *Ferromicrobium acidophilum*

Unlike the three previously described acidophilic iron-oxidizing mesophiles that are autotrophic, a unicellular iron-oxidizing mesophile that is obligately heterotrophic has been reported (Johnson and Roberto, 1997; Hallberg et al., 2006). It requires the presence of yeast extract to grow on ferrous iron. Its physiological temperature range is <20–40°C (optimum 37°C). Under limited aeration, it also has the ability to reduce ferric to ferrous iron. The GC ratio in its DNA is 51–55 mol%. The type strain is T-23 and it has been tentatively named *Ferromicrobium acidophilum* (Johnson and Roberto, 1997; private communication, 2000).

16.4.2.6 Strain CCH7

Another acidophilic, heterotrophic, iron-oxidizing bacterium, designated CCH7, was isolated by Johnson et al. (1993). It grew in a pH range of 2–4.4 (optimum pH 3). It formed streamers (filaments) >100 μm long when growing in liquid medium. The streamers broke up into motile cells and short filaments in later stages of growth. The authors concluded that this organism did not use Fe^{2+} as an energy source because addition of Fe^{2+} to the organic growth medium did not stimulate its growth. It produced a sheath like that of the *Sphaerotilus–Leptothrix* group (see Section 16.4.7.2), but its GC content of 62 mol% was lower than that of the *Sphaerotilus–Leptothrix* group, and it was acidophilic instead of neutrophilic. Other streamer-forming organisms in acid mine drainage in North Wales have been described by Hallberg et al. (2006).

16.4.3 Domain Bacteria: Thermophiles

16.4.3.1 *Sulfobacillus thermosulfidooxidans*

Sulfobacillus thermosulfidooxidans was isolated by Golovacheva and Karavaiko (1978) and further described by Norris (1990). It is a gram-positive, nonmotile, spore-forming (most strains), rod-shaped bacterium with tapered ends (0.6–0.8 μm × 1–1.3 μm, sometimes as long as 6 μm). The GC content of its DNA is 53.6–53.9 mol%. Its temperature range for growth is 28–60°C, with an optimum ~50°C, making it a *moderate thermophile*. It grows autotrophically on Fe(II), S^0, or metal sulfides as energy source and heterotrophically in the absence of an appropriate inorganic energy source (Golovacheva and Karavaiko, 1978). Autotrophic growth is stimulated by air enriched in CO_2 (Norris, 1997). It cannot use sulfate as a source of sulfur and consequently must be supplied with reduced sulfur (Norris and Barr, 1985). Autotrophic growth is stimulated by a trace (0.01–0.05%) of yeast extract (Golovacheva and Karavaiko, 1978), which causes the cells to increase significantly in size (Norris, 1997), but 0.1% yeast extract is inhibitory. Its pH range for growth is 1.9–3; the optimum range is 1.9–2.4 (Golovacheva and Karavaiko, 1978). A strain of this organism was formerly designated BC1 (Norris, 1997).

16.4.3.2 *Sulfobacillus acidophilus*

Sulfobacillus acidophilus is closely related to *S. thermosulfidooxidans*. A strain of *S. acidophilus* was previously designated as ALV (Norris et al., 1988). These two *Sulfobacillus* species differ in their GC content, with *S. thermosulfidooxidans* exhibiting a ratio of 48–50 mol% and *S. acidophilus* 55–57 mol% (Norris, 1997). Unlike the cells of *S. thermosulfidooxidans*, the cells of *S. acidophilus* do not increase significantly in size when growing on ferrous iron in the presence of a trace of yeast extract or when growth is heterotrophic in the absence of iron. Both species can appear elongated or in chains during autotrophic growth on mineral sulfides at acidities near maximum tolerance. In laboratory culture, *S. acidophilus* oxidizes elemental sulfur more readily than *S. thermosulfidooxidans*, whereas the latter oxidizes iron and mineral sulfides more readily (see Norris, 1997).

16.4.3.3 *Acidimicrobium ferrooxidans*

Acidimicrobium (Amicro.) ferrooxidans is another moderately thermophilic bacterium that grows autotrophically when oxidizing ferrous iron in air (Norris, 1997). Two strains have been studied: TH3 (Norris et al., 1980) and ICP (Clark and Norris, 1996). The DNA of strain TH3 exhibits a GC ratio of 68 mol%, and that of strain ICP 67.3 mol% (Clark and Norris, 1996). Strain TH3 can exhibit a filamentous growth habit (Clark and Norris, 1996). Growth of *Amicro. ferrooxidans* is not stimulated in air enriched with CO_2. It is less tolerant to ferric iron accumulation than *Sulfobacillus* spp. The ribulose 1,5-bisphosphate carboxylase of *Amicro. ferrooxidans* has an amino acid sequence that shares a high degree of similarity with that of *Acidithiobacillus ferrooxidans* (Norris, 1997).

16.4.4 DOMAIN ARCHAEA: MESOPHILES

16.4.4.1 *Ferroplasma acidiphilum*

Ferroplasma acidiphilum is an iron-oxidizing, chemolithotrophic, acidophilic microorganism whose cells are delimited by only a single peripheral membrane (Golyshina et al., 2000). Its pH optimum for growth is 1.7 (pH range 1.3–2.2). It requires a trace of yeast extract (0.02%) in its growth medium. At least some strains can also grow chemoorganotrophically on yeast extract or one of the several sugars, and they can grow chemomixotrophically on ferrous iron and yeast extract or sugars with temperature optima between 35 and 42°C (Dopson et al., 2004).

16.4.4.2 *Ferroplasma acidarmanus*

Ferroplasma acidarmanus (Figure 16.5; Barker et al., 1998) resembles *F. acidiphilum* but has a pH optimum of 1.2 (pH range 0–2.5) and grows three times as fast as the latter at optimum pH. It has been found in biofilms in pyritic sediments in the drainage tunnels at the Iron Mountain pyrite mine in California (Edwards et al., 2000). Like *F. acidiphilum*, it requires 0.02% yeast extract in its medium for autotrophic growth. It is also able to grow chemoorganotrophically with yeast extract or one of the several sugars as sole energy source and chemomixotrophically in media containing ferrous iron and yeast extract or one of the several sugars (Edwards et al., 2000; Dopson et al., 2004).

16.4.5 DOMAIN ARCHAEA: THERMOPHILES

16.4.5.1 *Acidianus brierleyi*

Acidianus brierleyi (formerly *Sulfolobus brierleyi*) (Segerer et al., 1986) is an extremely thermophilic acidophile growing in a temperature range of 55–90°C (optimum 70–75°C) and a pH range of 1–5 (optimum ~3). It is a non-spore-forming, nonmotile, pleomorphic organism (1–1.5 μm in diameter) (Brierley and Brierley, 1973; Brierley and Murr, 1973; McClure and Wyckoff, 1982; Segerer et al., 1986; Figure 16.6A). The GC ratio of its DNA is 31 mol% (Segerer and Stetter, 1998). This organism contains no peptidoglycan in its cell wall, nor does it feature an outer membrane as in gram-negative bacteria (Berry and Murr, 1980). Instead, the cells are surrounded by a protein layer called the S layer

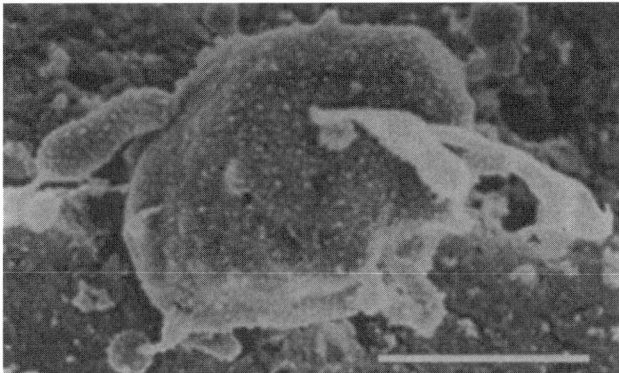

FIGURE 16.5 A cryoscanning electron photomicrograph of *Ferroplasma acidarmanus* isolate fer1 (bar = 500 nm). The cells were in late log growth phase when they were prepared for cryoscanning electron microscopy as described by Barker et al. (1998). The cells were viewed with a Hitachi S900 scanning electron microscope operated at 2 kV on a Gatan cryostage. Irregularly sized and shaped cellular protrusions are inferred to be the budding sections of the cell. (From Edwards KJ, Bond PL, Gihring TM, Banfield JF, *Science*, 287, 1796–1799, 2000. With permission from the American Association for the Advancement of Science.)

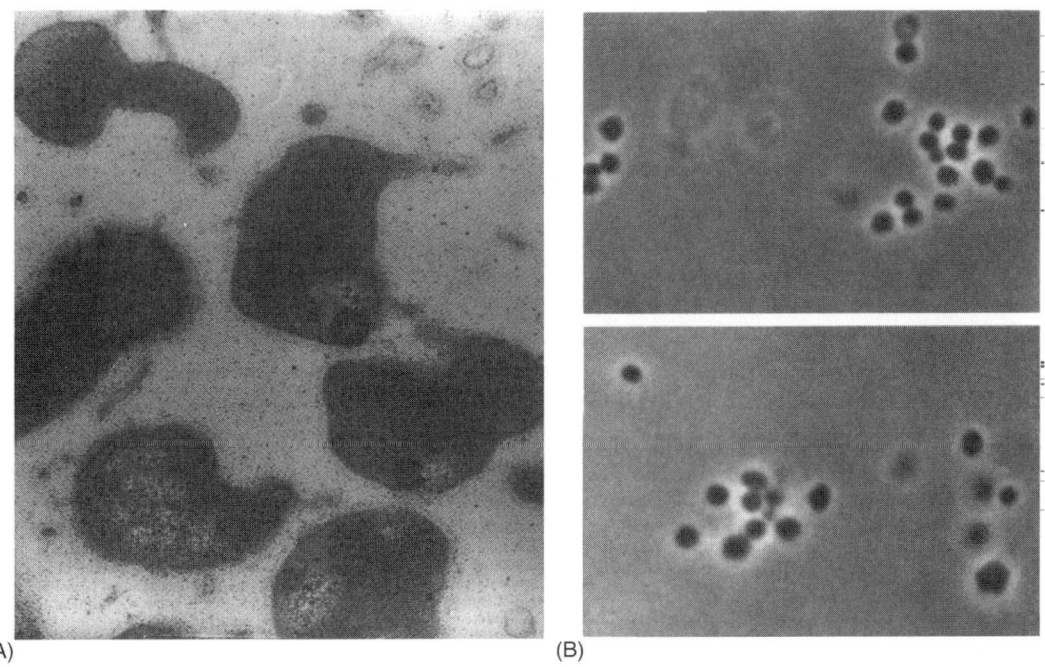

(A)   (B)

FIGURE 16.6 Iron- and sulfur-oxidizing archaea. (A) *Acidianus* (formerly *Sulfolobus*) sp. (×28,760). (Courtesy of Brierley CL, Brierley JA, *Can. J. Microbiol.*, 19, 183–188, 1973.) (B) *Sulfolobus acidocaldarius* (×3540). (Reproduced from Brock TD, Brock KM, Belly RT, Weiss RI, *Arch. Mikrobiol.*, 84, 54–68, 1972. With permission from Springer Science and Business Media.)

(Taylor et al., 1982). It can grow autotrophically on ferrous iron or S^0 as energy source. Autotrophic growth is stimulated by the addition of 0.2% yeast extract. CO_2 is not assimilated by a Calvin–Benson cycle but more likely by a reverse TCA cycle (Atlas, 1997; see also Chapter 6). It can grow heterotrophically on yeast extract at elevated concentrations.

16.4.5.2 *Sulfolobus acidocaldarius*

Sulfolobus acidocaldarius (Figure 16.6B) is an extremely acidophilic, thermophilic, chemolitho-trophic microorganism that has the capacity to use ferrous iron oxidation as a source of energy (Brock et al., 1972, 1976). It closely resembles *Acidianus brierley* morphologically and physiologi-cally. However, it differs from *A. brierleyi* on a genotypic basis (Segerer and Stetter, 1998). The GC ratio of its DNA is 60–68 mol%.

16.4.6 DOMAIN BACTERIA: NEUTROPHILIC IRON OXIDIZERS

16.4.6.1 Unicellular Bacteria

Bona fide unicellular, microaerobic, lithoautotrophic, neutrophilic iron oxidizers were discovered only relatively recently. *Sideroxydans lithotrophicus* strain ES-1 was isolated from Fe floc-contain-ing groundwater in Michigan, and strain RL-1 from a wetland in Michigan (Emerson and Moyer, 1997; Emerson, 2000; Emerson et al., 2008). Marine strains JV-1 and PV-1 were isolated from the Loihi hydrothermal vent site (Emerson and Moyer, 2002). Methods of isolation for such organisms have been described by Emerson and Weiss (2004) and Emerson and Floyd (2005). Other diverse marine, neutrophilic iron-oxidizing microorganisms have been described by Edwards et al. (2004). Anaerobic neutrophilic iron oxidizers are described in Section 16.5.

Freshwater neutrophilic iron-oxidizing bacteria are commonly encountered in iron plaque on root systems of most wetland and submerged aquatic plants together with ferric iron-reducing bacteria (Weiss et al., 2006).

16.4.7 APPENDAGED BACTERIA

16.4.7.1 *Gallionella ferruginea*

Gallionella ferruginea is a mesophilic organism that consists of a bean-shaped cell and a lateral stalk of twisted bundles of fibers (Figure 16.7; see also figures in Hanert, 1981; Ghiorse, 1984). It is currently assigned to the domain Bacteria; Proteobacteria; β-Proteobacteria; Nitrosomonadales; Gallionellaceae; but its genome has not been sequenced as yet. This organism was first described by Ehrenberg (1836). Cholodny (1924) was the first to recognize that the bean-shaped cell on the lateral stalk was an integral part of this organism. The stalk may branch dichotomously, each branch carrying a single bean-shaped cell at its tip. The stalk is usually anchored to a solid surface. The stalk may be heavily encrusted with ferric hydroxide. The cells, which may form one or two polar flagella, may detach from their stalk, swim away as swarmers, seek a new site for attachment, and develop a stalked growth habit.

Gallionella has been described as a gradient organism, that is, it grows best under low oxygen tension (0.1–1 mg of O_2 per liter) and in an E_h range of +200 to +320 mV (Kucera and Wolfe, 1957; Hanert, 1981). Its optimal growth temperature is 20°C (Hallbeck and Pedersen, 1990), although in nature the growth of some strains has been observed up to 47°C (Hanert, 1981). It prefers a pH range of 6–7.6 (Hanert, 1981). Its low-oxygen tension requirement explains as to why this organism can catalyze Fe^{2+} oxidation at neutral pH. Under these partially reduced conditions, ferrous iron autoxidizes only slowly (Wolfe, 1964).

Hallbeck and Pedersen (1990) studied a strain of *Gallionella ferruginea* in the laboratory whose cells were free living and motile with a single flagellum and no stalk when growing exponentially under aerobic gradient conditions. Stalks began to form only in stationary phase when the cell population exceeded 6×10^5 mL^{-1} at a pH >6. No stalks were formed when ferrous iron did not autoxidize under microaerobic conditions (e.g., pH 6.5, E_h <−40 mV). The investigators suggested that the stalk formation by *G. ferruginea* protects against "increasing reducing capacity of ferrous iron as it becomes unstable in an environment that becomes oxidizing."

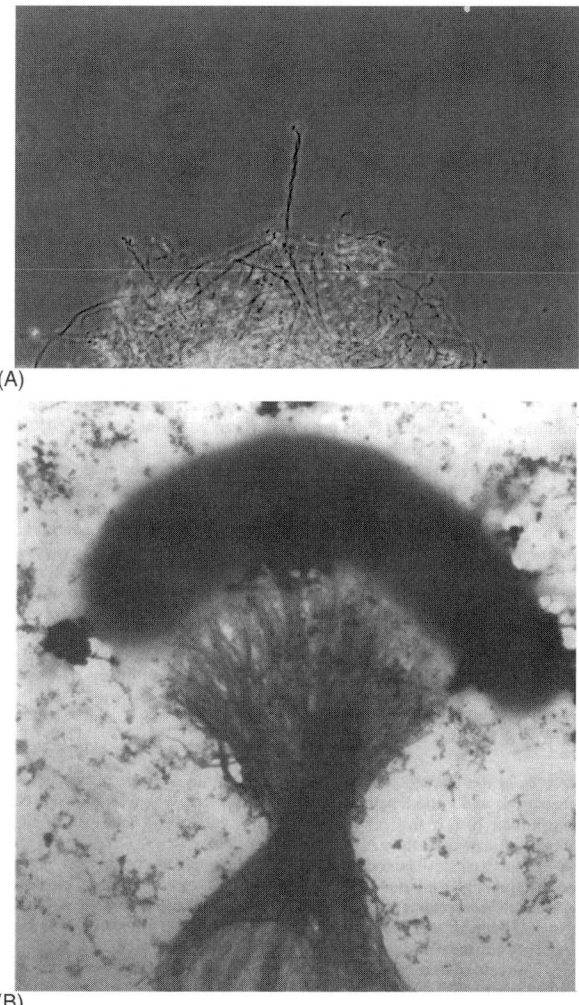

(A)

(B)

FIGURE 16.7 (A) Phase contrast photomicrograph of a tangle of *Gallionella ferruginea*. Note the small cell at the tip of the twisted stalk projecting from the tangle ($\times$576). (B) Electron photomicrograph of unstained and unshadowed *Gallionella* cell showing fibrillar nature of lateral twisted stalk ($\times$65,160). (Courtesy of Ghiorse WC, Department of Microbiology, Cornell University, Ithaca, NY, USA.)

Using environmental scanning electron microscopy, Hallberg and Ferris (2004) studied the nature of iron mineralization associated with *Gallionella* stalks. They demonstrated minute flaky crystallites of iron oxide with a ratio of Fe to O of ~0.67 consistent with the formula Fe_2O_3 for hematite inside the stalk fibers, which was believed by them to be the site of the formation of crystallites. Thermodynamically, the crystallite formation probably depended on CO_2 as well as oxygen, according to the authors. The authors also inferred that different forms of iron oxyhydroxides including goethite, which was observed by them to cover older stalks of *Gallionella*, probably form chiefly as a result of inorganic processes.

Gallionella can grow autotrophically and mixotrophically. Hallbeck and Pedersen (1991) demonstrated that their organism obtained all their carbon from CO_2 when growing in a mineral salts medium in an aerobic gradient with ferrous sulfide as energy source. The same investigators showed that glucose, fructose, and sucrose could meet part of the energy requirement and part or all of the carbon requirement of the organism, depending on the concentrations of the sugars. Hanert (1968) previously claimed that *Gallionella* does not grow without oxidizable iron in the medium. It may well be that strictly autotrophic as well as facultatively autotrophic strains exist. This organism appears to fix CO_2 via ribulose 1,5-bisphosphate carboxylase/oxidase (Lütters and Hanert, 1989; Hallbeck and Pedersen, 1991).

According to Lütters-Czekalla (1990), *Gallionella ferruginea* strain BD was able to grow using sulfide and thiosulfate as energy sources and electron donors but not elemental sulfur or tetrathionate at the interface of the oxidizing and reducing zones of a microgradient culture. Addition of organic carbon did not stimulate growth. Under these growth conditions, this organism did not form the characteristic stalk, which is formed when growing on Fe(II). Lütters-Czekalla (1990) took this to indicate that the stalk is a product of iron oxidation. Her culture nevertheless excreted a significant amount of an unidentified extracellular polymeric material. The strain of *G. ferruginea* isolated by Hallbeck (1993) was not able to grow with sulfide or thiosulfate as sole source of energy and reducing power.

Isolation and propagation methods for *Gallionella* were summarized by Hanert (1981). They are adaptations of the method originally described by Kucera and Wolfe (1957), using the medium formulated by them. The medium consists of NH_4Cl, 0.1%; K_2HPO_4, 0.05%; and $MgSO_4$, 0.02%, with freshly prepared ferrous sulfide suspension making up 10% of the total volume of the medium. Tap water or natural water is used in making up the medium because distilled water apparently does not supply a required component. A small amount of CO_2 is bubbled through the salts solution prior to the addition of ferrous sulfide. The medium is placed in test tubes, which are stoppered to prevent the loss of CO_2. A redox gradient is established in the culture medium as oxygen from the air in the headspace diffuses into the medium and reacts with some of the FeS. When growing in this medium, *Gallionella* occupies the lower two-thirds of the volume above the ferrous sulfide. This organism will not grow anaerobically, whether nitrate is added or not.

Owing to the complex growth habit of *Gallionella* under some culture conditions, determination of its growth rate can be a problem. Individual development of stalked organisms can be followed quantitatively in microculture by microscopically following stalk elongation and twisting (Hanert, 1974a). In other words, growth is measured in terms of increase in mass of the organism. An elongation rate of 40–50 $\mu m\ h^{-1}$ in the first generation has been obtained. This is two to four times as fast as that in a natural environment. Stalks do not elongate further after three or four divisions of the apical cell. Stalk lengthening occurs at the tip where the apical cell is attached. The stalk twists as it lengthens, owing to the rotation of the apical cell. The rotation occurs at a constant rate. In a natural environment, the rate of growth of *Gallionella* is measured in terms of the rate of attachment to a solid surface such as a submerged microscope slide and stalk elongation and is shown in the relationship (Hanert, 1973)

$$V_t = (b_v l_v)\ (t^2)/2 \qquad (16.16)$$

where b_v is the average rate of attachment, l_v the average rate of stalk elongation, and t the length of the growth period, which should not be longer than 10 h if this relationship is to hold. V_t is a measure of the amount of growth at time t.

Growth can also be followed by determining viable counts by a most probable number method and total counts by a direct counting method with an epifluorescence microscope after staining the cells with acridine orange (Hallbeck and Pedersen, 1990). These latter methods have to be employed for measuring growth when no stalks are formed.

The rate of iron oxidation by *Gallionella* in the natural environment may be measured by submerging a microscope slide at a site of *Gallionella* development for a desired length of time, then removing the slide, and measuring the amount of iron deposited on it (Hanert, 1974b). The quantity of iron deposited can be measured in terms of the amount of iron per unit surface area of the slide that was submerged and on which iron was laid down.

Another species of *Gallionella* was discovered ~10 years ago, named *G. capsiferriformans* strain ES-2, from groundwater in Michigan (Emerson and Moyer, 1997; http://www.genomesonline.org/DBs/goldtable.txt). It has also been assigned to the β-Proteobacteria. It was originally described as a unicellular organism and closely related to *Sideroxydans lithotrophicus* strain ES-1 (Emerson, 2000).

16.4.7.2 Sheathed, Encapsulated, and Wall-Less Iron Bacteria

Other mesophilic bacteria that have been associated with ferrous iron oxidation at circumneutral pH include sheathed bacteria, such as *Sphaerotilus*, *Leptothrix* spp. (Figure 16.8), *Crenothrix polyspora*, *Clonothrix* sp., and *Lieskeella bifida*, and some encapsulated bacteria such as the Siderocapsaceae group. Many of these are more likely iron-depositing bacteria, that is, they bind preoxidized iron at their cell surface (Ghiorse, 1984; Ghiorse and Ehrlich, 1992).

Ultrastructural examination of the sheath of *Leptothrix discophora* SP-6 showed it to be a tubular structure of condensed fibrils (6.5 nm in diameter) overlain by a somewhat diffuse capsular layer (Emerson and Ghiorse, 1993a). The fibrillar part of the sheath was anchored by bridges to the outer membrane of the gram-negative cells in the sheath. The capsular layer had a net negative charge. Purified sheaths contained 34–35% polysaccharide consisting of a 1:1 mixture of uronic acids and galactosamine, 23–25% protein enriched in cysteine, 8% lipid, and 4% inorganic ash. The cysteine in the sheath protein is thought to be important in the maintenance of the integrity of the sheath (Emerson and Ghiorse, 1993b).

Leptothrix spp., which sometimes have been classed with *Sphaerotilus* (Pringsheim, 1949; Stokes, 1954; Hoehnl, 1955), have been examined on several occasions for enzymatic iron oxidation. Winogradsky (1888) first reported that *L. ochracea* (probably *L. discophora*, according to Cholodny, 1926) could oxidize ferrous iron. He found that he could grow this organism in hay infusion only if he added ferrous carbonate. The iron was oxidized and deposited in the sheath of the organism. He inferred from this observation that the organism was an autotroph. Molisch (1910) and Pringsheim (1949) disagreed with Winogradsky's conclusion about iron oxidation by *L. ochracea*, believing that the organism merely deposited autoxidized iron in its sheath. However, Lieske (1919) confirmed Winogradsky's observations of growth on ferrous carbonate in very dilute organic solution and suggested that the organism might be mixotrophic. Cholodny (1926), Sartory and Meyer (1947), and Praeve (1957) also made observations similar to those of Winogradsky and Lieske. Most claims in the past for enzymatic iron oxidation by *Leptothrix* were based mainly on the growth requirement for ferrous iron in dilute medium and the oxidation of the ferrous iron during growth. However, Praeve (1957) also showed a stimulation of oxygen uptake by the organism when Fe^{2+} was present as the only exogenous, oxidizable substrate in Warburg respiration experiments. Significantly, he found that the empty sheaths were unable to take up oxygen on Fe^{2+}. Most recently, de Vrind-de Jong et al. (1990) reported iron-oxidizing activity in spent medium from a culture of *L. discophora* SS-1. Corstjens et al. (1992) related this iron-oxidizing activity to a 150 kDa protein. It behaved like an enzyme and was not produced by a spontaneous mutant strain that lacked iron-oxidizing activity. The factor was distinct from the manganese-oxidizing protein excreted by the wild-type SS-1 strain described by Adams and Ghiorse (1987).

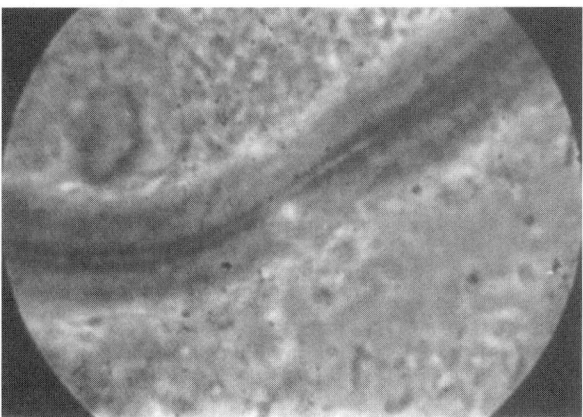

FIGURE 16.8   *Leptothrix* sp. Portion of the sheathed organism (×3460). (Courtesy of Arcuri EJ)

Enzymatic iron oxidation by all other sheathed bacteria is at most presumptive, based on gross morphological similarities with *Leptothrix* and the observation of oxidized iron deposits on their sheaths. It is quite likely that most or all of these organisms merely deposit preoxidized iron on their sheaths.

Dubinina (1978a,b) has reported that *Leptothrix pseudoochracea*, *Metallogenium*, and *Arthrobacter siderocapsulatus* oxidized Fe^{2+} with metabolically produced H_2O_2 through catalysis by catalase of the organism.

A wall-less bacterium, *Mycoplasma laidlawii* (now known as *Acholeplasma laidlawii*), has been reported to oxidize ferrous iron (Balashova and Zarvazin, 1972). This organism was cultured in a salt-free meat–peptone medium containing iron wire or powder. Ferric iron was formed during active growth and, in part, precipitated on the cells of the organism. Addition of catalase was found to depress ferric oxide production, suggesting that H_2O_2 played a role in the oxidation process. It is interesting that in this instance catalase did not accelerate iron oxidation as was found with other organisms by Dubinina (1978a). It is not clear from the report whether other enzymes played a direct role in the oxidation of iron in this instance. It seems that different mechanisms of enzymatic iron oxidation may exist among neutrophilic bacteria.

16.5 ANAEROBIC OXIDATION OF FERROUS IRON

16.5.1 PHOTOTROPHIC OXIDATION

Although until fairly recently bacterial oxidation of ferrous iron was generally assumed to require oxygen as terminal electron acceptor, exceptions that occur at circumneutral pH have been found. Diverse anaerobic photosynthetic bacteria, including strains *Rhodomicrobium*, *Rhodopseudomonas (Rhps.)*, *Rhodobacter*, *Rhodovulum*, *Chlorobium* (Figure 16.9), and *Thiodictyon*, are able to oxidize Fe(II) to Fe(III) oxides in the light. Jiao et al. (2005) studied a genetically tractable variant of *Rhps. palustris*, strain TIE-1, in which they demonstrated the need for two genetically determined components in a functioning phototrophic Fe(II)-oxidizing system, using transposon mutagenesis. Jiao and Newman (2007) demonstrated that the three-gene *pio* operon in this organism is essential for

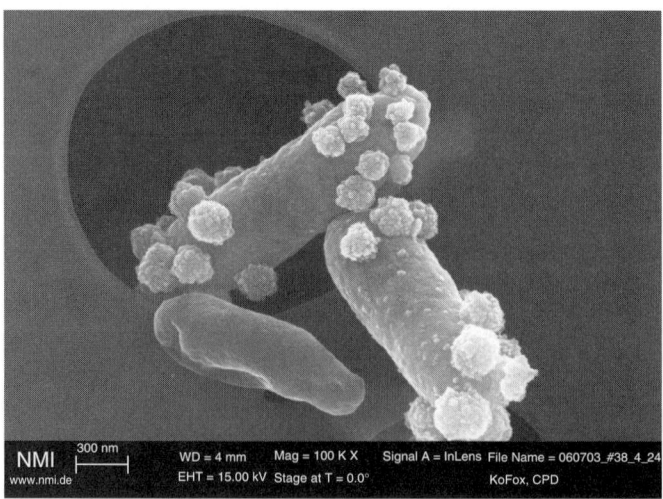

FIGURE 16.9 Scanning electron micrograph of critical point dried cells of the phototrophic iron-oxidizing bacterium *Chlorobium ferrooxidans* sp. strain KoFox. The particles attached to the surface of the cell are iron minerals precipitated during iron oxidation. Image taken by Sebastian Schädler, Claus Burkhardt, and Andreas Kappler at the Natural and Medical Science Institute at the University of Tuebingen (NMI), using the InLens Detector of a Zeiss Crossbeam Leo 1540 XB microscope. Working distance = 4 mm; acceleration voltage = 15 kV. (Courtesy of Kappler A)

growth by phototrophic Fe(II) oxidation, whereas Croal et al. (2007) showed that the three-gene *fox* operon enhances phototrophic Fe(II) oxidation in *Rhodobacter capsulatus* SB1003. All of these bacteria except *R. capsulatus* SB1003 can use the reducing power generated in the oxidation of Fe(II) to Fe(III) oxide in CO_2 assimilation (fixation) (Ehrenreich and Widdel, 1994; Widdel et al., 1993; Caiazza et al., 2007). Caiazza et al. (2007) have now shown that *R. capsulatus* SB1003 can grow photoheterotrophically by oxidizing Fe(II)-citrate to Fe(III)-citrate under anaerobic conditions and converting the citrate of Fe(III)-citrate to acetoacetate, whose carbon is then assimilated. Fe(II)-NTA can also be used in a similar fashion. No iron oxidation was observed anaerobically in the dark or light in the absence of the bacteria. Furthermore, no growth by these bacteria occurred in the test medium in the absence of added iron.

Widdel et al. (1993) suggested that anaerobic oxidation of ferrous to ferric iron by phototrophic bacteria could have contributed to the early stages of BIFs that arose in Archean times.

16.5.2 CHEMOTROPHIC OXIDATION

In addition to anaerobic iron-oxidizing phototrophs, respirers that oxidize Fe(II) anaerobically have been described. Straub et al. (1996) found that enrichment cultures from town ditches in Bremen, Germany, and from brackish water lagoons, as well as some denitrifying isolates were able to oxidize iron anaerobically using nitrate as terminal electron acceptor. Some were able to use ferrous iron as exclusive electron donor and grew lithotrophically, and some others used acetate concurrently with ferrous iron as electron donors and thus grew mixotrophically. The nitrate was reduced to dinitrogen in all instances. Ammonia production from nitrate was not detected. *Thiobacillus denitrificans* was also found to be able to reduce nitrate with ferrous iron (Straub et al., 1996). The following equation summarizes the biooxidation of ferrous iron by nitrate in these observations (Straub et al., 1996):

$$10FeCO_3 + 2NO_3^- + 24H_2O \rightarrow 10Fe(OH)_3 + N_2 + 10HCO_3^- + 8H^+ \qquad (16.17)$$

Benz et al. (1998) detected anaerobic, nitrate-coupled ferrous iron oxidation in culture enrichments with sediments from freshwater, brackish water, and marine water. They isolated a strain labeled HidR2, which was a motile, non-spore-forming, gram-negative rod that oxidized ferrous iron in the presence of 0.2–1.1 mM acetate as cosubstrate and nitrate as terminal electron acceptor at pH 7.2 and 30°C. The ferrous iron served as energy source. Although the authors considered this strain as heterotrophic, its growth with ferrous iron as energy source indicates that it grows mixotrophically under these conditions. This organism is capable of anaerobic growth on acetate in the absence of ferrous iron but in the presence of nitrate. It is also capable of aerobic growth in the presence of low concentrations of acetate with O_2 as terminal electron acceptor. In a more recent study, Straub et al. (2004) reported that the three earlier iron-oxidizing, nitrate-reducing isolates, BrG1, BrG2, and BrG3, obtained from freshwater sediments, affiliated with *Acidovorax*, *Aquabacterium*, and *Thermomonas*, respectively, which belong to the β- and γ-subgroups of the Proteobacteria.

Weber et al. (2006b) isolated a lithoautotrophic culture, strain 2002, from an anoxic, highly reduced sediment core taken from freshwater Campus Lake at Southern Illinois University in Carbondale, Illinois. The isolate grew autotrophically while reducing Fe(II) anaerobically with nitrate as terminal electron acceptor. Its autotrophy was facultative, because it was able to grow with some simple organic carbon compounds, such as acetate, propionate, and so on, as sole carbon and energy source. On acetate, it was able to grow with nitrate, nitrite, nitrous oxide, or oxygen as terminal electron acceptor. This organism was detected in significant numbers throughout the sediment core in a range of ~10^3–10^4 g^{-1} of sediment. It was closely related to the common soil bacterium *Chromobacterium violaceum* in the β-Proteobacteria.

A hyperthermophilic archeon, *Ferroglobus placidus*, has also been found to oxidize ferrous iron anaerobically with nitrate as electron acceptor (Hafenbrandl et al., 1996). It can grow lithoautotrophically or heterotrophically between 65 and 95°C (optimum 85°C) and at pH 7. The cells require 0.5–4.5% NaCl (~2% optimum) in their growth medium to prevent their lysis. The organism reduces the nitrate chiefly to nitrite, but during prolonged incubation the nitrite is converted to NO and NO_2. However, no N_2, N_2O, or NH_3 is formed. The following reaction best describes the overall oxidation of ferrous iron by this organism (Hafenbrandl et al., 1996, amended):

$$2FeCO_3 + NO_3^- + 5H_2O \rightarrow 2Fe(OH)_3 + NO_2^- + 2HCO_3^- + 2H^+ \qquad (16.18)$$

The bacterial processes described in this section can be considered to be part of an anaerobic iron cycle. For more information, the reader is referred to two recent reviews (Kappler and Straub, 2005; Weber et al., 2006).

16.6 IRON(III) AS TERMINAL ELECTRON ACCEPTOR IN BACTERIAL RESPIRATION

In nature, ferric iron may be microbiologically reduced to ferrous iron. The ferric iron that is reduced by microbes may be in solution or insoluble. Examples of insoluble forms are amorphous oxides or hydroxides and minerals such as limonite (FeOOH), goethite ($Fe_2O_3 \cdot H_2O$), hematite (Fe_2O_3), and so forth. As in the case of iron oxidation, this reduction may be enzymatic or nonenzymatic. Enzymatic ferric iron reduction may manifest itself as a form of respiration, mostly anaerobic, in which ferric iron serves as a *dominant* or *exclusive* terminal electron acceptor, or it may accompany fermentation in which ferric iron serves as a supplementary, as opposed to dominant or exclusive, terminal electron acceptor. Both of these ferric iron-reducing processes are forms of *dissimilatory iron reduction*.

When ferric iron is reduced during uptake or incorporation into specific cellular components, the process represents *assimilatory iron reduction*. Relatively large quantities of iron are consumed in dissimilatory reduction, whereas only very small quantities are consumed in assimilatory reduction. In assimilatory reduction, the ferric iron when acquired at circumneutral pH is usually complexed by siderophores and may be reduced in this complexed form or after release from the ligand in the cell envelope (discussed earlier). Dissimilatory and assimilatory reductases have been reviewed by Schröder et al. (2003). The emphasis in the following sections is on dissimilatory iron reduction.

16.6.1 BACTERIAL FERRIC IRON REDUCTION ACCOMPANYING FERMENTATION

For some time, ferric iron has been known to influence fermentative metabolism of bacteria as a result of its ability to act as terminal electron acceptor. Roberts (1947) showed a change in fermentation balance when comparing the action of *Bacillus polymyxa* on glucose anaerobically in the presence and absence of ferric iron (Table 16.4). The ferric iron in these experiments was supplied as freshly precipitated, dialyzed ferric hydroxide suspension obtained in a reaction of ferric chloride and an excess of potassium hydroxide. The suspension had a pH of 7.8. The ferric iron seemed to act as a supplementary electron acceptor in fermentation, and in this way it changed the relative quantities of certain products formed from glucose. Thus, in the presence of ferric iron, less H_2, CO_2, and 2,3-butylene glycol but more ethanol were formed in either *organic* or *synthetic* medium than in the absence of iron. Also, more glucose was consumed in the presence of iron than in its absence in either medium.

Bromfield (1954a) showed that besides *Bacillus polymyxa*, growing cultures of *B. circulans* are also able to reduce ferric iron. Depending on the medium, he found that even *Escherichia freundii*, *Aerobacter* (now *Enterobacter*) sp., and *Paracolobactrum* (now probably *Citrobacter*) could do so. However, from his results he inferred that the reduction of iron was not directly involved in the oxidation of the substrate (energy source), which is at variance with resting cells with his results (Bromfield, 1954b). He found that completely anaerobic conditions were not required to observe ferric iron reduction by bacteria. But when the level of aeration of the cultures was increased, ferrous iron became reoxidized because of autoxidation.

TABLE 16.4
Fermentation Balances for *Bacillus polymyxa* Growing in Two Different Media in the Presence and Absence of Ferric Hydroxide

Products	Synthetic Medium[a] (mol/100 mol Glucose)		Organic Medium[b] (mol/100 mol Glucose)	
	$-Fe(OH)_3$	$+Fe(OH)_3$	$-Fe(OH)_3$	$+Fe(OH)_3$
CO_2	199	170	186	178
H_2	51	31	53	33
HCOOH	11	12	9	12
Lactic acid	17	19	14	7
Ethanol	72	82	78	94
Acetoin	0.5	1	1	2
2,3-Butylene glycol	64	51	49	44
Acetic acid	0	0	0	0
Iron reduced	—	42	—	61
Glucose fermented (mg/100 mL)	1029	2333	1334	2380
C recovery (%)	112.1	101.8	98.8	97.2
O/R index	1.06	1	1.06	1.03

[a] Glucose, 2.4%; asparagine, 0.5%; K_2HOP_4, 0.08%; KH_2PO_4, 0.02%; KCl, 0.02%; $MgSO_4 \cdot 7H_2O$, 0.5%.
[b] Glucose, 2.5%; peptone, 1%; K_2HPO_4, 0.08%; KH_2PO_4, 0.02%; KCl, 0.02%; $MgSO_4 \cdot 7H_2O$, 0.5%.
Note: Incubation was for 7 days at 35°C.
Source: From Roberts JL, *Soil Sci.*, 63, 135–140, 1947. With permission.

Bromfield (1954b) also showed that washed cells of *Bacillus circulans*, *B. megaterium*, and *Enterobacter aerogenes* reduced ferric iron of several ferric compounds ($FeCl_3$, $Fe(OH)_3$, $Fe(lactate)_2$) in the presence of electron donors such as glucose, succinate, and malate. He was able to inhibit reduction by first boiling the cells or adding chloroform or toluene to the reaction mixture, but he did not observe inhibition with either azide or cyanide. He interpreted his findings to indicate that ferric iron reduction was associated with dehydrogenase activity. However, he thought that the reduction of insoluble ferric iron (e.g., ferric hydroxide) could have occurred only in the presence of a complexing agent. From more recent studies, it has become clear that although some complexing agents may speed up the rate of reduction, as did α,α-dipyridyl in his experiments and those of De Castro and Ehrlich (1970), nitriloacetic acid in experiments of Lovley and Woodward (1996), or an electron shuttle such as anthraquinone-2,6-disulfonate in experiments of Lovley et al. (1996, 1998) and Nevin and Lovley (2000), these agents are not essential in the majority of instances.

Some other bacteria have since been shown to be able to reduce ferric iron in conjunction with a fermentative process. They include aerobes, such as *Pseudomonas* spp. and *Vibrio* spp., and anaerobes such as *Clostridium* spp. and *Bacteroides hypermegas* (see review by Lovley, 1987). In several of these instances, the iron-reducing activity was viewed as a sink for excess reducing power from which the organism could not derive useful energy in its oxidation (see reviews by Lovley, 1987, 1991). However, this explanation holds only if it can be demonstrated that iron reduction in these instances is not accompanied by energy conservation, as was reported to be the case with *Clostridium beijerinckii* (Dobbin et al., 1999). *Pseudomonas ferrireductans* (now *Shewanella putrefaciens* strain 200) contains both constitutive and inducible ferric iron reductases. The constitutive enzyme is involved in ferric iron respiration (in which energy is conserved), and the inducible enzyme, produced at lower oxygen tension, is involved in electron scavenging without energy conservation (electron sink) (Arnold et al., 1986a,b).

Interestingly, Pollock et al. (2007) recently isolated a novel alkaliphilic, halotolerant *Bacillus* sp. strain SFB from salt flat sediments of Soap Lake, State of Washington, which unlike the strain of *B. polymyxa* studied by Roberts (1947), for instance, also had a capacity to grow anaerobically in Luria broth using Fe(III) as terminal electron acceptor in a process of anaerobic respiration (see later discussion in this section). Some unidentified component(s) in the broth must have served as electron donor(s).

Even some fungi have been implicated in Fe(III) reduction (Ottow and von Klopotek, 1969). However, their ability to reduce ferric iron is not likely to involve anaerobic respiration but instead either assimilatory iron reduction or the production of one or more metabolic products that act as a chemical reductant of the ferric iron.

16.6.2 Ferric Iron Respiration: Early History

Some typical heterotrophic and autotrophic ferric iron respirers are listed in Table 16.5. The entries in this table show that the ability to use ferric iron as terminal electron acceptor is spread among various members of the domains Bacteria and Archaea. These members include strictly anaerobic and facultative organisms, the latter of which can grow aerobically as well as anaerobically. In general, all of them reduce Fe(III) only anaerobically, but a few exceptions are known (Short and Blakemore, 1986; De Castro and Ehrlich, 1970; Brock and Gustafson, 1976). The electron donors used by the heterotrophic Fe(III) respirers include a wide range of organic compounds as well as H_2 (Lovley and Lonergan, 1990; Lovley et al., 1989a,b; Coates et al., 1999). The organic compounds include substances as simple as acetate and lactate and as complex as palmitate and some aromatic compounds. Different Fe(III) reducers utilize different types of these compounds. Some of these organisms are unable to degrade their organic electron donor completely, usually accumulating acetate (Lovley et al., 1989b). For Fe(III) to serve as terminal electron acceptor in anaerobic respiration by iron(III) reducers at circumneutral pH, it may be in the form of a soluble complex formed with citrate, nitrilotriacetate, or some other ligand (Lovley and Woodward, 1996). However, various iron reducers have also been shown to attack insoluble forms of Fe(III) such as crystalline goethite (Roden and Zacchara, 1996; Nevin and Lovley, 2000). Ferrous iron, which is the product of Fe(III) reduction, when adsorbed at the surface of an iron oxide such as goethite, interferes with microbial attack of the iron oxide. Its removal from the iron oxide surface promotes the reduction of the oxide (Roden and Urrutia, 1999; Roden et al., 2000).

As Table 16.5 shows, some autotrophs such as *Acidithiobacillus thiooxidans, A. ferrooxidans, Leptospirillum ferroxidans, Sulfolobus (sflb.) spp., Sulfobacillus acidophilus, S. thermosulfidooxidans,* and *Acidimicrobium ferrooxidans* can also respire with Fe(III) as terminal electron acceptor using S^0 as electron donor (Bridge and Johnson, 1998; Brock and Gustafson, 1976; Pronk et al., 1992; Sugio et al., 1992b). *A. thiooxidans* can bring about this reduction aerobically because the ferrous iron it produces at acid pH (~pH 2.5) does not autoxidize readily. However, *A. ferrooxidans* accumulates Fe^{2+} only anaerobically because aerobically it reoxidizes the Fe^{2+}. *sflb. acidocaldarius* can form Fe^{2+} microaerophilically at 70°C because under limited oxygen availability it does not reoxidize Fe^{2+}. *Amicro. ferrooxidans, S. acidophilus,* and *S. thermosulfidooxidans* also reduce Fe(III) to Fe^{2+} under oxygen limitation, but they perform this reduction best with organic electron donors (Bridge and Johnson, 1998). According to Corbett and Ingledew (1987), some growing cultures of *A. ferrooxidans* appear to use a branched pathway when oxidizing sulfur aerobically in the presence of ferric iron in which electrons from sulfite pass to iron(III) or O_2 via a cytochrome bc_1 complex. This reaction can be summarized as follows:

$$SO_3^{2-} + H_2O \rightarrow SO_4^{2-} + 2H^+ + 2e \tag{16.19}$$

$$2Fe^{3+} + 2e \rightarrow 2Fe^{2+} \tag{16.20}$$

$$0.5O_2 + 2H^+ + 2e \rightarrow H_2O \tag{16.21}$$

TABLE 16.5
Examples of Fe(III)-Respiring Bacteria

Organism	References
a. Heterotrophs: strict anaerobes	
Bacillus infernos	Boone et al. (1995)
Clostridium spp.	See discussion in text
Desulfobulbus propionicus	Lonergan et al. (1996)
Desulfovibrio desulfuricans	Coleman et al. (1993)
Desulfuromonas acetoxidans	Roden and Lovley (1993)
Desulfuromusa bakii	Lonergan et al. (1996)
Desulfuromusa succinoxidans	Lonergan et al. (1996)
Desulfuromusa kysingii	Lonergan et al. (1996)
Ferribacterium limnieticum	Cummings et al. (1999)
Geobacter chapellii	Lonergan et al. (1996)
Geobacter hydrogenophilus	Lonergan et al. (1996)
Geobacter metallireducens GS-15	Lovley et al. (1993)
Geobacter sulfurreducens	Caccavo et al. (1994)
Geospirillum barnesii[a]	Laverman et al. (1995) and Lonergan et al. (1996)
Geothrix fermentans	Coates et al. (1999)
Geovibrio ferrireducens	Caccavo et al. (1996)
Pyrobaculum islandicum	Huber et al. (1987) and Kashefi and Lovley (2000)
Pelobacter acetylinicus	Lonergan et al. (1996)
Pelobacter carbinolicus	Lonergan et al. (1996)
Pelobacter propionicus	Lonergan et al. (1996)
Pelobacter ventianus	Lonergan et al. (1996)
b. Heterotrophs: facultative aerobes	
Aeromonas hydrophila	Knight and Blakemore (1998)
Bacillus spp.	See discussion in text
Ferrimonas balearica	Rosselló-Mora et al. (1995)
Pseudomonas sp.	Balashova and Zavarzin (1979)
Shewanella alga	Caccavo et al. (1992)
Shewanella oneidensis MR-1	Myers and Nealson (1988), Lovley et al. (1989a), and Venkateswaran et al. (1999)
Shewanella putrefaciens	Obuekwe and Westlake (1982) and Arnold et al. (1986a,b)
Shewanella sp.	Rosselló-Mora et al. (1994)
c. Autotrophs	
Acidimicrobium ferrooxidans	Bridge and Johnson (1998)
Acidithiobacillus ferrooxidans	Brock and Gustafson (1976) and Pronk et al. (1992)
Acidithiobacillus thiooxidans	Brock and Gustafson (1976)
Leptospirillum ferrooxidans[b]	Sugio et al. (1992b)
Sulfobacillus acidophilus	Bridge and Johnson (1998)
Sulfobacillus thermosulfidooxidans	Bridge and Johnson (1998)
Sulfolobus spp.	Brock and Gustafson (1976)

[a] Now *Sulfurospirillum barnessi*.
[b] No growth when reducing Fe(III) with sulfur (Sugio et al., 1992b).

The sulfite is a metabolic intermediate in the oxidation of S^0. Its formation from sulfur involves an oxygenation and thus requires oxygen (see Chapter 19).

Acidithiobacillus ferrooxidans AP10-3, unlike the strain of *A. ferrooxidans* used by Brock and Gustafson (1976), reduces ferric iron both aerobically and anaerobically with sulfur with an enzyme system that includes a sulfide:Fe(III) and sulfite:Fe(III) oxidoreductase (Sugio et al., 1989, 1992a). Sugio et al. (1992b) found that some other strains of *A. ferrooxidans* and *Leptospirillum ferrooxidans*, which was not previously known to oxidize sulfur (S^0), also possess this enzyme system. In this process, the bacteria transform elemental sulfur to sulfide in the presence of reduced glutathione (GSH). The reaction involved in the oxidation of sulfur with Fe(III) by *A. ferrooxidans* AP19-3 can be summarized as follows:

$$S^0 + 2GSH \rightarrow H_2S + GSSG \tag{16.22}$$
$$\text{Glutathione disulfide}$$

$$H_2S + 6Fe^{3+} + 3H_2O \rightarrow SO_3^{2-} + 6Fe^{2+} + 8H^+ \tag{16.23}$$

$$SO_3^{2-} + 2Fe^{3+} + H_2O \rightarrow SO_4^{3-} + 2Fe^{2+} + 2H^+ \tag{16.24}$$

Growth by *A. ferrooxidans* AP19-3 in the presence of Fe(III) and elemental sulfur occurs only aerobically (Sugio et al., 1988a,b). Growth on sulfur by *L. ferrooxidans* has not been observed so far (Sugio et al., 1992b). It seems to conserve energy only from ferrous iron oxidation, not from sulfur oxidation by ferric iron. Chemical oxidation of sulfur intermediates by ferric iron in the periplasm of *A. ferrooxidans* AP19-3 has also been noted (Sugio et al., 1985). Sulfite oxidation by ferric iron in *A. ferrooxidans* AP19-3 is viewed as a mechanism of detoxification because sulfite is toxic if allowed to accumulate (Sugio et al., 1988b).

Dissimilatory iron reduction in the form of anaerobic respiration has now been recognized as an important means of mineralization of organic matter in environments where sulfate or nitrate occurs in amounts insufficient to sustain sulfate or nitrate respiration, respectively (Lovley, 1987, 1991; Nealson and Saffarini, 1994). The process can operate with various organic acids, including volatile fatty acids, and with aromatic compounds as electron donors. It can displace methanogenesis by outcompeting for H_2 and acetate (Lovley, 1991).

16.6.3 Metabolic Evidence for Enzymatic Ferric Iron Reduction

Most of the early evidence for dissimilatory ferric iron reduction rested on observations with growing cultures. Troshanov (1968, 1969), following up on Bromfield's earlier observations, demonstrated that *Bacillus circulans*, *B. mesentericus*, *B. cereus*, *B. centrosporus*, *B. mycoides*, *B. polymyxa*, *Pseudomonas liquefaciens*, and *Micrococcus* sp., which he isolated from sediment from several lakes in the Karelian peninsula in the former USSR, could reduce ferric iron to varying degrees. He found that all of his cultures that reduced ferric iron could also reduce manganese(IV), but the reverse was not true. The effect of oxygen on iron(III) reduction in his experiments depended on the culture tested by him. Some organisms, such as *B. circulans*, reduced iron more readily microaerobically, whereas others, including *B. polymyxa*, did not. Troshanov noted that the form in which the iron was presented to his cultures affected the rate of reduction. Insoluble ferric iron in bog ore was reduced more slowly than soluble $FeCl_3$. Cultures also varied in their ability to reduce insoluble ferric iron. He found that *B. circulans* actively reduced bog iron ore, whereas *B. polymyxa* did not.

Similar findings were made independently with soil bacteria by Ottow (1969a, 1971), Ottow and Glathe (1971), Hammann and Ottow (1974), Munch et al. (1978), and Munch and Ottow (1980). Support for the notion that ferric iron reduction in these instances was enzymatic was gained from the observation that nitrate and chlorate were able to inhibit ferric iron reduction reversibly by the members of the genera *Enterobacter* and *Bacillus* as well as *Pseudomonas* and *Micrococcus* (Ottow, 1969b, 1970a). Because all of these organisms were known to possess dissimilatory nitrate reductase, which catalyzes the reduction of nitrate and chlorate (Pichinoty, 1963), it was inferred that iron reductase is the same enzyme as nitrate reductase in these organisms. However, this was not observed with all iron reducers, because some other bacteria (e.g., *B. pumilus, B. sphaericus, Clostridium saccharobutylicum*, and *C. butyricum*) that lack nitrate reductase activity were nevertheless able to reduce ferric iron (Ottow, 1970a; Munch and Ottow, 1977). The investigators inferred that these organisms must possess another kind of iron reductase. They supported this inference with observations with mutants that lacked nitrate reductase (Nit⁻), which they derived from wild-type strains possessing the enzyme (Nit⁺). Iron reduction with these Nit⁻ strains was found to be insensitive to inhibition by nitrate or chlorate (Ottow, 1970a). Most of the Nit⁻ mutants reduced iron less rapidly than their wild-type parent, but a Nit⁻ mutant of *B. polymyxa* reduced iron more intensely than its wild-type parent. Also consistent with the inference that some dissimilatory iron reducers use an enzyme other than nitrate reductase in Fe(III) reduction, Ottow and Ottow (1970) noted that the size of the soil microflora capable of reducing iron is usually greater than that of the microflora capable of reducing nitrate. The list of inhibitors of bacterial ferric iron reduction has been extended to include permanganate, dichromate, sulfite, thiosulfate, and the redox dyes methylene blue, indochlorophenol, and phenazine methosulfate when *Shewanella* (formerly *Pseudomonas*) *putrefaciens* strain 200 is the test organism (Obuekwe and Westlake, 1982).

Nitrate inhibition of ferric iron reduction need not be due to competitive inhibition of iron reductase (nitrate reductase) (Obuekwe et al., 1981). Nitrate was found to stimulate ferric iron reduction by *Shewanella putrefaciens* during short-term incubation but to depress it during long-term incubation. The inhibitory effect of nitrate in this instance was related to the chemical oxidation of ferrous iron to ferric iron by nitrate (see also discussion by Sørensen, 1987). Nevertheless, the authors found that when *S. putrefaciens* was preinduced by nitrate, the resultant strain reduced Fe(III) faster than the uninduced strain, supporting the notion that nitrate reductase can catalyze ferric iron reduction by this organism. Obuekwe and Westlake (1982) explained this effect as merely reflecting a better physiological state of induced cells than uninduced cells. Nitrate may also act as a noncompetitive inhibitor of ferric iron reduction, as in *Staphylococcus aureus* (Lascelle and Burke, 1978).

Other early evidence of the enzymatic nature of Fe(III) reduction by some microbes was the observation by De Castro and Ehrlich (1970) that a cell extract from marine *Bacillus* strain 29A, whose intact cells actively reduced ferric iron, could reduce iron(III) in the mineral limonite. This activity was partially destroyed by heating and inhibited by mercuric chloride and *p*-mercuribenzoate. Lascelle and Burke (1978) detected ferric iron reduction by a membrane fraction from *Staphylococcus aureus*, which could also reduce nitrate. They found evidence for the involvement of a branched electron transport pathway in ferric iron reduction in their organisms. By use of selective inhibitors, they showed that nitrate received electrons via a cytochrome *b*-requiring branch, whereas ferric iron received electrons via a branch that originated ahead of cytochrome *b*. Nitrate was thought to inhibit ferric iron reduction by *Staph. aureus* because nitrate accepts electrons more readily than ferric iron in this system. Obuekwe (1986) demonstrated that the ferric reductase in *Shewanella putrefaciens* strain 200 was inducible and ferric iron reduction in intact cells was inhibited by sodium amytal, 2-*n*-heptyl-4-hydroxyquinoline-*N*-oxide (HQNO), and sodium cyanide, suggesting the involvement of a cytochrome pathway in transferring electrons from donor to ferric iron. However, he was unable to demonstrate ferric iron-reducing activity with cell membranes or periplasmic or cytoplasmic fractions from the cells. *S. putrefaciens* 200R grown separately on Fe(III), Mn(IV), U(VI), SO_3^{2-}, or $S_2O_3^{2-}$ as terminal electron acceptor under microaerobic or anaerobic conditions was able to reduce Fe(III). This was not the case when this culture was grown separately on O_2, NO_3^-, NO_2^-, or trimethylamine *N*-oxide under otherwise similar conditions (Blakeney et al., 2000). A kinetic study of ferric iron reduction by *S. putrefaciens* strain 200 grown anaerobically on Fe(III) indicated that the nature of the soluble ferric iron species determines the reaction rate (Arnold et al., 1986b).

16.6.4 Ferric Iron Respiration: Current Status

The most direct evidence for bacterial Fe(III) respiration has come from the studies with *Geobacter metallireducens* (strain GS-15), *G. sulfurreducens, Shewanella oneidensis* MR-1, two cultures of

S. putrefaciens, strains 200 and ATCC 8071, and *Desulfuromonas acetoxidans* (Obuekwe et al., 1981; Myers and Nealson, 1988; Lovley and Phillips, 1988a; Lovley et al., 1989a; Caccavo et al., 1994). *G. metallireducens*, an obligate anaerobe, was isolated from freshwater sediment. *G. sulfurreducens*, also an obligate anaerobe, was isolated from a hydrocarbon-contaminated ditch. *S. oneidensis*, a facultative anaerobe, was isolated from sediment of Lake Oneida, New York, whereas *S. putrefaciens* strain 200, also a facultative anaerobe, was isolated from a Canadian oil pipeline. All of these isolates are able to use Fe(III) and Mn(IV) anaerobically as terminal electron acceptors for growth.

Our most complete understanding of the enzymatic mechanisms of anaerobic respiration with Fe(III) and Mn(IV) come from the studies with *Shewanella oneidensis* MR-1 and *Geobacter sulfurreducens*. *S. oneidensis* MR-1 as well as *S. putrefaciens* can use H_2 (Figure 16.10) as well as formate, lactate, and pyruvate as electron donors in the anaerobic reduction of Fe(III), but lactate and pyruvate are only incompletely oxidized to acetate and CO_2 (Lovley et al., 1989a; Myers and Nealson, 1988, 1990). Myers and Nealson (1990) found that *S. putrefaciens* strain MR-1 was renamed as *S. oneidensis* strain.

MR-1 by Venkateswaran et al. (1999) conserved energy when reducing ferric iron anaerobically with lactate as electron donor. They demonstrated respiration-linked proton translocation, which was inhibited completely by 20 μM carbonylcyanide *m*-chlorophenylhydrazone and partially to completely by 50 μM 2-*n*-heptyl-4-hydroxyquinoline-*N*-oxide.

When grown anaerobically, *Shewanella oneidensis* MR-1 has 80% of its cytochrome complement (mostly *c*-type cytochrome) in the outer membrane of its cell envelope, with the rest (*c*- and *b*-type cytochromes) being mostly or entirely associated with the plasma membrane. When grown aerobically, the same organism contains the major portion of its cytochrome complement in its plasma membrane, which, according to many previous studies, is the more common location of cytochromes in other bacteria, except for periplasmic *c*-type cytochrome (Myers and Myers, 1992a).

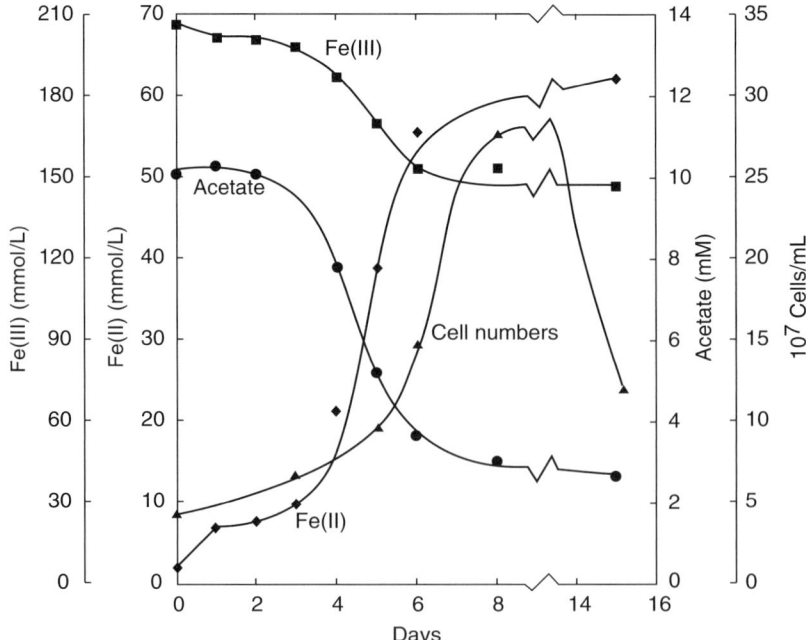

FIGURE 16.10 Reduction of ferric iron by acetate through mediation of anaerobic bacterial strain GS-15 (*Geobacter metallireducens*). This family of curves illustrates the reduction of oxalate-extractable Fe(III) to Fe(II) at the expense of acetate consumption during growth of the culture in FWA medium containing amorphic Fe(III) oxide. (Reproduced from Lovley DR, Phillips EJP, *Appl. Environ. Microbiol.*, 54, 1472–1480, 1988a. With permission.)

Slightly more than 50% of formate-dependent ferric reductase activity of anaerobically grown *S. oneidensis* MR-1 was found to be associated with the outer membrane, and the rest with the plasma membrane. The membranes of aerobically grown cells were devoid of this activity (Myers and Myers, 1993). Most of the formate-dehydrogenase activity was soluble (Myers and Myers, 1993), suggesting a possible periplasmic location, as with fumarate reductase in anaerobically grown cells of this organism (Myers and Myers, 1992b). Addition of nitrate, nitrite, fumarate, or trimethylamine *N*-oxide as alternative electron acceptor did not inhibit ferric reductase activity. NADH was able to replace formate as electron donor in experiments with a membrane fraction containing ferric reductase (Myers and Myers, 1993).

Outer membrane cytochromes encoded in genes *omcA* and *omcB* in *Shewanella oneidensis* MR-1 are required for the reduction of Mn(IV) oxide, which is insoluble. However, these cytochromes are not involved in the reduction of some soluble electron acceptors, including ferric citrate, nitrate, and thiosulfate (Myers and Myers, 2001), although they are required for the reduction of electron-shuttling compound 2,6-anthraquinone disulfonate (Lies et al., 2005). As in the reduction of Mn(IV) oxide, the studies of Myers and Myers support a model in which outer membrane proteins are also involved in the reduction of insoluble iron oxides by *S. oneidensis* MR-1 under anaerobic conditions. Lower et al. (2001) obtained direct evidence for the involvement of surface contact between *S. putrefaciens* MR-1 and goethite by atomic force electron microscopy during attack of the goethite under anaerobic conditions. Their results support a model in which the contact enables electron transfer at the cell–mineral interface by means of a 150 kDa iron reductase. In *S. putrefaciens* strain 200, a 91 kDa heme-containing protein in the outer membrane seems to be involved in anaerobic electron transfer to Fe(III) and Mn(IV) oxides (DiChristina et al., 2002).

Geobacter sulfurreducens strain PCA is able to use H_2 or acetate as sole electron donor in Fe(III) reduction, whereas *G. metallireducens* can use acetate, lactate, and several other organic compounds, such as propionate, butyrate, benzoate, or phenol, as sole electron donors, but not H_2. Figure 16.11 summarizes diagrammatically how H_2, acetate, and lactate are used in the reduction of FeOOH.

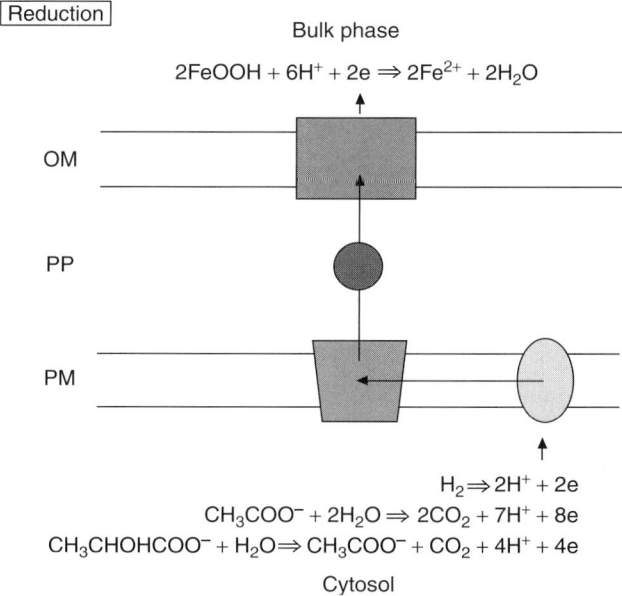

FIGURE 16.11 Path of electrons from H_2, acetate, or lactate to FeOOH in *Geobacter sulfurreducens*, *Shewanella oneidensis*, and some other anaerobically respiring gram-negative bacteria. OM, outer membrane; PP, periplasm; PM, plasma membrane; oval in PM, complex I; trapezoid in PM, complex III; circle in PP; and rectangle in OM represent different *c*-type cytochromes. The arrows indicate the direction of electron flow.

Elemental sulfur, Co(III)-EDTA, fumarate, and malate were found to be able to serve as alternative electron acceptors for *G. sulfurreducens*, but not Mn(IV) or U(VI). *G. sulfurreducens* cannot reduce nitrate, sulfate, sulfite, or thiosulfate with acetate as electron donor (Caccavo et al., 1994).

Lovley (2000) proposed a model of an Fe(III)-reductase system in *Geobacter sulfurreducens* that consists of a 41 kDa cytochrome located in the outer membrane, which conveys electrons to bulk-phase Fe(III) at the exterior of the outer membrane. The cytochrome receives electrons from NADH in a sequence of redox reactions involving NADH-dehydrogenase complex and an 89 kDa cytochrome in the plasma membrane, and a 9 kDa cytochrome in the periplasm (Figure 16.10). An alternative model has been proposed in which an extracellular 9.6 kDa *c*-type cytochrome released by *G. sulfurreducens* acts as an electron shuttle between the cell and the bulk-phase Fe(III). This model is based on the finding of such a cytochrome in the spent medium of *G. sulfurreducens* cultures (Seeliger et al., 1998). However, strong evidence against this model makes it unlikely (Lloyd et al., 1999).

Magnuson et al. (2000) detected a 300 kDa Fe^{3+}-reductase complex in *Geobacter sulfurreducens* of which 67% was located in the outer membrane and the rest in the cytoplasmic membrane. The protein complex included a 90 kDa *c*-type cytochrome and flavin adenine dinucleotide (FAD) and had an E_h of −100 mV. It was reduced by NADH but not by H_2. Its Fe(III)-reducing activity was inhibited by rotenone, myxthozole, quinacrine, or *p*-chloromercuribenzoate but not by inhibitors that act on the ubiquinone:cytochrome *c* oxidoreductase.

Kaufmann and Lovley (2001) found a soluble Fe(III) reductase with an ability to reduce soluble Fe(III) (e.g., Fe(III)-nitrilotriacetic acid) using NADPH but not NADH as electron donor in *Geobacter sulfurreducens*. The physiological role of this reductase in the organism is uncertain but unlikely to be in assimilatory Fe(III) reduction because, as pointed out by the authors, *G. sulfurreducens* inhabits anaerobic environments likely to harbor a plentiful supply of Fe(II). The authors also question as to whether a dissimilatory role of the enzyme is likely because the enzyme protein is expressed at the same level in the organism when growing in the presence of fumarate as in the presence of Fe(III).

Geobacter metallireducens has been shown to reduce amorphous iron oxide to magnetite (Fe_3O_4), but it does not readily reduce crystalline iron oxides (Lovley and Phillips, 1986b). Roden and Urrutia (1999) showed later that removal of ferrous iron from the surface of crystalline iron oxide was necessary for ready attack of the iron oxide by the organism. The organism appears to use chemotaxis to locate insoluble Fe(III) oxide substrates (Childers et al., 2002). *G. metallireducens* can use acetate (Figure 16.10), ethanol, butyrate, and propionate as electron donors in Fe(III) respiration. It oxidizes these compounds to CO_2 in this process (Lovley et al., 1993). In studies with intact cells of *G. metallireducens*, Gorby and Lovley (1991) found that the transfer of reducing power (electrons) from an appropriate electron donor to ferric iron included a plasma-membrane-bound electron transport system that involved *b*- but not *c*-type cytochrome and Fe(III) reductase, because the system was not inhibited by the electron transport inhibitors 2-*n*-heptyl-4-hydroxyquinoline-*N*-oxide, sodium azide, and sodium cyanide. *G. metallireducens* oxidizes acetate via the TCA cycle (see Chapter 6; Champine and Goodwin, 1991; Tang et al., 2007). Champine et al. (2000) proposed that energy conservation occurred in the electron transfer from NADH to oxidized menaquinone and from reduced cytochrome c_7, a low-potential cytochrome (≤−91 mV), to oxidized terminal oxidase (Champine et al., 2000).

As previously mentioned, *Shewanella putrefaciens* strain 200 has been found to form a constitutive and an inducible Fe(III) reductase. The latter is induced only anaerobically. Both reductases are responsible for rapid Fe(III) reduction under anaerobic conditions. The constitutive reductase will reduce Fe(III) at low O_2 tension, but the rate of Fe(III) reduction is very slow. Apparently, competition with O_2 is responsible for the slow rate of reduction. A branched respiratory pathway is postulated in which one branch leads to O_2 and the other to Fe(III) (Arnold et al., 1990).

Not all Fe(III) reductases function in dissimilatory iron reduction. The iron reductase activity detected in *Pseudomonas aeruginosa* by Cox (1980) evidently represents an assimilatory iron-reductase system because it reduced only complexed iron such as ferripyrochelin and ferric citrate (pyrochelin and citrate are known siderophores). The ferric reductase had a cytoplasmic location, whereas the ferripyrochelin reductase was located in the periplasm and cytoplasm. These activities appear to be linked to pyridine nucleotide, although reduced GSH was also able to serve as electron donor. Similarly, the Fe(III) reduction observed with *Escherichia coli* K12 (Williams and Poole, 1987) involved an assimilatory iron reductase. Like *P. aeroginosa*, *E. coli* K12 reduced ferric citrate directly with reduced pyridine nucleotide as electron donor without the involvement of the respiratory chain in the plasma membrane of the organism. Adenosine 5'-triphosphate (ATP) and cyanide were found to stimulate ferric citrate reduction, possibly forming complexes with Fe(III).

16.6.5 ELECTRON TRANSFER FROM CELL SURFACE OF A DISSIMILATORY FE(III) REDUCER TO FERRIC OXIDE SURFACE

Dissimilatory Fe(III) reducers may transfer electrons from a *c*-type cytochrome-containing enzyme complex in the outer membrane (see Section 16.6.4) to the surface of a ferric oxide in one of the several ways. In *Shewanella oneidensis* MR-1, this electron transfer process is complex and includes involvement of several key cytochromes to different extents (Bretschger et al., 2007). One way may involve physical contact between the cell and the surface of the ferric oxide. Recently, direct electron transfer has been shown *in vitro* between OmcA and hematite (Xiong et al., 2006). Lower et al. (2007) have obtained direct evidence for the formation of specific bonds between hematite and cytochromes MtrC and OmcA in the outer membrane of *S. oneidensis* MR-1 that are likely to be involved in electron transfer during anaerobic respiration with hematite as terminal electron acceptor. It remains to be demonstrated whether *in vivo* electron transfer between such outer membrane cytochromes and insoluble metal oxides is direct or mediated by small molecules. For further discussion, see two recent reviews, one by Gralnick and Newman (2007) and the other by Shi et al. (2007).

Another way may involve contact with the ferric oxide surface via special fimbriae or pili that act as conductors of electrons or nanowires. They have been shown to be formed by *Geobacter sulfurreducens* (Reguera et al., 2005, 2006) and *Shewanella oneidensis* MR-1 (Gorby et al., 2006). The pili of *G. sulfurreducens* may also have a nonconductive role (Reguera et al., 2007). If direct physical contact between the Fe(III)-reducing cell and a ferric oxide is not essential or is not involved in the reduction of Fe(III) of an Fe(III) oxide, an Fe(III) chelator or an electron shuttle may be involved (Lovley, 2000; Newman and Kolter, 2000; Royer et al., 2002; von Canstein et al., 2008). Nevin and Lovley (2002) found that *Geothrix (Gthr.) fermentans* produced a quinone-type compound that served as an electron shuttle in the reduction of Fe(III) oxide when the organism was not in direct contact with Fe(III) oxide. However, although an artificial chelator, nitrilotriacetic acid, was found to stimulate Fe(III) oxide reduction by *G. metallireducens* (Lovley and Woodward, 1966), no chelator or electron shuttle was found to be produced by this organism (Nevin and Lovley, 2000). Feinberg and Holden (2006) found that *Pyrobaculum aerophilum*, a hyperthermophilic archaeon, appeared to employ an as yet unidentified electron shuttle instead of a chelator when reducing poorly crystalline Fe(III) oxide in the absence of direct contact with the oxide. von Canstein et al. (2008) found that growing cultures of *S. oneidensis* MR-1 and some other strains of *Shewanella* released flavin mononucleotide into the medium, which could function as an electron shuttle in the reduction of poorly crystalline Fe(III) oxide. Nongrowing cells of *Shewanella* in the stationary phase released riboflavin, which also could serve as an electron shuttle in the reduction of the Fe(III) oxide. Flavin mononucleotide (FMN) did not serve as a shuttle in the reduction of the soluble compounds such as ferric citrate or fumarate. Independently, Marsili et al. (2008) detected riboflavin and riboflavin 5'-phosphate in biofilms of *S. oneidensis* MR-1 and *Shewanella* sp. MR-4 using electrochemical techniques and demonstrated that these molecules promoted electron transfer to electrodes as well as Fe(III) and Mn(IV) oxy(hydr)oxides.

16.6.6 BIOENERGETICS OF DISSIMILATORY IRON REDUCTION

As proton translocation has been demonstrated during Fe(III) reduction by *Shewanella oneidensis* MR-1 (Myers and Nealson, 1990), it can be inferred that this organism conserves energy chemiosmotically in this process. The strains of *S. putrefaciens* studied by Obuekwe (1986), Arnold et al. (1986a,b), and Lovley et al. (1989a) probably conserve energy by the same mechanism as *S. oneidensis* MR-1. At least they grow anaerobically when Fe(III) is the only terminal electron acceptor but not in its absence or in the absence of an alternative acceptor. *Geobacter metallireducens* most likely conserves energy chemiosmotically in Fe(III) respiration because, as previously mentioned, electrons from a suitable energy source are transported to Fe(III) by a membrane-bound system that includes *b*-type cytochrome and an Fe(III) reductase (Gorby and Lovley, 1991).

Free energy calculations show that when acetate is the electron donor (reductant) and Fe^{3+} is the electron acceptor (oxidant), the standard free energy change at pH 7 (ΔG_r^0) is -193.4 kcal mol^{-1} (-808.4 kJ mol^{-1}). This value is close to that when O_2 is the electron acceptor (-201.8 kcal mol^{-1} or -843.5 kJ mol^{-1}). However, when $Fe(OH)_3$ is the oxidant of acetate, the standard free energy change at pH 7 is only -5.5 kcal mol^{-1} (-23 kJ mol^{-1}) (Ehrlich, 1993). This indicates that undissolved iron oxides, oxyhydroxides, or hydroxides are poor electron acceptors at neutral or alkaline pH and do not allow for the magnitude of biochemical energy conservation that the dissolved Fe(III) does. The iron oxides are better acceptors and more effective for biochemical energy conservation at pH values slightly <7. Nevertheless, conditions exist in nature in which minerals containing ferric iron serve effectively as electron acceptors without ferric iron first being dissolved.

16.6.7 FERRIC IRON REDUCTION AS ELECTRON SINK

As previously mentioned, it has been suggested that not all iron(III)-reducing bacteria gain energy in the process. Ghiorse (1988) and Lovley (1991) proposed that a number of bacteria reduce ferric iron merely to dispose of excess reducing power via secondary respiratory pathways without conserving energy or the iron reduction that was observed with these organisms was part of their iron assimilation process. These interpretations ignore the possibility that an absence of growth stimulation under anaerobic conditions when an organism reduces Fe(III) with stoichiometric release of extracellular Fe^{2+} is not necessarily an indication that energy is not conserved in the reduction. The Fe(III) reduction process may simply yield insufficient energy to support growth and may have to be accompanied by an additional respiratory or fermentative energy-yielding process to meet the full energy demand of the cell. In other words, Fe(III) reduction is an ancillary source of energy for these particular organisms and may be important for survival rather than growth *per se*. Recently, these ideas have been explored in the context of Fe(III) reduction by *Pseudomonas* species mediated by redox-active metabolites (Price-Whelan et al., 2006).

The observation of Lovley et al. (1995) that *Pelobacter carbinolicus* can catabolize 2,3-butanediol with the production of ethanol and acetate by fermentation as well as reduction of Fe(III) or S^0 may provide new insight into this puzzle. The authors noted that the ratio of ethanol to acetate produced from 2,3-butanediol was significantly greater in the absence of Fe(III) than in its presence. Because *P. cabinolicus* lacks *c*-type cytochromes, it may be using Fe(III) merely as electron sink in this case. However, the authors also found that the organism can grow on H_2 as the sole energy source in a medium in which acetate is the sole carbon source (it cannot use acetate as an energy source) and Fe(III) is the only terminal electron acceptor. Under these conditions, the organism must respire and conserve energy chemiosmotically. This raises the question of whether when growing on 2,3-butanediol, the organism uses Fe(III), when present, merely as an electron sink and does not conserve energy from its reduction.

16.6.8 REDUCTION OF FERRIC IRON BY FUNGI

Some fungi seem to be able to reduce ferric iron. Lieske reported such a phenomenon (see Starkey and Halvorson, 1927). It was also observed by Ottow and von Klopotek (1969). They noted reduction of ferric iron in hematite (Fe_2O_3) by *Alternaria tenuis*, *Fusarium oxysporum*, and *F. solani*, all of which possessed an inducible nitrate reductase. Fungi tested in this study, which were incapable of reducing nitrate, were also incapable of reducing ferric iron. Ottow and von Klopotek concluded from their findings that fungi that possess nitrate reductase can reduce ferric iron as well as nitrate, like some bacteria they had studied. Their conclusion, however, raises a question as to where in the fungi the nitrate reductase is located to be able to act on insoluble hematite and whether this represents dissimilatory or assimilatory iron reduction. The assimilatory reduction seems more likely. In the simultaneous presence of nitrate and hematite, it is possible that nitrite produced from the nitrate by these fungi reduces Fe(III) in hematite nonenzymatically, as in the case of *Shewanella putrefaciens* (Obuekwe et al., 1981).

16.6.9 TYPES OF FERRIC COMPOUNDS ATTACKED BY DISSIMILATORY IRON(III) REDUCTION

The ease with which bacteria reduce ferric iron depends in part on the form in which they encounter it. In the case of insoluble forms, the order of decreasing solubilization in one study (Ottow, 1969b) was $FePO_4 \cdot 4H_2O$ > $Fe(OH)_3$ > lepidochrosite (γ-FeOOH) > goethite (α-FeOOH) > hematite (α-Fe_2O_3) when *Bacillus polymyxa*, *B. sphaericus*, and *Pseudomonas aeruginosa* were tested with glucose as electron donor. The order was inverted with respect to $FePO_4 \cdot 4H_2O$ and $Fe(OH)_3$ when *Enterobacter* (*Aerobacter*) *aerogenes* and *B. mesentericus* were tested. In another study (De Castro and Ehrlich, 1970), using glucose as electron donor, marine *Bacillus* 29A was found to solubilize larger amounts of iron from limonite and goethite than from hematite. In initial stages of the reduction, the order of decreasing activity was goethite > limonite > hematite. The bacterial iron-solubilizing activity in these experiments was enhanced by the addition of phenosafranin.

Bacillus 29, the wild-type parent of strain 29A, did not reduce ferric iron extensively when it occurred in ferromanganese nodules from the deep sea, although it reduced the Mn(IV) oxide (Ehrlich et al., 1973). Because of the insoluble nature of the iron(III) in the nodules, the aerobic conditions, and in some cases the circumneutral pH at which these observations were made, it seems unlikely that the absence of measurable iron reduction was due to the manganese inhibition phenomenon described by Lovley and Phillips (1988b). More likely it was related to the significantly greater amount of free energy available from Mn(IV) oxide reduction than from ferric oxide reduction (see Ehrlich, 1993). In a study of anaerobic reduction of pedogenic iron oxides by *Clostridium butyricum*, Munch and Ottow (1980) found that amorphous oxides were more readily attacked than crystalline oxides. This was also observed by Lovley and Phillips (1986b) in anaerobic experiments with an enrichment culture from Potomac River sediment tested on various forms of Fe(III) oxides.

Various species of *Acidiphilium*, including *A. acidophilus* and strain SJH, are able to reduce Fe(III) anaerobically (Johnson and Bridge, 2002). In a study with *Acidiphilium* SJH, Bridge and Johnson (2000) found that the order of decreasing Fe(III) reduction was dissolved Fe^{3+} > amorphous $Fe(OH)_3$ > magnetite > goethite = natrojarosite > akaganite > jarosite > hematite (no significant dissolution). In this case the minerals were not attacked directly by the organism. The authors found that the bacterial reduction of the iron in the minerals depended on the abiotic mobilization of Fe(III) by the acidity of the medium, which was pH 2 initially. The abiotic Fe(III) mobilization was enhanced by bacterial reduction of the solubilized Fe(III), which resulted in product removal in the abiotic dissolution step.

Although magnetite had been thought to be resistant to microbial reductive attack, its anaerobic reductive dissolution to Fe^{2+} by two strains *of Shewanella putrefaciens* using glucose and lactate as electron donors has now been observed (Kostka and Nealson, 1995). Although not mentioned by Hilgenfeldt (2000), the extensive dissolution of biogenic magnetite observed by him in the surface sediments of the Benguela upwelling system may have been due to the activity of Fe(III)-reducing bacteria. Magnetite, as well as ferrihydrite, goethite, and hematite, has also been shown to be reduced during growth of *Desulfovibrio desulfuricans* strain G-20 in laboratory culture (Li et al., 2006). In the presence of added sulfate, the rate of Fe(III) reduction of these minerals was accelerated owing to an interplay of respiratory and chemical Fe(III)-reducing reactions and the timing of Fe(III) respiration and sulfate reduction in the culture.

The anaerobic reduction of ferric iron that replaces some of the Al in the octahedral sheets of the clay smectite using lactate and formate as electron donors by *Shewanella oneidensis* MR-1 and some unidentified bacteria was observed by Stucki et al. (1987) and Kostka et al. (1996). This microbial action has also been shown with three strains of *Pseudomonas fluorescens* and a strain of *P. putida* (Ernstsen et al., 1998). Kashefi et al. (2008) reported growth of some thermophilic and hyperthermophilic Fe(III) reducers using Fe(III) in ferruginous smectite as the sole electron acceptor. Interestingly, mobilization of reduced iron (Fe^{2+}) in the clay was either very limited or not detected, depending on the study. Stucki and Roth (1977) proposed that the reduction proceeds in two steps. In the first step, an increase in layer charge without structural change occurs during initial reduction of some Fe(III), and in the second step, constant layer charge is maintained as a result of elimination of structural OH accompanied by a decrease of the iron in the octahedral sheet during subsequent reduction of additional Fe(III). The reduction of Fe^{3+} in the octahedral sheet of smectite results in an increase in net surface charge of the mineral (Stucki et al., 1984) and a decrease in its swellability (Stucki et al., 2000). The mechanism whereby the bacteria inject electrons into the smectite to reduce the structural iron remains to be elucidated.

As oxides of ferric iron are highly insoluble at near-neutral pH, the question arises as to how bacteria can attack these compounds. The tendency of the oxides to dissociate into soluble species in water at near-neutral pH is simply too small to explain the phenomenon. Physical contact between the active cell and the surface of the oxide particle appears essential. This was first demonstrated by Munch and Ottow (1982) in experiments in which they placed iron oxide inside a dialysis bag whose pores did not permit the passage of bacterial cells into the bag but permitted ready diffusion of culture medium with its dissolved inorganic and organic chemical species. No iron reduction occurred when bacteria (*Clostridium butyricum* or *Bacillus polymyxa*) were placed in the medium outside the bag, but it occurred when both the bacteria and iron oxide were present inside the bag. Interestingly, metabolites, including acids, produced by the bacteria were not able to dissolve significant amounts of oxide. Furthermore, the lowered E_h (redox potential) generated by the bacteria, whether inside or outside the dialysis bag, also did not result in the reduction of ferric iron in the oxide. Nevin and Lovley (2000) concluded that some bacteria mobilized ferric iron in oxide minerals as ferrous iron by direct attack via attachment to the mineral surface. The model for the Fe(III)-reductase system in *Geobacter sulfurreducens* (see Lovley, 2000), previously discussed, illustrates how direct attack by bacteria at the surface of a ferric oxide particle could occur.

16.7 NONENZYMATIC OXIDATION OF FERROUS IRON AND REDUCTION OF FERRIC IRON BY MICROBES

16.7.1 NONENZYMATIC OXIDATION

Many different kinds of microorganisms can promote the oxidation of ferrous iron indirectly (i.e., nonenzymatically). They can accomplish this by affecting the redox potential of the environment. This means they release metabolically formed oxidants that have the ability to oxidize ferrous iron into their surroundings. They can also accomplish this by generating a pH >5 at which Fe^{2+} is oxidized by oxygen in the air (*autoxidation*). Among the first to recognize indirect iron oxidation by microbes were Harder (1919), Winogradsky (1922), and Cholodny (1926). Starkey and Halvorson (1927) demonstrated indirect oxidation of iron by bacteria in laboratory experiments. They explained their findings in terms of reaction kinetics involving the oxidation of ferrous iron and the solubilization of ferric iron by acid. From their work it can be inferred that any organism that raises the pH of a medium by forming ammonia from protein-derived material (Reaction 16.25) or consuming salts of organic acids (Reaction 16.26) can promote ferrous iron oxidation in aerated medium.

$$RCHCOOH + H_2O + 0.5O_2 \rightarrow RCCOOH + NH_4^+ + OH^- \tag{16.25}$$

$$\underset{NH_2}{|} \qquad\qquad \underset{O}{\|}$$

$$\text{Amino acid} \qquad\qquad \text{Keto acid}$$

$$\underset{\text{Lactate}}{C_3H_5O_3^-} + 3O_2 \rightarrow 3CO_2 + 2H_2O + OH^- \tag{16.26}$$

A more specialized case of indirect microbial iron oxidation is the one associated with cyanobacterial and algal photosynthesis. The photosynthetic process may promote ferrous iron oxidation by creating conditions that favor autoxidation in two ways by raising (1) the pH of the waters in which they grow (bulk phase) and (2) the oxygen level in the waters around them. The pH rise is explained by the following equations:

$$2HCO_3^- \rightarrow CO_3^{2-} + CO_2(g) + H_2O \tag{16.27}$$

$$CO_3^{2-} + H_2O \rightarrow HCO_3^- + OH^- \tag{16.28}$$

Reaction 16.27 is promoted by CO_2 assimilation in oxygenic photosynthesis, and the overall reaction may be summarized as follows:

$$CO_2 + H_2O \xrightarrow[\text{Light}]{\text{Chl}} CH_2O + O_2 \tag{16.29}$$

Reaction 16.29 also yields much of the oxygen needed for autoxidation of the ferrous iron. Its genesis causes a rise of E_h because of increased oxygen saturation or even supersaturation of the water around the cyanobacterial or algal cells.

Ferrous iron may be protected from chemical oxidation at elevated pH and E_h by chelation with gluconate, lactate, or oxalate, for example. In that instance, bacterial breakdown of the ligand will free the ferrous iron, which then autoxidizes to ferric iron. This has been demonstrated in the laboratory with a *Pseudomonas* and a *Bacillus* strain (Kullmann and Schweisfurth, 1978). These cultures do not derive any energy from the iron oxidation but rather from the oxidation of the ligand.

The production of ferric iron from the oxidation of ferrous iron at pH values >5 usually leads to precipitation of iron. But the presence of chelating agents such as humic substances, citrate, and others can prevent the precipitation. Unchelated ferric iron tends to hydrolyze at higher pH values and may form compounds such as ferric hydroxide.

$$Fe^{3+} + 3H_2O \rightarrow Fe(OH)_3 + 3H^+ \tag{16.30}$$

Ferric hydroxide is relatively insoluble and will settle out of suspension. It may crystallize and dehydrate, forming FeOOH, goethite ($Fe_2O_3 \cdot H_2O$), or hematite (Fe_2O_3). In excess base, $Fe(OH)_3$ may, however, form water-soluble, charged entities such as $Fe(OH)_2O^-$ or $Fe(OH)O_2^{2-}$ because $Fe(OH)_3$ has amphoteric properties.

16.7.2 Nonenzymatic Reduction

Starkey and Halvorson (1927) tried to explain ferric iron reduction in nature as an indirect effect of microbes. They argued that with a drop in pH and lowering in oxygen tension caused by microbes, ferric iron would be changed to ferrous iron according to the following relationship:

$$4Fe^{2+} + O_2 + 10H_2O \leftrightarrow 4Fe(OH)_3 + 8H^+ \tag{16.31}$$

in which Fe(III) was considered to be an insoluble phase $(Fe(OH)_3)$. From this chemical equation they derived the relationship

$$[Fe^{2+}] = K[H^+]^2/[O_2]^{1/4} \qquad (16.32)$$

But as the experiments of Munch and Ottow (1982) convincingly showed, this mode of Fe(III) reduction is usually not very significant in nature. This was also shown by Lovley et al. (1991). Iron oxide minerals are relatively stable in the absence of oxygen when a strong reducing agent is not present. In the presence of such reducing agents, however, chemical Fe(III) reduction does occur. For instance, H_2S produced by sulfate-reducing bacteria may reduce ferric to ferrous iron before precipitating ferrous sulfide (Berner, 1962).

$$2FeOOH + 3H_2S \xrightarrow{\text{pH } 7-9} 2FeS + S^0 + 4H_2O \qquad (16.33)$$

$$2FeOOH + 3H_2S \xrightarrow{\text{pH } 4} FeS + FeS_2 + 4H_2O \qquad (16.34)$$

Marine bacteria that disproportionate elemental sulfur into H_2S and sulfate according to the reaction

$$4S^0 + 4H_2O \rightarrow 3H_2S + SO_4^{2-} + 2H^+ \qquad (16.35)$$

have been shown to reduce Fe(III) and Mn(IV) chemically under anaerobic conditions (Thamdrup et al., 1993).

Formate produced by a number of bacteria (e.g., *Escherichia coli*) can reduce Fe(III).

$$2Fe^{3+} + HCOOH \rightarrow 2Fe^{2+} + 2H^+ + CO_2 \qquad (16.36)$$

Some other metabolic products can also act as reductant of ferric iron (see discussion by Ghiorse, 1988). In general, reduction is favored by acid pH.

With the discovery that some Fe(III)-reducing organisms can reduce humic substances (Lovley et al., 1996, 1998; Scott et al., 1998), it has been proposed that the microbially reduced humic substances may reduce Fe(III) in soil abiologically, with the hydroquinone groups of the humics acting as electron shuttle between cells and extracellular Fe(III) (Lovley, 2000).

16.8 MICROBIAL PRECIPITATION OF IRON

16.8.1 ENZYMATIC PROCESSES

The clearest example of enzymatic Fe(III) precipitation is that associated with *Gallionella ferruginea*. The ferric iron (FeOOH) produced by this organism in its oxidation of ferrous iron is deposited in its stalk fibrils by a mechanism that is still unclear at this time (see discussion by Ghiorse, 1984; Ghiorse and Ehrlich, 1992; Hallbeck, 1993). James and Ferris (2004) showed that *G. ferruginea* and *Leptothrix ochracea* played a significant role in the oxidation of Fe(II) and the subsequent precipitation of the resultant Fe(III) as ocherous bacteriogenic iron oxide in the water at the mouth of a neutral groundwater spring draining into Ogilvie Creek near the Deep River, Ontario, Canada. This resulted in a progressive decrease in ferrous and total iron concentration and increase in redox potential and dissolved oxygen with distance from the spring.

Enzymatic iron reduction by other organisms can result in the precipitation of magnetite (Fe_3O_4) (Lovley et al., 1987) and Fe(II) as siderite $(FeCO_3)$ (Coleman et al., 1993; see Chapter 9).

16.8.2 NONENZYMATIC PROCESSES

Some bacteria may deposit ferric oxides formed by them nonenzymatically on their cells. This non-enzymatic oxidation may involve ligand destruction of iron chelates (Aristovskaya and Zavarzin, 1971). Such organisms include sheathed bacteria such as *Leptothrix* spp. (Figure 16.8), *Siderocapsa* (Figure 16.12), *Naumannniella, Ochrobium, Siderococcus, Pedomicrobium, Herpetosyphon, Seliberia* (Figure 16.13), *Toxothrix* (Krul et al., 1970), *Acinetobacter* (MacRae and Celo, 1975), and *Archangium*. The precipitation on the cells probably involves acid exopolymer (glycocalyx) in most if not all instances (see Section 16.8.3). Neutrophilic bacteria have been implicated to play a role in the formation of iron stalactites in caves (Kasama and Murakami, 2001). The authors attributed Fe ferrihydrite precipitation in this instance to the reaction of preformed ferrihydrite with bacterial exopolysaccharides and not to bacterial Fe(II) oxidation.

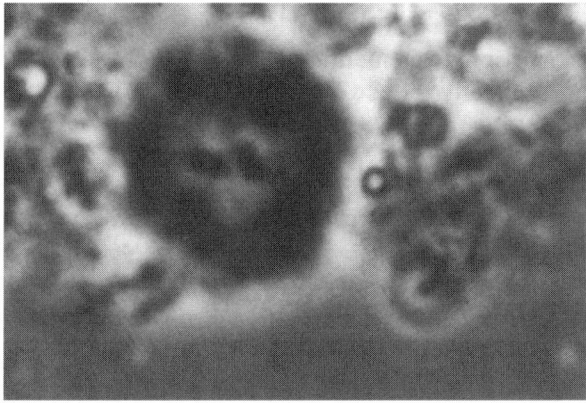

FIGURE 16.12 *Siderocapsa geminata*, Skuja (1956) (×7000). Specimen from filtered water from Pluss See, Schleswig-Holstein, Germany. Note the capsule surrounding the pair of bacterial cells. (Courtesy of Ghiorse WC, Schmidt W-D, Department of Microbiology, Cornell University, Ithaca, NY, USA.)

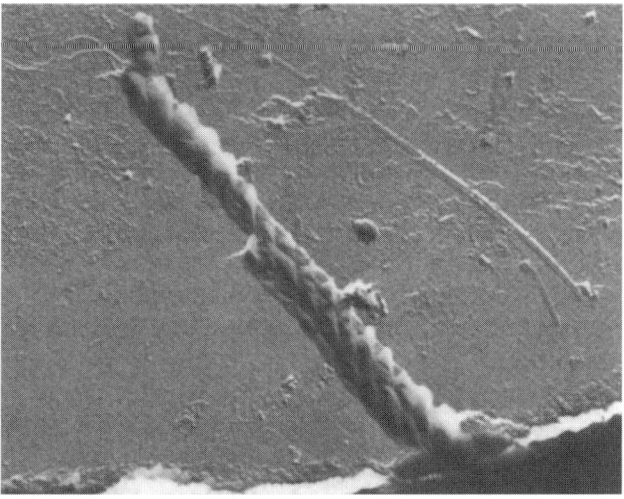

FIGURE 16.13 *Seliberia* sp. from forest pond neuston (×8200). (Courtesy of Ghiorse WC, Department of Microbiology, Cornell University, Ithaca, NY, USA.)

16.8.3 Bioaccumulation of Iron

Most or all of the bacteria listed in Section 16.8.2 may also accumulate ferric iron produced either enzymatically or nonenzymatically by other organisms in the bulk phase (Ferris, 2005). Usually, the iron is passively collected on the cell surface by reacting with acidic exopolymer, which exposes negative charges. The exopolymer may be organized in the form of a sheath, a capsule, or slime (see discussion by Ghiorse, 1984). Some protozoans such as *Anthophysa*, *Euglena* (Mann et al., 1987), *Bikosoeca*, and *Siphomonas* are also known to deposit iron on their cells.

16.9 CONCEPT OF IRON BACTERIA

A survey of the literature on bacteria that interact with iron shows that the term *iron bacteria* is differently defined by different authors. Some have included in this term any bacteria that precipitate iron, whether by active (enzymatic) oxidation or by passive accumulation of ferric oxides or hydroxides about their cells, whether they possess cellular structures with specific affinity for ferric iron or not. Starkey (1945) suggested that the term *iron bacteria* is best reserved for those bacteria that oxidize iron enzymatically.

Bacteria that passively accumulate iron should be called *iron accumulators* among which *specific* and *nonspecific* accumulators should be recognized as subgroups.

Magnetotactic bacteria occupy a special position among iron bacteria (Blakemore, 1982) (Figures 6.14A and 16.14B; Balkwill et al., 1980; Gorby et al., 1988). They comprise some motile rods, spirilla, and cocci with crystalline inclusions with magnetic properties. The inclusions consist of magnetite or greigite, or rarely both. The inclusions are formed by uptake of complexed ferric iron, which is then transformed into magnetite (Fe_3O_4) or greigite (Fe_3S_4) by mechanisms that in the case of magnetite seem to involve reduction and partial reoxidation of iron (Frankel et al., 1983; Bazylinski and Frankel, 2000). A possible clue to how the magnetite in *magnetosomes* arises may reside in the observations of formation of cytoplasmic iron granules of mixed valence in *Shewanella putrefaciens* CN32, which is not, however, a magnetotactic bacterium (Glasauer et al., 2007). Magnetite synthesis in *Magnetospirillum magnetotacticum* appears to involve the enzymes NADH-Fe(III) oxidoreductase, Fe(II)-nitrite oxidoreductase, and nitrate reductase (Fukumori, 2004). Magnetite-containing magnetotactic bacteria have been found in both freshwater and marine environments, whereas greigite-containing bacteria have been found only in marine environments (see Simmons et al., 2004). The magnetite or greigite is deposited as crystals in membrane-bound structures called magnetosomes (Figure 16.13) (Blakemore, 1982; Bazylinski, 1999). The magnetite and greigite crystals are single-domain magnets. The distribution of magnetotactic bacteria with magnetite- or greigite-containing magnetosomes, both of which are present in the water column of seasonally stratified Salt Pond in Woods Hole, Massachusetts, differed and was found to be affected by depth and oxygen and sulfide concentration (Simmons et al., 2004). The greigite-containing organisms were most numerous in the region where low concentrations of sulfide were present near the aerobic–anaerobic interface, whereas the magnetite-containing organisms were most numerous just above this zone.

Until recently it was believed that the magnetite- and greigite-containing magnetosomes enable the respective bacteria containing them to align with the Earth's magnetic field, which is inclined downward in the northern hemisphere and upward in the southern hemisphere, to assist them in locating their preferred habitat, which is a partially reduced environment. Because, by convention, the magnetic field direction is defined in terms of the north-seeking end of a compass needle, north-seeking magnetic polarity in magnetotactic bacteria in the northern hemisphere has been thought to assist them passively in their downward movement from an oxygen-rich environment to a more oxygen-poor environment. As a corollary, it was predicted that magnetotactic bacteria in the southern hemisphere must have a polarity opposite to that of bacteria in the northern hemisphere. Indeed, the existence of such bacteria in the southern hemisphere has been demonstrated (Blakemore, 1982; Bazylinski, 1999).

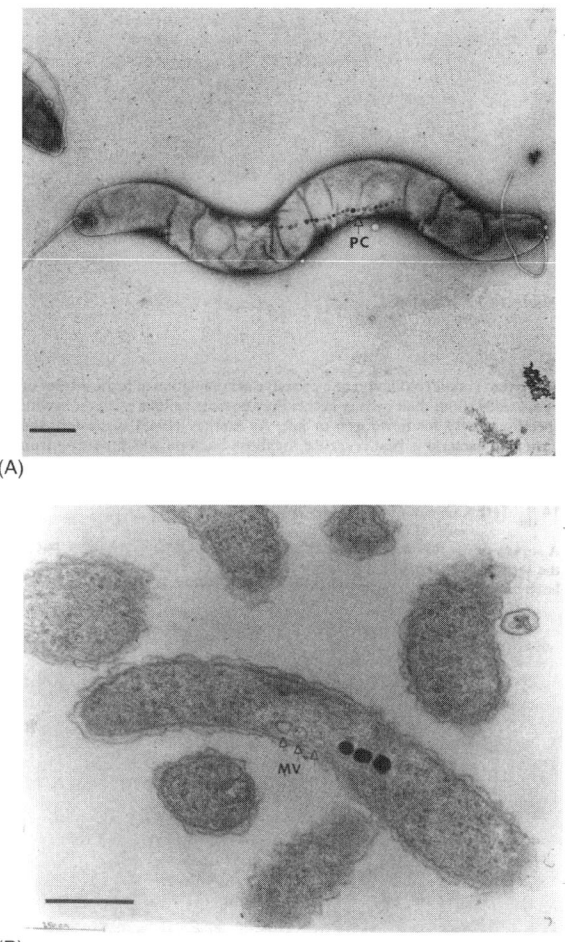

FIGURE 16.14 *Aquaspirillum magnetotacticum.* (A) Transmission electron micrograph of a negatively stained cell showing an electron-dense particle chain (PC) of magnetosomes (bar = 500 nm). (Reproduced from Balkwill DL, Maratea D, Blakemore RP, *J. Bacteriol.*, 141, 1399–1408, 1980. With permission.) (B) Transmission electron micrograph of a thin section showing trilaminate structure of the membranous vesicles (MV), which lie along the same axis as complete magnetosomes (bar = 250 nm). (Reproduced from Gorby YA, Beveridge TJ, Blakemore RP, *J. Bacteriol.*, 170, 834–841, 1988. With permission.)

The response of magnetotactic bacteria to a magnetic field is either polar or axial. A polar response is a strictly magnetotactic one in which the bacteria swim in a preferred direction by responding solely to the magnetic field (magnetotaxis), whereas an axial response is a magnetotactic or aerotactic one in which the bacteria swim back and forth in both directions by responding to the magnetic field and the oxygen gradient (aerotaxis). When the polarity of a magnetic field applied to a suspension of magnetotactic bacteria is reversed, polarly responding bacteria reverse direction, whereas axially responding bacteria rotate 180° but continue to swim back and forth (Frankel et al., 1997). A particular bacterial response is species specific. So far, polarly responding magnetotactic bacteria have been found to be much more common than axially responding ones.

Unexpectedly, recent observations have revealed the presence of significant numbers of both north-seeking and south-seeking magnetotactic bacteria in Salt Pond in Woods Hole, Massachusetts, which is inconsistent with our current understanding of the significance of magnetosomes in magnetotactic bacteria (Simmons et al., 2006). This raises a fundamental question about the adaptive value of magnetotaxis in magnetotactic bacteria to which a satisfactory answer eludes us at this time. Current molecular genetic studies of magnetotactic bacteria are laying the experimental groundwork necessary to resolve this question in the future (see recent review by Komeili, 2007).

Upon the death of magnetotactic bacteria with magnetite magnetosomes, the magnetite crystals in them are liberated and in nature may become incorporated in sediment. It has been suggested that remanent magnetism detected in some rocks may be due to magnetite residues from magnetotactic bacteria (Blakemore, 1982; Karlin et al., 1987). However, magnetite is also formed extracellularly by some nonmagnetotactic bacteria (Lovley et al., 1987; Jiao et al., 2005) (see Section 16.10).

16.10   SEDIMENTARY IRON DEPOSITS OF PUTATIVE BIOGENIC ORIGIN

Sedimentary iron deposits, many quite extensive, may feature iron in the form of oxides, sulfides, or carbonates. Microbes appear to have participated in the formation of many of these deposits (Konhauser, 1997). Their participation can be inferred by (1) the presence in some deposits of fossilized microbes with imputed iron-oxidizing or iron-accumulating potential when they were alive, (2) the presence of living iron-oxidizing or iron-accumulating bacteria in currently forming deposits of ferric oxide (Fe_2O_3) or FeOOH, (3) the presence of Fe(III)-reducing bacteria in currently forming magnetite (Fe_3O_4) deposits, or (4) the presence of sulfate-reducing bacteria in currently forming iron sulfide deposits. The probable environmental conditions prevailing at the time of formation of a biogenic iron deposit may be inferred from microfossils associated with it that resemble modern microorganisms for which environmental requirements for growth in their natural habitat are known. In this section, only iron oxide deposits are considered. Biogenic iron sulfide formation will be discussed in Chapter 20. Biogenic iron carbonate (siderite) formation was briefly examined in Chapter 9.

It has been thought that some mineral deposits may have been initiated by passive adsorption of a particular metal ion by one or more components of the cell surface of certain bacteria acting as a template and subsequent chemical reaction of the adsorbed ion with an appropriate counterion to form the nucleus in the formation of a mineral composed of the corresponding metal compound (Beveridge, 1989; Chan et al., 2004). Rancourt et al. (2005), however, obtained some evidence from a comparative study of hydrous ferric oxide (HFO) formation in the presence and absence of nonmetabolizing *Bacillus subtilis* or *B. licheniformis* that the bacterial surface did not act as a template in the formation of HFO by passively adsorbing Fe(III) or Fe(II) from which HFO was then formed. Instead, they found evidence that the HFO formed by coprecipitation in which redox functional groups of bacterial cell wall components may have inhibited some reactions that lead to an observable difference in biotically and abiotically formed HFO.

Among the most ancient iron deposits in the formation of which microbes may have played a central role are the BIFs that arose mostly in the Precambrian, in the time interval roughly between 3.3 and 1.8 eons ago. The major deposits were formed between 2.2 and 1.9 eons ago (see discussion by Cloud, 1973; Lundgren and Dean, 1979; Walker et al., 1983; Nealson and Myers, 1990). These formations have been found in various parts of the world, and in many places they are extensive enough to be economically exploitable as iron ore. They are characterized by alternating layers rich in chert, a form of silica (SiO_2), and layers rich in iron minerals such as hematite (Fe_2O_3), magnetite (Fe_3O_4), the iron silicate chamosite, and even siderite ($FeCO_3$). Ferric iron predominates over ferrous iron in the iron-rich layers. Average thickness of the layers is 1–2 cm, but they may be thinner or thicker. Because the most important BIFs were formed during that part of the Precambrian when the atmosphere around the Earth changed from reducing or nonoxidizing to oxidizing as a result of the emergence and incremental dominance of oxygenic photosynthesis, Cloud (1973) argued that the alternating layers in BIFs reflected episodic deposition of iron oxides, involving seasonal, annual, or longer-period cycles. In the reducing or nonoxidizing atmosphere of the Archean, iron in the Earth's crust was mostly in the ferrous form and thus could act as a scavenger of oxygen initially produced by oxygenic photosynthesis. The scavenging reaction involved autoxidation of the ferrous iron. Because autoxidation of iron is a very rapid reaction at circumneutral pH and higher, the supply of ferrous iron could have been periodically depleted. Further oxygen scavenging then must have had to wait until the supply of ferrous iron was replenished by leaching from rock or hydrothermal emissions (Holm, 1987). Alternatively, if much of the ferrous iron for scavenging O_2 was formed by ferric iron-respiring bacteria while they consumed organic carbon produced by seasonally growing photosynthesizers, periodic depletion of organic carbon rather than ferrous iron could have been the basis for the iron-poor layers (Nealson and Myers, 1990). Bacterial reduction of oxidized iron could also explain the origin of magnetite-rich layers of BIFs; Lovely et al. (1987) and Lovley and Phillips (1988a) found in their laboratory experiments that *Geobacter metallireducens* precipitated magnetite when reducing Fe^{3+}. Previously, it had been proposed that the magnetite in the iron-rich layers could have resulted from partial reduction of hematitic iron by organic carbon from biological activity (Perry et al., 1973). For a recent discussion on the potential significance of microbial Fe(III) reduction in BIF deposition, see Konhauser et al. (2005).

An ongoing process of iron deposition involving microbial mats consisting of cyanobacteria in a hot spring may provide another model for the origin of some BIFs (Pierson and Parenteau, 2000). Pierson and Parenteau observed three kinds of mats in Chocolate Pots Hot Springs in Yellowstone National Park. The first kind consisted of *Synechocystis* and *Chloroflexus* and the second of *Pseudoanabaena* and *Mastigocladus*. Both of these were present at 48–54°C. The third kind of mat consisted mostly of *Oscillatoria* and occurred at 36–45°C. The iron minerals occurred below the upper 0.5 mm of the mats, having become entrapped in the biofilm formed by the cyanobacteria. Gliding motility seemed to facilitate the encrustation process of the cyanobacteria. The iron in the water bathing the mats was mostly ferrous. Oxygenic photosynthetic activity of the mats appeared to promote the ferrous iron oxidation (Pierson et al., 1999), leading to the formation of amorphous iron oxides that are the source of the iron deposit. The ultimate development of the mineral deposit appeared to involve the death and decay of iron-encrusted cells.

If iron was indeed the scavenger of the oxygen of early oxygenic photosynthesis, it means that the sediments became increasingly more oxidizing relative to the atmosphere owing to the accumulation of the oxidized iron (Walker, 1987). Only when free oxygen began to accumulate would the atmosphere have become oxidizing relative to the sediment. Chert was deposited in BIFs because no silicate-depositing microorganisms (e.g., diatoms, radiolarians) had yet evolved. As increasing amounts of oxygen were generated by a growing population of oxygenic photosynthesizers, ferrous iron reserves became largely depleted, permitting the oxygen content of the atmosphere to increase to levels that were associated with an oxidizing atmosphere. This in turn would have restricted Fe(III) reduction to the remaining and ever-shrinking oxygen-depleted (anaerobic) environments. Although an extensive array of microfossils has been found in the cherty layers of various BIFs (Figure 16.15) (Gruner, 1923; Cloud and Licari, 1968), not all sedimentologists agree that the origin of BIFs depended on biological activity (see discussion by Walker et al., 1983).

Some microbiologists who have found that some purple photosynthetic bacteria can oxidize Fe(II) anaerobically by using it in place of reduced sulfur in the reduction of CO_2 have suggested that BIF formation may have been initiated before oxygenic photosynthesis had evolved (Hartmann, 1984; Widdel et al., 1993; Kappler et al., 2005). Anbar and Holland (1992) previously suggested that BIFs may have begun to form in anoxic Precambrian oceans as a result of abiotic photochemical oxidation of Mn^{2+} to a manganese(IV) oxide such as birnessite, which then served as oxidant in the oxidation of ferrous hydroxide complexes to γ-FeOOH or amorphous ferric hydroxide. Because various biological and abiotic pathways can generate the same ferric (hydr)oxide products, distinguishing which process was involved in BIF formation for any given interval in Earth history is a challenge. Multiple approaches must be used to address this question, but one that has gained traction in recent years is measuring Fe isotope fractionation in the iron minerals in BIFs (Johnson et al., 2008). Although it is not yet clear whether Fe isotopes can be used to distinguish different pathways from one another (Anbar, 2004), as more is understood about the mechanisms of the reactions (both biological and abiotic), Fe isotope fractionation determinations applied to BIFs may become more informative.

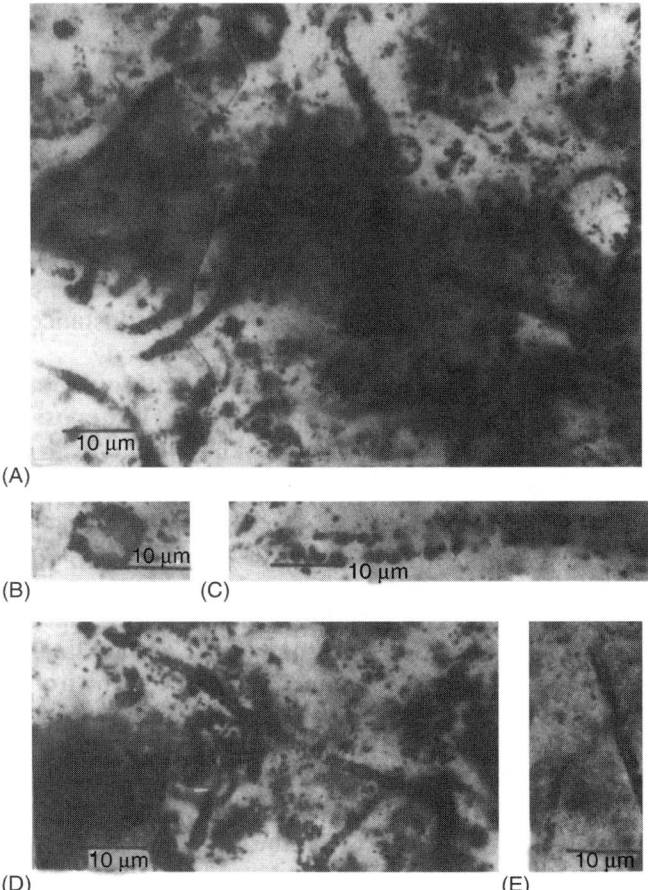

FIGURE 16.15 Gunflint microbiota in the stromatolites of the Biwabik Iron Formation, Corsica Mine, Minnesota. (A) *Gunflintia* filaments and *Huroniospora spheroides* replaced by hematite are abundant in some of the dark laminations in the stromatolitic rocks at the Corsica Mine. (B) *Huroniospora*, Corsica Mine. (C) Wide filaments, Corsica Mine. (D, E) *Gunflintia* filaments, Corsica Mine. (Reproduced from Cloud PE, Jr., Licari GR, *Proc. Natl. Acad. Sci.*, 61, 779–786, 1968.)

Among modern iron deposits in whose formation microorganisms are or have been participating are ocher deposits, bog and similar iron ores, and others. Formation of ocher deposits consisting of amorphous iron oxides (Ivarson and Sojak, 1978) is a common observation in field drains. *Gallionella*, sometimes associated with *Leptothrix*, is usually the causal organism. Hanert (1974b) measured rates of ocher deposition in field drains in terms of ferric iron deposition on submerged microscope slides. He found 4, 8.8, and 20.2 µg of iron deposited per square centimeter in 1, 2, and 3 days, respectively. Biogenic ocher deposits were also reported to be forming in a bay (the caldera of Santorini) of the Greek island Palaea Kameni in the Aegean Sea (Puchelt et al., 1973; Holm, 1987). *G. ferruginea* has been identified as the source of this ocher. The ferrous iron that is oxidized by *Gallionella* in this instance is of hydrothermal origin.

Bog iron deposition has been observable in the Pine Barrens of southern New Jersey (Crerar et al., 1979; Madsen et al., 1986). The depositional process seems to depend on several kinds of bacteria: *Acidithiobacillus ferrooxidans*, *Leptothrix ochracea*, *Crenothrix polyspora*, *Siderocapsa geminata*, and an iron-oxidizing *Metallogenium* (Walsh and Mitchell, 1972). According to Crerar et al. (1979), the iron is oxidized and precipitated by the iron-oxidizing bacteria from acid surface waters, which exhibit a pH range of 4.3–4.5 in summer. The oxidized iron accumulates as limonite (FeOOH)-impregnated sands and silts. The source of the iron is glauconite and to a much lesser extent pyrite in underlying sedimentary formations, from which it is released into groundwater—whether with microbial help has not yet be ascertained. The iron is brought to the surface by groundwater base flow, which feeds local streams. Biocatalysis of iron oxidation seems essential to account for the rapid rate of iron oxidation in the acid waters. However, *A. ferrooxidans* is probably least important because it is only infrequently encountered, perhaps because the environmental pH is above its optimum. Dissolved ferric iron in the streams in the Pine Barrens can be photoreduced (Madsen et al., 1986).

Trafford et al. (1973) described an example in which *Acidithiobacillus ferrooxidans* appeared responsible for ocher formation in field drains from pyritic soils. It is surprising that no jarosite was reported to be formed under these conditions (Silver et al., 1986).

16.11 MICROBIAL MOBILIZATION OF IRON FROM MINERALS IN ORE, SOIL, AND SEDIMENTS

Bacteria and fungi are able to mobilize significant quantities of iron from minerals in ore, soil, and sediment in which iron is a major constituent. The minerals include carbonates, oxides, and sulfides. Attack of iron sulfide will be discussed in Chapter 20, and attack of carbonates was briefly discussed in Chapter 9. In this chapter the stress is on iron oxides.

As indicated in Section 15.5, iron reduction has been observed for some time with bacteria from soil and aquatic sources. In recent years, a more systematic study of the importance of this activity in soils and sediments has been undertaken (see, for instance, review of activity in marine sediments by Burdige, 1993). Ferric iron reduction (iron respiration) can be an important form of anaerobic respiration in environments where nitrate or sulfate is present in insufficient quantities as terminal electron acceptor and where methanogenesis is not occurring (Lovley, 1987, 1991, 2000; Sørensen and Jørgensen, 1987; Lowe et al., 2000). Under these conditions, it can make an important contribution to anaerobic mineralization of organic carbon. Indeed, the presence of bioreducible ferric iron can inhibit sulfate reduction and methanogenesis when electron donors such as acetate or hydrogen are limiting (Lovley and Phillips, 1986a, 1987). Caccavo et al. (1992) found a facultative anaerobe, *Shewanella alba* BrY, that oxidized hydrogen anaerobically using Fe(III) as electron acceptor in the Great Bay Estuary, New Hampshire. Evidence that some ferric iron-reducing bacteria can use hydrogen as electron donor was first reported by Balashova and Zavarzin (1979).

The form in which ferric iron occurs can determine whether bacterial iron respiration will occur or not. In sediments from a freshwater site in the Potomac River estuary in Maryland, only amorphous FeOOH present in the upper 4 cm was bacterially reducible. Mixed Fe(II)–Fe(III) compounds that were present in the deeper layer were not attacked (Lovley and Phillips, 1986b). Local E_h and pH conditions can determine the form in which the iron produced in the bioreduction of FeOOH will occur. A mixed culture from sediment of Contrary Creek in central Virginia produced magnetite (Fe_3O_4) in the laboratory when the culture medium was allowed to go alkaline (pH 8.5 in the absence of glucose), but it produced Fe^{2+} when the medium was allowed to go acidic (pH 5.5 in the presence of glucose) (Bell et al., 1987). E_h values dropped from a range between 0 and -100 mV to less than -200 mV at the end of the experiments. A growing pure culture of the strict anaerobe, *Geobacter metallireducens* strain GS-15, produced magnetite from amorphous iron oxide directly with acetate as reductant at pH 6.7–7 and 30–35°C (Lovley and Phillips, 1988a; Lovley et al., 1987; see also Section 15.5).

The phenomenon of *gleying* in soil has come to be associated with bacterial reduction of iron oxides. It is a process that occurs under anaerobic conditions that may result from waterlogging. The affected soil becomes sticky and takes on a gray or light greenish-blue coloration (Alexander, 1977). Although once attributed to microbial sulfate reduction, this was later considered not to be the primary cause of gleying (Bloomfield, 1950). Waterlogged soil has been observed to bleach before sulfate reduction is detectable (Takai and Takamura, 1966). Bloomfield (1951) suggested that gleying was due at least in part to the action of products of plant decomposition, although he had earlier demonstrated gleying under artificial conditions in a sugar-containing medium that was allowed to ferment. Presently, microbial reduction of iron in which ferric iron is reduced by anaerobic bacterial respiration is favored as an explanation of gleying (Ottow, 1969a, 1970b, 1971). The reaction that iron undergoes in the microbial reduction has been summarized as follows (Ottow, 1971):

$$\text{Energy source} \xrightarrow[\text{Dehydrogenase}]{\text{Substrate}} e + H^+ + ATP + \text{end products} \tag{16.37}$$

$$Fe(OH)_3 + 3H^+ + e \rightarrow Fe^{2+} + 3H_2O \tag{16.38}$$

$$2Fe(OH)_3 + Fe^{2+} + 2OH^- \rightarrow \underset{\substack{\text{Gray-green mixed} \\ \text{oxide of gley}}}{Fe_3(OH)_8} \tag{16.39}$$

16.12   MICROBES AND IRON CYCLE

Microbial transformations of iron play an important role in the cycling of iron in nature (Figure 16.16) (Slobodkin et al., 2004). Weathering of iron-containing minerals in rocks, soils, and sediments introduces iron into the cycle. This weathering action is promoted partly by bacterial and fungal action and partly by chemical activity (Bloomfield, 1953a,b; Lovley, 2000). In many cases, the microbial

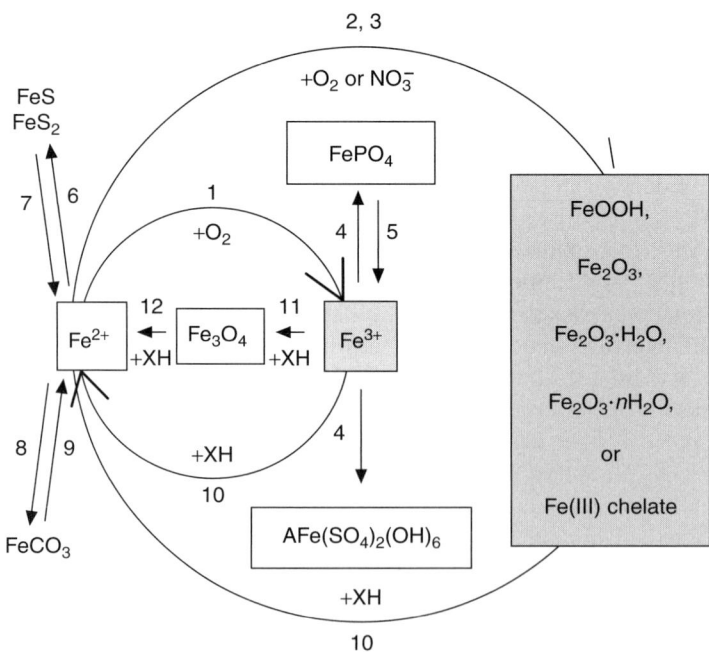

FIGURE 16.16   The iron cycle. Numerical code: 1, microbially at acid pH; 2, microbially at neutral pH in the dark with O_2 when O_2 tension is low, or anaerobically in the dark with NO_3^-, or anaerobically in light (anoxygenic photosynthesis); 3, chemically at neutral pH when O_2 tension is high; 4, chemically; 5, +H^+, microbially; 6, +H_2S, often of microbial origin; 7, +O_2, microbially or chemically; 8, +CO_3^{2-}; 9, +H^+, microbially or chemically; 10, microbially or chemically; e.g., microbially by *Geobacter metallireducens* GS-15 (Based on Lovley DR, Phillips EJP, *Appl. Environ. Microbiol.*, 54, 1472–1480, 1988a), *Desulfovibrio* spp. (Based on Lovley DR, Roden EE, Phillips EJP, Woodward JC, *Arch. Microbiol.*, 159, 336–344, 1993), *Archeoglobus fulgidus*, for instance (Based on Slobodkin AI, Chistyakova NI, Rusakov VS, *Microbiology*, 73, 469–473, 2004); 12, microbially by *Shewanella putrefaciens*.

action involves the interaction of the minerals with metabolic end products from microbes (Berthelin and Kogblevi, 1972, 1974; Berthelin et al., 1972; Berthelin et al., 1974). The mobilized iron, if ferrous, may be biologically or nonbiologically oxidized to ferric iron at a pH >5 under anaerobic or partially or fully aerobic conditions. In biological Fe(II) oxidation, light utilization may or may not be involved depending on whether or not the iron is the electron donor in a bacterial photosynthetic process (see Section 15.3). At a pH <4, ferrous iron is oxidized mainly biologically. The oxidation of ferrous iron may be followed by immediate precipitation of the ferric iron as a hydroxide, oxide, phosphate, or sulfate. If natural, soluble organic ligands abound, the ferric iron may be converted to soluble complexes and be dispersed from its site of formation. It may also be complexed by humic substances. In podzolic soils (spodosols) in temperate climates, complexed iron may be transported by groundwater from the upper A horizon to the B horizon. In hot, humid climates, ferric iron more likely precipitates at the site of its formation after release from iron-containing soil minerals, owing to intense microbial activity that rapidly and fairly completely mineralizes available organic matter including ligands, which prevents extensive formation of soluble ferric iron complexes. The iron precipitates tend to cement soil particles together in a process known as *laterization.* Aluminum hydroxide liberated in the weathering process may also be precipitated and contribute to laterization (Brooks and Kaplan, 1972, p. 75; Merkle, 1955, p. 294; see also Chapter 11).

A bacterially promoted iron cycle appears to operate in the rhizosphere of wetland plants (Weiss et al., 2003). Weiss et al. (2003) detected aerobic, neutrophilic, lithotrophic Fe(II) oxidizers as well as acetate-utilizing Fe(III) reducers associated with roots covered by iron plaque. The concentration of Fe(II) oxidizers and Fe(III) reducers in the rhizosphere was significantly higher than that in the bulk soil. The Fe(II) oxidizers are thought to be the major contributor to iron plaque formation. Because the Fe(II) oxidizers are microaerobic whereas the Fe(III) reducers are generally anaerobic, Weiss et al. (2003) suggest that these two kinds of organisms inhabit different niches in the rhizosphere. Microaerobic conditions prevail closest to the roots, which release some oxygen into the rhizosphere (Christensen et al., 1994). In another study, in a laboratory batch experiment, a lithotrophic Fe(II) oxidizer, strain BrT, isolated from the rhizosphere of *Typha latifola,* accounted for 18–53% of total iron oxidation, whereas in a continuous flow experiment it accounted for 62% of total iron oxidation (Neubauer et al., 2002). The investigators suggested that consumption of the limited amount of oxygen available in the rhizosphere by neutrophilic Fe(II)-oxidizing bacteria must influence other biogeochemical cycles in the wetlands where this occurs.

In freshwater environments, mobilization of iron may be prominent in sediments when the hypolimnion is oxygen-depleted. In sediment from Blelham Tarn in the English Lake District, bacteria occur that mineralize organic matter by iron respiration (Jones et al., 1983, 1984). Formation of Fe(II) by ferric iron reduction by a *Vibrio* of the bacterial flora in the sediment was stimulated by the addition of malate and inhibited by nitrate, MnO_2, and Mn_2O_3. H_2 was able to serve as a reductant (energy source) for the organism. Mineralization of the organic matter by iron respiration can be very important for the carbon cycle in some aquatic reducing environments (Lovley and Phillips, 1986a, 1987, 1988a). Reoxidation of mobilized ferrous iron by *Gallionella* has been reported to occur in the metalimnion of Hortlandsstemma, a small eutrophic lake in Norway where the dissolved oxygen concentration was only 0.005 mmol L^{-1} at a pH of 6.5 (Heldal and Tumyr, 1983). At higher oxygen concentrations, the iron would have been autoxidized.

The microbial genesis of iron sulfides, including pyrite, at near-neutral pH and the biological oxidation of iron sulfides at acid pH make very important contributions to the cycling of iron in environments such as salt marshes and sulfide ore deposits, respectively (see Chapter 20).

16.13 SUMMARY

Some bacteria can oxidize ferrous iron enzymatically with the generation of useful energy. Much of the evidence for this comes from the study of acidophilic iron oxidizers such as *Acidithiobacillus ferrooxidans, Leptospirillum ferrooxidans, Sulfolobus* spp., *Acidianus brierleyi,* and *Sulfobacillus thermosulfidooxidans*. The first of these has been most extensively studied. The reason the most convincing evidence of microbial iron oxidation has come from the study of acidophiles is that ferrous iron is least susceptible to autoxidation below pH 5. Some bacteria growing at circumneutral pH can also oxidize ferrous iron enzymatically, but only under partially reduced conditions. The stalked bacterium *Gallionella ferruginea* and some other circumneutrophilic iron-oxidizing bacteria use the energy gained from this oxidation chemolithotrophically, whereas the sheathed bacteria *Leptothrix* spp. may use it mixotrophically. *Leptothrix* may, in addition, precipitate preexistent ferric iron in the environment passively on their sheath. A few chemoautotrophic and photoautotrophic bacteria are now known that can oxidize Fe^{2+} anaerobically using nitrate as terminal electron acceptor. The chemoautotrophs use the ferrous iron as a source of energy and reducing power, whereas the photoautotrophs use it only as source of reducing power in the assimilation of CO_2.

Ferrous iron can also be oxidized nonenzymatically by microorganisms when they raise the redox potential and pH of the environment to levels that favor autoxidation. The rise in redox potential in the bulk phase results from the buildup of oxidants from their metabolism. The rise in pH is the consequence of photosynthesis, ammonia production, or the consumption of organic acids and their salts, some of which may chelate ferrous iron.

Ferric iron precipitation need not involve iron oxidation but rather the destruction of ferric iron chelates. Naturally occurring chelators that may solubilize extensive amounts of ferric iron include microbially produced oxalate, citrate, humic acids, and tannins. The resultant ferric chelates are relatively stable in solution unless the chelators are degraded (mineralized) by microbes. When such degradation occurs, ferric iron precipitation will result if the prevailing pH is circumneutral. A number of bacteria have this degradative capacity. Ferric iron may be locally concentrated by adsorption to cell surfaces. Various microorganisms, including bacteria and protozoa, have been found to have this capacity.

Microbial formation and accumulation of hydrous iron oxides in aqueous environments may cause concurrent accumulation of other heavy metal ions by coprecipitation with or adsorption to the hydrous iron oxides (Gunkel, 1986). The adsorbed metals may be remobilized by reduction of the iron oxides or acidification without reduction of the ferric adsorbent (Tipping et al., 1986).

Ferric iron can be enzymatically reduced to ferrous iron with a suitable electron donor. Nitrate reductase A is one enzyme that was implicated in early studies. However, at least one other enzyme appeared to be independently active in some nitrate reducers. Both bacteria and fungi have been implicated in ferric iron reduction by nitrate reductase. Currently, various facultative and strictly anaerobic bacteria have been found that can use ferric iron as terminal electron acceptor under anaerobic conditions. In at least two gram-negative bacteria involved, *Shewanella oneidensis* MR-1 and *Geobacter sulfurreducens*, the ferric reductase appears to be a *c*-type cytochrome or a protein associated with it in the outer membrane. Many of the ferric iron reducers are heterotrophs that mineralize the organic carbon used by them as electron donor, whereas others degrade this carbon only incompletely. Some can use H_2 as electron donor. A few chemoautotrophs have been found to be able to use ferric iron in place of oxygen as terminal electron acceptor. The ferric iron-respiring bacteria can play an important role in promoting the carbon cycle anaerobically.

Microbial ferric iron reduction is not always enzymatic. Some ferric iron reduction can be the result of reduction by products of microbial metabolism such as H_2S or formate, or secondary metabolites.

In nature, both oxidative and reductive reactions involving iron and brought about by microbes play important roles in the iron cycle. They affect the mobility of iron as well as its local accumulation. The formation of some sedimentary iron deposits has been attributed directly to microbial activity. The formation of ocher and bog- and lake-iron ore has been associated with bacterial iron oxidation. Gleying has been associated with bacterial iron reduction, as has the formation of magnetite and siderite. Ferric iron respiration can be a more important mechanism of carbon mineralization in some anaerobic environments than sulfate respiration.

REFERENCES

Adams LF, Ghiorse WC. 1987. Characterization of extracellular Mn^{2+}-oxidizing activity and isolation of Mn^{2+}-oxidizing protein from *Leptothrix discophora* SS-1. *J Bacteriol* 169:1279–1285.

Agate AD, Vishniac W. 1970. Transport characteristics of phospholipids from thiobacilli. *Bacteriol Proc* GP14:50.

Ahonen L, Tuovinen OH. 1989. Microbial oxidation of ferrous iron at low temperature. *Appl Environ Microbiol* 55:312–316.

Aleem MIH, Lees H, Nicholas DJD. 1963. Adenosine triphosphate-dependent reduction of nicotinamide adenine dinucleotide by ferro-cytochrome *c* in chemoautotrophic bacteria. *Nature* (London) 200:759–761.

Alexander M. 1977. *Introduction to Soil Microbiology*. 2nd ed. New York: Wiley, p. 377.

Anbar AD. 2004. Iron stable isotopes: Beyond biosignatures. *Earth Planet Sci Lett* 217:223–236.

Anbar AD, Holland HD. 1992. The photochemistry of manganese and the origin of banded iron formations. *Geochim Cosmochim Acta* 56:2595–2603.

Anderson KJ, Lundgren DG. 1969. Enzymatic studies of the iron-oxidizing bacterium, *Ferrobacillus ferrooxidans*: Evidence for a glycolytic pathway and Krebs cycle. *Can J Microbiol* 15:73–79.

Appia-Ayme C, Guiliani N, Ratouchniak J, Bonnefoy V. 1999. Characterization of an operon encoding two *c*-type cytochromes, an aa3-type cytochrome oxidase, and rusticyanin in *Thiobacillus ferrooxidans* ATCC 33020. *Appl Environ Microbiol* 65:4781–4787.

Arcuri, EJ.

Aristovskaya TV, Zavarzin GA. 1971. Biochemistry of iron in soil. In: McLaren AD, Skujins J, eds. *Soil Biochemistry*, Vol 2. New York: Marcel Dekker, pp. 385–408.

Arkesteyn GJMW. 1979. Pyrite oxidation of *Thiobacillus ferrooxidans* with special reference to the sulfur moiety of the mineral. *Antonie v Leeuwenhoek* 45:423–435.

Arkesteyn GJMW, DeBont JAM. 1980. *Thiobacillus acidophilus*: A study of its presence in *Thiobacillus ferrooxidans* cultures. *Can J Microbiol* 26:1057–1065.

Arnold RG, DiChristina TJ, Hoffmann MR. 1986a. Inhibitor studies of dissimilative Fe(III) reduction by *Pseudomonas* sp. 200 ("*Pseudomonas ferrireducens*"). *Appl Environ Microbiol* 52:281–289.

Arnold RG, Hoffmann MR, DiChristina TJ, Picardal FW. 1990. Regulation of dissimilatory Fe(III) reduction activity in *Shewanella putrefaciens*. *Appl Environ Microbiol* 56:2811–2817.

Arnold RG, Olson TM, Hoffmann MR. 1986b. Kinetics and mechanism of dissimilative Fe(III) reduction by *Pseudomonas* sp. 200. *Biotech Bioeng* 28:1657–1671.

Atlas RM. 1997. *Principles of Microbiology*. 2nd ed. Boston, MA: WCB McGraw-Hill.

Avakyan AA, Karavaiko GI. 1970. Submicroscopic organization of *Thiobacillus ferrooxidans*. *Mikrobiologiya* 39:855–861 (Engl. transl., pp. 744–751).

Baas Becking LGM, Parks GS. 1927. Energy relations in the metabolism of autotrophic bacteria. *Physiol Rev* 7:85–106.

Balashova VV, Vedina IYa, Markosyan GE, Zavarzin GA. 1974. *Leptospirillum ferrooxidans* and aspects of its autotrophic growth. *Mikrobiologiya* 43:581–585 (Engl. transl., pp. 491–494).

Balashova VV, Zavarzin GA. 1972. Iron oxidation by *Mycoplasm laidlawii*. *Mikrobiologiya* 41:909–911 (Engl. transl., pp. 808–810).

Balashova VV, Zavarzin GA. 1979. Anaerobic reduction of ferric iron by hydrogen bacteria. *Mikrobiologiya* 48:773–778 (Engl. transl., pp. 635–639).

Balkwill DL, Maratea D, Blakemore RP. 1980. Ultrastructure of a magnetotactic bacterium. *J Bacteriol* 141:1399–1408.

Barker WW, Welch SA, Chu S, Banfield JF. 1998. Experimental observations of the effects of bacteria in aluminosilicate weathering. *Am Mineral* 83:1551–1563.

Barros MEC, Rawlings DE, Woods DR. 1984. Mixotrophic growth of the *Thiobacillus ferrooxidans* strain. *Appl Environ Microbiol* 47:593–595.

Bazylinski DA. 1999. Synthesis of the bacterial magnetosome: The making of a magnetic personality. *Int Microbiol* 2:71–80.

Bazylinski DA, Frankel RB. 2000. Biologically controlled mineralization of magnetic iron minerals by magnetotactic bacteria. In: Lovley DR, ed. *Environmental Microbe-Metal Interactions*. Washington, DC: ASM Press, pp 109–144.

Beck JV. 1960. A ferrous-iron oxidizing bacterium. I. Isolation and some general physiological characteristics. *J Bacteriol* 79:502–509.

Beck JV, Elsden SR. 1958. Isolation and some characteristics of an iron-oxidizing bacterium. *J Gen Microbiol* 19:i.

Bell PE, Mills AL, Herman JS. 1987. Biogeochemical conditions favoring magnetite formation during anaerobic iron reduction. *Appl Environ Microbiol* 53:2610–2616.

Benz M, Brune A, Schink B. 1998. Anaerobic and aerobic oxidation of ferrous iron at neutral pH by chemoheterotrophic nitrate-reducing bacteria. *Arch Microbiol* 169:159–165.

Berner RA. 1962. Experimental studies of the formation of sedimentary iron sulfides. In: Jensen ML, ed. *Biogeochemistry of Sulfur Isotopes*. New Haven, CT: Yale University Press, pp. 107–120.

Berry VK, Murr LE. 1980. Morphological and ultrastructural study of the cell envelope of thermophilic and acidophilic microorganisms as compared to *Thiobacillus ferrooxidans*. *Biotech Bioeng* 22:2543–2555.

Berthelin J, Dommergues Y, Boymond D. 1972. Rôle de produits du métabolisme microbien dans la solubilization de minéraux d'une arène granitique. *Rev Ecol Biol Sol* 9:397–406.

Berthelin J, Kogblevi A. 1972. Influence de la sterilization partielle sur la solubilization microbienne des minéraux dans les sols. *Rev Ecol Biol* 9:407–419.

Berthelin J, Kogblevi A. 1974. Influence de l'engorgement sur l'altération microbienne des minéraux dans les sols. *Rev Ecol Biol* 11:499–509.

Berthelin J, Kogblevi A, Dommergues Y. 1974. Microbial weathering of a brown forest soil: Influence of partial sterilization. *Soil Biol Biochem* 6:393–399.

Beveridge TJ. 1989. Metal ion and bacteria. In: Beveridge TJ, Doyle RJ, eds. *Metal Ions and Bacteria*. New York: Wiley, pp. 1–29.

Blake R II, Waskowsky J, Harrison AP Jr. 1992. Respiratory components in acidophilic bacteria that respire on iron. *Geomicrobiol J* 10:173–192.

Blake RC, Shute EA. 1987. Respiratory enzymes of *Thiobacillus ferrooxidans*. *J Biol Chem* 262:14983–14989.

Blake RC II, Shute EA. 1994. Respiratory enzymes of *Thiobacillus ferrooxidans*. Kinetic properties of an acid-stable iron:rusticyanin oxidoreductase. *Biochemistry* 33:9220–9228.

Blakemore RP. 1982. Magnetotactic bacteria. *Annu Rev Microbiol* 36:217–238.

Blakeney MD, Moulaei T, DiChristina TJ. 2000. Fe(III) reduction activity and cytochrome content of *Shewanella putrefaciens* grown on ten compounds as sole terminal electron acceptor. *Microbiol Res* 155:87–94.

Blaylock BA, Nason A. 1963. Electron transport systems of the chemoautotroph *Ferrobacillus ferrooxidans*. I. Cytochrome *c*-containing iron oxidase. *J Biol Chem* 238:3453–3462.

Bloomfield C. 1950. Some observations on gleying. *J Soil Sci* 1:205–211.

Bloomfield C. 1951. Experiments on the mechanism of gley formation. *J Soil Sci* 2:196–211.

Bloomfield C. 1953a. A study of podzolization. Part I. The mobilization of iron and aluminum by Scots pine needles. *J Soil Sci* 4:5–16.

Bloomfield C. 1953b. A study of podzolization. Part II. The mobilization of iron and aluminum by the leaves and bark of *Agathis australis* (Kauri). *J Soil Sci* 4:17–23.

Boone DR, Liu Y, Zhao Z-J, Balkwill DL, Drake GR, Stevens TO, Aldrich HC. 1995. *Bacillus infernus* sp. nov., an Fe(III)- and Mn(IV)-reducing anaerobe from the deep terrestrial subsurface. *Int J Syst Bacteriol* 45:441–448.

Bretschger O, Obraztsova A, Sturm CA, Chang IS, Gorby YA, Reed SB, Culley DE, Reardon CL, Barua S, Romine MF, Zhou J, Beliaev AS, Bouhenni R, Saffarini D, Mansfeld F, Kim B-H, Fredrickson JK, Nealson KH. 2007. Current production and metal oxide reduction by *Shewanella oneidensis* MR-1 wild type and mutants. *Appl Environ Microbiol* 73:7003–7012.

Bridge TAM, Johnson DB. 1998. Reduction of soluble iron and reductive dissolution of ferric iron containing minerals by moderately thermophilic iron-oxidizing bacteria. *Appl Environ Microbiol* 64:2181–2186.

Bridge TAM, Johnson DB. 2000. Reductive dissolution of ferric iron minerals by *Acidiphilium* SJH. *Geomicrobiol J* 17:193–206.

Brierley CL, Brierley JA. 1973. A chemoautotrophic microorganism isolated from an acid hot spring. *Can J Microbiol* 19:183–188.

Brierley CL, Murr LE. 1973. Leaching: Use of a thermophilic and chemoautotrophic microbe. *Science* 179:488–489.

Brock TD, Brock KM, Belly RT, Weiss RL. 1972. *Sulfolobus*: A new genus of sulfur-oxidizing bacteria living at low pH and high temperature. *Arch Microbiol* 84:54–68.

Brock TD, Cook S, Peterson S, Mosser JL. 1976. Biogeochemistry and bacteriology of ferrous iron oxidation in geothermal habitats. *Geochim Cosmochim Acta* 40:493–500.

Brock TD, Gustafson J. 1976. Ferric iron reduction by sulfur- and iron-oxidizing bacteria. *Appl Environ Microbiol* 32:567–571.

Bromfield SM. 1954a. The reduction of iron oxide by bacteria. *J Soil Sci* 5:129–139.

Bromfield SM. 1954b. Reduction of ferric compounds by soil bacteria. *J Gen Microbiol* 11:1–6.

Brooks RR, Kaplan IR. 1972. Biogeochemistry. In: *The Encyclopedia of Geochemistry and Environmental Sciences*. Encycl Earth Sci Ser, Vol IVA. New York: Van Nostrand Reinhold, pp. 74–82.

Brown KA, Ratledge C. 1975. Iron transport in *Mycobacterium smegmatis*: Ferrimycobactin reductase (NAD(P) H:ferrimycobactin oxidoreductase), the enzyme releasing iron from its carrier. *FEBS Lett* 53:262–266.

Bruscella P, Appia-Ayme C, Levcán G, Ratouchniak J, Jedlicki E, Holmes DS, Bonnefoy V. 2007. Differential expression of two bc_1 complexes in the strict acidophilic chemolithoautotrophic bacterium *Acidithiobacillus ferrooxidans* suggests a model for their respective roles in iron and sulfur oxidation. *Microbiology* (Reading) 153:102–110.

Buchanan RE, Gibbons NE, eds. 1974. Bergey's manual of determinative bacteriology, 8th ed. *FEBS Lett* 53:262–266.

Burdige DJ. 1993. The biogeochemistry of manganese and iron reduction in marine sediments. *Earth-Sci Rev* 35:249–284.

Caccavo F Jr, Blakemore RP, Lovley DR. 1992. A hydrogen-oxidizing, Fe(III)-reducing microorganism from the Great Bay Estuary, New Hampshire. *Appl Environ Microbiol* 58:3211–3216.

Caccavo F Jr, Coates JD, Rossello-Mara RA, Ludwig W, Schleifer KH, Lovley DR, McInerney MJ. 1996. *Geovibrio ferrireducens*, a phylogenetically distinct dissimilatory Fe(III)-reducing bacterium. *Arch Microbiol* 165:370–376.

Caccavo F Jr, Lonergan DJ, Lovley DR, Davis M, Stolz JF, McInerney MJ. 1994. *Geobacter sulfurreducens* sp. nov., hydrogen-, and acetate-oxidizing dissimilatory metal-reducing microorganism. *Appl Environ Microbiol* 60:3752–3759.

Caiazza NC, Lies DP, Newman DK. 2007. Phototrophic Fe(II) oxidation promotes organic carbon acquisition by *Rhodobacter capsulatus* SB1003. *Appl Environ Microbiol* 73:6150–6158.

Champine JE, Goodwin S. 1991. Acetate catabolism in the dissimilatory iron-reducing isolate GS-15. *J Bacteriol* 173:2704–2706.

Champine JE, Underhill B, Johnston JM, Lilly WW, Goodwin S. 2000. Electron transfer in the dissimilatory iron-reducing bacterium *Geobacter metallireducens*. *Anaerobe* 6:187–196.

Chan CS, De Stasio G, Welch SA, Girasole M, Frazer BH, Nesterova MV, Fakra S, Banfield JF. 2004. Microbial polysaccharides template assembly of nanocrystal fibers. *Science* 303:1656–1658.

Childers SE, Ciufo S, Lovley DR. 2002. *Geobacter metallireducens* accesses insoluble Fe(III) oxide by chemotaxis. *Nature* (London) 416:767–769.

Cholodny N. 1924. Zur Morphologie der Eisenbakterien *Gallionella* and *Spirophyllum*. *Ber deutsch Bot Ges* 42:35–44.

Cholodny N. 1926. Die Eisenbakterien. Beiträge zu einer Monographie. Jena, Germany: von Gustav Fischer.

Christensen PB, Revsbech NP, Sand-Jensen K. 1994. Microsensor analysis of oxygen in the rhizosphere of the aquatic macrophyte *Littorella uniflora* (L.) Ascherson. *Plant Physiol* 105:847–852.

Clark DA, Norris PR. 1996. *Acidimicrobium ferrooxidans* gen. nov., sp. nov.: Mixed culture ferrous iron oxidation by *Sulfolobus* species. *Microbiology* (Reading) 142:785–790.

Cloud PE Jr. 1973. Paleoecological significance of the banded-iron formations. *Econ Geol* 68:1135–1143.

Cloud PE Jr, Licari GR. 1968. Microbiotas of the banded iron formations. *Proc Natl Acad Sci USA* 61:779–786.

Coates JD, Ellis DJ, Gaw CV, Lovley DR. 1999. *Geothrix fermentans*, gen. nov., sp. nov., a novel Fe(III)-reducing bacterium from a hydrocarbon contaminated aquifer. *Int J Syst Bacteriol* 49:1615–1622.

Cobley JG, Haddock BA. 1975. The respiratory chain of *Thiobacillus ferrooxidans*: The reduction of cytochromes by Fe^{2+} and the preliminary characterization of rusticyanin, a novel "blue" copper protein. *FEBS Lett* 60:29–33.

Coleman ML, Hedrick DB, Lovley DR, White DC, Pye K. 1993. Reduction of Fe(III) in sediments by sulfate-reducing bacteria. *Nature* (London) 361:436–438.

Corbett CM, Ingledew WJ. 1987. Is Fe^{3+}/Fe^{2+} cycling an intermediate in sulfur oxidation by Fe^{2+}-grown *Thiobacillus ferrooxidans*? *FEMS Microbiol Lett* 41:1–6.

Corstjens PLAM, de Vrind JPM, Westbroek P, de Vrind-de Jong EW. 1992. Enzymatic iron oxidation by *Leptothrix discophora*: Identification of an iron-oxidizing protein. *Appl Environ Microbiol* 58:450–454.

Cox CD. 1980. Iron reductases from *Pseudomonas aeruginosa*. *J Bacteriol* 141:199–204.

Cox JC, Boxer DH. 1978. The purification and some properties of rusticyanin, a blue copper protein involved in iron(II) oxidation from *Thiobacillus ferrooxidans*. *Biochem J* 174:497–502.

Cox JC, Boxer DH. 1986. The role of rusticyanin, a blue copper protein, in the electron transport chain of *Thiobacillus ferrooxidans* grown on iron or thiosulfate. *Biotech Appl Biochem* 8:269–275.

Crerar DA, Knox GW, Means JL. 1979. Biogeochemistry of bog iron in the New Jersey Pine Barrens. *Chem Geol* 24:111–135.

Croal LR, Gralnick JA, Malasarn D, Newman DK. 2004. The genetics of geochemistry. *Ann Rev Genetics* 38:175–202.

Croal LR, Jiao Y, Newman DK. 2007. The *fox* operon from *Rhodobacter* strain SW2 promotes phototrophic Fe(II) oxidation in *Rhodobacter capsulatus* SB1003. *J Bacteriol* 189:1774–1782.

Cummings DE, Caccavo F Jr, Spring S, Rosenzweig RF. 1999. *Ferribacterium limneticum*, gen. nov., sp. nov., an Fe(III)-reducing microorganism from mining impacted freshwater lake sediments. *Arch Microbiol* 171:183–188.

de Castro AF, Ehrlich HL. 1970. Reduction of iron oxide minerals by a marine *Bacillus*. *Antonie van Leeuwenhoek* 36:317–327.

de Vrind-de Jong EW, Corstjens PLAM, Kempers ES, Westbroek P, de Vrind JPM. 1990. Oxidation of manganese and iron by *Leptothrix discophora*: Use of *N,N,N',N'*-tetramethyl-*p*-phenylenediamine as an indicator of metal oxidation. *Appl Environ Microbiol* 56:3458–3462.

Dhungana S, Crumbliss AL. 2005. Coordination chemistry and redox processes in siderophore-mediated transport. *Geomicrobiol J* 22:87–89.

DiChristina TJ, Moore CM, Haller CA. 2002. Dissimilatory Fe(III) and Mn(IV) reduction by *Shewanella putrefaciens* requires *ferE*, a homolog of *pulE (gspE)* type II protein secretion gene. *J Bacteriol* 184:142–151.

Din GA, Suzuki I, Lees H. 1967a. Ferrous iron oxidation by *Ferrobacillus ferrooxidans*. Purification and properties of Fe^{2+}-cytochrome *c* reductase. *Can J Biochem* 45:1523–1546.

Din GA, Suzuki I, Lees H. 1967b. Carbon dioxide and phosphoenolpyruvate carboxylase in *Ferrobacillus ferrooxidans*. *Can J Microbiol* 13:1413–1419.

Dobbin PS, Carter JP, San Juan G-SC, von Hobe M, Powell AK, Richardson DJ. 1999. Dissimilatory Fe(III) reduction by *Clostridium beijerinckii* isolated from freshwater sediment using Fe(III) maltol enrichment. *FEMS Microbiol Lett* 176:131–138.

Dopson M, Baker-Austin C, Hind M, Bowman JP, Bond PL. 2004. Characterization of *Ferroplasma* isolates and *Ferroplasma acidarmanus* sp. nov., extreme acidophiles from acid mine drainage and industrial bioleaching environments. *Appl Environ Microbiol* 70:2079–2088.

Drobner E, Huber H, Stetter KO. 1990. *Thiobacillus ferrooxidans*, a facultative hydrogen oxidizer. *Appl Environ Microbiol* 56:2911–2923.

Dubinina GA. 1978a. Mechanism of the oxidation of divalent iron and manganese by iron bacteria growing at neutral pH of the medium. *Mikrobiologiya* 47:591–599 (Engl. transl., pp. 471–478).

Dubinina GA, 1978b. Functional role of bivalent iron and manganese oxidation in *Leptothrix pseudoochracea*. *Mikrobiologiya* 47:783–789 (Engl. transl., pp. 631–639).

Dugan PR, Lundgren DG. 1965. Energy supply for the chemoautotroph *Ferrobacillus ferrooxidans*. *J Bacteriol* 89:825–834.

Edwards KJ, Bach W, McCollum TM, Rogers DR. 2004. Neutrophilic iron-oxidizing bacteria in the oceans: Their habitats, diversity, and roles in mineral deposition, rock alteration, and biomass production in the deep-sea. *Geomicrobiol J* 21:393–404.

Edwards KJ, Bond PL, Gihring TM, Banfield JF. 2000. An Archaeal iron-oxidizing extreme acidophile important in acid mine drainage. *Science* 287:1796–1799.

Ehrenberg CG. 1836. Vorläufige Mitteilungen über das wirkliche Vorkommen fossiler Infusorien und ihre gross Verbreitung. *Poggendorf's Ann Phys Chem* 38:213–227.

Ehrenreich A, Widdel F. 1994. Anaerobic oxidation of ferrous iron by purple bacteria, a new type of phototrophic metabolism. *Appl Environ Microbiol* 60:4517–4526.

Ehrlich HL. 1993. Bacterial mineralization of organic carbon under anaerobic conditions. In: Bollag J-M, Stotzky G, eds. *Soil Biochemistry*, Vol 8. New York: Marcel Dekker, pp. 219–247.

Ehrlich HL, Yang SH, Mainwaring JD. 1973. Bacteriology of manganese nodules. VI. Fate of copper, nickel, cobalt, and iron during bacterial and chemical reduction of the manganese (IV). *Z Allg Mikrobiol* 13:39–48.

Elbehti A, Brasseur G, Lemesle-Meunier D. 2000. First evidence for existence of an uphill electron transfer through the bc_1 and NADH-Q oxidoreductase complexes of the acidophilic obligate chemolithotrophic ferrous iron-oxidizing bacterium *Thiobacillus ferrooxidans*. *J Bacteriol* 182:3602–3606.

Emerson D. 2000. Microbial oxidation of Fe(II) and Mn(II) at circumneutral pH. In: Lovley DR, ed. *Environmental Microbe-Metal Interactions*. Washington, DC: ASM Press, pp. 31–52.

Emerson D, Floyd MM. 2005. Enrichment and isolation of iron-oxidizing bacteria at neutral pH. *Meth Enzymol* 397:112–123.

Emerson D, Ghiorse WC. 1993a. Ultrastructure and chemical composition of the sheath of *Leptothrix discophora* SP-6. *J Bacteriol* 175:7808–7818.

Emerson D, Ghiorse WC. 1993b. Role of disulfide bonds in maintaining the structural integrity of the sheath of *Leptothrix discophora* SP-6. *J Bacteriol* 175:7819–7827.

Emerson D, Moyer C. 1997. Isolation and characterization of novel iron-oxidizing bacteria that grow at circumneutral pH. *Appl Environ Microbiol* 63:4784–4792.

Emerson D, Moyer CL. 2002. Neutrophilic Fe-oxidizing bacteria are abundant at the Loihi Seamount hydrothermal vents and play a major role in Fe oxide deposition. *Appl Environ Microbiol* 68:3085–3093.

Emerson D, Rentz JA, Plaia T. 2008. *Sideroxydans lithotrophicus*, gen. nov., spec. nov., and *Gallionella capsiferriformans* sp. nov., oxygen-dependent ferrous iron oxidizing bacteria that grow at circumneutral pH. *Int Syst Evol* (in preparation).

Emerson D, Weiss JV. 2004. Bacterial iron oxidation in circumneutral freshwater habitats: Findings from the field and laboratory. *Geomicrobiol J* 21:405–414.

Ernst JG, Winkelmann G. 1977. Enzymatic release of iron from sideramines in fungi. NADH:sideramine oxidoreductase in *Neurospora crassa*. *Biochim Biophys Acta* 500:27–41.

Ernstsen V, Gates WP, Stucki JW. 1998. Microbial reduction of structural iron in clays: A renewable resource of reduction capacity. *J Environ Qual* 27:761–766.

Essén SA, Johnsson A, Bylund D, Pedersen K, Lundström US. 2007. Siderophore production by *Pseudomonas stutzeri* under aerobic and anaerobic conditions. *Appl Environ Microbiol* 73:5857–5864.

Feinberg LF, Holden JF. 2006. Characterization of dissimilatory Fe(III) versus NO_3^- reduction in the hyperthermophilic archaeon *Pyrobaculum aerophilum*. *J Bacteriol* 188:525–531.

Ferris FG. 2005. Biogeochemical properties of bacteriogenic iron oxides. *Geomicrobiol J* 22:79–85.

Fischer J, Quentmeister A, Kostka S, Kraft R, Friedrich CG. 1996. Purification and characterization of the hydrogenase from *Thiobacillus ferrooxidans*. *Arch Microbiol* 165:289–296.

Frankel RB, Bazylinski DA, Johnson MS, Taylor BL. 1997. Magneto-aerotaxis in marine coccoid bacteria. *Biophysical J* 73:994–1000.

Frankel RB, Papaefthymiou GC, Blakemore RP, O'Brien W. 1983. Fe_3O_4 precipitation in magnetotactic bacteria. *Biochim Biophys Acta* 763:147–159.

Fry IV, Lazaroff N, Packer L. 1986. Sulfate dependent iron oxidation by *Thiobacillus ferrooxidans*: Characterization of a new EPR detectable electron transport component on the reducing side of rusticyanin. *Arch Biochem Biophys* 246:650–654.

Fukumori Y. 2004. Enzymes for magnetite synthesis in *Magnetospirillum magnetotacticum*. In: Bäuerlein E, ed. *Biomineralization*. 2nd ed. Weinheim, Germany: Wiley-VCH, pp. 75–90.

Fukumori Y, Yano T, Sato A, Yamanaka T. 1988. Fe(II)-oxidizing enzyme purified from *Thiobacillus ferrooxidans*. *FEMS Microbiol Lett* 50:169–172.

Gale NL, Beck JV. 1967. Evidence for the Calvin cycle and hexose monophosphate pathway in *Thiobacillus ferrooxidans*. *J Bacteriol* 94:1052–1059.

Gelder RJ. 1999. Complex lessons of iron uptake. *Nature* (London) 400:815–816.

Ghiorse WC. Department of Microbiology, Cornell University, Ithaca, NY, USA.

Ghiorse WC. 1984. Biology of iron- and manganese-depositing bacteria. *Annu Rev Microbiol* 38:515–550.

Ghiorse WC. 1988. Microbial reduction of manganese and iron. In: Zehnder AJB, ed. *Biology of Anaerobic Microorganisms*. New York: Wiley, pp. 305–331.

Ghiorse WC, Ehrlich HL. 1992. Microbial biomineralization of iron and manganese. In: Skinner HCW, Fitzpatrick RW, eds. *Biomineralization. Process of Iron and Manganese. Modern and Ancient Environments*. Catena Suppl 21. Cremlingen, Germany: Catena, pp. 75–99.

Ghiorse, WC, Schmidt, W-D. Department of Microbiology, Cornell University, Ithaca, NY, USA.

Gibson F, Magrath DI. 1969. The isolation and characterization of a hydroxamic acid (aerobactin) formed by *Aerobacter aerogenes* 62-1. *Biochim Biophys Acta* 192:175–184.

Glasauer S, Langley S, Boyanov M, Lai G, Kemner K, Beveridge TJ. 2007. Mixed-valence cytoplasmic iron granules are linked to anaerobic respiration. *Appl Environ Microbiol* 73:993–996.

Golovacheva RS, Karavaiko GI. 1978. A new genus of thermophilic spore-forming bacteria, *Sulfobacillus*. *Mikrobiologiya* 47:815–822 (Engl. transl., pp. 658–665).

Golyshina OV, Pivovarova TA, Karavaiko GI, Kondrat'eva TF, Moore ERB, Abraham WR, Lunsdorf H, Timmis KN, Yakimov MM, Golyshin PM. 2000. *Ferroplasma acidiphilum* gen. nov., spec. nov., an acidophilic member of the *Ferroplasmaceae* fam. nov., comprising a distinct lineage of Archaea. *Int J Syst Evol Microbiol* 50:997–1006.

Gómez JM, Cantero D, Johnson DB. 1999. Comparison of the effects of temperature and pH on iron oxidation and survival by *Thiobacillus ferrooxidans* (type strain) and a "*Leptospirillum ferrooxidans*"-like isolate. In: Amils R, Ballester A, eds. *Biohydrometallurgy and the Environment toward the Mining of the 21st Century. Part A*. Amsterdam: Elsevier, pp. 689–696.

González-Toril E, Gómez N, Izabal R, Amils R, Marín I. 1999. Comparative genomic characterization of iron oxidizing bacteria isolated from the Tinto River. In: Amils R, Ballester A, eds. *Biohydrometallurgy and the Environment toward the Mining of the 21st Century. Part B*. Amsterdam: Elsevier, pp. 149–157.

Gorby YA, Beveridge TJ, Blakemore RP. 1988. Characterization of the bacterial magnetosome membrane. *J Bacteriol* 170:834–841.

Gorby YA, Lovley DR. 1991. Electron transport in the dissimilatory iron reducer, GS-15. *Appl Environ Microbiol* 57:867–870.

Gorby YA, Vanina S, McLean JS, Rosso KM, Moyles D, Dohnalkova A, Beveridge TJ, Chang IS, Kim BH, Kim KS, Culley DE, Reed SB, Romine MF, Saffarini DA, Hill EA, Shi L, Elias DA, Kennedy DW, Pinchuk G, Watanabe K, Ishii S, Logan B, Nealson KH, Fredrickson JK. 2006. Electrically conductive bacterial nanowires produced by *Shewanella oneidensis* strain MR-1 and other microorganisms. *Proc Natl Acad Sci USA* 103:11358–11363.

Gralnick JA, Newman DK. 2007. Extracellular respiration. *Mol Microbiol* 65:1–11.

Gruner JW. 1923. Algae, believed to be Archean. *J Geol* 31:146–148.

Guay R, Silver M. 1975. *Thiobacillus acidophilus* sp. nov. Isolation and some physical characteristics. *Can J Microbiol* 21:281–288.

Gunkel G. 1986. Studies of the fate of heavy metals in lakes. I. The role of ferrous oxidizing bacteria in coprecipitation of heavy metals. *Arch Hydrobiol* 105:489–515.

Hafenbrandl D, Keller M, Dirmeier R, Rachel R, Rossnagel P, Burggrag S, Huber H, Stetter KO. 1996. *Ferroglobus placidus* gen. nov., spec. nov., a novel hyperthermophilic archeum that oxidizes Fe^{2+} at neutral pH under anoxic conditions. *Arch Microbiol* 166:308–314.

Hallbeck L. 1993. On the biology of the iron-oxidizing and stalk-forming bacterium *Gallionella ferruginea*. PhD Thesis. University of Göteborg, Sweden.

Hallbeck L, Pedersen K. 1990. Culture parameters regulating stalk formation and growth rate of *Gallionella ferruginea*. *J Gen Microbiol* 136:1675–1680.

Hallbeck L, Pedersen K. 1991. Autotrophic and mixotrophic growth of *Gallionella ferruginea*. *J Gen Microbiol* 137:2657–2661.

Hallberg R, Ferris FG. 2004. Biomineralization by *Gallionella*. *Geomicrobiol J* 21:325–330.

Hallberg KB, Coupland K, Kimura S, Johnson DB. 2006. Macroscopic streamer growths in acidic, metal-rich mine waters in north Wales consisting of novel and remarkably simple bacterial communities. *Appl Environ Microbiol* 72:2022–2030.

Hammann R, Ottow JCG. 1974. Reductive dissolution of Fe_2O_3 by saccharolytic Clostridia and *Bacillus polymyxa* under anaerobic conditions. *Z Pflanzenernhr Bodenk* 137:108–115.

Hanert H. 1968. Untersuchungen zur Isolierung, Stoffwechselphysiologie und Morphologie von *Gallionella ferruginea* Ehrenberg. *Arch Mikrobiol* 60:348–376.

Hanert H. 1973. Quantifizierung der Massenentwiclung des Eisenbakteriums *Gallionella ferruginea* unter natürlichen Bedignungen. *Arch Mikrobiol* 88:225–243.

Hanert H. 1974a. Untersuchungen zur individuellen Entwicklungskinetik von *Gallionella ferruginea* in statischer Mikrokultur. *Arch Mikrobiol* 96:59–74.

Hanert H. 1974b. In situ Untersuchungen zur Analyse und Intensität der Eisen(III)-Fällung in Dränungen. *Z Kulturtech Flurberein* 15:80–90.

Hanert HH. 1981. The genus *Gallionella*. In: Starr MP, Stolp H, Trüper HG, Balows A, Schlegel HG, eds. *The Prokaryotes. A Handbook on Habitats, Isolation, and Identification of Bacteria*. Berlin: Springer, pp. 509–515.

Harder EC. 1919. *Iron Depositing Bacteria and Their Geologic Relations*. US Geol Surv Prof Pap 113.

Harrison AP Jr. 1981. *Acidiphilium cryptum* gen. nov., spec. nov., heterotrophic bacterium from acidic mineral environments. *Int J Syst Bacteriol* 31:327–332.

Harrison AP Jr. 1982. Genomic and physiological diversity amongst strains of *Thiobacillus ferrooxidans* and genomic comparison with *Thiobacillus thiooxidans*. *Arch Microbiol* 131:68–76.

Harrison AP Jr. 1983. Genomic and physiological comparisons between heterotrophic thiobacilli and *Acidiphilium cryptum, Thiobacillus versutus* sp. nov., and *Thiobacillus acidophilus* nom. rev. *Int J Syst Bacteriol* 33:211–217.

Harrison AP Jr. 1984. The acidophilic thiobacilli and other acidophilic bacteria that share their habitat. *Annu Rev Microbiol* 38:265–292.

Harrison AP Jr. 1986. The phylogeny of iron-oxidizing bacteria. In: Ehrlich HL, Holmes DS, eds. *Workshop on Biotechnology for the Mining, Metal-Refining and Fossil Fuel Processing Industries*. Biotech Bioeng Symp 16. New York: Wiley, pp. 311–317.

Harrison AP Jr., Norris RP. 1985. *Leptospirillum ferrooxidans* and similar bacteria: Some characteristics and genomic diversity. *FEMS Microbiol Lett* 30:99–102.

Hart A, Murrell JC, Poole RK, Norris PR. 1991. An acid-stable cytochrome in iron oxidizing *Leptospirillum ferrooxidans. FEMS Microbiol Lett* 81:89–94.

Hartmann H. 1984. The evolution of photosynthesis and microbial mats. A speculation on banded iron formations. In: Cohen Y, Castenholz RW, Halvorson HO, eds. *Microbial Mats: Stromatolites*. New York: Alan R. Liss, pp. 451–453.

Heldal M, Tumyr O. 1983. *Gallionella* from metalimnion in a eutrophic lake: Morphology, X-ray energy dispersive microanalysis of apical cells and stalks. *Can J Microbiol* 29:303–308.

Hersman LE, hung A, Maurice PA, Forsythe JH. 2000. Siderophore production and iron reduction by *Pseudomonas mendocina* in response to iron deprivation. *Geomicrobiol J* 17:261–273.

Hilgenfeldt K. 2000. Diagenetic dissolution of biogenic magnetite in surface sediments of the Benguela upwelling system. *Int J Earth Sci* 88:630–640.

Hoehnl G. 1955. Ein Beitrag zur Physiologie der Eisenbakterien. *Vom Wasser* 22:176–193.

Holm NG. 1987. Biogenic influence on the geochemistry of certain ferruginous sediments of hydrothermal origin. *Chem Geol* 63:45–57.

Holmes DS, Lobos JH, Bopp LH, Welch GC. 1983. Setting up a genetic system de novo for studying the acidophilic thiobacillus *T. ferrooxidans*. In: Rossi G, Torma AE, eds. *Recent Progress in Biohydrometallurgy*. Iglesias, Italy: Associazione Mineraria Sarda, pp. 541–554.

Holmes DS, Yates JR, Schrader J. 1989. Mobile repeated DNA sequences in *Thiobacillus ferrooxidans* and their significance for biomining. In: Norris PR, Kelly DP, eds. *Biohydrometallurgy*. Kew Surrey, U.K.: Science & Technology Letters, pp. 153–160.

Holmes DS, Zhao H-L, Levican G, Ratouchniak J, Bonnefoy V, Varela P, Jedlicki E. 2001. ISAFe I, and ISL3 family insertion sequence from *Acidithiobacillus ferrooxidans* ATCC 19859. *J Bacteriol* 183:4323–4329.

Huber H, Stetter KO. 1989. *Thiobacillus prosperus* sp. nov. represents a new genus of halotolerant metal-mobilizing bacteria isolated from a marine geothermal field. *Arch Microbiol* 151:479–485.

Huber R, Kristjansson JK, Stetter KO. 1987. *Pyrobaculum* gen. nov., a new genus of neutrophilic, rod-shaped archaeobacteria from continental solfatara growing optimally at 100°C. *Arch Microbiol* 149:95–101.

Hutchinson M, Johnstone KI, White D. 1966. Taxonomy of the acidophilic Thiobacilli. *J Gen Microbiol* 44:373–381.

Ingledew WJ. 1982. *Thiobacillus ferrooxidans*. The bioenergetics of an acidophilic chemolithotroph. *Biochim Biophys Acta* 683:89–117.

Ingledew WJ. 1986. Ferrous iron oxidation by *Thiobacillus ferrooxidans*. In: Ehrlich HL, Holmes DS, eds. *Workshop on Biotechnology for the Mining, Metal-Refining and Fossil Fuel Processing Industries*. Biotech Bioeng Symp 16. New York: Wiley, pp. 23–33.

Ingledew WJ, Cox JC, Halling PJ. 1977. A proposed mechanism for energy conservation during Fe^{2+} oxidation by *Thiobacillus ferrooxidans*: Chemiosmotic coupling to net H^+ influx. *FEMS Microbiol Lett* 2:193–197.

Ingledew WJ, Houston A. 1986. The organization of the respiratory chain of *Thiobacillus ferrooxidans*. *Biotech Appl Biochem* 8:242–248.

Ivanov MV, Lyalikova NN. 1962. Taxonomy of iron-oxidizing bacilli. *Mikrobiologiya* 31:468–469 (Engl. transl., pp. 382–383).

Ivarson KC, Sojak M. 1978. Microorganisms and ochre deposition in field drains of Ontario. *Can J Soil Sci* 58:1–17.

James RE, FG. 2004. Evidence for microbial-mediated iron oxidation at a neutrophilic groundwater spring. *Chem Geol* 212:301–311.

Jiao Y, Newman DK. 2007. The *pio* operon is essential for phototrophic Fe(II) oxidation in *Rhodopseudomonas palustris* TIE-1. *J Bacteriol* 189:1765–1773.

Jiao Y, Kappler A, Croal LR, Newman DK. 2005. Isolation and characterization of a genetically tractable photoautotrophic Fe(II)-oxidizing bacterium, *Rhodopseudomonas palustris* strain TIE-1. *Appl Environ Microbiol* 71:4487–4496.

Johnson CM, Beard BL, Klein C, Beukes NJ, Roden EE. 2008. Iron isotopes constrain biologic and abiologic processes in banded iron formation genesis. *Geochim Cosmochim Acta* 72:151–169.

Johnson DB, Bridge TAM. 2002. Reduction of ferric iron by acidophilic heterotrophic bacteria: Evidence for constitutive and inducible enzyme systems in *Acidophilium* spp. *J Appl Microbiol* 92:315–321.

Johnson DB, Ghauri MA, Said MF. 1993. Isolation and characterization of an acidophilic, heterotrophic bacterium capable of oxidizing ferrous iron. *Appl Environ Microbiol* 58:1423–1428.

Johnson DB, Kelso WI. 1983. Detection of heterotrophic contaminants in cultures of *Thiobacillus ferrooxidans* and their elimination by subculturing in media containing copper sulfate. *J Gen Microbiol* 129:2969–2972.

Johnson DB, Macvicarl JHM, Rolfe S. 1987. A new solid medium for isolation and enumerations of *Thiobacillus ferrooxidans* and acidophilic heterotrophic bacteria. *J Microbiol Methods* 17:9–18.

Johnson DB, McGinness S. 1991. A highly efficient and universal solid medium for growing mesophilic and moderately thermophilic iron-oxidizing acidophilic bacteria. *J Microbiol Methods* 13:113–122.

Johnson DB, Roberto FF. 1997. Heterotrophic acidophiles and their roles in bioleaching of sulfide minerals. In: Rawlings DE, ed. *Biomining: Theory, Microbes and Industrial Processes*. Berlin: Springer, pp. 259–279.

Johnson DB, Said MF, Ghauri MA, McGinness S. 1989. Isolation of novel acidophiles and their potential use in bioleaching operations. In: Salley J, McCready RGL, Wichlacz PL, eds. *Biohydrometallurgy. Proceedings of International Symposium*, Jackson Hole, WY, Aug 13–18, 1989. CANMET SP89-10, Ottawa, ON: Canada Centre for Mineral and Energy Technology, pp. 403–414.

Jones CA, Kelly DP. 1983. Growth of *Thiobacillus ferrooxidans* on ferrous iron in chemostat culture: Influence of product and substrate inhibition. *J Chem Tech Biotechnol* 33B:241–261.

Jones JG, Gardener S, Simon BM. 1983. Bacterial reduction of ferric iron in a stratified eutrophic lake. *J Gen Microbiol* 129:131–139.

Jones JG, Gardener S, Simon BM. 1984. Reduction of ferric iron by heterotrophic bacteria in lake sediments. *J Gen Microbiol* 130:45–51.

Kamalov MR. 1967. Oxidation of divalent iron by *Thiobacillus ferrooxidans* culture in the presence of chloride and sulfate ions. *Izv Akad Nauk Kaz SSR Ser Biol* 5:47–50.

Kamalov MR. 1972. Adaptation of *Thiobacillus ferrooxidans* cultures to increased amounts of copper, zinc, and molybdenum in an acid medium. *Izv Akad Nauk Kaz Ser Biol* 10:39–44.

Kappler A.

Kappler A, Pasquero C, Konhauser KO, Newman DK. 2005. Deposition of banded iron formations by photo-autotrophic Fe(II)-oxidizing bacteria. *Geology* 33:865–868.

Kappler A, Straub KL. 2005. Geomicrobiological cycling of iron. *Rev Mineral Geochem* 59:85–108.

Karavaiko GI, Avakyan AA. 1970. Mechanism of *Thiobacillus ferrioxidans* multiplication. *Mikrobiologiya* 39:950–952.

Karlin R, Lyle M, Heath GR. 1987. Authigenic magnetite formation in suboxic marine sediments. *Nature* (London) 326:490–493.

Kasama T, Murakami T. 2001. The effect of microorganisms on Fe precipitation rates at neutral pH. *Chem Geol* 180:117–128.

Kashefi K, Lovley DR. 2000. Reduction of Fe(III), Mn(IV), and toxic metals at 100°C by *Pyrobaculum islandicum*. *Appl Environ Microbiol* 66:1050–1056.

Kashefi K, Shelobolina ES, Elliorr WC, Lovley DR. 2008. Growth of thermophilic and hyperthermophilic Fe(III)-reducing microorganisms on a ferruginous smectite as the sole electron acceptor. *Appl Environ Microbiol* 74:251–258.

Kaufmann F, Lovley DR. 2001. Isolation and characterization of a soluble NADPH-dependent Fe(III) reductase from *Geobacter sulfurreducens*. *J Bacteriol* 183:4468–4476.

Kelly DP, Jones CA. 1978. Factors affecting metabolism and ferrous iron oxidation in suspensions and batch cultures of *Thiobacillus ferrooxidans*: Relevance to ferric iron leach solution generation. In: Murr LE, Torma AE, Brierley JA, eds. *Metallurgical Applications of Bacterial Leaching and Related Microbiological Phenomena*. New York: Academic Press, pp. 19–44.

Kelly DP, Tuovinen OH. 1972. Recommendation that the names *Ferrobacillus ferrooxidans* Leathen and Braley and *Ferrobacillus sulfooxidans* Kinsel be recognized as synonymous of *Thiobacillus ferrooxidans* Temple and Colmer. *Int J Syst Bacteriol* 22:170–172.

Kelly DP, Wood AP. 2000. Reclassification of some species of *Thiobacillus* to the newly designated genera *Acidithiobacillus* gen. nov., *Halothiobacillus* gen. nov., and *Thermithiobacillus* gen. nov. *Int J Syst Evol Microbiol* 50:511–516.

Kinsel N. 1960. New sulfur oxidizing iron bacterium: *Ferrobacillus sulfooxidans* sp. n. *J Bacteriol* 80:628–632.

Knight V. Blakemore R. 1998. Reduction of diverse electron acceptors by *Aeromonas hydrophila*. *Arch Microbiol* 169:239–248.

Komeili A. 2007. Molecular mechanisms of magnetosome formation. *Ann Rev Biochem* 76:351–366.

Konhauser KO. 1997. Bacterial iron biomineralization in nature. *FEMS Microbiol Rev* 20:315–326.

Konhauser KO, Newman DK, Kappler A. 2005. The potential significance of microbial Fe(III) reduction during deposition of Precambrian banded iron formations. *Geobiology* 3:167–177.

Kostka JE, Nealson KH. 1995. Dissolution and reduction of magnetite by bacteria. *Environ Sci Technol* 29:2535–2540.

Kostka JE, Stucki JW, Nealson KH, Wu J. 1996. Reduction of structural Fe(III) in smectite by a pure culture of *Shewanella putrefaciens* strain MR-1. *Clays Clay Miner* 44:522–529.

Kovalenko TV, Karavaiko GI, Piskunov VP. 1982. Effect of Fe^{2+} ions on the oxidation of ferrous iron by *Thiobacillus ferrooxidans* at various temperatures. *Mikrobiologiya* 51:156–160 (Engl. transl., pp. 142–146).

Krul IM, Hirsch P, Staley JT. 1970. *Toxothrix trichogenes* (Mol) Berger and Bingmann: The organism and its biology. *Antonie van Leeuwenhoek* 36:409–420.

Kucera K-H, Wolfe RS. 1957. A selective enrichment method for *Gallionella ferruginea*. *J Bacteriol* 74:344–349.

Kullmann K-H, Schweisfurth R. 1978. Eisenoxydierende, stäbchenförmige Bakterien. II. Quantitative Untersuchungen zum Stoffwechsel und zur Eisenoxydation mit Eisen(II)-oxalat. *Z Allg Mikrobiol* 18:321–327.

Kusano T, Sugawara K, Inoue C, Takeshima T, Numata M, Shiratori T. 1992. Electro-transformation of *Thiobacillus ferrooxidans* with plasmids containing a *mer* determinant as the selective marker by electroporation. *J Bacteriol* 174:6617–6623.

Kuznetsov SI, Romanenko VI. 1963. *The Microbiological Study of Inland Waters*. Moscow: Izvddatel'stvo Akad Nauk SSSR.

Landesman J, Duncan DW, Walden CC. 1966. Iron oxidation by washed cell suspensions of the chemoautotroph *Thiobacillus ferrooxidans*. *Can J Microbiol* 12:25–33.

Lascelle J, Burke KA. 1978. Reduction of ferric iron by L-lactate and DL-glycerol-3-phosphate in membrane preparations from *Staphylococcus aureus* and interactions with the nitrate reductase system. *J Bacteriol* 134:585–589.

Laverman AM, Blum JS, Schaefer JK, Phillips EJP, Lovley DR, Oremland RS. 1995. Growth of strain SES-3 with arsenate and other diverse electron acceptors. *Appl Environ Microbiol* 61:3556–3561.

Lazaroff N. 1963. Sulfate requirement for iron oxidation by *Thiobacillus ferrooxidans*. *J Bacteriol* 85:78–83.

Lazaroff N. 1977. The specificity of the anionic requirement for iron oxidation by *Thiobacillus ferrooxidans*. *J Gen Microbiol* 101:85–91.

Lazaroff N. 1983. The exclusion of D_2O from the hydration sphere of $FeSO_4 \cdot 7H_2O$ oxidized by *Thiobacillus ferrooxidans*. *Science* 222:1331–1334.

Lazaroff N, Melanson L, Lewis E, Santoro N, Pueschel C. 1985. Scanning electron microscopy and infrared spectroscopy of iron sediments formed by *Thiobacillus ferrooxidans*. *Geomicrobiol J* 4:231–268.

Lazaroff N, Sigal W, Wasserman A. 1982. Iron oxidation and precipitation of ferric hydroxysulfates by resting *Thiobacillus ferrooxidans* cells. *Appl Environ Microbiol* 43:924–938.

Leathen WW, Kinsel NA, Braley SA Jr. 1956. *Ferrobacillus ferrooxidans*: A synthetic autotrophic bacterium. *J Bacteriol* 72:700–704.

Lees H, Kwok SC, Suzuki I. 1969. The thermodynamics of iron oxidation by ferrobacilli. *Can J Microbiol* 15:43–46.

Li Y-L, Vali H, Yang J, Phelps TJ, Zhang CL. 2006. Reduction of iron oxides enhanced by a sulfate-reducing bacterium and biogenic H_2S. *Geomicrobiol J* 23:103–117.

Lies DP, Hernandez ME, Kappler A, Mielke RE, Gralnick JA, Newman DK. 2005. *Shewanella oneidensis* MR-1 uses overlapping pathways for iron reduction at a distance and by direct contact under conditions relevant for biofilms. *Appl Environ Microbiol* 71:4414–4426.

Lieske R. 1919. Zur Ernährungsphysiologie der Eisenbakterien. *Z Bakt Parasitenk Infektionskr Hyg Abt II* 49:413–425.

Lloyd JR, Blunt-Harris EL, Lovley DR. 1999. The periplasmic 9.6 kDa *c*-type cytochrome is not an electron shuttle to Fe(III). *J Bacteriol* 181:7647–7649.

Lobos JH, Chisholm TE, Bopp LH, Holmes DS. 1986. *Acidiphilium organovorum* sp. nov., an acidophilic heterotroph isolated from a *Thiobacillus ferrooxidans* culture. *Int J Syst Bacteriol* 36:139–144.

Lonergan DJ, Jenter HL, Philips EJP, Schmidt TM, Lovley DR. 1996. Phylogenetic analysis of dissimilatory Fe(III)-reducing bacteria. *J Bacteriol* 178:223–1333.

Lovley DR. 1987. Organic matter mineralization with the reduction of ferric iron: A review. *Geomicrobiol J* 5:375–399.

Lovley DR. 1991. Dissimilatory Fe(III) and Mn(IV) reductions. *Microbiol Rev* 55:259–287.

Lovley DR. 2000. Fe(III) and Mn(IV) reduction. In: Lovley DR, ed. *Environmental Microbe-Metal Interactions*. Washington, DC: ASM Press, pp. 3–30.

Lovley DR, Badaeker MJ, Lonergan DJ, Cozzarelli IM, Phillips EJP, Siegel DI. 1989b. Oxidation of aromatic contaminants coupled to microbial iron reduction. *Nature* (London) 339:297–299.

Lovley DR, Coates JD, Blunt-Harris EL, Phillips EJP, Woodword JC. 1996. Humic substances as electron acceptors for microbial respiration. *Nature* (London) 382:445–448.

Lovley DR, Fraga JL, Blunt-Harris EL, Hayes LA, Phillips EJP, Coates JD. 1998. Humic substances as a mediator for microbially catalyzed metal reduction. *Acta Hydrochim Hydrobiol* 26:152–157.

Lovley DR, Giovannoni SJ, White DC, Champine JE, Phillips EJP, Gorby YA, Goodwin S. 1993. *Geobacter metallireducens* gen. nov., sp. nov., a microorganism capable of coupling the complete oxidation of organic compounds to the reduction of iron and other metals. *Arch Microbiol* 159:336–344.

Lovley DR, Lonergan DJ. 1990. Anaerobic oxidation of toluene, phenol, and *p*-cresol by the dissimilatory iron-reducing organism, GS-15. *Appl Environ Microbiol* 56:1858–1864.

Lovley DR, Phillips EJP. 1986a. Organic matter mineralization with the reduction of ferric iron in anaerobic sediments. *Appl Environ Microbiol* 51:683–689.

Lovely DR, Phillips EJP. 1986b. Availability of ferric iron for microbial reduction in bottom sediments of the freshwater tidal Potomac River. *Appl Environ Microbiol* 52:751–757.

Lovley DR, Phillips EJP. 1987. Competitive mechanisms for inhibition of sulfate reduction in sediments. *Appl Environ Microbiol* 53:2636–2641.

Lovley DR, Phillips EJP. 1988a. Novel mode of microbial energy metabolism: Organic carbon oxidation coupled to dissimilatory reduction of iron or manganese. *Appl Environ Microbiol* 54:1472–1480.

Lovley DR, Phillips EJP. 1988b. Manganese inhibition of microbial iron reduction in anaerobic sediments. *Geomicrobiol J* 6:145–155.

Lovley DR, Phillips EJP, Lonergan DJ. 1989a. Hydrogen and formate oxidation coupled to dissimilatory reduction of iron and manganese by *Alteromonas putrefaciens*. *Appl Environ Microbiol* 55:700–706.

Lovley DR, Phillips EJP, Lonergan DJ. 1991. Enzymatic versus nonenzymatic mechanisms for Fe(III) reduction in aquatic sediments. *Environ Sci Technol* 25:1062–1067.

Lovley DR, Phillips EJP, Lonergan DJ, Widman PK. 1995. Fe(III) and S^0 reduction by *Pelobacter carbinolicus*. *Appl Environ Microbiol* 61:2132–2138.

Lovley DR, Roden EE, Phillips EJP, Woodward JC. 1993. Enzymic iron and uranium reduction by sulfate-reducing bacteria. *Mar Geol* 113:41–53.

Lovley DR, Stolz JF, Nord GL, Phillips EJP. 1987. Anaerobic production of magnetite by a dissimilatory iron-reducing microorganism. *Nature* (London) 330:252–254.

Lovley DR, Woodward JC. 1996. Mechanism for chelator stimulation of microbial Fe(III) oxide reduction. *Chem Geol* 132:19–24.

Lowe KL, DiChristina TJ, Toychoudhury AN, van Cappellen P. 2000. Microbiological and geochemical characterization of microbial Fe(III) reduction in salt marsh sediments. *Geomicrobiol J* 17:163–178.

Lower BH, Shi L, Yongsunthon R, Doubray TC, McCready DE, Lower SK. 2007. Specific bonds between an iron oxide surface and outer membrane cytochromes MtrC and OmcA from *Shewanella oneidensis* MR-1. *J Bacteriol* 189:4944–4952.

Lower SK, Hochella MR Jr, Beveridge TJ. 2001. Bacterial recognition of mineral surfaces: Nanoscale interactions between *Shewanella* and α-FeOOH. *Science* 292:1360–1363.

Lundgren DG, Department of Biology, Syracause University, Syracuse, NY, USA.

Lundgren DG, Andersen KJ, Remsen CC, Mahoney RP. 1964. Culture, structure, and physiology of the chemoautotroph *Ferrobacillus ferrooxidans*. *Dev Indust Microbiol* 6:250–259.

Lundgren DG, Dean W. 1979. Biogeochemistry of iron. In: Trudinger PA, Swaine DJ, eds. *Biogeochemical Cycling of Mineral-Forming Elements*. Amsterdam: Elsevier, pp. 211–251.

Lütters S, Hanert HH. 1989. The ultrastructure of chemolithoautotrophic *Gallionella ferruginea* and *Thiobacillus ferrooxidans* as revealed by chemical fixation and freeze-etching. *Arch Microbiol* 151:245–251.

Lütters-Czekalla S. 1990. Lithoautotrophic growth of the iron bacterium *Gallionella ferruginea* with thiosulfate or sulfide as energy source. *Arch Microbiol* 154:417–421.

Lyalikova NN. 1958. A study of chemosynthesis in *Thiobacillus ferrooxidans*. *Mikrobiologiya* 27:556–559.

Maciag WJ, Lundgren DG. 1964. Carbon dioxide fixation in the chemoautotroph, *Ferrobacillus ferrooxidans*. *Biochim Biophys Res Commun* 17:603–607.

Mackintosh ME. 1978. Nitrogen fixation by *Thiobacillus ferrooxidans*. *J Gen Microbiol* 105:215–218.

MacRae IC, Celo JS. 1975. Influence of colloidal iron on the respiration of a species of the genus *Acinetobacter*. *Appl Microbiol* 29:837–840.

Madsen EL, Morgan MD, Good RE. 1986. Simultaneous photoreduction and microbial oxidation of iron in a stream in the New Jersey pinelands. *Limnol Oceanogr* 31:832–838.

Magnuson TS, Hodges-Myerson AL, Lovley DR. 2000. Characterization of a membrane-bound NADH-dependent Fe^{3+}-reductase from dissimilatory Fe^{3+}-reducing bacterium *Geobacter sulfurreducens*. *FEMS Microbiol Lett* 185:205–211.

Mann H, Tazaki T, Fyfe WS, Beveridge TJ, Humphrey R. 1987. Cellular lepidocrocite precipitation and heavy-metal sorption in *Euglena* sp. (unicellular alga): Implications for biomineralization. *Chem Geol* 63:39–43.

Manning HL. 1975. New medium for isolating iron-oxidizing and heterotrophic acidophilic bacteria from acid mine drainage. *Appl Microbiol* 30:1010–1016.

Mansch R, Sand W. 1992. Acid-stable cytochromes in ferrous iron oxidizing cell-free preparations from *Thiobacillus ferrooxidans*. *FEMS Microbiol Lett* 92:83–88.

Mao MWH, Dugan PR, Martin PAW, Tuovinen OH. 1980. Plasmid DNA in chemoorganotrophic *Thiobacillus ferrooxidans* and *T. acidophilus*. *FEMS Microbiol Lett* 8:121–125.

Marchlewitz B, Hasche D, Schwartz W. 1961. Untersuchungen über das Verhalten von Thiobakterien gegen-über Schwermetallen. *Z Allg Mikrobiol* 1:179–191.

Markosyan GE. 1972. A new iron-oxidizing bacterium *Leptospirillum ferrooxidans* gen. et sp. nov. *Bio Zh Arm* 25:26.

Marsili E, Baron DB, Shikhare ID, Coursolle D, Gralnick JA, Bond DR. 2008. *Shewanella* secretes flavins that mediate extracellular electron transfer. *Proc Natl Acad Sci* 105:3968–3973.

Martin PAW, Dugan PR, Tuovinen OH. 1981. Plasmid DNA in acidophilic, chemolithotrophic thiobacilli. *Can J Microbiol* 27:850–853.

Mason J, Kelly DP. 1988. Mixotrophic and autotrophic growth of *Thiobacillus acidophilus*. *Arch Microbiol* 149:317–323.

Mason J, Kelly DP, Wood AP. 1987. Chemolithotrophic and autotrophic growth of *Thermothrix thiopara* and some thiobacilli on thiosulfate and polythionates, and a reassessment of the growth yields of *Thx. thiopara* in chemostat culture. *J Gen Microbiol* 133:1249–1256.

McClure MA, Wyckoff RWG. 1982. Ultrastructural characteristics of *Sulfolobus acidocaldarius*. *J Gen Microbiol* 128:433–437.

Merkle FG. 1955. Oxidation-reduction processes in soils. In: Bear FE, ed. *Chemistry of the Soil*. New York: Reinhold, pp. 200–218.

Mishra AK, Roy P. 1979. A note on the growth of *Thiobacillus ferrooxidans* on solid medium. *J Appl Bacteriol* 47:289–292.

Mishra AK, Roy P, Roy Mahapatra SS. 1983. Isolation of *Thiobacillus ferrooxidans* from various habitats and their growth pattern on solid medium. *Curr Microbiol* 8:147–152.

Molisch H. 1910. *Die Eisenbacterien*. Jena: Gustav Fischer.

Munch JC, Hillebrand TH, Ottow JCG. 1978. Transformation in the Fe_o/Fe_d ratio of pedogenic iron oxides affected by iron-reducing bacteria. *Can J Soil Sci* 58:475–486.

Munch JC, Ottow JCG. 1977. Modelluntersuchungen zum Mechanismus der bakteriellen Eisenreduktion in hydromorphen Böden. *Z Pflanzenernhr Bodenk* 140:549–562.

Munch JC, Ottow JCG. 1980. Preferential reduction of amorphous crystalline iron oxides by bacteria activity. *Soil Sci* 129:1–21.

Munch JC, Ottow JCG. 1982. Einfluss von Zellkontakt und Eisen(III)-oxidform auf die bakterielle Eisenreduktion. *Z Pflanzenernhr Bodenk* 145:66–77.

Myers CR, Myers JM. 1992a. Localization of cytochromes to the outer membrane of anaerobically grown *Shewanella putrefaciens* MR-1. *J Bacteriol* 174:3429–3438.

Myers CR, Myers JM. 1992b. Fumarate reductase is a soluble enzyme in anaerobically grown *Shewanella putrefaciens* MR-1. *FEMS Microbiol Lett* 98:13–20.

Myers CR, Myers JM. 1993. Ferric reductase is associated with the membranes of anaerobically grown *Shewanella putrefaciens* MR-1. *FEMS Microbiol Lett* 108:15–22.

Myers CR, Nealson KH. 1988. Bacterial manganese reduction and growth with manganese oxide as sole elec-tron acceptor. *Science* 240:1319–1321.

Myers CR, Nealson KH. 1990. Respiration-linked proton translocation coupled to anaerobic reduction of manganese(IV) and iron(III) in *Shewanella putrefaciens* MR-1. *J Bacteriol* 172:6236–6238.

Myers JM, Myers CR. 2001. Role of outer membrane cytochromes OmcA and OmcB of *Shewanella putrefa-ciens* MR-1 in reduction of manganese dioxide. *Appl Environ Microbiol* 67:260–269.

Nealson KH, Myers CR. 1990. Iron reduction by bacteria: A potential role in the genesis of banded iron forma-tions. *Am J Sci* 290-A:35–40.

Nealson KH, Saffarini D. 1994. Iron and manganese in anaerobic respiration: Environmental significance, physiology, and regulation. *Annu Rev Microbiol* 48:311–343.

Neilands JB, ed. 1974. *Microbial Iron Metabolism.* New York: Academic Press.

Neubauer SC, Emerson D, Megonigal JP. 2002. Life at the energetic edge: Kinetics of circumneutral iron oxidation by lithotrophic iron-oxidizing bacteria isolated from the wetland-plant rhizosphere. *Appl Environ Microbiol* 68:3988–3995.

Nevin KP, Lovley DR. 2000. Lack of production of electron-shuttling compounds or solubilization of Fe(III) during reduction of insoluble Fe(III) oxide by *Geobacter metallireducens. Appl Environ Microbiol* 66:2248–2251.

Nevin KP, Lovley DR. 2002. Mechanisms of accessing insoluble Fe(III) oxide during dissimilatory Fe(III) reduction by *Geothrix fermentans. Appl Environ Microbiol* 68:2294–2299.

Newman DK, Kolter R. 2000. A role for excreted quinones in extracellular electron transfer. *Nature* (London) 405:94–97.

Niemelä SI, Sivalä C, Luoma T, Tuovinen OH. 1994. Maximum temperature limits for acidophilic, mesophilic bacteria in biological leaching systems. *Appl Environ Microbiol* 60:3444–3446.

Norris PR. 1990. Acidophilic bacteria and their activity in mineral sulfide oxidation. In: Ehrlich HL, Brierley CL, eds. *Microbial Mineral Recovery.* New York: McGraw-Hill, pp. 3–27.

Norris PR. 1997. Thermophiles and bioleaching. In: Rawlings DE, ed. *Biomining: Theory, Microbes and Industrial Processes.* Berlin: Springer, pp. 247–258.

Norris PR, Barr DW. 1985. Growth and iron oxidation by acidophilic moderate thermophiles. *FEMS Microbiol Lett* 28:221–224.

Norris PR, Barr DW, Hinson D. 1988. Iron and mineral oxidation by acidophilic bacteria: Affinities for iron and attachment to pyrite. In: Norris PR, Kelly DP, eds. *Biohydrometallurgy.* Kew Surrey, U.K.: Science & Technology Letters, pp. 43–59.

Norris PR, Brierley JA, Kelly DP. 1980. Physiological characteristics of two facultatively thermophilic mineral-oxidizing bacteria. *FEMS Microbiol Lett* 7:119–122.

Norris PR, Marsh RM, Lindstrom EB. 1986. Growth of mesophilic and thermophilic bacteria on sulfur and tetrathionate. *Biotechnol Appl Biochem* 8:313–329.

Obuekwe CO. 1986. Studies on the physiological reduction of ferric iron by *Pseudomonas* sp. *Microbiol Lett* 33:81–97.

Obuekwe CO, Westlake DWS. 1982. Effect of reducible compounds (potential electron acceptors) on reduction of ferric iron by *Pseudomonas* sp. *Microbiol Lett* 19:57–62.

Obuekwe CO, Westlake DWS, Cook FD. 1981. Effect of nitrate on reduction of ferric iron by a bacterium isolated from crude oil. *Can J Microbiol* 27:692–697.

Ottow JCG. 1969a. The distribution and differentiation of iron-reducing bacteria in gley soils. *Zentralbl Baketeriol Parasitenkd Infektionskr Hyg Abt 2* 123:600–615.

Ottow JCG. 1969b. Der Einfluss von Nitrat, Chlorat, Sulfat, Eisenoxydform und Wachsbedingungen auf das Ausmass der bakteriellen Eisenreduktion. *Z Pflanzenernhr Bodenk* 124:238–253.

Ottow JCG. 1970a. Selection, characterization and iron-reducing capacity of nitrate reductaseless (nit⁻) mutants of iron-reducing bacteria. *Z Allg Mikrobiol* 10:55–63.

Ottow JCG. 1970b. Bacterial mechanism of gley formation in artificially submerged soil. *Nature* (London) 225:103.

Ottow JCG. 1971. Iron reduction and gley formation by nitrogen-fixing Clostridia. *Oecologia* 6:164–175.

Ottow JCG, Glathe H. 1971. Isolation and identification of iron-reducing bacteria from gley soils. *Soil Biol Biochem* 3:43–55.

Ottow JCG, Ottow H. 1970. Gibt es eine Korrelation zwischen der eisenreduzierenden und nitratreduzierenden Flora des Bodens? *Zentralbl Bakteriol Parasitenkd Inffektionskr Hyg Abt 2* 124:314–318.

Ottow JCG, von Klopotek A. 1969. Enzymatic reduction of iron oxide by fungi. *Appl Microbiol* 18:41–43.

Pandey A, Bringel F, Meyer J-M. 1994. Iron requirement and search for siderophores in lactic acid bacteria. *Biotechnol* 40:735–739.

Peng J-B, Yan W-M, Bao XZ. 1994. Expression of heterogenous arsenic resistance genes in the obligately autotrophic biomining bacterium *Thiobacillus ferrooxidans. Appl Environ Microbiol* 60:2653–2656.

Perry EC Jr., Tan FC, Morey GB. 1973. Geology and stable isotope geochemistry of Biwabic Iron Formation, northern Minnesota. *Econ Geol* 68:1110–1125.

Pichinoty F. 1963. Récherches sur la nitrate réductase d'*Aerobacter aerogenes. Ann Inst Past* (Paris) 104:394–418.

Pierson BK, Parenteau MN. 2000. Phototrophs in high iron microbial mats: Microstructure of mats in iron-depositing hot springs. *FEMS Microbiol Ecol* 32:181–196.

Pierson BK, Parenteau MN, Griffin BM. 1999. Phototrophs in high-iron concentration microbial mats: Physiological ecology of phototrophs in an iron-depositing hot spring. *Appl Environ Microbiol* 65:5474–5483.

Pivovarova TA, Markosyan GE, Karavaiko GI. 1981. Morphogenesis and fine structure of *Leptospirillum ferrooxidans*. *Mikrobiologiya* 50:482–486 (Engl. transl., pp. 339–344).

Pollack JR, Neilands JB. 1970. Enterobactin, an iron transport compound. *Biochem Biophys Res Commun* 38:989–992.

Pollock J, Weber KA, Lack J, Achenbach LA, Mormile MR, Coates JD. 2007. Alkaline iron(III) reduction by a novel alkaliphilic, halotolerant *Bacillus* sp. isolated from salt flat sediments of Soap Lake. *Appl Microbiol Biotechnol* 77:927–934.

Praeve P. 1957. Untersuchungen über die Stoffwechselphysiologie des Eisenbakteriums *Leptothrix ochracea* Kützing. *Arch Mikrobiol* 27:33–62.

Pretorius IM, Rawlings DE, O'Neill EF, Jones WA, Kirby R, Woods DR. 1987. Nucleotide sequence of the gene encoding the nitrogenase iron protein of *Thiobacillus ferrooxidans*. *J Bacteriol* 169:367–370.

Pretorius IM, Rawlings DE, Woods DR. 1986. Identification and cloning of *Thiobacillus ferrooxidans* structural *nif*-genes in *Escherichia coli*. *Gene* 45:59–65.

Price-Whelan A, Dietrich L, Newman DK. 2006. Rethinking "secondary" metabolism, physiological roles for phenazine antibiotics. *Nat Chem Biol* 2:71–78.

Pringsheim EG. 1949. The filamentous bacteria *Sphaerotilus*, *Leptothrix* and *Cladothrix* and their relation to iron and manganese. *Philos Trans R Soc Lond B Biol Sci* 233:453–482.

Pronk JT, de Bruyn JC, Bos P, Kuenen JG. 1992. Anaerobic growth of *Thiobacillus ferrooxidans*. *Appl Environ Microbiol* 58:2227–2230.

Pronk JT, Johnson DB. 1992. Oxidation and reduction of iron by acidophilic bacteria. *Geomicrobiol J* 10:149–171.

Pronk JT, Liem K, Bos P, Kuenen JG. 1991b. Energy transduction by anaerobic ferric iron respiration in *Thiobacillus ferrooxidans*. *Appl Environ Microbiol* 57:2063–2068.

Pronk JT, Meijer WM, Hazeu W, van Dijken JP, Bos P, Kuenen JG. 1991a. Growth of *Thiobacillus ferrooxidans* on formic acid. *Appl Environ Microbiol* 57:2057–2062.

Puchelt H, Schock HH, Schroll E, Hanert H. 1973. Rezente marine Eisenerze aus Santorin, Griechenland. *Geol Rundschau* 62:786–812.

Rancourt DG, Thibault P-J, Mavrocordatos D, Lamarche G. 2005. Hydrous ferric oxide precipitation in the presence of nonmetabolizing bacteria: Constraints on the mechanism of a biotic effect. *Geochim Cosmochim Acta* 69:553–577.

Rankama K, Sahama TG. 1950. *Geochemistry*. Chicago, IL: University of Chicago Press, pp. 657–676.

Rawlings DE, Kusano T. 1994. Molecular genetics of *Thiobacillus ferrooxidans*. *Microbiol Rev* 58:39–55.

Rawlings DE, Pretorius IM, Woods DR. 1986. Expression of *Thiobacillus ferrooxidans* plasmid function and the development of genetic systems for the thiobacilli. In: Ehrlich HL, Holmes DS, eds. *Workshop on Biotechnology for the Mining, Metal-Refining and Fossil Fuel Processing Industries*. Biotech Bioeng Symp 16. New York: Wiley, pp. 281–287.

Razzell WE, Trussell PC. 1963. Isolation and properties of an iron-oxidizing *Thiobacillus*. *J Bacteriol* 85:595–603.

Reid RT, Live DH, Faulkner DJ, Butler A. 1993. A siderophore from a marine bacterium with an exceptional ferric ion affinity constant. *Nature* (London) 366:455–458.

Remsen C, Lundgren DG. 1966. Electron microscopy of the cell envelope of *Ferrobacillus ferrooxidans* prepared by freeze-etching and chemical fixation techniques. *J Bacteriol* 92:1765–1771.

Reguera G, McCarthy KD, Mehta T, Nicoll JS, Tuominen MR, Lovley DR. 2005. Extracelluar electron transfer via microbial nanowires. *Nature* (London) 435:1098–1101.

Reguera G, Nevin KP, Nicoll JS, Covalla SF, Woodward TL, Lovley DR. 2006. Biofilm and nanowire production lead to increased current in microbial fuel cells. *Appl Environ Microbiol* 72:7345–7348.

Reguera G, Pollina RB, Nicoll JS, Lovley DR. 2007. Possible nonconductive role of *Geobacter sulfurreducens* pilus nanowires in biofilm formation. *J Bacteriol* 189:2125–2127.

Roberts JL. 1947. Reduction of ferric hydroxide by strains of *Bacillus polymyxa*. *Soil Sci* 63:135–140.

Roden EE, Lovley DR. 1993. Dissimilatory Fe(III) reduction by the marine microorganism *Desulfuromonas acetoxidans*. *Appl Environ Microbiol* 59:734–742.

Roden EE, Urrutia MM. 1999. Ferrous iron removal promotes microbial reduction of crystalline iron(III) oxides. *Environ Sci Technol* 33:1847–1853.

Roden EE, Urrutia MM, Mann CJ. 2000. Bacterial reductive dissolution of crystalline Fe(III) oxide in continuous-flow column reactors. *Appl Environ Microbiol* 66:1062–1065.

Roden EE, Zachara JM. 1996. Microbial reduction of crystalline iron(III) oxides: Influence of oxide surface area and potential for cell growth. *Environ Sci Technol* 30:1618–1628.

Rosenberg H, Young IG. 1974. Iron transport in the enteric bacteria. In: Neilands JB, ed. *Microbial Iron Metabolism: A Comprehensive Treatise*. New York: Academic Press, pp. 67–82.

Rosselló-Mora RA, Caccavo F Jr., Osterlehner K, Springer N, Spring S, Schüler D, Ludwig W, Amann R, Vanncanney M, Schleifer KH. 1994. Isolation and taxonomic characterization of a halotolerant, facultative iron-reducing bacterium. *Syst Appl Microbiol* 17:569–573.

Rosselló-Mora RA, Ludwig W, Kämpfer P, Amann R, Schleifer K-H. 1995. *Ferrimonas balearica* gen. nov., spec. nov., a new marine facultative Fe(III)-reducing bacterium. *Syst Appl Microbiol* 18:196–202.

Royer RA, Burgos WD, Fisher AS, Unz RF, Dempsey BA. 2002. Enhancement of biological reduction of hematite by electron shuttling and Fe(II) complexation. *Environ Sci Technol* 36:1939–1946.

Sartory A, Meyer J. 1947. Contribution á l'étude du métabolisme hydrocarboné des bactéries ferrigineuses. *CR Acad Sci* (Paris) 225:541–542.

Schnaitman CA, Korczinski MS, Lundgren DG. 1969. Kinetic studies of iron oxidation by whole cells of *Ferrobacillus ferrooxidans*. *J Bacteriol* 99:552–557.

Schnaitman CA, Lundgren DG. 1965. Organic compounds in the spent medium of *Ferrobacillus ferrooxidans*. *Can J Microbiol* 11:23–27.

Schrader J, Holmes DS. 1988. Phenotypic switching of *Thiobacillus ferrooxidans*. *J Bacteriol* 170:3915–3923.

Schröder I, Johnson E, de Vries S. 2003. Microbial ferric iron reductases. *FEMS Microbiol Rev* 27:427–447.

Scott DT, McKnight DM, Blunt-Harris EL, Kolesar SE, Lovley DR. 1998. Quinone moieties act as electron acceptors in the reduction of humic substances by humics-reducing microorganisms. *Environ Sci Technol* 32:2984–2989.

Seeger M, Jerez CA. 1993. Phosphate starvation-induced changes in *Thiobacillus ferrooxidans*. *FEMS Microbiol Lett* 108:35–42.

Seeliger S, Cord-Ruwish R, Schink B. 1998. A periplasmic and extracellular *c*-type cytochrome of *Geobacter sulfurreducens* acts as a ferric iron reductase and as an electron carrier to other acceptors or to partner bacteria. *J Bacteriol* 180:3686–3691.

Segel IH. 1975. *Enzyme Kinetics: Behavior and Analysis or Rapid Equilibrium and Steady-State Enzyme Systems*. New York: Wiley.

Segerer A, Neuner A, Kristjansson JK, Stetter KO. 1986. *Acidianus infernus*, gen. nov., spec. nov.: Facultatively aerobic, extremely acidophilic thermophilic sulfur-metabolizing archaeabacteria. *Arch Microbiol* 36:559–564.

Segerer AH, Stetter KO. 1998. 29. The order sulfolobales. In: Dworkin M, ed. *The Prokaryotes*. Electronic version. Berlin: Springer.

Selkov E, Overbeek R, Kogan Y, Chu L, Vonstein V, Holmes D, Silver S, Haselkorn R, Fonstein M. 2000. Functional analysis of gapped microbial genomes: Amino acid metabolism of *Thiobacillus ferrooxidans*. *Proc Natl Acad Sci USA* 97:3509–3514.

Shafia F, Brinson KR, Heinzman MW, Brady JM. 1972. Transition of chemolithotroph *Ferrobacillus ferrooxidans* to obligate organotrophy and metabolic capabilities of glucose-grown cells. *J Bacteriol* 111:56–65.

Shafia F, Wilkinson RF. 1969. Growth of *Ferrobacillus ferrooxidans* on organic matter. *J Bacteriol* 97:256–260.

Shi L, Squier TC, Zachara JM, Fredrickson JK. 2007. Respiration of metal (hydr)oxides by *Shewanella* and *Geobacter*, a key role for multihaem *c*-type cytochromes. *Mol Microbiol* 65:12–20.

Short KA, Blakemore RP. 1986. Iron respiration-driven proton translocation in aerobic bacteria. *J Bacteriol* 167:729–731.

Silver M, Ehrlich HL, Ivarson KC. 1986. Soil mineral transformation mediated by soil microbes. In: Huang PM, Schnitzer M, eds. *Interaction of Soil Minerals with Natural Organics and Microbes*. SSSA Special Publication No. 17. Madison, WI: Soil Science Society of America, pp. 497–519.

Silverman MP, Lundgren DG. 1959a. Studies on the chemoautotrophic iron bacterium *Ferrobacillus ferrooxidans*. I. An improved medium and a harvesting procedure for securing high cell yields. *J Bacteriol* 77:642–647.

Silverman MP, Lundgren DG. 1959b. Studies on the chemoautotrophic iron bacterium *Ferrobacillus ferrooxidans*. II. Manometric studies. *J Bacteriol* 78:326–331.

Simmons SL, Bazylinski DA, Edwards KJ. 2006. South-seeking magnetotactic bacteria in the Northern Hemisphere. *Science* 311:371–374.

Simmons SL, Sievert SM, Frankel RB, Bazylinski DA, Edwards KJ. 2004. Spatiotemporal distribution of marine magnetotactic bacteria in a seasonally stratified coastal salt pond. *Appl Environ Microbiol* 70:6230–6239.

Skuja 1956. Max-Planck-Institute for Limnology, Plön, Germany.

Slobodkin AI, Chistyakova NI, Rusakov VS. 2004. High temperature microbial sulfate reduction can be accompanied by magnetite formation. *Microbiology* 73:469–473.

Sørensen J. 1987. Nitrate reduction in marine sediment: Pathways and interactions with iron and sulfur cycling. *Geomicrobiol J* 5:401–421.

Sørensen J, Jørgensen BB. 1987. Early diagenesis in sediments from Danish coastal waters: Microbial activity and Mn-Fe-S geochemistry. *Geochim Cosmochim Acta* 51:1583–1590.

Starkey RL. 1945. Precipitation of ferric hydrate by iron bacteria. *Science* 102:532–533.

Starkey RL, Halvorson HO. 1927. Studies on the transformations of iron in nature. II. Concerning the importance of microorganisms in the solution and precipitation of iron. *Soil Sci* 24:381–402.

Starr MP, Stolp H, Trüper HG, Balows A, Schlegel HG, eds. 1981. *The Prokaryotes: A Handbook on Habitats, Isolation, and Identification of Bacteria*. Berlin: Springer.

Stevens CJ, Dugan PR, Tuovinen OH. 1986. Acetylene reduction (nitrogen fixation) by *Thiobacillus ferrooxidans*. *Biotechnol Appl Biochem* 8:351–359.

Stokes JL. 1954. Studies on the filamentous sheathed iron bacterium *Sphaerotilus natans*. *J Bacteriol* 67:278–291.

Straub KL, Benz M, Schink B, Widdel F. 1996. Anaerobic, nitrate-dependent microbial oxidation of ferrous iron. *Appl Environ Microbiol* 62:1458–1460.

Straub KL, Schönhuber WA, Buchholz-Cleven BEE, Schink B. 2004. Diversity of ferrous iron-oxidizing, nitrate-reducing bacteria and their involvement in oxygen-dependent iron cycling. *Geomicrobiol J* 21:371–378.

Stucki JW, Golden DC, Roth CB. 1984. Effects of reduction and reoxidation of structural iron(III) in smectites. *Clays Clay Miner* 32:350–356.

Stucki JW, Komadel P, Wilkinson HT. 1987. Microbial reduction of structural iron(III) in smectites. *Soil Sci Soc Am J* 51:1663–1665.

Stucki JW, Roth CB. 1977. Oxidation-reduction mechanism for structural iron in nontronite. *Soil Sci Soc Am J* 41:808–814.

Stucki JW, Wu J, Gan H, Komadel P, Banin A. 2000. Effects of iron oxidation state and organic cations on dioctahedral smectite hydration. *Clays Clay Miner* 48:290–298.

Sugio T, Anzai Y, Tano T, Imai K. 1981. Isolation and some properties of an obligate and a facultative iron-oxidizing bacterium. *Agric Biol Chem* 45:1141–1151.

Sugio T, Domatsu C, Munakata O, Tano T, Imai K. 1985. Role of a ferric iron reducing system in sulfur oxidation by *Thiobacillus ferrooxidans*. *Appl Environ Microbiol* 49:1401–1406.

Sugio T, Hirose T, Lii-Zhen Y, Tano T. 1992a. Purification and some properties of sulfite:ferric ion oxidoreductase from *Thiobacillus ferrooxidans*. *J Bacteriol* 174:4189–4192.

Sugio T, Katagiri T, Inagaki K, Tano T. 1989. Actual substrate for elemental sulfur oxidation by sulfur:ferric iron oxidoreductase purified from *Thiobacillus ferrooxidans*. *Biochim Biophys Acta* 973:250–256.

Sugio T, Katagiri T, Moriyama M, Zhen YL, Inagaki K, Tano T. 1988b. Existence of a new type of sulfite oxidase which utilizes ferric ions as an electron acceptor in *Thiobacillus ferrooxidans*. *Appl Environ Microbiol* 54:153–157.

Sugio T, Wada K, Mori M, Inagaki K, Tano T. 1988a. Synthesis of an iron-oxidizing system during growth of *Thiobacillus ferrooxidans* on sulfur-basal salts medium. *Appl Environ Microbiol* 54:150–152.

Sugio T, White KJ, Shute E, Choate D, Blake RC II. 1992b. Existence of a hydrogen sulfide:ferric ion oxidoreductase in iron-oxidizing bacteria. *Appl Environ Microbiol* 58:431–433.

Tabita R, Lundgren DG. 1971a. Utilization of glucose and the effect of organic compounds on the chemolithotroph *Thiobacillus ferrooxidans*. *J Bacteriol* 108:328–333.

Tabita R, Lundgren DG. 1971b. Heterotrophic metabolism of the chemolithotroph *Thiobacillus ferrooxidans*. *J Bacteriol* 108:334–342.

Tait GH. 1975. The identification and biosynthesis of siderochromes formed by *Micrococcus denitrificans*. *Biochem J* 146:191–204.

Takai Y, Takamura T. 1966. The mechanism of reduction in water-logged paddy soil. *Folia Microbiol* 11:304–315.

Tang YJ, Chakraborty R, García Martín H, Chu J, Hazen TC, Keasling JD. 2007. Flux analysis of central metabolic pathways in *Geobacter metallireducens* during reduction of soluble Fe(III)-nitrilotriacetic acid. *Appl Environ Microbiol* 73:3859–3864.

Taylor KA, Deatherage JF, Amos LA. 1982. Structure of the S-layer of *Sulfolobus acidocaldarius*. *Nature* (London) 299:840–842.

Temple KL, Colmer AR. 1951. The autotrophic oxidation of iron by a new bacterium: *Thiobacillus ferrooxidans*. *J Bacteriol* 62:605–611.

Thamdrup B, Finster K, Wuergler Hansen J, Bak F. 1993. Bacterial disproportionation of elemental sulfur coupled to chemical reduction of iron or manganese. *Appl Environ Microbiol* 99:101–108.

Tikhonova GV, Lisenkova LL, Doman NG, Skulachev VP. 1967. Electron transport pathways in *Thiobacillus ferrooxidans*. *Mikrobiologiya* 32:725–734 (Engl. transl., pp. 599–606).

Tipping E, Thompson DW, Ohnstad MJ, Hetherington NB. 1986. Effects of pH on the release of metals from naturally occurring oxides of manganese and iron. *Environ Technol Lett* 7:109–114.

Trafford BD, Bloomfield C, Kelso WI, Pruden G. 1973. Ochre formation in field drains in pyritic soils. *J Soil Sci* 24:453–460.

Troshanov EP. 1968. Iron- and manganese-reducing microorganisms in ore-containing lakes of the Karelian isthmus. *Mikrobiologiya* 37:934–940 (Engl. transl., pp. 786–791).

Troshanov EP. 1969. Conditions affecting the reduction of iron and manganese by bacteria in the ore-bearing lakes of the Karelian isthmus. *Mikrobiologiya* 38:634–643 (Engl. transl., pp. 528–535).

Tuovinen OH, Kelly DP. 1973. Studies on the growth of *Thiobacillus ferrooxidans*. I. Use of membrane filters and ferrous iron agar to determine viable numbers, and comparison with $^{14}CO_2-$ fixation and iron oxidation measures of growth. *Arch Mikrobiol* 88:285–298.

Tuovinen OH, Kelly DP. 1974. Studies on the growth of *Thiobacillus ferrooxidans*. II. Toxicity of uranium to growing cultures and tolerance conferred by mutation, other metal cations and EDTA. *Arch Microbiol* 95:153–164.

Tuovinen OH, Nicholas DJD. 1977. Transition of *Thiobacillus ferrooxidans* KG-4 from heterotrophic growth on glucose to autotrophic growth on ferrous-iron. *Arch Microbiol* 114:193–195.

Tuovinen OH, Niemal SI, Gyllenberg HG. 1971. Tolerance of *Thiobacillus ferrooxidans* to some metals. *Antonie van Leeuwenhoek* 37:489–496.

Tuttle JH, Dugan PR. 1976. Inhibition of growth, iron, and sulfur oxidation by *Thiobacillus ferrooxidans* by simple organic compounds. *Can J Microbiol* 22:719–730.

Unz RF, Lundgren DG. 1961. A comparative nutritional study of three chemoautotrophic bacteria: *Ferrobacillus ferrooxidans, Thiobacillus ferrooxidans* and *Thiobacillus thiooxidans*. *Soil Sci* 92:302–313.

Vartanian SE, Cowart RE. 1999. Extracellular iron reductases: Identification of a new class of enzymes by siderophore-producing microorganisms. *Arch Biochem Biophys* 364:75–82.

Venkateswaran K, Moser DP, Dollhopf ME, Lies DP, Saffarini DA, McGregor BJ, Ringelberg DB, White DC, Nishijima M, Sano H, Burghardt J, Stackebrand E, Nealson KH. 1999. Polyphasic taxonomy of the genus *Shewanella* and description of *Shewanella oneidensis* sp. nov. *Int J Syst Bacteriol* 49:705–724.

Vernon LP, Mangum JH, Beck JV, Shafia FM. 1960. Studies on a ferrous-ion-oxidizing bacterium. II. Cytochrome composition. *Arch Biochem Biophys* 88:227–231.

von Canstein H, Ogawa J, Shimizu S, Lloyd JR. 2008. Secretion of flavins by *Shewanella* species and their role in extracellular electron transfer. *Appl Environ Microbiol* 74:615–623.

Vorreiter L, Madgwick JC. 1982. The effect of sodium chloride on bacterial leaching of low-grade copper ore. *Proc Australas Inst Min Metall* 284:63–66.

Walker JCG. 1987. Was the Archean biosphere upside down? *Nature* (London) 329:710–712.

Walker JCG, Klein C, Schidlowski M, Schopf JW, Stevenson DJ, Walter MR. 1983. Environmental evolution of the Archean-Early Proterozoic Earth. In: Schopf JW, ed. *Earth's Earliest Biosphere: Its Origin and Evolution*. Princeton, NJ: Princeton University Press, pp. 260–290.

Walsh F, Mitchell R. 1972. An acid-tolerant iron-oxidizing *Metallogenium*. *J Gen Microbiol* 72:369–376.

Wandersman C, Delepelaire P. 2004. Bacterial iron sources from siderophores to hemophores. *Ann Rev Microbiol* 88:611–647.

Wang Y, Newman DK. 2008. Redox reactions of phenazine antibiotics with ferric (hydr)oxides and molecular oxygen. *Env Sci Technol* 42:2380–2386.

Weber KA, Achenbach LA, Coates JD. 2006a. Microorganisms pumping iron: Anaerobic microbial iron oxidation and reduction. *Nat Rev Microbiol* 4:L752–L764.

Weber KA, Pollock J, Cole KA, O'Connor SM, Achenbach LA, Coates JD. 2006b. Anaerobic nitrate-dependent iron(II) bio-oxidation by a novel lithoautotrophic Betaproteobacterium, strain 2002. *Appl Environ Microbiol* 72:686–694.

Weiss JV, Emerson D, Backer SM, Megonigal JP. 2006. Enumeration of Fe(II)-oxidizing and Fe(III)-reducing bacteria in the root zone of wetland plants: Implications for a rhizosphere iron cycle. *Biogeochemistry* 64:77–96.

Wichlacz PL, Unz RF, Langworthy TA. 1986. *Acidiphilium angustum* sp. nov., *Acidiphilium facilis* sp. nov., and *Acidiphilium rubrum* sp. nov.: Acidophilic heterotrophic bacteria isolated from acid coal mine drainage. *Int J Syst Bacteriol* 36:197–201.

Widdel F, Schnell S, Heising S, Ehrenreich A, Assmus B, Schink B. 1993. Ferrous oxidation by anoxygenic phototrophic bacteria. *Nature* (London) 362:834–836.

Williams HD, Poole RK. 1987. Reduction of iron(III) by *Escherichia coli* K12: Lack of involvement of the respiratory chains. *Curr Microbiol* 15:319–324.

Winogradsky S. 1888. Über Eisenbakterien. *Bot Ztg* 46:261–276.

Winogradsky S. 1922. Eisenbakterien als Anorgoxydanten. *Zentralbl Bakteriol Parasitenkd Infektionskr Hyg Abt 2* 57:1–21.

Wolfe RS. 1964. Iron and manganese bacteria. In: Heukelakian H, Dondero N, eds. *Principles and Applications in Aquatic Microbiology*. New York: Wiley, pp. 82–97.

Xiong Y, Shi L, Chen B, Mayer MJ, Lower BH, Londer Y, Saumyaditya B, Hochella MF, Fredrickson JK, Squier TC. 2006. High-affinity binding and direct electron transfer to solid metals by the *Shewanella oneidensis* MR-1 outer membrane c-type cytochrome OmcA. *J Am Chem Soc* 128:13978–13979.

Yarzábal A, Appia-Ayme C, Ratouchniak J, Bonnefoy V. 2004. Regulation of the expression of the *Acidithiobacillus ferrooxidans rus* operon encoding two cytochromes *c*, a cytochrome oxidase and rusticyanin. *Microbiology* (Reading) 150:2113–2123.

Yarzábal A, Brasseur G, Bonnefoy V. 2002b. Cytochromes *c* of *Acidithiobacillus ferrooxidans*. *FEMS Microbiol Lett* 209:189–195.

Yarzábal A, Brasseur G, Ratouchniak J, Lund K, Lemesle-Meunier D, DeMoss JA, Bonnefoy V. 2002a. The high-molecular-weight cytochrome c Cyc2 of *Acidithiobacillus ferrooxidans* is an outer membrane protein. *J Bacteriol* 184:313–317.

Yates JR, Holmes DS. 1987. Two families of repeated DNA sequences in *Thiobacillus ferrooxidans*. *Bacteriol* 169:1861–1870.

Yates MG, Nason A. 1966a. Enhancing effect of nucleic acids and their derivatives in the reduction of cytochrome *c* by ferrous ions. *J Biol Chem* 241:4861–4871.

Yates MG, Nason A. 1966b. Electron transport systems of the chemoautotroph *Ferrobacillus ferrooxidans*. II. Purification and properties of a heat-labile iron- cychrome *c* reductase. *J Biol Chem* 241:4872–4880.

Zavarzin GA. 1972. Heterotrophic satellite of *Thiobacillus ferrooxidans*. *Mikrobiologiya* 41:369–370 (Engl. transl., pp. 323–324).

17 Geomicrobiology of Manganese

17.1 OCCURRENCE OF MANGANESE IN EARTH'S CRUST

The abundance of manganese in the Earth's crust has been estimated to be 0.1% (Alexandrov, 1972, p. 670). The element is, therefore, only 1/50th as plentiful as iron in this part of the Earth. Its distribution in the crust is by no means uniform. In soils, for instance, its concentration can range from 0.002 to 10% (Goldschmidt, 1954). An average concentration in freshwater has been reported to be 8 μg kg^{-1} (Bowen, 1979). Concentrations slightly in excess of 1 mg kg^{-1} can be encountered in anoxic hypolimnia of some lakes. In seawater, an average concentration has been reported to be 0.2 μg kg^{-1} (Bowen, 1979), but concentrations more than three orders of magnitude greater can be encountered near active hydrothermal vents at midocean spreading centers.

Manganese is found as a major or minor component in more than 100 naturally occurring minerals (Bureau of Mines, 1965, p. 556; Post, 1999). Major accumulations of manganese occur in the form of oxides, carbonates, and silicates. Among the oxides, psilomelane [Ba,Mn^{2+},(Mn^{4+})$_8$O$_{16}$(OH)$_4$], birnessite (δMnO$_2$), pyrolusite and vernadite (MnO$_2$), manganite (MnOOH), hausmannite (Mn$_3$O$_4$), and todorokite [(Mn^{2+},Mg^{2+},Ba^{2+},Ca^{2+},K$^+$,Na$^+$)$_2$(Mn^{4+})$_5$O$_{12}$·3H$_2$O] are important examples. Among the carbonates, rhodochrosite (MnCO$_3$) is important, and among the silicates, rhodonite (MnSiO$_3$) and braunite [Mn,Si$_2$O$_3$] are important. The oxides, carbonates, and silicates of manganese originated mostly as secondary authigenic minerals formed by reprecipitation of dissolved manganese.

Minerals that contain manganese as a minor constituent include ferromagnesian minerals such as pyroxenes and amphiboles (Trost, 1958), and micas such as biotite (Lawton, 1955, p. 59). These are all of igneous origin.

17.2 GEOCHEMICALLY IMPORTANT PROPERTIES OF MANGANESE

Manganese is one of the elements in the first transition series in the periodic table, which includes, in order of increasing atomic number from 21 to 29, the elements Sc, Ti, V, Cr, Mn, Fe, Co, Ni, and Cu. Electronically, these elements differ mostly in the degree to which their d orbitals are filled (Drew, 1972). Their increasing oxidation states are attributed to removal of $4s$ and $3d$ electrons (Sienko and Plane, 1966).

Manganese can exist in the oxidation states 0, +2, +3, +4, +6, and +7. However, in nature only the +2, +3, and +4 oxidation states are commonly found. Of the three naturally occurring oxidation states, only manganese in the +2 oxidation state can occur as a free ion in aqueous solution. In this oxidation state it may also occur in soluble inorganic or organic complexes. Manganese in the +3 oxidation state can occur in aqueous solution only when it is complexed. The free +3 ion tends to disproportionate into the +2 and +4 oxidation states:

$$2Mn^{3+} + 2H_2O \Leftrightarrow Mn^{2+} + MnO_2 + 4H^+ \tag{17.1}$$

Citrate, pyrophosphate, and pyoverdin are examples of effective ligands for Mn^{3+}. Of these, pyoverdin, a yellowish-green fluorescent pigment formed by some *Pseudomonas* species, forms the strongest complex with Mn(III) (the stability constant for pyoverdin from *P. putida* strain MnB1 is log K = 44.6 ± 0.5). Reaction of the Mn(III) complex with ascorbate releases the manganese as Mn^{2+} (Parker et al., 2004).

The +4 oxides of manganese are insoluble in water. They have amphoteric properties that account for their affinity for various cations, especially for heavy metal ions such as Co^{2+}, Ni^{2+}, and Cu^{2+}. Mn(IV) oxides have long been known as scavengers of metallic cations (Geloso, 1927; Goldberg, 1954). They are frequently associated with ferric iron in nature. Although the natural occurrence of soluble Mn(III) in complexed ionic form in aqueous systems has been considered rare until now, a recent study of the oxygen-poor interface (suboxic) zones ($O_2 < 3$ µM, $H_2S < 0.2$ µM) in the Black Sea and Chesapeake Bay has revealed a significant presence of this form of manganese, where it is likely an important catalyst of chemical cycles involving sulfide and sulfur, and manganese(II) and (IV) (Trouwborst et al., 2006; Johnson, 2006).

Manganous ion is more stable than ferrous ion at similar pH and E_h. Based on equilibrium computations, manganese should exist predominantly as Mn^{2+} below pH 5.5 and 3.8×10^{-4} atm of CO_2, and as Mn(IV) above pH 5.5 if the E_h is ~800 mV and Mn^{2+} activity is 0.1 ppm. At an E_h of 500 mV and below, Mn^{2+} at an activity as high as 10 ppm may predominate at pH values up to pH 7.8–8.0 (Hem, 1963). Although in theory, 0.1 ppm of Mn^{2+} ions in aqueous solution should readily autoxidize when exposed to air at pH values above 4, they actually do not do so until the pH exceeds 8. Apart from Mn^{2+} concentration and E_h effects, one explanation for this resistance of Mn^{2+} ions to autoxidation is the high energy of activation requirement for the reaction (Crerar and Barnes, 1974). Another explanation is that the Mn^{2+} may be extensively complexed and thereby stabilized by such inorganic ions as Cl^-, SO_4^{2-}, and HCO_3^- (Hem, 1963; Goldberg and Arrhenius, 1958) or by organic compounds such as amino acids, humic acids, and others (Graham, 1959; Hood, 1963; Hood and Slowey, 1964).

17.3   BIOLOGICAL IMPORTANCE OF MANGANESE

Manganese is an important trace element in biological systems. It is essential in microbial, plant, and animal nutrition. It is required as an activator by a number of enzymes such as isocitric dehydrogenase or malic enzyme and may replace Mg^{2+} ion as an activator, for example, in enolase (Mahler and Cordes, 1966). It is also required in oxygenic photosynthesis, where it functions in the production of oxygen from water by photosystem II (Klimov, 1984). In Section 17.5, the ability of Mn(II) to serve as an energy source for some bacteria will be discussed, and in Section 17.6, the ability of Mn(III) and Mn(IV) to serve as terminal electron acceptors in respiration by some bacteria will be examined. As in the case of iron, the most important geomicrobial interactions with manganese are those that lead to precipitation of dissolved manganese in an insoluble phase (immobilization) or solubilization of insoluble forms of manganese (mobilization). The reactions frequently but not always involve oxidations or reductions of manganese.

17.4   MANGANESE-OXIDIZING AND MANGANESE-REDUCING BACTERIA AND FUNGI

17.4.1   MANGANESE-OXIDIZING BACTERIA AND FUNGI

Jackson (1901a,b) and Beijerinck (1913) were the first to describe the existence of manganese-oxidizing bacteria. Since their discovery, a significant number of different kinds of bacteria, many of which are taxonomically unrelated, have been reported to oxidize manganese. Among these, some promote the oxidation enzymatically, others nonenzymatically. With still other bacteria, it is as yet unclear whether they oxidize manganese enzymatically or nonenzymatically. To date, all known bacterial manganese oxidizers seem to belong to the domain Bacteria; none have so far been found that belong to the domain Archaea. They include gram-positive and gram-negative forms and are represented by spore-forming and non-spore-forming rods, cocci, vibrios, spirilla, and sheathed and appendaged bacteria. No *bona fide* autotrophic manganese oxidizers have so far been identified, although two unconfirmed claims of autotrophy have appeared in the bacteriological literature (Ali and Stokes, 1971; Kepkay and Nealson, 1987). Caspi et al. (1996) reported finding a cryptic ribulose 1,5-bisphosphate carboxylase/oxygenase gene in a marine, gram-negative α-Proteobacterium, strain S185-9A1, which did not couple CO_2 fixation to Mn^{2+} oxidation. In most instances, growth in the presence of manganese is either mixotrophic (manganese oxidation supplies some or all of the energy needed by the organism, but its carbon source is organic) or heterotrophic (manganese oxidation furnishes no useful energy; carbon and energy are derived from organic carbon). Table 17.1 lists examples of bacteria that have been shown to oxidize manganese enzymatically or nonenzymatically. They have been detected in very diverse environments such as *desert varnish* on rock surfaces, in soil, in the water column and sediments of freshwater lakes and streams, and in ocean water and sediments, including on and in ferromanganese concretions on the ocean floor.

Some mycelium-forming fungi have also been found to promote manganese oxidation, at least under laboratory conditions. In most instances, this oxidation is probably nonenzymatic and due to interaction with a metabolic product from fungi (e.g., hydroxy acid) or a fungal cell component.

TABLE 17.1
Some Bacteria that Oxidize Manganese

A. Attack dissolved Mn^{2+} enzymatically
  1. Derive useful energy
    Marine strains SSW_{22}, S_{13}, HCM-41, and E_{13} (all are gram-negative rods) (Ehrlich, 1983, 1985; Ehrlich and Salerno, 1990; unpublished results)
    Hyphomicrobium manganoxidans (Eleftheriadis, 1976)
    Pseudomonas strain S-36 (Kepkay and Nealson, 1987)
  2. Do not derive useful energy
    Arthrobacter siderocapsulatus (Dubinina, 1978a,b)
    Leptothrix discophora (Adams and Ghiorse, 1987)
    Leptothrix pseudoochracea (Dubinina, 1978a,b)
    Metallogenium (Dubinina, 1978a,b)
    Strain FMn-1 (Zindulis and Ehrlich, 1983)
  3. Not known if able to derive useful energy
    Aeromonas sp. (Nealson, 1978)
    Arthrobacter B (Bromfield, 1956; Bromfield and David, 1976)
    Arthrobacter citreus (Dubinina and Zhdanov, 1975)
    Arthrobacter globiformis (Dubinina and Zhdanov, 1975)
    Arthrobacter simplex (Dubinina and Zhdanov, 1975)
    Citrobacter freundii E_4 (Douka, 1977)
    Flavobacterium (Nealson, 1978)
    Hyphomicrobium T37 (Tyler and Marshall, 1967b)
Pedomicrobium (Aristovskaya, 1961; Ghiorse and Hirsch, 1979)
    Pseudomonas E_1 (Douka, 1977)
    Pseudomonas putida GB-1 (Tebo et al., 1997)
    Pseudomonas putida Mn-1 (Tebo et al., 1997)
    Pseudomonas spp. (Zavarzin, 1962; Nealson, 1978)
B. Attack Mn^{2+} prebound to Mn(IV) oxide or some clays enzymatically
  1. Derive useful energy
    Arthrobacter 37 (Ehrlich, 1969; Arcuri, 1978)
    Oceanospirillum (Ehrlich, 1976, 1978a)
    Marine strain CFP-11 (Ehrlich, 1983, 1985; Ehrlich and Salerno, 1990)
C. Attack Mn^{2+} nonenzymatically
    Pseudomonas manganoxidans (Jung and Schweisfurth, 1979; but see also Tebo et al., 1997)
    Streptomyces sp. (Bromfield, 1978, 1979)
    Bacillus SG-1 (de Vrind et al., 1976b ; but see also Van Wassbergen et al., 1993)

However, in the case of some fungi, such as the white rot fungus *Phanerochaete chrysosporium*, oxidation may be the result of an extracellular Mn(II)-dependent peroxidase that oxidizes Mn(II) to Mn(III) with the consumption of H_2O_2 (Glenn and Gold, 1985; Glenn et al., 1986). This reaction is similar to one observed with peroxidase in plant extracts (horseradish, turnip) (Kenten and Mann, 1949, 1950). The Mn(III) is stabilized as a complex {Mn^{3+}}, for example, lactate complex, which may then react with an organic compound (YH) such as a lignin component that reduces the Mn(III) back to Mn(II) and is itself oxidized (Figure 17.1; Glenn et al., 1986; Paszczynski et al., 1986). In this system, the rereduction of Mn(III) to Mn(II) makes manganese merely an electron shuttle in the peroxidase system and has no geochemical significance insofar as manganese redistribution in the environment is concerned. However, Kenten and Mann (1949) proposed that at a very low H_2O_2 concentration, oxidized manganese may accumulate because reduction of the oxidized manganese would be negligible. In the peroxidase M2 reaction in the absence of YH, Glenn et al. (1986) observed the formation of a brown precipitate that was a manganese oxide (tentatively identified as MnO_2, but could have been Mn_2O_3 or Mn_3O_4) in a laboratory experiment in which H_2O_2 was slowly diffused into a weakly buffered solution of enzyme and Mn(II).

Höfer and Schlosser (1999) found a laccase, an extracellularly active enzyme produced by *Trametes versicolor*, which oxidized Mn^{2+} to Mn^{3+} in the presence of pyrophosphate in a pure culture of the fungus. O_2 was the terminal electron acceptor and not H_2O_2 as in the case of manganese peroxidase. The enzyme is one of the multicopper oxidases. Subsequently, Schlosser and Höfer (2002) found that a laccase from the fungus *Stropharia rugosoannulata* oxidized Mn^{2+} to Mn^{3+} in the presence of oxalate or malonate accompanied by H_2O_2 production. No H_2O_2 was formed in this reaction when pyrophosphate replaced oxalate or malate. The H_2O_2 resulted from *abiotic* decomposition of the oxalate or malonate by biotically formed Mn^{3+}. By contrast, Miyata et al. (2006) found that a laccase produced by anamorphic ascomycete fungi isolated from streambed pebbles formed manganese(IV) oxide (δ-MnO_2) identified by x-ray diffraction (XRD) and x-ray absorption near-edge structure spectroscopy (XANES). Growing cultures produced the manganese oxide in

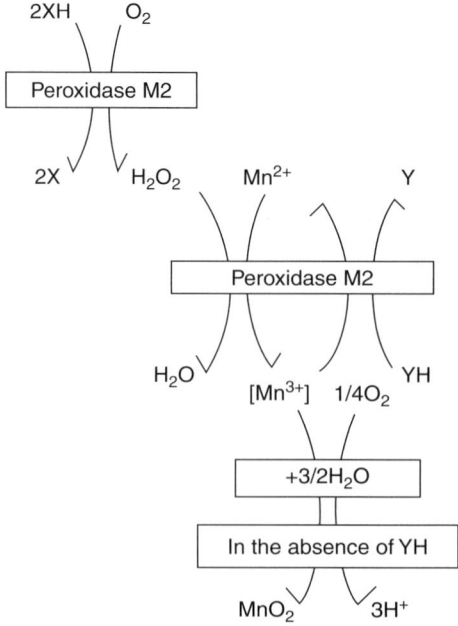

FIGURE 17.1 Interactions of extracellular peroxidase M2 from *Phanerochaete chrysosporium* with manganese. (Based on description by Glenn JK et al., *Arch. Biochem. Biophys.*, 251, 688–696, 1986; Paszczynski et al., 1986.)

an acetate-based medium with 8.8 μM Mn(II) buffered at pH 7.0 with HEPES. One of the laccase-producing fungi was affiliated to the *Xylariales* and three others with the *Pleosporales*.

17.4.2 Manganese-Reducing Bacteria and Fungi

A number of different, taxonomically unrelated bacteria have been found to reduce manganese either enzymatically or nonenzymatically. The bacteria that reduce manganese enzymatically often do so as a form of respiration in which oxidized manganese serves as terminal electron acceptor and is reduced to Mn(II). Some bacteria can reduce the oxidized manganese aerobically or anaerobically, whereas others can reduce it only anaerobically. The manganese-reducing bacteria include aerobes and strict and facultative anaerobes. In some cases, manganese may be reduced to satisfy a nutritional need for soluble Mn(II) (see de Vrind et al., 1986a; also discussion by Ehrlich, 1987) or to scavenge excess reducing power, as in some cases of NO_3^- reduction (Robertson et al., 1988) and ferric reduction (Ghiorse, 1988; Lovley, 1991).

Bacteria that reduce manganese oxides include gram-positive and gram-negative forms, with representatives among rods, cocci, and vibrios—all belonging to the domain Bacteria. The majority of the bacteria studied to date that can respire with manganese(IV) oxide as terminal electron acceptor use reduced carbon as electron donor (reductant), but a few types can also use H_2 anaerobically. Like manganese oxidizers, they have been found in very diverse environments, including soil and deep subsurface, freshwater, and marine habitats. Most of the manganese-reducing bacteria described to date do not seem to have the ability to oxidize Mn(II) as well, but a few exceptions have been reported. Representative examples of Mn reducers are listed in Table 17.2.

TABLE 17.2
Some Mn(IV)-Reducing Bacteria

A. Gram-positive bacteria, aerobic or facultative, reduce Mn(IV) aerobically and anaerobically
   Arthrobacter strain B (Bromfield and David, 1976)
   Bacillus 29 (Trimble and Ehrlich, 1968)
   Bacillus SG-1 (de Vrind et al., 1986a)
   Bacillus circulans (Troshanov, 1968, 1969)
   Bacillus (now *Paenibacillus*) *polymyxa* (Troshanov, 1968, 1969)
   Bacillus mesentericus (Troshanov, 1968, 1969)
   Bacillus mycoides (Troshanov, 1968, 1969)
   Bacillus cereus (Troshanov, 1968, 1969)
   Bacillus centrosporus (Troshanov, 1968, 1969)
   Bacillus filaris (Troshanov, 1968, 1969)
   Bacillus GJ33 (Ehrlich et al., 1973)
   Coccus 32 (Trimble and Ehrlich, 1968)
B. Gram-negative bacteria, anaerobic
   Geobacter metallireducens (formerly strain GS-15) (Lovley and Phillips, 1988a)
   Geobacter sulfurreducens (Caccavo et al., 1994)
C. Gram-negative bacteria, facultative, reduce Mn(IV) only anaerobically
   Shewanella (formerly *Alteromonas*) *oneidensis* (formerly *putrefaciens*) (Myers and Nealson, 1988a)
D. Gram-negative bacteria, aerobes, reduce Mn(IV) aerobically and anaerobically
   Strain BIII 88 (Ehrlich, 1993a)
E. Gram-negative bacteria, aerobes, reduce Mn(IV) aerobically
   Pseudomonas liquefaciens (Troshanov, 1968, 1969)
   Acinetobacter calcoaceticus (Karavaiko et al., 1986)[a]
   Strain BIII 32 (Ehrlich, 1980)[a]
   Strain BIII 41 (Ehrlich, 1980)[a]

[a] May also reduce Mn(IV) anaerobically, but not specifically tested.

Some mycelial fungi have been found to reduce manganese oxides, at least under laboratory conditions. As in the case of Mn(II) oxidation by fungi, reduction of manganese oxides by them must be nonenzymatic in most cases, although experimental proof for any specific mechanism is mostly lacking. Nonenzymatic reduction of manganese oxides by fungi is best explained in terms of formation of metabolic products by them, which act as reductants.

17.5   BIOOXIDATION OF MANGANESE

Like iron oxidation, manganese oxidation by microbes may be enzymatic or nonenzymatic. However, unlike enzymatic iron oxidation, it has not been reported to occur at very acidic pH. Enzymatic iron oxidation in air-saturated solutions or solutions close to air saturation is favored by very acidic pH because at that pH iron does not autoxidize at a significant rate. Enzymatic manganese oxidation in solution under similar air saturation levels is not favored at pH values much lower than neutrality. This is because in contrast to iron, the standard free energy change (ΔG^0) when manganese is oxidized with oxygen as electron acceptor decreases steadily until it assumes a positive value near pH 1.0 (see standard free energy values at pH 0 and 7 listed by Ehrlich, 1978a). One instance of nonenzymatic manganese biooxidation at a pH of ~5.0 has been documented (Bromfield, 1978, 1979).

Among the first to report on bacterial manganese oxidation were Jackson (1901a,b) and Beijerinck (1913). Beijerinck suggested that it may be associated with autotrophic growth. Lieske (1919) and Sartory and Meyer (1947) suggested that manganese oxidation may also be associated with mixotrophic growth. Either way, an enzymatic process was implied. The first experiments to demonstrate Mn^{2+} oxidation among resting (nongrowing) cells were performed by Bromfield (1956). He showed that cells from stationary growth phase of *Arthrobacter* strain B (formerly called *Corynebacterium* strain B) oxidized Mn^{2+} in a 0.005% $MnSO_4 \cdot 4H_2O$ solution at 40°C but not above 45°C. The oxidation was inhibited by copper and mercury salts and by azide and cyanide. In a later study, Bromfield (1974) showed that the composition of organic substrate and its concentration affected manganese oxidation by *Arthrobacter* strain B. Bromfield and David (1976) reported that cells of *Arthrobacter* strain B rapidly adsorbed Mn^{2+} ions from solution as well as oxidizing them. The adsorbed but unoxidized Mn^{2+} could be displaced by the addition of 5 mM Cu^{2+}. Manganese oxidation kinetics of the organism are shown in Figure 17.2. Bromfield and David (1976) also found that this organism could reduce oxidized manganese. Most studies to date leave the impression that enzymatic manganese oxidation is restricted mainly to bacteria. Exceptional instances of manganese oxidation by some fungi are discussed in Section 17.4.

17.5.1   Enzymatic Manganese Oxidation

On a physiological basis, various reports of enzymatic manganese(II) oxidation by bacteria suggest that not all of them use the same mechanism. However on a molecular basis, some degree of commonality among four cultures—two belonging to the genus *Pseudomonas*, one to *Leptothrix*, and one to *Bacillus*—may exist. All four seem to employ a multicopper oxidase in the oxidation (Brouwers et al., 2000a). Another common feature of these cultures is that they deposit the oxidized manganese on their cells, sheath, or spore, respectively (Okazaki et al., 1997; Mulder, 1972; Rosson and Nealson, 1982). Although these molecular studies have led some researchers to suggest that this oxidase is common to all manganese oxidizers, physiological considerations of various other manganese-oxidizing bacteria described in the literature raise doubt about a universal manganese-oxidizing mechanism. In what follows, a *physiological classification* is presented based on what is known about these bacteria to date. For the majority of bacteria, a phylogenetic, molecular analysis is lacking. The classification contains three major groups, at least one of which can be further divided into several subgroups (Table 17.3).

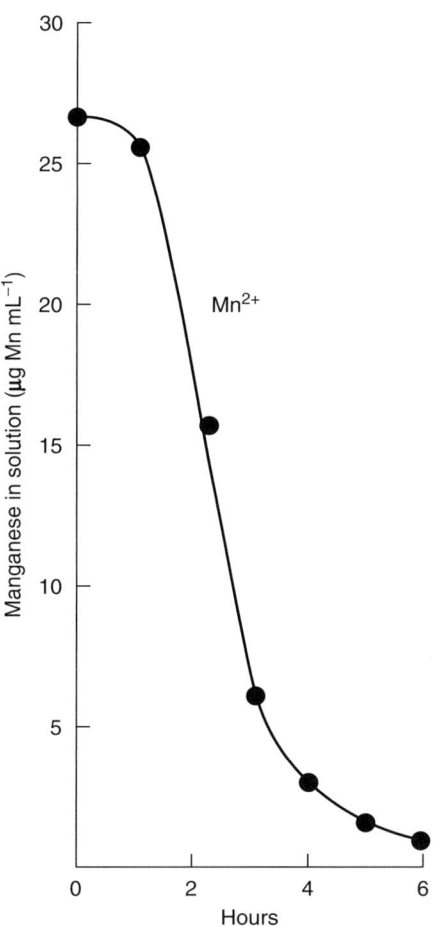

FIGURE 17.2 Oxidation of 0.5 mM Mn^{2+} by a suspension of *Arthrobacter* strain B (formerly *Corynebacterium* strain B) at pH 7.0. The mixture (80 mL) contained 4 mg cells per milliliter and 27 μg Mn per milliliter. (From Bromfield SM, David DJ, *Soil. Biol. Biochem.*, 8, 37–43, 1976, Elsevier Limited. With permission.)

TABLE 17.3
Enzymatic Mechanisms in Manganese-Oxidizing Bacteria

Group I manganese-oxidizers oxidize dissolved manganese (Mn^{2+}) with O_2 as terminal electron acceptor

  Subgroup Ia: Conserve energy from Mn^{2+} oxidation.

  Subgroup Ib: Do not conserve energy from Mn^{2+} oxidation. They use oxidized soluble factor in periplasm as oxidant of Mn^{2+}. The resultant reduced factor is reoxidized in cytoplasm and returned to periplasm.

  Subgroup Ic: Do not conserve energy from Mn^{2+} oxidation. This subgroup includes some sheathed bacteria and spores of *Bacillus* SG-1 and a strain of *B. megaterium*.

Group II manganese-oxidizers oxidize Mn^{2+} with O_2 as terminal electron acceptor when the Mn^{2+} is prebound to a suitable binding agent such as Mn(IV) oxide, ferromanganese, montmorillonite, or kaolinite. Group II oxidizers conserve energy from the oxidation.

Group III manganese-oxidizers oxidize Mn^{2+} with H_2O_2 as oxidant catalyzed by catalase.

17.5.2 GROUP I MANGANESE OXIDIZERS

Group I manganese oxidizers include those bacteria that oxidize dissolved Mn(II) species, for example, Mn^{2+}, using oxygen as terminal electron acceptor. Some bacteria in this group have the capacity to gain useful energy from this reaction (Subgroup Ia), whereas others do not (Subgroup Ib). The overall oxidation reaction can be summarized as follows:

$$Mn^{2+} + 0.5O_2 + H_2O \rightarrow MnO_2 + 2H^+ \tag{17.2}$$

In some gram-negative bacteria in Group Ia whose manganese-oxidizing system has been examined to date, Mn^{2+} appears to be oxidized in or at their cell envelope external to their plasma membrane. Until now, Ehrlich's proposal has been that the oxidation in gram-negative organisms occurred in their periplasm (Figure 17.3A; Ehrlich, 1990, 1996a, 2002). The discovery of cytochrome Cyc2 in the outer membrane of *Acidithiobacillus ferrooxidans* (gram-negative bacteria) responsible for oxidation of Fe^{2+} at the external surface of the outer membrane (see Chapter 16) suggests a better model to explain the oxidation of Mn^{2+}. It now seems more likely that in gram-negative bacteria, Mn(II) oxidation by these organisms occurs at the outer surface of their outer membrane (Figure 17.3 B). Indeed, in *Pseudomonas putida* strain MnB1 and GB-1, both gram-negative, the manganese oxidase, a protein complex that includes a multicopper oxidase appears to reside in the outer membrane, and Mn^{2+} oxidation occurs at the cell surface (Okazaki et al., 1997; Brouwers et al., 2000a). The product of Mn(II) oxidation by *P. putida* strain MnB1 has been characterized as a poorly crystalline layer-type oxide with hexagonal symmetry comparable to δ-MnO_2 and a poorly crystalline hexagonal birnessite (Villalobos et al., 2003; see also Toner et al., 2005).

In the case of some gram-positive bacteria examined to date (Ehrlich and Zapkin, 1985), Mn^{2+} may be oxidized on the external surface of the plasma membrane or in the wall region consisting of peptidoglycan and teichoic or teichuronic acid polymer, which Hobot et al. (1984) and Beveridge and Kadurugamuwa (1996) consider as a periplasmic equivalent, or external to it (Ehrlich, 2007).

17.5.2.1 Subgroup Ia

In bacteria of Subgroup Ia, Mn^{2+} oxidation can be coupled to ATP synthesis (energy conservation), as has been demonstrated directly with everted and hybrid membrane vesicles produced by sonication from a gram-negative marine bacterium, strain SSW_{22} (Ehrlich, 1983; Ehrlich and Salerno, 1990), or by uncoupling of ATP synthesis in this organism with 2,4-dinitrophenol (Ehrlich and Salerno, 1988, 1990). Synthesis of ATP in strain SSW_{22} appears to be very tightly coupled to manganese oxidation because added ADP stimulated manganese oxidation by everted and hybrid vesicles from the organism (Ehrlich and Salerno, 1990). A periplasmic factor is required for Mn^{2+} oxidation by membranes from strain SSW_{22} (Clark and Ehrlich, 1988; Clark, 1991). If Mn^{2+} is oxidized by an oxidoreductase at the outer surface of the outer membrane of this gram-negative organism, this periplasmic factor likely functions in an electron transfer process from the outer membrane to the electron transport system (ETS) of the plasma membrane, as illustrated in Figure 17.3B.

The working models in Figures 17.3A and 17.3B assume that chemiosmosis is the underlying energy-transducing mechanism. In the model in Figure 17.3A, the manganese-oxidizing half-reaction occurs external to the plasma membrane in the periplasm. It also assumes that oxygen reduction by electrons removed in the Mn^{2+}-oxidizing half-reaction occurs on the inner surface of the plasma membrane. Proton generation by the manganese-oxidizing half-reaction and proton pumping linked to the passage of electrons from manganese to oxygen via the ETS in the plasma membrane result in a proton gradient across the membrane (outside acidic relative to the cytoplasm). This gradient contributes to the proton motive force (PMF) needed for ATP synthesis via ATP synthase. ADP stimulation of Mn^{2+} oxidation by everted and hybrid membrane vesicles from strain SSW_{22} is explained by a mass action effect involving vectorial and scalar proton consumption in ATP synthesis. The everted and hybrid vesicles appeared to become freely permeable to Mn^{2+} in their preparation.

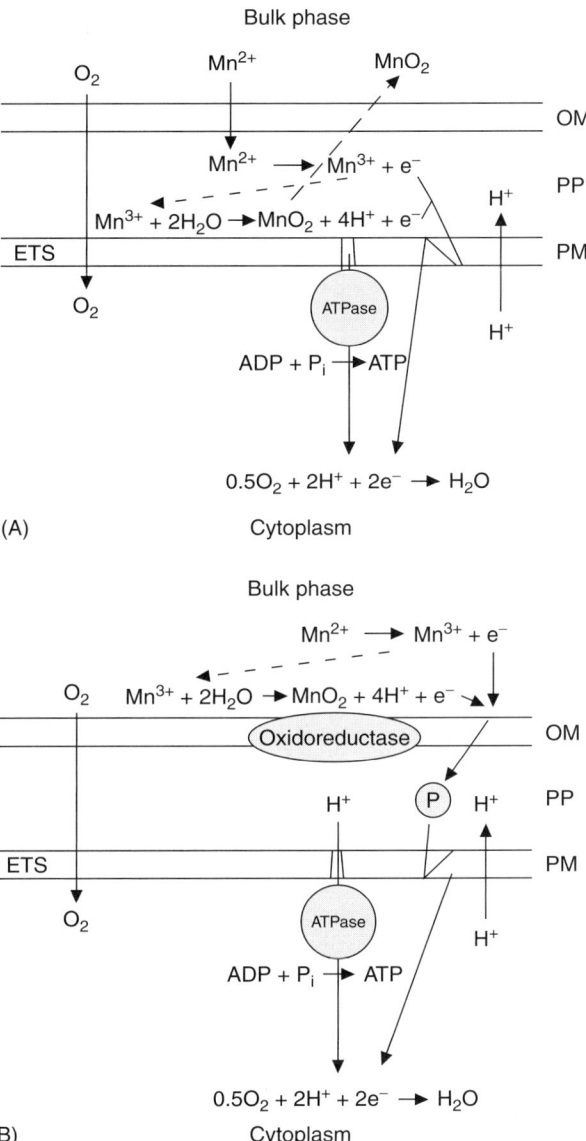

FIGURE 17.3 (A) A model proposed by Ehrlich (2002) of Mn^{2+} oxidation with energy conservation by Group Ia bacteria, in which manganese oxidation occurs in the periplasm (PP) of gram-negative bacteria, whereas oxygen reduction occurs at the inner surface of the plasma membrane (PM). Protons pumped into the PP from the cytosol by components of the electron transport system (ETS) in the PM (details not shown) together with protons from the oxidative half-reaction involving Mn^{2+} contribute to the proton gradient (ΔpH) across the membrane, which together with the transmembrane gradient ($\Delta\Psi$) contribute to ATP synthesis. (B) New proposal for a model of Mn^{2+} oxidation with energy conservation by gram-negative Group Ia bacteria in which manganese oxidation occurs at the outer surface of the outer membrane (OM) at sites where a Mn^{2+}-oxidoreductase complex is located. Electrons removed by the oxidoreductase complex are transferred to the ETS in the PM via one or more electron transfer agents (P) in the periplasm (PP). Oxygen reduction occurs at the inner surface of the PM. Protons pumped into the PP from the cytosol by key components of the ETS in the PM (details not shown) result in the proton gradient (ΔpH) across the membrane, which together with the transmembrane potential ($\Delta\Psi$) contribute to ATP synthesis.

In the model in Figure 17.3B, the Mn^{2+}-oxidizing complex (oxidoreductase) is located in the outer membrane with the reactions directly involved in the Mn^{2+} oxidation taking place on the outer surface of the membrane. This eliminates the necessity for a mechanism to export MnO_2 produced in the Mn^{2+} oxidation from the periplasm to the bulk phase. In the model in Figure 17.3A, it is assumed that the MnO_2 produced in the oxidation in the periplasm is colloidal and somehow passes from the periplasm through the outer membrane into the bulk phase by an unknown mechanism. In the model in Figure 17.3B, the MnO_2 formed is still assumed to be colloidal because when strain SSW_{22} is grown in streaks on a Mn^{2+}-containing agar slants, the MnO_2 formed is not deposited in the streaks but appears in the agar surrounding them (Ehrlich, 1999). In the model in Figure 17.3B, Clark's periplasmic factor (Clark and Ehrlich, 1988; Clark, 1991), as previously mentioned, is viewed as the agent(s) involved in the transfer of the electrons from the enzyme complex at the outer membrane surface to the plasma membrane.

Some gram-positive bacteria have also been found to oxidize Mn^{2+} in a process in which some energy appears to be conserved (Ehrlich and Zapkin, 1985). Because gram-positive bacteria do not possess an outer membrane, the location of the Mn(II) oxidoreductase must be different from that in gram-negative bacteria. A location of the oxidoreductase at the outer surface of the plasma membrane is possible, but it is also possible that the oxidoreductase is located at the surface of the cell wall and that electron-transferring structures within the wall transfer the electrons to the plasma membrane (Ehrlich, 2007). Nothing definitive is known as yet about this aspect of Mn(II)-oxidation by gram-positive bacteria in Group I, Subgroup Ia.

For thermodynamic reasons, the biooxidation of Mn(II) to Mn(IV) in the two models in Figure 17.3 is presented in two half-reactions, with Mn(III) as an intermediate:

$$Mn^{2+} \rightarrow \{Mn^{3+}\} + e \quad (E_0' = 0.84\,V) \tag{17.3a}$$

$$\{Mn^{3+}\} + 2H_2O \rightarrow MnO_2 + 4H^+ + e \quad (E_0' = 0.08\,V) \tag{17.3b}$$

where $\{Mn^{3+}\}$ represents bound Mn(III), probably in an enzyme complex. Some spectroscopic confirmation of such a two-step reaction involving a manganese-oxidizing multicopper oxidase in the exosporium of spores of marine *Bacillus* SG-1 (see Section 17.5.2.4) has been obtained (Webb et al., 2005). Reaction 17.3b is the one that drives ATP synthesis because its E_0' is low enough to permit entry of the electron into the electron transport chain in the plasma membrane at the level of a cytochrome bc_1 complex, as suggested in the experiment. Although oxidation of Mn^{2+} to MnO_2 in a single step (Reaction 17.2) by removal of two electrons is thermodynamically possible ($\Delta G_0' = -16.3$ kcal or -68.1 kJ) (Luther, 2005), its E_0' is too high ($+0.46$ V) to allow for ATP synthesis coupled to electron transport. Experimental evidence for a Mn(III) intermediate in bacterial oxidation of Mn(II) to Mn(IV) was recently demonstrated (Webb et al., 2005). When coupled to the reductive half-reaction for O_2,

$$0.5O_2 + 2H^+ + 2e \rightarrow H_2O \tag{17.3c}$$

the value for $\Delta G_0'$ for the oxidation of Mn^{2+} to $\{Mn^{3+}\}$ is about $+1.21$ kcal ($+5.1$ kJ) and the oxidation of $\{Mn^{3+}\}$ to MnO_2, about -33.8 kcal (-141 kJ) (for further discussion see Ehrlich, 1999).

It is possible that enzymatic oxidation of Mn(II) to Mn(IV) in a single two-electron transfer step (Luther, 2005) is performed by some manganese oxidizers that detoxify Mn(II) by oxidizing it to Mn(IV) without conserving energy from the oxidation. This has yet to be investigated.

On the basis of a report by Kepkay et al. (1984), *Pseudomonas* strain S-36 should be assigned to Subgroup Ia. This organism binds and oxidizes Mn^{2+} and appears to derive energy from the oxidation of manganese. It may also use some of this energy to fix CO_2 (Kepkay and Nealson, 1987). However, Nealson et al. (1988) did not rule out the possibility, even though they

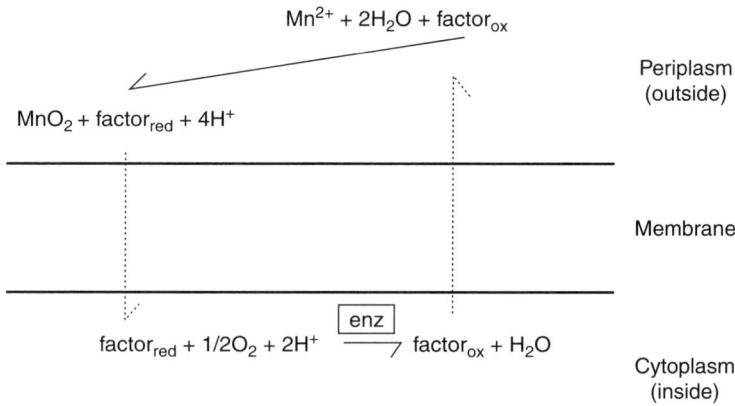

FIGURE 17.4 Current model for manganese oxidation by a Group Ib bacterium. The electron transport system of the plasma membrane is not directly involved. Manganese oxidation is the result of interaction with an as-yet-unidentified factor in the periplasm, which is reduced by Mn^{2+}. The reduced factor is reoxidized in the cytosol. Whether the protons from the manganese oxidation in the periplasm, if formed as shown, contribute to ATP synthesis is not known.

considered it remote, that the manganese stimulated carbon utilization by a mechanism other than oxidation.

17.5.2.2 Subgroup Ib

Subgroup Ib includes bacteria in which manganese oxidation is not directly coupled to ATP synthesis (Figure 17.4). The only documented example is a gram-negative rod (strain FMn1) isolated from a body of freshwater (Tomhannock Reservoir, Troy, New York; Zapkin and Ehrlich, 1983) that is thought to oxidize Mn^{2+} in its periplasm (Zindulis and Ehrlich, 1983), although oxidation at the outer surface of the outer membrane instead cannot be ruled out. The oxidation involves an as-yet-unidentified factor that behaves like an oxidant and not like an enzyme (Zindulis and Ehrlich, 1983). The oxidized form of the factor is reduced by Mn^{2+}, and the reduced factor, after passing into the cytoplasm, is then enzymatically reoxidized with O_2 as terminal electron acceptor. Results from inhibitor studies indicate that the ETS in the plasma membrane is not involved (Zindulis and Ehrlich, 1983).

17.5.2.3 Subgroup Ic

Subgroup Ic includes sheathed bacteria of the genus *Leptothrix*. Early work with *L. discophora* led to contradictory findings. Johnson and Stokes (1966) found that nongrowing (resting) cell suspensions of a strain of *L. discophora* (also called *Sphaerotilus discophorus* in some of the literature) oxidized Mn^{2+} if the culture had been pregrown in a medium supplemented with manganese but not if pregrown in medium without manganese supplementation. A subsequent study with a different strain of *L. discophora*, however, failed to confirm inducibility of manganese-oxidizing ability in the organism (Van Veen, 1972). A difference in the strains might have been the cause of this difference in results. In another strain of *L. discophora*, manganese-oxidizing activity was found to be associated with a particulate cell fraction that sediments at 48,000 × g. It appeared to be coupled to cytochrome oxidase because it was inhibited by 10^{-5} M cyanide and 10^{-4} M azide, but coupling to *b*- and *c*-type cytochrome was not observed (Hogan, 1973; Mills and Randles, 1979). *L. discophora* has been reported to be able to grow autotrophically and mixotrophically on Mn(II) as an energy source (Ali and Stokes, 1971), but these observations were not confirmed (Hajj and Makemson, 1976).

More recent studies with a freshly isolated strain of *Leptothrix discophora* (SS-1; ATCC 43182) called the earlier findings into question. Strain SS-1, which has lost the ability to form a sheath but continues to be able to oxidize Mn^{2+} (Adams and Ghiorse, 1986), releases Mn^{2+}-oxidizing protein, which has a molecular weight of 100,000–110,000, into liquid culture medium. The protein has a polysaccharide moiety linked to it. The manganese-oxidizing activity of the protein exhibits a pH optimum ~7.5 and a temperature optimum of 25°C. The activity is inhibited by cyanide, o-phenanthroline, and $HgCl_2$, but not by azide. The oxidized manganese becomes associated with the protein and can be removed by reduction with ascorbate, thereby restoring the manganese-oxidizing activity of the protein (Adams and Ghiorse, 1987; Boogerd and de Vrind, 1987). Corstjens et al. (1997) identified the *mofA* gene in *L. discophora* SS-1, which codes for a manganese-oxidizing protein that includes copper in its structure and is related to multicopper oxidases. Addition of Cu^{2+} to the culture medium was found to stimulate production of the manganese-oxidizing protein in growing cultures of *L. discophora* SS-1 although it decreased cell yield (Brouwers et al., 2000b). The excretion of a manganese-oxidizing protein by *Leptothrix* spp. was first reported by Mulder (1972) and Van Veen (1972).

Adams and Ghiorse (1988) reported that the excreted protein acts as a catalyst in association with an acidic exopolymer in oxidizing Mn(II) to Mn(III). The formation of Mn in the +3 oxidation state yields only ~10 kcal mol^{-1} at pH 7.0 (assuming hausmannite as the product) under otherwise standard conditions compared to the oxidation of Mn(II) to Mn(IV), which yields ~16 kcal mol^{-1} at pH 7 under otherwise standard conditions. Excess concentrations of Mn^{2+} inhibit the growth of *Leptothrix discophora* SS-1, probably because the manganese oxide that it deposits on the cells in the absence of a sheath deprives it of essential iron by binding it (Adams and Ghiorse, 1985).

17.5.2.4   Subgroup Id

Subgroup Id currently accommodates marine *Bacillus* SG-1 (Nealson and Ford, 1980). The vegetative cells of this organism do not oxidize Mn(II), but its dormant spores do (Rosson and Nealson, 1982; de Vrind et al., 1986b; Francis and Tebo, 1999). The formation of the spores by the vegetative cells of the bacillus is enhanced by solid surfaces in dilute seawater medium (Kepkay and Nealson, 1982). The spores bind and oxidize Mn^{2+} (Rosson and Nealson, 1982). The Mn^{2+} is bound by a protein component in the exosporium (de Vrind et al., 1986b; Francis and Tebo, 1999). The protein is actually a complex of several smaller proteins. The complex has a molecular mass of 205 kDa and may be a glycoprotein (Tebo et al., 1988), but according to Francis and Tebo (1999), these results have been difficult to reproduce. An operon, labeled *mnx*, consisting of genes *mnxA* to *mnxF* on the chromosome of *Bacillus* SG-1, codes for factors required for manganese oxidation by the spores (Van Waasbergen et al., 1993, 1996). The gene *mnxG* codes for a protein that belongs to the multicopper oxidase family. The product of gene *mnxC* may be a redox-active protein involved in the oxidation of Mn^{2+}, but it may also be involved in Cu^{2+} transport into the cell (Tebo et al., 1997). Addition of Cu^{2+} to the growth medium stimulates manganese oxidation by the spores (Van Waasbergen et al., 1996), consistent with the notion that the oxidation involves a multicopper oxidase. The initial product of manganese oxidation by the spores was first thought to be Mn_3O_4 (hausmannite) (Hastings and Emerson, 1986) but was later identified as Mn(IV) oxide (10Å manganate) (Mandernack et al., 1995). The direct formation of Mn(IV) oxide from Mn(II) was confirmed in an application of XANES analysis of the spores when oxidizing Mn^{2+}. Contrary to earlier observations by Bargar et al. (2000), experimental evidence for intermediate Mn(III) formation during Mn(II) oxidation by the spores was obtained by Webb et al. (2005). Francis and Tebo (1999) previously suggest that the Mn_3O_4 detected on spores in early work may have been the product of Mn(II) oxidation catalyzed abiotically by enzymatically formed Mn(IV) oxide on the spore surface and thus would be the product of secondary reaction, previously suggested by de Vrind et al. (1986b) and Mann et al. (1988), but this interpretation is brought into question by the observation of Webb et al. (2005). Because the products of manganese oxidation by the spores remain bound to the spore surface, it remains unclear whether the spores can sustain continuous enzymatic manganese oxidation.

Spores produced by a strain of *Bacillus megaterium* (strain BC1), which was isolated from a microbial mat at Laguna Figueroa, Baja California, can also oxidize Mn^{2+}. They are thus similar to those formed by *Bacillus* SG-1 (Gong-Collins, 1986).

Although vegetative cells of *Bacillus* SG-1 cannot oxidize Mn^{2+}, de Vrind et al. (1986a) discovered that they are able to reduce MnO_2 at low oxygen tension. The electron donor in these organisms is an unidentified intracellular compound; externally supplied glucose or succinate did not act as electron donor. MnO_2 reduction involved a branched, membrane-bound electron transport pathway in which oxygen at normal tension competed with MnO_2 as terminal electron acceptor. The MnO_2-reducing activity in this organism is thought to supply needed Mn^{2+} for sporulation in environments where the supply of Mn^{2+} is limited or absent but where manganese oxide is available (de Vrind et al., 1986a).

The oxidation of Mn(II) by spores but not by vegetative cells of members of the genus *Bacillus* is by no means a universal phenomenon with spore formers. Ehrlich and Zapkin (1985) and Vojak et al. (1984) isolated bacilli whose oxidation of Mn(II) depended on vegetative cells, in the latter instance cells that were in the process of sporulating. These organisms should be assigned to Subgroup Ia based on knowledge to date.

17.5.2.5 Uncertain Subgroup Affiliations

Subgroup assignment of *Pseudomonas putida* strains MnB1 and GB-1, both gram-negative, is problematic. These two cultures do show some affinity to Subgroup Ia based on characterizations to date (DePalma, 1993; Okazaki et al., 1997; figure 7 in Tebo et al., 1997; Brouwers et al., 2000a). However, it is presently unclear whether these strains can conserve energy from manganese oxidation. Their manganese oxidase appears to be a protein complex that includes multicopper oxidase (Brouwers et al., 1999) and appears to reside in the outer membrane because Mn^{2+} oxidation occurs at the cell surface (Okazaki et al., 1997; Brouwers et al., 2000a). *P. putida* strain MnB1 appears to form Mn(III) as an intermediate in its oxidation of Mn(II) to Mn(IV) (Toner et al., 2005).

Similarly problematic is the subgroup assignment of strain SD-21, a marine α-Proteobacterium, isolated from surface sediment in San Francisco Bay, which is able to oxidize Mn(II) to Mn(III,IV) oxides. The Mn(II)-oxidizing activity of this organism has been attributed to two proteins, one 250 kDa and the other 150 kDa (Francis et al., 2001).

Subgroup assignment within Group I of the manganese-oxidizing *Pseudomonas* strain E1 and *Citrobacter freundii* strain E4, isolated from manganese concretions found in alfisol soil in West Peleponnesus, Greece (Douka, 1977, 1980; Douka and Vaziourakis, 1982), is also problematic. It was previously assigned to Subgroup Id (Ehrlich, 1996a), but based on the current definition of the other subgroups, the information on these organisms is insufficient for subgroup assignment. Manganese oxidation by the *Pseudomonas* and *Citrobacter* strains was demonstrated with intact cells and cell extracts. The activity was heat-sensitive and inhibited by $HgCl_2$. Cell extracts exhibited a temperature optimum at 34°C. Whether these bacteria can derive energy from manganese oxidation remains to be determined.

17.5.3 GROUP II MANGANESE OXIDIZERS

Group II of manganese oxidizers includes those bacteria that oxidize Mn^{2+} only when it is prebound to one of several solids external to the cell. Like Group I bacteria, they use oxygen as terminal electron acceptor. At least some bacteria using this mechanism can gain useful energy from the oxidation. The oxidative reaction catalyzed by these bacteria when Mn^{2+} is bound to hydrated Mn(IV) oxide such as $MnO_2 \cdot H_2O$ (H_2MnO_3) can be summarized as follows:

$$MnMnO_3 + 0.5O_2 + 2H_2O \rightarrow 2H_2MnO_3 \qquad (17.4a)$$

The $MnMnO_3$ results from a reaction in which Mn^{2+} displaces the protons of H_2MnO_3:

$$Mn^{2+} + H_2MnO_3 \rightarrow MnMnO_3 + 2H^+ \tag{17.4b}$$

Reaction 17.4a is catalyzed by the bacteria on the surface of the $MnMnO_3$ particles and is the rate-limiting step in the overall oxidation of Mn(II) to Mn(IV) in this process. Reaction 17.4b is very rapid and not catalyzed by the bacteria. It is oxygen-independent. A study by Pecher et al. (2003) offers support for such a reaction sequence in ferromanganese nodule development.

One similarity in the Mn(II) oxidation mechanism of bacteria assigned to Group II organisms and to bacteria assigned to Group I, Subgroup Ia has already been noted. Members of both groups are able to couple ATP synthesis to Mn^{2+} oxidation by using the respiratory chain in their plasma membrane (Ehrlich, 1976; Arcuri and Ehrlich, 1979; Ehrlich and Salerno, 1990). However, Group II organisms oxidize Mn^{2+} only if prebound to solid substrates such as Mn(IV) oxide (see Reaction 17.4b earlier), ferromanganese, or clays such as montmorillonite or kaolinite, but not to illite (Arcuri, 1978; Arcuri and Ehrlich, 1979; Ehrlich, 1982, 1984). In the case of oxidation of Mn^{2+} bound to clays, pretreatment of the clays overnight with ferric chloride in the absence of added Mn^{2+} was essential for whole cells to oxidize manganese(II) that was subsequently allowed to be bound to clays (Ehrlich, 1982). Such ferric-chloride pretreatment was not necessary when cell extracts instead of whole cells were used, but the amount of Mn(II) oxidized in 4 h was less without than with ferric-chloride pretreatment. The iron was not a factor required for bacterial activity because when ferric chloride and manganese(II) were added simultaneously to the reaction mixture, the bacteria did not oxidize the manganese(II). The iron seems to play a steric role in making bound Mn^{2+} more accessible to oxidation by the cells or cell extract.

A model for oxidation of Mn^{2+} by gram-negative Group II organisms originally proposed by Ehrlich (1990, 1996a, 2002) is shown in Figure 17.5A. As in the original model for Mn^{2+} oxidation in Group I, Subgroup Ia organisms (Figure 17.3A), the oxidation in the original model of Group II organisms was previously thought to occur in the periplasm of gram-negative bacteria or the periplasmic equivalent of gram-positive bacteria. The model did not explain how the Mn^{2+} bound to an external solid like MnO_2 with which the cell had to be in contact was transferred into the periplasm for subsequent oxidation. It also did not explain how Mn(IV) oxide formed in the periplasm as a result of Mn^{2+} oxidation was exported into the bulk phase. As previously pointed out in connection with bacterial oxidation of unbound Mn^{2+}, the discovery of cytochrome Cyc2 in the outer membrane of *Acidithiobacillus ferrooxidans* (gram-negative bacteria) responsible for oxidation of Fe^{2+} at the external surface of the outer membrane (see Chapter 16) suggested a better model to explain the oxidation of Mn^{2+} bound to MnO_2, ferromanganese, or certain clays, which is shown in Figure 17.5B. In this model, the need for dissociation of Mn^{2+} from the solid to which it is bound and from which it is transferred into the periplasm is eliminated. The oxidation of the Mn^{2+} takes place external to the periplasm, catalyzed by a specific enzyme complex (oxidoreductase) in the outer membrane whose catalytic site is exposed at the outer surface of the membrane. The electrons removed in the oxidation of bound Mn^{2+} are transferred to a periplasmic factor shown essential to Mn^{2+} oxidation by Group II organisms (Arcuri and Ehrlich, 1979, 1980). This periplasmic factor then transfers the electrons it acquired to the ETS in the plasma membrane of the organisms at an as-yet-unspecified location for transfer to the terminal electron acceptor, O_2, accompanied by chemiosmotic ATP synthesis. The Mn(IV) oxide formed as a result of the oxidation of Mn^{2+} becomes part of the MnO_2 (or the ferromanganese or clay) to which the Mn^{2+} was bound. As in organisms of Group I, Subgroup Ia, the oxidation of Mn^{2+} to Mn(IV) oxide is assumed to proceed in two steps, of which only the second step yields enough energy to be conserved.

Like Group I, Subgroup Ia, Group II includes gram-positive bacteria (e.g., *Arthrobacter* 37 (Ehrlich, 1963, 1966, 1968)) that can oxidize Mn(II), presumably with an ability to conserve some of the energy from this oxidation. Unlike gram-positive organisms in Group I, Subgroup Ia, the location of the Mn(II) oxidoreductase is less likely the plasma membrane and more likely the surface of the

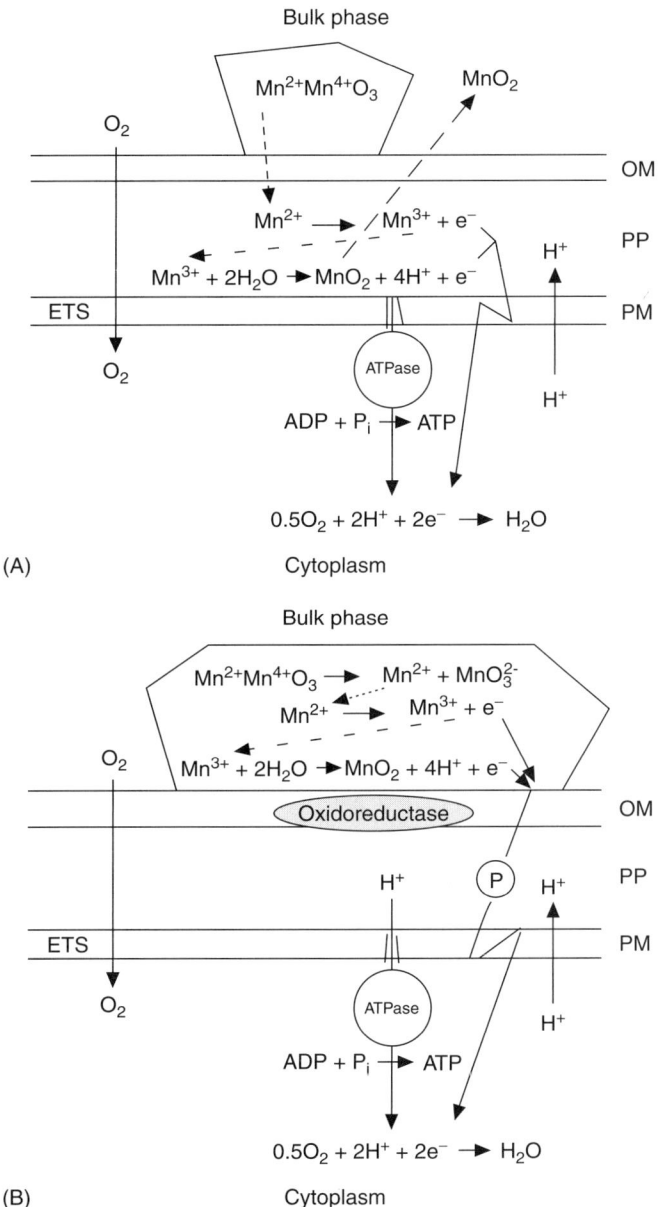

FIGURE 17.5 (A) A model proposed by Ehrlich (2002) of oxidation of Mn(II) bound to manganese(IV) oxide (MnO_2) with energy conservation by Group II bacteria. In this model, Mn^{2+} oxidation occurs in the periplasm (PP) of gram-negative bacteria and oxygen reduction occurs at the inner surface of the plasma membrane (PM). Note that this model is similar to that for Group Ia bacteria (Figure 17.3A), except that the Mn^{2+} derives from the Mn(II) bound to Mn(IV) oxide with which the cell is in direct contact. The mechanism by which the Mn^{2+} enters the PP is unspecified. (B) New proposal for a model of oxidation of Mn^{2+} bound to MnO_2 ($Mn^{+2}Mn^{+4}O_3$) with energy conservation by gram-negative Group II bacteria in which oxidation of the bound Mn^{2+} occurs at the outer surface of the outer membrane (OM) at sites where a Mn^{2+}-oxidoreductase capable of oxidizing bound Mn^{2+} in direct contact with the $Mn^{+2}Mn^{+4}O_3$ particles is located. Electrons removed from the bound Mn^{2+} are transferred across the PP by electron transfer agents (P) to the electron transport system (ETS) in the PM. Oxygen reduction occurs at the inner surface of the PM. Protons pumped into the PP from the cytosol by key components of the ETS in the PM (details not shown) result in the proton gradient (ΔpH) across the membrane, which together with the transmembrane potential ($\Delta\Psi$) contribute to ATP synthesis.

cell wall because the Mn^{2+} in these instances is bound to MnO_2 or ferromanganese (Mn^{2+} bound to clay has not yet been tested with these gram-positive organisms). Electron-conducting components within the cell wall, if they exist, may transfer the electron to the plasma membrane (see discussion by Ehrlich, 2007), but no evidence in support of such a mechanism exists at this time.

17.5.4 GROUP III MANGANESE OXIDIZERS

Group III manganese oxidizers include those bacteria that oxidize dissolved Mn(II) with H_2O_2 using catalase as the enzyme that promotes this reaction. The reaction can be summarized as follows:

$$Mn^{2+} + H_2O_2 \rightarrow MnO_2 + 2H^+ \tag{17.5}$$

Group III bacteria so far include *Metallogenium*, *Leptothrix pseudoochracea*, and *Siderobacter capsulatus* (Dubinina, 1978a). These organisms appear to generate H_2O_2 faster in their metabolism than their catalase can destroy it by the reaction

$$2H_2O_2 \rightarrow 2H_2O + O_2 \tag{17.6}$$

In the presence of Mn^{2+} at neutral pH, their catalase can function as a peroxidase and use H_2O_2 as an oxidant to generate MnO_2 according to Reaction 17.5 (Dubinina, 1978b). The reaction can be reproduced in the laboratory by generating H_2O_2 from glucose oxidation by glucose oxidase and then causing the H_2O_2 to oxidize Mn^{2+} with a commercial preparation of catalase (Dubinina, 1978a). The oxidation product was identified as a Mn(IV) compound, but because it was complexed with pyrophosphate it could have been a Mn(III) compound. Fe^{2+} can replace Mn^{2+} in this reaction (Dubinina, 1978b). At acid pH, H_2O_2 can reduce MnO_2 to Mn^{2+} chemically according to the reaction

$$MnO_2 + H_2O_2 + 2H^+ \rightarrow Mn^{2+} + 2H_2O + O_2 \tag{17.7}$$

This reaction proceeds without catalase.

17.5.5 NONENZYMATIC MANGANESE OXIDATION

Autoxidation of Mn^{2+} is promoted when environmental conditions feature an E_h greater than +500 mV, a pH >8, and a Mn^{2+} concentration >0.01 ppm. These pH and E_h limits are much narrower than those required for autoxidation of iron (see Chapter 16). Nonenzymatic manganese oxidation may also be promoted through production of one or more metabolic end products, which cause chemical oxidation of Mn^{2+}. According to Soehngen (1914), a large number of microorganisms can cause such manganese oxidation in the presence of hydroxycarboxylic acids such as citrate, lactate, malate, gluconate, or tartrate. He indicated that metabolic utilization of a part of such acids, or more exactly their salts, causes a rise in pH of the culture medium (if unbuffered), and that when the pH turns alkaline, residual hydroxycarboxylic acid catalyzes the oxidation of Mn^{2+}. Hydroxycarboxylic acids can be formed by both bacteria and fungi. In apparent agreement, Van Veen (1973) found that with *Arthrobacter* 216, hydroxycarboxylic acids are required for Mn^{2+} oxidation. However, he felt that other microorganisms with more specific manganese-oxidizing capacity play a more important role in manganese oxidation in soil.

Another example of nonenzymatic manganese oxidation is furnished by an actinomycete, *Streptomyces* sp., found in Australian soil. In the laboratory it causes Mn^{2+} to be oxidized in a pH range of 5–6.5 when growing in soil agar (Bromfield, 1978, 1979). The actinomycete produces a water-soluble extracellular compound that is responsible for the oxidation. The oxidized manganese is thought to protect the organism from inhibition by Mn^{2+} ions (Bromfield, 1978).

Pseudomonas manganoxydans has been reported to form a Mn^{2+}-oxidizing protein that oxidizes Mn(II) intracellularly and is consumed in the process (Jung and Schweisfurth, 1979). On the basis of this observation, the authors concluded that the protein is not an enzyme. They also found that this oxidation did not require oxygen. It proceeded optimally at pH 7.0. The formation of the protein depended on cessation of growth by the culture at the end of its exponential growth phase but did not require added Mn^{2+} in the medium to stimulate its synthesis (Jung and Schweisfurth, 1979). Later, study of this culture by DePalma (1993) and Okazaki et al. (1997) led to its reclassification as *P. putida* strain MnB1. They also demonstrated that the strain did possess a manganese-oxidizing enzyme system (see Section 17.5.2.5).

Zygospores of the alga *Chlamydomonas* from soil have been reported to become encrusted with Mn(IV) oxide, presumably through oxidation of Mn^{2+} (Schulz-Baldes and Lewin, 1975), but how manganese came to be oxidized was not explained. It could have been a case of passive accumulation of manganese oxide (see Section 17.7).

17.6 BIOREDUCTION OF MANGANESE

Microbial reduction of oxidized manganese can be enzymatic or nonenzymatic, as in the case of microbial oxidation of manganese. The occurrence of microbial reduction of manganese oxide was suggested as far back as the 1890s by Adeny (1894), who found that manganese dioxide, formed when permanganate was added to sewage, was reduced to manganous carbonate. He attributed this reduction to the action of bacteria and thought it analogous to bacterial nitrate reduction. Mann and Quastel (1946) reported the microbial reduction of biogenically formed manganese oxide in a soil-perfusion column when it was fed with glucose solution. They were able to inhibit this reduction by feeding the column with azide. They suggested that the oxides of manganese acted as hydrogen acceptors. Troshanov (1968) isolated a number of Mn(IV)-reducing bacteria from reduced horizons of several Karalian lakes (former USSR). His isolates included *Bacillus circulans, B. polymyxa, B. mesentericus, B. mycoides, B. cereus, B. centrosporus, B. filaris, Pseudomonas liquefaciens,* and *Micrococcus* sp., among others.

Some of these strains were able to reduce both manganese and iron oxides, but strains that reduced only manganese oxides were encountered most frequently. Nitrate did not inhibit Mn(IV) reduction the way it did Fe(III) reduction by *Bacillus circulans, B. polymyxa, B mesentericus,* and *Pseudomonas liquefaciens* (Troshanov, 1969). Anaerobic conditions stimulated Mn(IV) reduction by *B. centrosporus, B. mycoides, B. filaris,* and *B. polymyxa,* but not by *B. circulans, B. mesentericus,* or *Micrococcus albus.* Not all carbohydrates tested were equally good sources of reducing power for a given organism (Troshanov, 1969). Although a significant part of the Mn(IV) seemed to be reduced enzymatically, some of it appeared to be reduced chemically because some reduction of oxidized manganese in lake ores by glucose and xylose and some organic acids (e.g., acetic and butyric acids) was observed at pH 4.3–4.4 under sterile conditions (Troshanov, 1968). Such reaction of glucose and other organic reducing agents with Mn(IV) oxides is, however, dependent on the form of the oxide; hydrous Mn(IV) oxides react readily, whereas crystalline MnO_2 does not (see Ehrlich, 1984; Nealson and Saffarini, 1994).

Bromfield and David (1976) reported that *Arthrobacter* strain B, which can oxidize Mn(II) aerobically, is able to reduce manganese oxides at lowered oxygen tension. Ehrlich (1988) reported that several gram-negative bacteria that can oxidize Mn(II) and thiosulfate aerobically and reduce tetrathionate anaerobically (Tuttle and Ehrlich, 1986) can also reduce MnO_2 aerobically and anaerobically.

Clearly, a variety of microbes have the capacity to reduce oxidized manganese. Some bacteria can reduce Mn(IV) oxide aerobically and anaerobically (Ehrlich, 1966; Di-Ruggiero and Gounot, 1990; Troshanov, 1968). Others reduce it only anaerobically (Burdige and Nealson, 1985; Lovley and Phillips, 1988a; Myers and Nealson, 1988a). Among these latter bacteria, some, like *Shewanella* (formerly *Alteromonas, Pseudomonas*) *putrefaciens,* are facultative but nevertheless reduce Mn(IV) oxide only anaerobically (Myers and Nealson, 1988a), whereas others, like *Geobacter metallire-ducens* (formerly strain GS-15), are strict anaerobes and reduce MnO_2 only anaerobically (Lovley and Phillips, 1988a). Only one member of the Archaea has so far been found to reduce Mn(IV) oxide, namely, *Pyrobaculum islandicum* (Kashefi and Lovley, 2000).

17.6.1 ORGANISMS CAPABLE OF REDUCING MANGANESE OXIDES ONLY ANAEROBICALLY

Bacterial reduction of manganese oxides, such as MnO_2, that proceeds only anaerobically has been well documented (Burdige and Nealson, 1985; Lovley and Phillips 1988a; Lovley and Goodwin 1988; Myers and Nealson, 1988a; Lovley et al., 1989, 1993b). Organisms that are facultative aerobes but that reduce Mn(IV) oxide only anaerobically in an anaerobic respiratory process include *Shewanella* (formerly *Alteromonas, Pseudomonas*) *oneidensis* (formerly *putrefaciens*) MR-1 (Myers and Nealson, 1988a), *Bacillus polymyxa* D1 and *Bacillus* MBX1 (Rusin et al., 1991a,b), and *Pantoea agglomerans* SP1 (Francis et al., 2000). Organisms that are strict anaerobes and reduce Mn(IV) oxides in an anaerobic respiratory process include *Geobacter metallireducens* (formerly strain GS-15) (Lovley and Phillips, 1988b), *G. sulfurreducens* (Caccavo et al., 1994), *Geothrix fermentans* (Coates et al., 1999), *Pyrobaculum islandicum* (Kashefi and Lovley, 2000), *Sulfurospirillum barnesii* SES-3 (Laverman et al., 1995), *Desulfovibrio desulfuricans, Desulfomicrobium baculatum, Desulfobacterium autotrophicum*, and *Desulfuromonas acetoxidans* (Lovley and Phillips, 1994a). All but one of these organisms, *P. islandicum*, belong to the domain Bacteria. All are quite versatile in regard to the electron donors that they can utilize for MnO_2 reduction. For instance, *G. metalli-reducens* can use butyrate, propionate, lactate, succinate and acetate, and several other compounds, which are completely oxidized to CO_2 (Lovley and Phillips, 1988a, Lovley, 1991; Myers et al., 1994; Langenhoff et al., 1997). *S. putrefaciens*, on the other hand, can utilize lactate and pyruvate as electron donors but oxidizes these only to acetate. It can, however, use H_2 and formate as electron donors whereas *G. metallireducens* cannot (Lovley et al., 1989). Unlike *G. metallireducens, G. sul-furreducens* can use H_2 as electron donor (Caccavo et al., 1994). As described in Chapter 16, these organisms are also not restricted to MnO_2 as terminal electron acceptor anaerobically. Depending on which organism is considered, Fe(III), uranyl ion, chromate, nitrate, iodate, elemental sulfur, sulfite, sulfate, thiosulfate, fumarate, and glycine may serve. In the case of *S. putrefaciens*, oxygen, nitrate, and fumarate inhibited MnO_2 reduction, but sulfate, sulfite, molybdate, nitrite, or tungstate did not (Myers and Nealson, 1988a).

Anaerobic utilization of acetate as electron donor in reduction of MnO_2 is proof that the process is a form of anaerobic respiration. This is because acetate is unfermentable except in the special case of acetotrophic methanogenesis. Because acetate can also be respired in air where oxygen can serve as terminal electron acceptor, it is interesting to note that gram-negative marine strain SSW_{22} can reduce MnO_2 with acetate, succinate, and glucose aerobically as well as anaerobically to the same degree (Ehrlich, 1988; Ehrlich et al., 1991; Ehrlich, unpublished data). This raises a question about differences in electron transport pathways between those organisms in which electron transport from acetate to MnO_2 is blocked by oxygen and those in which it is not. In at least some instances it may have to do with the nature of the manganese(IV) reductase, or genetic control (repression). As discussed in Section 17.6.2, in marine pseudomonad BIII 88, which can reduce MnO_2 aerobically and anaerobically, the manganese reductase is postulated to act on Mn(III) produced by reaction between cell-surface-bound Mn^{2+} and MnO_2 with which the cell surface is in physical contact. In *Shewanella oneidensis* MR-1, *c*-type cytochrome that occurs in the outer membrane of anaerobically but not aerobically grown cells is involved in MnO_2 reduction as it is in dissimilatory Fe(III) reduction (Myers and Myers, 1992, 1993, 2000). Indeed, Myers and Myers (2001) have presented strong evidence that cytochromes OcmA (Myers and Myers, 1997, 1998) and OcmB in the outer membrane of this organism are required for reduction of MnO_2 but not for alternative *soluble* electron acceptors (see also discussion in Chapter 16). Direct evidence of electron transfer under anaerobic conditions between a putative 150 kD protein in the outer membrane of *S. oneidensis*

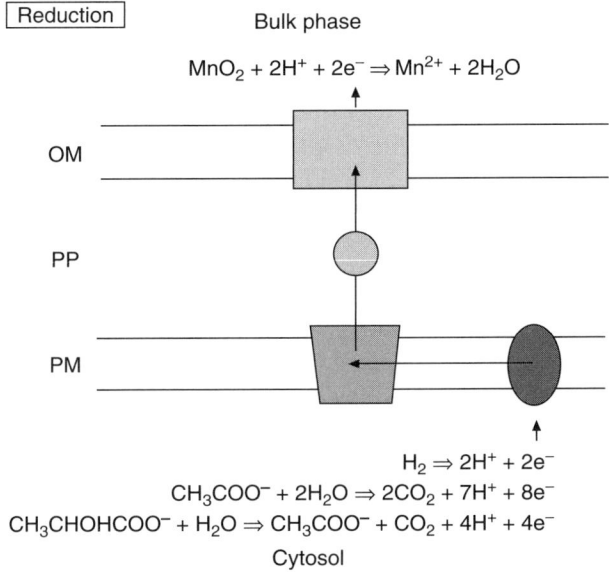

| Reduction | Bulk phase |

$$MnO_2 + 2H^+ + 2e^- \Rightarrow Mn^{2+} + 2H_2O$$

OM

PP

PM

$$H_2 \Rightarrow 2H^+ + 2e^-$$
$$CH_3COO^- + 2H_2O \Rightarrow 2CO_2 + 7H^+ + 8e^-$$
$$CH_3CHOHCOO^- + H_2O \Rightarrow CH_3COO^- + CO_2 + 4H^+ + 4e^-$$

Cytosol

FIGURE 17.6 Path of electrons from H_2, acetate, or lactate to MnO_2 in *Geobacter sulfurreducens*, *Shewanella oneidensis*, and some other anaerobically respiring gram-negative bacteria. OM, outer membrane; PP, periplasm; PM, plasma membrane. The oval in PM represents complex I. The trapezoid in PM represents complex III. The circle in the PP and rectangle in the OM represent different *c*-type cytochromes. The arrows indicate the direction of electron flow.

MR-1 and the surface of goethite (α-FeOOH) with which it is in contact has been reported by Lower et al. (2001). Indeed, the complexity of electron transfer in anaerobic respiration on insoluble metal oxides has become very apparent from a recent study with *S. oneidensis* MR-1 (Bretschger et al., 2007). Cytochrome c_3 may be involved in MnO_2 reduction by *Desulfovibrio* spp., as it was shown to be in the reduction of chromate (Lovley and Phillips, 1994b) and of uranium(VI) (Lovley et al., 1993a). A special sensitivity of certain reduced cytochromes to autoxidation may contribute to an explanation, at least in part, of why some facultative bacteria can reduce MnO_2 only anaerobically.

As to how electrons may be transferred from the *c*-type cytochrome enzyme complex in the outer membrane of gram-negative organisms to MnO_2 or other Mn(IV) oxide minerals, the same considerations as for transfer of electrons from the cell-surface to Fe(III) oxides apply (see discussion in Chapter 16, Section 16.6).

Figure. 17.6 shows a general model for the electron transport pathway to MnO_2 in gram-negative bacteria like the obligate anaerobes *Geobacter* spp. and *Shewanella oneidensis* MR-1, which are capable of dissimilatory MnO_2 reduction, based on currently available information.

17.6.2 REDUCTION OF MANGANESE OXIDES BY ORGANISMS CAPABLE OF REDUCING MANGANESE OXIDES AEROBICALLY AND ANAEROBICALLY

A number of different bacterial cultures have been isolated from marine and freshwater environments that can reduce Mn(IV) oxides aerobically and anaerobically. As in the case of enzymatic Mn(II) oxidation by bacteria, all the marine cultures studied to date do not employ the same electron transport pathway in Mn(IV) reduction (Ehrlich, 1980, 1993b). Less is known about the pathways of Mn(IV) reduction under aerobic and anaerobic conditions by freshwater isolates than by the marine isolates. Growing cultures of marine *Bacillus* 29 are able to reduce MnO_2 aerobically and anaerobically using metabolites formed from glucose as electron donors. However, a component of the ETS involved in MnO_2 reduction in this culture is only aerobically inducible (Trimble and Ehrlich, 1968). Both Mn^{2+} and MnO_2 can serve as inducers. Because of the aerobic requirement for induction, uninduced *Bacillus* 29 cannot reduce MnO_2 anaerobically unless an artificial electron carrier such as ferricyanide is added (Ehrlich, 1966; Trimble and Ehrlich, 1968, 1970). Once initiated by a growing culture of *Bacillus* 29, MnO_2 reduction is quickly inhibited by addition of $HgCl_2$ to a final concentration of 10^{-3} M. During MnO_2 reduction with glucose as electron donor, the concentrations of pyruvate and lactate formed by the organism from glucose were greater in the presence than in the absence of MnO_2 by day 11 (Table 17.4). However, calculations from the data in Table 17.4 show that on the basis of 100 nmol of glucose consumed, *Bacillus* 29 formed 7.9 nmol pyruvate with MnO_2 but only 3.1 nmol without it by day 11. Yet, it formed 10.4 nmol lactate with MnO_2 and 24.2 nmol without it per 100 nmol glucose consumed in the same duration. The organism consumed more glucose in the presence of MnO_2 than in its absence (Table 17.4). Glucose consumption and total product formation qualitatively similar to those with growing cultures of *Bacillus* 29 occurred with marine coccus 32 (Table 17.4) (Trimble and Ehrlich, 1968).

Aerobic MnO_2 reduction has also been studied with resting cell cultures and with cell extracts. In the case of *Bacillus* 29, uninduced resting cells were able to reduce MnO_2 only in the presence of dilute ferri- or ferrocyanide, whereas induced resting cells were able to reduce MnO_2 without ferri- or ferrocyanide. The ferri- or ferrocyanide was substituted for a missing component in the ETS to MnO_2 in uninduced cells (Ehrlich, 1966; Trimble and Ehrlich, 1970; Ghiorse and Ehrlich, 1976). Substrate level concentrations of ferrocyanide can serve as electron donor in place of glucose in the reduction of MnO_2 by *Bacillus* 29 (Ghiorse and Ehrlich, 1976). Mann and Quastel (1946) had previously observed that pyocyanin could serve as electron carrier in MnO_2 reduction by bacteria, but they believed that such a carrier was essential for such reduction to occur.

Induction of the missing electron transport component in *Bacillus* 29 involves protein synthesis because inhibitors of protein synthesis prevent its induction by manganese (Trimble and Ehrlich, 1970). Reduction of MnO_2 by cell extracts from *Bacillus* 29 proceeds in the same way as with resting cells (Ghiorse and Ehrlich, 1976). Extracts from cells previously grown without added $MnSO_4$ do not reduce MnO_2 in the absence of added small amounts of ferri- or ferrocyanide, whereas extracts from cells previously grown in the presence of added $MnSO_4$ do. Results from spectrophotometric and inhibitor studies of the electron transport components in the plasma membrane of *Bacillus* 29 led initially to the inference that electron transfer from electron donors to MnO_2 in induced cells involved a branched pathway. It included flavoproteins, *c*-type cytochrome, and two metalloproteins, one an MnO_2 reductase and the other a conventional oxidase (Figure 17.7A). The existence of two metalloproteins was initially postulated because MnO_2-reducing activity was stimulated by sodium azide at 1.0 and 10 mM concentrations. The oxidase was presumed to be more azide sensitive than the MnO_2 reductase (Ghiorse and Ehrlich, 1976). Subsequent observations, however, led to the conclusion that the azide stimulation in *Bacillus* 29 was the result of inhibition of catalase activity with a resultant reduction of some MnO_2 by accumulating H_2O_2 (see Reaction 17.7; Figure 17.7B) (Ghiorse, 1988). This reduction by metabolically produced H_2O_2 is similar to that previously reported by Dubinina (1978a,b) with *Leptothrix pseudoochracea*, *Arthrobacter siderocapsulatus*, and *Metallogenium*.

The organization of the MnO_2-reducing system in *Bacillus* 29 is far from universal in manganese(IV)-reducing bacteria. In other marine MnO_2-reducing bacteria so far examined, electron transport appears to involve more conventional electron transport carriers and enzymes, as suggested by electron transport inhibitor studies. These inhibitor studies, moreover, indicated that the pathway appears not to be exactly the same in each organism tested (Ehrlich, 1980, 1993b). H_2O_2 appears to play no significant role in MnO_2 reduction by these organisms because 1 mM azide did not stimulate MnO_2 reduction. Ghiorse (1988) suggested that in *Bacillus* 29, MnO_2 reduction may serve as a means of disposing of excess reducing power without energy coupling. In strains involving more conventional electron transport components for the reduction of MnO_2, such reduction in air may be a source of supplement energy.

TABLE 17.4
Chemical and Physical Changes during Growth[a] of *Bacillus* 29 and Coccus 32 in the Presence and Absence of MnO_2

Culture	Time (days)	Glucose Consumed (nmol mL^{-1})		Manganese Released (nmol mL^{-1})		Pyruvate Produced (nmol mL^{-1})		Lactate Produced (nmol mL^{-1})		E_h (mV)		pH		Cells (mL^{-1}) ($\times 10^7$)	
		$+MnO_2$	$-MnO_2$	$-MnO_2$	$+MnO_2$	$+MnO_2$	$-MnO_2$	$+MnO_2$	$-MnO_2$	$+MnO_2$	$-MnO_2$	$+MnO_2$	$-MnO_2$	$+MnO_2$	$-MnO_2$
29	0	0	0		0	0	0	0	0	469	474	7.3	7.2	1.6	1.6
	1	1720	655		25	43	32	461	449	528	482	6.6	6.7	3.8	4.0
	3	3060	1239		240	190	153	1635	1612	502	512	5.4	5.6	4.5	4.8
	4	3686	2342		420	438	330	1430	1955	573	514	5.7	5.5	4.5	5.0
	7	6120	3635		1060	705	243	2080	1387	615	594	5.7	5.5	4.6	5.2
	8	8750	5286		1260	879	330	2040	1705	617	589	5.5	4.9	4.8	5.0
	11	11893	7281		1830	936	225	2430	1761	602	589	5.8	5.1	5.0	5.0
32	0	0	0		0	0	0	0	0	426	426	7.7	7.7	2.5	2.5
	1	480	0		0	50	52	0	0	394	419	7.0	6.9	4.6	5.8
	2	1440	1440		0	84	75	9	5	519	490	6.8	6.5	6.2	7.0
	3		130		130	138	79			491	496	6.6	6.4	6.2	7.0
	7	3000	2650		490	1025	575	174	68	484	495	5.7	4.7	6.0	4.8
	9	4900	3125		780	1650	980	250	124	514	549	5.4	4.5	5.8	4.2
	11	5550	3600		1020	2625	1120	250	123	475	509	5.6	4.3	5.4	4.8

[a] Culture medium: For *Bacillus* 29, 0.48% glucose and 0.048% peptone in seawater; for coccus 32, 0.60% glucose and 0.048% peptone in seawater. One gram portions of reagent grade MnO_2 were added to 20 mL of medium as needed. Incubation was at 25°C.

Source: From Trimble RB, Ehrlich HL, *Appl. Microbiol.*, 16, 695–702, 1968. With permission.

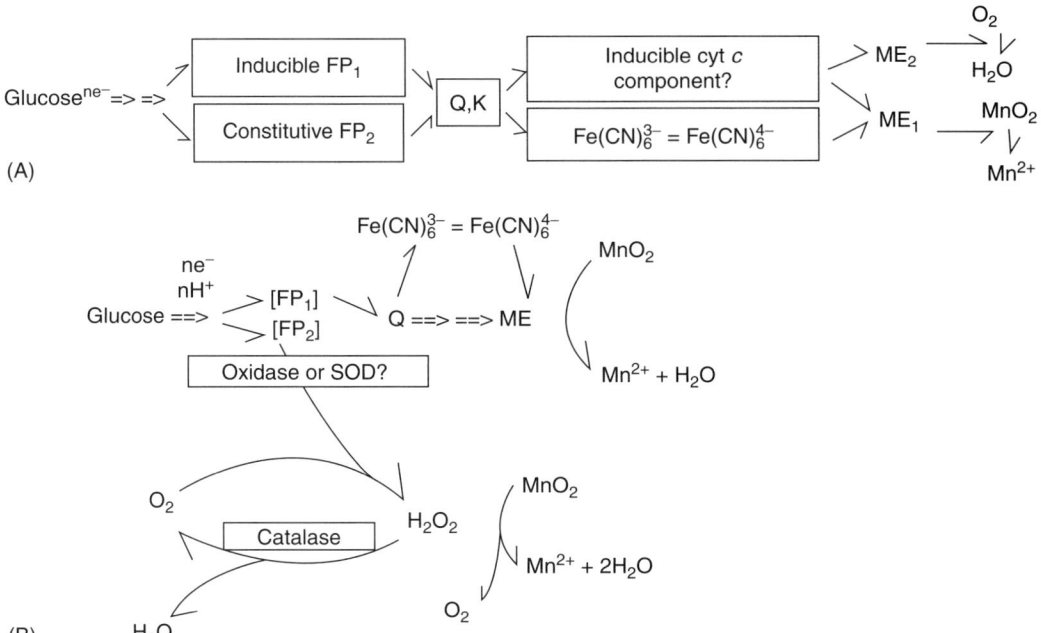

FIGURE 17.7 Electron transfer pathways in MnO_2 reduction by *Bacillus* 29. (A) Original proposal. FP_1, flavoprotein 1; FP_2, flavoprotein 2; Q,K, coenzyme Q; ME_1, metalloprotein 1 (MnO_2 reductase); ME_2, metalloprotein 2 (an oxidase such as cytochrome oxidase). (Redrawn from Ghiorse WC, Ehrlich HL, *Appl. Environ. Microbiol.*, 31, 977–985, 1976. With permission.) (B) Amended proposal. FP_1, flavoprotein; FP_2, glucose oxidase; SOD, superoxide dismutase (converts superoxide radicals formed during oxygen reduction to H_2O_2 and O_2); Q, coenzyme Q; ME, metalloprotein (MnO_2 reductase). (Modified from Ghiorse WC, *Biology of Anaerobic Microorganisms*, Wiley, New York, 305–331, 1988.)

The question of how MnO_2 can partially replace oxygen as terminal electron acceptor for some aerobic bacteria needs to be answered. *Bacillus* 29 was found to consume less oxygen in the presence of MnO_2 than in its absence when respiring on glucose (Trimble, 1968; Ehrlich, 1981). MnO_2 reduction in *Bacillus* GJ33 was accelerated in rotary-shake culture in air at 200 rpm but depressed at 300 rpm (Ehrlich, 1988). This can be explained thermodynamically (Ehrlich, 1987). Considering that the standard free energy change at pH 7.0 ($\Delta G^{0'}$) for the half-reaction

$$MnO_2 + 4H^+ + 2e \rightarrow Mn^{2+} + 2H_2O \qquad (17.8)$$

is only -18.5 kcal (-77.3 kJ), and that of the half-reaction

$$0.5O_2 + 2H^+ + 2e \rightarrow H_2O \qquad (17.9)$$

at pH 7.0 it is -37.6 kcal (-157 kJ), it might be concluded that oxygen should be the better electron acceptor and therefore MnO_2 should not depress its consumption. However, the free energy values above are for standard conditions at pH 7.0 (1 M solute concentrations, except for H^+, and 1 atm for gases). These conditions do not apply to laboratory experiments or in nature. Oxygen concentration in pure water at atmospheric pressure and 25°C is $<10^{-2.89}$ M, whereas the concentration of Mn(IV) in MnO_2 that the bacteria encounter suspended in water is orders of magnitude >1 M because MnO_2 is insoluble in water. Because of its insolubility, the bacteria must be in physical contact with the MnO_2 (see discussion by Ghiorse, 1988). The specific mechanism by which *aerobic* bacteria that respire with MnO_2 as one of the terminal electron acceptors are able to transfer electrons to insoluble MnO_2 has still not been unraveled. A proposal for one possible mechanism is described later.

The ability to reduce MnO_2 was found to be inducible in all marine isolates able to carry out the reduction aerobically and anaerobically, as tested by Ehrlich (1973). Despite differences in the electron transport pathway from donor to recipient, the overall reaction involving MnO_2 reduction appears to be similar in all marine cultures studied by Ehrlich and collaborators (Ehrlich et al., 1972; Ehrlich, 1973). Considering glucose as electron donor, the overall reaction may be summarized as follows:

$$\text{Glucose} \xrightarrow{\text{bacteria}} ne^- + nH^+ + \text{end products} \tag{17.10}$$

$$0.5nMnO_2 + ne^- + nH^+ \xrightarrow[\text{or uninduced bacteria + ferri- or ferrocyanide}]{\text{induced bacteria}} 0.5nMn(OH)_2 \tag{17.11}$$

$$0.5nMn(OH)_2 + nH^+ \rightarrow 0.5nMn^{2+} + nH_2O \tag{17.12}$$

The reason for representing the direct product of MnO_2 reduction by the bacteria as $Mn(OH)_2$ rather than Mn^{2+} is that in resting-cell experiments or in experiments with cell extract that were run for only 3–4 h, it was necessary to acidify the reaction mixture to about pH 2 on completion of incubation to bring Mn^{2+} into solution. No manganese was solubilized by acidification at time 0 from the MnO_2 employed in these experiments. The acidification was not necessary in growth experiments in which acid production from glucose by the bacteria or complexation by medium constituents brought Mn^{2+} into solution (Trimble and Ehrlich, 1968).

The enzyme directly responsible for MnO_2 reduction by organisms that do it aerobically and anaerobically has so far not been isolated and characterized. In the case of *Acinetobacter calcoaceticus*, assimilatory nitrate reductase has been implicated (Karavaiko et al., 1986). MnO_2 reduction by cell extract from this organism was found to be inhibited by NH_4^+ and NO_3^-. The reduction was stimulated if the cells had been cultured in the presence of added molybdenum or if molybdate was added to the reaction mixture in which MnO_2 was being reduced with extract from cells grown in medium without added molybdenum (Karavaiko et al., 1986). MnO_2-reducing activity by *A. calcoaceticus* appears to be assisted by the simultaneous production of organic acids by the organism (Yurchenko et al., 1987).

Ehrlich (1993a,b) has proposed a model (Figure 17.8) to explain how gram negative marine pseudomonad strain BIII 88 is able to reduce insoluble MnO_2 either aerobically or anaerobically using acetate as electron donor. This model requires that the bacteria be in direct contact with MnO_2 particles. Mn^{2+} bound in the outer cell membrane enters into a disproportionation reaction with MnO_2 at the site of attachment to the particle (Reaction 17.1). The oxidation state of the bound Mn^{2+} is thereby raised to Mn^{3+} and that of the reacting Mn(IV) is lowered to Mn^{3+}. Reducing power removed by the bacteria from the electron donors (acetate in the example) is transferred by enzymes and electron carriers to the Mn^{3+} in the outer cell membrane, reducing it to Mn^{2+}. Most of this reduced Mn^{2+} is released into the bulk phase, but some remains bound in the outer cell membrane to continue the disproportionation reaction between MnO_2 of the particle and the Mn^{2+} bound in the outer cell membrane. The model is based on the following experimental observations with marine pseudomonad BIII 88 (Ehrlich, 1993a).

The plasma membrane of marine pseudomonad BIII 88 contains a respiratory system that includes *b* and *c* cytochromes. Aerobically, MnO_2 reduction with acetate by intact cells was stimulated by electron transport inhibitors antimycin A and 2-heptyl-4-hydroxyquinoline *N*-oxide (HQNO), suggesting a branched respiratory pathway to the terminal electron acceptors MnO_2 and O_2, the branch to O_2 being more sensitive to the inhibitors than the branch to MnO_2. Anaerobically, both inhibitors caused a decrease in MnO_2 reduction. Oxidative phosphorylation uncouplers 2,4-dinitrophenol

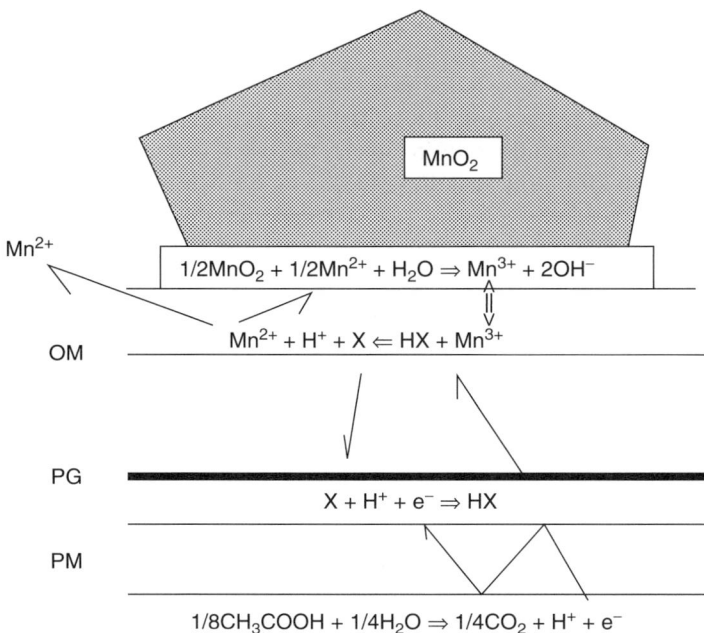

FIGURE 17.8 Schematic representation of a model explaining the transfer of reducing power (electrons) across the interface between the cell surface of marine pseudomonad BIII 88 and the surface of an MnO_2 particle with which the bacterium is in physical contact. OM, outer cell membrane; PG, peptidoglycan layer; PM, plasma membrane; X, hypothetical carrier of reducing power in the cell envelope. (From Ehrlich HL, *Biohydrometallurgical Technologies,* Vol II, The Minerals, Metals and Materials Society, Warrendale, PA, 1993b. With permission.)

and carbonyl cyanide *m*-chlorophenylhydrazone (CCCP) stimulated MnO_2 reduction at critical concentrations aerobically and anaerobically, indicating that energy is conserved in MnO_2 reduction. Induced cells contain significantly more manganese in their cell envelope than uninduced cells. The extent of the MnO_2-reducing activity was strongly correlated with the manganese concentration in the cell envelope. Cell envelopes of marine bacterial strains that could oxidize Mn^{2+} but not reduce MnO_2 or that could neither oxidize Mn^{2+} nor reduce MnO_2 contained less manganese and exhibited marginal or no MnO_2-reducing activity.

This model helps to explain why pseudomonad BIII 88, unlike *Geobacter metallireducens* or *Shewanella oneidensis*, which reduce MnO_2 and insoluble Fe(III) oxides only anaerobically, can reduce MnO_2 aerobically as well as anaerobically. It depends on the stability of dissolved Mn^{2+} in air at circumneutral and weakly acidic pH in the absence of a catalyst, as opposed to Fe^{2+}, which would rapidly autoxidize.

17.6.3 Bacterial Reduction of Manganese(III)

Because Mn(III)-containing minerals such as hausmannite exist, their possible function as terminal electron acceptors in bacterial respiration in nature needs to be considered. Complexed Mn(III) that can be abiologically formed in a reaction between MnO_2 and Mn^{2+} in the presence of a suitable ligand must also be considered as a potential terminal electron acceptor. Very limited evidence of such bacterial use of Mn(III) has been obtained to date. Gottfreund and Schweisfurth (1983) reported the reduction of Mn(III) pyrophosphate to Mn(II) by bacterial strain Red 16 using glucose, glycerol, or fructose as electron donor under aerobic conditions and glucose, fructose, glycerol, lactate, succinate, or acetate under anaerobic conditions. However, strain

Red 16 could not grow on acetate, either aerobically or anaerobically. The authors proposed that this organism produces Mn(III) in complexed form as a detectable intermediate in the reduction of Mn(IV) to Mn(II). Another study (Gottfreund et al., 1985) indicated the existence of bacteria that need not attack Mn(IV) to form Mn(III) but can produce a ligand that binds preexistent Mn(III). The Mn(III) was believed by the investigators to be derived by nonenzymatic reduction of Mn(IV). The bacteria then reduced the chelated Mn(III) to Mn(II). The authors suggested that Mn(III) is an intermediate in Mn(IV) reduction as well as in Mn(II) oxidation. However, whether Mn(III) is an obligatory intermediate in manganese reduction has yet to be clearly established (see Ehrlich, 1988, 1993a,b). The study of Mn(III) as an intermediate is fraught with difficulty because of the ease with which Mn(II) can react with some form of Mn(IV) in a disproportionation reaction to form Mn(III) in the presence of a ligand for Mn(III) such as pyrophosphate (Ehrlich, 1964).

More recently, Kostka et al. (1995) demonstrated the reduction of Mn(III) pyrophosphate by *Shewanella oneidensis* MR-1 using formate or lactate as electron donor. The reduction required anaerobic conditions and was inhibited by formaldehyde, tetrachlorosalicylanilide, CCCP, and HQNO but not at all by antimycin A. The alternative electron acceptor nitrate inhibited Mn(III) pyrophosphate reduction fully under aerobic conditions but only slightly under anaerobic conditions. Acetate, when it was the electron donor, was completely oxidized to CO_2. Energy derived from the reduction of Mn(III) pyrophosphate was able to drive protein synthesis. Abiotically, Mn(III) pyrophosphate is reduced rapidly by Fe(II) and sulfide. Either of these reductants may be of microbial origin.

Considering insoluble forms of Mn(III), Larsen et al. (1998) demonstrated anaerobic manganite (MnOOH) reduction by a growing culture of *Shewanella putrefaciens* MR-4 in a lactate-containing medium. Reduction was optimal at pH 7 and 26°C. Cell contact was required for the reduction. The reduction was inhibited by $HgCl_2$ (no concentration given) and by 0.1 mM CCCP and oxygen, but not by 0.2 mM cyanide or 0.05 mM HQNO.

Bacterial reduction of other Mn(III)-containing minerals seems not to have been reported so far but probably can occur.

17.6.4 NONENZYMATIC REDUCTION OF MANGANESE OXIDES

Some bacteria and most of those fungi that reduce Mn(IV) oxides such as MnO_2 reduce them indirectly (nonenzymatically). A likely mechanism of reaction is the production of metabolic products that are strong enough reductants for Mn(IV) oxides. *Escherichia coli*, for instance, produces formic acid from glucose, which is capable of reacting nonenzymatically with MnO_2:

$$MnO_2 + HCOO^- + 3H^+ \rightarrow Mn^{2+} + CO_2 + 2H_2O \qquad (17.13)$$

Pyruvate, which is a metabolic product of some bacteria, can also react nonenzymatically with manganese oxides at acid pH (Stone, 1987). Sulfate reducers produce H_2S from sulfate, and *Shewanella putrefaciens* can produce H_2S and sulfite from thiosulfate and Fe^{2+} from Fe(III) by anaerobic respiration. The resultant H_2S, sulfite, and Fe^{2+} can readily reduce some manganese oxides nonenzymatically (Burdige and Nealson, 1986; Mulder 1972; Lovley and Phillips, 1988b; Myers and Nealson, 1988b; Nealson and Saffarini, 1994).

Many fungi produce oxalic acid, which can also reduce MnO_2 nonenzymatically (Stone, 1987):

$$MnO_2 + HOOCCOO^- + 3H^+ \rightarrow Mn^{2+} + 2CO_2 + 2H_2O \qquad (17.14)$$

Because the electron transport mechanism in fungi, which are eukaryotic organisms, is located in mitochondrial membranes and not in the plasma membrane as in prokaryotic cells, fungi cannot be expected to reduce MnO_2 enzymatically (Ehrlich, 1978a), except by extracellular enzymes produced by them (see Section 17.4). MnO_2 reduction in acid solution under anoxic conditions by the reaction

$$MnO_2 + 2H^+ \rightarrow Mn^{2+} + 0.5O_2 + H_2O \qquad (17.15)$$

is not very likely because of the requirement for a high energy of activation.

The fact that MnO_2 is reduced indirectly by fungi can be readily demonstrated on a glucose-containing agar medium into which MnO_2 has been incorporated (Schweisfurth, 1968; Tortoriello, 1971). Fungal colonies growing on such a medium develop a halo (clear zone) around them in which MnO_2 has been dissolved (reduced). Because enzymes do not work across a spatial separation between them and their substrate, the formation of the halo can best be explained in one of two ways, which are not necessarily mutually exclusive. One way involves excretion by the fungus of a reducing compound, which then reacts with MnO_2. The other way involves acidification of the medium by the fungus to a pH range at which a residual medium constituent such as glucose reduces MnO_2.

17.7   BIOACCUMULATION OF MANGANESE

Just as some microorganisms that accumulate ferric oxide exist, so do microorganisms that accumulate manganese oxide. In a number of instances, the same organism may accumulate either iron or manganese oxide or both. As with iron oxide, the organisms deposit manganese oxides on the surface of their cell envelope or on some structure surrounding it, such as a slime layer or a sheath, but not intracellularly (see reviews by Ghiorse, 1984; Ghiorse and Ehrlich, 1992; Ehrlich, 1999). In some instances the accumulation may represent the product of oxidation of Mn^{2+} by the organism; in others, it may represent accumulation of manganese or iron oxides that were formed by other organisms or abiotically. Prokaryotes that accumulate manganese oxides include sheathed bacteria such as *Leptothrix* (Figure 16.8), budding and appendaged bacteria such as *Metallogenium* (Figures 17.9A and 17.9B), *Pedomicrobium* (Figure 17.10; Ghiorse and Hirsch, 1979; Ghiorse and Ehrlich, 1992), *Hyphomicrobium* (Figure 17.11), *Planctomyces* (Schmidt et al., 1981, 1982), *Caulococcus*, and *Kuznetsova*, bacteria with capsules such as *Siderocapsa* (Figure 16.12) and *Naumanniella*, and Mn^{2+}-oxidizing fungi (Figure 17.12). Mn(II) oxidation by *Pedomicrobium* sp. ACM 3067 has been shown to depend on a membrane fraction produced by cell lysis in a French press at 110 mPa, and in intact cells it is associated with Mn(II) adsorption by the cells (Larson et al., 1999). A Cu-dependent enzyme appears to be involved in the oxidation. As previously mentioned (see Section 17.5), the spores but not the vegetative cells of *Bacillus* SG-1 deposit manganese oxide at the level of the exosporium. The formation of this oxide from Mn^{2+} is catalyzed by a multicopper oxidase in the exosporium.

Mineral deposition around microbial cells raises an interesting question that begs for an answer. Although bacteria with stalks or sheaths generally deposit manganese oxide only on these structures and can thus readily escape from being trapped in such mineralizations by detaching from the stalk or by migrating out of the sheath, how do cells of nonappendaged and nonsheathed bacteria that become completely encased in such mineralization escape from it to multiply further? Are they permanently trapped in the mineralization and prevented from escaping it? It is possible that, in nature, permanent entrapment in a solid coating of manganese or iron oxide of nonappendaged and nonsheathed bacteria is preventable as long as their growth rate exceeds the mineral-oxide deposition rate (see also Ehrlich, 1999).

Bioaccumulation of manganese in soil has been shown with certain bacteria, including actinomycetes, and with fungi isolated from soil of the Chiatura manganese biogeochemical province of the former Georgian SSR. They accumulated 0.04–0.3 mg of Mn per gram dry weight of biomass in Czapek's medium unsupplemented with manganese, and 9.2–101 mg g^{-1} biomass in Czapek's medium supplemented with 0.1% Mn (Letunova et al., 1978). The organisms analyzed in this study

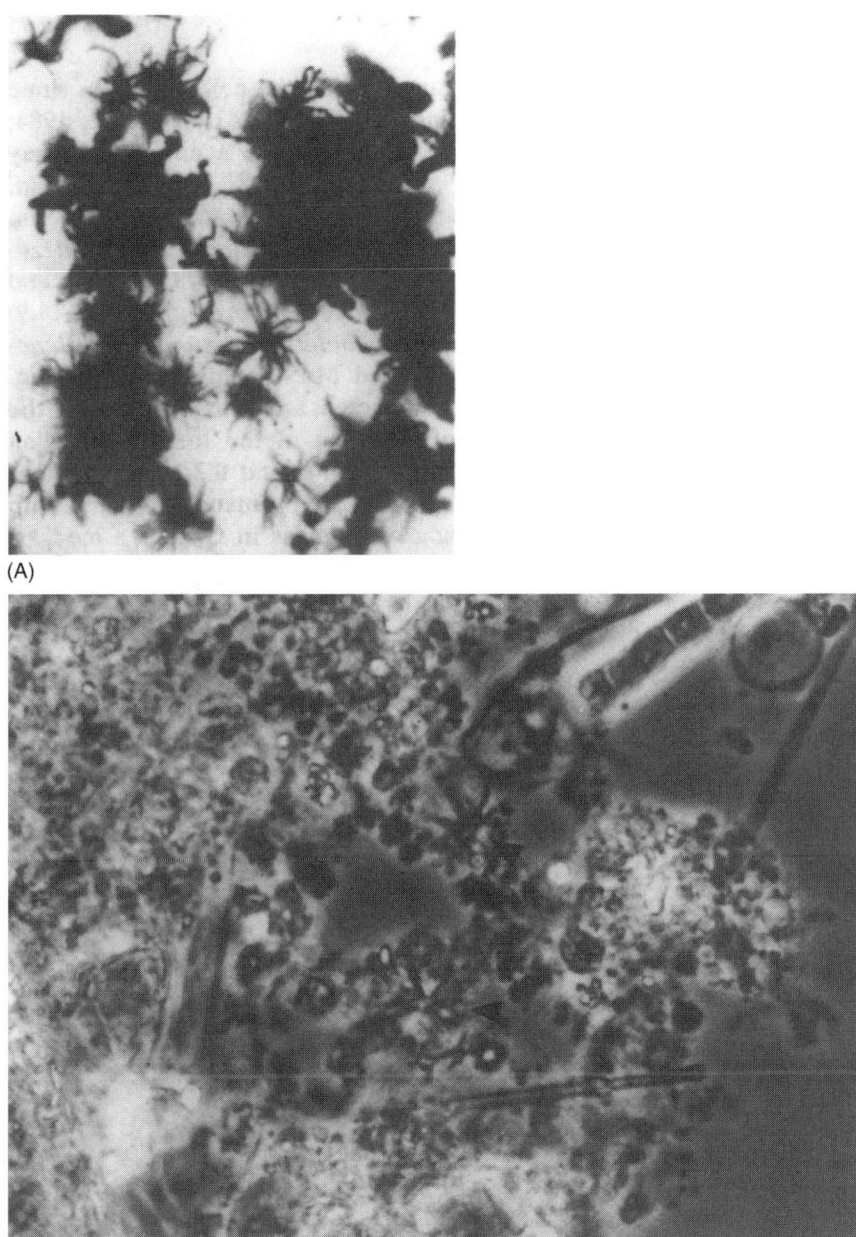

(A)

(B)

FIGURE 17.9 *Metallogenium*. (A) Microaccretions of *Metallogenium* from Lake Ukshezero in a split pedoscope after 50 days; growth in central part of microzone (×909). (From Perfil'ev BV, Gabe DR, *Applied Capillary Microscopy. The Role of Microorganisms in the Formation of Iron Manganese Deposits*, Consultants Bureau, New York, 1965. With permission.) (B) *Metallogenium*, filtered from Pluss See, Schleswig-Holstein, Germany (×2450). (Courtesy of Ghiorse WC, Schmidt W-D, Dept. of Microbiology, Cornell University, Ithaca, NY, USA.)

included *Bacillus megaterium, Actinomyces violaceus, A. indigocolor, Cladosporium herbarium*, and *Fusarium oxysporum*. The manganese concentration in the soil samples from which these organisms were isolated ranged from 1.05–296 g kg^{-1}. Development of the actinomycetes and fungi but not the bacteria appeared to be stimulated by high manganese levels in their native habitat.

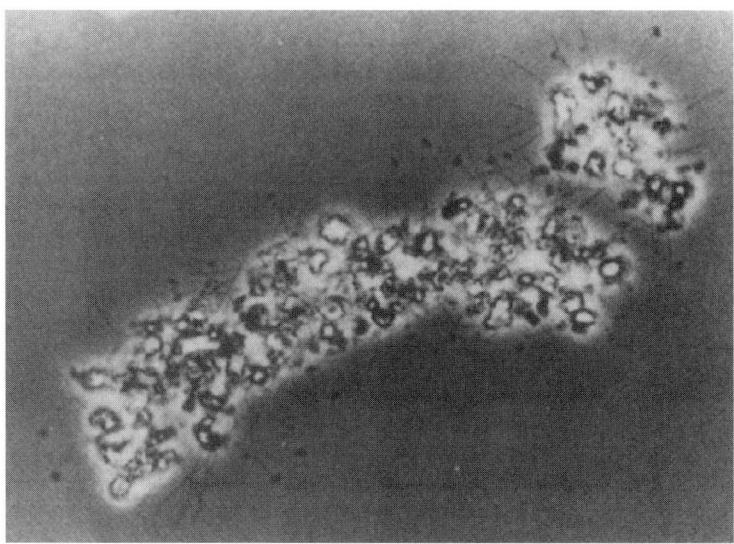

FIGURE 17.10 *Pedomicrobium* in association with manganese oxide particles (×2800). (Courtesy of Ghiorse WC, Dept. of Microbiology, Cornell University, Ithaca, NY, USA.)

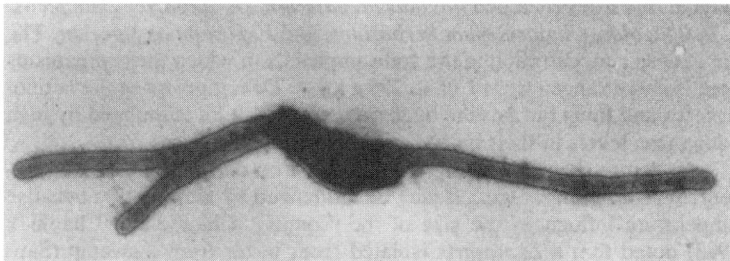

FIGURE 17.11 Manganese-oxidizing *Hyphomicrobium* sp. isolated from a Baltic Sea iron–manganese crust (×15,600). (Courtesy of Ghiorse WC, Dept. of Microbiology, Cornell University, Ithaca, NY, USA.)

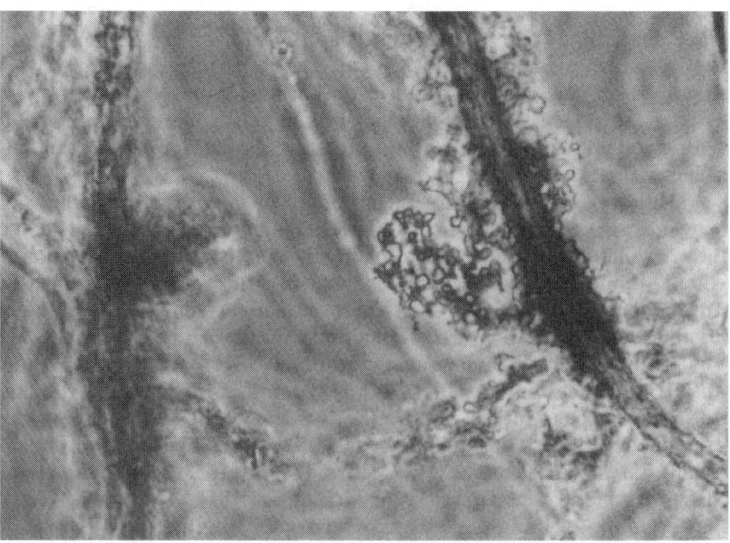

FIGURE 17.12 Manganese oxide deposit on fungal hyphae (×900). (Courtesy of Schweisfurth R)

From the standpoint of a natural environment, bacterial manganese or iron accumulation may be transient. With *Leptothrix* spp., it may be influenced by temperature because temperature influences the size of the biomass within its growth range. Ghiorse and Chapnik (1983) noted that a *Leptothrix* strain isolated from water from a swamp (Sapsucker Woods, Ithaca, New York) grew optimally between 20 and 30°C in the laboratory. This correlated with observations that *Leptothrix* and particulate iron and manganese oxide were most abundant in the surface water of the swamp when the temperature in the surface water was in the range of 20–30°C. Outside this temperature range, significantly less particulate iron and manganese were found in these waters.

17.8 MICROBIAL MANGANESE DEPOSITION IN SOIL AND ON ROCKS

17.8.1 SOIL

Manganese is an essential micronutrient for plants. Lack of sufficiently available manganese in soil can lead to manganese deficiency in plants, which may manifest itself in the form of grey-speck disease, for instance. Such lack of availability of manganese may be due to a lack of its mobility in a soil and can be the result of activity of manganese-oxidizing microbes.

Microorganisms can have a profound effect on the distribution of manganese in soil. Oxidizers of Mn(II) will tend to immobilize it and make it less available or unavailable by converting it to an insoluble oxide. Mobile forms of manganese include Mn^{2+} ions, complexes of Mn^{2+}, especially complexes such as those of humic and tannic acids, complexes of Mn^{3+}, and colloidal MnO_2. Immobile forms of manganese include Mn^{2+} adsorbed to clays or manganese or ferromanganese oxides, $Mn(OH)_2$, and insoluble salts of Mn(II) such as $MnCO_3$, $MnSiO_3$, and MnS, and various Mn(III) and Mn(IV) oxides. Oxidized manganese can accumulate as concretions in some soils, often accompanied by iron and various other trace elements (Drosdoff and Nikiforoff, 1940; Taylor et al., 1964; Taylor and McKenzie, 1966; McKenzie 1967, 1970). The formation of these nodules can be microbially mediated (Drosdoff and Nikiforoff, 1940; Douka 1977) (see also Section 17.5). The extent of insolubility of various immobile manganese compounds is governed by prevailing pH and E_h conditions (Collins and Buol, 1970). Reducing conditions at circumneutral or alkaline pH are especially favorable for the stability of Mn(III) and Mn(IV) oxides. Agriculturally, the most important forms of insoluble manganese include the oxides of Mn(IV) and mixed oxides such as $MnO \cdot MnO_2$ (also written as Mn_2O_3) and $2MnO \cdot MnO_2$ or $MnO \cdot Mn_2O_3$ (also written as Mn_3O_4). The stability of these oxides may also be affected by the presence of iron (Collins and Buol, 1970). Ferrous iron generated by Fe(III)-reducing bacteria can chemically reduce manganese(IV) oxides (Lovley and Phillips, 1988b; Myers and Nealson, 1988b).

Soil chemists have distinguished between the various forms of manganese in soil in an empirical fashion through the use of various extraction methods (see Robinson, 1929; Sherman et al., 1942; Leeper, 1947; Reid and Miller, 1963; Bromfield and David, 1978). Thus, Sherman et al. (1942) measured, in successive steps, water-soluble manganese by extracting a soil sample with distilled water, then exchangeable manganese by extracting the residue with 1 N ammonium acetate at pH 7.0, and finally, easily reducible manganese by extracting the residue from the previous step with 1 N ammonium acetate containing 0.2% hydroquinone. Other investigators have used different extraction reagents to measure exchangeable and easily reducible manganese (Robinson, 1929; Leeper, 1947).

The soil percolation experiment by Mann and Quastel (1946) clearly showed the role that microbes can play in immobilizing manganese in soil. These investigators showed that when they continuously perfused nonsterile soil columns with 0.02 M $MnSO_4$ solution, the manganese was progressively retained in the columns by being transformed into oxide paralleled by a progressive disappearance of manganous manganese in the column effluent. The oxidized state of the manganese was demonstrated by reacting it *in situ* with hydroxylamine reagent (a reducing agent). This reaction releases manganese from the columns in soluble form. The cause of the oxidation of the perfused manganese in the columns was inferred to have been microbial activity because poisons such as chloretone, sodium iodoacetate, and sodium azide inhibited it.

Manganese oxidation by soil microbes can also be demonstrated in the laboratory by the method of Gerretsen (1937). It involves preparation of petri plates of agar mixed with unsterile soil. A central core is removed from the agar and replaced with a sterile sandy soil mixture containing 1% $MnSO_4$. As Mn^{2+} diffuses from the sandy soil mixture into the agar, some developing bacteria and fungal colonies accumulate brown precipitate and manganese oxide in and around them as a result of enzymatic or nonenzymatic manganese oxidation if the pH of the medium is held below 8 (Figure 17.13). Controls in which the soil is antiseptically treated before adding it to the agar will not show evidence of manganese oxidation. This method was used effectively by Leeper and Swaby (1940) in demonstrating the presence of Mn^{2+}-oxidizing microbes in Australian soil.

Manganese oxidation in soil can also be studied in deep cultures of a soil agar mixture. Uren and Leeper (1978) studied manganese oxidation in cultures consisting of 85 mL of 1% soil agar in a 150-mL jar, set up like a Gerretsen plate. The oxidation seemed to occur preferentially at reduced oxygen levels because oxidation was most intense at 10–17 mm below the agar surface when air was passed over the surface. Above this zone, sparse manganese oxide was deposited only around bits of organic matter from the soil and around fungal hyphae. Raising the agar concentration in the soil medium to 2% caused the intense zone of oxidation to appear 4–11 mm below the surface. When the jar was sealed with an air space above the agar, an 8 mm thick oxidation zone developed from the surface downward. The investigators found that the factor that controlled the position of the zone of

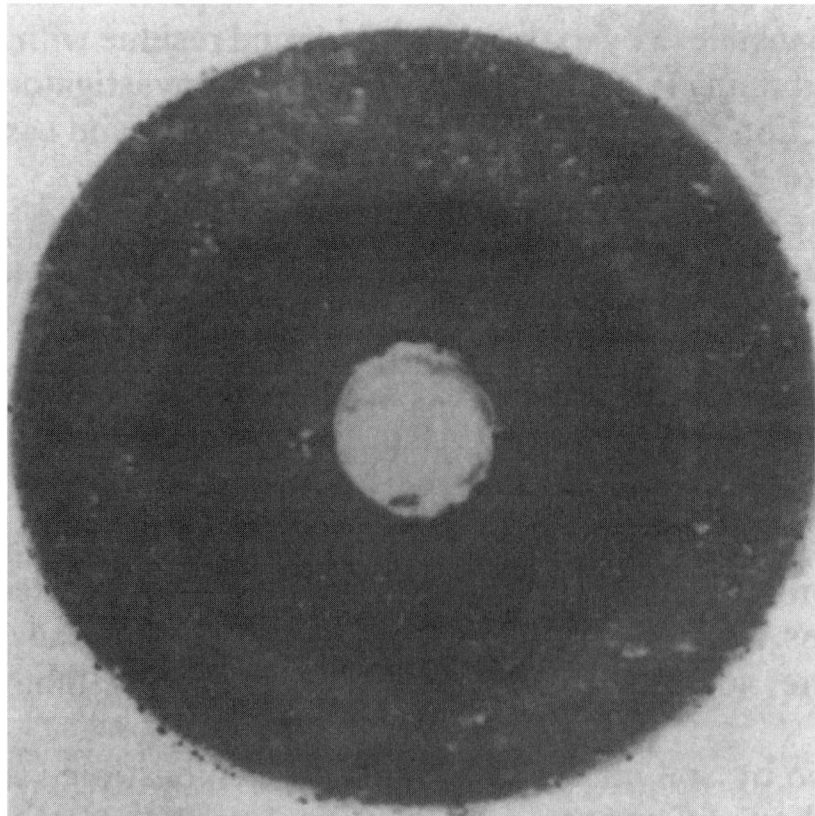

FIGURE 17.13 Gerretsen plate by method modified by Leeper and Swaby (1940), showing manganese oxide deposition (dark halo) as a result of microbial growth around the central $MnSO_4$-containing agar plug. Initial pH 7.3, final pH 6.7. (From Leeper EW, Swaby RJ, *Soil. Sci.,* 49, 163–169, 1940. With permission.)

intense manganese oxidation in the soil column was available CO_2 because aeration at the surface with a mixture of 97% N_2, 2% O_2, and 1% CO_2 permitted development of an oxidation zone in the top 1 cm whereas a gas mixture of 98% N_2 and 2% O_2 did not. The observed oxidation was due to microbial activity because incorporation of azide to a final concentration of 1 mM into the soil agar inhibited manganese oxidation. Microbial manganese-oxidizing activity was noted even in soils as acidic as pH 5.0 (Uren and Leeper, 1978; Sparrow and Uren, 1987).

In situ manganese oxidation by soil microbes has been observed with pedoscopes (Perfil'ev and Gabe, 1965). This apparatus consists of one or more glass capillaries with optically flat sides for direct microscopic observation of capillary content. Inserting a sterilized pedoscope into the soil enables development of soil microbes in the lumen of the capillaries under soil conditions. Periodic removal of the pedoscope and examination under a microscope permits visual assessment of the developmental progress of the organisms and the deposition of manganese oxide, if any (see Chapter 7). Manganese-oxidizing and manganese-depositing organisms detected for the first time by use of a pedoscope include *Metallogenium personatum*, *Kuznetsova polymorpha*, *Caulococcus manganifer*, and others (Perfil'ev and Gabe, 1965, 1969).*

Application of the above techniques and others has provided ample evidence of the important role that microbes can play in immobilizing manganese in the soil by oxidizing it. The method of Gerretsen was adapted to show the role microbes play in manganese oxidation in Australian soil (Leeper and Swaby, 1940; Uren and Leeper, 1978; Sparrow and Uren, 1987) and in soils of South Sakhalin (former USSR) (Ten Khak-mun, 1967). Other investigations have demonstrated the presence of manganese-oxidizing microbes (bacteria and fungi) in soils in various parts of the world using various methods. These include studies by Timonin (1950a,b), Timonin et al. (1972), Bromfield and Skerman (1950), Bromfield (1956, 1978), Aristovskaya (1961), Aristovskaya and Parinkina (1963), Perfil'ev and Gabe (1965), Schweisfurth (1969, 1971), Van Veen (1973), and others. Among the bacteria identified as active agents of Mn(II) oxidation in these studies, *Arthrobacter* or *Corynebacterium*, *Pedomicrobium*, *Pseudomonas*, and *Metallogenium* were the most frequently mentioned, but a number of other unrelated genera were also identified as being active.

17.8.2 ROCKS

Manganese oxides are sometimes found in thin, brown-to-black veneers and iron oxides in orange veneers (each up to 100 μm thick) covering rock surfaces in some semiarid and arid regions of the world. In the brown-to-black coatings, manganese and iron oxides are major components (20–30%) along with clay (~60%) and various trace elements (Potter and Rossman, 1977; Dorn, 1991). These manganese- and iron-rich coatings are known as *desert varnish* or *rock varnish*. Manganese-rich coatings have been detected on some rocks in the Sonoran and Mojave Deserts (North America),

---

* The status of the genus *Metallogenium* is presently uncertain. Although isolated and cultured repeatedly in the former Soviet Union by a number of investigators (Perfil'ev and Gabe, 1965; Zavarzin, 1961, 1964; Mirchink et al., 1970; Dubinina, 1970, 1978b, 1984), living cultures that oxidize manganese have generally not been successfully isolated outside Russia. An exception is Herschel (1999), who isolated *Metallogenium* sp. (probably *M. personatum*) from a reservoir in Nordrhein-Westfalen, Germany, and subcultured it in continuous culture in a chemostat. Structures resembling *Metallogenium* were collected by membrane filtration from waters of several lakes outside Russia, but these structures proved nonviable. Indeed, they did not contain any trace of nucleic acid, protein, or membrane lipid (Klaveness, 1977; Gregory et al., 1980; Schmidt and Overbeck, 1984; Maki et al., 1987). Structures interpreted as *Metallogenium symbioticum*, which is said to grow in obligate association with some fungi and bacteria (Zavarzin, 1961; 1964; Dubinina, 1984), may be fibrous manganese produced nonenzymatically by the fungus (Schweisfurth, 1969; Schweisfurth and Hehn, 1972; Schmidt and Overbeck, 1984) as, for instance, in manganese oxidation by exopolymers produced by an arthroconidial anamorph of a basidiomycete (Emerson et al., 1989). It is possible that the starlike structures (arais) associated with *Metallogenium* coenobia (Balashova and Dubinina, 1989) are nonliving appendages from which the actual cell that gave rise to the appendages has become detached before or during sample collection. Like the stalks of *Gallionella*, the "appendages" may be structures impregnated with manganese oxide. A more recent observation of arais showing "secondary coccoid body formation" as part of a manganese-depositing biofilm on the inner walls of a water pipe lends support to this view (Sly et al., 1988).

Negev Desert (Middle East), the Gibson and Victoria Deserts (Western Australia), and the Gobi Desert (Asia). Although they are very likely the result of microbial activity (Krumbein, 1969; Dorn and Oberlander, 1981; Taylor-George et al. 1983; Hungate et al., 1987), this view is not universally shared. Krumbein (1969) and Krumbein and Jens (1981) implicated cyanobacteria and fungi in the formation of desert varnish in the Negev Desert, whereas Dorn and Oberlander (1981) implicated *Pedomicrobium*- and *Metallogenium*-like bacteria in desert varnish from Desert Valley in the Mojave Desert. The latter investigators produced desert varnish-like deposits in the laboratory with their bacterial isolates. Dorn (1991) is of the opinion that the following conditions favor bacterial participation in manganese-rich varnish formation: (1) moist rock surfaces, (2) rock surfaces of low nutrient content that favor bacterial manganese oxidation by mixotrophs, and (3) rock surfaces exhibiting circumneutral pH.

Adams et al. (1992) suggested that in the formation of iron-rich varnish, siderophores, produced by bacteria on rock surfaces where the varnish forms, mobilize the ferric iron from wind-borne dust or rock surfaces. The mobilized iron is then concentrated as iron oxide or oxyhydroxide on the cell walls of bacteria on the rock surfaces. Upon the death of the iron-coated bacteria, the iron oxide becomes part of the varnish.

Microscopic examination of desert varnish from the Sonoran Desert revealed the presence of fungi (dematiaceous hyphomycetes) and bacteria. Manganese oxidizers in samples from this source included *Arthrobacter, Micrococcus, Bacillus, Pedomicrobium*, and possibly *Geodermatophilus* (Taylor-George et al., 1983). Active respiration but little CO_2 fixation was detected in the varnish. The respiration was attributed mainly to the fungi. The absence of significant CO_2 fixation indicated an absence of photosynthetic, that is, cyanobacterial or algal, activity. Varnish formation on Sonoran Desert rocks has been postulated to be the result of formation of fungal microcolonies on rock surfaces, which trap wind-borne clay and other mineral particles. Upon death of the fungi, intermittent, moisture-dependent weathering of minerals is thought to lead to microbial and abiotic concentration of mobilized manganese through oxidation. Dead fungal biomass could provide nutrients to manganese-accumulating bacteria at this stage. Further development of fungal microcolonies and their coalescence was postulated to lead ultimately to the formation of continuous films of varnish. Besides contributing to rock weathering, the fungal mycelium in the model seems to serve as an *anchor* for Mn oxide formed at least in part by bacteria (Taylor-George et al., 1983). The MnO_2 in the varnish may act as a screen against ultraviolet radiation and may thus serve a protective function (Taylor-George et al., 1983). Examination of the bacterial flora of desert varnish from the Negev Desert yielded a range of bacterial isolates similar but not identical to those from varnish of the Sonoran Desert. Of 79 bacterial isolates, 74 oxidized manganese under laboratory conditions. They were assigned to the genera *Bacillus, Geodermatophilus, Arthrobacter*, and *Micrococcus* (Hungate et al., 1987). The prevalence of manganese oxidizers in desert varnish samples from geographically very distinct desert sites further supports the idea that they play a role in its formation (Hungate et al., 1987).

17.8.3 ORES

It has been suggested that some sedimentary manganese ore deposits are of biogenic origin. This is based in part on the observation of structures in the ore that have been identified as microfossils by the discoverers. Examination of manganese ore from the Groote Eylandt deposits in Australia revealed the presence of algal (cyanobacterial?) stromatolite structures, cyanobacterial oncolites, coccoid microfossils, and microfossils enclosed in metalcolloidal oxides of manganese (Ostwald, 1981). In this deposit, the microbes are viewed as being the main cause of manganese oxide formation and accretion, with subsequent nonbiological diagenetic changes leading to the ultimate form of the deposit. Some Precambrian deposits and a Cretaceous–Paleocene manganese deposit (Seical deposit on the island of Timor) have been attributed to the activity of *Metallogenium* because coenobial structures resembling modern forms of this organism as described by Russian investigators have been found by microscopic study of sections of the respective ores (Crerar et al., 1979). Similar observations were made by Shternberg (1967) on samples of Oligocene Chiatura manganese deposit and the Paleozoic Tetri-Tsarko manganese deposit.

17.9 MICROBIAL MANGANESE DEPOSITION IN FRESHWATER ENVIRONMENTS

The first reports on manganese-oxidizing microorganisms in freshwater environments appeared at the beginning of the twentieth century. They include those of Neufeld (1904), Molisch (1910), and Lieske (1919) as cited by Moese and Brantner (1966), Thiel (1925), von Wolzogen-Kühr (1927), Zappfe (1931), and Sartory and Meyer (1947). The organisms found by these investigators were usually detected in sediments, organic debris, or manganiferous crusts. More recent investigations dealt with microbial manganese oxidation in springs, lakes, and water distribution systems.

17.9.1 BACTERIAL MANGANESE OXIDATION IN SPRINGS

Precipitation of manganese hydroxides (presumably hydrous manganese oxide), which upon aging became transformed into pyrolusite and birnessite in a mineral spring near Komaga-dake on the island of Hokkaido, Japan, was attributed to the activity of manganese-oxidizing bacteria, including sheathed bacteria (Hariya and Kikuchi, 1964). The bacteria were detected by filtration of water from the spring through ordinary filter paper and membrane filters and by incubating samples of water in the laboratory. The water contained (in milligrams per liter) Mn^{2+}, 4.75; K^+, 16; Na^+, 128; Ca^{2+}, 101; Mg^{2+}, 52; Cl^-, 156; SO_4^{2-}, 481; HCO_3^-, 0; and CO_2, 11; but only traces of Fe^{2+} and Fe^{3+}. The temperature of the water was 23°C, and its pH was 6.8. During incubation of water samples in the laboratory, manganese precipitated progressively between 20 and 50 days accompanied by a fall in pH, a gradual rise in E_h after 20 days, and an abrupt rise in E_h after 45 days. Only 1.18 mg of dissolved Mn^{2+} per liter was left after 50 days.

Another instance of bacterial manganese oxidation associated with a spring was reported from Squalicum Creek Valley near Bellingham, Washington, DC (Mustoe, 1981). In this example, a black soil surrounded the spring over an area 5 m × 25 m and to a depth of 30 cm. It contained 43% MnO_2 and 20–30% iron oxide (classified as Fe_2O_3). Two pseudomonad strains that oxidized Mn^{2+} and Fe^{2+} when growing on medium containing soil organic matter were isolated from it. The oxides were deposited extracellularly. The isolates were considered to be a cause of manganese and iron deposition in soil.

Still another instance of bacterial manganese oxidation in springwater and sediment was observed at Ein Feshkha on the western shore of the Dead Sea, Israel (Ehrlich and Zapkin, 1985). Gram-positive spore-forming and non-spore-forming rods and gram-negative rods were isolated from water and sediment samples from drainage channels from the spring, which oxidized Mn^{2+} enzymatically. They were Group I type (see Section 17.5). It was postulated that at least some of the manganese oxide formed by the bacteria in the springwater was carried in drainage water from the spring to the Dead Sea and became incorporated as laminations in calcareous crusts found along the shore of the lake.

17.9.2 BACTERIAL MANGANESE OXIDATION IN LAKES

Evidence of bacterial manganese oxidation has been observed in Lake Punnus-Yarvi on the Karelian Peninsula in the former USSR (Sokolova-Dubinina and Deryugina, 1967a,b). The lake is of glacial origin, oligotrophic, and slightly stratified thermally. It is 7 km long, up to 1.5 km wide, and up to 14 m deep. It contains significant amounts of dissolved manganese (0.02–1.4 mg L^{-1}) and iron (0.7–1.8 mg L^{-1}) only in its deep waters. The lake is fed by two rivers and 24 streams, which drain surrounding swamps. The manganese in the lake is supplied by surface water and groundwater drainage containing 0.2–0.8 mg Mn L^{-1} and 0.4–2 mg Fe L^{-1}. Most of the iron and manganese in the lake are deposited on the northwestern banks of Punnus-Yoki Bay situated at the outflow from the lake into the Punnus-Ioki River in a deposit 5–7 cm thick at a water depth of up to 5–7 m. The deposit includes hydrogoethite (nFeO · nH$_2$O), wad (MnO$_2$ · nH$_2$O), and psilomelane (mMnO · MnO$_2$). The iron content of the deposit ranges from 18 to 60% and the manganese content from 10 to 58%. The deposit also includes 5–16% SiO$_2$ and Al, Ba, and Mg in amounts ranging from 0.3 to 0.7%. The bacterium held responsible for manganese oxidation was *Metallogenium*, which was found in all parts of the lake.

Manganese deposition was demonstrated in the laboratory by incubating sediment samples at 8°C for several months. Dark brown compact spots of manganese oxide were then detected, which revealed no characteristic bacterial structures by examination with a light microscope but did show by electron microscopy the presence of *Metallogenium* that closely resembled *Metallogenium* fossils in Chiatura manganese ore (Dubinina and Deryugina, 1971). *Metallogenium* was cultured from lake samples on manganese acetate agar. No *Metallogenium* cultures were obtained from sediment samples from parts of the lake where no microconcretions of manganese oxide were found. *Metallogenium* development was also studied in pedoscopes, which showed progressive encrustation with manganese oxide. Manganese deposition in the lake occurred in a redox potential range of +435 to +720 mV and a pH range of 6.3–7.1. It was concluded from data gathered by other investigators that ore formation may begin at an E$_h$ as low as +230 mV and a pH of 6.5. At Mn^{2+} concentrations below 10 mg L^{-1}, autoxidation was not considered likely. Manganese oxide deposition attributed to *Metallogenium* activity was also noted in some other Karelian lakes where a steady supply of dissolved manganese occurred, redox conditions were favorable, and bacterial reducing processes did not occur (Sokolova-Dubinina and Deryugina, 1968). More recently, *Metallogenium* was reported to occur in sediment samples from Lake Geneva (Lac Léman) (Jaquet et al., 1982) and was implicated in manganese-oxidizing activity in Lake Constance (Stabel and Kleiner, 1983).

In Oneida Lake, near Syracuse, New York, manganese deposition has been studied for many years because of the occurrence of ferromanganese concretions on the sediment surface of shallow, well-oxygenated central areas of the lake (Figures 17.14A, 17.4B, and 17.4C). The lake has a surface area of 207 km^2 and varies in depth, its average depth being 6.8 m (Dean and Ghosh, 1978). It never exhibits seasonal stratification. Wind agitation keeps it well aerated to all depths when it is not frozen, although deeper water may have oxygen concentrations only 50% of saturation (Dean and Ghosh, 1978). The ferromanganese concretions appear as crusts around rocks at the edge of shoals where the water is <4.3 m in depth (Dean, 1970), but some have also been recovered from hard sediments at depths of 8 m (Chapnick et al., 1982). The crust may take on a saucer shape, concave upward (Dean, 1970). Many exhibit coarse concentric banding of alternating zones rich in manganese and rich in iron (Moore, 1981). At other times the concretions have been described as flattened, disk-shaped structures. Their growth rate as determined by natural-radioisotope analysis has been estimated to vary between >1 mm per 100 years at some periods and no growth at others (Moore et al., 1980).

The origin of ferromanganese concretions in Oneida Lake has been a puzzle. Hypotheses of abiogenesis (Ghosh, 1975) and biogenesis (Gillette, 1961; Dean, 1970; Dean and Ghosh, 1978; Dean and Greeson, 1979; Chapnick et al., 1982) have been offered at various times. Gillette (1961) and Chapnick et al. (1982) favor a direct bacterial role, whereas Dean (1970) and Dean and Ghosh (1978) favor an indirect role played by algae in the lake. Gillette (1961) recognized bacterial cells in pulverized concretions by microscopic examination and showed that some of the isolates precipitated iron. He speculated that bacterial deposition of iron and manganese in the lake could be the result of degradation of organic complexes but also suggested that iron may be precipitated by autoxidation with oxygen generated in the photosynthesis of algae. Dean and Ghosh (1978) and Dean and Greeson (1979) favor a primary role for cyanobacteria and algae as producers of chelators for iron and manganese and accumulators of the chelates. These notions are supported by recent observations that cyanobacteria in the lake belonging to *Microcystis* sp. generate microenvironments with strongly alkaline conditions (pH >9) in their mats as a result of photosynthetic activity (Richardson et al., 1988) (see Equations 16.27 and 16.28 for a chemical explanation). Such alkaline conditions promote

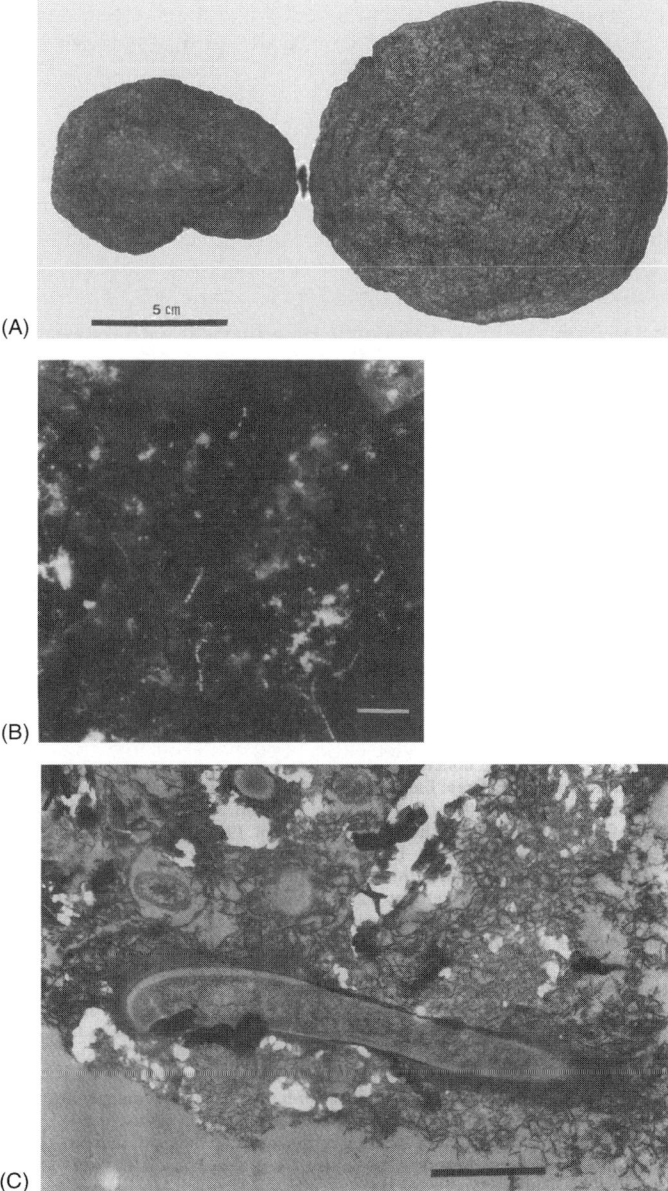

FIGURE 17.14 Ferromanganese concretions from the *nodule-rich* area in Oneida Lake, central New York State. (A) Two concretions: the one on the right is typical of the flat *pancake* variety; the one on the left is a rock with 2–3 cm ferromanganese oxide crust. (B) Epifluorescent light photomicrograph of a portion of the surface of the concretion on the left in (A) stained with 0.01% acridine orange. The sample was removed from the surface and then viewed under ultraviolet epi-illumination. The brightly fluorescent filamentous and coccoid bacteria are present in the ferromanganese-mineralized biofilm on the surface. Scale bar = 10 μm. (C) Transmission electron photomicrograph of an ultrathin section of a piece of material from a concretion similar to those shown in (A). The sample was fixed in glutaraldehyde followed by osmium tetroxide and uranyl acetate, dehydrated in ethanol, and embedded in epoxyresin. A portion of a filamentous bacterial cell ~0.6 μm in diameter is surrounded by a dark-colored complex composed of polymer-ferromanganese oxide. Note the smaller encrusted coccoid cell in the upper left surrounded by a clear zone produced by shrinkage during preparation for electron microscopy. The wispy dendritic black material is Fe–Mn oxide. Scale bar = 1 μm. (From Ghiorse WC, Ehrlich HL, *Biomineralization: Process of Iron and Manganese. Modern and Ancient Environments*. Catena Suppl 21, Catena, Cremlingen, Germany, 1992. With permission.)

autoxidation of Mn^{2+}. The oxides become entrapped in the mats. Upon death, the cyanobacterial and algal biomass is believed to carry the oxidized iron and manganese to the lake bottom to be released on the decay of the biomass and somehow incorporated into concretions. Chapnick et al. (1982) demonstrated that bacteria in the bottom water of the lake are able to catalyze the oxidation of Mn^{2+} dissolved in it. They also noted some binding of Mn^{2+} without oxidation. Analysis of the manganese budget for the lake supports the notion that most of the dissolved manganese (95%) becomes incorporated into the nodules. Chapnick et al. (1982) suggested that cyanobacteria and algae help in transporting the manganese from the surface waters to the nodule-forming regions where bacteria participate in the oxidation of Mn^{2+} in the bottom waters to an as-yet-undetermined extent. Bottom water movement carries nodules or fragments of nodule to the surface of deeper anoxic sediments, where they are buried and undergo reduction with remobilization of manganese. *Shewanella* (formerly *Alternaria*) *putrefaciens*, now renamed *S. oneidensis*, is one organism found in this sediment that has been demonstrated in laboratory experiments to be an effective reducer of Mn(IV) oxide under anaerobic conditions (Myers and Nealson, 1988a). Deepwater circulation reintroduces the dissolved manganese into the water column, and the cycle of manganese transformations is repeated. Manganese that is lost in the lake outflow (50 tons per year) is more than replaced by manganese influx (75 tons per year) into the lake (Dean et al., 1981).

Another lake in which ferromanganese concretions have been found is Lake Charlotte, Nova Scotia, first studied by Kindle (1932) and later by Harriss and Troup (1970). The concretions occur in a shallow region of the lake that Kindle (1935) named Concretion Cove. It receives runoff from soil rich in metals and organics. Kindle (1932) thought that an indirect mechanism was operating in which removal of CO_2 from the lake water through photosynthesis by diatoms caused a rise in pH that promoted autoxidation of Mn^{2+} to MnO_2. Later investigations suggested more direct bacterial involvement (Kepkay, 1985a,b). Using a special device called a peeper (Burdige and Kepkay, 1983) emplaced into sediment in Concretion Cove, Kepkay (1985a,b) demonstrated bacterially dependent oxidation of Mn(II) in the dark, that is, without direct participation of algal photosynthesis. He also demonstrated the binding of nickel, and to a lesser extent, copper that was enhanced by microbial oxidation of manganese. In experiments in which filter membranes with 0.2 μm pore size were submerged in Lake Charlotte water, a succession of bacterial cell types were observed: cocci developed into rods, suggesting *Arthrobacter*. The coccus-to-rod transformation coincided with a peak in oxygen consumption. CO_2 fixation was detectable as the rods started to bind Mn and Fe to their cell surface. If this metal binding involved, at least in part, oxidation, especially that of manganese, autotrophic growth at the expense of manganese could have been taking place at that point (Kepkay et al., 1986).

Sediment samples from the experimental site in Lake Charlotte where concretions are found revealed the presence of microscopic precipitates of manganese and iron (Kepkay, 1985a,b). How these precipitates are related to the concretions has yet to be elucidated. In places where macrophytes (*Eriocaulon sepangulare*) grew in the Concretion Cove area, bacterial manganese oxidation resulted in the formation of finely dispersed oxide within the sediments. In places where macrophytes were absent, ferromanganese concretions were noted at the sediment surface. The macrophyte root system appeared to influence manganese oxidation by somehow promoting downward movement of manganous manganese in the sediments. Manganese oxidation was most active 1–3 cm below the sediment surface, where the macrophytes grew, but also occurred at the sediment surface where macrophytes were absent (Kepkay, 1985c).

Ferromanganese concretions have been reported in other lakes in North America. Examples include Great Lake, Nova Scotia, and Mosque Lake, Ontario (Harriss and Troup, 1970); Lake Ontario (Cronan and Thomas, 1970); Lake George, New York (Schöttle and Friedman, 1971); and Lake Michigan (Rossman and Callender, 1968). Typical compositions of the concretions are listed in Table 17.5. In general, lake concretions form in areas where the sedimentation rate is low. Growth rates may be very slow, as in Lake Ontario (0.015 mm per year), or more rapid, as in Mosque Lake (1.5 mm per year).

TABLE 17.5
Composition of Some Lake Ores (Values in Percent)

Mn	Fe	Si	Al	Source	Reference
36.08 (MnO_2)	13.74 (Fe_2O_3), 7.70 (FeO)	12.75 (SiO_2)	12.50 (Al_2O_3)	Ship Harbor Lake	Kindle (1932)
13.4–15.4 (Mn)	19.5–27.5 (Fe)	4–10 (Si)	0.7–0.95 (Al)	Oneida Lake	Dean (1970)
31.7–35.9 (Mn)	14.2–20.9	—	—	Great Lake	Harriss and Troup (1970)
15.7 (Mn)	39.8–40.2 (Fe)	—	—	Mosque Lake	Harriss and Troup (1970)
17.0 (Mn)	20.6 (Fe)	—	—	Lake Ontario	Cronan and Thomas (1970)
3.57 (Mn)	33.2	—	—	Lake George	Schöttle and Friedman (1971)
0.89–22.2 (MnO)	1.34–60.8 (FeO)			Lake Michigan	Rossman and Callender (1968)

Not all geologists have been convinced that microbes play a role in the formation of the concretions. Varentsov (1972), for instance, explained the formation of the nodules in Eningi-Lampi Lake, Karelia, Russia, in terms of chemosorption and autocatalytic action.

One instance of possible transient manganite (γMnOOH) formation in an artificial lake was reported by Greene and Madgwick (1991). The lake was created behind a tailings dam in the mining operation of the Mary Kathleen Mine, North Queensland, Australia. In it, bacteria occur in association with the microalga *Chlamydomonas acidophilus* that oxidize Mn^{2+} to form disordered γMnO_2 via manganite under laboratory conditions (Greene and Madgwick, 1991). An unidentified component of the algal cells stimulated the oxidation process. Intact cells, however, were most effective as stimulants of the oxidation process. This process is evidently an example in which Mn(III) oxide appears to be an intermediate in Mn(IV) oxide formation.

17.9.3 BACTERIAL MANGANESE OXIDATION IN WATER DISTRIBUTION SYSTEMS

In 1962, a case of microbial manganese precipitation in a water pipeline connecting a reservoir with a filtration plant of the waterworks in the city of Trier, Germany, was reported (Schweisfurth and Mertes, 1962). The accumulation of precipitate in the pipes caused a loss of water pressure in the line. The sediment in the pipe had a dark brown to black coloration and was rich in manganese but relatively poor in iron content. Microscopic examination of the sediment revealed the presence of cocci and rods after removal of MnO_2 with 10% oxalic acid solution. Sheathed bacteria were found only on the rubber seams, and then always in association with other bacteria. Evidence for fungal mycelia and streptomycetes was also found. In culture experiments, only gram-negative rods and fungi were detected. Chemical examination of the reservoir revealed that the manganese concentration in the bottom water was between 0.25 and 0.5 mg L^{-1} during most of 1960, except in September and October, when it ran as high as 6 mg L^{-1}. The peak in manganese concentration in the bottom waters was correlated with water temperature, which reached its peak at about the same time as the manganese concentration in the water. The two feed streams into the reservoir did not contribute large amounts of manganese, only ~0.05 mg L^{-1}. The major source of the manganese could have been the manganiferous minerals of the reservoir basin and surrounding watershed.

Similar observations were made in some pipelines connecting water reservoirs with hydroelectric plants in Tasmania, Australia (Tyler and Marshall, 1967a). In this instance, pipelines from Lake

King William were found to have heavy deposits of manganese oxide, whereas those from Great Lake did not. The deposits in pipes leading from Lake King William accumulated to a maximum thickness of 7 mm in 6–12 months. The manganic oxide deposition process was reproduced in the laboratory in a recirculation apparatus (Tyler and Marshall, 1967a). With water from Lake King William, a deposition of brown manganic oxide formed at the edge of coverslips after 24 h and for 6 days thereafter. Subsequent addition of $MnSO_4$ to the water caused further deposition after 6 days. By contrast, only traces of deposit developed with Great Lake water in similar experiments. The difference in manganese deposition from the two lake waters was explained in terms of the difference in the manganese content. Lake King William's water contained 0.01–0.07 ppm of manganese, whereas Great Lake water contained only 0.001–0.013 ppm. In the laboratory, inoculation of Great Lake water with some Lake King William water did not promote the oxidation unless $MnSO_4$ was also added. It was also found that if Lake King William water was autoclaved or was treated with azide (10^{-3} M final concentration), manganic oxide deposition was prevented, suggesting participation of a microbial agent in the reaction. Inoculation of autoclaved Lake King William water with unautoclaved water caused resumption of the oxidation. The dominant organism involved in the oxidation appeared to be one identified as a *Hyphomicrobium* sp. Sheathed bacteria and fungi and possibly *Metallogenium symbioticum* were also encountered in platings from pipeline deposits, but only at low dilutions. It was not resolved whether *Hyphomicrobium* sp. oxidized Mn^{2+} enzymatically or nonenzymatically (Tyler and Marshall, 1967b). After the publication of the original studies of manganese oxide pipeline deposits in Tasmania, Tyler (1970) found *Hyphomicrobium* in pipeline deposits in other parts of the world. It was not always the only manganese-oxidizing organism present, however.

In a seasonal study of North Pine Dam, a body of freshwater located 25 km west of Brisbane, Australia, Johnson et al. (1995) noted that microbial manganese oxidation prevailed in the lower epilimnion between November and May (Australian summer). Abiotic manganese oxidation became more noticeable in late summer and autumn, when it represented up to 25% of the total. Microbial oxidation required a minimum temperature of 19°C and was optimal at 30°C. The oxidation state of the oxidized manganese ranged from 2.55 to 3.25 in the epilimnion and was 3.59 in a hypolimnion sample after a few days of aerobic incubation. It was highest when microbial activity was greatest.

Micronodule formation in biofilms formed by *Pseudomonas* spp. on the surface of polyvinyl chloride (PVC) and high-density polyethylene (HDPE) pipe material after 2 weeks of incubation was reported by Murdoch and Smith (1999). The nodules had an average diameter of 10 μm. Most featured a ~2 μm hole at their center. They were rich in manganese. The role of microbes in the micronodule-forming process remains to be elucidated.

17.10 MICROBIAL MANGANESE DEPOSITION IN MARINE ENVIRONMENTS

Manganese-oxidizing bacteria have been detected in various parts of the marine environment, including the water column and surface sediments in estuaries, continental shelves and slopes, and abyssal depths. They appear to play an integral part in the manganese cycle in the sea.

Manganese is unevenly distributed in the marine environment. It occurs in greater quantities on and in sediments than in the seawater and in greater quantities in seawater than in the biomass (see Table 17.6 for manganese distribution in the Pacific Ocean; Poldervaart, 1955; Mero, 1962; Bowen, 1966). In some parts of the oceans, a significant portion of manganese at the sediment interface at abyssal depths is concentrated in ferromanganese concretions (nodules) and crusts. The concentration of manganese in surface seawater from the Pacific Ocean has been reported to fall in the range of 0.3–3.0 nmol kg^{-1} (16.4–164.8 ng kg^{-1}) (Klinkhammer and Bender, 1980; Landing and Bruland, 1980). The concentration in bottom water is generally less than that in surface water. Klinkhammer and Bender (1980) found it to be one-fourth or less at some stations in the Pacific Ocean. Manganese concentrations in surface waters over the continental slope near the mouth of major rivers such as the Columbia River on the coast of the state of Washington, DC, may be as high as 5.24 nmol kg^{-1} (164.8 ng kg^{-1}) (Jones and Murray, 1985). Exceptionally high manganese concentrations occur

---

TABLE 17.6
Manganese Budget for the Pacific Ocean

Total Mn (as MnO) in sediments	1.4×10^{15} t
Total Mn (as MnO) in nodules	3.1×10^{11} t (170 times)[a]
Total Mn (as MnO) in seawater	1.8×10^{9} t
Total Mn (as MnO) in biomass	1×10^{7} t (0.0055 times)[a]

[a] Relative to manganese in seawater.

Source: Poldervaart A, in *Crust of the Earth. A Symposium. Special Paper 62*, Geological Society of America, Boulder, 1955, 119–144; Mero JL, *Econ Geol.*, 57, 747–767, 1962; and Bowen HJM, *Trace Elements in Biogeochemistry*, Academic Press, London, 1966.

---

around active hydrothermal vents on mid-ocean spreading centers. Manganese concentrations in hydrothermal solution as high as 1002 µmol kg^{-1} (55.05 mg kg^{-1}) have been reported (Von Damm et al., 1985). Although diluted as much as 8500 times within tens of meters away from a vent source when the hydrothermal solution mixes with bottom water, elevated manganese concentrations may be encountered in plumes extending 1 km or more from the vent source (Baker and Massoth, 1986), and in some instances for hundreds of kilometers.

The dominant oxidation state of manganese in seawater is +2 despite the alkaline pH of seawater (7.5–8.3) (Park, 1968) and its E_h of +430 mV (ZoBell, 1946). The stability of the divalent manganese in seawater is attributable to its complexation by chloride ions (Goldberg and Arrhenius, 1958), by sulfate and bicarbonate ions (Hem, 1963), and by organic substances such as amino acids (Graham, 1959).

In water from suboxic zones of the Black Sea and Chesapeake Bay, complexed Mn^{3+} was recently reported to be the prevailing manganese species (up to 100% of total manganese; 5 µmol maximum), as previously mentioned in Section 17.2 (Trouwborst et al., 2006). Although uncomplexed Mn^{3+} in aqueous solution rapidly disproportionates into Mn^{2+} and MnO_2 (see Equation 17.1), this reaction is prevented by the complexation of Mn(III). In the laboratory, pyrophosphate and desferrioxamine have been found to be effective ligands. In nature, specific effective ligands have yet to be identified. Trouwborst et al. (2006) predicted that Mn(III) should be ubiquitous at all oxic and anoxic interfaces in water columns and sediment in the environment.

The oxidation state of manganese in marine sediments and ferromanganese concretions is mainly +4 based on samples taken in the Pacific Ocean, north of the equator and in the Central Indian Ocean (Kalhorn and Emerson, 1984; Murray et al., 1984; Pattan and Mudholkar, 1990). A primary source of manganese in the oceans is that injected by hydrothermal solutions at vents on mid-ocean spreading centers. Magma and volcanic exhalations also make a contribution. Other contributions are from continental runoff and from aeolian sources (windblown dust).

Manganese at the low concentrations found in most of the seawater is important biologically as a micronutrient but is too dilute to serve as energy source without prior concentration. The manganese assimilated by marine organisms can be viewed as transiently immobilized and is returned to the available manganese pool upon death of the organisms.

17.10.1 MICROBIAL MANGANESE OXIDATIONS IN BAYS, ESTUARIES, INLETS, THE BLACK SEA, ETC.

Thiel (1925) detected manganese-oxidizing bacteria in marine mud collected in Woods Hole, Massachusetts. He also found that in anaerobic culture, sulfate-reducing bacteria precipitated $MnCO_3$. Krumbein (1971) reported the presence of heterotrophic bacteria and fungi in the Bay of Biscay and in the Heligoland Bight (North Sea) by culturing sediment samples collected at these locations. In the Bay of Biscay, the organisms were present in all samples collected in the uppermost millimeters of sediment at water depths of 13–180 m, but few (in only one of several samples taken) between 280 and 540 m. In plate counts from sediment and water samples from the estuary of the River Tamar and the English Channel, Vojak et al. (1985a) found that 11–85% of the total bacterial population formed colonies representing manganese-oxidizing bacteria. The plating was done on an $MnSO_4$-containing culture medium prepared in seawater (3% sea salt) with peptone and yeast extract. Vojak et al. (1985a) identified the manganese-oxidizing colonies by applying berbelin blue reagent (Krumbein and Altmann, 1973), which is turned blue by oxidized manganese. No obvious correlation between total number of bacteria, proportion of manganese oxidizers, and particulate load or salinity was noted, nor were any seasonal trends with regard to distribution of manganese oxidizers observed. Manganese oxidation rates in water from Tamar Estuary amended with 2 mg L^{-1} of Mn^{2+} were 3.62 µg L^{-1} h^{-1} in freshwater containing 30 mg L^{-1} suspended matter and 0.7 µg L^{-1} h^{-1} in saline water (23‰ salinity) with the same amount of suspended matter. The rate of manganese oxidation was proportional to particulate load up to 100 mg L^{-1}. Oxidation was depressed in the presence of metabolic inhibitors such as chloramphenicol (100 µg mL^{-1}). The oxidation had a temperature optimum of 30°C, which was above the *in situ* temperature (13.5°C) when the water sample was taken (Vojak et al., 1985b).

Saanich Inlet, on the southeast side of Vancouver Island, British Columbia, Canada, presents an interesting natural setting to study bacterial Mn^{2+} oxidation. As described by Anderson and Devol (1973) and Richards (1965) (see also Tebo and Emerson, 1985), it is a fjord having a maximum water depth of 220 m and features a sill at its mouth at a depth of 70 m. Water behind the sill develops a chemocline in late winter and summer, becoming anoxic below 130 m. The anoxic condition prevails for about 6 months, after which it is displaced by dense oxygenated water pushed over the sill as a result of strong coastal upwelling. Bacterial Mn^{2+}-oxidizing activity was measured just above the O_2 and H_2S interface and found to be O_2 limited and temperature dependent. Excess concentration of Mn^{2+} inhibited its oxidation (Emerson et al., 1982; Tebo and Emerson, 1985). Removal of manganese from solution involved both binding to particulates (bacterial cells and inorganic aggregates) and oxidation of manganese. The oxidation state of particulate manganese in the water samples upon collection was in the range of 2.3–2.7, suggesting possible *in situ* formation of Mn(III) and mixed Mn(II)–Mn(IV) oxides (Emerson et al, 1982). Mn(II) binding and oxidation was accompanied by Co(II) binding in the chemocline (Tebo et al., 1984).

In Framvaren Fjord, Norway, evidence of bacterial Mn^{2+} oxidation has also been observed above the chemocline, which is located in the euphotic zone at a depth of 17 m and persists all year (Jacobs et al., 1985). The manganese oxidation rate measured as O_2-dependent ^{54}Mn binding was greater than in Saanich Inlet (Tebo et al., 1984).

Microbially catalyzed Mn(II) oxidation has been observed in the suboxic zone in the Black Sea (Tebo, 1991; Schippers et al., 2005). This oxidation is not anaerobic but is driven by lateral intrusions of O_2 into this zone (Schippers et al., 2005). No anaerobic Mn(II) oxidizers were detected in enrichment-culture experiments with samples from the study site (Schippers et al., 2005).

17.10.2 Manganese Oxidation in Mixed Layer of Ocean

A comparison of the manganese-oxidizing activity in the mixed layer (waters above the thermocline) of the Sargasso Sea (Atlantic Ocean) at the Bermuda Atlantic Time Series Station (BATS) and two stations in the Pacific (0°N, 140°W and 9°N, 147°W) in 1991 (Moffett, 1997) led to the following conclusions. Particulate manganese formation in the Sargasso Sea is chiefly the result of microbial oxidation throughout the year. In the Pacific, no microbial oxidation was measured above 175 m. The formation of particulate manganese (oxide) was inhibited in the presence of azide, indicating biological involvement. The particulates were not dissolved by ascorbate. Whereas in the Pacific, light stimulated nonoxidative biological $^{54}Mn^{2+}$ uptake suggesting phytoplankton involvement in the Sargasso Sea, light inhibited oxidative biological uptake. Interestingly, Moffett and Ho (1996) found Mn^{2+}-oxidizing bacteria in Waquoit Bay, Massachusetts, that also have a capacity to oxidize Co^{2+} to Co^{3+}. A common enzyme appears to be involved. In the Sargasso Sea, by contrast, they found that Co was taken up by particulates nonoxidatively in a light-dependent process, suggesting phytoplankton involvement. In this body of water, Co uptake is thus completely decoupled from Mn uptake into particulates. Tebo and Lee (1993) previously reported Co^{2+} oxidation by spores of marine *Bacillus* SG-1. Very recently, however, Murray et al. (2007) found that spores of *Bacillus* SG-1 oxidized Co(II) indirectly when Co(II) and Mn^{2+} occur together. The mechanism seems to involve the enzymatic oxidation of Mn^{2+} to Mn(IV) oxide by the spores followed by abiotic oxidation of Co(II) to Co(III) by the Mn(IV) oxide. They explained previous findings of enzymatic Co(II) oxidation by the spores as most likely having been due to low-level contamination of the Co(II) substrate.

17.10.3 MANGANESE OXIDATION ON OCEAN FLOOR

Manganese oxides can be found in large quantities in concretions (nodules) or crusts on the ocean floor at great distances from hydrothermal discharges, where the rate of sedimentation is low (Figure 17.15) (Margolis and Burns, 1976). Such concretions have been found in all the oceans of

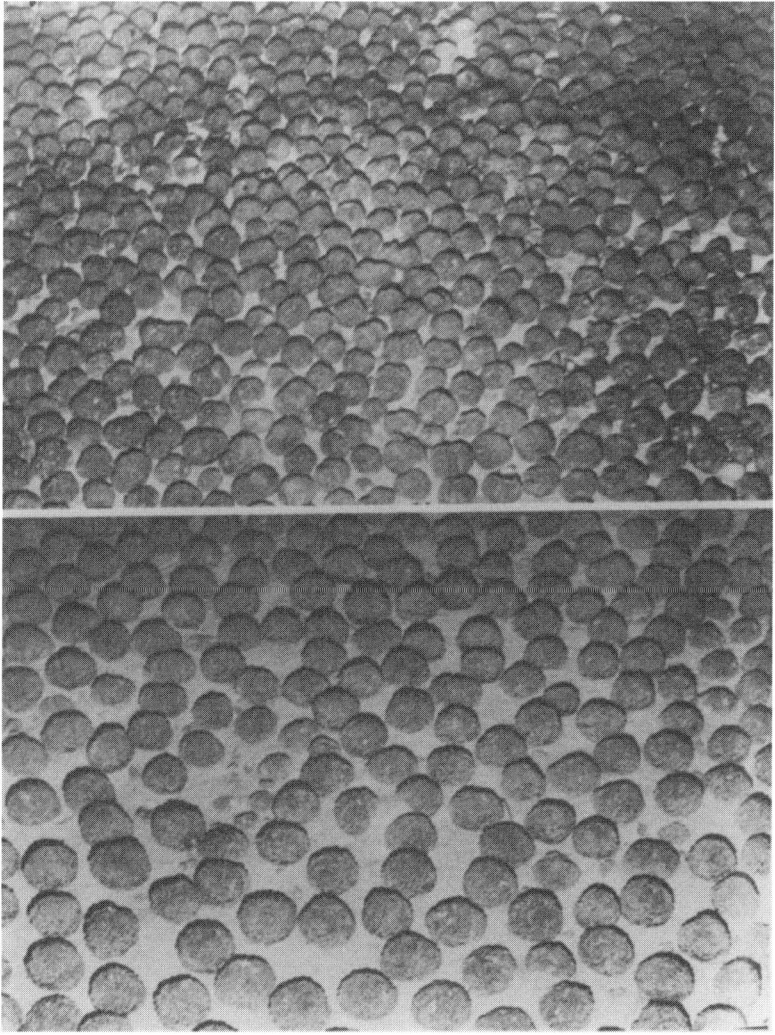

FIGURE 17.15 A bed of ferromanganese nodules on the ocean floor at a depth of 5292 m in the southwest Pacific Ocean at 43°01′S and 139°37′W. Nodules may range in size from <1 to 25 cm in diameter. Average size has been given as 3 cm. (Courtesy of Woods Hole Oceanographic Institute, Woods Hole, MA, USA.)

the world (Horn et al., 1972). They may cover vast areas of the ocean floor, as on some parts of the Pacific Ocean floor, or be distributed in patches. Typical composition of such nodules is given in Table 17.7. The chemical components of a nodule are not evenly distributed throughout its mass (Sorem and Foster, 1972). When examined in cross section, nodules are seen to have developed around a nucleus, which may be a foraminiferal test, a piece of pumice, a shark's tooth, a piece of coral, an ear bone from a whale, an older nodule fragment, etc.

The oxidation state of the manganese in deep-sea nodules is mainly +4 (Murray et al., 1984; Piper et al., 1984; Pattan and Mudholkar, 1990). Manganese occurs as todorokite, birnessite, and δMnO_2 (disordered birnessite, according to the nomenclature of Burns et al., 1974). The Mn(IV) oxides in the nodule have a strong capacity to scavenge cations (Crerar and Barnes, 1974; Ehrlich et al., 1973; Loganathan and Burau, 1973; Varentsov and Pronina, 1973), particularly Mn^{2+} and other cations of transition metals of the first series. The nodules thus serve as concentrators of divalent manganese and some other ions in seawater and are therefore able to furnish Mn(II) in sufficient concentration to serve as an energy source to Mn(II)-oxidizing bacteria that live on nodules. This makes ferromanganese nodules a selective habitat for these bacteria and other organisms that may depend on them directly or indirectly. Electron microscopic examination and culture tests have shown the presence of various kinds of bacteria on the surface and within nodules (Figure 17.16;

TABLE 17.7
Average Concentration of Some Major Constituents of Manganese Nodules from the Pacific Ocean (Percentage by Weight)

Mn	24.2	Mg	2.7
Fe	14.0	Na	2.6
Co	0.35	Al	2.9
Cu	0.53	Si	9.4
Ni	0.99	LOI[a]	25.8

[a] LOI, loss on ignition.
Source: Mero JL, *Econ. Geol.*, 57, 747–767, 1962.

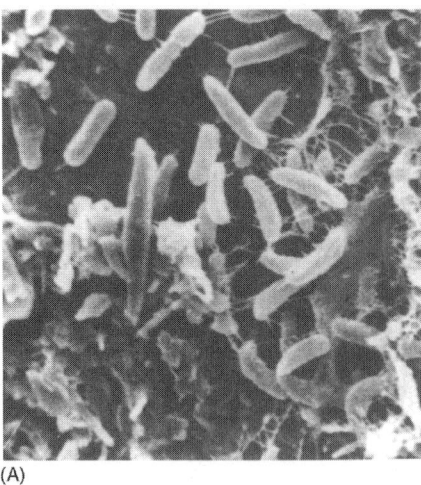
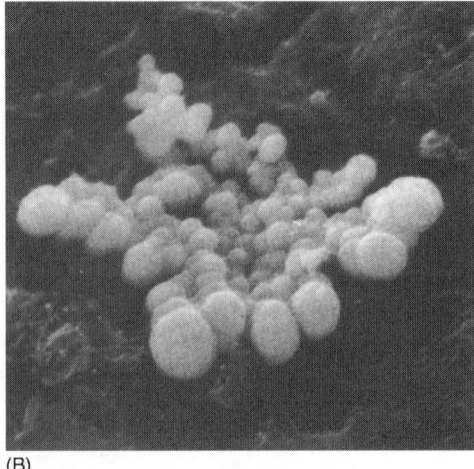

(A)   (B)

FIGURE 17.16 Scanning electron photomicrographs of bacteria attached to the surface of ferromanganese nodules from Blake Plateau, off the southern Atlantic coast of the United States. Note the slime strands anchoring the rod-shaped bacteria to the nodule surface. (Reproduced from LaRock PA, Ehrlich HL, *Microb. Ecol.*, 2, 84–96, 1975. With permission of Springer Science and Business Media.)

LaRock and Ehrlich, 1975; Burnett and Nealson, 1981; Ehrlich et al., 1972). Their numbers have been found to range from hundreds to tens of thousands per gram of nodule, as determined by plate counts on seawater-nutrient agar at 14–18°C and atmospheric pressure (Ehrlich et al., 1972; see also Sorokin, 1972). These numbers are probably underestimates because the organisms grow in clumps or microcolonies on or in a nodule from which they cannot be easily dislodged. Indeed, it is necessary to plate suspended, crushed nodule material rather than the washings of nodule fragments to make the counts.

The microbial population on nodules includes three types of bacteria when considered in terms of the action on manganese compounds (Ehrlich et al., 1972; Ehrlich 2000). The three types are Mn(II) oxidizers (Group II, see Section 17.5), Mn(IV) reducers (Section 17.6), and a group that can neither oxidize Mn(II) nor reduce Mn(IV). In the nodules examined by Ehrlich et al. (1972), the Mn(IV) reducers were most numerous. These findings must not, however, be interpreted to mean that the Mn(IV) in the nodules examined was necessarily undergoing active reduction. Neither the Mn(II) oxidizers nor the Mn(IV) reducers are dependent on respective forms of manganese to grow. Thus, it is quite conceivable that the nodules were undergoing net manganese accretion at the time of collection. A plentiful supply of Mn(II) and oxygen as well as a deficit in organic carbon are needed to favor Mn(II) oxidation, whereas an adequate supply of oxidizable organic carbon is needed to favor Mn(IV) reduction. The bacteria that neither oxidize nor reduce manganese may play an important role in keeping the level of organic carbon low, thereby favoring the Mn(II)-oxidizing activity by the Mn(II)-oxidizing bacteria. The inability of Sorokin (1972) to culture Mn(II)-oxidizing bacteria from his nodule specimens may have had to do with the culturing method and the method of detection of manganese oxidation he employed.

Most of the organisms found on nodules are gram-negative rods (Ehrlich et al., 1972), although gram-positive bacilli, micrococci, and *Arthrobacter* have been isolated from them (Ehrlich, 1963). A curious and as-yet-unexplained finding has been that a significantly greater number of isolates recovered from the central Pacific than those from the eastern Pacific were unable to grow in freshwater media.

All Mn(II)-oxidizing bacterial cultures isolated from nodules and associated sediments by Ehrlich and associates are of the Group II type (see Section 17.5). They oxidize Mn^{2+} only if it is first bound to Mn(IV) oxide, ferromanganese, or certain ferric-iron coated sediment particles (Ehrlich, 1978b, 1982, 1984). Support for a reaction sequence consisting of Mn(II) sorption followed by oxidation of the sorbed Mn(II) to Mn(IV) via a Mn(III) intermediate oxidation state was recently offered by a study by Pecher et al. (2003). As previously mentioned, Mn(IV) oxides at neutral to alkaline pH act as scavengers of cations, including Mn^{2+} ions. Given that Mn(IV) oxide is the product of manganese oxidation, the oxidation process generates new scavenging sites, and in this way keeps nodules growing. The scavenging action of Mn(IV) oxides probably explains how other cationic constituents get incorporated into nodules:

$$H_2MnO_3 + X^{2+} \rightarrow XMnO_3 + H^+ \tag{17.16}$$

where X^{2+} represents a divalent cation (e.g., Cu^{2+}, Co^{2+}, Ni^{2+}) (Goldberg, 1954; Ehrlich et al., 1973; Loganathan and Burau, 1973; Varentsov and Pronina, 1973). Iron, if incorporated in this fashion, is probably picked up as Fe(III).

If the mechanism by which the Mn(II)-oxidizing bacteria form Mn(IV) oxide is according to the model proposed in Figure 17.5B, it may be directly incorporated into preexistent Mn(IV) oxide, or, if formed on a clay particle, it may form the seed particle to which subsequently formed Mn(IV) oxide is added in the oxidation of scavenged Mn(II).

The growth rate of manganese nodules in the deep sea is reportedly very slow. Ku and Broecker (1969) and Kadko and Burckle (1980), for instance, reported rates ranging from 1 to 10 mm per 10^6 years, based on radioisotope dating methods. Reyss et al. (1982) found a nodule in the Peru Basin whose rate of growth they estimated radiometrically to have been 168 ± 24 mm per 10^6 years. These estimates assume a constant rate of growth. Heye and Beiersdorf, (1973) found variable growth rates varying from 0 to 15.1 mm per 10^6 years. In other words, the nodules these investigators examined did not grow at constant rates. This suggests that conditions must not be continually favorable for nodule growth and that quiescent periods, and perhaps even periods of nodule degradation, very likely aided by the Mn(IV)-reducing bacteria, must intervene between growth periods. Bender et al. (1970) estimated manganese accumulation rates from five nodule specimens as ranging from 0.2 to 1.0 mg cm^{-2} per 1000 years. If manganese incorporation and therefore nodule growth is an intermittent process, actual growth rates may be somewhat faster than above estimates from radiodating, but probably by no more than an order of magnitude. Rates of manganese oxidation with bacteria isolated from nodules determined under optimal conditions in the laboratory (15°C, hydrostatic pressure of ~1 atm) were many orders of magnitude faster than any estimated *in situ* rates of manganese incorporation into nodules (and therefore nodule growth rates). By contrast, growth rates of nodule bacteria and bacterial manganese oxidation rates are greatly slowed in laboratory experiments by applying *in situ* hydrostatic pressure (e.g., 300–500 atm) and low temperature (e.g., 4°C) (Ehrlich, 1972). Hence, the slow *in situ* growth rates of nodules are not a sufficient argument against microbial participation in nodule genesis. Conditions appear not to be continually favorable for growth.

Other biogenic mechanisms of nodule growth have been evoked in the past. Butkevich (1928) tried to explain the growth in terms of iron precipitation by *Gallionella* and *Persius marinus*, the latter newly isolated by him but not otherwise known, which he found associated with brown deposits of the Petchora and White Seas. However, Sorokin (1971) concluded that *Gallionella* does not occur in deep-sea nodules, nor could he find evidence for the presence of *Metallogenium*. The only types of bacteria that he detected in nodules were some other heterotrophs. The Petchora and White Seas are, of course, marginal seas and untypical of the open ocean, which may explain Butkevich's findings. Novozhilova and Berezina (1984) did report finding *Gallionella* in samples from the Atlantis Deep I in the Red Sea. They found *Metallogenium*, *Caulococcus*, *Siderococcus*, and *Naumaniella* besides some other types of bacteria in samples from the northwestern Indian Ocean. Thus, these terrestrial types of bacteria can be isolated from marine settings. It is, however, very likely that the stations from which they collected their samples were subject to continental influences that contributed these bacteria. Graham (1959) proposed that manganese nodules form as a result of bacterial destruction of organic complexes of Mn(II) in seawater, liberating Mn^{2+} ions, which then autoxidize and precipitate, the oxides collecting on solids such as foraminiferal tests, which become the nuclei around which manganese nodules develop. According to Graham, other trace metals could be deposited in a similar manner. He felt that amino acids or peptides were the complexing agents of Mn^{2+} that were attacked by the bacteria. He detected amino acid-like material in nodules from Blake Plateau. Graham and Cooper (1959) were able to recover foraminiferal tests from Blake Plateau, among which arenaceous ones were coated with a veneer of manganese-rich material containing Cu, Ni, Co, and Fe. Because the manganese-rich coating was on the surface of the tests, the authors reasoned that the manganese was deposited on foraminifera after their death. They found calcareous foraminiferal tests to be free of this deposit. They implicated protein-rich coating on arenaceous tests as responsible for providing a habitat for organisms (bacteria?) that remove trace element chelators from seawater.

Kalinenko et al. (1962) visualized the origin of ferromanganese nodules in terms of bacterial colonies growing on ooze particles. The bacteria on these ooze particles were thought to mineralize the organic matter coating the particles. The authors assumed that in the process, manganese and iron oxides are formed and deposited on the colonies together with other trace elements. Slime formed by the bacteria was assumed to cause the deposits to agglutinate, producing micronodules. The investigators studied this process in laboratory experiments by observing bacterial development on glass slides introduced into oozes from the bottom of several Indian Ocean stations. They described the bacteria as the living cement that holds the nodules together.

Ghiorse (1980) examined the surface layers of ferromanganese concretions from the Baltic Sea by scanning and transmission electron microscopy and by light microscopy. He noted filamentous structures (possibly fungal hyphae) and single cells, microcolonies, or aggregates of bacteria on the outer surface and some rods and cocci within a matrix of polymerlike material, and occasionally sheathed bacteria. He also noted budding and star-shaped bacteria and bacteria with internal membranes (possibly cyanobacteria, purple photosynthetic bacteria, or methanotrophic bacteria). The bacterial assemblage described by Ghiorse reflects a not unexpected terrestrial influence on the development of the Baltic Sea concretions and suggests an involvement of some different microbes in the growth of these nodules.

Burnett and Nealson (1983) examined the surface of some fragments from North Pacific Ocean nodules by energy-dispersive x-ray analysis. They found regions of high manganese and iron concentrations that outlined microorganismlike objects on the nodule surface. They inferred from this observation that the microorganisms were involved in accretion or removal of the metals because they would have expected the Mn and Fe to have been evenly distributed over the surface if a purely physicochemical process had been involved.

Greenslate (1974a) observed manganese deposition in microcavities of planktonic debris, especially from diatoms, and proposed that such deposition was the beginning of nodule growth. He also found remains of shelter-building organisms such as benthic foraminifera on nodules, which became encrusted and ultimately buried in the nodule structure. He proposed that the skeletal remains provided a framework on which manganese and other components were deposited, perhaps with the help of bacterial action (Greenslate, 1974b). Others have since reported evidence of traces of such organisms on nodules (Fredericks-Jantzen et al., 1975; Bignot and Dangeard, 1976; Dugolinsky et al., 1977; Harada, 1978; Riemann, 1983) (Figures 17.17A, 17.17B, and 17.17C).

The finding that benthic foraminifera and other protozoa have grown and may presently be growing on ferromanganese nodules is of significance in explaining the role of Mn(II)-oxidizing bacteria in nodule growth. Because these foraminifera and other protozoa are *phagotrophic* (i.e., they prey on living bacterial cells), they probably feed on the Mn(II)-oxidizing bacteria, among other types on the nodules. To maintain this food supply, uneaten Mn(II)-oxidizing bacteria must therefore continue to multiply. Thus, Mn(II)-oxidizing bacteria may play a dual role: (1) to aid in manganese accretion to nodules and (2) to serve as food for phagotrophic protozoa. A somewhat different interpretation of the interrelationship of bacteria, protozoa, and tube-building polychaete worms resident on nodules was offered by Riemann (1983). He found that rhizopods (Protozoa) on the nodules accumulated large volumes of fecal pellets (stercomata) containing biogenic and mineral particles (manganese oxides, etc.) in their tests as a result of feeding on bacteria, detritus, and organic primary film on the nodule surface. He proposed that manganese oxides ingested by the protozoa are at least partially reduced, released, and reprecipitated as a result of reoxidation, in part by bacterial activity on the nodule surface. He viewed polychaete worms feeding on the rhizopods as further aiding in the concentration of the manganese–oxide-containing stercomata.

Although the hypotheses of Graham, Greenslate, and to some extent Kalinenko et al. may well have some bearing on the initiation of formation of some nodules, it seems doubtful that they apply to the major growth phase of nodules. Graham and Kalinenko et al. assumed that most of the Mn^{2+} in seawater is organically complexed. According to them, it is incorporated into nodules as preformed Mn(IV) oxide that resulted from autoxidation of the Mn^{2+} freed from organic complexes through ligand degradation by microbes. Incorporation of Mn(IV) oxides into nodules is also a part of the assumption by Riemann (1983), but his hypothesis does not absolutely depend on it. Sorokin (1972) also assumed that the role of heterotrophic bacteria on nodules in nodule genesis was one of digesting organic manganese complexes, releasing Mn^{2+}. However, he postulated that the Mn^{2+} becomes bound to the surface of the nodules and is then abiotically oxidized. Filter-feeding benthic invertebrates that are resident on the nodules were viewed by Sorokin as the source of the organically complexed manganese. None of these proposals give recognition to the fact that most

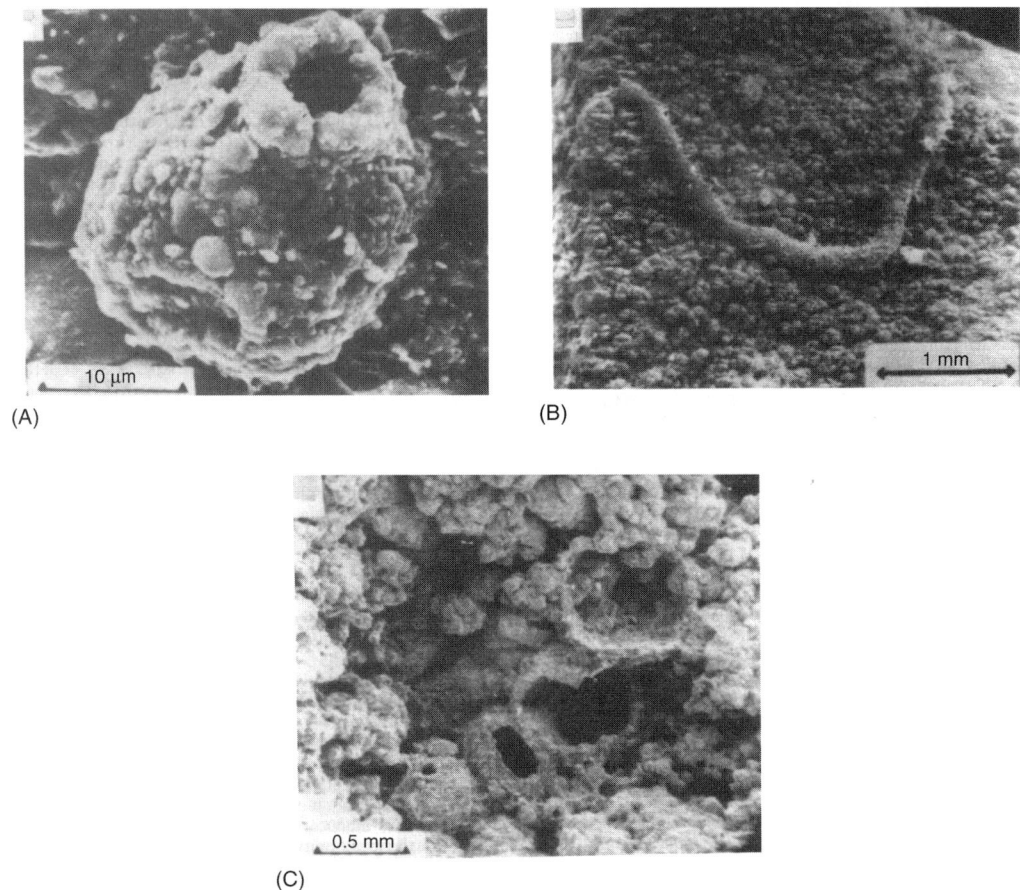

(A)   (B)

(C)

FIGURE 17.17   Benthic, test-forming protozoans inhabiting the surface of ferromanganese nodules. (A) Fresh remains of chambered encrusting protozoans on a nodule surface, showing siliceous biogenic material used in test construction. (B) Surface of a nodule with a partial test of *Saccorhiza ramosa*, the most common and longest of any tubular agglutinating foraminifera yet found on nodules. (C) Test of unidentified form composed almost entirely of manganese micronodules. (From Dugolinsky BK, Margolis SV, Dudley WC, *J. Sediment. Petrol.*, 47, 428–445, 1977. With permission.)

manganese in seawater, as previously pointed out, exists as inorganic complexes from which Mn^{2+} is readily adsorbed by the nodules.

17.10.4   MANGANESE OXIDATION AROUND HYDROTHERMAL VENTS

A special type of biological community driven by geothermal rather than radiant energy from the sun has been discovered around hydrothermal vents situated on mid-ocean spreading centers. Because sunlight does not reach the ocean depths where these communities exist, and because introduction of nutrients by thermal convection from overlying surface waters does not furnish sufficient nutrients, the ecosystem depends on primary producers consisting of chemolithotrophic bacteria. The consumers in the community depend on these lithotrophs, directly at the first trophic level, and indirectly at other trophic levels. Indeed, at the first trophic level, many of the consumers have established an intimate symbiotic relationship with some primary producers (see Chapter 20 and discussions by Jannasch and Wirsen, 1979; Jannasch, 1984; and Jannasch and Mottl, 1985). The available energy sources for primary producers (chemolithotrophic bacteria) vary, depending on the output of the vents. Potential energy sources include reduced forms of sulfur (H_2S, S^0, $S_2O_3^{2-}$), H_2, NH_4^+, NO_2^-, Mn^{2+}, CH_4, and CO (Jannasch and Mottle, 1985), and iron sulfide. Although ferrous iron is frequently a major constituent of hydrothermal fluid from black smokers (see later), it is least likely to serve as an energy source because of its great tendency to rapidly precipitate as iron sulfide at the mouth of a vent when the hydrothermal solution meets cold, oxygenated bottom water. Most of any iron that escapes sulfide precipitation quickly autoxidizes. Available electron acceptors in solution for bacteria in the vent community include O_2, NO_3^-, S^0, SO_4^{2-}, CO_2, and Fe^{3+} (Jannasch and Mottl, 1985), and very likely Mn(IV) oxide.

The hydrothermal solution results from deep penetration (as much as 1–3 km) of seawater into the basalt beneath the ocean floor at mid-ocean spreading centers. When a magma chamber under-lies the region of seawater penetration, the heat diffusing from the chamber (the temperature may be in excess of 350°C) and the high hydrostatic pressure exerted at this site cause the seawater to react with the basalt. Seawater sulfate may be chemically reduced to sulfide by ferrous iron in the basalt (a rare example of chemical sulfate reduction in nature), bicarbonate may become reduced to methane, and Mg^{2+} from seawater may react with silica in the basalt to form talc and protons according to the reaction (Shanks et al., 1981)

$$Fe^{2+} + 2Mg^{2+} + 4H_2O + 4SiO_2 \rightarrow 6H^+ + FeMg_2Si_4O_{10}(OH)_2 \tag{17.17}$$

and render the altered seawater acidic. The acidity (protons) aids in leaching base metals (Mn^{2+}, Cu^{2+}, Zn^{2+}, Co^{2+}, etc.) from the basalt. Acidity may also be generated in the formation of calcium compounds from basalt constituents and seawater calcium (Bischoff and Rosenbauer, 1983; Seyfried and Janecky, 1985). As a result, the altered seawater (hydrothermal solution) becomes loaded with H_2S, Fe^{2+}, Mn^{2+}, Cu^{2+}, Zn^{2+}, etc. (Jannasch and Mottle, 1985).

Vents on the mid-ocean spreading centers from which the hydrothermal solution enters the ocean are differentiated between the so-called white smokers and black smokers. *White smokers* emit gases and a hydrothermal solution that has a temperature between 6 and 23°C at the mouth of their vents. *Black smokers* emit gases and a hydrothermal solution that has a temperature near 350°C at the mouth of their vents. In the case of white smokers, the upward-moving hydrothermal solution meets downward-moving cold seawater and mixes with it before emerging from the vents. As a result of this mixing with seawater, the hydrothermal solution loses a significant portion of its metal charge through precipitation in the rock stratum in which this mixing occurs. Consequently, when this mixed solution emerges from the vent, it still contains a significant amount of H_2S and Mn^{2+} but little iron. In the case of black smokers, the upward-moving hydrothermal solution does not encoun-ter significant amounts of downward-moving cold seawater and is little altered when it issues from vents. On emergence and mixing with cold seawater, voluminous amounts of black iron, copper, and zinc sulfides precipitate around the mouth of a black smoker vent and may be deposited in the form of a chimney along with anhydrite ($CaSO_4$) (Figure 17.18). Much of the iron in the hydrothermal solution precipitates as sulfide, but that which does not precipitates as iron oxide (FeOOH). A major portion of the manganese (Mn^{2+}) remains in solution and may be carried a considerable distance (tens of kilometers) before it is precipitated.

Although H_2S appears to be the energy source most widely used by the primary producers of vent communities at mid-ocean spreading centers, Mn^{2+}, because of its relative stability in seawater, could also be a potential energy source. Indeed, Mn^{2+}-oxidizing bacteria (Group I, Subgroup Ia, see Section 17.5) have been isolated from water samples collected around a white smoker (Mussel Bed Vent) on the Galapagos Rift (Ehrlich, 1983) and black smokers on the East Pacific Rise at 21°N (Ehrlich, 1985) and 10°N (Ehrlich, unpublished results). Mandernack (1992) and Mandernack and Tebo (1993) measured moderate-to-high microbial Mn removal (oxidizing activity) at the source of Mn plumes on a hydrothermal vent field on the Galapagos Rift at 50–100 m above the mouth of the vent and ~10 km distance. This study included Mussel Bed Vent. Similarly, Mandernack measured low-to-high microbial Mn removal (oxidizing activity) at the source of Mn plumes on the Juan de

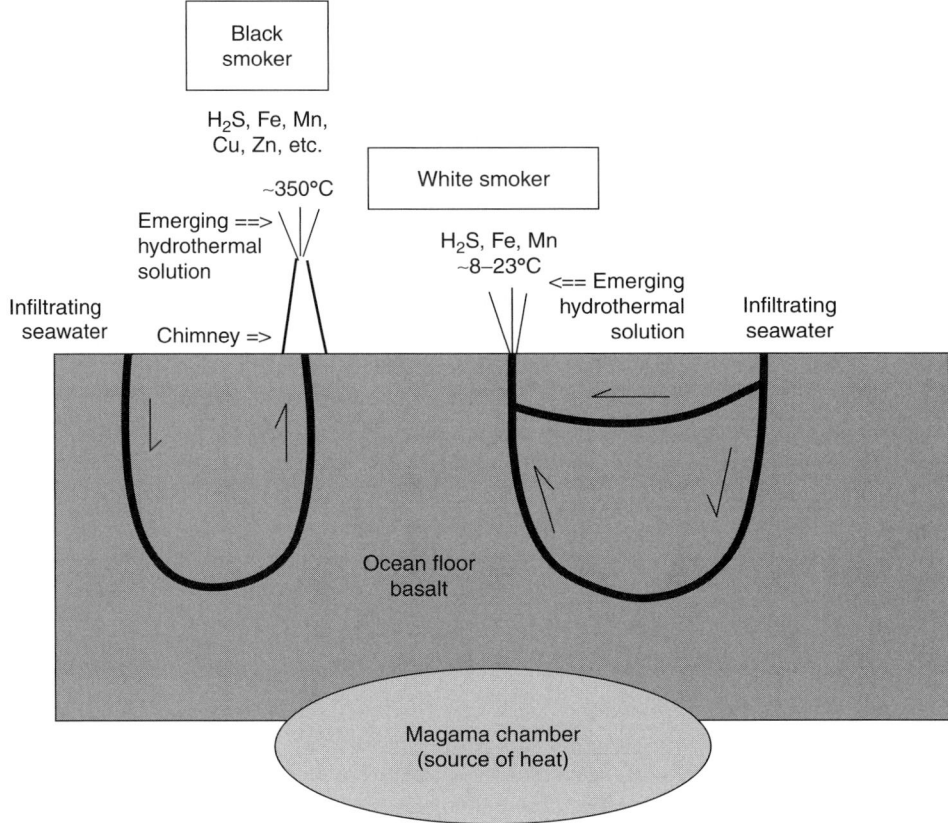

FIGURE 17.18 Schematic representation of the origin of hydrothermal solution from black smoker and white smoker vents at mid-ocean rifts. (Adapted from Jannasch HW, Mottle MJ, *Science*, 229, 717–725, 1985.)

Fuca hydrothermal vent field on Endeavor Ridge at 50–200 m above the mouth of the vent and ~10 km distance.

Two bacterial isolates from around Mussel Bed Vent and three from around the vent at 21°N on the East Pacific Rise have been studied in some detail (Ehrlich, 1983, 1985). They are all gram-negative bacteria and, depending on the isolate, have the shape of short or curved rods or spirilla (Figures 17.19A and 17.19B). The isolates from around the white smoker grew in a temperature range of 4–37°C, and those from around the black smoker in a temperature range of 5–45°C. None grew in broth prepared in freshwater. The two isolates from Mussel Bed Vent also did not grow in nutrient broth prepared with 3% NaCl. Both isolates from Mussel Bed Vent and two of the isolates from the vent at 21°N, East Pacific Rise, are examples of Group I, Subgroup Ia (see Section 17.5) manganese oxidizers, and the third, isolated from 21°N, East Pacific Rise, is an example of a Group II manganese oxidizer (see Section 17.5). The manganese-oxidizing system was inducible in all isolates, even in the Group II manganese oxidizer. This is in contrast to all Group II manganese oxidizers isolated from ferromanganese nodules and associated sediments in which the manganese-oxidizing system was constitutive. All five vent isolates appear to be able to obtain energy from $Mn(II)$ oxidation. This conclusion was reached on the basis of inhibition studies of manganese oxidation with electron transport inhibitors and in the case of the Mussel Bed Vent isolates by direct determination of ATP synthesis coupled with Mn^{2+} oxidation (Ehrlich, 1983, 1985). In the Mussel Bed Vent isolates, ATP synthesis was very tightly coupled with Mn^{2+} oxidation, and growth in batch culture showed significant stimulation by Mn^{2+} in initial stages, which may be attributable to mixotrophy (Ehrlich and Salerno, 1988, 1990).

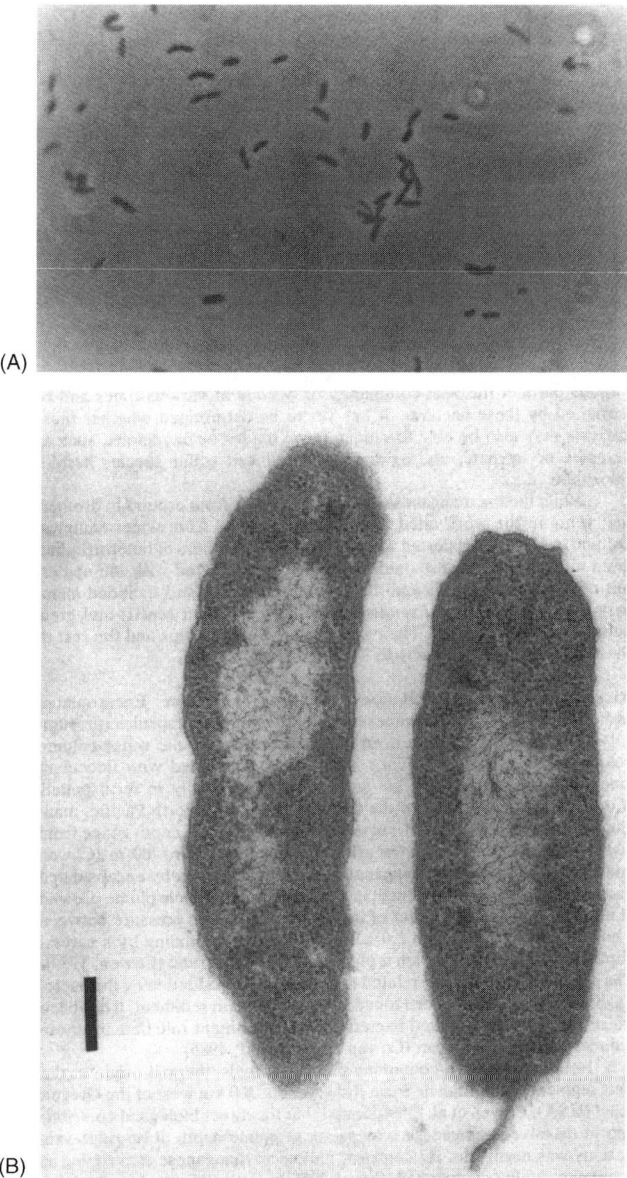

FIGURE 17.19   Bacterial strain SSW$_{22}$ isolated from a seawater sample from near a hydrothermal vent on the Galapagos Rift. (A) Typical morphology in a seawater medium containing tryptone, dextrose, yeast extract, and MnSO$_4$. Note range of cell size, typical of growth in this medium (×1500). (B) Longitudinal sections of SSW$_{22}$ viewed by electron microscopy (×32,250). (Courtesy of Ghiorse WC, Dept. of Microbiology, Cornell University, Ithaca, NY, USA.)

The two isolates from around Mussel Bed Vent were metabolically very versatile. In addition to being able to oxidize Mn^{2+}, they were also able to reduce MnO$_2$ aerobically and anaerobically with glucose, succinate, and acetate as electron donors (Ehrlich et al., 1991; Ehrlich, unpublished data). Their MnO$_2$ reductase was inducible. They were also able to oxidize thiosulfate aerobically and reduce tetrathionate anaerobically (Tuttle and Ehrlich, 1986). The ability to use this variety of electron donors and acceptors suggests that these organisms are opportunists that will use whatever energy source or terminal electron acceptor is in plentiful supply. Mn^{2+} emitted by the vents, thiosulfate, and tetrathionate formed in partial chemical or biological oxidation of H$_2$S emitted by the vents, or MnO_2 resulting from biological or chemical oxidation of Mn^{2+} can all be expected to occur in various parts of the vent community or beyond at various times. Each can be exploited by these bacteria. It is yet to be determined whether these bacteria may also be able to reduce ferric oxides or oxyanions, such as arsenate or selenate, and oxidize reduced sulfur species besides thiosulfate.

Although the manganese-oxidizing cultures from around hydrothermal vents in the work cited earlier were obtained from water samples, Durand et al. (1990) reported finding significant numbers of heterotrophic, gram-negative manganese-oxidizing bacteria associated with the epidermis of polychaete worms and their tubes from vent sites. The bacteria included members of *Aeromonas* and *Pseudomonas*. They oxidized Mn(II) and grew better at 40°C than at 20°C. Their significance to the worms and the rest of the vent community remains to be established.

17.10.5 BACTERIAL MANGANESE PRECIPITATION IN SEAWATER COLUMN

Encapsulated bacteria with iron and manganese precipitates in their capsules (presumably their glycocalyx) have been detected in the oceanic water column below 100 m (Figure 17.20). They were found associated with flocculent amorphous aggregates (marine snow) and occasionally in fecal pellets (Cowen and Silver, 1984). In the eastern subtropical North Pacific, manganese deposits on bacterial capsules were absent in a depth range from 100 to 700 m but became increasingly noticeable below 700 m (Cowen and Bruland, 1985). Such manganese-scavenging activity by encapsulated bacteria was also very prominent in a hydrothermal vent plume (Cowen et al., 1986). A positive effect of increased hydrostatic pressure between 1 and 200 bar (1.01×10^2 to 2.03×10^4 kPa) on Mn^{2+} binding by a natural population of bacteria in such a plume has been observed (Cowen, 1989). The pressure effect may be related to exopolymer production by the bacteria. Because bacterial capsules have been found to be abundant in sediment, it has been suggested that encapsulated bacteria play a prominent role in manganese sedimentation in the ocean (Cowen and Bruland, 1985).

In the case of Mn-containing plumes of hydrothermal origin at the cleft segment of the Juan de Fuca Ridge, ~300 km west of the Oregon coast, Cowen et al. (1990) found that the direct biological

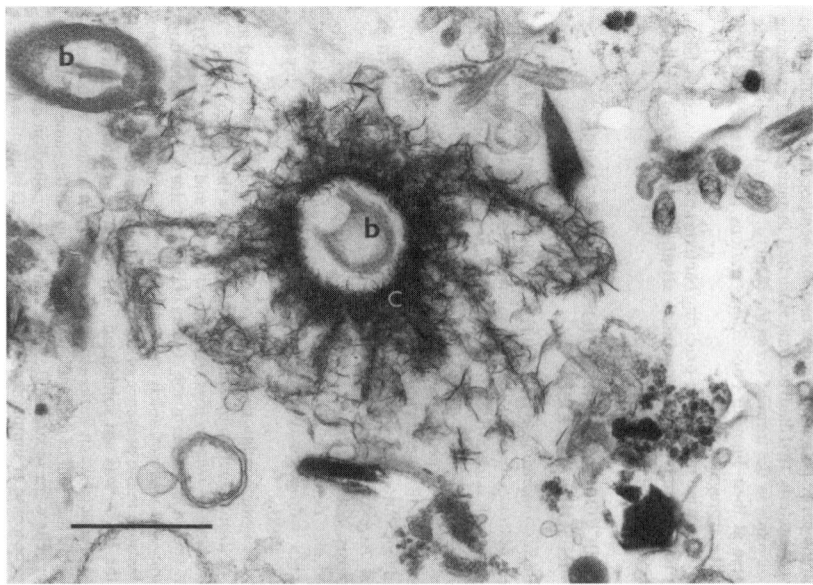

FIGURE 17.20 Marine bacteria encapsulated in manganese oxide. Transmission electron photomicrographs of bacterial cell (b) and bacteria capsule (c). Scale bar = 1 μm. (Reprinted from Cowen JP, Massoth GJ, Baker ET, *Nature*, 322, 169–171, 1986. With permission of Macmillan Magazines Ltd.)

contribution to dissolved-manganese scavenging at plume depth at stations on the ridge axis was negligible. By contrast, biological manganese scavenging at nonplume depths contributed almost 50% to the total process. At most off-ridge axis stations, most of the dissolved-manganese scavenging was almost totally biological at plume depth. Manganese scavenging rates from a vent plume ranged from 1.7 to 3.4 mM m^{-2} per year. In Gorda Ridge vent plumes, Cowen et al. (1998) found that the ratio of encapsulated Mn-precipitating bacteria to total bacteria was greater in a day-old plume than in older plumes. The higher ratio in recently formed plumes may have been due to resuspension of encapsulated bacteria in disturbed sediment, whereas the lower ratio in older plumes was due to aggregation and particle settlement.

Tambiev and Demina (1992) found dissolved manganese to be much more rapidly converted to particulate manganese in the southern trough of the Guaymas Basin, Gulf of California (Sea of Cortez) than in the caldera of the Axial Mountain on the Juan de Fuca Ridge. They suggested that the rapid conversion of dissolved manganese in the Guaymas Basin involved participation of bacteria, and that benthopelagic zooplankton grazed on bacteria that had become enriched in manganese, excreting manganese oxide in the form of vernadite (MnO_2). They found particulate manganese at their Juan de Fuca stations to be associated with bacterialike aggregates.

Despite the foregoing report of manganese oxide deposition on bacterial cell envelopes, it should be stressed that of the marine manganese-oxidizing bacteria from various sources studied in various laboratories to date, only some accumulated significant amounts of oxidized manganese on their cell envelope (see discussion by Ehrlich, 1999).

17.11   MICROBIAL MOBILIZATION OF MANGANESE IN SOILS AND ORES

17.11.1   Soils

Because Mn(III) and Mn(IV) oxides are water-insoluble, their manganese is not directly available as a micronutrient to plants and soil microbes, which have a need for it in their nutrition. The manganese in these oxides must be reduced to Mn(II) to be available nutritionally. Mn^{2+} that is sorbed by the oxides as well as by clays may become available by ion exchange. Reduction of manganese oxides may be brought about by bacterial respiration in which Mn(IV) or Mn(III) oxide replaces oxygen as terminal electron acceptor. As explained earlier, this process is not always dependent on anaerobiosis, although bacteria that reduce MnO_2 exclusively under anaerobic conditions exist and may dominate in a given environment. The electron donors are usually organic compounds, but hydrogen or formate may serve under anaerobic conditions (Section 17.6; Lovley and Goodwin, 1988).

Under some conditions, Mn(IV) oxides in soil can be solubilized as a result of their reduction by elemental sulfur (S^0) or thiosulfate $S_2O_3^{2-}$ in the presence of *Thiobacillus* (now *Acidithiobacillus*) *thiooxidans* (Vavra and Frederick, 1952). This was shown in soil perfusion studies in the laboratory. The production of sulfuric acid was not solely responsible for the solubilization of the MnO_2 because in the presence of acid, but in the absence of any reduced sulfur, the MnO_2 was not solubilized. Although the greatest quantity of MnO_2 was reduced when *T. thiooxidans* cells were in direct contact with MnO_2, slightly more than half of the MnO_2 was reduced under the same conditions when the cells were separated from the MnO_2 by a collodion membrane. This may be a case of simultaneous direct and indirect reduction of MnO_2, the indirect reduction being the result of chemical reaction with partially reduced sulfur species generated in the bacterial oxidation of sulfur. A field study in Indiana, the United States, confirmed that S^0 and $S_2O_3^{2-}$ can mobilize fixed manganese in an agriculturally manganese-deficient soil (Garey and Barber, 1952).

As previously pointed out, ferrous iron and H_2S produced in microbial Fe(III) and sulfate respiration, respectively, can chemically reduce Mn(IV) oxides in soil and sediments (Burdige and Nealson, 1986; Lovley and Phillips, 1988b; Myers and Nealson, 1988b; Nealson and Saffarini, 1994). Reduction of manganese oxides of microbial origin in soil can also be brought about chemically by some constituents of root exudates as from oats and vetch, especially at acidic pH (Bromfield, 1958a,b). Bacteria are apparently not needed to promote Mn(IV) reduction by active root exudate components (Bromfield, 1958a). Malate is an active root exudate component that has been shown to reduce Mn(IV) oxide chemically (Jauregui and Reisenauer, 1982). However, in view of the widespread ability of various types of bacteria to reduce MnO_2 and their widespread occurrence, bacterial reduction of oxides of Mn(IV) in soil is probably a more important factor in remobilizing fixed manganese in nature.

Enumeration of viable manganese reducers in soil by culture methods is an incompletely solved problem. In the past, attempts have been made to use heterotroph-supporting agar medium containing hydrous manganese oxide for differential plate counts (Schweisfurth, 1968; Tortoriello, 1971). MnO_2-reducing colonies have been identified as those that develop a clear halo around them, suggesting reduction of the manganese oxide in the area of the halo. The absence of manganese oxide in the halo can be confirmed by applying berbelin blue or leucocrystal violet reagent to the colony (Krumbein and Altmann, 1973; Kessick et al., 1972). Although the clear halo unquestionably indicates manganese reduction, enzymatic manganese reduction is not the only possible explanation for its development. Manganese reductase of a bacterial cell, insofar as it is not an extracellular enzyme, cannot work over a distance, meaning that the cell must be in physical contact with any manganese oxide whose reduction the enzyme is to catalyze. Thus, MnO_2 reduction by a bacterial colony that produces a halo around itself in an agar medium may have been due to indirect enzymatic action (via an extracellular electron shuttle) and abiotic action by a metabolically produced reductant or by a medium ingredient. If the medium contains glucose, the manganese oxide reduction may merely have been the result of reaction with residual glucose at acidic pH generated by the colony. An experimentally verifiable model for the action of an electron shuttle is demonstrable with ferricyanide at low concentration as a shuttle proxy in an agar medium containing dispersed MnO_2 powder. The nutrient that serves as the source of reducing power to the microbes with which the medium is inoculated must be one that cannot reduce MnO_2 abiotically under the experimental conditions and not give rise to a metabolic product(s) that could serve as reductant of the MnO_2 in the agar medium. If those conditions are met, a manganese oxide-free halo around a colony from which the MnO_2 particles have disappeared would indicate reduction of the MnO_2 to Mn^{2+} by the ferricyanide shuttle. In that case, the bacteria in the colony reduce the ferricyanide to ferrocyanide, which then diffuses through the medium to an MnO_2 particle and reduces its MnO_2, the ferrocyanide thereby becoming reoxidized to ferricyanide. The ferricyanide then diffuses back to the bacterial colony where it is rereduced to ferrocyanide, and the process is repeated.

Ultimately, molecular biological techniques will be introduced into the enumeration procedure for Mn(III,VI) reducers.

17.11.2 ORES

The autotroph *Acidithiobacillus thiooxidans*, when acting on elemental sulfur, has been shown to be able to extract manganese from ores (Imai and Tano, 1967; Ghosh and Imai, 1985a,b; Kumari and Natarajan, 2001). Imai and Tano leached an ore containing MnO_2 (10.6%), Fe_2O_3 (25%), SiO_2 (55%), MgO (5.23%), and traces of Ca, Al, and S. Ground ore was suspended at a concentration (pulp density) of 3% in a culture medium containing K_2HPO_4 (0.4%), $MgSO_4$ (0.03%), $CaCl_2$ (0.02%), $FeSO_4$ (0.001%), $(NH_4)_2SO_4$ (0.2%), and S^0 (1%). Addition of FeS or $FeSO_4$ stimulated both the growth of *A. thiooxidans* and solubilization of manganese, whereas the addition of ZnS stimulated only the solubilization of manganese; $Fe_2(SO_4)_3$ was without effect. The reducing activity of partially reduced sulfur species such as thiosulfate or sulfite produced by *A. thiooxidans* in the oxidation of elemental sulfur to sulfuric acid was probably the basis for the solubilization of MnO_2 in the ore.

Ghosh and Imai (1985a,b) leached manganese from MnO_2 (>90% pure, −400 mesh size) with *Acidithiobacillus ferrooxidans* and *A. thiooxidans* in medium containing the mineral salts mixture of 9K medium (see Chapter 16) with 1% elemental sulfur but without ferrous sulfate (Silverman and Lundgren, 1959). As with *A. thiooxidans* alone, leaching was probably the result of production of partially reduced, soluble sulfur species, such as SO_3^{2-}, shown to be produced from S^0 by *A. ferrooxidans* strain AP19-3 (Sugio et al., 1988a,b), which then reduced MnO_2 to Mn^{2+}. *A. ferrooxidans* was able to leach manganese from MnO_2 when S^0 in the medium was replaced by chalcocite (Cu_2S) or covellite (CuS) ground to -100 mesh size. Increasing the pulp density (particle concentration) of MnO_2 from 0.5 to 5.0% increased the amount of manganese leached from MnO_2 but decreased the amount of Cu leached from chalcocite or covellite (Ghosh and Imai, 1985b).

Heterotrophic bacterial leaching of manganese ores was first attempted by Perkins and Novielli (1962). They isolated a *Bacillus* that was able to solubilize manganese from a variety of ores in organic culture medium that contained molasses as an ingredient. The mineralogy of the ores they tested was not reported, and thus it is not clear whether the manganese in the ore was mobilized by microbial reduction or merely by solvent action of some metabolic products formed by the bacteria from the molasses in the medium. Heterotrophic leaching that did involve reduction of manganese oxides was reported by Trimble and Ehrlich (1968), Ehrlich et al. (1973), Agate and Deshpande (1977), Mercz and Madgwick (1982), Holden and Madgwick (1983), Kozub and Madgwick (1983), Babenko et al. (1983), Silvero (1985), Rusin et al. (1991a,b), and others.

17.12   MICROBIAL MOBILIZATION OF MANGANESE IN FRESHWATER ENVIRONMENTS

Detection of manganese-reducing bacteria in some of the Karelian lakes (former USSR) led to the inference that manganese reduction is occurring in these lakes (Sokolova-Dubinina and Deryugina, 1967a; Troshanov, 1968). In Lake Punnus-Yarvi, *Bacillus circulans, B. polymyxa,* and an unidentified non-spore-forming rod were thought to be involved in the formation of rhodochrosite ($MnCO_3$) from bacterially generated manganese oxide. The manganese-reducing activity that led to $MnCO_3$ formation was found to occur on the shoreward side of a depression in Punnus Ioki Bay and was related to the bacterial degradation of plant debris originating from plant life along the shore of the lake. The manganese oxides appeared to act as terminal electron acceptors. CO_2 from the mineralization of organic matter contributed the carbonate. Limited bacterial sulfate reduction was thought to assist the process by helping to maintain reducing conditions. As much as 5% $MnCO_3$ has been found in the sediment at the active site (see also Chapter 9).

Lovley and Phillips (1988a) demonstrated in the laboratory the reduction of MnO_2 with acetate by the action of *Geobacter metallireducens* under anaerobic conditions, leading to the production of $MnCO_3$ identified as rhodochrosite. *G. metallireducens* was isolated from Potomac River sediment. In this instance, manganese was mobilized and at least a portion reimmobilized.

Microbial Mn(IV)-reducing activity has been detected in sediments of Blelham Tarn in the English Lake District. A malate-fermenting *Vibrio* was isolated that could use Mn(IV), Fe(III), and NO_3^- as terminal electron acceptors (Jones et al., 1984). In laboratory experiments, this organism exhibited a 20% greater molar growth yield anaerobically on Mn(IV) oxide than on Fe(III). Mn(IV) oxide and NO_3^- inhibited reduction of iron(III) by the organism. Similar inhibition of iron(III) reduction was observed when a sample of lake sediment was emended with Mn_2O_3 (Jones et al., 1983).

Davison et al. (1982) reported microbial manganese reduction in the deeper sediment of Esthwaite Water (U.K.). It was intense during winter months when manganese oxides accumulated transiently in the sediment. In the summer months, manganese oxides generated in the water column were reduced in the hypolimnion before reaching the sediment. Myers and Nealson (1988a) found a *Shewanella* (formerly *Alteromonas, Pseudomonas*) *putrefaciens* (now *oneidensis*), isolated from Oneida Lake (New York) that, although a facultative aerobe, reduced Mn(IV) oxide only anaerobically with a variety of organic electron donors. It is probably one of the organisms responsible for recycling oxidized manganese in the lake. Gottfreund et al. (1983) detected manganese-reducing bacteria in groundwater to which they attributed Mn(III)-reducing ability.

17.13   MICROBIAL MOBILIZATION OF MANGANESE
IN MARINE ENVIRONMENTS

Manganese(IV)-reducing bacteria have been isolated from seawater, marine sediments, and deep-sea ferromanganese concretions (nodules). To date, all isolates tested in the laboratory have been heterotrophs or mixotrophs that can use one or more of the following electron donors: glucose, lactate, succinate, acetate, or the inorganic donor H_2. In 1988, *in situ* observations suggested that anaerobically respiring thiobacilli and some other bacteria in anaerobic sediment at the edge of a salt marsh on Skidaway Island, Georgia, catalyze the reduction of Mn(IV) oxides that are in contact with sulfide in a *solid phase* (e.g., FeS). The Mn(IV) oxide was thought to act as electron acceptor while the sulfide was oxidized to sulfate. The process was inferred from experiments in which microbial reduction of Mn(IV) oxide by sulfide was inhibited by azide and 2,4-dinitrophenol (Aller and Rude, 1988). A process resembling this has since been demonstrated in the laboratory with elemental sulfur as electron donor (Lovley and Phillips, 1994a). Organisms capable of catalyzing this reaction include *Desulfovibrio desulfuricans, Desulfomicrobium baculatum, Desulfobacterium autotrophicum, Desulfuromonas acetoxidans*, and *Geobacter metallireducens*. Stoichiometric transformation according to the reaction

$$S^0 + 3MnO_2 + 4H^+ \rightarrow SO_4^{2-} + 3Mn^{2+} + 2H_2O \qquad (17.18)$$

was demonstrated with *D. desulfuricans*. Fe(III) could not replace MnO_2 as terminal electron acceptor in this reaction. This activity is not only of significance for the marine manganese cycle but also presents an important mechanism by which sulfate can be regenerated from reduced forms of sulfur anaerobically in the dark in the marine sulfur cycle.

Hydrogen sulfide (H_2S) produced anaerobically by sulfate-reducing bacteria has been shown to be able to reduce Mn(IV) oxide nonenzymatically, with S^0 being the chief product of sulfide oxidation (Burdige and Nealson, 1986). In addition, H_2S produced by bacterial disproportionation of elemental sulfur into H_2S and SO_4^{2-} has been shown to reduce Mn(IV) oxide chemically. Indeed, this reaction with Mn(IV) oxide [or Fe(III)] appears to be thermodynamically essential in promoting continued bacterial disproportionation of the elemental sulfur (Thamdrup et al., 1993).

Most manganese-reducing bacterial isolates from marine environments studied to date have been aerobes that can reduce MnO_2 aerobically or anaerobically, but some evidence for strictly anaerobic reduction has been obtained (see reviews by Ehrlich, 1987; Burdige 1993). Caccavo et al. (1992) reported the isolation of a facultative anaerobe, strain BrY (*Shewanella alga* BrY), from Great Bay Estuary, New Hampshire, that reduced Mn(IV) oxide only anaerobically with hydrogen or lactate. The extent of *in situ* activity of marine MnO_2-reducing bacteria has so far not been estimated in any part of the marine environment.

Because Mn(IV) oxides are good scavengers of trace metals such as Cu, Co, and Ni, it is noteworthy that bacterial reduction of the manganese oxide in pulverized ferromanganese nodules in laboratory experiments resulted in solubilization of these metals along with manganese (Figure 17.21; Ehrlich et al., 1973). Ni and Co solubilization was absolutely dependent on Mn(IV) reduction. Cu release, on the contrary, was only partially dependent on it, being initially mobilized by complexation with peptone in the culture medium. Only in later stages did Cu mobilization show a direct dependence on bacterial action. This finding suggests that Cu may be more loosely bound in the nodule structure than Ni or Co. The need for bacterial action in Cu release appears to arise only when it is encapsulated by Mn(IV) oxides and not in direct contact with the solvent. Also noteworthy in these experiments was the observation that only negligible amounts of iron were mobilized, although the Mn(IV)-reducing organisms used in these studies also had the capacity to reduce limonite and goethite. Whether this apparent inability to solubilize iron was due to an inability of the bacteria to reduce it in the nodules or to either chemical reoxidation of any Fe(II) produced from Fe(III) oxides in the nodules by remaining Mn(IV) oxide in the nodules or to

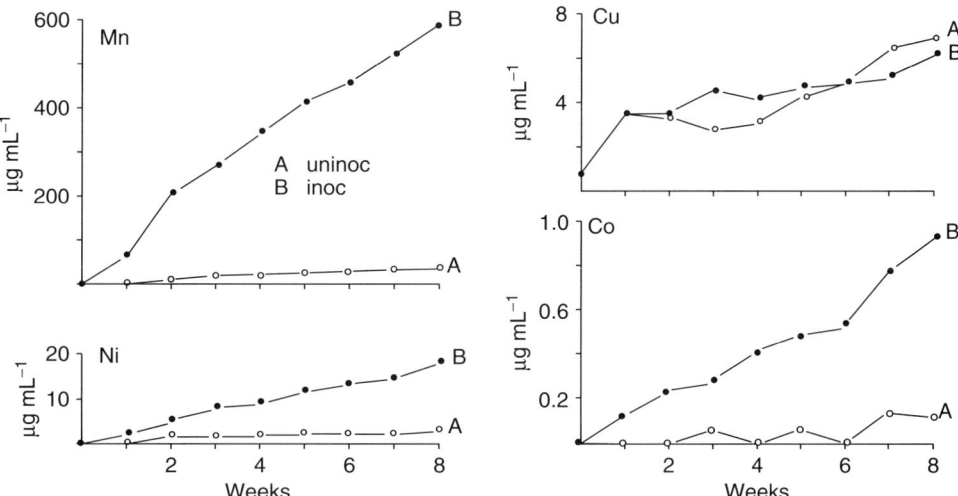

FIGURE 17.21 Manganese, copper, nickel, and cobalt release from ferromanganese nodule substance by *Bacillus* GJ33 in seawater containing 1% glucose and 0.05% peptone. A, uninoculated; B, inoculated. (Reproduced from Ehrlich HL, Yang SH, Mainwaring JD, Jr., *Z. Allg. Mikrobiol.*, 13, 39–48, 1973, Copyright Wiley-VCH Verlag GmbH & Co. KGaA. With permission.)

immediate autoxidation and precipitation after release from the nodules is not known. However, on the basis of standard free energy yields at pH 7.0 ($\Delta G^{0'}$) from the reduction of iron oxide minerals, the first explanation is more plausible. The mobilization of the scavenged trace metals by bacterial ferromanganese reduction may be of nutritional importance in the ecology of the marine environment where manganese oxides such as ferromanganese occur.

In laboratory experiments, mobilization of Co contained in ferromanganese nodules could also be achieved by action of microbially produced siderophorelike phenolic compounds (Mukherjee et al., 2003). The Co-mobilizing action of the siderophores was believed to be due to their reaction with the iron in the nodules in this instance.

17.14 MICROBIAL MANGANESE REDUCTION AND MINERALIZATION OF ORGANIC MATTER

As discussed earlier (see Section 17.6), enzymatic manganese reduction by bacteria is often a form of respiration that can occur aerobically and anaerobically, depending on the type of organism and on prevailing environmental conditions. Organic electron donors in the bacterial process can be any of a variety of different compounds. Even under anaerobic conditions, they may be completely degraded to CO_2 by a single species of organism (Lovley et al., 1993b), but complete degradation often requires the successive action of two or more species (see Lovley et al., 1989). Thus, manganese(IV) oxide respiration can be viewed environmentally as a form of mineralization of organic compounds (Ehrlich, 1993a,b; Nealson and Saffarini, 1994). This form of organic carbon mineralization is unlike that by bacterial sulfate reduction and most forms of iron and nitrate respiration because it can occur readily both aerobically and anaerobically if the appropriate organisms are present. However, it is probably only anaerobically that manganese respiration has significant impact on the carbon cycle, and then only if no other competing forms of anaerobic respiration occur (Lovley and Phillips, 1988b; Ehrlich, 1987). Thamdrup et al. (2000) found that in the Black Sea shelf sediments, dissimilatory Mn reduction was the most important means of organic carbon mineralization in the surface layer down to ~1 cm depth, whereas dissimilatory sulfate reduction was the exclusive carbon mineralization process below this depth. The Mn respiration was accompanied by $MnCO_3$ formation.

17.15 MICROBIAL ROLE IN MANGANESE CYCLE IN NATURE

It must be inferred from the widespread occurrence of manganese-oxidizing and manganese-reducing microorganisms in terrestrial, freshwater, and marine environments that they play an important role in the geochemical cycle of manganese. At neutral pH, under aerobic conditions, manganese-oxidizing microbes clearly are more important in immobilizing manganese in soils and sediments than are iron-oxidizing microbes in immobilizing Fe(II) in view of the relative resistance of Mn(II) to autoxidation at pH values below 8 in contrast to Fe(II) (see Diem and Stumm, 1984; see also Chapter 16). The manganese-reducing bacteria are important at neutral pH because manganese oxides under reducing conditions in the absence of strong reducing agents such as H_2S or Fe^{2+} are relatively stable. As Figure 17.22 shows, manganese oxidation reactions generally lead to manganese fixation because most Mn(III) and all Mn(IV) oxidation products are insoluble. Gottfreund and Schweisfurth suggested in 1983 that soluble Mn(III) complexes may be formed in microbial manganese oxidation, which is also suggested by recent observations in the Black Sea and Chesapeake Bay by Trouwborst et al. (2006). By contrast, reduction of Mn(III) and Mn(IV) oxides generally leads to solubilization of manganese. Under some reducing conditions, the solubilized

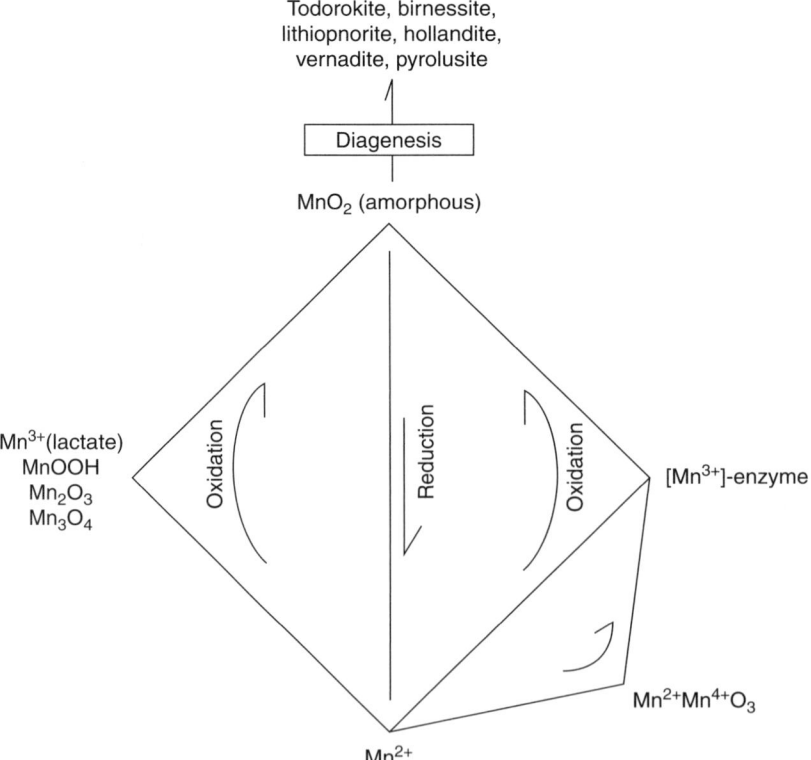

FIGURE 17.22 The manganese cycle. The oxidation reactions on the left side of the diagram involve a dismutation (disproportionation) reaction of Mn(III), whereas the oxidation reactions on the right side of the diagram involve direct oxidation of enzyme-bound Mn(III). Mn(IV) reduction may or may not involve Mn(III) as an intermediate. Bacterial oxidation of dissolved complexed Mn(III) and bacterial reduction of complexed Mn(III) and Mn(III) oxide does occur, but these reactions are not indicated in the diagram.

manganese may, however, precipitate as $MnCO_3$ (see Sokolova-Dubinina and Deryugina, 1967a; Lovley and Phillips, 1988a; see also Section 17.9).

The effect on the manganese geochemical cycle of predation of bacteria that have accumulated Mn(III/IV) oxides on their cell surface as a result of Mn(II) oxidation by phagotrophic protozoa was recently investigated using a model system. The prey in this system was the bacterium *Leptothrix discophora* SS-1 and the predator was the protozoan *Tetrahymena thermophila* (Zeiner et al., 2006). The overall results showed that consumption by the protozoan predator of the bacteria with manganese oxide on their surface did not change the distribution of manganese between solid and solution phases in a major way, suggesting that in nature, this activity should not be expected to affect the manganese cycle significantly.

Studies of biogeochemical aspects of the manganese cycle in soil and aquatic environments are still relatively few. One reason for this is the difficulty in methodology. Two approaches are being taken. One is the reconstruction of the process in laboratory experiments. Thus, manganese and iron reduction and oxidation cycles in an acidic (pH 5.6) flooded coastal prairie soil of southwest Louisiana were reproduced in the laboratory under controlled pH and E_h conditions (Patrick and Henderson, 1981). The results showed that the rate of redox potential change could affect the oxidation and reduction rates differentially if the change was too rapid. Some Mn(IV) reduction in soils may be caused nonmicrobiologically by interaction with nitrite of microbial origin in which Mn(IV) may be reduced to Mn(II) or Mn(III), depending on the nitrite concentration relative to Mn(IV) oxide (Bartlett, 1981). At high MnO_2/nitrite ratios, Mn(III) has been found to predominate.

$$NO_2^- + MnO_2 + 2H^+ \rightarrow Mn^{2+} + NO_3^- + H_2O \qquad (17.19a)$$

$$NO_2^- + 2MnO_2 \rightarrow NO_3^- + Mn_2O_3 \qquad (17.19b)$$

The complete reduction to Mn^{2+} is favored by acidic conditions, whereas the reduction to Mn_2O_3 is independent of pH. Microcosm experiments with surface sediment from Long Island Sound led Hulth et al. (1999) to speculate on biochemical coupling of anaerobic lithotrophic nitrification $(NH_4^+ \rightarrow NO_3^-)$ to MnO_2 reduction $(MnO_2 \rightarrow Mn^{2+})$.

Luther et al. (1997) studied denitrification (dinitrogen formation) in marine sediments resulting from interaction of MnO_2 with NH_3 and organic N, and Mn^{2+} with NO_3^-. These reactions are not sensitive to the presence of O_2 in air. According to Luther et al. (1997), denitrification due to oxidation of NH_3 and organic nitrogen by MnO_2 can predominate over nitrification $(NH_3 \rightarrow NO_3^-)$ in continental margin sediments and may account for 90% of the nitrogen formed in this environment. What role, if any, microbes play in these denitrifications remains to be elucidated. Observations by Vandenabeele et al. (1995) on stimulation of manganese removal from sand filters in water purification plants by nitrate suggest that bacteria can play a role.

The second approach to studying biogeochemical aspects of the manganese cycle is *in situ* investigations (Nealson et al., 1988). Some examples were mentioned in Section 17.9 (Lake Punnus-Yarvi, Lake Oneida, Saanich Inlet, Framvaren Fjord). In the studies of Oneida Lake, Saanich Inlet, and Framvaren Fjord, microbial Mn(II) oxidation was demonstrated by manganese removal in the presence and absence of cell poisons from water samples collected at various depths and spiked with $^{54}Mn^{2+}$. The metabolic poisons included sodium azide, penicillin G, and tetracycline·HCl. A critical evaluation of poisons of potential use as inhibitors of manganese oxidation in environmental samples was made by Rosson et al. (1984). The removal of manganese reflected both binding of manganese(II) to particulates (bacterial cells and organic and inorganic aggregates) and oxidation of Mn(II). Binding was distinguished partially from oxidation by use of formaldehyde. The oxidation state of particulate manganese in the water samples upon collection was in the range of 2.3–2.7, suggesting possible *in situ* formation of Mn(III) and mixed Mn(II)-Mn(IV) oxides (Emerson et al., 1982). An even better assessment of manganese oxidation rate can be obtained by measuring Mn(II) removal from solution in the presence and absence of oxygen instead of using metabolic poisons (Tebo and Emerson, 1985). In the absence of oxygen,

Mn(II) removal should be due only to surface binding. The difference between total Mn(II) removal in air and Mn(II) removal in the absence of oxygen should represent the amount of manganese that was oxidized.

An important question in the biogeochemical cycle of manganese is the oxidation state of the manganese in the first detectable product of biological and chemical oxidation of Mn(II). Diem and Stumm (1984) and Tipping et al. (1984, 1985) found that Mn(IV) oxide was formed microbiologically as the first recognizable oxidation product in manganese-containing water samples. Tipping et al. (1984, 1985) found that the oxide they recovered resembled vernadite (MnO_2). It had an oxidation state of 3.5. By contrast, Hastings and Emerson (1986) concluded from experiments with spores of marine *Bacillus* SG-1 that Mn^{2+} was transformed to amorphous Mn(III) oxide that recrystallized to hausmannite (Mn_3O_4 or $MnO_{1.33}$), especially in an excess of Mn(II) (Mann et al., 1988). However, subsequent work by Mandernack (1992) and Mandernack et al. (1995) suggested that the immediate product of manganese oxidation by the spores is Mn(IV) oxide. Nevertheless, a study by Pecher et al. (2003) indicates that a form of Mn(III) is an intermediate state in Mn(IV) oxide formation by the spores. The oxidation is caused by an exosporium protein of the *Bacillus* SG-1 spores (Tebo et al., 1988; Mann et al., 1988; Francis and Tebo, 1999). The first stages of the Mn(II) oxidation by the exosporium protein may be enzymatic, but once the spore is covered by the insoluble manganese oxide, further oxidation is abiotic (de Vrind et al., 1986b; Mann et al., 1988).

Because the first oxidation product of abiotic Mn^{2+} oxidation was found to be Mn_3O_4,

$$3Mn^{2+} + 3H_2O + 0.5O_2 \rightarrow Mn_3O_4 + 6H^+ \tag{17.20}$$

which subsequently may disproportionate at a very slow rate to MnO_2 and Mn^{2+} under appropriate conditions of pH,

$$Mn_3O_4 + 4H^+ \rightarrow MnO_2 + 2Mn^{2+} + 2H_2O \tag{17.21}$$

(Hem, 1981; Hem and Lind, 1983; Hem et al., 1982; Murray et al., 1985), the following interpretation can be offered for contrasting biological and chemical oxidation results to date. When Mn^{2+} oxidation is enzymatically catalyzed by a manganese oxidase system with energy conservation in intact cells, the reaction is a two-step sequence in which the first step involves enzymatic oxidation of Mn^{2+} to Mn(III), with the Mn(III) existing as a bound (e.g., enzymatically bound) intermediate $\{Mn^{3+}\}$:

$$Mn^{2+} + H^+ + 0.25O_2 \rightarrow \{Mn^{3+}\} + 0.5H_2O \quad (\Delta G^{0\prime} = +0.61\,\text{kcal or} +2.5\,\text{kJ}) \tag{17.22}$$

A very recent study by Webb et al. (2005) (see also Tebo et al., 2005) presented evidence for the formation of Mn(III) intermediates in the oxidation of Mn^{2+} by a multicopper oxidase in spores of marine *Bacillus* SG-1. Reaction 17.22 is followed by rapid enzymatic oxidation to the final, and at the same time, first detectable product, namely a form of Mn(IV) oxide, here written as MnO_2:

$$\{Mn^{3+}\} + 1.5H_2O + 0.25O_2 \rightarrow MnO_2 + 3H^+ \quad (\Delta G^{0\prime} = -16.9\,\text{kcal or} -70.6\,\text{kJ}) \tag{17.23}$$

A free energy of formation value of -35 kcal is estimated for $\{Mn^{3+}\}$ (bound Mn^{3+}). In this reaction sequence, all Mn^{2+} that is oxidized is transformed to MnO_2. It is Reaction 17.23 from which the energy that the cell conserves is derived (see also Ehrlich, 1999).

When Mn^{2+} is biologically oxidized by a nonenzymatic process, it is probably oxidized according to the abiotic mechanism in which Mn(III) oxide such as Mn_3O_4 is the first detectable product of oxidation, and MnO_2, if it forms at all, is formed very slowly by disproportionation of Mn_3O_4.

The major difference between the reactions catalyzed by manganese oxidase on the one hand and by abiotic oxidation on the other is that the manganese oxidase prevents disproportionation of Mn(III) by catalyzing its complete conversion to Mn(IV) oxide at a rapid rate, whereas in the abiotic reaction Mn_3O_4 is the first detectable product, and Mn(IV) oxide, if it forms at all, is formed very slowly by abiotic disproportionation, with the result that only one-third of the reacting Mn(III) is converted to Mn(IV), the other two-thirds being reconverted to Mn(II) (Reaction 17.21). Environmentally it is thus possible to associate rapidly formed, fresh Mn(IV) oxide deposits with biological catalysis and Mn(III) oxide deposits with abiotic reactions (Ehrlich, 1996b).

17.16 SUMMARY

Some bacteria can oxidize manganous manganese enzymatically or nonenzymatically. Enzymatically, they may oxidize dissolved Mn^{2+} with O_2, catalyzed by an inducible or a constitutive oxidase, or they may oxidize it with metabolically produced H_2O_2, catalyzing the manganese oxidation with catalase. They may oxidize Mn^{2+} to MnO_2, ferromanganese, montmorillonite- or kaolinite-type clay, or some marine clay with O_2 catalyzed, depending on the organism, by a constitutive or inducible oxidase system. At least some of the oxidase-catalyzed reactions yield useful energy to the bacteria. Some fungi can oxidize Mn^{2+} with an extracellular peroxidase or with laccase. Oxidation of Mn^{2+} in solution occurs in soil, freshwater, and marine environments. Oxidation Mn^{2+} prebound to a suitable sorbent has so far been demonstrated only with bacteria from marine environments.

Some bacteria and fungi may promote nonenzymatic manganese oxidation by utilization of hydroxycarboxylic acids, leading to a rise in pH above neutrality followed by oxidation of Mn(II) catalyzed by residual hydroxycarboxylic acid. Others may produce other metabolic end products, which then act as oxidants of Mn(II). Still other microbes may raise the pH by deamination of organic nitrogen compounds such as amino acids to a range where Mn(II) autoxidizes. Some cyanobacteria and algae can promote Mn^{2+} autoxidation by raising the pH of their immediate surroundings photosynthetically. Some bacteria and fungi can precipitate preformed oxidized manganese through sorption to their cell surface or to extracellular slime.

Some bacteria can reduce Mn(IV) oxides to manganous manganese [Mn(II)] with suitable electron donors. This reduction may be aerobic or anaerobic, depending on the bacterial strain and environmental conditions. Such reduction has been demonstrated with bacteria from soil, freshwater, and marine environments. Some bacteria and fungi can also reduce Mn(IV) oxides nonenzymatically with metabolic end products such as H_2S and Fe^{2+}, or oxalate.

Manganese oxidizing or manganese reducing microorganisms play an important role in immobilization and mobilization, respectively, of manganese in soil. In some anaerobic environments, manganese-reducing bacteria may play an important role in mobilizing fixed, nutritionally unavailable manganese and in mineralization of organic matter. Fixed manganese in soil can be concentrated in concretions or in arid environments as desert varnish. Because manganese is required in plant nutrition, microbial manganese-oxidizing and manganese-reducing activity is ecologically very significant.

Manganese-oxidizing and manganese-reducing bacteria can also play a significant role in the manganese cycle in freshwater and marine environments. Manganese-oxidizing microbes may contribute to the accumulation of manganese oxides on and in sediments. The oxides they form may sometimes be deposited as concretions formed around a nucleus, such as a sediment grain, a pebble, or a dead biological structure (e.g., mollusk shell, coral fragment, or other debris). Conversely, manganese-reducing microorganisms may mobilize oxidized or fixed manganese, releasing it into the aqueous phase. In reducing freshwater environments in the presence of abundant plant debris, the microbial reduction of manganese oxides may lead to the formation of manganous carbonate, a different form of fixed manganese.

Ferromanganese nodules on parts of the ocean floor are inhabited by bacteria. Some of these bacteria can oxidize Mn(II) associated with the nodules, whereas others can reduce the oxidized manganese of the nodules. Together with benthic foraminifera, which can be expected to feed on the bacteria, some of which accumulated manganese oxides on their tests and tubes, the manganese(II) oxidizers appear to constitute a biological system that contributes to nodule formation.

REFERENCES

Adams JB, Palmer F, Staley JT. 1992. Rock weathering in deserts: Mobilization and concentration of ferric iron by microorganisms. *Geomicrobiol J* 10:99–114.

Adams LF, Ghiorse WC. 1985. Influence of manganese on the growth of a sheathless strain of *Leptothrix discophora*. *Appl Environ Microbiol* 49:556–562.

Adams LF, Ghiorse WC. 1986. Physiology and ultrastructure of *Leptothrix discophora* SS-1. *Arch Microbiol* 145:126–135.

Adams LF, Ghiorse WC. 1987. Characterization and extracellular Mn^{2+}-oxidizing activity and isolation of an Mn^{2+}-oxidizing protein from *Leptothrix discophora* SS-1. *J Bacteriol* 169:1279–1285.

Adams LF, Ghiorse WC. 1988. Oxidation state of Mn in the Mn oxide produced by *Leptothrix discophora* SS-1. *Geochim Cosmochim Acta* 52:2073–2076.

Adeny WE. 1894. On the reduction of manganese peroxide in sewage. *Proceedings of the Royal Dublin Society*, Chapter 27, pp. 247–251.

Agate AD, Deshpande HA. 1977. Leaching of manganese ores using *Arthrobacter* species. In: Schwaartz W, ed. *Conference Bacterial Leaching. Gesellschaft für Biotechnologische Forschung, mbH, Braunschweig-Stöckheim*. Weinheim, Germany: Verlag Chemie, pp. 243–250.

Alexandrov EA. 1972. Manganese: Element and geochemistry. In: Fairbridge RW, ed. *Encyclopedia of Geochemistry and Environmental Sciences. Encyclopedia of Earth Science Series,* Vol IVA. New York: Van Nostrand Reinhold, pp. 670–671.

Ali H, Stokes JL. 1971. Stimulation of heterotrophic and autotrophic growth of *Sphaerotilus discophorus* by manganous ions. *Antonie van Leeuwenhoek* 37:519–528.

Aller RC, Rude PD. 1988. Complete oxidation of solid phase sulfides by manganese and bacteria in anoxic marine sediments. *Geochim Cosmochim Acta* 52:751–765.

Anderson JJ, Devol AH. 1973. Deep water renewal in Saanich Inlet, an intermittently anoxic basin. *Estuar Coast Mar Sci* 1:1–10.

Arcuri EJ. 1978. Identification of the cytochrome complements of several strains of marine manganese oxidizing bacteria and their involvement in manganese oxidation. PhD Thesis. Troy, New York: Rensselaer Polytechnic Institute.

Arcuri EJ, Ehrlich HL. 1979. Cytochrome involvement in Mn(II) oxidation by two marine bacteria. *Appl Environ Microbiol* 37:916–923.

Arcuri EJ, Ehrlich HL. 1980. Electron transfer coupled to Mn(II) oxidation in two deep-sea Pacific Ocean isolates: In: Trudinger PA, Walter MR, Ralph BJ, eds. *Biogeochemistry of Ancient and Modern Environments*. Canberra, Australia: Australian Academy of Science, pp. 339–344.

Aristovskaya TV. 1961. Accumulation of iron in breakdown of organo-mineral humus complexes by microorganisms. *Dokl Akad Nauk SSSR* 136:954–957.

Aristovskaya TV, Parinkina OM. 1963. A new soil microorganism *Seliberia stellata* n. gen., n. sp. *Akad Nauk SSSR Ser Biol* 218:49–56.

Babenko YS, Doligh LM, Serebryana MZ. 1983. Characteristics of the bacterial breakdown of primarily oxidized manganese ores of the Nikopol deposit. *Mikrobiologiya* 52:851–856.

Baker ET, Massoth GJ. 1986. Hydrothermal plume measurements. A regional perspective. *Science* 234:980–982.

Balashova VV, Dubinina GA. 1989. The ultrastructure of *Metallogenium* in axenic culture. *Mikrobiologiya* 58:841–846 (English translation, pp. 681–685).

Bargar JR, Tebo BM, Villinski JE. 2000. In situ characterization of Mn(II) oxidation by spores of the marine *Bacillus* sp. strain SG1. *Geochim Cosmochim Acta* 64:2775–2778.

Bartlett RJ. 1981. Nonmicrobial nitrite-to-nitrate transformation in soil. *Soil Sci Am J* 45:1054–1058.

Beijerinck MW. 1913. Oxydation des Mangancarbonates durch Bakterien und Schimmelpilze. *Fol Microbiol Holländ Beitr Gesamt Mikrobiol Delft* 2:123–134.

Bender ML, Ku T-L, Broecker WS. 1970. Accumulation rates of manganese in pelagic sediments and nodules. *Earth Planet Sci Lett* 8:143–148.

Beveridge TJ, Kadurugamuwa JL. 1996. Periplasm, periplasmic spaces, and their relation to bacterial wall structure: Novel secretion of selected periplasmic proteins from *Pseudomonas aeruginosa*. *Microb Drug Resist* 2:1–8.

Bignot G, Dangeard L. 1976. Contribution à l'étude de la fraction biogène des nodules polymétalliques des fonds océanique actuels. *CR Somm Soc Geol Fr* 3:96–99.

Bischoff JL, Rosenbauer RJ. 1983. A note on the chemistry of seawater in the range of 350°–500°C. *Geochim Cosmochim Acta* 47:139–144.

Boogerd FC, de Vrind JPM. 1987. Manganese oxidation by *Leptothrix discophora*. *J Bacteriol* 169:489–494.

Bowen HJM. 1966. *Trace Elements in Biogeochemistry*. London: Academic Press.

Bowen HJM. 1979. *Environmental Chemistry of the Elements*. London: Academic Press.

Bretschger O, Obraztsova A, Sturm CA, Chang IS, Gorby YA, Reed SB, Culley DE, Reardon CL, Barua S, Romine MF, Zhou J, Beliaev AS, Bouhenni R, Saffarini D, Mansfeld F, Kim B-H, Fredrickson JK, Nealson KH. 2007. Current production and metal oxide reduction by *Shewanella oneidensis* MR-1 wild type and mutants. *Appl Environ Microbiol* 73:7003–7012.

Bromfield SM. 1956. Oxidation of manganese by soil microorganisms. *Aust J Biol Sci* 9:238–252.

Bromfield SM. 1958a. The properties of a biologically formed manganese oxide, its availability to oats and its solution by root washings. *Plant Soil* 9:325–337.

Bromfield SM. 1958b. The solution of γ-MnO$_2$ by substances released from soil and from roots of oats and vetch in relation to manganese availability. *Plant Soil* 10:147–160.

Bromfield SM. 1974. Bacterial oxidation of manganous ion as affected by organic substrate concentration and composition. *Soil Biol Biochem* 6:383–392.

Bromfield SM. 1978. The oxidation of manganous ion under acid conditions by an acidophilous actinomycete from acid soil. *Aust J Biol Soil Res* 16:91–100.

Bromfield SM. 1979. Manganous ion oxidation at pH values below 5.0 by cell-free substances from *Streptomyces* sp. cultures. *Soil Biol Biochem* 11:115–118.

Bromfield SM, David DJ. 1976. Sorption and oxidation of manganous ions and reduction of manganese oxide by cell suspensions of a manganese oxidizing bacterium. *Soil Biol Biochem* 8:37–43.

Bromfield SM, David DJ. 1978. Properties of biologically formed manganese oxide in relation to soil manganese. *Aust J Soil Res* 16:79–89.

Bromfield SM, Skerman VDB. 1950. Biological oxidation of manganese in soils. *Soil Sci* 69:337–348.

Brouwers GJ, de Vrind JPM, Corstjens PLAM, Cornelis P, Baysse C, de Vrind-de Jong EW. 1999. *CumA*, a gene encoding a multicopper oxidase, is involved in Mn^{2+}-oxidation in *Pseudomonas putida* GB-1. *Appl Environ Microbiol* 65:1762–1768.

Brouwers GJ, Vijgenboom E, Corstjens PLAM, de Vrind JPM, de Vrind-de Jong EW. 2000a. Bacterial Mn^{2+} oxidizing systems and multicopper oxidases: An overview of mechanisms and functions. *Geomicrobiol J* 17:1–24.

Brouwers GJ, Corstjens PLAM, de Vrind JPM, Verkammen A, De Kuyper M, de Vrind-de Jong EW. 2000b. Stimulation of Mn^{2+} oxidation in *Leptothrix discophora* SS-1 by Cu^{2+} and sequence analysis of the region flanking the gene encoding putative multicopper oxidase MofA. *Geomicrobiol J* 17:25–33.

Burdige DJ. 1993. The biogeochemistry of manganese and iron reduction in marine sediments. *Earth-Sci Rev* 35:249–284.

Burdige DJ, Kepkay PE. 1983. Determination of bacterial manganese oxidation rates in sediments using an in situ dialysis technique. I. Laboratory studies. *Geochim Cosmochim Acta* 47:1907–1916.

Burdige DJ, Nealson KH. 1985. Microbiological manganese reduction by enrichment cultures from coastal marine sediments. *Appl Environ Microbiol* 50:491–497.

Burdige DJ, Nealson KH. 1986. Chemical and microbiological studies of sulfide-mediated manganese reduction. *Geomicrobiol J* 4:361–378.

Bureau of Mines. 1965. *Mineral Facts and Problems. Bull. 630*. Washington, DC: Bureau of Mines, U.S. Department of the Interior.

Burnett BR, Nealson KH. 1981. Organic films and microorganisms associated with manganese nodules. *Deep-Sea Res* 28A:637–645.

Burnett BR, Nealson KH. 1983. Energy dispersive X-ray analysis of the surface of a deep-sea ferromanganese nodule. *Mar Geol* 53:313–329.

Burns RG, Burns VM, Smig W. 1974. Ferromanganese mineralogy: Suggested terminology of the principal manganese oxide phases. *Abstr Annu Meet Geol Soc Am*.

Butkevich VS. 1928. The formation of marine manganese deposits and the role of microorganisms in the latter. *Wissenschaft Meeresinst Ber* 3:7–80.

Caccavo F, Jr., Blakemore RP, Lovley DR. 1992. A hydrogen-oxidizing, Fe(III)-reducing microorganism from the Great Bay Estuary, New Hampshire. *Appl Environ Microbiol* 58:3211–3216.

Caccavo F, Jr., Lonergan DJ, Lovley DR, Davis M, Stolz SF, McInerney MJ. 1994. *Geobacter sulfurreducens* sp. nov., a new hydrogen- and acetate-oxidizing dissimilatory metal-reducing microorganism. *Appl Environ Microbiol* 60:3752–3759.

Caspi R, Haygood MG, Tebo BM. 1996. Unusual ribulose-1,5-bisphosphate carboxylase/oxygenase genes from a marine manganese oxidizing bacterium. *Microbiology* (Reading) 142:2549–2559.

Chapnick SD, Moore WS, Nealson KH. 1982. Microbially mediated manganese oxidation in a freshwater lake. *Limnol Oceanogr* 27:1004–1014.

Clark TR. 1991. Manganese-oxidation by SSW_{22}, a hydrothermal vent bacterial isolate. PhD Thesis. Troy, New York: Rensselaer Polytechnic Institute.

Clark TR, Ehrlich HL. 1988. Manganese oxidation by cell fractions from a bacterial hydrothermal vent isolate. *Abstr Annu Meet Am Soc Microbiol* K-152, p. 232.

Coates JD, Ellis DJ, Gaw CV, Lovley DR. 1999. *Geothrix fermentans* gen nov., sp. nov., a novel Fe(III)-reducing bacterium from a hydrocarbon contaminated aquifer. *Int J Syst Bacteriol* 49:1615–1622.

Collins BF, Buol SW. 1970. Effects of fluctuations in the E_h-pH environment on iron and/or manganese equilibria. *Soil Sci* 110:111–118.

Corstjens PLAM, de Vrind JPM, Goosen T, de Vrind-de Jong EW. 1997. Identification and molecular analysis of the *Leptothrix discophora* SS-1 mofA gene, a gene putatively encoding a manganese oxidizing protein with copper domains. *Geomicrobiol J* 14:91–108.

Cowen JP. 1989. Positive effect on manganese binding by bacteria in deep-sea hydrothermal plumes. *Appl Environ Microbiol* 55:764–766.

Cowen JP, Bertram MA, Baker ET, Feely RA, Massoth GJ, Summit M. 1998. Geomicrobial transformation of manganese in Gorda Ridge event plumes. *Deep-Sea Res II* 45:2713–2737.

Cowen JP, Bruland KW. 1985. Metal deposits associated with bacteria: Implications for Fe and Mn marine biogeochemistry. *Deep-Sea Res* 32:253–272.

Cowen JP, Massoth GJ, Baker ET. 1986. Bacterial scavenging of Mn and Fe in a mid- to far-field hydrothermal particle plume. *Nature* (London) 322:169–171.

Cowen JP, Massoth GJ, Feely RA. 1990. Scavenging rates of dissolved manganese in a hydrothermal vent plume. *Deep-Sea Res* 37:1619–1637.

Cowen JP, Silver MW. 1984. The association of iron and manganese with bacteria on marine microparticulate material. *Science* 224:1340–1342.

Crerar DA, Barnes HL. 1974. Deposition of deep-sea manganese nodules. *Geochim Cosmochim Acta* 38:279–300.

Crerar DA, Fischer AG, Plaza CL. 1979. *Metallogenium* and biogenic deposition of manganese from Precambrian to recent time. In: *Geology and Geochemistry of Manganese,* Vol III. Budapest: Publishing House of the Hungarian Academy of Sciences, pp. 285–303.

Cronan DS, Thomas RL. 1970. Ferromanganese concretions in Lake Ontario. *Can J Earth Sci* 7:1346–1349.

Davison W, Woof C, Rigg E. 1982. The dynamics of iron and manganese in a seasonally anoxic lake: Direct measurement of fluxes using sediment traps. *Limnol Oceanogr* 27:987–1003.

Dean WE. 1970. Fe-Mn oxide crusts in Oneida Lake, New York. *Proceedings of the 13th Conference on Great Lakes Research*, Ann Arbor, MI: International Association of Great Lakes Research, pp. 217–226.

Dean WE, Ghosh SK. 1978. Factors contributing to the formation of ferromanganese nodules in Oneida Lake, New York, *J Res US Geol Surv* 6:231–240.

Dean WE, Greeson PE. 1979. Influences of algae on the formation of freshwater ferromanganese nodules, Oneida Lake, New York. *Arch Hydrobiol* 86:181–192.

Dean WE, Moore WS, Nealson KH. 1981. Manganese cycles and the origin of manganese nodules, Oneida Lake, New York. *Chem Geol* 34:53–64.

DePalma SR. 1993. Manganese oxidation by *Pseudomonas putida*. PhD Dissertation. Cambridge, MA: Harvard University (as cited by Brouwers et al., 2000a).

de Vrind JPM, Boogerd FC, de Vrind-de Jong EW. 1986a. Manganese reduction by a marine *Bacillus* species. *J Bacteriol* 167:30–34.

de Vrind JPM, de Vrind-de Jong EW, de Vogt J-WH, Westbroek P, Boogerd FC, Rosson R. 1986b. Manganese oxidation by spore coats of marine *Bacillus* sp. *Appl Environ Microbiol* 52:1096–1100.

Diem D, Stumm W. 1984. Is dissolved Mn^{2+} being oxidized by O_2 in the absence of Mn-bacteria or surface catalysts? *Geochim Cosmochim Acta* 48:1571–1573.

Di-Ruggiero J, Gounot AM. 1990. Microbial manganese reduction by bacterial strains isolated from aquifer sediments. *Microb Ecol* 20:53–63.

Dorn RI. 1991. Rock varnish. *Am Sci* 79:542–553.

Dorn RI, Oberlander TM. 1981. Microbial origin of desert varnish. *Science* 213:1245–1247.

Douka CE. 1977. Study of bacteria from manganese concretions. Precipitation of manganese by whole cells and cell-free extracts of isolated bacteria. *Soil Biol Biochem* 9:89–97.

Douka CE. 1980. Kinetics of manganese oxidation by cell-free extracts of bacteria isolated from manganese concretions from soil. *Appl Environ Microbiol* 39:74–80.

Douka CE, Vaziourakis CD. 1982. Enzymatic oxidation of manganese by cell-extracts of bacteria from manganese concretions from soil. *Folia Microbiol* 27:418–422.

Drew IM. 1972. Atomic number and periodic table. In: Fairbridge RW, ed. *The Encyclopedia of Geochemistry and Environmental Sciences. Encyclopedia Earth Science Series*, Vol IVA. New York: Van Nostrand Reinhold, pp. 43–48.

Drosdoff M, Nikiforoff CC. 1940. Iron-manganese concretions in Dayton soils. *Soil Sci* 49:333–345.

Dubinina GA. 1970. Untersuchungen über die Morphologie von *Metallogenium* und die Beziehung zu *Mycoplasma*. *Z Allg Mikrobiol* 10:309–320.

Dubinina GA. 1978a. Functional role of bivalent iron and manganese oxidation in *Leptothrix pseudoochracea*. *Mikrobiologiya* 47:783–789 (English translation, pp. 631–636).

Dubinina GA. 1978b. Mechanism of oxidation of divalent iron and manganese by iron bacteria growing at neutral pH of the medium. *Mikrobiologiya* 47:591–599 (English translation, pp. 471–478).

Dubinina GA, 1984. Infection of prokaryotic and eukaryotic microorganisms with *Metallogenium*. *Curr Microbiol* 11:349–356.

Dubinina GA, Deryugina ZP. 1971. Electron microscope study of iron-manganese concretions from Lake Punnus-Yarvi. *Dokl Akad Nauk SSSR* 201:714–716 (English translation, pp. 738–740).

Dubinina GA, Zhdanov AV. 1975. Recognition of the iron bacteria *Siderocapsa* as *Arthrobacter siderocapsulatus* n. sp. *Int J Syst Bacteriol* 25:340–350.

Dugolinsky BK, Margolis SV, Dudley WC. 1977. Biogenic influence on the growth of manganese nodules. *J Sediment Petrol* 47:428–445.

Durand P, Prieur D, Jeanthon C, Jacq E. 1990. Occurrence and activity of heterotrophic manganese oxidizing bacteria associated with Alvinellids (polychaetous annelids) from a deep hydrothermal vent site on the East Pacific Rise. *CR Acad Sci* (Paris) 310 (Ser III):273–278.

Ehrlich HL. 1963. Bacteriology of manganese nodules. I. Bacterial action on manganese in nodule enrichments. *Appl Microbiol* 11:15–19.

Ehrlich HL. 1964. Microbial transformation of minerals. In: Heukelakian H, Dondero N, eds. *Principles and Applications in Aquatic Microbiology*. New York: Wiley, pp. 43–60.

Ehrlich HL. 1966. Reactions with manganese by bacteria from ferromanganese nodules. *Dev Ind Microbiol* 7:43–60.

Ehrlich HL. 1968. Bacteriol of manganese nodules. II. Manganese oxidation by cell-free extract from a manganese nodule bacterium. *Appl Microbiol* 16:197–202.

Ehrlich HL. 1972. Response of some activities of ferromanganese nodule bacteria to hydrostatic pressure. In: Colwell RR, Morita RY, eds. *Effect of the Ocean Environment on Microbial Activities*. Baltimore, MD: University of Park Press, pp. 208–221.

Ehrlich HL. 1973. The biology of ferromanganese nodules. Determination of the effect of storage by freezing on the viable flora, and a check on the reliability of the results from a test to identify MnO_2-reducing cultures. In: *Interuniversity Program of Research of Ferromanganese Deposits on the Ocean Floor. Phase I Rep, Seabed Assessment Program, International Decade of Ocean Exploration*. Washington, DC: National Science Foundation, pp. 217–219.

Ehrlich HL. 1976. Manganese as an energy source for bacteria. In: Nriagu JO, ed. *Environmental Biogeochemistry,* Vol 2. Ann Arbor, MI: Ann Arbor Science Publishers, pp. 633–644.

Ehrlich HL 1978a. Inorganic energy sources for chemolithotrophic and mixotrophic bacteria. *Geomicrobiol J* 1:65–83.

Ehrlich HL 1978b. Conditions for bacterial participation in the initiation of manganese deposition around marine sediment particles. In: Krumbein WE, ed. *Environmental Biogeochemistry and Geomicrobiology*, Vol 3. *Methods, Metals, and Assessment*. Ann Arbor, MI: Ann Arbor Science Publishers, pp. 839–845.

Ehrlich HL. 1980. Bacterial leaching of manganese ores. In: Trudinger PA, Walter MR, Ralph BJ, eds. *Biogeochemistry of Ancient and Modern Environments*. Canberra, Australia: Australian Academy of Science, pp. 609–614.

Ehrlich HL. 1981. Microbial oxidation and reduction of manganese as aids in its migration in soil. In: Colloques Internationaux du C.N .R.S. No. 301: Migrations Organo-Minérales dans les Sols Temperés. Paris: Editions du CNRS, pp. 209–213.

Ehrlich HL. 1982. Enhanced removal of Mn^{2+} from seawater by marine sediments and clay minerals in the presence of bacteria. *Can J Microbiol* 28:1389–1395.

Ehrlich HL. 1983. Manganese oxidizing bacteria from a hydrothermally active area on the Galapagos Rift. *Ecol Bull* (Stockholm) 35:357–366.

Ehrlich HL. 1984. Different forms of bacterial manganese oxidation. In: Strohl WR, Tuovinen OH, eds. *Microbial Chemoautotrophy*. Columbus, OH: Ohio State University Press, pp. 47–56.

Ehrlich HL. 1985. Mesophilic manganese-oxidizing bacteria from hydrothermal discharge areas at 21° north on the East Pacific Rise. In: Caldwell DE, Brierley JA, Brierley CL, eds. *Planetary Ecology*. New York: Van Nostrand Reinhold, pp. 186–194.

Ehrlich HL. 1987. Manganese oxide reduction as a form of anaerobic respiration. *Geomicrobiol J* 5: 423–431.

Ehrlich HL. 1988. Bioleaching of manganese by marine bacteria. In: Durand G, Bobichon L, Florent J, eds. *Proceedings of the 8th International Biotechnology Symposium*, Vol 2. Paris: Société Française Microbiologie, pp. 1094–1105.

Ehrlich HL. 1990. *Geomicrobiology*, 2nd ed. rev. New York: Marcel Dekker.

Ehrlich HL. 1993a. Electron transfer from acetate to the surface of MnO_2 particles by a marine bacterium. *J Ind Microbiol* 12:121–128.

Ehrlich HL. 1993b. A possible mechanism for the transfer of reducing power to insoluble mineral oxide in bacterial respiration. In: Torma AE, Apel ML, Brierley CL, eds. *Biohydrometallurgical Technologies*, Vol II. Warrendale, PA: The Minerals, Metals and Materials Society, pp. 415–422.

Ehrlich HL. 1996a. *Geomicrobiology*. 3rd ed. rev. New York: Marcel Dekker.

Ehrlich HL. 1999. Microbes as geologic agents: Their role in mineral formation. *Geomicrobiol J* 16:135–153.

Ehrlich HL. 2000. Ocean manganese nodules: Biogenesis and bioleaching possibilities. *Miner Metall Process* 17:121–128.

Ehrlich HL. 2002. *Geomicrobiology*, 4th ed. rev. New York: Marcel Dekker.

Ehrlich HL. 2008. Are gram-positive bacteria capable of electron transfer across their cell wall without an externally available electron shuttle? *Geobiology* 6:220–224.

Ehrlich HL, Ghiorse WC, Johnson GL II. 1972. Distribution of microbes in manganese nodules from the Atlantic and Pacific Oceans. *Dev Ind Microbiol* 13:57–65.

Ehrlich HL, Graham LA, Salerno JS. 1991. MnO_2 reduction by marine Mn^{2+} oxidizing bacteria from around hydrothermal vents at a mid-ocean spreading center and from the Black Sea. In: Berthelin J, ed. *Diversity of Environmental Biogeochemistry*. Amsterdam: Elsevier, pp. 217–224.

Ehrlich HL, Salerno JC. 1988. Stimulation by ADP and phosphate of Mn^{2+} oxidation by cell extracts and membrane vesicles of induced Mn-oxidizing bacteria. *Abstr Annu Meet Am Soc Microbiol* K-151:231.

Ehrlich HL, Salerno JC. 1990. Energy coupling in Mn^{2+} oxidation by a marine bacterium. *Arch Microbiol* 154:12–17.

Ehrlich HL, Yang SH, Mainwaring JD, Jr. 1973. Bacteriology of manganese nodules. VI. Fate of copper, nickel, cobalt and iron during bacterial and chemical reduction of the manganese(IV). *Z Allg Mikrobiol* 13:39–48.

Ehrlich HL, Zapkin MA. 1985. Manganese-rich layers in calcareous deposits along the western shore of the Dead Sea may have a bacterial origin. *Geomicrobiol J* 4:207–221.

Eleftheriadis DK. 1976. Mangan- und Eisenoxydation in Mineral- and Termal-Quellen-Mikrobiologie, Chemie, und Geochimie. PhD Thesis. Saarbrücken, West Germany: University of Saarlandes.

Emerson D, Garen RE, Ghiorse WC. 1989. Formation of *Metallogenium*-like structures by a manganese-oxidizing fungus. *Arch Microbiol* 151:223–231.

Emerson S, Kalhorn S, Jacobs L, Tebo BM, Nealson KH, Rosson RA. 1982. Environmental oxidation rate of manganese(II): Bacterial catalysis. *Geochim Cosmochim Acta* 46:1073–1079.

Francis CA, Co E-M, Tebo MB. 2001. Enzymatic manganese(II) oxidation by a marine α-Proteobacterium. *Appl Environ Microbiol* 67:4024–4029.

Francis CA, Obraztsova AY, Tebo BM. 2000. Dissimilatory metal reduction by the facultative anaerobe *Pantoea agglomerans* SP1. *Appl Environ Microbiol* 66:543–548.

Francis CA, Tebo BM. 1999. Marine *Bacillus* spores as catalysts for oxidative precipitation and sorption of metals. *J Mol Microbiol Biotechnol* 1:71–78.

Fredericks-Jantzen CM, Herman H, Herley P. 1975. Microorganisms on manganese nodules. *Nature* (London) 258:270.

Garey CL, Barber SA. 1952. Evaluation of certain factors involved in increasing manganese availability with sulfur. *Soil Sci Soc Am Proc* 16:173–175.

Geloso M. 1927. *Ann Chim* 7:113–150.

Gerretsen FC. 1937. Manganese deficiency of oats and its relation to soil bacteria. *Ann Bot* 1:207–230.

Ghiorse WC. 1980. Electron microscopic analysis of metal-depositing microorganisms in surface layers of Baltic Sea ferromanganese concretions. In: Trudinger PA, Walter MR, Ralph BJ, eds. *Biogeochemistry of Ancient and Modern Environments*. Canberra, Australia: Australian Academy of Science, pp. 345–354.

Ghiorse WC. 1984. Biology of iron- and manganese-depositing bacteria. *Annu Rev Microbiol* 38:515–550.

Ghiorse WC. 1988. Microbial reduction of manganese and iron. In: Zehnder JB, ed. *Biology of Anaerobic Microorganisms*. New York: Wiley, pp. 305–331.

Ghiorse WC, Chapnik SD. 1983. Metal-depositing bacteria and the distribution of manganese and iron in swamp waters. *Ecol Bull* (Stockholm) 35:367–376.

Ghiorse WC, Ehrlich HL. 1976. Electron transport components of the MnO_2 reductase system and the location of the terminal reductase in a marine bacillus. *Appl Environ Microbiol* 31:977–985.

Ghiorse WC, Ehrlich HL. 1992. Microbial biomineralization of iron and manganese. In: Skinner HCW, Fitzpatrick RW, eds. *Biomineralization. Process of Iron and Manganese. Modern and Ancient Environments*. Catena Suppl 21. Cremlingen, Germany: Catena Verlag, pp. 75–99.

Ghiorse WC, Hirsch P. 1979. An ultrastructural study of iron and manganese deposition associated with extra-cellular polymers of *Pedomicrobium*-like budding bacteria. *Arch Microbiol* 123:213–226.

Ghosh J, Imai K. 1985a. Leaching of manganese dioxide by *Thiobacillus ferrooxidans* growing on elemental sulfur. *J Ferment Technol* 63:259–262.

Ghosh J, Imai K. 1985b. Leaching mechanism of manganese dioxide by *Thiobacillus ferrooxidans*. *J Ferment Technol* 63:295–298.

Ghosh SK. 1975. Origin and geochemistry of ferromanganese nodules in Oneida Lake, New York. *Diss Abstr Int* 36:4905-B.

Gillette NJ. 1961. Oneida Lake Pancakes. *New York State Conservationist* 41(April–May):21.

Glenn JK, Akileswaram L, Gold MH. 1986. Mn(II) oxidation is the principal function of the extracellular Mn-peroxidase from *Phanerochaete chrysosporium*. *Arch Biochem Biophys* 251:688–696.

Glenn JK, Gold MH. 1985. Purification and characterization of an extracellular Mn(II)-dependent peroxidase from the lignin degrading basidiomycete *Phanerochaete chrysosporium*. *Arch Biochem Biophys* 242:329–341.

Goldberg ED. 1954. Marine geochemistry. I. Chemical scavengers of the sea. *J Geol* 62:249–265.

Goldberg ED, Arrhenius GOS. 1958. Chemistry of Pacific pelagic sediments. *Geochim Cosmochim Acta* 13:153–212.

Goldschmidt VM. 1954. *Geochemistry*. Oxford, U.K.: Clarendon Press, pp. 621–642.

Gong-Collins E. 1986. A euryhaline, manganese- and iron-oxidizing *Bacillus megaterium* from a microbial mat at Laguna Figueroa, Baja California, Mexico. *Microbios* 48:109–126.

Gottfreund J, Schmitt G, Schweisfurth R. 1983. Chemische und mikrobiologische Untersuchungen and Komplexverbindungen des Mangans. *Mitteilgn Deutsch Bodenkundl Gesellsch* 38:325–330.

Gottfreund J, Schmitt G, Schweisfurth R. 1985. Wertigkeitswechsel von Manganspecies duch Bakterien in Nährlösungen und in Lockergestein. *Landwitsch Forschung* 38:80–86.

Gottfreund J, Schweisfurth R. 1983. Mikrobiologische Oxidation und Reduktion von Manganspecies. *Fresenius Z Allg Chem* 316:634–638.

Graham JW. 1959. Metabolically induced precipitation of trace elements from sea water. *Science* 129:1428–1429.

Graham JW, Cooper S. 1959. Biological origin of manganese-rich deposits on the sea floor. *Nature* (London) 183:1050–1051.

Greene AC, Madgwick JC. 1991. Microbial formation of manganese oxides. *Appl Environ Microbiol* 57:1114–1120.

Greenslate J. 1974a. Manganese and biotic debris associations in some deep-sea sediments. *Science* 186:529–531.

Greenslate J. 1974b. Microorganisms participate in the construction of manganese nodules. *Nature* (London) 249:181–183.

Gregory E, Perry RS, Staley JT. 1980. Characterization, distribution, and significance of *Metallogenium* in Lake Washington. *Microb Ecol* 6:125–140.

Hajj J, Makemson J. 1976. Determination of growth of *Sphaerotilus discophorus* in the presence of manganese. *Appl Environ Microbiol* 32:699–702.

Harada K. 1978. Micropaleontologic investigation of Pacific manganese nodules. *Mem Fac Sci Kyoto Univ Ser Geol Mineral* 45:111–132.

Hariya T, Kikuchi T. 1964. Precipitation of manganese by bacteria in mineral springs. *Nature* (London) 202:416–417.

Harriss RC, Troup AG. 1970. Chemistry and origin of freshwater ferromanganese concretions. *Limnol Oceanogr* 15:702–712.

Hastings D, Emerson S. 1986. Oxidation of manganese by spores of a marine bacillus: Kinetic and thermodynamic considerations. *Geochim Cosmochim Acta* 50:1819–1824.

Hem JD. 1963. Chemical equilibria and rate of manganese oxidation. U.S. Geological Survey Water Supply Paper 1667-A.

Hem JD. 1981. Rates of manganese oxidation in aqueous systems. *Geochim Cosmochim Acta* 45: 1369–1374.

Hem JD, Lind CJ. 1983. Nonequilibrium models for predicting forms of precipitated manganese oxides. *Geochim Cosmochim Acta* 47:2037–2046.

Hem JD, Roberson E, Fournier RB. 1982. Stability of β-MnOOH and manganese oxide deposition from springwater. *Water Resour Res* 18:563–570.

Herschel A. 1999. Das Verhalten des Mikroorganismus *Metallogenium* spec. in einer kontinuierlichen Kultur. *Limnologica* 29:105–200

Heye D, Beiersdorf H. 1973. Radioaktive und magnetische Untersuchungen an Manganknollen zur Ermittlung der Wachstumsgeschwindigkeit bzw. Zur Altersbestimmung. *Z Geophysik* 39:703–726.

Hobot JA, Carlemalm E, Villigier W, Kellenberger E. 1984. Periplasmic gel: New concept resulting from the reinvestigation of bacterial cell envelope ultrastructure by new methods. *J Bacteriol* 160:143–152.

Höfer C, Schlosser D. 1999. Novel enzymatic oxidation of Mn^{2+} to Mn^{3+} catalyzed by a fungal laccase. *FEBS Lett* 451:186–190.

Hogan VC. 1973. Electron transport and manganese oxidation in *Leptothrix discophorus*. PhD Thesis. Columbus, OH: Ohio State University.

Holden PJ, Madgwick JC. 1983. Mixed culture bacterial leaching of manganese dioxide. *Proc Australas Inst Min Metall* 286(June):61–63.

Hood DW. 1963. Chemical oceanography. *Oceanogr Mar Biol Annu Rev* 1:129–155.

Hood DW, Slowey JF. 1964. Texas A and M Univ Progr Rept, Proj 276, AEC Contract No. AT-(40-1)-2799.

Horn DR, Horn BM, Delach MN. 1972. Distribution of ferromanganese deposits in the world ocean. In: Horn DR, ed. *Ferromanganese Deposits on the Ocean Floor. The Office of the International Decade of Ocean Exploration.* Washington, DC: National Science Foundation, pp. 9–17.

Hulth S, Aller RC, Gilbert F. 1999. Coupled anoxic nitrification/manganese reduction in marine sediments. *Geochim Cosmochim Acta* 63:49–66.

Hungate B, Danin A, Pellerin NB, Stemmler J, Kjellander P, Adams BJ, Staley JT. 1987. Characterization of manganese oxidizing ($Mn^{II}{\rightarrow}Mn^{IV}$) bacteria from Negev Desert rock varnish: Implications in desert varnish formation. *Can J Microbiol* 33:939–943.

Imai K, Tano T. 1967. Leaching of manganese by *Thiobacillus thiooxidans*. *Hakko Kyokaishi* 25:166–167.

Jackson DD. 1901a. A new species of *Crenothrix* (*C. manganifera*). *Trans Am Microsc Soc* 23:31–39.

Jackson DD. 1901b. The precipitation of iron, manganese, and aluminum by bacterial action. *J Soc Chem Ind* 21:681–684.

Jacobs L, Emerson S, Skei J. 1985. Partitioning and transport of metals across an O_2/H_2S interface in a permanently anoxic basin: Framvaren Fjord, Norway. *Geochim Cosmochim Acta* 49:1433–1444.

Jannasch HW. 1984. Microbial processes at deep sea hydrothermal vents. In: Rona PA, Bostrom K, Laubier L, Smith KL, eds. *Hydrothermal Processes at Seafloor Spreading Centers.* New York: Plenum Press, pp. 677–709.

Jannasch HW, Mottl MJ. 1985. Geomicrobiology of deep-sea hydrothermal vents. *Science* 229:717–725.

Jannasch HW, Wirsen CO. 1979. Chemosynthetic primary production at East Pacific seafloor spreading centers. *Bioscience* 29:592–598.

Jaquet JM, Nembrini G, Garcia J, Vernet JP. 1982. The manganese cycle in Lac Léman, Switzerland: The role of *Metallogenium*. *Hydrobiology* 91:323–340.

Jauregui MA, Reisenauer HM. 1982. Dissolution of oxides of manganese and iron by root exudates components. *Soil Sci Soc Am* 46:314–317.

Johnson AH, Stokes JL. 1966. Manganese oxidation by *Sphaerotilus discophorus*. *J Bacteriol* 91:1543–1547.

Johnson D, Chiswell B, O'Halloran K. 1995. Microorganisms and manganese cycling in a seasonally stratified freshwater dam. *Water Res* 29:2739–2745.

Johnson KS. 2006. Manganese redox chemistry revisited. *Science* 313:1896–1897.

Jones CJ, Murray JW. 1985. The geochemistry of manganese in the northeast Pacific Ocean off Washington. *Limnol Oceanogr* 30:81–92.

Jones JG, Gardener S, Simon BM. 1983. Bacterial reduction of ferric iron in a stratified eutrophic lake. *J Gen Microbiol* 129:131–139.

Jones JG, Gardener S, Simon BM. 1984. Reduction of ferric iron by heterotrophic bacteria in lake sediments. *J Gen Microbiol* 130:45–51.

Jung WK, Schweisfurth R. 1979. Manganese oxidation by an intracellular protein of a *Pseudomonas* species. *Z Allg Mikrobiol* 19:107–115.

Kadko D, Burckle LH. 1980. Manganese nodule growth rates determined by fossil diatom dating. *Nature* (London) 287:725–726.

Kalhorn S, Emerson S. 1984. The oxidation state of manganese in surface sediments of the deep sea. *Geochim Cosmochim Acta* 48:897–902.

Kalinenko VO, Belokopytova OV, Nikolaeva GG. 1962. Bacteriogenic formation of iron-manganese concretions in the Indian Ocean. *Okeanologiya* 11:1050–1059 (English translation).

Karavaiko G, Yuchenko VA, Remizov VI, Klyushnikova TM. 1986. Reduction of manganese dioxide by cell-free *Acinetobacter calcoaceticus* extracts. *Mikrobiologiya* 55:709–714 (English translation, pp. 553–558).

Kashefi K, Lovley DR. 2000. Reduction of Fe(III), Mn(IV), and toxic metals at 100°C by *Pyrobaculum islandicum*. *Appl Environ Microbiol* 66:1050–1056.

Kenten RH, Mann PJG. 1949. The oxidation of manganese by plant extracts in the presence of hydrogen peroxide. *Biochem J* 45:225–263.

Kenten RH, Mann PJG. 1950. The oxidation of manganese by peroxidase systems. *Biochem J* 50:67–73.

Kepkay PE. 1985a. Kinetics of microbial manganese oxidation and trace metal binding in sediments: Results from an in situ dialysis technique. *Limnol Oceanogra* 30:713–726.

Kepkay PE. 1985b. Microbial manganese oxidation and trace metal binding in sediments: Results from an in situ dialysis technique. In: Caldwell DE, Brierley JA, Brierley CL, eds. *Planetrary Ecology*. New York: Van Nostrand Reinhold, pp. 195–209.

Kepkay PE. 1985c. Microbial manganese oxidation and nitrification in relation to the occurrence of macrophyte roots in lacustrine sediments. *Hydrobiology* 128:135–142.

Kepkay PE, Burdige DJ, Nealson KH. 1984. Kinetics of bacterial manganese binding and oxidation in chemostat. *Geomicrobiol J* 3:245–262.

Kepkay PE, Nealson KH. 1982. Surface enhancement of sporulation and manganese oxidation by a marine bacillus. *J Bacteriol* 151:1022–1026.

Kepkay PE, Nealson KH. 1987. Growth of a manganese oxidizing *Pseudomonas* sp. in continuous cultures. *Arch Microbiol* 148:3–67.

Kepkay PE, Schwinghamer P, Willar T, Bowen AJ. 1986. Metabolism and metal binding by surface colonizing bacteria: Results of microgradient measurements. *Appl Environ Microbiol* 51:163–170.

Kessick MA, Vuceta J, Morgan JJ. 1972. Spectrophotometric determination of oxidized manganese with leuco crystal violet. *Environ Sci Technol* 6:642–644.

Kindle EM. 1932. Lacustrine concretions of manganese. *Am J Sci* 24:496–504.

Kindle EM. 1935. Manganese concretions in Nova Scotia lakes. *Roy Soc Can Trans Sec IV* 29:163–180.

Klaveness D. 1977. Morphology, distribution and significance of the manganese-accumulating microorganisms *Metallogenium* in lakes. *Hydrobiologia* 56:25–33.

Klimov V V. 1984. Charge separation in photosystem II reaction centers. The role of phaeophytin and manganese. In: Sybesina C, ed. *Advances in Photosynthetic Research,* Vol 1. Proc Int Congr Photosynth 6th. The Hague: Nijhoff, pp. 131–138.

Klinkhammer GP, Bender ML. 1980. The distribution of manganese in the Pacific Ocean. *Earth Planet Sci Lett* 46:361–384.

Kostka JE, Luther GW III, Nealson KH. 1995. Chemical and biological reduction of Mn(III) pyrophosphate complexes: Potential importance of dissolved Mn(III) as an environmental oxidant. *Geochim Cosmochim Acta* 59:885–894.

Kozub JM, Madgwick JC. 1983. Microaerobic microbial manganese dioxide leaching. *Proc Autralas Inst Min Metall* 288(December):51–54.

Krumbein WE. 1969. Ueber den Einfluss der Mikroflora auf die exogene Dynamik (Verwitterung und Krustenbildung) *Geol Rundsch* 58:333–365.

Krumbein WE. 1971. Manganese oxidizing fungi and bacteria in recent shelf sediments of the Bay of Biscay and the North Sea. *Naturwissenschaften* 58:56–57.

Krumbein WE, Altmann HJ. 1973. A new method for the detection and enumeration of manganese-oxidizing and reducing microorganisms. *Helgoländer Wissenschaftliche Meeresuntersuchungen* 25:347–356.

Krumbein WE, Jens K. 1981. Biogenic rock varnishes of the Negev Desert (Israel): An ecological study of iron and manganese transformation by cyanobacteria and fungi. *Oecologia* (Berlin) 50:25–38.

Ku TL, Broecker WS. 1969. Radiochemical studies on manganese nodules of deep-sea origin. *Deep-Sea Res Oceanogr Abstr* 16:625–635.

Kumari A, Natarajan KA. 2001. Bioleaching of ocean manganese nodules in the presence of reducing agents. *Eur J Miner Process Environ Protect* 1:10–24.

Landing WM, Bruland K. 1980. Manganese in the North Pacific. *Earth Planet Sci Lett* 49:45–56.

Langenhoff AAM, Brouwers-Ceiler DL, Engelberting JHL, Quist JJ, Wolkenfeldt JGPN, Zehnder AJB, Schraa G. 1997. Microbial reduction of manganese coupled to toluene oxidation. *FEMS Microbiol Ecol* 22:119–127.

LaRock PA, Ehrlich HL. 1975. Observations of bacterial microcolonies on the surface of ferromanganese nodules from Blake Plateau by scanning electron microscopy. *Microb Ecol* 2:84–96.

Larsen I, Little B, Nealson KH, Ray R, Stone A, Tian J. 1998. Manganite reduction by *Shewanella putrefaciens* MR-4. *Ann Mineral* 83:1564–1572.

Larson EI, Sly LI, McEwan AG. 1999. Manganese(II) adsorption and oxidation by whole cells and a membrane fraction of *Pedomicrobium* sp. ACM 3067. *Arch Microbiol* 171:257–264.

Laverman AM, Switzer Blum J, Schaefer JK, Philips EJP, Lovley DR, Oremland RS. 1995. Growth of strain SES-3 with arsenate and other diverse electron acceptors. *Appl Environ Microbiol* 61:3556–3561.

Lawton K. 1955. Chemical composition of soils. In: Bear FE, ed. *Chemistry of the Soil*. New York: Reinhold, pp. 53–54.

Leeper GW. 1947. The forms and reactions of manganese in the soil. *Soil Sci* 63:79–94.

Leeper GW, Swaby RJ. 1940. The oxidation of manganous compounds by microorganisms in the soil. *Soil Sci* 49:163–169.

Letunova SV, Ulubekova MV, Shcherbakov VI. 1978. Manganese concentration by microorganisms inhabiting soils of the manganese biogeochemical province of the Georgian SSR. *Mikrobiologiya* 47:332–337 (English translation, pp. 273–278).

Lieske R. 1919. Zur Ernährungsphysiologie der Eisenbakterien. *Zentralbl Bakteriol Parasitenk Infektionskr Hyg Abt II* 49:413–425.

Loganathan P, Burau RG. 1973. Sorption of heavy metal ions by hydrous manganese oxide. *Geochim Cosmochim Acta* 37:1277–1293.

Lovley DR. 1991. Dissimilatory Fe(III) and Mn(IV) reduction. *Microbiol Rev* 55:259–287.

Lovley DR, Giovannoni SJ, White DC, Champine JE, Phillips EJP, Gorby YA, Goodwin S. 1993b. *Geobacter metallireducens* gen. nov. sp. nov., a microorganism capable of coupling the complete oxidation of organic compounds to the reduction of iron and other metals. *Arch Microbiol* 159:336–344.

Lovley DR, Goodwin S. 1988. Hydrogen concentrations as an indicator of the predominant terminal electron-accepting reactions in aquatic sediments. *Abstr Annu Meet Am Soc Microbiol* NK-30:249.

Lovley DR, Phillips EJP. 1988a. Novel mode of microbial energy metabolism: Organic carbon oxidation coupled to dissimilatory reduction of iron and manganese. *Appl Environ Microbiol* 54:1472–1480.

Lovley DR, Phillips EJP. 1988b. Manganese inhibition of microbial iron reduction in anaerobic sediments. *Geomicrobiol J* 6:145–155.

Lovley DR, Phillips EJP. 1994a. Novel process for anaerobic sulfate production from elemental sulfur by sulfate-reducing bacteria. *Appl Environ Microbiol* 60:2394–2399.

Lovley DR, Phillips EJP. 1994b. Reduction of chromate by *Desulfovibrio vulgaris* and its c_3 cytochrome. *Appl Environ Microbiol* 60:726–728.

Lovley DR, Phillips EJP, Lonergan DJ. 1989. Hydrogen and formate oxidation coupled to dissimilatory reduction of iron and manganese by *Alteromonas putrefaciens*. *Appl Environ Microbiol* 55:700–706.

Lovley DR, Widman PK, Woodward JC, Phillips JP. 1993a. Reduction of uranium by cytochrome c_3 of *Desulfovibrio vulgaris*. *Appl Environ Microbiol* 59:3572–3576.

Lower SK, Hochella MF, Jr., Beveridge TJ. 2001. Bacterial recognition of mineral surfaces: Nanoscale interactions between *Shewanella* and α-FeOOH. *Science* 292:1360–1363.

Luther GW III. 2005. Manganese(II) oxidation and Mn(IV) reduction in the environment—Two one-electron transfer steps versus a single two-electron step. *Geomicrobiol J* 22:195–203.

Luther GW III, Sunby B, Lewis BL, Brendel PJ, Silverberg N. 1997. Interactions of manganese with the nitrogen cycle: Alternative pathways to dinitrogen. *Geochim Cosmochim Acta* 61:4043–4052.

Mahler HR, Cordes EH. 1966. *Biological Chemistry*. New York: Harper & Row.

Maki JS, Tebo BM, Palmer FE, Nealson KH, Staley JR. 1987. The abundance and biological activity of manganese-oxidizing bacteria and *Metallogenium*-morphotypes in Lake Washington. *FEMS Microbiol Ecol* 45:21–29.

Mandernack KW. 1992. Oxygen isotopic, mineralogical, and field studies. PhD Thesis. San Diego, CA: University of California.

Mandernack KW, Post J, Tebo BM. 1995. Manganese mineral formation by bacterial spores of the marine *Bacillus*, strain GS-1: Evidence for the direct oxidation of Mn(II) to Mn(IV). *Geochim Cosmochim Acta* 59:4393–4408.

Mandernack KW, Tebo BM. 1993. Manganese scavenging and oxidation at hydrothermal vents and in vent plumes. *Geochim Cosmochim Acta* 57:3907–3923.

Mann PJG, Quastel JH. 1946. Manganese metabolism in soils. *Nature* (London) 158:154–156.

Mann S, Sparks NHC, Scott GHE, de Vrind-de Jong EW. 1988. Oxidation of manganese and formation of Mn_3O_4 (haumannite) by spore coats of a marine *Bacillus* sp. *Appl Environ Microbiol* 54:2140–2143.

Margolis JV, Burns RG. 1976. Pacific deep-sea manganese nodules: Their distribution, composition, and origin. *Annu Rev Earth Planet Sci* 4:229–263.

McKenzie RM. 1967. The sorption of cobalt by manganese minerals in soils. *Aust J Soil Res* 5:235–246.

McKenzie RM. 1970. The reaction of cobalt with manganese minerals. *Aust J Soil Res* 8:97–106.

Mercz TI, Madgwick JC. 1982. Enhancement of bacterial manganese leaching by microalgal growth products. *Proc Australas Inst Min Metall* 238(September):43–46.

Mero JL. 1962. Ocean-floor manganese nodules. *Econ Geol* 57:747–767.

Mills VH, Randles CI. 1979. Manganese oxidation in *Sphaerotilus discophorus* particles. *J Gen Appl Microbiol* 25:205–207.

Mirchink TG, Zaprometova KM, Zvyagintsev DG. 1970. Satellite fungi of manganese oxidizing bacteria. *Mikrobiologiya* 39:379–383 (English translation, pp. 327–330).

Miyata N, Maruo K, Tani KY, Tsuno H, Seyama H, Soma M, Iwahori K. 2006. Production of biogenic manganese oxides by anamorphic ascomycete fungi isolated from streambed pebbles. *Geomicrobiol J* 23:63–73.

Moese JR, Brantner H. 1966. Mikrobiologische Studien an manganoxydierenden Bakterien. *Zentralbl Bakt Parasitenk Infektionskr Hyg Abt II* 120:480–495.

Moffett JW. 1997. The importance of microbial Mn oxidation in the upper ocean: a comparison of the Sargassso Sea and equatorial Pacific. *Deep-Sea Res I*, 44:1277–1291.

Moffett JW, Ho J. 1996. Oxidation of cobalt and manganese in seawater is a common microbially catalyzed pathway. *Geochim Cosmochim Acta* 60:3415–3424.

Molisch H. 1910. *Die Eisenbakterien*. Jena, Germany: Gustav Fischer Verlag.

Moore WS. 1981. Iron-manganese banding in Oneida Lake ferromanganese nodules. *Nature* (London) 292:233–235.

Moore WS, Dean WE, Krishnaswami S, Borole DV. 1980. Growth rates of manganese nodules in Oneida Lake, New York. *Earth Planet Sci Lett* 46:191–200.

Mukherjee A, Raichur AM, Modak JM, Natarajan KA. 2003. Solubilization of cobalt from ocean nodules at neutral pH—a novel bioprocess. *J Industr Microbiol Biotechnol* 30:606–612.

Mulder GE. 1972. Le cycle biologique tellurique et aquatique du fer et du manganèse. *Rev Ecol Biol Sol* 9:321–348.

Murdoch F, Smith PG. 1999. Formation of manganese micro-nodules on water pipeline materials. *Water Res* 33:2893–2895.

Murray JW, Balistrieri LS, Paul B. 1984. The oxidation state of manganese in marine sediments and ferromanganese nodules. *Geochim Cosmochim Acta* 48:1237–1247.

Murray JW, Dillard JG, Giovanoli R, Moers H, Stumm W. 1985. Oxidation of Mn(II): Intitial mineralogy, oxidation state and ageing. *Geochim Cosmochim Acta* 49:463–470.

Murray KJ, Webb, SM, Bargar JR, Tebo BM. 2007. Indirect oxidation of Co(II) in the presence of marine Mn(II)-oxidizing bacterium *Bacillus* sp. strain SG-1. *Appl Environ Microbiol* 73:6905–6909.

Mustoe GE. 1981. Bacterial oxidation of manganese and iron in a modern cold spring. *Geol Soc Am Bull Part 1*, 92:147–153.

Myers CR, Alatalo LJ, Myer JM. 1994. Microbial potential for the anaerobic degradation of simple aromatic compounds in sediments of the Milwaukee harbor, Green Bay, and Lake Erie. *Environ Toxicol Chem* 13:461–471.

Myers CR, Myers JM. 1992. Localization of cytochromes to the outer membrane of anaerobically grown *Shewanella putrefaciens* MR-1. *J Bacteriol* 174:3429–3438.

Myers CR, Myers JM. 1993. Ferric reductase is associated with the membranes of *Shewanella putrefaciens* MR-1. *Microbiol Lett* 108:15–22.

Myers CR, Myers JM. 1997. Outer membrane cytochromes of *Shewanella putrefaciens* MR-1: Spectral analyses, and purification of the 83-kDa c-type cytochrome. *Biochim Biophys Acta* 1326:307–318.

Myers CR, Nealson KH. 1988a. Bacterial manganese reduction and growth with manganese oxide as the sole electron acceptor. *Science* 240:1319–1321.

Myers CR, Nealson KH. 1988b. Microbial reduction of manganese oxides: Interaction with iron and sulfur. *Geochim Cosmochim Acta* 52:2727–2732.

Myers JM, Myers CR. 1998. Isolation and sequence of *omcA*, a gene encoding a decaheme outer membrane cytochrome c of *Shewanella putrefaciens*. *Biochim Biophys Acta* 1873:237–251.

Myers JM, Myers CR. 2000. Role of the tetraheme cytochrome Cym A in anaerobic electron transport in cells of *Shewanella putrefaciens* MR-1 in reduction of manganese dioxide. *J Bacteriol* 182:67–75.

Myers JM, Myers CR. 2001. Role for outer membrane cytochromes OmcA and OmcB of *Shewanella putrefaciens* MR-1 in reduction of manganese dioxide. *Appl Environ Microbiol* 67:260–269.

Nealson KH. 1978. The isolation and characterization of marine bacteria which catalyze manganese oxidation. In: Krumbein WE, ed. *Environmental Biogeochemistry, Vol 3. Methods, Metals and Assessment.* Ann Arbor, MI: Ann Arbor Science Publishers, pp. 847–858.

Nealson KH, Ford J. 1980. Surface enhancement of bacterial manganese oxidation: Implications for aquatic environments. *Geomicrobiol J* 2:21–37.

Nealson KH, Saffarini D. 1994. Iron and manganese in anaerobic respiration: Environmental significance, physiology, regulation. *Annu Rev Microbiol* 48:311–343.

Nealson KH, Tebo BM, Rosson RA. 1988. Occurrence and mechanism of microbial oxidation of manganese. *Adv Appl Microbiol* 33:279–318.

Neufeld CA. 1904. Z. Uners. Nahrungs-Genussmitt 7:478 (as cited by Moese and Brantner, 1966).

Novozhilova MI, Berezina FS. 1984. Iron- and manganese-oxidizing microorganisms in grounds in the northwestern part of the Indian Ocean and the Red Sea. *Mikrobiologiya* 53:129–136 (English translation, pp. 106–112).

Okazaki M, Sugita T, Shimizu M, Ohode Y, Iwamoto K, de Vrind-de Jong EW, de Vrind JPM, Corstjen PLAM. 1997. Partial purification and characterization of manganese oxidizing factors of *Pseudomonas fluorescens* GB-1. *Appl Environ Microbiol* 63:4793–4799.

Ostwald J. 1981. Evidence for a biogeochemical origin of the Groote Eylandt manganese ores. *Econ Geol* 76:556–567.

Park PK. 1968. Seawater hydrogen-ion concentration: Vertical distribution. *Science* 162:357–358.

Parker DL, Sposito G, Tebo BM. 2004. Manganese(III) binding to a pyoverdine siderophore produced by a manganese(II)-oxidizing bacterium. *Geochim Cosmochim Acta* 68:4809–4820.

Paszczynski A, Huynh V-B, Crawford R. 1986. Comparison of lignase-1 and peroxidase-M2 from the white rot fungus *Phanerochaete chrysosporium*. *Arch Biochem Biophys* 244:750–765.

Patrick WH, Jr., Henderson RE. 1981. Reduction and reoxidation cycles of manganese and iron in flooded soil and in water solution. *Soil Sci Soc Am J* 45:855–859.

Pattan JN, Mudholkar AV. 1990. The oxidation state of manganese in ferromanganese nodules and deep-sea sediments from the Central Indian Ocean. *Chem Geol* 85:171–181.

Pecher K, McCubbery D, Kneedler E, Rothe J, Bargar J, Meigs G, Cox L, Nealson K, Tonner B. 2003. Quantitative charge state analysis of manganese biominerals in aqueous suspension using scanning transmission X-ray microscopy (STXM). *Geochim Cosmochim Acta* 67:1089–1098.

Perfil'ev BV, Gabe DR. 1965. The use of the microbial-landscape method to investigate bacteria which concentrate manganese and iron in bottom deposits. In: Perfil'ev BV, Gabe DR, Gal'perina AM, Rabinovich VA, Saponitskii AA, Sherman EE, Troshanov EP, eds. *Applied Capillary Microscopy. The Role of Microorganisms in the Formation of Iron Manganese Deposits.* New York: Consultants Bureau, pp. 9–54.

Perfil'ev BV, Gabe DR. 1969. *Capillary Methods of Studying Microorganisms.* English translation by J Shewan. Toronto, Canada: University of Toronto Press.

Perkins EC, Novielli F. 1962. Bacterial leaching of manganese ores. *US Bur Mines Rep Invest* 6102.

Piper DZ, Basler JR, Bischoff JL. 1984. Oxidation state of marine manganese nodules. *Geochim Cosmochim Acta* 48:2347–2355.

Poldervaart A. 1955. Chemistry of the Earth's crust. In: Poldervaart A, ed. *Crust of the Earth. A Symposium. Special Paper 62.* Boulder, CO: Geological Society of America, pp. 119–144.

Post JE. 1999. Manganese oxide minerals: Crystal structures and economic and environmental significance. *Proc Natl Acad Sci USA* 96:3447–3454.

Potter RM, Rossman GR. 1977. Desert varnish: The importance of clay minerals. *Science* 196:1446–1448.

Reid ASJ, Miller MH. 1963. The manganese cycle. II. Forms of soil manganese in equilibrium with solution manganese. *Can J Soil Sci* 43:250–259.

Reyss JL, Marchig V, Ku TL. 1982. Rapid growth of a deep-sea manganese nodule. *Nature* (London) 295:401–403.

Richards FA. 1965. Anoxic basins and fjords. In: Riley PG, Skirrow G, eds. *Chemical Oceanography*, Vol 1. New York: Academic Press, pp. 611–641.

Richardson LL, Aguilar C, Nealson KH. 1988. Manganese oxidation in pH and O_2 environments produced by phytoplankton. *Limnol Oceanogr* 33:352–363.

Riemann F. 1983. Biological aspects of deep-sea manganese nodule formation. *Oceanol Acta* 6:303–311.

Robertson LA, Van Niel EWJ, Torremans RAM, Kuenen GJ. 1988. Simultaneous nitrification and deni-trification in aerobic chemostat cultures of *Thiosphaera pantotropha*. *Appl Environ Microbiol* 54:2821–2818.

Robinson WO. 1929. Detection and significance of manganese dioxide in the soil. *Soil Sci* 27:335–349.

Rossman R, Callender E. 1968. Manganese nodules in Lake Michigan. *Science* 162:1123–1124.

Rosson RA, Nealson KH. 1982. Manganese binding and oxidation by spores of a marine bacillus. *J Bacteriol* 151:1027–1034.

Rosson RA, Tebo BM, Nealson KH. 1984. Use of poisons in determination of microbial manganese binding rates in seawater. *Appl Environ Microbiol* 47:740–745.

Rusin PA, Quintana L, Sinclair NA, Arnold RG, Oden KL. 1991a. Physiology and kinetics of manganese-reducing *Bacillus polymyxa* strain D1 isolated from manganiferous silver ore. *Geomicrobiol J* 9:13–25.

Rusin PA, Sharp JE, Arnold RG, Sinclair NA. 1991b. Enhanced recovery of manganese and silver from refrac-tory ores. In: Smith RW, Mishra M, eds. *Minerals Bioprocessing*. Warrendale, PA: The Minerals, Metals and Materials Society, pp. 207–218.

Sartory A, Meyer J. 1947. Contribution à l'étude du métabolisme hydrocarboné des bactéries ferrugineuses. *CR Acad Sci* (Paris) 225:541–542.

Schippers A, Neretin LN, Lavik G, Leipe T, Pollehne F. 2005. Manganese(II) oxidation driven by lateral oxy-gen intrusions in the western Black Sea. *Geochim Cosmochim Acta* 69:2241–2252.

Schlosser D, Höfer C. 2002. Laccase-catalyzed oxidation of Mn^{2+} in the presence of natural Mn^{3+} chelators as a novel source of extracellular H_2O_2 production and its impact on manganese peroxidase. *Appl Environ Microbiol* 68:3514–3521.

Schmidt JM, Overbeck J. 1984. Studies of iron bacteria from Lake Pluss (West Germany). I. Morphology, fine structure and distribution of *Metallogenium* sp. and *Siderocapsa germinata*. *Z All Mikrobiol* 24:324–339.

Schmidt JM, Sharp WP, Starr MP. 1981. Manganese and iron encrustations and other features of *Planctomyces crassus* Hortobagyi 1965, morphotype Ib of the *Blastocaulis-Planctomyces* group of budding and append-aged bacteria, examined by electron microscopy and X-ray micro-analysis. *Curr Microbiol* 5:241–246.

Schmidt JM, Sharp WP, Starr MP. 1982. Metallic oxide encrustations of the nonprothecate stalks of naturally occurring populations of *Planktomyces bekfii*. *Curr Microbiol* 7:389–394.

Schöttle M, Friedman GM. 1971. Fresh water iron-manganese nodules in Lake George, New York. *Geol Soc Am Bull* 82:101–110.

Schulz-Baldes A, Lewin RA. 1975. Manganese encrustation of zygospores of a *Chlamydomonas* (Chlorophyta: Volvocales). *Science* 188:1119–1120.

Schweisfurth R. 1968. Untersuchungen über manganoxidierende und –reduzierende Mikroorganismen. *Mitt Int Verein Limnol* 14:179–186.

Schweisfurth R. 1969. Manganoxidierende Pilze. *Zentralbl Bakt Parasitenk Infektionskr Hyg Abt I Orig* 212:486–491.

Schweisfurth R. 1971. Manganoxidierende Pilze. I. Vorkommen, Isolierung und mikroskopische Untersuchungen. *Z Allg Mikrobiol* 11:415–430.

Schweisfurth R, Hehn Gv. 1972. Licht- und elektronenmikroskopische Untersuchungen sowie Kulturversuche zum *Metallogenium*-Problem. *Zentralbl Bakt Parasitenk Infektionskr Hyg I Abt Orig* A220:357–361.

Schweisfurth R, Mertes R. 1962. Mikrobiologische und chemische Untersuchungen über Bildung and Bekämpfung von Manganschlammablagerung einer Druckleitung für Talsperrenwasser. *Arch Hyg Bakteriol* 146:401–417.

Seyfried WE, Jr., Janecky DR. 1985. Heavy metal and sulfur transport during subcritical and supercritical hydrothermal alteration of basalt: Influence of fluid pressure and basal composition and crystallinity. *Geochim Cosmochim Acta* 49:2545–2560.

Shanks WC III, Bischoff JL, Rosenbauer RJ. 1981. Seawater sulfate reduction and sulfur isotope fraction-ation in basaltic systems: Interaction of seawater with fayalite and magnetite at 200–350°C. *Geochim Cosmochim Acta* 45:1977–1995.

Sherman GD, McHargue JS, Hodgkiss WS. 1942. Determination of active manganese in soil. *Soil Sci* 54:253–257.

Shternberg LE. 1967. Biogenic structures in manganese ores. *Mikrobiologiya* 36:710–712 (English translation, pp. 595–597).

Sienko MJ, Plane RA. 1966. *Chemistry: Principles and Properties*. New York: McGraw-Hill.

Silvero CM. 1985. Microaerobic manganese dioxide reduction. *NSTA Technol J*, October–December, 51–63.

Silverman MP, Lundgren DG. 1959. Studies on the chemoautotrophic iron bacterium *Ferrobacillus ferroox-idans*. I. An improved medium and a harvesting procedure for securing high cell yields. *J Bacteriol* 77:642–647.

Sly LI, Hodkinson MC, Arunpairojana V. 1988. Effect of water velocity on the early development of manganese-depositing biofilm in a drinking-water distribution system. *FEMS Microbiol Ecol* 53:175–186.

Soehngen NL. 1914. Umwandlung von Manganverbindungen unter dem Einfluss mikrobiologischer Prozesse. *Zentralbl Bakt Parasitenk Infektionskr Hyg II abt* 40:545–554.

Sokolova-Dubinina GA, Deryugina ZP. 1967a. On the role of microorganisms in the formation of rhodochrosite in Punnus-Yarvi lake. *Mikrobiologiya* 36:535–542 (English translation, pp. 445–451).

Sokolova-Dubinina GA, Deryugina ZP. 1967b. Process of iron-manganese concretion formation in Lake Punnus-Yarvi. *Mikrobiologiya* 36:1066–1076 (English translation, pp. 892–900).

Sokolova-Dubinina GA, Deryugina ZP. 1968. Influence of limnetic environmental conditions on microbial manganese ore formation. *Mikrobiologiya* 37:143–153 (English translation, pp. 123–128).

Sorem RK, Foster AR. 1972. Internal structure of manganese nodules and implications in beneficiation. In: Horn DR, ed. *Ferromanganese Deposits on the Ocean Floor*. Washington, DC: Office of the International Decade of Ocean Exploration, National Science Foundation, pp. 167–181.

Sorokin YI. 1971. Microflora of iron-manganese concretions. *Mikrobiologiya* 40:563–566 (English translation, pp. 493–495).

Sorokin YI. 1972. Role of biological factors in the sedimentation of iron, manganese, and cobalt and in the formation of nodules. *Oceanology* (Oekonologiya) 12:1–11.

Sparrow LA, Uren NC. 1987. Oxidation and reduction of Mn in acid soils: Effect of temperature and soil pH. *Soil Biol Biochem* 19:143–148.

Stabel HH, Kleiner J 1983. Endogenic flux of manganese to the bottom of Lake Constance. *Arch Hydrobiol* 98:307–316.

Stone AT. 1987. Microbial metabolites and the reductive dissolution of manganese oxides: Oxalate and pyruvate. *Geochim Cosmochim Acta* 51:919–925.

Sugio T, Katagiri T, Moriyama M, Zhen YL, Inagaki K, Tano T. 1988a. Existence of a new type of sulfide oxidase which utilizes ferric iron as an electron acceptor in *Thiobacillus ferrooxidans*. *Appl Environ Microbiol* 54:153–157.

Sugio T, Tsujitsu Y, Hirayama K, Inagaki K, Tano T. 1988b. Mechanism of tetravalent manganese reduction with elemental sulfur. *Agric Microbiol Chem* 52:185–190.

Tambiev SB, Demina LS. 1992. Biogeochemistry and fluxes of manganese and some other metals in regions of hydrothermal activities (Axial Mountain, Juan de Fuc Ridge and Guaymas Basin, Gulf of California). *Deep-Sea Res* 39:687–703.

Taylor RM, McKenzie RM. 1966. The association of trace elements with manganese minerals in Australian soils. *Aust J Soil Res* 4:29–39.

Taylor RM, McKenzie RM, Norrish K. 1964. The mineralogy and chemistry of manganese in some Australian soils. *Aust J Soil Res* 2:235–248.

Taylor-George S, Palmer F, Staley JT, Borns DJ, Curtiss B, Adams JB. 1983. Fungi and bacteria involved in desert varnish formation. *Microb Ecol* 9:227–245.

Tebo BM, Emerson S. 1985. Effect of oxygen tension, Mn(II) concentration, and temperature on the microbially catalyzed Mn(II) oxidation rate in a marine fjord. *Appl Environ Microbiol* 50:1268–1273.

Tebo BM, Ghiorse WC, van Waasbergen LG, Siering PL, Caspi R. 1997. Bacterial mineral formation: Insights into manganese(II) oxidation from molecular genetic and biochemical studies. *Rev Mineral* 35:225–266.

Tebo BM, Johnson HA, McCarthy JK, Templeton AS. 2005. Geomicrobiology of manganese(II) oxidation. *Trends Microbiol* 13:421–428.

Tebo BM, Lee Y. 1993. Microbial oxidation of cobalt. In: Torma AE, Wey JE, Lakshmanan VI, eds. *Biohydrometallurgical Technologies,* Vol 1. *Bioleaching Processes*. Warrendale, PA: The Minerals, Metals and Materials Society, pp. 695–704.

Tebo BM, Mandernack K, Rosson RA. 1988. Manganese oxidation by a spore coat or exosporium protein from spores of a manganese(II) oxidizing marine bacillus. *Abstr Annu Meet Am Soc Microbiol* I-121:201.

Tebo BM, Nealson KH, Emerson S, Jacobs L. 1984. Microbial mediation of Mn(II) and Co(II) precipitation at the O_2/H_2S interfaces in two anoxic fjords. *Limnol Oceanogr* 29:1247–1258.

Tebo MB. 1991. Manganese(II) oxidation in the suboxic zone of the Black Sea. *Deep-Sea Res* 38:S883–S905.

Ten Khak-mun. 1967. Iron- and manganese-oxidizing microorganisms in soils of South Sakhalin. *Mikrobiologiya* 36:337–344 (English translation, pp. 276–281).

Thamdrup B, Finster K, Wuergler Hansen J, Bak F. 1993. Bacterial disproportionation of elemental sulfur coupled to chemical reduction of iron or manganese. *Appl Environ Microbiol* 59:101–108.

Thamdrup B, Rosselló-Mora R, Amann R. 2000. Microbial manganese and sulfate reduction in Black Sea shelf sediments. *Appl Environ Microbiol* 66:2888–2897.

Thiel GA. 1925. Manganese precipitated by microorganisms. *Econ Geol* 20:301–310.

Timonin MI. 1950a. Soil microflora and manganese deficiency. *Trans 4th Int Congr Soil Sci* 3:97–99.

Timonin MI. 1950b. Soil microflora in relation to manganese deficiency. *Sci Agric* 30:324–325.

Timonin MI, Illman WI, Hartgering T. 1972. Oxidation of manganous salts of manganese by soil fungi. *Can J Microbiol* 18:793–799.

Tipping E, Jones JG, Woof C. 1985. Lacustrine manganese oxides: Mn oxidation states and relationships to Mn depositing bacteria. *Arch Hydrobiol* 105:161–175.

Tipping E, Thompson DW, Davidson W. 1984. Oxidation products of Mn(II) oxidation in lake waters. *Chem Geol* 44:359–383.

Toner B, Fakra S, Villalobos M, Warwick T, Sposito G. 2005. Spatially resolved characterization of biogenic manganese oxide production within a bacterial biofilm. *Appl Environ Microbiol* 71:1300–1310.

Tortoriello RC. 1971. Manganic oxide reduction by microorganisms in fresh water environments. PhD Thesis. Troy, New York: Rensselaer Polytechnic Institute.

Trimble RB. 1968. MnO_2-reduction by two strains of marine ferromanganese nodule bacteria. MS Thesis. Troy, New York: Rensserlaer Polytechnic Institute.

Trimble RB, Ehrlich HL. 1968. Bacteriology of manganese nodules. III. Reduction of MnO_2 by two strains of nodule bacteria. *Appl Microbiol* 16:695–702.

Trimble RB, Ehrlich HL. 1970. Bacteriology of manganese nodules. IV. Induction of an MnO_2-reductase system in *Bacillus*. *Appl Microbiol* 19:966–972.

Troshanov EP. 1968. Iron- and manganese-reducing microorganisms in ore-containing lakes of the Karelian Isthmus. *Mikrobiologiya* 37:934–940 (English translation, pp. 786–791).

Troshanov EP. 1969. Conditions affecting the reduction of iron and manganese by bacteria in the ore-bearing lakes of the Karelian Isthmus. *Mikrobiologiya* 38:634–643 (English translation, pp. 528–535).

Trost WR. 1958. *The Chemistry of Manganese Deposits. Mines Branch Res Rep R8*. Ottawa, Canada: Department of Mines and Technical Surveys.

Trouwborst RE, Clement BG, Tebo BM, Glazer BT, Luther GW III. 2006. Soluble Mn(III) in suboxic zones. *Science* 313:1955–1957.

Tuttle JH, Ehrlich HL. 1986. Coexistence of inorganic sulfur metabolism and manganese oxidation in marine bacteria. *Abstr Annu Meet Am Soc Microbiol* 1–21:168.

Tyler PA. 1970. Hyphomicrobia and the oxidation of manganese in aquatic ecosystems. *Antonie van Leeuwenhoek* 36:567–578.

Tyler PA, Marshall KC. 1967a. Microbial oxidation of manganese in hydro-electric pipelines. *Antonie van Leeuwenhoek* 33:171–183.

Tyler PA, Marshall KC. 1967b. Hyphomicrobia: A significant factor in manganese problems. *J Am Water Works Assoc* 59:1043–1048.

Uren NC, Leeper GW. 1978. Microbial oxidation of divalent manganese. *Soil Biol Biochem* 10:85–87.

Vandenabeele J, De Beer D, Germonpré R, Van de Sande R, Verstraete W. 1995. Influence of nitrate on manganese removing microbial consortia from sand filters. *Water Res* 29:579–587.

Van Veen WL. 1972. Factors affecting the oxidation of manganese by *Sphaerotilus discophorus*. *Antonie van Leeuwenhoek* 38:623–626.

Van Veen WL. 1973. Biological oxidation of manganese in soils. *Antonie van Leeuwenhoek* 39:657–662.

Van Waasbergen LG, Hildebrand M, Tebo BM. 1996. Identification and characterization of a gene cluster involved in manganese oxidation by spores of the marine *Bacillus* sp. SG-1. *J Bacteriol* 178:3517–3530.

Van Waasbergen LG, Hoch JA, Tebo BM. 1993. Genetic analysis of the marine manganese-oxidizing *Bacillus* sp. SG-1: Protoplast transformation, Tn*917* mutagenesis, and identification of chromosomal loci involved in manganese oxidation. *J Bacteriol* 175:7594–7603.

Varentsov IM. 1972. Geochemical studies on the formation of iron-manganese nodules and crusts in recent basins. I. Eningi-Lampi Lake, Central Karelia. *Usta Mineral Petrogr Szeged* 20:363–381.

Varentsov IM, Pronina NV. 1973. On the study of mechanisms of iron-manganese ore formation in recent basins: The experimental data on nickel and cobalt. *Mineral Depos* (Berlin) 8:161–178.

Vavra P, Frederick L. 1952. The effect of sulfur oxidation on the availability of manganese. *Soil Sci Soc Am Proc* 16:141–144.

Villalobos M, Toner B, Bargar J, Sposito G. 2003. Characterization of the manganese oxide produced by *Pseudomonas putida* strain MnB1. *Geochim Cosmochim Acta* 67:2649–2662.

Vojak PWL, Edwards C, Jones MV. 1984. Manganese oxidation and sporulation by estuarine *Bacillus* species. *Microbios* 41:39–47.

Vojak PWL, Edwards C, Jones MV. 1985a. A note on the enumeration of manganese-oxidizing bacteria in estuarine water and sediment samples. *J Appl Microbiol* 59:375–379.

Vojak PWL, Edwards C, Jones MV. 1985b. Evidence for microbiological manganese oxidation in the River Tamar estuary, South West England. *Estuar Coast Shelf Sci* 20:661–671.

Von Damm KL, Edmond JM, Grant B, Measures CI, Walden B, Weiss RF. 1985. Chemistry of submarine hydrothermal solutions at 21°N, East Pacific Rise. *Geochim Cosmochim Acta* 49:2197–2220.

Von Wolzogen-Kühr CAH. 1927. Manganese in waterworks. *J Am Water Works Assoc* 18:1–31.

Webb SM, Dick GJ, Bargar JR, Tebo BM. 2005. Evidence for the presence of Mn(III) intermediates in the bacterial oxidation of Mn(II). *Proc Natl Acad Sci USA* 102:5558–5563.

Woods Hole Oceanographic Institute, Woods Hole, MA.

Yurchenko VA, Karavaiko GI, Remizov VI, Klyushnikova TM, Tarusin AD. 1987. Role of the organic acids produced by *Acinetobacter calcoaceticus* in manganese leaching. *Prikl Biokhim Mikrobiol* 23:404–412.

Zappfe C. 1931. Deposition of manganese. *Econ Geol* 26:799–832.

Zapkin MA, Ehrlich HL. 1983. A comparison of manganese oxidation by growing and resting cells of a freshwater bacterial isolate, strain FMn1. *Z Allg Mikrobiol* 23:447–455.

Zavarzin GA. 1961. Symbiotic culture of a new microorganism oxidizing manganese. *Mikrobiologiya* 30:393–395 (English translation, pp. 343–345).

Zavarzin GA. 1962. Symbiotic oxidation of manganese by two species of *Pseudomonas*. *Mikrobiologiya* 31:586–588 (English translation, pp. 481–482).

Zavarzin GA. 1964. *Metallogenium symbioticum*. *Z Allg Mikrobiol* 4:390–395.

Zeiner CA, Lion LW, Shuler ML, Ghiorse WC, Hay A. 2006. Cycling of biogenic Mn-oxides in a model microbial predator-prey system. *Geomicrobiol J* 23:37–43.

Zindulis J, Ehrlich HL. 1983. A novel Mn^{2+}-oxidizing enzyme system in a freshwater bacterium. *Z Allg Mikrobiol* 23:457–465.

ZoBell CE. 1946. *Marine Microbiology*. Waltham, MA: Chronica Botanica.

18 Geomicrobial Interactions with Chromium, Molybdenum, Vanadium, Uranium, Polonium, and Plutonium

18.1 MICROBIAL INTERACTION WITH CHROMIUM

18.1.1 OCCURRENCE OF CHROMIUM

Chromium is not a very plentiful element in the Earth's crust, but it is nevertheless fairly widespread. Its average crustal abundance of 122 ppm (Fortescue, 1980) is less than that of manganese. Average concentrations in rocks range from 4 to 90 mg kg^{-1}, in soil ~70 mg kg^{-1}, in freshwater ~1 μg kg^{-1}, and in seawater ~0.3 μg kg^{-1} (Bowen, 1979). Its chief mineral occurrence is chromite, in which the chromium has an oxidation state of +3. It is a spinel whose end members are MgCr$_2$O$_4$ and FeCr$_2$O$_4$. The chromium in this mineral can be partially replaced by Al or Fe. Chromite is of igneous origin. Other chromium minerals of minor occurrence include eskolite (Cr$_2$O$_3$), daubréelite (FeS·Cr$_2$S$_3$), crocoite (PbCrO$_4$), uvarovite, which is also known as garnet [Ca$_3$Cr$_2$(SiO$_4$)$_3$], and others (Smith, 1972).

18.1.2 CHEMICALLY AND BIOLOGICALLY IMPORTANT PROPERTIES

The element chromium is a member of the first transition series of elements in the periodic table together with scandium, titanium, vanadium, manganese, iron, cobalt, nickel, copper, and zinc. The chief oxidation states of chromium are 0, +2, +3, and +6. A +5 oxidation state is also known, which appears to be of significance in at least some biochemical reductions of Cr(VI) (Shi and Dalal, 1990; Suzuki et al., 1992). The geomicrobially important oxidation states are +3 and +6.

Chromium in the hexavalent state (+6) is very toxic, in part because of its high solubility as chromate (CrO$_4^{2-}$) and dichromate (Cr$_2$O$_7^{2-}$) in the physiological pH range. At high enough concentrations, Cr(VI) can be mutagenic and carcinogenic. Chromium in the trivalent state is less toxic, in part probably because it is less soluble in this oxidation state at physiological pH. At neutral pH, Cr^{3+} tends to precipitate as a hydroxide [Cr(OH)$_3$] or a hydrated oxide. Chromate and dichromate are strong oxidizing agents.

Chromium has been reported to be nutritionally essential in trace amounts (Miller and Neathery, 1977; Mertz, 1981). Its metabolic function in biochemical terms is still unclear.

As Cr$_2$(SO$_4$)$_3$, K$_2$CrO$_4$, or K$_2$Cr$_2$O$_7$, chromium is inhibitory to growth of bacteria at appropriate concentrations (Forsberg, 1978; Wong et al., 1982; Bopp et al., 1983). When taken into the cell, hexavalent chromium can act as a mutagen in prokaryotes and eukaryotes (Nishioka, 1975; Petrelli and DeFlora, 1977; Venitt and Levy, 1974) and as carcinogen in animals (Gruber and Jennette, 1978; Sittig, 1985). In bacteria, chromate can be taken into the cell via the sulfate uptake system, which involves active transport (Ohtake et al., 1987; Silver and Walderhaug, 1992). Whether dichromate is transported by the same system is not clear. Bacterial resistance to Cr(VI) has been observed (Bader et al., 1999). In some instances, at least, the genetic trait is plasmid-borne (Summers and Jacoby,

1978; Bopp et al., 1983; Cervantes, 1991; Silver and Walderhaug, 1992). In *Pseudomonas fluorescens* LB300, resistance to chromate (CrO_4^{2-}) was found to be due to decreased chromate uptake (Bopp, 1980; Ohtake et al., 1987). In *P. ambigua*, chromate resistance was attributed to formation of a thickened cell envelope that reduced permeability of Cr(VI) and to an ability to reduce Cr(VI) to Cr(III) (Horitsu et al., 1987).

The basis for resistance to dichromate ($Cr_2O_7^{2-}$) and chromite (Cr^{3+}) has not been clearly established. The resistance mechanism for dichromate need not be the same as for chromate, because *Pseudomonas fluorescens* LB300, which is resistant to chromate, is much more sensitive to dichromate (Bopp, 1980; Bopp et al., 1983). Some bacteria have an ability to accumulate chromium. In at least some cases, the accumulation may be due to adsorption (Coleman and Paran, 1983; Johnson et al., 1981; Marques et al., 1982).

18.1.3 MOBILIZATION OF CHROMIUM WITH MICROBIALLY GENERATED LIXIVIANTS

Acidithiobacillus thiooxidans and *A. ferrooxidans* have been found to solubilize only a limited amount of chromium contained in the mineral chromite (Cr_2O_3) with sulfuric acid generated by the oxidation of sulfur (Ehrlich, 1983). Similarly, acid produced during iron oxidation by *A. ferrooxidans* was able to solubilize only limited amounts of chromium from chromite (Wong et al., 1982). On the contrary, chromium can be successfully leached from some solid industrial wastes with biologically formed sulfuric acid (Bosecker, 1986).

18.1.4 BIOOXIDATION OF CHROMIUM(III)

No observations of enzymatic oxidation of Cr(III) to Cr(VI) have been reported. However, nonenzymatic oxidation of Cr(III) to Cr(VI) may occur in soil environments, where biogenic (or abiogenic) Mn(III) or Mn(IV)- oxides may oxidize Cr(III) to Cr(VI) (Bartlett and James, 1979). These interactions can be summarized as

$$Cr^{3+} + 3MnOOH + H^+ \rightarrow CrO_4^{2-} + 3Mn^{2+} + 2H_2O \tag{18.1a}$$

$$2Cr^{3+} + 3MnO_2 + 2H_2O \rightarrow 2CrO_4^{2-} + 3Mn^{2+} + 4H^+ \tag{18.1b}$$

Such oxidation can be detrimental if the Cr(VI) produced reaches a toxic level. Similar observations were made by Chen et al. (1997) and by Kozuh et al. (2000). The latter emphasized that Cr(III) oxidation by Mn(IV) is favored by low organic matter concentration and high concentration of Mn(IV) oxides.

18.1.5 BIOREDUCTION OF CHROMIUM(VI)

A number of bacterial species have been shown to reduce Cr(VI) to Cr(III) (Romanenko and Koren'kov, 1977; Horitsu et al., 1978; Lebedeva and Lyalikova, 1979; Shimada, 1979; Kvasnikov et al., 1985; Gvozdyak et al., 1986; Wang et al., 1989; Ishibashi et al., 1990; Shen and Wang, 1993; Llovera et al., 1993; Lovley and Phillips, 1994; Gopalan and Veeramani, 1994; Garbisu et al., 1998; Philip et al., 1998; Wani et al., 2007). They include *Achromobacter eurydice, Aeromonas dechromatica, Agrobacterium radiobacter* strain EPS-916, *Arthrobacter* spp., *Bacillus subtilis, B. cereus, B. coagulans, Desulfovibrio vulgaris* (Hildenborough) ATCC 29579, *Escherichia coli* K-12 and ATCC 33456, *Enterobacter cloacae* HO1, *Flavobacterium devorans, Sarcina flava, Micrococcus roseus, Pseudomonas* spp., *Shewanella putrefaciens* (now *Sh. oneidensis*) MR-1, and *Burkholderia cepacia* MCMB-821. It is unclear whether all these strains reduce Cr(VI) enzymatically. Sulfate-reducing bacteria can reduce chromate with the H_2S they produce from sulfate (Bopp, 1980), but

D. vulgaris can do it enzymatically as well (Lovley and Phillips, 1994). Arias and Tebo (2003) observed, however, that sulfate-reducing bacteria in general are inhibited at elevated Cr(VI) concentrations so that chromate reduction by biogenic H_2S is likely to be significant only at low Cr(VI) concentrations in the environment. *T. ferrooxidans* can reduce dichromate with partially reduced sulfur species it forms during the oxidation of elemental sulfur (Sisti et al., 1996). Of those bacteria that reduce Cr(VI) enzymatically, some of the facultative strains reduce it only anaerobically, whereas others will do it aerobically and anaerobically. Many bacterial strains reduce Cr(VI) as a form of respiration, but at least one (*P. ambigua* G-1) reduces it as a means of detoxification (Horitsu et al., 1987). Kwak et al. (2003) showed that the Cr(VI) reductase of *P. ambigua* G-1 is homologous with the nitroreductase of strains KCTC of *Vibrio harveyi* and strain DH5α of *E. coli*. Marsh et al. (2000) explored some of the factors that affect biological chromate reduction in microcosms of sandy aquifer material. They found that biological reduction occurred only in light-colored sediment, abiotic reduction being observed in black, claylike sediment. The pH optimum for Cr(VI) reduction in this sediment was 6.8. Although they detected two temperature optima, at 22 and 50°C, the lower optimum probably represented that of the dominant bacterial group in view of the ambient *in situ* temperature of the sediment. The presence of oxygen in the sediment was inhibitory as was the addition of nitrate, but not of selenate or ferrous iron. The presence of Cr(VI) prevented a loss of sulfate or production of Fe(II). Molybdate, an inhibitor of sulfate reduction, inhibited Cr(VI) reduction only at concentrations 40 times that of Cr(VI). Bromoethanesulfonic acid, an inhibitor of methanogenesis, strongly inhibited Cr(VI) reduction at 20 mM concentration, but only slightly inhibited it at concentrations between 0.2 and 2.0 mM.

Pseudomonas fluorescens LB 300, isolated from the upper Hudson River (New York State), can reduce chromate aerobically with glucose or citrate as electron donor (Bopp and Ehrlich, 1988; DeLeo and Ehrlich, 1994). Conditions under which aerobic reduction has been studied include batch culturing with shaking at 200 rpm and continuous culturing with stirring and forced aeration (DeLeo and Ehrlich, 1994). The organism converts chromate to Cr^{3+} in batch culture when growing in glucose–mineral salts solution (Vogel–Bonner [VB] medium) and in continuous culture (chemostat) when growing in a citrate–yeast extract–tryptone solution buffered with phosphate (Figure 18.1). Anaerobically, *P. fluorescens* strain LB300 was found to reduce chromate only when growing with acetate as energy source (electron donor). Furthermore, although *P. fluorescens* LB300 will reduce chromate aerobically at an initial concentration as high as 314 μg mL^{-1}, anaerobically it reduces chromate only at a concentration below 50 μg mL^{-1} (Bopp and Ehrlich, 1988; DeLeo and Ehrlich, 1994). Other bacteria that can reduce chromate aerobically and anaerobically include *Escherichia coli* ATCC 33456 (Shen and Wang, 1993), *Agrobacterium radiobacter* EPS-916 (Llovera et al., 1993), and *Burkholderia cepacia* MCMB-821 (Wani et al., 2007). Reduction of chromate by *E. coli* ATCC 33456 is, however, partially repressed by oxygen through uncompetitive inhibition (Shen and Wang, 1993). Reduction of chromate by resting cells of *A. radiobacter* EPS-916 proceeded initially at similar rates aerobically and anaerobically but subsequently slowed significantly in air (Llovera et al., 1993). *P. putida* PRS2000 reduces chromate aerobically more rapidly than anaerobically (Ishibashi et al., 1990). *Pseudomonas* sp. strain C7 has so far been tested only aerobically (Gopalan and Veeramani, 1994). *Burholderia cepacia* MCMB-821 reduces chromate equally efficiently aerobically and anaerobically with lactose as electron donor at pH 9—the organism being an alkaliphile (Wani et al., 2007).

By contrast, *Pseudomonas dechromaticans*, *P. chromatophila*, *Enterobacter cloacae* OH1, and *Desulfovibrio vulgaris* reduce Cr(VI) only anaerobically with organic electron donors, or H_2 in the case of *D. vulgaris* (Romanenko and Koren'kov, 1977; Lebedeva and Lyalikova, 1979; Komori et al., 1989; Lovley and Phillips, 1994). Except for *E. cloacae* OH1, these organisms cannot use glucose as reductant. *P. dechromatican* and *P. dechromatophila* appear to be able to reduce chromate and dichromate.

Cell extracts of *Pseudomonas fluorescens* LB300 reduce chromate with added glucose or NADH (Figure 18.2). One or more plasma membrane components appear to be required (Bopp and Ehrlich,

1988). *Enterobacter cloacae* HO1 also uses a membrane-bound respiratory system to reduce chromate, but it functions only under anaerobic conditions (Wang et al., 1991). By contrast, most of the chromate-reducing activity in *Escherichia coli* ATCC 33456 appears to be soluble, that is, it does not involve plasma membrane components, but is mediated by NADH (Shen and Wang, 1993). The chromate-reducing activity of *P. putida* PRS2000 also does not depend on plasma membrane components. It mediates reduction via NADH or NADPH (Ishibashi et al., 1990). *Desulfovibrio vulgaris* ATCC 29759 uses its cytochrome c_3 as its Cr(VI) reductase coupled to hydrogenase when using H_2 as reductant (Lovley and Phillips, 1994). Ackerley et al. (2004) detected soluble Cr(VI)-reducing activity in aerobically growing strains of *P. putida* and *E. coli* associated with soluble bacterial flavoproteins. The flavoprotein from *P. putida*, labeled ChrR, and that from *E. coli*, labeled YieF, were dimers. They tested the activity of the two reductases with NADH as electron donor. Unlike the YieF dimer, the ChrR dimer generated a flavin semiquinone transiently and reactive oxygen species

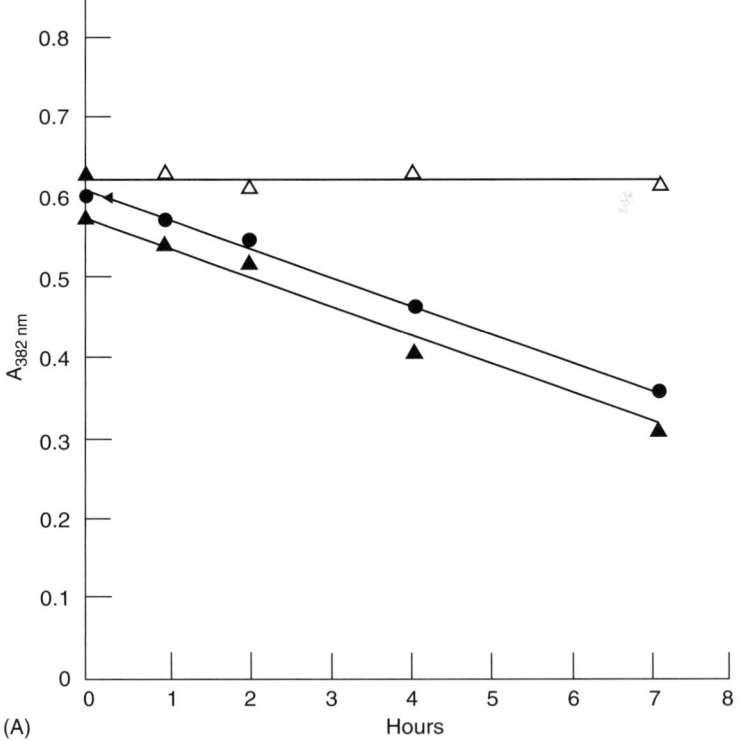

(A)

FIGURE 18.1   Chromate reduction by resting and growing cells of chromate-resistant *Pseudomonas fluorescens* LB300. (A) Resting cells grown with and without chromate. (△) Chromate-grown cells in the absence of electron donor (results were the same for cells grown without chromate and assayed in the absence of electron donor; no chromate reduction was observed; (●) chromate-grown cells with 0.5% (w/v) glucose; (▲) cells grown without chromate and assayed with 0.5% (w/v) glucose. Chromate was not reduced by spent medium from either chromate-grown cells or cells grown without chromate or by assay buffer containing either 0.25 or 0.5% glucose. Chromate concentration was measured as absorbance at 328 nm after cell removal. (B) Growing cells in VB broth at an initial K_2CrO_4 concentration of 40 µg mL^{-1}; growth of the culture was measured photometrically as turbidity at 600 nm; chromate concentration was measured as absorbance at 382 nm after first removing cells from replicate samples by centrifugation followed by filtration. ((A) and (B): From Bopp LH, Ehrlich HL, *Arch. Microbiol.*, 150, 426–431, 1988. With permission of Springer Science and Business media.) (C) Chromate reduction by cells growing in citrate-chromate medium in a chemostat at a dilution rate of 1.17 mL h^{-1}; (△) chromate concentration in uninoculated reactor; (○) chromate concentration in inoculated reactor; (□) cell concentration in inoculated reactor. ((C): From DeLeo PC, Ehrlich HL, *Appl. Microbiol. Biotechnol.*, 40, 756–759, 1994. With permission of Springer Science and Business media.)

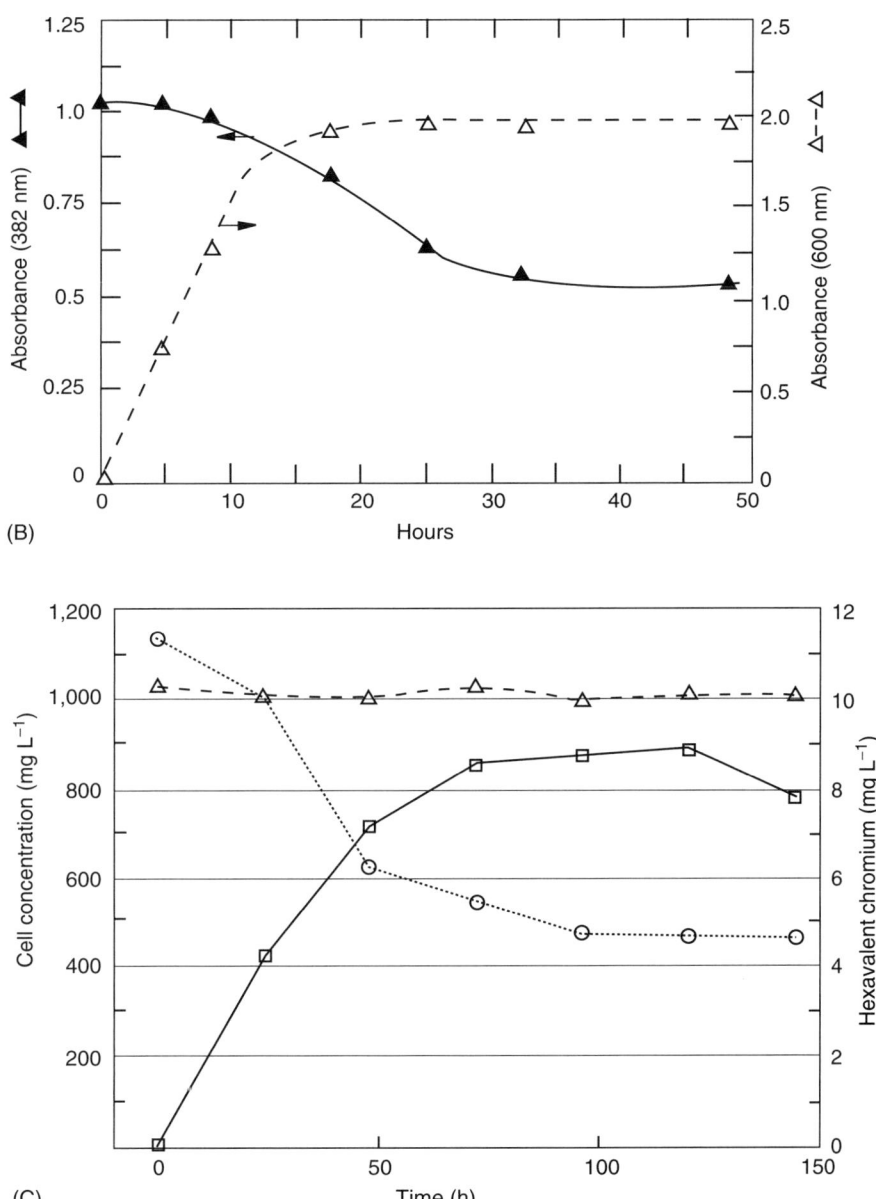

(B)

(C)

FIGURE 18.1 (Continued)

diminished over time, which suggested that Cr(V) may be an intermediate in Cr(VI) reduction by this reductase. Because the YieF dimer did not generate flavin semiquinone and used only 25% of the electrons from NADH for reduction of reactive oxygen species, Ackerley et al. (2004) suggest that in reducing Cr(VI), this reductase may transfer three electrons to Cr(VI) and one to reactive oxygen species for every two NADH oxidized.

In anaerobically grown *Shewanella putrefaciens* (now *Sh. oneidensis*) MR-1, chromate reductase activity is associated with the cytoplasmic membrane (Myers et al., 2000). Although this organism is facultative, it reduces chromate only anaerobically. Both formate and NADH but not L-lactate or NADPH can serve as electron donors to the Cr(VI) reductase system, which includes a multicomponent electron transport system. Some of the activity is inducible in cells grown anaerobically with

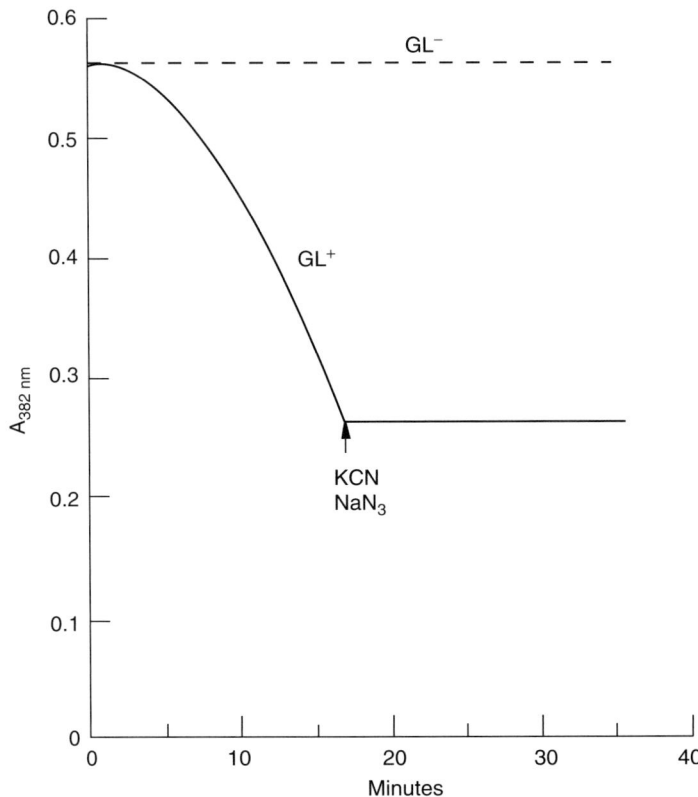

FIGURE 18.2 Chromate reduction by cell extract from *Pseudomonas fluorescens* LB300. GL$^-$, without added glucose; GL$^+$, with glucose from time 0 and KCN and NaN$_3$ at time indicated. Chromate concentration was measured as absorbance at 382 nm after cell removal. (From Bopp LH, Ehrlich HL, *Arch. Microbiol.*, 150, 426–431, 1988. With permission of Springer Science and Business Media.)

lactate as electron donor and nitrate or fumarate as electron acceptor, but this activity is inhibited by nitrite (Viamajala et al., 2002). Although Cr(III) is the usual product of respiratory Cr(VI) reduction by *Sh. oneidensis* and other unrelated Cr(VI) reducers, one case in which Cr(II) accompanied Cr(III) in Cr(VI) respiration by *Sh. oneidensis* MR-1 has been reported (Daulton et al., 2007). The Cr(II) in this instance was concentrated near the cytoplasmic membrane.

18.1.6   *In Situ* Chromate Reducing Activity

In situ rates of microbial Cr(VI)-reducing activity are generally not readily available in the literature, although such measurements would be useful because a number of different bacteria possess the ability to reduce Cr(VI). On the contrary, Wang and Shen (1997) examined rate parameters for a number of pure cultures under laboratory conditions. For example, they reported that the half-maximal Cr(VI) reducing velocity constant K_s, in mg Cr(VI) L^{-1}, is 5.43 for *Bacillus subtilis* (aerobic), 19.2 for *Desulfovibrio vulgaris* ATCC 29579 (anaerobic), 8.64 for *Escherichia coli* ATCC 33456 (anaerobic), 641.9 for *Pseudomonas ambigua* G-1 (aerobic), and 5.55 for *P. fluorescens* LB300 (aerobic), based on Monod kinetics. Natural levels of Cr(VI) in most environments can be expected to be low. However, anthropogenic pollution can cause very significant elevation in environmental chromium concentrations.

Lebedeva and Lyalikova (1979) isolated a strain of *Pseudomonas chromatophila* from the effluent of a chromite mine in Yugoslavia that contained the mineral crocoite (PbCrO$_4$) in its oxidation zone. At this site, the organism clearly found a plentiful source of Cr(VI). The isolate was shown to use a range of carbon and energy sources anaerobically for chromate reduction. These included ribose, fructose, benzoate, lactate, acetate, succinate, butyrate, glycerol, and ethylene glycol but not glucose or hydrogen. Crocoite reduction with lactate as electron donor was demonstrated anaerobically in the laboratory.

18.1.7 Applied Aspects of Chromium(VI) Reduction

The first practical application of microbial Cr(VI) reduction as a bioremediation process was explored by Russian investigators. They presented evidence that indicated that bacterial chromate reduction can be harnessed in wastewater and sewage treatment to remove chromate (Romanenko et al., 1976; Pleshakov et al., 1981; Serpokrylov et al., 1985; Simonova et al. 1985). The process also has potential application in treatment of tannery and, especially, electroplating wastes and *in situ* bioremediation. In the case of tannery and electroplating waste treatment, prior dilution of the waste may be necessary to bring the Cr(VI) concentration into a range tolerated by the Cr(VI)-reducing bacteria.

Extensive research in the use of *Enterobacter cloacae* HO1 in bioremediation of Cr(VI)-containing wastewaters was performed in Japan (Ohtake et al., 1990; Komori et al., 1990a,b; Yamamoto et al., 1993; Fuji et al., 1994).

18.2 MICROBIAL INTERACTION WITH MOLYBDENUM

18.2.1 Occurrence and Properties of Molybdenum

Molybdenum is an element of the second transition series in the periodic table. In mineral form, it occurs extensively as molybdenite (MoS_2). The minerals wulfenite ($PbMoO_4$) and powellite ($CaMoO_4$) are often associated with the oxidation zone of molybdenite deposits (Holliday, 1965). Molybdite (MoO_3) is another molybdenum-containing mineral that may be encountered in nature. The abundance of Mo has been reported to be 2–4 g t^{-1} in basaltic rock, 2.3 g t^{-1} in granitic rock, and 0.001–0.005 g t^{-1} in ocean waters (Enzmann, 1972).

The oxidation states in which molybdenum can exist include 0, +2, +3, +4, +5, and +6. Of these, the +4 and +6 states are the most common, but the +5 state is of biological significance. Molybdenum oxyanions of the +6 oxidation state tend to polymerize the complexity of the polymers depending on the pH of the solution (Latimer and Hildebrand, 1942).

Molybdenum is a biologically important trace element. A number of enzymes feature it in their structure, for example, nitrogenase, nitrate reductase (Brock and Madigan, 1991), sulfite reductase, and arsenite oxidase (Anderson et al., 1992). Molybdate is an effective inhibitor of bacterial sulfate reduction (Oremland and Capone, 1988).

18.2.2 Microbial Oxidation and Reduction

Molybdenite (MoS_2) is aerobically oxidizable as an energy source by *Acidianus brierleyi* (Brierley and Murr, 1973) (see also Chapter 20) with the formation of molybdate and sulfate. *Acidithiobacillus ferrooxidans* can also oxidize molybdenite, but it is poisoned by the resulting molybdate (Tuovinen et al., 1971) unless the molybdate is rendered insoluble, for instance by reaction with Fe^{3+}. Sugio et al. (1992) reported that *Thiobacillus* (now *Acidithiobacillus*) *ferrooxidans* AP-19-3 contains an enzyme that oxidizes molybdenum blue (Mo^{5+}) to molybdate (Mo^{6+}). They purified the molybdenum oxidase and found it to be an enzyme complex that included cytochrome oxidase as an important component. The function of molybdenum oxidase in the organism remains unclear in view of its sensitivity to molybdate.

Molybdate was first shown to be reduced by *Sulfolobus* sp. by Brierley and Brierley (1982). In a more detailed study, molybdate was shown to be reduced microbially to molybdenum blue (containing Mo^{5+}) by *Acidithiobacillus ferrooxidans* using sulfur as electron donor (Sugio et al., 1988). The enzyme that reduced the Mo^{6+} was identified as sulfur:ferric ion oxidoreductase. Molybdate has also been shown to be reduced anaerobically to molybdenum blue by *Enterobacter cloacae* strain 48 using glucose as electron donor (Ghani et al., 1993). The reduction appears to be mediated via NAD and *b*-type cytochrome. Other bacteria reported to be able to reduce Mo(VI) include *Pseudomonas guillermondii*, *Micrococcus* sp., and *Desulfovibrio desulfuricans* (reviewed by Lloyd, 2003).

18.3 MICROBIAL INTERACTION WITH VANADIUM

Vanadium belongs to the first transition series of elements in the periodic table. In mineral form, it often occurs in complex forms such as patronite (a complex sulfide), roscoelite (a vanadium mica), vanadinite (a lead vanadate), and carnotite (a hydrous potassium uranium vanadate) (DeHuff, 1965). Its average abundance (milligrams per kilogram) in granites is 72, in basalts 270, and in soil 90. Its average concentration (nanograms per cubic meter) in freshwater is 0.0005 and in seawater 0.0025 (Bowen, 1979).

Vanadium occurs in the oxidation states of 0, +2, +3, +4, and +5. Pentavalent vanadium in solution occurs as the oxyanion VO_3^- (vanadate) and is colorless. Tetravalent vanadium in solution occurs as VO^{2+} and is deep blue. Trivalent vanadium (V^{3+}) forms a green solution, and divalent vanadium (V^{2+}) a violet solution (Dickerson et al., 1979).

As a trace element in prokaryotes, vanadium has been found to occur in place of molybdenum in certain nitrogenases (Brock and Madigan, 1991) (see also Chapter 13). It also occurs in oxygen-carrying blood pigment of ascidian worms.

18.3.1 BACTERIAL OXIDATION OF VANADIUM

Five different bacteria have been reported to be able to reduce vanadate. The first three are *Veillonella* (*Micrococcus*) *lactilyticus*, *Desulfovibrio desulfuricans*, and *Clostridium pasteurianum*, which were shown by Woolfolk and Whiteley (1962) to be able to reduce vanadate to vanadyl with hydrogen using a hydrogenase,

$$VO_3^- + H_2 \rightarrow VO(OH) + OH^- \tag{18.2}$$

The fourth and fifth organisms are new isolates assigned to the genus *Pseudomonas* (Yurkova and Lyalikova, 1990; Lyalikova and Yurkova, 1992). One of these, isolated from a waste stream from a ferrovanadium factory, was named *P. vanadiumreductans*, and the other, isolated from seawater in Kraternaya Bay, Kuril Islands, *P. issachenkovii*. Both are gram-negative, motile, non-spore-forming rods that can grow as facultative chemolithotrophs and facultative anaerobes. Anaerobically, chemolithotrophic growth was observed with H_2 and CO as alternative energy sources, CO_2 as carbon source, and vanadate as terminal electron acceptor. However, the organisms can also grow organotrophically under anaerobic conditions with glucose, maltose, ribose, galactose, lactose, arabinose, lactate, proline, histidine, threonine, and serine as carbon and energy sources. *P. issachenkovii* can also use asparagine as carbon and energy source. Vanadate reduction by the organism involved transformation of pentavalent vanadium to tetravalent and trivalent vanadium. The tetravalent oxidation state was identified in the medium by development of a blue color and the trivalent state by formation of a black precipitate and by its reaction with tairon reagent. An equation describing the overall reduction of vanadium by lactate in these experiments was presented by the authors as

$$2NaVO_3 + NaC_3H_5O_3 \rightarrow V_2O_3 + NaC_2H_3O_2 + NaHCO_3 + NaOH \tag{18.3}$$

It accounts for the alkaline pH developed by the medium during growth that started at pH 7.2. Antipov et al. (2000) found that molybdenum- and molybdenum cofactor-free nitrate reductases of *P. issachenkovii* appear to mediate the vanadate reduction. Homogeneous membrane-bound nitrate reductase from the organism reduced vanadate with NADH as electron donor. In a medium containing both nitrate and vanadate, the organism reduced nitrate before vanadate. A vanadium mineral similar to sherwoodite was detected in cultures reducing vanadium, suggesting that these bacteria may play a role in epigenetic vanadium mineral formation (Lyalikova and Yurkova, 1992).

More recently, *Shewanella oneidensis* MR-1 was shown to be able to use vanadate as sole electron acceptor anaerobically, that is, it was able to respire anaerobically on lactate, pyruvate, formate, fumarate, Fe(III), or citrate using vanadate as terminal electron acceptor (Carpentier et al., 2003, 2005). It reduces vanadate [V^V] to vanadyl [V^{IV}] (Carpentier et al., 2003). The reduction was inhibited by 2-heptyl-4-hyroxyquinoline *N*-oxide and antimycin A, and proton translocation associated with vanadate reduction was abolished by the protonophores dinitrophenol and carbonyl cyanide *m*-chlorophenuylhydrazone (Carpentier et al., 2005).

Geobacter metallireducens has also been shown to be able to grow anaerobically with acetate as electron donor and vanadate as terminal electron acceptor (Ortiz-Bernad et al., 2004a). It reduced V^V to V^{IV}, which subsequently precipitated. Microprobe analysis of the precipitate suggested that it could have been vanadyl phosphate. In a bioremediation study of groundwater contaminated with uranium(VI) and vanadium(V), acetate stimulation of Geobacteraceae caused precipitation of the vanadium as a result of its reduction (Ortiz-Bernad et al., 2004a).

Vanadium(V) can also be reduced nonenzymatically by bacteria. An example is the reduction of vanadium(V), at concentrations up to 5 mM, to vanadium(IV) by *Acidithiobacillus thiooxidans*, using elemental sulfur as its energy source. The vanadium reduction in this instance is brought about by partially reduced sulfur intermediates produced by the organism during oxidation of the elemental sulfur (Briand et al., 1996).

18.4   MICROBIAL INTERACTION WITH URANIUM

18.4.1   OCCURRENCE AND PROPERTIES OF URANIUM

Uranium is one of the naturally occurring radioactive elements. Its abundance in the Earth's crust is only 0.0002%. It is found in more than 150 minerals, but the most important are the igneous minerals pitchblende and coffinite and the secondary mineral carnotite (Baroch, 1965). It is found in small amounts in granitic rocks (4 4 mg kg^{-1}) and in even smaller amounts in basalt (0.43 mg kg^{-1}). In freshwater it has been reported in concentrations of 0.0004 mg kg^{-1} and in seawater, 0.0032 mg kg^{-1} (Bowen, 1979).

Uranium can exist in the oxidation states 0, +3, +4, +5, and +6 (Weast and Astle, 1982). The +4 and +6 oxidation states are of greatest significance microbiologically. In nature, the +4 oxidation state usually manifests itself in insoluble forms of uranium, for example, UO_2. The +6 oxidation state predominates in nature in soluble, and hence mobile, form, for example, UO_2^{2+} (Haglund, 1972). In radioactive decay of an isotopic uranium mixture, alpha, beta, and gamma radiation are emitted, but the overall rate of decay is very slow because the dominant isotopes have very long half-lives (Stecher, 1960). This slow rate of decay probably accounts for the ability of bacteria to interact with uranium species without experiencing lethal radiation damage.

18.4.2   MICROBIAL OXIDATION OF U(IV)

Acidithiobacillus ferrooxidans has been shown to oxidize tetravalent U^{4+} to hexavalent UO_2^{2+} in a reaction that yields enough energy to enable the organism to fix CO_2. Nevertheless, experimental demonstration of growth of *A. ferrooxidans* with U^{4+} as sole energy source has not succeeded to date. *Thiobacillus acidophilus* (now renamed *Acidiphilium acidophilum*) was also found to oxidize

U^{4+} but without energy conservation (DiSpirito and Tuovinen, 1981, 1982a,b) (see also Chapter 20). Recently, the autotroph, *Thiobacillus denitrificans*, was shown to promote oxidation of U(IV) oxide, uraninite, anaerobically in the presence of nitrate in an in-laboratory glove-box experiment (Beller, 2005). Although U(IV) oxidation was accompanied by significant nitrate reduction, the amount of nitrate consumed significantly exceeded the amount of U(IV) oxidized. The excess nitrate reduction was attributable to biooxidation of H_2 in the atmosphere in the glove box (Beller, 2005).

18.4.3 MICROBIAL REDUCTION OF U(IV)

A number of organisms have been shown to be able to reduce hexavalent uranium (UO_2^{2+}) to tetravalent uranium (UO_2). The first demonstration was with *Veilonella* (*Micrococcus*) *lactilyticus* using H_2 as electron donor under anaerobic conditions (Woolfolk and Whiteley, 1962). Much more recently, some other bacteria were shown to be able to reduce U(VI) to U(IV) anaerobically. They are the facultative organism *Shewanella putrefaciens* (Lovley et al., 1991) and *Sh. alga* strain BrY (Caccavo et al., 1992) and the strict anaerobe *Geobacter metallireducens* strain GS-15 (Lovley et al., 1991, 1993a), *Desulfovibrio desulfuricans* (Lovley and Phillips, 1992), *D. vulgaris* (Lovley et al., 1993b), and *Desulfovibrio* sp. (Pietzsch et al., 1999). In *Desulfovibrio*, cytochrome c_3 appears to be the U(VI) reductase (Figure 18.3) (Lovley et al., 1993b). The electron donors used by these organisms may be organic or, in some instances, H_2. Although *G. metallireducens* and *Sh. putrefaciens* can gain energy from the process of U(VI) reduction (Lovley et al., 1991), *D. desulfuricans* and *D. vulgaris* are unable to do so (Lovley and Phillips, 1992; Lovley et al., 1993c). *Desulfovibrio* sp., on the contrary, is able to gain energy from the process (Pietzsch et al., 1999).

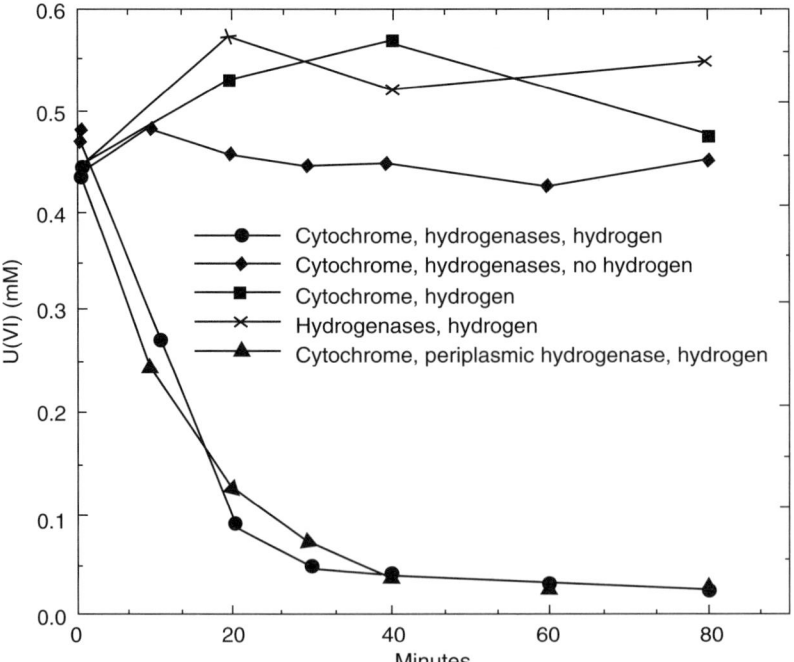

FIGURE 18.3 Reduction of U(VI) by electron transfer from H_2 to U(VI) via hydrogenase and cytochrome c_3. As noted, pure periplasmic hydrogenase or a protein fraction containing two hydrogenases was used. (From Lovley DR, Widman PK, Woodward JC, Phillips JEP, *Appl. Environ. Microbiol.*, 59, 3572–3576, 1993b. With permission.)

Insofar as the study of electron transport pathways in U(VI) and Cr(VI) reduction under anaerobic conditions by *Shewanella oneidensis* MR-1 is concerned, Bencheikh-Latmani et al. (2005), using whole-genome DNA microarrays, showed that U(VI)-reducing conditions caused upregulation of 121 genes and Cr(VI)-reducing conditions 83 genes. Genes whose products are known to enable the use of the alternative electron acceptors fumarate, dimethyl sulfoxide, Mn(IV), or soluble Fe(III) were upregulated under both U(VI)- and Cr(VI)-reducing conditions. Mutant studies confirmed that several genes, whose products are known to be involved in ferric citrate reduction, are also involved in the reduction of both U(VI) and Cr(VI) by *Sh. oneidensis* MR-1. The genes are *mtrA*, *mtrB*, *mtrC*, and *menC*. However, genes coding for efflux pumps were upregulated under Cr(VI)-reducing conditions but not U(VI)-reducing conditions. These findings by Bencheikh-Latmani et al. (2005) show that the same electron transport pathway or parts thereof may be involved in the reduction of U(VI) and Cr(VI) and several other terminal electron acceptors.

The isolation of some of the organisms from freshwater and marine sediments suggests that this microbial activity may play or may have played a significant role in the immobilization of uranium and its accumulation in sedimentary rock. The first evidence in support of such immobilization was obtained by Gorby and Lovley (1992) in experiments with groundwater amended with 0.4 or 1.0 mM uranyl acetate and 30 mM $NaHCO_3$ and inoculated with *Geobacter metallireducens* GS-15, which produced a black precipitate from the dissolved UO_2^{2+}. The black precipitate was identified as uraninite (UO_2). Frederickson et al. (2000) studied U(VI) reduction by *Shewanella putrefaciens* CN32 in the presence of goethite ($Fe_2O_3 \cdot H_2O$). They found that besides enzymatic reduction of the U(VI) by *Sh. putrefaciens* CN32, Fe(II) that had been sorbed to goethite was able to reduce U(VI) abiotically, as was the humic analog anthraquinone-2,6-disulfonate (AQDS).

18.4.4 BIOREMEDIATION OF URANIUM POLLUTION

Uranium(VI)-contaminated aquifer groundwater can be bioremediated by reduction of the U(VI) to insoluble U(IV) under various conditions. Thus in a field experiment, the uranium level in a contaminated aquifer in Rifle, Colorado, United States, was lowered to <0.18 µM from a range of 0.4 to 1.4 µM within 50 days by injection of 1–3 mM acetate per day (Anderson et al., 2003). Initial loss of uranium from the groundwater was attributed to the activity of *Geobacter* spp., which can reduce U(VI) enzymatically as well as by producing Fe(II) from reduction of Fe(III). But as acetate injection continued, *Geobacter* spp. was gradually replaced by sulfate reducers, stimulated by the injected acetate, which they used as carbon and energy source, and by using sulfate in the groundwater as terminal electron acceptor, and causing the uranium concentration in the groundwater to increase because the particular sulfate reducers that became selectively enriched were incapable of reducing U(VI) or promoting its reduction. It appears that sustained bioremediation of a uranium-contaminated aquifer by *Geobacter* spp. would benefit from maintenance of conditions that favor the activity of *Geobacter* spp. over sulfate-reducers. It also appears that only U(VI) dissolved in the groundwater at the Rifle, Colorado, aquifer site is amenable to bioremediation. U(VI) bound to the sediment in the aquifer was not microbially reducible (Ortiz-Bernad et al., 2004b).

Suzuki et al. (2005) have presented evidence from shallow water sediment from an open pit of a uranium mine in Washington State, United States, that showed U(VI) immobilization by bioreduction to U(IV) oxide, in which both Fe(III)- and sulfate-reducers, members of the Geobacteraceae and Desulfovibrionacea, respectively, were implicated. Both families of organisms were previously known to contain members capable of U(VI) reduction.

Nevin et al. (2003) demonstrated uranium bioremediation in a high-salinity subsurface sediment from an aquifer associated with uranium mine tailings at Shiprock, New Mexico, United States, by stimulation with acetate, which enriched populations of microorganisms related to *Pseudomonas* and *Desulfosporosinus* species.

Shelobolina et al. (2004) obtained evidence for potential U(VI) anaerobic bioreduction at low pH in nitrate- and U(VI)-contaminated subsurface sediment from the Natural and Accelerated

Bioremediation Research Field Center in Oak Ridge, Tennessee, United States. *Salmonella sub-terranean* sp. nov., a gram-negative, motile rod capable of using O_2, NO_3^-, $S_2O_3^{2-}$, fumarate, and malate as terminal electron acceptors and of reducing U(VI) in cell suspension was isolated from the sediment.

18.5   BACTERIAL INTERACTION WITH POLONIUM

Polonium is a radioactive element that occurs naturally in association with uranium and thorium minerals. Different isotopes of polonium are produced in the decay of ^{238}U, ^{235}U, and ^{232}Th. Of these isotopes, ^{210}Po, which originates from the decay of ^{238}U, has the longest half-life (138.4 days) (Lietzke, 1972). Its known oxidation states are +2 and +4 (see LaRock et al., 1996).

Polonium-210 can be an environmental pollutant arising by release from uranium-containing phosphorite, which is commercially exploited for its phosphate and phosphogypsum, which is a by-product in the manufacture of phosphoric acid from phosphorite (see LaRock et al., 1996).

Sulfate-reducing bacteria that were found by LaRock et al. (1996) to be associated with phospho-gypsum are able to mobilize Po contained in the phosphogypsum. In laboratory experiments, this mobilization required that the dissolved sulfate concentration in the bulk phase was below 10 µM. Above this sulfate concentration, enough H_2S was produced to coprecipitate the mobilized Po as a metal sulfide (LaRock et al., 1996). The release of polonium apparently depended on the reduction of the sulfate in the gypsum. Aerobic bacteria were also able to mobilize Po in the phosphogypsum. The mechanism in this instance may involve Po complexation by ligands produced by the active organisms (LaRock et al., 1996).

Immobilization of Po by at least one aerobic bacterial isolate has also been reported (Cherrier et al., 1995). The Po was taken into the cell by a mechanism that appeared to differ from that of sulfate uptake. However, its partitioning after uptake paralleled that of sulfur among cell compo-nents that included cell envelope, cytoplasm, and cytoplasmic protein. Most polonium and sulfur were detected in the cytoplasmic fraction. In nature, such immobilization of Po must be considered transient if upon death of these cells the Po becomes redissolved in the bulk phase.

18.6   BACTERIAL INTERACTION WITH PLUTONIUM

Recently enzymatic reduction of plutonium(IV) to the more soluble plutonium(III) under anaerobic conditions was definitively demonstrated with *Geobacter metallireducens* GS-15 and *Shewanella oneidensis* MR-1 using freshly precipitated, amorphous Pu(IV)(OH)$_4$ and soluble Pu(IV)(EDTA) (Boukhalfa et al., 2007). Both organisms reduced the Pu(IV)(EDTA) and amorphous Pu(IV)(OH)$_4$ in the presence of EDTA, forming Pu(III)(EDTA). In the absence of a complexing ligand such as EDTA, *Sh. oneidensis* MR-1 produced only minor amounts of Pu(III) and *G. metallireducens* pro-duced little or no Pu(III). The Pu(IV) reduction did not support growth of either organism.

The only previous observation suggesting bacterial interaction with plutonium involved a study with *Bacillus polymyxa* and *B. circulans* (Rusin et al., 1994). However, reduction of Pu(IV) to Pu(III) by the bacteria was only inferred but not directly demonstrated.

18.7   SUMMARY

Enzymatic oxidation of Cr(III) by bacteria has not been demonstrated. Nonenzymatic oxidation of Cr(III), which is dependent on biogenic (bacterial, fungal) formation of Mn(IV), which then oxi-dizes Cr(III) to Cr(VI) chemically, may occur in soil.

Aerobic and anaerobic reduction of Cr(VI) by bacteria has been demonstrated. The process is in many instances a form of respiration. Various organic electron donors may serve, but not all act equally well aerobically and anaerobically. Chromate and dichromate are not necessarily reduced equally well. The ability to reduce chromate does not always correlate with chromate tolerance.

Although chromite is not very susceptible to leaching by acid formed by acidophiles such as *Acidithiobacillus ferrooxidans* and *A. thiooxidans*, chromium in some solid inorganic industrial wastes can be leached by sulfuric acid formed by *A. thiooxidans* in sulfur oxidation.

Cr(VI) reduction to Cr(III) is beneficial ecologically because it lowers chromium toxicity. At equivalent concentrations, Cr(VI) is more toxic than Cr(III). Cr(III) also tends to precipitate as a hydroxo compound around neutrality, the pH range at which all known Cr(VI) reducers operate.

Bacteria have been discovered that can enzymatically oxidize Mo(IV) and Mo(V) to Mo(VI) in air. Other bacteria have been found that can enzymatically reduce Mo(VI) to Mo(V), some aerobically, others anaerobically.

Vanadate (VO_3^-) has been found to be reduced anaerobically to vanadyl [VO(OH)] by a number of bacteria. At least two of these organisms can use vanadate as terminal electron acceptor during chemolithotrophic growth with H_2 as electron donor.

Tetravalent uranium can be oxidized to hexavalent uranium, an oxidation that serves as a source of energy to *Acidithiobacillus ferrooxidans*, although it does not support its growth as sole energy source. Anaerobically, hexavalent uranium can be reduced to tetravalent uranium by a number of bacteria using either H_2 or one of a variety of organic electron donors.

The bacterial oxidations and reductions of Cr, Mo, V, and U can play an important role in their mobilization and immobilization in soils and sediments.

Polonium can be mobilized or immobilized by bacteria, but whether redox reactions involving Po are involved remains unknown. Bacterial reduction of plutonium(IV) to plutonium(III) can occur but does not support growth of the responsible organisms.

REFERENCES

Ackerley DF, Gonzalez CF, Park CH, Blake R II, Keyhan M, Matin A. 2004. Chromate-reducing properties of soluble flavoproteins from *Pseudomonas putida* and *Escherichia coli*. *Appl Environ Microbiol* 70:873–882.

Anderson GL, Williams J, Hille R. 1992. The purification and characterization of arsenite oxidase from *Alcaligenes faecalis*, a molybdenum-containing hydroxylase. *J Biol Chem* 267:23674–23682.

Anderson RT, Vrionis HA, Ortiz-Bernad I, Resch CT, Long PE, Dayvault R, Karp K, Marutzky S, Metzler DR, Peacock A, White DC, Lowe M, Lovley DR. 2003. Stimulating the *in situ* activity of *Geobacter* species to remove uranium from the groundwater of a uranium-contaminated aquifer. *Appl Environ Microbiol* 69:5884–5891.

Antipov AN, Lyalikova NN, Khijniak TV, L'vov NP. 2000. Vanadate reduction by molybdenum free dissimilatory nitrate reductase from vanadate-reducing bacteria. *IUBMB Life* 50:39–42.

Arias YM, Tebo BM. 2003. Cr(VI) reduction by sulfidogenic and nonsulfidogenic microbial consortia. *Appl Environ Microbiol* 69:1847–1853.

Bader JL, Gonzalez G, Goodell PC. 1999. Chromium-resistant bacterial populations from a site heavily contaminated with hexavalent chromium. *Water Air Soil Pollut* 109:263–276.

Baroch CT. 1965. Uranium. In: *Mineral Facts and Problems. Bulletin 630*. Washington, DC: Bureau of Mines, U.S. Department of the Interior, pp. 1007–1037.

Bartlett RJ, James BR. 1979. Behavior of chromium in soils: III. Oxidation. *J Environ Qual* 8:31–25.

Beller HR. 2005. Anaerobic, nitrate-dependent oxidation of U(IV) oxide minerals by the chemolithoautotrophic bacterium *Thiobacillus denitrificans*. *Appl Environ Microbiol* 71:2170–2174.

Bencheikh-Latmani R, Williams SM, Haucke L, Criddle CS, Wu L, Zhou J, Tebo BM. 2005. Global transcriptional profiling of *Shewanella oneidensis* MR-1 during Cr(VI) and U(VI) reduction. *Appl Environ Microbiol* 71:7453–7460.

Bopp LH. 1980. Chromate resistance and chromate reduction in bacteria. PhD Thesis. Troy, New York: Rensselaer Polytechnic Institute.

Bopp LH, Chakrabarty AM, Ehrlich HL. 1983. Chromate resistance plasmid in *Pseudomonas fluorescens*. *J Bacteriol* 155:1105–1109.

Bopp LH, Ehrlich HL. 1988. Chromate resistance and reduction in *Pseudomonas fluorescens* strain LB300. *Arch Microbiol* 150:426–431.

Bosecker K. 1986. Bacterial metal recovery and detoxification of industrial waste. In: Ehrlich HL, Holmes DS, eds. *Workshop on Biotechnology for the Mining, Metal-Refining and Fossil Fuel Processing Industries. Biotechnology and Bioengineering Symposium 16.* New York: Wiley, pp. 105–120.

Boukhalfa H, Icopini GA, Reilly SD, Neu MP. 2007. Plutonium(IV) reduction by the metal-reducing bacteria *Geobacter metallireducens* GS15 and *Shewanella oneidensis* MR1. *Appl Environ Microbiol* 73:5897–5903.

Bowen HJM. 1979. *Environmental Chemistry of the Elements.* London: Academic Press.

Bierley CL, Brierley JA. 1982. Anaerobic reduction of molybdenum by *Sulfolobus* species. *Zentralbl Bakteriol Hyg I Abt Orig* C3:289–294.

Briand L, Thomas H, Donati E. 1996. Vanadium(V) reduction by *Thiobacillus thiooxidans* cultures on elemental sulfur. *Biotechnol Lett* 18:505–508.

Brierley CL, Murr LE. 1973. A chemoautotrophic and thermophilic chemoautotrophic microbe. *Science* 179:488–490.

Brock TD, Madigan MT. 1991. *Biology of Microorganisms.* 6th ed. Englewood Cliffs, NJ: Prentice-Hall.

Caccavo F, Jr, Blakemore RP, Lovley DR. 1992. A hydrogen-oxidizing, Fe(III) reducing microorganism from the Great Bay Estuary, New Hampshire. *Appl Environ Microbiol* 58:3211–3216.

Carpentier W, De Smet L, Van Beeumen J, Brigé A. 2005. Respiration and growth of *Shewanella oneidensis* MR-1 using vanadate as the sole electron acceptor. *J. Bacteriol* 187:3293–3301.

Carpentier W, Sandra K, De Smet I, Brigé A, De Smet L, Van Beeumen J. 2003. Microbial reduction and precipitation of vanadium by *Shewanella oneidensis. Appl Environ Microbiol* 69:3636–3639.

Cervantes C. 1991. Bacterial interactions with chromium. *Antonie van Leeuwenhoek* 59:229–233.

Chen Y, Chen YY, Lin Q, Hu Z, Hu H, Wu J. 1997. Factors affecting Cr(III) oxidation by manganese oxides. *Pedosphere* 7:185–192.

Cherrier J, Burnett WC, LaRock PA. 1995. Uptake of polonium and sulfur by bacteria. *Geomicrobiol J* 13:103–115.

Coleman RN, Paran JH. 1983. Accumulation of hexavalent chromium by selected bacteria. *Environ Technol Lett* 4:149–156.

Daulton TL, Little BJ, Jones-Meehan J, Blom DA, Allard LF. 2007. Microbial reduction of chromium from the hexavalent to the divalent state. *Geochim Cosmochim Acta* 71:556–565.

DeHuff GL. 1965. Vanadium. In: *Mineral Facts and Problems. Bulletin 630.* Washington, DC: Bureau of Mines, U.S. Department of the Interior, pp. 1039–1053.

DeLeo PC, Ehrlich HL. 1994. Reduction of hexavalent chromium by *Pseudomonas fluorescens* LB300 in batch and continuous culture. *Appl Microbiol Biotechnol* 40:756–759.

Dickerson RE, Gray HB, Haight GP, Jr. 1979. *Chemical Principles.* 3rd ed. Menlo Park, CA: Benjamin/Cummings.

DiSpirito AA, Tuovinen OH. 1981. Oxygen uptake couples with uranous sulfate oxidation by *Thiobacillus ferrooxidans* and *T. acidophilus. Geomicrobiol J* 2:275–291.

DiSpirito AA, Tuovinen OH. 1982a. Uranous ion oxidation and carbon dioxide fixation by *Thiobacillus ferrooxidans. Arch Microbiol* 133:28–32.

DiSpirito AA, Tuovinen OH. 1982b. Kinetics of uranium and ferrous iron oxidation by *Thiobacillus ferrooxidans. Arch Microbiol* 133:33–37.

Ehrlich HL. 1983. Leaching chromite ore and sulfide matte with dilute sulfuric acid generated by *Thiobacillus thiooxidans* from sulfur. In: Rossi G, Torma AE, eds. *Recent Progress in Biohydrometallurgy.* Iglesias, Italy: Associazione Mineraria Sarda, pp. 19–42.

Enzmann RD. 1972. Molybdenum: Element and geochemistry. In: Fairbridge RW, ed. *The Encyclopedia of Geochemistry and Environmental Sciences.* Encyclopedia of Earth Science Series, Vol. IVA. New York: Van Nostrand Reinhold, pp. 753–759.

Forsberg CW. 1978. Effect of heavy metals and other trace elements on the fermentative activity of the rumen flora and growth of functionally important rumen bacteria. *Can J Microbiol* 24:298–306.

Fortescue JAC. 1980. *Environmental Geochemistry.* New York: Springer.

Frederickson JK, Zachara JM, Kennedy DW, Duff MC, Gorby YA, Li S-MW, Krupka KM. 2000. Reduction of U(VI) in goethite (α-FeOOH) suspensions by a dissimilatory metal-reducing bacterium. *Geochim Cosmochim Acta* 64:3085–3098.

Fuji E, Tsuchida T, Urano K, Ohtake H. 1994. Development of a bioreactor system for the treatment of chromate wastewater using *Enterobacter cloacae* HO1. *Water Sci Technol* 30:235–243.

Garbisu C, Alkorta I, Llama MJ, Serra JL. 1998. Aerobic chromate reduction by *Bacillus subtilis. Biodegradation* 9:133–141.

Ghani B, Takai M, Hisham NZ, Kishimoto N, Ismail AKM, Tano T, Sugio T. 1993. Isolation and characterization of a Mo^{6+}-reducing bacterium. *Appl Environ Microbiol* 59:1176–1180.

Gopalan R, Veeramani H. 1994. Studies on microbial chromate reduction by *Pseudomonas* sp. in aerobic continuous suspended culture. *Biotechnol Bioeng* 43:471–476.

Gorby YA, Lovley DR. 1992. Enzymatic uranium precipitation. *Environ Sci Technol* 26:205–207.

Gruber JE, Jennette KW. 1978. Metabolism of the carcinogen chromate by rat liver microsomes. *Biochem Biophys Res Commun* 82:700–706.

Gvozdyak PI, Mogilevich NF, Ryl'skit AF, Grishchenko NI. 1986. Reduction of hexavalent chromium by collection strains of bacteria. *Mikrobiologiya* 55:962–965 (English translation, pp. 770–773).

Haglund DS. 1972. Uranium: Element and geochemistry. In: Fairbridge RW, ed. *The Encyclopedia of Geochemistry and Environment Sciences*. Encyclopedia of Earth Science Series, Vol. IVA. New York: Van Nostrand Reinhold, pp. 1215–1222.

Holliday RW. 1965. Molybdenum. In: *Mineral Facts and Problems. Bulletin 630*. Washington, DC: Bureau of Mines, U.S. Department of the Interior, pp. 595–606.

Horitsu H, Futo S, Myazawa Y, Ogai S, Kawai K. 1987. Enzymatic reduction of hexavalent chromium by hexavalent chromium tolerant *Pseudomonas ambigua* G-1. *Biol Chem* 51:2417–2420.

Horitsu H, Nishida H, Kato H, Tomoyeda M. 1978. Isolation of potassium chromate-tolerant bacterium and chromate uptake by the bacterium. *Agric Biol Chem* 42:2037–2043.

Ishibashi Y, Cervantes C, Silver S. 1990. Chromium reduction by *Pseudomonas putida*. *Appl Environ Microbiol* 56:2268–2270.

Johnson I, Flower N, Loutit MW. 1981. Contribution of periphytic bacteria to the concentration of chromium in the crab *Helice crassa*. *Microb Ecol* 7:245–252.

Komori K, Rivas A, Toda K, Ohtake H. 1990a. Biological removal of toxic chromium using an *Enterobacter cloacae* strain that reduces chromate under anaerobic conditions. *Biotechnol Bioeng* 35:951–954.

Komori K, Rivas A, Toda K, Ohtake H. 1990b. A method for removal of toxic chromium using dialysis-sac cultures of the chromate reducing strain of *Enterobacter cloacae*. *Appl Microbiol Biotechnol* 33:117–119.

Komori K, Wang P-C, Toda K, Ohtake H. 1989. Factors affecting chromate reduction in *Enterobacter cloacae* HO1. *Appl Microbiol Biotechnol* 31:567–570.

Kozuh N, Stupar J, Gorenc B. 2000. Reduction and oxidation processes of chromium in soils. *Environ Sci Technol* 34:112–119.

Kvasnikov EI, Stepnyuk VV, Klyushnikova TM, Serpokrylov NS, Simonova GA, Kasatkina TP, Panchenko LP. 1985. A new chromium-reducing, gram-variable bacterium with mixed type of flagellation. *Mikrobiologiya* 54:83–88 (English translation, pp. 69–75).

Kwak YH, Lee DS, Kim HB. 2003. *Vibrio harveyi* nitroreductase is also a chromate reductase. *Appl Environ Microbiol* 69:4390–4395.

LaRock PA, Huyn J-H, Boutelle S, Burnett WC, Hull CD. 1996. Bacterial mobilization of polonium. *Geochim Cosmochim Acta* 60:4321–4328.

Latimer WM, Hildebrand JH. 1942. *Reference Book of Inorganic Chemistry*. Revised edition. New York: Macmillan, pp. 357–360.

Lebedeva EV, Lyalikova NN. 1979. Reduction of crocoite by *Pseudomonas chromatophila* nov. sp. *Mikrobiologiya* 48:517–522.

Lietzke MH. 1972. Polonium: Element and geochemistry. In: Fairbridge RW, ed. *The Encyclopedia of Geochemistry and Environmental Sciences*. Encyclopedia of Earth Science Series, Vol. IVA. New York: Van Nostrand Reinhold, pp. 962–964.

Llovera S, Bonet R, Simon-Pujol MD, Congregado F. 1993. Chromate reduction by resting cells of *Agrobacterium radiobacter* EPS-916. *Appl Environ Microbiol* 59:3516–3518.

Lloyd JR. 2003. Microbial reduction of metals and radionuclides. *FEMS Microbiol Rev* 27:411–425.

Lovley DR, Giovannoni SJ, White DC, Champine JE, Phillips EJP, Gorby YA, Goodwin S. 1993a. *Geobacter metallireducens* gen. nov. sp. nov., a microorganism capable of coupling the complete oxidation of organic compounds to the reduction of iron and other metals. *Arch Microbiol* 159:336–344.

Lovley DR, Phillips EJP. 1992. Reduction of uranium by *Desulfovibrio desulfuricans*. *Appl Environ Microbiol* 58:850–856.

Lovley DR, Phillips EJP. 1994. Reduction of chromate by *Desulfovibrio vulgaris* and its c_3 cytochrome. *Appl Environ Microbiol* 60:726–728.

Lovley DR, Phillips EJP, Gorby YA, Landa ER. 1991. Microbial reduction of uranium. *Nature* (London) 350:413–416.

Lovley DR, Roden EE, Phillips EJP, Woodward JC. 1993c. Enzymatic iron and uranium reduction by sulfate-reducing bacteria. *Mar Geol* 113:41–53.

Lovley DR, Widman PK, Woodward JC, Phillips JEP. 1993b. Reduction of uranium by cytochrome c_3 of *Desulfovibrio vulgaris*. *Appl Environ Microbiol* 59:3572–3576.

Lyalikova NN, Yurkova NA. 1992. Role of microorganisms in vanadium concentration and dispersion. *Geomicrobiol J* 10:15–26.

Marques AM, Tomas MJE, Congregado F, Simon-Pujol MD. 1982. Accumulation of chromium by *Pseudomonas aeruginosa*. *Microbios Lett* 21:143–147.

Marsh TL, Leon NM, McInerney MJ. 2000. Physicochemical factors affecting chromate reduction by aquifer materials. *Geomicrobiol J* 17:291–303.

Mertz W. 1981. The essential trace elements. *Science* 213:1332–1338.

Miller WJ, Neathery MW. 1977. Newly recognized trace mineral elements and their role in animal nutrition. *BioScience* 27:674–679.

Myers CR, Carstens BP, Antholine WE, Myers JM. 2000. Chromium(VI) reductase activity is associated with the cytoplasmic membrane of anaerobically grown *Shewanella putrefaciens* MR-1. *J Appl Microbiol* 88:98–106.

Nevin KP, Finnernan KT, Lovley DR. 2003. Microorganisms associated with uranium bioremediation in a high-salinity subsurface sediment. *Appl Environ Microbiol* 69:3672–3675.

Nishioka H. 1975. Mutagenic activities of metal compounds in bacteria. *Mutat Res* 31:185–190.

Ohtake H, Cervantes C, Silver S. 1987. Decreased chromate uptake in *Pseudomonas fluorescens* carrying a chromate resistance plasmid. *J Bacteriol* 169:3853–3856.

Ohtake H, Fuji E, Toda K. 1990. Reduction of toxic chromate in an industrial effluent by use of a chromate-reducing strain of *Enterobacter cloacae*. *Environ Technol* 11:663–668.

Oremland RS, Capone DG. 1988. Use of "specific" inhibitors in biogeochemistry and microbial ecology. *Adv Microb Ecol* 10:285–383.

Ortiz-Bernad I, Anderson RT, Vrionis HA, Lovley DR. 2004a. Vanadium respiration by *Geobacter metallireducens*: Novel strategy for *in situ* removal of vanadium from groundwater. *Appl Environ Microbiol* 70:3091–3095.

Ortiz-Bernad I, Anderson RT, Vrionis HA, Lovley DR. 2004b. Resistance of solid-phase U(VI) to microbial reduction during *in situ* bioremediation of uranium-contaminated groundwater. *Appl Environ Microbiol* 70:7558–7560.

Petrelli FL, DeFlora S. 1977. Toxicity and mutagenicity of hexavalent chromium on *Salmonella typhimurium*. *Appl Environ Microbiol* 33:805–809.

Philip L, Iyengar L, Venkobachar C. 1998. Cr(VI) reduction by *Bacillus coagulans* isolated from contaminated soils. *J Environ Eng* 124:1165–1170.

Pietzsch K, Hard BC, Babel W. 1999. A *Desulfovibrio* sp. capable of growing by reducing U(VI). *J Basic Microbiol* 39:365–372.

Pleshakov VD, Koren'kov VN, Zhukov IM, Serpokrylov NS, Lemper IA, Pankrova II. 1981. Biochemical removal of chromium(VI) compounds from wastewaters. USSR Patent SU 835,978. June 7, 1981.

Romanenko VI, Koren'kov VN. 1977. A pure culture of bacteria utilizing chromates and bichromates as hydrogen acceptors in growth under anaerobic conditions. *Mikrobiologiya* 46:414–417 (English translation, pp. 329–332).

Romanenko VI, Kuznetsov SI, Koren'kov VN. 1976. Koren'kov method for biological purification of wastewater. USSR Patent SU 521,234. July 15, 1976.

Rusin PA, Quintana L, Brainard JR, Strietelmeier BA, Tait CD, Ekberg SA, Palmer PD, Newton TW, Clark DI. 1994. Solubilization of plutonium hydrous oxide by iron-reducing bacteria. *Environ Sci Technol* 28:1686–1690.

Serpokrylov NS, Zhukov IM, Simonova GA, Kvasnikov EI, Klyushnikova TM, Kasatkina TP, Kostyukov VP. 1985. Biological removal of chromium(VI) compounds from wastewater by anaerobic microorganisms in the presence of an organic substrate. USSR Patent SU 1,198,020. December 1985.

Shelobolina ES, Sullivan SA, O'Neill KR, Nevin KP, Lovley DR. 2004. Isolation, characterization, and U(VI)-reducing potential of a facultatively anaerobic, acid-resistant bacterium from low-pH, nitrate- and U(VI)-contaminated subsurface sediment and description of *Salmonella subterranean* sp. nov. *Appl Environ Microbiol* 70:2959–2965.

Shen H, Wang Y-T. 1993. Characterization of enzymatic reduction of hexavalent chromium by *Escherichia coli* ATCC 33456. *Appl Environ Microbiol* 59:3771–3777.

Shi X, Dalal NS. 1990. One-electron reduction of chromate by NADPH-dependent glutathione reductase. *J Inorg Biochem* 40:1–12.

Shimada A. 1979. Effect of six-valent chromium on growth and enzyme production of chromium-resistant bacteria. *Abst, Annu Meet Am Soc Microbiol* Q[H]25:238.

Silver S, Walderhaug M. 1992. Gene regulation of plasmid- and chromosome-determined inorganic ion transport in bacterial. *Microbiol Rev* 56:195–228.

Simonova GA, Serpokrylov NS, Tokareva LL. 1985. Adaptation characteristics of a culture of *Aeromonas dechromaticans* KS-11 in microbial removal of hexavalent chromium from water. *Izv Sev Kavk Nauhn Tsentra Vyssh Shk Estestv Nauki* 4:89–91.

Sisti F, Allegretti P, Donati E. 1996. Reduction of dichromate by *Thiobacillus ferrooxidans*. *Biotechnol Lett* 18:1477–1480.

Sittig M. 1985. *Handbook of Toxic and Hazardous Chemicals and Carcinogens*. Park Ridge, NJ: Noyes.

Smith CH. 1972. Chromium: Element and geochemistry. In: Fairbridge RW, ed. *The Encyclopedia of Geochemistry and Environmental Sciences*. Encyclopedia of Earth Science Series, Vol. IVA. New York: Van Nostrand Reinhold, pp. 167–170.

Stecher PG, ed. 1960. *The Merck Index of Chemicals and Drugs*. 7th ed. Rahway, NJ: Merck, p. 1493.

Sugio T, Hirajama K, Inagaki K, Tanaka H, Tano T. 1992. Molybdenum oxidation by *Thiobacillus ferrooxidans*. *Appl Environ Microbiol* 58:1768–1771.

Sugio T, Tsujita Y, Katagiri T, Inagaki K, Tano T. 1988. Reduction of Mo^{6+} with elemental sulfur by *Thiobacillus ferrooxidans*. *J Bacteriol* 170:5956–5959.

Summers AO, Jacoby GA. 1978. Plasmid-determined resistance to boron and chromium compounds in *Pseudomonas aeruginosa*. *Anitmicrob Agents Chemother* 13:637–640.

Suzuki T, Miyata N, Horitsu H, Kawai K, Takamizawa K, Tai Y, Okazaki M. 1992. NAD(P)H-dependent chromium(VI) reductase of *Pseudomonas ambigua* G-1: A Cr(V) intermediate is formed during the reduction of Cr(VI) to Cr(III). *J Bacteriol* 174:5340–5345.

Suzuki Y, Kelly SD, Kemner KM, Banfield JF. 2005. Direct microbial reduction and subsequent preservation of uranium in natural near-surface sediment. *Appl Environ Microbiol* 71:1790–1797.

Tuovinen OH, Niemalä SI, Gyllenberg HG. 1971. Tolerance of *Thiobacillus ferrooxidans* to some metals. *Antonie van Leeuwenhoek* 37:489–496.

Venitt S, Levy LS. 1974. Mutagenicity of chromates in bacteria and its relevance to carcinogenesis. *Nature* (London) 250:493–495.

Viamajala S, Peyton BM, Apel WA, Petersen JN. 2002. Chromate reduction in *Shewanella oneidensis* MR-1 is an inducible process associated with anaerobic growth. *Biotechnol Prog* 18:290–295.

Wang P, Mori T, Komori K, Sasatsu M, Toda K, Ohtake H. 1989. Isolation and characterization of an *Enterobacter cloacae* strain that reduces hexavalent chromium under anaerobic conditions. *Appl Environ Microbiol* 55:1665–1669.

Wang P-C, Toda K, Ohtake H, Kusaka I, Yabe I. 1991. Membrane-bound respiratory system of *Enterobacter cloacae* strain HO1 grown anaerobically with chromate. *FEMS Microbiol Lett* 78:11–16.

Wang Y-T, Shen H. 1997. Modeling Cr(VI) reduction by pure bacterial cultures. *Water Res* 31:727–732.

Wani R, Kodam KM, Gawai KR, Dhakephalkar PK. 2007. Chromate reduction by *Burkholdia cepacia* MCMB-821, isolated from the pristine habitat of alkaline crater lake. *Appl Microbiol Biotechnol* 75:627–632.

Weast RC, Astle MJ. 1982. *CRC Handbook of Chemistry and Physics*. 63rd ed. Boca Raton, FL: CRC Press, pp. B44–B45.

Wong C, Silver M, Kushner DJ. 1982. Effects of chromium and manganese on *Thiobacillus ferrooxidans*. *Can J Microbiol* 28:536–544.

Woolfolk CA, Whiteley HR. 1962. Reduction of inorganic compounds with molecular hydrogen by *Micrococcus lactilyticus*. *J. Bacteriol* 84:647–658.

Yamamoto K, Kato J, Yano I, Ohtake H. 1993. Kinetics and modeling of hexavalent chromium reduction in *Enterobacter cloacae*. *Biotechnol Bioeng* 41:129–133.

Yurkova NA, Lyalikova NN. 1990. New vanadate-reducing facultative chemolithotrophic bacteria. *Mikrobiologiya* 59:968–975 (English translation, pp. 672–677).

19 Geomicrobiology of Sulfur

19.1 OCCURRENCE OF SULFUR IN EARTH'S CRUST

The abundance of sulfur in the Earth's crust has been reported to be ~520 ppm by Goldschmidt (1954) and 880 ppm by Wedepohl (1984). It is thus one of the more common elements in the biosphere. Its concentration (as total sulfur) in rocks, including igneous and sedimentary rocks, can range from 270 to 2400 ppm (Bowen, 1979). The average total sulfur concentrations in freshwater and seawater are ~3.7 and 905 ppm, respectively (Bowen, 1979). In field soils in humid, temperate regions, the total sulfur concentration may range from 100 to 1500 ppm, of which 50–500 ppm is soluble in weak acid or water (Lawton, 1955). Most of the sulfur in soil of pastureland in humid to semiarid climates is organic, whereas that in drier soils is contained in gypsum ($CaSO_4 \cdot 2H_2O$), epsomite ($MgSO_4 \cdot 7H_2O$), and in lesser amounts in sphalerite ($ZnSO_4$), chalcopyrite ($CuFeS_2$), and pyrite or marcasite (FeS_2) (Freney, 1967).

19.2 GEOCHEMICALLY IMPORTANT PROPERTIES OF SULFUR

Inorganic sulfur occurs most commonly in the -2, 0, $+2$, $+4$, and $+6$ oxidation states (Roy and Trudinger, 1970). Table 19.1 lists geochemically important forms and their various oxidation states. In nature, the -2, 0, and $+6$ oxidation states are the most common, represented by sulfide, elemental sulfur, and sulfate, respectively. However, in some environments (e.g., chemoclines in aquatic environments, and in some soils and sediments), some other mixed oxidation states (e.g., thiosulfate, tetrathionate) may also occur, though in lesser amounts.

Some abiotic reactions involving elemental sulfur may have a geomicrobiological impact. For instance, elemental sulfur (usually written as S^0, although it is really S_8 because it consists of eight sulfur atoms in a ring) can react reversibly with sulfite to form thiosulfate (Roy and Trudinger, 1970):

$$S^0 + SO_3^{2-} \Leftrightarrow S_2O_3^{2-} \tag{19.1}$$

The forward reaction is favored by neutral to alkaline pH, whereas the back reaction is favored by acid pH. Thiosulfate is very unstable in aqueous solution below pH 5. In stable form it can be readily oxidized or reduced by various bacteria.

Elemental sulfur also reacts with sulfide, forming polysulfides (Roy and Trudinger, 1970), a reaction that can play an important role in elemental sulfur metabolism:

$$S_8^0 + HS^- \rightarrow HSS_n^- + S_{8-n}^0 \tag{19.2}$$

The value of n may equal 2, 3, or higher up to 8. It is related to the sulfide concentration.

Polythionates, starting with trithionate ($S_3O_6^{2-}$), may be viewed as disulfonic acid derivatives of sulfanes (Roy and Trudinger, 1970). They may be formed as by-products in the oxidation of sulfide and sulfur to sulfate and in disproportionation reactions. Polythionates can also be metabolized by some microorganisms.

TABLE 19.1
Geomicrobially Important Forms of Sulfur and Their Oxide States

Compound	Formula	Oxidation State(s) of Sulfur
Sulfide	S^{2-}	-2
Polysulfide	S_n^{2-}	$-2, 0$
Sulfur[a]	S_8	0
Hyposulfite (dithionite)	$S_2O_4^{2-}$	$+3$
Sulfite	SO_3^{2-}	$+4$
Thiosulfate[b]	$S_2O_3^{2-}$	$-1, +5$
Dithionate	$S_2O_6^{2-}$	$+4$
Trithionate	$S_3O_6^{2-}$	$-2, +6$
Tetrathionate	$S_4O_6^{2-}$	$-2, +6$
Pentathionate	$S_5O_6^{2-}$	$-2, +6$
Sulfate	SO_4^{2-}	$+6$

[a] Occurs in an octagonal ring in crystalline form.
[b] Outer sulfur has an oxidation state of -1; the inner sulfur has an oxidation state of $+5$.

19.3 BIOLOGICAL IMPORTANCE OF SULFUR

Sulfur is an important element for life. In the cell it is especially important in stabilizing protein structure and in the transfer of hydrogen by enzymes in redox metabolism. For some prokaryotes, reduced forms of sulfur can serve as sources of energy and reducing power. For other prokaryotes, oxidized forms, especially sulfate but also elemental sulfur, can serve as terminal electron acceptors in anaerobic respiration. Geomicrobiologically, it is these oxidation and reduction reactions involving sulfur and sulfur compounds that are especially important.

19.4 MINERALIZATION OF ORGANIC SULFUR COMPOUNDS

As part of the carbon cycle, microbes degrade organic sulfur compounds such as the amino acids cysteine, cystine, methionine; agar-agar (a sulfuric acid ester of a linear galactan); tyrosine-O-sulfate; and so on. Desulfurization is usually the first step in the degradation. Desulfurization of cysteine by bacteria may occur anaerobically by the reaction (Freney, 1967; Roy and Trudinger, 1970)

$$HSCH_2CHCOOH + H_2O \xrightarrow[\text{desulfhydrase}]{\text{cysteine}} H_2S + NH_3 + CH_3COCOOH \qquad (19.3)$$
$$| $$
$$NH_2$$

or by the reaction (Freney, 1967; Roy and Trudinger, 1970)

$$HSCH_2CHCOOH \underset{-H_2O}{\overset{+H_2O}{\rightleftarrows}} HOCH_2CHCOOH + H_2S \qquad (19.4)$$
$$|\qquad\qquad \text{serine} \qquad\qquad |$$
$$NH_2 \qquad \text{sulfhydrase} \qquad NH_2$$

The sulfur of cysteine may also be released aerobically as sulfate. The reaction in that case is not completely certain and may differ with different types of organisms (Freney, 1967; Roy and Trudinger, 1970). Although alanine-3-sulfinic acid [$HO_2SCH_2CH(NH_2)COOH$] has been postulated as a key intermediate by some workers, others have questioned it, at least for rat liver mitochondria (Wainer, 1964, 1967).

Methionine is decomposed by extracts of *Clostridium sporogenes* or *Pseudomonas* sp. as follows (Freney, 1967):

$$\text{Methionine} \rightarrow \alpha\text{-ketobutyrate} + CH_3SH + NH_3 \tag{19.5}$$

19.5   SULFUR ASSIMILATION

Inorganic sulfur is usually assimilated into organic compounds through utilization of sulfate by plants and most microorganisms. One possible pathway of assimilation in bacteria is the reduction of sulfate to sulfide and its subsequent reaction with serine to form cysteine, as in *Salmonella typhimurium* (see Freney, 1967, p. 239)*

$$ATP + SO_4^{2-} \xrightarrow{\text{ATP sulfurylase}} APS + PP_i \tag{19.6}$$

$$ASP + ATP \xrightarrow{\text{APS kinase}} PAPS + ADP \tag{19.7}$$

$$PAPS + 2e \xrightarrow{\text{PAPS reductase}} SO_3^{2-} + PAP \tag{19.8}$$

$$SO_3^{2-} + 7H^+ + 6e \xrightarrow{\text{SO}_3^{2-} \text{ reductase}} HS^- + 3H_2O \tag{19.9}$$

$$HS^- + serine \xrightarrow{\text{cysteine synthase}} cysteine + H_2O \tag{19.10}$$

This reaction sequence has also been found in *Bacillus subtilis, Aspergillus niger, Micrococcus aureus,* and *Enterobacter aerogenes* (Roy and Trudinger, 1970). Reaction 19.10 may be replaced by the sequence:

$$Serine + acetyl \sim SCoA \longrightarrow \underset{\substack{| \\ CHNH_2 \\ | \\ COOH \\ \textit{O}\text{-acetylserine}}}{CH_2OOCCH_3} + CoASH \tag{19.11}$$

$$\textit{O}\text{-Acetylserine} + H_2S \rightarrow cysteine + acetate + H^+ \tag{19.12}$$

This latter sequence has been observed in *Escherichia coli* and *S. typhimurium* (Roy and Trudinger, 1970). The reduction of sulfate to *active thiosulfate* and its incorporation into serine from cysteine is also possible for some organisms, such as *E. coli* (Freney, 1967).

Sulfate reduction occurs not only as part of an *assimilatory* process but also as a *dissimilatory* or *respiratory* process. The latter occurs in the great majority of known instances only in special anaerobic bacteria. The majority of dissimilatory sulfate-reducers known to date are members of the domain Bacteria, but at least two sulfate-reducers, *Archeoglobus fulgidus* (Stetter et al., 1987; Stetter, 1988) and *A. profundus* (Burggraf et al., 1990), have been shown to belong to the domain Archaea. Geomicrobiologically, it is the dissimilatory sulfate-reducers that are important.

Almost without exception, assimilatory sulfate reduction does not consume more sulfate than is needed for assimilation, so excess sulfide is not produced. The only known exceptions are a strain of *Bacillus megaterium* (Bromfield, 1953) and *Pseudomonas zelinskii* (Shturm, 1948). Unlike

---

* APS, adenosine 5′-phosphatosulfate; PAPS, adenosine 3′-phosphate-5′-sulfatophosphate; PPi, inorganic pyrophosphate; PAP, adenosine 3′,5′-diphosphate.

dissimilatory sulfate reduction, assimilatory sulfate reduction is thus a form of transitory sulfur immobilization involving small amounts of sulfur per cell; the fixed sulfur is returned to the sulfur cycle upon the death of the organism that assimilated it.

19.6 GEOMICROBIALLY IMPORTANT TYPES OF BACTERIA THAT REACT WITH SULFUR AND SULFUR COMPOUNDS

19.6.1 OXIDIZERS OF REDUCED SULFUR

Most geomicrobiologically important microorganisms that oxidize reduced forms of sulfur in relatively large quantities are prokaryotes. They include representatives of the domains Bacteria and Archaea. They comprise aerobes, facultative organisms, and anaerobes and are mostly obligate or facultative autotrophs or mixotrophs. Among the aerobes in the domain Bacteria, one of the most important groups terrestrially is that of the *Thiobacillaceae* (Table 19.2). This group includes obligate and facultative autotrophs as well as mixotrophs. Among aerobes in the domain Archaea, one

TABLE 19.2
Some Aerobic Sulfur-Oxidizing Bacteria[a,b]

Autotrophic	Mixotrophic	Heterotrophic
Acidithiobacillus albertensis[c]	*Pseudomonas* spp.	*Beggiatoa* spp.
Acidithiobacillus caldus[c]	*Thiobacillus intermedius*	*Thiobacillus perometabolis*
Acidithiobacillus ferrooxidans[c]	*Thiobacillus organoparus*	
Acidithiobacillus thiooxidans[c]	*Thiobacillus versutus*[d]	
Acidianus brierleyi[e]		
Alicyclobacillus disulfidooxidans[f,g]		
Alicyclobacillus tolerans[f,g]		
Beggiatoa alba MS-81-6		
Sulfolobus acidocaldarius[e]		
Thermothrix thiopara[f]		
Thiobacillus denitrificans[h]		
Thiobacillus neapolitanus		
Thiobacillus novellus[f]		
Thiobacillus tepidarius		
Thiobacillus thioparus		

[a] A more complete survey of aerobic sulfur-oxidizing bacteria can be found in Balows et al. (1992) and Dworkin (2001).

[b] All members of the domain Bacteria in this table are gram-negative except for *Alicyclobacillus disulfidooxidans and A. tolerans*.

[c] Formerly assigned to the genus *Thiobacillus* (see Kelly and Wood, 2000).

[d] Can also grow autotrophically and heterotrophically.

[e] Archeon.

[f] *Alicyclobacillus disulfidooxidans* formerly known as *Sulfobacillus disulfidooxidans* and *Alicyclobacillus tolerans* formerly known as *Sulfobacillus thermosulfidooxidans* subsp. *thermotolerans* (see Karavaiko et al., 2005).

[g] Facultative autotroph.

[h] Facultative anaerobe.

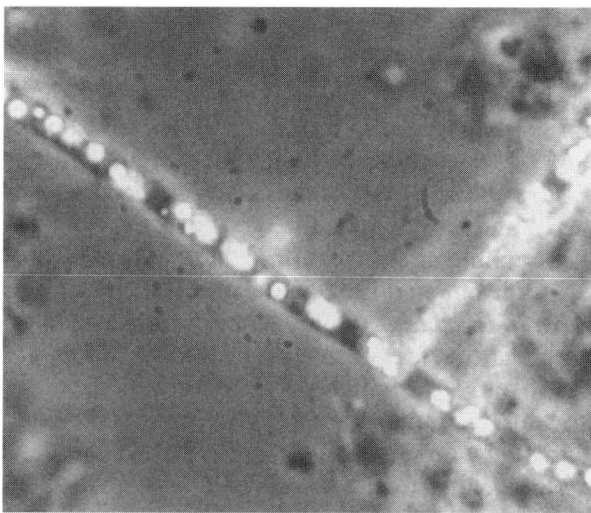

FIGURE 19.1   Trichome of *Beggiatoa* with sulfur granules in pond-water enrichment (×5240). (Courtesy of Arcuri EJ)

of the most widely studied groups consists of the genera *Sulfolobus* and *Acidianus* (Table 19.2). Another group in the domain Bacteria that oxidizes sulfide and is of some importance in freshwater and marine environments is the family *Beggiatoaceae* (Figure 19.1). Most cultured members of the group use hydrogen sulfide mixotrophically or heterotrophically. In the latter instance, they employ H_2S oxidation as protection against metabolically produced H_2O_2 in the absence of catalase (Kuenen and Beudecker, 1982; Nelson and Castenholz, 1981), but at least one marine strain, *Beggiatoa alba* MS-81-6, can grow autotrophically (Nelson and Jannasch, 1983). Other hydrogen sulfide oxidizers found in aquatic environments include *Thiovulum* (autotrophic) (e.g., Wirsen and Jannasch, 1978), *Achromatium*, *Thiothrix*, *Thiobacterium* (LaRiviere and Schmidt, 1981), and *Thiomicrospira* (Kuenen and Tuovinen, 1981). Of all these groups, only the thiobacilli produce sulfate directly without accumulating elemental sulfur when oxidizing H_2S at normal oxygen tension. The other groups accumulate sulfur (S^0), which they may oxidize to sulfate when the supply of H_2S is limited or depleted.

Among members of the domain Bacteria, *Thiobacillus thioparus* oxidizes S^0 slowly to H_2SO_4. It is inhibited as the pH drops below 4.5. *Halothiobacillus halophilus* (formerly *T. halophilus*) is another neutrophilic, but extremely halophilic, obligate chemolithotroph that oxidizes elemental sulfur to sulfate (Wood and Kelly, 1991). By contrast, *Acidithiobacillus thiooxidans*, *A. albertensis* (formerly *T. albertis*) (Bryant et al., 1983), and *A. ferrooxidans* readily oxidize elemental sulfur to sulfuric acid. Being acidophilic, they may drop the pH as low as 1.0 in batch culture. All these organisms are strict autotrophs.

The Archaea *Sulfolobus* spp. and *Acidianus* spp. are also able to oxidize elemental sulfur to sulfuric acid. Both the genera are extremely thermophilic. *S. acidocaldarius* will oxidize sulfur between 55 and 85°C (70–75° optimum) in a pH range of 0.9–5.8 (pH 2–3 optimum) (Brock et al., 1972; Shivvers and Brock, 1973). The organisms are facultative autotrophs. *Acidianus* (formerly *Sulfolobus*) *brierleyi* has traits similar to those of *S. acidocaldarius* but can also reduce S^0 anaerobically with H_2 and has a different GC (guanine + cytosine) content (31 versus 37 mol%) (Brierley and Brierley, 1973; Segerer et al., 1986).

Moderately thermophilic bacteria capable of oxidizing sulfur have also been observed. Several were incompletely characterized. Some were isolated from sulfurous hot springs, others from ore deposits. One of these has been described as a motile rod capable of forming endospores in plectridia. It is a facultative autotroph capable of oxidizing various sulfides and organic compounds besides elemental sulfur. It was named *Thiobacillus thermophilica* Imshenetskii (Egorova and Deryugina, 1963). Another is an aerobic, gram-positive, facultative thermophile capable of sporulation, which is able to oxidize not only elemental sulfur but also Fe^{2+} and metal sulfides mixotrophically. It was originally named *Sulfobacillus thermosulfidooxidans* subsp. *thermotolerans* strain K1 (Golovacheva and Karavaiko, 1978; Bogdanova et al., 1990), and later renamed *Alicyclobacillus tolerans* (Karavaiko et al., 2005). Still another is a gram-negative, facultatively autotrophic *Thiobacillus* sp. capable of growth at 50 and 55°C with a pH optimum of 5.6 (range 4.8–8) (Williams and Hoare, 1972). Other thermophilic thiobacilluslike bacteria have been isolated that can grow on thiosulfate at 60 and 75°C and a pH of 7.5, and still others that can grow at 60 and 75°C and at pH 4.8 (LeRoux et al., 1977). A moderately thermophilic acidophile, *Acidithiobacillus caldus* (formerly *T. caldus*), with an optimum growth temperature of ~45°C was isolated by Hallberg and Lindström (1994) and found capable of oxidizing S^{2-}, S^0, SO_3^{2-}, $S_2O_3^{2-}$, and $S_4O_6^{2-}$ (Hallberg et al., 1996).

A number of heterotrophs have been reported to be able to oxidize reduced sulfur in the form of elemental sulfur, thiosulfate, and tetrathionate. They include bacteria and fungi. A diverse group of heterotrophic thiosulfate-oxidizing bacteria have been detected in marine sediments and around hydrothermal vents (Teske et al., 2000). Many bacteria that oxidize elemental sulfur oxidize it to thiosulfate, whereas others oxidize thiosulfate to sulfuric acid (Guittonneau, 1927; Guittonneau and Keiling, 1927; Grayston and Wainright, 1988; see also Roy and Trudinger, 1970, pp. 248–249). Some marine pseudomonads and others can gain useful energy from thiosulfate oxidation by using it as a supplemental energy source (Tuttle et al., 1974; Tuttle and Ehrlich, 1986).

Two examples of facultatively anaerobic sulfur oxidizers in the domain Bacteria are *Thiobacillus denitrificans* (e.g., Justin and Kelly, 1978) and *Thermothrix thiopara* (Caldwell et al., 1976; Brannan and Caldwell, 1980), the former a mesophile and the latter a thermophile. The genome sequence of *T. denitrificans* has been unraveled (Beller et al., 2006). Anaerobically both organisms use nitrate as terminal electron acceptor and reduce it to oxides of nitrogen and dinitrogen, with nitrite being an intermediate product. They can use sulfur in various oxidation states as an energy source. *T. denitrificans* is an obligate autotroph whereas *Thx. thiopara* is a facultative autotroph.

The strictly anaerobic sulfur oxidizers are represented by photosynthetic purple and green bacteria (Pfennig, 1977) and certain cyanobacteria (Table 19.3). Some of these bacteria may also

---

TABLE 19.3
Some Anaerobic Sulfur-Oxidizing Bacteria[a]

Photolithotrophs	Chemolithotrophs
Chromatinum spp.	*Thermothrix thiopara*[b,c]
Chlorobium spp.	*Thiobacillus denitrificans*[c]
Ectothiorhodospira spp.	
Rhodopseudomonas spp.[b]	
Chloroflexus aurantiacus[b]	
Oscillatoria sp.[c]	
Lyngbya spp.[c]	
Aphanothece spp.[c]	
Microcoleus spp.[c]	
Phormidium spp.[c]	

[a] For a more complete description of anaerobic sulfur-oxidizing bacteria, see Starr et al. (1981), Holt (1984), and Dworkin (2001).

[b] Facultatively autotrophic.

[c] Facultatively anaerobic.

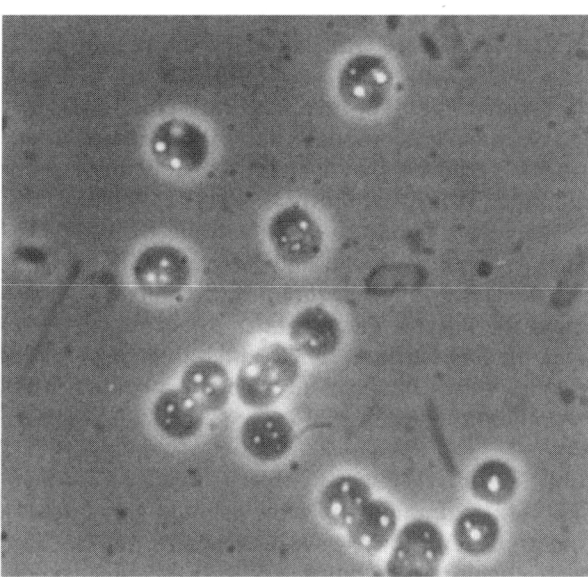

FIGURE 19.2 Unidentified purple sulfur bacteria (probably *Chromatium* sp.) in an enrichment culture (×5385). Note conspicuous sulfur granules in the spherical cells.

grow aerobically but without oxidizing reduced sulfur compounds. The purple sulfur bacteria (Chromataceae) (Figure 19.2) are obligate anaerobes that oxidize reduced sulfur, especially H_2S, by using it as a source of reducing power for CO_2 assimilation. Despite the terminology, several purple nonsulfur bacteria (Rhodospirillaceae) can also grow autotrophically on H_2S as a source of reducing power for CO_2 assimilation, but for the most part they tolerate only low concentrations of sulfide, in contrast to purple sulfur bacteria. In the laboratory, purple nonsulfur bacteria can also grow photoheterotrophically, using reduced carbon as a major carbon source. Most purple sulfur bacteria when growing on H_2S oxidize it to S^0, which they deposit intracellularly (Figure 19.2), but *Ectothiorhodospira* spp. are an exception in depositing it extracellularly. Under conditions of H_2S limitation, these strains oxidize the sulfur they accumulate further to sulfate. Among the purple nonsulfur bacteria, *Rhodopseudomonas palustris* and *R. sulfidophila* do not form elemental sulfur as an intermediate from H_2S but oxidize sulfide directly to sulfate (Hansen and van Gemerden, 1972; Hansen and Veldkamp, 1973). In contrast, *Rhodospirillum rubrum*, *R. capsulata*, and *R. spheroides* form elemental sulfur from sulfide, which they deposit extracellularly (Hansen and van Gemerden, 1972). *R. sulfidophila* differs from most purple nonsulfur bacteria in being more tolerant of high concentrations of sulfide.

Green sulfur bacteria (Chlorobiaceae) are strictly anaerobic photoautotrophs that oxidize H_2S by using it as a source of reducing power in CO_2 fixation. They deposit the sulfur (S^0) they produce *extracellularly*. Under H_2S limitation, they oxidize the sulfur further to sulfate. At least a few strains of *Chlorobium limicola* forma *thiosulfatophilum* do not accumulate sulfur but oxidize H_2S directly to sulfate (Ivanov, 1968, p. 137; Paschinger et al., 1974). Many photosynthetic anaerobic bacteria can also use thiosulfate as electron donor in the place of hydrogen sulfide.

Filamentous gliding green bacteria (Chloroflexacea) grow photoheterotrophically under anaerobic conditions, but at least some can also grow photoautotrophically with H_2S as electron donor under anaerobic conditions (Brock and Madigan, 1988).

A few filamentous cyanobacteria, including some members of the genera *Oscillatoria*, *Lyngbya*, *Aphanothece*, *Microcoleus*, and *Phormidium*, which are normally classified as oxygenic photoautotrophs, can grow photosynthetically under anaerobic conditions with H_2S as a source of reducing power (Cohen et al., 1975; Garlick et al., 1977). They oxidize H_2S to elemental sulfur and deposit it extracellularly. In the dark, they can re-reduce the sulfur they produce using internal reserves of polyglucose as reductant (Oren and Shilo, 1979). At this time there is no evidence that these organisms can oxidize the sulfur they produce anaerobically further to sulfate under H_2S limitation.

19.6.2 REDUCERS OF OXIDIZED FORMS OF SULFUR

19.6.2.1 Sulfate Reduction

A geomicrobially and geochemically very important group of bacteria are the sulfate-reducers. Most of the known examples that have been cultured belong to the domain Bacteria, but at least two belong to Archaea, namely, *Archeoglobus fulgidus* (Stetter et al., 1987; Speich and Trüper, 1988) and *A. profundus* (Burggraf et al., 1990). More than three decades ago, the sulfate-reducers were thought to be represented by only three genera of the Bacteria, *Desulfovibrio*, *Desulfotomaculum* (originally classified as *Clostridium* because of its ability to form endospores), and *Desulfomonas*. These organisms are very specialized nutritionally in that among organic energy sources they can use only lactate, pyruvate, fumarate, malate, and ethanol. Furthermore, none of these organisms are able to degrade their organic energy sources beyond acetate (Postgate, 1984), with the result that at the time the importance of the sulfate-reducers in anaerobic mineralization of organic matter in sulfate-rich environments was unappreciated. After 1976, this restricted view of sulfate-reducers changed rapidly with the discovery of a sulfate-reducer, *Desulfotomaculum acetoxidans* (Widdel and Pfennig, 1977, 1981), which is able to oxidize acetate anaerobically to CO_2 and H_2O with sulfate. Subsequently, a wide variety of other sulfate-reducers were discovered that differed in the nature of the energy sources they were capable of utilizing. Various aliphatic, aromatic, and heterocyclic compounds were found to be attacked, and in many cases completely mineralized, each by a specific

---

TABLE 19.4
Some Sulfate-Reducing Bacteria in the Domain Bacteria[a]

Heterotrophs	Autotrophs[b]
Desulfovibrio desulfuricans[c,d]	*Desulfovibrio baarsii*
Desulfovibrio vulgaris	*Desulfobacter hydrogenophilus*
Desulfovibrio gigas	*Desulfosarcina variabilis*
Desulfovibrio fructosovorans	*Desulfonema limicola*
Desulfovibrio sulfodismutans	
Desulfomonas pigra	
Desulfotomaculum nigrificans	
Desulfotomaculum acetoxidans	
Desulfotomaculum orientis[d]	
Desulfobacter postgatei	
Desulfolobus propionicus	
Desulfobacterium phenolicum[e]	
Desulfobacterium indolicum[f]	
Desulfobacterium catecholicum[g]	

[a] For a more detailed description of sulfate-reducers, see Pfennig et al. (1981), Postgate (1984), and Dworkin (2001).
[b] Autotrophic growth on H_2 and CO_2.
[c] Some strains can grow mixotrophically on H_2 and CO_2 and acetate.
[d] At least one strain can grow autotrophically on H_2 and CO_2.
[e] Bak and Widdel (1986b).
[f] Bak and Widdel (1986a).
[g] Szewzyk and Pfennig (1987).

group of sulfate-reducers (e.g., Pfennig et al., 1981; Imhoff-Stuckle and Pfennig, 1983; Braun and Stolp, 1985; Bak and Widdel, 1986a,b; Szewzyk and Pfennig, 1987; Platen et al., 1990; Zellner et al., 1990; Qatabi et al., 1991; Schnell and Schink, 1991; Boopathy and Daniels, 1991; Aeckersberg et al., 1991; Tasaki et al., 1991, 1992; Kuever et al., 1993; Rueter et al., 1994; Janssen and Schink, 1995; Rees et al., 1998; Londry et al., 1999; Meckenstock et al., 2000). Some of the newly discovered sulfate-reducers were also found to the able to use H_2 as an energy source. Most require an organic carbon source, but a few can grow autotrophically on hydrogen. Table 19.4 presents a list of some of the different kinds of sulfate-reducers in the domain Bacteria and their nutritional ranges. Although most sulfate-reducers discovered to date are mesophiles, thermophilic types are also known now (e.g., Pfennig et al., 1981; Zeikus et al., 1983; Stetter et al., 1987; Burggraf et al., 1990). At least one moderate psychrophile (growth optimum 10–19°C) has been described (Isaksen and Teske, 1996) and named *Desulforhopalus vacuolatus* gen. nov., spec. nov. It was isolated from sediment in Kysing Fjord, Denmark, at 10°C. It grows anaerobically on propionate, lactate, and alcohols as energy and carbon sources with sulfate as terminal electron acceptor. It can also grow heterotrophically on H_2 as an energy source and acetate as a carbon source. Its cells are vacuolated.

Morphologically, sulfate-reducers are a very varied group including cocci, sarcinae, rods, vibrios (Figure 19.3), spirilla, and filaments. The representatives in the domain Bacteria are of a gram-negative cell type.

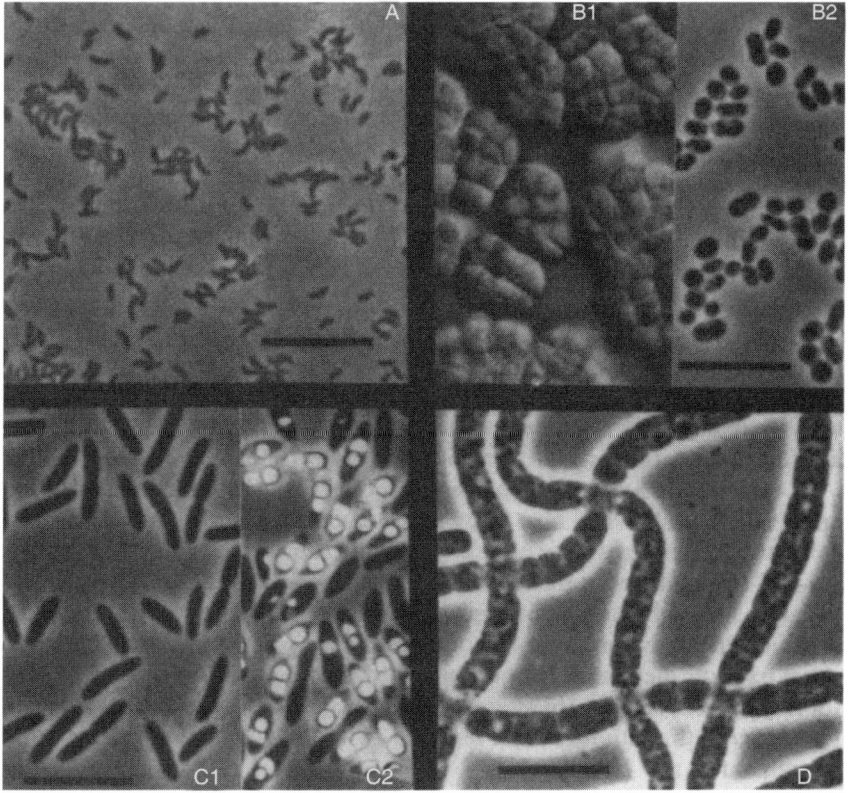

FIGURE 19.3   Sulfate-reducing bacteria. (A) *Desulfovibrio desulfuricans* (phase contrast). (B1, B2) *Desulfosarcina variabilis*: (B1) sarcina packets (interference contrast); (B2) free-living cells (phase contrast). (C1, C2) *Desulfotomaculum acetoxidans*: (C1) vegetative cells (phase contrast); (C2) cells with spherical spores and gas vacuoles (phase contrast). (D) *Desulfonema limicola* (phase contrast). (From Pfennig N, Widdel F, Trüper HG. *The Prokaryotes: A Handbook on Habitats, Isolation, and Identification of Bacteria*, Vol I. Springer, Berlin, Copyright 1981. With kind permission of Springer Science and Business Media.)

The two sulfate-reducers in the domain Archaea that were discovered by Stetter et al. (1987) and Burggraf et al. (1990) are extremely thermophilic, anaerobic, gram-negative, irregularly shaped cocci. *Archeoglobus fulgidus* was found to grow naturally in a hydrothermal system at temperatures between 70 and 100°C in the vicinities of Vulcano and Stufe di Nerone, Italy. Under laboratory conditions, the cultures grow anaerobically in marine mineral salts medium supplemented with yeast extract. In this medium they produce a large amount of hydrogen sulfide and some methane. Thiosulfate, but not elemental sulfur, can act as alternative electron acceptor. Energy sources include hydrogen and some simple organic molecules as well as glucose, yeast extract, and other more complex substrates. Cells contain a number of compounds such as 8-OH-5-deazaflavin and methanopterin previously found only in methanogens, which are also members of the domain Archaea, but 2-mercaptoethanesulfonic acid and factor F430, which are found in methanogens, were absent (Stetter et al., 1987).

Archeoglobus profundus was isolated from the Guaymas hot vent area (Gulf of California, also known as the Sea of Cortez). It grows anaerobically at temperatures between 65 and 95°C (optimum 82°C) in a pH range of 4.5–7.5 at an NaCl concentration in the range of 0.9–3.6%. Unlike *A. fulgidus*, it is an obligate mixotroph that requires H_2 as an energy source. Its organic carbon requirement can be satisfied by acetate, lactate, pyruvate, yeast extract, meat extract, peptone, or acetate-containing crude oil. As for *A. fulgidus*, sulfate, thiosulfate, and sulfate can serve as terminal electron acceptors for growth. Although S^0 is reduced by resting cells, it inhibits growth (Burggraf et al., 1990).

The presence of as-yet-unidentified, extremely thermophilic sulfate-reducers was detected in hot deep-sea sediments at the hydrothermal vents of the tectonic spreading center of Guaymas Basin (Sea of Cortez or Gulf of California). Sulfate-reducing activity was measurable between 100 and 110°C (optimum 103–106°C). The responsible organisms are probably examples of Archaea (Jørgensen et al., 1992).

19.6.2.2   Sulfite Reduction

Some *Clostridia* have been found to be strong sulfite reducers. However, they cannot reduce sulfate in a dissimilatory mode. An example is *Clostridium pasteurianum* (McCready et al., 1975; Laishley and Krouse, 1978). The geomicrobial significance of this trait is not clear because sulfite normally does not occur free in the environment in significant amounts in the absence of pollution from human activities. An exception is around volcanic vents that may emit SO_2. This gas, which may also be formed when sulfur-containing fossil fuels are combusted, forms H_2SO_3 when it dissolves in water. *Shewanella putrefaciens* from the Black Sea is another organism that cannot reduce sulfate but can reduce thiosulfate, sulfite, and elemental sulfur to sulfide (Perry et al., 1993). The organism represents 20–50% of the bacterial population in the suboxic zone of the Black Sea, where its ability to reduce potentially oxidized sulfur compounds appears to play a significant role in the sulfur cycle.

19.6.2.3   Reduction of Elemental Sulfur

A number of different bacteria have been found to be able to use elemental sulfur anaerobically as terminal electron acceptor, reducing the sulfur to H_2S (Pfennig and Biebl, 1981). Some of these bacteria can grow autotrophically on sulfur using H_2 or methane as an energy source. They are generally thermophilic members of the domain Archaea (Stetter et al., 1986). Others grow heterotrophically on elemental sulfur, using organic carbon as complex as sugars or as simple as acetate as electron donors (Belkin et al., 1986; Pfennig and Biebl, 1976) to reduce the sulfur in their energy metabolism. This group includes members of both the Bacteria and the Archaea. A few, but by no means all, strains of sulfate-reducing bacteria of the genus *Desulfovibrio* have the ability to use sulfur in the place of sulfate as terminal electron acceptor for growth, but strains of *Desulfotomaculum* and *Desulfomonas* are unable to do so (Biebl and Pfennig, 1977).

Some fungi can also reduce sulfur to sulfide with an electron donor such as glucose (e.g., Ehrlich and Fox, 1967). As expected, the activity is greater anaerobically than aerobically.

19.7 PHYSIOLOGY AND BIOCHEMISTRY OF MICROBIAL OXIDATION OF REDUCED FORMS OF SULFUR

19.7.1 SULFIDE

19.7.1.1 Aerobic Attack

Many aerobic bacteria that oxidize sulfide are obligate or facultative chemosynthetic autotrophs (chemolithoautotrophs). When growing in the autotrophic mode, they use sulfide as an energy source to assimilate CO_2. Most of them oxidize the sulfide to sulfate, regardless of the level of oxygen tension (e.g., *Acidithiobacillus thiooxidans* (London and Rittenberg, 1964)). However, some like *Thiobacillus thioparus* form elemental sulfur (S^0) if the pH of their milieu is initially alkaline and the rH_2* is 12, that is, if the milieu is partially reduced due to an oxygen tension below saturation. Thus, *T. thioparus* T5, isolated from a microbial mat, produces elemental sulfur in continuous culture in a chemostat under conditions of oxygen limitation. In this case, small amounts of thiosulfate together with even smaller amounts of tetrathionate and polysulfide are also formed (van den Ende and van Gemerden, 1993). In batch culture under oxygen limitation, *T. thioparus* has been observed to produce initially a slight increase in pH followed by a drop to ~7.5 in 4 days and a rise in rH_2 to ~20 (Sokolova and Karavaiko, 1968). The reaction leading to the formation of elemental sulfur can be summarized as

$$HS^- + 0.5O_2 + H^+ \rightarrow S^0 + H_2O \tag{19.13}$$

Wirsen et al. (2002) described an autotrophic, microaerophilic sulfide-oxidizing organism from a coastal marine environment, *Candidatus* Arcobacter sulfidicus that produced filamentous elemental sulfur. It fixed CO_2 via the reductive tricarboxylic acid cycle rather than the Calvin–Benson–Bassham cycle (Hügler et al., 2005). The organism was able to fix nitrogen (Wirsen et al., 2002).

Under conditions of high oxygen tension (at or near saturation), *Thiobacillus thioparus* will oxidize soluble sulfide all the way to sulfate (London and Rittenberg, 1964; Sokolova and Karavaiko, 1968; van den Ende and van Gemerden, 1993):

$$HS^- + 2O_2 \rightarrow SO_4^{2-} + H^+ \tag{19.14}$$

Thiovulum sp. is another example of a member of the domain Bacteria that oxidizes sulfide to sulfur under reduced oxygen tension (Wirsen and Jannasch, 1978).

London and Rittenberg (1964) (see also Vishniac and Santer, 1957) suggested that the intermediate steps in the oxidation of sulfide to sulfate involved

$$4S^{2-} \rightarrow 2S_2O_3^{2-} \rightarrow S_4O_6^{2-} \rightarrow SO_3^{2-} + S_3O_6^{2-} \rightarrow 4SO_3^{2-} \rightarrow SO_4^{2-} \tag{19.15}$$

However, this reaction sequence does not explain the formation of elemental sulfur at reduced oxygen tension. Unless this occurs by way of a specialized pathway, which seems doubtful, a more attractive model of the pathway that explains both the processes, the formation of S^0 and SO_4^{2-}, in a unified way is the one proposed by Roy and Trudinger (1970) (see also Suzuki, 1999; Suzuki et al., 1994; Yamanaka, 1996):

$$S^{2-} \rightarrow x \rightarrow SO_3^{2-} \rightarrow SO_4^{2-}$$
$$\updownarrow$$
$$S^o$$
$$\tag{19.16}$$

Here X represents a common intermediate in the oxidation of sulfide and elemental sulfur to sulfite. Roy and Trudinger visualized X as a derivative of glutathione or a membrane-bound thiol. It may also be a representative of the *intermediate sulfur* described by Pronk et al. (1990). The scheme of

---

* $rH_2 = -\log[H_2] = (E_h/0.029) + 2pH$, because $E_h = -0.029 \log[H_2] + 0.058 \log[H^+]$.

Roy and Trudinger permits integration of a mechanism for elemental sulfur oxidation into a unified pathway for oxidizing reduced forms of sulfur. Hallberg et al. (1996) found this mechanism consistent with the action of *Acidithiobacillus caldus* on reduced forms of sulfur.

Sorokin (1970) questioned the sulfide-oxidizing ability of thiobacilli, believing that they oxidize only thiosulfate resulting from chemical oxidation of sulfide by oxygen and that any elemental sulfur formed by thiobacilli from sulfide is due to the chemical interaction of bacterial oxidation products with S^{2-} and $S_2O_3^{2-}$, as previously proposed by Nathansohn (1902) and Vishniac (1952). This view is not accepted today. Indeed, Vainshtein (1977) and others have presented clear evidence to the contrary. More recently, Nübel et al. (2000) showed that hyperthermophilic, microaerophilic, chemolithotrophic *Aquifex aeolicus* VF5 oxidizes sulfide to elemental sulfur using a membrane-bound electron transport pathway that conveys electrons from the oxidation of sulfide to oxygen. The pathway includes a quinone pool, a cytochrome bc_1 complex, and cytochrome oxidase.

19.7.1.2 Anaerobic Attack

Most bacteria that oxidize sulfide anaerobically are photosynthetic autotrophs (Chlorobacteriaceae, Chromataceae, some Rhodospirillaceae, and a few Cyanobacteria), but a few, like the facultative anaerobes *Thiobacillus denitrificans* and *Thermothrix thiopara*, are chemosynthetic autotrophs (chemolithoautotrophs). In the presence of nonlimiting concentrations of sulfide, most photosynthetic autotrophs oxidize sulfide to elemental sulfur, using the reducing power from this reaction in the assimilation of CO_2. However, some exceptional organisms exist that never form elemental sulfur (see Section 19.6). When elemental sulfur is formed, it is usually accumulated intracellularly by purple sulfur bacteria and extracellularly by green sulfur bacteria and cyanobacteria. Elemental sulfur accumulated extracellularly by *Chlorobium* appears to be readily available to the cell that formed it but not to other individuals in the population of the same organism or to other photosynthetic bacteria that can oxidize elemental sulfur. The sulfur is apparently attached to the cell surface (van Gemerden, 1986). Recent study by environmental scanning electron microscopy suggests that the extracellularly deposited sulfur is associated with spinae on the cell surface (Douglas and Douglas, 2000). Spinae are helically arrayed proteins, which form a hollow tube protruding from the cell surface (Easterbrook and Coombs, 1976). Details of the biochemistry of sulfide oxidation by the photosynthetic autotrophs remain to be explored.

The chemosynthetic autotroph *Thiobacillus denitrificans* can oxidize sulfide to sulfate anaerobically with nitrate as terminal electron acceptor. As the sulfide is oxidized, nitrate is reduced via nitrite to nitric oxide (NO), nitrous oxide (N_2O), and dinitrogen (N_2) (Baalsrud and Baalsrud, 1954; Milhaud et al., 1958; Peeters and Aleem, 1970; Adams et al., 1971; Aminuddin and Nicholas, 1973). Acetylene has been found to cause accumulation of sulfur rather than sulfate in gradient culture of a strain of *T. denitrificans* using nitrous oxide as terminal electron acceptor. In the absence of acetylene, the gradient culture, unlike a batch culture, did not even accumulate sulfur transiently. It was suggested that acetylene prevents the transformation of S^0 to SO_3^{2-} in this culture (Daalsgaard and Bak, 1992). Polysulfide ($S_{n-1}SH^-$), but not free sulfur, appears to be an intermediate in sulfide oxidation to sulfate by this organism (Aminuddin and Nicholas, 1973). The polysulfide appears to be oxidized to sulfite and thence to sulfate (Aminuddin and Nicholas, 1973, 1974a,b). The reaction sequence is like that proposed in Reaction 19.18.

Oil field brine from the Coleville oil field in Sasketchewan, Canada, has yielded two microaerophilic strains of bacteria, one (strain CVO) resembling *Thiomicrospira denitrificans* and the other (strain FWKO B) resembling *Arcobacter*. Both of these strains can oxidize sulfide anaerobically with nitrate as terminal electron acceptor (Gevertz et al., 2000). Each can grow autotrophically, but strain CVO can also use acetate in the place of CO_2. Strain CVO produces elemental sulfur or sulfate, depending on the sulfide concentration, while reducing nitrate or nitrite to dinitrogen. Strain FWKO B produces only sulfur and reduces nitrate only to nitrite.

19.7.1.3 Oxidation of Sulfide by Heterotrophs and Mixotrophs

Hydrogen sulfide oxidation is not limited to autotrophs. Most known strains of *Beggiatoa* (Figure 19.1) grow mixotrophically or heterotrophically on sulfide. In the former instance, the organisms derive energy from oxidation of the H_2S. In the latter, they apparently use sulfide oxidation to eliminate metabolically formed hydrogen peroxide in the absence of catalase (Burton and Morita, 1964) (see also Section 19.6). *Beggiatoa* deposits sulfur granules resulting from sulfide oxidation in its cells external to the cytoplasmic membrane in invaginated, double-layered membrane pockets (Strohl and Larkin, 1978; see also discussion by Ehrlich, 1999). The sulfur can be further oxidized to sulfate under sulfide limitation (Pringsheim, 1967). At least one strain of *Beggiatoa* has proven to be autotrophic. It was isolated from the marine environment (Nelson and Jannasch, 1983; see also Jannasch et al., 1989, and Section 19.6.1). The heterotrophs *Sphaerotilus natans* (prokaryote, domain Bacteria), *Alternaria*, and yeast (eukaryotes, fungi) have also been reported to oxidize H_2S to elemental sulfur (Skerman et al., 1957a,b). Whether these organisms derive useful energy from this oxidation has not been established.

19.7.2 ELEMENTAL SULFUR

19.7.2.1 Aerobic Attack

Elemental sulfur may be enzymatically oxidized to sulfuric acid by certain members of the Bacteria and Archaea. The overall reaction may be written as

$$S^0 + 1.5O_2 + H_2O \rightarrow H_2SO_4 \tag{19.17}$$

Cell extract of *Acidithiobacillus thiooxidans*, to which catalytic amounts of glutathione were added, oxidized sulfur to sulfite (Suzuki and Silver, 1966). In the absence of formaldehyde as trapping agent for sulfite, thiosulfate was recovered in the reaction mixture (Suzuki, 1965), but this was an artifact resulting from the chemical reaction of sulfite with residual sulfur (Suzuki and Silver, 1966) (see Reaction 19.10). Sulfite was also shown to be accumulated when sulfur was oxidized by *A. thiooxidans* in the presence of the inhibitor 2-*n*-heptyl-4-hydroxyquinoline *N*-oxide (HQNO), which has been shown to inhibit sulfite oxidation. The stoichiometry when the availability of sulfur was limited was 1 mol sulfite accumulated per mole each of sulfur and oxygen consumed (Suzuki et al., 1992). A sulfur-oxidizing enzyme in *Thiobacillus thioparus* used glutathione as a cofactor to produce sulfite (Suzuki and Silver, 1966). The enzyme in both organisms contained nonheme iron and was classed as an oxygenase. The mechanism of sulfur oxidation is consistent with the model described in Reaction 19.16. The glutathione in this instance forms a polysulfide (compound *X* in Reaction 19.16) with the substrate sulfur, which is then converted to sulfite by the introduction of molecular oxygen. This reaction appears not to yield useful energy to the cell. Sulfur oxidation to sulfite that does not involve oxygenase but an oxidase with a potential for energy conservation has also been considered. Some experimental evidence supports such a mechanism (see Pronk et al., 1990).

19.7.2.2 Anaerobic Oxidation of Elemental Sulfur

Few details are known as yet as to how elemental sulfur is oxidized in anaerobes, especially photosynthetic autotrophs. *Thiobacillus denitrificans* appears to follow the reaction sequence in Reaction 19.16 except that oxidized forms of nitrogen substitute for oxygen as terminal electron acceptor.

Acidithiobacillus ferrooxidans has the capacity to oxidize elemental sulfur anaerobically using ferric iron as terminal electron acceptor (Brock and Gustafson, 1976; Corbett and Ingledew, 1987). The anaerobic oxidation yields enough energy to support growth at a doubling time of ~24 h (Pronk et al., 1991, 1992).

19.7.2.3 Disproportionation of Sulfur

Anaerobic marine enrichment cultures consisting predominantly of slightly curved bacterial rods have been shown to contain chemolithotrophic bacteria that were able to grow on sulfur by disproportionating it into H_2S and SO_4^{2-}, but only in the presence of sulfide scavengers such as FeOOH, $FeCO_3$, or MnO_2 (Thamdrup et al., 1993; see also Janssen et al., 1996). The disproportionation reaction can be summarized as

$$4S^0 + 4H_2O \rightarrow SO_4^{2-} + 3H_2S + 2H^+ \qquad (19.18)$$

Added ferrous iron scavenges the sulfide by forming FeS, whereas added MnO_2 scavenges sulfide in a redox reaction in which MnO_2 is reduced to Mn^{2+} by the sulfide, producing SO_4^{2-}, with S^0 a probable intermediate. The scavenging action is needed to propel the reaction in the direction of sulfur disproportionation. In the disproportionation reaction, three pairs of electrons from one atom of sulfur are transferred via an as-yet-undefined electron transport pathway to three other atoms of sulfur, generating H_2S in Reaction 19.18. The sulfur atom yielding the electrons is transformed into sulfate. The transfer of the three pairs of electrons is the source of the energy conserved by the organism for growth and reproduction. This sulfur disproportionation reaction is similar to the one that has been observed under laboratory conditions with the photolithotrophic green sulfur bacteria *Chlorobium limicola* subspecies *thiosulfaticum* and *Chl. vibrioforme* under an inert atmosphere in the light in the absence of CO_2. To keep the reaction going, the H_2S produced had to be removed by continuous flushing with nitrogen (see Trüper, 1984b).

A study of sulfur isotope fractionation as a result of sulfur disproportionation by enrichment cultures from Åarhus Bay, Denmark, and other sediment sources revealed that the sulfide produced may be depleted in ^{34}S by as much as 7.3–8.6‰ and the corresponding sulfate produced may be enriched by as much as 12.6–15.3‰. When sulfur disproportionation is coupled to oxidation in nature, it can lead to the formation of sedimentary sulfides that are more depleted in ^{34}S than would be sulfides that are generated through bacterial sulfate reduction alone (Canfield and Thamdrup, 1994).

19.7.3   Sulfite Oxidation

19.7.3.1   Oxidation by Aerobes

Sulfite may be oxidized by two different mechanisms, one of which includes substrate-level phosphorylation whereas the other does not, although both yield useful energy through oxidative phosphorylation by the intact cell (see, e.g., review by Wood, 1988). In substrate-level phosphorylation, sulfite reacts oxidatively with adenylic acid (AMP) to give adenosine 5′-sulfatophosphate (APS):

$$SO_3^{2-} + AMP \xrightarrow{\text{APS reductase}} APS + 2e \qquad (19.19)$$

The sulfate of APS is then exchanged for phosphate:

$$APS + P_i \xrightarrow{\text{ADP sulfurylase}} ADP + SO_4^{2-} \qquad (19.20)$$

ADP can then be converted to ATP as follows:

$$2ADP \xrightarrow{\text{adenylate kinase}} ATP + AMP \qquad (19.21)$$

Hence the oxidation of 1 mol of sulfite yields 0.5 mol of ATP formed by substrate-level phosphorylation. However, most energy conserved as ATP is gained from shuttling electrons in Reaction 19.19 through the membrane-bound electron transport system to oxygen (Davis and Johnson, 1967).

A number of thiobacilli appear to use an AMP-independent sulfite oxidase system (Roy and Trudinger, 1970, p. 214). These systems do not all seem to be alike. The AMP-independent sulfite oxidase of autotrophically grown *Thiobacillus novellus* may use the following electron transport pathway (Charles and Suzuki, 1966):

$$SO_3^{2-} \rightarrow \text{cytochrome } c \rightarrow \text{cytochrome oxidase} \rightarrow O_2 \qquad (19.22)$$

The sulfite oxidase of *T. neapolitanus* can be pictured as a single enzyme complex that may react either with sulfite and AMP in an oxidation that gives rise to APS and sulfate or with sulfite and water followed by oxidation to sulfate (Roy and Trudinger, 1970). The enzyme complex then transfers the reducing power that is generated to oxygen. Sulfite-oxidizing enzymes that do not require the presence of AMP have also been detected in *Acidithiobacillus thiooxidans*, *T. denitrificans*, and *T. thioparus*. *T. concretivorus* (now considered a strain of *A. thiooxidans*) was reported to shuttle electrons from SO_3^{2-} oxidation via the following pathway to oxygen (Moriarty and Nicholas, 1970):

$$SO_3^{2-} \rightarrow (\text{flavin?}) \rightarrow \text{coenzyme } Q_8 \rightarrow \text{cytochrome } b \rightarrow \text{cytochrome } c \rightarrow \text{cytochrome } a_1 \rightarrow O_2 \quad (19.23)$$

The archaeon *Acidianus ambivalens* appears to possess both an ADP-dependent and an ADP-independent pathway. The former occurs in the cytosol, whereas the latter is membrane associated (Zimmermann et al., 1999).

19.7.3.2 Oxidation by Anaerobes

Thiobacillus denitrificans is able to form APS reductase (Bowen et al., 1966) that is not membrane bound, as well as a membrane-bound AMP-independent sulfite oxidase (Aminuddin and Nicholas, 1973, 1974a,b). Both enzyme systems appear to be active in anaerobically grown cells (Aminuddin, 1980). The electron transport pathway under anaerobic conditions terminates in cytochrome *d*, whereas under aerobic conditions it terminates in cytochromes aa_3 and *d*. Nitrate but not nitrite acts as electron acceptor anaerobically when sulfite is the electron donor (Aminuddin and Nicholas, 1974b).

19.7.4 THIOSULFATE OXIDATION

Most chemosynthetic autotrophic bacteria that can oxidize sulfur can also oxidize thiosulfate to sulfate. The photosynthetic, autotrophic, purple and green sulfur bacteria and some purple nonsulfur bacteria oxidize thiosulfate to sulfate as a source of reducing power for CO_2 assimilation (e.g., Trüper 1978; Neutzling et al., 1985). However, the mechanism of thiosulfate oxidation is probably not the same in all these organisms. The chemosynthetic, aerobic autotrophic *Thiobacillus thioparus* will transiently accumulate elemental sulfur outside its cells when growing in excess thiosulfate in batch culture but only sulfate when growing in limited amounts of thiosulfate. *T. denitrificans* will do the same anaerobically with nitrate as terminal electron acceptor (Schedel and Trüper, 1980). The photosynthetic purple bacteria may also accumulate sulfur transiently, but some green sulfur bacteria (Chlorobiaceae) do not (see discussion by Trüper, 1978). Several of the purple nonsulfur bacteria (Rhodospirillaceae) when growing photoautotrophically with thiosulfate do not accumulate sulfur in their cells (Neutzling et al., 1985). Some mixotrophic bacteria oxidize thiosulfate only to tetrathionate.

Thiosulfate is a reduced sulfur compound with sulfur in a mixed valence state. Current evidence indicates that the two sulfurs are covalently linked, the outer or sulfane sulfur of $S:SO_3^{2-}$ having a valence of -1 and the inner or sulfone sulfur having a valence of $+5$. An older view was that the sulfane sulfur had a valence of -2 and the sulfone sulfur $+6$.

Charles and Suzuki (1966) proposed that when thiosulfate is oxidized, it is first cleaved according to the reaction

$$S_2O_3^{2-} \rightarrow SO_3^{2-} + S^0 \qquad (19.24a)$$

The sulfite is then oxidized to sulfate

$$SO_3^{2-} + H_2O \rightarrow SO_4^{2-} + 2H^+ + 2e \qquad (19.24b)$$

and the sulfur is oxidized to sulfate via sulfite as previously described:

$$S^0 \rightarrow SO_3^{2-} \rightarrow SO_4^{2-} \qquad (19.24c)$$

Alternatively, thiosulfate oxidation may be preceded by a reduction reaction, resulting in the formation of sulfite from sulfone and sulfide from the sulfane sulfur:

$$S_2O_3^{2-} + 2e \rightarrow S^{2-} + SO_3^{2-} \qquad (19.25)$$

These products are then each oxidized to sulfate (Peck, 1962). In the latter case, it is conceivable that sulfur could accumulate transiently by the mechanisms suggested by Reaction 19.16, but sulfur could also result from asymmetric hydrolysis of tetrathionate resulting from direct oxidation of thiosulfate (see Roy and Trudinger (1970) for detailed discussion)

$$2S_2O_3^{2-} + 2H^+ + 0.5O_2 \rightarrow S_4O_6^{2-} + H_2O \qquad (19.26)$$

$$S_4O_6^{2-} + OH^- \rightarrow S_2O_3^{2-} + S^0 + HSO_4^- \qquad (19.27)$$

The direct oxidation reaction may involve the enzymes thiosulfate oxidase and thiosulfate cytochrome c reductase, a thiosulfate activating enzyme (Aleem, 1965; Trudinger, 1961). The thiosulfate oxidase may use glutathione as coenzyme (see summary by Roy and Trudinger (1970) and Wood (1988)).

Thiosulfate may also be cleaved by the enzyme rhodanese, which is found in most sulfur-oxidizing bacteria. For instance, it can transfer sulfane sulfur to acceptor molecules such as cyanide to form thiocyanate. This enzyme may also play a role in thiosulfate oxidation. In anaerobically growing *Thiobacillus denitrificans* strain RT, for instance, rhodanese initiates thiosulfate oxidation by forming sulfite from the sulfone sulfur, which is then oxidized to sulfate. The sulfane sulfur accumulates transiently as elemental sulfur outside the cells, and when the sulfone sulfur is depleted, the sulfane sulfur is rapidly oxidized to sulfate (Schedel and Trüper, 1980). In another strain of *T. denitrificans*, however, thiosulfate reductase rather than rhodanese catalyzes the initial step of thiosulfate oxidation, and both the sulfane and sulfone sulfur are attacked concurrently (Peeters and Aleem, 1970). *T. versutus* (formerly *Thiobacillus* A_2) seems to oxidize thiosulfate to sulfate by a unique pathway (Lu and Kelly, 1983) that involves a thiosulfate multienzyme system that has a periplasmic location (Lu, 1986). No free intermediates appear to be formed from either the sulfane or the sulfone sulfur of thiosulfate.

Pronk et al. (1990) summarized the evidence that supports a model in which *Acidithiobacillus ferrooxidans*, *A. thiooxidans*, and *Acidiphilium acidophilum* oxidize thiosulfate by forming tetrathionate in an initial step:

$$2S_2O_3^{2-} \rightarrow S_4O_6^{2-} + 2e \qquad (19.28)$$

This is followed in the model by a series of hydrolytic and oxidative steps whereby tetrathionate is transformed into sulfate with transient accumulation of intermediary sulfur from sulfane-monosulfinic acids (polythionates). Thiosulfate dehydrogenase from *Aph. acidophilum*, which catalyzes the oxidation of thiosulfate to tetrathionate, was purified and partially characterized by Meulenberg et al. (1993).

19.7.4.1   Disproportionation of Thiosulfate

It has been demonstrated experimentally that some bacteria, like *Desulfovibrio sulfodismutans*, can obtain energy anaerobically by disproportionating thiosulfate into sulfate and sulfide (Bak and Cypionka, 1987; Bak and Pfennig, 1987; Jørgensen, 1990a,b):

$$S_2O_3^{2-} + H_2O \rightarrow SO_4^{2-} + HS^- + H^+ \quad (-19.1\,\text{kcal mol}^{-1}\ \text{or} -79.8\,\text{kJ mol}^{-1}) \quad (19.29)$$

The energy from this reaction enables the organisms to assimilate carbon from a combination of CO_2 and acetate. Energy conservation by thiosulfate disproportionation seems, however, paradoxical if the oxidation state of the sulfane sulfur is –2 and that of the sulfone sulfur is +6, as formerly believed, because no redox reaction would be required to generate a mole of sulfate and sulfide each per mole of thiosulfate. A solution to this paradox has been provided by the report of Vairavamurthy et al. (1993), which demonstrated spectroscopically that the charge density of the sulfane sulfur in thiosulfate is really –1 and that of the sulfone sulfur is +5. Based on this finding, the formation of sulfide and sulfate by disproportionation of thiosulfate would indeed require a redox reaction. Another organism that has been shown to be able to disproportionate thiosulfate is *Desulfotomaculum thermobenzoicum* (Jackson and McInerney, 2000). The addition of acetate to the growth medium stimulated thiosulfate disproportionation by this organism. Thiosulfate disproportionation has also been observed with *Desulfocapsa thiozymogenes* (Janssen et al., 1996).

Desulfovibrio desulfodismutans can also generate useful energy from the disproportionation of sulfite and dithionite to sulfide and sulfate (Bak and Pfennig, 1987). The overall reaction for sulfite disproportionation is

$$4SO_3^{2-} + H^+ \rightarrow 3SO_4^{2-} + HS^- \quad (14.1\,\text{kcal mol}^{-1}\ \text{or} -58.9\,\text{kJ mol}^{-1}) \quad (19.30)$$

and that for dithionite disproportionation is

$$4S_2O_4^{2-} + 4H_2O \rightarrow 5SO_4^{2-} + 3HS^- + 5H^+ \quad (-32.1\,\text{kcal mol}^{-1}\ \text{or} -134.0\,\text{kJ mol}^{-1}) \quad (19.31)$$

D. sulfodismutans can also grow on lactate, ethanol, propanol, and butanol as energy sources and sulfate as terminal electron acceptor, like typical sulfate-reducers, but growth is slower than by disproportionation of partially reduced sulfur compounds. Bak and Pfennig (1987) suggested that from an evolutionary standpoint, *D. sulfodismutans*-type sulfate-reducers could be representative of the progenitors of typical sulfate-reducers.

Perry et al. (1993) suggested that *Shewanella putrefaciens* MR-4, which they isolated from the Black Sea, disproportionates thiosulfate into either sulfide and sulfite or elemental sulfur and sulfite. They never detected any sulfate among the products in these reactions. These disproportionations are, however, endogonic (+7.39 and +3.84 kcal mol^{-1} or +30.98 and 16.10 kJ mol^{-1} at pH 7, 1 atm, and 25°C, respectively). Perry and coworkers suggested that in *S. putrefaciens* MR-4 these reactions must be coupled to exogonic reactions such as carbon oxidation.

Thiosulfate disproportionation seems to play a significant role in the sulfur cycle in marine environments (Jørgensen, 1990a). In Kysing Fjord (Denmark) sediment, thiosulfate was identified as a major intermediate product of anaerobic sulfide oxidation that was simultaneously reduced to sulfide, oxidized to sulfate, and disproportionated to sulfide and sulfate. This occurred at a rapid rate as reflected by a small thiosulfate pool. The metabolic fate of thiosulfate in these experiments was determined by adding differentially labeled ^{35}S-thiosulfate and following the consumption of the thiosulfate and the isotopic distribution in sulfide and sulfate formed from the sulfane and sulfone sulfur atoms of the labeled thiosulfate over time in separate experiments. According to Jørgensen (1990a), the disproportionation reaction can explain the observed large difference in $^{34}S/^{32}S$ in sulfate and sulfide in the sediments. These findings were extended to anoxic sulfur transformations in further experiments with Kysing Fjord sediments and in new experiments with sediments from

Braband Lake, Åarhus Bay, and Aggersund by Elsgaard and Jørgensen (1992). They showed a significant contribution made by thiosulfate disproportionation in anaerobic production of sulfate from sulfide. Addition of nitrate stimulated anoxic oxidation of sulfide to sulfate. Addition of iron in the form of lepidocrocite (FeOOH) caused partial oxidation of sulfide with the formation of pyrite and sulfur and precipitation of iron sulfides.

19.7.5 TETRATHIONATE OXIDATION

Although bacterial oxidation of tetrathionate has been reported, the mechanism of oxidation is still not certain (see Roy and Trudinger, 1970; Kelly, 1982). It may involve disproportionation and hydrolysis reactions. A more detailed scheme was described by Pronk et al. (1990), which was already mentioned above in Section 19.7.4 in connection with thiosulfate oxidation.

19.7.6 COMMON MECHANISM FOR OXIDIZING REDUCED INORGANIC SULFUR COMPOUNDS IN DOMAIN BACTERIA

Friedrich et al. (2001) suggested that the mechanism for oxidizing inorganic reduced sulfur compounds by aerobic and anaerobic sulfur-oxidizing bacteria, including anoxygenic phototrophic bacteria, have certain common features. Their suggestion is based on molecular comparisons of the *Sox* genes and the proteins they encode between those in *Paracoccus pantotrophus* and those in other bacteria capable of oxidizing inorganic reduced sulfur compounds. The *Sox* enzyme system in the archeon *Sulfolobus sulfataricus* appears to differ from that in Bacteria on the basis of genomic analysis.

19.8 AUTOTROPHIC AND MIXOTROPHIC GROWTH ON REDUCED FORMS OF SULFUR

19.8.1 ENERGY COUPLING IN BACTERIAL SULFUR OXIDATION

All evidence to date indicates that to conserve biochemically useful energy, chemosynthetic autotrophic and mixotrophic bacteria that oxidize reduced forms of sulfur feed the reducing power (electrons) into a plasma membrane-bound electron transport system whether oxygen, nitrate, or nitrite is the terminal electron acceptor (Peeters and Aleem, 1970; Sadler and Johnson, 1972; Aminuddin and Nicholas, 1974b; Loya et al., 1982; Lu and Kelly, 1983; Smith and Strohl, 1991; Kelly et al., 1993; also see review by Kelly, 1982). However, the components of the electron transport system in the plasma membrane, that is, the cytochromes, quinones, and nonheme iron proteins, are not identical in all organisms. Whatever the electron transport chain makeup in the plasma membrane, it is the oxidation state of a particular sulfur compound being oxidized, or more exactly the midpoint potential of its redox couple at physiological pH, that determines the entry point into the electron transport chain of the electrons removed during the oxidation of the sulfur compound. Thus, the electrons from elemental sulfur are generally thought to enter the transport chain at the level of a cytochrome bc_1 complex or equivalent. Actually, as pointed out earlier, the first step in the oxidation of sulfur to sulfate can be the formation of sulfite by an oxygenation involving direct interaction with oxygen without involvement of the cytochrome system. Only in the subsequent oxidation of sulfite to sulfate is the electron transport system directly involved starting at the level of the cytochrome bc_1 complex or equivalent. Also, as discussed earlier, sulfite may be oxidized by an AMP-dependent or AMP-independent pathway. In either case, electrons are passed into the electron transport system at the level of a cytochrome bc_1 complex or equivalent. In the AMP-dependent pathway, most of the energy coupling can be assumed to be chemiosmotic, that is, on average 1 or 2 mol of ATP can be formed per electron pair passed to oxygen by the electron transport system, but in addition 0.5 mL of ATP can be formed via substrate-level phosphorylation (Reactions 19.19 and 19.20). By contrast, only 1 or 2 mol of ATP can be formed on average per electron pair passed to oxygen by the AMP-independent pathway.

Chemiosmosis is best explained if it is assumed that the sulfite oxidation half-reaction occurs at the exterior of the plasma membrane (in the periplasm)

$$SO_3^{2-} + H_2O \rightarrow SO_4^{2-} + 2H^+ + 2e \qquad (19.32)$$

and the oxygen reduction half-reaction on the inner surface of the plasma membrane (cytoplasmic side)

$$0.5O_2 + 2H^+ + 2e \rightarrow H_2O \qquad (19.33)$$

In *Thiobacillus versutus*, a thiosulfate-oxidizing, multienzyme system has been located in the periplasm (Lu, 1986).

The pH gradient resulting from sulfite oxidation and any proton pumping associated with electron transport in the plasma membrane together with any electrochemical gradient provide the proton motive force for ATP generation via F_0F_1 ATP synthase. Proton translocation during thiosulfate oxidation has been observed in *Thiobacillus versutus* (Lu and Kelly, 1988). Involvement of energy coupling via chemiosmosis is also indicated for *T. neapolitanus* using thiosulfate as energy source. The evidence for this is (1) inhibition of CO_2 uptake by the uncouplers carbonyl cyanide *m*-chlorophenylhydrazone (CCCP) and carbonylcyanide *p*-trifluoromethoxyphenylhydrazone (FCCP) and (2) an increase in transmembrane electrochemical potential and CO_2 uptake in response to nigericin (Holthuijzen et al., 1987).

19.8.2 REDUCED FORMS OF SULFUR AS SOURCES OF REDUCING POWER FOR CO₂ FIXATION BY AUTOTROPHS

19.8.2.1 Chemosynthetic Autotrophs

Reduced sulfur is not only an energy source but also a source of reducing power for chemosynthetic autotrophs that oxidize it. Because the midpoint potential for pyridine nucleotides is much reduced compared to that for reduced sulfur compounds that could serve as potential electron donors, *reverse electron transport* from the electron-donating sulfur substrate to pyridine nucleotide is required (see Chapter 6). Electrons must travel up the electron transport chain, that is, against the redox gradient, to NADP with consumption of ATP providing the needed energy. This applies to both aerobes and anaerobes that use nitrate as terminal electron acceptor (denitrifiers).

19.8.2.2 Photosynthetic Autotrophs

In purple sulfur and nonsulfur bacteria, reverse electron transport, a light-independent sequence, is used to generate reduced pyridine nucleotide (NADPH) using ATP from photophosphorylation to provide the needed energy. In green sulfur bacteria as well as cyanobacteria, light-energized-electron transport is used to generate NADPH (Stanier et al., 1986) (see also discussion in Chapter 6).

19.8.3 CO₂ FIXATION BY AUTOTROPHS

19.8.3.1 Chemosynthetic Autotrophs

Insofar as studied, thiobacilli (domain Bacteria) generally fix CO_2 by the Calvin–Benson–Bassham cycle (see Chapter 6), that is, by means of the ribulose 1.5-bisphosphate carboxylase pathway. In at least some thiobacilli, the enzyme is detected in both the cytosol and the cytoplasmic polyhedral bodies called carboxysomes (Shively et al., 1973). The carboxysomes appear to contain no other enzymes and may represent a means of regulating the level of carboxylase activity in the cytosol (Beudecker et al., 1980, 1981; Holthuijzen et al., 1986a,b). *Sulfolobus* (domain Archaea) assimilates CO_2 via a reverse, that is, a reductive tricarboxylic acid cycle (see Brock and Madigan, 1988), like green sulfur bacteria (domain Bacteria) (see Chapter 6).

19.8.3.2 Photosynthetic Autotrophs

Purple sulfur bacteria, purple nonsulfur bacteria, and those cyanobacteria capable of anoxygenic photosynthesis fix CO_2 by the Calvin–Benson–Bassham cycle, that is, via the ribulose 1,5-bisphosphate carboxylate pathway, when growing photoautotrophically on reduced sulfur (see Chapter 6). Green sulfur bacteria, however, use a reverse, that is, a reductive tricarboxylic acid cycle mechanism (Stanier et al., 1986). However, *Chloroflexus aurantiacus*, a filamentous green nonsulfur bacterium, uses a 3-hydroxypropionate cycle (see discussion in Chapter 6).

19.8.4 Mixotrophy

19.8.4.1 Free-Living Bacteria

Some sulfur-oxidizing chemosynthetic autotrophs can also grow mixotrophically (e.g., Smith et al., 1980). Among oxidizers of reduced sulfur, *Thiobacillus versutus* is a good model for studying autotrophy, mixotrophy, and heterotrophy. It can even grow anaerobically on nitrate (e.g., Wood and Kelly, 1983; Claassen et al., 1987). The organism can use each of these forms of metabolism depending on medium composition (see review by Kelly, 1982). Another more recently studied example is *Acidiphilium acidophilum* growing on tetrathionate (Mason and Kelly, 1988).

Thiobacillus intermedius, which grows poorly as an autotroph in a thiosulfate–mineral salts medium, grows well in this medium if it is supplemented with yeast extract, glucose, glutamate, or other organic additive (London, 1963; London and Rittenberg, 1966). The organic matter seems to repress the CO_2-assimilating mechanism in this organism but not its ability to generate energy from thiosulfate oxidation (London and Rittenberg, 1966). *T. intermedius* also grows well heterotrophically in a medium containing glucose with yeast extract or glutathione but not in a glucose–mineral salts medium without thiosulfate (London and Rittenberg, 1966). It needs thiosulfate or organic sulfur compounds because it cannot assimilate sulfate (Smith and Rittenberg, 1974). A nutritionally similar organism is *T. organoparus*, an acidophilic, facultatively heterotrophic bacterium, which resembles *Acidiphilium acidophilum* (Wood and Kelly, 1978). *T. organoparus* was first isolated from acid mine water in copper deposits in Alaverdi (former Armenian SSR). It was found to grow autotrophically and mixotrophically with reduced sulfur compounds (Markosyan, 1973).

Thiobacillus perometabolis cannot grow at all autotrophically in thiosulfate–mineral salts medium but requires the addition of yeast extract, casein hydrolysate, or an appropriate single organic compound to utilize thiosulfate as an energy source (London and Rittenberg, 1967). Growth on yeast extract or casein hydrolysate is much less luxuriant in the absence of thiosulfate.

Some marine pseudomonads, which are ordinarily considered to grow heterotrophically, have been shown to grow mixotrophically on reduced sulfur compounds (Tuttle et al., 1974). Growth of the cultures on yeast extract was stimulated by the addition of thiosulfate. The bacteria oxidized it to tetrathionate. The growth stimulation by thiosulfate oxidation manifested itself in increased organic carbon assimilation. A number of other heterotrophic bacteria, actinomycetes, and filamentous fungi are also able to oxidize thiosulfate to tetrathionate (Trautwein, 1921; Starkey, 1934; Guittonneau and Keiling, 1927), but whether the growth of any of these is enhanced by this oxidation is unknown at this time. Even if it is not, these organisms may play a role in promoting the sulfur cycle in soil (Vishniac and Santer, 1957).

19.8.5 Unusual Consortia

Very unusual consortia involving invertebrates and autotrophic sulfide-oxidizing bacteria have been discovered in submarine hydrothermal vent communities (Jannasch, 1984; Jannasch and Mottl, 1985; Jannasch and Taylor, 1984). Vestimentiferan tube worms (*Riftia pachyptila*), which grow around the submarine vents, especially white smokers, lack a mouth and digestive tract, and harbor special organelles in their body cavity called collectively a trophosome. These organelles when viewed in section under a transmission electron microscope are seen to consist of tightly packed bacteria (Cavanaugh et al., 1981). Metabolic evidence indicates that these are chemosynthetic, autotrophic bacteria (Felbeck, 1981; Felbeck et al., 1981; Rau, 1981; Williams et al., 1988). Some bacteria have been cultured from the trophosomes, but whether any of them are the important symbionts of the worm remains to be established (Jannasch and Taylor, 1984). The bacteria in the trophosomes appear to be autotrophic sulfur-oxidizing bacteria that share the carbon they fix with the worm. The worm absorbs sulfide (HS^-) and oxygen from the water through a special organ at its anterior end consisting of a tentacular plume attached to a central supporting obturaculum (Jones, 1981; Goffredi et al., 1997), and transmits these via its circulatory system to the trophosome. The blood of the worm contains hemoglobin for reversible binding of oxygen and another special protein for reversible binding of sulfide. The latter protein prevents reaction of sulfide with the hemoglobin and its consequent destruction (Arp and Childress, 1983; Powell and Somero, 1983). The bound hydrogen sulfide and oxygen are released at the site of the trophosome.

Somewhat less intimate consortia around hydrothermal vents are formed by giant clams and mussels (*Mollusca*) with autotrophic sulfide-oxidizing bacteria. The bacteria in these instances reside not in the gut of the animals but on their gills (see Jannasch and Taylor, 1984, for discussion; also Rau and Hedges, 1979). These looser consortia involving autotrophic sulfide-oxidizing bacteria and mollusks are not restricted to hydrothermal vent communities but also occur in shallow water environments rich in hydrogen sulfide (Cavanaugh, 1983).

19.9 ANAEROBIC RESPIRATION USING OXIDIZED FORMS OF SULFUR AS TERMINAL ELECTRON ACCEPTORS

19.9.1 REDUCTION OF FULLY OR PARTIALLY OXIDIZED SULFUR

Various forms of oxidized sulfur can serve as terminal electron acceptor in the respiration of some bacteria under anaerobic conditions. The sulfur compounds include sulfate, thiosulfate, and elemental sulfur, among others.

19.9.2 BIOCHEMISTRY OF DISSIMILATORY SULFATE REDUCTION

A variety of strictly anaerobic bacteria respire using sulfate as terminal electron acceptor. Many are taxonomically quite unrelated and include members of the domains Bacteria and Archaea (see Section 19.6.2). Insofar as is now known, the mechanism by which they reduce sulfate follows a very similar but not necessarily identical pattern in all. As presently understood, the enzymatic reduction of sulfate requires an initial activation by ATP to form adenine phosphatosulfate and pyrophosphate:

$$SO_4^{2-} + ATP \xrightarrow{\text{ATP sulfurylase}} APS + PP_i \tag{19.34}$$

In members of the genus *Desulfovibrio*, the pyrophosphate (PP_i) is hydrolyzed to inorganic phosphate (P_i), which helps to pull the reaction in the direction of APS:

$$PP_i + H_2O \xrightarrow{\text{pyrophosphatase}} 2P_i \tag{19.35}$$

The energy in the anhydride bond of pyrophosphate is thus not available to *Desulfovibrio*. By contrast, this energy is conserved by members of the genus *Desulfotomaculum*. They do not hydrolyze the pyrophosphate but use it as a substitute for ATP (Liu et al., 1982). This also has the effect of pulling Reaction 19.34 in the direction of APS.

Unlike in assimilatory sulfate reduction, APS, once formed, is reduced directly to sulfite and adenylic acid (AMP):

$$APS + 2e \xrightarrow{\text{APS reductase}} SO_3^{2-} + AMP \tag{19.36}$$

The APS reductase, unlike PAPS reductase, does not require NADP as a cofactor but, like PAPS reductase, contains bound flavine adenine dinucleotide (FAD) and iron (for further discussion see, for instance, Peck, 1993).

The subsequent details in the reduction of sulfite to sulfide have not been fully agreed upon. One line of experimental evidence suggests a multistep process involving trithionate and thiosulfate as intermediates (Kobiyashi et al., 1969; modified by Akagi et al., 1974; Drake and Akagi, 1978):

$$3HSO_3^{2-} + 2H^+ + 2e \xrightarrow{\text{bisulfite reductase}} S_3O_6^{2-} + 2H_2O + OH^- \tag{19.37}$$

$$S_3O_6^{2-} + H^+ + 2e \xrightarrow{\text{trithionate reductase}} S_2O_3^{2-} + HSO_3^- \tag{19.38}$$

$$S_2O_3^{2-} + 2H^+ + 2e \xrightarrow{\text{thiosulfate reductase}} HS^- + HSO_3^- \tag{19.39}$$

In most *Desulfovibrio* cultures, the bisulfite reductase seems to be identical to desulfoviridin (Kobiyashi et al., 1972; Lee and Peck, 1971). In *D. desulfuricans* strain Essex 6, both a soluble and a membrane-bound activity amounted to 90% of the total. Unlike the soluble activity, the membrane-bound activity could be coupled to hydrogenase and cytochrome c_3. Sulfide was the main product of reduction with this enzyme (Steuber et al., 1994). In *D. desulfuricans* strain Norway 4, which lacks desulfoviridin, desulforubidin appears to be the bisulfite reductase (Lee et al., 1973), and in *D. thermophilus* it has been identified as desulfofuscidin (Fauque et al., 1990). In *Desulfotomaculum nigrificans*, a carbon monoxide–binding pigment, called P582 by Trudinger (1970), accounts for bisulfite reducing activity, which according to Akagi et al. (1974), leads to the formation of trithionate, with thiosulfate and sulfide as endogenous side products. An F_0F_1-ATP synthase involved in energy conservation from sulfate reduction by *D. vulgaris* has been identified (Ozawa et al., 2000).

Inducible sulfite reduction has also been observed with *Clostridium pasteurianum*, an anaerobic bacterium that is not a dissimilatory sulfate-reducer. It can reduce sulfite to sulfide. In the absence of added selenite, whole cells do not release detectable amounts of trithionate or thiosulfate when reducing sulfite, but in the presence of selenite they do. Selenite was found to inhibit thiosulfate reductase but not trithionate reductase in whole cells, but inhibited both in cell extracts (Harrison et al., 1980). A purified sulfite reductase from *C. pasteurianum* produced sulfide from sulfite. It was also able to reduce NH_2OH, SeO_3^{2-}, and NO_2^- but did not reduce trithionate or thiosulfate (Harrison et al., 1984). Several physical and chemical properties of this enzyme differed from those of bisulfite reductases in sulfate-reducers. Its role in *C. pasteurianum* may be in detoxification when excess bisulfite is present (Harrison et al., 1984). Peck (1993) referred to the enzymes involved in the transformation of bisulfite to sulfide collectively as bisulfite reductase. Distinct sulfite reductase, trithionite reductase, and thiosulfate reductase were also identified by Peck and LeGall (1982). However, at the time they did not visualize a major role for these enzymes in sulfite reduction to sulfide.

Chambers and Trudinger (1975) questioned whether the trithionate pathway of sulfite reduction is the major pathway of *Desulfovibrio* spp. They found that results of experiments with isotopically labeled $^{35}SO_4^{2-}$, $^{35}SSO_3^{2-}$, and $S^{35}SO_3^{2-}$ could not be reconciled with the trithionate pathway but were more consistent with a pathway involving the assimilatory kind of sulfite reductase. Their view was supported by Peck and LeGall (1982). However, Vainshtein et al. (1981), after reinvestigating this problem, concluded that the findings of Chambers and Trudinger (1975) were the result of using a heavy cell concentration at a limiting bisulfite concentration that did not permit transient accumulation of thiosulfate. LeGall and Fauque (as cited by Fauque et al., 1991) concluded in 1988 that a direct pathway from sulfite to sulfide is used by *Desulfovibrio* and the trithionate pathway by *Desulfotomaculum*. It appears that our understanding of the details of sulfite reduction to sulfide by sulfate-reducing bacteria remains incomplete at this time.

19.9.3 SULFUR ISOTOPE FRACTIONATION

Sulfate-reducing bacteria can distinguish between ^{32}S and ^{34}S isotopes of sulfur; that is, they can bring about isotope fractionation (Harrison and Thode, 1957; Jones and Starkey, 1957). Both the isotopes are stable. The ^{32}S isotope of sulfur is the most abundant (average 95.1%) and the ^{34}S isotope is the next most abundant (average 4.2%). The $^{32}S/^{34}S$ ratio of unfractionated natural sulfur ranges between 21.3 and 23.2. Meteoritic sulfur has a $^{32}S/^{34}S$ ratio of 22.22. Because this ratio appears to be relatively constant from sample to sample, it is often used as a reference standard against which to compare sulfur isotope ratios of other materials that may be either enriched or depleted in ^{34}S. Isotope fractionation by microbes is the result of preferential attack of ^{32}S over ^{34}S as a consequence of which the sulfur in the product of the attack becomes enriched in ^{32}S. Although early work based on observations with *Desulfovibrio desulfuricans* suggested that sulfur isotope fractionation by sulfate-reducers proceeded more readily under conditions of slow growth (Tables 19.5 and 19.6) (Jones and Starkey, 1957, 1962), recent results, albeit with a variety of sulfate-reducers that did not include *D. desulfuricans*, did not confirm this observation (Detmers et al., 2001). The nature of the electron donor may affect the degree of isotope fractionation (Kemp and Thode, 1968), but the temperature range for growth does not (Böttcher et al., 1999).

The extent of isotope fractionation is calculated in terms of $\delta^{34}S$ values expressed in parts per thousand (‰):

$$\delta^{34}S = \frac{^{34}S/^{32}S \text{ sample} - {}^{34}S/^{32}S \text{ meteoritic standard}}{^{34}S/^{32}S \text{ meteoritic standard}} \times 1000 \qquad (19.40)$$

TABLE 19.5
Sulfide Production and Fractionation of Stable Isotopes of Sulfur by *Desulfovibrio desulfuricans* Cultivated at 28°C (Rapid Growth)

Sample Number	Incubation Period (h)[a]	Sulfide S in PbS (mg)	Sulfate Reduced[b] (%)	Number of Isotope Determinations	$\delta^{34}S$
1	44	996	6.3	4	−5.4
2	8	2168	20.0	4	−4.9
3	4	1931	32.2	2	−3.1
4	5	1448	41.4	2	−3.1
5	4	1394	50.2	2	5.4
6	3	1248	58.1	2	−5.4
7	9	317	60.1	2	−6.7
8	14	191	61.3	2	−8.9
9	41	103	62.0	2	−9.8
10	68	115	62.7	2	−12.9
11	59.5	387	65.2	2	−7.2
12	25.5	901	70.9	2	−3.1
13	7	615	74.8	2	−0.5
14	6	474	77.8	2	+0.9
15	24	856	83.2	2	+0.5
16	43	106	83.9	2	−4.9

[a] Periods were calculated from the time sulfide first appeared in the culture substrate. This was 60 h after the medium was inoculated.

[b] The initial sulfate S in the substrate was 3943 ppm.

Source: Adapted from Jones GE, Starkey RL, *Appl. Microbiol.*, 5, 111–118, 1957. With permission.

TABLE 19.6
Sulfide Production and Fractionation of Stable Isotopes of Sulfur by *Desulfovibrio desulfuricans* Cultivated at Low Temperatures (Slow Growth)

Sample Number	Incubation Period (h)[a]	Sulfide S in PbS (mg)	Sulfate Reduced (%)	$\delta^{34}S$
17	200	21.2	4.4	−22.1
18	142	65.8	4.5	−25.9
19	120	112.7	8.3	−25.9
20	120	167.8	10.0	−24.2
21	96	174.6	13.1	−24.2
22	120	180.8	16.0	−22.9
23	144	134.5	16.8	−21.6
24	120	102.0	18.4	−19.5

[a] Periods were calculated from the time sulfide first appeared in the culture substrate; this was 18 h after the medium was inoculated.

Source: Adapted from Jones GE, Starkey RL, *Appl. Microbiol.*, 5, 111–118, 1957. With permission.

Harrison and Thode (1957) (see also Hoefs, 1997) proposed that the S–O bond breakage was the rate-controlling reaction in bacterial sulfate reduction (i.e., reduction of APS to sulfite and AMP; Reaction 19.36) that is responsible for the isotope fractionation phenomenon.

Dissimilatory sulfate reduction is not the only process that may lead to sulfur isotope fractionation. Sulfite reduction by *Desulfovibrio*, *Saccharomyces cerevisiae* (Kaplan and Rittenberg, 1962, p. 81), and *Clostridium pasteurianum* (Laishley and Krouse, 1978), sulfide release from cysteine by *Proteus vulgaris* (Kaplan and Rittenberg, 1962, 1964), and assimilatory sulfate reduction by *Escherichia coli* and *S. cerevisiae* (Kaplan and Rittenberg, 1962) can also lead to sulfur isotope fractionation.

Sulfur isotope fractionation has also been observed when thiosulfate is reduced by *Desulfovibrio desulfuricans* (Smock et al., 1998). In this instance it was found that the depletion in ^{34}S of the H_2S formed was 10‰ with respect to the total sulfur in thiosulfate. However, sulfane (outer) and sulfone (inner) sulfur of thiosulfate contributed differently to the overall fractionation, the sulfone sulfur contributing 15.4‰ and the sulfane sulfur 5.0‰. Although S–O bond cleavage of sulfone sulfur is thought to have contributed significantly to the observed fractionation, Smock et al. (1998) suggested that other factors such as thiosulfate uptake, sulfonate activation, intracellular concentration, or the physiological state of the cells could influence the observed isotope effect.

Isotopic analysis of sulfur minerals in nature has helped in deciding whether biogenesis was involved in their accumulation. Any given deposit must, however, be sampled at a number of locations, because isotope enrichment values ($\delta^{34}S$) generally fall in a narrow or a wide range. Abiogenic $\delta^{34}S$ values generally fall in a narrow range and usually have a positive (+) sign, whereas biogenic values tend to fall in a wide range and have a negative (−) sign.

19.9.4 Reduction of Elemental Sulfur

Elemental sulfur can be used anaerobically as terminal electron acceptor in bacterial respiration or as an electron sink for disposal of excess reducing power. The product of S^0 reduction in either case is sulfide. Polysulfide may be an intermediate in respiration (Schauder and Müller, 1993; Fauque et al., 1991). Some members of both Bacteria and Archaea can respire on sulfur (Schauder and Kröger, 1993; Ma et al., 2000; Bonch-Osmolovskaya, 1994). Examples of Bacteria include *Desulfuromonas acetoxidans*, *Desulfovibrio gigas*, and some other sulfate-reducers (Pfennig and Biebl, 1976; Biebl and Pfennig, 1977; Fauque et al., 1991). Examples of Archaea include *Pyrococcus furiosus* (Schicho et al., 1993), *Pyrodictium* (Stetter, 1983, 1985), *Pyrobaculum* (Huber et al., 1987), and *Acidianus*.

Organisms that use S^0 reduction as an electron sink include *Thermotoga* spp. in the domain Bacteria, and *Thermoproteus, Desulfurococcus,* and *Thermofilum* in the domain Archaea (Jannasch et al., 1988a,b). These organisms are fermenters that dispose in this way excess of H_2 they produce, which would otherwise inhibit their growth (Bonch-Osmolovskaya et al., 1990; Janssen and Morgan, 1992; Bonch-Osmolovskaya, 1994). It is possible that these organisms can salvage some energy in the disposal of H_2 (e.g., Schicho et al., 1993). Some fungi, for example, *Rhodotorula* and *Trichosporon* (Ehrlich and Fox, 1967), can also reduce sulfur to H_2S with glucose as electron donor. This is probably not a form of respiration.

The energy source for the sulfur-respiring Archaea is sometimes hydrogen and methane but more often organic molecules such as glucose and small peptides, whereas that for Bacteria may be simple organic compounds (e.g., ethanol, acetate, propanol) or more complex organics. In the case of *Desulfuromonas acetoxidans* (domain Bacteria), an electron transport pathway including cytochromes appears to be involved (Pfennig and Biebl, 1976). When acetate is used as an energy source, oxidation proceeds anaerobically by way of the tricarboxylic acid cycle (see Chapter 6). The oxalacetate required for initiation of the cycle is formed by carboxylation of pyruvate, which arises from carboxylation of acetate (Gebhardt et al., 1985). Energy is gained in the oxidation of isocitrate and 2-ketoglutarate. Membrane preparations were shown to oxidize succinate using S^0 or NAD as electron acceptor by an ATP-dependent reaction. Similar membrane preparations reduced fumarate to succinate with H_2S as electron donor by an ADP-independent reaction. Menaquinone mediated hydrogen transfer. Protonophores and uncouplers of phosphorylation inhibited reduction of S^0 but not fumarate. The compound 2-*n*-nonyl-4-hydroxyquinoline *N*-oxide inhibited electron transport to S^0 and fumarate. Together these observations support the notion that S^0 reduction in *D. acetoxidans* involves a membrane-bound electron transport system and the ATP is formed chemiosmotically, that is, by oxidative phosphorylation, when growing on acetate (Paulsen et al., 1986).

The hyperthermophilic Archaea, *Thermoproteus tenax* and *Pyrobaculum islandicum*, growing on S^0 and glucose or casamino acids in the case of the former and on peptone in the case of the latter, mineralized their carbon substrates completely. They produced CO_2 and H_2S in a ratio of 1:2 using the tricarboxylic acid cycle (Selig and Schönheit, 1994).

19.9.5 REDUCTION OF THIOSULFATE

Growth and growth yield of some members of the anaerobic and thermophilic and hyperthermophilic *Thermotogales* were shown to be stimulated in the presence of thiosulfate (Ravot et al., 1995). The test organisms included *Fervidobacterium islandicum, Thermosipho africanus, Thermotoga maritime, T. neapolitana,* and *Thermotoga* sp. SERB 2665. The last named was isolated from an oil field. All reduced thiosulfate to sulfide. The *Thermotogales* in this group are able to ferment glucose among various energy-yielding substrates. Thiosulfate, like sulfur (see, e.g., Janssen and Morgan, 1992), appears to serve as an electron sink by suppressing H_2 accumulation in the fermentation of glucose, for instance. This accumulation has an inhibitory effect on the growth of these organisms. The biochemical mechanism by which they reduce thiosulfate remains to be elucidated. *Pyrobaculum islandicum* is able to mineralize peptone by way of the tricarboxylic acid cycle, using thiosulfate as terminal electron acceptor, producing CO_2 and H_2S in a ratio of 1:1 (Selig and Schönheit, 1994).

19.9.6 TERMINAL ELECTRON ACCEPTORS OTHER THAN SULFATE, SULFITE, THIOSULFATE, OR SULFUR

A few sulfate-reducers can grow with Fe(III) (Coleman et al., 1993; Lovley et al., 1993), nitrate, nitrite (McCready et al., 1983; Keith and Herbert, 1983; Seitz and Cypionka, 1986), fumarate (Miller and Wakerley, 1966), or, in the case of *Desulfomonile tiedje*, chloroaromatics (DeWeerd et al., 1990, 1991) as terminal electron acceptors. A few strains of *Desulfovibrio* can even grow on pyruvate or fumarate without an external terminal electron acceptor by generating H_2 as one of the metabolic end products (Postgate, 1952, 1963). *Desulfovibrio gigas* and a few strains of *D. desulfuricans* grow on fumarate by disproportionating it. They reduce a portion of the fumarate to succinate and oxidize the remainder to malate and acetate (Miller and Wakerley, 1966).

When Fe(III) serves as terminal electron acceptor, it may be reduced to $FeCO_3$ (siderite) or Fe^{2+}. When NO_3^- and NO_2^- are the terminal electron acceptors, they are reduced to ammonia (nitrate ammonification). When fumarate is the terminal electron acceptor, it is reduced to succinate. When a chloroaromatic compound such as 3-chlorobenzoate is the terminal electron acceptor, it is reductively dechlorinated to benzoate and chloride. The number of different sulfate-reducing bacteria that are able to substitute any of the terminal electron acceptors for sulfate has yet to be systematically explored. The fact that some sulfur reducers can avail themselves of such *substitute* terminal electron acceptors may explain why the presence of such organisms can be detected in environments, like most soils, in which the natural sulfate, sulfite, thiosulfate, or sulfur concentration is very low.

19.9.7 OXYGEN TOLERANCE OF SULFATE-REDUCERS

In general, sulfate-reducers are strict anaerobes, yet they have shown limited oxygen tolerance (Wall et al., 1990; Abdollahi and Wimpenny, 1990; Marshall et al., 1993). Indeed, *Desulfovibrio desulfuricans*, *D. vulgaris*, *D. desulfodismutans*, *Desulfobacterium autotrophicum*, *Desulfolobus propionicus*, and *Desulfococcus multivorans* have shown an ability to use oxygen as terminal electron acceptor, that is, to respire microaerophilically (<10 μM dissolved O_2) without being able to grow under these conditions (Dilling and Cypionka, 1990; Baumgartner et al., 2001).

Some evidence has been presented in support of aerobic sulfate reduction by bacteria, none of which have been obtained in pure culture until relatively recently (Canfield and Des Marais, 1991; Jørgensen and Bak, 1991; Fründ and Cohen, 1992). A chemostat study of a coculture of *Desulfovibrio oxyclinae* and *Marinobacter* strain MB isolated from a mat from Solar Lake in the Sinai Peninsula, showed *D. oxyclinae* is able to grow slowly on lactate in the presence of air and the concurrent absence of sulfate or thiosulfate. The lactate is oxidized to acetate by *D. oxyclinae* (Krekeler et al., 1997; Sigalevich and Cohen, 2000; Sigalevich et al., 2000a). *Marinobacter* strain MB is a facultatively aerobic heterotroph. When grown on lactate in the presence of sulfate in a chemostat supplied with oxygen after an initial anaerobic growth phase, a pure culture of *D. oxyclinae* tended to form clumps after ~149 h of exposure to oxygen (Sigalevich et al., 2000b). Such clumps were not formed in coculture with *Marinobacter* strain MB (Sigalevich et al., 2000b). The clumping may represent a defense mechanism against exposure to oxygen for sulfate-reducing bacteria in general because the interior of clumps >3 μm in size will become anoxic.

19.10 AUTOTROPHY, MIXOTROPHY, AND HETEROTROPHY AMONG SULFATE-REDUCING BACTERIA

19.10.1 AUTOTROPHY

Although the ability of *Desulfovibrio desulfuricans* to grow autotrophically with hydrogen (H_2) as an energy source had been previously suggested, experiments by Mechalas and Rittenberg (1960) failed to demonstrate it. Seitz and Cypionka (1986), however, obtained autotrophic growth of *D. desulfuricans* strain Essex 6 with hydrogen, but the growth yield was less when sulfate was the terminal electron acceptor. Better yields were obtained with nitrate or nitrite as terminal electron acceptor, presumably because the latter two acceptors did not need to be activated by ATP, which is a requirement for sulfate reduction. Nitrate and nitrite are reduced to ammonia by *Desulfovibrio* (McCready et al., 1983; Mitchell et al., 1986; Keith and Herbert, 1983). *Desulfotomaculum orientis* also has the ability to grow autotrophically with hydrogen as energy source and sulfate, thiosulfate, or sulfite as terminal electron acceptors (Cypionka and Pfennig, 1986). Under optimal conditions, better growth yields were obtained with this organism than had been reported for *D. desulfuricans* (12.4 versus 9.4 g of dry cell mass per mole of sulfate reduced). This may be explainable on the basis that *Desulfotomaculum* can utilize inorganic pyrophosphate generated in sulfate activation as an energy source whereas *Desulfovibrio* cannot. *Desulfotomaculum orientis* gave better growth yields when thiosulfate or sulfite was the terminal electron acceptor than when sulfate was. The organism excreted acetate that was formed as part of its CO_2 fixation process (Cypionka and Pfennig, 1986). The acetate may have been formed via the activated acetate pathway in which it is formed directly from two molecules of CO_2 as is the case in methanogens and homoacetogens (see Chapters 6 and 22), and as has now been shown to occur in *D. baarsii*, which can also grow with hydrogen and sulfate (Jansen et al., 1984) and in *Desulfobacterium autotrophicum* (Schauder et al., 1989). *Desulfobacter hydrogenophilus*, by contrast, assimilates CO_2 by a reductive tricarboxylic acid cycle when growing autotrophically with H_2 as energy source and sulfate as terminal electron acceptor (Schauder et al., 1987). Other sulfate-reducers that are able to grow autotrophically on hydrogen as energy source and sulfate as terminal electron acceptor include *Desulfonema limicola*, *Desulfonema ishimotoi*, *Desulfosarcina variabilis* (Pfennig et al., 1981; Fukui et al., 1999), and *Desulfobacterium autotrophicum* (Schauder et al., 1989).

19.10.2 MIXOTROPHY

Desulfovibrio desulfuricans has been shown to grow mixotrophically with any one of several different compounds as sole energy source, including hydrogen, formate, and isobutanol. The carbon in the organic energy sources was not assimilated. It was derived instead from substances as complex as yeast extract or as simple as acetate or acetate and CO_2. Sulfate was the terminal electron acceptor in all instances (Mechalas and Rittenberg, 1960; Sorokin, 1966a–d, Badziong and Thauer, 1978; Badziong et al., 1978; Brandis and Thauer, 1981). A strain of *D. desulfuricans* used by Sorokin (1966a) was able to derive as much as 50% of its carbon from CO_2 when it grew on hydrogen as energy source and acetate and CO_2 as carbon source, whereas on lactate and CO_2 it derived only 30% of its carbon from CO_2. Badziong et al. (1978), using a different strain of *Desulfovibrio*, found that 30% of its carbon was derived from CO_2 when it grew on hydrogen and acetate and CO_2.

Member of some other genera of sulfate-reducing bacteria can also grow mixotrophically on hydrogen and acetate and CO_2 (Pfennig et al., 1981). In all the instances, ATP is generated chemiosmotically from hydrogen oxidation in the periplasm.

19.10.3 HETEROTROPHY

The great majority of autotrophic sulfate-reducers can grow heterotrophically with sulfate as terminal electron acceptor. In general, sulfate-reducers specialize with respect to the carbon or energy source they can utilize (see Section 19.6) (see also Pfennig et al., 1981). When acetate serves as energy source, it may be completely oxidized anaerobically via the tricarboxylic acid, as in the case of *Desulfobacter postgatei* (Brandis-Heep et al., 1983; Gebhardt et al., 1983; Höller et al., 1987). More commonly, however, sulfate-reducers oxidize acetate by reversal of the active-acetate-synthesis pathway (Schauder et al., 1986). Assimilation of acetate most likely involves carboxylation to pyruvate. ATP synthesis in the heterotrophic mode of sulfate reduction, insofar as it is understood, is mainly by oxidative phosphorylation (chemiosmotically) involving transfer of hydrogen abstracted from an organic substrate into the periplasm followed by its oxidation (Odom and Peck, 1981; but see also Odom and Wall, 1987; Kramer et al., 1987). In the case of lactate, this hydrogen transfer from the cytoplasm to the periplasm across the plasma membrane appears to be energy-driven (Pankhania et al., 1988). Some ATP may be formed by substrate-level phosphorylation.

19.11   BIODEPOSITION OF NATIVE SULFUR

19.11.1   TYPES OF DEPOSITS

Deposits of elemental sulfur of biogenic origin, and in most cases of abiogenic origin, have resulted from the oxidation of H_2S:

$$H_2S + 0.5O_2 \rightarrow S^0 + H_2O \tag{19.41}$$

In some fumaroles, sulfur may also be formed abiogenically through the interaction of H_2S with SO_2:

$$2H_2S + SO_2 \rightarrow 3S^0 + 2H_2O \tag{19.42}$$

Most known native sulfur deposits are not of volcanic origin. Indeed, only 5% of the known reserves are the result of volcanism (Ivanov, 1968, p. 139). Biogenic sulfur accumulation in sedimentary deposits may originate syngenetically or epigenetically. In syngenetic formation, sulfur is deposited contemporaneously with the enclosing host rock or sediment during its sedimentation. In epigenetic formation, sulfur is laid down in cracks and fissures of preformed host rock. This sulfur may originate from a diagenetic process in which a sulfate component of the host rock is converted to sulfur, or it may involve the conversion of dissolved sulfate or sulfide to sulfur in a solution percolating through cracks and fissures of host rock. Syngenetic sulfur deposits are generally formed in limnetic environments, whereas epigenetic sulfur deposits tend to form in terrestrial environments. If the source of elemental sulfur is sulfate, the microbial transformation to sulfur is a two-stage process. The first stage involves dissimilatory sulfate reduction to sulfide (elemental sulfur is not an intermediate in the process), and the second stage involves oxidation of the sulfide to elemental sulfur under limited oxygen availability or anaerobically.

19.11.2   EXAMPLES OF SYNGENETIC SULFUR DEPOSITION

19.11.2.1   Cyrenaican Lakes, Libya, North Africa

A typical example of contemporaneous syngenetic sulfur deposition is found in the sediments of the Cyrenaican lakes Ain ez Zauia, Ain el Rabaiba, and Ain el Braghi. The origin of the sulfur in these lakes was first studied by Butlin and Postgate (1952) and Butlin (1953). The extensive native sulfur in the lake sediments makes up as much as half of the silt. The lake waters have a strong odor of hydrogen sulfide and are opalescent, owing to a fine suspension of sulfur crystals. A fourth lake in the general area, called Ain amm el Gelud, also contains sulfuretted water but shows no evidence of sulfur in the sediments. Ain ez Zauia was the most thoroughly studied by Butlin and Postgate. It is made up of two adjacent basins, 55×30 and 90×70 m in expanse, respectively, and no deeper than 1.5 m. Other characteristics of the lake are summarized in Tables 19.7 and 19.8. The water in the lake is introduced by warm springs (Butlin, 1953). The border of Ain ez Zauia as well as those of the other two lakes with sulfur deposits featured a characteristic red carpetlike gelatinous material that extended several yards into the shallow water in some places. The underside of this red gelatinous material harbored a green and black material. Some of the red material was found floating in the water in the form of red bulbous formations. The red material was a massive growth of the photosynthetic purple sulfur bacterium *Chromatium*, and the green material consisted of a growth of the green photosynthetic sulfur bacterium *Chlorobium*. Many sulfate-reducing bacteria were also detected in the lakes. From these qualitative observations, Butlin and Postgate (1952) inferred that sulfate-reducers were responsible for producing hydrogen sulfide from the sulfate in the water, using as carbon and energy source some of the organic carbon produced by the photoautotrophic bacteria. Their model can be visualized as a cycle in which the photosynthetic bacteria oxidize the hydrogen sulfide produced by the sulfate-reducing bacteria to elemental sulfur while, at the same time, assimilating CO_2 photosynthetically in the process. The sulfate-reducing bacteria, in turn, use some of the fixed carbon produced by the photosynthetic bacteria to reduce sulfate in the lake to sulfide.

TABLE 19.7
Physical Characteristics of Lake Ain ez Zauni

Surface area	7950 m^2
Maximum depth	1.5 m
Surface temperature	30°C
Bottom temperature	32°C
Air temperature	16°C
Sulfur production per year	100 t

Source: Ivanov MV, *Microbiological Processes in the Formation of Sulfur Deposits*, Israel Program for Scientific Translations, U.S. Department of Agriculture and National Science Foundation, Washington, DC, 1968.

TABLE 19.8
Chemical Composition of the Waters of Lake Ain ez Zauni

H$_2$S in surface water	15–20 mg L^{-1}
H$_2$S in bottom water	108 mg L^{-1}
Total solids	25.25 g L^{-1}
Ca	1179 mg L^{-1}
Mg	336 mg L^{-1}
Na	7636 mg L^{-1}
K	320 mg L^{-1}
NH$_3$	8 mg L^{-1}
Cl	13,520 mg L^{-1}
HCO$_3$	145 mg L^{-1}
SO$_4$	1848 mg L^{-1}
NO$_3$	3 mg L^{-1}
SiO$_3$	70 mg L^{-1}

Source: Ivanov MV, *Microbiological Processes in the Formation of Sulfur Deposits*. Israel Program for Scientific Trans-lations, U.S. Department of Agriculture and National Science Foundation, Washington, DC, 1968.

Butlin and Postage recognized that some of the hydrogen sulfide in the lake water could undergo autoxidation to form sulfur, but they considered this process unimportant because lake Ain amm el Gelud, which contains sulfuretted water, contains an insignificant amount of sulfur in its sediment and also lacks noticeable growth of photosynthetic bacteria. Butlin and Postgate were able to reconstruct an artificial system in the laboratory with pure and mixed cultures of sulfate-reducing and photosynthetic sulfur bacteria that reproduced the processes they postulated for sulfur deposition in the Cyrenaican lakes. Significantly, however, they found it best to supplement their artificial lake water with 0.1% sodium malate to achieve good sulfur production. This led to questions about the correctness of their model for biogenesis of sulfur in the Cyrenaican lakes.

Ivanov (1968) pointed out that Butlin and Postgate's model did not account for all the carbon needed for sulfur production from sulfate in the Cyrenaican lakes. He argued that a cyclical mechanism in which the photosynthetic sulfur bacteria produce the organic carbon with which the sulfate-reducing bacteria reduce sulfate to sulfide, which the photosynthetic sulfur bacteria then turn into S^0, suffers from carbon limitation. He showed that each turn of the cycle produces only one-fourth or less of the hydrogen sulfide that was produced in the just preceding cycle. As a result, the photosynthetic sulfur bacteria produce only one-fourth or less fixed carbon in each succeeding cycle. This is best illustrated by the reactions

$$2CO_2 + 4H_2S \rightarrow 2(CH_2O) + 4S^0 + 2H_2O \tag{19.43}$$

$$2(CH_2O) + SO_4^{2-} + 2H^+ \rightarrow H_2S + 2H_2O + 2CO_2 \tag{19.44}$$

Reaction 19.43 illustrates the photosynthetic reaction, and Reaction 19.44 illustrates sulfate reduction. It is seen that to produce the organic carbon (CH_2O) needed to reduce sulfate (Reaction 19.43), four times as much H_2S is consumed as is produced in sulfate reduction (Reaction 19.44). Ivanov (1968) therefore argued that most of the sulfide turned into sulfur by the photosynthetic sulfur bacteria is introduced into the lake by the warm springs feeding it and does not result from sulfate reduction. He noted that Butlin and Postgate (Butlin, 1953) had actually demonstrated that many artesian wells in the area contained sulfuretted water with sulfate-reducing bacteria. Ivanov did not consider, however, the possibility that these wells might also inject H_2 into the lakes that sulfate-reducers could employ either autotrophically or mixotrophically as an alternative energy source and as a reductant of sulfate in a carbon-sparing action.

Ivanov (1968) also suggested that a portion of the sulfur in the lake may be produced by nonphotosynthetic sulfur bacteria and by autoxidation. No matter what the source of the H_2S is, biogenesis of the sulfur in the Cyrenaican lakes has been confirmed on the basis of stable isotope analysis (Macnamara and Thode, 1951; Harrison and Thode, 1958; Kaplan et al., 1960). Figure 19.4 summarizes the different biological reactions by which sulfur may be generated in the Cyrenaican lakes.

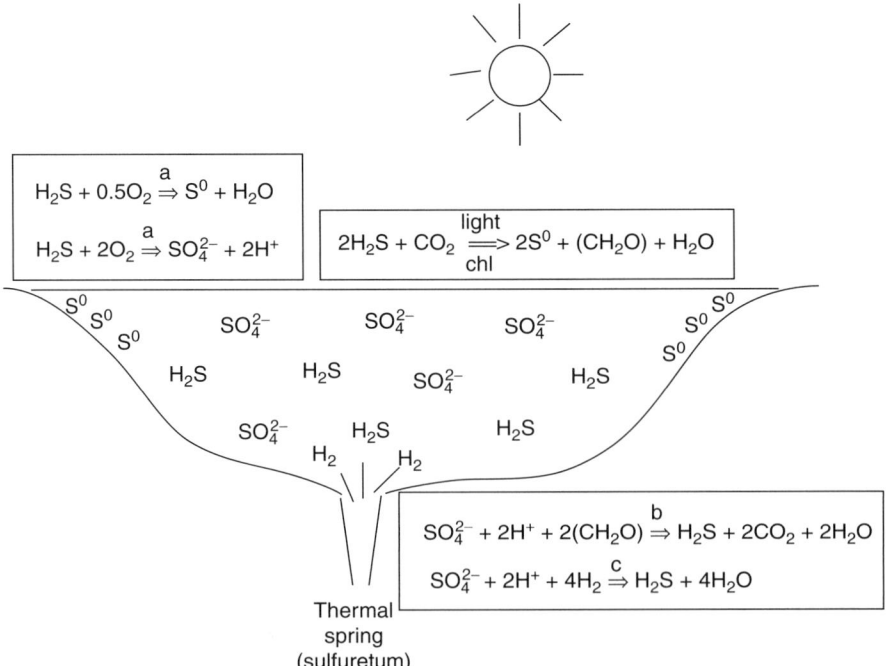

FIGURE 19.4 Summary of microbial reactions that can account for the formation of elemental sulfur in the Cyrenaican lakes, Libya. (a) Caused by thiobacilli in the oxidizing zones of the lake (suggested by Ivanov, 1968) or by autoxidation; (b) caused by sulfate-reducers such as *Desulfovibrio desulfuricans*; and (c) caused by autotrophic sulfate-reducers such as those listed in Table 19.4 (not reported by either Butlin and Postgate (1952) or Ivanov (1968)). chl = bacteriochlorophyll in green or purple bacteria.

19.11.2.2 Lake Senoye

Lake Sernoye is located in the Kuibishev Oblast in the central Volga region of Russia. It is an artificial, relatively shallow reservoir fed by the Sergievsk sulfuretted springs (Ivanov, 1968). The water output of these springs is ~6000 m^3 per day. The waters contain 83–86 mg of H_2S per liter and have a pH of 6.7. The water temperature in summer ranges ~8°C. The lake drains into Molochni Creek. The waters that enter Molochni Creek are reported to be opalescent owing to suspended native sulfur in them. The sulfur originates from the oxidation of H_2S in the lake. Much of the lake sediment contains ~0.5% native sulfur, but some sediment contains as much as 2–5%. The lake freezes over in winter, at which time no significant oxidation of H_2S occurs. This fact is reflected in the stratified occurrence of sulfur in the lake sediment. Pure sulfur crystals, which are paragenetic with calcite crystals, have been found in some sediment cores (Sokolova, 1962). Most of the sulfur in the lake is deposited around the sulfuretted springs. At these locations, masses of purple and green sulfur bacteria are seen. Impression smears have shown the presence of *Chromatium* and large numbers of rod-shaped bacteria, which on culturing reveal themselves to be mostly thiobacilli. A study of the H_2S oxidation in Lake Sernoye waters around the springs using $Na_2^{35}S$ revealed that the microflora of the lake made a significant contribution (>59%). The study differentially measured chemical and biological sulfide oxidation in the dark as well as biological, light-dependent sulfide oxidation. About the same amount of native sulfur was precipitated in both the dark and the light. These results suggested that most of the H_2S in the lake that is biologically oxidized to native sulfur is attacked by thiobacilli, in particular *Thiobacillus thioparus* (Sokolova, 1962). The photosynthetic bacteria appeared to oxidize H_2S for the most part directly to sulfate. They were found to be of the type that is physiologically like *Chromatium thiosulfatophilum*. An average dark production of native sulfur during the summer months has been estimated to be 150 kg per day (Ivanov, 1968).

19.11.2.3 Lake Eyre

Lake Eyre in Australia represents another locality in which evidence of syngenetic sulfur deposition promoted by bacteria has been noted. In shallow water on the southern bank of this lake, sulfur nodules have been found by Bonython (see Ivanov, 1968, pp. 146–150). The nodules are oval to spherical and are usually covered with crusts of crystalline gypsum on the outside while being cavernous on the inside (Baas Becking and Kaplan, 1956). Their composition includes the following (in percent by dry weight): $CaSO_4$, 34.8; S^0, 62–63; $NaCl$, 0.8; Fe_2O_3, 0.45; $CaCO_3$, 0.32; organic carbon, 1.8; and moisture, 7.54 (Baas Becking and Kaplan, 1956). Most nodules as well as the water and muds in the lake were found to harbor active sulfate-reducing bacteria and thiobacilli (Baas Becking and Kaplan, 1956). Flagellates and cellulolytic, methane-forming, and other bacteria were also found to abound. Baas Becking and Kaplan (1956) at first proposed that the nodules were forming all about the time of observation with photosynthetic flagellates providing organic carbon that cellulolytic bacteria converted into a form that can be used by sulfate-reducing bacteria for the reduction of sulfate in gypsum in the surrounding sedimentary rock. The resultant H_2S was then thought to be subject to chemical and biological oxidation by thiobacilli to elemental sulfur. The nodule structure was seen to result from the original dispersion of the gypsum in septaria in which the gypsum is replaced by elemental sulfur. The difficulty with this model of the origin of the sulfur concretions, as Ivanov (1968) argues, is that the present oxidation–reduction potential of the ecosystem is +280 to +340 mV, which is too high for intense sulfate reduction. Sulfate reduction usually occurs at redox potentials no higher than around –110 mV.

Radiodating of some sulfur nodules from Lake Eyre has shown them to be 19,600 year old (Baas Becking and Kaplan, 1956). The $^{32}S/^{34}S$ ratios of the sulfur and the gypsum of the outer crust of the sulfur nodules were found to be very similar (22.40–22.56 and 22.31–22.53, respectively), whereas that of the gypsum of the surrounding rock was found to be 22.11. This clearly suggests that the gypsum of the nodule crust is a secondary formation, biogenically produced through the oxidation of the sulfur in the nodules, which was itself biogenically produced in Quaternary time, the age of the surrounding sedimentary deposit. It is possible, therefore, that the sulfur in the nodules derived from H_2S microbiologically generated from the primary gypsum deposit and released into the water and converted to elemental sulfur by organisms like photosynthetic bacteria or *Beggiatoa* residing on the surface of the concretions.

19.11.2.4 Solar Lake

An example of a lake in which sulfur is produced biogenically but not permanently deposited in the sediment is Solar Lake in the Sinai on the western shore of the Gulf of Aqaba. It is a small hypersaline pond (7000 m^2 surface area, 4–6 m depth), which has undergone very extensive limnological investigation (Cohen et al., 1977a–c). It is tropical and has a chemocline (O_2/H_2S interface) and a thermocline that is inverted in winter (i.e., the hypolimnion is warmer than the epilimnion). The chemocline, which is 0–10 cm thick and located at a depth of 2–4 m below the surface, undergoes diurnal migration over a distance of 20–30 cm. The chief cause of this migration is the activity of the cyanobacteria *Oscillatoria limnetica* and *Microcoleus* sp., whose growth extends from the epilimnion into the hypolimnion. Sulfate-reducing bacteria, including a *Desulfotomaculum acetoxidans* type, near the bottom in the anoxic hypolimnion generate H_2S from SO_4^{2-} in the lake water. Some of this H_2S migrates upward to the chemocline. During early daylight hours, H_2S in the chemocline and below is oxidized to elemental sulfur by anaerobic photosynthesis of the cyanobacteria. After they have depleted the H_2S available to them, the cyanobacteria switch to aerobic photosynthesis, generating O_2. Thus, during daylight hours, the chemocline gradually drops. After dark, when all photosynthesis by the cyanobacteria has ceased, H_2S generated by the sulfate-reducers builds up in the chemocline together with H_2S generated by the cyanobacteria when they reduce the S^0 they formed earlier with polyglucose they stored from oxygenic photosynthesis. Some of the S^0 is also reduced by bacteria such as *Desulfuromonas acetoxidans*. Thus, during dark hours, the chemocline rises. The cycle is repeated with break of day.

Some thiosulfate is found in the chemocline during daylight hours, formed primarily by chemical oxidation of sulfide. This thiosulfate is reduced in the night hours by biological and chemical means. Sulfur therefore undergoes a cyclical transformation in the lake such that elemental sulfur does not accumulate to a significant extent. The major driving force of the sulfur cycle is sunlight. (See Jørgensen et al. (1979a,b) for further details of the sulfur cycle in this lake.)

A more recent study of cyanobacterial mats at the oxygen chemocline in Solar Lake revealed the presence of sulfate-reducers in the upper 4 mm of the mat. Significant presence of members of the *Desulfovibrionaceae*, *Desulfobacteriaceae*, and *Desulfonema*-like bacteria was observed by using appropriate rRNA probes (Minz et al., 1999a,b).

19.11.2.5 Thermal Lakes and Springs

An example of syngenetic sulfur deposition in a thermal lake in which bacteria appear to play a role has been found in Lake Ixpaca, Guatemala. This body of water is a crater lake that is supplied with H_2S by *solfataras* (fumarole hot springs that yield sulfuretted waters), which discharge water at a temperature of 87–85°C (Ljunggren, 1960). The water in the lake has a temperature in the range 29–32°C. The H_2S concentration of the lake water was reported to be 0.10–0.18 g L^{-1}. Some of it is oxidized to native sulfur, rendering the water of the lake opalescent. A portion of this sulfur settles out and is incorporated into the sediment. Another smaller portion of the sulfur is oxidized to sulfuric acid, acidifying the lake to a pH of 2.27. The sulfate content of the lake water was reported in the range from 0.46 to 1.17 g L^{-1}. The sulfuric acid in the lake water is very corrosive. It transforms igneous minerals such as pyroxenes and feldspars into clay minerals (e.g., pickingerite). Ljunggren (1960) found an extensive presence of *Beggiatoa* in the waters associated with the sediments of this lake. He implicated this microorganism in the conversion of H_2S into native sulfur. He suggested that microorganisms could also have a role in the formation of the sulfuric acid, as they do in some hot springs in Yellowstone National Park in the United States (Brock, 1978; Ehrlich and

Schoen, 1967; Schoen and Ehrlich, 1968; Schoen and Rye, 1970). More recent studies of solfataras elsewhere led to the discovery of a number of thermophiles, mostly in the domain Archaea, such as *Acidianus* (Brierley and Brierley, 1973; Segerer et al., 1986), *Sulfolobus* (Brock et al., 1972), *Thermoproteus* (Zillig et al., 1981), *Thermofilum pendens* (Zillig et al., 1983), *Sulfurococcus yellowstonii* (Karavaiko et al., 1994), and *Desulfurococcus* (Zillig et al., 1982), but also occasional members of the domain Bacteria, such as *Thermothrix thiopara* (Caldwell et al., 1976), which have a capacity to either oxidize H_2S, S^0, or thiosulfate or reduce S^0.

Studies of some hot springs in Yellowstone National Park (Brock, 1978) showed that in most, the H_2S emitted in their discharge appears to be chemically oxidized to sulfur. A major exception is Mammoth Hot Springs in which H_2S is biochemically oxidized to sulfur as deduced from sulfur isotope fractionation studies. Physiological evidence for bacterial H_2S oxidation was also obtained at Boulder Springs in the park. Unlike most of the springs, its water has a pH in the range of 8–9. Here oxidation can occur at a temperature as high as 93°C (80–90°C, optimum). The bacteria are mixotrophic in this case, and are able to use H_2S or other reduced sulfur compounds as energy source and organic matter as carbon source (Brock et al., 1971). Further study of the oxidation of elemental sulfur to sulfuric acid in the acid hot springs in the park revealed that *Acidithiobacillus thiooxidans* was responsible at temperatures <55°C and *Sulfolobus acidocaldarius* at temperatures between 55 and 85°C (Fliermans and Brock, 1972; Mosser et al., 1973). Almost all sulfur oxidation in the hot acid springs and hot acid soils was biochemical, because sulfur appeared to be stable in the absence of bacterial activity (Mosser et al., 1973).

Sulfolobus acidocaldarius consists of spherical cells that frequently form lobes and lack peptidoglycan in their cell wall (Brierley, 1966; Brock et al., 1972) (Figure 16.5). The organism is acidophilic (optimum pH 2–3; pH range 0.9–5.8) and thermophilic (temperature optimum 70–75°C; range 55–80°C). It has a guanine–cytosine ratio of 60–69 mol%. The genomic sequence of strain DSM639 has been determined (Chen et al., 2005). In growing cultures in the laboratory, the growth rate parallels the oxidation rate of elemental sulfur crystals. The presence of yeast extract in the medium was found to partially inhibit sulfur oxidation but not growth. The growth rates of *S. acidocaldarius* in several hot springs in Yellowstone National Park exhibit steady-state doubling times on the order of 10–20 h in the water of small springs having volumes of 20–2000 L, and on the order of 30 days in large springs with 1×10^6 L volumes. Doubling times during exponential growth in the water of artificially drained springs were on the order of a few hours (Mosser et al., 1974).

In effluent channels from alkaline hot springs in Yellowstone National Park where the temperature does not exceed 70°C, a bacterium called *Chloroflexus aurantiacus* was discovered (Brock, 1978;

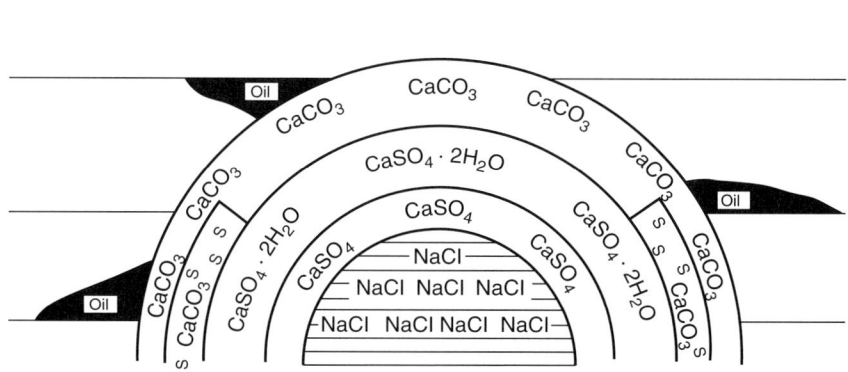

FIGURE 19.5 Diagrammatic representation of a salt dome. (After Ivanov MV, *Microbiological Processes in the Formation of Sulfur Deposits*, Israel Program for Scientific Translations, U.S. Department of Agriculture and National Science Foundation, Washington, DC, 1968.)

Pierson and Castenholz, 1974). It is characterized as a gliding, filamentous (0.5–0.7 μm in width, variable in length), phototrophic bacterium with a tendency to form orange mats below, and to a lesser extent above, thin layers of cyanobacteria such as *Synechococcus* (Doemel and Brock, 1974, 1977). Its photosynthetic pigments include bacteriochlorophylls *a* and *c*, and β- and γ-carotene. The pigments occur in *Chlorobium*-type vesicles. Anaerobically, the organism can grow photoautotrophically in the presence of sulfide and bicarbonate (Madigan and Brock, 1975), but it can also grow photoheterotrophically with yeast extract and certain other organic supplements. Aerobically, the organism is capable of heterotrophic growth in the dark. Although it shows some physiological resemblance to Rhodospirillaceae (purple nonsulfur bacteria) and even greater resemblance to Chlorobiaceae (green sulfur bacteria), phylogenetically *Chloroflexus* is related to neither (Brock and Madigan, 1988).

The mats formed by *Chloroflexus* in some hot springs in Yellowstone National Park may be models for the formation of ancient stromatolites. They often incorporate detrital silica in the form of siliceous sinter from the geyser basins and in time are transformed into structures recognizable as stromatolites.

Sulfate in the mats may be reduced to sulfide by sulfate-reducers below the upper 3 mm, and this sulfide can be converted to elemental sulfur by *Chloroflexus* in the mats (Doemel and Brock, 1976). This is another example in which below 70°C at least some elemental sulfur in hot springs or their effluent may be of biogenic origin.

A study of bacterial mats in hot springs in southwestern Iceland revealed the presence of 14 bacterial types and 5 archaeal types in mats loaded with precipitated sulfur formed in sulfide-rich water at 60–80°C. Mats formed in low-sulfide springs at 65–70°C were dominated by *Chloroflexus* (Skirnisdottir et al., 2000).

19.11.3 EXAMPLES OF EPIGENETIC SULFUR DEPOSITS

19.11.3.1 Sicilian Sulfur Deposits

An example of epigenetic sulfur deposition in which microbes must have played a role is found on the volcanic Mediterranean island of Sicily. Isotopic studies (Jensen, 1968) showed that the elemental sulfur in these deposits is significantly enriched in ^{32}S relative to associated sulfate. This finding signifies that the sulfur could not have originated from volcanic activity but must have been the result of microbial sulfate reduction in evaporite deposits that originated from Mediterranean seawater. The biological agents must have been a consortium of dissimilatory sulfate-reducing bacteria that reduced the sulfate in the evaporite to sulfide and chemosynthetic sulfide-oxidizing bacteria that formed elemental sulfur from the sulfide. Abiotic sulfide oxidation could also have contributed to sulfur formation. Organic carbon, if used in the microbial reduction of the sulfate, came presumably from organic detritus in the sediment (algal or other remains).

19.11.3.2 Salt Domes

Another example of biogenic native sulfur of epigenetic origin in a sedimentary environment is that associated with salt domes (Figure 19.5) such as those found on the Gulf Coast of the United States and Mexico (northern and western shores of the Gulf Coast, including those of Texas, Louisiana, and Mexico) (see, e.g., Ivanov, 1968, pp. 92ff.; Martinez, 1991). Such salt domes reside directly over a central plug consisting of 90–95% rock salt (NaCl), 5–10% anhydrite ($CaSO_4$), and traces of dolomite ($CaMgCO_3$), barite ($BaSO_4$), and celestite ($SrSO_4$). Petroleum may be entrapped in peripheral deformations. The domes consist mainly of anhydrite topped by calcite ($CaCO_3$), which may also have exploitable petroleum associated with it. Between the calcite and anhydrite exists a zone containing gypsum ($CaSO_4 \cdot 2H_2O$), calcite, and anhydrite relicts. Elemental sulfur is associated with the calcite in this intermediate zone. The salt domes originated from evaporite, which formed in a period between the late Paleozoic (230–280 million years ago) and the Jurassic (Middle Mesozoic, 135–180 million years ago). Current theory (see Strahler, 1977) proposes that the salt domes on the Gulf Coast began as beds of evaporite along the continental margins of the newly emergent Atlantic Ocean ~180 million years ago. The evaporite derived from hypersaline waters with the aid of heat emanating from underlying magma reservoirs in these tectonically active areas. As the ocean basin broadened as a result of continental drift, and as the continental margins became more defined, turbidity currents began to bury the evaporite beds under ever thicker layers of sediment. Ultimately, these sediment layers became so heavy that they forced portions of the evaporite, which has plastic properties, upward as fingerlike salt plugs through ever younger sediment strata. As these plugs intruded into the groundwater zone, they lost their more water-soluble constituents, particularly the rock salt, leaving behind relatively insoluble constituents, especially anhydrite, which became the cap rock. In time, some of the anhydrite was converted to more soluble gypsum, and some was dissolved away. At that point, bacterial sulfate reduction is thought to have begun, lasting perhaps for a period of 1 million years. The active bacteria were most likely introduced into the dome structures from the native flora in the groundwater. The organic carbon needed for bacterial sulfate reduction is thought to have derived from adjacent petroleum deposits.

When the biological contribution to elemental sulfur formation in the salt domes was first recognized (Jones et al., 1956), the only known sulfate-reducer was *Desulfovibrio desulfuricans*, which uses lactate, pyruvate, and in most instances, malate as energy sources, which it oxidizes to acetate with sulfate as terminal electron acceptor. For *D. desulfuricans* to have used petroleum constituents as a carbon source, it would have had to depend on other bacteria that could transform these constituents, preferably anaerobically, into carbon substrates it could metabolize.

A more recent observation indicated the existence of a sulfate-reducer that can use methane (CH_4) as a source of energy (Panganiban and Hanson, 1976; Panganiban et al., 1979). Methane is a major gaseous constituent associated with petroleum deposits.

Even more recent studies revealed the existence of sulfate-reducers with an ability to use some short-chain saturated aliphatic (including chain lengths of C_8–C_{16}) or aromatic hydrocarbons, or heterocyclic compounds, many of which they are able to mineralize (see Section. 19.6) (see also Aeckersberg et al., 1991; Rueter et al., 1994). Many sulfate-reducers have also been shown to be able to use H_2 as their energy source. H_2 occurs in detectable amounts in oil wells. Thus, it is not necessary to postulate the past existence of a complex assemblage of anaerobic fermentative bacteria that converted petroleum hydrocarbons into energy sources for sulfate-reducing bacteria in salt domes.

Whatever sulfate-reducing bacteria were active in the salt domes, they produced not only H_2S but also CO_2. The CO_2 arose from fermentation, anaerobic respiration, and mineralization of the organic carbon consumed for energy conservation in sulfate reduction. The H_2S was subsequently oxidized biologically or chemically to native sulfur, whereas the CO_2 was extensively precipitated as carbonate (secondary calcite) with the calcium from the anhydrite or gypsum attacked by the sulfate-reducing bacteria:

$$CaSO_4 + 2(CH_2O) \rightarrow CaS + 2CO_2 + 2H_2O \qquad (19.45)$$

$$CaS + CO_2 + H_2O \rightarrow CaCO_3 + H_2S \qquad (19.46)$$

$$H_2S + 0.5O_2 \rightarrow S^0 + H_2O \qquad (19.47)$$

Mineralogical and isotopic study has shown that the native sulfur and secondary calcite are physically associated in the cap rock (paragenetic). The isotopic enrichment of the sulfur and secondary calcite indicates a biological origin (Jones et al., 1956; Thode et al., 1954). The sulfur exhibits enrichment with respect to ^{32}S, and the secondary calcite with respect to ^{12}C (see discussion by Ivanov, 1968). Although the enrichment of ^{12}C in the secondary calcite was originally attributed to degradation of petroleum hydrocarbons by sulfate-reducers, it seems more likely that it might have been due, at least in part if not entirely, to anaerobic oxidation of ^{12}C-enriched biogenic methane by sulfate-reducers, as apparently happened in a native sulfur deposit at Machéw, Poland (Böttcher and Parafiniuk, 1998).

A recent study by Detmers et al. (2001) showed that sulfur isotope fractionation was greater when the sulfate-reducers completely oxidized (mineralized) the carbon they consumed than when they oxidized it incompletely. The degree of sulfur isotope fractionation appears to be affected by the metabolic pathway and the regulation of transmembrane transport (Detmers et al., 2001).

19.11.3.3   Gaurdak Sulfur Deposit

Epigenetic sulfur deposition in a mode somewhat similar to that associated with the salt domes in the United States took place in the Gaurdak Deposit in Turkmenistan (Ivanov, 1968). This deposit resides in rock of Upper Jurassic age and was probably emplaced in the Quaternary as plutonic waters picked up organic carbon from the Kugitang Suite containing bituminous limestone and sulfate from the anhydrite-carbonate rocks of the Gaurdak Suite. Sulfate-reducing bacteria that entered the plutonic waters reduced the sulfate in it to H_2S with the help of the organic carbon derived from the bituminous material. *Thiobacillus thioparus* oxidized the H_2S to S^0 at the interface where the plutonic water encountered infiltrating oxygenated surface water. Where the elemental sulfur presently encounters oxygenated water, intense bio-oxidation of the sulfur to sulfuric acid has been noted, causing secondary paragenetic calcite, formed during sulfate reduction to H_2S in the initial phase of sulfur genesis, to be transformed into secondary gypsum. The bacteria, *T. thioparus* and *Acidithiobacillus thiooxidans*, have been found in significant numbers in sulfuretted waters in the sulfur deposits with paragenetic (secondary) calcite and in acidic sulfur deposits with secondary gypsum, respectively. Elemental sulfur, therefore, appears to be deposited and degraded in Gaurdak formation at the present time.

19.11.3.4   Shor-Su Sulfur Deposit

Another example of epigenetic microbial sulfur deposition is the Shor-Su Deposit in the northern foothills of the Altai mountain range in the southern corner of the West Siberian Plain. Here an extensive, folded sedimentary formation of lagoonal origin and mainly of Paleocene and Cretaceous age contains major sulfuretted regions in lower Paleocene strata (Bukhara and Suzak) of the second anticline and to a lesser extent in Quaternary conglomerates (Figure 19.6) (see Ivanov, 1968, pp. 33–34).

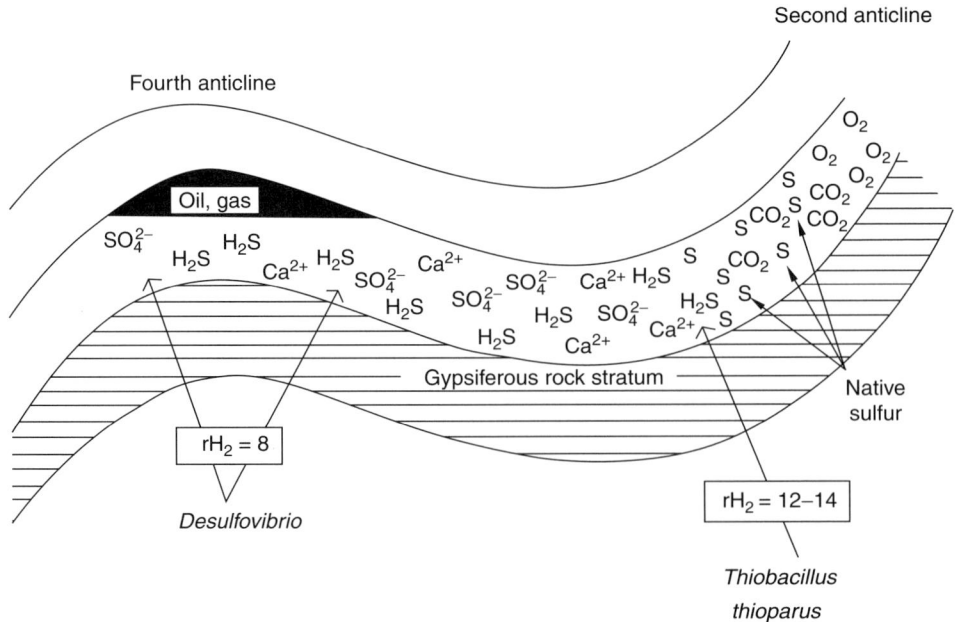

FIGURE 19.6 Diagrammatic representation of essential features of the Shor-Su formation. (After Ivanov MV, *Microbiological Processes in the Formation of Sulfur Deposits*, Israel Program for Scientific Translations, U.S. Department of Agriculture and National Science Foundation, Washington, DC, 1968.)

The native sulfur of the main deposits occurs in heavily broken rock surrounded by gypsified rock. It contains some relict gypsum lenses. It is enclosed in a variety of cavernous rock and associated with calcite and celestite in the Bukhara stratum and in cavities and slit like caves in the Suzak stratum. Petroleum and natural gas deposits are associated with the fourth anticline. The second anticline contains most of the sulfur. The two anticlines are hydraulically connected. One basis of this claim of hydraulic connection between the two anticlines is that their pore waters are chemically very similar in composition. Sulfate-reducing bacteria occur in the plutonic waters that flow through the permeable strata from the fourth to the second anticline. It is believed that these bacteria have been reducing the sulfate that the plutonic waters picked up from dissolution of some of the gypsum and anhydrite in the surrounding rock. The bacteria are presumed to have been using petroleum hydrocarbons or derivatives from them as a source of energy (reducing power) and carbon for the process. The presence of sulfate-reducing bacteria has been reported in waters of the second anticline and in rock in which sulfur occurs (Ivanov, 1968). These bacteria were demonstrated to be able to reduce sulfate under *in situ* conditions at a measurable rate (0.009–0.179 mg H_2S L^{-1} d^{-1}). Native sulfur has been forming where rising plutonic water has been mixing with downward-seeping oxygenated surface water. In this zone of mixing of the two waters, *Thiobacillus thioparus* was detected and shown to oxidize H_2S from the plutonic water to native sulfur.

Measurements have shown that sulfate reduction predominates where plutonic waters carry sulfate derived from surrounding gypsiferous rock and organic matter derived from associated petroleum. The waters at these sites in the deposit have an rH_2 that often is below 8, indicating strong reducing conditions. The H_2S is transported by moving plutonic waters to a region in the second anticline where it encounters aerated surface waters. The waters have an rH_2 ~12–14 (16.5 maximum). In this environment, *Thiobacillus thioparus* is favored. It causes conversion of H_2S into S^0. Where the rH_2 exceeds 16.5 owing to extensive exposure to surface water, as in outcroppings of the western conglomerate of the Shor-Su, the sulfur is undergoing extensive oxidation by *Acidithiobacillus thiooxidans*. The pH is found to drop from neutrality to less than 1 where the bacteria are most active. Although unknown then, it is possible that at very low pH hyperacidophilic archaea may be active, as at Iron Mountain, California (Edwards et al., 2000). The sulfur in the main strata began to be laid down in the Quaternary according to Ivanov (1968). Deposition continues to the present day. For this reason, events in the geologic past can be reconstructed from current observations of bacterial distribution and activity in the Quaternary strata of the Shor-Su (Ivanov, 1968).

19.11.3.5 Kara Kum Sulfur Deposit

Spatially, a somewhat different mechanism of epigenetic sulfur deposition has been recognized in the Kara Kum Deposit north of Ashkhabad in Turkmenistan (Ivanov, 1968). Sulfate-reducing bacteria and H_2S-oxidizing bacteria have also been playing a role in native sulfur formation at this site. However, sulfate reduction has been taking place in a different stratum from that involving sulfur deposition, implying that these two activities are spatially separated. The H_2S has been transported to another site before conversion to native sulfur. Hence, paragenetic (secondary) calcite is not found associated with the native sulfur of this deposit, and consequently sulfuric acid formed at the sites of outcropping of the sulfur deposit cannot form gypsum but reacts with sandstone, liberating aluminum and iron, which are precipitated as oxides in a more neutral environment.

19.12 MICROBIAL ROLE IN SULFUR CYCLE

As the foregoing discussion shows, microbes play an important role in inorganic as well as organic sulfur transformations (Trüper, 1984a). Figure 19.7 shows how these various biological interactions fit into the sulfur cycle in soil, sediment, and aquatic environments. Although some of these transformations such as aerobic oxidation of H_2S or S^0 may proceed partly by an abiotic route, albeit often significantly more slowly than by a biotic route, at least two other transformations, the anaerobic oxidation of H_2S and S^0 to sulfuric acid and the reduction of sulfate to H_2S, do not proceed

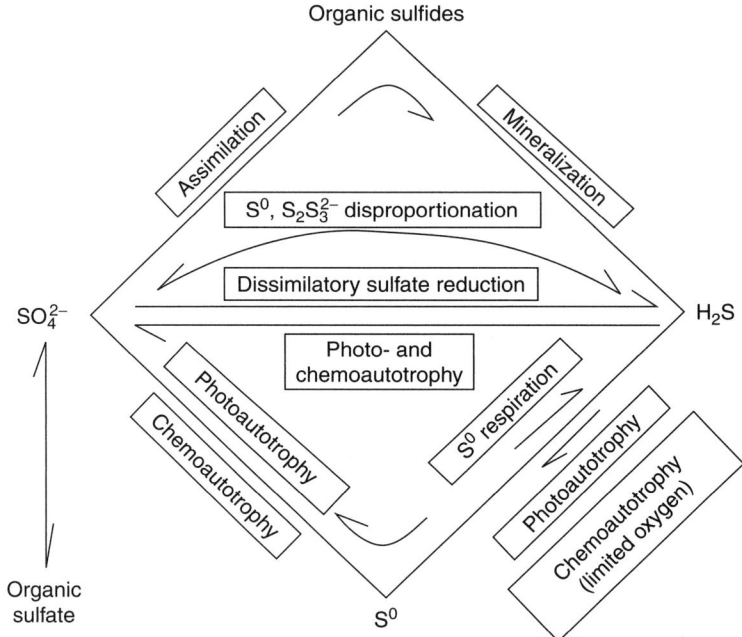

FIGURE 19.7   The sulfur cycle.

readily abiotically at atmospheric pressure in the temperature range that prevails at the Earth's surface. Sulfate reduction is now recognized to be an important mechanism of anaerobic mineralization of organic carbon in anaerobic estuarine and other coastal environments where plentiful sulfate is available from seawater (Skyring, 1987). Geochemically, sulfur-oxidizing and sulfur-reducing bacteria are important catalysts in the sulfur cycle in the biosphere.

19.13   SUMMARY

Sulfur, which occurs in organic and inorganic forms in nature, is essential to life. Different organisms may assimilate it in organic or inorganic form. Plants and many microbes normally take it up as sulfate. Microbes are important in mineralizing organic sulfur compounds in soil and aqueous environments. The biogeochemistry of organic sulfur mineralization as well as the synthesis of organic sulfur compounds has been studied in some detail.

Inorganic sulfur may exist in various oxidation states in nature, most commonly as sulfide (−2), elemental sulfur (0), and sulfate (+6). Thiosulfate and tetrathionate, each with sulfur in mixed oxidation states, may also occur in significant amounts in some environments. Some microbes in soil and water play an important role in the interconversion of these oxidation states. These include several different groups in the domain Bacteria (even some cyanobacteria under special conditions) and some different groups in the domain Archaea. Among members of the Bacteria that oxidize reduced forms of sulfur are chemolithotrophs, anoxygenic and oxygenic (cyanobacterial) photolithotrophs, mixotrophs, and heterotrophs. Most chemolithotrophs and mixotrophs use oxygen as oxidant, but a few chemolithotrophs can substitute nitrate or ferric iron when oxygen is absent. Some chemolithotrophs, such as *Thiobacillus thioparus*, can oxidize H_2S to S^0 under partially reduced conditions, but form H_2SO_4 under fully oxidizing conditions. The anoxygenic photolithotrophic bacteria (purple and green bacteria) oxidize H_2S to S^0 or H_2SO_4 to generate reducing power for CO_2 fixation or ATP synthesis. Certain cyanobacteria oxidize H_2S to S^0 in the absence of oxygen for generating energy and reducing power for CO_2 fixation. Various chemolithotrophs and mixotrophs can oxidize S^0 to H_2SO_4 aerobically in neutral or acid environment. Sulfur oxidation has been noted in mesophilic and thermophilic environments, in the latter instance at temperatures exceeding 100°C in some cases. Thiosulfate is readily oxidized by some chemolithotrophs, mixotrophs, and heterotrophs. Some marine pseudomonads have been shown to use it as a supplemental energy source, oxidizing it to tetrathionate. Some bacteria can conserve energy by disproportionating elemental sulfur, dithionite, sulfite, or thiosulfate to sulfide and sulfate under anaerobic conditions.

Oxidized forms of sulfur may be reduced by various microorganisms. Elemental sulfur is reduced to H_2S with or without energy conservation by some anaerobic members of the Bacteria and Archaea. Among the Bacteria that conserve energy are *Desulfuromonas acetoxidans*, *Desulfovibrio gigas*, and some other sulfate-reducing bacteria. Among the Archaea that conserve energy are *Pyrococcus furiosus*, *Pyrodictium*, and *Acidianus*. Two fungi, *Rhodotorula* and *Trichosporon*, have also been found to be able to reduce S^0 to H_2S, but probably without energy conservation.

Sulfate may be reduced in sulfate respiration (dissimilatory sulfate reduction) by a number of specialized bacteria. Most known species are members of the domain Bacteria, but at least two species are known among the members of the domain Archaea. Geologically, this microbial activity is of major importance because under natural conditions at the Earth's surface, sulfate cannot be reduced by purely chemical means due to the high energy of activation required by the process. Sulfate is reduced aerobically by various microbes and plants, but only in small amounts without any extracellular accumulation of H_2S (assimilatory sulfate reduction). The mechanisms of dissimilatory and assimilatory sulfate reduction differ biochemically.

Some reducers and oxidizers of sulfur and its compounds can distinguish between stable isotopes ^{32}S and ^{34}S and can bring about isotope fractionation. Geologically, this is useful, for example, in determining whether ancient sulfur deposits were formed biogenically or abiogenically.

Contemporary biogenic sulfur deposition involving sulfate-reducing bacteria and aerobic and anaerobic sulfide-oxidizing bacteria have been identified in several lacustrine environments. These represent syngenetic deposits. Bacterial oxidation of elemental sulfur to sulfuric acid in certain hot springs has also been reported.

Ancient epigenetic native sulfur deposits of microbial origin have been identified in salt domes and other geological formations associated with hydrocarbon (petroleum) deposits in various parts of the world. The native sulfur in these instances arose from bacterial reduction of sulfate derived from anhydrite or gypsum followed by bacterial oxidation to elemental sulfur under partially reduced conditions. On full exposure to air, some of the elemental sulfur is presently being oxidized to sulfuric acid by bacteria.

Less spectacular oxidative and reductive transformations of sulfur occur in soil, where they play an important role in the maintenance of soil fertility.

REFERENCES

Abdollahi H, Wimpenny JWT. 1990. Effects of oxygen on the growth of *Desulfovibrio desulfuricans*. *J Gen Microbiol* 136:1025–1030.

Adams CA, Warnes GM, Nicholas DJD. 1971. A sulfite-dependent nitrate reductase from *Thiobacillus denitrificans*. *Biochim Biophys Acta* 235:398–406.

Aeckersberg F, Bak F, Widdel F. 1991. Anaerobic oxidation of saturated hydrocarbons to CO_2 by a new type of sulfate-reducing bacterium. *Arch Microbiol* 156:5–14.

Akagi JM, Chan M, Adams V. 1974. Observations on the bisulfite reductase (P582) isolated from *Desulfotomaculum nigrificans*. *J Bacteriol* 120:240–244.

Aleem MIH. 1965. Thiosulfate oxidation and electron transport in *Thiobacillus novellas*. *J Bacteriol* 90:95–101.

Aminuddin M. 1980. Substrate level versus oxidative phosphorylation in the generation of ATP in *Thiobacillus denitrificans*. *Arch Microbiol* 128:19–25.

Aminuddin M, Nicholas DJD. 1973. Sulfide oxidation linked to the reduction of nitrate to nitrite in *Thiobacillus denitrificans*. *Biochim Biophys Acta* 325:81–93.

Aminuddin M, Nicholas DJD. 1974a. An AMP-independent sulfite oxidase from *Thiobacillus denitrificans*. *J Gen Microbiol* 82:103–113.

Aminuddin M, Nicholas DJD. 1974b. Electron transfer during sulfide to sulfite oxidation in *Thiobacillus denitrificans*. *J Gen Microbiol* 82:115–123.

Arcuri EJ. Pasonian Springs, VA, USA.

Arp AJ, Childress JJ. 1983. Sulfide binding by the blood of the hydrothermal vent tube worm *Riftia pachyptila*. *Science* 219:295–297.

Baalsrud K, Baalsrud KS. 1954. Studies on *Thiobacillus denitrificans*. *Arch Mikrobiol* 20:34–62.

Baas-Becking LGM, Kaplan IR. 1956. The microbiological origin of the sulfur nodules of Lake Eyre. *Trans Roy Soc Aust* 79:62–65.

Badziong W, Thauer RK. 1978. Growth yields and growth rates of *Desulfovibrio vulgaris* (Marburg) growing on hydrogen plus sulfate and hydrogen plus thiosulfate as sole energy sources. *Arch Microbiol* 117:209–214.

Badziong W, Thauer RK, Zeikus JG. 1978. Isolation and characterization of *Desulfovibrio* growing on hydrogen plus sulfate as the sole energy source. *Arch Microbiol* 116:41–49.

Bak F, Cypionka H. 1987. A novel type of energy metabolism involving fermentation of inorganic sulfur compounds. *Nature* (London) 326:891–892.

Bak F, Pfennig N. 1987. Chemolithotrophic growth of *Desulfovibrio sulfodismutans* sp. nov. by disproportionation of inorganic sulfur compounds. *Arch Microbiol* 147:184–189.

Bak F, Widdel F. 1986a. Anaerobic degradation of indolic compounds by sulfate-reducing enrichment cultures, and description of *Desulfobacterium indolicum* gen. nov., spec. nov. *Arch Microbiol* 146:170–176.

Bak F, Widdel F. 1986b. Anaerobic degradation of phenol and phenol derivatives by *Desulfobacterium phenolicum* sp. nov. *Arch Microbiol* 146:177–180.

Balows A, Trüper HG, Dworkin M, Harder W, Schleifer K-H, eds. 1992. *The Prokaryotes. 2nd ed. A Handbook on the Biology of Bacteria; Ecophysiology, Isolation, Identification, Applications*. New York: Springer.

Baumgartner A, Redenius I, Kranzoch J, Cypionka H. 2001. Periplasmic oxygen reduction by *Desulfovibrio* species. *Arch Microbiol* 176:306–309.

Belkin S, Wirsen CO, Jannasch HW. 1986. A new sulfur-reducing, extremely thermophilic eubacterium from a submarine thermal vent. *Appl Environ Microbiol* 51:1180–1185.

Beller HR, Chain PSG, Letain TE, Chakicherla A, Larimer FW, Richardson PM, Coleman MA, Wood AP, Kelly DP. 2006. The genome sequence of the obligately chemolithoautotrophic, facultatively anaerobic bacterium *Thiobacillus denitrificans*. *J Bacteriol* 188:1473–1488.

Beudecker RF, Cannon GC, Kuenen JG, Shively JM. 1980. Relations between D-ribulose-1,5-bisphosphate carboxylase, carboxysomes, and CO_2 fixing capacity in the obligate chemolithotroph *Thiobacillus neapolitanus* grown under different limitations in the chemostat. *Arch Microbiol* 124:185–189.

Beudecker RF, Codd GA, Kuenen JG. 1981. Quantification and intracellular distribution of ribulose-1,5-biphosphate carboxylase in *Thiobacillus neapolitanus*, as related to possible functions of carboxysomes. *Arch Microbiol* 129:361–367.

Biebl H, Pfennig N. 1977. Growth of sulfate-reducing bacteria with sulfur as electron acceptor. *Arch Microbiol* 112:115–117.

Bogdanova TI, Tsaplina IA, Sayakin DD, Karavaiko GI. 1990. Morphology and cytology of '*Sulfobacillus thermosulfidooxidans* subsp. *thermotolerans*' bacterium. *Mikrobiologiya* 60:577–586.

Bonch-Osmolovskaya EA. 1994. Bacterial sulfur reduction in hot vents. *FEMS Microbiol Rev* 15:65–77.

Bonch-Osmolovskaya EA, Miroshnichenko ML, Kostrikina NA, Chernych NA, Zavarzin GA. 1990. *Thermoproteus uzoniensis* sp. nov., a new extremely thermophilic archaebacterium from Kamchatka continental hot springs. *Arch Microbiol* 154:556–559.

Boopathy R, Daniels L. 1991. Isolation and characterization of a furfural degrading sulfate-reducing bacterium from an anaerobic digester. *Curr Microbiol* 23:327–332.

Böttcher ME, Parafiniuk J. 1998. Methane-derived carbonates in a native sulfur deposit: Stable isotope and trace elements discrimination related to the transformation of aragonite to calcite. *Isot Environ Health Stud* 34:177–190.

Böttcher ME, Sievert SM, Kuever J. 1999. Fractionation of sulfur isotopes during dissimilatory reduction of sulfate by a thermophilic gram-negative bacterium at 60°C. *Arch Microbiol* 172:125–128.

Bowen HJM. 1979. *Environmental Chemistry of the Elements*. London, UK: Academic Press.

Bowen TJ, Happold FC, Taylor BF. 1966. Studies on adenosine 5′-phosphosulfate reductase from *Thiobacillus denitrificans*. *Biochim Biophys Acta* 118:566–576.

Brandis A, Thauer RK. 1981. Growth of *Desulfovibrio* species on hydrogen and sulfate as sole energy source. *J Gen Microbiol* 126:249–252.

Brandis-Heep A, Gebhardt NA, Thauer RK, Widdel F, Pfennig N. 1983. Anaerobic acetate oxidation to CO_2 by *Desulfobacter postgatei*. 1. Demonstration of all enzymes required for the operation of the citric acid cycle. *Arch Microbiol* 136:222–229.

Brannan DK, Caldwell DE. 1980. *Thermothrix thiopara*: Growth and metabolism of a newly isolated thermophile capable of oxidizing sulfur and sulfur compounds. *Appl Environ Microbiol* 40:211–216.

Braun M, Stolp H. 1985. Degradation of methanol by a sulfate reducing bacterium. *Arch Microbiol* 142:77–80.

Brierley CL, Brierley JA. 1973. A chemoautotrophic and thermophilic microorganism isolated from an acid hot spring. *Can J Microbiol* 19:183–188.

Brierley JA. 1966. Contribution of chemoautotrophic bacteria to the acid thermal waters of the Geyser Springs group in Yellowstone Park. PhD Thesis. Montana State University, Bozeman.

Brock TD. 1978. *Thermophilic Microorganisms and Life at High Temperatures*. New York: Springer.

Brock TD, Brock KM, Belly RT, Weiss RL. 1972. *Sulfolobus*: A new genus of sulfur-oxidizing bacteria living at low pH and high temperature. *Arch Microbiol* 84:54–68.

Brock TD, Brock ML, Bott TL, Edwards MR. 1971. Microbial life at 90°C: The sulfur bacteria of Boulder Springs. *J Bacteriol* 107:303–314.

Brock TD, Gustafson J. 1976. Ferric iron reduction by sulfur- and iron-oxidizing bacteria. *Appl Environ Microbiol* 32:567–571.

Brock TD, Madigan MT. 1988. *Biology of Microorganisms*. 5th ed. Englewood Cliffs, NJ: Prentice Hall.

Bromfield SM. 1953. Sulfate reduction in partially sterilized soil exposed to air. *J Gen Microbiol* 8:378–390.

Bryant RD, McGroarty KM, Costerton JW, Laishley EJ. 1983. Isolation and characterization of a new acidophilic *Thiobacillus* species (*T. albertis*). *Can J Microbiol* 29:1159–1170.

Burggraf S, Jannasch HW, Nicholaus B, Stetter KO. 1990. *Archeoglobus profundus* sp. nov., represents a new species within the sulfate-reducing archaebacteria. *Syst Appl Microbiol* 13:24–28.

Burton SD, Morita RY. 1964. Effect of catalase and cultural conditions on growth of *Beggiatoa*. *J Bacteriol* 88:1755–1761.

Butlin K. 1953. The bacterial sulfur cycle. *Research* 6:184–191.

Butlin KR, Postgate JR. 1952. The microbiological formation of sulfur in the Cyrenaican Lakes. In: Cloudsley-Thompson J. ed. *Biology of the Deserts*. London, UK: Institute of Biology, pp. 112–122.

Caldwell DE, Caldwell SJ, Laylock JP. 1976. *Thermothrix thiopara* gen et spec nov, a facultatively anaerobic, facultative chemolithotroph living at neutral pH and temperatures. *Can J Microbiol* 22:1509–1517.

Canfield DE, Des Marais DJ. 1991. Aerobic sulfate reduction in microbial mats. *Science* 251:1471–1473.

Canfield DE, Thamdrup B. 1994. The production of ^{34}S-depleted sulfide during bacterial disproportionation of elemental sulfur. *Science* 266:1973–1975.

Cavanaugh CM. 1983. Symbiotic chemoautotrophic bacteria in marine invertebrates from sulfide-rich habitats. *Nature* (London) 302:58–61.

Cavanaugh CM, Gardiner SL, Jones ML, Jannasch HW, Waterbury JB. 1981. Prokaryotic cells in the hydrothermal vent tube worm *Riftia pachyptila* Jones: Possible chemoautotrophic symbionts. *Science* 213:340 342.

Chambers LA, Trudinger PA. 1975. Are thiosulfate and trithionate intermediates in dissimilatory sulfate reduction? *J Bacteriol* 123:36–40.

Charles AM, Suzuki I. 1966. Mechanism of thiosulfate oxidation by *Thiobacillus novellus*. *Biochim Cosmochim Acta* 128:510–521.

Chen L, Brügger K, Skovgaard M, Redder P, She Q, Torarinsson E, Greve B, Awayez M, Zibat A, Klenk H-P, Garrett RA. 2005. The genome of *Sulfolobus acidocaldarius*, a model organism of the *Crenarchaeota*. *J Bacteriol* 187:4992–4999.

Claassen PAM, van den Heuvel MHMJ, Zehnder AJB. 1987. Enzyme profiles of *Thiobacillus versutus* after aerobic and denitrifying growth: Regulation of isocitrate lyase. *Arch Microbiol* 147:30–36.

Cohen Y, Krumbein WE, Goldberg M, Shilo M. 1977a. Solar Lake (Sinai). 1. Physical and chemical limnology. *Limnol Oceanogr* 22:597–608.

Cohen Y, Krumbein WE, Shilo M. 1977b. Solar Lake (Sinai). 2. Distribution of photosynthetic microorganisms and primary production. *Limnol Oceanogr* 22:609–620.

Cohen Y, Krumbein WE, Shilo M. 1977c. Solar Lake (Sinai). 4. Stromatolitic cyanobacterial mats. *Limnol Oceanogr* 22:635–656.

Cohen Y, Padan E, Shilo M. 1975. Facultative anoxygenic photosynthesis in the cyanobacterium *Oscillatoria limnetica*. *J Bacteriol* 123:855–861.

Coleman ML, Hedrick DB, Lovley DR, White DC, Pye K. 1993. Reduction of Fe(III) in sediments by sulphate-reducing bacteria. *Nature* (London) 361:436–438.

Corbett CM, Ingledew WJ. 1987. Is Fe^{3+}/Fe^{2+} cycling an intermediate in sulfur oxidation by *Thiobacillus ferrooxidans*? *FEMS Microbiol Lett* 41:1–6.

Cypionka H, Pfennig N. 1986. Growth yields of *Desulfotomaculum orientis* with hydrogen in chemostat culture. *Arch Microbiol* 143:396–399.

Daalsgaard T, Bak F. 1992. Effect of acetylene on nitrous oxide reduction and sulfide oxidation in batch and gradient cultures of *Thiobacillus denitrificans*. *Appl Environ Microbiol* 58:1601–1608.

Davis EA, Johnson EJ. 1967. Phosphorylation coupled to the oxidation of sulfide and 2-mercaptoethanol in extracts of *Thiobacillus thioparus*. *Can J Microbiol* 13:873–884.

Detmers J, Brüchert V, Habicht KS, Kuever J. 2001. Diversity of sulfur isotope fractionations by sulfate-reducing prokaryotes. *Appl Environ Microbiol* 67:888–894.

DeWeerd KA, Concannon F, Suflita JM. 1991. Relationship between hydrogen consumption, dehalogenation, and the reduction of sulfur oxyanions by *Desulfomonile tiedjei*. *Appl Environ Microbiol* 57:1929–1934.

DeWeerd KA, Mandelco L, Tanner RS, Woese CR, Suflita JM. 1990. *Desulfomonile tiedje* gen. nov. spec. nov., a novel anaerobic dehalogenating, sulfate-reducing bacterium. *Arch Microbiol* 154:23–30.

Dilling W, Cypionka H. 1990. Aerobic respiration in sulfate reducing bacteria. *FEMS Microbiol Lett* 71:123–128.

Doemel WN, Brock TD. 1974. Bacterial stromatolites. Origin of laminations. *Science* 184:1083–1085.

Doemel WN, Brock TD. 1976. Vertical distribution of sulfur species in benthic algal mats. *Limnol Oceanogr* 21:237–244.

Doemel WN, Brock TD. 1977. Structure, growth and decomposition of laminated algal-bacterial mats in alkaline hot springs. *Appl Environ Microbiol* 34:433–452.

Douglas S, Douglas DD. 2000. Environmental scanning electron microscopy studies of colloidal sulfur deposition in a natural microbial community from a cold sulfide spring near Ancaster, Ontario, Canada. *Geomicrobiol J* 17:275–289.

Drake HL, Akagi JM. 1978. Dissimilatory reduction of bisulfite by *Desulfovibrio vulgaris*. *J Bacteriol* 136:916–923.

Dworkin M, editor-in-chief. 2001. *The Prokaryotes. Electronic version*. New York: Springer.

Easterbrook KB, Coombs RW. 1976. Spinin: The subunit protein of bacterial spinae. *Can J Microbiol* 23:438–440.

Edwards KJ, Bond PL, Gihring TM, Banfield JF. 2000. A new iron-oxidizing, extremely acidophilic arachaea is implicated in acid mine drainage generation. *Science* 287:1796–1799.

Egorova AA, Deryugina ZP. 1963. The spore forming thermophilic thiobacterium: *Thiobacillus thermophilica* Imschenetskii nov. spec. *Mikrobiologiya* 32:439–446.

Ehrlich GG, Schoen R. 1967. Possible role of sulfur-oxidizing bacteria in surficial acid alteration new hot springs. US Geol Survey Prof Paper 575C, pp. C110–C112.

Ehrlich HL. 1999. Microbes as geologic agents: Their role in mineral formation. *Geomicrobiol J* 16:135–153.

Ehrlich HL, Fox SI. 1967. Copper sulfide precipitation by yeast from acid mine-waters. *Appl Microbiol* 15:135–139.

Elsgaard L, Jørgensen BB. 1992. Anoxic transformations of radiolabeled hydrogen sulfide in marine and freshwater sediments. *Geochim Cosmochim Acta* 56:2425–2435.

Fauque G, LeGall J, Barton LL. 1991. Sulfate-reducing and sulfur-reducing bacteria. In: Shively JM, Barton LL, eds. *Variations in Autotrophic Life*. London, UK: Academic Press, pp. 271–337.

Fauque G, Lino AR, Czechowski M, Kang L, DerVertanian DV, Moura JJG, LeGall J, Mora I. 1990. Purification and characterization of bisulfite reductase (desulfofuscidin) from *Desulfovibrio thermophilus* and its complexes with exogenous ligands. *Biochim Biophys Acta* 1040:112–118.

Felbeck H, 1981. Chemoautotrophic potential of the hydrothermal vent tube worm, *Riftia pachyptila* Jones (Vestimentifera). *Science* 213:336–338.

Felbeck H, Childress JJ, Solmero GN. 1981. Calvin–Benson cycle and sulfide oxidation enzymes in animals from sulfide-rich habitats. *Nature* (London) 293:291–293.

Fliermans CB, Brock TD. 1972. Ecology of sulfur-oxidizing bacteria in hot acid soils. *J Bacteriol* 111:343–350.

Freney JR. 1967. Sulfur-containing organics. In: McLaren AD, Petersen GH, eds. *Soil Biochemistry*. New York: Marcel Dekker, pp. 229–259.

Friedrich CG, Rother D, Bardischewsky F, Quentmeier A, Fischer J. 2001. Oxidation of reduced inorganic sulfur compounds by bacteria. Emergence of a common mechanism? *Appl Environ Microbiol* 67:2873–2882.

Fründ C, Cohen Y. 1992. Diurnal cycles of sulfate reduction under oxic conditions in cyanobacterial mats. *Appl Environ Microbiol* 58:70–77.

Fukui M, Teske A, Assmus F, Muyzer G, Widdel F. 1999. Physiology, phylogenetic relationships, and ecology of filamentous sulfate-reducing bacteria (genus *Desulfonema*). *Arch Microbiol* 172:193–203.

Garlick S, Oren A, Padan E. 1977. Occurrence of facultative anoxygenic photosynthesis among filamentous and unicellular cyanobacteria. *J Bacteriol* 129:623–629.

Gebhardt NA, Linder D, Thauer RK. 1983. Anaerobic oxidation of CO_2 by *Desulfobacter postgatei*. 2. Evidence from ^{14}C-labelling studies for the operation of the citric acid cycle. *Arch Microbiol* 136:230–233.

Gebhardt NA, Thauer RK, Linder D, Kaulfers P-M, Pfennig N. 1985. Mechanism of acetate oxidation to CO_2 with elemental sulfur in *Desulfuromonas acetoxidans*. *Arch Microbiol* 141:392–398.

Gevertz D, Telang AJ, Voordrouw G, Jenneman GE. 2000. Isolation and characterization of strains CVO and FWKO B, two novel nitrate-reducing, sulfide-oxidizing bacteria isolated from oil field brine. *Appl Environ Microbiol* 66:2491–2501.

Goffredi SK, Childress JJ, Desaulniers NT, Lallier FH. 1997. Sulfide acquisition by the vent worm *Riftia pachyptila* appears to be via uptake of HS^-, rather than H_2S. *J Exp Biol* 200 (pt 20):2609–2616.

Goldschmidt VM. 1954. *Geochemistry*. Oxford, UK: Clarendon Press, pp. 621–642.

Golovacheva RS, Karavaiko GI. 1978. *Sulfobacillus*, a new genus of thermophilic sporeforming bacteria. *Mikrobiologiya* 47:815–822 (Engl transl, pp. 658–665).

Grayston SJ, Wainright M. 1988. Sulfur oxidation by soil fungi including species of mycorrhizae and wood-rotting basidiomycetes. *FEMS Microbiol Ecol* 53:1–8.

Guittonneau G. 1927. Sur l'oxidation microbienne du soufre au cours de l'ammonisation. *CR Acad Sci* (Paris) 184:45–46.

Guittonneau G, Keiling J. 1927. Sur la solubilisation du soufre élémentaire et la formation des hyposulfides dans une terre riche en azote organique. *CR Acad Sci* (Paris) 184:898–901.

Hallberg KB, Dopson M, Lindström EB. 1996. Reduced sulfur compound oxidation by *Thiobacillus caldus*. *J Bacteriol* 178:6–11.

Hallberg KB, Lindström EB. 1994. Characterization of *Thiobacillus caldus*, sp. nov., a moderately thermophilic acidophile. *Microbiology* (Reading) 140:3451–3456.

Hansen TA, van Gemerden H. 1972. Sulfide utilization by purple sulfur bacteria. *Arch Microbiol* 86:49–56.

Hansen TA, Veldkamp H. 1973. *Rhodopseudomonas sulfidophila* nov. spec., a new species of the purple non-sulfur bacteria. *Arch Microbiol* 92:45–58.

Harrison AG, Thode H. 1957. The kinetic isotope effect in the chemical reduction of sulfate. *Trans Faraday Soc* 53:1–4.

Harrison AG, Thode H. 1958. Mechanism of the bacterial reduction of sulfate from isotope fractionation studies. *Trans Faraday Soc* 54:84–92.

Harrison G, Curle C, Laishley EJ. 1984. Purification and characterization of an inducible dissimilatory type of sulfite reductase from *Clostridium pasteurianum*. *Arch Microbiol* 138:172–178.

Harrison GI, Laishley EJ, Krouse HR. 1980. Stable isotope fractionation by *Clostridium pasteurianum*. 3. Effect of SeO_3^- on the physiology of associated sulfur isotope fractionation during SO_3^{2-} and SO_4^{2-} reduction. *Can J Microbiol* 26:952–958.

Hoefs J. 1997. *Stable Isotope Geochemistry*. 4th ed. Berlin: Springer.

Holt JG, ed. 1984. *Bergey's Manual of Systematic Bacteriology*. Vol 1. Baltimore, MD: Wiliams & Wilkins.

Holthuijzen YA, van Breemen JFL, Konings WN, van Bruggen EFJ. 1986a. Electron microscopic studies of carboxysomes of *Thiobacillus neapolitanus*. *Arch Microbiol* 144:258–262.

Holthuijzen YA, van Breemen JFL, Kuenen JG, Konings WN. 1986b. Protein composition of the carboxysomes of *Thiobacillus neapolitanus*. *Arch Microbiol* 144:398–404.

Holthuijzen YA, Van Dissel-Emiliani FFM, Kuenen JG, Konings WN. 1987. Energetic aspects of CO_2 uptake in *Thiobacillus neapolitanus*. *Arch Microbiol* 147:285–290.

Huber R, Kristjansson JK, Stetter KO. 1987. *Pyrobaculum* gen. nov., a new genus of neutrophilic, rod-shaped archaebacteria from continental solfataras growing optimally at 100°C. *Arch Microbiol* 149:95–101.

Hügler M, Wirsen CO, Fuchs G, Taylor CD, Sievert SM. 2005. Evidence for autotrophic CO_2 fixation via the reductive tricarboxylic acid cycle by members of the ε subdivision of Proteobacteria. *J. Bacteriol* 187:3020–3027.

Imhoff-Stuckle D, Pfennig N. 1983. Isolation and characterization of a nicotinic acid-degrading sulfate-reducing bacterium, *Desulfococcus niacini* sp. nov. *Arch Microbiol* 136:194–198.

Isaksen MF, Teske A. 1996. *Desulforhopalus vacuolatus* gen. nov., spec. nov., a new moderately psychrophilic sulfate-reducing bacterium with gas vacuoles isolated from a temperate estuary. *Arch Microbiol* 166:160–168.

Ivanov MV. 1968. *Microbiological Processes in the Formation of Sulfur Deposits.* Israel Program for Scientific Translations. Washington, DC. US Department of Agriculture and National Science Foundation.

Jackson BE, McInerney MJ. 2000. Thiosulfate disproportionation by *Desulfotomaculum thermobenzoicum. Appl Environ Microbiol* 66:3650–3653.

Jannasch HW. 1984. Microbial processes at deep-sea hydrothermal vents. In: Rona PA, Bostrom K, Laubier L, Smith KL Jr, eds. *Hydrothermal Processes at Sea Floor Spreading Centers.* New York: Plenum Press, pp. 677–709.

Jannasch HW, Huber R, Belkin S, Stetter KO. 1988b. *Thermotoga neapolitana* sp. nov. of the extremely thermophilic, eubacterial genus *Thermotoga. Arch Microbiol* 150:103–104.

Jannasch HW, Mottl MJ. 1985. Geomicrobiology of deep-sea hydrothermal vents. *Science* 229:717–725.

Jannasch HW, Nelson DC, Wirsen CO. 1989. Massive natural occurrence of unusually large bacteria (*Beggiatoa* sp.) at a hydrothermal deep-sea vent site. *Nature* (London) 342:834–836.

Jannasch HW, Taylor CD. 1984. Deep-sea microbiology. *Annu Rev Microbiol* 38:487–514.

Jannasch HW, Wirsen CO, Molyneaux SJ, Langworthy TA. 1988a. Extremely thermophilic fermentative archaebacteria of the genus *Desulfurococcus* from deep-sea hydrothermal vents. *Appl Environ Microbiol* 54:1203–1209.

Jansen K, Thauer RK, Widdel F, Fuchs G. 1984. Carbon assimilation pathways in sulfate reducing bacteria. Formate, carbon dioxide, carbon monoxide, and acetate assimilation by *Desulfovibrio baarsii. Arch Microbiol* 138:257–262.

Janssen PH, Morgan HW. 1992. Heterotrophic sulfur reduction by *Thermotoga* sp. strain FjSS3B1. *FEMS Microbiol Lett* 96:213–218.

Janssen PH, Schink B. 1995. Metabolic pathways and energetics of the acetone-oxidizing sulfate-reducing bacterium, *Desulfobacterium cetonicum. Arch Microbiol* 163:188–194.

Janssen PH, Schuhmann A, Bak F, Liesack W. 1996. Disproportionation of inorganic sulfur compounds by the sulfate-reducing bacterium *Desulfocapsa thiozymogenes* gen. nov., spec. nov. *Arch Microbiol* 166:184–192.

Jensen ML. 1968. Isotopic geology and the origin of Gulf Coast and Sicilian sulfur deposits. In: International Conference on Saline Deposits. 1962. Special Paper 88, Boulder, CO: Geological Society of America, pp. 526–536.

Jones GE, Starkey RL. 1957. Fractionation of stable isotopes of sulfur by microorganisms and their role in deposition of native sulfur. *Appl Microbiol* 5:111–118.

Jones GE, Starkey RL. 1962. Some necessary conditions for fractionation of stable isotopes of sulfur by *Desulfovibrio desulfuricans.* In: Jensen ML, ed. *Biogeochemistry of Sulfur Isotopes.* NSF Symposium. New Haven, CT: Yale University Press, pp. 61–79.

Jones GE, Starkey RL, Feely HW, Kulp JL. 1956. Biological origin of native sulfur in salt domes of Texas and Louisiana. *Science* 123:1124–1125.

Jones ML. 1981. *Riftia pachyptila* Jones: Observations on the vestimentiferan worm from the Galápagos Rift. *Science* 213:333–336.

Jørgensen BB. 1990a. A thiosulfate shunt in the sulfur cycle of marine sediments. *Science* 249:152–154.

Jørgensen BB. 1990b. The sulfur cycle of freshwater sediments: Role of thiosulfate. *Limnol Oceanogr* 35:1329–1342.

Jørgensen BB, Bak F. 1991. Pathways of microbiology of thiosulfate transformations and sulfate reduction in marine sediment (Kattegat, Denmark). *Appl Environ Microbiol* 57:847–856.

Jørgensen BB, Isaksen MF, Jannasch HW. 1992. Bacterial sulfate reduction above 100°C in deep-sea hydrothermal vent sediments. *Science* 258:1756–1757.

Jørgensen BB, Kuenen JG, Cohen Y. 1979a. Microbial transformations of sulfur compounds in a stratified lake (Solar Lake, Sinai). *Limnol Oceanogr* 24:799–822.

Jørgensen BB, Revsbech NP, Blackburn H, Cohen Y. 1979b. Diurnal cycle of oxygen and sulfide microgradient and microbial photosynthesis in a cyanobacterial mat sediment. *Appl Environ Microbiol* 38:46–58.

Justin P, Kelly DP. 1978. Growth kinetics of *Thiobacillus denitrificans* in anaerobic and aerobic chemostat culture. *J Gen Microbiol* 107:123–300.

Kaplan IR, Rafter TA, Hulston JR. 1960. Sulfur isotopic variations in nature. Part 8. Applications to some biochemical problems. *NZ J Sci* 3:338–361.

Kaplan IR, Rittenberg SC. 1962. The microbiological fractionation of sulfur isotopes. In: Jensen MLL, ed. *Biogeochemistry of Sulfur Isotopes.* NSF Symposium. New Haven, CT: Yale University Press, pp. 80–93.

Kaplan IR, Rittenberg SC. 1964. Microbiological fractionation of sulfur isotopes. *J Gen Microbiol* 34:195–212.

Karavaiko GI, Bogdanova TI, Tourova TP, Kondrat'eva TF, Tsalpina IA, Egorova MA, Karsil'nikova EN, Zakharchuk LM. 2005. Reclassification of 'Sulfobacillus thermosulfidooxidans subsp. thermotolerans' strain K1 as Alicyclobacillus tolerans sp. nov. and Sulfobacillus disulfidooxidans Dufresne et al. 1996 as Alicyclobacillus disulfidooxidans comb. nov., and emended description of the genus Alicyclobacillus. Int J Syst Evol Microbiol 55:941–947.

Karavaiko GI, Golyshina OV, Troitskii AV, Valicho-Roman KKM. Golovacheva RS, Pivovarova TA. 1994. Sulfurococcus yellowstonii sp. nov., a new species of iron- and sulfur-oxidizing thermoacidophilic archaebacteria. Mikrobiologiya 63:668–682 (Engl transl, 379–387).

Keith SM, Herbert RA. 1983. Dissimilatory nitrate reduction by a strain of Desulfovibrio desulfuricans. FEMS Microbiol Lett 18:55–59.

Kelly DP. 1982. Biochemistry of the chemolithotrophic oxidation of inorganic sulfur. Phil Trans Roy Soc Lond B 298:499–528.

Kelly DP, Lu W-P, Poole PK. 1993. Cytochromes in Thiobacillus tepidarius and the respiratory chain involved in the oxidation of thiosulfate and tetrathionate. Arch Microbiol 160:87–95.

Kelly DP, Wood AP. 2000. Reclassification of some species of Thiobacillus to the newly designated genera Acidithiobacillus gen. nov., Halobacillus gen. nov. and Thermithiobacillus gen. nov. Int J Syst Evol Microbiol 50:511–516.

Kemp ALW, Thode HG. 1968. The mechanism of the bacterial reduction of sulfate and sulfite from isotope fractionation studies. Geochim Cosmochim Acta 32:71–91.

Kobiyashi KS, Tashibana S, Ishimoto M. 1969. Intermediary formation of trithionate in sulfite reduction by a sulfate-reducing bacterium. J Biochem (Tokyo) 65:155–157.

Kobiyashi KS, Tukahashi E, Ishimoto M. 1972. Biochemical studies on sulfate-reducing bacteria. XI. Purification and some properties of sulfite reductase, desulfoviridin. J Biochem (Tokyo) 72:879–887.

Kramer JF, Pope DH, Salerno JC. 1987. Pathways of electron transfer in Desulfovibrio. In: Kim CH, Tedeschi H, Diwan JJ, Salerno JC, eds. Advances in Membrane Biochemistry and Bioenergetics. New York: Plenum Press, pp. 249–258.

Krekeler D, Sigalevich P, Teske A, Cypionka H, Cohen Y. 1997. A sulfate-reducing bacterium from the oxic layer of a microbial mat from Solar Lake (Sinai), Desulfovibrio oxyclinae sp. nov. Arch Microbiol 167:369–375.

Kuenen JG, Beudecker RF. 1982. Microbiology of thiobacilli and other sulfur-oxidizing autotrophs, mixotrophs, and heterotrophs. Phil Trans Roy Soc Lond B 298:473–497.

Kuenen JG, Tuovinen OH. 1981. The genera Thiobacillus and Thiomicrospira. In: Starr MP, Stolp H, Trüper HG, Balows A, Schlegel H, eds. The Prokaryotes: A Handbook of Habitats, Isolation and Identification of Bacteria. Berlin, Germany: Springer, pp. 1023–1036.

Kuever J, Kulmer J, Jannsen S, Fischer U, Blotevogel K-H. 1993. Isolation and characterization of a new sporeforming sulfate-reducing bacterium growing by complete oxidation of catechol. Arch Microbiol 159:282–288.

Laishley EJ, Krouse HR. 1978. Stable isotope fractionation by Clostridium pasteurianum. 2. Regulation of sulfite reductase by sulfur amino acids and their influence on sulfur isotope fractionation during SO_3^{2-} and SO_4^{2-} reduction. Can J Microbiol 24:716–724.

LaRiviere JWM, Schmidt K. 1981. Morphologically conspicuous sulfur-oxidizing bacteria. In: Starr MP, Stolp H, Trüper HG, Balows A, Schlegel H. eds. The Prokaryotes: A Handbook of Habitats, Isolation and Identification of Bacteria. Berlin, Germany: Springer, pp.1037–1048.

Lawton K. 1955. Chemical composition of soils. In: Bear FE, ed. Chemistry of the Soil. New York: Reinhold, pp. 53–84.

Lee JP, Peck HD Jr. 1971. Purification of the enzyme reducing bisulfite to trithionate from Desulfovibrio gigas and its identification as desulfoviridin. Biochem Biophys Res Commun 45:583–589.

Lee JP, Yi CS, LeGall J, Peck HD. 1973. Isolation of a new pigment, desulfoviridin, from Desulfovibrio desulfuricans (Norway strain) and its role in sulfite reduction. J Bacteriol 115:453–455.

LeRoux N, Wakerley DS, Hunt SD. 1977. Thermophilic thiobacillus-type bacteria from Icelandic thermal areas. J Gen Microbiol 100:197–201.

Liu C-L, Hart N, Peck HD Jr. 1982. Inorganic pyrophosphate: Energy source for sulfate-reducing bacteria of the genus Desulfotomaculum. Science 217:363–364.

Ljunggren P. 1960. A sulfur mud deposit formed through bacterial transformation of fumarolic hydrogen sulfide. Econ Geol Bull Soc Econ Geol 55:531–538.

London J. 1963. Thiobacillus intermedius nov. sp. a novel type of facultative autotroph. Arch Microbiol 46:329–337.

London J, Rittenberg SC. 1964. Path of sulfur in sulfide and thiosulfate oxidation by thiobacilli. *Proc Natl Acad Sci USA* 52:1183–1190.

London J, Rittenberg SC. 1966. Effects of organic matter on the growth of *Thiobacillus intermedius*. *J Bacteriol* 91:1062–1069.

London J, Rittenberg SC. 1967. *Thiobacillus perometabolis* nov. sp., a non-autotrophic thiobacillus. *Arch Microbiol* 59:218–225.

Londry KL, Suflita JM, Tanner RS. 1999. Cresol metabolism by the sulfate-reducing bacterium *Desulfotomaculum* sp. strain Groll. *Can J Microbiol* 45:458–463.

Lovley DR, Roden EE, Phillips EJP, Woodward JC. 1993. Enzymatic iron and uranium reduction by sulfate-reducing bacteria. *Mar Geol* 113:41–53.

Loya S, Yanofsky SA, Epel BL. 1982. Characterization of cytochromes in lithotrophically and organotrophically grown cells of *Thiobacillus* A_2. *J Gen Microbiol* 128:2371–2378.

Lu W-P. 1986. A periplasmic location for the bisulfite-oxidizing multienzyme system from *Thiobacillus versutus*. *FEMS Microbiol Lett* 34:313–317.

Lu W-P, Kelly DP. 1983. Purification and some properties of two principal enzymes of the thiosulfate-oxidizing multienzyme system from *Thiobacillus* A_2. *J Gen Microbiol* 129:3549–3562.

Lu W-P, Kelly DP. 1988. Respiration-driven proton translocation in *Thiobacillus versutus* and the role of the periplasmic thiosulfate-oxidizing enzyme system. *Arch Microbiol* 149:297–302.

Ma K, Weiss R, Adams MWW. 2000. Characterization of hydrogenase II from the hyperthermophilic archeon *Pyrococcus furiosus* and assessment of its role in sulfur reduction. *J Bacteriol* 182:1864–1871.

Macnamara J, Thode H. 1951. The distribution of ^{34}S in nature and the origin of native sulfur deposits. *Research* 4:582–583.

Madigan MT, Brock TD. 1975. Photosynthetic sulfide oxidation by *Chloroflexus aurantiacus*, a filamentous, photosynthetic, gliding bacterium. *J Bacteriol* 122:782–784.

Markosyan GE. 1973. A new mixotrophic sulfur bacterium developing in acidic media, *Thiobacillus organoparus* sp. n. *Dokl Akad Nauk SSSR Ser Biol* 211:1205–1208.

Marshall C, Frenzel P, Cypionka H. 1993. Influence of oxygen on sulfur reduction and growth of sulfate-reducing bacteria. *Arch Microbiol* 159:168–173.

Martinez JD. 1991. Salt domes. *Am Sci* 79:420–431.

Mason J, Kelly DP. 1988. Mixotrophic and autotrophic growth of *Thiobacillus acidophilus* on tetrathionate. *Arch Microbiol* 149:317–323.

McCready RGL, Gould WD, Cook FD. 1983. Respiratory nitrate reduction by *Desulfovibrio* sp. *Arch Microbiol* 135:182–185.

McCready RGL, Laishley EJ, Krouse HR. 1975. Stable isotope fractionation by *Clostridium pasteurianum*. 1. ^{34}S/^{32}S: Inverse isotope effects during SO_4^{2-} and SO_3^{2-} reduction. *Can J Microbiol* 21:235–244.

Mechalas BJ, Rittenberg SC. 1960. Energy coupling in *Desulfovibrio desulfuricans*. *J Bacteriol* 80:501–507.

Meckenstock RU, Annweiler E, Michaelis W, Richnow HH, Schink B. 2000. Anaerobic naphthalene degradation by a sulfate-reducing enrichment culture. *Appl Environ Microbiol* 66:2743–2747.

Meulenberg R, Pronk JT, Hazeu W, van Dijken JP, Frank J, Bos P, Kuenen JG. 1993. Purification and partial characterization of thiosulfate dehydrogenase from *Thiobacillus acidophilus*. *J Gen Microbiol* 139:2033–2039.

Milhaud G, Aubert JP, Millet J. 1958. Role physiologique du cytochrome C de la bactérie chemieautotrophe *Thiobacillus denitrificans*. *CR Acad Sci* (Paris) 246:1766–1769.

Miller JDA, Wakerley DS. 1966. Growth of sulfate-reducing bacteria by fumarate dismutation. *J Gen Microbiol* 43:101–107.

Minz D, Fishbain S, Green SJ, Muyzer G, Cohen Y, Rittman BE, Stahl DA. 1999a. Unexpected population distribution in a microbial mat community: Sulfate-reducing bacteria localized to the highly oxic chemocline in contrast to eukaryotic preference for anoxia. *Appl Environ Microbiol* 65:4659–4665.

Minz D, Flax JL, Green SJ, Muyzer G, Cohen Y, Wagner M, Rittman BE, Stahl DA. 1999b. Diversity of sulfate-reducing bacteria in oxic and anoxic regions of a microbial mat characterized by comparative analysis of dissimilatory sulfite reductase genes. *Appl Environ Microbiol* 65:4666–4671.

Mitchell GJ, Jones JG, Cole JA. 1986. Distribution and regulation of nitrate and nitrite reduction by *Desulfovibrio* and *Desulfotomaculum* species. *Arch Microbiol* 144:35–40.

Möller D, Schauder R, Fuchs G, Thauer RK. 1987. Acetate oxidation to CO_2 via a citric acid cycle involving an ATP-citrate lyase: A mechanism for the synthesis of ATP via substrate-level phosphorylation in *Desulfobacter postgatei* growing on acetate and sulfate. *Arch Microbiol* 148:202–207.

Moriarty DJW, Nicholas DJD. 1970. Electron transfer during sulfide and sulfite oxidation by *Thiobacillus concretivorus*. *Biochim Biophys Acta* 216:130–138.

Mosser JL, Bohlool BB, Brock TD. 1974. Growth rates of *Sulfolobus acidocaldarius* in nature. *J Bacteriol* 118:1075–1081.

Mosser JL, Mosser AG, Brock TD. 1973. Bacterial origin of sulfuric acid in geothermal habitats. *Science* 179:1323–1324.

Nathansohn A. 1902. Über eine neue Gruppe von Schwefelbakterien und ihren Stoffwechsel. *Mitt Zool Sta Neapel* 15:655–680.

Nelson DC, Castenholz RW. 1981. Use of reduced sulfur compounds by *Beggiatoa* sp. *J Bacteriol* 147:140–154.

Nelson DC, Jannasch HW. 1983. Chemoautotrophic growth of a marine *Beggiatoa* in sulfide-gradient cultures. *Arch Microbiol* 136:262–269.

Neutzling O, Pfleiderer C, Trüper HG. 1985. Dissimilatory sulfur metabolism in phototrophic "non-sulfur" bacteria. *J Gen Microbiol* 131:791–798.

Nübel T, Klughammer C, Huber R, Hauska G, Schütz M. 2000. Sulfide:quinone oxidoreductase in membranes of the hyperthermophilic bacterium *Aquifex aeolicus* (VF5). *Arch Microbiol* 173:233–244.

Odom JM, Peck HD Jr. 1981. Hydrogen cycling as a general mechanism for energy coupling in the sulfate-reducing bacteria, *Desulfovibrio* sp. *FEMS Microbiol Lett* 12:47–50.

Odom JM, Wall JD. 1987. Properties of a hydrogen-inhibited mutant of *Desulfovibrio desulfuricans* ATCC 27774. *J Bacteriol* 169:1335–1337.

Oren A, Shilo M. 1979. Anaerobic heterotrophic dark metabolism in the cyanobacterium *Oscillatoria limnetica*: Sulfur respiration and lactate fermentation. *Arch Microbiol* 122:77–84.

Ozawa K, Meikari T, Motohashi K, Yoshida M, Akutsu H. 2000. Evidence for the presence of an F-type ATP synthase involved in sulfate respiration in *Desulfovibrio vulgaris*. *J Bacteriol* 182:2200–2206.

Panganiban AT, Hanson RS. 1976. Isolation of a bacterium that oxidizes methane in the absence of oxygen. *Abstr Annu Meet Am Soc Microbiol* 159:121.

Panganiban AT, Patt TE, Hart W, Hanson RS. 1979. Oxidation of methane in the absence of oxygen in lake water samples. *Appl Environ Microbiol* 37:303–309.

Pankhania IP, Sporman AM, Hamilton WA, Thauer RK. 1988. Lactate conversion to acetate, CO_2, and H_2 in cell suspensions of *Desulfovibrio vulgaris* (Marburg): Indications for the involvement of an energy driven reaction. *Arch Microbiol* 150:26–31.

Paschinger H, Paschinger J, Gaffron H. 1974. Photochemical disproportionation of sulfur into sulfide and sulfate by *Chlorobium limicola* forma *thiosulfatophilum*. *Arch Microbiol* 96:341–351.

Paulsen J, Kröger A, Thauer RK. 1986. ATP-driven succinate oxidation in the catabolism of *Desulfuromonas acetoxidans*. *Arch Microbiol* 144:78–83.

Peck HD Jr. 1962. Symposium on metabolism of inorganic compounds. V. Comparative metabolism of inorganic sulfur compounds in microorganisms. *Bacteriol Rev* 26:67–94.

Peck HD Jr. 1993. Bioenergetic strategies of the sulfate-reducing bacteria. In: Odom JM, Singleton Jr, eds. *The Sulfate-Reducing Bacteria: Contemporary Perspectives*. New York: Springer, pp. 41–76.

Peck HD Jr., LeGall J. 1982. Biochemistry of dissimilatory sulfate reduction. *Phil Trans Roy Soc Lond B* 298:443–466.

Peeters T, Aleem MIH. 1970. Oxidation of sulfur compounds and electron transport in *Thiobacillus denitrificans*. *Arch Microbiol* 71:319–330.

Perry KA, Kostka JE, Luther GW III, Nealson KH. 1993. Mediation of sulfur speciation by a Black Sea facultative anaerobe. *Science* 259:801–803.

Pfennig N. 1977. Phototrophic green and purple bacteria: A comparative, systematic survey. *Annu Rev Microbiol* 31:275–290.

Pfennig N, Biebl H. 1976. *Desulfuromonas acetoxidans* gen nov. and sp. nov., a new anaerobic, sulfur-reducing, acetate-oxidizing bacterium. *Arch Microbiol* 110:3–12.

Pfennig N, Biebl H. 1981. The dissimilatory sulfur-reducing bacteria. In: Starr MP, Stolp H, Trüper HG, Balows A, Schlegel H, eds. *The Prokaryotes: A Handbook of Habitats, Isolation and Identification of Bacteria*. Vol 1. Berlin, Germany: Springer, pp. 941–947.

Pfennig N, Widdel F, Trüper HG. 1981. The dissimilatory sulfate-reducing bacteria. In: Starr MP, Stolp H, Trüper HG, Balows A, Schlegel H, eds. *The Prokaryotes: A Handbook of Habitats, Isolation and Identification of Bacteria*. Vol 1. Berlin, Germany: Springer, pp. 926–940.

Pierson BK, Castenholz RW. 1974. A photosynthetic gliding filamentous bacterium of hot springs, *Chloroflexus aurantiacus* gen. nov. and spec. nov. *Arch Microbiol* 100:5–24.

Platen H, Temmes A, Schink B. 1990. Anaerobic degradation of acetone by *Desulfurococcus biacutus* spec. nov. *Arch Microbiol* 154:355–361.

Postgate JR. 1952. Growth of sulfate reducing bacteria in sulfate-free media. *Research* 5:189–190.

Postgate JR. 1963. Sulfate-free growth of *Cl. nigrificans*. *J Bacteriol* 85:1450–1451.

Postgate JR. 1984. *The Sulfate-Reducing Bacteria*. 2nd ed. Cambridge, UK: Cambridge University Press.

Powell MA, Somero GN. 1983. Blood components prevent sulfide poisoning of respiration of the hydrothermal vent tube worm *Riftia pachyptila*. *Science* 219:297–299.

Pringsheim EG. 1967. Die Mixotrophie von *Beggiatoa*. *Arch Mikrobiol* 59:247–254.

Pronk JT, De Bruyn JC, Bos P, Kuenen JG. 1992. Anaerobic growth of *Thiobacillus ferrooxidans*. *Appl Environ Microbiol* 58:2227–2230.

Pronk JT, Liem K, Bos P, Kuenen JG. 1991. Energy transduction by anaerobic ferric iron respiration in *Thiobacillus ferrooxidans*. *Appl Environ Microbiol* 57:2063–2068.

Pronk JT, Meulenberg R, Hazeu W, Bos P, Kuenen JG. 1990. Oxidation of reduced inorganic sulfur compounds by acidophilic thiobacilli. *FEMS Microbiol Rev* 75:293–306.

Qatabi AI, Nivière V, Garcia JL. 1991. *Desulfovibrio alcoholovorans* sp. nov., a sulfate-reducing bacterium able to grow on glycerol, 1,2- and 1,3-propanol. *Arch Microbiol* 155:143–148.

Rau GH. 1981. Hydrothermal vent clam and tube worm $^{13}C/^{12}C$: Further evidence of nonphotosynthetic food sources. *Science* 213:338–340.

Rau GH, Hedges JI. 1979. Carbon-13 depletion in a hydrothermal vent mussel: Suggestion of a chemosynthetic food source. *Science* 203:648–649.

Ravot G, Magot M, Fardeau ML, Patel BKC, Prensier G, Egan A, Garcia JL, Ollivier B. 1995. *Thermotoga elfii* sp. nov., a novel thermophilic bacterium from an African oil-producing well. *Int J Syst Bacteriol* 45:308–314.

Rees GN, Harfoot CG, Sheehy AJ. 1998. Amino acid degradation by the mesophilic sulfate-reducing bacterium *Desulfobacterium vacuolatum*. *Arch Microbiol* 169:76–80.

Roy AB, Trudinger PA. 1970. *The Biochemistry of Inorganic Compounds of Sulfur*. Cambridge, UK: Cambridge University Press.

Rueter P, Rabus R, Wilkes H, Aeckersberg F, Rainey FA, Jannasch HW, Widdel F. 1994. Anaerobic oxidation of hydrocarbons in crude oil by new types of sulfate-reducing bacteria. *Nature* (London) 372:455–458.

Sadler MH, Johnson EJ. 1972. A comparison of the NADH oxidase electron transport system of two obligately chemolithotrophic bacteria. *Biochim Biophys Acta* 283:167–179.

Schauder R, Eikmanns B, Thauer RK, Widdel F, Fuchs G. 1986. Acetate oxidation to CO_2 in anaerobic bacteria via a novel pathway not involving reactions of ther citric acid cycle. *Arch Microbiol* 145:162–172.

Schauder R, Kröger A. 1993. Bacterial sulfur respiration. *Arch Microbiol* 159:491–497.

Schauder R, Müller E. 1993. Polysulfide as a possible substrate for sulfur-reducing bacteria. *Arch Microbiol* 160:377–382.

Schauder R, Preuss A, Jetten M, Fuchs G. 1989. Oxidative and reductive acetyl CoA/carbon monoxide dehydrogenase pathway in *Desulfobacterium autotrophicum*. *Arch Microbiol* 151:84–89.

Schauder R, Widdel F, Fuchs G. 1987. Carbon assimilation pathways in sulfate-reducing bacteria. II. Enzymes of a reductive citric acid cycle in autotrophic *Desulfobacter hydrogenophilus*. *Arch Microbiol* 148:218–225.

Schedel M, Trüper HG. 1980. Anaerobic oxidation of thiosulfate and elemental sulfur in *Thiobacillus denitrificans*. *Arch Microbiol* 124:205–210.

Schicho RN, Ma K, Adams MWW, Kelly RM. 1993. Bioenergetics of sulfur reduction in the hyperthermophilic archeon *Pyrococcus furiosus*. *J Bacteriol* 175:1823–1830.

Schnell S, Schink B. 1991. Anaerobic aniline degradation via reductive deamination of a 4-aminobenzoyl~CoA in *Desulfobacterium anilini*. *Arch Microbiol* 155:183–190.

Schoen R, Ehrlich GG. 1968. Bacterial origin of sulfuric acid in sulfurous hot springs. *23rd Int Geol Congr* 17:171–178.

Schoen R, Rye RO. 1970. Sulfur isotope distribution in solfataras, Yellowstone National Park. *Science* 170:1082–1084.

Segerer A, Neuner A, Kristiansson JK, Stetter KO. 1986. *Acidianus infernos* gen. nov., sp. nov., and *Acidianus brierleyi* comb. nov.: Facultative aerobic, extremely acidophilic thermophilic sulfur-metabolizing archaebacteria. *Int J Syst Bacteriol* 36:559–564.

Seitz H-J, Cypionka H. 1986. Chemolithotrophic growth of *Desulfovibrio desulfuricans* with hydrogen coupled to ammonification of nitrate or nitrite. *Arch Microbiol* 146:63–67.

Selig M, Schönheit P. 1994. Oxidation of organic compounds to CO_2 with sulfur or thiosulfate as electron acceptor in the anaerobic hyperthermophilic archaea *Thermoproteus tenax* and *Pyrobaculum islandicum* proceeds via the citric acid cycle. *Arch Microbiol* 162:286–294.

Shively JM, Ball F, Brown DH, Saunders RE. 1973. Functional organelles in prokaryotes: Polyhedral inclusions (carboxysomes) of *Thiobacillus neapolitanus*. *Science* 182:584–586.

Shivvers DW, Brock TD. 1973. Oxidation of elemental sulfur by *Sulfolobus acidocaldarius*. *J Bacteriol* 114:706–710.

Shturm LD. 1948. Sulfate reduction by facultative aerobic bacteria. *Mikrobiologiya* 17:415–418.

Sigalevich P, Baev MV, Teske A, Cohen Y. 2000a. Sulfate reduction and possible aerobic metabolism of the sulfate-reducing bacterium *Desulfovibrio oxyclinae* in a chemostat coculture with *Marinobacter* sp. strain MB under exposure of increasing oxygen concentrations. *Appl Environ Microbiol* 66:5013–5018.

Sigalevich P, Cohen Y. 2000. Oxygen-dependent growth of sulfur-reducing bacterium *Desulfovibrio oxyclinae* in coculture with *Marinobacter* sp. strain MB in an aerated sulfate-dependent chemostat. *Appl Environ Microbiol* 66:5019–5023.

Sigalevich P, Meshorer E, Helman Y, Cohen Y. 2000b. Transition from anaerobic to aerobic growth conditions for the sulfate-reducing bacterium *Desulfovibrio oxyclinae* results in flocculation. *Appl Environ Microbiol* 66:5005–5012.

Skerman VBD, Dementyeva G, Carey B. 1957a. Intracellular deposition of sulfur by *Sphaerotilus natans*. *J Bacteriol* 73:504–512.

Skerman VBD, Dementyeva G, Skyring GW. 1957b. Deposition of sulfur from hydrogen sulfide by bacteria and yeasts. *Nature* (London) 179:742.

Skirnisdottir S, Hreggvidsson GD, Hjörleifsdottir S, Marteinsson VT, Petursdottir SK, Holst O, Qrist jansson JK. 2000. Influence of sulfide and temperature on species composition and community structure of hot spring microbial mats. *Appl Environ Microbiol* 66:2835–2841.

Skyring GW. 1987. Sulfur reduction in coast ecosystems. *Geomicrobiol J* 5:295–374.

Smith AL, Kelly DP, Wood AP. 1980. Metabolism of *Thiobacillus* A_2 grown under autotrophic, mixotrophic, and heterotrophic conditions in a chemostat culture. *J Gen Microbiol* 121:127–138.

Smith DW, Rittenberg SC. 1974. On the sulfur-source requirement for growth of *Thiobacillus intermedius*. *Arch Microbiol* 100:65–71.

Smith DW, Strohl WR. 1991. Sulfur-oxidizing bacteria. In: Shively JM, Barton LL, eds. *Variations in Autotrophic Life*. London, UK: Academic Press, pp. 121–146.

Smock AM, Böttcher ME, Cypionka H. 1998. Fractionation of sulfur isotopes during thiosulfate reduction by *Desulfovibrio desulfuricans*. *Arch Microbiol* 169:460–463.

Sokolova GA. 1962. Microbiological sulfur formation in Sulfur Lake. *Mikrobiologiya* 31:324–327 (Engl transl, pp. 264–266).

Sokolova GA, Karavaiko GI. 1968. *Physiology and Geochemical Activity of Thiobacilli*. Sprinfield, VA: US Department of Commerce. Clearinghouse Fed Tech Info (Engl transl).

Sorokin YI. 1966a. Role of carbon dioxide and acetate in biosynthesis of sulfate-reducing bacteria. *Nature* (London) 210:551–552.

Sorokin YI. 1966b. Sources of energy and carbon for biosynthesis by sulfate-reducing bacteria. *Mikrobiologiya* 35:761–766 (Engl transl, pp. 643–647).

Sorokin YI. 1966c. Investigation of the structural metabolism of sulfate-reducing bacteria with [14]C. *Mikrobiologiya* 35:967–977 (Engl transl, pp. 806–814).

Sorokin YI. 1966d. The role of carbon dioxide and acetate in biosynthesis in sulfate reducing bacteria. *Dokl Akad Nauk SSSR* 168:199.

Sorokin YI. 1970. The mechanism of chemical and biological oxidation of sodium, calcium, and iron sulfides. *Mikrobiologiya* 39:253–258 (Engl transl, pp. 220–224).

Speich N, Trüper HG. 1988. Adenylylsulfate reductase in a dissimilatory sulfate-reducing archaebacterium. *J Gen Microbiol* 134:1419–1425.

Stanier RY, Ingraham JL, Wheelis ML, Painter PR. 1986. *The Microbial World*. 5th ed. Englewood Cliffs, NJ: Prentice Hall.

Starkey RL. 1934. The production of polythionates from thiosulfate by microorganisms. *J Bacteriol* 28:387–400.

Stetter KO. 1985. Thermophilic archaebacteria occurring in submarine hydrothermal areas. In: Caldwell DE, Brierley JA, Brierley CL, eds. *Planetary Ecology*. New York: Van Nostrand Reinhold, pp. 320–332.

Stetter KO. 1988. *Archaeoglobus fulgidus* gen. nov., sp. nov.: A new taxon of extremely thermophilic archaebacteria. *Syst Appl Microbiol* 10:172–173.

Stetter KO, Koenig H, Stackebrandt E. 1983. *Pyrodictium* gen. nov., a new genus of submarine disk-shaped sulfur-reducing archaebacteria growing optimally at 105°C. *Syst Appl Microbiol* 4:535–551.

Stetter KO, Lauerer G, Thomm M, Neuner A. 1987. Isolation of extremely thermophilic sulfate reducers: Evidence for a novel branch of archaebacteria. *Science* 236:822–824.

Stetter KO, Segerer A, Zillig W, Huber G, Fiala G, Huber R, Koenig H. 1986. Extremely thermophilic sulfur-metabolizing archaebacteria. *Syst Appl Microbiol* 7:393–397.

Steuber J, Cypionka H, Kroneck PMH. 1994. Mechanism of dissimilatory sulfite reduction by *Desulfovibrio desulfuricans*: Purification of membrane-bound sulfite reductase and coupling with cytochrome c_3 and hydrogenase. *Arch Microbiol* 162:255–260.

Strahler AN. 1977. *Principles of Physical Geology*. New York: Harper & Row.

Strohl WR, Larkin JM. 1978. Enumeration, isolation, and characterization of *Beggiatoa* from freshwater sediments. *Appl Environ Microbiol* 36:755–770.

Suzuki I. 1965. Oxidation of elemental sulfur by an enzyme system of *Thiobacillus thiooxidans*. *Biochim Biophys Acta* 104:359–371.

Suzuki I. 1999. Oxidation of inorganic sulfur compounds: Chemical and enzymatic reactions. *Can J Microbiol* 45:97–105.

Suzuki I, Chan CW, Takeuchi TL. 1992. Oxidation of elemental sulfur to sulfite by *Thiobacillus thiooxidans* cells. *Appl Environ Microbiol* 58:3767–3769.

Suzuki I, Chan CW, Takeuchi TL. 1994. Oxidation of inorganic sulfur compounds by Thiobacilli. In: Alpers CN, Blowes DW, eds. *Environmental Geochemistry of Sulfide Oxidation*. ACS Symposium 550. Washington, DC: American Chemical Society, pp. 60–67.

Suzuki I, Silver M. 1966. The initial product and properties of the sulfur-oxidizing enzyme of thiobacilli. *Biochim Biophys Acta* 122:22–33.

Szewzyk R, Pfennig N. 1987. Complete oxidation of catechol by a strictly anaerobic sulfate-reducing *Desulfobacterium catecholicum* sp. nov. *Arch Microbiol* 147:163–168.

Tasaki M, Kamagata Y, Nakamura K, Mikami E. 1991. Isolation and characterization of a thermophilic benzoate-degrading sulfate-reducing bacterium, *Desulfotomaculum thermobenzoicum* sp. nov. *Arch Microbiol* 155:348–352.

Tasaki M, Kamagata Y, Nakamura K, Mikami E. 1992. Utilization of methoxylated benzoates and formation of intermediates by *Desulfotomaculum thermobenzoicum* in the presence and absence of sulfate. *Arch Microbiol* 157:209–212.

Teske A, Brinkhoff T, Muyzer G, Moser DP, Rethmeier J, Jannasch HW. 2000. Diversity of thiosulfate-oxidizing bacteria from marine sediments and hydrothermal vents. *Appl Environ Microbiol* 66:3125–3133.

Thamdrup B, Finster K, Hansen JW, Bak F. 1993. Bacterial disproportionation of elemental sulfur coupled to chemical reduction of iron and manganese. *Appl Environ Microbiol* 59:101–108.

Thode HG, Wanless K, Wallouch R. 1954. The origin of native sulfur deposits from isotopic fractionation studies. *Geochim Cosmochim Acta* 5:286–298.

Trautwein K. 1921. Beitrag zur Physiologie und Morphologie der Thionsäurebakterien. *Zentralbl Bakteriol Parasitenk Infektionskr Hyg Abt II* 53:513–548.

Trudinger PA. 1961. Thiosulfate oxidation and cytochromes in *Thiobacillus* X. 2. Thiosulfate oxidizing enzyme. *Biochem J* 78:680–686.

Trudinger PA. 1970. Carbon monoxide-reacting pigment from *Desulfotomaculum nigrificans* and its possible relevance to sulfite oxidation. *J Bacteriol* 104:158–170.

Trüper HG. 1978. Sulfur metabolism. In: Clayton RK, Sistrom WR, eds. *The Photosynthetic Bacteria*. New York: Plenum Press, pp. 677–690.

Trüper HG. 1984a. Microorganisms and the sulfur cycle. In: Müller A, Krebs B , eds. *Sulfur, Its Significance for Chemistry, for the Geo-, Bio, and Cosmosphere and Technology*, Vol 5. Amsterdam: Elsevier, pp. 351–365.

Trüper HG. 1984b. Phototrophic bacteria and the sulfur metabolism. In: Müller A, Krebs B, eds. *Sulfur, Its Significance for Chemistry, for the Geo-, Bio, and Cosmosphere and Technology*, Vol 5. Amsterdam: Elsevier, pp. 367–382.

Tuttle JH, Ehrlich HL. 1986. Coexistence of inorganic sulfur metabolism and manganese oxidation in marine bacteria. *Abstr Annu Meet Am Soc Microbiol* I-21:168.

Tuttle JH, Holmes PE, Jannasch HW. 1974. Growth rate stimulation of marine pseudomonads by thiosulfate. *Arch Microbiol* 99:1–14.

Vainshtein MB. 1977. Oxidation of hydrogen sulfide by thionic bacteria. *Mikrobiologiya* 46:1114–1116 (Engl transl, pp 898–899).

Vainshtein MB, Matrosov AG, Baskunov VB, Zyakun AM, Ivanov MV. 1981. Thiosulfate as in intermediate product of bacterial sulfate reduction. *Mikrobiologiya* 49:855–859 (Engl transl, pp. 672–675).

Vairavamurthy A, Manowitz B, Luther GW III, Jeon Y. 1993. Oxidation state of sulfur in thiosulfate and implications for anaerobic energy metabolism. *Geochim Cosmochim Acta* 57:1619–1623.

van den Ende FP, van Gemerden H. 1993. Sulfide oxidation under oxygen limitation by a *Thiobacillus thioparus* isolated from a marine microbial mat. *FEMS Microbiol Ecol* 13:69–78.

van Gemerden H. 1986. Production of elemental sulfur by green and purple sulfur bacteria. *Arch Microbiol* 146:52–56.

Vishniac W. 1952. The metabolism of *Thiobacillus thioparus*. I. The oxidation of thiosulfate. *J Bacteriol* 64:363–373.

Vishniac W, Santer M. 1957. The thiobacilli. *Bacteriol Rev* 21:195–213.

Wainer A. 1964. The production of sulfate from cysteine without the formation of free cysteinesulfinic acid. *Biochem Biophys Res Commun* 16:141–144.

Wainer A. 1967. Mitochondrial oxidation of cysteine. *Biochim Biophys Acta* 141:466–472.

Wall JD, Rapp-Giles BJ, Brown MF, White JA. 1990. Response of *Desulfovibrio desulfuricans* colonies to oxygen stress. *Can J Microbiol* 36:400–408.

Wedepohl KH. 1984. Sulfur in the Earth's crust, its origin and natural cycle. In: Müller A, Krebs B, eds. *Sulfur, Its Significance for Chemistry, for the Geo-, Bio-, and Cosmosphere and Technology*, Vol 5. Amsterdam: Elsevier, pp. 39–54.

Widdel F, Pfennig N. 1977. A new anaerobic, sporing, acetate-oxidizing, sulfate-reducing bacterium, *Desulfotomaculum* (emend) *acetoxidans*. *Arch Microbiol* 112:119–122.

Widdel F, Pfennig N. 1981. Sporulation and further nutritional characteristics of *Desulfotomaculum acetoxidans*. *Arch Microbiol* 129:401–402.

Williams CD, Nelson DC, Farah BA, Jannasch HW, Shively JM. 1988. Ribulose bisphosphate carboxylase of the prokaryotic symbiont of a hydrothermal vent tube worm: Kinetics, activity, and gene hybridization. *FEMS Microbiol Lett* 50:107–112.

Williams RD, Hoare DS. 1972. Physiology of a new facultative autotrophic thermophilic *Thiobacillus*. *J Gen Microbiol* 70:555–566.

Wirsen CO, Jannasch HW. 1978. Physiological and morphological observations on *Thiovulum* sp. *J Bacteriol* 136:765–774.

Wirsen CO, Sievert SM, Cavanaugh CM, Molyneaux SJ, Ahmad A, Taylor LT, DeLong EF, Taylor CD. 2002. Characterization of an autotrophic sulfide-oxidizing marine *Arcobacter* sp. that produces filamentous sulfur. *Appl Environ Microbiol* 68:316–325.

Wood AP, Kelly DP. 1978. Comparative radiorespirometric studies of glucose oxidation in three facultative heterotrophic thiobacilli. *FEMS Microbiol Lett* 4:283–286.

Wood AP, Kelly DP. 1983. Autotrophic, mixotrophic and heterotrophic growth with denitrification by *Thiobacillus* A$_2$ under anaerobic conditions. *FEMS Microbiol Lett* 16:363–370.

Wood AP, Kelly DP. 1991. Isolation and characterization of *Thiobacillus halophilus* sp. nov., a sulfur-oxidizing autotrophic eubacterium from a Western Australian hypersaline lake. *Arch Microbiol* 156:277–280.

Wood P. 1988. Chemolithotrophy. In: Anthony C, ed. *Bacterial Energy Transduction*. London: Academic Press, pp. 183 230.

Yamanaka T. 1996. Mechanism of oxidation of inorganic electron donors in autotrophic bacteria. *Plant Cell Physiol* 37:569–574.

Zeikus JG, Swanson MA, Thompson TE, Ingvosen K, Hatchikian EC. 1983. Microbial ecology of volcanic sulfidogenesis: Isolation and characterization of *Thermodesulfobacterium commune* gen. nov. and spec. nov. *J Gen Microbiol* 129:1159–1169.

Zellner G, Kneifel H, Winter J. 1990. Oxidation of benzaldehydes to benzoic acid derivatives by three *Desulfovibrio* strains. *Appl Environ Microbiol* 56:2228–2233.

Zillig W, Gierl A, Schreiber G, Wunderl S, Janekovic D, Stetter KO, Klenk HP. 1983. The archaebacterium *Thermofilum pendens*, a novel genus of the thermophilic, anaerobic sulfur respiring Thermoproteales. *Syst Appl Microbiol* 4:79–87.

Zillig W, Stetter KO, Prangishvilli D, Schäfer S, Janekovic D, Holz I, Palm P. 1982. Desulfurococcaceae: The second family of the extremely thermophilic, anaerobic, sulfur respiring Thermoproteales. *Zentralbl Bakt Hyg I Abt Orig C* 3:304–317.

Zillig W, Tu J, Holz I. 1981. Thermoproteales: A third order of thermoacidophilic archaebacteria. *Nature* (London) 293:85–86.

Zimmermann P, Laska S, Kletzin A. 1999. Two modes of sulfite oxidation in the extremely thermophilic and acidophilic archeon *Acidianus ambivalens*. *Arch Microbiol* 172:76–82.

20 Biogenesis and Biodegradation of Sulfide Minerals at Earth's Surface

20.1 INTRODUCTION

Sulfate-reducing bacteria play an important role in some sedimentary environments in the formation of certain sulfide minerals, especially iron pyrite (FeS_2). Other microbes play an even more pervasive role in the oxidation of a wide range of metal sulfides in some soils and sediments or exposed at rock surfaces, regardless of the mode of origin of these minerals. The oxidative microbial activity is being industrially exploited in the extraction of metals from some metal sulfide ores. Currently, the bioextractable sulfidic ores of commercial interest include those of copper, nickel, zinc, and cobalt. At least one of the kinds of bacteria capable of leaching the metal in sulfidic ores is also capable of leaching uranium from the nonsulfidic ore uraninite (UO_2). Although gold in sulfidic ores is not commercially bioextracted, microbial pretreatment (*biobeneficiation*) of such gold ores to remove interfering pyrite and arsenopyrite impurities is now being practiced on a commercial scale. The pyrites in these ores encapsulate the gold, making it inaccessible to a chemical extractant such as aqueous cyanide or thiourea. When cyanide is used as extractant, the pyrites cause excessive consumption of it, resulting in the formation of cyanide complexes with the iron and sulfur components of pyrite, that is, ferro- and ferricyanide and thiocyanate, from none of which the cyanide is readily recoverable. A great potential exists for industrial bioextraction of a variety of other metal sulfide ores.

In the metals industry, a widely used term for metal bioextraction from ores is *bioleaching*. In this chapter, we examine ore biogenesis and biomobilization, including bioleaching, in some detail. Table 20.1 lists metal sulfide minerals of geomicrobial interest.

20.2 NATURAL ORIGIN OF METAL SULFIDES

20.2.1 HYDROTHERMAL ORIGIN (ABIOTIC)

Most metal sulfides, including those of commercial interest, are of igneous origin. Current theory explaining their formation invokes plate tectonics, which has played and is playing a central role in their formation. Terrestrial deposits of *porphyry copper ore* (small crystals of copper sulfides richly dispersed in host rock) are thought to have originated as a result of subduction of oceanic crust that had become somewhat enriched in copper by hydrothermal activity at mid-ocean spreading centers. Subsequent formation of terrestrial deposits of porphyry sulfide ores from subducted oceanic crust is thought to have involved the following successive steps: (1) remelting of the subducted oceanic crust, (2) rising of the resultant magma, (3) release of water with fracturing of incipient rock and formation of hydrothermal solution containing hydrogen sulfide during progressive partial cooling of the magma, and finally (4) re-formation of copper and other metal sulfides by crystallization of the cooling magma and from reaction of H_2S in the hydrothermal solution with metal constituents in the cooled magma in the fractured rock (see Strahler, 1977; Bonatti, 1978; Tittley, 1981).

TABLE 20.1
Metal Sulfides of Geomicrobial Interest

Mineral or Synthetic Compound	Formula	References
Antimony trisulfide	Sb_2S_3	Silver and Torma (1974), Torma and Gabra (1977)
Argentite	Ag_2S	Baas Becking and Moore (1961)
Arsenopyrite	FeAsS	Ehrlich (1964)
Bornite	Cu_5FeS_4	Cuthbert (1962), Bryner et al. (1954)
Chalcocite	Cu_2S	Bryner et al. (1954), Ivanov (1962), Razzell and Trussell (1963), Sutton and Corrick (1963, 1964), Fox (1967), Nielsen and Beck (1972)
Chalcopyrite	$CuFeS_2$	Bryner and Anderson (1957)
Cobalt sulfide	CoS	Torma (1971)
Covellite	CuS	Bryner et al. (1954), Razzell and Trussell (1963)
Digenite	CU_9S_5	Baas Becking and Moore (1961), Nielsen and Beck (1972)
Enargite	$3Cu_2S \cdot As_2S_5$	Ehrlich (1964)
Galena	PbS	Silver and Torma (1974)
Gallium sulfide	Ga_2S_3	Torma (1978)
Marcasite, pyrite	FeS_2	Leathen et al. (1953), Silverman et al. (1961)
Millerite	NiS	Razzell and Trussell (1963)
Molybdenite	MoS_2	Bryner and Anderson (1957), Bryner and Jameson (1958), Brierley and Murr (1973)
Orpiment	As_2S_3	Ehrlich (1963a)
Nickel sulfide	NiS	Torma (1971)
Pyrrhotite	Fe_4S_5	Freke and Tate (1961)
Sphalerite	ZnS	Ivanov et al. (1961), Ivanov (1962), Malouf and Prater (1961)
Tetrahedrite	$Cu_8Sb_2S_7$	Bryner et al. (1954)

The enrichment of the surficial deposits of metal sulfide in and on the oceanic crust has occurred and is occurring in hydrothermally active regions at seafloor spreading centers (mid-ocean ridges) at depths of 2500–2699 m. Examples of such sites are the eastern Pacific Ocean at the Galapagos Rift and the East Pacific Rise (Ballard and Grassle, 1979; Corliss et al., 1979) and the Atlantic Ocean at the Mid-Atlantic Ridge (Klinkhammer et al., 1985). Metal sulfide deposits are evident on the seafloor where some hydrothermal vents (*black smokers*; see Chapters 2 and 17) discharge brine solution that has a temperature ~350°C and is metal-laden and charged with H_2S. Metal sulfides such as chalcopyrite ($CuFeS_2$) and sphalerite (ZnS) precipitate around the mouth of these vents as the brine meets cold seawater and are often deposited in the form of hollow tubes (chimneys). The hydrothermal solution discharged by these vents originated from seawater that penetrated deep into porous volcanic rock (basalt) at the mid-ocean spreading centers to depths as great as 10 km below the seafloor (Bonatti, 1978). As this water penetrated ever deeper into the rock, it absorbed heat diffusing away from underlying magma chambers and was subjected to increasing hydrostatic pressure. This caused the seawater to react with the basalt and pick up various metal species and hydrogen sulfide. The reactions responsible for these seawater modifications include, among others, the interaction of magnesium in the seawater with the rock to form new minerals with an accompanying release of acid (H^+) (Seyfried and Mottl, 1982). The acid leaches metals from the basalt (Edmond et al., 1982; Marchig and Grundlach, 1982). H_2S is formed by the reduction of the sulfate in seawater and sulfur in basalt by ferrous iron released from the basalt (Shanks et al., 1981; Mottl et al., 1979; Styrt et al., 1981). As long as the hydrothermal solution is subjected to high temperature and pressure in the basalt, metal sulfides are prevented from precipitating.

A quantitatively more significant deposition of metal sulfides occurs within the upper oceanic crust associated with white smokers. Here hot, metal-charged hydrothermal brine rising from the lower crust meets and mixes with cold seawater that penetrated the upper crust. The mixing of the two solutions in the upper crust results in partial cooling of the solution and consequent precipitation of metal sulfides in the upper crust. This contrasts with the precipitation of metal sulfides associated with black smokers, which occurs external to the crust around the mouth of the vents and becomes deposited mostly in the walls of the vent chimneys. The brine emerging from the vents of white smokers is depleted in some base metals but still contains major quantities of iron, manganese, and hydrogen sulfide. It is much cooler than the hydrothermal solution issuing from the vents of black smokers. Figure 17.17 shows diagrammatically the origin of the hydrothermal solution and metal sulfides associated with black and white smokers at mid-ocean spreading centers.

A study of bioalteration of sulfur and mineral sulfide samples deployed and incubated under conditions prevailing in the vicinity of a seafloor hydrothermal vent systems (main Endeavor segment of the Juan de Fuca Ridge axis, Pacific Ocean) revealed ready colonization by Bacteria but not Archaea (Edwards et al., 2003). Elemental sulfur appeared to be most readily attacked. Extensive Fe-oxide accumulation on Fe-containing minerals suggested activity of neutrophilic iron-oxidizing bacteria to the investigators.

20.2.2   SEDIMENTARY METAL SULFIDES OF BIOGENIC ORIGIN

Among sedimentary metal sulfides of biogenic origin, iron sulfides are the most common. They are usually associated with reducing zones in sedimentary deposits in estuarine environments, which have a plentiful supply of sulfate. The presence of sulfate is important, because the formation of these metal sulfides is usually the result of an interaction of iron compounds with H_2S that originated from bacterial reduction of the sulfate under anaerobic conditions at these sites. The interaction of the H_2S with the iron compounds leads to the formation of iron pyrite (FeS_2). Whether amorphous sulfide (FeS), mackinawite (FeS), and greigite (Fe_3S_4) are intermediates in the formation of the pyrite depends on prevailing environmental conditions (Schoonen and Barnes, 1991a,b; Luther, 1991). In at least one salt marsh (Great Sippewissett Marsh, Massachusetts) where pyrite forms, the pore waters were found to be undersaturated with respect to these compounds (Jørgensen, 1977; Fenchel and Blackburn, 1979; Howarth, 1979; Berner, 1984; Giblin and Howarth, 1984; Howarth and Merkel, 1984). Rapid and extensive microbial pyrite formation has been observed in salt marsh peat on Cape Cod, Massachusetts (Howarth, 1979). Pyrite formation from biogenic H_2S has also been noted in organic-rich sediments at the Peru Margin of the Pacific Ocean (Mossmann et al., 1991), in Long Island Sound off the Atlantic coast in Connecticut and New York (Westrich and Berner, 1984), along the Danish coast (Thode-Andersen and Jørgensen, 1989), and in two seepage lakes, Gerritsfles and Kliplo, and two moorland ponds in the Netherlands (Marnette et al., 1993).

In many sedimentary environments, pyrite does not represent a permanent sink for iron because the pyrite may be subject to seasonal reoxidation as conditions in the environment change from reducing to oxidizing (Luther et al., 1982; Giblin and Howarth, 1984; King et al., 1985; Giblin, 1988).

Active growth of marsh grass may draw oxygen into the sediment by evapotranspiration (Giblin, 1988). Of all the biogenic sulfide formed in these environments, only a portion is consumed in the formation of pyrite and other metal sulfides. The rest is reoxidized as it enters the oxidizing zones (Jørgensen, 1977). This oxidation may be biological or abiological (Fenchel and Blackburn, 1979).

Nonferrous sulfide deposits of sedimentary origin, especially biogenic ones, appear to be relatively rare. They are generally thought to have formed syngenetically. The metals in question were precipitated by hydrogen sulfide of hydrothermal origin (abiotic formation) or of microbial origin and then buried in contemporaneously formed sediment. The limiting conditions for sedimentary sulfide formation by bacteria as calculated by Rickard (1973) require a minimum of 0.1% carbon (dry weight) and an enriched source of metals such as a hydrothermal solution if more than 1% metal is to be deposited. More recent studies of microbial sulfate reduction revealed, however, that a significant amount of reducing power for sulfate reduction may be furnished by hydrogen (H_2), which would lower the requirement for organic carbon correspondingly (Nedwell and Banat, 1981; see also Section 19.9).

Examples of nonferrous sedimentary sulfide deposits, which may have been biogenically formed, include the Permian Kupferschiefer of Mansfeld in Germany (Love, 1962; Stanton, 1972, p. 1139), Black Sea sediments (Bonatti, 1972, p. 51), the Roan Antelope Deposit in Zambia and Katanga (Africa) (Cuthbert, 1962; Stanton, 1972, p. 1139), the Zechstein Deposit in southwestern Poland (Serkies et al., 1967), and the deposits in Pernatty Lagoon (Australia) (Lambert et al., 1971). By contrast, the sulfide deposit in the Pine Point Pb–Zn property in Northwest Territories, Canada, was abiotically formed (Powell and MacQueen, 1984). $\delta^{34}S$ analyses of the metal sulfides in this deposit suggest that the sulfide resulted from a reaction between bitumen and sulfate at elevated temperature and pressure.

As an example of ongoing nonferrous sulfide biodeposition, the following observation at the Piquette Pb–Zn deposit in Tennyson, Wisconsin, must be cited. At this site, investigators examined a flooded tunnel in carbonate rock and found the presence of biofilms in which aerotolerant members of sulfate-reducing bacteria of the family Desulfobacteriaceae were precipitating sphalerite (ZnS) at a pH between ~7.2 and 8.6. The sphalerite accumulated in the biofilm in aggregates of particles that had a diameter of 2–5 nm (Labrenz et al., 2000).

Although most instances of metal sulfide biogenesis in nature are associated with bacterial sulfate reduction, at least one case of biogenesis of galena has been attributed to the aerobic mineralization of organic sulfur compounds by *Sarcina flava* Bary (Dévigne, 1968a,b, 1973). The *Sarcina* strain was isolated from earthy concretions between crystals of galena in an accumulation in a karstic pocket located in the lead–zinc deposit of Djebel Azered, Tunisia. In laboratory experiments, the organism was shown to produce PbS from Pb^{2+} bound to sulfhydryl groups of sulfur-containing amino acids in peptone.

20.3   PRINCIPLES OF METAL SULFIDE FORMATION

Metal sulfides in nature result from an interaction between an appropriate metal ion and biogenically or abiogenically formed sulfide ion:

$$M^{2+} + S^{2-} \rightarrow MS \tag{20.1}$$

The source of the sulfide in the reaction determines whether a metal sulfide is considered to be of biogenic or abiogenic origin. In the case of biogenic sulfide, it does not matter whether the sulfide resulted from bacterial sulfate reduction (Chapter 19) or from bacterial mineralization of organic sulfur-containing compounds (Dévigne, 1968a,b, 1973). Because of their relative insolubility, the metal sulfides form readily at ambient temperatures and pressures. Table 20.2 lists solubility products for a few common simple sulfide compounds.

The following calculations will show that relatively low concentrations of metal ions, typical in some lakes, will form metal sulfides by reacting with low concentrations of H_2S. The ionic activities

---

TABLE 20.2
Solubility Products for Some Metal Sulfides

CdS	1.4×10^{-28}	FeS	1×10^{-19}	NiS	3×10^{-21}
Bi_2S_3	1.6×10^{-72}	PbS	3.4×10^{-28}	Ag_2S	1×10^{-51}
CoS	7×10^{-23}	MnS	5.6×10^{-16}	SnS	8×10^{-29}
Cu_2S	2.5×10^{-50}	Hg_2S	1×10^{-45}	ZnS	1.2×10^{-23}
CuS	8.5×10^{-45}	HgS	3×10^{-53}	H_2S	1.1×10^{-7}
				HS^-	1×10^{-15}

Source: Latimer WM, Hildebrand JH, *Reference Book of Inorganic Chemistry*. Rev ed., Macmillan, New York, 1942; Weast RC, Astle MJ, *CRC Handbook of Chemistry and Physics*. 63rd ed., CRC Press, Boca Raton, FL, 1982.

---

in these calculations are taken as approximately equal to concentration because of the low concentrations involved. The following examines the case of amorphous iron sulfide (FeS) formation.

The ionization constant for FeS is

$$[Fe^{2+}][S^{2-}] = 10^{-19} \tag{20.2}$$

The ionization constant for H_2S is

$$[S^{2-}] = 10^{-21.96}[H_2S]/[H^+]^2 \tag{20.3}$$

This relationship is derived from the constant for the dissociation of H_2S into HS^- and H^+,

$$[HS^-][H^+]/[H_2S] = 10^{-6.96} \tag{20.4}$$

and the constant for the dissociation of HS^- into S^{2-} and H^+,

$$[S^{2-}][H^+]/[HS^-] = 10^{-15} \tag{20.5}$$

Substituting Equation 20.3 into Equation 20.2, the following relationship is obtained:

$$[Fe^{2+}] = [H^+]^2/[H_2S] \times 10^{-19}/10^{-21.96} = [H^+]^2/[H_2S] \times 10^{21.96} \tag{20.6}$$

Assuming that the bottom water of a lake contains ~34 mg of H_2S L^{-1} (i.e., 10^{-3} M) at pH 7, ~5.08 μg Fe^{2+} L^{-1} (i.e., $10^{-7.04}$ M) will be precipitated as FeS by 3.4 mg of hydrogen sulfide per liter (i.e., 10^{-4} M). The remaining H_2S will ensure reducing conditions, which will keep the iron in the ferrous state. Because ferrous sulfide is one of the most soluble sulfides, metals whose sulfides have even smaller solubility products will require even less sulfide for precipitation. In the excess of sulfide, the FeS would probably be transformed into FeS_2, which is more stable than FeS.

20.4 LABORATORY EVIDENCE IN SUPPORT OF BIOGENESIS OF METAL SULFIDES

20.4.1 BATCH CULTURES

Metal sulfides have been generated in laboratory experiments using H_2S from bacterial sulfate reduction. Miller (1949, 1950) reported that sulfides of Sb, Bi, Co, Cd, Fe, Pb, Ni, and Zn were formed in a lactate-containing broth culture of *Desulfovibrio desulfuricans* to which insoluble salts of selected metals had been added. For instance, he found that bismuth sulfide was formed on addition of $(BiO_2)_2CO_3 \cdot H_2O$, cobalt sulfide on addition of $2CoCO_3 \cdot 3Co(OH)_2$, lead sulfide on addition of $2PbCO_3 \cdot Pb(OH)_2$ or $PbSO_4$, nickel sulfide on addition of $NiCO_3$ or $Ni(OH)_2$, and zinc sulfide on addition of $2ZnCO_3 \cdot 3Zn(OH)_2$. The metal salt reactants were added as insoluble compounds to minimize metal toxicity for *D. desulfuricans*. Metal ion toxicity depends in part on the solubility of the metal compound from which the ion derives. Obviously, for a metal sulfide to be formed from another metal compound that is relatively insoluble, the metal sulfide must be even more insoluble than the source compound of the metal. Miller was not able to demonstrate copper sulfide formation from malachite $[CuCO_3 \cdot Cu(OH)_2]$, probably because malachite was too insoluble relative to copper sulfides in the medium. Miller (1949) also showed that with addition of Cd or Zn ions to the culture medium, the yield of total sulfide produced from sulfate by the bacteria in batch culture was greater than in the absence of the added metal ions. This was because the uncombined sulfide itself becomes toxic to sulfate-reducers at high enough concentration.

Baas Becking and Moore (1961) also undertook a study of biogenesis of sulfide minerals. Like Miller, they worked with batch cultures of sulfate-reducing bacteria. The bacteria they employed were *Desulfovibrio desulfuricans* and *Desulfotomaculum* sp. (which they called *Clostridium desulfuricans*). They grew them in lactate or acetate medium containing steel wool. The steel wool in the medium was meant to serve as a source of hydrogen for the bacterial reduction of sulfate. The hydrogen resulted from corrosion of the steel wool by the spontaneous reaction,

$$Fe^0 + 2H_2O \rightarrow H_2 + Fe(OH)_2 \tag{20.7}$$

The H_2 was then used by the sulfate-reducers in the formation of hydrogen sulfide,

$$4H_2 + SO_4^{2-} + 2H^+ \rightarrow H_2S + 4H_2O \tag{20.8}$$

The media were saline to simulate marine (near-shore and estuarine) conditions under which the investigators thought the reactions are likely to occur in nature. They formed ferrous sulfide from strengite ($FePO_4$) and from hematite (Fe_2O_3). They also formed covellite (CuS) from malachite $[CuCO_3 \cdot Cu(OH)_2]$; argentite (Ag_2S) from silver chloride (Ag_2Cl_2) and from silver carbonate ($AgCO_3$); galena (PbS) from lead carbonate ($PbCO_3$) and from lead hydroxycarbonate $[PbCO_3 \cdot Pb(OH)_2]$; and sphalerite (ZnS) from smithsonite ($ZnCO_3$). All mineral products were identified by x-ray powder diffraction studies. Baas Becking and Moore (1961) were unable to form cinnabar (HgS) from mercuric carbonate ($HgCO_3$), probably owing to the toxicity of the Hg^{2+} ion. They were also unable to form alabandite (MnS) from rhodochrosite ($MnCO_3$), or bornite (Cu_5FeS_4) or chalcopyrite ($CuFeS_2$) from a mixture of cuprous oxide (Cu_2O) or malachite and hematite and lepidochrosite. They succeeded in forming covellite from malachite where Miller (1950) failed, probably because they performed their experiment in a saline medium (3% NaCl) in which Cl^- could complex Cu^{2+}, thereby increasing the solubility of Cu^{2+}. The starting materials that were the source of metal were all relatively insoluble, as in Miller's experiments. Baas Becking and Moore found that in the formation of covellite and argentite, native copper and silver were respective intermediates that disappeared with continued bacterial H_2S production.

Leleu et al. (1975) synthesized ZnS by passing H_2S produced by unnamed strains of sulfate-reducing bacteria through a solution of $ZnSO_4$. In one experiment, biogenic H_2S formation and ZnS precipitation by the biogenic H_2S occurred in separate vessels. In a second experiment, biogenesis of H_2S and precipitation of ZnS occurred in the same vessel at an initial $ZnSO_4$ concentration in the culture medium of 10^{-2} M. The ZnS formed under either experimental condition was identified as a sphalerite–wurtzite mixture by powder x-ray diffraction. The presence of Zn directly in the culture medium caused a lag in H_2S production, which was not observed when H_2S was generated in a separate vessel.

20.4.2 COLUMN EXPERIMENT: MODEL FOR BIOGENESIS OF SEDIMENTARY METAL SULFIDES

The relatively high toxicity of many of the heavy metals for sulfate-reducing bacteria has been used as an argument that these organisms could not have been responsible for metal sulfide precipitation in nature (Davidson, 1962a,b). However, in a sedimentary environment, metal ions will be mostly adsorbed to sediment particles such as clays or complexed by organic matter (Hallberg, 1978), which lessens their toxicity. Such adsorbed or complexed ions are still capable of reacting with sulfide and precipitating as metal sulfides, as was shown experimentally by Temple and LeRoux (1964). They constructed a column in which clay or ferric hydroxide slurry carrying adsorbed Cu^{2+}, Pb^{2+}, and Zn^{2+} ions was separated by an agar plug from an underlying liquid culture of sulfate-reducers actively generating hydrogen sulfide in saline medium. They also tested clay that was carrying Fe^{3+} in this setup. They found that, in time, bands of precipitate formed in the agar plug separating the slurry of metal-carrying adsorbent from the culture of sulfate-reducing bacteria (Figure 20.1). The bands formed as upward-diffusing sulfide ion species and downward diffusing, desorbed metal ion species encountered each other in the agar. Differential desorption of metal ions from the adsorbent and differential diffusion in the agar accounted for the discrete banding of the various sulfides. These results demonstrate that biogenesis of relatively large amounts of sulfides in a sedimentary environment is possible, even in the presence of relatively large amounts of metal ions. The main requirement is that the metal ions are in a nontoxic form (e.g., adsorbed or complexed) or combined in the form of insoluble mineral oxide, carbonate, or sulfate. As Temple (1964) pointed out, syngenetic microbial production of metal sulfide in nature is possible. Restrictions on the process,

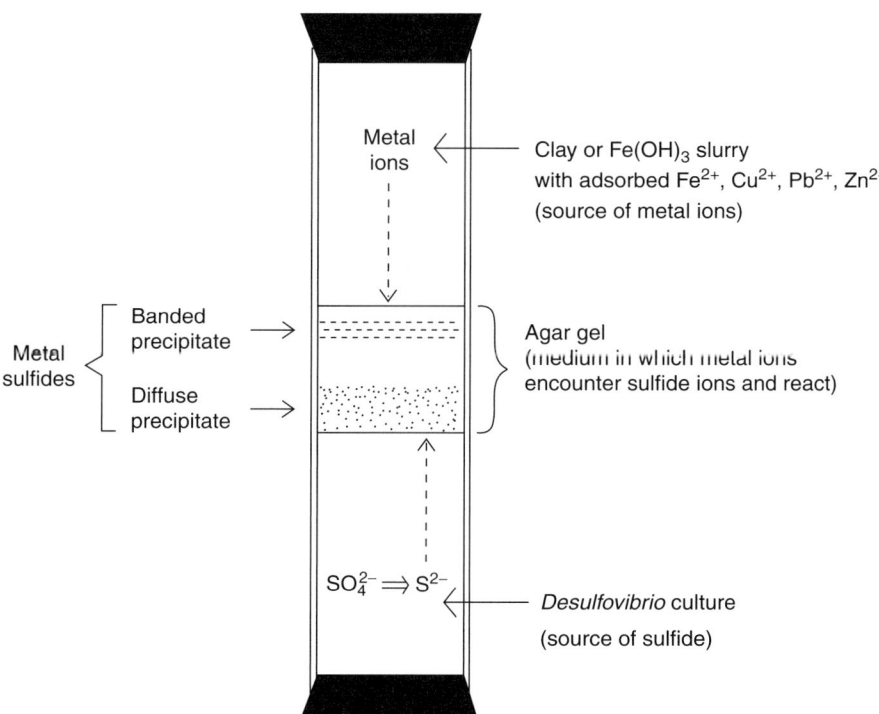

FIGURE 20.1 Temple and LeRoux column modeling a sedimentary environment in which sulfate-reducers can precipitate metal sulfides with sulfide they produce. The adsorbents, clay or $Fe(OH)_3$ slurry, control the concentration of metal ions in solution, and the plug of agar gel prevents physical contact of the sulfate-reducers with metal ions. In nature, sediments can act as adsorbents of metal ions. They keep the metal ion concentration in the interstitial water at a level that does not poison the sulfate-reducers.

according to him, are not metal toxicity but free movement of the bacterially generated sulfide and a need for metal-enriched zones in the sedimentary environment. On a biochemical basis, Temple suspected that microbial sulfate reduction evolved in the Precambrian. Subsequent stable isotope analyses of samples representing the early Precambrian in South Africa indicated that extensive biogenic sulfate reduction occurred at least 2350 million years ago (Cameron, 1982).

20.5   BIOOXIDATION OF METAL SULFIDES

Regardless of whether they are of abiogenic or biogenic origin, metal sulfides in nature may be subject to microbial oxidation. This may take the form of *direct* or *indirect* interaction (Silverman and Ehrlich, 1964). In direct interaction, the microbes oxidize a metal sulfide in physical contact with the mineral surface. In indirect interaction, the microbes usually generate an oxidant (commonly ferric iron from ferrous iron) in the bulk phase. The oxidant then attacks the metal sulfide. In most instances, the metal is solubilized as a metal ion by either direct or indirect mode of oxidation. The biooxidation of galena (PbS) is an exception because the mobilized metal reacts with sulfate ion, which is generated during the oxidation of the lead sulfide and which is also present in the bulk phase from other sources, to form insoluble lead sulfate ($PbSO_4$). Some microbes can mobilize metals in metal sulfides in an indirect mode by generating ligands, which may also be acids. These mobilize the metals by forming soluble metal complexes.

20.5.1   ORGANISMS INVOLVED IN BIOOXIDATION OF METAL SULFIDES

A number of different acidophilic, iron-oxidizing bacteria have been detected at sites where metal sulfide oxidation is occurring (Norris, 1990; Rawlings, 1997b; Okibe et al., 2003; Mousavi et al., 2005). The most important of these have been identified as the mesophiles *Acidithiobacillus ferrooxidans*, *Leptospirillum ferrooxidans*, *Ferroplasma acidiphilum*, and *F. acidarmanus*; the moderate thermophiles *Alicyclobacillus tolerans* (formerly *Sulfobacillus thermosulfidooxidans*), and *Acidimicrobium ferrooxidans*; and extreme thermophiles *Sulfolobus* spp., and *Acidianus brierleyi* (formerly *Sulfolobus brierleyi*). All are autotrophs, and all but *F. acidarmanus* grow best in a pH range of ~1.5–2.5. *F. acidarmanus*, a recent discovery, grows at a pH as low as 0 (optimum pH 1.2) at a temperature of ~40°C. It was isolated from pyrite surfaces of the ore body at Iron Mountain, California, and has been described by Edwards et al. (2000a) as a cell-wall-lacking, iron-oxidizing autotroph. *F. acidiphilum*, also a recent discovery, and a close relative of *F. acidarmanus*, was isolated from a bioleaching pilot plant (Golyshina et al., 2000). It grows in a pH range of 1.3–2.2 (optimum pH 1.7) in a temperature range of 15–45°C. An organism closely related to *F. acidiphilum* was isolated from a bioleaching operation by Mintek in South Africa.

   Acidithiobacillus ferrooxidans, *Leptospirillum ferrooxidans*, *Alicyclobacillus tolerans*, and *Acidimicrobium ferrooxidans* are members of the domain Bacteria (Norris, 1997). *Sulfolobus* spp., *Acidianus brierleyi*, *Ferroplasma acidarmanus*, and *F. acidiphilum* are members of the domain Archaea. Although *L. ferrooxidans* and *Acidimicrobium ferrooxidans* oxidize Fe^{2+} and pyrite, they do not oxidize reduced forms of sulfur as *Acidithiobacillus ferrooxidans* and *Alicyclobacillus tolerans* do. This seems to suggest that *L. ferrooxidans* and *Acidimicrobium ferrooxidans* can promote metal sulfide oxidation only by generating Fe^{3+} from dissolved Fe^{2+}, which then oxidizes metal sulfide abiotically. However, because of a structural feature possessed by both *Acidithiobacillus ferrooxidans* and *L. ferrooxidans*, both organisms may also be able to oxidize metal sulfides by attacking them directly. The common structural feature is exopolymer (EPS) secreted by the cells that contains bound iron (Gehrke et al., 1995, 1998; Sand et al., 1997; Harneit et al., 2006). Barreto et al. (2005) have identified the gene cluster involved in EPS formation in *Acidithiobacillus ferrooxidans*. The EPSs enable attachment to sulfide mineral surfaces. In addition, as will be explained later, the iron in the EPS may serve as an electron shuttle or conductor for conveying electrons from the oxidative half-reaction of metal sulfides to the electron transport system in the plasma membrane of the cells via cytochrome in the outer membrane, such as cytochrome Cyc2 in *Acidithiobacillus ferrooxidans*, and specific electron carriers in the periplasm, such as rusticyanin and cytochrome Cyc1 in *Acidithiobacillus ferrooxidans* (see Figure 16.4B). It remains to be determined if *Acidimicrobium ferrooxidans* forms EPS with bound iron and transfers electrons from its outer membrane to its plasma membrane by a mechanism similar to that in *Acidithiobacillus ferrooxidans*.

Although *Acidithiobacillus ferrooxidans, Sulfolobus* spp., and *Acidianus brierleyi* are autotrophs, growth of the two archaea in this group of three is stimulated by a trace of yeast extract in laboratory culture. In the absence of dissolved ferrous iron or reduced forms of sulfur in the medium, all three organisms can use appropriate metal sulfides as energy sources. Depending on the oxidation state of the metal moiety in the metal sulfide, both the metal and the sulfide may serve as energy sources. For example, in the oxidation of chalcocite (Cu_2S), *Acidithiobacillus ferrooxidans* can use the energy from Cu(I) oxidation for CO_2 fixation (Nielsen and Beck, 1972) (see also further discussion in Section 20.5.2). Cell extracts from *Acidithiobacillus ferrooxidans* have been prepared that catalyze the oxidation of cuprous copper in Cu_2S but not of elemental sulfur (Imai et al., 1973). The oxidation is not inhibited by quinacrine (atebrine). It needs the addition of a trace of iron for proper activity. The effect of traces of iron on metal sulfide oxidation had been previously noted in experiments in which the addition of 9 mg of ferrous iron per liter of medium stimulated metal sulfide oxidation by whole cells of *Acidithiobacillus ferrooxidans* (Ehrlich and Fox, 1967).

Acidithiobacillus ferrooxidans can use NH_4^+ and some amino acids as nitrogen sources (see Sugio et al., 1987; see also Chapter 16). At least some strains are able to fix nitrogen (Mackintosh, 1978; Stevens et al., 1986).

Acidithiobacillus ferrooxidans is very versatile in attacking metal sulfides. It has been reported to oxidize arsenopyrite ($FeS_2 \cdot FeAs_2$ or FeAsS), bornite (Cu_5FeS_4), chalcocite (Cu_2S), chalcopyrite ($CuFeS_2$), covellite (CuS), enargite ($3Cu_2S \cdot As_2S_5$), galena (PbS), millerite (NiS), orpiment (As_2S_3), pyrite (FeS_2), marcasite (FeS_2), sphalerite (ZnS), stibnite (Sb_2S_3), and tetrahedrite ($Cu_8Sb_2S_7$) (see Silverman and Ehrlich, 1964; Wang et al., 2007). In addition, the oxidation of gallium sulfide, pyrrhotite, and synthetic preparations of CoS, NiS, and ZnS by *Acidithiobacillus ferrooxidans* has been reported (Torma, 1971, 1978; Pinka, 1991; Bhatti et al., 1993). The mode of attack of any of these minerals may be direct, indirect, or both.

Although not as exhaustively tested as *Acidithiobacillus ferrooxidans*, the archaea *Acidianus brierleyi* and *Sulfolobus* sp. can also oxidize a variety of metal sulfides including pyrite, marcasite, arsenopyrite, chalcopyrite, NiS, and probably CoS (Brierley, 1978a,b, 1982; Brierley and Murr, 1973; Dew et al., 1999; Wang et al., 2007). Unlike *Acidithiobacillus ferrooxidans*, *Acidianus brierleyi* can oxidize molybdenite in the absence of added iron (Brierley and Murr, 1973) because molybdate ion is less toxic to *Acidianus brierleyi* than to *Acidithiobacillus ferrooxidans* (Tuovinen et al., 1971).

20.5.2 Direct Oxidation

According to the concept of direct oxidation of susceptible metal sulfides as defined by Silverman and Ehrlich (1964), the crystal lattice of such sulfides is attacked through enzymatic oxidation. To accomplish this, the microbes have to be in intimate contact with the mineral they attack. Evidence for rapid attachment of *Acidithiobacillus ferrooxidans* to mineral surfaces of chalcopyrite particles ($CuFeS_2$) has been presented by McGoran et al. (1969) and Shrihari et al. (1991); to covellite particles by Pogliani et al. (1990); to galena crystals by Tributsch (1976); to pyrite crystals by Bennett and Tributsch (1978), Rodriguez-Leiva and Tributsch (1988), Mustin et al. (1992), Murthy and Natarajan (1992), and Edwards et al. (1998); and to pyrite/arsenopyrite-containing auriferous ore by Norman and Snyman (1988).

Bacterial attachment to mineral sulfide surfaces appears not to be random but to occur at specific sites and even specific crystal faces. Some evidence suggests that direct microbial attack is initiated at sites of crystal imperfections. Selective attachment of *Acidithiobacillus ferrooxidans* or *Sulfolobus acidocaldarius* to newly exposed pyrite crystals in coal is very rapid, that is, ~90% complete in 2–5 min (Badigian and Myerson, 1986; Chen and Skidmore, 1987, 1988). Although the details of how microbes attack crystal lattices of metal sulfides once they have attached are in most respects not yet understood, a collective model is that bacterial cells possessing this ability act as catalytic conductors in transferring electrons from cathodic areas on crystal surfaces of a metal sulfide via an electron transport system in the cell envelope to oxygen (Figure 20.2). The model assumes the existence of a special electron transport system, which in gram-negative bacteria involves components of the outer membrane, the periplasm, and the plasma membrane. Indeed, the bacteria can be viewed as cathodic extensions. They benefit from this process by coupling energy conservation (ATP synthesis) to it.

The mere spontaneous dissociation of a mineral to yield oxidizable ion species in solution that *Acidithiobacillus ferrooxidans* can attack is too small in the case of minerals that are very insoluble in acid solution. For example, covellite (CuS), in which the only oxidizable constituent is the sulfide, has a solubility constant of $10^{-44.07}$ (Table 20.2). *Acidithiobacillus ferrooxidans* is able to oxidize this mineral at pH 2 (see later in this section). Simple calculations show that at equilibrium at pH 2 in water, the dissociation of CuS will only generate a concentration of HS^- equal to $10^{-15.53}$ M and a concentration of H_2S equal to $10^{-13.06}$. This is insufficient for sulfide oxidation by *Acidithiobacillus ferrooxidans* because the most recent K_s value for sulfide oxidase in intact cells of this organism has

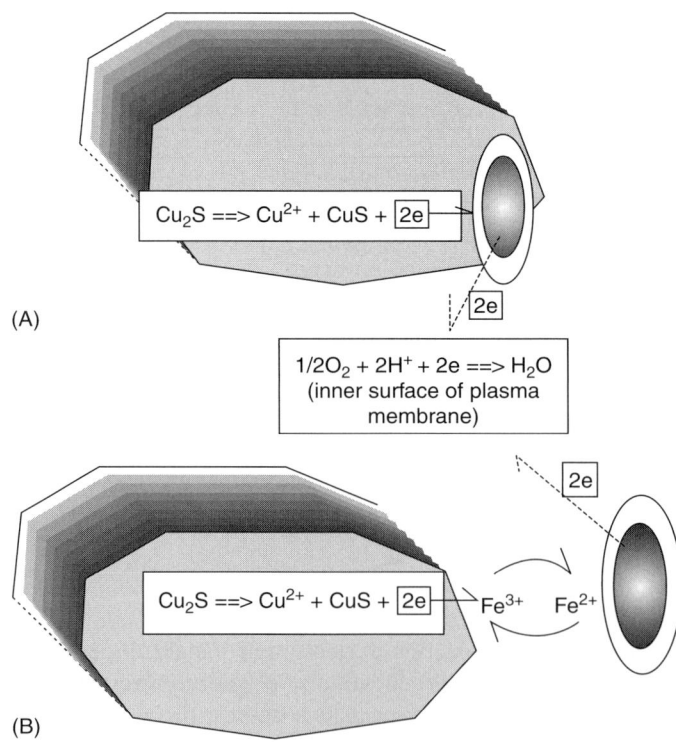

FIGURE 20.2 Schematic representation of direct and indirect oxidation of a particle of Cu_2S by *Acidithiobacillus ferrooxidans*. (A) Direct oxidation. In this model the bacterial cell, attached to the surface of a Cu_2S particle, acts essentially as a conductor of electrons it removes in the oxidation of Cu(I) of Cu_2S and transfers to oxygen. Not shown is the mechanism by which the electrons cross the interface between the particle surface and the cell surface. For a possible mechanism, for this electron transfer, see Chapter 16, Figure 16.4B and discussion in Section 20.5 of this chapter. (B) Indirect oxidation. In this model, planktonic (unattached) bacterial cells generate and regenerate the oxidant (Fe^{3+}) by oxidizing Fe^{2+} in the bulk phase. The iron acts as a shuttle that carries electrons from the oxidation of Cu(I) of Cu_2S to the bacterium, which transfers them to oxygen.

been reported to be $10^{-5.30}$ (Pronk et al., 1990). K_s values are a measure of the substrate concentration at which a reaction velocity catalyzed by intact cells is half-maximal (Michaelis–Menten kinetics). (See footnote in Section 16.4. K_s for a cell is equivalent to K_m for an individual enzyme.) Thus the mere dissociation of CuS into Cu^{2+}, HS^-, and H_2S cannot furnish nearly enough sulfide substrate to sustain its oxidation by *Acidithiobacillus ferrooxidans* at a reasonable velocity, regardless of whether HS^- or H_2S or both are the actual substrate for sulfide oxidase. Because *Acidithiobacillus ferrooxidans* can oxidize covellite in the absence of added iron, it must be in direct contact with a mineral to attack it under that condition. The need for direct contact in covellite oxidation by *Acidithiobacillus ferrooxidans* in the absence of added Fe^{2+} was demonstrated experimentally by Pogliani et al. (1990). By contrast, *Acidithiobacillus thiooxidans* was shown by Donati et al. (1995) to promote covellite oxidation only with the addition to the medium of Fe(III) or by Fe(II) autoxidized to Fe(III). The oxidation of covellite by Fe(III) generates S^0 (and possibly small amounts of other partially reduced and dissolved sulfur species) as first described by Sullivan (1930),

$$CuS + 2Fe^{3+} \rightarrow Cu^{2+} + S^0 + 2Fe^{2+} \tag{20.9}$$

As pointed out in Chapter 19, *Acidithiobacillus thiooxidans* cannot oxidize Fe^{2+}.

On the contrary, if we consider a more soluble sulfide mineral such as ZnS, which has a solubility constant of $10^{-22.9}$, calculations similar to those for CuS show that at pH 2, ZnS dissociation will yield $10^{-4.95}$ M HS^- and $10^{-1.47}$ H_2S (Table 20.2). These concentrations of HS^- and H_2S are more than sufficient to satisfy the K_s of $10^{-5.30}$ for sulfide oxidase in *Acidithiobacillus ferrooxidans* and to permit its growth without direct attack of ZnS at the mineral surface. Indeed, it has been shown that *Acidithiobacillus thiooxidans*, which is unable to oxidize Fe^{2+}, will readily promote the dissolution of ZnS at pH 2 (Pistorio et al., 1994). The relative solubility of PbS in acid solution also explains why Garcia et al. (1995) found that *Acidithiobacillus thiooxidans* promoted dissolution of PbS (galena). The solubility constant of this metal sulfide is $10^{-27.5}$, a little smaller than that of ZnS ($10^{-22.9}$) but significantly larger than that of CuS ($10^{-44.07}$). At pH 2, PbS dissociates to yield $10^{-7.25}$ M HS^- and $10^{-4.77}$ M H_2S (Table 20.2).

The exact nature of the interaction between a sulfide mineral surface and the *Acidithiobacillus ferrooxidans* cell surface, on which the enzyme-catalyzed oxidation of the mineral depends, has become a little clearer in the past few years. Previously, Ingledew proposed that iron bound in the cell envelope of *Acidithiobacillus ferrooxidans* served as an electron shuttle that conveys electron from an external electron donor across the outer membrane to electron carriers in the periplasm of the cell (see Ingledew, 1986; Ehrlich, 2000). Alternatively, Tributsch (1999) proposed that the ferric iron bound in the EPS at the cell surface of *Acidithiobacillus ferrooxidans* in contact with a mineral surface generated elemental sulfur according to Reaction 20.9. This sulfur was then supposed to be oxidized by *Acidithiobacillus ferrooxidans* by a known reaction (see Chapter 19). However, this proposal assumes that EPS-bound iron(III) is as strong an oxidant as ferric iron in the bulk phase.

It has now been demonstrated by Yarzábal et al. (2002a,b) that the outer membrane of *Acidithiobacillus ferrooxidans* contains a high-molecular weight *c*-type cytochrome Cyc2 that has the capacity to promote the oxidation of Fe^{2+} to Fe^{3+} at the outer surface of the outer membrane. Cytochrome Cyc2 in the outer membrane then conveys the electrons it removed from Fe^{2+} to the multicopper oxidase rusticyanin and from it to a low-molecular weight *c*-type cytochrome Cyc1, both located in the periplasm. Cytochrome Cyc1 then passes the electrons it receives to aa_3 cytochrome oxidase in the plasma membrane, which passes them to O_2 accompanied by energy conservation via ATP synthase (see also Section 16.4 and Figure 16.4B). Although it remains to be experimentally demonstrated, it seems reasonable to assume that cytochrome Cyc2 or another *c*-type cytochrome in the outer membrane is involved in conveying electrons from the oxidation of a metal sulfide with which *Acidithiobacillus ferrooxidans* is in physical contact, to oxygen either directly or via iron bound in the EPS of the cells acting as an electron shuttle between the mineral and cytochrome Cyc2 in the outer membrane, as explained later.

Precedents for electron transfer between an outer cell surface and a mineral surface with which it is in contact exist among bacteria that are involved in the reduction of insoluble electron acceptors such as ferric oxide or MnO_2 (see Chapters 16 and 17). The electrons in these instances travel in a direction opposite to that when an oxidation of an insoluble electron donor, like metal sulfide or a substrate like Fe^{2+}, at the outer surface of the outer membrane is involved.

Gehrke et al. (1995, 1998) and Sand et al. (1997) have proposed that the iron bound in the EPS around the cells of *Acidithiobacillus ferrooxidans* mediates attack of the sulfur moiety in pyrite. They view bulk-phase iron as inducing EPS formation and enabling attachment to pyrite. However, they do not differentiate unbound, bulk-phase iron and iron bound (complexed) by EPS of *Acidithiobacillus ferrooxidans*. Bulk-phase ferric iron when interacting with metal sulfide mineral behaves as a chemical reactant, which is consumed in the oxidation of the metal sulfide. EPS-bound iron should be viewed as an *electron shuttle*, which is reversibly reduced and oxidized during electron transfer from the metal sulfide being oxidized to an acceptor molecule of the cell (e.g., cytochrome Cyc2). This follows from the following consideration. The standard reduction potential for the Fe^{3+}/Fe^{2+} couple is $+777$ mV, whereas that of the $Fe(CN)_6^{3-}/Fe(CN)_6^{4-}$ in 0.01 N NaOH is $+460$ mV (Weast and Astle, 1982), and that for oxidized cytochrome c/reduced cytochrome c is $+0.245$ mmV (Lehninger, 1975, p. 479). Although no reduction potential for EPS-bound iron is available, its value is likely to lie between that for uncomplexed iron and the porphyrin-bound iron in cytochrome. EPS-bound iron would therefore be a significantly weaker oxidant and function better as an electron shuttle.

Although the mechanism by which *Acidithiobacillus ferrooxidans* promotes direct oxidative attack of metal sulfides is becoming clearer, that by which other bacteria known to promote such oxidative attack remains to be clarified. Some may only be able to promote such oxidation by indirect attack, but others are likely to be able to promote it by both direct and indirect attack. Because *Acidithiobacillus ferrooxidans* is a gram-negative organism, other gram-negative organisms, such as *Leptospirillum ferrooxidans* for instance, may use the same mode of direct attack, but not gram-positive organisms or archaea, such as *Alicyclobacterium tolerans* and *Acidianus brierleyi*, respectively, that may be capable of it. This is because the envelope structure of gram-positive bacteria and archaea is very different from that of gram-negative bacteria (see Chapter 6).

Evidence for enzymatic attack of synthetic covellite (CuS) by noting inhibition of oxygen consumption and Cu^{2+} and SO_4^{2-} ion production by *Acidithiobacillus ferrooxidans* in the presence of the enzyme inhibitor trichloroacetate (8 mM) was obtained by Rickard and Vanselow (1978). In the case of CuS, only the sulfide moiety of the mineral is attacked because the metal moiety is already fully oxidized. The oxidation of the mineral probably proceeds in two steps (Fox, 1967):

$$CuS + 0.5O_2 + 2H^+ \xrightarrow{\text{Bacteria}} Cu^{2+} + S^0 + H_2O \qquad (20.10)$$

$$S^0 + 1.5O_2 + H_2O \xrightarrow{\text{Bacteria}} SO_4^{2-} + 2H^+ \qquad (20.11)$$

By contrast, *Thiobacillus thioparus* promotes covellite oxidation only after autoxidation of the mineral to $CuSO_4$ and S^0 (similar to Reaction 20.10 but in the absence of bacterial catalysis; Rickard and Vanselow, 1978). It is the bacterial catalysis of the oxidation of S^0 to sulfate that helps the reaction by removing a product of the autoxidation of CuS.

In some instances, both an oxidizable metal moiety and the sulfide moiety may be attacked by separate enzymes, for example, in the case of chalcopyrite ($CuFeS_2$) (assuming the Fe of chalcopyrite to have an oxidation state of $+2$; Duncan et al., 1967; Shrihari et al., 1991). Although Duncan and coworkers reported Fe and S to be simultaneously attacked, Shrihari et al. (1991) found that iron-grown *Acidithiobacillus ferrooxidans* oxidized the sulfide sulfur of chalcopyrite by direct attack before oxidizing ferrous iron in solution to ferric iron. When the dissolved ferric iron attained a significant concentration, it promoted chemical oxidation of residual chalcopyrite. The overall reaction by *Acidithiobacillus ferrooxidans* may be written as follows:

$$4CuFeS_2 + 17O_2 + 4H^+ \xrightarrow{\text{\textit{Acidithiobacillus ferrooxidans}}} 4Cu^{2+} + 4Fe^{3+} + 8SO_4^{2-} + 2H_2O \qquad (20.12)$$

$$4Fe^{3+} + 12H_2O \rightarrow 4Fe(OH)_3 + 12H^+ \qquad (20.13)$$

$$4CuFeS_2 + 17O_2 + 10H_2O \xrightarrow{\text{\textit{Acidithiobacillus ferrooxidans}}} 4Cu^{2+} + 4Fe(OH)_3 + 8SO_4^{2-} + 8H^+ \qquad (20.14)$$

Reaction 20.14 is the sum of Reactions 20.12 and 20.13.

In other cases of direct attack, the oxidizable metal moiety may be oxidized before the sulfide, as in the example of chalcocite (Cu_2S) oxidation (Fox, 1967; Nielsen and Beck, 1972) (Figure 20.2):

$$Cu_2S + 0.5O_2 + 2H^+ \xrightarrow{\text{\textit{Acidithiobacillus ferrooxidans}}} Cu^{2+} + CuS + H_2O \qquad (20.15)$$

$$CuS + 0.5O_2 + 2H^+ \xrightarrow{\text{\textit{Acidithiobacillus ferrooxidans}}} Cu^{2+} + S^0 + H_2O \qquad (20.16)$$

$$S^0 + 1.5O_2 + H_2O \xrightarrow{\text{\textit{Acidithiobacillus ferrooxidans}}} SO_4^{2-} + 2H^+ \qquad (20.17)$$

Digenite (Cu_9S_5) can be an intermediate in the formation of CuS from Cu_2S (Nielsen and Beck, 1972).

Although in the laboratory it is possible to demonstrate exclusive direct oxidation by *Acidithiobacillus ferrooxidans* of certain nonferrous metal sulfides by using iron-free mineral in an iron-free culture medium, in nature these conditions never occur. This is because nonferrous metal sulfides are always accompanied by pyrites in ore deposits. Thus, in nature, direct and indirect oxidation usually occurs together. It is prevailing environmental conditions that determine the extent to which each mode contributes to overall oxidation of a metal sulfide. Pyrite oxidation presents a special problem in applying the concepts of direct and indirect attack in bacterial leaching and is treated in Section 20.5.4.

20.5.3 INDIRECT OXIDATION

In indirect biooxidation of metal sulfides, a major role of the bacteria is the generation of a lixiviant that chemically oxidizes the sulfide ore. This lixiviant is ferric iron (Fe^{3+}). It is a major *consumable reactant* in the oxidation of the metal sulfide. It may be generated initially from dissolved ferrous iron (Fe^{2+}) at pH values of 3.5–5.0 by *Metallogenium* in a mesophilic temperature range (Walsh and Mitchell, 1972a). At pH values below 3.5, ferric iron may be generated from Fe^{2+} by bacteria such as *Acidithiobacillus ferrooxidans* and *Leptospirillum ferrooxidans* Markosyan (see Balashova et al., 1974) in a mesophilic temperature range and by *Sulfolobus* spp., *Acidianus brierleyi*, *Alicyclobacillus tolerans*, and others in a thermophilic temperature range (Brierley, 1978b; Brierley and Brierley, 1973; Brierley and Murr, 1973; Brierley and Lockwood, 1977; Brierley et al., 1978; Balashova et al., 1974; Golovacheva and Karavaiko, 1979; Harrison and Norris, 1985; Pivovarova et al., 1981; Segerer et al., 1986). In addition, ferric iron can be generated from iron pyrites (e.g., FeS_2) by indirect attack by *Acidithiobacillus ferrooxidans* and other iron-oxidizing acidophiles. At least some of these organisms may also generate Fe^{3+} from pyrites by direct attack. In whichever way it is formed, ferric iron in acid solution acts as an oxidant of the metal sulfides in indirect attack (Sullivan, 1930; Ehrlich and Fox, 1967):

$$MS + 2Fe^{3+} \rightarrow M^{2+} + S^0 + 2Fe^{2+} \qquad (20.18)$$

where M may be any metal in an appropriate oxidation state, which does not always have to be +2 (e.g., $Cu(I)$ in Cu_2S). A central role of *Acidithiobacillus ferrooxidans* in an indirect oxidation process is to regenerate Fe^{3+} from the Fe^{2+} formed in Reaction 20.18. It should be noted that in *chemical* as opposed to *biochemical* oxidation, the sulfide of a mineral is mostly oxidized to elemental sulfur (S^0) (pyrite is an exception). Further oxidation to sulfuric acid (H_2SO_4) is very slow but is likely to be greatly accelerated by microorganisms such as *Acidithiobacillus thiooxidans, Acidithiobacillus ferrooxidans, Sulfolobus* spp., and *Acidianus brierleyi* but not by *L. ferrooxidans*. Elemental sulfur may form a film on the surface of metal sulfide crystals in chemical oxidation and interfere with further chemical oxidation of the residual metal sulfide. The chemical oxidation of metal sulfides must occur in acid solution below pH 5 to keep enough ferric iron in solution. In nature, the needed acid may be formed chemically through autoxidation of sulfur and other partially reduced forms of sulfur, but much more likely biologically through bacterial oxidation of the sulfur. The acid may also form as a result of autoxidation or biooxidation of ferrous iron or pyrite. In ferrous iron biooxidation, the acid forms as follows (autoxidation proceeds by the same reaction, but without bacterial participation, and much more slowly):

$$2Fe^{2+} + 0.5O_2 + 2H^+ \xrightarrow{\text{Bacteria}} 2Fe^{3+} + H_2O \qquad (20.19)$$

$$Fe^{3+} + 3H_2O \rightarrow Fe(OH)_3 + 3H^+ \qquad (20.20)$$

Because this reaction normally occurs in the presence of sulfate, the ferric hydroxide may convert to the more insoluble jarosite, especially in the presence of *Acidithiobacillus ferrooxidans* and probably other acidophilic iron-oxidizers (Lazaroff, 1983; Lazaroff et al., 1982, 1985; Carlson et al., 1992):

$$A^+ + 3Fe(OH)_3 + 2SO_4^{2-} \rightarrow AFe_3(SO_4)_2(OH)_6 + 3OH^- \qquad (20.21)$$

where A^+ may represent Na^+, K^+, NH_4^+, or H_3O^+ (Duncan and Walden, 1972). The formation of jarosite decreases the ratio of protons produced per iron oxidized from 2:1 to 1:1. In pyrite oxidation, the acid forms as a result of Reactions 20.22 through 20.24 (see later). Reaction 20.22 proceeds by autoxidation if the process is indirect. Like sulfur, jarosite may also form on the surface of metal sulfide crystals and block further oxidation.

20.5.4 PYRITE OXIDATION

Pyrite oxidation by *Acidithiobacillus ferrooxidans* represents a special case in which direct and indirect oxidation of the mineral cannot be readily separated because ferric iron is always a product. Experimentally, Mustin et al. (1992) recognized four phases in the leaching of pyrite by *Acidithiobacillus ferrooxidans* in a stirred reactor. The first phase, which lasted ~5 days under the experimental conditions, featured a measurable decrease in planktonic (*unattached*) bacteria. The small amount of dissolved ferric iron added with the inoculum reacted with some of the pyrite.

The second phase, which also lasted ~5 days, featured the start of pyrite dissolution with oxidation of its iron and sulfur, but with sulfur being preferentially oxidized. Planktonic bacteria multiplied exponentially and the pH began to drop.

The third phase, which lasted ~10 days, featured a significant increase in dissolved ferric iron, the ferrous iron concentration remaining low. Both iron and sulfur in the pyrite were being oxidized at high rates. However, the rate of sulfur oxidation decreased with time relative to iron oxidation, the ratio of sulfate to ferric iron becoming stoichiometric by day 18. Planktonic bacteria continued to increase exponentially, and the pH continued to drop. The surface of the pyrite crystals began to show evidence of corrosion cracks.

In the fourth and last phase, which lasted ~25 days, the dissolved Fe(III)/Fe(II) ratio decreased slightly, iron and sulfur in the pyrite continued to be strongly oxidized, and the planktonic bacteria reached a stationary phase. At the same time, the pH continued to drop to 1.3 by the 45th day. The surface of the pyrite particles now showed easily recognizable square or hexagonal corrosion pits. Mustin and coworkers followed pH and electrochemical (redox) changes during the entire experiment of pyrite oxidation.

Classically, direct bacterial oxidation of iron pyrite (FeS_2) has been described by the overall reaction

$$FeS_2 + 3.5O_2 + H_2O \xrightarrow{\text{Attached \textit{Acidithiobacillus ferrooxidans}}} Fe^{2+} + 2H^+ + 2SO_4^{2-} \qquad (20.22)$$

The dissolved ferrous iron generated in this reaction is further oxidized by planktonic bacteria according to the overall reaction

$$2Fe^{2+} + 0.5O_2 + 2H^+ \xrightarrow{\text{Planktonic \textit{Acidithiobacillus ferrooxidans}}} 2Fe^{3+} + H_2O \qquad (20.23)$$

The resultant ferric iron then causes chemical oxidation of residual pyrite according to the reaction

$$FeS_2 + 14Fe^{3+} + 8H_2O \rightarrow 15Fe^{2+} + 2SO_4^{2-} + 16H^+ \qquad (20.24)$$

whereby Fe^{2+} is regenerated. The oxidation of Fe^{2+} by the planktonic bacteria becomes the rate-controlling reaction in this model of pyrite oxidation according to Singer and Stumm (1970), which makes biooxidation of pyrite mostly an indirect process. This is because in the absence of the iron-oxidizing bacteria the rate of Fe^{2+} oxidation is very slow at acid pH. According to a study by Moses et al. (1987), Fe^{3+} is the preferred chemical oxidant of pyrite over O_2, suggesting that an organism like *Acidithiobacillus ferrooxidans* will facilitate initiation of pyrite oxidation by generating Fe^{3+} from pyrite (Reactions 20.22 and 20.23).

Schippers and Sand (1999) viewed the role of *Acidithiobacillus ferrooxidans* in the oxidation of pyrite to be only indirect, as defined in this book, and they extended this concept to bacterial oxidation of other metal sulfides. In their model, *Acidithiobacillus ferrooxidans* catalyzes the oxidation of the dissolved Fe^{2+} in the bulk phase that results from the chemical oxidation of pyrite by Fe^{3+},

$$FeS_2 + 6Fe^{3+} + 3H_2O \rightarrow S_2O_3^{2-} + 7Fe^{2+} + 6H^+ \qquad (20.25)$$

in which thiosulfate is formed as a product (see Rimstidt and Vaughn, 2003 and Descostes et al., 2004 for chemical pyrite oxidation). They proposed furthermore that the thiosulfate is subsequently oxidized by the ferric iron generated by *Acidithiobacillus ferrooxidans*,

$$S_2O_3^{2-} + 8Fe^{3+} + 5H_2O \rightarrow 2SO_4^{2-} + 8Fe^{2+} + 10H^+ \qquad (20.26)$$

ferric iron being the oxidant in this reaction. At the same time, the thiosulfate in their model is oxidized by *Acidithiobacillus ferrooxidans* itself,

$$S_2O_3^{2-} + 2O_2 + H_2O \xrightarrow{\text{\textit{Acidithiobacillus ferrooxidans}}} 2SO_4^{2-} + 2H^+ \qquad (20.27)$$

In the bacterial oxidation, oxygen rather than Fe^{3+} is the terminal electron acceptor (oxidant). It is unclear, however, whether thiosulfate should exist at all at the acid pH (in the range of $<$2–3.5) at which these reactions would occur. Thiosulfate is known to decompose into elemental sulfur and sulfite (SO_3^{2-}) below a pH of 4–5 (see Roy and Trudinger, 1970, p. 18; also Rimstidt and Vaughn, 2003, p. 879).

Sand et al. (1995) believe that direct oxidation of pyrite or any other sulfide mineral by *Acidithiobacillus ferrooxidans* does not occur. Thus, their model of bacterial metal sulfide oxidation relies on an abiotic attack by Fe^{3+}, the role of the bacteria being the regeneration of Fe^{3+} from the Fe^{2+} formed in the chemical oxidation of the metal sulfide (Reaction 20.23). Their general model suggests that *Acidithiobacillus ferrooxidans* cannot oxidize any metal sulfide in the absence of iron in the bulk phase. This is, however, contrary to previous observations (see experiments by S.I. Fox summarized by Ehrlich, 1978, p. 71; Nielsen and Beck, 1972; Pogliani et al., 1990). In the case of pyrite oxidation, it seems impossible to distinguish between direct and indirect bacterial mechanisms in the absence of any suppression of the chemical action of ferric iron produced in pyrite oxidation. The ferric iron will always be a product of pyrite oxidation, no matter whether pyrite is attacked directly or indirectly. As commented earlier, Sand and collaborators' model does not distinguish between bulk-phase (dissolved) ferric iron and EPS-bound and cell-wall-bound (complexed) iron of *Acidithiobacillus ferrooxidans* in terms of reactivity.

Edwards et al. (2001) approached the question of direct versus indirect action by determining what effect *Acidithiobacillus ferrooxidans* and *Ferroplasma acidarmanus* had on pitting of the mineral surface of pyrite, marcasite, and arsenopyrite in a mineral salt medium. They used scanning electron microscopy to make their assessments. They found that extensive pitting on the mineral surface occurred in the absence of bacteria with added ferric chloride but not without it. In the presence of bacteria, pitting depended on the kind of bacteria and their attachment, the kind of mineral, and probably other experimental conditions. Thus, *Acidithiobacillus ferrooxidans* produced cell-sized and cell-shaped dissolution pits on pyrite but not on marcasite or arsenopyrite. *F. acidarmanus* produced such pits on pyrite and arsenopyrite but not on marcasite. However, individual cells were found in shallow pits on marcasite. The investigators came to the conclusion that overall sulfide dissolution in their experiments was dominated by reaction of a given mineral with bulk-phase Fe^{3+} and not by reaction at the cell/mineral surface interface. Nevertheless, they did not rule out the possibility of a cell/mineral surface interface reaction. If it did occur, they believed it to have been of minor importance in pit formation in these experiments.

If the bound (complexed) iron acts as an electron shuttle in the sense of Ingledew (1986), Moses et al. (1987) and Rawlings et al. (1999) suggest that the bound iron in bacterial cells when attached to a pyrite surface may accept the electrons from the half-reactions describing the initial steps in corrosion of pyrite:

$$FeS_2 + 2H_2O \rightarrow Fe(OH)_2S_2 + 2H^+ + 2e \qquad (20.28a)$$

and

$$FeS_2 + 8H_2O \rightarrow Fe^{3+} + 2SO_4^{2-} + 16H^+ + 15e \qquad (20.28b)$$

In this way, the bacteria take the place of ferric iron in attacking pyrite. Because *Leptospirillum ferrooxidans* also features bound iron in its EPS, it may thus also be able to attack pyrite by the direct mechanism when attached to a pyrite surface.

Because iron pyrites usually accompany other metal sulfides in nature, iron pyrite oxidation is an important source of acid for the oxidizing reactions of nonferrous metal sulfides, especially those that consume acid, for example, the oxidation of chalcocite (Reactions 20.15 through 20.17). In some cases the host rock in which metal sulfides, including pyrites, are contained may itself consume acid and thus raise the pH of the environment high enough to cause extensive precipitation of ferric iron and thereby prevent oxidation of metal sulfide by it (Ehrlich, 1977, pp. 149–150).

The foregoing discussion of direct and indirect metal sulfide oxidation dealt mainly with *Acidithiobacillus ferrooxidans*, which is capable of the oxidation of both ferrous iron and reduced sulfur. Investigations need to be extended to *Leptospirillum ferrooxidans*, which oxidizes ferrous iron readily but cannot oxidize reduced forms of sulfur. Yet, *L. ferrooxidans* has been reported to be the dominant member of the leaching microflora in many heap/dump leaching operations (Sand et al., 1992, 1993; Asmah et al., 1999; Bruhn et al., 1999). The ability to oxidize ferrous iron but not reduced forms of sulfur might suggest that *L. ferrooxidans* oxidizes metal sulfides only in an indirect mode.

However, like *Acidithiobacillus ferrooxidans, L. ferrooxidans* produces EPS that has an affinity for iron (Gehrke et al., 1995; Sand et al., 1997). If this bound iron can act as an electron shuttle that transfers the electrons to a cytochrome in the outer membrane of *L. ferrooxidans*, as it may in *Acidithiobacillus ferrooxidans*, the organism should also be capable of oxidizing a metal sulfide in a direct mode. As previously pointed out, investigation of the mode(s) of action in metal sulfide oxidation also needs to be extended to the other microorganisms that have been found active in metal sulfide oxidation. Most of these can oxidize both Fe^{2+} and reduced forms of sulfur.

20.6 BIOLEACHING OF METAL SULFIDE AND URANINITE ORES

20.6.1 METAL SULFIDE ORES

When metal sulfide ore bodies are exposed to moisture and air during mining activities, the ore mineral may begin to undergo gradual oxidation, which may be accelerated by native microorganisms, especially acidophilic iron-oxidizers. Groundwater passing through a zone of ore oxidation will pick up soluble products of the oxidation and will issue from the site as acid mine drainage (AMD), which contains the metal solubilized in the oxidation. Mining companies harness the microbial oxidizing activity that mobilizes metals in some sulfidic ores on an industrial scale as an economic means of metal extraction. Such metal bioleaching may be applied to rubblized ore *in situ* (McCready, 1988), in ore heaps or dumps, or to crushed ore in special reactors. Initially bioleaching was used commercially only with low-grade portions of an ore and with ore tailings, but with more recent improvements in the process it is now also used in treating high-grade ore and ore concentrates.

Low-grade sulfide ores generally contain metal values at concentrations below 0.5% (w/w). Their extraction by smelting after milling and subsequent ore enrichment (ore beneficiation) by flotation is uneconomic because of an unfavorable gangue/metal ratio. Years of experience have shown that for most efficient bioleaching, ore heaps should be constructed to heights limited to tens of feet to avoid slumping. Ore heaps may consist of waste rock or mine tailings, which are by-products of mining that still contain traces of recoverable metal values, but nowadays may also consist of high-grade ore. The lixiviant can be water, acidified water, or spent acidic leach solution (barren solution) containing ferric sulfate from a previous leaching cycle. It is applied in a fine spray onto ore heaps and dumps (Figure 20.3A). The spraying avoids waterlogging of the heaps and dumps, which would exclude needed oxygen (air). Simultaneous diffusion of oxygen into the ore in a heap or dump being leached is important because the microbial leaching process is aerobic. If oxygen does not reach all parts of a heap or dump because its concentration in the leach solution is insufficient to meet the demand for microbial metal sulfide oxidation, anaerobic conditions will develop in those parts that are not reached by oxygen. In such regions, a microbial flora, mostly heterotrophs, has been shown to develop that includes bacteria that reduce ferric to ferrous iron and others that reduce sulfate to H_2S (Fortin et al., 1995; Johnson and Roberto, 1997; Fortin and Beveridge, 1997). The Fe-reducers lower the ferric iron concentration available for chemical oxidation of metal sulfides in the anoxic zone. The sulfate-reducers cause metal species mobilized in the oxidized zones to reprecipitate as sulfides. In some heap-leaching operations, perforated pipes are placed in strategic positions within the heap during its construction to optimize access of air to deeper portions of the heap.

The lixiviant solution applied to the ore makes possible the growth and multiplication of appropriate acidophilic iron-oxidizers and the oxidation of pyrite, chalcopyrite, and nonferrous metal sulfides in the ore. Initially, solutions issuing from the heaps or dumps may be recirculated without any treatment, but ultimately, as microbial and chemical activities continue, the solution in the heaps and dumps becomes charged with dissolved metal values, and after issuing from the heaps or dumps it is collected as *pregnant solution* in special sumps. Pregnant solution frequently harbors a variety of microorganisms, including autotrophic and heterotrophic bacteria, fungi, and protozoa, despite its very acidic pH and high metal content (Ehrlich, 1963b; Baker and Banfield, 2003). Indeed, fungi

(A)

(B)

FIGURE 20.3   Bioleaching of copper from sulfidic copper ores. (A) Top of a leach dump showing corrosion-resistant pipes and hose for watering the dump with barren solution. The dark patches represent moistened areas in which oxidation has occurred. Note the thin streams of solution issuing from the hoses in the distance. (B) Launder used in recovering copper from pregnant solution from the leach dumps by cementation with sponge iron. The copper recovered in this way has to be purified by smelting. Currently, the preferred method of copper recovery from pregnant solution is by electrolysis (electrowinning) because it yields a pure product. (Courtesy of Duval Corporation.)

and protozoa have now been identified in acid mine solution in a pH range from 0.8 to 1.38 in the Richmond Mine at Iron Mountain in California (Baker et al., 2004). Some of the protozoa may harbor prokaryotic endosymbionts affiliated with a Rickettsiales lineage (Baker et al., 2003). When the concentration of desired metal values in the pregnant solution in a sump is high enough, they are stripped from the solution. This may be accomplished in one of several ways. A formerly widely used method for copper separation involved treatment of pregnant solution with sponge iron (Fe^0) in a specially constructed basin called a launder (Figure 20.3B). The sponge iron precipitated the copper by *cementation* in a process involving the reaction

$$Cu^{2+} + Fe^0 \rightarrow Cu^0 + Fe^{2+} \tag{20.29}$$

The copper metal formed in this way was very impure and required further refinement by smelting.

After the metal value is stripped from a pregnant solution, the metal-depleted solution is called *barren solution*. It can be recirculated as lixiviant in the heap-leaching operation. However, when cementation was used to strip the copper from pregnant solution, the resultant barren solution was significantly enriched in Fe^{2+}. In many instances it was enriched in ferrous iron to such an extent that some of this iron had to be removed before the solution could be reintroduced in the leach heaps or dumps. Without removal of the excess ferrous iron, there was a danger of excessive jarosite formation upon its oxidation in the ore heaps or dumps being leached. The jarosite could precipitate on the ore mineral surfaces, interfering with further oxidation, and could also clog the drainage channels in the ore heaps and dumps. Plugging and impeding of the leach process would be the result. Excess iron removal from acid barren solution was best accomplished by biooxidation in shallow lagoons called *oxidation ponds*. In these ponds, acidophilic iron-oxidizers present in the barren solution from previous leaching cycles promoted the oxidation of the ferrous iron with concomitant acidification. A significant portion of the oxidized iron precipitated as basic ferric sulfates, including jarosite, in the oxidation ponds. When reintroduced into the heap or dump, the residual iron in the treated barren solution, which was mostly ferric, caused indirect leaching of the metal sulfides in the ore.

The acid in the recirculated barren solution that entered the heaps and dumps caused relatively rapid weathering of the host rock (gangue) of the ore, resulting in liberation of aluminum and aluminosilicates. This weathering is important in exposing occluded metal sulfide crystals to the lixiviant and to the bacteria active in the leaching process. The liberated aluminum could ultimately be separated as $Al(OH)_3$ by neutralizing the pregnant solution (Zimmerley et al., 1958; Moshuyakova et al., 1971), but this has not been done in practice. The recovered $Al(OH)_3$ could be subsequently used in the manufacture of aluminum metal. High acidity of the lixiviant may also play an important role in preventing metal ions formed during leaching from being adsorbed by the host rock (gangue) (Ehrlich, 1977; Ehrlich and Fox, 1967).

A currently preferred method of recovering metal values from pregnant solution involves *electrowinning* when the pregnant solution contains only one major metal value, or *solvent extraction* when several different metal values are present, followed by electrowinning of each of the separated metal values. Electrowinning is an electrolytic process in which the metal to be recovered is deposited on a cathode made of the same metal. The anode is usually made of carbon. The metal product of electrowinning is usually of high purity and normally does not need further refining. These metal separation processes have the advantage of not raising the ferrous iron concentration in barren solution. However, recovery of metal values from pregnant solution by solvent extraction can introduce reagents into the resultant barren solution that are inhibitory to the bacteria involved in leaching and may have to be removed before recirculation of the barren solution in heaps or dumps.

The acidophilic iron-oxidizing bacteria developing in an ore-leaching process play a dual role in solubilizing metal values, as already explained. They generate acidic ferric sulfate lixiviant by attacking pyrite and by reoxidizing Fe^{2+}, and they also attack the mineral sulfides directly, as previously discussed. It is usually not possible to assess quantitatively the extent to which these organisms are involved in direct and indirect oxidation if they are capable of both.

The interior temperature of some ore dumps or heaps, especially if they are not well ventilated, can rise as high as 70–80°C. Lyalikova (1960) observed that the heating can be accelerated by bacterial action. The heating is due to the fact that metal sulfide oxidation is an exothermic process. Such a temperature rise is unfavorable for the growth of mesophilic bioleaching bacteria, such as *Acidithiobacillus ferrooxidans* and *Leptospirillum ferrooxidans*, when it occurs in a heap or a dump interior. Microbial activity in oxidation ponds for removal of excess iron in barren solution is not affected by the heat generated in the heaps and dumps.

In the 1960s, the observation of interior heating of ore heaps and dumps during leaching suggested to some that the leaching process in dumps and heaps of metal sulfide ores is mostly abiotic. This view changed after the discovery of thermophiles capable of promoting bioleaching of metal sulfides. *Acidithiobacillus ferrooxidans* and *Leptospirillum ferrooxidans*, which are unable to live at temperatures in excess of 37–40°C, are succeeded by acidophilic, iron-oxidizing thermophiles in the interior of leach dumps and heaps where the temperature has risen to the thermophilic range (Norris, 1997). These microorganisms, which catalyze reactions similar to those of *Acidithiobacillus ferrooxidans*, operate optimally at the higher temperatures. Thus, contrary to earlier views, leaching in all parts of a heap or dump is most likely biological. It has also been suggested that the thermophiles may be responsible for regulating the internal temperature of active leach heaps and dumps (Murr and Brierley, 1978).

The primary copper mineral in ore bodies of magmatic hydrothermal origin is chalcopyrite ($CuFeS_2$). This mineral tends to be somewhat refractory to chemical (abiotic) leaching by acidic ferric sulfate compared to the secondary copper sulfide minerals chalcocite (Cu_2S) and covellite (CuS). The oxidation of chalcopyrite in nature can thus be significantly enhanced by *Acidithiobacillus ferrooxidans* and *Acidianus brierleyi* (Razzell and Trussell, 1963; Brierley, 1974). Ferric iron, when present in excess of 1000 ppm, has been found to inhibit chalcopyrite oxidation (Duncan and Walden, 1972; Ehrlich, 1977), probably because it precipitates as jarosite or adsorbs to the surface of chalcopyrite crystals and prevents further oxidation. Bacteria themselves may interfere by generating excess ferric iron from ferrous iron that precipitates or is adsorbed by residual chalcopyrite.

A typical leach cycle in which copper is recovered by either cementation or by electrowinning is represented in Figure 20.4 (For further discussion of bacterial leaching see Brierley, 1978a, 1982; Lundgren and Malouf, 1983; Rawlings, 1997a, 2002; Rohwerder et al., 2003; Olson et al., 2003).

In practice, the leaching of Pb from PbS through oxidation by *Acidithiobacillus ferrooxidans* can present a special problem because the oxidation product, $PbSO_4$, is relatively insoluble. As oxidation of PbS proceeds, $PbSO_4$ is likely to accumulate on the crystal surface and block further access to PbS by the bacteria, dissolved Fe^{3+} if present, and oxygen. In the laboratory this problem is largely eliminated when *Acidithiobacillus ferrooxidans* oxidizes PbS in batch culture with agitation (Silver and Torma, 1974) or in a large volume of lixiviant in a stirred continuous-flow reactor (Ehrlich, 1988).

Metal sulfide-oxidizing bacteria are naturally associated with metal sulfide-containing deposits, including ore bodies, bituminous coal seams, and the like. They usually exist in a consortium when ore from such deposits is bioleached (Bruhn et al., 1999). Therefore, heap, dump, and *in situ* leaching operations do not require inoculation with active bacteria, although the leaching process may be improved by it (Brierley et al., 1995). Reactor leaching operations, on the contrary, greatly benefit from inoculation with a strain selected for enhanced activity. The inoculum has to be massive for either leaching process in order to outgrow the organisms naturally present on the ore. The ore cannot be sterilized on an industrial scale.

Under natural conditions, growth and activity of the leaching organisms may be limited by one or more environmental factors. These include limited access to an energy source (metal sulfide

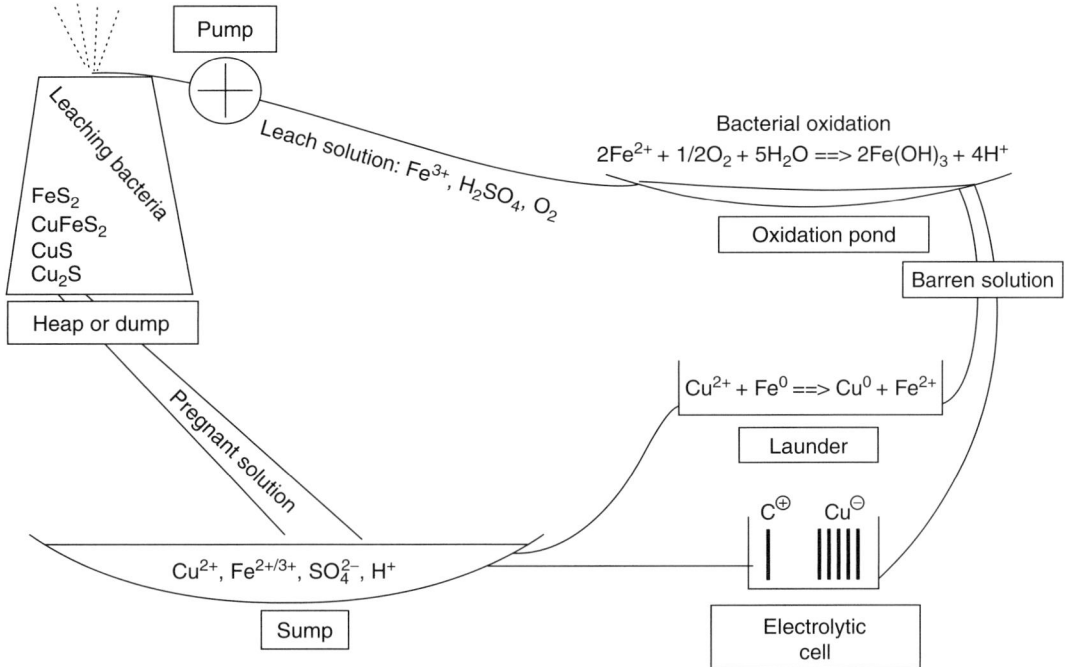

FIGURE 20.4 Schematic representation of a bioleach circuit for heap or dump leaching of copper sulfide ore. In addition to copper recovery from pregnant solution by cementation with sponge iron in a launder or by electrowinning, copper can also be recovered by solvent extraction in combination with electrowinning when the pregnant solution contains two or more base metals.

crystal), limited nitrogen source, unfavorable temperature, and limited access to air or moisture (Ehrlich and Fox, 1967; Brock, 1975; Ahonen and Tuovinen, 1989, 1992). It should be noted, however, that *Acidithiobacillus ferrooxidans* is capable of anaerobic growth with H_2, formate, or S^0 as electron donor and Fe(III) as terminal electron acceptor (see Chapter 16).

20.6.2 URANINITE LEACHING

The principles of bioleaching have also been applied on a practical scale to the leaching of uraninite ores, especially if the ores are low grade. The process may involve dump, heap, or *in situ* leaching (Wadden and Gallant, 1985; McCready and Gould, 1990). *Acidithiobacillus ferrooxidans* is one organism that has been harnessed for this process. Its action in this instance is chiefly indirect by generating an oxidizing lixiviant, acid ferric sulfate, which oxidizes U(IV) in the uraninite ore to soluble U(VI). The overall key reactions leading to uranium mobilization from uraninite can be summarized as follows:

$$2Fe^{2+} + 0.5O_2 + 2H^+ \xrightarrow{\text{\textit{Acidithiobacillus ferrooxidans}}} 2Fe^{3+} + H_2O \qquad (20.30)$$

Some of the resultant ferric iron hydrolyzes and generates acid:

$$Fe^{3+} + 3H_2O \rightarrow Fe(OH)_3 + 3H^+ \qquad (20.31)$$

The ultimate product of Fe^{3+} hydrolysis will more likely be a basic ferric sulfate such as jarosite rather than ferric hydroxide. The consequence of jarosite formation is that the net yield of acid (protons) will be less than in Reaction 20.31 (see Reaction 20.21). The remaining dissolved ferric iron can then react abiologically with the uraninite to form uranyl ions:

$$UO_2 + 2Fe^{3+} \rightarrow UO_2^{2+} + 2Fe^{2+} \tag{20.32}$$

Acidithiobacillus ferrooxidans will reoxidize the ferrous iron from this reaction, thus maintaining the process without the need for continual external resupply of ferric iron to the system. The dissolved uranium may be recovered from solution through concentration by ion exchange.

The acidophilic iron-oxidizing bacteria are often naturally associated with the ore body if it contains pyrite or another form of iron sulfide. In an absence of pyrite in depleted mines, the growth of these bacteria may be stimulated for *in situ* leaching by intermittent spraying of $FeSO_4$-nutrient-enriched solution onto the floors, walls, and mud of mine stopes (Zajic, 1969). If the leachate becomes anoxic and its pH rises, sulfate-reducing bacteria may develop in it. These bacteria can reprecipitate UO_2 as a result of reaction of UO_2^{2+} with H_2S:

$$UO_2^{2+} + H_2S \rightarrow UO_2 + 2H^+ + S^0 \tag{20.33}$$

Some observations have indicated that *Acidithiobacillus ferrooxidans* can enzymatically catalyze the oxidation of U(IV) to U(VI), but it has not yet been shown to grow under these conditions in the laboratory (DiSpirito and Tuovinen, 1981, 1982a,b; see also Chapter 18). Living cells accumulate significantly less uranium than dead cells and bind it chiefly in their cell envelope (DiSpirito et al., 1983). Despite this capacity of *Acidithiobacillus ferrooxidans* to oxidize (IV) directly, bioleaching of uraninite is believed to involve chiefly the indirect pathway.

Uranium bioleaching is another example of a natural process that is artificially stimulated. Under natural conditions, the reaction described here must occur on a very limited scale and thus only result in slow mobilization of uranium.

UO_2^{2+} in drainage from uranium mines can be microbiologically precipitated by its reduction under anaerobic conditions to insoluble UO_2. Some sulfate-reducing bacteria are capable of this reaction (Lovley et al., 1993). *Geobacter metallireducens* and *Shewanella putrefaciens* are two other bacterial species capable of precipitating U(IV) under anaerobic conditions (Gorby and Lovley, 1992). The electron donors used by the bacteria in these reactions are usually organic compounds but can be H_2 in the case of some sulfate-reducers and *Shewanella*. This activity can be useful in remediating uranium-containing mine drainage, although only at a circumneutral pH.

20.6.3 MOBILIZATION OF URANIUM IN GRANITIC ROCKS BY HETEROTROPHS

Acidithiobacillus ferrooxidans is not the only organism capable of uranium mobilization. Heterotrophic microorganisms such as some members of the soil microflora and bacteria from granites or mine waters (*Pseudomonas fluorescens, P. putida*, and *Achromobacter*) can mobilize uranium in granitic rocks, ore, and sand by weathering that results from mineral interaction with organic acids and chelators produced by the microorganisms (Magne et al., 1973, 1974; Zajic, 1969). Magne et al. found experimentally that the addition of thymol to percolation columns of uraniferous material fed with glucose solution selected a microbial flora whose efficiency in uranium mobilization was improved by greater production of oxalic acid. The authors suggested that in nature, phenolic and quinoid compounds of plant origin can serve the role of thymol. They also reported that microbes can precipitate uranium by digestion of soluble uranium complexes (Magne et al., 1974), that is, by microbial destruction of the organic moiety that complexes the uranium. These observations may explain how in nature uranium in granitic rock may be mobilized by bacteria and reprecipitated and concentrated elsewhere under the influence of other microbial activity.

20.6.4 STUDY OF BIOLEACHING KINETICS

A number of studies have been published on the kinetics of bioleaching of metal sulfides under controlled conditions in the presence of ferrous iron. They include studies by Boon et al. (1995), Hansford (1997), Hansford and Vargas (1999), Crundwell (1995, 1997), Nordstrom and Southam (1997), Driessens et al. (1999), Fowler and Crundwell (1999), and Howard and Crundwell (1999). In these studies the assumption was made or the inference drawn that bioleaching of metal sulfides proceeds in only one mode. The existence of separate direct and indirect modes that may occur concurrently in the same leaching operation appears to have been rejected. The authors did not consider that the experimental conditions used in their experiments happened to favor strongly an indirect mode of leaching over a direct mode. However, Fowler and Crundwell (1999) do assign oxidation of S^0 that appears at the surface of ZnS during leaching to attached *Acidithiobacillus ferrooxidans* cells.

20.6.5 INDUSTRIAL VERSUS NATURAL BIOLEACHING

Industrial bioleaching of metal sulfide ores harnesses naturally occurring microbiological processes by creating selective and optimized conditions that allow leaching to occur at fast rates. In the absence of human intervention, the same processes occur only at very slow rates in highly localized situations, contributing in the case of sulfide ore to a very slow, gradual change from reduced to oxidized ore. This accounts for the relative stability of undisturbed ore bodies.

20.7 BIOEXTRACTION OF METAL SULFIDE ORES BY COMPLEXATION

Some metal sulfide ores cannot be oxidized by acidophilic iron-oxidizing bacteria because they contain too great an amount of acid-consuming constituents in the host rock (gangue). The metals in such ores may be amenable to extraction by some microorganisms such as fungi (Burgstaller and Schinner, 1993). Wenberg et al. (1971) reported the isolation of the fungus *Penicillium* sp. from a mine-tailings pond of the White Pine Copper Co. in Michigan, which produced unidentified metabolites that could mobilize copper from sedimentary ores of the White Pine deposit in Czapek's broth containing sucrose, $NaNO_3$, and cysteine, methionine, or glutamic acid. *Acidithiobacillus ferrooxidans* could not be employed for leaching of this ore because of the presence of significant quantities of calcium carbonate that would neutralize the required acid. Similar findings were reported by Hartmannova and Kuhr (1974), who found that not only *Penicillium* sp. but also *Aspergillus* sp. (e.g., *A. niger*) were active in producing complexing agents that leached copper. More recently, Mulligan and Galvez-Cloutier (2000) demonstrated mobilization of copper in an oxidized mining residue by *A. niger* in a sucrose–mineral salts medium. The chief mobilizing agents produced by the fungus were gluconic and citric acids, which can act as acidulants as well as ligands of metal ions.

In some of their experiments, Wenberg et al. (1971) grew their fungus in the presence of copper ore (sulfide or native copper minerals with basic gangue constituents). The addition of some citrate to the medium lowered the toxicity of the extracted copper when the fungus was grown in the presence of the ore. They obtained better results when they grew the fungus in the absence of the ore and then treated the ore with the spent medium from the fungus culture. The principle of action of the fungi in all the above-cited experiments is similar to that involved in a study by Kee and Bloomfield (1961), who noted the dissolution of the oxides of several trace elements (e.g., ZnO, PbO_2, MnO_2, CoO, Co_2O_3) with anaerobically fermented plant material (Lucerne and Cocksfoot). The principle of action is also similar to that employed in the experiments by Parès (1964a,b,c), in which *Serratia marcescens, Bacillus subtilis*, *B. sphaericus,* and *B. firmus* solubilized copper and some other metals that were associated with laterites and clays. The organisms formed ligands that extracted the metals from the ores by forming complexes with them, which were more stable than the original insoluble form of the metals in the ores. This type of reaction can be formulated as follows:

$$MA + HCh \rightarrow MCh + H^+ + A^- \tag{20.34}$$

where MA is a metal salt (mineral), HCh a ligand (chelating agent), MCh the resultant metal chelate, and A^- the counter ion of the original metal salt, which would be S^{2-} in the case of metal sulfides. The S^{2-} may undergo chemical or bacterial oxidation. The use of carboxylic acids in industrial leaching of ores has been proposed as a general process (Chemical Processing, 1965).

20.8 FORMATION OF ACID COAL MINE DRAINAGE

When bituminous coal seams that contain pyrite inclusions are exposed to air and moisture during mining, the pyrites undergo oxidation, leading to the formation of AMD. With the onset of pyrite oxidation, acidophilic iron-oxidizing thiobacilli become readily detectable in the drainage (Leathen et al., 1953). *Acidithiobacillus thiooxidans* also makes an appearance. When *Acidithiobacillus ferrooxidans* is involved, pyrite biooxidation proceeds by the reactions previously described (Reactions 20.19 through 20.24). *Acidithiobacillus thiooxidans*, which cannot oxidize ferrous iron, probably oxidizes elemental sulfur (S^0) and other partially reduced sulfur species, which may form as intermediates in pyrite oxidation, to sulfuric acid (Mustin et al., 1992, 1993) (Reaction 20.17). The chief products of pyrite oxidation are thus sulfuric acid and basic ferric sulfate (jarosite and amorphous basic ferric sulfates). Streams that receive this mine drainage may exhibit pH values ranging from 2 to 4.5 and sulfate ion concentrations ranging from 1,000 to 20,000 mg L^{-1} but a nondetectable ferrous iron concentration (Lundgren et al., 1972). Walsh and Mitchell (1972b) proposed that a *Metallogenium*-like organism that they isolated from AMD may be the dominant iron-oxidizing organism attacking the pyrite in exposed pyrite-containing coal seams until the pH drops below 3.5. After that, the more acid-tolerant *Acidithiobacillus ferrooxidans* and probably *Leptospirillum ferrooxidans*, which was unknown to Walsh and Mitchell, take over. The *Metallogenium*-like organism does not, however, appear to be essential to lower the pH in a pyritic environment to make it favorable for the acidophilic iron-oxidizers. *Acidithiobacillus ferrooxidans* itself may be capable of doing that, at least in pyrite-containing coal or overburden (Kleinmann and Crerar, 1979). It may accomplish this by initial direct attack of pyrite (see Reaction 20.22), creating an acid microenvironment from which the organism and the acid it generates spread.

An early study of microbial succession in coal spoil under laboratory conditions was carried out by Harrison (1978). He constructed an artificial coal spoil by heaping a homogeneous mixture of 1 part crushed, sifted coal and 2 parts shale and 8 parts subsoil from the overburden of a coal deposit into a mound 50 cm in diameter and 25 cm high on a plastic tray. The mound was inoculated with 20 L of an emulsion of acid soil, drainage water, and mud from a spoil from an old coal strip mine, which was poured on the bottom of the plastic tray. The inoculum was absorbed by the mound and migrated upward, presumably by capillary action. Evaporation losses during the experiment were made up by periodic additions of distilled water to the free liquid on the tray.

Initial samples taken at the base of the mound yielded evidence of the presence of heterotrophic bacteria. These bacteria were dominant and reached a population density of ~10^7 cells g^{-1} within 2 weeks. After 8 weeks, heterotrophs were still dominant, although the pH had dropped from 7 to 5. Between 12 and 20 weeks, the population decreased by about an order of magnitude, coinciding with a slight decrease in pH to just below 5 caused by a burst of growth by sulfur-oxidizing bacteria, which then died off progressively. Thereafter, the heterotrophic population increased again to just below 10^7 g^{-1}. In samples from near the summit of the mound, heterotrophs predominated for the first 15 weeks but then decreased dramatically from 10^6 to 10^2 cells g^{-1}, concomitant with a drop in pH to 2.6. The pH drop was correlated with a marked rise in population density of sulfur- and iron-oxidizing autotrophic bacteria (*Acidithiobacillus thiooxidans* and *Acidithiobacillus ferrooxidans*), the former dominating briefly over the latter in the initial weeks. Protozoans, algae, an arthropod, and a moss were also noted, mostly at the higher pH values. The presence of protozoa in coal mine drainage was earlier reported by Lackey (1938) and Joseph (1953) and later by Johnson and Rang (1993). *Metallogenium* of the type of Walsh and Mitchell (1972a) was not seen.

The sulfur-oxidizing bacteria were assumed to be making use of elemental sulfur resulting from the oxidation of pyrite by ferric sulfate:

$$FeS_2 + Fe_2(SO_4)_3 \rightarrow 3FeSO_4 + 2S^0 \qquad (20.35)$$

More specifically, as Mustin et al. (1992, 1993) indicated, the sulfur may arise as a result of anodic reactions at the surface of pyrite crystals. It is also possible that at least some of the sulfur arises indirectly from the reduction of microbial sulfate in anaerobic zones of the coal spoil. The reduction of sulfate would yield H_2S, which then becomes the energy source for thiobacilli such as *Thiobacillus thioparus* that oxidize it to elemental sulfur at the interface of the oxidized and reduced zones, provided the ambient pH is not too far below neutrality. Anaerobic bacteria were not sought in this study.

After 7 weeks of incubation, a mineral efflorescence developed on the surface of the mound. It consisted mainly of sulfates of Mg, Ca, Na, Al, and Fe. The magnesium sulfate was in the form of the hexahydrate rather than epsomite. The leached metals were derived from the coal, but magnesium was also leached from the overburden material. Harrison's study, which was reported in 1978 before the introduction of the techniques of molecular biology to microbial ecology, deserves to be repeated using molecular techniques. It could yield new insights into the microbial succession in the development of AMD from coal spoils. Baker and Banfield (2003) have reviewed recent findings from studies of microbial diversity in AMD that relied on DNA analysis rather on culture techniques. The findings hint at a complex series of microbial successions that must have taken place over time as AMD generation evolved.

Darland et al. (1970) isolated thermophilic, acidophilic *Thermoplasma acidophilum* from a coal refuse pile that had become self-heated. The organism was described as lacking a true bacterial cell wall and to resemble mycoplasmas, however, it is an archeon. Its optimum growth temperature was 59°C (range 45–62°C) and its optimum pH for growth was between 1 and 2 (pH range 0.96–3.5). The organism is a heterotroph growing readily in a medium of 0.02% $(NH_4)_2SO_4$, 0.05% $MgSO_4$, 0.025% $CaCl_2 \cdot 2H_2O$, 0.3% KH_2PO_4, 0.1% yeast extract, and 1.0% glucose at pH 3. Its relation to the coal environment and its contribution, if any, to the AMD problem needs to be clarified.

20.8.1 NEW DISCOVERIES RELATING TO ACID MINE DRAINAGE

A fairly recent study of abandoned mines at Iron Mountain, California, resulted in a startling discovery insofar as the generation of AMD is concerned, at least at this mine. The ore body at Iron Mountain at the time it was mined contained various metal sulfides and was a source of Fe, Cu, Ag, and Au. A significant part of the iron was in the form of pyrite. The drainage currently coming from the abandoned mine workings contains varying amounts of these metals as well as Cd and is very acidic. A survey of the distribution of *Acidithiobacillus ferrooxidans* and *Leptospirillum ferrooxidans* in solutions from a pyrite deposit in the Richmond Mine, seepage from a tailings pile, and AMD storage tanks outside this mine was undertaken (Schrenk et al., 1998). It revealed that *Acidithiobacillus ferrooxidans* occurred in slime-based communities at pH >1.3 at temperatures below 30°C, whereas *L. ferrooxidans* was abundant in subsurface slime-based communities and also occurred in planktonic form at pH values in the range of 0.3–0.7 between 30 and 50°C (Figure 20.5). *Acidithiobacillus ferrooxidans* appeared to affect precipitation of ferric iron but seemed to have a minor role in acid generation. Neither *Acidithiobacillus ferrooxidans* nor *L. ferrooxidans* was thought to exert a direct catalytic effect on metal sulfide oxidation, but they were thought to play an active role in generating ferric iron as an oxidizing agent (Schrenk et al., 1998). Microbiological investigation of underground areas (drifts) in the Richmond Mine revealed the presence of Archaea

FIGURE 20.5 A field site in an abandoned stope within the Richmond Mine at Iron Mountain in northern California, showing streamers of bacterial slime in the foreground. A pH electrode is visible on the right. A fishing line in the center secures a white bottle with a filter lid that contains sulfide mineral samples to study their fate *in situ*. (Courtesy of Thomas Gihring, Department of Oceanography, Florida State University, Tallahassee.)

as well as *Acidithiobacillus ferrooxidans* and *L. ferrooxidans* (Edwards et al., 1999a), but only the Archaea and *L. ferrooxidans* were associated with acid-generating sites (Edwards et al., 1999b). The proportions in which they were detected varied with the site and the season of the year. Members of the Bacteria were most abundant during winter months when Archaea were nearly undetectable. The reverse was found in summer and fall months, when Archaea represented ~50% of the total population. The authors correlated these population fluctuations with rainfall and conductivity (dissolved solids), pH, and temperature of the mine water. As noted above, *Acidithiobacillus ferrooxidans* was the least pH-tolerant and was less temperature-tolerant than *L. ferrooxidans*. As already mentioned, *Acidithiobacillus ferrooxidans* was absent at acid-generating sites in the mine (Edwards et al., 1999b). Any attachment of iron-oxidizing bacteria was restricted to pyrite phases. Based on dissolution rate measurements, the investigators found that attached and planktonic species contributed to similar extents to acid release. Among the Archaea, a newly discovered microbe, *Ferroplasma acidarmanus*, grew in slime streamers on the pyrite surfaces (Edwards et al., 2000a). It constituted up to 85% of the total communities in the slimes and sediments that were examined. In laboratory study, it was found to be extremely acid-tolerant. It was able to grow at pH 0 and exhibited a pH optimum at 1.2 (see also Section 20.5.4). Its cells lack a wall. It belongs to the Archaean order Thermoplasmales (Edwards et al., 2000a) and is a close relative of *Ferroplasma acidophilum* (Golyshina et al., 2000). On the basis of studies to date, Edwards et al. (2001) believe that *F. acidarmanus* and *F. acidophilum* promote pyrite oxidation by generating the oxidant Fe^{3+} from Fe^{2+} (indirect mechanism).

Using molecular phylogenetic techniques, examination of a ~1 cm thick slime on finely disseminated pyrite ore collected in the Richmond Mine revealed the presence of a variety of bacterial types (Bond et al., 2000b). Predominant ones were *Leptospirillum* species representing 71% of the clones recovered. *Acidimicrobium*-related species, including *Ferromicrobium acidophilus*, were detected. Archaea were represented by the family of Ferroplasmaceae. They included organisms closely related to *Ferroplasma acidophilum* and *F. acidarmanus* and organisms with an affinity to *Thermoplasma acidophilum*. Also detected were members of the δ-subdivision of sulfate- and metal-reducers. These findings indicate the presence of microniches in the slime, some supporting aerobic (oxidizing) activity, others anaerobic (reducing) activity. The makeup of microbial communities at environmentally different sites in the Richmond Mine were found to differ quantitatively and to some degree qualitatively (Bond et al., 2000a).

The interesting studies at Iron Mountain relegate the position of *Acidithiobacillus ferrooxidans* in AMD formation to the periphery of the remaining ore body at the present time. This position appears to be dictated by prevailing environmental conditions inside and outside the mine. This may not always have been the position of *Acidithiobacillus ferrooxidans*. Its current position may be the result of a species succession in microbial community development that started when the pyrite and other metal sulfides in the ore body first began to be oxidized upon exposure to air and water as mining proceeded. In AMD arising from the weathering of As- and pyrite-rich mine tailings at an abandoned mining site in Carnoulès, France, *Acidithiobacillus ferrooxidans* was readily detectable along with many as-yet-uncultured organisms (Duquesne et al., 2003; Bruneel et al., 2006). It was also readily detected in addition to *Ferroplasma acidiphilum* and some other organisms in the Tinto River, southwestern Spain (González-Toril et al., 2003). Thus, microbial succession is likely to have taken place over time in the formation of AMD in the Iron Mountain deposit (Edwards et al., 2000b), and is strongly suggested in a study of a tank bioleaching process at the Mintek plant in South Africa in which thermoplasmales appeared (Okibe et al., 2003).

The findings from Iron Mountain raise the question of whether similar changes have occurred or are taking place in the generation of AMD from bituminous coal mines where the pyrite in the coal is more dispersed than in the Iron Mountain ore body. Clearly, AMD formation from bituminous coal mines needs to be reinvestigated. The question of microbial succession in the development of AMD in the mining of metal sulfide ore bodies also needs to be investigated. The evolution of microbial diversity over time in AMD cannot be readily deduced from analysis of a single sample late in the process.

20.9 SUMMARY

Metal sulfides may occur locally at high concentrations, in which case they constitute ores. Although most nonferrous sulfides are formed abiogenically through magmatic and hydrothermal processes, a few sedimentary deposits are of biogenic origin. More important, some sedimentary ferrous sulfide accumulations are biogenically formed. The microbial role in biogenesis of any of the sulfide deposits is the generation of H_2S, usually from the bacterial reduction of sulfate, but in a few special cases possibly from the mineralization of organic sulfur compounds. Because metal sulfides are relatively water insoluble, spontaneous reaction of metal ions with the biogenic sulfide proceeds readily. Biogenesis of specific metal sulfide minerals has been demonstrated in the laboratory. These experiments require relatively insoluble metal compounds as starting materials to limit the toxicity of the metal ions to the sulfate-reducing bacteria. In nature, adsorption of the metal ions by sediment components serves a similar function in lowering the concentration below their toxic levels for sulfate-reducers.

Metal sulfides are also subject to oxidation by bacteria such as *Acidithiobacillus ferrooxidans, Leptospirillum ferrooxidans, Sulfolobus* spp., *Acidianus brierleyi*, and others. The bacterial action may involve direct oxidative attack of the crystal lattice of a metal sulfide or indirect oxidative attack by generation of lixiviant (acid ferric sulfate), which oxidizes the metal sulfide chemically. The indirect mechanism is of primary importance in the solubilization of uraninite (UO_2). Microbial oxidation of metal sulfides is industrially exploited in extracting metal values from low-grade metal sulfide ore and uraninite and has been tested successfully on some high-grade ore and ore concentrates. In bituminous coal seams that are exposed as a result of mining activity, pyrite oxidation by these bacteria is an environmentally deleterious process; it is the source of AMD.

REFERENCES

Ahonen L, Tuovinen OH. 1989. Microbiological oxidation of ferrous iron at low temperatures. *Appl Environ Microbiol* 55:312–316.

Ahonen L, Tuovinen OH. 1992. Bacterial oxidation of sulfide minerals in column leaching experiments at suboptimal temperatures. *Appl Environ Microbiol* 58:600–606.

Asmah RH, Bosompem KM, Osei YD, Rodriguez FK, Addy ME, Clement C, Wilson MD. 1999. Isolation and characterization of mineral oxidizing bacteria from the Obuasi gold mining site, Ghana. In: Amils R, Ballester A, eds. *Biohydrometallurgy and the Environment Toward the Mining of the 21st Century, Part A.* Amsterdam: Elsevier, pp. 657–662.

Baas Becking LGM, Moore D. 1961. Biogenic sulfides. *Econ Geol* 56:259–272.

Badigian RM, Myerson AS. 1986. The adsorption of *Thiobacillus ferrooxidans* on coal surfaces. *Biotech Bioeng* 28:467–479.

Baker BJ, Banfield JF. 2003. Microbial communities in acid mine drainage. *FEMS Microbiol Ecol* 44:139–152.

Baker BJ, Hugenholtz P, Dawson SC, Banfield JF. 2003. Extremely acidophilic protists from acid miner drainage host *Rickettsiales*-lineage endosymbionts that have intervening sequences in the 16S rRNA genes. *Appl Environ Microbiol* 69:5512–5518.

Baker BJ, Lutz MA, Dawson SC, Bond PL, Banfield JF. 2004. Metabolically active eukaryotic communities in extremely acidic mine drainage. *Appl Environ Microbiol* 70:6264–6271.

Balashova VV, Vedenina IYa, Markosyan GE, Zavarzin GA. 1974. The autotrophic growth of *Leptospirillum ferrooxidans*. *Mikrobiologiya* 43:581–585 (English translation, pp. 491–494).

Ballard RD, Grassle JF. 1979. Strange world without sun. Return to oases of the deep. *Natl Geogr Mag* 156:680–703.

Barreto M, Jedlicki E, Holmes DS. 2005. Identification of a gene cluster for the formation of extracellular polysaccharide precursors in the chemolithoautotroph *Acidithiobacillus ferrooxidans*. *Appl Environ Microbiol* 71:2902–2909.

Bennett JC, Tributsch H. 1978. Bacterial leaching patterns of pyrite crystal surfaces. *J Bacteriol* 134:310–317.

Berner RA. 1984. Sedimentary pyrite formation: An update. *Geochim Cosmochim Acta* 48:605–615.

Bhatti TM, Gigham JM, Carlson L, Tuovinen OH. 1993. Mineral products of pyrrhotite oxidation by *Thiobacillus ferrooxidans*. *Appl Environ Microbiol* 59:1984–1990.

Bond PL, Druschel GK, Banfield JF. 2000a. Comparison of acid mine drainage microbial communities in physically and geochemically distinct ecosystems. *Appl Environ Microbiol* 66:4962–4971.

Bond PL, Smriga SP, Banfield JF. 2000b. Phylogeny of microorganisms populating a thick, subaerial, predominantly lithotrophic biofilm at an extreme acid mine drainage site. *Appl Environ Microbiol* 66:3842–3849.

Bonnatti E. 1972. Authigenesis of minerals—marine. In: Fairbridge RW, ed. *The Encyclopedia of Geochemistry and Environmental Sciences. Encyclopedia Earth Science Series, Vol IVA.* New York: Van Nostrand Reinhold, pp. 48–56.

Bonnatti E. 1978. The origin of metal deposits in the oceanic lithosphere. *Sci Am* 238:54–61.

Boon M, Hansford GS, Heijnen JJ. 1995. The role of bacterial ferrous iron oxidation in the bio-oxidation of pyrite. In: Vargas T, Jerez CA, Wiertz JV, Toledo H, eds. *Biohydrometallurgical Processing, Vol 1.* Santiago, Chile: University of Chile, pp. 153–163.

Brierley CL. 1974. Leaching. Use of a high-temperature microbe. In: Aplan FF, Mckinney WA, Pernichele AD, eds. *Solution Mining Symposium. Proceedings of the 103rd AIME Annual Meet*, Dallas, TX, February 25–27, pp. 461–469.

Brierley CL. 1978a. Bacterial leaching. *CRC Crit Rev* 6:107–262.

Brierley CL. 1982. Microbiological mining. *Sci Am* 247:44–53, 150.

Brierley CL, Brierley JA. 1973. A chemoautotrophic and thermophilic microorganism isolated from an acid hot spring. *Can J Microbiol* 19:183–188.

Brierley CL, Murr LE. 1973. Leaching: Use of a thermophilic chemoautotrophic microbe. *Science* 179:488–490.

Brierley JA. 1978b. Thermophilic iron-oxidizing bacteria found in copper leaching dumps. *Appl Environ Microbiol* 36:523–525.

Brierley JA, Lockwood SJ. 1977. The occurrence of thermophilic iron-oxidizing bacteria in a copper leaching system. *FEMS Microbiol Lett* 2:163–165.

Brierley JA, Norris PR, Kelly DP, LeRoux NW. 1978. Characteristics of a moderately thermophilic and acidophilic iron-oxidizing *Thiobacillus*. *Eur J Appl Microbiol Biotechnol* 5:291–299.

Brieley JA, Wan RY, Hill DL, Logan TC. 1995. Biooxidation-heap pretreatment technology for processing lower grade refractory gold ores. In: Vargas T, Jerez CA, Wiertz JV, Toledo H, eds. *Biohydrometallurgical Processing*. Vol 1. Santiago, Chile: University of Chile, pp. 253–262.

Brock TD. 1975. Effect of water potential on growth and iron oxidation by *Thiobacillus ferrooxidans*. *Appl Microbiol* 29:495–501.

Bruhn DF, Thompson DN, Noah KS. 1999. Microbial ecology assessment of a mixed copper oxide/sulfide dump leach operation. In: Amils R, Ballester A, eds. *Biohydrometallurgy and the Environment Towards the Mining of the 21st Century, Part A*. Amsterdam: Elsevier, pp. 799–808.

Bruneel O, Duran R, Casiot C, Elbaz-Poulichet F, Personné J-C. 2006. Diversity of microorganisms in Fe-As-rich acid mine drainage waters at Carnoulès, France. *Appl Environ Microbiol* 72:551–556.

Bryner LC, Anderson R. 1957. Microorganisms in leaching sulfide minerals. *Ind Eng Chem* 49:1721–1724.

Bryner LC, Beck JV, Davis BB, Wilson DG. 1954. Microorganisms in leaching sulfide minerals. *Ind Eng Chem* 46:2587–2592.

Bryner LC, Jameson AK. 1958. Microorganisms in leaching sulfide minerals. *Appl Microbiol* 6:281–287.

Burgstaller W, Schinner F. 1993. Leaching of metals with fungi. *J Biotechnol* 27:91–116.

Cameron EM. 1982. Sulfate and sulfate reduction in early Precambrian oceans. *Nature* (London) 296:145–148.

Carlson L, Lindström EB, Hallberg KB, Tuovinen OH. 1992. Solid-phase products of bacterial oxidation of arsenical pyrite. *Appl Environ Microbiol* 58:1046–1049.

Chemical Processing. 1965. Carboxylic acids in metal extraction. *Chem Proc* 11:24–25.

Chen C-Y, Skidmore DR. 1987. Langmuir adsorption isotherm for *Sulfolobus acidocaldarius* on coal particles. *Biotech Lett* 9:191–194.

Chen C-Y, Skidmore DR. 1988. Attachment of *Sulfolobus acidocaldarius* cells in coal particles. *Biotechnol Progr* 4:25–30.

Corliss JB, Dymond J, Gordon LI, Edmond JM, von Herzen RP, Ballard RD, Green K, Williams D, Bainbridge A, Crane K, van Andel TH. 1979. Submarine thermal springs on the Galapagos Rift. *Science* 203:1073–1083.

Crundwell FK. 1995. Mathematical modeling and optimization of bacterial leaching plants. In: Vargas T, Jerez CA, Wiertz JV, Toledo H, eds. *Biohydrometallurgical Processing, Vol 1*. Santiago, Chile: University of Chile, pp. 437–446.

Crundwell FK. 1997. Physical chemistry of bacterial leaching. In: Amils R, Ballester A, eds. *Biohydrometallurgy and the Environment Toward the Mining of the 21st Century, Part A*. Amsterdam: Elsevier, pp. 13–26.

Cuthbert ME. 1962. Formation of bornite at atmospheric temperature and pressure. *Econ Geol* 57:38–41.

Darland G, Brock TD, Sansonoff W, Conti SF. 1970. A thermophilic, acidophilic mycoplasma isolated from a coal refuse pile. *Science* 170:1416–1418.

Davidson CF. 1962a. The origin of some strata-bound sulfide ore deposits. *Econ Geol* 57:265–274.

Davidson CF. 1962b. Further remarks on biogenic sulfides. *Econ Geol* 57:1134–1137.

Descostes M, Vitorge P, Beaucaire C. 2004. Pyrite dissolution in acidic media. *Geochim Cosmochim Acta* 68:4559–4569.

Dévigne J-P. 1968a. Précipitation du sulfure de plomb par un micrococcus tellurique. *CR Acad Sci* (Paris) 267:935–937.

Dévigne J-P. 1968b. Une bactérie saturnophile, *Sarcina flava* Bary 1887. *Arch Inst Pasteur* (Tunis) 45:341–358.

Dévigne J-P. 1973. Une métallogenèse microbienne probable en milieux sédimentaires: Celle de la galena. *Cah Geol* 89:35–37.

Dew DW, van Buuren C, McEwan K, Bowker C. 1999. Bioleaching of base metal sulfide concentrates: A comparison of mesophile and thermophile bacterial cultures. In: Amils R, Ballester A, eds. *Biohydrometallurgy and the Environment Toward the Mining of the 21st Century, Part A*. Amsterdam: Elsevier, pp. 229–238.

DiSpirito AA, Talnagi JW, Jr., Tuovinen OH. 1983. Accumulation and cellular distribution of uranium in *Thiobacillus ferrooxidans*. *Arch Microbiol* 135:250–253.

DiSpirito AA, Tuovinen OH. 1981. Oxygen uptake coupled with uranous sulfate oxidation by *Thiobacillus ferrooxidans* and *T. acidophilus*. *Geomicrobiol J* 2:275–291.

DiSpirito AA, Tuovinen OH. 1982a. Uranous ion oxidation and carbon dioxide fixation by *Thiobacillus ferrooxidans*. *Arch Microbiol* 133:28–32.

DiSpirito AA, Tuovinen OH. 1982b. Kinetics of uranous and ferrous ion oxidation by *Thiobacillus ferrooxidans*. *Arch Microbiol* 133:33–37.

Donati E, Curutchet G, Pogliani C, Tedesco P. 1995. Bioleaching of covellite by individual or combined cultures of *Thiobacillus ferrooxidans* and *Thiobacillus thiooxidans*. In: Vargas T, Jerez CA, Wiertz JV, Toledo H, eds. *Biohydrometallurgical Processing, Vol 1*. Santiago, Chile: University of Chile, pp. 91–99.

Driessens YPM, Fowler TA, Crundwell FK. 1999. A comparison of the bacterial and chemical leaching of sphalerite at the same solution conditions. In: Amils R, Ballester A, eds. *Biohydrometallurgy and the Environment Toward the Mining of the 21st Century, Part A*. Amsterdam: Elsevier, pp. 201–208.

Duncan DW, Landesman J, Walden CC. 1967. Role of *Thiobacillus ferrooxidans* in the oxidation of sulfide minerals. *Can J Microbiol* 13:397–403.

Duncan DW, Walden CC. 1972. Microbial leaching in the presence of ferric iron. *Dev Ind Microbiol* 13:66–75.

Duquesne K, Lebrun S, Casiot C, Bruneel O, Personné J-C, Leblanc M, Elbaz-Poulichet F, Morin G, Bonnefoy V. 2003. Immobilization of arsenite and ferric iron by *Acidithiobacillus ferrooxidans* and its relevance to acid mine drainage. *Appl Environ Microbiol* 69: 6165–6173.

Edmond JM, Von Damm KL, McDuff RE, Measures CI. 1982. Chemistry of hot springs on the East Pacific Rise and their effluent dispersal. *Nature* (London) 297:187–191.

Edwards KJ, Bond PL, Druschel GK, McGuire MM, Hamers RJ, Banfield JF. 2000b. Geochemical and biological aspects of sulfide mineral dissolution: Lessons from Iron Mountain, California. *Chem Geol* 169:383–397.

Edwards KJ, Bond PL, Gihring TM, Banfield JF. 2000a. An archaeal iron-oxidizing extreme acidophile important in acid mine drainage. *Science* 287:1796–1799.

Edwards KJ, Gihring TM, Banfield JF. 1999a. Seasonal variations in microbial populations and environmental conditions in an extreme acid mine drainage environment. *Appl Environ Microbiol* 65:3627–3632.

Edwards KJ, Goebell BM, Rodgers RM, Schrenk MO, Gihring TM, Cardona MM, Hu B, McGuire MM, Hamers RJ, Pace NR. 1999b. Geomicrobiology of pyrite (FeS_2) dissolution: Case study at Iron Mountain, California. *Geomicrobiol J* 16:155–179.

Edwards KJ, Hu B, Hamers RJ, Banfield JF. 2001. A new look at patterns on sulfide minerals. *FEMS Microbiol Ecol* 34:197–206.

Edwards KJ, McCollum TM, Konishi H, Buseck PR. 2003. Seafloor bioalteration of sulfide minerals: Results from in situ incubation studies. *Geochim Cosmochim Acta* 67:2843–2856.

Edwards KJ, Schrenk MO, Hamers R, Banfield JF. 1998. Mineral oxidation of pyrite: Experiments using microorganisms from an extreme acidic environment. *Am Mineral* 83:1444–1453.

Ehrlich HL. 1963a. Bacterial action on orpiment. *Econ Geol* 58:991–994.

Ehrlich HL. 1963b. Microorganisms in acid mine drainage from a copper mine. *J Bacteriol* 86:350–352.

Ehrlich HL. 1964. Bacterial oxidation of arsenopyrite and enargite. *Econ Geol* 59:1306–1312.

Ehrlich HL. 1977. Bacterial leaching of low-grade copper sulfide ore with different lixiviants. In: Schwartz E, ed. *Conference Bacterial Leaching. Gesellschaft für Biotechnologische Forschung mbH, Braunschweig-Stöckheim*. Weinstein, Germany: Verlag Chemie, pp. 145–155.

Ehrlich HL. 1978. Inorganic energy sources for chemolithotrophic and mixotrophic bacteria. *Geomicrobiol J* 1:65–83.

Ehrlich HL. 1988. Bioleaching of silver from a mixed sulfide ore in a stirred reactor. In: Norris PR, Kelly DP, eds. *Biohydrometallurgy*. Kew Surrey U.K.: Science and Technology Letters, pp. 223–231.

Ehrlich HL. 2000. Past, present, and future in biohydrometallurgy. *Hydrometallurgy* 59:127–134.

Ehrlich HL, Fox SI. 1967. Environmental effects on bacterial copper extraction from low-grade copper sulfide ores. *Biotech Bioeng* 9:471–485.

Fenchel T, Blackburn TH. 1979. *Bacteria and Mineral Cycling*. London: Academic Press.

Fortin D, Beveridge TJ. 1997. Microbial sulfate reduction within sulfidic mine tailings: Formation of diagenetic Fe sulfides. *Geomicrobiol J* 14:1–21.

Fortin D, Davis B, Southam G, Beveridge TJ. 1995. Biogeochemical phenomena induced by bacteria within sulfidic mine tailings. *J Ind Microbiol* 14:178–185.

Fowler TA, Crundwell FK. 1999. Leaching zinc sulfide by *Thiobacillus ferrooxidans*: Bacterial oxidation of the sulfur product layer increases the rate of zinc sulfide dissolution at high concentrations of ferrous ions. *Appl Environ Microbiol* 65:5285–5292.

Fox SI. 1967. Bacterial oxidation of simple copper sulfides. PhD Thesis. Rensselaer Polytechnic Institute, Troy, New York.

Freke AM, Tate D. 1961. The formation of magnetic iron sulfide by bacterial reduction of iron solutions. *J Biochem Microbiol Technol Eng* 3:29–39.

Garcia O, Jr., Bigham JM, Tuovinen OH. 1995. Oxidation of galena by *Thiobacillus ferrooxidans* and *Thiobacillus thiooxidans*. *Can J Microbiol* 41:508–516.

Gehrke T, Hallmann R, Sand W. 1995. Importance of exopolymers from *Thiobacillus ferrooxidans* and *Leptospirillum ferrooxidans* for bioleaching. In: Vargas T, Jerez CA, Wiertz JV, Toledo H, eds. *Biohydrometallurgical Processing*, Vol 1. Santiago, Chile: University of Chile, pp. 1–11.

Gehrke T, Telegdi J, Thierry D, Sand W. 1998. Importance of extracellular polymeric substances from *Thiobacillus ferrooxidans* for bioleaching. *Appl Environ Microbiol* 64:2743–2747.

Giblin AE. 1988. Pyrite formation in marshes during early diagenesis. *Geomicrobiol J* 6:77–97.

Giblin AE, Howarth RW. 1984. Porewater evidence for a dynamic sedimentary iron cycle in salt marshes. *Limnol Oceanogr* 29:47–63.

Golovacheva RS, Karavaiko GI. 1979. A new genus of thermophilic spore-forming bacteria, *Sulfobacillus*. *Mikrobiologiya* 47:815–822 (English translation, pp. 658–665).

Golyshina OV, Pivovarova TA, Karavaiko GI, Kondrat'eve TF, Moore RB, Abraham W-R, Lunsdorf H, Timmis KN, Yakimov MM, Golyshin PN. 2000. *Ferroplasma acidophilum* gen nov., sp. nov., an acidophilic, autotrophic, ferrous-iron-oxidizing, cell-wall-lacking, mesophilic member of the Ferroplasmaceae, fam. nov., comprising a distinct lineage of Archaea. *Int J Syst Evol Microbiol* 50:997–1006.

González-Toril E, Llobet-Brossa E, Casamayor EO, Amann R, Amils R. 2003. Microbial ecology of an extreme acidic environment, the Tinto River. *Appl Environ Microbiol* 69:4853–4865.

Gorby YA, Lovley DR. 1992. Enzymatic uranium precipitation. *Environ Sci Technol* 26:205–207.

Hallberg RO. 1978. Metal-organic interactions at the redoxcline. In: Krumbein WE, ed. *Environmental Biogeochemistry and Geomicrobiology, Vol 3. Methods, Metals, and Assessment*. Ann Arbor, MI: Ann Arbor Science, pp. 947–953.

Hansford GS. 1997. Recent developments in modeling kinetics of bioleaching. In: Rawlings DE, ed. *Biomining: Theory, Microbes and Industrial Processes*. Berlin, Germany: Springer, pp. 153–175.

Hansford GS, Vargas T. 1999. Chemical and electrochemical basis for bioleaching processes. In: Amils R, Ballester A, eds. *Biohydrometallurgy and the Environment Toward the Mining of the 21st Century, Part A*. Amsterdam: Elsevier, pp. 13–26.

Harneit K, Göksel A, Kock D, Klock J-H, Gehrke T, Sand W. 2006. Adhesion to metal sulfide surfaces by cells of *Acidithiobacillus ferrooxidans, Acidithiobacillus thiooxidans* and *Leptospirillum ferrooxidans*. *Hydrometallurgy* 83:245–254.

Harrison AP, Jr. 1978. Microbial succession and mineral leaching in an artificial coal spoil. *Appl Microbiol* 36:861–869.

Harrison AP, Jr., Norris PR. 1985. *Leptospirillum ferrooxidans* and similar bacteria: Some characteristics and genomic diversity. *FEMS Microbiol Lett* 30:99–102.

Hartmannova V, Kuhr I. 1974. Copper leaching by lower fungi. *Rudy* 22:234–238.

Howard D, Crundwell FK. 1999. A kinetic study of the leaching of chalcopyrite with *Sulfolobus metallicus*. In: Amils R, Ballester A, eds. *Biohydrometallurgy and the Environment Toward the Mining of the 21st Century. Part A, Bioleaching, Microbiology*. Amsterdam: Elsevier, pp. 209–217.

Howarth RW. 1979. Pyrite: Its rapid formation in a salt marsh and its importance in ecosystem metabolism. *Science* 203:49–51.

Howarth RW, Merkel S. 1984. Pyrite formation and the measurement of sulfate reduction in salt marsh sediments. *Limnol Oceanogrf* 29:598–608.

Imai KH, Sakaguchi H, Sugio T, Tano T. 1973. On the mechanism of chalcocite oxidation by *Thiobacillus ferrooxidans*. *J Ferment Technol* 51:865–870.

Ingledew WJ. 1986. Ferrous iron oxidation by *Thiobacillus ferrooxidans*. In: Ehrlich HL, Holmes DS, eds. *Biotechnology for the Mining, Metal Refining, and Fossil Fuel Processing Industries. Biotechnol Bioeng Symp 16*. New York: Wiley, pp.23–33.

Ivanov VI. 1962. Effect of some factors on iron oxidation by cultures of *Thiobacillus ferrooxidans*. *Mikrobiologiya* 31:795–799 (English translation, pp. 645–648).

Ivanov VI, Nagirnyak FI, Stepanov BA. 1961. Bacterial oxidation of sulfide ores. I. Role of *Thiobacillus ferrooxidans* in the oxidation of chalcopyrite and sphalerite. *Mikrobiologiya* 30:688–692.

Johnson BS, Rang L. 1993. Effects of acidophilic protozoa on populations of metal-mobilizing bacteria during the leaching of pyritic coal. *J Gen Microbiol* 139:1417–1423.

Johnson BS, Roberto FF. 1997. Heterotrophic acidophiles and their roles in bioleaching of sulfide minerals. In: Rawlings DE, ed. *Biomining: Theory, Microbes and Industrial Processes*. Berlin, Germany: Springer, pp. 259–279.

Jørgensen BB. 1977. The sulfur cycle of a coastal marine sediment (Limfjorden, Denmark). *Limnol Oceanogr* 22:814–832.

Joseph JM. 1953. Microbiological study of acid mine water: Preliminary report. *Ohio J Sci* 53:123–127.

Kee NS, Bloomfield C. 1961. The solution of some minor element oxides by decomposing plant materials. *Geochim Cosmochim Acta* 24:206–225.

King GM, Howes BL, Dacey JWH. 1985. Short-term end-products of sulfate reduction in a salt marsh: Formation of acid volatile sulfides, elemental sulfur, and pyrite. *Geochim Cosmochim Acta* 49:1561–1566.

Kleinmann RLP, Crerar DA. 1979. *Thiobacillus ferrooxidans* and the formation of acidity in simulated coal mine environments. *Geomicrobiol J* 1:373–388.

Klinkhammer GP, Rona P, Greaves M, Elderfeld, H. 1985. Hydrothermal manganese plumes in the Mid-Atlantic Ridge Rift Valley. *Nature* (London) 314:727–731.

Labrenz M, Druschel GK, Thomsen-Ebert T, Gilbert B, Welxh SA, Kemner KM, Logan GA, Summons RE, De Stasio G, Bond PL, Lai B, Kelly SD, Banfield JG. 2000. Formation of sphalerite (ZnS) deposits in natural biofilms of sulfate-reducing bacteria. *Science* 290:1744–1747.

Lackey JB. 1938. The flora and fauna of surface water polluted by acid mine drainage. *Publ Health Rep* 53:1499–1507.

Lambert IB, McAndrew J, Jones HE. 1971. Geochemical and bacteriological studies of the cupriferous environment at Pernatty Lagoon, South Australia. *Aust Inst Min Met Proc* 240:15–23.

Latimer WM, Hildebrand JH. 1942. *Reference Book of Inorganic Chemistry*. Rev ed. New York: Macmillan.

Lazaroff N. 1983. The exclusion of D_2O from the hydration sphere of $FeSO_4 \cdot 7H_2O$ oxidized by *Thiobacillus ferrooxidans*. *Science* 222:1331–1334.

Lazaroff N, Melanson L, Lewis E, Santoro N, Pueschel C. 1985. Scanning electron microscopy and infrared spectroscopy of iron sediments formed by *Thiobacillus ferrooxidans*. *Geomicrobiol J* 4:231–268.

Lazaroff N, Sigal W, Wasserman A. 1982. Iron oxidation and precipitation of ferric hydroxysulfates by resting *Thiobacillus ferrooxidans* cells. *Appl Environ Microbiol* 43:924–938.

Leathen WW, Braley SA, McIntyre LD. 1953. The role of bacteria in the formation of acid from certain sulfuritic constituents associated with bituminous coal. II. Ferrous iron oxidizing bacteria. *Appl Microbiol* 1:65–68.

Lehninger AL. 1975. *Biochemistry*. 2nd ed. New York: Worth.

Leleu MT, Gulgalski T, Goni J. 1975. Synthèse de wurtzite par voie bactérienne. *Mineral Deposita* (Berlin) 10:323–329.

Love LG. 1962. Biogenic primary sulfide of the Permian Kupferschiefer and marl slate. *Econ Geol* 57:350–366.

Lovley DR, Roden EE, Phillips EJP, Woodward JC. 1993. Enzymatic iron and uranium reduction by sulfate-reducing bacteria. *Mar Geol* 113:41–53.

Lundgren DG, Malouf EE. 1983. Microbial extraction and concentration of metals. *Adv Biotechnol Proc* 1:223–249.

Lundgren DG, Vestal JR, Tabita FR. 1972. The microbiology of mine drainage pollution. In: Mitchell R, ed. *Water Pollution Microbiology*. New York: Wiley-Interscience, pp. 69–88.

Luther GW III, 1991. Pyrite synthesis via polysulfide compounds. *Geochim Cosmochim Acta* 55:2839–2849.

Luther GW III, Giblin A, Howarth RW, Ryans RA. 1982. Pyrite and oxidized iron mineral phases formed from pyrite oxidation in salt marsh and estuarine sediments. *Geochim Cosmochim Acta* 46:2665–2669.

Lyalikova NN. 1960. Participation of *Thiobacillus ferrooxidans* in the oxidation of ores in pyrite beds of the Middle Ural. *Mikrobiologiya* 29:382–387.

Mackintosh ME. 1978. Nitrogen fixation by *Thiobacillus ferrooxidans*. *J Gen Microbiol* 105:215–218.

Magne R, Berthelin J, Dommergues Y. 1973. Solubilisation de l'uranium dans les roches par des bactéries n'appartenant pas au genre *Thiobacillus*. *CR Acad Sci* (Paris) 276:2625–2628.

Magne R, Berthelin JR, Dommergues Y. 1974. Solubilisation et insolubilisation de l'uranium des granites par des bactéries heterotroph. In: *Formation of Uranium Ore Deposits*. Vienna, Austria: International Atomic Energy Commission, pp. 73–88.

Malouf EE, Prater JD. 1961. Role of bacteria in the alteration of sulfide minerals. *J Metals* 13:353–356.

Marchig V, Grundlach H. 1982. Iron-rich metalliferous sediments on the East Pacific Rise: Prototype of undifferentiated metalliferous sediments on divergent plate boundaries. *Earth Planet Sci Lett* 58:361–382.

Marnette ECL, Van Breemen N, Hordijk KA, Cappenberg TE. 1993. Pyrite formation in two freshwater systems in the Netherlands. *Geochim Cosmochim Acta* 57:4165–4177.

McCready RGL. 1988. Progress in the bacterial leaching of metals in Canada. In: Norris PR, Kelly DP, eds. *Biohydrometallurgy*. Kew, Surrey, U.K.: Science and Technology Letters, pp. 177–195.

McCready RGL, Gould WD. 1990. Bioleaching of uranium. In: Ehrlich HL, Brierley CL, eds. *Microbial Mineral Recovery*. New York: McGraw-Hill, pp. 107–125.

McGoran CJM, Ducan DW, Walden CC. 1969. Growth of *Thiobacillus ferrooxidans* on various substrates. *Can J Microbiol* 15:135–138.

Miller LP. 1949. Stimulation of hydrogen sulfide production by sulfate-reducing bacteria. *Boyce Thompson Inst Contr* 15:467–474.

Miller LP. 1950. Formation of metal sulfides through the activities of sulfate reducing bacteria. *Boyce Thompson Inst Contr* 16:85–89.

Moses CO, Nordstrom DK, Herman JS, Mills AL. 1987. Aqueous pyrite oxidation by dissolved oxygen and by ferric iron. *Geochim Cosmochim Acta* 51:1561–1571.

Moshuyakova SA, Karavaiko GI, Shchetinina EV. 1971. Role of *Thiobacillus ferrooxidans* in leaching nickel, copper, cobalt, iron, aluminum, magnesium, and calcium from ores of copper-nickel deposits. *Mikrobiologiya* 40:1100–1107 (English translation, pp. 659–969).

Mossman JR, Aplin AC, Curtis CD, Coleman ML. 1991. Geochemistry of inorganic and organic sulfur in organic-rich sediments from the Peru Margin. *Geochim Cosmochim Acta* 55:3581–3595.

Mottl MJ, Holland HD, Corr RF. 1979. Chemical exchange during hydrothermal alteration of basalt by seawater. II. Experimental results for Fe, Mn, and sulfur species. *Geochim Cosmochim Acta* 43:869–884.

Mousavi SM, Yaghmaei S, Vossoughi M, Jafari A, Hoseini SA. 2005. Comparison of bioleaching ability of two native mesophilic and thermophilic bacteria on copper recovery from chalcopyrite concentrate in an airlift bioreactor. *Hydrometallurgy* 80:139–144.

Mulligan CN, Galvez-Cloutier R. 2000. Bioleaching of copper mining residues by *Aspergillus niger*. *Water Sci Technol* 41:255–262.

Murr LE, Brierley JA. 1978. The use of large-scale test facilities in studies of the role of microorganisms in commercial leaching operations. In: Murr LE, Torma AE, Brierley JA, eds. *Metallurgical Applications of Bacteria*. New York: Academic Press, pp. 491–520.

Murthy KSN, Natarajan KA. 1992. The role of surface attachment of *Thiobacillus ferrooxidans* on the oxidation of pyrite. *Miner Metall Process*, 9:20–24.

Mustin C, Berthelin J, Marion P, de Donato P. 1992. Corrosion and electrochemical oxidation of a pyrite by *Thiobacillus ferrooxidans*. *Appl Environ Microbiol* 58:1175–1182.

Mustin C, de Donato P, Berthelin J. 1993. Surface oxidized species, a key factor in the study of bioleaching processes. In: Torma AE, Wey JE, Lakshmanan VI, eds. *Biohydrometallurgical Technologies, Vol 1*, Bioleaching Process. Warrendale, PA: The Minerals, Metals and Materials Society, pp. 175–184.

Myers CR, Myers JM. 1997. Outer membrane cytochromes of *Shewanella putrefaciens* MR-1: Spectral analysis, and purification of the 83-kDa c-type cytochrome. *Biochim Biophys Acta* 1326:307–318.

Myers JM, Myers CR. 1998. Isolation and sequence of omcA, a gene encoding dodecaheme outer membrane cytochrome c of *Shewanella putrefaciens* MR-1, and detection of omcA homologs in other strains of *S. putrefaciens*. *Biochim Biophys Acta* 1373:237–251.

Myers JM, Myers CR. 2001. Role of outer membrane cytochromes OmcA and OmcB of *Shewanella putrefaciens* MR-1 in reduction of manganese dioxide. *Appl Environ Microbiol* 67:260–269.

Nedwell DB, Banat IM. 1981. Hydrogen as an electron donor for sulfate-reducing bacteria in slurries of salt marsh sediment. *Microb Ecol* 7:305–313.

Nielsen AM, Beck JV. 1972. Chalcocite oxidation and coupled carbon dioxide fixation by *Thiobacillus ferrooxidans*. *Science* 175:1124–1126.

Nordstrom DK, Southam G. 1997. Geomicrobiology of sulfide mineral oxidation. *Rev Mineral* 35:361–390.

Norman PG, Snyman CP. 1988. The biological and chemical leaching of auriferous pyrite/arsenopyrite flotation concentrate: A microscopic examination. *Geomicrobiol J* 6:1–10.

Norris PR. 1990. Acidophilic bacteria and their activity in mineral sulfide oxidation. In: Ehrlich HL, Brierley CL, eds. *Microbial Mineral Recovery*. New York: McGraw-Hill, pp. 3–27.

Norris PR. 1997. Thermophiles and bioleaching. In: Rawlings DE, ed. *Biomining: Theory, Microbes and Industrial Processes*. Berlin, Germany: Springer, pp. 3–27.

Okibe N, Gericke M, Hallberg KB, Johnson DB. 2003. Enumeration and characterization of acidophilic microorganisms isolated from a pilot plant stirred-tank bioleaching operation. *Appl Environ Microbiol* 69:1936–1943.

Olson GJ, Brierley JA, Brierley CL. 2003. Bioleaching review part B: Progress in bioleaching: applications of microbial processes by the minerals industries. *Appl Microbiol Biotechnol* 63:249–257.

Parès Y. 1964a. Intervention des bactéries dans la solubilisation du cuivre. *Ann Inst Pasteur* (Paris) 107:132–135.

Parès Y. 1964b. Action de *Serratia marcescens* dans le cycle biologique des métaux. *Ann Inst Pasteur* (Paris) 107:136–141.

Parès Y. 1964c. Action d'*Agrobacterium tumefaciens* dans la mise en solution de l'or. *Ann Inst Pasteur* (Paris) 107:141–143.

Pinka J. 1991. Bacterial oxidation of pyrite and pyrrhotite. *Erzmetall* 44:571–573.

Pistorio M, Curutchet G, Donati E, Tedesco P. 1994. Direct zinc sulfide bioleaching by *Thiobacillus ferrooxidans* and *Thiobacillus thiooxidans*. *Biotechnol Lett* 16:419–424.

Pivovarova TA, Markosyan GE, Karavaiko GI. 1981. *Mikrobiologiya* 50:482–486 (English translation, pp. 339–344).

Pogliani C, Curutchet G, Donati E, Tedesco PH. 1990. A need for direct contact with particle surfaces in the bacterial oxidation of covellite in the absence of a chemical lixiviant. *Biotechnol Lett* 12:515–518.

Powell TG, MacQueen RW. 1984. Precipitation of sulfide ores and organic matter: Sulfate reduction at Pine Point, Canada. *Science* 224:63–66.

Pronk JT, Meulenberg R, Hazeu W, Bos P, Kuenen JG. 1990. Oxidation of reduced inorganic sulfur compounds by acidophilic thiobacilli. *FEMS Microbiol Rev* 75:293–306.

Rawlings DE. 1997a. *Biomining: Theory, Microbes and Industrial Processes*. Berlin, Germany: Springer.

Rawlings DE. 1997b. Mesophilic, autotrophic bioleaching bacteria: Description, physiology and role. In: Rawlings DE, ed. *Biomining: Theory, Microbes and Industrial Processes*. Berlin, Germany: Springer, pp. 229–245.

Rawlings DE. 2002. Heavy metal mining using microbes. *Annu Rev Microbiol* 56:65–91.

Rawlings DE, Tributsch H, Hansford GS. 1999. Reasons why *Leptospirillum*-like species rather than *Thiobacillus ferrooxidans* are the dominant iron-oxidizing bacteria in many commercial processes for the biooxidation of pyrite and related ores. *Microbiology* (Reading) 145:5–13.

Razzell WE, Trussell PC. 1963. Microbiological leaching of metallic sulfides. *Appl Microbiol* 11:105–110.

Rickard DT. 1973. Limiting conditions for synsedimentary sulfide ore formation. *Econ Geol* 68:605–617.

Rickard PAD, Vanselow DG. 1978. Investigation into the kinetics and stoichiometry of bacterial oxidation of covellite (CuS) using a polarographic oxygen probe. *Can J Microbiol* 24:998–1003.

Rimstidt JD, Vaughn DJ. 2003. Pyrite oxidation: A state-of-the-art assessment of the reaction mechanism. *Geochim Cosmochim Acta* 67:873–880.

Rodriguez-Leiva M, Tributsch H. 1988. Morphology of bacterial leaching patterns by *Thiobacillus ferrooxidans* on synthetic pyrite. *Arch Microbiol* 149:401–405.

Rohwerder T, Gehrke T, Kinzler K, Sand W. 2003. Bioleaching review part A: Progress in bioleaching: fundamentals and mechanisms of bacterial metal sulfide oxidation. *Appl Microbiol Biotechnol* 63:239–248.

Roy AB, Trudinger PA. 1970. *The Biochemistry of Inorganic Compounds of Sulphur*. Cambridge, U.K.: Cambridge University Press.

Sand W, Gehrke T, Hallmann R, Rohde K, Sabotke B, Wentzien S. 1993. In-situ bioleaching of metal sulfides: The importance of *Leptospirillum ferrooxidans*. In Torma AE, Wey JE, Lashmanan VL. *Biohydrometallurgical Technologies, Vol 1*. Warrendale, PA: The Minerals, Metals and Materials Society, pp. 15–27.

Sand W, Gehrke T, Hallmann R, Schippers A. 1995. Sulfur chemistry, biofilm, and the (in)direct attack mechanism: A critical evaluation of bacterial leaching. *Appl Microbiol Biotechnol* 43:961–966.

Sand W, Gehrke T, Jozsa P-G, Schippers A. 1997. Novel mechanism for bioleaching of metal sulfides. In: *Biotechnology Comes of Age. International Biohydrometallurgy Symposium IBS97 BIOMINE 97. Conference Proceedings*. Glenside, South Australia: Australasian Mineral Foundation, pp. QP2.1–QP2.10.

Sand W, Rohde K, Sabotke B, Zenneck C. 1992. Evaluation of *Leptospirillum ferrooxidans* for leaching. *Appl Environ Microbiol* 58:85–92.

Schippers A, Sand W. 1999. Bacterial leaching of metal sulfides proceeds by two indirect mechanisms via thiosulfate or via polysulfides and sulfur. *Appl Environ Microbiol* 65:319–321.

Schoonen MAS, Barnes HL. 1991a. Reactions forming pyrite and marcasite from solution. 1. Nucleation of FeS$_2$ below 100°C. *Geochim Cosmochim Acta* 55:1495–1504.

Schoonen MAS, Barnes HL. 1991b. Reactions forming pyrite and marcasite from solution. II. Via FeS precursors below 100°C. *Geochim Cosmochim Acta* 55:1505–1514.

Schrenk MO, Edwards KJ, Goodman RM, Hamers RJ, Banfield JF. 1998. Distribution of *Thiobacillus ferrooxidans* and *Leptospirillum ferrooxidans*: Implications for generation of acid mine drainage. *Science* 279:1519–1522.

Segerer A, Neuner A, Kristjansson JK, Stetter KO. 1986. *Acidianus infernus* gen nov. sp. nov., and *Acidianus brierleyi* comb. nov.: Facultative aerobic, extremely acidophilic thermophilic sulfur-metabolizing archaebacteria. *Arch Microbiol* 36:559–564.

Serkies J, Oberc J, Idzikowski A. 1967. The geochemical bearings of the genesis of Zechstein copper deposits in southwest Poland as exemplified by the studies on the Zechstein of the Leszczyna syncline. *Chem Geol* 2:217–2232.

Seyfried WE, Jr., Mottl MJ. 1982. Hydrothermal alteration of basalt by seawater under seawater-dominated conditions. *Geochim Cosmochim Acta* 46:985–1002.

Shanks WC III, Bischoff JL, Rosenauer RJ. 1981. Seawater sulfate reduction and sulfur isotope fractionation in basaltic systems: Interaction of seawater with fayalite and magnetite at 200–350°C. *Geochim Cosmochim Acta* 45:1977–1981.

Shrihari RK, Ghandi KS, Natarajan KA. 1991. Role of cell attachment in leaching chalcopyrite mineral by *Thiobacillus ferrooxidans*. *Appl Microbiol Biotechnol* 36:278–282.

Silver M, Torma AE. 1974. Oxidation of metal sulfides by *Thiobacillus ferrooxidans* grown on different substrates. *Can J Microbiol* 20:141–147.

Silverman MP, Ehrlich HL. 1964. Microbial formation and degradation of minerals. *Adv Appl Microbiol* 6:153–206.

Silverman MP, Rogoff MH, Wender I. 1961. Bacterial oxidation of pyritic materials in coal. *Appl Microbiol* 9:491–496.

Singer PC, Stumm W. 1970. Acid mine drainage: The rate-determining step. *Science* 167:1121–1123.

Stanton RL. 1972. Sulfides in sediments. In: Fairbridge RW, ed. *The Encyclopedia of Geochemistry and Environmental Sciences. Encyclopedia Earth Science Series, Vol IVA*. New York: Van Nostrand Reinhold, pp. 1134–1141.

Stevens CJ, Dugan PR, Tuovinen OH. 1986. Acetylene reduction (nitrogen fixation) by *Thiobacillus ferrooxidans*. *Biotechnol Appl Biochem* 8:351–359.

Strahler AN. 1977. *Principles of Physical Geology*. New York: Harper & Row.

Styrt MM, Brackman AJ, Holland HD, Clark BC, Pisutha-Arnold V, Eldridge CS, Ohmoto H. 1981. The mineralogy and isotopic composition of sulfur in hydrothermal sulfide/sulfate deposits on the East Pacific Rise, 21°N latitude. *Earth Planet Sci Lett* 53:382–390.

Sugio T, Tanijiri S, Fukuda K, Yamargo K, Inagaki K, Tano T. 1987. Utilization of amino acids as sole source of nitrogen by obligate chemoautotroph *Thiobacillus ferrooxidans*. *Agric Biol Chem* 51:2229–2236.

Sullivan JD. 1930. Chemistry of leaching covellite. Technical Paper 487. Washington, DC: U.S. Department of Commerce, Bureau of Mines.

Sutton JA, Corrick JD. 1963. Microbial leaching of copper minerals. *Mining Eng* 15:37–40.

Sutton JA, Corrick JD. 1964. Bacteria in mining and metallurgy: Leaching selected ores and minerals; experiments with *Thiobacillus ferrooxidans*. *Rept Invest RI 5839*. Washington, DC: Bureau of Mines, U.S. Department of the Interior.

Temple KL. 1964. Syngenesis of sulfide ores. An explanation of biochemical aspects. *Econ Geol* 59:1473–1491.

Temple KL, LeRoux N. 1964. Syngenesis of sulfide ores: Desorption of adsorbed metal ions and the precipitation of sulfides. *Econ Geol* 59: 647–655.

Thode-Andersen S, Jørgensen BB. 1989. Sulfate reduction and the formation of ^{35}S-labeled FeS, FeS$_2$, and S^0 in coastal marine sediments. *Limnol Oceanogr* 34:793–806.

Tittley SR. 1981. Porphyry copper. *Am Sci* 69:632–638.

Torma AE. 1971. Microbial oxidation of synthetic cobalt, nickel and zinc sulfides by *Thiobacillus ferrooxidans*. *Rev Can Biol* 30:209–216.

Torma AE. 1978. Oxidation of gallium sulfides by *Thiobacillus ferrooxidans*. *Can J Microbiol* 24:888–891.

Torma AE, Gabra GG. 1977. Oxidation of stibnite by *Thiobacillus ferrooxidans*. *Antonie van Leeuwenhoek* 43:1–6.

Tributsch H. 1976. The oxidative disintegration of sulfide crystals by *Thiobacillus ferrooxidans*. *Naturwissenschaften* 63:88.

Tributsch H. 1999. Direct versus indirect bioleaching. In: Amils R, Ballester A, eds. *Biohydrometallurgy and the Environment Toward the Mining of the 21st Century, Part A*. Amsterdam: Elsevier, pp. 51–60.

Tuovinen OH, Niemelä SI, Gyllenberg HG. 1971. Tolerance of *Thiobacillus ferrooxidans* to some metals. *Antonie van Leeuwenhoek* 37:489–496.

Wadden D, Gallant A. 1985. The in-place leaching of uranium at Denison Mines. *Can J Metall Quart* 24:127–134.

Walsh F, Mitchell R. 1972a. An acid-tolerant iron-oxidizing *Metallogenium*. *J Gen Microbiol* 72:369–376.

Walsh F, Mitchell R. 1972b. The pH-dependent succession of iron bacteria. *Environ Sci Technol* 6:809–812.

Wang H, Bigham JM, Tuovinen OH. 2007. Oxidation of marcasite and pyrite by iron-oxidizing bacteria and archaea. *Hydrometallurgy* 88:127–131.

Weast RC, Astle MJ. 1982. *CRC Handbook of Chemistry and Physics*. 63rd ed. Boca Raton, FL: CRC Press.

Wenberg GM, Erbisch FH, Volin M. 1971. Leaching of copper by fungi. *Trans Soc Min Eng AIME* 250:207–212.

Westrich JT, Berner RA. 1984. The role of sedimentary organic matter in bacterial sulfur reduction: The G model tested. *Limnol Oceanogr* 29:236–249.

Yarzábal A, Brasseur G, Bonnefoy V. 2002b. Cytochromes *c* of *Acidithiobacillus ferrooxidans*. *FEMS Microbiol Lett* 209:189–195.

Yarzábal A, Brasseur G, Ratouchniak J, Lund K, Lemesle-Meunier D, DeMoss JA, Bonnefoy V. 2002a. The high-molecular-weight cytochrome *c* Cyc2 of *Acidithiobacillus ferrooxidans* is an outer membrane protein. *J Bacteriol* 184:313–317.

Zajic JE. 1969. *Microbial Biogeochemistry*. New York: Academic Press.

Zimmerley SR, Wilson DG, Prater JD. 1958. Cyclic leaching process employing iron oxidizing bacteria. U.S. Patent 2,829,964.

21 Geomicrobiology of Selenium and Tellurium

21.1 OCCURRENCE IN EARTH'S CRUST

The elements selenium and tellurium, like sulfur, belong to group VI of the periodic table. All three have some properties in common, but selenium and tellurium, especially the latter, have some metallic attributes, unlike sulfur. Selenium and tellurium are much less abundant than sulfur in the Earth's crust. Selenium amounts to only 0.05–0.14 ppm (Rapp, 1972, p. 1080) and tellurium to 10^{-5}–10^{-2} ppm (Lansche, 1965). Both are associated with metal sulfides in nature and occur in distinct minerals, for example, ferroselite, ($FeSe_2$), challomenite ($CuSeO_3 \cdot 2H_2O$), hessite (Ag_2Te), and tetradymite (Bi_2Te_2S). Selenium occurs in small amounts in various soils in concentrations in the range of 0.01–100 ppm. High concentrations are associated with arid, alkaline soils that contain some free $CaCO_3$ (Rosenfeld and Beath, 1964).

21.2 BIOLOGICAL IMPORTANCE

Some plants, such as *Astragalus* spp. and *Stanleya*, can accumulate large amounts of selenium in the form of organic selenium compounds. However, not all forms of selenium in soil are available for assimilation by these plants.

Selenium is required nutritionally as a trace element by at least some microorganisms, plants, and animals, including human beings (Stadtman, 1974; Miller and Neathery, 1977; Combs and Scott, 1977; Patrick, 1978; Mertz, 1981). It has been found to be an essential component of the enzyme glutathione peroxidase in mammalian red blood corpuscles (Rotruck et al., 1973). The enzyme catalyzes the reaction

$$2GSH + H_2O_2 \rightarrow GSSG + 2H_2O \tag{21.1}$$

Selenium has also been found to be essential together with molybdenum in the structure of formate dehydrogenase in the bacteria *Escherichia coli*, *Clostridium thermoaceticum*, *C. sticklandii*, and *Methanococcus vannielii* among others (Pinsent, 1954; Lester and DeMoss, 1971; Shum and Murphy, 1972; Andreesen and Ljungdahl, 1973; Enoch and Lester, 1972; Stadtman, 1974) and with tungsten in the formate dehydrogenase of *C. thermoaceticum* when grown in the presence of tungsten instead of molybdenum (Yamamoto et al., 1983). The enzyme catalyzes the reaction

$$HCOOH + NAD^+ \rightarrow CO_2 + NADH + H^+ \tag{21.2}$$

Selenium has also been found essential to protein A of glycine reductase in *Clostridia* (Stadtman, 1974), an enzyme that catalyzes the reaction

$$\underset{NH_2}{CH_2COOH} + R(SH)_2 + P_i + ADP \quad CH_3COOH + NH_3 + R\diagdown\!\!\overset{S}{\underset{S}{\big|}} + ATP \tag{21.3}$$

For this reason, *C. purinolyticum* exhibits an absolute requirement for selenium in its growth medium for fermentation of glycine (Duerre and Andreesen, 1982).

No biological requirement for tellurium has been observed up to now.

21.3 TOXICITY OF SELENIUM AND TELLURIUM

Both selenium and tellurium are toxic when present in excess, but the minimum toxic doses vary depending on the organism. As mentioned in Section 21.2, some plants accumulate selenium to the extent of 1.5–2 g kg^{-1} dry weight of tissue (Stadtman, 1974). They usually grow in arid environments with unusually high concentrations of selenium in the soil. In the Kesterson National Wildlife Refuge in California, where extensive selenium intoxication of wild animals has been observed, selenate concentrations of 1.8–18 µM (0.14–1.4 ppm) have been reported, contrasted with normal concentrations of ~1.3–21.1 nM in the San Joaquin River, ~0.4–1.3 nM in the Sacramento River, and <0.2 nM in San Francisco Bay, all located in California (Zehr and Oremland, 1987). These normal concentrations are below minimum inhibitory concentrations (MICs) of selenate [Se(VI)] for three selenium-sensitive strains of bacteria from the same general area in California. Their MICs were found to range from 0.78 to 1.56 mM for selenate and from 1.56 to 25 mM for selenite [Se(IV)]. Selenium-resistant bacteria from the Kesterson National Wildlife Refuge exhibited MICs of 50 to >200 mM selenate and selenite (Burton et al., 1987). By contrast, both selenium-resistant and selenium-sensitive organisms from these same sites in California exhibited MICs for tellurate in the range of 0.03–1 mM and for tellurite in the range of 0.03–4 mM (Burton et al., 1987).

Selenium and tellurium resistance appears to be regulated by different genes. In *Escherichia coli*, tellurium resistance appears to be mediated by the arsenical ATPase efflux pump. The genetic determinants for this pump reside on resistance plasmid R773 (Turner et al., 1992). Higher forms of life appear to be relatively more sensitive to Se than bacteria, although they require Se as a nutritional trace element. Biochemically, Se toxicity appears to be the result of superoxide or H_2O_2 production in excess of antioxidant production by a cell. A similar mechanism may be the basis for Te toxicity (see references 31, 35, and 41 cited by Guzzo and Dubow, 2000).

21.4 BIOOXIDATION OF REDUCED FORMS OF SELENIUM

Some inorganic forms of selenium have been reported to be oxidizable by microorganisms. *Micrococcus selenicus* isolated from mud (Breed et al., 1948), a rod-shaped bacterium isolated from soil and thought to be autotrophic (Lipman and Waksman, 1923), and a purple bacterium (Sapozhnikov, 1937) were observed to oxidize Se^0 to SeO_4^{2-}. A strain of *Bacillus megaterium* from topsoil in a river alluvium was found to oxidize Se^0 to SeO_3^{2-} and traces of SeO_4^{2-}. Red selenium was more readily attacked than gray selenium (Sarathchandra and Watkinson, 1981). Dowdle and Oremland (1998) observed elemental selenium oxidation in soil slurries that was inhibited by autoclaving the slurry or by addition of formalin, azide, 2,4-dinitrophenol, or the antibiotics chloramphenicol + tetracycline or cycloheximide + nystatin. Se^0 oxidation in the slurries was enhanced by addition of sulfide, acetate, or glucose, suggesting that sulfur-oxidizing autotrophs and heterotrophs were involved in the oxidation.

Acidithiobacillus ferrooxidans has been shown to oxidize copper selenide (CuSe) to cupric copper (Cu^{2+}) and elemental selenium (Se^0) (Torma and Habashi, 1972). The reaction may be written as

$$CuSe + 2H^+ + 0.5O_2 \rightarrow Cu^{2+} + Se^0 + H_2O \tag{21.4}$$

21.5 BIOREDUCTION OF OXIDIZED SELENIUM COMPOUNDS

Various inorganic selenium compounds have been found to be reduced anaerobically by some microorganisms. Crude cell extract of *Micrococcus lactilyticus* (also known as *Veillonella lactilyticus*) has been shown to reduce selenite but not selenate to Se^0, and Se^0 to HSe^-. The reductant was hydrogen (H_2) (Woolfolk and Whiteley, 1962). Cell extracts from strains of *Desulfovibrio desulfuricans* and *Clostridium pasteurianum* were also found to reduce selenite with hydrogen. The enzyme hydrogenase mediated electron transfer from hydrogen in these reactions (Woolfolk and Whiteley,

1962). A variety of other bacteria, actinomycetes, and fungi have been shown to reduce selenate and selenite to Se^0 (Bautista and Alexander, 1972; Lortie et al., 1992; Stolz and Oremland, 1999; Tomei et al., 1992; Zalokar, 1953). The bacteria include *Pseudomonas stutzeri*, *Wolinella succinogenes*, and *Micrococcus* sp. *Acidithiobacillus ferrooxidans* is able to reduce the red form of Se^0 to H$_2$Se anaerobically, albeit in small amounts (Bacon and Ingledew, 1989).

A relatively recently discovered bacterium, *Thauera selenatis*, can grow anaerobically with selenate or nitrate as terminal electron acceptor (Macy et al., 1993; Rech and Macy, 1992). In the absence of nitrate, it reduces selenate to selenite (DeMoll-Decker and Macy, 1993). The reductases for selenate and nitrate in this organism are distinct enzymes with different pH optima. Thus in contrast to the response of a selenate-reducing enrichment culture (Steinberg et al., 1992), nitrate does not inhibit selenate reduction by *T. selenatis*. Indeed, when present together, both selenate and nitrate are reduced simultaneously, with selenate reduced to elemental selenium (DeMoll-Decker and Macy, 1993). The selenate reductase in this organism, which catalyzes the reduction of selenate to selenite, is found in the periplasm, whereas its nitrate reductase, which catalyzes the reduction of nitrate to nitrite, is found in its cytoplasmic membrane (Rech and Macy, 1992), Selenate reductase is a metalloprotein containing Mo, Fe, acid-labile sulfur, and a cytochrome *b* subunit (Schroeder et al., 1997). Nitrite reductase is found in the periplasm of *T. selenatis* and plays a role in selenite reduction, besides catalyzing nitrite reduction (DeMoll-Decker and Macy, 1993). This explains why *T. selenatis* produces elemental selenium in the presence of nitrate, but selenite in its absence. Selenite does not support growth of *T. selenatis* (DeMoll-Decker and Macy, 1993).

A selenite reductase enzyme has been obtained from the fungus *Candida albicans* (Falcone and Nickerson, 1963; Nickerson and Falcone, 1963). It reduces selenite to Se^0. A characterization of the enzyme has shown that it requires a quinone, a thiol compound (e.g., glutathione), a pyridine nucleotide (NADP), and an electron donor (e.g., glucose 6-phosphate) for activity. Electron transfer between NADP and quinone is probably mediated by flavin mononucleotide in this system. It is possible that this enzyme is part of an assimilatory SeO$_4^{2-}$ and SeO$_3^{2-}$ reductase system. How this enzyme compares with that in *T. selenatis* remains to be established.

Sulfurospirillum barnesii (formerly *Geospirillum barnesii*, also called strain SES-3) (Oremland et al., 1994; Stolz et al., 1999) is another bacterium that can reduce selenate to elemental selenium. Cells of this organism grew with lactate as carbon and energy source and selenate as terminal electron acceptor, which was reduced to selenite. As with *Thauera selenatis*, resting cells of *Ssp. barnesii* but not growing cells were able to reduce selenite to Se^0 (Oremland et al., 1994). One important difference between *Ssp. barnesii* and *T. selenatis* is that *Ssp. barnesii* is able to use a much wider range of reducible anions as terminal electron acceptors than *T. selenatis* (Stolz and Oremland, 1999). *Ssp. barnesii* can reduce selenate and nitrate simultaneously whether pregrown on selenate or nitrate, consistent with the observation that selenate reductase is constitutive in this organism (Oremland et al., 1999).

Two newly discovered selenate reducers, both gram-positive bacteria, are *Bacillus arsenicoselenatis* and *B. selenitireducens* (Switzer Blum et al., 1998). The first forms spores but the second does not. Both were isolated from anoxic muds from Mono Lake, California, which is alkaline, hypersaline, and arsenic-rich. *B. arsenicoselenatis* reduces selenate to selenite whereas *B. selenitireducens* reduces selenite to elemental selenium as forms of anaerobic respiration. In coculture, the two strains together can reduce selenate to elemental selenium. Both strains can reduce arsenate as well as selenate (Switzer Blum et al., 1998). *Sulfurospirillum barnesii* and *B. arsenicoselenatis* can reduce selenate and nitrate simultaneously, but unlike the selenate reductase in *Ssp. barnesii*, that in *B. arsenicoselenatis* is not constitutive because it does not appear in nitrate-grown cells (Oremland et al., 1999). Therefore, in order for *B. arsenicoselenatis* to reduce selenate and nitrate simultaneously, it has to be grown in the presence of a mixture of the two electron acceptors.

A moderately halophilic selenate reducer was isolated from Dead Sea (Israel) sediment. It reduced selenate to selenite and elemental selenium. It is a gram-negative organism and has been named *Selenihalanaerobacter shriftii* (Switzer Blum et al., 2001). When it respires on glycerol or glucose, it forms acetate and CO_2. Nitrate and trimethylamine N-oxide could serve as alternative electron acceptors, but reduced forms of sulfur, nitrite, arsenate, fumarate, or dimethylsulfoxide could not.

All previously mentioned selenate- and selenite-reducing bacteria belong to the domain Bacteria. Recently, a hyperthermophilic member of the domain Archaea capable of respiring organotrophically on selenate was isolated from a hot spring near Naples, Italy (Huber et al., 2000) (see also Chapter 14). Its name is *Pyrobaculum arsenaticum*. It reduces selenate to elemental selenium. Previously isolated *P. aerophilum* (Völkl et al., 1993) was found capable of respiring organotrophically on selenate and selenite and autotrophically on selenate with H_2 as electron donor (Huber et al., 2000). Elemental selenium was the reduction product.

In most studies of bacterial reduction of selenate and selenite, elemental selenium (red form), when formed, is usually found to be a major, if not the only, product. This is noteworthy because sulfate and sulfite cannot be directly reduced to S^0 but are reduced to H_2S without intermediate formation of S^0. Yet selenium and sulfur are members of the same chemical family. The implication is that enzymatic mechanisms of reduction for oxidized forms of these two elements are different. To date, none of the true selenate respirers have been found capable of sulfate respiration, which could be related to the significantly higher energy yield in selenate respiration ($\Delta G'$, -15.53 kcal mol^{-1} e^{-1}) than in sulfate respiration ($\Delta G'$, 0.10 kcal mol^{-1} e^{-1}) (Newman et al., 1998). It must be noted, however, that *Desulfovibrio desulfuricans* subsp. *aestuarii* has been found to reduce nanomolar but not millimolar quantities of selenate to selenite (Zehr and Oremland, 1987). Sulfate inhibited reduction of selenate, suggesting but not proving that the mechanism of sulfate and selenate reduction in this case may be a common one. As Zehr and Oremland (1987) pointed out, when sulfate is being reduced to H_2S in the absence of selenate, some of the H_2S formed may subsequently reduce biogenically formed selenite chemically to Se^0. They found that in nature, the sulfate reducer can reduce selenate only if the ambient sulfate concentration is <4 mM. Hockin and Gadd (2003) found that in mixed biofilms, *Desulfomicrobium norvegicum* could reduce selenite that diffused into the biofilm with H_2S it produced anaerobically by reduction of sulfate, resulting in the formation of S^0 and Se^0. This reaction was abiotic and can be formulated as follows:

$$3HS^- + SeO_4^{2-} + 5H^+ \rightarrow 3S^0 + Se^0 + 4H_2O \tag{21.5}$$

The sulfur and selenium precipitated within the biofilm as nanometer-sized selenium–sulfur granules. By contrast, *Sulfurospirillum barnesii*, *Bacillus selenitireducens*, and *Selenihalanaerobacter shriftii* can form nanospheres consisting exclusively of Se^0 when reducing selenite enzymatically (Oremland et al., 2004).

Whereas selenate and selenite reduction by the previously described organisms resulted in extracellular deposition of Se^0, intracellular deposition of Se^0 has been observed with some other organisms. *Chromatium vinosum* can deposit Se^0 intracellularly as a result of an interaction of H_2Se, which is produced by *Desulfovibrio desulfuricans* in selenate reduction in coculture with *Chr. vinosum*. The Se^0 is stored in the form of globules in the *Chr. vinosum* cells (Nelson et al., 1996). *Rhodobacter spheroides* deposited red Se^0 in or on its cells when it reduced selenate and selenite (Van Fleet-Stadler et al., 2000). *Ralstonia metallidurans* CH34 can reduce selenite to red Se^0, which it stores in its cytoplasm and occasionally in its periplasm (Roux et al., 2001).

21.5.1 OTHER PRODUCTS OF SELENATE AND SELENITE REDUCTION

In *Escherichia coli*, a significant portion of selenite reduced during glucose metabolism is deposited as Se^0 on its cell membrane but not in its cytoplasm (Gerrard et al., 1974), and another portion is incorporated as selenide in organic compounds such as selenomethionine (Ahluwalia et al., 1968). Some soil microbes reduce selenate or selenite to dimethylselenide [$(CH_3)_2Se$] at elevated selenium concentrations (Kovalskii et al., 1968; Fleming and Alexander, 1972; Alexander, 1977; Doran and Alexander, 1977). Other volatile selenium compounds may also be formed, their relative quantities depending on reaction conditions (Reamer and Zoller, 1980). The compounds include dimethyl diselenide [$(CH_3)_2Se_2$] and dimethyl selenone [$(CH_3)_2SeO_2$].

Some fungi have been found to be effective in forming methylated selenium compounds (Barkes and Fleming, 1974). *Alternaria alternata* isolated from seleniferous water from a sample series collected from evaporation ponds at the Kesterson Reservoir, Lost Hills, and Peck Ranch in California formed dimethylselenide more rapidly from selenate and selenite than from selenium sulfide (SeS_2) or various organic Se compounds. Methionine, a known biochemical methyl donor, and methylcobalamin, a known methyl carrier in biochemical transmethylation, stimulated dimethylselenide formation by the fungus (Thompson-Eagle et al., 1989). Crude cell extracts and a supernatant fraction from the fungus *Pichia guillermondi* after centrifugation at 144,000g reduced selenite but not selenate (Bautista and Alexander, 1972). In a mechanism proposed by Reamer and Zoller (1980), all methylated forms of selenium arise by methylation of selenite and subsequent reductions and, where needed, by additional methylation of the methylated products. Dimethylselenone is viewed as a precursor of dimethylselenide, whereas methylselenide [$(CH_3)SeH$] and $(CH_3)SeOH$ are viewed as precursors of dimethyldiselenide.

The archaeon *Methanococcus voltae* has been found to be able to use dimethylselenide [$(CH_3)_2Se$] as a source of Se required for its growth. Demethylation in this instance involved a corrinoid protein and two methyltransferases (Niess and Klein, 2004).

21.5.2   SELENIUM REDUCTION IN THE ENVIRONMENT

Bacterial reduction of selenate and selenite has been detected *in situ*, especially in environments with significant soluble selenium. In the Kesterson National Wildlife Refuge in California, Maiers et al. (1988) reported that 4% of water samples, 92% of sediment samples, and 100% of the soil samples they collected exhibited microbial selenium reduction. Of 100 mg selenate per liter, up to 75% was reduced to red Se^0, the rest to selenite. In the interstitial water of core samples from a wastewater evaporation pond in Fresno, California, selenate removal was stimulated by H_2 and the addition of acetate, and inhibited by O_2, NO_3^-, MnO_2, CrO_4^{2-}, and WO_4^{2-}, but not by the addition of SO_4^{2-}, MoO_4^{2-}, or FeOOH (Oremland et al., 1989). At other sites in California and also in Nevada, Steinberg and Oremland (1990) found measurable selenate-reducing activity in surficial sediment samples from bodies of freshwater to waters with salinities of 250 g L^{-1} but not 320 g L^{-1}. Nitrate, nitrite, molybdate, and tungstate added separately to samples from the agricultural drains were inhibitory to different extents. Sulfate partially inhibited the reduction of selenate in a sample from a freshwater site but not in one from a site with water having a salinity of 60 g L^{-1}. These differences are likely reflections of differences in the type of selenate reducers in the different samples and therefore of differences in mechanisms of selenate reduction. Additional studies in the agricultural drainage region of western Nevada revealed a selenate turnover rate of 0.04–1.8 h^{-1} at ambient Se oxyanion concentrations (13–455 nM). Rates of removal of selenium oxyanions ranged from 14 to 155 µmol m^{-2} per day (Oremland et al., 1991). The formation of elemental Se has a potential for selenium immobilization in soil and sediment under anaerobic conditions. Owing to the possibility of Se^0 reoxidation under aerobic conditions, remobilization may occur. However, such reoxidation has been found to be a slow process compared to microbial selenate reduction (Dowdle and Oremland, 1998).

A recent study of selenate-respiring bacteria detected in culture enrichments with sediments from water bodies in Chennai, India, and New Jersey, United States, revealed the presence of a diverse group of bacteria classifiable with *Gammaproteobacteria*, *Deltaproteobacteria*, *Deferribacteres*, and *Chrysiogenetes* (Narasingarao and Häggblom, 2007).

Methylation of selenium in aquatic environments has also been observed (e.g., Chau et al., 1976; Frankenberger and Karlson, 1992, 1995). This activity has a potential for Se removal from polluted soils and waters. Ranjard et al. (2002) demonstrated transmethylation of different forms of selenium with bacterial thiopurine methyltransferase in a strain of *Escherichia coli*, DH10B, acting on selenate, selenite, (methyl)selenocysteine, and selenomethionine.

Ecologically, anaerobic reduction of selenate and selenite to selenium (Se^0) represents a respiratory, energy-conserving process in some microorganisms and serves to detoxify the immediate environment for all organisms as long as anaerobic conditions are maintained. Selenium volatilization serves as a permanent detoxification process in water, soils, and sediments, and can occur aerobically, although in at least one instance it was more effective anaerobically (Frankenberger and Karlson, 1995).

21.6   SELENIUM CYCLE

The existence of a selenium cycle in nature was suggested by Shrift (1964). However, some of the details of this cycle are still obscure. The ultimate source of selenium must be igneous rocks, but whether microbes play a role in mobilizing the selenium from selenium-containing minerals is unknown. Similarly, little is known about the role that microbes play in mobilizing selenium in soil and sediment. Such an activity, when it occurs, is of great importance in understanding and controlling selenium pollution, as has occurred, for instance, in the Kesterson National Wildlife Refuge in California. The source of selenium in that case appears to be the drainage of irrigation water applied to farmland in the San Joaquin Valley. The irrigation water leached the selenium from the soil. This drainage has been collecting in the wildlife refuge. Different processes in selenium cycling in wetlands include redox reactions involving selenium, methylation and volatilization of selenium, organic and inorganic complexation of selenium, precipitation and dissolution of Se-containing minerals, and sorption and desorption of ionic species of selenium (Masscheleyn and Patrick, 1993). The known biochemical steps of a selenium cycle are shown in Figure 21.1.

21.7   BIOOXIDATION OF REDUCED FORMS OF TELLURIUM

Microbial oxidation of reduced forms of tellurium has so far not been reported. This may mean that this process does not occur in nature, but it is more likely that so far it has not been sought by investigators. Its geomicrobial importance is likely to be limited because the natural occurrence of tellurium is much rarer than that of selenium (see Section 21.1).

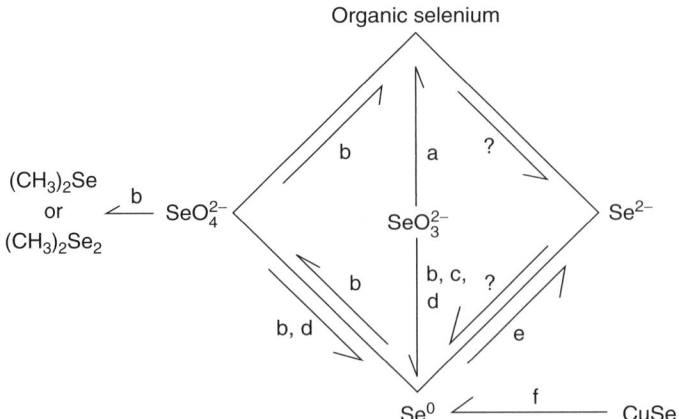

FIGURE 21.1   The selenium cycle. (a) *Escherichia coli*, (b) bacteria, (c) actinomycetes, (d) fungi, (e) *Micrococcus lactilyticus*, (f) *Acidithiobacillus ferrooxidans*. (See also Doran JW, Alexander M, *Appl Environ Microbiol*, 33, 31–37, 1977.)

21.8 BIOREDUCTION OF OXIDIZED FORMS OF TELLURIUM

Microbial reduction of tellurates and tellurites to elemental tellurium (Te^0) and dimethyltelluride [$(CH_3)_2Te$] has been reported (Woolfolk and Whiteley, 1962; Silverman and Ehrlich, 1964; Nagai, 1965; Bautista and Alexander, 1972; Trutkoet al., 2000; Klonowska et al., 2005; Csotonyi et al., 2006; Baesman et al., 2007). Trutko et al. (2000) presented evidence that in some gram-negative bacteria the respiratory chain was involved in tellurite reduction. The tellurite was reduced to tellurium crystallites, which appeared in the periplasmic space or on the outer or inner surface of the plasma membrane. The makeup of the respiratory chain differed to some extent among the different bacterial cultures tested. Klonowska et al. (2005) found that although *Shewanella oneidensis* was able to reduce selenite and tellurite anaerobically, the electron transport pathway to the two electron acceptors diverged upstream from tetracytochrome *c*, CymA.

Baesman et al. (2007) found that *Bacillus selenitireducens* and *Sulfurospirillum barnesii* produced Te^0 in the form of nanocrystals when respiring on tellurate or tellurite. With *B. selenitireducens*, Te^0 was deposited in the form of nanorods on the surface of the cells, which subsequently formed clusters, called *shards*, and rosettes. Nanorods also appeared in the bulk phase. Some crystals in the form of nanorods that aggregated into shard-like nanocrystals also formed inside the cells. *Ssp. barnesii* deposited Te^0 as irregularly shaped nanospheres (~20 nm diameter) frequently attached to the cell surface, which coalesced into larger clusters (500–1000 nm diameter). Nanospheres were also observed inside the cells. A question arises whether the difference in morphologies of the Te^0 nanocrystals is somehow related to a difference in the gram-staining properties of these two organisms, *B. selenitireducens* being gram-positive whereas *Ssp. barnesii* being gram-negative. This difference in gram reactivity may reflect a difference in organization of their respective electron transport systems, as is probably the case, for instance, in Mn(II)-oxidation and Mn(IV)-reduction by gram-positive and gram-negative bacteria (see Sections 17.5 and 17.6).

The fungus *Penicillium* sp. has been found to produce $(CH_3)_2Te$ from several inorganic tellurium compounds, provided only that reducible selenium compounds were also present (Fleming and Alexander, 1972). The amount of dialkyltelluride formed was related to the relative concentrations of Se and Te in the medium. Microbial reduction of oxidized forms of tellurium may represent detoxification reactions rather than a form of respiration, but this needs further investigation.

21.9 SUMMARY

Selenium, although a very toxic element, is nutritionally required by some bacteria, plants, and animals. Microorganisms have been described that can oxidize reduced selenium compounds. At least one, *Acidithiobacillus ferrooxidans*, can use selenide in the form of CuSe as a sole source of energy, oxidizing the compound to elemental selenium (Se^0) and Cu^{2+}. Oxidized forms of inorganic selenium compounds can be reduced by microorganisms, including members of the domains Bacteria and Archaea, and Fungi. Selenate and selenite may be reduced to one or more of the following: Se^0, H_2Se, dimethylselenide [$(CH_3)_2Te$], dimethyl diselenide [$(CH_3)_2Se_2$], and dimethyl selenone [$(CH_3)_2SeO_2$]. The reductions are enzymatic and in some bacteria represent a form of respiration. The microbial interactions with various forms of selenium contribute to a selenium cycle in nature. Microbial selenate and selenite reduction to elemental selenium in soil and sediment is a form of selenium immobilization that is potentially reversible. Microbial selenate and selenite reduction to volatile forms of selenium in soil, sediment, and water columns of bodies of water is a form of selenium removal that is permanent.

Tellurium occurs in such low concentrations in nature that it does not seem geomicrobially important. Nevertheless, microbial reduction of tellurate and tellurite to elemental tellurium (Te^0) and dimethyltelluride [$(CH_3)_2Te$] has been observed. Microbial oxidation of tellurides has so far not been reported.

REFERENCES

Ahluwalia GS, Saxena YR, Williams HH. 1968. Quantitative studies on selenite metabolism in *Escherichia coli*. *Arch Biochem Biophys* 124:79–84.

Alexander M. 1977. *Introduction to Soil Microbiology*. 2nd ed. New York: Wiley.

Andreesen JR, Ljungdahl LG. 1973. Formate dehydrogenase of *Clostridium thermoaceticum*: Incorporation of selenium-75, and the effects of selenite, molybdate, and tungstate on the enzyme. *J Bacteriol* 116:869–873.

Bacon M, Ingledew WJ. 1989. The reductive reactions of *Thiobacillus ferrooxidans* on sulfur and selenium. *FEMS Microbiol Lett* 58:189–194.

Baesman SM, Bullen TD, Dewald J, Zhang D, Curran S, Islam FS, Beveridge TJ, Oremland RS. 2007. Formation of tellurium nanocrystals during anaerobic growth of bacteria that use Te oxyanions as respiratory electron acceptors. *Appl Environ Microbiol* 73:2135–2143.

Barkes L, Fleming RW. 1974. Production of dimethylselenide gas from inorganic selenium by eleven fungi. *Bull Environ Contam Toxicol* 12:308–311.

Bautista EM, Alexander M. 1972. Reduction of inorganic compounds by soil microorganisms. *Soil Sci Soc Am Proc* 36:918–920.

Breed RS, Murray EGD, Smith NR. 1948. *Bergey's Manual of Determinative Bacteriology*. 6th ed. Baltimore, MD: Williams & Wilkins.

Burton GA Jr., Giddings TH, DeBrine P, Fall R. 1987. High incidence of selenite-resistant bacteria from a site polluted with selenium. *Appl Environ Microbiol* 53:185–188.

Chau YK, Wong PTS, Silverberg BA, Luxon PL, Bengert GA. 1976. Methylation of selenium in the aquatic environment. *Science* 192:1130–1131.

Combs GF Jr., Scott ML. 1977. Nutritional interrelationships of vitamin E and selenium. *BioScience* 27:467–473.

Csotonyi JT, Stackebrandt E, Yurkov V. 2006. Anaerobic respiration on tellurate and other metalloids in bacteria from hydrothermal vent fields in the eastern Pacific Ocean. *Appl Environ Microbiol* 72:4950–4956.

DeMoll-Decker H, Macy JM. 1993. The periplasmic nitrate reductase of *Thauera selenatis* may catalyze the reduction of selenite to elemental selenium. *Arch Microbiol* 160:241–247.

Doran JW, Alexander M. 1977. Microbial transformation of selenium. *Appl Environ Microbiol* 33:31–37.

Dowdle PR, Oremland RS. 1998. Microbial oxidation of elemental selenium in soil slurries and bacterial cultures. *Environ Sci Technol* 32:3749–3755.

Duerre P, Andreesen JR. 1982. Selenium-dependent growth and glycine fermentation by *Clostridium purinolyticum*. *J Gen Microbiol* 128:1457–1466.

Enoch HG, Lester RL. 1972. Effects of molybdate, tungstate, and selenium compounds on formate dehydrogenase and other enzymes in *Escherichia coli*. *J Bacteriol* 110:1032–1040.

Falcone G, Nickerson WJ. 1963. Reduction of selenite by intact yeast cells and cell-free preparations. *J Bacteriol* 85:754–762.

Fleming RW, Alexander M. 1972. Dimethyl selenide and dimethyl telluride formation by a strain of *Penicillium*. *Appl Microbiol* 24:424–429.

Frankenberger WT, Karlson U. 1992. Dissipation of soil selenium by microbial volatilization. In: Adriano DC, ed. *Biogeochemistry of Trace Metals*. Boca Raton, FL: Lewis Publishers, pp. 365–381.

Frankenberger WT, Karlson U. 1995. Soil management factors affecting volatilization of selenium from dewatered sediments. *Geomicrobiol J* 12:265–277.

Gerrard TL, Telford JN, Williams HH. 1974. Detection of selenium deposits in *Escherichia coli* by electron microscopy. *J Bacteriol* 119:1057–1060.

Guzzo J, Dubow MS. 2000. A novel selenite- and tellurite-inducible gene in *Escherichia coli*. *Appl Environ Microbiol* 66:4972–4978.

Hockin SL, Gadd GM. 2003. Linked redox precipitation of sulfur and selenium under anaerobic conditions by sulfate-reducing bacterial biofilms. *Appl Environ Microbiol* 69:7063–7072.

Huber R, Sacher M, Vollmann A, Huber H, Rose D. 2000. Respiration of arsenate and selenate by hyperthermophilic archaea. *Syst Appl Microbiol* 23:305–314.

Klonowska A, Heulin T, Vermeglio A. 2005. Selenite and tellurite reduction by *Shewanella oneidensis*. *Appl Microbiol* 71:5607–5609.

Kovalskii VV, Ermakov VV, Letunova SV. 1968. Geochemical ecology of microorganisms in soils with different selenium content. *Mikrobiologiya* 37:122–139.

Lansche AM. 1965. Tellurium. In: *Mineral Facts and Problems*. Washington, DC: Bureau of Mines, Department of the Interior, pp. 935–939.

Lester RL, DeMoss JA. 1971. Effects of molybdate and selenite on formate and nitrate metabolism in *Escherichia coli*. *J Bacteriol* 105:1006–1014.

Lipman JG, Waksman SA. 1923. The oxidation of selenium by a new group of autotrophic microorganisms. *Science* 57:60.

Lortie L, Gould WD, Rajan S, McCready RGL, Cheng J-J. 1992. Reduction of selenate and selenite to elemental selenium by a *Pseudomonas stutzeri* isolate. *Appl Environ Microbiol* 58:4042–4044.

Macy JM, Rech S, Auling G, Dorsch M, Stackebrandt E, Sly L. 1993. *Thauera selenatis* gen. nov. sp. nov., a member of the beta-subclass of Proteobacteria with a novel type of anaerobic respiration. *Int J Syst Bacteriol* 43:135–142.

Maiers DT, Wichlacz PL, Thompson DL, Bruhn DF. 1988. Selenate reduction by bacteria from a selenium-rich environment. *Appl Environ Microbiol* 54:2591–2593.

Masscheleyn PH, Patrick WH Jr. 1993. Biogeochemical processes affecting selenium cycling in wetlands. *Environ Toxicol Chem* 12:2235–2243.

Mertz W. 1981. The essential trace elements. *Science* 213:1332–1338.

Miller WJ, Neathery MW. 1977. Newly recognized trace mineral elements and their role in animal nutrition. *BioScience* 27:674–679.

Nagai S. 1965. Differential reduction of tellurite by growing colonies of normal yeasts and respiration deficient mutants. *J Bacteriol* 90:220–222.

Narasingarao P, Häggblom MM. 2007. Identification of anaerobic selenate-respiring bacteria from aquatic sediments. *Appl Environ Microbiol* 73:3519–3527.

Nelson DC, Casey WH, Sison JD, Mack EE, Ahmad A, Pollack JS. 1996. Selenium uptake by sulfur-accumulating bacteria. *Geochim Cosmochim Acta* 60:3531–3539.

Newman DK, Ahmann D, Morel FMM. 1998. A brief review of microbial arsenate respiration. *Geomicrobiol J* 15:255–268.

Nickerson WJ, Falcone G. 1963. Enzymatic reduction of selenite. *J Bacteriol* 85:763–771.

Niess UM, Klein A. 2004. Dimethylselenide demethylation is an adaptive response to selenium deprivation in the archeon *Methanococcus voltae*. *J Bacteriol* 186:3640–3648.

Oremland RS, Herbel MJ, Switzer Blum J, Langley S, Beveridge TJ, Ajayan PM, Sutto T, Ellis AV, Curran S. 2004. Structural and spectral features of selenium nanospheres produced by Se-respiring bacteria. *Appl Environ Microbiol* 70:52–60.

Oremland RS, Hollibaugh JT, Maest AS, Presser TS, Miller LB, Cuthberson CW. 1989. Selenate reduction to elemental selenium by anaerobic bacteria in sediments and culture: Biogeochemical significance of a novel sulfate-independent respiration. *Appl Environ Microbiol* 55:2333–2343.

Oremland RS, Steinberg NA, Presser TS, Miller LG. 1991. In situ bacterial selenate reduction in the agricultural drainage system of western Nevada. *Appl Environ Microbiol* 57:615–617.

Oremland RS, Switzer Blum J, Burns Bindi A, Dowdle PR, Herbel M, Stolz JF. 1999. Simultaneous reduction of nitrate and selenate by cell suspensions of selenium-respiring bacteria. *Appl Environ Microbiol* 65:4385–4392.

Oremland RS, Switzer Blum J, Culbertson CW, Visscher PT, Miller LG, Dowdle P, Strohmaier FE. 1994. Isolation, growth, and metabolism of an obligately anaerobic, selenate-respiring bacterium, strain SES-3. *Appl Environ Microbiol* 60:3011–3019.

Patrick R. 1978. Effects of trace metals in the aquatic ecosystem. *Am Sci* 66:185–191.

Pinsent J. 1954. The need for selenite and molybdate in the coli-aerogenes group of bacteria. *Biochem J* 57:10–16.

Ranjard L, Prigent-Combaret C, Nazaret S, Cournoyer B. 2002. Methylation of inorganic and organic selenium by the bacterial thiopurine methyltransferase. *J Bacteriol* 184:3146–3149.

Rapp G Jr. 1972. Selenium: Element and geochemistry. In: Fairbridge RW, ed. *The Encyclopedia of Geochemistry and Environmental Sciences*. Encyclopedia of Earth Science Series, Vol. IVA. New York: Van Nostrand Reinhold, pp. 1079–1080.

Reamer DC, Zoller WH. 1980. Selenium biomethylation products from soil and sewage. *Science* 208:500–502.

Rech S, Macy JM. 1992. The terminal reductases from selenate and nitrate respiration in *Thauera selenatis* are two distinct enzymes. *J Bacteriol* 174:7316–7320.

Rosenfeld I, Beath OA. 1964. *Selenium, Geobotany, Biochemistry, Toxicity, and Nutrition*. New York: Academic Press.

Rotruck JT, Pope AL, Ganther HE, Swanson AB, Hafman DG, Hoekstra WG. 1973. Selenium: Biochemical role as a component of glutathione peroxidase. *Science* 179:588–590.

Roux M, Sarret G, Pignot-Paintrand I, Fontecave M, Coves J. 2001. Mobilization of selenite by *Ralstonia metallidurans* CH34. *Appl Environ Microbiol* 67:769–773.

Sapozhnikov DI. 1937. The substitution of selenium for sulfur in the photoreduction of carbonic acid by purple sulfur bacteria. *Mikrobiologiya* 6:643–644.

Sarathchandra SU, Watkinson JH. 1981. Oxidation of elemental selenium to selenite by *Bacillus megaterium*. *Science* 211:600–601.

Schroeder I, Rech S, Krafft T, Macy JM. 1997. Purification and characterization of the selenate reductase from *Thauera selenatis*. *J Biol Chem* 272:23765–23768.

Shrift A. 1964. Selenium in nature. *Nature* (London) 201:1304–1305.

Shum AD, Murphy JC. 1972. Effects of selenium compounds on formate metabolism and coincidence of selenium-75 incorporation and formic dehydrogenase activity in cell-free preparations of *Escherichia coli*. *J Bacteriol* 110:447–449.

Silverman MP, Ehrlich HL. 1964. Microbial formation and degradation of minerals. *Adv Appl Microbiol* 6:153–206.

Stadtman RC. 1974. Selenium biochemistry. *Science* 183:915–922.

Steinberg NA, Blum JS, Hochstein L, Ormland RS. 1992. Nitrate is a preferred electron acceptor for growth of freshwater selenate-respiring bacteria. *Appl Environ Microbiol* 58:426–428.

Steinberg NA, Oremland RS. 1990. Dissimilatory selenate reductase potentials in a diversity of sediment types. *Appl Environ Microbiol* 56:3550–3557.

Stolz JF, Ellis DJ, Switzer Blum J, Ahmann D, Oremland RS, Lovley DR. 1999. *Sulfurospirillum barnesii* sp. nov. and *Sulfurospirillum arsenophilus* sp. nov., new members of the *Sulfurospirillum* clade of the ε-Proteobacteria. *Int J Syst Bacteriol* 49:1177–1180.

Stolz JF, Oremland RS. 1999. Bacterial respiration of arsenic and selenium. *FEMS Microbiol Rev* 23:615–627.

Switzer Blum J, Burns Bindi A, Buzzelli J, Stolz JG, Oremland RS. 1998. *Bacillus arsenicoselenatis*, sp. nov., and *Bacillus selenitireducens*, sp. nov.: Two haloalkaliphiles from Mono Lake, California that respired oxyanions of selenium and arsenic. *Arch Microbiol* 171:19–30.

Switzer Blum J, Stolz JF, Oren A, Oremland RS. 2001. *Selenihalanaerobacter shriftii* gen nov., spec. nov., a halophilic anaerobed from Dead Sea sediments that respire selenate. *Arch Microbiol* 175:208–219.

Thompson-Eagle ET, Frankenberger WT Jr., Karlson U. 1989. Volatilization of selenium by *Alternaria alternata*. *Appl Environ Microbiol* 55:1406–1413.

Tomei FA, Barton LL, Lemansky CL, Zocco TG. 1992. Reduction of selenate and selenite to elemental selenium by *Wolinella succinogenes*. *Can J Microbiol* 38:1328–1333.

Torma AE, Habashi F. 1972. Oxidation of copper(II) selenide by *Thiobacillus ferrooxidans*. *Can J Microbiol* 18:1780–1781.

Trutko SM, Akimenko VK, Suzina NE, Anisimova LA, Shlyapnikov MG, Baskunov BP, Duda VI, Boronin AM. 2000. Involvement of the respiratory chain of gram-negative bacteria in the reduction of tellurite. *Arch Microbiol* 173:178–186.

Turner RJ, Hou Y, Weiner JH, Taylor DE. 1992. The arsenical ATPase efflux pump mediates tellurite resistance. *J Bacteriol* 174: 3092–3094.

Van Fleet-Stadler V, Chasteen TG, Pickering IJ, George GN, Prince RC. 2000. Fate of selenate and selenite metabolized by *Rhodobacter sphaeroides*. *Appl Environ Microbiol* 66:4849–4853.

Völkl P, Huber R, Drobner E, Rachel R, Burggraf S, Trincone A, Stetter KO. 1993. *Pyrobaculum aerophilum* sp. nov., a novel nitrate-reducing hyperthermophilic archaeum. *Appl Environ Microbiol* 59:2918–2926.

Woolfolk CA, Whiteley HR. 1962. Reduction of inorganic compounds with molecular hydrogen by *Micrococcus lactilyticus*. I. Stoichiometry with compounds of arsenic, selenium, tellurium, transition and other elements. *J Bacteriol* 84:647–658.

Yamamoto I, Saiki T, Liu S-M, Ljungdahl LG. 1983. Purification and properties of NADP-dependent formate dehydrogenase from *Clostridium thermoaceticum*, a tungstate-selenium protein. *J Biol Chem* 258:1826–1832.

Zalokar M. 1953. Reduction of selenite by *Neurospora*. *Arch Biochem Biophys* 44:330–337.

Zehr JP, Oremland RS. 1987. Reduction of selenate to selenide by sulfate-respiring bacteria: Experiments with cell suspensions and estuarine sediments. *Appl Environ Microbiol* 53:1365–1369.

22 Geomicrobiology of Fossil Fuels

22.1 INTRODUCTION

Although much of the organic carbon in the biosphere is continually recycled, a very significant amount has become trapped in special sedimentary formations, where it is inaccessible to mineralization by microbes until it becomes reexposed to water through natural causes or human intervention. Microbial mineralization of such reexposed organic carbon also depends on the access to suitable terminal electron acceptors, that is, oxygen in air in the case of aerobes, and inorganic electron acceptors in the case of facultative or anaerobic microbes. The trapped organic carbon exists in various forms. The degree of its chemically reduced state is related to the length of time it has been trapped and any secondary changes that it has undergone during this time. Some of this trapped carbon has value as a fuel, a source of energy for industrial and other human activity, and is exploited for this purpose. Because of the great age of this material, it is known as *fossil fuel*. The remainder of the trapped carbon is chiefly kerogen and bitumen, some of which can be converted to fuel by human intervention. Fossil fuels include methane gas, natural gas (which is largely methane), petroleum, oil shale, coal, and peat. They are generally considered to have had a microbial origin (Ourisson et al., 1984).

22.2 NATURAL ABUNDANCE OF FOSSIL FUELS

A major portion of the total carbon at the Earth's surface is in the form of carbonate (Figure 22.1). It represents a major sink for carbon. The other sink is the trapped organic carbon that is not directly accessible for microbial mineralization. The carbonate carbon is not an absolute sink unless it is deeply buried because it is in a steady-state relationship with dissolved carbonate/bicarbonate and atmospheric CO_2, which in turn are in a steady-state relationship with organic carbon in living and dead biomass. The passage of carbon from one compartment into another is under biological control (Figure 22.1; Fenchel and Blackburn, 1979).

22.3 METHANE

Methane at atmospheric pressure and ambient temperature is a colorless, odorless, and flammable gas. Because its autoignition temperature is 650°C, it does not catch fire spontaneously. It is sparingly soluble in water (3.5 mL per 100 mL of water) but readily soluble in organic solvents, including liquid hydrocarbons. It may have an abiotic or biogenic origin. Biogenic accumulations of methane may occur in nature when it is formed in consolidated sediment from which it cannot readily escape. In some deep-ocean sediments under conditions of high pressure and low temperature, methane accumulations are found in the form of methane hydrates (Kvenvolden, 1988; Haq, 1999). A special, mixed community of certain members of the Archaea and Bacteria living very close to methane hydrate has been detected in the forearc basin of the Nankai Trough off the east coast of Japan by Reed et al. (2002) and in solid gas hydrates in the Gulf of Mexico by Mills et al., 2005.

Bioformation of methane comes about when organic matter in the sediment is undergoing anaerobic microbial breakdown in the absence of significant quantities of alternative terminal electron

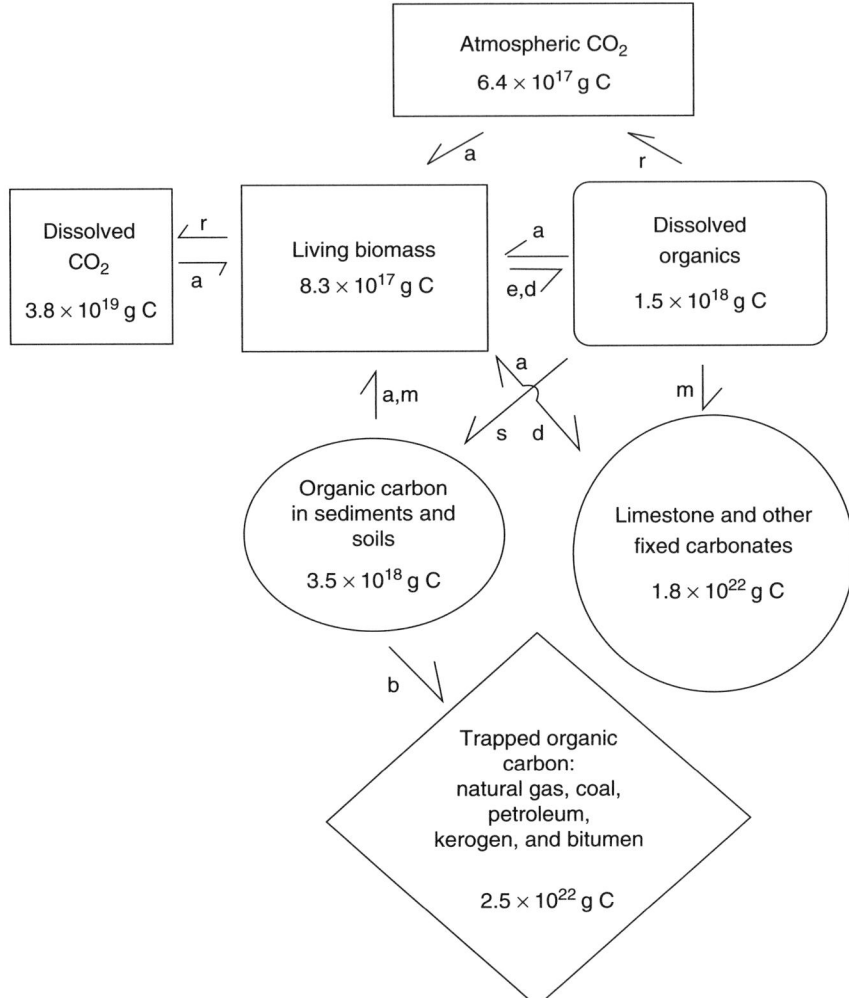

FIGURE 22.1 Microbial and physical processes contributing to carbon transfer among different compartments in the biosphere. a, Microbial assimilation; b, burial; d, decomposition; e, excretion; m, microbial mineralization; r, respiration; s, sedimentation. (Quantitative estimates from Fenchel T, Blackburn TH, *Bacteria and Mineral Cycling*, Academic Press, London, U.K., 1979; Bowen HJM, *Environmental Chemistry of the Elements*, Academic Press, London, U.K., 1979.)

acceptors such as nitrate, Fe(III), Mn(IV), or sulfate. If gas pressure due to methane builds up sufficiently in anaerobic lake or coastal sediment, it may escape in the form of large gas bubbles that break at the water surface to release their methane into the atmosphere (Martens, 1976; Zeikus, 1977). In marshes, escaping methane may be ignited (by biogenic phosphene?) to burn as the so-called will-o'-the-wisps.

Many of the methane accumulations on Earth are of biogenic origin. Methane may occur in association with peat, coal, and oil deposits, or independent of them. That which occurs in association with coal and oil was probably microbially generated in the early stages of their formation, although some may have been formed abiotically in later diagenetic phases. Methane associated with coal deposits can be the cause of serious mine explosions when accidentally ignited. Such methane is called *coal damp* by coal miners.

Biogenic methane formation is a unique biochemical process that appears to have arisen very early in the evolution of life. Indeed, the methanogenesis that results from the microbial reduction of CO_2 by H_2 may represent the first process or one of the first processes on Earth that autotrophs have harnessed for energy conservation (see Chapter 3).

22.3.1 METHANOGENS

All methane-forming bacteria, that is, *methanogens*, are members of the domain Archaea. As a group, they are very diverse phylogenetically (Jones et al., 1987; Boone et al., 1993). They also show great diversity morphologically, existing as rods, spirilla, cocci, and sarcinae (Figure 22.2). The feature that they share in common is that of being strict anaerobes that form methane as a product of their energy-generating metabolism (respiration and fermentation). The large majority of them are obligate or facultative autotrophs. The majority can get their energy from the reduction of carbon dioxide with hydrogen or its equivalent (formate or CO), whereas they obtain their carbon exclusively by assimilating carbon dioxide. A few, like *Methanosarcina* (formerly *Methanothrix*) *Soehngenii* (Zehnder et al., 1980; Huser et al., 1982) and *Methanosaeta* (formerly *Methanothrix*) *concilii* (Patel, 1984), are heterotrophs that use acetate as energy and carbon source, and for this

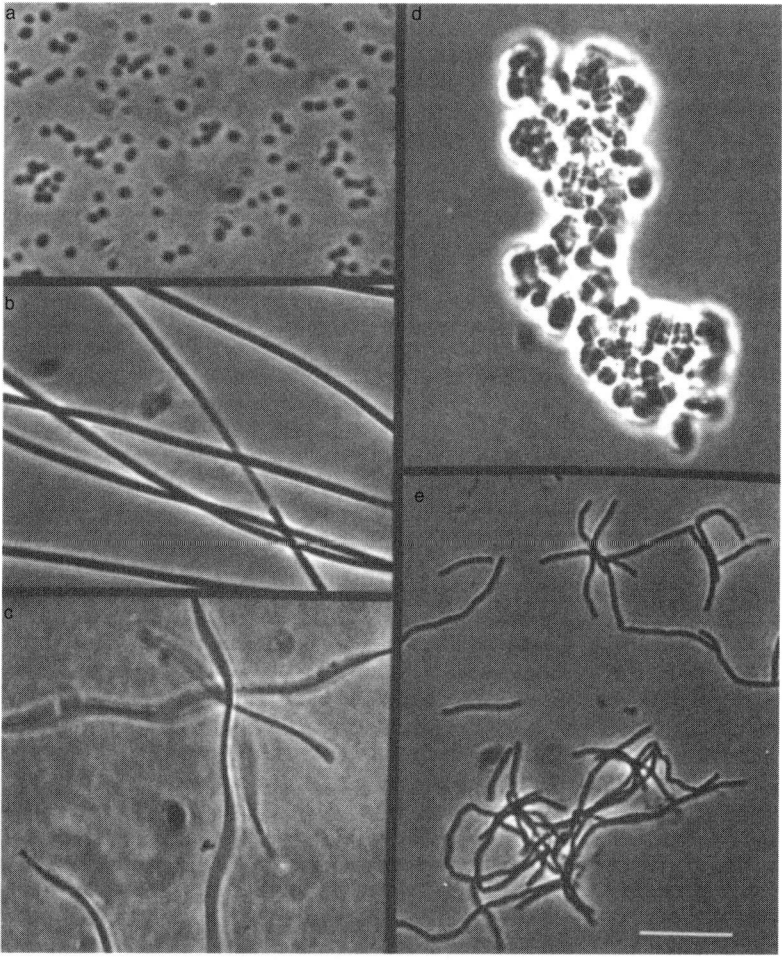

FIGURE 22.2 Morphologies of different methanogens. (a) *Methanocullens bourgensis* (formerly, *Methanogenium bourgense*), (b) *Methanosaeta* (formerly *Methanothrix*) *concilii*, (c) *Methanospirillum hungatei* OGC 16, (d) *Methanosarcina barkeri* OGC 35, (e) *Methanobacterium formicicum* OGC 55. Scale mark represents 10 µm and applies to all panels. (Courtesy of Boone DR, Wills R)

reason are known as *acetotrophic* or *aceticlastic* methanogens. At least one methanogen can use S^0 in addition to CO_2 as the terminal electron acceptor (Stetter and Gaag, 1983). As a group, most methanogens are nutritionally restricted to the following energy sources: H_2, CO, HCOOH, methanol, methylamines, and acetate (Atlas, 1997; Brock and Madigan, 1988). But exceptions exist. Widdel reported that a freshwater strain of *Methanospirillum* and a strain of *Methanogenium* were each able to grow on 2-propanol and 2-butanol as well as on H_2 and formate. Zellner et al. (1989) found that *Methanobacterium palustre* was able to grow on 2-propanol as an energy source as well as on H_2 and formate. It was able to oxidize but not grow on 2-butanol. In 1990, Zellner et al. reported that *Methanogenium liminatans* can use 2-propanol, 2-butanol, and cyclopentanol as energy sources in addition to H_2 and formate (Zellner et al., 1990). Finster et al. (1992) found that strain MTP4 can use methanediol and dimethylsulfide as well as methylamines, methanol, and acetate as energy sources. Yang et al. (1992) found that *Methanococcus voltae*, *M. maripaludis*, and *M. vannielii* can each use pyruvate as an energy source in the absence of H_2.

For methanogens to be able to draw on the wide range of oxidizable carbon compounds that may be available in their environment but that cannot be metabolized by them directly, they associate with heterotrophic fermenters or anaerobic respirers that do not completely mineralize their organic energy sources (see, e.g., Jain and Zeikus, 1989; Sharak Genthner et al., 1989; Grbic-Galic, 1990). To optimize access to the microbially generated energy sources and the electron acceptor CO_2 that these methanogens need, some of them form intimate consortia (*syntrophic associations*) with other anaerobic bacteria that can furnish them with these energy sources (H_2, acetate) and CO_2 through their metabolic end products (see, e.g., Bochem et al., 1982; MacLeod et al., 1990; McInerney et al., 1979; Winter and Wolfe, 1979, 1980; Zinder and Koch, 1984; Wolin and Miller, 1987). Frequently, the metabolites that are the basis for these syntrophic associations are not readily detectable when all the members of the consortium are growing together in mixed culture. This is because the metabolites are consumed as quickly as they are formed. When hydrogen is the metabolite, the process is called interspecies hydrogen transfer (Wolin and Miller, 1987).

Among the most widely recognized genera of methanogens are *Methanobacterium*, *Methanothermobacter*, *Methanobrevibacterium*, *Methanococcus*, *Methanomicrobium*, *Methanogenium*, *Methanospirillum*, *Methanosarcina*, *Methanoculleus*, and *Methanosaeta* (see Brock and Madigan, 1988; Bhatnagar et al., 1991; Boone et al., 1993; Atlas, 1997). Methanogens may be mesophilic or thermophilic. They are found in diverse anaerobic habitats (Zinder, 1993), including some marine environments such as salt marsh sediments (Oremland et al., 1982; Jones et al., 1983b), coastal sediments (Gorlatov et al., 1986; Sansone and Martens, 1981), anoxic basins (Romesser et al., 1979), geothermally heated seafloor (Huber et al., 1982), hydrothermal vent effluent on the East Pacific Rise (Jones et al., 1983a), sediment effluent channel of the Crystal River Nuclear Power Plant (Florida; Rivard and Smith, 1982), lakes (Deuser et al., 1973; Jones et al., 1982; Giani et al., 1984), soils (Jakobsen et al., 1981), desert environments (Worakit et al., 1986), solfataric fields (Wildgruber et al., 1982; Zabel et al., 1984), oil deposits (Nazina and Rozanova, 1980; Rubinshtein and Oborin, 1986; Stetter et al., 1993), the digestive tract of insects and higher animals, especially ruminants and herbivores (Breznak, 1982; Brock and Madigan, 1988; Wolin, 1981; Zimmerman et al., 1982; Atlas, 1997), and as endosymbionts (van Bruggen et al., 1984; Fenchel and Finlay, 1992). Thus, despite their obligately anaerobic nature, methanogens are fairly ubiquitous.

Methanogens play an important but not exclusive role in anaerobic mineralization of organic carbon compounds in soil and aquatic environments, especially freshwater sediments (Wolin and Miller, 1987). In marine sediments, where methanogens have to share hydrogen and acetate as sources of energy with sulfate-reducing bacteria, they tend to be outcompeted by the sulfate reducers because of the latter's higher affinity for hydrogen and acetate (Abrams and Nedwell, 1978; Kristjansson et al., 1982; Schönheit et al., 1982; Robinson and Tiedje, 1984). Thus, in many estuarine or coastal anaerobic muds, sulfate-reducing activity and methanogenesis occur usually in spatially separated zones in the sediment profile, with the zone exhibiting sulfate-reducing activity overlying the zone exhibiting methanogenesis (e.g., Martens and Berner, 1974; Sansone and Martens, 1981).

Recent evidence indicates that some sulfate-reducing bacteria can also use methane as electron donor (Section 22.3.5.2).

Under two special circumstances, methanogenesis and sulfate reduction can be compatible in an anaerobic marine environment. One circumstance is the existence of an excess supply of a shared energy source (H_2 or acetate; Oremland and Taylor, 1978). The other circumstance is one where sulfate reducers and methanogens use different energy sources, namely, products of decaying plant material and methanol or trimethylamine, respectively (Oremland et al., 1982). In anaerobic freshwater sediments and soils where sulfate, nitrate, ferric oxide, and manganese(IV) oxide concentrations are very low, methanogenesis is usually the dominant mechanism of organic carbon mineralization. Yet, even here, certain sulfate-reducing bacteria may grow in the same niche as methanogens (e.g., Koizumi et al., 2003). Indeed, they may form a consortium with them. In the absence of sulfate, these sulfate reducers ferment suitable organic carbon with the production of H_2, which the methanogens then use in their energy metabolism to form methane (e.g., Bryant et al., 1977).

A few methanogens, in particular, *Methanosarcina barkeri* grown with H_2/CO_2 or methanol, have the ability to reduce Fe(III) in place of CO_2 (van Bodegom et al., 2004). This ability may explain in part the inhibition of methanogenesis in soil and sediment by Fe(III).

22.3.2 METHANOGENESIS AND CARBON ASSIMILATION BY METHANOGENS

22.3.2.1 Methanogenesis

One kind of autotrophic methane formation represents a form of anaerobic respiration in which hydrogen (H_2) is the electron donor and CO_2 is the terminal electron acceptor, with the CO_2 being transformed to CH_4 according to the overall reaction

$$4H_2 + CO_2 \rightarrow CH_4 + 2H_2O \tag{22.1}$$

This reaction is exothermic and yields energy ($\Delta G^0 = -33$ kcal or -137.9 kJ) that can be used by the organism to do metabolic work.

In a few instances, secondary alcohols were found to serve as electron donors, with CO_2 as the terminal electron acceptor. The CO_2 was therefore the source of the methane formed (Widdel, 1986; Zellner et al., 1989). In these reactions, the alcohols were replacing H_2 as the reductant of CO_2. At least one instance is known in which ethanol served as the electron donor for methane formation from CO_2, the ethanol being oxidized to acetate (Frimmer and Widdel, 1989):

$$2CH_3CH_2OH + HCO_3^- \rightarrow 2CH_3COO^- + CH_4 + H_2O + H^+$$

$$(\Delta G^0 = -27.8 \text{ kcal/mol } CH_4 \text{ or } -116.3 \text{ kJ/mol } CH_4) \tag{22.2}$$

The organism in this instance was a nonautotrophic methanogen, *Methanogenium organophilum* growing in a medium containing 0.05% tryticase peptone and 0.05% yeast extract as nitrogen sources among other ingredients (Frimmer and Widdel, 1989).

Some methanogens can form methane by a disproportionation reaction, that is, by fermentation, in which a portion of the substrate molecule acts as the electron donor (energy source) and the rest as the electron acceptor. For example, they can produce methane from carbon monoxide, formic acid, methanol, acetate, and methylamines without H_2 as the electron donor (Brock and Madigan, 1988; Atlas, 1997; Mah et al., 1978; Smith and Mah, 1978; Zeikus, 1977),

$$4HCOOH \rightarrow CH_4 + 3CO_2 + 2H_2O \quad (\Delta G^0 = -35 \text{ kcal or } -146.3 \text{ kJ}) \tag{22.3}$$

$$4CH_3OH \rightarrow 3CH_4 + CO_2 + 2H_2O \quad (\Delta G^0 = -76 \text{ kcal or } 317.7 \text{ kcal}) \tag{22.4}$$

$$CH_3COOH \rightarrow CH_4 + CO_2 \quad (\Delta G^0 = -9 \text{ kcal or } -37.6 \text{ kJ}) \tag{22.5}$$

$$4CH_3NH_2 + 2H_2O \rightarrow 3CH_4 + CO_2 + 4NH_3 \quad (\Delta G^0 = -75 \text{ kcal or } -313.5 \text{ kJ}) \tag{22.6}$$

$$4CO + 2H_2O \rightarrow CH_4 + 3CO_2 \quad (\Delta G^0 = -44.5 \text{ kcal or } -186 \text{ kJ}) \tag{22.7}$$

Although methanogenesis from acetate by the disproportionation reaction of *aceticlastic* methanogens (Reaction 22.5) is fairly common during anaerobic degradation of organic matter in the absence of a plentiful external supply of external electron acceptors such as Fe(III), Mn(IV), or sulfate, a consortium of anaerobic acetate oxidizers like *Clostridium* spp., which generate H_2 and CO_2 from the acetate, and *hydrogenotrophic* methanogens like *Methanomicrobium* or *Methanobacterium* may form methane in the absence of aceticlastic methanogens (Karakashev et al., 2006).

Some methanogens can form methane from pyruvate by disproportionation (Yang et al., 1992). Resting cells of *Methanococcus* spp. grown in a pyruvate-containing medium under an N_2 atmosphere were shown to transform pyruvate to acetate, methane, and CO_2 according to the following stoichiometry (Yang et al., 1992):

$$4CH_3COCOOH + 2H_2O \rightarrow 4CH_3COOH + 3CO_2 + CH_4 \quad (\Delta G^0 = -74.9 \text{ kcal or } 313.1 \text{ kJ}) \tag{22.8}$$

This stoichiometry is attained if the organism oxidatively decarboxylates pyruvate:

$$4CH_3COCOOH + 4H_2O \rightarrow 4CH_3COOH + 4CO_2 + 8(H) \tag{22.8a}$$

and uses the reducing power [8(H)] to reduce one-fourth of the CO_2 to CH_4:

$$CO_2 + 8(H) \rightarrow CH_4 + 2H_2O \tag{22.8b}$$

Bock et al. (1994) found that a spontaneous mutant of *Methanosarcina barkeri* could grow by fermenting pyruvate to methane and CO_2 with the following stoichiometry:

$$CH_3COCOOH + 0.5H_2O \rightarrow 1.25CH_4 + 1.75CO_2 \tag{22.9}$$

To achieve this stoichiometry, the authors proposed the following mechanism based on known enzyme reactions in methanogens. Pyruvate is oxidatively decarboxylated to acetyl~SCoA and CO_2:

$$CH_3COCOOH + CoASH \rightarrow CH_3CO\sim SCoA + CO_2 + 2(H) \tag{22.9a}$$

The available reducing power [2(H)] from this reaction is then used to reduce one-fourth of the CO_2 formed to methane:

$$0.25CO_2 + 2(H) \rightarrow 0.25CH_4 + 0.5H_2O \tag{22.9b}$$

and the acetyl~SCoA is decarboxylated to methane and CO_2:

$$CH_3CO\sim SCoA + H_2O \rightarrow CH_4 + CO_2 + CoASH \tag{22.9c}$$

The standard free energy yield at pH 7 (ΔG^0) was calculated to be -22.9 kcal mol^{-1} or -96 kJ mol^{-1} of methane produced (Bock et al., 1994).

Although Reactions 22.1 through 22.7 look very disparate, they share a common metabolic pathway (Figure 22.3). The reason methanogens differ with respect to the methane-forming reactions they can perform is that not all of them possess the same key enzymes that permit entry of particular methanogenic substrates into the common pathway (Vogels and Visser, 1983; Zeikus et al., 1985; Stanier et al., 1986; Brock and Madigan, 1988; Atlas, 1997). The pathway involves stepwise reduction of carbon from the +4 to the −4 oxidation state via bound formyl, methylene, and methyl carbon. The operation of the methane-forming pathway requires some unique coenzyme and carrier molecules (Table 22.1). Coenzyme M (2-mercaptoethylsulfonate) is unique to methanogens and

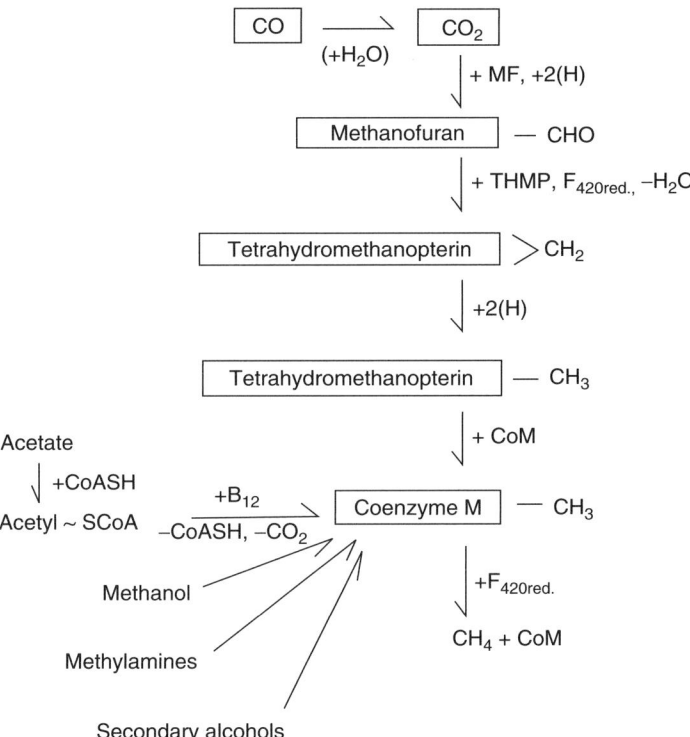

FIGURE 22.3 Pathways of methanogenesis from CO, CO_2, acetate, methanol, secondary alcohols, and methylamines. (THMP, Tetrahydromethanoptein.)

TABLE 22.1
Unusual Coenzymes in Methanogens

Coenzyme[a]	Function
Methanofuran	CO_2 reduction factor in first step of methanogenesis
Methanopterin (coenzyme F_{342})	Formyl and methene carrier in methanogenesis
Coenzyme M (2-mercaptoethane sulfonate)	Methyl carrier in methanogenesis
Coenzyme F_{430}	Hydrogen carrier for reduction of methyl coenzyme M
Coenzyme F_{420} (nickel-containing tetrapyrrole)	Mediates electron transfer between hydrogenase or formate and NADP, reductive carboxylation of acety1~CoA, and succinyl~CoA

[a] For structures of these coenzymes, see Brock and Madigan (1988) and Blaut et al. (1992).

may be used to identify them as methane formers. The large majority of methanogens synthesize this molecule *de novo*.

Methanogenic reactions with hydrogen as the electron donor that utilize formic or acetic acid, methanol, or methylamines as electron acceptors instead of CO_2 such as

$$3H_2 + HCOOH \rightarrow CH_4 + 2H_2O \quad (\Delta G^0 = -42 \text{ kcal or } -175.6 \text{ kJ}) \tag{22.10}$$

$$4H_2 + CH_3COOH \rightarrow 2CH_4 + 2H_2O \quad (\Delta G^0 = -49 \text{ kcal or } 204.8 \text{ kJ}) \tag{22.11}$$

$$H_2 + CH_3OH \rightarrow CH_4 + H_2O \quad (\Delta G^0 = -26.9 \text{ kcal or } -112.4 \text{ kJ}) \tag{22.12}$$

$$H_2 + CH_3NH_2 \rightarrow CH_4 + NH_3 \quad (\Delta G^0 = -9 \text{ kcal or } -37.6 \text{ kJ}) \tag{22.13}$$

are not known to occur.

New evidence suggests that Reaction 22.1 can occur abiotically in the presence of a nickel–iron alloy under hydrothermal conditions (e.g., 200–400°C, 50 MPa), conditions met in parts of the oceanic crust, as reported, for instance, by Horita and Berndt (1999).

22.3.3 BIOENERGETICS OF METHANOGENESIS

As an anaerobic respiratory process, methane formation is performed to yield useful energy to the cell. Evidence to date indicates that adenosine 5-triphosphate (ATP) is generated by chemiosmotic energy-coupling metabolism (e.g., Mountford, 1978; Doddema et al., 1978, 1979; Blaut and Gottschalk, 1984; Sprott et al., 1985; Gottschalk and Blaut, 1990; Blaut et al., 1990, 1992; Müller et al., 1993; Atlas, 1997; Li et al., 2006). The chemiosmotic coupling mechanism seems to involve pumping of protons or sodium ions across the plasma membrane, depending on the methanogen. Membrane-associated electron transport constituents required in chemiosmotic energy conservation involving proton coupling in *Methanosarcina strain* Göl include reduced factor F_{420} dehydrogenase, an unknown electron carrier, cytochrome b, and heterodisulfide reductase (see Blaut et al., 1992). The heterodisulfide consists of coenzyme M covalently linked to 7-mercaptoheptanoyl-threonine by a disulfide bond (Blaut et al., 1992). A proton-translocating ATPase associated with the membrane catalyzes ATP synthesis in this organism.

An example of a methanogen that employs sodium ion coupling is *Methanococcus voltae* (Dybas and Konisky, 1992; Chen and Konisky, 1993). It appears to employ Na^+-translocating ATPase that is insensitive to proton translocation inhibitors. A scheme for pumping sodium ions from the cytoplasm to the periplasm that depends on membrane-bound methyl transferase was proposed by Blaut et al. (1992). *Methanosarcina acetivorans* is another example of a methanogen that employs sodium ion coupling in its ATP synthesis (Li et al., 2006).

22.3.4 CARBON FIXATION BY METHANOGENS

When methanogens grow autotrophically, their carbon source is CO_2. The mechanism by which they assimilate CO_2 is different from that of most autotrophs (Simpson and Whitman, 1993). Most autotrophs in the domain Bacteria use the pentose diphosphate pathway (Calvin-Benson-Bassham cycle). Among the exceptions are green sulfur bacteria, which use a reverse tricarboxylic acid cycle; *Chloroflexus aurantiacus*, which uses bicyclic CO_2-fixation involving 3-hydroxypropionate as a key intermediate; and the methane-oxidizing bacteria, which use either the hexulose monophosphate or the serine pathway (see Section 22.3.7). In methanogens, as in homoacetogens (see Chapter 6) and some sulfate-reducing bacteria (see Chapter 19), the chief mechanism of carbon assimilation is by reduction of one of the two molecules of CO_2 to methyl carbon and the second to a formyl carbon followed by the coupling of the formyl carbon to the methyl carbon to form acetyl~SCoA (see Figure 6.8 in Chapter 6). To form the important metabolic intermediate pyruvate, they next carboxylate the acetyl~SCoA reductively. All other cellular constituents are then synthesized from pyruvate and may utilize incomplete reductive or oxidative tricarboxylic acid cycles (Simpson and Whitman, 1993).

Examination of the genome sequence of *Methanocaldococcus* (*Methanococcus*) *jannaschii* and *Methanosarcina acetivorans* has revealed the presence of genes for a ribulose 1,5-bisphosphate carboxylase/oxidase (Finn and Tabita, 2003), but these organisms do not use the Calvin-Benson-Bassham cycle for primary CO_2 fixation (see introduction of the article by Finn and Tabita, 2004, and Sprott et al., 1993).

22.3.5 MICROBIAL METHANE OXIDATION

22.3.5.1 Aerobic Methanotrophy

Methane can be used as a primary energy source by a number of aerobic bacteria. Some of these are obligate *methanotrophs*; others are facultative (Higgins et al., 1981; see also Theisen and Murrell, 2005). Methane is also oxidized by some yeasts (Higgins et al., 1981). Except for the anaerobic methanotrophic consortia described in Section 22.3.5.2, most, if not all, known methanotrophs are aerobes. Examples of obligate methanotrophs are *Methylomonas, Methylococcus, Methylobacter, Methylosinus,* and *Methylocystis* (Figure 22.4). All are gram-negative and feature intracytoplasmic membranes. On the basis of the organization of these membranes, each obligate methanotroph can be assigned to one of two types (Davies and Whittenbury, 1970). Members of *Type I* have stacked membranes, whereas members of *Type II* have paired membranes concentric with the plasma membrane and forming vesiclelike or tubular structures (Figure 22.5). Facultative methanotrophs feature internal membranes with an appearance like those of the *Type II* obligate methanotrophs. All methanotrophs can also use methanol as primary energy source, but not all methanol oxidizers can grow on methane as primary energy source. Methanol oxidizers that cannot oxidize methane are called *methylotrophs.*

Recently, the surprising discovery was made that the filamentous, sheathed bacteria *Crenothrix polyspora* Cohn 1870 and *Clonothrix fusca* Roze 1896 have methanotrophic ability (Stoecker et al., 2006; Vigliotta et al., 2007). *C. polyspora* was found to have an unusual methane monooxygenase (Stoecker et al., 2006). *C. fusca* was shown to possess an elaborate internal membrane system to oxidize and assimilate methane (Vigliotta et al., 2007).

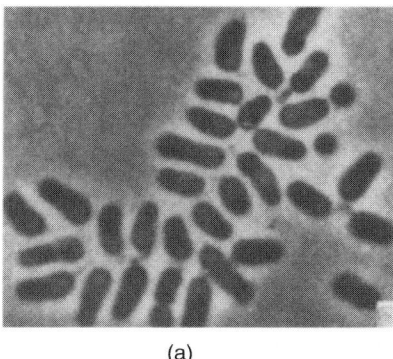   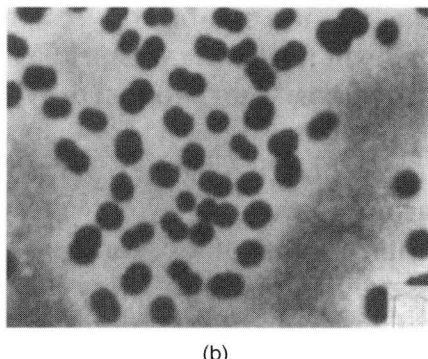

(a)   (b)

FIGURE 22.4 Aerobic methane-oxidizing bacteria (methanotrophs) (×19,000). (a) *Methylosinus trichosporium* in rosette arrangement. Organisms are anchored by visible holdfast material. (b) *Methylococcus capsulatus*. (From Whittenbury R, Phillips KC, Wilkinson JF, *J. Gen. Microbiol.*, 61, 205–218, 1970. With permission.)

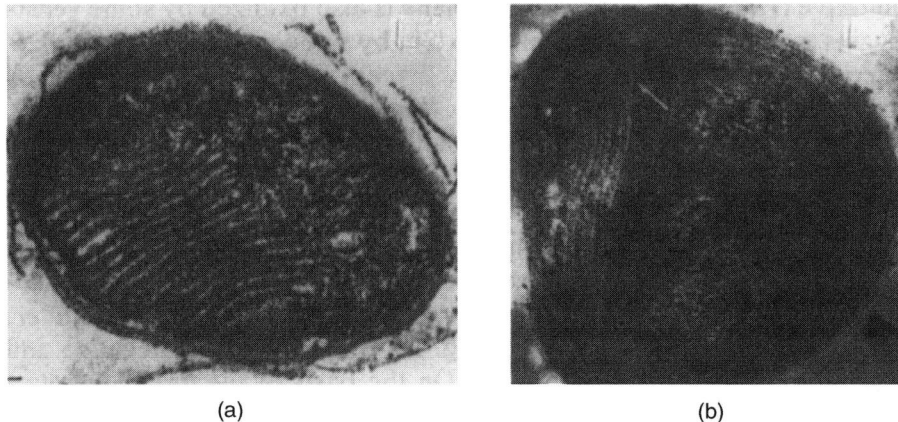

(a)   (b)

FIGURE 22.5 Fine structure of methane-oxidizing bacteria (×80,000). (a) Section of *Methylococcus* (subgroup *minimus*) showing Type I membrane system. (b) Peripheral arrangement of membranes in *Methylosinus* (subgroup *sporium*) characteristic of Type II membrane systems. (From Davies SL, Whittenbury R, *J. Gen. Microbiol.*, 61, 227–232, 1970. With permission.)

Methanotrophs are important for the carbon cycle in returning the carbon of methane, which is always generated anaerobically, to the reservoir of CO_2 (e.g., Vogels, 1979). Obligate methanotrophs are found mainly at aerobic/anaerobic interfaces in soils and aquatic environments that are crossed by methane (e.g., Alexander, 1977; Reeburgh, 1976; Ward and Brock, 1978; Sieburth et al., 1987; Hyun et al., 1997; Horz et al., 2002; Inagaki et al., 2004; Sundh et al., 2005) and also in coal and petroleum deposits (Ivanov et al., 1978; Kuznetsov et al., 1963). Some methanotrophs are also important intracellular symbionts in mussels from marine hydrocarbon seeps and in other benthic invertebrates that encounter methane in their habitat. Cavanaugh et al. (1987) found evidence of the presence of such symbiotic methanotrophs in the epithelial cells of the gills of some mussels from reducing sediments at hypersaline seeps at abyssal depths in the Gulf of Mexico at the Florida Escarpment. MacDonald et al. (1990) made similar observations in mussels that occurred in a large bed surrounding a pool of hypersaline water rich in methane at a depth of 650 m on the continental slope south of Louisiana. Transmission electron microscopic examination by Cavanaugh et al. (1987) showed that the symbionts feature typical intracytoplasmic membranes of Type I methanotrophs. They possess the key enzymes associated with methane oxidation in the group (see Section 22.3.6). The basis for the symbiosis between the invertebrate host and the methanotrophs is the sharing of fixed carbon derived from methane taken up by the host and metabolized by the methanotroph (Childress et al., 1986). This activity is similar to that of the symbiotic H_2S-oxidizing bacteria in some invertebrates of hydrothermal vent communities, which share with their host the carbon they fix from CO_2 taken up by the host (see Chapter 20). Whether the source of the methane issuing from Gulf of Mexico seeps is biogenic or abiogenic is unclear at this time. The invertebrate fauna in the vicinity of the hydrocarbon seeps on the Louisiana slope in the Gulf of Mexico features intracellular methanotrophic bacterial endosymbionts only in mussels. Vestimentiferan worms and the clam species in these locations feature autotrophic sulfur-metabolizing symbionts (Brooks et al., 1987). Interestingly, the mussel *Bathymodiolus* sp. from a methane seep on the Gabon continental margin in the southeast Atlantic hosts not only methane-oxidizing symbionts on its gills but also sulfide-oxidizing symbionts (Duperron et al., 2005). The methane-oxidizing symbionts in this mussel are associated with the basal region of the gill epithelium, whereas the sulfide-oxidizing symbionts are associated with the apical region. Some animals at the Oregon subduction zone have also formed associations with methanotrophs that enable them to feed on methane (Kulm et al., 1986).

22.3.5.2 Anaerobic Methanotrophy

Methane may also be anaerobically oxidized by some microbes. Most of the evidence for this activity derives from the study of marine environments, although in at least two instances, evidence was obtained from two freshwater lakes, one being Lake Mendota, Madison, Wisconsin, United States, and the other being Plußsee, Germany. Initially, sulfate-reducing bacteria were implicated in this oxidation on the assumption that some of them can use methane as electron donor for the reduction of sulfate (e.g., Oremland and Taylor, 1978; Panganiban et al., 1979; Reeburgh, 1980; Martens and Klump, 1984; Iversen and Jørgensen, 1985; Gal'chenko et al., 1986; Henrichs and Reeburgh, 1987; Ward et al., 1987). A presumably pure, sulfate-reducing culture with methane-oxidizing ability was obtained from sediment in Lake Mendota (Panganiban and Hanson, 1976; Panganiban et al., 1979), but it was incompletely characterized. A consortium of a methanogen and a sulfate reducer, which together oxidize methane anaerobically to CO_2, has been detected in the anoxic zone in Plußsee (Eller et al., 2005). A similar relationship probably exists in the upper sediments (0–20 cm) of Lake Biwa, Japan, as described by Koizumi et al. (2003).

As to marine environments, Thomsen et al. (2001), who studied anaerobic sediments from Aarhus Bay, Denmark, suggested that an archaeal sulfate reducer with methane-oxidizing capacity could have been active in their samples but were unable to rule out other possibilities. Hoehler et al. (1994, 1998), Hansen et al. (1998), Niewöhner et al. (1998), Hinrichs et al. (1999), Boetius et al. (2000), and Pancost et al. (2000) suggested that anaerobic methane oxidation may involve a consortium of methanogens and sulfate reducers. Biogeochemical evidence in support of anaerobic methane oxidation was presented by Schouten et al. (2003).

It was an observation by Orphan et al. (2001a,b) that clarified what kind of organisms were responsible for the anaerobic methane oxidation. By means of rRNA gene and lipid analysis of anoxic methane seep sediments from the California continental margin, they implicated a consortium of *Methanosarcinales* and *Desulfosarcinales* in anaerobic methane oxidation. Orphan et al. (2002) recognized involvement of at least two archaeal groups: ANME-1 and ANME-2. Group ANME-1, unlike ANME-2, was often encountered in monospecific aggregates or single filaments without a bacterial partner closely related to *Desulfosarcina*. Teske et al. (2002) found evidence for the occurrence of anaerobic methanotrophic communities that showed affiliation to the ANME-1 and ANME-2 groups in the hydrothermal sediments in the Guaymas Basin. Niemann et al. (2006) identified an additional archaeal group, ANME-3, in the Haakon Mosby Mud Volcano, Barents Sea. In this group, *Methanococcoides* and *Methanolobus* are partnered with *Desulfobulbus*. Nauhaus et al. (2002) demonstrated *in vitro*, anaerobic methane oxidation at the expense of sulfate reduction by a consortium of members of the *Methanosarcinales* and the *Desulfosarcina-Desulfococcus* cluster in marine sediments associated with a gas hydrate deposit at the hydrate ridge. Treude et al. (2005) quantified anaerobic oxidation of methane (AOM) and sulfate reduction occurring along the upwelling region at the Chilean continental margin. They measured AOM rates between 7 and 1124 mmol m^{-2} a^{-1}, individual measurements at various depths showing the highest rates at a depth of 800 m. Their data suggested that in the sulfate–methane transition zone, sulfate reduction was mainly coupled to anaerobic methane oxidation. They associated the high methane turnover rates they found with elevated input of organic matter that was converted to methane in this region. They did not determine phylogenetically which methanogens and sulfate reducers were active in the methane turnover. Kallmeyer and Boetius (2004) reported that AOM in hydrothermal sediments from the Guaymas Basin was observable from 35 to 90°C but accounted for less than 5% of observable sulfate reduction. Sulfate reduction in these sediments was maximal between 60 and 95°C. Treude et al. (2007) detected methane as well as CO_2 consumption in methanotrophic microbial mats from gas seeps of the anoxic Black Sea. The anaerobic methanotrophy involved differentially distributed members of the ANME-1 and ANME-2 groups.

Ecologically interesting is the finding by Lösekann et al. (2007) of the coexistence of aerobic and anaerobic microbial methanotrophs in the Haakon Mosby Mud Volcano—site of a prominent methane seep. The aerobic methanotrophs dominated the active volcano center, whereas the anaer-obes (ANME-3) dominated in the sediment 2–3 cm below a *Beggiatoa* mat that occurred in a circle at the rim of the center of the mud volcano.

In anaerobic methane oxidation by consortia of methanogens and sulfate reducers, the methano-gens are thought to oxidize methane anaerobically by reversing the methanogenic reaction involv-ing CO_2 and H_2 (Reaction 22.1):

$$CH_4 + 2H_2O \rightarrow CO_2 + 4H_2 \tag{22.14}$$

Sulfate reducers would immediately consume the H_2 formed in Reaction 22.14:

$$HSO_4^- + 4H_2 \rightarrow HS^- + 4H_2O \tag{22.15}$$

It is difficult to visualize Reaction 22.14 as a simple reversal of the energy-yielding Reaction 22.1 that would enable hydrogen-oxidizing methanogens to grow. Reaction 22.14 should consume energy if it uses the same enzymes that are used in the formation of methane from CO_2 and H_2. Zehnder and Brock (1979) did observe such a reaction with a methanogenic culture under laboratory condi-tions, but it was very weak. Reaction 22.14 as written may merely represent an overall reaction of a unique enzymatic pathway in those methanogens that have an ability to oxidize methane anaerobi-cally as well as to form it. Although the combination of Reactions 22.14 and 22.15 is inconsistent with an incompatibility of methanogenesis and sulfate reduction (Section 22.3.1), it would explain the absence of very low concentrations of methane in sulfate-reducing zones overlying methano-genic zones. It is noteworthy that the results from a genomic study of methane-oxidizing Archaea from deep-sea sediments were consistent with an anaerobic reverse-methanogenesis hypothesis (Hallam et al., 2004).

Unlike the members of anaerobic methane-oxidizing marine consortia, which appear to be in physical contact, the members of the anaerobic methane-oxidizing consortium in Plußsee are not in direct physical contact, as determined by the staining technique involving catalyst-amplified-reported-deposition, fluorescence *in situ* hybridization (CARD-FISH; Pernthaler et al., 2002).

22.3.6 BIOCHEMISTRY OF METHANE OXIDATION IN AEROBIC METHANOTROPHS

Obligate aerobic methanotrophs can use methane, methanol, and methylamines as energy sources by oxidizing them to CO_2, H_2O, and NH_3, respectively. When methane is the energy source, the following steps are involved in its oxidation:

$$CH_4 \xrightarrow{0.5O_2} CH_3OH \xrightarrow{0.5O_2} HCHO \xrightarrow{0.5O_2} HCOOH \xrightarrow{0.5O_2} CO_2 + H_2O \tag{22.16}$$

The first step in this reaction sequence is catalyzed by a monooxygenase that causes the direct introduction of an atom of molecular oxygen into the methane molecule (Anthony, 1986). This step is generally considered not to yield useful energy to the cell. A report by Sokolov (1986), however, suggests the contrary, at least with *Methylomonas alba* BG8 and *Methylosinus trichospo-rium* OB3b. Because monooxygenase requires pyridine nucleotide ($NADH + H^+$) in its catalytic process to provide electrons for reduction of one of the two oxygen atoms in O_2 to H_2O (the other oxygen atom is introduced into methane to form methanol), a proton motive force is generated in the electron transfer from the reduced pyridine nucleotide, which the cell may be able to couple to ATP synthesis. The enzyme that catalyzes methanol oxidation is methanol dehydrogenase, which in *Methylococcus thermophilus*, as in other methanotrophs, does not use pyridine nucleotide as the cofactor (Anthony, 1986; Sokolov et al., 1981). Instead, the enzyme contains pyrroloquinone (+90 mV) as its prosthetic group, which feeds electrons from methanol into the electron transport chain. The formaldehyde resulting from methanol oxidation is oxidized to formate (Reaction 22.16;

see Roitsch and Stolp, 1985). The oxidation of formaldehyde to formate may involve a pyridine nucleotide–linked dehydrogenase or a pyrroloquinone-linked dehydrogenase (Stanier et al., 1986). Whichever the mechanism of formaldehyde oxidation, the reducing power is fed into the electron transport system for energy generation. The pyridine nucleotide–coupled formate dehydrogenase oxidizes formate to CO_2 and H_2O and feeds electrons mobilized by the oxidation into the electron transport system via the pyridine nucleotide to generate energy. ATP synthesis in methanotrophs appears to be mainly or entirely by chemiosmosis (Anthony, 1986).

It should be noted that the methane monooxygenase of methanotrophs is not a very specific enzyme. It can also catalyze NH_3 oxidation (O'Neill and Wilkinson, 1977). In this instance, the monooxygenase hydroxylates ammonia to NH_2OH. Ammonia-oxidizing autotrophic bacteria can similarly oxidize methane to methanol (Jones and Morita, 1983). However, just as ammonia oxidizers cannot grow on methane as the energy source, methanotrophs cannot grow on ammonia as the energy source. This is because they lack the enzyme sequences for methanol or hydroxylamine oxidation, respectively.

22.3.7 CARBON ASSIMILATION BY AEROBIC METHANOTROPHS

All autotrophically grown methanotrophs of Types I and II assimilate some carbon (up to 30%) in the form of CO_2 (Romanovskaya et al., 1980). The enzyme involved in the fixation appears to be phosphoenolpyruvate (PEP) carboxylase. The mechanism of fixation of the remaining carbon depends on the methanotroph type. Both types assimilate it at the formaldehyde oxidation state of carbon. Members of Type I fix this carbon via an assimilatory, cyclic ribulose monophosphate pathway (Figure 22.6a), whereas members of Type II fix it via a cyclic serine pathway (Figure 22.6b). In the ribulose monophosphate pathway, 3-phosphoglyceraldehyde is the key

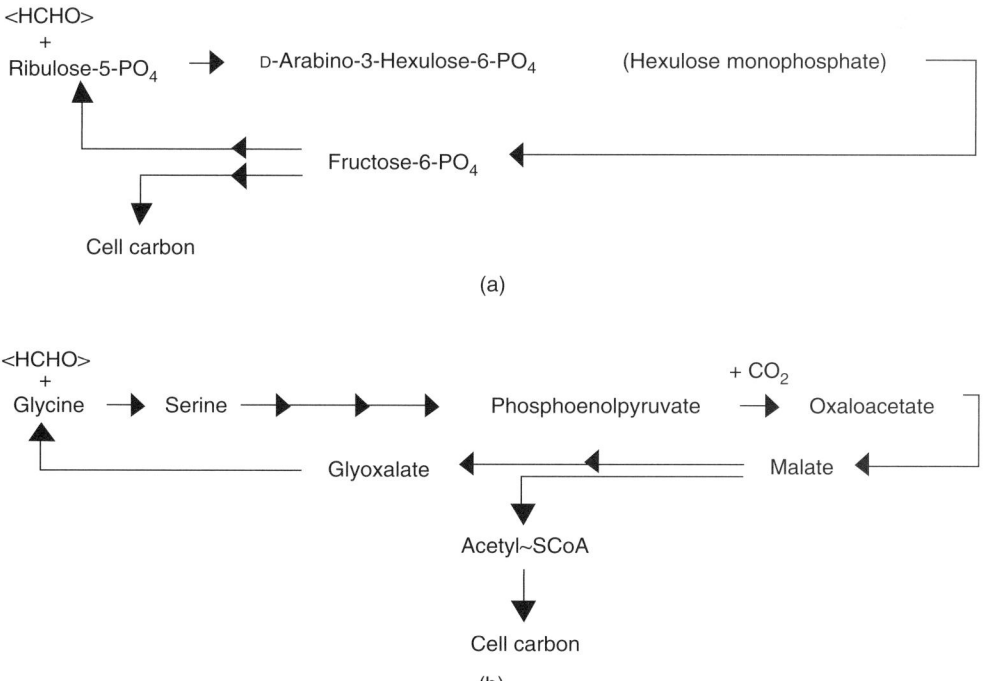

FIGURE 22.6 Alternative pathways of formaldehyde-carbon assimilation in methanotrophs. (a) The hexulose monophosphate (HuMP) pathway used by Type I methanotrophs. (b) The serine pathway used by Type II methanotrophs.

intermediate in carbon assimilation, whereas in the serine pathway it is acetyl~SCoA (Brock and Madigan, 1988; Stanier et al., 1986; Gottschalk, 1986; Atlas, 1997). Reducing power for assimilation derives from methane dissimilation and may require reverse electron transport to generate needed NADPH + H$^+$. Methylotrophs generally use the serine pathway for carbon assimilation from C_1 compounds.

Obligate methanotrophs are positioned somewhere between typical autotrophs and heterotrophs in the carbon-assimilating mechanism (Quayle and Ferenci, 1978).

22.3.8 POSITION OF METHANE IN CARBON CYCLE

For a sufficient pool of biologically available carbon to be maintained in the biosphere, carbon has to be continually recycled from organic to inorganic carbon. This is accomplished both aerobically and anaerobically. Quantitatively, the aerobic process has been thought to make the greater contribution, but the anaerobic process is far from negligible; indeed, with newer knowledge about the extent to which the deep subsurface is inhabited by microbes, it is probably quite significant. Henrichs and Reeburgh (1987) have estimated that the rate of anaerobic mineralization in sediment equals approximately the rate of burial of organic carbon.

Anaerobic mineralization involves fermentation processes coupled with anaerobic forms of respiration, including dissimilatory nitrate reduction (nitrate respiration, denitrification, nitrate ammonification) (e.g., Sørensen, 1987), iron and manganese respiration (Lovley, 1987, 1993; Lovley and Phillips, 1988; Myers and Nealson, 1988; Ehrlich, 1993), sulfate respiration (Skyring, 1987), and methanogenesis (Wolin and Miller, 1987; Young and Frazer, 1987). Each of these forms of respiration is dominant where the respective terminal electron acceptor is dominant and other environmental conditions are optimal. In some instances, more than one form of anaerobic respiration may occur simultaneously in the same general environment, provided there is no competition for the same electron donor or other growth-limiting substance (see Ehrlich, 1993).

In anoxic soils (paddy spoils) or anoxic freshwater or marine sediments where extensive methanogenesis occurs, a small portion of the methane is oxidized to CO_2 without the benefit of oxygen. A larger portion of the methane that is not trapped escapes into an oxidizing environment and is extensively oxidized to CO_2 by aerobic methanotrophs (Higgins et al., 1981). A small amount of methane may be used as energy and carbon source by special marine invertebrates that have formed a symbiotic relationship with specific methanotrophs. Any methane that is not bio-oxidized or otherwise combusted or trapped in natural sedimentary reservoirs escapes this biological attack and enters the atmosphere, where it may be chemically oxidized in the troposphere (Vogels, 1979). The various paths for methanogenesis and methane oxidation are summarized in Figure 22.7.

22.4 PEAT

22.4.1 NATURE OF PEAT

Although peat and coal are two different substances, their modes of origin have included common initial steps. Indeed, the formation of peat may have been an intermediate step in the formation of coal. Peat is a form of organic soil or histosol. It is mostly derived from plant remains that have accumulated in marshes and bogs (Figure 22.8). According to Francis (1954), these remains have come from (1) sphagnum, grasses, and heather, yielding high moor peat; (2) reeds, grasses, sedges, shrubs, and bushes, yielding low moor peat; (3) trees, branches, and debris of large forests in low-lying wet ground, yielding forest peat; or (4) plant debris accumulated in swamps, yielding sedimentary or lake peat. In all these instances, plant growth outstripped the decay of the plant remains, ensuring a continual supply of new raw material for peat formation.

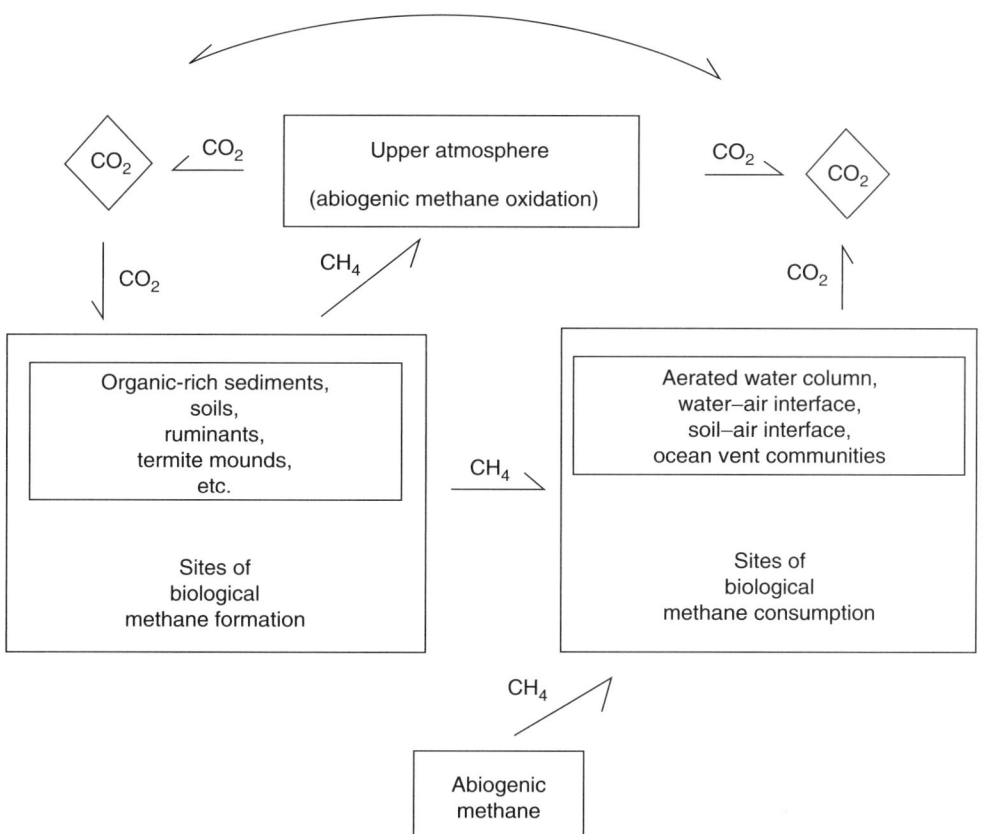

FIGURE 22.7   The methane cycle, emphasizing methanogenesis and methanotrophy.

FIGURE 22.8   Peat. (a) Section of a ditch near Vestburg showing light sphagnum peat over dark peat: (1) living sphagnum, (2) sphagnum peat, (3) shrub remains, (4) sedge rootstock, (5) pond lily rootstocks, (6) laminated peat. (b) View of surface near ditch, showing corresponding vegetative zones: (1) shrub zone, (2) grass zone, (3) sedge zone, (4) pond lily zone. (From Davis CA, Plate XII from Annual Report, Geological Survey of Michigan, 1907. With permission.)

22.4.2 ROLES OF MICROBES IN PEAT FORMATION

Initially, the plant remains may have undergone attack by some of their own enzymes but soon were attacked by fungi, which degraded the relatively stable polymers such as cellulose, hemicellulose, and lignin. Bacteria degraded the more easily oxidizable substances and the breakdown products of fungal activity that were not consumed by the fungi themselves. Fungal activity continued for as long as the organisms had access to air, but as they and their remaining substrate became buried and conditions became anaerobic, bacterial fermentation and anaerobic respiration, including methanogenesis, set in and continued until arrested by accumulation of inhibitory (toxic) wastes, lack of sufficient moisture, depletion of suitable electron acceptors for anaerobic respiration (i.e., nitrate, sulfate, Fe(III), Mn(IV), and carbon dioxide), and other factors (Francis, 1954; Kuznetsov et al., 1963; Rogoff et al., 1962). Combined with these limiting factors, the overwhelming accumulation of organic debris must also be taken into account. It was more than the system could handle before conditions for continued mineralizing activity became unfavorable.

The uppermost aerobic layers of peat may sometimes harbor a viable microflora even today, indicating that peat formation may be occurring at the present time (Kuznetsov et al., 1963). Indeed, viable anaerobic bacteria and actinomycetes have even been detected in the deeper layers of some peats (Rogoff et al., 1962; Zvyagintsev et al., 1993; Metje and Frenzel, 2005). Besides members of the domain Bacteria, methanogenic Archaea are also found associated with peat (e.g., Chan et al., 2002). Differences in methane production in various peats are a reflection of the differences in their origin with respect to source materials and environmental conditions (e.g., Yavitt and Lang, 1990; Brown and Overend, 1993; Galand et al., 2005).

During its formation, peat becomes enriched in lignin, ulmins, and humic acids. The first of these compounds is a relatively stable polymer of woody tissue, and the second and third compounds are complex materials that resulted from the incomplete breakdown of plant matter, including lignin. Peat also contains other compounds that are relatively resistant to microbial attack, such as resins and waxes from cuticles, stems, and spore exines of the peat-forming plants. Compared to the C, H, O, N, and S contents of the original undecomposed plant material, peat is slightly enriched in carbon, nitrogen, and sulfur, but depleted in oxygen and sometimes hydrogen (Francis, 1954). This enrichment in C, N, and S over oxygen may be explained in part by the volatilization of products formed from the less resistant components by microbial attack and in part by the buildup of residues (resins, waxes, lignin) that have a relatively low oxygen content owing to their hydrocarbon-like and aromatic properties. The plant origin of peat is still clearly visible on examination of its structure.

22.5 COAL

22.5.1 NATURE OF COAL

Coal has been defined by Francis (1954) as "a compact, stratified mass of mummified plants, which have been modified chemically in varying degrees, interspersed with smaller amounts of inorganic matter." Peat can be distinguished from coal in chemical terms by its much lower carbon content (51–59% dry weight) and higher hydrogen content (5.6–6.1% dry weight) compared to coal (carbon, 75–95% dry weight; hydrogen 2.0–5.8% dry weight) (Francis, 1954, p. 295). The average carbon content of typical wood has been given as 49.2% (dry weight), and its average hydrogen content as 6.1% (dry weight) (Francis, 1954). Coalification can thus be seen to have resulted in an enrichment in carbon and a slight depletion in hydrogen of the substance that gives rise to coal. Coal is generally found buried below layers of sedimentary strata (*overburden*). Its geologic age is generally advanced. Significant deposits formed in the Upper Paleozoic between 300 and 210 million years ago, in the Mesozoic between 180 and 100 million years ago, and in the Tertiary between 60 and 2.5 million years ago. Peats, however, have generally developed from about 1 million years ago up to the present.

FIGURE 22.9 A section of the Pittsburgh coal seam in the Safety Research coal mine of the U.S. Bureau of Mines in Pittsburgh, Pennsylvania (United States). Although not visible in this black-and-white photograph, extensive brown iron stains were present on portions of the face of the coal seam shown here. The stains are evidence of acid mine-drainage emanating from the fracture at the upper limit of the coal seam. The acid drainage resulted from microbial oxidation of exposed iron pyrite inclusions in the seam. (Courtesy of the U.S. Bureau of Mines.)

Coal is classified by rank. According to the American Society for Testing and Materials (ASTM) classification system, four major classes are recognized (Bureau of Mines, 1965). They are—starting with the least developed coal—lignitic coal, subbituminous coal, bituminous coal, and anthracite coal. Lignitic coal (lignite) resembles peat structurally and has the highest moisture and lowest carbon content (59.2–72.3% dry weight) (Francis, 1954, p. 335) as well as has the lowest heat value of any of the coals. This coal formed during the Tertiary times. Subbituminous coal has a slightly lower moisture content and higher carbon content (72.3–80.4% dry weight). It also has a higher heat content than lignitic coal. Bituminous coal (Figure 22.9) has a carbon content ranging from 80.4 to 90.9% (dry weight) and a high heat value. Subbituminous and bituminous coals are mostly of the Paleozoic and the Mesozoic age. Anthracite coals have very low moisture content, few volatiles, and a high carbon content (92.9–94.7% dry weight). They are of Paleozoic age. Cannel coal is a special type of bituminous coal that was derived mainly from wind-blown spores and pollen rather than from woody plant tissue.

22.5.2 ROLE OF MICROBES IN COAL FORMATION

As mentioned earlier, coal deposits developed at special periods in geologic time. In these periods, the climate, landscape features, and biological activity were favorable. Large amounts of plant debris accumulated in swamps or shallow lakes, because owing to warm, moist climatic conditions, plant growth was very profuse and provided a continual supply of plant debris. At times, the accumulated debris would become buried, covered by clay and sand under water, before a new layer of plant debris accumulated. As more sediment was deposited, subsidence followed and is believed to have played an integral part in the formation of coal deposits (Francis, 1954).

Bacteria and fungi are generally believed to have had an important role in coalification only in the initial stages. Their role was similar to that in peat formation. They destroyed the easily metabolizable substances, such as sugars, amino acids, and volatile acids, in a short time and degraded the more stable polymers, such as cellulose, hemicellulose, lignins, waxes, and resins, more slowly. Many of the latter were degraded only very incompletely before microbial activity ceased for reasons similar to those in peat formation (Section 22.4.2). Hyphal remains, sclerotia, and fungal spores have been identified in some coal remains. Initial microbial attack is believed to have been aerobic and mainly fungal. Later attack, mostly by bacteria, occurred under progressively more anaerobic conditions. Tauson believed, however, that in coal formation, the anaerobic phase was abiological (as cited by Kuznetsov et al., 1963, pp. 79–81). Conversion of the residue from microbial activity (peat?) to coal is presently believed to have been due to physical and chemical agencies of an unidentified nature, but probably involved heat and pressure, which resulted in the loss of volatile components.

22.5.3 COAL AS MICROBIAL SUBSTRATE

Coal is not a very suitable nutrient to support microbial growth. According to early views, this is because coal contains inhibitory substances ("antibiotics") that may suppress it. These antibiotics have been thought to be associated with the waxy and resinous part of coal, extractable with methanol (Rogoff et al., 1962). In the first culture experiments with coal slurries, only marginal bacterial growth was obtainable, the limiting factors being the presence of inhibitory substances and lack of assimilable nutrients (Koburger, 1964). Growth of *Escherichia freundii* and *Pseudomonas rathonis* in such slurries improved when coal was first treated with H_2O_2. In much more recent experiments with run-of-the-mine bituminous coal from Pennsylvania, in which coal particles in a size range 0.5–13 mm were wetted in glass columns with air-saturated distilled water at 10- to 14-day intervals, a bacterial community did develop. It consisted mainly of autotrophic bacteria (iron and sulfur oxidizers) and, to a lesser extent, heterotrophic bacteria. Progressive acid production by the autotrophs was thought to limit the development of heterotrophs. Observed changes in the rate of acetate metabolism may be a reflection of microbial succession among the heterotrophs (Radway et al., 1987, 1989). The bacteria in these experiments lived at the expense of impurities in the coal, such as pyrite.

Two basidiomycete fungi, including *Trametes versicolor* (also known as *Polysporus versicolor* and *Coriolus versicolor*) and *Poria manticola*, have been shown to grow directly on crushed lignite coal as well as in minimal lignite-noble agar medium (Cohen and Gabriele, 1982). With time, the cultures growing on the lignite exuded a black liquid that was a product of lignite attack. Infrared spectra of the exudates gave an indication that conjugated aromatic rings from the lignite had been structurally modified. It must be stressed that the coal in this case was lignite, a low-rank coal, and not bituminous coal. Other fungi, including *Paecilomyces*, *Penicillium* spp., *Phanerochaete chrysosporium*, *Candida* sp., and *Cunninghamella* sp., as well as *Streptomyces* sp. (actinomycete), have also been shown to grow on and degrade lignite and, in some cases, even bituminous coal (see Cohen et al., 1990; Stewart et al., 1990). In the case of *Trametes versicolor*, Cohen et al. (1987) and Pyne et al. (1987) at first attributed lignite solubilization to a protein secreted by the fungus that had polyphenol oxidase activity (syringalda-zinc oxidase; Pyne et al., 1987). Subsequently, Cohen et al. (1990) and Fredrickson et al. (1990) reported that a ligand produced by the fungus and identified as ammonium oxalate (Cohen et al., 1990) was the real solubilizing agent. The oxalate acted as a siderophore by removing iron from the test substrate leonardite (oxidized lignin). This later conclusion seems a little puzzling because it implies that iron(III) plays a central role in holding the leonardite structure together and that no modification of the organic skeleton of leonardite takes place. More recently, Hölker et al. (2002) implicated extracellular esterases and phenolic oxidases (probably laccases) in the solubilization of lignite by *Trichoderma atroviride*.

White-rot fungi, of which *Polysporus versicolor* and *Phanerochaete chrysosporium* are examples, are to produce two kinds of extracellular enzymes: lignin peroxidase and manganese-dependent peroxidase, both of which catalyze lignin attack (e.g., Paszczynski et al., 1986). Indeed, Stewart et al. (1990) reported that *P. chrysosporium* degraded lignite and bituminous coal from Pennsylvania, albeit weakly. Some bacteria also can form such extracellular enzymes that are active on low-rank coals (Crawford and Gupta, 1993). Moreover, dibenzothiophene-degrading aerobic bacteria have been found that are able to break down part of the carbon framework of liquefied bituminous coal (suspension of pulverized coal), and in the process, remove some of the sulfur bound in the framework (Stoner et al., 1990). It would therefore seem reasonable that such types of enzymes play a role in lignite attack. Depending on the coal, it is probably a combination of enzymatic and nonenzymatic processes that leads to the transformation of low-rank coals. The interested reader is referred to Crawford (1993) for further details on these processes.

Lignin and lignin derivatives can also be biologically attacked under anaerobic conditions (Young and Frazer, 1987). It would be of interest to study this action on lignite.

22.5.4 Microbial Desulfurization of Coal

Bituminous coal may contain significant amounts of sulfur in inorganic (pyrite, marcasite, elemental sulfur, sulfate) or organic form. The total sulfur can range from 0.5 to 11% (Finnerty and Robinson, 1986). The proportion of pyritic and organic sulfur in the total sulfur depends on the source of the coal. Iron pyrite or marcasite (FeS_2) originated in some bituminous coal seams during their formation. Some or all of this iron disulfide may well have been formed biogenically, in which case it is a reflection of the anaerobic biogenic phase of coalification, representative of sulfate respiration (Chapters 19 and 20) and also the mineralization of organic sulfur compounds such as sulfur-containing amino acids. A model system illustrating how pyrite may have formed in coal is provided by the sulfur transformations presently occurring in the formation of Everglades peat (Casagrande and Siefert, 1977; Altschuler et al., 1983). The source of the sulfur in this case is organic.

The presence of pyritic sulfur in coal lowers its commercial value because on combustion of such coal, air pollutants such as SO_2 are generated. Pyritic sulfur in coal can be removed in various ways (Bos and Kuenen, 1990; Blazquez et al., 1993). In all cases, the coal must first be pulverized to expose the pyrite. The pyrite particles can be separated by differential flotation, in which pyrite flotation is suppressed. Flotation suppression can be achieved chemically or biologically. In the latter case, *Acidithiobacillus ferrooxidans*, which attaches rapidly and selectively to pyrite particles, is the suppression agent (Pooley and Atkins, 1983; Bagdigian and Myerson, 1986; Townsley et al., 1987). However, pyrite can also be removed by oxidizing it. This can be accomplished by the action of pyrite-oxidizing bacteria such as *Acidithiobacillus ferrooxidans*, *Sulfolobus* spp., *Acidianus* spp., and *Metallosphaera* (Dugan, 1986; Andrews et al., 1988; Merrettig et al., 1989; Larsson et al., 1990; Baldi et al., 1992; Clark et al., 1993).

Bituminous coal may also contain organically bound sulfur. Its presence is undesirable because on combustion, it too contributes to air pollution. Microbiological methods continue to be explored to remove it, but progress has been slow (Dugan, 1986; Finnerty and Robinson, 1986; Crawford and Gupta, 1990; Mormile and Atlas, 1988; Van Afferden et al., 1990; Stoner et al., 1990; Omori et al., 1992; Olson et al., 1993; Izumi et al., 1994). Dibenzothiophene is used as a model structure for organically bound sulfur in coal. Because the sulfur is likely to occur in different types of compounds bound in the structure of bituminous coal, whose degradation may require different enzyme systems, it may be that no one organism in nature can attack the whole range of these substances. Genetic engineering is being applied to find a solution to this problem. In any approach taken to remove organic sulfur from coal, it is important to find a way to remove the sulfur without significant loss of carbon, which would lower the caloric value of the coal.

22.6 PETROLEUM

22.6.1 Nature of Petroleum

Petroleum is a mixture of aromatic and aliphatic hydrocarbons and various heterocyclics including oxygen-, nitrogen-, and sulfur-containing compounds. The aliphatic hydrocarbons include gaseous ones of the paraffinic series such as methane, ethane, propane, and butane, besides longer-chain, nongaseous ones. Some of the heterocyclic compounds, such as porphyrin derivatives, may contain metals such as vanadium and nickel bound in their structure. Petroleum accumulations are found in some folded, porous, sedimentary rock strata, such as limestone or sandstone, or in other fractured rocks such as fissured shale of igneous rock. In petroleum geology, these rock formations are collectively known as reservoir rocks. The age of reservoir rocks may range from Late Cambrian (500 million years) to the Pliocene (1–13 million years). Very extensive petroleum reservoirs are found in rocks of the Tertiary age (70 million years; North, 1985).

Most petroleum derived mainly from planktonic debris that was deposited on the floor of depressions of shallow seas and ultimately buried under heavy layers of sediment, deposited perhaps by turbidity currents. Over geologic time, the trapped organic matter became converted to petroleum and natural gas (chiefly methane; North, 1985).

Many theories have been advanced to explain the origin of petroleum and associated natural gas (see, e.g., Beerstecher, 1954; Robertson, 1966; North, 1985). None of these have been fully accepted. Some theories invoke heat or pressure or both as agents that promoted abiological conversion of planktonic residues to the hydrocarbons and other constituents of petroleum and natural gas. The source of the heat has been viewed as the natural radioactivity of the Earth's interior but was more likely heat diffusing from magma chambers underlying tectonically active areas. Other theories have invoked inorganic catalysis with or without the influence of heat and pressure and with or without prior acid or alkaline hydrolysis. Still other theories have proposed that petroleum represents a residue of naturally occurring hydrocarbons in the planktonic remains after all other components have been biologically destroyed. It has even been proposed that biological agents produced the hydrocarbons by aerobic or anaerobic reduction of fatty acids, proteins or amino acids, carbohydrates, carotenoids, sterols, glycerol, chlorophyll, and lignin–humus complexes, together with appropriate decarboxylations and deaminations. A theory has also been put forward that methane, formed biogenically from planktonic debris, then became polymerized under high temperature and pressure in the possible presence of inorganic catalysts (see, e.g., Mango, 1992).

Alternatively, it was theorized that bacteria modified the planktonic material to substances closely resembling petroleum components, which were then converted to petroleum and natural gas constituents by heat and pressure. Abiotically formed methane could also have been a source and could have been polymerized as proposed for biogenic methane. Chemical reaction of methane with liquid hydrocarbons has been noted in the laboratory at high temperature and pressure (1000 atm, 150–259°C; Gold et al., 1986). At a hydrothermal mound area in the southern rift of the Guaymas Basin in the Gulf of California (Sea of Cortez), organic matter appears to be actively and abiotically transformed into petroliferous substances including gasoline-range aliphatic and aromatic hydrocarbons (Simoneit and Lonsdale, 1982; Didyk and Simoneit, 1989). This site may also be a source of methane used by methanotrophic consortia (see Section 22.3).

22.6.2 Role of Microbes in Petroleum Formation

At present, it is generally thought that bacteria played a role in the initial stages of petroleum formation, but what this role was remains obscure, except in the case of methane formation. ZoBell (1952, 1963) suggested that the planktonic debris was fermented, leading to compounds enriched in hydrogen and depleted in oxygen, sulfur, and phosphorus. Davis (1967, p. 23) visualized microbial processes not unlike those in peat formation, involving aerobic attack of the sedimented planktonic debris followed by anaerobic activity after initial burial. This activity may have included hydrolytic, decarboxylating, deaminating, and sulfate-reducing reactions, resulting in the accumulation of marine humus, that is, stabilized organic matter. Progressively, deeper burial led to the cessation of microbial activity and to compaction, accompanied by evolution of small amounts of hydrocarbon substance plus petroleum precursors. The biotic reactions were followed by an abiotic phase of very long duration during which the microbially produced precursors were transformed under the influence of heat and pressure into the range of hydrocarbons associated with petroleum. This sequence of biotic and abiotic reactions is supported by observations on light hydrocarbon formation in marine sediments (Hunt, 1984; Hunt et al., 1980). Clays could have catalytically promoted further chemical reductions of petroleum precursors.

22.6.3 Role of Microbes in Petroleum Migration in Reservoir Rock

As hydrocarbons accumulated during petroleum formation, the more volatile compounds generated increasing pressure, which helped force the more liquid components through porous rock (sandstone, limestone, fractured rock) to anticlinal folds. The hydrocarbons were trapped in the apex of these folds below a stratum of impervious rock to form a petroleum reservoir. We tap such reservoirs today for our petroleum supply. The migration of petroleum from the source rock to the reservoir rock was probably helped by groundwater movement and by the action of natural detergents such as fatty acid soaps and other surface-active compounds of microbial origin. ZoBell (1952) suggested that bacteria themselves may help to liberate oil from rock surfaces at the site of its formation and thereby promote its migration by dissolving carbonate and sulfate minerals to which oil may adhere and by generating CO_2, whose gas pressure could help force the migration of petroleum. Bacterially produced methane may lower the viscosity of petroleum liquid by dissolving in it and thus help migration (ZoBell, 1952). The important microbial contributions to petroleum formation thus come during the initial action on the source material (planktonic biomass) and the final stages by promoting migration of petroleum from the source rock to the reservoir rock. In addition, sulfate-reducing bacteria may play a role in the sealing of an oil deposit in reservoir rock by deposition of secondary $CaCO_3$ at the interface between the oil-bearing stratum and the stratal waters (Ashirov and Sazanova, 1962; Davis, 1967; Kuznetsov et al., 1963) (see also Chapter 9).

Viable bacteria have been detected in brines associated with petroleum reservoirs to which access was gained by drilling. They were assigned to three specific groups. One included the sulfate reducers *Desulfovibrio desulfuricans* and *Desulfotomaculum nigrificans* (Kuznetsov et al., 1963; Nazina and Rozanova, 1978). The second group included the methanogens *Methanobacterium mazei*, *Sarcina methanica* (now *Methanosarcina methanica*), and *Methanobacterium omelianskii* (now *Methanobacterium* MOH). The third group included the phototroph *Rhodopseudomonas palustris* (Rozanova, 1971), which is capable of anaerobic respiration in the dark (Madigan and Gest, 1978; Yen and Marrs, 1977). The petroleum-associated brines may be connate seawater whose mineral content was somewhat altered through contact with enclosing rock strata (Chapter 5). Such brines may be low in sulfate but high in chlorides and not very conducive to microbial growth. Sulfate-containing groundwaters and alkaline carbonate waters, on mixing with the brines, can provide a milieu suitable for the active sulfate-reducing bacteria (Table 22.2). These waters furnish needed

---

TABLE 22.2
Composition of Petroleum-Associated Brines
(Percentage Equivalents)

Cl^-	7.4–49.90	Ca^{2+}	0.33–11.02
SO_4^{2-}	0.03–10.06	Mg^{2+}	0.04–4.70
CO_3^{2-}	0.03–42.2	K^+ and Na^+	34.28–49.34

Source: After Kuznetsov SI, Ivanov MV, Lyalikova NN in *Introduction to Geological Microbiology*, McGraw-Hill, New York, 1963, p. 17.

---

moisture and, in the case of sulfate-reducing bacteria, the terminal electron acceptor sulfate for their respiration. By the current understanding of methanogenesis (see Section 22.3), the methane bacteria in these brines probably rely mainly on H_2 for energy and CO_2 as terminal electron acceptor and as the source of carbon. The sulfate-reducing bacteria may obtain their energy source directly from petroleum, as has been claimed by some, but may also depend on other bacteria to produce what they need as carbon and energy sources (Ivanov, 1967; Chapters 19 and 20). If the plutonic waters associated with an oil reservoir are hydraulically connected with infiltrating surface waters, it is also possible that carbon and energy sources for the sulfate reducers derive at least in part from products formed by aerobic bacteria in oxidizing strata (Jobson et al., 1979). The observation of Panganiban and Hanson (1976) that at least one sulfate reducer can use methane for energy and acetate for carbon makes a petroleum reservoir a not impossible direct source of an energy substrate for sulfate reducers. This notion is also supported by the more recent discovery of a sulfate reducer that uses saturated hydrocarbon as an energy source (Aeckersberg et al., 1991).

22.6.4 Microbes in Secondary and Tertiary Oil Recovery

When a petroleum reservoir is first tapped for commercial exploitation, the initial oil is recovered by being forced to the surface by gas pressure from the volatile components of the oil and by pumping. This action, however, recovers only part of the total oil in a reservoir. To recover additional oil, a reservoir may be flooded with water by injection to force out additional oil (*secondary oil recovery*). Even secondary oil recovery will yield no more than 30–40% of the oil (North, 1985). To recover even part of the remaining oil, which is more viscous than the previously extracted oil, *tertiary* or *enhanced oil recovery* treatment is necessary. This may involve a thermal method (e.g., steam injection) to reduce viscosity (e.g., North, 1985) or other chemical or physical methods (see, e.g., Orr and Taber, 1984). Alternatively, it may involve biological methods such as generation of surface-active agents of microbial origin to facilitate mobility of the oil, or it may involve the generation of gas pressure by fermentation to force movement of the oil, or it may involve a combination of both processes (McInerney and Westlake, 1990; Tanner et al., 1991; Adkins et al., 1992; Van Hamme et al., 2003; Brown, 2007). Microbially enhanced oil recovery (MEOR) may also involve the promotion of selective plugging by microbes of high-permeability zones in oil reservoirs to increase volumetric sweep efficiency and the microscopic oil displacement efficiency (Raiders et al., 1989; McInerney and Westlake, 1990; Brown, 2007). Volumetric sweep efficiency refers to the ability of injected water to recover oil from less permeable zones.

Water injection into oil reservoirs stimulates microbial activity by sulfate reducers as well as methanogens and fermentative bacteria (Rozanova, 1978; Nazina et al., 1985; Belyaev et al., 1990a,b). Because the injected water carries oxygen, hydrocarbon-oxidizing bacteria have also been detected (Belyaev et al., 1990a,b). Indeed, a succession of organisms may occur, with the aerobic hydrocarbon oxidizers and others producing the substrates (e.g., acetate and higher fatty acids) for fermenting and sulfate-reducing bacteria, and the fermenting bacteria producing hydrogen for use by methanogens and sulfate reducers (Nazina et al., 1985). In the Bondyuzh oil field (former Tatar SSR), maximal numbers of aerobic hydrocarbon-oxidizing bacteria were found at the interface of injection and stratal waters. The destructive effect of these bacteria on petroleum, in one case in which this was studied, appeared to be limited by the salt concentration of the stratal waters (Gorlatov and Belyaev, 1984).

For tertiary oil recovery that involves the use of surface-active agents of microbial origin, xanthan gums of the bacterium *Xanthomonas campestris* or glucan polymers of fungi such as *Sclerotium*, *Stromantinia*, and *Helotium* (Compere and Griffith, 1978) may be generated in separate processes and then injected into an oil well to facilitate oil recovery. As an alternative approach, appropriate organisms may be introduced into the oil well to produce surface-active agents *in situ* (Finnerty et al., 1984). Promotion of tertiary oil recovery may also involve injection of dilute molasses solution into an oil well and subsequent injection of a gas-producing culture such as *Clostridium acetobutylicum*, which ferments the molasses to the solvents acetone, butanol, and ethanol and large amounts of CO_2 and H_2. The solvents help to lower the viscosity of the oil, and the gases provide pressure to move the oil (Yarbrough and Coty, 1983).

22.6.5 REMOVAL OF ORGANIC SULFUR FROM PETROLEUM

As in the case of coal, petroleum that contains significant amounts of organic sulfur may not be usable as a fuel because of air pollution by the volatile sulfur compounds such as SO_2 that result from the combustion of this fuel. The feasibility of removing this sulfur microbiologically has been actively explored (e.g., Foght et al., 1990; Van Hamme et al., 2003). The experimental approach being taken is similar to that in the investigations to remove organic sulfur from coal. The model substance to evaluate microbial desulfurization of petroleum is also dibenzothiophene. The best microbial agents for any industrially applicable process are those that remove sulfur withour significant concomitant oxidation of the carbon to CO_2. *Rhodococcus erythropolis* is an example of such an organism (e.g., Izumi et al., 1994), and *Rhodococcus* sp. strain ECRD-1 (Grossman et al., 2001) is another.

22.6.6 MICROBES IN PETROLEUM DEGRADATION

When natural petroleum reservoirs become industrially exploited, constituents in the oil, whether the oil is still in its reservoir or removed from it, become susceptible to microbial attack. This attack may be aerobic or anaerobic. An extensive literature has accumulated around this subject, largely because such microbial attack can be used in the management of oil pollution. A useful, relatively recent review is one by Van Hamme et al. (2003). Only the major principles of microbial petroleum degradation will be discussed here.

Although hydrocarbon oxidation by microbes was once considered a strictly aerobic process because the initial attack usually involves an oxygenation, clear evidence now exists for anaerobic degradation of some oil constituents as well. Simakova et al. (1968) found that methane and high-paraffinaceous oil were more intensely attacked aerobically than anaerobically. They detected very similar products, including fatty acids of high and low molecular weight, amino acids, alcohols, and aldehydes under either condition. However, hydroxyacids were found only under aerobic conditions. This work showed that in this instance, petroleum degradation is much slower anaerobically than aerobically.

Old claims, to which reference has already been made, that sulfate-reducing bacteria in oil well brines are able to derive energy and carbon from petroleum constituents, especially methane, can be found in the literature (Davis, 1967, p. 243; Davis and Yarbrough, 1966; Panganiban et al., 1976) (see also Section 22.3). It was also suggested that even if a sulfate reducer was not able to attack hydrocarbons themselves, satellite organisms might be able to convert such compounds to products that could be used by sulfate reducers (Dutova, 1962). Kuznetsova and Gorlenko (1965) reported anaerobic attack of hydrocarbons by a strain of *Pseudomonas* in a mineral salts medium with petroleum as the only carbon source. The bacterial population in these experiments increased a millionfold maximally; the redox potential dropped from +40 to –110 mV. Similarly, Kvasnikov et al. (1973) obtained growth of *Clostridium* (*Bacillus*, now *Paenifacillus*) *polymyxa* anaerobically with *n*-alkanes as the sole source of carbon.

Ward and Brock (1978) observed very slow anaerobic conversion of [1-^{14}C]hexadecane added to reducing sediments and bottom water from Lake Mendota, Wisconsin, United States. They reported that 13.7% of the [1-^{14}C]hexadecane in the sediment was converted into $^{14}CO_2$ and ^{14}C-containing cell carbon in 375 h of incubation. Aerobically, the hexadecane was degraded much more rapidly in the same sediment samples. Zehnder and Brock (1979) found that methanogens are able to oxidize small amounts of the methane they formed anaerobically. The methane oxidation mechanism in these organisms seems to differ from the methane-forming mechanism. The slow rate of anaerobic hydrocarbon degradation, when it occurs, helps to explain in part why petroleum has remained preserved over eons of time. However, prolonged periods of an absence of degradative activity of any kind, probably due to a lack of moisture, must have been a more important factor in petroleum preservation. However, a case of *in situ* microbial conversion of petroleum into bitumen has been attributed to Alberta (Canada) oil sands on the basis of laboratory simulation (Rubinstein et al., 1977).

22.6.7 CURRENT STATE OF KNOWLEDGE OF AEROBIC AND ANAEROBIC PETROLEUM DEGRADATION BY MICROBES

A variety of bacteria and fungi are able to metabolize hydrocarbons (Atlas, 1981, 1984, 1988; Atlas and Bartha, 1998; So et al., 2003; Johnson and Hyman, 2006). Some examples are listed in Table 22.3. The mode of attack of hydrocarbons by microorganisms depends on the kind of organism involved and the environmental conditions. Aerobically, alkanes may be attacked monoterminally to form an alcohol by oxygenation (Doelle, 1975; Atlas, 1981; Gottschalk, 1986), which may subsequently be oxidized to a corresponding carboxylic acid (Reactions 22.16):

$$RCH_2CH_3 \xrightarrow{+0.5O_2} RCH_2CH_2OH \xrightarrow{-2H} RCH_2CHO \xrightarrow{+H2O,-2H} RCH_2COOH \quad (22.17)$$

The carboxylic acid may be oxidized to acetate, which may then be oxidized to CO_2 and H_2O.

Alkanes may also be monoterminally attacked to form a ketone (Fredricks, 1967) or a hydroperoxide (Stewart et al., 1959). They may also be attacked diterminally (Doelle, 1975). For instance, *Pseudomonas aeruginosa* can attack 2-methylhexane at either end of the carbon chain, forming a

---

TABLE 22.3
Examples of Microorganisms Capable of Aerobic Hydrocarbon Metabolism

Organism	Substrate	Mode of Attack	References[a]
Pseudomonas oleovorans	Octane	Desaturation	Abbott and Hou, 1973
Ps. putida GPo-1	Propane, butane	Oxidation	Johnson and Hyman, 2006
Ps. Fluorescens and *Ps. aeruginosa*	Aromatic hydrocarbons	Oxidation	Van der Linden Thijsse, 1965
Azoarcus evansii	Benzoate	Oxidation	Gescher et al., 2002
Nocardia salmonicolor	Hexadecane	Desaturation	Abbott and Casida, 1968
Yeasts		NS	Ahearn et al., 1971
Trichosporon	*n*-Paraffins	NS	Barna et al., 1970
Arthrobacter	*n*-Alkane	Oxidation	Klein et al., 1968
	Aromatics	Oxidation	Stevenson, 1967
Mycobacterium	Butane	NS	Nette et al., 1965
		Oxidation	Phillips and Perry, 1974
Brevibacterium eythrogenes	Alkane	Oxidation	Pirnik et al., 1974
Nocardia	Mono- and dicyclic hydrocarbons	Oxidation	Raymond et al., 1967
Cladosporium	*n*-Alkane	NS	Teh and Lee, 1973
		Oxidation	Walker and Cooney, 1973
Graphium	Ethane	Oxidation	Volesky and Zajic, 1970

[a] For more recent reviews, see Assinder and Williams (1990), Cerniglia (1984), and Atlas (1984).
Note: NS = not specified.

mixture of 5-methylhexanoic acid and 2-methylhexanoic acid (Foster, 1962). Furthermore, alkanes may be desaturated terminally or subterminally, forming alkenes (Chouteau et al., 1962; Abbott and Casida, 1968). Subterminal desaturation may proceed as follows:

$$\text{Hexadecane} \rightarrow \begin{matrix} \text{8-hexadecene} \\ \text{7-hexadecene} \\ \text{6-hexadecene} \end{matrix} \qquad (22.18)$$

Alkenes may be attacked by forming epoxides, which may then be further metabolized (Abbott and Hou, 1973). Diols may be formed in the process. In all the foregoing processes, atmospheric oxygen acts as a terminal electron acceptor and as the source of oxygen in oxygenation.

Alkane hydroxylases, which fall into different classes in terms of composition and cofactor requirements, location in the cell, and substrate range, play an important role in microbial degradation of petroleum hydrocarbons, chlorinated hydrocarbons, and other related compounds. Their occurrence in different microbes and their mode of action have been recently reviewed by van Beilen and Funhoff (2007).

Until about 1990, anaerobic attack of alkanes and alkenes was believed possible only if these compounds carried one or more substituents, in particular, halogens. For instance, dichloromethane (CH_2Cl_2) was shown to be degraded anaerobically. The initial attack in this case did not involve an oxygenation but instead a dehalogenation by a consortium of two different bacterial strains (Braus-Stromeyer et al., 1993):

$$CH_2Cl_2 + H_2O \rightarrow (HCHO) + 2HCl \qquad (22.19)$$

This reaction was followed by an oxidation of the formaldehyde-like intermediate (HCHO) to formic acid. The formic acid was subsequently converted to acetate in an acetogenic reaction (Braus-Stromeyer et al., 1993).

In the case of tetra- and trichloroethylene, anaerobic dechlorination by a reductive process has been observed. In this process, the chlorinated hydrocarbon serves as a terminal electron acceptor. Complete bacterial dechlorination of tetra- and trichloroethylene was observed by Freedman and Gossett (1989), Ensley (1991), and Maymó-Gatell et al. (1999). De Bruin et al. (1992) observed the formation of ethane by reductive transformation of tetrachloroethene.

A clear demonstration that a saturated, unsubstituted alkane can be mineralized by anaerobic bacteria was presented by Aeckersberg et al. (1991) (see also So et al., 2003). These investigators isolated a sulfate-reducing bacterium, HxD3, from the precipitate of an oil–water separator in an oil field near Hamburg, Germany. This organism mineralized hexadecane, using sulfate as oxidant (terminal electron acceptor). The nature of the initial attack of the hexadecane was not elucidated, but hydroxylation was probably involved (see review by van Beilen and Funhoff, 2007). The overall reaction of hexadecane mineralization was consistent with the following stoichiometry (Aeckersberg et al., 1991):

$$C_{16}H_{34} + 12.25SO_4^{2-} + 1.5H^+ \rightarrow 16HCO_3^- + H_2O \qquad (22.20)$$

Ehrenreich et al. (2000) demonstrated that three distinct types of denitrifying bacteria were able to oxidize alkanes anaerobically, and Kropp et al. (2000) found that a sulfate-reducing enrichment culture was able to mineralize n-dodecane.

A wide range of aromatic compounds can be aerobically and anaerobically degraded by microbes (Tables 22.3 and 22.4). *Aerobic scission* of the ring structure of the aromatic compounds involves oxygenation either between adjacent oxygenated carbon atoms (*ortho*-fission) or adjacent to one of them (*meta*-fission; Dagley, 1975). In the case of benzene, aerobic degradation of the ring structure involves initial hydroxylation catalyzed by a mixed function or monooxygenase to form catechol

TABLE 22.4
Bacteria Capable of Anaerobic Metabolism of Aromatic and Heterocyclic Hydrocarbons

Organism	Substrate	Mode of Attack	Reference[a]
Rhodopseudomonas palustris	Benzoate, hydroxybenzoate	Reductive ring cleavage	Dutton and Evans, 1969
Desulfobacterium phenolicum	Phenol and derivatives	Anaerobic degradation	Bak and Widdel, 1986a
Desulfobacterium indolicum	Indolic compounds	Anaerobic degradation	Bak and Widdel, 1986b
Desulfobacterium catecholicum	Catechol	Anaerobic degradation	Szewzyk and Pfennig, 1987
Desulfococcus niacini	Nicotinic acid	Anaerobic degradation	Imhoff-Stuckle and Pfennig, 1983

[a] For reviews, see Evans and Fuchs (1988), Reineke and Knackmuss (1988), and Higson (1992).

followed by the action of dioxygenase to cleave the catechol ring to form *cis,cis*-muconate by *ortho*-fission. This product can then be degraded enzymatically in several steps to acetate (Doelle, 1975). The acetate is then oxidized to CO_2 and H_2O. Napthalene, anthracene, and phenanthrene and their derivatives can be degraded by a similar mechanism by attacking each ring in succession (Doelle, 1975). In some instances, benzene derivatives are attacked by *meta*-fission instead of *ortho*-fission as in the previous examples (Doelle, 1975).

Anaerobic scission of an aromatic ring structure involves ring saturation, hydration, and dehydrogenation (Evans and Fuchs, 1988; Colberg, 1990; Grbic-Galic, 1990). Some aromatic hydrocarbons such as benzoate can also be biodegraded anaerobically by photometabolism of certain *Rhodospirillaceae* (purple nonsulfur bacteria; Table 22.4). Ring cleavage is by hydration of pimelate (Dutton and Evans, 1969). Biodegradation of benzene can result in significant carbon and hydrogen isotopic fractionation with enrichment factors ranging from –1.9 to –3.6‰ for carbon and –29 to –79‰ for hydrogen (Mancini et al., 2003).

Chlorinated aromatics have been shown to be completely degradable anaerobically by a consortium of bacteria (e.g., Sharak Genthner et al., 1989; Colberg, 1990; Grbic-Galic, 1990). In one kind of consortium, *Desulfomonile tiedje* DCB-1 was found to act on a chlorinated hydrocarbon by using it as a terminal electron acceptor in a respiratory process that includes dechlorination (Shelton and Tiedje, 1984; Dolfing, 1990; Mohn and Tiedje, 1990, 1991). Subsequent mineralization of the dechlorinated aromatic product depends on other anaerobic organisms in the consortium.

The ability of an organism to attack hydrocarbons aerobically does not necessarily mean that it can use such compounds as the sole source of carbon and energy. Many cases are known in which hydrocarbons are oxidized in a process known a *cooxidation*, wherein another compound, which may be quite unrelated, is the carbon and energy source but which somehow permits the simultaneous oxidation of the hydrocarbon. Examples are the oxidation of ethane to acetic acid, the oxidation of propane to propionic acid and acetone, and the oxidation of butane to butanoic acid and methyl ethyl ketone by *Pseudomonas methanica* growing on methane, as first shown by Leadbetter and Foster (1959). Methane is the only hydrocarbon on which this organism can grow. Another example is the oxidation of alkylbenzenes by a strain of *Micrococcus cerificans* growing on *n*-paraffins (Donos and Frankenfeld, 1968). Other examples have also been summarized by Horvath (1972).

Chain length and branching of aliphatic hydrocarbons can affect microbial attack. *In situ* observations revealed rapid microbial degradation of pristane and phytane (Atlas and Cerniglia, 1995). Some bacteria that attack alkanes of chain lengths C_8–C_{20} may not be able to attack alkanes of chain lengths C_1–C_6, whereas others cannot grow on alkanes of chain lengths greater than C_{10} (Johnson, 1964). Fungi that can grow on alkanes of chain lengths up to C_{34} are known. It has also been noted that certain placements of methyl or propyl groups in the alkane carbon chain lessen or prevent utilization of the compounds (McKenna and Kallio, 1964).

22.6.8 USE OF MICROBES IN PROSPECTING FOR PETROLEUM

Prospecting for petroleum through the detection of hydrocarbon-utilizing microorganisms has been explored (e.g., Brown, 2007). The scientific basis for this method is the detection of microseepage of petroleum or some of its constituents, especially the more volatile components, in the ground over-lying a deposit using the presence of any hydrocarbon-utilizing microbes as indicators. One method involves enriching soil, sediment, and water samples from a suspected seepage area for microbes that can metabolize gaseous hydrocarbons and demonstrating hydrocarbon consumption (Davis, 1967). An enrichment medium consisting of a mineral salts solution with added volatile hydro-carbon (ethane, propane, butane, isobutene, *but not methane*) is satisfactory. Methane-oxidizing bacteria are poor indicators in petroleum prospecting because methane can occur in the absence of petroleum deposits and, moreover, some methane-oxidizing bacteria are unable to oxidize other aliphatic hydrocarbons. Bacteria that can oxidize ethane and longer-chain hydrocarbons, however, provide presumptive evidence for a hydrocarbon seep and an underlying petroleum reservoir (Davis, 1967; Brown, 2007). It is assumed that ethane and propane formed in anaerobic fermentation are produced in quantities too small to select for a hydrocarbon-utilizing microflora. Likely organisms active in soil enrichments from hydrocarbon seeps may include *Mycobacterium paraffinicum* and *Streptomyces* spp.

Hydrocarbon enrichment cultures may be prepared using ^{14}C-labeled hydrocarbon. This allows the easy identification of the activity of hydrocarbon-oxidizing bacteria in water and sedi-ments (Caparello and LaRock, 1975). With this method, the hydrocarbon-oxidizing potential of a sample can be correlated with the hydrocarbon burden of the environment from which the sample came.

22.6.9 MICROBES AND SHALE OIL

North (1985) described *oil shales* as bituminous, nonmarine limestones, or marlstones contain-ing kerogens. *Tar sands* are consolidated or unconsolidated rock coated with bituminous material (North, 1985). *Bitumens* are solid hydrocarbons that are soluble in organic solvents and fusible below ~150°C. Kerogens are insoluble in organic solvents. They are intermediate products in the diagenetic transformation of organic matter in sediments and are considered a precursor in petro-leum formation. As in the formation of peat, coal, and petroleum, fungi and bacteria probably played a role in the early stages of transformation of the source material (mostly terrestrial biomass). Later stages involved physicochemical processes. However, in the case of oil sand bitumens of Alberta, the origin appears to be a partial biodegradation of petroleum, leaving behind the high-viscosity compounds (Rubinshtein et al., 1977).

Bitumen and kerogen can be converted to a petroleum-like substance by heat treatment (e.g., retorting; North, 1985). Their separation from host rock, especially if it is limestone, can be facili-tated if the limestone is dissolved. This can be achieved, at least on a laboratory scale, by acid-forming microbes (e.g., sulfuric acid from S^0 oxidation by *Acidithiobacillus thiooxdans*; Meyer and Yen, 1976).

Although raw shale is considered relatively resistant to microbial attack, some reports indicate otherwise. Both aerobic and anaerobic attacks have been observed, but the anaerobic attack pro-ceeded at a much slower rate (Roffey and Norqvist, 1991; Wolf and Bachofen, 1991; Ait-Langomazino et al., 1991). Hydrogenated shale S metabolized by some gram-negative organisms (*Alcaligenes* and *Pseudomonas* or *Pseudomonas*-like organisms; Westlake et al., 1976).

22.7 SUMMARY

Not all carbon in the biosphere is continually being recycled. Some is trapped as organic carbon in special sedimentary formations, where it is inaccessible to microbial attack. The forms in which the trapped carbon appears are methane, peat, coal, petroleum, bitumen, and kerogen.

Most methane in sedimentary formations is of biogenic origin. It may occur by itself or in association with peat, coal, or petroleum deposits. Its biogenic formation is a strictly anaerobic process involving methanogenic bacteria that may reduce CO_2 with H_2 or a few other hydrogen donors, or it may involve disproportionation (fermentation) of formate, methanol, methylamines, secondary alcohols, or acetate in the absence of H_2. The autotrophic methanogens use a mechanism for CO_2 assimilation that involves reduction of one CO_2 to methyl carbon and a second CO_2 to formyl carbon and then coupling the two to form acetate. The acetate is subsequently carboxylated to form pyruvate—the key intermediate for forming building blocks for all the cell constituents. Acetotrophic methanogens obtain their carbon from part of the acetate they consume. Methanogenesis occurs mostly mesophilically and thermophilically.

Methane may be oxidized and assimilated by a special group of microorganisms called methanotrophs. This process is generally aerobic, although clear evidence of anaerobic methane oxidation now exists. At least one of the organisms responsible for anaerobic methane oxidation by itself appears to be a sulfate reducer. Limited anaerobic methane oxidation by a methanogen has also been observed. In some anaerobic niches in the marine environment, significant anaerobic methane oxidation has been shown to occur that is caused by consortia involving a methanogen (e.g., *Methanosarcina*) and a sulfate reducer (e.g., *Desulfosarcina*). Carbon assimilation by aerobic methanotrophs may be via a hexulose monophosphate pathway or a serine pathway, in each case involving integration of the carbon at the oxidation level of formaldehyde that is produced as an intermediate in methane oxidation. In addition, both types of methanotrophs derive some of their carbon from CO_2.

Peat is the result of partial biodegradation of plant remains accumulating in marshes and bogs. Aerobic attack by enzymes in the plant debris and by fungi and some bacteria initiates the process. It is followed by anaerobic attack by bacteria during burial resulting from continual sedimentation until inhibited by accumulating wastes, lack of sufficient moisture, and so on. A viable microbial flora can usually be detected in peat even though the peat may have formed over a geologically extended period. Coal is thought to have formed like peat, except that in the advanced stages, as a result of deeper burial, it was subject to physical and chemical influences that converted the peat to coal. Different ranks of coal exist, which differ from one another largely in carbon and moisture content and in heat value, and are a reflection of the maturity of the coal. It is questionable whether coal itself harbors an indigenous microbial flora.

Bituminous coal has pyrite or marcasite associated with it as inclusions. Upon exposure to air and moisture during mining, this iron disulfide becomes subject to attack by acidophilic, iron-oxidizing thiobacilli, sulfur-oxidizing bacteria, and Archaea, and is the source of acid mine drainage.

Whereas peat and coal are derived from terrestrial plant matter, petroleum and associated natural gas (mostly methane) are derived from phytoplankton remains that accumulated in depressions of shallow seas and became gradually buried in sediment. Microbial attack altered these remains biochemically until complete burial by accumulating sediment stopped the organisms. In tectonically active areas, the buried and biochemically altered organic matter became subject to further alteration by heat from magmatic activity and pressure from the weight of overlying sediment. The chemical alterations may have been catalyzed by clay minerals. The final products of these transformations were petroleum hydrocarbons. At least some of the natural gas associated with petroleum may represent biogenic methane formed in the initial stages of plankton debris fermentation. At its site of formation, petroleum is highly dispersed. As a result of gas (natural gas, CO_2) and hydrostatic pressure as well as lubrication of the surfaces of the sediment matrix by bacteria and some of their metabolic products, matured petroleum may be forced to migrate through pervious sediment strata until it is collected in a trap such as the apex of an anticlinal fold. It is such petroleum-filled traps that constitute commercially exploitable petroleum reservoirs. Sulfate-reducing bacteria may assist in trapping petroleum in reservoirs by laying down impervious calcite layers at the interface between the trapped petroleum and groundwater. This calcite may, however, also interfere with petroleum recovery.

At least some petroleum hydrocarbons can be oxidized in air by certain bacteria and fungi. Anaerobic bacterial attack of unsubstituted alkanes as well as chlorinated alkanes and of a wide range of aromatic hydrocarbons has also been demonstrated. Hydrocarbon-utilizing microorganisms may be used as indicators in prospecting for petroleum.

REFERENCES

Abbott BJ, Casida LE Jr. 1968. Oxidation of alkanes to internal monoalkenes by a *Nocardia*. *J Bacteriol* 96:925–930.

Abbott BJ, Hou CT. 1973. Oxidation of 1-alkanes to 1,2epoxy-alkanes by *Pseudomonas oleovorans*. *Appl Microbiol* 26:86–91.

Abrams JW, Nedwell DB. 1978. Inhibition of methanogenesis by sulfate reducing bacteria competing for transferred hydrogen. *Arch Microbiol* 117:89–92.

Adkins JP, Tanner RS, Udegbunam EO, McInerney MJ, Knapp RM. 1992. Microbially enhanced oil recovery from unconsolidated limestone cores. *Geomicrobiol J* 10:77–86.

Aeckersberg F, Bak F, Widdel F. 1991. Anaerobic oxidation of saturated hydrocarbons to CO_2 by a new type of sulfate-reducing bacterium. *Arch Microbiol* 156:5–14.

Ahearn DG, Meyers SP, Standard PG. 1971. The role of yeasts in the decomposition of oils in marine environments. *Dev Ind Microbiol* 12:126–134.

Ait-Langomazino N, Sellier R, Jouquet G, Trescinski M. 1991. Microbial degradation of bitumen. *Experientia* 47:533–539.

Alexander M. 1977. *Introduction to Soil Microbiology*. 2nd ed. New York: Wiley.

Altschuler ZS, Schnepfe MM, Silber CC, Simon FO. 1983. Sulfur diagnosis in Everglades peat and origin of pyrite in coal. *Science* 221:221–227.

Andrews G, Darroch M, Hansson T. 1988. Bacterial removal of pyrite from concentrated coal slurries. *Biotech Bioeng* 32:813–820.

Anthony C. 1986. Bacterial oxidation of methane and methanol. *Adv Microb Physiol* 27:113–210.

Ashirov KB, Sazanova IV. 1962. Biogenic sealing of oil deposits in carbonate reservoirs. *Mikrobiologiya* 31:680–683 (Engl. transl., pp. 555–557).

Assinder SJ, Williams PA. 1990. The TOL plasmids: Determinants of the catabolism of toluene and the xylenes. *Adv Microb Physiol* 31:1–69.

Atlas RM. 1981. Microbial degradation of petroleum hydrocarbons: An environmental perspective. *Microbiol Rev* 45:180–209.

Atlas RM. 1984. *Petroleum Microbiology*. New York: McGraw-Hill.

Atlas RM. 1988. *Microbiology. Fundamentals and Applications*. 2nd ed. New York: Macmillan.

Atlas RM. 1997. *Principles of Microbiology*. 2nd ed. Boston, MA: WCB McGraw-Hill.

Atlas RM, Bartha R. 1998. *Microbial Ecology. Fundamentals and Applications*. 4th ed. Menlo Park, CA: Benjamin Cummings.

Atlas RM, Cerniglia CE. 1995. Bioremediation of petroleum pollutants. *BioScience* 45:332–338.

Bagdigian RM, Myerson AS. 1986. The adsorption of *Thiobacillus ferrooxidans* on coal surfaces. *Biotech Bioeng* 28:467–479.

Bak F, Widdel F. 1986a. Anaerobic degradation of phenol and phenol derivatives by *Desulfobacterium phenolicum* sp. nov. *Arch Microbiol* 146:177–180.

Bak F, Widdel F. 1986b. Anaerobic degradation of indolic compounds by sulfate-reducing enrichment cultures, and description of *Desulfobacterium indolicum* gen. nov. sp. nov. *Arch Microbiol* 146:170–176.

Baldi F, Clark T, Pollack SS, Olson GJ. 1992. Leaching pyrites of various reactivities by *Thiobacillus ferrooxidans*. *Appl Environ Microbiol* 58:1853–1856.

Barna PK, Bhagat SD, Pillai KR, Singh HD, Branah JN, Iyengar MS. 1970. Comparative utilization of paraffins by a *Trichosporon* species. *Appl Microbiol* 20:657–661.

Beerstecher E. 1954. *Petroleum Microbiology*. Houston, TX: Elsevier.

Belyaev SS, Rozanova EP, Borzenkov IA, Charakhch'yan IA, Miller YuM, Sokolov My, Ivanov MV. 1990a. Characteristics of microbiological processes in a water-flooded oilfield in the middle Ob' region. *Mikrobiologiya* 59:1075–1081 (Engl. transl., pp. 754–759).

Belyaev SS, Borzenkov IA, Milekhina EI, Charakhch'yan IA, Ivanov MV. 1990b. Development of microbiological processes in reservoirs of the Romashkino oilfield. *Mikrobiologiya* 59:1118–1126 (Engl. transl., pp. 786–792).

Bhatnagar L, Jain MK, Zeikus JG. 1991. Methanogenic bacteria. In: Shively JM, Barton LL, eds. *Variations in Autotrophic Life*. London, U.K.: Academic Press, pp. 251–270.

Blaut M, Gottschalk G. 1984. Protonmotive force-driven synthesis of ATP during methane formation from molecular hydrogen and formaldehyde or carbon dioxide in *Methanosarcina barkeri*. *FEMS Microbiol Lett* 24:103–107.

Blaut M, Peinemann S, Deppenmeier U, Gottschalk G. 1990. Energy transduction in vesicles of methanogenic strain Gö 1. *FEMS Microbiol Rev* 87:367–372.

Blaut M, Müller V, Gottschalk G. 1992. Energetics of methanogenesis studied in vesicular systems. *J Bioenerg Biomembr* 24:529–546.

Blazquez ML, Ballester A, Gonzalez F, Mier JL. 1993. Coal biodesulfurization. *Biorecovery* 2:155–157.

Bochem HP, Schoberth SM, Sprey B, Wengler P. 1982. Thermophilic biomethanation of acetic acid: Morphology and ultrastructure of granular consortium. *Can J Microbiol* 28:500–510.

Bock A-K, Prieger-Kraft A, Schönheit P. 1994. Pyruvate: A novel substrate for growth and methane formation in *Methanosarcina barkeri*. *Arch Microbiol* 161:33–46.

Boetius A, Ravenschlag K, Schubert CJ, Rickert D, Widdel F, Gieseke A, Amann R, Jørgensen BB, Witte U, Pfannkuche O. 2000. A marine microbial consortium apparently mediating anaerobic oxidation of methane. *Nature* (London) 407:623–626.

Boone DR, Whitman WB, Rouvière P. 1993. Diversity and taxonomy of methanogens. In: Ferry JG, ed. *Methanogenesis*. New York: Chapman & Hall, pp. 35–89.

Bos P, Kuenen JG. 1990. Microbial treatment of coal. In: Ehrlich HL, Brierley CL, eds. *Microbial Mineral Recovery*. New York: McGraw-Hill, pp. 343–377.

Bowen HJM. 1979. *Environmental Chemistry of the Elements*. London, U.K.: Academic Press.

Braus-Stromeyer SA, Hermann R, Cook AM, Leisinger T. 1993. Dichloromethane as the sole carbon source for an acetogenic mixed culture and isolation of a fermentative dichloromethane-degrading bacterium. *Appl Environ Microbiol* 59:3790–3797.

Breznak JA. 1982. Intestinal microbiota of termites and other xylophagus insects. *Annu Rev Microbiol* 36:323–343.

Brock TD, Madigan MT. 1988. *Biology of Microorganisms*. 5th ed. Englewood Cliffs, NJ: Prentice-Hall.

Brooks JM, Kennicutt MC II, Fisher CR, Macko SA, Cole K, Childress JJ, Bridigare RR, Vetter RD. 1987. Deep-sea hydrocarbon seep communities: Evidence for energy and nutritional carbon sources. *Science* 238:1138–1142.

Brown LR. 2007. The relationship of microbiology to the petroleum industry. *SIM News* 57:180–190.

Brown DA, Overend RP. 1993. Methane metabolism in raised bogs of northern wetlands. *Geomicrobiol J* 11:35–48.

Bryant MP, Campbell LL, Reddy CA, Crabill MR. 1977. Growth of *Desulfovibrio* in lactate or methanol media low in sulfate in association with H_2-utilizing methanogenic bacteria. *Appl Environ Microbiol* 33:1162–1169.

Bureau of Mines. 1965. Mineral Facts and Problems. Bulletin 630. Washington, DC: Bur Mines, US Department of the Interior.

Caparello DM, LaRock PA. 1975. A radioisotope assay for quantification of hydrocarbon biodegradation potential in environmental samples. *Microb Ecol* 2:28–42.

Casagrande D, Siefert K. 1977. Origins and sulfur in coal: Importance of the ester sulfate content in peat. *Science* 195:675–676.

Cavanaugh CM, Levering PR, Maki JS, Mitchell R, Lidstrom ME. 1987. Symbiosis of methylotrophic bacteria and deep-sea mussels. *Nature* (London) 325:346–348.

Cerniglia CE. 1984. Microbial metabolism of polycyclic aromatic hydrocarbons. *Adv Appl Microbiol* 30:31–71.

Chan OC, Wolf M, Hepperle D, Casper P. 2002. Methanogenic archaeal community in the sediment of an artificially partitioned acidic bog lake. *FEMS Microbiol Ecol* 42:119–129.

Chen W, Konisky J. 1993. Characterization of a membrane-associated ATPase from *Methanococcus voltae*, a methanogenic member of the Archaea. *J Bacteriol* 175:5677–5682.

Childress JJ, Fisher CR, Brooks JM, Kennicutt MC II, Bidigare R, Anderson AE. 1986. A methanotrophic marine molluscan (Bivalvia, Mytilidae) symbiosis: Mussel fueled by gas. *Science* 233:1306–1308.

Chouteau J, Azoulay E, Senez JC. 1962. Dégradation bactérienne des hydrocarbons paraffiniques. IV. Identification par spectrophotométrie infrarouge du hept-1-ene produit à partir du *n*-heptane par des suspensions nonproliférantes de *Pseudomonas aeruginosa*. *Bull Soc Chim Biol* 44:1670–1672.

Clark TR, Baldi F, Olson GJ. 1993. Coal depyritization by the thermophilic archeon *Metallosphaera sedula*. *Appl Environ Microbiol* 59:2375–2379.

Cohen MS, Gabriele PD. 1982. Degradation of coal by the fungi *Polysporus versicolor* and *Poria monticols*. *Appl Environ Microbiol* 44:23–27.

Cohen MS, Bowers WC, Aronson H, Gray ET Jr. 1987. Cell-free solubilization of coal by *Polysporus versicolor*. *Appl Environ Microbiol* 53:2840–2843.

Cohen MS, Feldman KA, Brown CS, Gray ET Jr. 1990. Isolation and identification of the coal-solubilizing agent produced by *Trametes versicolor*. *Appl Environ Microbiol* 56:3285–3291.

Colberg PJS. 1990. Role of sulfate in microbial transformations of environmental contaminants: Chlorinated aromatic compounds. *Geomicrobiol J* 8:147–165.

Compere Al, Griffith WL. 1978. Production of high viscosity glucans from hydrolyzed cellulosics. *Dev Ind Microbiol* 19:601–607.

Crawford DL, ed. 1993. *Microbial Transformations of Low Rank Coals*. Boca Raton, FL: CRC Press.

Crawford DL, Gupta RK. 1990. Oxidation of dibenzothiophene by *Cunninghamella elegans*. *Curr Microbiol* 21:229–231.

Crawford DL, Gupta RK. 1993. Microbial depolymerization of coal. In: Crawford DL, ed. *Microbial Transformations of Low Rank Coals*. Boca Raton, FL: CRC Press, pp. 65–92.

Dagley S. 1975. Microbial degradation of organic compounds in the biosphere. *Sci Am* 63:681–689.

Davies SL, Whittenbury R. 1970. Fine structure of methane and other hydrocarbon-utilizing bacteria. *J Gen Microbiol* 61:227–232.

Davis CA. 1907. *Peat: Essays on the Origin and Distribution in Michigan*. Lansing, MI: Wynkoop Hallenbeck Crawford, State Printers.

Davis JB. 1967. *Petroleum Microbiology*. Amsterdam: Elsevier.

Davis JB, Yarbrough HF. 1966. Anaerobic oxidation of hydrocarbons by *Desulfovibrio desulfuricans*. *Chem Geol* 1:137–144.

De Bruin WP, Kotterman MJ, Posthumus MA, Schraa G, Zehnder AJ. 1992. Complete biological reductive transformation of tetrachloroethene to ethane. *Appl Environ Microbiol* 58:1996–2000.

Deuser WG, Degens ET, Harvey GR. 1973. Methane in Lake Kivu: New data bearing on origin. *Science* 181:51–54.

Didyk BM, Simoneit BRT. 1989. Hydrothermal oil of Guaymas Basin and implications for petroleum formation mechanisms. *Nature* (London) 342:65–69.

Doddema HJ, Hutten TJ, van der Drift C, Vogels GD. 1978. ATP hydrolysis and synthesis by the membrane-bound ATP synthetase complexes of *Methanobacterium thermoautotrophicum*. *J Bacteriol* 136:19–23.

Doddema HJ, van der Drift C, Vogels GD, Veenhuis M. 1979. Chemiosmotic coupling in *Methanbacterium thermoautotrophicum*: Hydrogen-dependent adenosine 5'-triphosphate synthesis by subcellular particles. *J Bacteriol* 140:1081–1089.

Doelle HW. 1975. *Bacterial Metabolism*. 2nd ed. New York: Academic Press.

Dolfing J. 1990. Reductive dechlorination of 3-chlorobenzoate is coupled to ATP production and growth of anaerobic bacterium strain DCB-1. *Arch Microbiol* 153:264–266.

Donos JD, Frankenfeld JW. 1968. Oxidation of alkyl benzenes by a strain of *Micrococcus cerificans* growing on *n*-paraffins. *Appl Miccrobiol* 16:532–533.

Dugan PR. 1986. Microbiological desulfurization of coal and its increased monetary value. In: Ehrlich HL, Holmes DS, eds. *Workshop on Biotechnology for the Mining, Metal-Refining and Fossil Fuel Processing Industries*. Biotech Bioeng Symp 16. New York: Wiley, pp. 185–203.

Duperron S, Nadalig T, Caprais J-C, Sibuet M, Fiala-Médioni A, Amann R, Dubilier N. 2005. Dual symbiosis in a *Bathymodiolus* sp. mussel from a methane seep on the Gabon continental margin (Southeast Atlantic): 16S rRNA phylogeny and distribution of the symbionts in gills. *Appl Environ Microbiol* 71:1694–1700.

Dutova EN. 1962. The significance of sulfate-reducing bacteria in prospecting for oil as exemplified in the study of ground water in Cental Asia. In: Kuznetsov SI, ed. *Geologic Activity of Microorganisms*. New York: Consultants Bureau, pp. 76–78.

Dutton PL, Evans WC. 1969. The metabolism of aromatic compounds by *Rhodopseudomonas palustris*. A new reductive method of aromatic ring metabolism. *Biochem J* 113:525–536.

Dybas M, Konisky J. 1992. Energy transduction in the methanogen *Methanococcus voltae* is based on a sodium current. *J Bacteriol* 174:5575–5583.

Ehrenreich P, Behrends A, Harder J, Widdel F. 2000. Anaerobic oxidation of alkanes by newly isolated denitrifying bacteria. *Arch Microbiol* 173:58–64.

Ehrlich HL. 1993. Bacterial mineralization of organic carbon under anaerobic conditions. In: Bollag J-M, Stotzky G, eds. *Soil Biochemistry*, Vol 8. New York: Marcel Dekker, pp. 219–247.

Eller G, Känel L, Krüger M. 2005. Cooccurrence of aerobic and anaerobic methane oxidation in the water column of Lake Plußsee. *Appl Environ Microbiol* 71:8925–8928.

Ensley BD. 1991. Biochemical diversity of trichloroethylene metabolism. *Annu Rev Microbiol* 45:283–299.

Evans WC, Fuchs G. 1988. Anaerobic degradation of aromatic compounds. *Annu Rev Microbiol* 42:289–317.

Fenchel T, Blackburn TH. 1979. *Bacteria and Mineral Cycling*. London, U.K.: Academic Press.

Fenchel T, Finlay BJ. 1992. Production of methane and hydrogen by anaerobic ciliates containing symbiotic methanogens. *Arch Microbiol* 157:475–480.

Finn MW, Tabita FR. 2003. Synthesis of catalytically active form of III ribulose 1,5-bisphosphate carboxylase/oxidase in archaea. *J Bacteriol* 185:3049–3059.

Finn MW, Tabita FR. 2004. Modified pathway to synthesize ribulose 1,5-bisphosphate in the methanogenic Archaea. *J Bacteriol* 186:6360–6366.

Finnerty WR, Robinson M. 1986. Microbial desulfurization of fossil fuels: A review. In: Ehrlich HL, Holmes DS, eds. *Workshop on Biotechnology for the Mining, Metal-Refining and Fossil Fuel Processing Industries*. Biotech Bioeng Symp 16. New York: Wiley, pp. 205–221.

Finnerty WR, Singer ME, King AD. 1984. Microbial processes and the recovery of heavy petroleum. In: Meyer RF, Wynne JC, Olson JC, eds. *Future Heavy Crude Tar Sands*. 2nd Int Conf. New York: McGraw-Hill, pp. 424–429.

Finster K, Tanimoto Y, Bak F. 1992. Fermentation of methanediol and diemethylsulfide by a newly isolated methanogenic bacterium. *Arch Microbiol* 157:425–430.

Foght JM, Fedorak PM, Gray MR, Westlake DWS. 1990. Microbial desulfurization of petroleum. In: Ehrlich HL, Brierley CL, eds. *Microbial Mineral Recovery*. New York: McGraw-Hill, pp. 379–407.

Foster JW. 1962. Hydrocarbons as substrates for microorganisms. *Antonie v Leeuwenhoek* 28:241–274.

Francis W. 1954. *Coal: Its Formation and Composition*. London, U.K.: Edward Arnold.

Fredricks KM. 1967. Products of the oxidation of *n*-decane by *Pseudomonas aeruginosa* and *Mycobacterium rhodochrous*. *Antonie v Leeuwenhoek* 33:41–48.

Fredrickson JK, Stewart DL, Campbell JA, Powell MA, McMulloch M, Pyne JW, Bean RM. 1990. Biosolubilization of low-rank coal by *Trametes versicolor* siderophore-like product and complexing agents. *J Ind Microbiol* 5:401–406.

Freedman DL, Gossett JM. 1989. Biological reductive dechlorination of tetrachloroethylene and trichloroethylene under methanogenic conditions. *Appl Environ Microbiol* 55:2144–2151.

Frimmer U, Widdel F. 1989. Oxidation of ethanol by methanogenic bacteria. *Arch Microbiol* 152:479–483.

Galand PE, Fritze H, Conrad R, Yrjälä K. 2005. Pathways for methanogenesis and diversity of methanogenic Archaea in three boreal peatland ecosystems. *Appl Environ Microbiol* 71:2195–2198.

Gal'chenko VF, Gorlatov SN, Tokarev VG. 1986. Microbial oxidation of methane in Bering Sea sediments. *Mikrobiologiya* 55:669–673 (Engl. transl., pp. 526–530).

Gescher J, Zaar A, Mohamed M, Schagger H, Fuchs G. 2002. Genes coding for a new pathway of aerobic benzoate metabolism in *Azoaxcus evansii*. *J Bacteriol* 184:6301–6315.

Giani D, Giani L, Cohen Y, Krumbein WE. 1984. Methanogenesis in the hypersaline Solar Lake (Sinai). *FEMS Microbiol Lett* 25:219–224.

Gold T, Gordon BE, Streett W, Bilson E, Panaik P. 1986. Experimental study of the reaction of methane with petroleum hydrocarbons in geological conditions. *Geochim Cosmochim Acta* 50:2411–2418.

Gorlatov SN, Belyaev SS. 1984. The aerobic microflora of an oil field and its ability to destroy petroleum. *Mikrobiologiya* 53:843–849 (Engl. transl., pp. 701–706).

Gorlatov SN, Gal'chenko VF, Tokarev VG. 1986. Microbiological methane formation in deposits of the Bering Sea. *Mikrobiologiya* 55:490–495 (Engl. transl., pp. 380–385).

Gottschalk G. 1986. *Bacterial Metabolism*. 2nd ed. New York: Springer.

Gottschalk G, Blaut M. 1990. Generation of proton and sodium motive forces in methanogenic bacteria. *Biochem Biophys Acta* 1018:263–266.

Grbic-Galic D. 1990. Methanogenic transformation of aromatic hydrocarbons and phenols in groundwater aquifers. *Geomicrobiol J* 8:167–200.

Grossman MJ, Lee MK, Prince RC, Minak-Bernero V, George GN, Pickering IJ. 2001. Deep desulfurization of extensively hydrodesulfurized middle distillate oil by *Rhodococcus* sp. strain ECRD-1. *Appl Environ Microbiol* 67:1949–1952.

Hallam SJ, Putnam N, Preston CM, Detter JC, Rokshar D, Richardson PM, DeLong EF. 2004. Reverse methanogenesis: Testing the hypothesis with environmental genomics. *Science* 305:1457–1462.

Hansen LB, Finster K, Fossing H, Iversen N. 1998. Anaerobic methane oxidation in sulfur depleted sediments: Effects of sulfate and molybdate additions. *Aquat Microb Ecol* 14:195–204.

Henrichs SM, Reeburgh WS. 1987. Anaerobic mineralization of organic matter: Rates and the role of anaerobic processes in the oceanic carbon economy. *Geomicrobiol J* 5:191–237.

Haq BU. 1999. Methane in the deep blue sea. *Science* 285:543–544.

Higgins IJ, Best DJ, Hammond RC, Scott D. 1981. Methane-oxidizing microorganisms. *Microbiol Rev* 45:556–590.

Higson FK. 1992. Microbial degradation of biphenyl and its derivatives. *Adv Appl Microbiol* 37:135–164.

Hinrichs K-U, Hayes JM, Sylva SP, Brewer PG, DeLong EF. 1999. Methane-consuming archaebacteria in marine sediments. *Nature* (London) 398:802–805.

Hoehler TM, Alperin MJ, Albert DB, Martens CS. 1994. Field and laboratory studies of methane oxidation in an anoxic marine sediment: Evidence for a methanogen-sulfur reducer consortium. *Gobal Biogeochem Cycles* 8:451–463.

Hoehler TM, Alperin MJ, Albert DB, Martens CS. 1998. Thermodynamic control on hydrogen concentrations in anoxic sediments. *Geochim Cosmochim Acta* 62:1745–1756.

Hölker U, Schmiers H, Große S, Winkelhöfer M, Polsakiewicz M, Ludwig S, Dohse J, Höfer M. 2002. Solubilization of low-rank coal by *Trichoderma atroviride*: Evidence for the involvement of hydrolytic and oxidative enzymes by using ^{14}C-labelled lignite. *J Ind Microbiol Biotechnol* 28:207–212.

Horita J, Berndt ME. 1999. Abiogenic methane formation and isotope fractionation under hydrothermal conditions. *Science* 285;1055–1057.

Horz H-P, Raghubanshi AS, Heyer J, Kamman C, Conrad R, Dunfield PF. 2002. Activity and community structure of methane-oxidising bacteria in a wet meadow soil. *FEMS Microbiol Ecol* 41:247–257.

Horvath RS. 1972. Microbial catabolism and the degradation of organic compounds in nature. *Bacteriol Rev* 36:146–155.

Huber H, Thomm M, König G, Thies G, Stetter KO. 1982. *Methanococcus thermolithotrophicus*, a novel thermophilic lithotrophic methanogen. *Arch Microbiol* 132:47–50.

Hunt JM. 1984. Generation and migration of light hydrocarbons. *Science* 226:1265–1270.

Hunt JM, Whelan JK, Hue AY. 1980. Genesis of petroleum hydrocarbons in marine sediments. *Science* 209:403–404.

Huser BA, Wuhrmann K, Zehnder AJB. 1982. *Methanothrix soehngii* gen. nov. spec. nov. a new acetotrophic non-hydrogen-oxidizing methane bacterium. *Arch Microbiol* 132:1–9.

Hyun J–H, Bennison BW, LaRock PA. 1997. The formation of large aggregates at depth within the Louisiana hydrocarbon seep zone. *Microb Ecol* 33:216–222.

Imhoff-Stuckle D, Pfennig N. 1983. Isolation and characterization of a nicotinic acid degrading sulfate-reducing bacterium, *Desulfococcus niacini* sp. nov. *Arch Microbiol* 136:194–198.

Inagaki F, Tsunogai U, Suzuki M, Kosaka A, Machiyama H, Takai K, Nunora T, Nealson KH, Horikoshi K. 2004. Characterization of C_1-metabolizing prokaryotic communities in methane seep habitats at the Kuroshima Knoll, Southern Ryukyu Arc, by analyzing *pmoA, mmoX, mxaF, mcrA,* and 16S rRNA genes. *Appl Environ Microbiol* 70:7445–7455.

Ivanov MV. 1967. The development of geological microbiology in the U.S.S.R. *Microbiologiya* 36:849–859 (Engl. transl., pp. 751–722).

Ivanov MV, Nesterov AI, Nasaraev GB, Gal'chenko VF, Nazarenko AV. 1978. Distribution and geochemical activity of methanotrophic bacteria in coal mine waters. *Mikrobiologiya* 47:489–494 (Engl. transl., pp. 396–401).

Iversen N, Jørgensen BB. 1985. Anaerobic methane oxidation rates at the sulfate-methane transition in marine sediments from Kattegat and Skagerrak (Denmark). *Limnol Oceanogr* 30:944–955.

Izumi Y, Ohshiro T, Ogino H, Hine Y, Shimao M. 1994. Selective desulfurization of dibenzothiophene by *Rhodococcus erythropolis* D-1. *Appl Environ Microbiol* 60:223–226.

Jain MK, Zeikus JG. 1989. Bioconversion of gelatin to methane by a coculture of *Clostridium collagenovorans* and *Methanosarcina barkeri*. *Appl Environ Microbiol* 55:366–371.

Jakobsen P, Patrick WH Jr, Williams BG. 1981. Sulfide and methane formation in soils and sediments. *Soil Sci* 132:279–287.

Jobson AM, Cook FD, Westlake DWS. 1979. Interaction of aerobic and anaerobic bacteria in petroleum biodegradation. *Chem Geol* 24:355–365.

Johnson EJ, Hyman MR. 2006. Propane and *n*-butane oxidation by *Pseudomonas putida* GPo1. *Appl Environ Microbiol* 72:950–952.

Johnson EL, Hyman MR. 2006. Propane and *n*-butane oxidation by *Pseudomonas putida* GPo1. *Appl Environ Microbiol* 72:950–952.

Johnson MJ. 1964. Utilization of hydrocarbons by microorgansism. *Chem Ind* 36:1532–1537.

Jones JG, Simon BM, Gardener S. 1982. Factors affecting methanogenesis and associated anaerobic processes in the sediments of a stratified eutrophic lake. *J Gen Microbiol* 128:1–11.

Jones RD, Morita RY. 1983. Methane oxidation by *Nitrosococcus oceanus* and *Nitrosomonas europaea*. *Appl Environ Microbiol* 45:401–410.

Jones WJ, Leigh JA, Mayer F, Woese CR, Wolfe RS. 1983a. *Methanococcus jannaschii* sp. nov., an extremely thermophilic methanogen from a submarine hydrothermal vent. *Arch Microbiol* 136:254–261.

Jones WJ, Paynter MJB, Gupta R. 1983b. Characteristics of *Methanococcus maripaludis* sp. nov., a new methanogen from salt marsh sediment. *Arch Microbiol* 135:91–97.

Jones WJ, Nagle DP Jr, Whitman RB. 1987. Methanogens and the diversity of archaebacteria. *Microbiol Rev* 51:135–177.

Kallmeyer J, Boetius A. 2004. Effects of temperature and pressure on sulfate reduction and anaerobic oxidation of methane in hydrothermal sediment of Guaymas Basin. *Appl Environ Microbiol* 70:1231–1233.

Karakashev D, Batstone DJ, Trably E, Angelidaki I. 2006. Acetate oxidation is the dominant methanogenic parthway from acetate in the absence of *Methanosaetaceae*. *Appl Environ Microbiol* 72:5138–5141,

Klein DA, Davis JA, Casida LE. 1968. Oxidation of *n*-alkanes to ketones by an *Arthrobacter* species. *Antonie v Leeuwenhoek* 34:495–503.

Koburger JA. 1964. Microbiology of coal: Growth of bacteria in plain and oxidized coal slurries. 39th Annu Session of W Virginia Acad Sci Proc. *West Virginia Acad Sci* 36:26–30.

Koizumi Y, Takii S, Nishino M, Nakajima T. 2003. Vertical distributions of sulfate-reducing bacteria and methane-producing archaea quantified by oligonucleotide probe hybridization in the profundal sediment of a mesotrophic lake. *FEMS Microbiol Ecol* 44:101–108.

Kristjansson JK, Schönheit P, Thauer RK. 1982. Different K_s values for hydrogen of methanogenic and sulfate-reducing bacteria: An explanation for the apparent inhibition of methanogenesis by sulfate. *Arch Microbiol* 131:278–282.

Kropp KG, Davidova IA, Suflita JM. 2000. Anaerobic oxidation of *n*-dodecane by an enrichment reaction in a sulfate-reducing bacterial enrichment culture. *Appl Environ Microbiol* 66:5393–5398.

Kulm LD, Suess E, Moore JC, Carson B, Lewis BT, Ritger SD, Kadko DC, Thornburg TM, Embley RW, Rugh WD, Massoth GJ, Langseth MG, Cochrane GR, Scamman RL. 1986. Oregon subduction zone: Venting, fauna, and carbonates. *Science* 231:561–566.

Kuznetsov SI, Ivanov MV, Lyalikova NN. 1963. *Introduction to Geological Microbiology*. New York: McGraw-Hill.

Kuznetsova VA, Gorlenko VM. 1965. The growth of hydrocarbon-oxidizing bacteria under anaerobic conditions. *Prikl Biokhim Mikrobiol* 1:623–626.

Kvasnikov EI, Lipshits VV, Zubova NV. 1973. Facultative anaerobic bacteria of producing petroleum wells. *Mikrobiologiya* 42:925–930 (Engl. transl., pp. 823–827).

Kvenvolden KA. 1988. Methane hydrate—a major reservoir of carbon in the shallow geosphere. *Chem Geol* 71:41–51.

Larsson L, Olsson G, Holst O, Karlsson HT. 1990. Pyrite oxidation by thermophilic archaebacteria. *Appl Environ Microbiol* 56:697–701.

Leadbetter ER, Foster JW. 1959. Oxidation products formed from gaseous alkanes by the bacterium *Pseudomonas methanica*. *Arch Biochem Biophys* 82:491–492.

Li Q, Li L, Rejtar T, Lessner DJ, Karger BL, Ferry JG. 2006. Electron transport in the pathway of acetate conversion to methane in the marine archaeon *Methanosarciana acetivorans*. *J Bacteriol* 188:702–710.

Lösekann T, Knittel K, Nadalig T, Fuchs B, Niemann H, Boetius A, Amann R. 2007. Diversity and abundance of aerobic and anaerobic methane oxidizers at the Haakon Mosby Mud Volcano, Barents Sea. *Appl Environ Microbiol* 73:3348–3362.

Lovley DR. 1987. Organic matter mineralization with the reduction of ferric iron: A review. *Geomicrobiol J* 5:375–399.

Lovley DR. 1993. Dissimilatory metal reduction. *Annu Rev Microbiol* 47:263–290.

Lovley DR, Phillips EJP. 1988. Novel mode of microbial energy metabolism: Organic carbon oxidation coupled to dissimilatory reduction of iron and manganese. *Appl Environ Microbiol* 54:1472–1480.

MacDonald IR, Reilly JF, Guinasso NL Jr, Brooks JM, Carney RS, Bryant WA, Bright TJ. 1990. Chemosynthetic mussels at a brine-filled pockmark in the northern Gulf of Mexico. *Science* 248:1096–1099.

MacLeod FA, Guiot SR, Costerton JW. 1990. Layered structure of bacterial aggregates produced in an upflow anaerobic sludge bed and filter reactor. *Appl Environ Microbiol* 56:1598–1607.

Madigan MT, Gest H. 1978. Growth of a photosynthetic bacterium anaerobically in darkness, supported by "oxidant-dependent" sugar fermentation. *Arch Microbiol* 117:119–122.

Mah RA, Smith MR, Baresi L. 1978. Studies on an acetate-fermenting strain of *Methanosarcina*. *Appl Environ Microbiol* 35:1174–1184.

Mancini SA, Ulrich AC, Lacrampe-Couloume G, Sleep B, Edwards EA, Sherwood Lollar B. 2003. Carbon and hydrogen isotopic fractionation during anaerobic biodegradation of benzene. *Appl Environ Microbiol* 69:191–198.

Mango FD. 1992. Transition metal catalysis in the generation of petroleum and natural gas. *Geochim Cosmochim Acta* 56:3851–3854.

Martens CS. 1976. Control of methane sediment-water bubble transport by macroinfaunal irrigation in Cape Lookout Bight, North Carolina. *Science* 192:998–1000.

Martens CS, Berner RA. 1974. Methane production in the interstitial waters of sulfate-depleted marine sediments. *Science* 185:1167–1169.

Martens CS, Klump JV. 1984. Biogeochemical cycling in an organic-rich coastal marine basin. 4. An organic carbon budget for sediments dominated by sulfate reduction and methanogensis. *Geochim Cosmochim Acta* 48:1987–2004.

Maymó-Gatell X, Anguish T, Zinder SH. 1999. Reductive dechlorination of chlorinated ethenes and 1,2-dichloroethane by "*Dehalococcoides ethenogenes*" 195. *Appl Environ Microbiol* 65:3108–3113.

McInerney MJ, Westlake DWS. 1990. Microbially enhanced oil recovery. In: Ehrlich HL, Brierley CL, eds. *Microbial Mineral Recovery*. New York: McGraw-Hill, pp. 409–445.

McInerney MJ, Bryant MP, Pfennig N. 1979. Anaerobic bacterium that degrades fatty acids in syntrophic association with methanogens. *Arch Microbiol* 122:129–135.

McKenna EJ, Kallio RE. 1964. Hydrocarbon structure: Its effect on bacterial utilization of alkanes. In: Heukelakian H, Dondero N, eds. *Principles and Applications in Aquatic Microbiology*. New York: Wiley, pp. 1–14.

Merrettig U, Wlotzka P, Onken U. 1989. The removal of pyritic sulfur from coal by *Leptospirillum*-like bacteria. *Appl Microbiol Biotechnol* 31:626–628.

Metje M, Frenzel P. 2005. Effect of temperature on anaerobic ethanol oxidation and methanogenesis in acidic peat from a northern wetland. *Appl Environ Microbiol* 71:8191–8200.

Meyer WC, Yen TF. 1976. Enhanced dissolution of oil shale by bioleaching with thiobacilli. *Appl Environ Microbiol* 32:610–616.

Mills HJ, Martinez RJ, Story S, Sopbecky PA. 2005. Characterization of microbial community structure in Gulf of Mexico gas hydrates: Comparative analysis of DNA- and RNA-derived clone libraries. *Appl Environ Microbiol* 71:3235–3247.

Mohn WW, Tiedje JM. 1990. Strain DCB-1 conserves energy for growth from reductive dechlorination coupled to formate oxidation. *Arch Microbiol* 153:267–271.

Mohn WW, Tiedje JM. 1991. Evidence for chemiosmotic coupling of reductive dechlorination and ATP synthesis in *Desulfomonile tiedjei*. *Arch Microbiol* 157:1–6.

Mormile MR, Atlas RM. 1988. Mineralization of dibenzothiophene biodegradation products 3-hydroxy-2-formyl benzothiophene and dibenzothiophene sulfone. *Appl Environ Microbiol* 54:3183–3184.

Mountford DO. 1978. Evidence for ATP synthesis driven by a proton gradient in *Methanosarcina barkeri*. *Biochem Biophys Res Commun* 85:1346–1351.

Müller V, Blaut M, Gottschalk B. 1993. Bioenergetics of methanogenesis. In: Ferry JG, ed. *Methanogenesis*. New York: Chapman & Hall, pp. 360–406.

Myers CR, Nealson KH. 1988. Bacterial manganese reduction and growth with manganese oxide as the sole electron acceptor. *Science* 240:1319–1321.

Nauhaus K, Boetius A, Krüger M, Widdel F. 2002. *In vitro* demonstration of anaerobic oxidation of methane coupled to sulphate reduction in sediment from a marine gas hydrate area. *Environ Microbiol* 4:296–305.

Nazina TN, Rozanova EP. 1978. Thermophilic sulfate-reducing bacteria from oil strata. *Mikrobiologiya* 47:142–148 (Engl. transl., pp. 113–118).

Nazina TN, Rozanova EP. 1980. Ecological conditions for the distribution of methane producing bacteria in oil-containing strata of Apsheron. *Mikrobiologiya* 49:123–129 (Engl. transl., pp. 104–109).

Nazina TN, Rozanova EP, Kuznetsov SI. 1985. Microbial oil transformation processes accompanied by methane and hydrogen-sulfide formation. *Geomicrobiol J* 4:103–130.

Nette IG, Grechushkina NN, Rabotnova IL. 1965. Growth of certain mycobacteria in petroleum and petroleum products. *Prikl Biokhim Mikrobiol* 1:167–174.

Niemann H, Lösekann T, de Beer D, Ebert M, Nadalig T, Knittel K, Amann R, Santer EJ, Schlüter M, Klages M, Foucher JP, Boetius A. 2006. Novel microbial communities of the Haakon Mosby mud volcano and their roles of methane sink. *Nature* (London) 443:854–858.

Niewöhner C, Hensen C, Kasten S, Zabel M, Schulz HD. 1998. Deep sulfate reduction completely mediated by anaerobic methane oxidation in sediments of the upwelling area of Namibia. *Geochim Cosmochim Acta* 62:455–464.

North FK. 1985. *Petroleum Geology*. Boston, MA: Allen and Unwin.

Olson ES, Stanley DC, Gallagher JR. 1993. Characterization of intermediates in the microbial desulfurization of dibenzothiophene. *Energy Fuels* 7:159–164.

Omori T, Monna L, Saiki Y, Kodama T. 1992. Desulfurization of dibenzothiophene by *Corynebacterium* sp. strain SY1. *Appl Environ Microbiol* 58:911–915.

O'Neill JG, Wilkinson JF. 1977. Oxidation of ammonia by methane-oxidizing bacteria and the effects of ammonia on methane oxidation. *J Gen Microbiol* 100:407–412.

Oremland RS, Taylor BF. 1978. Sulfate reduction and methanogenesis in marine sediments. *Geochim Cosmochim Acta* 42:209–214.

Oremland RS, Marsh LM, Polcin S. 1982. Methane production and simultaneous sulfate reduction in anoxic, salt marsh sediments. *Nature* (London) 296:143–145.

Orphan VJ, House CH, Hinrichs K-U, McKeegan KD, DeLong EF. 2001a. Methane-consuming Archaea revealed by directly coupled isotopic and phylogenetic analysis. *Science* 293:484–487.

Orphan VJ, Hinrichs K-U, Ussler W III, Paull CK, Taylor LlT, Sylva SP, Hayes JM, DeLong EF. 2001b. Comparative analysis of methane-oxidizing Archaea and sulfate-reducing Bacteria in anoxic marine sediments. *Appl Environ Microbiol* 67:1922–1934.

Orphan VJ, House CH, Hinrichs K-U, McKeegan KD, DeLong EF. 2002. Multiple archaeal groups mediate methane oxidation in anoxic cold seep sediments. *Proc Natl Acad Sci* (Washington) 99:7663–7668.

Orr FM Jr, Taber JJ. 1984. Use of carbon dioxide in enhanced oil recovery. *Science* 224:563–569.

Ourisson G, Albrecht P, Rohmer M. 1984. Microbial origin of fossil fuels. *Sci Am* 251:44–51.

Pancost RD, Damsté JSS, de Lint S, van der Maarel MJEC, Gottschal JC, Medinaut Shipboard Scientific Party. 2000. Biomarker evidence for widespread anaerobic methane oxidation in Mediterranean sediments by a consortium of methanogenic Archaea and Bacteria. *Appl Environ Microbiol* 66:1126–1132.

Panganiban A, Hanson RS. 1976. Isolation of a bacterium that oxidizes methane in the absence of oxygen. Abstract, *Annu Meet Am Soc Microbiol*, 159:121.

Panganiban AT Jr, Patt TE, Hart W, Hanson RS. 1979. Oxidation of methane in the absence of oxygen in lake water samples. *Appl Environ Microbiol* 37:303–309.

Paszczynski A, Huynh V-B, Crawford R. 1986. Comparison of lignase-1 and peroxidase-M2 from the white-rot fungus *Phanerochaete chrysosporium*. *Arch Biochem Biophys* 244:750–785.

Patel GB. 1984. Characterization and nutritional properties of *Methanothrix concilii* sp. nov., an aceticlastic methanogen. *Can J Microbiol* 30:1383–1396.

Pernthaler A, Pernthaler J, Amann R. 2002. Fluorescence in situ hybridization and catalyzed reporter deposition for the identification of marine bacteria. *Appl Environ Microbiol* 68:3094–3101.

Phillips WE Jr, Perry JJ. 1974. Metabolism of *n*-butane and 3-butanone by *Mycobacterium vaccae*. *J Bacteriol* 120:987–989.

Pirnik MP, Atlas RM, Bartha R. 1974. Hydrocarbon metabolism by *Brevibacterium erythrogenes*: Normal and branched alkanes. *J Bacteriol* 119:868–878.

Pooley FD, Atkins AS. 1983. Desulfurization of coal using bacteria by both dump and process plant techniques. In: Rossi G, Torma AE, eds. *Recent Progress in Biohydrometallurgy*. Iglesias, Italy: Assoc Minerar Sarda, pp. 511–526.

Pyne JW Jr, Stewart DL, Fredrickson J, Wilson BW. 1987. Solubilization of leonardite by an extracellular fraction from *Coriolus versicolor*. *Appl Environ Microbiol* 53:2844–2848.

Quayle JR, Ferenci T. 1978. Evolutionary aspects of autotrophy. *Microbiol Rev* 42:251–273.

Radway J, Tuttle JH, Fendinger NJ, Means JC. 1987. Microbially mediated leaching of low-sulfur coal in experimental coal columns. *Appl Environ Microbiol* 53:1056–1063.

Radway J, Tuttle JH, Fendinger NJ. 1989. Influence of coal source and treatment upon indigenous microbial communities. *J Ind Microbiol* 4:195–208.

Raiders RA, Knapp RM, McInerney JM. 1989. Microbial selective plugging and enhanced oil recovery. *J Ind Microbiol* 4:215–230.

Raymond RL, Jamison VW, Hudson JO. 1967. Microbial hydrocarbon cooxidation. I. Oxidation of mono- and dicyclic hydrocarbons by soil isolates of the genus *Nocardia*. *Appl Microbiol* 15:857–865.

Reeburgh WS. 1976. Methane consumption in Cariaco Trench waters and sediments. *Earth Planet Sci Lett* 28:337–344.

Reeburgh WS. 1980. Anaerobic methane oxidation: Rate depth distribution in Skan Bay sediments. *Earth Planet Sci Lett* 47:345–352.

Reed DW, Fujita Y, Delwiche ME, Blackwelder DB, Sheridan PP, Uchida T, Colwell FS. 2002. Microbial communities from methane hydrate-bearing deep marine sediments in a forearc basin. *Appl Environ Microbiol* 68:3759–3770.

Reineke W, Knackmuss H-J. 1988. Microbial degradation of haloraromatics. *Annu Rev Microbiol* 42:263–287.

Rivard CJ, Smith PH. 1982. Isolation and characterization of a thermophilic marine methanogenic bacterium, *Methanogenium thermophilicum* sp. nov. *Int J Syst Bacteriol* 32:430–436.

Robertson R. 1966. The origins of petroleum. *Nature* (London) 212:1291–1295.

Robinson JA, Tiedje JM. 1984. Competition between sulfate-reducing and methanogenic bacteria for H_2 under resting and growing conditions. *Arch Microbiol* 137:26–32.

Roffey R, Norqvist A. 1991. Biodegradation of bitumen used for nuclear waste disposal. *Experientia* 47:539–542.

Rogoff MH, Wender I, Anderson RB. 1962. *Microbiology of Coal*. Info Circ 8057. Washington, DC: Bur Mines, US Dept of the Interior.

Roitsch T, Stolp H. 1985. Distribution of dissimilatory enzymes in methane and methanol oxidizing bacteria. *Arch Microbiol* 143:233–236.

Romanovskaya VA, Lyudvichenko ES, Kryshtab TP, Zhukov VG, Sokolov IG, Malashenko YuR. 1980. Role of exogenous carbon dioxide in metabolism of methane-oxidizing bacteria. *Mikrobiologiya* 49:687–693 (Engl. transl., pp. 566–571).

Romesser JA, Wolfe RS, Mayer F, Spiess E, Walther-Mauruschat A. 1979. *Methanogenium*, a new genus of marine methanogenic bacteria, and characterization of *Methanogenium cariaci* sp. nov. and *Methanogenium marisnigri* sp. nov. *Arch Microbiol* 121:147–153.

Rozanova EP. 1971. Morphology and certain physiological properties of purple bacteria from oil-bearing strata. *Mikrobiologiya* 40:152–157 (Engl. transl., pp. 134–138).

Rozanova EP. 1978. Sulfate reduction and water-soluble organic substances in a flooded oil reservoir. *Mikrobiologiya* 47:495–500 (Engl. transl., pp. 401–405).

Rubinshtein LM, Oborin AA. 1986. Microbial methane production in stratal waters of oilfields in the Perm area of the Cis-Ural region. *Mikrobiologiya* 55:674–678 (Engl. transl., pp. 530–534).

Rubinstein I, Strausz OP, Spyckerelle C, Crawford RJ, Westlake DW. 1977. The origin of oil and bitumens of Alberta: A chemical and a microbiological study. *Geochim Cosmochim Acta* 41:1341–1353.

Sansone FJ, Martens CS. 1981. Methane production from acetate and associated methane fluxes from anoxic coastal sediments. *Science* 211:707–709.

Schönheit P, Kristjansson JK, Thauer RK. 1982. Kinetic mechanism for the ability of sulfate reducers to out-compete methanogens for acetate. *Arch Microbiol* 132:285–288.

Schouten S, Wakeham SG, Hopmans EC, Sinninghe Damsté JS. 2003. Biogeochemical evidence that thermophilic Archaea mediate the anaerobic oxidation of methane. *Appl Environ Microbiol* 69:1680–1686.

Sharak Genthner BR, Price WA II, Pritchard PH. 1989. Anaerobic degradation of chloroaromatic compounds in aquatic sediments under a variety of enrichment conditions. *Appl Environ Microbiol* 55:1466–1477.

Shelton DR, Tiedje JM. 1984. Isolation and characterization of bacteria in an anaerobic consortium that mineralizes 3-chlorobenzoic acid. *Appl Environ Microbiol* 48:840–848.

Sieburth J McN, Johson PW, Eberhardt MA, Sieracki ME, Lidstron ME, Laux D. 1987. The first methane-oxidizing bacterium from the upper mixing layer of the deep ocean: *Methylomonas pelagica* sp. nov. *Curr Microbiol* 14:285–293.

Simakova TL, Kolesnik ZA, Strigaleva NV. 1968. Transformation of high-paraffinaceous oil by microorganisms under anaerobic and aerobic conditions. *Mikrobiologiya* 37:233–238 (Engl. transl., pp. 194–198).

Simoneit BRT, Londsdale PG. 1982. Hydrothermal petroleum in mineralized mounds at the seabed of Guaymas Basin. *Nature* (London) 295:198–202.

Simpson PG, Whitman WB. 1993. Anabolic pathways in methanogens. In: Ferry JG, ed. *Methanogenesis*. New York: Chapman & Hall, pp. 445–472.

Skyring GW. 1987. Sulfate reduction in coastal ecosystems. *Geomicrobiol J* 5:295–374.

Smith MR, Mah RA. 1978. Gowth and methane genesis by *Methanosarcina* strain 227 on acetate and methanol. *Appl Environ Micorbiol* 36:870–879.

So CM, Phelps CD, Young LY. 2003. Anaerobic transformation of alkanes to fatty acids by a sulfate-reducing bacterium, strain Hxd3. *Appl Environ Microbiol* 69:3892–3900.

Sokolov IG. 1986. Coupling of the process of electron transport to methane monooxygenasts with the translocation of protons in methane-oxidizing bacteria. *Mikrobiologiya* 55:715–722 (Engl. transl., pp. 559–565).

Sokolov IG, Malashenko YuR, Romanovskaya VA. 1981. Electron transport chain of thermophilic methane-oxidizing culture *Methylococcus thermophilus*. *Mikrobiologiya* 50:13–20 (Engl. transl., pp. 7–13).

Sørensen J. 1987. Nitrate reduction in marine sediment: Pathways and interactions with iron and sulfur cycling. *Geomicrobiol J* 5:401–421.

Sprott GE, Bird SE, MacDonald IJ. 1985. Proton motive force as a function of the pH at which *Methanobacterium bryantii* is grown. *Can J Microbiol* 31:1031–1034.

Sprott GD, Ekiel I, Patel G. 1993. Metabolic pathways in *Methanococcus jannaschii* and other methanogenic bacteria. *Appl Environ Microbiol* 59: 1092–1098.

Stanier RY, Ingraham JL, Wheelis ML, Painter PR. 1986. *The Microbial World*, 5th ed.. Englewood Cliffs, NJ: Prentice-Hall.

Stetter KO, Gaag G, 1983. Reduction of molecular sulfur by methanogenic bacteria. *Nature* (London) 305:309–311.

Stetter KO, Huber R, Boechl E, Kurr M, Eden RD, Fielder M, Cash H, Vance I. 1993. Hyperthermophilic archaea are thriving in deep North Sea and Alaskan oil reservoirs. *Nature* (London) 365:743–745.

Stevenson IL. 1967. Utilization of aromatic hydrocarbons by *Arthorbacter* spp. *Can J Microbiol* 13:205–211.

Stewart DL, Thomas BL, Bean RM, Fredrickson JK. 1990. Colonization and degradation of bituminous and lignite coals by fungi. *J Ind Microbiol* 6:53–59.

Stewart JE, Kallio RE, Stevenson DP, Jones AC, Schissler DO. 1959. Bacterial hydrocarbon oxidation. I. Oxidation of *n*-hexadecane by a gram-negative coccus. *J Bacteriol* 78:441–448.

Stoecker K, Bendinger B, Schöning B, Nielsen PH, Nielsen JL, Baranyi C, Toenshoff ER, Daims H, Wagner M. 2006. Cohn's *Crenothrix* is a filamentous methane oxidizer with an unusual methane monooxygenase. *Proc Natl Acad Sci* (Washington, DC) 103:2363–2367.

Stoner DL, Wey JW, Barrett KB, Jolley JG, Wright RB, Dugan PR. 1990. Modification of water-soluble coal-derived products by dibenzothiophene-degrading microorganisms. *Appl Environ Microbiol* 56:2667–2676.

Sundh I, Bastviken D, Tranvik LJ. 2005. Abundance, activity, and community structure of pelagic, methane-oxidizing bacteria in temperate lakes. *Appl Environ Microbiol* 71:6746–6752.

Szewzyk R, Pfennig N. 1987. Complete oxidation of catechol by the strictly anaerobic sulfate reducing *Desulfobacterium catecholicum* sp. nov. *Arch Microbiol* 147:163–168.

Tanner RS, Udegbunam EO, McInerney MJ, Knapp RM. 1991. Microbially enhanced oil recovery from carbonate reservoirs. *J Ind Microbiol* 9:169–195.

Teh JS, Lee KH. 1973. Utilization of *n*-alkanes by *Cladosporium resinae*. *Appl Microbiol* 25:454–457.

Teske A, Hinrichs K-U, Edgcomb V, de Vera Gomez A, Kysela D, Sylva SP, Sogin ML, Jannasch HW. 2002. Microbial diversity of hydrothermal sediments in the Guaymas Basin: Evidence for anaerobic methanotrophic communities. *Appl Environ Microbiol* 68:1994–2007.

Theisen AR, Murrell JC. 2005. Facultative methanotrophs revisited. *J Bacteriol* 187:4303–4305.

Thomsen TR, Finster K, Ramsing NB. 2001. Biogeochemical and molecular signatures of anaerobic methane oxidation in a marine sediment. *Appl Environ Microbiol* 67:1646–1656.

Townsley CC, Atkins AS, Davis AJ. 1987. Suppression of pyritic sulfur during flotation tests using the bacterium *Thiobacillus ferrooxidans*. *Biotech Bioeng* 30:1–8.

Treude T, Niggermann J, Kallmeyer J, Wintersteller P, Schubert CJ, Boetius A, Jøregensen BB. 2005. Anaerobic oxidation of methane and sulfate reduction along the Chilean continental margin. *Geochim Cosmochim Acta* 69:2767–2779.

Treude T, Orphan V, Knittel K, Gieseke A, House CH, Boetius A. 2007. Consummption of methane and CO_2 by methanotrophic microbial mats from gas seeps of the anoxic Black Sea. *Appl Environ Microbiol* 73:2271–2283.

Van Afferden M, Schacht S, Klein J, Trüper HG. 1990. Degradation of dibenzothiophene by *Brevibacterium* sp. DO. *Arch Microbiol* 153:324–328.

van Beilen JB, Funhoff EG. 2007. Alkane hydroxylases involved in microbial alkane degradation. *Appl Microbiol Biotechnol* 74:13–21.

van Bodegom PM, Scholten JCM, Stams AJM. 2004. Direct inhibition of methanogenesis by ferric iron. *FEMS Microbiol Ecol* 49:261–268.

Van Bruggen JJA, Zwart KB, van Assema RM, Stumm CK, Vogels GD. 1984. *Methanobacterium formicum*, an endosymbiont of the anaerobic ciliate *Metopus striatus* McMurrich. *Arch Microbiol* 139:1–7.

Van der Linden AC, Thijsse GJE. 1965. The mechanism of microbial oxidations of petroleum hydrocarbons. *Adv Enzymol* 27:469–546.

Van Hamme JD, Singh A, Ward OP. 2003. Recent advances in petroleum microbiology. *Microbiol Mol Biol Rev* 67:503–509.

Vigliotta G, Nurticati E, Carata E, Tredici SM, De Stefano M, Pontieri P, Massardo DR, Prati MV, De Bellis L, Alifano P. 2007. *Clonothrix fusca* Roze 1996, a filamentous, sheathed, methanotrophic γ-Proteobacterium. *Appl Environ Microbiol* 73:3556–3565.

Vogels GD. 1979. The global cycle of methane. *Antonie v Leeuwenhoek* 45:347–352.

Vogels CD, Visser CM. 1983. Interconnections of methanogenic and acetogenic pathways. *FEMS Microbiol Lett* 20:291–297.

Volesky B, Zajic JE. 1970. Ethane and natural gas oxidation by fungi. *Dev Ind Microbiol* 11:184–195.

Walker JD, Cooney JJ. 1973. Pathway of *n*-alkane oxidation in *Cladosporium resinae*. *J Bacteriol* 115:635–639.

Ward BB, Kilpatrick KA, Novelli PC, Scanton MI. 1987. Methane oxidation and methane fluxes in the ocean surface layer and deep anoxic layers. *Nature* (London) 327:226–229.

Ward DM, Brock TD. 1978. Anaerobic metabolism of hexadecane in sediments. *Geomicrobiol J* 1:1–9.

Westlake DWS, Belicek W, Jobson A, Cook FD. 1976. Microbial utilization of raw and hydrogenated shale oils. *Can J Microbiol* 22:221–227.

Whittenbury R, Phillips KC, Wilkinson JF. 1970. Enrichment isolation and some properties of methane-utilizing bacteria. *J Gen Microbiol* 61:205–218.

Widdel F. 1986. Growth of methanogenic bacteria in pure culture with 2-propanol and other alcohols as hydrogen donors. *Appl Envirom Microbiol* 51:1056–1062.

Wildgruber G, Thomm M, König H, Ober K, Ricchiuto T, Stetter KO. 1982. *Methanoplanus limicola*, a plate-shaped methanogen representing a novel family, the Methanoplanaceae. *Arch Microbiol* 132:31–36.

Winter J, Wolfe RS. 1979. Complete degradation of carbohydrate to carbon dioxide and methane by syntrophic cultures of *Acetobacter woodii* and *Methanosarcina barkeri*. *Arch Microbiol* 121:97–102.

Winter JU, Wolfe RS. 1980. Methane formation from fructose by syntrophic associations of *Acetobacterium woodii* and different strains of methanogens. *Arch Microbiol* 124:73–79.

Wolf M, Bachofen R. 1991. Microbial degradation of bitumen. *Experientia* 47:542–548.

Wolin MJ. 1981. Fermentation in the rumen and human large intestine. *Science* 213:1463–1468.

Wolin MJ, Miller TL. 1987. Bioconversion of organic carbon to CH_4 and CO_2. *Geomicrobiol J* 5:239–259.

Worakit S, Boone DR, Mah RA, Abdel-Samie M-E, El-Halwagi MM. 1986. *Methanobacterium alcaliphilum* sp. nov., an H_2-utilizing methanogen that grows at high pH values. *Int J Syst Bacteriol* 36:380–382.

Yang KY-L, Lapado J, Whitman WB. 1992. Pyruvate oxidation by *Methanococcus* spp. *Arch Microbiol* 158:271–275.

Yarbrough HF, Coty VF. 1983. Microbially enhanced oil recovery from the Upper Cretaceous Nacatoch formation, Union County, Arkansas. In: Donaldson EC, Clark JB, eds. *Proceedings of the International Conference on Microbial Enhancement of Oil Recovery* (Conf 8205140). Springfield, VA: NTIS, pp. 149–153.

Yavitt JB, Lang GE. 1990. Methane production in contrasting wetland sites: Response to organic-chemical components of peat and to sulfate reduction. *Geomicrobiol J* 8:27–46.

Yen HC, Marrs B. 1977. Growth of *Rhodopseudomonas capsulata* under anaerobic dark conditions with dimethy sulfoxide. *Arch Biochem Biophys* 181:411–418.

Young LY, Frazer AC. 1987. The fate of lignin and lignin-derived compounds in anaerobic environments. *Geomicrobiol J* 5:261–293.

Zabel HP, König H, Winter J. 1984. Isolation and characterization of a new coccoid methanogen, *Methanogenium tatii* spec. nov. from a solfataric field on Mount Tatio. *Arch Microbiol* 137:308–315.

Zehnder AJB, Brock TD. 1979. Methane formation and methane oxidation by methanogenic bacteria. *J Bacteriol* 137:420–432

Zehnder AJB, Huser BA, Brock TD, Wuhrmann K. 1980. Characterization of an acetate-carboxylating, non-hydrogen-oxidizing methane bacterium. *Arch Microbiol* 124:1–11.

Zeikus JG. 1977. The biology of methanogenic bacteria. *Bacteriol Rev* 41:514–541.

Zeikus JG, Kerby R, Krzycki JA. 1985. Single-carbon chemistry of acetogenic and methanogenic bacteria. *Science* 127:1167–1173.

Zellner G, Bleicher K, Braun E, Kneifel H, Tindall BJ, Conway de Marcario E, Winter J. 1989. Characterization of a new mesophilic, secondary alcohol-utilizing methanogen, *Methanobacterium palustre* spec nov. from a peat bog. *Arch Microbiol* 151:1–9.

Zellner G, Sleytr UB, Messner P, Kneifel H, Winter J. 1990. *Methanogenium liminatans* spec. nov., a new coccoid, mesophilic methanogen able to oxidize secondary alcohols. *Arch Microbiol* 153:287–293.

Zimmerman PR, Greenberg JP, Windiga SO, Crutzen PJ. 1982. Termites: A potentially large source of atmospheric methane, carbon dioxide and molecular hydrogen. *Science* 218:563–565.

Zinder SH. 1993. Physiological ecology of methanogens. In: Ferry JG, ed. *Methanogenesis*. New York: Chapman & Hall, pp. 128–206.

Zinder SH, Koch M. 1984. Non-aceticlastic methanogenesis from acetate: Acetate oxidation by a thermophilic syntrophic coculture. *Arch Microbiol* 138:263–272.

ZoBell CE. 1952. Part played by bacteria in petroleum formation. *J Sediment Petrol* 22:42–49.

ZoBell CE. 1963. The origin of oil. *Int Sci Technol* (August): 42–48.

Zvyagintsev DG, Zenova GM, Shirokykh IG. 1993. Distribution of actinomycetes with the vertical structure of peat bog ecosystems. *Mikrobiologiya* 62:548–555 (Engl. transl., pp. 339–342).

Glossary

Acetogen   A bacterial culture that forms acetic acid as the sole product in the reduction of CO_2 with hydrogen, or in the fermentation of a sugar such as glucose. Also called homoacetogen.

Acidophilic bacteria (acidophiles)   Bacteria that need an acid environment to grow.

Actinomycete   Mycelium-forming bacterium; found in soil and aquatic environments.

Adenosine 5′-diphosphate (ADP)   A phosphate anhydride that serves as acceptor of high-energy phosphate in metabolic energy conservation, or it may be a product in enzymatic phosphorylation of a metabolite with adenosine 5′-triphosphate. ADP contains one high-energy phosphate bond.

Adenosine phosphosulfate (APS)   APS and adenosine sulfatophosphate.

Adenosine 5′-triphosphate (ATP)   A phosphate anhydride that conserves metabolic energy. It contains two high-energy phosphate bonds.

Adenylic acid   Adenosine monophosphate (AMP) consisting of ribose with adenine attached to its C_1 and phosphate ester linked to its C_5.

Adipocere   A waxy substance from dead organic matter; a soap.

Adventitious organisms   Organisms introduced naturally from an adjacent habitat; they may or may not be able to survive in the new habitat.

Aerobe   An organism that lives in air and uses oxygen as terminal electron acceptor in its respiratory process.

Aerobic heterotroph   An organism that uses organic substances as carbon and energy sources in air (oxygen-respiring).

Agar (also **agar-agar**)   A polysaccharide heteropolymer derived from the walls of certain red algae and used for gelling bacteriological culture media.

Agar shake culture   A bacterial culture method in which the bacterial inoculum is completely mixed in melted agar medium in a test tube, after which the agar medium is allowed to solidify.

Allochthonous   Introduced from another place.

Aluminosilicate   A mineral containing a combination of aluminum and silicate.

Amictic lake   A lake that never turns over.

Ammonification   A biochemical process that releases amino nitrogen as ammonia from organic compounds such as proteins and amino acids.

Amorphous   Noncrystalline.

Amphibole   A ferromagnesian mineral with two infinite chains of silica tetrahedra linked to each other; the double chains are cross-linked by Ca, Mg, and Fe.

Amphoteric   Having both acidic and basic properties.

Anabolism   The part of metabolism that deals with synthesis and polymerization of biomolecules in an energy-consuming process.

Anaerobe   An organism that grows in the absence of oxygen; it may be oxygen-tolerant or oxygen-intolerant.

Anaerobic heterotrophy   A form of nutrition using organic energy and carbon sources in the absence of oxygen.

Anaerobic respiration   A respiratory process in which nitrate, sulfate, sulfur, carbon dioxide, Fe(III), Mn(IV), or some other externally supplied reducible inorganic or organic compounds substitute for oxygen as terminal electron acceptor.

Anhydrite   A calcium sulfate mineral, $CaSO_4$.

Anodic surface   A surface exhibiting a net positive charge.

Anticyclones   In oceanography, a small closed-current system of water derived from a surface current; it has a warm core surrounded by colder water, spinning in a clockwise rotation (see also Rings and Meddies).

Aragonite   A calcium carbonate mineral forming slender pointed crystals.

Archaea   (formerly **Archaebacteria, Archaeobacteria**) A domain of the prokaryotes that includes methanogens, *Sulfolobus, Acidianus, Halobacterium*, and *Thermoplasma*, among other genera, which have a unique cell envelope and plasma membrane structure that distinguishes it from the domain Bacteria (also called Eubacteria).

Archaeon   A member of the Archaea.

Aridisol   A mature desert soil.

Arroyo   A dried-up riverbed in a desert region through which water flows after a rainstorm; a wash.

Arsenopyrite   An iron–arsenic sulfide mineral, FeAsS.

Arsine AsH_3.

Ascomycetes   Fungi that deposit their sexual spores in sacs (asci), for example, *Neurospora*.

Asparagine   The amide of aspartic acid, a dicarboxylic amino acid.

Assimilation   Uptake and incorporation of nutrients by cells.

Asthenosphere   Upper portion of the Earth's mantle, which is thought to have a plastic consistency and upon which crustal plates flow.

Atmosphere   The gaseous envelope around the Earth.

Augite   A pyroxene type of mineral.

Authigenic   Formed *de novo* from dissolved species in the case of minerals.

Autochthonous   Generated in place; indigenous.

Autotroph   An organism capable of growth exclusively at the expense of inorganic nutrients.

Bacteria   A domain of prokaryotes, formerly called Eubacteria, which is distinct from the domain Archaea.

Bacterioneuston   The bacterial population located in a thin film at the air/water interface in a natural body of water.

Bacterioplankton   Unattached bacterial forms in an aqueous environment.

Bacterium   A prokaryotic single- or multicellular organism. Single cells may appear as rods, spheres, spirals, or other shapes.

Baltica   A former continent encompassing Russia, west of the Urals; Scandinavia; Poland; and northern Germany.

Banded iron formation (BIF)   A sedimentary deposit featuring alternating iron oxide-rich (Fe_2O_3 or Fe_3O_4) layers and iron oxide-poor cherty layers; thought to have originated in the Precambrian at the time of transition from a nonoxidizing to an oxidizing atmosphere resulting from the build-up of oxygen in it, but it may have begun to form much earlier.

Barium psilomelane   A complex manganese(IV) oxide.

Barophile   An organism capable of growth at elevated hydrostatic pressure.

Barren solution   Pregnant solution from an ore-leaching operation after its valuable metals have been removed.

Basaltic rock   Rock of volcanic origin showing very fine crystallization due to rapid cooling. Basalt is rich in pyroxenes and feldspars.

Basidiomycetes   Fungi that form sexual spores on basidia (club-shaped cells) and feature septate mycelia (e.g., mushrooms).

Benthic   Located at the bottom of a body of water.

Betaine $HOOCCH_2N^+(CH_3)_3$.

Binary fission   Cell division in which one cell divides into two cells of approximately equal size.

Bioherm   A large mineral aggregate of biological origin; a microbialite.

Bioleaching   A process whereby microbes extract metal values from ore by solubilizing them through oxidation, reduction, or complexation.

Biosphere The portion of the Earth inhabited by living organisms.

Birnessite A manganese(IV) oxide mineral, δMnO_2.

Bisulfite HSO_3^-.

Calcareous ooze A sediment having calcareous structures from foraminifera, coccolithophores, or other $CaCO_3$-depositing organisms as major constituents.

Calcite A calcium carbonate mineral with a rhombohedral structure.

Capillary culture method A method that employs a glass capillary with optically flat sides inserted into soil or sediment for culturing microbes from these sources *in situ*. Developing microbes in the capillaries may be observed directly under the microscope after withdrawal of the capillaries from the soil or sediment.

Catabolism The part of metabolism that involves degradation of nutrients and energy conservation from their oxidation.

Catalase An enzyme capable of catalyzing the reaction $H_2O_2 \rightarrow H_2O + 0.5O_2$; it can also catalyze the reduction of H_2O_2 with an organic hydrogen donor or inorganic electron donor.

Cathodic surface A surface exhibiting a net negative charge.

Celestite A strontium sulfate mineral, $SrSO_4$.

Cellulolytic Referring to cellulose hydrolysis, which may be enzymatic.

Centric geometry Cylindrical, in reference to diatoms.

Chalcopyrite A copper–iron sulfide mineral, $CuFeS_2$.

Chasmolithtic Living inside preformed pores, fissures, or cavities.

Chemocline A chemical gradient zone in a water column that separates a more dilute and less dense phase from a more concentrated phase.

Chemolithotroph An autotroph that derives energy from the oxidation of inorganic matter.

Chemostat A culture system permitting microbial growth under steady-state conditions.

Chlorophyll A light-harvesting and energy-transducing type of pigment of photosynthetic organisms.

Chloroplasts Photosynthetic organelles in eukaryotic cells.

Choline $HOCH_2CH_2N^+(CH_3)_3$.

Coccolithophore A chrysophyte alga whose surface is covered with $CaCO_3$ platelets (coccoliths).

Colony counting A method of enumerating viable, culturable microbes by counting colonies (visible aggregates) formed by them on or in agar medium in a petri dish or test tube.

Conjugation Unidirectional transfer of genetic information between prokaryotic cells that requires cell-to-cell contact.

Consortium An association of two or more different microbes that exhibit a metabolic interdependence.

Constitutive enzyme An enzyme that is always present in an active form in a cell, whether needed or not.

Contaminant An organism accidentally introduced during experimental manipulation of a habitat.

Continental drift Migration of continents on the Earth's surface as a result of crustal plate movements.

Continental margin The edge of a continent.

Continental rise Gently sloping seafloor at the base of a continental slope.

Continental shelf Gently sloping seafloor between the shore and the continental slope.

Continental slope Steeply sloping seafloor at the outer edge of the continental shelf.

Convergence The confluence of two water masses.

Cooxidation Simultaneous microbial oxidation of two compounds, which may be quite unrelated, only one of which supports growth.

Copiotroph A bacterium that requires a nutrient-rich environment.

Coriolis force An apparent force that seems to deflect a moving object to the right in the northern hemisphere and to the left in the southern hemisphere of the Earth.

Crustal plates   Portions of the Earth's crust, which have irregular shapes and sizes and which contact and interact with each other while floating on the asthenosphere.

Cyanobacteria   Oxygenic, photosynthetic members of the domain Bacteria, formerly known as blue-green algae.

Cysteine $HSCH_2CH(NH_2)COOH$.

Cytochrome system   An electron transport system used in biological oxidation (respiration) that includes iron porphyrin proteins called cytochromes.

Dehydrogenase   An enzyme that catalyzes removal or addition of hydrogen.

Denitrification   A process in which nitrate is reduced to dinitrogen (N_2), nitrous oxide, and nitric oxide.

Deoxyribonucleic acid (DNA)   A biopolymer consisting of purine and pyrimidine bases, deoxyribose, and phosphate and has genetic information encoded in it.

Desert varnish   A manganese- and iron-rich coating on a rock surface.

Desferrisiderophore   A siderophore that does not contain ferric iron complexed by it.

Deuteromycetes   Fungi that do not form sexual spores.

Diagenesis   A process of transformation or alteration of rocks or minerals.

Diatom   A chrysophyte alga grouped with the Bacillarophyceae, which is encased in a siliceous cell wall.

Diatomaceous ooze   A sediment having diatom frustules as a major constituent.

Dimethyl arsinate $CH_3AsO(OH)$ This acid form is also known as cacodylic acid.
$$\underset{CH_3}{\overset{|}{}}$$

Dimethylmercury $(CH_3)_2Hg$.

Dimictic lake   A lake that turns over twice a year.

Disproportionation reaction   A stoichiometric chemical reaction in which part of the reactant undergoes oxidation and the rest undergoes reduction, for example, $2H_2O_2 \rightarrow 2H_2O + O_2$.

Dithiothreitol $HSCH_2CH(OH)CH_2SH$.

Divergence   A separation of two water masses.

DNA   Deoxyribonucleic acid.

Dolomite   A $Ca(Mg)(CO_3)_2$ mineral.

Domain   In phylogeny, the highest level of grouping based on specific 16S rRNA sequences and on cell organization, for example, in the Prokaryotes, the domains of the Bacteria and the Archaea.

Dunite   An ultrabasic rock rich in the mineral olivine.

Dystrophic   Referring to waters with an oversupply of organic matter that is only incompletely decomposed because of an insufficiency of oxygen, phosphorus, and nitrogen.

Earth's core   The innermost portion of the Earth, consisting mostly of Fe and some Ni.

Earth's mantle   The portion of the Earth overlying the core, consisting mainly of O, Mg, and Si with lesser amounts of Fe, Al, Ca, and Na.

Enargite   A copper–arsenic sulfide mineral, Cu_3AsS_4.

Endolithic   Living inside rock (limestone) as a result of boring into it.

Endosymbiosis   A relationship in which cells live inside other cells for mutual benefit.

Enrichment culture   A culture method that selects for a desired organism(s) by providing special nutrients and physical conditions that favor its development; also known as selective culture method.

Entisol   An immature desert soil.

Epigenetic   Referring to emplacement of a mineral in cracks or fissures of preexisting rock.

Epilimnion   The portion of a stratified lake above the thermocline.

Epiphytes   Organisms attached to the surface of other living organisms or inanimate objects.

Eukaryotic cell   A cell with a true nucleus, mitochondria, and chloroplasts (if photosynthetic).

Euphotic zone   The part of a body of water that is penetrated by sunlight in sufficient quantity to permit photosynthesis.

Euryhaline   Capable of growth over a wide range of salinities.

Eutrophic   Referring to a nutrient-rich status of a body of natural water.

Facultative chemolithotroph   A bacterium that can grow heterotrophically or chemolithotrophically, depending on growth conditions.

Facultative microorganism   A microorganism capable of living aerobically or anaerobically.

Fauna   A term used in ecology to denote an assemblage of organisms that may include members of one or more of the following groups: Protozoa, invertebrates, and vertebrates, although protozoans do not belong to the Animalia in modern systematics.

Fecal pellet   Compacted fecal matter packaged in a membrane by the organism that excretes it.

Feldspar   A type of mineral consisting of anhydrous aluminosilicates of Na, K, Ca, and Ba.

Fermentation   A metabolic process of intramolecular oxidation/reduction operating without an externally supplied terminal electron acceptor; a biochemical disproportionation.

Ferrisiderophore   A siderophore that contains ferric iron complexed by it.

Flora   A term used in ecology to denote an assemblage of organisms that may include members of one or more of the following groups: prokaryotes, algae, fungi, and plants, even though the first three groups are not considered plants in modern systematics.

Fluorescence microscopy   A microscopy method making use of natural or artificial fluorescence of objects upon irradiation with UV light.

Foraminifera   Amoeboid protozoa that mostly form a calcareous test (shell) about them; some form tests by cementing sand grains or other inorganic particles to their cell surface (e.g., arenaceous foraminifera).

Fungi   Mycelial or, occasionally single-celled eukaryotic organisms, possessing a cell wall but no chloroplasts; yeasts, molds, mildews, and mushrooms are examples.

Galena   A lead sulfide mineral, PbS.

Gangue   A term of technical slang that refers to the host rock of an ore that encloses the metal-containing minerals of the ore.

Garnet   A silicate mineral of Ca, Mg, Fe, or Mn; it is hard and vitreous.

Generation time   The average time required for cell doubling.

Genotype   The genetic makeup of an organism, as distinguished from its behavioral properties (phenotype).

Geomicrobiology   The study of microbes and the role they have played and are playing in a number of fundamental geologic processes.

Gleying   An anaerobic process in some soils involving microbial reduction of ferric iron manifested by a color change from brownish to grayish and development of stickiness; often associated with water logging of soil.

Glutathione

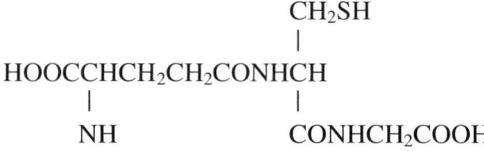

Goethite   An iron oxide mineral, $Fe_2O_3 \cdot H_2O$ or α-FeOOH.

Gondwana   A former continent encompassing Africa, South America, Australia, Antarctica, and India.

Gram-negative   Referring to a differential staining reaction of bacteria in which a counterstain, usually safranin, is retained by the cell.

Gram-positive   Referring to a differential staining reaction of bacteria in which the primary stain, crystal violet, is retained by the cell.

Grandiorite   A volcanic rock intermediate between granite and diorite, showing coarse crystallization.

Granite   Rock of volcanic origin showing coarse crystallization due to slow cooling of the magma from which it arose; granite is rich in quartz and feldspars.

Gravitational water   A film of water, which moves by gravity and responds to hydrostatic pressure and may freeze. It surrounds pellicular water.

GSH   Reduced glutathione.

GSSH   Oxidized glutathione.

Guyot   Flat-topped seamount.

Gypsum   A calcium sulfate mineral, $CaSO_4 \cdot 2H_2O$.

Halophile   A microbe that grows preferentially at a high salt concentration.

Hematite   An iron oxide mineral, Fe_2O_3.

Heterotroph   An organism requiring one or more organic nutrients for carbon and energy for growth.

Heulandite   A type of zeolite mineral.

Histosol   Organic soil.

Holozoic   Feeding on living cells; predatory.

Homeostatsis   Maintenance of a state of equilibrium.

Hornblende   A type of amphibole mineral.

Humic acid   A humus fraction that is acid- and alcohol-soluble.

Humus   In soil it is a mixture of substances derived from partial decomposition of plant, animal, and microbial remains and from microbial syntheses; in marine sediment of the open ocean, it is a mixture of substances derived from phytoplankton remains.

Hydrogenase   An enzyme catalyzing the reaction $H_2 \Leftrightarrow 2H^+ + 2e$.

Hydrosphere   The portion of the Earth's surface that is covered by water; it includes the oceans, lakes, rivers, and groundwater.

Hydrothermal solution   A hot, metal-laden solution generated by a reaction of water (e.g., seawater) with rock in the lithosphere in regions receiving heat from adjacent magma chamber(s).

Hygroscopic water   A thin film of water covering a soil particle, which never freezes or moves a liquid.

Hypersthene   A type of pyroxene mineral.

Hypha(e)   A branch of a mycelium; it is filamentous.

Hypolimnion   The portion of a lake located below the thermocline.

Hypophosphite HPO_2^{2-}.

Igneous rock   Rock of volcanic origin.

Illite   A group of mica-like clay minerals that have a three-layered structure like montmorillonite in which Al may substitute for Si and that contain significant amounts of Fe and Mg.

Indigenous organisms   Organisms native to a habitat.

Inducible enzyme   An enzyme that is formed by a cell only when needed.

Juvenile water   Water from within the Earth that had never before reached the Earth's surface.

Kaolinite   A type of clay $[Al_4SiO_{10}(OH)_8]$ featuring alternating aluminum oxide and tetrahedral silica sheets.

Karstic   Referring to a landscape with sinkholes or cavities due to local dissolution of limestone.

Kazakhstania   A former continent encompassing present-day Kazakhstan.

Labradorite   A type of feldspar mineral related to plagioclase.

Laterization   A soil transformation in which iron and aluminum oxides, silicates, and carbonates are precipitated, cementing soil particles together and thus destroying the porosity of the soil.

Laurasia   A former continent encompassing North America, Europe, and most of Asia.

Laurentia   A former continent encompassing most of North America, Greenland, Scotland, and the Chukotski Peninsula of eastern Russia.

Lentic waters   Static waters.

Lichen   A consortium involving an intimate association of a fungus and a green alga or a cyanobacterium.

Lignin   A heteropolymer of units of substituted phenylpropane derivatives; an abundant constituent of wood.

Limestone   A type of rock that is rich in $CaCO_3$.

Limonite   An amorphous iron oxide mineral, $FeOOH$ or $Fe_2O_3 \cdot nH_2O$.

Lithification   A process of rock formation by compaction or cementation of sediment.

Lotic waters   Flowing waters.

Macrofauna   The fauna excluding protozoa and microscopic invertebrates.

Magma   Molten rock beneath the Earth's surface.

Mannitol   A polyhydric alcohol, which may be formed by reduction of fructose or mannose.

Meddies   In oceanography, small closed-current systems of water whose core is more saline than the surrounding water and that exhibit clockwise rotation (*see also* Rings and Anticyclones).

Mercaptoethanol $HOCH_2CH_2SH$.

Mesophile   A microorganism capable of growth in a temperature range of 10–45°C (optimal range between 25 and 40°C).

Mesotrophic   Referring to a nutritional state of a natural body of water between oligotrophic and eutrophic.

Metabolism   Cellular biochemical activities collectively.

Metabolite   A metabolic reactant or product.

Metamorphic rock   Rock produced by alteration of igneous or sedimentary rock through action of heat and pressure.

Methanogen   A methane-forming bacterium (archeon).

Methanotroph   A methane-oxidizing bacterium.

Methylotroph   A methanol-oxidizing microbe, which can oxidize methanol but not methane.

Microaerophilic organism   An organism that requires a low concentration of oxygen.

Microcosm   An experimental setup that approximates important features of a natural environment, but can be manipulated on a small scale, for example, a soil or sediment percolation column.

Mineralization   In *microbial physiology*, the complete decomposition of an organic compound into CO_2 and H_2O, and if the corresponding elements are present in the organic compound, PO_4^{2-}, NO_3^-, or NH_4^+, and SO_4^{2-} or H_2S. In *mineralogy*, the formation of a mineral.

Mitochondria   Cytoplasmic organelle of eukaryotic cells in which respiration takes place by which energy is conserved through ATP synthesis.

Mixotroph   A bacterium that uses simultaneously organic and inorganic energy sources and/or inorganic or organic carbon sources.

Molybdenite   A molybdenum disulfide mineral, MoS_2.

Monomethyl arsinate $CH_3AsO(OH)$ in acid form.
$$|$$
$$OH$$

Monomictic lake   A lake that turns over once a year.

Montmorillonite   A type of clay mineral $[Al_2Si_4O_{10}(OH)_2 \cdot nH_2O]$ consisting of successive aluminum oxide sheets, each sandwiched between two sheets of silica tetrahedra.

Mutant   A strain that differs genetically in some way from the parent strain of the species.

Mycelium   A network of hyphae produced by most fungi and some bacteria.

Nepheline   A sodium aluminum silicate.

Nitrate ammonification   Anaerobic reduction of nitrate to ammonia via nitrite.

Nitrification A bacterial process in which ammonia is converted to nitrate autotrophically or heterotrophically; some fungi are also capable of heterotrophic nitrification.

Nitrogen fixation A bacterial process in which dinitrogen (N_2) is enzymatically reduced to ammonia.

Nucleic acid A biopolymer containing purines, pyrimidines, pentose, or deoxypentose and phosphate found in chromosomes, plasmids, ribosomes, plastids, and cytoplasm of cells.

Nucleotides Polymer units of nucleic acid consisting of a pyrine or pyrimidine and pentose or deoxypentose and phosphate.

Ocean eddies Collectively, oceanic, small, closed-current systems: rings, anticyclones, and meddies.

Ocean trench Deep cleft in the ocean floor; a site of subduction of an oceanic crustal plate below a continental plate.

Ochre An iron oxide ore, $FeOOH$.

Oligotrophic Referring to a nutrient-poor state in a natural body of water.

Olivine A mineral consisting of orthosilicate of magnesium and iron.

Organic soil A soil formed from accumulation of slow and incomplete decomposition of organic matter in a sedimentary environment.

Orogeny Mountain building.

Orpiment An arsenic sulfide mineral, As_2S_3.

Orthoclase A feldspar mineral.

Orthophosphate Monomeric phosphate, H_3PO_4.

Orthosilicate Monomeric silicate, H_4SiO_4.

Oxidative phosphorylation A process of ATP synthesis coupled to electron transport in respiration.

Oxisol A soil type in tropic and subtropic humid climates.

Pangaea A supercontinent including all major continents of today, existing from ~250 to 200 million years ago.

Panspermia Transfer of life in the form of spores from one world (universe, planet) to Earth.

PAPS 3′-Phosphoadenosine phosphosulfate; 3′-phosphoadenosine phosphatosulfate.

Pectinolytic Capable of enzymatic hydrolysis of pectin.

Pedoscope A system of glass capillaries with optical flat sides for insertion into soil and subsequent microscopic inspection for microbial development in the capillary lumen.

Pellicular water A film of water surrounding hygroscopic water that moves by intermolecular attraction and that may freeze.

Peloscope A system of capillaries with optically flat sides for insertion into sediment and subsequent microscopic inspection for microbial development in the capillary lumen.

Pennate geometry Symmetrical about a long and short axis, in reference to diatoms.

Peptone A mixture of peptides from a digest of beef muscle by pepsin; used in bacterial culture media.

Peridotite An igneous rock, rich in olivines by lacking feldspars.

Peroxidase An enzyme that catalyzes the reduction of H_2O_2 by an oxidizable organic molecule.

Phagotrophic Consuming whole cells by engulfment (phagocytosis).

Phenotype The observable characteristics of an organism; the physical expression of its genotype.

Phosphatase An enzyme that catalyzes the hydrolysis of phosphate esters.

Phosphine PH_3.

Phosphite HPO_3^{2-}.

Phosphorite A calcium phosphate mineral; apatite.

Photolithotroph An autotroph that derives its energy from sunlight.

Photophosphorylation A light-dependent process of ATP synthesis associated with photosynthesis.

Photosynthesis A metabolic process using energy from sunlight for the assimilation of carbon in the form of CO_2, HCO_3^-, or CO_3^{2-}.

Phycomycete Aquatic or terrestrial fungus whose vegetative mycelium shows no septation (e.g., *Rhizopus*).

Phytoplankton   Photosynthetic plankton.

Plankton   Free-floating biota in an aqueous habitat.

Plasmid   An extrachromosomal bit of genetic substance (DNA).

Plutonic water   Deep, anoxic underground water, likely containing significant amounts of sulfate and chloride.

Podzolic soil   A type of spodosol associated with humid, temperate climates; a naturally acidic forest soil.

Pregnant solution   A metal-laden effluent from an ore-leaching operation.

Primary producers   Organisms that transform (fix) CO_2 into organic carbon; include photo- and chemolithotrophs.

Prokaryotic cell   A cell lacking a true nucleus, mitochondria, and chloroplasts.

Proteolytic   Referring to enzymatic hydrolysis of proteins.

Psychrophile   A microorganism capable of growth in a temperature range from slightly below 0 to 20°C (optimum 15°C or below).

Psychrotolerant   Capable of surviving but not growing at a temperature in the psychrophilic range.

Psychrotroph   A microorganism capable of growth in a temperature range of 0–30°C (optimum ~25°C).

Pure culture   A microbial culture that consists of one and only one species or strain.

Purines   A group of organic bases having a purine ring structure in common.

Pyridine nucleotide   Nicotinamide adenine dinucleotide or nicotinamide adenine dinucleotide phosphate; a coenzyme capable of hydrogen transfer.

Pyrimidines   A group of organic bases having the pyrimidine ring structure in common.

Pyrite (iron)   An iron disulfide mineral, FeS_2.

Pyroxine   A ferromagnesian mineral with silica tetrahedra linked in single chains and cross-linked mainly by Ca, Mg, and Fe.

Quatzite   A metamorphic rock derived from sandstone.

Radiolarian ooze   A sediment having radiolarian tests as a major constituent.

Red-bed deposit   A sedimentary deposit rich in ferric oxide; first appeared after the atmosphere of the Earth became oxidizing.

Respiration   Biological oxidation utilizing an electron transport system that may operate with either oxygen of another external reducible inorganic or organic compound as terminal electron acceptor.

Reverse electron transport   The transfer of electrons by an electron transport system against the redox gradient, requiring the input of metabolic energy.

Rhizosphere   The zone in soil that surrounds the root system of a plant, where special environmental conditions may prevail as a result of root secretions and uptake of specific inorganic or organic substances. It represents a special habitat for some microbes.

Rhodanese   An enzyme capable of catalyzing the reaction $CN^- + S_2O_3^{2-} \rightarrow SCN^- + SO_3^{2-}$, and the reductive cleavage of $S_2O_3^{2-}$.

Rhodochrosite   A mineral form of $MnCO_3$.

Rhyolite   An igneous rock rich in plagioclase feldspar.

Ribonucleic acid (RNA)   Heteropolymer consisting of purine and pyrimidine bases, ribose, and phosphate. Different forms of RNA served as templates in protein synthesis (messenger RNA), as amino acid transfer RNA (to locate positions in peptide chains determined by messenger RNA), and as part of the structure of ribosomes.

Ribosome   Submicroscopic, intracellular particle that consists of different RNAs and proteins and is part of the protein synthesizing system of cells.

Ribulose bisphosphate carboxylase/oxygenase   An enzyme that catalyzes carboxylation or oxygenation of ribulose diphosphate in many autotrophic bacteria and in algae and plants.

Rings   In oceanography, as small closed-current systems, with a diameter as great as 300 km, a depth as great as 2 km, and a core of cold water surrounded by warmer water and rotating counterclockwise.

RNA   Ribonucleic acid.

Rock   Massive, solid inorganic matter, usually consisting of two or more intergrown minerals.

Rusticyanin   A copper-containing, periplasmic enzyme that is involved in Fe^{2+} oxidation in *Acidithiobacillus ferrooxidans*.

Saccharolytic   Capable of enzymatic hydrolysis or fermentation of sugars.

Salinity   A measure of the salt content of seawater based on its chlorinity.

Salt dome   The cap rock composed of anhydrite, gypsum, and calcite at the top of a salt plug; a geologic formation.

Sandstone   A rock formed from compacted and cemented sand.

Saponite   A montmorillonite type of clay in which Mg replaces Al.

Saprozoic   Feeding on dead organic matter.

Satellite microorganism   An organism not identical to the dominant organism in a mixed culture, which will give rise to distinctive colonies on appropriate solid medium.

Sclerotium   A vegetative, resting, food storage body in higher fungi, composed of a compact mass of hardened mycelium. Plural: sclerotia.

Sediment   Finely divided mineral and organic matter that has settled to the bottom in a body of water.

Sedimentary rock   Rock formed from compaction and cementation of sediment.

Seismic activity   Earth tremors.

Shale   A laminate sedimentary rock formed from mud or clay.

Sheath   In bacteriology, an organic tubular structure around some bacterial organisms.

Siderite   A mineral form of $FeCO_3$.

Siderophore   An organic iron-chelating substance produced by certain microbes.

Silica   Silicon dioxide; quartz and opal are examples.

Silicate   A salt of silicic acid; a mineral containing silicate.

Slime molds   A group of eukaryotic microorganisms that have a life cycle including a motile swarmer stage and an aggregational phase, which may be multinucleate, leading to the formation of a sessile fruiting body.

Sodium azide NaN_3, an inhibitor of cytochrome oxidase.

Soil horizon   A soil stratum as seen in a soil profile.

Soil profile   A vertical section through soil.

Solfatara   Fumarolic hot spring that yields sulfuretted water.

Spent culture medium   Culture medium after microbial growth has taken place in it.

Spodosol   Forest soil type in temperate climates.

Stenohaline   Capable of growth in a narrow range of salinities.

Stromatolite   A laminated structure formed from filamentous organisms that grew in mats that either entrapped inorganic detrital material or formed $CaCO_3$ deposits in which the organisms became embedded; the organisms in modern stromatolites are most commonly cyanobacteria; in Precambrian stromatolites, the organic remains have frequently disappeared owing to replacement by silica.

Subduction   A process in which the edge of an oceanic crustal plate slips under a continental plate manifested in the form of deep ocean trenches.

Substrate-level phosphorylation   A process of ATP synthesis involving high-energy phosphate bond formation on the substrate being oxidized.

Sulfate-reducing bacteria (SRB)   Bacteria that convert sulfate to sulfide as part of a respiratory process; include members of the domains Bacteria and Archaea.

Sulfhydryl compound   An organic compound with one or more –SH functional groups.

Superoxide dismutase   An enzyme that catalyzes the disproportionation of superoxide (O_2^-) into H_2O_2 and O_2.

Synergism   The interaction of two or more microorganisms, resulting in a reaction that none of the organisms could carry out alone.

Syngenetic   Referring to the deposition of an ore mineral contemporaneously with the enclosing sediment or rock.

Talc   A hydrous magnesium silicate mineral.

Tectonic activity   Interaction of crustal plates of the Earth.

Teichoic acid   A glycerol- or ribitiol-based polymeric constituent of the cell walls of gram-positive bacteria.

Tetrathionate $S_4O_6^{2-}$.

Thermocline   A zone in a water column with a steep temperature gradient.

Thermophiles   Bacteria that grow at temperatures >45°C; some have been shown capable of growth above the boiling point of water when under pressure.

Thiobacilli   Gram-negative rod-shaped bacteria, mostly chemolithotrophs, that can use H_2S, S^0, or $S_2O_3^{2-}$ as energy source.

Thiosulfate $S_2O_3^{2-}$.

Todorokite   A complex manganese(IV) oxide.

Transduction   A process of transfer of genetic information between bacteria involving a bacterial virus as the transmitting agent.

Transpiration   Loss of water by evaporation through the stomata (pores) of leaves.

Travertine   A porous limestone that may be formed by rapid $CaCO_3$ precipitation by cyanobacteria.

Tricarboxylic acid cycle   A cyclic sequence of biochemical reactions in which acetate is completely oxidized in one turn of the cycle.

Trithionate $S_3O_6^{2-}$.

Trophosome   A structure in the coelomic cavity of some vestimentiferan worms consisting of a mass of active symbiotic hydrogen sulfide-oxidizing bacteria, which share with the worm the carbon they fix chemoautotrophically.

Tundra soil   A soil type occurring at high northern latitudes.

Turbidity current   A strong ocean current of a sediment suspension; may exert scouring action as it moves over rock surfaces.

Ultramafic rock   An igneous rock, usually rich in olivine and pyroxenes.

Upwelling   An upward movement of a mass of deep, cold ocean water, which may bring nutrients (nitrate, phosphate) into surface waters.

Vermiculite   A micaceous mineral.

Wad   A complex manganese(IV) oxide.

Wadi   See Arroyo.

Water potential   A measure of water availability, for instance, in soil.

Weathering   A breakdown process of rock.

Wild type   A strain isolated from *the wild* as a reference strain for a process of interest.

Wollastonite   A calcium silicate mineral, $CaSiO_3$.

Zeolite   A hydrated silicate of aluminum containing alkali metals.

Zooplankton   Nonphotosynthetic plankton.

Zygospore   A sexual spore formed by certain algae and fungi.

Index

A

C